The Reticuloendothelial System
A COMPREHENSIVE TREATISE

Volume 3
Phylogeny and Ontogeny

The Reticuloendothelial System
A COMPREHENSIVE TREATISE

General Editors:
Herman Friedman, *University of South Florida, Tampa, Florida*
Mario Escobar, *Medical College of Virginia, Richmond, Virginia*
and
Sherwood M. Reichard, *Medical College of Georgia, Augusta, Georgia*

MORPHOLOGY
Edited by Ian Carr and W. T. Daems

BIOCHEMISTRY AND METABOLISM
Edited by Anthony J. Sbarra and Robert R. Strauss

PHYLOGENY AND ONTOGENY
Edited by Nicholas Cohen and M. Michael Sigel

IMMUNOPATHOLOGY
Edited by Noel R. Rose and Benjamin V. Siegel

PHYSIOLOGY
Edited by Sherwood M. Reichard and James P. Filkins

PHARMACOLOGY
Edited by John Hadden, Jack R. Battisto, and Andor Szentivanyi

IMMUNOLOGY
Edited by Joseph A. Bellanti and Herbert B. Herscowitz

CANCER
Edited by Herman Friedman and Ronald B. Herberman

HYPERSENSITIVITY
Edited by Peter Abramoff and S. Michael Phillips

INFECTION
Edited by John P. Utz and Mario R. Escobar

The Reticuloendothelial System
A COMPREHENSIVE TREATISE

Volume 3
Phylogeny and Ontogeny

Edited by
NICHOLAS COHEN
*The University of Rochester
School of Medicine and Dentistry
Rochester, New York*

and

M. MICHAEL SIGEL
*University of South Carolina School of Medicine
Columbia, South Carolina*

PLENUM PRESS • NEW YORK AND LONDON

Library of Congress Cataloging in Publication Data

Main entry under title:

The Reticuloendothelial system.

Includes bibliographical references and indexes.
CONTENTS: v. 1. Carr, I., Daems, W. T., and Lobo, A. Morphology. — —v. 3. Cohen, N. and Sigel, M. Phylogeny and ontogeny of the RES.
1. Reticulo-endothelial system. 2. Macrophages. I. Friedman, Herman, 1931- II. Escobar, Mario R. III. Reichard, Sherwood M. [DNLM: 1. Reticuloendothelial system. WH650 R437]
QP115.R47   591.2'95   79-25933
ISBN-13: 978-1-4684-4168-0   e-ISBN-13: 978-1-4684-4166-6
DOI: 10.1007/978-1-4684-4166-6

©1982 Plenum Press, New York
Softcover reprint of the hardcover 1st edition 1982

A Division of Plenum Publishing Corporation
233 Spring Street, New York, N.Y. 10013

All rights reserved

No part of this book may be reproduced, stored in a retrieval system, or transmitted in any form or by any means, electronic, mechanical, photocopying, microfilming, recording, or otherwise, without written permission from the Publisher

The American pioneers charted new directions. They cleared forests and allowed the sun to shine on their newly constructed homesteads. They tilled their fields, planted crops, nourished their children and watched them grow. Bill Hildemann is such a pioneer for he has mapped the wilderness of immunophylogeny, made it habitable, revealed its frontiers, and taught and encouraged its settlers. We dedicate this volume to him.

NICHOLAS COHEN
M. MICHAEL SIGEL

Contributors

ROBERT AUERBACH • Department of Zoology, University of Wisconsin, Madison, Wisconsin

CHARLES H. BIGGER • Immunogenetics Group, School of Medicine and Dental Research Institute, University of California, Los Angeles, California

DOMINIQUE BOILEDIEU • Départment de Biologie du Développement de l'Université de Bordeaux I et Centre de Morphologie Expérimentale du CNRS, Talence Cedex, France

MYRIN BORYSENKO • Department of Anatomy, Tufts University, School of Medicine, Boston, Massachusetts

KATHERINE A. COFFARO • Division of Natural Sciences, Thimann Laboratories, University of California, Santa Cruz, California

NICHOLAS COHEN • Department of Microbiology, Division of Immunology, University of Rochester School of Medicine and Dentistry, Rochester, New York

EDWIN L. COOPER • Department of Anatomy, School of Medicine, University of California, Los Angeles, California

C. ANWYL COOPER-WILLIS • Department of Microbiology and Immunology, University of Oregon Health Sciences Center, Portland, Oregon

MARVIN A. CUCHENS • Department of Microbiology, University of Mississippi Medical Center, Jackson, Mississippi

PIERRE DEPARIS • Laboratoire de Biologie Générale, Université Paul Sabatier, Toulouse Cedex, France

F. DIETERLEN-LIÈVRE • Institut d'Embryologie du CNRS et du Collège de France, Nogent-sur-Marne, France

CONTRIBUTORS

Louis Du Pasquier • Basel Institute for Immunology, Basel, Switzerland

Rashika El Ridi • Zoology Department, Faculty of Science, Cairo University, Cairo, Egypt

Thomas H. Ermak • Department of Physiology, University of California School of Medicine, San Francisco, California

Peter L. Ey • Department of Microbiology and Immunology, The University of Adelaide, Adelaide, South Australia

Thelma C. Fletcher • NERC Institute of Marine Biochemistry, Aberdeen, Scotland, U.K.

Bruce Glick • Poultry Science Department–MAFES, Mississippi State University, Mississippi State, Mississippi

Amiela Globerson • Department of Cell Biology, The Weizmann Institute of Science, Rehovot, Israel

William H. Hildemann • Department of Microbiology and Immunology, School of Medicine, and Immunogenetics Group, School of Medicine and Dental Research Institute, University of California, Los Angeles, California

John D. Horton • Department of Zoology, University of Durham, England

E. Houssaint • Institut d'Embryologie du CNRS et du Collège de France, Nogent-sur-Marne, France

André Jaylet • Laboratoire de Biologie Générale, Université Paul Sabatier, Toulouse Cedex, France

Charles R. Jenkin • Department of Microbiology and Immunology, The University of Adelaide, Adelaide, South Australia

Ian S. Johnston • Department of Microbiology and Immunology, School of Medicine, University of California, Los Angeles, California

F. Jotereau • Institut d'Embryologie du CNRS et du Collège de France, Nogent-sur-Marne, France

Richard D. Karp • Department of Biological Sciences, University of Cincinnati, Cincinnati, Ohio

N. Le Douarin • Institut d'Embryologie du CNRS et du Collège de France, Nogent-sur-Marne, France

MARGARET J. MANNING • Department of Biological Sciences, Plymouth Polytechnic, Plymouth, England

JOHN J. MARCHALONIS • Department of Biochemistry, Medical University of South Carolina, Charleston, South Carolina

C. MARTIN • Institut d'Embryologie du CNRS et du Collège de France, Nogent-sur-Marne, France

LARRY J. MCCUMBER • Department of Microbiology and Immunology, University of South Carolina School of Medicine, Columbia, South Carolina

VR. MUTHUKKARUPPAN • Department of Immunology, Madurai-Kamaraj University, Madurai, India

J. J. T. OWEN • Department of Anatomy, Medical School, Birmingham, England

NORMAN A. RATCLIFFE • Department of Zoology, University College of Swansea, Wales, U.K.

PHILIPPE ROCH • Département de Biologie du Développement de l'Université de Bordeaux I et Centre de Morphologie Expérimentale du CNRS, Talence Cedex, France

ANDREW F. ROWLEY • Department of Zoology, University College of Swansea, Wales, U.K.

M. MICHAEL SIGEL • Department of Microbiology and Immunology, University of South Carolina School of Medicine, Columbia, South Carolina

ROBERT TOMPKINS • Department of Biology, Tulane University, New Orleans, Louisiana

RICHARD J. TRAUGER • Department of Microbiology and Immunology, University of South Carolina School of Medicine, Columbia, South Carolina

JAMES B. TURPEN • Department of Biology, Pennsylvania State University, University Park, Pennsylvania

TEHILA UMIEL • Department of Cell Biology, The Weizmann Institute of Science, Rehovot, Israel

PIERRE VALEMBOIS • Département de Biologie du Développement de l'Université de Bordeaux I et Centre de Morphologie Expérimentale du CNRS, Talence Cedex, France

PIERSON J. VAN ALTEN • Department of Anatomy, University of Illinois at the Medical Center, Chicago, Illinois

E. PETER VOLPE • Department of Basic Medical Sciences, Mercer University School of Medicine, Macon, Georgia

ROY L. WALFORD • Department of Pathology, School of Medicine, University of California, Los Angeles, California

JULIA B. WALTERS • Department of Zoology, University College of Swansea, Wales, U.K.

GREGORY W. WARR • Department of Biochemistry, Medical University of South Carolina, Charleston, South Carolina

RICHARD H. WEINDRUCH • Department of Pathology, School of Medicine, University of California, Los Angeles, California

KEITH N. WHITE • Department of Zoology, University College of Swansea, Wales, U.K.

RICHARD K. WRIGHT • Department of Anatomy, School of Medicine, Center for the Health Sciences, University of California, Los Angeles, California

Foreword

This comprehensive treatise on the reticuloendothelial system is a project jointly shared by individual members of the Reticuloendothelial (RE) Society and biomedical scientists in general who are interested in the intricate system of cells and molecular moieties derived from these cells which constitute the RES. It may now be more fashionable in some quarters to consider these cells as part of what is called the mononuclear phagocytic system or the lymphoreticular system. Nevertheless, because of historical developments and current interest in the subject by investigators from many diverse areas, it seems advantageous to present in one comprehensive treatise current information and knowledge concerning basic aspects of the RES, such as morphology, biochemistry, phylogeny and ontogeny, physiology, and pharmacology as well as clinical areas including immunopathology, cancer, infectious diseases, allergy, and hypersensitivity. It is anticipated that by presenting information concerning these apparently heterogeneous topics under the unifying umbrella of the RES attention will be focused on the similarities as well as interactions among the cell types constituting the RES from the viewpoint of various disciplines. The treatise editors and their editorial board, consisting predominantly of the editors of individual volumes, are extremely grateful for the enthusiastic cooperation and enormous task undertaken by members of the biomedical community in general and especially by members of the American as well as European and Japanese Reticuloendothelial Societies. The assistance, cooperation, and great support from the editorial staff of Plenum Press are also valued greatly. It is hoped that this unique treatise, the first to offer a fully comprehensive treatment of our knowledge concerning the RES, will provide a unified framework for evaluating what is known and what still has to be investigated in this actively growing field. The various volumes of this treatise provide extensive in-depth and integrated information on classical as well as experimental aspects of the RES. It is expected that these volumes will serve as a major reference for day-to-day examination of various subjects dealing with the RES from many different viewpoints.

Herman Friedman
Mario R. Escobar
Sherwood M. Reichard

Introduction

During the last two decades immunology has truly come of age, with its own individuality, reaching the open stage where a number of eminent contributors from different but related backgrounds have felt justified in attempting to integrate and compare the immunologic or paraimmunologic responses of invertebrate and vertebrate host organisms. Although the steps in the evolution of the reticuloendothelial system (RES) in animals cannot be retraced with any certainty, the study of immunity in extant species has supported the concept of an orderly progression in immunologic development as one ascends the phylogenetic tree from unicellular organisms with a nucleus (eukaryotic) to man.

Phylogeny and Ontogeny, the third volume in a series on the RES, constitutes an essential part of the comprehensive treatise as it critically reviews the state of the art for each taxonomic group of invertebrates and scrutinizes the most current information on the lymphoreticular system of vertebrate animals. The carefully edited contents of this volume attest to the worldwide growth of basic immunology which is perhaps unique in the history of biology.

It is worth considering that although in many cases there is no structural or functional homology between the elements of host defense in invertebrates (e.g., antisomes) and those in vertebrate animals (e.g., antibodies), the development of these elements can be threaded along the evolutionary staircase by analogous similarities. Furthermore, immunologists studying mammals have already learned a great deal about many of the individual functional units in the immune system thanks to the detailed study of lower species that have a limited capacity for immune reactions, for example, in invertebrates, one can examine a primitive form of cellular immunity in the absence of specific antibodies.

We hope that the combined efforts of all the specialists who have contributed generously to this gigantic task and the talent of the editors of this volume to give it organization, unity and perspective will, at least, widen the horizon of classical immunologists by revealing to them the incredible teleological wisdom that is inherent to the evolutionary development of the RES.

Because of the uniqueness of the pioneering efforts by the editors of this volume, their conviction is expressed in the Editorial Comments section that follows. The volume editors wish to underline the importance attached to sources of information derived from pertinent research on species that evolved early in phylogeny. They believe that the knowledge acquired from that research may pave the way toward a better understanding of immunologic mechanisms operative in the more advanced or complex animals, and not conversely.

Herman Friedman
Mario R. Escobar
Sherwood M. Reichard

Editorial Comments

Higher vertebrates effortlessly perform what seem to be amazing feats of magic as they distinguish self- from modified-self antigens and respond to appropriate signals by manufacturing a panoply of antibody specificities and other effector molecules. We are just beginning to learn that some of the magic involves cellular and molecular on/off signals that exert profound regulatory and directional influences on the generation of immune responses. Only time and human ingenuity will reveal the extent of this biological wizardry.

Immunophylogeneticists attempt to answer when, how, and why this incredibly well integrated complexity came to be. Between the lines of many of the chapters of this volume, one can sense the heady experience of racing up or down what is hopefully a sturdy evolutionary staircase in pursuit of information about the very existence (let alone function) of ancestral lymphocytes, macrophages, and lymphoid organs in the myriad creatures that now inhabit our world. We have learned that ectothermic vertebrates display relatively minor variations on avian and mammalian themes of organization, specialization, and regulation. That the outlines of the mammalian reticuloendothelial systems are well delineated in primitive fishes and amphibians does not make them less worthy of serious scrutiny. Rather, it simply means that we must look elsewhere—to more primitive invertebrates—if we are to decipher the origins of the RES from extant organisms. And look we shall, since there is much to see. Although the diversity of specialized cells in simple invertebrates is certainly less than that of more advanced multicellular creatures, simple and complex multicellular animals alike, must perform, in a coordinated fashion, the universal physiological functions of respiration, digestion, reproduction, and self-protection if they are to survive. Thus, we cannot help but marvel at the abilities of these early metazoans to effectively carry out so many different defense reactions without the vertebrate attributes of antibody, suppressor and helper T cells, etc. We must continue to probe the uniqueness of these defense systems, for an understanding of how they work might help us to arrive at some understanding about the origins of immunity.

M. Michael Sigel
Nicholas Cohen

Preface

Concurrent with the recent veritable explosion of information about all facets of the mammalian immune system, has been the resurgence of interest in the problems posed by the phylogeny and the ontogeny of immunity. This volume reviews research on comparative and developmental aspects of components of the lymphoreticular system—components that are, by and large, responsible for protecting an organism in its antigenically hostile environment.

Each of this book's four sections has a distinctly different focus. In the first section (Chapters 1–9), a separate chapter is devoted to the primordial cellular defense systems of representatives from each taxonomic group of invertebrates. A perusal of these chapters devoted to sponges, coelenterates, worms of all sorts, molluscs, arthropods, echinoderms, and protochordates reveals unique features of organization and specialization as well as a common denominator—the phagocytic cell. This section, which is indeed a book unto itself, is both introduced by and concluded by a chapter in which defense systems of all invertebrates are viewed from a broad biological and biochemical perspective.

The first four chapters of the second section (Chapters 10–14) of this volume are definitive reviews of the structure and function of the lymphoreticular systems of fishes, amphibians, reptiles, and birds. Chapter 14 addresses the critical issue of the molecular basis of self/nonself discrimination of all ectothermic vertebrates. There are two omissions in the second section. First, the structure and function of the RES of mammals has not been covered; this is, of course, the purview of the nine other volumes of this treatise. Second, data bearing on the nonlymphoid components of the RES of ectothermic vertebrates are not presented in any depth. This omission is not an oversight. Rather, it is simply the result of our almost complete ignorance about these cellular entities in creatures other than birds and mammals.

The comparative theme that unites the first and second sections (and provides the rationale for our presumption in using the word, phylogeny, in the title of this volume) is continued in the second half of the book. In the third section (Chapters 15–17), the ontogeny of cells (primarily lymphocytes) of the lymphoreticular systems of amphibians, birds, and mammals is reviewed. In the fourth section (Chapters 18–21), the ontogeny of immunity in representatives of these three vertebrate classes is addressed. A final chapter discusses aging of the RES as an important and continual developmental event. Each of these chapters not only provides a state of the art summary but it reveals a wealth of new, often unique, models with which to study immunological problems that are not restricted to mammals. The second half of this book is also characterized by certain noteworthy omissions. On the one hand, these again reflect our desire to avoid redundancy in view of the abundant well reviewed literature on the ontogeny of immunity of mammals; on the other hand, they speak to our ignorance about the ontogeny of the immune system and immunity of fishes and reptiles. Fortunately, there are shimmerings on the comparative and developmental immunology horizon* that suggest that by the time this treatise will need to be revised and updated, Volume 3 will contain several new chapters, as well as many new references.

Nicholas Cohen
M. Michael Sigel

---

*See grouped papers on fish immunology in *Aspects of Developmental and Comparative Immunology I*. (J. B. Solomon, ed.) Pergamon Press, Oxford, 1981.

Contents

1. **Invertebrate Defense Systems: An Overview**

    EDWIN L. COOPER

    1. Introduction   1
        1.1. Why Study Invertebrate Immunity?   1
        1.2. Three Speculations Defined   2
        1.3. Three Steps in Immune Response Evolution   2
    2. Evidence for Immunogenesis   3
        2.1. Evolution of the Immune System   3
        2.2. Relationship between the Evolution of Immunity and Cancer   4
        2.3. Increasing Leukocyte Diversity and Specialization in Chordates   5
    3. The Animal Kingdom and Immunity   6
        3.1. Taxonomic Restrictions   6
        3.2. Adaptive Radiation, Convergence, and Divergence   6
        3.3. Homology and Analogy   7
        3.4. Basic Units of the Immune System   7
        3.5. Adaptive Immunity   8
    4. The Receptor Problem   9
        4.1. Recognition   9
        4.2. Receptors for Self/Non-self in Protostomes   9
        4.3. Humoral Analysis in a Deuterostome   10
        4.4. Evolution of Genes Coding for Hemolymph Components in Invertebrates   11
        4.5. Cell Surface Receptors   11
        4.6. Are There Immunologically Relevant Receptors on Cells from Annelids, Arthropods, and Molluscs?   12

5. Evidence for Cell Division and Other Coelomocyte Activity in Annelids 13
   5.1. Results with Mitogens 13
   5.2. Transplantation Antigens 14
   5.3. Chemotaxis 15
6. What Do Comparisons between Invertebrate and Vertebrate Defense and Immune Reactions Reveal? 15
   6.1. Introduction 15
   6.2. Invertebrate Immunology Defined 16
   6.3. Contrasts by Phases and Events Using Insects as Models 18
   6.4. Significance of Invertebrate Humoral Immunity 19
   6.5. Comparison of Invertebrate and Vertebrate Immune Strategies 21
7. Impact of Cellular Immunity in Annelids, Arthropods (Protostomes) and Echinoderms (Deuterostomes) 22
   7.1. Introduction 22
   7.2. Cytotoxicity and Other Aggressive Reactions Mediated by Invertebrate Immunocytes 22
8. Evolution of the Invertebrate Immune Response Capacity in Relation to the Major Histocompatibility Complex (MHC) 25
   8.1. Introduction 25
   8.2. Class I Characteristics 25
9. Paradigms for Self/Non-self Recognition Based upon Phagocytosis, the MHC, and V Genes 27
   9.1. Introduction 27
   9.2. Stereospecific Recognition Factors and Phagocytosis 27
   9.3. V Genes and MHC 29
References 30

2. **Cellular Defense Systems of the Porifera**

   IAN S. JOHNSTON and WILLIAM H. HILDEMANN

   1. Sponges and Sponge Cell Types 37
   2. General Defense Mechanisms 39
   3. Defense against Predators and Competitors 40
   4. Defense against Pathogens and Foreign-Body Invasion 41
   5. Interaction with Symbionts and Commensals 43
   6. Maintaining Tissue and Genetic Integrity 44
      6.1. Sponge Cell Aggregation 44
      6.2. Sponge Grafting 47
   References 54

3. **Cellular Defense Systems of the Coelenterata**

   CHARLES H. BIGGER and WILLIAM H. HILDEMANN

   1. Introduction 59

2. Cellular Defense   64
   2.1. Defense against Macroscopic Animals   64
   2.2. Pathogen and Parasite Defenses   68
   2.3. Defense against Cancerous or Foreign Tissue   75
3. Summary   82
References   83

4. Cellular Defense Systems of the Platyhelminthes, Nemertea, Sipunculida, and Annelida

PIERRE VALEMBOIS, PHILIPPE ROCH, and DOMINIQUE BOILEDIEU

1. Taxonomy of the Platyhelminthes, Nemertea, Sipunculida, and Annelida   89
2. The Leukocytes of Worms: Their Structure and General Properties   94
   2.1. Leukocytes of Platyhelminths and Nemerteans   95
   2.2. Leukocytes of Sipunculid Worms   96
   2.3. Leukocytes of Annelids   99
   2.4. Physiological Properties of Worm Leukocytes   101
3. The Graft Response in Platyhelminthes, Nemertea, Sipunculida, and Annelida   110
   3.1. Behavior of Incompatible Grafts   111
   3.2. Arguments for Specific Recognition and Immunological Memory in Nemerteans and Earthworms   114
   3.3. Cellular Aspects of the Host Response to Grafts   116
   3.4. Genetic Basis of Incompatibility   120
4. *In Vitro* Cytopathic Activities of Leukocytes and Body Fluid   123
   4.1. Cytotoxicity in Sipunculids and Earthworms   123
   4.2. Protective Activities of the Body Fluid   127
References   132

5. Cellular Defense Systems of the Mollusca

THELMA C. FLETCHER and C. ANWYL COOPER-WILLIS

1. Introduction   141
2. Recognition of Self and Non-Self   141
3. Pathogens of Molluscs   143
4. Host Resistance and Susceptibility   145
5. Cellular Defense Mechanisms   146
   5.1. Encapsulation   147
   5.2. Nacrezation   151
   5.3. Phagocytosis   151
   5.4. Pinocytosis   157
6. Relationships of Cellular Defense Mechanisms within the Mollusca and Comparison with Vertebrate Systems   158
References   159

6. **Cellular Defense Systems of the Arthropoda**

   NORMAN A. RATCLIFFE, KEITH N. WHITE, ANDREW F. ROWLEY, and JULIA B. WALTERS

   1. General Introduction   167
   2. Chelicerates   168
      2.1. Introduction   168
      2.2. Merostomates   168
      2.3. Arachnids   174
      2.4. Pycnogonids   176
   3. Crustaceans   176
      3.1. Introduction   176
      3.2. Circulatory System   177
      3.3. Crustacean Diseases   178
      3.4. Cells of the Defense System   178
      3.5. Cellular Defense Reactions   187
      3.6. Recognition of Foreignness   197
   4. Onychophorans and Myriapods   200
      4.1. Introduction   200
      4.2. Circulatory System   201
      4.3. Diseases   201
      4.4. Cells of the Defense System   201
      4.5. Cellular Defense Reactions   205
   5. Insects   206
      5.1. Introduction   206
      5.2. Circulatory System   207
      5.3. Insect Diseases   207
      5.4. Insect Host Defenses   208
      5.5. Cells of the Defense System   208
      5.6. Cellular Defense Reactions   215
   References   236

7. **Cellular Defense Systems of the Echinodermata**

   RICHARD D. KARP and KATHERINE A. COFFARO

   1. Introduction   257
   2. Coelomic Cell Types   258
      2.1. Asteroidea   258
      2.2. Echinoidea   260
      2.3. Holothuroidea   263
      2.4. Crinoidea   266
      2.5. Ophiuroidea   267
   3. Origin of the Coelomocytes   267
   4. Defense Reactions   268
      4.1. Phagocytosis   268

    4.2. Clotting Reaction   271
    4.3. Cell-Mediated Immune Responsiveness   272
 5. Conclusions   279
 References   280

8. **Cellular Defense Systems of the Protochordata**

    RICHARD K. WRIGHT and THOMAS H. ERMAK

 1. Introduction   283
 2. Ascidian Internal Defense System: Components, Origin, and Ontogeny   286
    2.1. Ascidian Blood Cells   286
    2.2. Blood Cell Origins   293
 3. Ascidian Internal Defense System: Reactions   297
    3.1. Serum and Tissue Fluid Factors   297
    3.2. Inflammatory Responses   300
    3.3. Histoincompatibility Responses   307
 4. Conclusions and Summary   314
 References   316

9. **Molecular Basis of Self/Non-self Discrimination in the Invertebrata**

    PETER L. EY and CHARLES R. JENKIN

 1. Introduction   321
 2. Primitive Metazoa (Porifera)   323
 3. Protostomia   325
    3.1. Mollusca   325
    3.2. Arthropoda   350
 4. Deuterostomia   360
    4.1. Echinodermata   360
    4.2. Protochordata   365
 5. Discussion   379
 References   382

10. **RES Structure and Function of the Fishes**

    LARRY J. MCCUMBER, M. MICHAEL SIGEL, RICHARD J. TRAUGER, and MARVIN A. CUCHENS

 1. Introduction   393
 2. Morphology of the Lymphoreticular System of Fish   394
    2.1. Organs   394
    2.2. Cells   397
 3. Functions of the Lymphoreticular System of Fish   403
    3.1. Phagocytosis   403

3.2. Graft Rejection   404
3.3. Interferon   405
3.4. Cytotoxicity   406
3.5. Cell Cooperation and Antibody Production: Lymphocyte Heterogeneity   409
References   417

11. RES Structure and Function of the Amphibia

MARGARET J. MANNING and JOHN D. HORTON

1. Introduction   423
2. Thymus   423
   2.1. Structure   423
   2.2. Function   430
3. Spleen   432
   3.1. Structure   432
   3.2. Function   433
4. Kidney   440
5. Lymphomyeloid Nodes   441
6. Gut-Associated Lymphoid Tissue   443
   6.1. Lymphoepithelial Tissue   443
   6.2. Nodular Gut-Associated Lymphoid Tissue   446
7. Bone Marrow   447
8. Liver   448
9. Blood   449
10. Discussion: Phylogenetic Considerations   450
References   451

12. RES Structure and Function of the Reptilia

VR. MUTHUKKARUPPAN, MYRIN BORYSENKO, and RASHIKA EL RIDI

1. Introduction   461
2. Lymphoid Organs and Cells   462
   2.1. Survey of Reptilian Lymphoid Tissues and Organs   462
   2.2. Histology of Thymus, Spleen, and GALT   463
   2.3. Development of Thymus, Spleen, and GALT   466
   2.4. Changes in Spleen Histology following Antigenic Stimulation   468
   2.5. Ultrastructure of Immunocompetent Cells   470
   2.6. Lymphocyte Heterogeneity   474
3. Humoral Immune Responses   475
   3.1. Characteristics of Antibody Response to Various Antigens   475
   3.2. Immunoglobulin Classes   478
   3.3. Anaphylaxis   479
   3.4. Rosette-Forming Cells and Plaque-Forming Cells   479

3.5. Route and Dose–Response Kinetics   481
3.6. Helper Function   482
3.7. Immune Response to Haptens   484
4. Cell-Mediated Immune Responses   486
  4.1. Transplantation Reactions   486
  4.2. *In Vitro* Correlate of Cell-Mediated Immunity   486
  4.3. Graft-vs.-Host Reactions   487
  4.4. Delayed Hypersensitivity Reactions   488
  4.5. CMI Response to Sheep Erythrocytes and Hapten–Carrier Complex   488
  4.6. Normal Lymphocyte Transfer Reaction   489
5. Role of Thymus, Spleen, and GALT   490
  5.1. Role of Thymus in Immunogenesis   490
  5.2. Role of Spleen and GALT in Immune Responses   494
6. The Effects of Nutrition and Environment on the Immune System   496
  6.1. The Influences of Malnutrition and Housing Mode on Immunity   496
  6.2. Lymphoid Tissues and Cells in the Different Seasons   497
  6.3. Humoral Immune Responses in the Different Seasons   500
7. Concluding Remarks   502
References   503

13. RES Structure and Function of the Aves

BRUCE GLICK

1. Introduction   509
2. Morphological and Structural Studies   509
  2.1. Bursa of Fabricius   509
  2.2. Thymus   515
  2.3. Spleen   517
  2.4. Cecal Tonsil   517
  2.5. Harderian Gland   521
  2.6. Lymph Nodules   522
3. Functional Studies   523
  3.1. Fc Receptors   523
  3.2. Kinetics of Lymphocytes in Reticuloendothelial Structures   524
  3.3. Immunoglobulin Synthesis   525
  3.4. Phagocytosis   527
  3.5. Subsets of T Cells   528
  3.6. Lymphokines   529
  3.7. Harderian Gland   530
References   532

14. **Molecular Basis of Self/Non-self Discrimination in the Ectothermic Vertebrates**

    GREGORY W. WARR and JOHN J. MARCHALONIS

    1. Introduction   541
    2. Immunoglobulins   543
        2.1. Immunoglobulins of Lower Vertebrates   546
        2.2. IgM of Fish   546
        2.3. IgM of Amphibians and Reptiles   551
        2.4. Non-IgM Immunoglobulins of Amphibians and Reptiles   551
        2.5. Immunoglobulin Light Chains   551
        2.6. The Antigen-Combining Site of Immunoglobulins of Ectothermic Vertebrates   553
    3. Antigen Receptors and the Cellular Basis of the Immune Response   555
    4. Immunoglobulin as Antigen Receptor on Lymphocytes   557
    References   562

15. **Ontogeny of Amphibian Hemopoietic Cells**

    JAMES B. TURPEN, NICHOLAS COHEN, PIERRE DEPARIS, ANDRÉ JAYLET, ROBERT TOMPKINS, and E. PETER VOLPE

    1. Introduction   569
    2. Ventral Blood Islands and Erythropoiesis   570
    3. Embryonic Origin of Lymphocytes during Normal Amphibian Development   572
        3.1. Ontogeny of Thymocytes: Studies with Anurans   572
        3.2. Ontogeny of Thymocytes: Studies with Urodeles   575
    4. Embryonic Origin of Other Hemopoietic Cells during Normal Anuran Development   578
        4.1. Hepatic Hemopoiesis   579
        4.2. Pronephric Hemopoiesis   579
        4.3. A Dorsally Located Hemopoietic Precursor Cell Compartment   580
    5. Discussion and Conclusions   583
    References   586

16. **Ontogeny of Avian Lymphocytes**

    N. LE DOUARIN, F. JOTEREAU, E. HOUSSAINT, C. MARTIN, and F. DIETERLEN-LIÈVRE

    1. Introduction   589
    2. Ontogeny of the Thymic Primordium Studied in the Quail–Chick Chimera System   590
        2.1. Contribution of the Endoderm and Mesenchyme   590

2.2. Chronology of Homing of Hemopoietic Stem Cells to the Embryonic Thymus   593
  2.3. Mechanisms Controlling the Immigration of LPC into the Thymus   595
  2.4. Chemotactic Hypothesis and Potentialities of LPC   597
3. Ontogeny of Bursa of Fabricius   600
  3.1. Seeding by Extrinsic LPC   600
  3.2. Developmental Relationships between Endoderm and Mesenchyme in Bursal Ontogeny   602
4. Embryonic Origin of LPC   604
  4.1. Demonstration of Intraembryonic Stem Cells in Yolk Sac Chimeras   605
  4.2. The Para-aortic Foci   609
5. Conclusion   612
References   613

17. Ontogeny of Mammalian Lymphocytes

J. J. T. OWEN

1. Introduction   617
2. Hemopoietic Stem Cells and the Ontogeny of Mammalian Lymphocytes   618
3. Mammalian Lymphocytes: Functional Types and Their Ontogeny   619
4. The Thymus and T-Lymphocyte Ontogeny   619
5. Ontogenetic Events in the Maturation of Thymic Epithelium   620
6. Ontogenetic Events in the Maturation of Thymic Lymphocytes   623
7. Mammalian B-Lymphocyte Ontogeny   625
8. Pre-B Cells   626
9. Newly Differentiated B Lymphocytes in Fetal Tissues   626
10. The Expression of V-Region Genes during Primary B-Lymphocyte Differentiation in Fetal Organs   627
11. The Expression of C-Region Genes during Primary B-Lymphocyte Differentiation in Fetal Organs   628
12. General Summary   629
References   629

18. Ontogeny of Immunological Functions in Amphibians

LOUIS DU PASQUIER

1. Introduction   633
2. Antibody Production   634
  2.1. Differentiation of B Cells   634
  2.2. Onset of Immunoglobulin Production and of Specific Antibody Responses   636

2.3. Thymus Dependency of the Antibody Response 639
3. Allogeneic Recognition 640
   3.1. Differentiation of the T Lymphocyte 640
   3.2. Early Appearance of T-Cell Function 641
   3.3. Appearance of Histocompatibility Antigens 643
   3.4. Thymus Dependency of MLR, Graft Rejection, and PHA Responsiveness 644
   3.5. Metamorphosis: Its Effect on Allogeneic Recognition 645
4. Self Tolerance 649
   4.1. Evidence for Anti-adult Self Reactivity in Amphibians 649
   4.2. Possible Influence on the Antibody Repertoire 651
5. General Conclusions 652
References 652

19. Ontogeny of RES Function in Birds

PIERSON J. VAN ALTEN

1. Introduction 659
2. Development of Competence of Endocytosis and Clearance 661
   2.1. Uptake of Colloidal Materials 661
   2.2. Uptake of Microorganisms 663
   2.3. Clearance of Foreign Materials from the Vascular System 663
   2.4. The Role of Opsonins in Clearance and Uptake 665
3. RES Function in Immunologically Deficient Chickens 667
   3.1. Clearance of Antigens and Colloids following Bursectomy 667
   3.2. Alteration in the Nonimmune Opsonins following Bursectomy 670
4. RES Functions of the Epithelium Covering the Lymphoid Follicles of the Bursa of Fabricius 674
   4.1. Morphology and Function of the Bursal Epithelium 674
   4.2. Bursal Function as a Peripheral Lymphoid Organ in Localized Immunity 675
   4.3. Bursal Function as a Peripheral Lymphoid Organ in Systemic Immunity 656
   4.4. Epithelial–Lymphoid Interaction in the Bursa and Consequent Functioning as Both a Central and a Peripheral Lymphoid Organ 678
5. Conclusions 680
References 681

20. Ontogeny of Cellular Immune Reactivity in the Mouse

ROBERT AUERBACH, AMIELA GLOBERSON, and TEHILA UMIEL

1. Introduction 687

2. Embryonic Effector Organs   688
   2.1. The Embryonic Yolk Sac: First Effector Organ of the Immune System   688
   2.2. The Embryonic Thymus as an Effector Organ   692
   2.3. The Embryonic Liver and the Development of T-Cell-Type Immune Functions   695
3. Ontogeny of Suppressor Cells   698
   3.1. Embryonic Liver as a Source of Cells Capable of Suppressing *in Vitro* Immune Reactions   698
   3.2. Other Embryonic Organs as Sources of Suppressor Cells   699
   3.3. *In Vivo* Function of Embryonic Suppressor Cells   701
   3.4. Developmental Considerations   701
   3.5. The Possible Relevance of Embryonic Liver-Derived Suppressor Cells for the Establishment of Tolerance to Self   703
4. General Discussion   704
References   706

21. Aging and Functions of the RES

RICHARD H. WEINDRUCH and ROY L. WALFORD

1. Introduction   713
2. Immunological Theories of Aging   713
   2.1. Immune Decline, Autoimmunity, and Aging   713
   2.2. Role of the Major Histocompatibility Complex   715
3. Organismic Level: Immunological Aspects of Mortality and the Diseases of Aging   716
4. Organ Level   718
5. Cellular/Subcellular Level   720
   5.1. Stem Cells   720
   5.2. Macrophages   721
   5.3. B Cells   722
   5.4. T Cells   723
   5.5. Biochemical Changes in Lymphoid Cells with Age   731
   5.6. Autoreactivity/Tolerance   732
6. Methods of Decelerating Immunological Aging   733
   6.1. Body Temperature Reduction   733
   6.2. Hormone and Drug Therapy   734
   6.3. Immunological Reconstitution   735
   6.4. Dietary Restriction   735
7. Concluding Comments   737
References   737

Index   749

Invertebrate Defense Systems
An Overview

EDWIN L. COOPER

1. INTRODUCTION

1.1. WHY STUDY INVERTEBRATE IMMUNITY?

Invertebrate immunity, a vast, still relatively unexplored, but exciting area, offers many reasons for sustained analysis. First, we will continue to reveal immune reactions as biological phenomena characteristic of all living species. Second, we are interested in mechanisms—the quest for cause-and-effect relationships such as the possible evolutionary pressures causing immunogenesis. Third, we can search for origins of vertebrate immune responses by experimenting with ancestral animals. A fourth, perhaps more practical reason is to question the utility of any immune system—this we do by comparing immune capabilities, whether primitive or advanced, invertebrate or vertebrate. In other words, the question of why the immune system evolved to its present status in humans can be answered only by understanding less complex immune systems.

Invertebrates respond to numerous foreign antigens by recognition, phagocytosis, encapsulation, and specific cellular immunity. Invertebrates also produce humoral substances that inactivate infectious microorganisms. In all instances, offending and possibly pathogenic material is eliminated. While the capacity to synthesize specific antibodies is apparently a characteristic exclusive to vertebrates, all other features of vertebrate responses seem to be shared by the invertebrates, perhaps in a somewhat attenuated form (Figure 1). This chapter will expand upon these generalizations, by specifically reviewing pertinent models and several unsolved problems related to the evolution of immunity. These include: (1) invertebrate defense reactions, (2) phylogeny, (3) affinities among invertebrates, vertebrates, and chordate immunity (Uhlenbruck and Steinhausen, 1977; Wright and Cooper, 1976), and (4) comparative immunology

---

EDWIN L. COOPER • Department of Anatomy, School of Medicine, University of California, Los Angeles, California 90024. Supported in part by Research Grants HD 09333-06 and AI 15976-02.

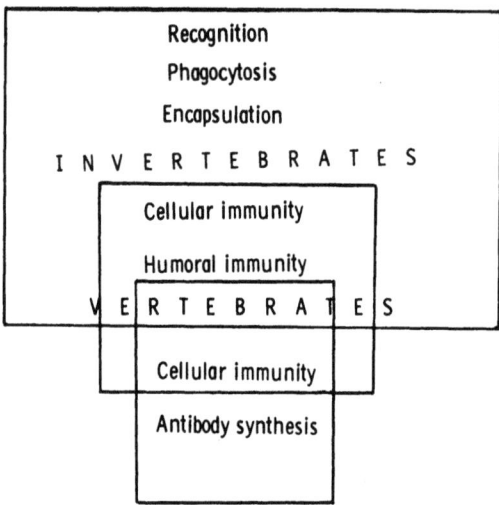

FIGURE 1. Overlapping characteristics of invertebrate and vertebrate immune responses built on primordial mechanisms of recognition, phagocytosis, and encapsulation. Whereas both animal groups share cellular and humoral immune mechanisms, antibody synthesis is exclusively a vertebrate attribute.

(Cooper, 1974a, 1976a, 1977a,b,c; Hildemann and Benedict, 1975; Manning and Turner, 1976; Marchalonis, 1976, 1977; Wright and Cooper, 1976). Invertebrate immunity is one of the important by-products resulting from ferment in our new discipline, comparative immunology.

1.2. THREE SPECULATIONS DEFINED

It is useful to speculate about the nature of those evolutionary pressures that were responsible for development of an immune system. A major pressure would be the need to maintain homeostasis in the face of threats of extinction by external pathogens. A second, selective force might have been the need to maintain the internal milieu as a safeguard against altered self components, i.e., the need for an internal surveillance and regulatory system to guard against the development of cancer or degenerative diseases. It is also of heuristic value to speculate that development of the vertebrate immune system probably paralleled an increasing diversity of leukocyte types (Figure 2).

1.3. THREE STEPS IN IMMUNE RESPONSE EVOLUTION

The immune system's ancestral cell existed among primitive invertebrates. This cell probably recognized foreign material or antigens, and responded by phagocytosis. Phagocytic cells have been conserved throughout the phylogenetic scale as macrophages and granulocytes; probably mast cells are also related to this ancestral cell. Thus, the phagocytic cell and its function is common to and pervades all phylogenetic levels of immunity. It is the first step.

The second evolutionary step probably occurred when coelomate invertebrates developed diverse leukocyte types, some of which are considered to be

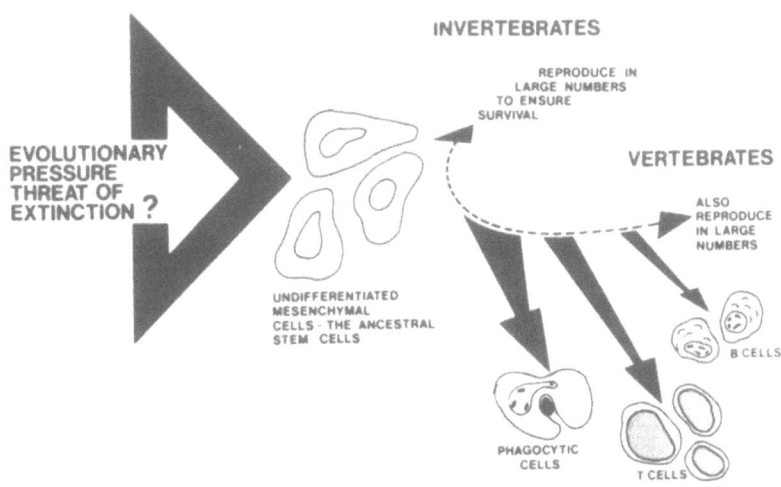

FIGURE 2. Evolutionary pressure (threat of extinction?) leading to the modification of ancestral cells. Further differentiation produced phagocytic cells which persisted throughout the animal kingdom. T cells and B cells evolved later.

precursors of lymphocytes (Wright, 1976). Functionally, invertebrate leukocyte responses may be related to vertebrate *cell-mediated immune responses* (e.g., graft rejection). This effector response appears to be the only one mediated by invertebrate leukocytes in which specificity and memory resemble the vertebrate condition. Production of humoral substances, such as structurally heterogeneous agglutinins, constitutes another form of immunity that evolved among coelomate invertebrates. Effectors of humoral immunity seem to show no structural homology with vertebrate antibodies. Moreover, effectors of invertebrate humoral immunity appear unique. Whether they represent a point where vertebrates and invertebrates share common features (e.g., the production of lymphokines; Waksman, 1979) remains to be determined. The relationship between cellular and humoral immunity in invertebrates is also unclear.

The third event in the evolution of immunity was the capacity for antibody synthesis. It appeared first among the most primitive vertebrates, the cyclostome fish. Increased complexity of antibody structure and diversity reflects a phylogenetic progression from fish to amphibians to mammals.

2. EVIDENCE FOR IMMUNOGENESIS

2.1. EVOLUTION OF THE IMMUNE SYSTEM

Cells basic to the immune system acquired structural properties necessary for recognition and memory as they evolved during immunogenesis from earlier cell types. Those evolutionary pressures causing the immune system to develop are unknown, but defense and immunity presumably evolved at least to prevent the threat of extinction and to preserve individuality of self as distinct from non-self (Acton *et al.*, 1972; Cooper, 1974a; Cushing, 1977; Duprat-Châteaureynaud *et al.*, 1970; Hildemann and Cooper, 1970, 1977; Hildemann and Uhlenbruck, 1974; Kolb, 1977; Panijel and Liacopoulos, 1972; Rothenberg, 1978; Theodor and Acton, 1974).

We analyze the *evolution* of the immune system by studying mechanisms from the comparative viewpoint (Cooper, 1975a,b, 1977c; Du Pasquier and Cooper, 1974; Hildemann and Benedict, 1975; Solomon and Horton, 1977). We should also stress the diversity of defense and immune reactions throughout the animal kingdom (Hildemann and Clem, 1971). Time spans that refer to past geological eras are important to comparative immunologists, and immunological phenomena that occur in animals today, are the supposed missing links in approaching immunogenesis during evolution. Thus, comparative immunological analyses of immune systems of extant species are imperative since their immune systems may reflect those of the extinct (Cooper, 1976a; Jones, 1977; Manning and Turner, 1976). Clearly, the fossil record is essentially of no use in immunology since ancestors are obviously unavailable for direct examination.

2.2. RELATIONSHIP BETWEEN THE EVOLUTION OF IMMUNITY AND CANCER

Cancer in invertebrates, fish, amphibians, and reptiles has been the subject of two reviews (Balls and Ruben, 1976; Dawe and Harshbarger, 1969). Much controversy and speculation once centered around the question of whether cancer in fact exists in animals other than birds and mammals. Such speculations have been similar to those concerning immunity in more primitive creatures. Now, while comparative immunology is beginning to keep pace with the parent discipline, comparative oncology still lags behind. The problem of immunity and neoplasia in invertebrates, just as in vertebrates, has a central characteristic—namely the capacity to distinguish between self and non-self material, in essence, the capacity for specific *recognition*. Exogenous infectious material introduced experimentally into invertebrates is sequestered and does induce a state of rapid immunity, but foreign cells from within (assuming they are present in invertebrates) in the form of neoplasia, may not be reacted against by the immune system. This remains to be determined. If one assumes a cause-and-effect relationship, then animals would possess cell-mediated immunity as defense against neoplasia. Thus, animals with immune systems can be expected to have neoplasia, and immunological deficiency caused by breakdown of the immune system may allow neoplasms to develop. Studying both neoplasia and immunity in primitive animals such as molluscs (snails, clams), where there are more data on both phenomena, is important for understanding how these two phenomena evolved (Cooper, 1976b) (Figure 3).

Some form of immunity with both specific and nonspecific components exists among the invertebrates. The molluscs and the arthropods are the only groups, however, where there are clear-cut cases of known neoplastic growths

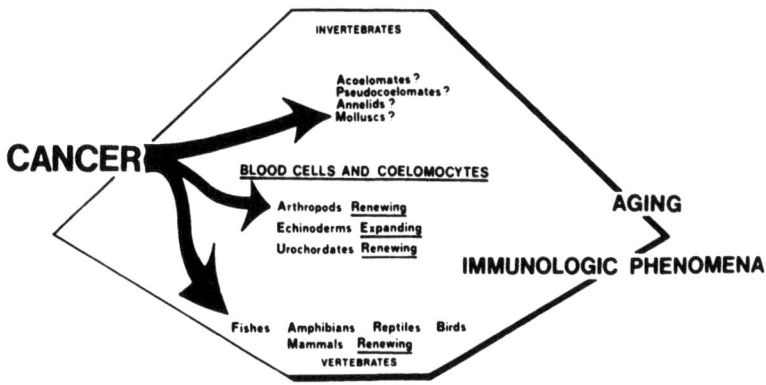

FIGURE 3. Cancer develops in arthropods and vertebrates where there are renewing coelomocyte and blood cell populations. Although molluscs develop cancer, the characteristics of their blood cells have not been reported. Urochordates possess renewing populations, but there are as yet no reports of cancer. The situation is unclear in annelids although benign tumors develop.

with characteristics similar to those in mammals (Figure 3). According to Weiss, the major problem remaining to be resolved is whether there is a causal association between the vigor and kind of invertebrate immune reactions and their capacity to resist the abnormal growths analogous to or homologous to vertebrate neoplasia (Weiss, 1976). Perhaps it is safe to conclude simply that neither neoplasia nor immunity is a purely vertebrate attribute.

2.3. INCREASING LEUKOCYTE DIVERSITY AND SPECIALIZATION IN CHORDATES

Studying vertebrate immunity by examining first the protostome line, where there is more information, fosters a restricted view of the immune system's origin (Cooper, 1977c). By broadening our analyses of immune capacities of *all* animals, we can understand more fully the extensiveness of immune responses. Thus, whether experimental animals be protostome or deuterostome is irrelevant if we wish to view immunity *solely as a biological phenomenon*.

Comparative analyses (such as those in endocrinology, behavior, physiology, embryology, biochemistry) need not require as their goal, the understanding of human immunity through what are viewed (often mistakenly) as simpler models. Yet, from the more practical viewpoint there is also the contrary, teleological and more urgent view for some investigators that the vertebrate immune system, especially that of mammals, is the real aim of all immunologists' work. Thus, this aim seeks to emphasize a more restricted study of deuterostome rather than protostome invertebrates. Vertebrates are generally believed to have evolved from primitive chordates which share certain characteristics with vertebrates, including: (1) a notochord or its equivalent, (2) a dorsal nerve cord, and (3) gill slits. A link to the immune system is found in the presence of lymphocytelike cells in close proximity to the gill region of tunicates, a position homologous to that of the vertebrate thymus (Ermak, 1976). According to Wright (1976), the vertebrate lymphocyte evolved from this early chordate ancestor. This suggests homology, whereas a consideration of lymphocytelike cells in protostomes underscores the concept of analogy (Cooper, 1976f).

3. THE ANIMAL KINGDOM AND IMMUNITY

3.1. TAXONOMIC RESTRICTIONS

When we describe the immune responses of different animal groups, we are often forced to use such terms as "primitive" and "advanced," or perhaps "lower" and "higher," or "simple" and "specialized." Such classifications should not imply, however, that the immune systems of some animals are "better" than others. Thus, a particular animal's immune response capacity should be analyzed on the one hand in relation to its position within the phylogenetic scale, and on the other, in relation to something higher. Those from protozoans through flatworms may be considered primitive, and the annelids, molluscs, and arthropods as advanced. These three advanced groups are also classified as *protostomes*, due to certain embryological developmental patterns. A second group of advanced invertebrates, the deuterostomes, includes the echinoderms and tunicates. Tunicates, like vertebrates, are chordates. Deuterostomes probably diverged from the main protostome line at a point in phylogenesis after the flatworms. Despite their prevailing phylogenetic position dictated by any of several taxonomic schemes, the fact remains that all animals have evolved immune responses built on a basic plan (whose limits are still being defined) that functions: (1) to recognize what is non-self and (2) to eliminate it without detriment to self.

3.2. ADAPTIVE RADIATION, CONVERGENCE, AND DIVERGENCE

General concepts seek to explain the evolution of animals and, in turn, the immune system. A common leukocyte ancestor evolved into a variety of leukocytes by *adaptive radiation* (Cooper, 1976b). In its first form, the leukocyte ancestor probably appeared as an ameboid wandering cell which differentiated, specialized, and evolved into monocytes, granulocytes, mast cells, and lymphocytes (perhaps not necessarily in that order). As a second step, lymphocytes also developed the capacity to divide after contact with antigen, ensuring amplification of the defense system and perhaps immunological memory, characteristics of vertebrate adaptive immune reactions. Based upon structural and functional studies, lymphocyte populations may have evolved from a common lymphocyte precursor among the amphibians (Cooper, 1976e). These lines may have adapted by *divergent evolution* in different ways to distributions in varied organ environments, producing T and B lymphocytes. The evolution of similar sets of lymphocyte characteristics by groups of lymphocytes from quite different evolutionary ancestries is known as *convergent evolution*. The structural and functional development of lymphocytes in mammals and lymphocytelike cells in advanced invertebrates may be an example of convergent evolution (Cohen, 1975; Cooper, 1973a,b, 1974b, 1975a,b, 1976a,b,c,d). Of course, these speculations require substantiation.

3.3. HOMOLOGY AND ANALOGY

Descriptions of the evolution of immune systems must draw distinctions between resemblances—*homology* and *analogy*. Homology may refer to leukocytes or cell products in different animals which share structural similarities due to inheritance of features from a common ancestor. Homologous leukocytes, their sources (blood-cell-producing organs), or their products (agglutinins) may resemble each other closely or they may be superficially quite distinct. Whether superficially alike or distinct, homologous entities share certain basic similarities in structure and function referable to their common ancestor. When cells or their products such as agglutinins in different animals (whether invertebrate or vertebrate) are functionally similar and have no common ancestor, this may be referred to as *analogy*, nonhomologous, but functionally similar. Analogous entities often have little gross structural resemblance to each other. To take an extreme example, the coelomic cavity of the earthworm, an invertebrate, and the bone marrow cavity of tetrapod vertebrates are both storehouses for and production centers for leukocytes. Both entities probably have no common ancestor but functionally, they produce and contain all types of leukocytes. Other analogous organs such as the white body of the octopus and the lymphoid aggregations of vertebrates may have a superficial structural resemblance (Cooper, 1976f). The function they have in common is the production of leukocytes.

3.4. BASIC UNITS OF THE IMMUNE SYSTEM

Cells of the immune system found in the circulating blood of a typical vertebrate are *granulocytes, lymphocytes,* and *monocytes*. All of these cell types have structural and perhaps even functional counterparts among the invertebrates (e.g., echinoderms and tunicates) and primitive vertebrates (Cuénot, 1891, 1897; Ermak, 1976; Freeman, 1970; Geddes, 1880; George, 1926, 1939; Hetzel, 1965; Holland *et al.*, 1965; Jordan, 1938; Jordan and Reynolds, 1933; Kindred, 1929; Kollmann, 1908; Overton, 1966; Pérès, 1943; Prenant, 1922; Roch *et al.*, 1975; Simpson, 1960; Stang-Voss, 1974; Vethamany and Fung, 1972; Wright and Cooper, 1975, 1976). The phagocytic cell, represented by monocytes or tissue and organ macrophages of vertebrates, is probably the only blood cell that has persisted since the protozoans and as a unique unit, functioning in phagocytosis, since the primitive metazoans. When food-getting and immunity became separate functions, cells resembling granulocytes evolved. Lymphocytes

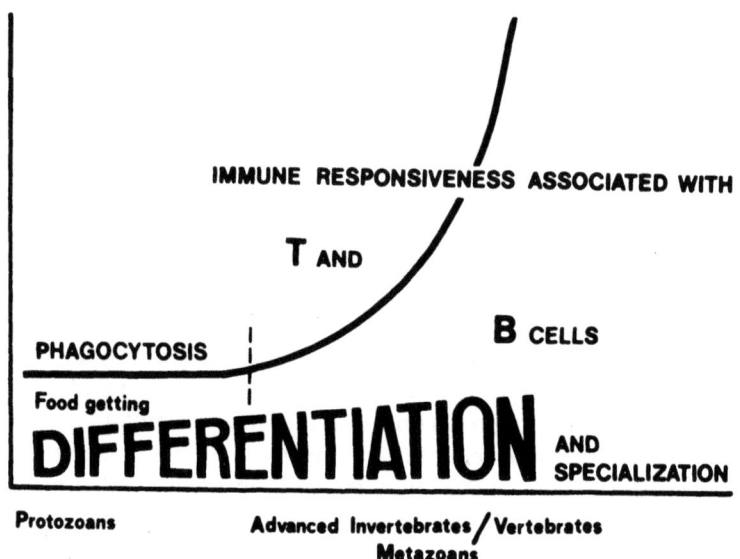

FIGURE 4. Phagocytosis was first associated with food-getting and defense in protozoans but as differentiation and specialization increased, in metazoans and advanced invertebrates, T cells appeared; B cells appeared later in vertebrates.

evolved when there was a further separation of food-getting, nonspecific immunity, and specific immunity, producing T and B subtypes (Figure 4).

3.5. ADAPTIVE IMMUNITY

Adaptive immunity is traditionally defined based on vertebrate cellular and humoral responses, supporting the erroneous assumption by classical immunologists that invertebrates possess no form of adaptive immunity. Phylogenetically, cell-mediated immunity, or some variant, seems to be the earliest of specific immune responses, and, therefore, common to all multicellular animals. Graft rejection in multicellular invertebrates is characterized by the infiltration into such transplants (*in vivo* assay) of leukocytes strongly resembling those of vertebrates. Although debatable, thought now focuses on the possibility that the lymphocytelike cells of certain advanced invertebrates may present T-cell precursors (Garland, 1978; Warr, 1978). A second view of adaptive immunity contends that humoral immunity, specifically, is an attribute peculiar to vertebrates. Although invertebrates synthesize humoral substances, such products are not antibodies. Antibodies or immunoglobulins can be cell surface receptors for antigens, but invertebrate humoral products (such as agglutinins) have not yet been clearly defined as receptors, although excellent work is beginning to define them as such (Renwrantz and Cheng, 1977; Roch and Valembois, 1978).

Unlike vertebrate antibody, humoral substances of invertebrates still remain to be clearly distinguished and defined. Progress is slow since the conceptual and more often the technical approaches to understanding synthesis of humoral mediators in invertebrates are still somewhat limited. Despite the debate concerning precursor T cells (cell-mediated immunity) and the question of substances as receptors (humoral immunity), a definition of adaptive immunity should encompass both vertebrate and invertebrate characteristics. Thus, there should be flexibility to assume that any animal reaches a stage when certain characteristics of its leukocytes or their by-products (humoral substances) are altered after antigenic stimulation triggers a receptor.

4. THE RECEPTOR PROBLEM

4.1. RECOGNITION

Whether dealing with simpler animals or the advanced protostomes and deuterostomes, or vertebrates, the ability to distinguish between self and non-self, i.e., recognition, is a universal attribute requiring a receptor. Indeed, recognition can be traced to the simplest protozoans, where transplantation of cell organelles causes incompatibility reactions as basic to immunity as primitive irritability is to the nervous system. *Quasi-immunorecognition* is the capacity for recognition of non-self tissue followed by incompatibility (Hildemann and Reddy, 1973). Regardless of the test system, recognition is probably the most basic event in the immune response, and continued analysis at all levels is essential to understanding fundamental mechanisms. The first two examples supporting the presumed existence of receptors (in invertebrates) (see Sections 4.2 and 4.3) are worth reviewing for two reasons. First, there is the protostome versus the deuterostome line, and second, both putative cell receptors are related to humoral substances, which recalls questions raised in Section 3.5. In other words, will invertebrate receptors turn out to be humoral products just as are the vertebrate leukocyte receptors? Finally, the conclusions reached concerning these receptors are quite similar in general even though the animals and assays were different.

4.2. RECEPTORS FOR SELF/NON-SELF IN PROTOSTOMES

Leukocytes of animals must recognize antigens, and to do this cells must possess receptors. (There is the alternative but less popular view that cells recognize self and all else is therefore non-self.) To test the assumption, Sloan *et al.* (1975) asked the important question: do crustaceans have recognition molecules, i.e., receptors, enabling them to recognize self, to distinguish it from various non-self foreign proteins, and to eliminate the latter from the circulation? Crayfish and lobsters can readily clear certain soluble antigens (i.e., mammalian serum and albumins) from the circulation but the specificity of elimination is questionable. This contrasts with the situation in higher species. If receptors are present, regardless of whether they are cell-bound or free in body fluids, they may be finite in number. To test whether putative receptors are saturable and whether they possess some degree of specificity, crayfish (*Procambarus clarkii*) were injected with radiolabeled proteins: self (crayfish hemocyanin) and non-self (bovine serum albumin and human γ-globulin). The rate of elimination was markedly different. Equilibration of self proteins occurred within 10 min but essentially no elimination (< 20%) was apparent after 24 hr. In contrast, non-self proteins were eliminated in significant amounts (~ 80%) within a 2-hr period. After recognition, clearance occurred, and the remaining proteins were concentrated in the gills. The crayfish were then injected with a mixture of labeled and unlabeled protein. In excess, the unlabeled protein always prevented the elimination of the same labeled protein. If the excess unlabeled protein was different from the labeled protein, there was no effect on elimination:

$$\text{protein A}^{\text{labeled}} + \text{protein A}^{\text{unlabeled}} \rightarrow \text{defective elimination}$$

$$\text{protein A}^{\text{labeled}} + \text{protein B}^{\text{unlabeled}} \rightarrow \text{no effect on elimination}$$

These results suggest that the crayfish has naturally occurring receptors or recognition molecules for at least three groups of foreign proteins (albumins, γ-globulins, and hemocyanins). Clearance and blocking of binding by competitive elimination do not precisely locate these putative receptors. However, their demonstration provides strong evidence that the functional analog of vertebrate receptors is present in invertebrates.

4.3. HUMORAL ANALYSIS IN A DEUTEROSTOME

The nature of recognition is crucial to several of the body's systems since effector function is first triggered by some stimulus acting on a cell equipped with receptors. Central to the immune system is the need to understand the nature of receptors and to define recognition in more precise terms. One recent approach by Marchalonis and Warr (1978) has been the development of a sensitive binding assay which detected and allowed the isolation of IgG and IgA molecules as well as IgM in mammalian sera. Immunoglobulinlike molecules have not been found in the hemolymph of any species more primitive than cyclostomes. In the tunicate *Pyura stolonifera*, however, the hemolymph does contain proteins characterized by charge heterogeneity that also possess DNP-binding as well as SRBC-binding molecules. These molecules consist of a single subunit of approximate molecular weight 65,000–70,000, and by using two-dimensional electrophoretic separations, preliminary evidence indicates a molecule or structure resembling immunoglobulins in heterogeneity.

These molecules require further study to determine if tunicate hemolymph components are, in fact, related to immunoglobulin, but at least two characteristics are worth mentioning. First, the subunit resembles the μ chain on polyacrylamide gel electrophoresis (PAGE); IgM is the primitive immunoglobulin which is found in all vertebrates. Second, analyses of the T-cell receptor using anti-idiotype sera suggest that this molecule consists of immunoglobulin heavy chains whose mass is comparable to that of the tunicate DNP-binding protein subunit. According to Marchalonis and Warr, lymphocytes possessing properties of T cells preceded the emergence of antibody-secreting B cells (a conclusion similar to that reached in Section 3.5). That tunicates might express a primitive μ-like chain considered to be ancestral to all subsequent immunoglobulin polypeptide chains is a possibility.

4.4. EVOLUTION OF GENES CODING FOR HEMOLYMPH COMPONENTS IN INVERTEBRATES

We know more concerning the genes that coded for antibody than for those genes that may have coded for invertebrate humoral substances (Hood *et al.*, 1975). According to one hypothesis, existing similarities between the hemagglutinins (molecular weight, behavior in electrophoresis) suggest that they: (1) had common origins; (2) may have been the first primitive receptor units; (3) were localized on the surface of immunocytes; (4) serve as recognition units or receptors for antigens. Speculation centers around a precursor gene that coded for a molecule of about 20,000 molecular weight. Later, through fusion, contiguous gene duplication, and possible translocation, genes evolved, coding for hemolymph molecules of about 69,000 molecular weight (e.g., in the spiny lobster). Such genes could have undergone considerable mutation, ultimately producing molecules with diverse functions (e.g., hemolysins, bactericidins, and hemagglutinins). As an alternative hypothesis, genes coding for invertebrate hemolymph factors (e.g., hemagglutinins) may have been lost when vertebrates evolved. For example, a hemagglutinin occurs in the lamprey, one of the most primitive jawless fish; its hemagglutinin differs from its immunoglobulin, a case supporting this second hypothesis (Kubo *et al.*, 1973).

4.5. CELL SURFACE RECEPTORS

Cell surface receptors among imvertebrates have been approached in a unique way. Amirante and Mazzalai (1978) have analyzed the hemagglutinins in hemocytes of the cockroach *Leucophaea maderae* L. and established for the first time that two anti-rabbit erythrocyte hemagglutinins from the hemolymph appear on the cell membrane and in the cytoplasm of numerous hemocytes. After isolation and purification, one hemagglutinin had a sedimentation constant of 18.4 S and electrophoretic mobility in the albumin range, whereas the other moved in the β_1-globulin range. Hemocytes treated with FITC rabbit antihemagglutinin antiserum revealed, by a marked and specific green fluorescence, the presence of hemagglutinins on the plasma membrane, in the cytoplasm, and in cytoplasmic vacuoles of granular hemocytes and spherule cells. The 18.4 S hemagglutinin, as revealed by anti-18.4 S-β_1, was found exclusively in the cytoplasm, mainly in the perinuclear regions. The β_1 was localized on the cell membrane, in the cytoplasm, as well as in the vacuoles. These two components within the hemolymph are probably released by hemocytes which participate in humoral reactions and may also be involved in cellular immunological reactions by acting as receptors for foreign material.

4.6. ARE THERE IMMUNOLOGICALLY RELEVANT RECEPTORS ON CELLS FROM ANNELIDS, ARTHROPODS, AND MOLLUSCS?

Earthworm (*Lumbricus terrestris*) coelomocytes were first shown to form spontaneous rosettes with sheep erythrocytes (E) (Cooper, 1973b). Since then, this observation has been repeated by Toupin and Lamoureaux (1976a) and more recently by Cooper and Perussia (in preparation). The most recent results suggest that earthworm coelomocytes, particularly the basophil (light microscopy) or lymphocytic coelomocyte (electron microscopy), show a preferential attachment to SRBC which can be blocked by adding human 7 S and 19 S immunoglobulin (EA), but partially restored by sensitizing EA with human complement (EAC). Earthworm coelomocytes also form rosettes with chicken, mouse, human, or ox erythrocytes. McKay and Jenkin (1970) and Rabinovitch and De Stefano (1970) showed rosette formation by hemocytes in the arthropods *Parachaeraps bicarinatus* (crayfish) and *Galleria mellonella* L. (wax moth), suggesting that binding is a characteristic common to the leukocytes of advanced invertebrates.

Anderson (1976) extended these studies on rosette formation by using two other insect species. *Prodenia eridania* macrophages express nonspecific receptors for foreign particles, such as human and sheep erythrocytes, aldehyde-treated erythrocytes, and latex beads. However, they lack Fc and C3 receptors commonly encountered on macrophages and certain lymphocytes of higher animals. Cytochalasin B reversibly inhibits rosette formation, whereas treatment of macrophages with vinblastine or colchicine has no effect, suggesting that normal microfilament function is required for rosette formation; however, microtubules apparently play no important role.

Macrophages from the insect *Spodoptera eridania* possess membrane receptors for unmodified avian and mammalian erythrocytes with which they form spontaneous rosettes (Anderson, 1977). Rosette formation occurs in the absence of serum proteins and divalent cations, and individual macrophages bear receptors for several erythrocyte types; however, the level of naturally occurring hemagglutinins against a particular test erythrocyte is not correlated with macrophage reactivity against it. Unlike mammalian macrophages, neuraminidase treatment of either hemocytes or erythrocytes causes no marked enhancement of binding. Pretreatment of macrophages or erythrocytes with cytochalasin B causes reversible inhibition of rosetting, probably by interfering with normal microfilament function, suggesting that optimal binding occurs when membranes are functioning normally on both macrophages and erythrocytes. Colchicine and vinblastine do not influence rosetting; therefore, microtubules are probably not involved in erythrocyte binding.

In earthworms, *Eisenia foetida* and *Lumbricus terrestris*, receptors for Con A have been shown (Valembois *et al.*, this volume). Roch *et al.* (1975) and Roch and Valembois (1978), using fluorescent and radioactive Con A, demonstrated Con A receptors on earthworm leukocyte membranes. Inhibition of binding by methyl-K-mannose, methyl-K-glucose, and D-fructose supported the suggestion that Con A can bind to specific sites. Differences of labeling between two types of leukocytes also support the view that physiological heterogeneity exists among leukocyte populations (Roch and Valembois, 1977). After 30 min at room temperature, patches of Con A-binding-site complexes were formed and capping occurred after 1 hr at room temperature in 30% of the small, round cells which represented 20% of the nonadherent leukocytes. The Con A molecules were then internalized by endocytosis of the cap membrane. In a preliminary collaborative study, Cooper (1979) confirmed the existence of Con A receptors on *Lumbricus* coelomocytes and the binding of FITC-labeled agglutinin from *Helix pomatia* to coelomocyte surfaces after neuraminidase treatment. This latter lectin is known to be specific for human T lymphocytes. The lectin from *Helix* was described with respect to its six specific combining groups (Hammarström and Kabat, 1971). It is also known to interact specifically with human T lymphocytes (Hammarström *et al.*, 1973).

The question of receptors has been investigated in molluscs. There is some evidence that hemolymph proteins interact with antigenic sites perhaps like vertebrate antibodies. Renwrantz and Cheng (1977) have reported the attachment of erythrocytes to the surface of hemocytes of *H. pomatia* by agglutinins which probably serve to link non-self material to the phagocytes.

5. EVIDENCE FOR CELL DIVISION AND OTHER COELOMOCYTE ACTIVITY IN ANNELIDS

5.1. RESULTS WITH MITOGENS

In addition to the existence of receptors, leukocytes of invertebrates must also be capable of division following specific stimulation by antigen if the hypothesis is correct that development of specific clones is a necessary component of adaptive immunity (Burnet, 1974). Vertebrate T lymphocytes transform specifically in the presence of PHA and Con A; in the case of PHA, for example, transformation is first triggered by a specific binding between *N*-acetylgalactosamine-like sugars and PHA. Roch *et al.* (1975) and Toupin and Lamoureaux (1976a) have demonstrated thymidine uptake by cultures of earthworm coelomocytes stimulated with PHA and Con A, and Toupin and Lamoureaux (1976b) studied transformations of *Lumbricus* coelomocytes induced by PHA at various concentrations and during different periods of incubation. To determine which cells or cell associations were transformed following exposure to PHA, subpopulations of coelomocytes were separated into nonadhering and adhering cells. The adhering coelomocytes were also separated into trypsin-sensitive and trypsin-resistant, and each subpopulation alone or mixed together was assayed for its capacity to transform in the presence of PHA.

The coelomocytes of *L. terrestris* do transform in the presence of PHA without the presence of an adhering coelomocyte population. Of the adhering coelomocytes, only the trypsin-resistant ones (5–10%), having a high phagocytic activity, were capable of inducing the transformation of nonadhering coelomocytes. None of these coelomocyte subpopulations transformed alone. These results: (1) reinforce the existence in earthworms of a T-cell-like characteristic; (2) suggest receptors similar to those observed in higher vertebrates; and (3) confirm the cooperation of two cell types in worms to achieve this activity.

Whether these responses adequately reflect what may occur *in vivo* when a coelomocyte recognizes a foreign antigen by receptors is not clear. In summary, the responses to SRBC, PHA, Con A, and *H. pomatia* lectin: (1) either support the origin of T-lymphocyte properties from primitive invertebrates or (2) these properties are nonspecific or (3) these properties developed independently during evolution. The significance of these observations was recently supported by Garland (1978) in relation to T-cell origins but has been challenged by Warr (1978).

5.2. TRANSPLANTATION ANTIGENS

Lemmi (1975a,b) demonstrated the *in vivo* synthesis of DNA, by [^{3}H]thymidine incorporation into coelomocytes of xenografted *L. terrestris* at various times postgrafting, with a peak incorporation of 4 days. Roch *et al.* (1975) investigated *in vitro* DNA synthesis by *E. foetida* coelomocytes following wounding, allografting, and xenografting. Peak incorporation of [^{3}H]thymidine occurred 4 days after all three operations, but was greatest after allografting [with an estimated stimulation index (SI) of approximately 15], less after xenografting (SI ~ 5), and least after wounding (SI ~ 4). Responses following second grafts or wounds were more complex with a peak at day 2 for xenografts (SI ~ 4), at days 2 and 6 for allografts (SI ~ 4), and at day 6 for wounds (SI ~ 3). Using autoradiographic techniques, xenografting (particularly second-sets) stimulated DNA synthesis in more small coelomocytes (< 10 μm) than in larger cells (> 10 μm).

Subsequent extension of the autoradiographic work (Valembois and Roch, 1977) indicated that, at least in terms of DNA synthesis, stimulation of coelomocytes by both wounding and grafting is greater than that which occurs with mitogens. Four days after xenografting, 13% of all coelomocytes incorporated [^{3}H]thymidine *in vitro*, and after second xenografts less (6%) are labeled. After first wounds, 6.5% are labeled, and after second wounds, 4%. Small cells appear to be stimulated more by grafting than by wounding. In normal cells (no grafts or wounds), larger coelomocytes, characterized by abundant cytoplasm, a few free ribosomes, some polyribosomes, moderate amounts of endoplasmic reticulum and vesicular elements, and large pseudopods, incorporated [^{3}H]thymidine more frequently than did small coelomocytes characterized by a high nucleus/cytoplasm ratio and many free ribosomes. After second xenografts, however, the situation is reversed, with more smaller cells labeled than large ones. Coelomocytes do respond to mitogens, wounds, and grafts, responses that show some degree of specificity, confirming the earthworm's primitive immunological capacity.

5.3. CHEMOTAXIS

Coelomocytes of *L. terrestris* were found by Marks *et al.* (1979) to respond *in vitro* by directional migration to both bacterial and foreign tissue antigens. In the case of bacteria (*Aeromonas hydrophila* and *Staphylococcus epidermidis*), the magnitude of the chemotactic response was in direct proportion to the bacterial concentration. Responses to body wall tissues from *E. foetida, Pheretima* sp., and *Tenebrio molitor* (an arthropod) were found to be highest toward *E. foetida*, moderate toward *Pheretima* sp., and least toward *T. molitor*, an inverse ratio to phylogenetic relatedness. The *Eisenia* chemotactant was postulated to be a low-molecular-weight protein, since it was dialyzable and heat-labile. A migration-inhibition factor appeared to be present in *L. terrestris* body wall tissue, since placing both *E. foetida* and *L. terrestris* tissue together in the chemotactant chamber resulted in reduced migration from that found with *E. foetida* tissue alone. Ninety-two to ninety-four percent of the responding coelomocytes were found to be neutrophils (type I granulocytes).

6. WHAT DO COMPARISONS BETWEEN INVERTEBRATE AND VERTEBRATE DEFENSE AND IMMUNE REACTIONS REVEAL?

6.1. INTRODUCTION

The subject of invertebrate immunity is presented in three different ways. First, invertebrates do possess immune reactions but their immune capabilities are different from those of vertebrates, as argued by Chadwick and Aston (1978) and Nappi (1978). Second, Bang (1975) suggested that comparisons must be made between homologous or analogous reactions. Reactions which are considered to be *homologous* are those presumed to have evolved in vertebrates directly from invertebrates, in contrast to *analogous* reactions which are similar in function but have evolved separately; they, therefore, may have made use of completely different molecular constituents. Bang's interpretation is supported by evidence concerning lytic activity in *Sipunculus* which destroys ciliates and is considered to be analogous to certain lytic activity in vertebrates. He cautions, however, against assuming that lysis in vertebrates evolved from *Sipunculus*. Finally, Bang, in citing the variety of molecules of invertebrates which agglutinate erythrocytes and bacteria, and lyse other foreign cells, has proposed the term *antisomes*, "so that they may be discussed, worked with and compared, but not confused with *antibodies*." Third, there are those who argue against immune phenomena in all animals other than vertebrates due to differences in interpretation of phenomena.

6.2. INVERTEBRATE IMMUNOLOGY DEFINED

6.2.1. "Paraimmunology" vs. Independent Evolution

Invertebrate immunology, according to an extreme view, was once referred to as "pseudoimmunology" or "paraimmunology" by Uhlenbruck and Steinhausen (1977). In their assumption, some doubt is cast on the presence of any "immune" or defense characteristic of the "paraimmunological" systems of invertebrates. In another sense, "paraimmunology" facilitates a useful, symbiotic coexistence between a host and an "antigen" or "parasite." By contrast, and more in line with immunology, however, they assume, like many other com-

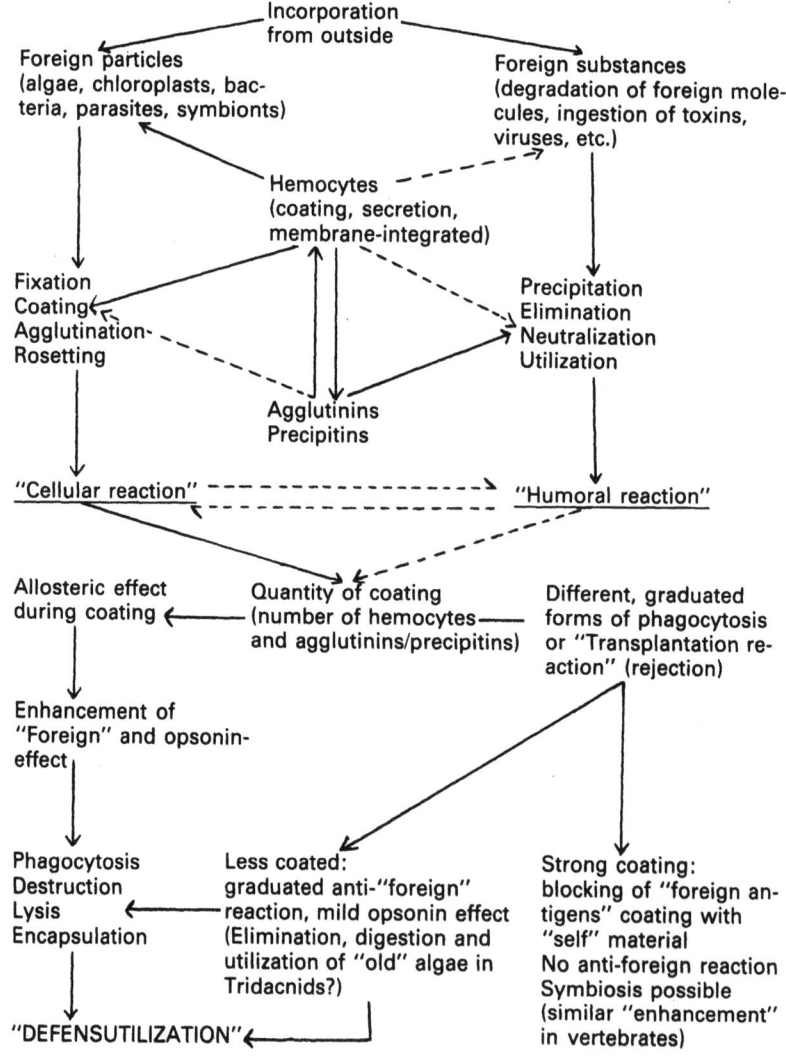

FIGURE 5. Steps leading to "defensutilization." Taken from Uhlenbruck and Steinhausen (1977).

parative immunologists, that the origins and roots of vertebrate immunology may become apparent among invertebrates (Figure 5), in which their immune systems, although primitive, may be the direct precursors of the vertebrate system. Yet, they warn that this seems unlikely, since no antibodylike structure has been detected biochemically in invertebrate "antibodylike substances" such as agglutinins and precipitins. Going further, they contend that only *certain* components of invertebrate recognition systems comprise vertebrate immunological defense mechanisms (e.g., cellular, delayed hypersensitivity). Finally, they have proposed that invertebrate and vertebrate immune systems are entirely independent. That is, they developed separately with different functions.

6.2.2. Examples of Independent Evolution in *Limulus*, Bivalve Clams, and Sponges

A critical review of the contributions from Uhlenbruck's school reveals three main, best investigated "immune" systems of invertebrates that support this hypothesis of independent evolution (Uhlenbruck and Steinhausen, 1977). The first system encompasses the agglutinins and precipitins that occur in the hemolymph of invertebrates. The classical example is the antineuraminyl agglutinin (limulin) of *Limulus polyphemus*, the horseshoe crab. Another example are the tridacnins, the high-molecular-weight (~ 500,000) antigalactose agglutinins/precipitins in the hemolymph of the bivalve clams, the tridacnids. The second system, not necessarily related to the first, occurs independently in some invertebrates in the sexual apparatus of these animals: the best known are the "proteins" of snail albumin glands and eggs. The third well-studied system involves "receptor-specific proteins" and can be observed in different sponges as recognition factors. These factors can also be demonstrated on different invertebrate "immune" cells and hemocytes by rosette formation, adherence phenomena or agglutination, and transplantation techniques (Vaith *et al.*, 1979a,b).

6.2.3. Support for Independent Evolution Based on Studies Using Insects

A large body of information relevant to establishing an accepted definition of invertebrate immunity is based on studies of insects. Insects are capable of mounting certain immune responses analogous to defense mechanisms of higher animals, but there is still little understanding of insect immunity including its molecular and cellular components (Chadwick and Aston, 1978). Future work should exploit current technology with one major aim toward fractionating components involved in recognition, probably the key to understanding immune responses. Chadwick and Aston (1978), leading workers in insect immunity, echoed this when they wrote: "Solving precisely what recognizes and what is recognized may lead to a clarification of how effector substances arise, what they are and how they work. It seems unlikely that insects have such a diverse repertoire of specific receptor substances characteristic of immune responses in higher animals." They further stated that "although investigations of insect immunity will probably always be influenced by mammalian immunity, the field has much to offer on its own merit. It should not serve only as a vehicle to further our understanding of the evolution of the mammalian system." As Cooper (1977a) pointed out, "immunity is as ubiquitous as respiration, reproduction, behavior, etc. and understanding its condition in all animals will broaden and unify our knowledge. It is our belief that in the next decade insects may serve as tools for such unification."

6.3. CONTRASTS BY PHASES AND EVENTS USING INSECTS AS MODELS

Chadwick and Aston (1978) consider it highly unlikely that protective immune responses of insects are homologous with those of mammals. In agreement with the views of Bang (1975), Chadwick and Aston assert that since insects are not in a direct evolutionary line with mammals, the work and interpretations emanating from their laboratory "make no claim for interpretations in the light of homologies but instead they search for analogies." Chadwick and Aston insist, however, from the functional viewpoint, that "the end result is the same: preservation of *self* and disposal of *non-self*." The essential difference between insects and mammals (and vertebrates) lies in the interval between the introduction of the foreign agent and its elimination. After considerable productive work, Chadwick and Aston have suggested several characteristics of insect immunity comparable to, but which can also be contrasted with, features of immunity in higher animals (Table 1).

In their model of insect immunity, Chadwick and Aston proposed three phases (Figure 6). Phase I is inductive, thus stimulating Phase II; this in turn creates a hostile environment within the insect. Phase III is the final effector period in which the induced mechanisms (insect immunity) prevent death. The initial event, Phase I, is *recognition* of the triggering agent or pathogen. During this phase, carbohydrate moieties may be important and breakdown of bacteria necessary. Cross-reactivity may exist and sharing of specificities due to molecular configuration may result in limited specificity. The inductive event of Phase I

TABLE 1. COMPARISON OF FEATURES OF INDUCED PROTECTIVE IMMUNITY IN INSECTS AND HIGHER ANIMALS[a]

Feature	Insects	Higher animals	Apparent analogy
Development	Hours	Days–weeks	Apparent
Duration	Few days	Variable	?
Specificity	Conflicting evidence	Specific	Not apparent
Memory	Second stimulus has no booster effect	Memory involved	None
Antibacterial activity in serum	Many positive reports	Present	Apparent
Phagocytosis	Variable	Present	In some cases
Antibody production	None	Present	None

[a] Taken from Chadwick and Aston (1978).

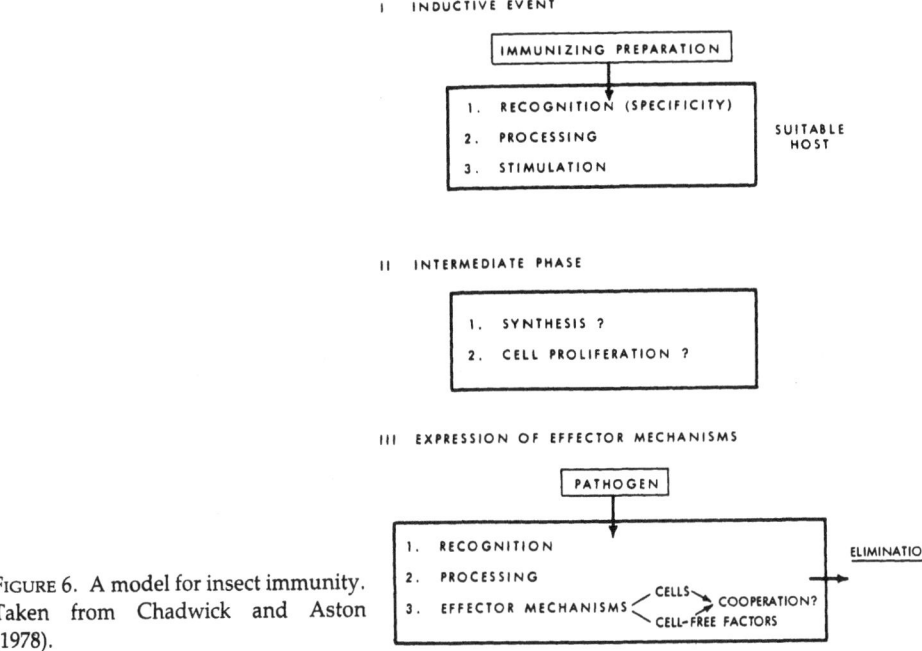

FIGURE 6. A model for insect immunity. Taken from Chadwick and Aston (1978).

triggers Phase II which may lead to alterations in protein levels and profiles, activities of certain enzymes, and changes in cell numbers. The resulting products may be synthesized either *de novo* or constitutively; and there is evidence for the former (Boman *et al.*, 1974; Faye *et al.*, 1975; Rasmussen and Boman, 1977).

6.4. SIGNIFICANCE OF INVERTEBRATE HUMORAL IMMUNITY

6.4.1. Possible Homologies and Comparison with Vertebrate Complement

Humoral immunity in invertebrates involves biologically active molecules that seem to occur naturally or that may be induced, against bacteria and heterologous erythrocytes (Dales, 1979; Hardy *et al.*, 1977). By virtue of their lytic or agglutinating properties, these molecules can act against antigens responsible for their induction. It is only in this respect (i.e., reaction against antigens), however, that such substances resemble vertebrate antibodies. A resolution of this question of homology requires more detailed, primary sequence data on the molecular structure of invertebrate humoral components and a comparison with antibodies. Despite important structural differences between invertebrate humoral substances and vertebrate antibody, both groups of molecules possess the common property of providing a means of inactivating offending pathogens by immune defense mechanisms.

In invertebrates, antigens stimulate antibody synthesis, and the resulting antibody binds to antigen, thus forming complexes. Under certain conditions, these antigen–antibody complexes can bind to complement that system of proteins that mediates a number of reactions such as cell lysis, chemotaxis, agglutination, and phagocytosis. Antigen–antibody complexes may also bind to certain cells, triggering an array of inflammatory reactions. The complement system alone can be activated if the need for antigen and antibody is bypassed by the alternative system, perhaps the oldest phylogenetically. In some invertebrates, there are hemolymph activities that resemble activities mediated by the terminal components of the vertebrate complement system. Complementlike activity represents, for now, the most striking similarity between invertebrate and vertebrate humoral immune components (Acton et al., 1972).

6.4.2. Heterogeneity

Similarities of hemagglutinins between species exist and heterogeneity within species has been suspected, but the first information reporting different hemagglutinins (heterogeneity) within the same lobster species was provided by Hall and Rowlands (1974a,b). Heterogeneity also exists within individuals, important in immunology because of implications to antigen-binding activity. Lobsters have two multiple agglutinins and possibly two independent agglutinins (L Ag-1 and L Ag-2), after purification from hemolymph by ammonium sulfate precipitation, Pevikon block electrophoresis, and gel chromatography. Each has a unique molecular size, electrophoretic mobility, and binding specificity. The molecular weight of L Ag-1 exceeds that of 19 S antibody and that of L Ag-2, 11 S. Antigen specificity is present since L Ag-1 reacts with human and mouse erythrocytes, whereas L Ag-2 reacts only with mouse erythrocytes. Both agglutinins are heat labile (56°C for 15 min), require calcium ions for activity, and are inactive after treatment with trypsin or after reduction and alkylation. These agglutinins dissociate into subunits of equal size (molecular weight 55,000) in 6 M urea, indicating a subunit structure joined by noncovalent bonds.

Hemagglutinin binding properties have been analyzed by inhibition studies using simple saccharides, enzyme treatment of erythrocytes, absorption of agglutinins, and microagglutination preparations. L Ag-2 contains an N-acetylgalactosamine (GalNAc) site which binds to GalNAc residues present on mouse and hamster erythrocytes. L Ag-1 contains an N-acetylneuraminic acid (NeuNAc) site which binds to NeuNAc residues on human erythrocytes; L Ag-1 also binds to mouse erythrocytes via the NeuNAc site or via a second binding site on the L Ag-1 molecule. That heterogeneity exists enlarges our view of the invertebrates' repertoire with respect to possible diversity of responses.

We have accepted in the past, without much solid evidence of molecular structure, that invertebrates recognize antigens and react against them by means of some kind of recognition molecule or receptor. Ideally, such receptors should show heterogeneity reflecting different polymeric forms of an identical subunit(s). This flexibility in structure would enable recognition units or receptors to react with or to bind to relatively large numbers of different antigens with unique specificities. The demonstration of heterogeneity in lobster hemagglutinin is profoundly important, therefore, to studies of immunoevolution. We also need to know the nature of the cells that synthesize such agglutinins (i.e., are they leukocytes?) (see description of work by Amirante and Mazzalai, this chapter). The agglutinins could be found on the surface of leukocytes after synthesis and prior to release in the hemolymph. Although lobsters are arthropods (protostomes) and not in the group (deuterostomes) from which vertebrates arose, the demonstration of hemagglutinin heterogeneity provides strong information of much theoretical importance to understanding how immunity evolved.

6.5. COMPARISON OF INVERTEBRATE AND VERTEBRATE IMMUNE STRATEGIES

Invertebrates do have the capacity for immune reactions since they often produce macromolecules or develop cell-mediated responses which protect them from foreign substances found in the environment (Cunningham, 1978). According to Cunningham, most invertebrates rely on conventional evolutionary strategy for survival when confronted with entirely new antigens (e.g., mutant pathogenic microorganisms) so that among the invertebrate species, there is wide variation and resistant individuals are selected. Vertebrates, by contrast, are more complicated and appear to have a more elaborate network mechanism which enables each *individual* to learn continuously to react against a variety of new and different antigens. This flexibility increases strategy and exploits the evolutionary potential of an internal population of lymphocytes capable of proliferation. As with bacterial or protozoan populations, this increases the chance for variants. Useful variants are selected by antigens in concordance with the clonal selection theory. Both vertebrates and invertebrates do share one common feature, the principle of adaptation to a changing antigenic environment, but the significant difference is the "unit" which survives or succumbs. In the case of invertebrates and depending upon their immune capability, it is the whole *individual*, whereas in vertebrates it is the *lymphocyte*.

In his comparison of mechanisms in invertebrates and vertebrates, Cunningham focused on the following five parameters: (1) *effector molecules*; (2) *specific cells*; (3) *adaptive functions*; (4) *response to unexpected stimuli*; (5) *possible genetic mechanisms*. In almost every instance, these comparisons revealed very limited potential among invertebrates, whereas the characteristics were uniformly positive or high for vertebrates. Cunningham's analysis is useful in pointing out the great gaps in information, even within any single invertebrate phylum. There is, however, a major flaw in that his generalities are based upon limited evidence which has been derived from limited phyla. Nevertheless, his views are important for heuristic reasons. His conclusion suggests no "true" immune system among invertebrates of the vertebrate kind; however, he stresses that invertebrate immunity still should be vigorously analyzed as an important biological discipline aimed at unearthing invertebrate recognition mechanisms of fundamental importance throughout the animal kingdom. Vertebrate immunity is considered a late evolutionary *frill*.

7. IMPACT OF CELLULAR IMMUNITY IN ANNELIDS, ARTHROPODS (PROTOSTOMES), AND ECHINODERMS (DEUTEROSTOMES)

7.1. INTRODUCTION

Comparative immunologists have used two broad approaches to study the evolution of immune mechanisms. Based upon studies using arthropods (protostomes), the greatest amount of information has focused on humoral immune responses. There is still the need, however, to extend invertebrate humoral immune mechanisms to integrate more classical definitions of immunity. Among the annelids and echinoderms where more information exists on cell-mediated responses, there is a closer link to vertebratelike reactions. Earthworms and starfish have been transplanted with autografts, allografts, and xenografts, revealing that coelomocytes in both groups make specific immunological distinction between self and non-self by migrating in and destroying foreign transplants. In both annelids and echinoderms, short-term memory is demonstrable after regrafting previously grafted hosts (Hostetter and Cooper, 1972, 1973, 1974; Karp, 1976; Valembois and Roch, 1977). This phenomenon of specific immunological memory suggests that effector coelomocytes do increase in number in response to a second antigenic challenge, but how they increase is a question of immense importance to comparative immunological theory.

Heightened coelomocyte numbers that result from a response to a second encounter with antigen can be explained by at least two mechanisms in earthworms. First, coelomocytes, sensitive to an initial antigenic encounter, may be *recruited* from throughout the host or from specific coelomocytopoietic or generation sites (Cooper, 1976b). In contrast to recruitment, coelomocytes could increase by *reproduction*; once a coelomocyte is challenged by an antigen, assuming the presence of an appropriate receptor, they would then be triggered to divide after a second challenge—clonal expansion. Support of either mechanism, *recruitment* or *reproduction*, would be important in linking what is now only a tenuous relationship between invertebrate and vertebrate cellular reactions (Marks *et al.*, 1979). However, specific reproduction of coelomocytes in response to antigen in invertebrates would provide evidence that elements of the *clonal selection theory* evolved from a Darwinian approach to immunological diversification.

7.2. CYTOTOXICITY AND OTHER AGGRESSIVE REACTIONS MEDIATED BY INVERTEBRATE IMMUNOCYTES

7.2.1. Cytotoxicity in Sipunculids

Graft rejection analyzed throughout the animal kingdom in major taxonomic groups, has been refined at the cellular level and assays now utilize single effector cell reactions against target cells to determine their capacities for mediating cell-killing reactions. This refinement of analysis has been used in such groups as sipunculids and arthropods (cytotoxicity), and in echinoderms (allo- and xenoaggression), and in each instance, the results have brought us closer to understanding, at the cellular level, the gross manifestations of graft rejection mediated by leukocytes. It is still problematical whether these effector cells are equivalent in all respects to lymphocytes.

Lymphoid effector cells of vertebrates are able to exert *in vitro* cytotoxic activity against xenogeneic, allogeneic, and even syngeneic, non-self target cells. Specific cytotoxicity results when donors of effector cells have previously been sensitized against antigen-carrying target cells; however, natural cytotoxicity mediated by effector cells from nonsensitized animals has also been demonstrated. A natural cytotoxic reaction, analogous to those of rodents, was successfully demonstrated in sipunculid worms by Boiledieu and Valembois (1977a,b). They used three species: *Sipunculus nudus* Linne, *Siphonosoma arcassonense* Cuenot, and *Phascolosoma vulgare* Blainville. *Siphonosoma* and *Sipunculus* were collected from two sandy beaches near the Bassin d'Arcachon about 6 km apart, whereas *Phascolosoma* and other *Sipunculus* were obtained from the Marine Biological Institute, Roscoff, France.

Leukocytes served as effector cells and erythrocytes as targets, revealing optimal cytotoxic activity, as in vertebrates, at a 25 : 1 leukocyte : erythrocyte ratio. Cytotoxicity, measured by trypan blue dye exclusion, always occurred when killer leukocytes reacted against xenogeneic target erythrocytes, or when allogeneic effectors and targets originated from remote stations. Allogeneic effector targets derived from nearby stations (6 km apart) only elicited cytotoxicity in one of three cases. Allogeneic cytotoxicity was not observed between *Sipunculus nudus* from the same station but it occurred in one case out of two between *Siphonosoma arcassonense* from the same station. Cytotoxicity could be specifically reduced after injecting target cells several times *in vivo* due perhaps to lack of sufficient numbers of effector cells *in vivo*. The significance of this blocked response is not clear but it suggests wider response possibilities for invertebrate leukocytes heretofore unrecognized.

7.2.2. Arthropods

Vertebrate lymphocytes and macrophages recognize and destroy foreign tissue cells, including those of tumors; how the capacity for recognition and effector cytotoxicity evolved is unclear. Since *in vivo* recognition of histocompatibility differences has hardly been studied in arthropods, recent work using them has focused on their capacity for *in vivo* cytotoxicity, a manifestation of graft rejection at the cellular level. Hemocytes from the freshwater crayfish, *Parachaeraps bicarinatus*, can kill vertebrate tumor cell targets as was shown by Tyson and Jenkin (1974). They determined whether monolayers of crayfish hemocytes were cytotoxic for ^{51}Cr-labeled Ehrlich ascites tumor cells (EAT), a tumor derived from mice. When ^{51}Cr-labeled EAT cells were added either to Leighton tubes containing crayfish hemocytes, or to control tubes containing no hemocytes, EAT cells were killed rapidly by crayfish hemocytes during the first 90 min. Recognition of the tumor cells appears to be mediated via trypsin-labile receptors on the membrane of hemocytes since enzyme treatment did not destroy the response.

7.2.3. Molluscan Reactions against Parasites

One of the most exciting recent demonstrations concerning cytotoxicity involves parasites (Lie and Heyneman, 1977). As an extension of this work, Bayne (1977) has reported on studies demonstrating the capacity to propagate a type of small cell (minicyte) from stimulated *Biomphalaria glabrata*. For initiating *in vitro* cytotoxic responses, minicyte cultures have been derived from tissues of infected snails. In this instance, the minicyte will voraciously attack and kill *Schistosoma mansoni* sporocysts by "inserting a finger" of cytoplasm into the sporocyst integument. Although relatively rapid (it occurs within 4 hr of the encounter), the mechanism of killing the parasite is unknown; killer cells then detach and move away from the target. According to Bayne, the presence of schistosome targets causes the proliferation of potential killer cells, a response quite different from that of normal snail hemocytes. This important work confirms the existence of cytotoxic responses in another invertebrate group and opens the door for studies in comparative immunoparasitology.

7.2.4. Alloaggression in Echinoderms

7.2.4a. Asteroidea–Echinoidea. There have been attempts to understand cell recognition in an *in vivo* model obviating the need for culture media and *in vitro* procedures. To show adverse host responses, Reinisch and Bang (1971) injected *Arbacia* (sea urchin) coelomocytes (about half of which are deeply pigmented) into *Asterias* (sea star). This caused an abrupt drop in circulating amebocytes in the hosts. Amebocytes from *Arbacia* adhered to and were phagocytosed by *Asterias* (host) amebocytes, which always clumped in regions of the body wall (papulae). The reciprocal experiment (i.e., injection of *Asterias* cells into *Arbacia*) elicited no cell clumping, nor was it followed by a drop in circulating amebocytes. Thus, they concluded that recognition by *Asterias* of intact foreign cells evokes a defense mechanism distinct from and unlike its reaction to allogeneic cells. In fact, this echinoderm response closely resembles the type of self/non-self cell recognition without memory demonstrable in vertebrates, except that in starfish allogeneic combinations, observations could have been longer, ensuring the observation of possible incompatibilities.

7.2.4b. Asteroidea–Birds and Mammals. Another *in vivo* approach, considerably more drastic involving distant xenogeneic combinations, has been developed in the sea star by Leclerc *et al.* (1977a,b), who have been interested in whether axial organ cells could express the same physiological activity in a distant phylogenetic environment. They collected the starfish *Asterias rubens* from along the Atlantic coastline near the Baie d'Arcachon. As sources of cells, they used fragments of brachial digestive organ, ovocytes, and the axial organ, a glandular organ whose cells evoke immune reactions suggestive of delayed hypersensitivity. Cell suspensions of axial organ injected subcutaneously into irradiated mice induced a characteristic angiogenesis reaction whose size and intensity appear to be dependent upon the number of injected cells. Axial organ cells also cause splenomegaly in 16-day-old chick embryos, but no significant angiogenesis or splenomegaly was induced by heat-killed axial organ cells, ovocytes, or brachial digestive organ cells.

Induced angiogenesis supports their hypothesis that the axial organ represents an ancestral primary lymphoid organ. In contrast to starfish axial organ cells, coelomocytes of *A. rubens* induced no significant angiogenesis reaction after subcutaneous injections into irradiated mice. After separating axial organ cells into adherent (by attachment to a plastic surface) and nonadherent cells, the nonadherent population, consisting mainly of small-lymphocyte-like cells, appeared to be the effector cells. Prendergast and Suzuki (1970) also found SSF, a protein extracted from starfish coelomocytes which has lymphokine activity, to be especially effective in inducing delayed hypersensitivity. Although the cell–host combinations are widely disparate, the implications are peculiar but promising, especially since embryogenesis is similar in echinoderms and protochordates, the latter being vertebrate ancestors.

8. EVOLUTION OF THE INVERTEBRATE IMMUNE RESPONSE CAPACITY IN RELATION TO THE MAJOR HISTOCOMPATIBILITY COMPLEX (MHC)

8.1. INTRODUCTION

The MHC is composed of a cluster of gene loci occupying a single chromosomal region, whose products govern certain immunobiological functions (Klein, 1977). These include: (1) *induction of B-cell differentiation* leading to humoral antibody production detectable by standard serological methods; (2) *induction of T-cell differentiation* after blast transformation producing cytotoxic effector lymphocytes; (3) *regulation of immune responses* to various antigens; and (4) *complement* biosynthesis and activation. The MHC has been divided into subregions and these subregions, in turn, into classes (Klein, 1977). Three different classes (I, II, III) have been designated based upon differences in biochemical properties, phenotypic expression, and function. Whether such a complex exists in any or all invertebrates is unknown (although it is suggested in tunicates), but the concept is worthy of conceptual consideration.

8.2. CLASS I CHARACTERISTICS

8.2.1. General Considerations

In mammals, Class I regions occur in pairs with its members closely linked but separated by other loci. With regard to Class I evolution, Schreffler *et al.* (1971) suggested that members of a pair originated from an ancestral gene by gene duplication. The products of Class I regions are glycoproteins whose monomers have a molecular weight of 45,000, and they are noncovalently associated with a single chain of β_2-microglobulin. Class I antigens are expressed on the plasma membrane of all adult nucleated cells, early embryonic cells, and some neoplasias. Mature T and B lymphocytes and macrophages possess the highest concentration of Class I molecules which will induce the production of humoral antibodies detectable by serological methods.

8.2.2. Phylogeny

8.2.2a. Introduction. How the MHC evolved has been difficult to assess for a number of important reasons, only fully appreciated by those few who have assiduously remained dedicated to comparative immunology. First, modern immunology, unlike other disciplines, is more concerned with mammals, particularly humans, and not so much the evolution of responses revealed by mechanisms in primitive systems. Second, interpretations are often difficult since those attracted to the phylogeny of immune mechanisms are usually zoologists first; by contrast, those interested in immunity emphasizing mammals are more highly specialized immunologists. Finally, the husbandry of primitive species is often not well worked out in relation to immunization schedules since most of these animals are "exotic." Since histoincompatibility reactions can be demonstrated in all metazoan invertebrates, prominent gaps can now be filled by finer genetic and immunochemical analyses.

8.2.2b. Histoincompatibility Reactions in Invertebrates. Histoincompatibility reactions in invertebrates are demonstrable in multicellular organisms. The incompatibility expressed in protozoans is assumed to signify cytoincompatibility, a term I introduced to designate incompatibility reactions peculiar to these one-cell animals (Cooper, 1977c). Recognition of non-self occurs in sponge cells (specific cell aggregation, phylum Porifera) and in embryonic tissues, recalling the original definition that Class I antigens are expressed in varying degrees on plasma membranes of all cells. *Acute aggressive reactions* have been described in Cnidaria and *encapsulation* in Arthropoda and Mollusca. These histoincompatibility reactions occur when tissues of two genetically different individuals come into contact and fusion is prevented. In contrast, histoincompatibility reactions also result from *slow aggression* of one tissue against another, as in sponges and coelenterates. These strengths of reaction are surely referable to a genetic control system yet to be fully defined.

Primitive allograft reactions have been studied most extensively in advanced invertebrates, the Annelida (earthworms), Echinodermata (sea stars), and Tunicata (sea squirts). Newer results continue to emerge from experiments using these groups and, recently, from a less primitive group, the ribbon worms (Nemertea). The rejection process in all species is mediated by leukocytes, is transferable adoptively to nonsensitized recipients (earthworms), and is characterized by a memory response. In line with other investigations, graft rejection is probably controlled by at least one histocompatibility locus. There are no reports on genetic control of allograft reactions in echinoderms nor its transfer by adoptive means, although the full array of leukocytes similar to those of vertebrates is involved. Incompatibility reactions in tunicates are not clearly explained although fusion or lack of it between various colonies of the genus *Botryllus* may be controlled by a single genetic system with multiple alleles, perhaps the earliest homolog of the MHC. Although genetic controls have been more elaborately studied after analyses of histoincompatibility reactions in coelenterates (*Hydractinia echinata*, *Hydra*, and gorgonians), such reactions still require more study if we are to understand their evolutionary relevance to allograft responses involving leukocytes, specificity, and memory. Isolated but intense studies on one phylogenetic group are only a beginning, leaving still other groups to be analyzed.

9. PARADIGMS FOR SELF/NON-SELF RECOGNITION BASED UPON PHAGOCYTOSIS, THE MHC, AND V GENES

9.1. INTRODUCTION

Roots of the immune system are grounded in *recognition*, known since the time of Metchnikoff; his observations were significant enough to cause a split, giving immunology its major cellular and humoral subdivisions. Additional significance is attached to his work since it stressed that cellular immunocompetence evolved among certain metazoan invertebrates, well before the additional vertebrate capacity to produce immunoglobulin antibodies. What comparative immunologists have done since his pioneering and prescient work is to extend his observations by showing, in addition to the capacity for recognition, that reactions against foreign components evolved still earlier, even against allogeneic combinations in protozoans. Once again, however, we are left with how cell surface recognition units and perhaps even primordial immunoglobulins or portions of their structure evolved among the "advanced invertebrates" (i.e., echinoderms and tunicates). The essential machinery is there since, clearly, these animals possess diverse types of leukocytes, even small-lymphocytelike cells. Comparative immunologists are moving away from (albeit slowly) the pure analysis of phagocytosis alone, by predicting the existence of and even searching for receptors.

9.2. STEREOSPECIFIC RECOGNITION FACTORS AND PHAGOCYTOSIS

Kolb (1977) suggested that during evolution from Protozoa to Metazoa, cells must have acquired the ability to adhere to cells of the same species. Allogeneic adherence is mediated by stereospecific recognition factors (receptors for self determinants) even in the most primitive metazoans (e.g., sponges). He has postulated that the same recognition system is used to enable phagocytes to free the organism from pathogens. Binding together of neighboring cells, via self-determinant receptors for self interaction, to a phagocyte leads to blockade of phagocytic activity. No blocking will occur when cell-to-cell contact does not lead to specific recognition (Figures 7,8). Foreign cells do not bind to receptors for self and thus will be destroyed. Lack of specific self recognition may be due

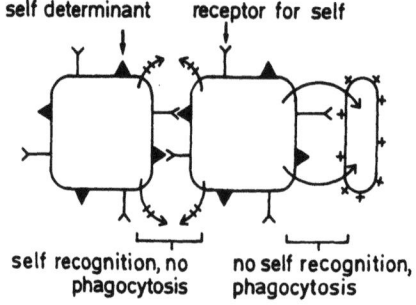

FIGURE 7. Direct recognition of foreign cells. Phagocytic activity of cells in a tissue is blocked when stereospecific contact is established. A foreign particle does not fit into the receptor for self and thus phagocytosis will occur. Taken from Kolb (1977).

other physical parameters that may impair close contact between host and foreign cell surface.

Whereas Kolb (1977) seemed to have focused on the first event (recognition) in his model, Rothenberg (1978) has offered another model stressing phagocytosis, which I interpret as secondary (Figure 9). The first premise relating to self recognition involved the ability of a phagocytic cell to distinguish self from non-self, a basic critical phenomenon necessary for the survival of all multicellular organisms. In higher animals, the nature of recognition units which discriminate between self and non-self is usually defined in terms of antigen–antibody complementarity. Thus, immunogens are recognized by immunoglobulin receptors on B lymphocytes, but we are not so clear about receptors on T cells. However, since no molecular structures that can preferentially recognize have as yet been found in invertebrates, and if phagocytosis of foreign bodies is non-

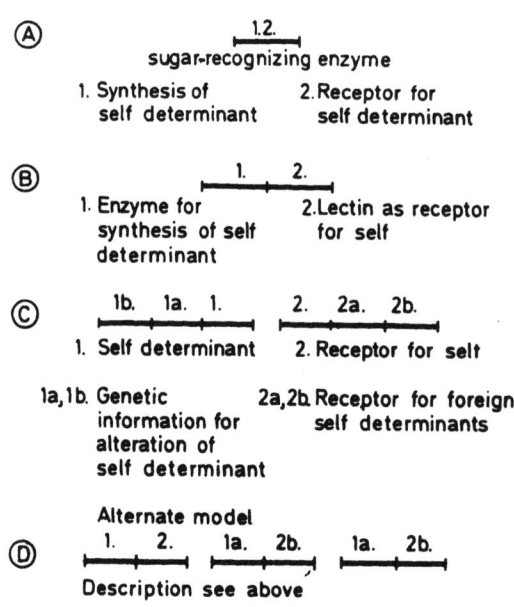

FIGURE 8. Genetic model for the evolution of self determinants and receptors for self. (A) Control of recognition by one gene. The receptor for self is a sugar-recognizing enzyme. It synthesizes the self determinant on its own cell surface and binds to neighboring cell surfaces. (B) Control of recognition by two genes. One gene is responsible for the synthesis of the self determinant; a second gene codes for the receptor for self (which is derived from the former gene product but is devoid of enzymatic activity). (C, D) Control of recognition by a multigene complex. Two models of a multigene complex coding for the self-recognition system and additional receptor systems which evolved by gene duplication and mutation. Taken from Kolb (1977).

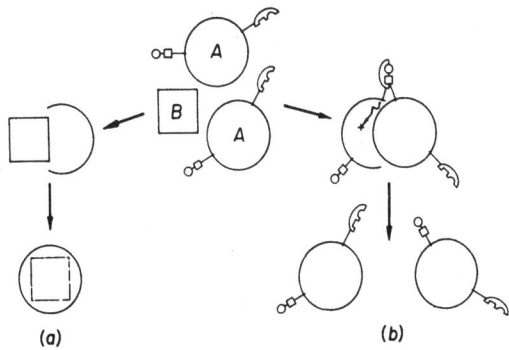

FIGURE 9. A lock-and-key model describing how primitive multicellular organisms, containing a distinct population of phagocytic cells, regulate self phagocytosis. (a) A lytic or phagocytic response is initiated by strain A coming into contact with strain B. This process continues, without restraint, until B is engulfed by A. (b) Phagocytosis or lysis is stimulated whenever an A cell comes into contact with an autologous cell. Here the same physical laws that initiated the interaction of A and B apply, but after the process has begun, the self-recognition units of two A cells become activated—their interaction inhibits any lytic or phagocytic action. Consequently, the cells may return to their resting state. Thus, by assuming that the initiation of phagocytic or lytic activity is a nonspecific process, we conclude that autologous phagocytic or lytic activity is regulated by a self-recognition system. Taken from Rothenberg (1978).

either to a lack of corresponding self determinants on foreign cells or, on the other hand, to differences in electric charge, charge pattern, surface tension, or specific, then how do these very potent and highly destructive cells avoid destroying self tissue?

Phagocytosis has been and is still the most actively studied defense function among the invertebrates, yet many of the approaches have hardly gone further nor become more imaginative than those during Metchnikoff's time and of early 20th century comparative immunologists. Recently, at least in theory, two hypothetical models accounting for the basis of self/non-self recognition with applications to phagocytosis have been proposed. According to Parish (1977), recognition factors are considered to be random associations of different host glycosyltransferase enzymes with specificities and binding properties for diverse sugar acceptors. Such recognition factors are incapable of recognizing self, since the self sugars or acceptors to which they could bind would be blocked by glycosylation. Two other activities, cytophilic and opsonic, may occur if polymerization of transferases into the recognition factor was initiated by adding a protein which also possessed an acceptor site on the hemocyte. Although this model is somewhat simple and awaits investigation, it does provide for a range of specificities, ensures that self-reactive recognition factors are not generated, and can withstand a high degree of genetic polymorphism without the appearance of self reactivity.

9.3. V GENES AND MHC

Jerne (1971) proposed a hypothesis linking the generation of antibody diversity with the: (1) genetic determination of immune responsiveness and (2) with the alloaggressive properties of the immune system. The genetics of immune responsiveness has revealed the importance of the histocompatibility pattern of an animal since it restricts the range of capabilities in antibody formation. Alloaggression refers to the remarkable capacity of an animal's immune system to react most strongly against the histocompatibility antigens of members of the same species. According to this hypothesis, an animal's germ cells carry a set of V genes which determine the combining sites of antibodies directed against a complete set of a certain class of histocompatibility antigens of the species to which the animal belongs. Although antibody synthesis remains a vertebrate attribute, the evolutionary development of V genes in phylogeny is traced back to the requirements for cell-to-cell recognition in all metazoans. Self tissues are not destroyed, according to Rothenberg, due to regulation by a self marker based upon protein–carbohydrate complementarity. The biological counterparts of these hypothetical molecules may be the products of the MHC; the MHC and foreign-recognition systems are direct descendants of the self-recognition genes. Rothenberg's idea is certainly related to Jerne's views on somatic generation of immune recognition.

In conclusion, several hypotheses have been offered to account for *recognition, phagocytosis,* and *cell-mediated immunity* (alloaggression, according to Jerne). They all stress the possible existence of receptors and the capacity of invertebrate immunocytes to divide. Burnet (1974) suggested that some forms of recognition of the difference between self and non-self are evident in invertebrates. For comparative immunology, the most interesting future development could be a demonstration that invertebrate hemocytes (or some subgroup of these cells) are capable of differentiation into forms analogous to immunocytes, i.e., multiple clones of cells each with a distinctive pattern or range of patterns of steric reactivity. Burnet asserts the existence of phenomena suggestive of specific adaptive immune responses in coelomate invertebrates, but immunocytes with specific receptors must still be defined.

Finally, I am convinced of the necessity for studying immunity in all forms of life (including plants). The reasons are both practical and intellectual. On the practical side, knowledge of how exotic animals (excluding mammals) fight pathogens should do more to aid in preserving natural food sources of which invertebrates and certain primitive vertebrates form an impressive and significant part. On the intellectual side, immunology is still a young discipline badly in need of a unifying theme, concept, or molecule such as exists for genetics, embryology, and the other basic biological sciences. This need has not been satisfied by the *immunoglobulin molecule,* the *clonal selection theory,* or the *T and B cell concept.* GOD (generator of diversity) will not be recognized until there are more facts about immunity in primitive animals.

REFERENCES

Acton, R. T., Evans, E. E., Weinheimer, P. F., Cooper, E. L., Campbell, R. D., Prowse, R. H., Bizot, M., Stewart, J. E., and Fuller, G. M., 1972, *Invertebrate Immune Mechanisms,* MSS Information Corporation, New York.

Amirante, G. A., and Mazzalai, F. G., 1978, Synthesis and localization of hemagglutinins in hemocytes of the cockroach *Leucophaea maderae* L., *Dev. Comp. Immunol.* 2:735.

Anderson, R. S., 1976, Expression of receptors by insect macrophages, in: *Phylogeny of Thymus and Bone Marrow-Bursa Cells* (R. K. Wright and E. L. Cooper, eds.), Elsevier/North-Holland, Amsterdam.

Anderson, R. S., 1977, Rosette formation by insect macrophages: Inhibition by cytochalasin B, *Cell. Immunol.* **29**:331.

Balls, M., and Ruben, L. N., 1976, Phylogeny of neoplasia and immune reactions to tumors, in: *Comparative Immunology* (J. J. Marchalonis, ed.), pp. 167–208, Blackwell, Oxford.

Bang, F. B., 1975, Phagocytosis in invertebrates, in: *Invertebrate Immunity* (K. Maramorosch and R. Shope, eds.), pp. 137–151, Academic Press, New York.

Bayne, C. J., 1977, Molluscan immunobiology: The elevation of responses, in: *Developmental Immunobiology* (J. B. Solomon and J. D. Horton, eds.), pp. 67–74, Elsevier/North-Holland, Amsterdam.

Boiledieu, D., and Valembois, P., 1977a, Natural cytotoxic activity of sipunculid leukocytes on allogenic and xenogenic erythrocytes, *Dev. Comp. Immunol.* **1**:207.

Boiledieu, D., and Valembois, P., 1977b, The mechanism of leukocyte cytotoxicity studied by time-lapse microcinematography and its inhibition: An example of *in vitro* specific recognition in invertebrates, in: *Developmental Immunobiology* (J. B. Solomon and J. D. Horton, eds.), pp. 51–57, Elsevier/North-Holland, Amsterdam.

Boman, H. G., Nilsson-Faye, I., Paul, K., and Rasmuson, T., Jr., 1974, Insect immunity. I. Characteristics of an inducible cell-free antibacterial reaction in hemolymph of *Samia cynthia* pupae, *Infect. Immun.* **10**:136.

Burnet, F. M., 1974, Invertebrate precursors to immune responses, in: *Contemporary Topics in Immunobiology*, Vol. 4 (E. L. Cooper, ed.), pp. 13–24, Plenum Press, New York.

Chadwick, J. M., and Aston, W. P., 1978, An overview of insect immunity, in: *Animal Models of Comparative and Developmental Aspects of Immunity and Disease* (M. E. Gershwin and E. L. Cooper, eds.), pp. 1–14, Pergamon Press, Elmsford, N.Y.

Cohen, N., 1975, Phylogeny of lymphocyte structure and function, *Am. Zool.* **15**:119.

Cooper, E. L., 1973a, Earthworm coelomocytes: Role in understanding the evolution of cellular immunity. I. Formation of monolayers and cytotoxicity, in: *Proceedings, III International Colloquium on Invertebrate Tissue Culture* (J. Reháček, D. Blaskovic, and W. F. Hink, eds.), pp. 381–404, Publishing House of the Slovak Academy of Science, Bratislava.

Cooper, E. L., 1973b, Evolution of cellular immunity, in: *Non-specific Factors Influencing Host Resistance* (W. Braun and J. Ungar, eds.), pp. 11–23, Karger, Basel.

Cooper, E. L. (ed.), 1974a, Invertebrate immunology, in: *Contemporary Topics in Immunobiology*, Vol. 4, Plenum Press, New York.

Cooper, E. L., 1974b, Phylogeny of leukocytes: Earthworm coelomocytes *in vitro* and *in vivo*, in: *Lymphocyte Recognition and Effector Mechanisms*, pp. 155–162, Academic Press, New York.

Cooper, E. L., 1975a, Characteristics of CMI and memory in annelids, *Adv. Exp. Med. Biol.* **64**:127.

Cooper, E. L. (ed.), 1975b, Developmental immunology, *Am. Zool.* **15**:1.

Cooper, E. L., 1976a, *Comparative Immunology*, Prentice-Hall, Englewood Cliffs, N.J.

Cooper, E. L., 1976b, Immunity and neoplasia in mollusks, *Isr. J. Med. Sci.* **12**:479.

Cooper, E. L., 1976c, The earthworm coelomocyte: A mediator of cellular immunity, in: *Phylogeny of Thymus and Bone Marrow-Bursa Cells* (R. K. Wright and E. L. Cooper, eds.), pp. 9–18, Elsevier/North-Holland, Amsterdam.

Cooper, E. L., 1976d, Cellular recognition of allografts and xenografts in invertebrates, in: *Comparative Immunology* (J. J. Marchalonis, ed.), pp. 36–79, Blackwell, Oxford.

Cooper, E. L., 1976e, Evolution of blood cells, *Ann. Immunol. (Inst. Pasteur)* **127C**:817.

Cooper, E. L., 1976f, The earthworm coelomocyte: A mediator of cellular immunity, in: *Phylogeny of Thymus and Bone Marrow-Bursa Cells* (R. K. Wright and E. L. Cooper, eds.), Elsevier/North-Holland, Amsterdam.

Cooper, E. L., 1977a, Preface, in: *Comparative Pathobiology* (L. Bulla, Jr. and T. Cheng, eds.), pp. ix–xii, Plenum Press, New York.

Cooper, E. L. (ed.), 1977b, *Developmental and Comparative Immunology 1*, Pergamon Press, Elmsford, N.Y.

Cooper, E. L., 1977c, Evolution of cell mediated immunity, in: *Developmental Immunobiology* (J. B. Solomon and J. D. Horton, eds.), Elsevier/North-Holland, Amsterdam.

Cooper, E. L., 1979, Properties of coelomocyte membranes, in: *Protides of the Biological Fluids* (H. Peeters, ed.), pp. 581–584, Pergamon Press, Elmsford, N.Y.

Cuénot, L., 1891, Études sur le sang et les glands lymphatiques dans la série animale (2e parties: Invertébrés), *Arch. Zool. Exp. Gen.* **9**:13.

Cuénot, L., 1897, Les globules sanguins et les organes lymphoides des invertébrés, *Arch. Anat. Microsc. Morphol. Exp.* **1**:153.

Cunningham, A. J., 1978, A comparison of the immune strategy of vertebrates and invertebrates, *Dev. Comp. Immunol.* **2**:243.

Cushing, J., 1977, Immunology and the processes for evolution, *Dev. Comp. Immunol.* **1**:65.

Dales, R. P., 1979, Defense of invertebrates against bacterial infection, *J. R. Soc. Med.* **72**:688.

Dawe, C. J., and Harshbarger, J. C. (eds.), 1969, *A Symposium on Neoplasms and Related Disorders of Invertebrate and Lower Vertebrate Animals*, Natl. Cancer Inst. Monogr. **31**.

Du Pasquier, L., and Cooper, E. L., 1974, Primitive vertebrate immunology, *Prog. Immunol. II* **2**:297.

Duprat-Châteaureynaud, P., Du Pasquier, L., and Valembois, P., 1970, Colloque sur les réactions immunitaires chez les invertébrés, Laboratoire de Zoologie (Université de Bordeaux) et Centre de Morphologie Expérimentale du CNRS.

Ermak, T., 1976, Hematogenic tissues of tunicates, in: *Phylogeny of Thymus and Bone Marrow-Bursa Cells* (R. K. Wright and E. L. Cooper, eds.), pp. 45–56, Elsevier/North-Holland, Amsterdam.

Faye, I., Pye, A., Rasmuson, T., Jr., Boman, H. G., and Boman, I. A., 1975, Insect immunity. II. Simultaneous induction of antibacterial activity and selective synthesis of some hemolymph proteins in diapausing pupae of *Hyalophora cecropia* and *Samia cynthia*, *Infect. Immun.* **12**:1426.

Freeman, G., 1970, The reticuloendothelial system of tunicates, *J. Reticuloendothelial Soc.* **7**:183.

Garland, J. M., 1978, The T-cell paradigm: A philosophical view of immunology, *Dev. Comp. Immunol.* **2**:39.

Geddes, P., 1880, Observations sur le fluide périviscéral des oursins, *Arch. Zool. Ser.* **8**:483.

George, W. C., 1926, The histology of the blood of *Perophora viridis* (ascidian), *J. Morphol* **41**:31.

George, W. C., 1939, A comparative study of the blood of tunicates, *Q. J. Microsc. Sci.* **81**:391.

Hall, J. L., and Rowlands, D. T., Jr., 1974a, Heterogeneity of lobster agglutinins. I. Purification and physicochemical characterization, *Biochemistry* **13**:821.

Hall, J. L., and Rowlands, D. T., Jr., 1974b, Heterogeneity of lobster agglutinins. II. Specificity of agglutinin–erythrocyte binding, *Biochemistry* **13**:828.

Hammarström, S., and Kabat, E. A., 1971, Studies on specificity and binding properties of the blood group A reactive hemagglutinin from *Helix pomatia*, *Biochemistry* **10**:1684.

Hammarström, S., Hellstrom, U., Perlmann, P., and Dillner, M. L., 1973, A new surface marker on T lymphocytes of human peripheral blood, *J. Exp. Med.* **138**:1270.

Hardy, S. W., Fletcher, T. C., and Olafsen, J. A., 1977, Aspects of cellular and humoral defense mechanisms in the Pacific oyster, *Crassostrea gigas*, in: *Developmental Immunobiology* (J. B. Solomon and J. D. Horton, eds.), pp. 59–66, Elsevier/North-Holland, Amsterdam.

Hetzel, H. R., 1965, Studies on holothurian coelomocytes. II. The origin of coelomocytes and the formation of brown bodies, *Biol. Bull.* **128**:102.

Hildemann, W. H., and Benedict, A. A. (eds.), 1975, *Immunologic Phylogeny*, Adv. Exp. Med. Biol. **64**, Plenum Press, New York.

Hildemann, W. H., and Clem, L. W., 1971, Phylogenetic aspects of immunity, *Prog. Immunol.* **I**:1305.

Hildemann, W. H., and Cooper, E. L. (eds.), 1970, Phylogeny of transplantation reactions, *Transplant. Proc.* **2**:179.

Hildemann, W. H., and Cooper, E. L., 1977, Phylogeny, *Prog. Immunol.* **III**:138.

Hildemann, W. H., and Reddy, A. L., 1973, Phylogeny of immune responsiveness: Marine invertebrates, *Fed. Proc.* **32**:2188.

Hildemann, W. H., and Uhlenbruck, G., 1974, Invertebrate immunology, *Prog. Immunol. II* **2**:292.

Holland, N. D., Phillips, J. H., Jr., and Giese, A. C., 1965, An autoradiographic investigation of coelomocyte production in the purple sea urchin (*Strongylocentrotus purpuratus*), *Biol. Bull.* **128**:259.

Hood, L., Campbell, J. H., and Elgin, S. C. R., 1975, The organization, expression, and evolution of antibody genes and other multigene families, *Annu. Rev. Genet.* **9**:305.

Hostetter, R. K., and Cooper, E. L., 1972, Coelomocytes as effector cells in earthworm immunity, *Immunol. Commun.* **1**:155.

Hostetter, R. K., and Cooper, E. L., 1973, Cellular anamnesis in earthworms, *Cell. Immunol.* **9**:384.

Hostetter, R. K., and Cooper, E. L., 1974, Earthworm coelomocyte immunity, in: *Contemporary Topics in Immunobiology* (E. L. Cooper, ed.), Vol. 4, pp. 91–107, Plenum Press, New York.

Jerne, N. K., 1971, The somatic generation of immune recognition, *Eur. J. Immunol.* **1**:1.

Jones, K., 1977, The need for a comparative approach to immunobiology, *Dev. Comp. Immunol.* **1**:279.

Jordan, H. E., 1938, Comparative hematology, in: *Handbook of Hematology* (H. Downey, ed.), pp. 700–862, Harper & Row (Hoeber), New York.

Jordan, H., and Reynolds, B. D., 1933, The blood cells of the trematode, *Diplodiscus temperatus, J. Morphol.* **55**:119.

Karp, R. D., 1976, Specific immunoreactivity in echinoderms, in: *Phylogeny of Thymus and Bone Marrow-Bursa Cells* (R. K. Wright and E. L. Cooper, eds.), p. 27, Elsevier/North-Holland, Amsterdam.

Kindred, J. E., 1929, The leucocytes and leucocytopoietic organs of an oligochaete, *Pheretima indica* (Horst), *J. Morphol.* **47**:435.

Klein, J., 1977, Evolution and function of the major histocompatibility system: Facts and speculation, in: *The Major Histocompatibility System in Man and Animals* (D. Gotze, ed.), pp. 340–378, Springer-Verlag, Berlin.

Kolb, H., 1977, On the phylogenetic origin of the immune system: A hypothesis, *Dev. Comp. Immunol.* **1**:193.

Kollmann, M., 1908, Recherches sur les leucocytes et le tissue lymphoïde des Invertébrés, *Ann. Sci. Nat. Zool. Biol. Anim.* Ser. 9 **8**:1.

Kubo, R. T., Zimmerman, B., and Grey, H. M., 1973, Phylogeny of immunoglobulins, in: *The Antigens* (M. Sela, ed.), pp. 417–477, Academic Press, New York.

Leclerc, M., Redziniak, G., Panijel, J., and El Lababidi, M., 1977a, Reactions induced in vertebrates by invertebrate cell suspensions. I. Specific effects of sea star axial organ cells injection, *Dev. Comp. Immunol.* **1**:299.

Leclerc, M., Redziniak, G., Panijel, J., and El Lababidi, M., 1977b, Reactions induced in vertebrates by invertebrate cell suspensions. II. Non-adherent axial organ cells as effector cells, *Dev. Comp. Immunol.* **1**:311.

Lemmi, C. A., 1975a, Tissue graft rejection mechanisms in the earthworm *Lumbricus terrestris*: Specific induction of coelomocyte proliferation, *Diss. Abstr. Int. B* **10**:4960.

Lemmi, C. A., 1975b, Specific induction of coelomocyte proliferation in earthworms, *Anat. Rec.* **181**:409.

Lie, K. J., and Heyneman, D., 1977, Studies on resistance in snails: Interference by non-irradiated echinostome larvae with natural resistance to *Schistosoma mansoni* in *Biomphalaria glabrata, J. Invertebr. Pathol.* **29**:118.

Manning, M. J., and Turner, R. J., 1976, *Comparative Immunobiology*, p. 184, Wiley, New York.

Marchalonis, J. J. (ed.), 1976, *Comparative Immunology*, p. 470, Blackwell, Oxford.

Marchalonis, J. J., 1977, *Immunity in Evolution*, Arnold, London.

Marchalonis, J. J., and Warr, G. W., 1978, Phylogenetic origins of immune recognition: Naturally occurring DNP-binding molecules in chordate sera and hemolymph, *Dev. Comp. Immunol.* **2**:443.

Marks, D. H., Stein, E. A., and Cooper, E. L., 1979, Chemotactic attraction to foreign tissue, *Dev. Comp. Immunol.* **3**:277.

McKay, D., and Jenkin, C. R., 1970, Immunity in the invertebrates: The role of serum factors in phagocytosis of erythrocytes by haemocytes of the fresh-water crayfish, *Parachaeraps bicarinatus, Aust. J. Exp. Biol. Med. Sci.* **48**:139.

Nappi, A. J., 1978, Immune reactions of invertebrates to foreign materials, in: *Animal Models of Comparative and Developmental Aspects of Immunity and Disease* (M. E. Gershwin and E. L. Cooper, eds.), pp. 15–24, Pergamon Press, Elmsford, N.Y.

Overton, J., 1966, The fine structure of blood cells in the ascidian *Perophora viridis, J. Morphol.* **119**:305.

Panijel, J., and Liacopoulos, F. (eds.), 1972, *Phylogenic and Ontogenic Study of the Immune Response and Its Contribution to the Immunologic Theory*, INSERM, Paris.

Parish, C. R., 1977, Simple model for self–non-self discrimination in invertebrates, *Nature (London)* **261**:711.

Pérès, J. M., 1943, Recherches sur le sang et les organes neuraux des tuniciers, *Ann. Inst. Oceanogr. (Monaco)* **21**:229.

Prenant, M., 1922, Recherches sur le parenchyme des plathelminthes: Essai d'histologie comparée, in: *Librairie Octave Doin* (G. Doin, ed.), p. 174, Paris.

Prendergast, R. A., and Suzuki, M., 1970, Invertebrate protein stimulating mediators of delayed hypersensitivity, *Nature (London)* **227**:277.

Rabinovitch, M., and De Stefano, M. J., 1970, Interactions of red cells with phagocytes of wax-moth (*Galleria mellonella* L.) and mouse, *Exp. Cell Res.* **59**:272.

Rasmussen, T., and Boman, H. G., 1977, The assay and the specificity problem in insect immunity, in: *Developmental Immunobiology* (J. B. Solomon and J. D. Horton, eds.), pp. 83–90, Elsevier/North-Holland, Amsterdam.

Reinisch, C. L., and Bang, F. B., 1971, Cell recognition: Reaction of the sea star, *Asterias vulgaris*, to the injection of amoebocytes of sea urchin, *Arbacia punctulata*, *Cell. Immunol.* **2**:496.

Renwrantz, L. R., and Cheng, T. C., 1977, Identification of agglutinin receptors on hemocytes of *Helix pomatia*, *J. Invertebr. Pathol.* **29**:88.

Roch, P., and Valembois, P., 1977, Physiological heterogeneity and cellular differentiation of earthworm leukocytes studied by concanavalin A, in: *Developmental Immunobiology* (J. B. Solomon and J. D. Horton, eds.), Elsevier/North-Holland, Amsterdam.

Roch, P., and Valembois, P., 1978, Evidence for concanavalin A receptors and their redistribution on lumbricid leukocytes, *Dev. Comp. Immunol.* **2**:51.

Roch, P. H., Valembois, P., and Du Pasquier, L., 1975, Response of earthworm leukocytes to concanavalin A and transplantation antigens, *Adv. Exp. Med. Biol.* **64**:44.

Rothenberg, B. E., 1978, The self-recognition concept: An active function for the molecules of the MHC based on the complementary interaction of protein and carbohydrate, *Dev. Comp. Immunol.* **2**:23.

Schreffler, D. C., David, C. S., Passmore, H. C., and Klein, J., 1971, Genetic organization and evolution of the mouse H-2 region: A duplication model, *Transplant. Proc.* **3**:176.

Simpson, G. G., 1960, *The Meaning of Evolution: A Study of the History of Life and of Its Significance for Man*, p. 365, Yale University Press, New Haven.

Sloan, B., Yocum, C., and Clem, L. W., 1975, Recognition of self from non-self in crustaceans, *Nature (London)* **258**:521.

Solomon, J. B., and Horton, J. D. (eds.), 1977, *Developmental Immunobiology*, Elsevier/North-Holland, Amsterdam.

Stang-Voss, C., 1974, On the ultrastructure of invertebrate hemocytes: An interpretation of their role in comparative hematology, in: *Contemporary Topics in Immunobiology*, Vol. 4 (E. L. Cooper, ed.), p. 65, Plenum Press, New York.

Theodor, J. L., and Acton, R. T., 1974, Primitive recognition systems, *Prog. Immunol. II* **2**:287.

Toupin, J., and Lamoureaux, G., 1976a, Coelomocytes of earthworms: The T-cell-like rosette, *Cell. Immunol.* **26**:127.

Toupin, J., and Lamoureaux, G., 1976b, Mitogen responsiveness of *Lumbricus terrestris* coelomocytes, in: *Phylongeny of Thymus and Bone Marrow-Bursa Cells* (R. K. Wright and E. L. Cooper, eds.), pp. 19–25, Elsevier/North-Holland, Amsterdam.

Tyson, C. J., and Jenkin, C. R., 1974, The cytotoxic effect of haemocytes from the crayfish (*Parachaeraps bicarinatus*) on tumor cells of vertebrates, *Aust. J. Exp. Biol. Med. Sci.* **52**:915.

Uhlenbruck, G., and Steinhausen, G., 1977, Tridacnins: Symbiosis-profit or defense purpose?, *Dev. Comp. Immunol.* **1**:183.

Vaith, P., Müller, W. E. G., and Uhlenbruck, G., 1979a, On the role of D-glucuronic acid in the aggregation of cells from the marine sponge *Geodia cydonium*, *Dev. Comp. Immunol.* **3**:259.

Vaith, P., Uhlenbruck, G., Muller, W. E., and Holz, G., 1979b, Sponge aggregation factor and sponge hemagglutinin: Possible relationships between two different molecules, *Dev. Comp. Immunol.* **3**:399.

Valembois, P., and Roch, P., 1977, Identification par autoradiographie des leucocytes stimulés à la suite de plaies ou de greffes chez un ver de terre, *Biol. Cell.* **28**:81.

Vethamany, V. G., and Fung, M., 1972, The fine structure of coelomocytes of the sea urchin *Strongylocentrotus dröbachiensis*, *Can. J. Zool.* **50**:77.

Waksman, B. H., 1979, Overview: Biology of the lymphokines, in: *Biology of the Lymphokines* (S. Cohen, E. Pick, and J. J. Oppenheim, eds.), pp. 585–616, Academic Press, New York.

Warr, G. W., 1978, On T cells in invertebrates, *Dev. Comp. Immunol.* **2**:555.

Weiss, D. W., 1976, Central problems in tumor immunology (Introduction and Preface), *Isr. J. Med. Sci.* **12**:281.

Wright, R. K., 1976, Phylogenetic origin of the vertebrate lymphocyte and lymphoid tissue, in: *Phylogeny of Thymus and Bone Marrow-Bursa Cells* (R. K. Wright and E. L. Cooper, eds.), Elsevier/North-Holland, Amsterdam.

Wright, R. K., and Cooper, E. L., 1975, Immunological maturation in the tunicate *Ciona intestinalis*, *Am. Zool.* **15**:21.

Wright, R. K., and Cooper, E. L. (eds.), 1976, *Phylogeny of Thymus and Bone Marrow-Bursa Cells*, p. 325, Elsevier/North-Holland, Amsterdam.

Cellular Defense Systems of the Porifera

IAN S. JOHNSTON and WILLIAM H. HILDEMANN

1. SPONGES AND SPONGE CELL TYPES

Sponges are diploblastic acoelomate Metazoa. They are sedentary, filter-feeding animals which utilize a layer of flagellated cells to pump a unidirectional water current through themselves. They are found in freshwater, but more abundantly in marine habitats. Sponges have been persistent throughout geological time from the Precambrian to the Recent, with special success during the Paleozoic. They are apparently the most primitive multicellular animals on a phylogenetic scale ranked by morphological complexity, although the levels of physiological and biochemical complexity found in sponges easily measure up to the degree of sophistication found in so-called higher animals. The Porifera (sponges) and Coelenterata are related as two phyla representing distinct stocks, but stemming from a presumed common although presently unknown origin (Hyman, 1940).

A generalized demosponge body plan is illustrated in Figure 1. The body openings, chambers, and canals constitute the aquiferous system. Although these animals do not possess a coelom or an enclosed internal vascular system, the water current pumped through the aquiferous system subserves feeding, respiratory, and excretory functions. The aquiferous system and the outside body wall of a sponge are lined with a thin pavement epithelium or pinacoderm (Bagby, 1970), except in the pumping chambers which are lined with spherical or cuboidal flagellated choanocytes (Brill, 1973). The internal body space lined by choanocytes and pinacocytes is called the mesohyl, or alternatively the mesogloea or mesenchyme; it contains a loosely organized mass of independent cells and an extracellular matrix. There are great variations between sponges in the proportions of different types of mesohyl cells, and in the ratio between total

---

IAN S. JOHNSTON • Department of Biology, Northwestern College, Orange City, Iowa 51041.   WILLIAM H. HILDEMANN • Department of Microbiology and Immunology, School of Medicine, University of California, Los Angeles, California 90024. Research work and manuscript preparation supported by National Institutes of Health Grant AI 15705.

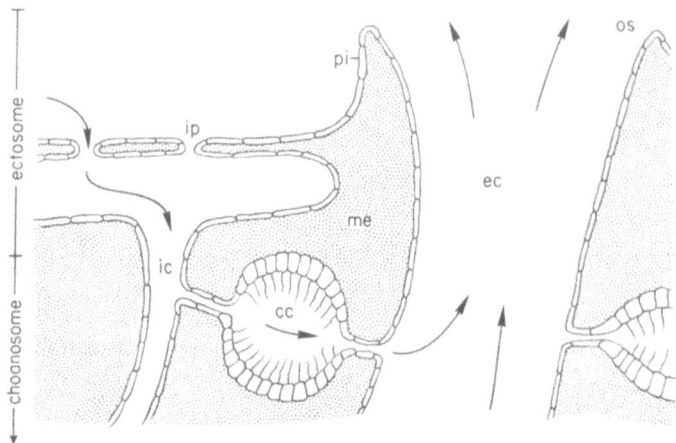

FIGURE 1. Schematic diagram of a section through a portion of a demosponge. The arrows indicate the direction of flow of water through the aquiferous system, beginning at the incurrent pores (ip) or ostia, traversing a subdermal space, down into an incurrent canal (ic), through a choanocyte chamber (cc) where the motive force is applied by flagellated choanocytes, out into an excurrent canal (ec), and finally out of the sponge via the osculum (os). The outer body wall and canals are lined with pinacocytes (pi). The stippled area represents the loosely organized mesohyl (me).

cellular volume and extracellular matrix volume. We have recently published an account of the basic morphology of the marine demosponge *Callyspongia diffusa* (Johnston and Hildemann, 1982).

The cells of the mesohyl have been characterized morphologically, but their functions are in some cases not well understood (Cheng *et al.*, 1968a; Levi, 1970). The mesohyl may contain skeletal structures ranging from a collagenous matrix, through hardened fibrous scaffolding structures composed of a form of collagen called spongin, and finally, fully mineralized spicules consisting of either silica or calcium carbonate. The cells presumed responsible for the elaboration of these structures are as follows: loose collagen is secreted by collencytes and lophocytes (Pavans de Ceccatty and Garrone, 1971); spongin is secreted by spongocytes (Garrone and Pottu, 1973); and the spicules are formed by sclerocytes (Simpson and Vaccaro, 1974). There are contractile cells or myocytes in the mesohyl, found principally around the oscula and ostia. Another group of mesohyl cells are those that contain distinct inclusions (e.g., Donadey and Vacelet, 1977). These include: the "gray cells," thought to be capable of synthesis and accumulation of glycogen and responsible for its transfer to sites of most intense metabolism (Boury-Esnault, 1977); the spherulous and rhabdiferous cells variously thought to be responsible for the synthesis of a sulfated acid mucopolysaccharide that is a part of a ground substance dispersed between the collagenous matrix of the mesohyl (Donadey and Vacelet, 1977); spherulous cells with chromolipid vesicles, observed close to the walls of excurrent canals and thought to serve an excretory function since they are able to migrate from the mesohyl into the excurrent water flume; and microgranular and globiferous cells which contain a variety of paracrystalline and granular deposits. All of these cells, which are described in detail by Levi (1970) and by Bergquist (1978, pp. 70–74), are thought to be terminally differentiated from another population of mesohyl cells called archeocytes.

Archeocytes tend to be large cells, often conspicuously filled with phagosomes and lysosomal structures. They are very mobile, and are most easily characterized by a large vesicular nucleus and prominent nucleolus. They are thought to be totipotent cells (Bergquist, 1978, p. 69) from which not only the mesohyl cells but all the cells of a sponge may be differentiated; hence, another name given to them is "polyblasts" (Paris, 1961). The term "amoebocyte" (e.g., Hyman, 1940) has been used to describe these cells, but this is somewhat misleading since many of the differentiated cells are also amebocytic, or they may be induced to be so. Besides their role in development and morphogenesis, the archeocytes are also presumed to be the major phagocytes carrying out digestive and excretory functions. Archeocytes have been called "the macrophage of the sponge" (Bergquist, 1978, p. 70), and they have been broadly implicated in the cellular defense mechanisms to be discussed in the rest of this chapter.

2. GENERAL DEFENSE MECHANISMS

The repertoire of instantaneous behavioral responses in adult sponges appears to be restricted to modification of water flow in the aquiferous system. An increase or decrease in overall pumping rate may correlate with the metabolic and nutritive requirements of the sponge. In some sponges, sudden transient reversal of water flow flushes debris clogging incurrent openings (cf. Reiswig, 1971a). The contractility of cells lining the incurrent and excurrent openings allows rapid closing of these apertures. This coordinated behavior, which is presumably subserved by chemo- and mechanoreceptors, may well prevent the access of unwanted materials or organisms into the sponge.

Morphogenetic response to irritants may involve drastic rerouting of the aquiferous systems, e.g., cellular reorganization permits the closing off of an osculum and its reconstruction elsewhere. Internal reorganization may represent another strategy for flushing debris if inhalant canals are transformed to exhalant canals. Reiswig (1971a) speculated that this was occurring during periods of cessation of pumping in *Tethya crypta* and *Verongia gigantea*. This plasticity of tissue organization is also apparent following surgical manipulations, in which case rapid recovery is noted during wound healing (Connes, 1966; Egami and Ishii, 1956). We have observed complete regeneration of a pinacoderm and oscula, on the cut surface of an excised 1-cm-diameter finger of *Callyspongia diffusa*, within 2 to 4 hr at 27°C.

Wound healing is apparently subserved by infiltration and large-scale migration of amebocytic archeocytes into the damaged area of the sponge (Paris, 1961; Evans and Curtis, 1979). In the freshwater sponge *Corvomeyenia carolinensis*, Harrison (1972) showed that pinacocytes became actively ameboid to initiate the regenerative process of wound healing.

3. DEFENSE AGAINST PREDATORS AND COMPETITORS

Sponges rely principally on biochemical interactions rather than on overt behavioral responses to discourage or damage predators and competitors. Sponge natural product chemistry is currently a fruitful and fascinating field (Faulkner, 1977), especially in the search for antimicrobials (see Section 4), but little is known about the biological role of many of the peculiar molecules discovered. One speculation is that some of these exotic molecules, being noxious chemical substances, may be highly effective in discouraging predation (Bakus and Green, 1974). For example, Dayton *et al.* (1974) described a lack of predation by starfish on a common antarctic sponge, *Cinachyra antarctica*, in a community where other similar sponges were heavily preyed upon. Bergquist (1978, p. 188) has subsequently demonstrated that several species of *Cinachyra* produce toxic terpenoid compounds that could mediate chemical defense against the starfish. Together with the presence of tough fibrous skeletons and spicules, a strategy of chemical defense may account for low levels of predation on tropical sponges by fish. Randall and Hartman (1968) discovered that only 11 of 212 species of West Indian reef and inshore fish were significant sponge predators (the criterion for significant predation was the presence of sponge material comprising 6% or more of the fishes' stomach contents) even though sponges comprised a large proportion of the total biomass present in such communities.

In benthic marine communities, especially in hard-bottom environments, space on which to live is a most important limiting resource. Therefore, sponges together with other sessile marine organisms have developed interference mechanisms to compete for this space. One such strategy, which is highly developed in plants, is the release of so-called "allelochemicals" into the immediate environment (cf. Muller, 1970). Such compounds may prevent overgrowth of a sponge by adjacent organisms (or by larval settling), either by altering behavior, by altering the direction of growth of the invading organism, or even by afflicting pathological effects (Jackson and Buss, 1975). There is currently no information on the cellular origins of endogenous toxins or of exogenously released allelochemicals in sponges, but Jackson and Buss have speculated that secretion together with mucus may provide a mechanism for concentration of allelochemicals in the immediate vicinity of a sponge without excessive dilution by water currents.

Some sponges can enter into relatively stable and complex epibiotic (i.e., one organism living on another) relationships with other species of sponge as well as other benthic organisms (Rützler, 1970; Sara, 1970). Intimate physical contact of xenogeneic tissues might be expected to cause antagonistic cytotoxic responses as is the case for some allogeneic interactions (see Section 6.2). The species specificity of some epibiotic associations implies that recognition processes are involved, but overt cellular antagonistic responses have yet to be described in epibiosis. Antagonism may either be suppressed in certain xenogeneic combinations or effectively blocked by inert extracellular secretions that prevent persistent cell-to-cell contact. An example of a physical barrier formed between adjacent individual sponges has been described as a result of allogeneic tissue contact in *Ephydatia fluviatilis* (Van de Vyver, 1970).

4. DEFENSE AGAINST PATHOGENS AND FOREIGN-BODY INVASION

Our understanding of the incidence, let alone the cause of sponge diseases is rather sparse, although a recent review by Lauckner (1980) brings together some of the relevant information. Shallow-water commercial cultivation of sponges was practiced in the Bahamas, Gulf of Mexico, and Caribbean in the first part of this century, until several bouts of "wasting disease" (Galtsoff *et al.*, 1939) destroyed most of the sponge stocks. One such episode was apparently caused by a water-current-borne infectious agent that was traced from the Bahamas to Tarpon Springs, Florida, and on to Teneriffe in British Honduras via the Gulf Stream Counter Current (Walton-Smith, 1941). The symptoms of the disease were foci of decay, first apparent in the deep choanosome and later spreading to the ectosome. The only detectable histopathological feature was the presence of fungallike filaments in and close to the affected areas of each sponge. Some of the infected sponges were able to stabilize themselves and prevent the spread of the disease by separating off a pocket of dead and dying tissue with a callouslike wall. This kind of tissue reaction is by no means restricted to this disease, but rather it is a fundamental strategy of wound healing (Paris, 1961; Connes, 1966). When large volumes of tissue are dying, the siliceous sponge *Tethya lyncurium* appears to isolate small masses of multipotent archeocytes out of the mass of decaying tissue (Connes, 1967), a strategy somewhat similar to gemmule formation in freshwater sponges.

Extracts of sponges, or even pieces of excised sponge tissue sometimes show specific antimicrobial activity against test organisms *in vitro* (see review by Burkholder, 1973). Antibiotics isolated from sponges are derived from the sponges themselves rather than from their internal bacterial flora (Jakowska and Nigrelli, 1960; see Section 5). The cellular origins of these antimicrobials are unknown but extractions of the deep choanosome are usually less reactive than those from the superficial ectosome (Connes, 1967). This correlates with the lower susceptibility of the ectosome to foci of disease decay and leads to the presumption that sponges produce compounds that are biologically active in suppressing the growth of otherwise pathogenic organisms. Intense interest in sponge natural product chemistry is encouraged by the fact that some of these compounds are highly effective against human pathogens; e.g., renierone, an isoquinoline quinone isolated from *Reniera* sp., has antibiotic activity against *Staphylococcus aureus*, *Bacillus subtilis*, and *Candida albicans* (McIntyre *et al.*, 1979).

The fundamental particle-feeding behavior of sponges is presumed to be subserved by phagocytosis (Reiswig, 1971b). Small bacteria-sized particles are apparently phagocytosed by choanocytes, whereas larger particles, unable to penetrate the openings of the choanocyte chambers, are captured by amebocytic cells found on the walls of the incurrent canals. Certain pinacocytes may also show phagocytic properties (Harrison, 1972) as do the archeocytes of the mesohyl. It is not surprising then that the first-line defense against foreign-body invasion involves widespread phagocytosis. Injection of India ink or carmine particles into the mesohyl of *Terpios zeteki* (Cheng et al., 1968b) resulted in: phagocytosis, "principally by archeocytes and to a lesser extent by collencytes"; the migration of particle-laden cells toward excurrent canals; the passage of these cells through the pinacoderm lining the excurrent canals; and the discharge of these cells from the sponge in the excurrent water flume. When xenogeneic biological materials such as human red blood cells and trematode cercariae were injected into this same sponge (Cheng et al., 1968c), smaller particles were phagocytosed and larger ones effectively encapsulated by archeocytes and collencytes. Intracellular digestion of phagocytosed red blood cells was evident. Both single archeocytes each enclosing a red cell, and small encapsulation complexes involving several archeocytes surrounding a foreign body, sometimes migrated into the excurrent canals and were thereby eliminated from the sponge. Among the different cell types present in a sponge, archeocytes appear to play the most conspicuous role in responses to foreign bodies.

Another strategy for the elimination of foreign bodies is said to involve the physical motion of spicules. Connes (1967) described the accumulation of spicules around a mass of bacteria and their subsequent migration to the surface of the sponge effectively "sweeping" the bacteria out.

Cheng et al. (1968b) noted that carmine and ink particles became clumped prior to their engulfment by archeocytes. This led to the speculation that sponges may synthesize a factor (or factors) responsible for extracellular clumping, which subsequently triggers mass phagocytosis and elimination. Agglutinating factors have been described from a variety of invertebrate animals, and although they are not thought to be produced by mechanisms equivalent to antibody production, they are claimed to be active in parasite immobilization, bactericidal and antiviral functions, and other defense functions (see symposium proceedings edited by Cohen, 1974).

While investigating the experimental reaggregation of dissociated sponge cells (see Section 6.1), Galtsoff (1929) noted a lytic and heteroagglutinating effect of extracts of one sponge on the cells of individuals of another sponge species. Such cytolytic effects were confirmed by Van de Vyver (1975) and the heteroagglutinating effects have also been noted by others (e.g., Spiegel, 1955). The latter effects were presumed to coincide with the ability of sponge extracts to cause the agglutination of red blood cells (MacLennan and Dodd, 1967). Purified hemagglutinins from sponges (of which the heteroagglutinins could be a subset) are lectins, each of which has a definable carbohydrate specificity with a binding site complementary to a monosaccharide (or disaccharide) terminal unit of an oligosaccharide chain. From *Axinella polypoides*, for example, Bretting and Kabat (1976) isolated two hemagglutinins with specificity for terminal nonreducing D-galactose, glycosidically linked $\beta1\rightarrow6$. These particular lectins are not present as cell-membrane-bound molecules and are therefore unlikely to participate in normal cell-to-cell contact and recognition. Instead, they are found in secretory vacuoles within spherulous cells of the mesohyl (Bretting and Konigsmann,

1979). In an unnamed species of *Axinella*, MacLennan (1974) described a similar galactose-specific component that had no effect when assayed for agglutination with conspecific sponge cells. More recently, a lectin isolated from *Geodia cydonium* has been shown to be totally unreactive with polysaccharide preparations from the same sponge (Vaith *et al.*, 1979b). These kinds of observations bear on the speculation that sponge agglutinins are specific for non-self markers (MacLennan, 1974) at least at the level of xenogeneic interactions. If this is so, and if the non-self markers happen to be present for example on pathogenic microorganisms, then this could provide the basis for at least the recognition phase of a foreign-body defense mechanism. Besides the agglutinins, there is another family of specific recognition molecules in sponges, called aggregation factors. These will be considered further in Section 6.1, but it is appropriate to mention that a fundamental distinction between the two is the binding specificity of aggregation factors for self markers at the levels of syngeneic and allogeneic interactions. It is noteworthy that "mucoid" cells from *Iricinia muscarum*, which might well have been an enriched population of spherulous cells, were specifically able to inactivate the aggregation factor of *G. cydonium* (Müller *et al.*, 1976a). Thus, there may be a potential for direct interaction between agglutinin and aggregation factors derived from heterologous sources.

5. INTERACTION WITH SYMBIONTS AND COMMENSALS

Sponges commonly contain a substantial and specific flora and fauna. The associated organisms consist of bacterial, cyanobacterial, and algal symbionts, commensal invertebrates, and even commensal vertebrates, e.g., fish (Tyler and Böhlke, 1972). Bacteria and cyanobacteria are found both free in the matrix of the mesohyl and also in vacuoles within specialized cells of the mesohyl, i.e., bacteriocytes and cyanocytes. In certain sponges, bacteria may occupy up to 40% of the mesohyl volume (Vacelet, 1975). Morphological data imply that only a limited and quite specific number of bacterial species are present in any given sponge. The associations are stable with no evidence of either seasonal or intraspecific variation (Wilkinson, 1978). Different types of bacteria, or even bacteria and cyanobacteria together, can coexist in the same cell vacuoles (Vacelet and Donadey, 1977). Occasionally, bacteria are seen in phagocytic vacuoles in archeocytes, and Wilkinson (1979) has especially noted the engulfment into secondary lysosomes of cyanobacteria infected with bdellovibriolike parasites. However, the bacterial symbionts are not thought to contribute significantly to the nutrition of the host sponge by their wholesale digestion, but rather by some kind of dynamic mutual exchange of metabolites (cf. Smith *et al.*, 1969). Nitrogen fixation in coral reef sponges that house cyanobacteria (Wilkinson and Fay, 1979) may represent a significant nutritional input not only to the sponges, but also to the whole reef ecosystem. The obligate nature of some of these associations is especially evident in the case of sponges that host cyanobacteria, which have an absolute requirement for light (Wilkinson and Vacelet, 1979).

Intracellular associations, between freshwater spongillids and symbiotic chlorellae, and between certain marine sponges and symbiotic dinoflagellates (zooxanthellae), are also presumed to confer mutual benefit by the exchange of photosynthetic products and other metabolites (Muscatine et al., 1967; Sara and Vacelet, 1973). Algal symbionts are found within cells that are morphologically equivalent to archeocytes (Cheng et al., 1968a).

In considering the associations so far described, there is an implicit assumption that the sponge is able to control the density and composition of its internal symbiont community. This could be carried out by cellular means, e.g., the engulfment and elimination of excess symbionts or of individuals of an inappropriate species, and/or by the action of specific antimicrobial factors (see Section 4).

A variety of animals covering most invertebrate phyla inhabit the larger canals of sponge aquiferous systems. A commensal infauna is dependent on the architecture and physiological properties of the sponge and therefore will show species specificity. The most common commensals are annelids and crustaceans, and Bergquist (1978, p. 197) mentions a single specimen of *Spheciospongia vesparia* from Florida that was found to contain over 16,000 alpheid shrimps.

Occasionally, an organism may irritate the host sponge and cause it to build a thickened wall around the chamber or canal housing the commensal. Such a reaction, seen for example between the sponge *T. lyncurium* and the amphipod *Leucothoe spinicarpa* (Connes, 1967), is subserved by an influx of amebocytic archeocytes. This outcome is very similar indeed to the walling off of internal parasites, as in the case of a parasitic copepod found in *Suberites domuncula* (Tuzet and Paris, 1964).

6. MAINTAINING TISSUE AND GENETIC INTEGRITY

6.1. SPONGE CELL AGGREGATION

H. V. Wilson's (1907) classical studies on "coalescence and regeneration" in sponges spawned a whole new research area in cell-to-cell recognition and adhesion. Wilson showed that whole sponges can reform from suspensions of their cells, and if cells from two species are intermingled, the resulting regenerated sponge individuals each consist of cells from only one species-type. These and subsequent observations on species- and strain-specific reaggregation of dissociated sponge cells have been applied to the interpretation of tissue and organ specificity in morphogenetic cell sorting during embryogenesis (e.g., Moscona, 1968; Curtis, 1974, 1978; Burger et al., 1975). Furthermore, the genetic integrity of an individual sponge and the tissue integrity of its overall body structure may well depend on the same recognition and adhesion properties that can be investigated during cell reaggregation. Therefore, a brief overview of cell behavior in aggregation is relevant to cellular defense mechanisms that promote the integrity of the individual. The molecular basis of sponge cell aggregation is certainly not yet fully understood (see reviews by Müller et al., 1978a; Burger et al., 1978).

At least three distinct stages are recognizable during cell reaggregation: (1) initial contact; (2) cellular adhesion; and (3) histiotypic rearrangement of cells with respect to one another to form new sponge individuals. Species-specific sorting-out, in mixtures of cells from different sponges, may complicate things at any or all stages of reaggregation, although the existence in some cases of transient bispecific aggregates (Sara, 1968) implies that it may sometimes be delayed until the third stage.

Humphreys (1963) isolated a discrete soluble molecular "aggregation factor" that was presumed to serve as an adhesive ligand between adjacent homospecific cells, i.e., as an intercellular bridge consisting either of a single symmetrical unit, or of two or more symmetrical units (e.g., Burger et al., 1975).

Since 1963, aggregation factors have been extensively characterized as either small proteins or large proteoglycans, depending on the biochemical techniques used to isolate them and the sponge being investigated. In G. cydonium, for example, reaggregation is mediated by an intercellular, high-molecular-weight particle (Müller and Zahn, 1973), which consists of an annular core, with a contour length of 350 nm, from which about 25 filamentous arms radiate. In addition to multivalent ligand activity (aggregation factor) which is presumed to reside in the arms, the particle also has the properties of a multiglycosyltransferase system (Müller et al., 1978b). Such complex particles are easily visualized at the level of resolution of the electron microscope when isolated from in vitro preparations (e.g., Humphreys et al., 1977) but they have eluded in situ observations at cell-to-cell adhesions (Evans and Bergquist, 1974; Kartha and Mookerjee, 1979). Aggregation factor associates specifically with a membrane-bound aggregation-receptor molecule (Müller et al., 1976b), also named "baseplate" (Weinbaum and Burger, 1973), and the association seems to be dependent on a lectin binding to a glycoprotein, although the polarity is not always clear (Vaith et al., 1979a). Very simply then, aggregation factor and baseplate are thought to be together responsible for specific cellular adhesion. But as already noted in some combinations, mixtures of cells from separate sponge species form initial aggregates that are at least transiently bispecific (see data compiled by MacLennan, 1970, Tables II and III). Furthermore, certain factors appear to stimulate aggregation of xenogeneic cells, leaving the impression that there may be some cross-specificity between aggregation factors from different sponges. Observations on the reaggregation of bispecific mixtures, or even of the effects of extracts of one sponge on the dissociated cells of another sponge, cannot simply be interpreted in terms of more-or-less specific interactions between aggregation factors and baseplates alone. Certainly, the "heteroagglutinins," which are thought to be independent of aggregation factors (see Section 4), will complicate these observations. Extracts from sponges very often inhibit the normal aggregation of allogeneic or xenogeneic cells (Curtis and Van de Vyver, 1971; McClay, 1974). While this could represent another manifestation of the heteroagglutinins, Curtis (1974) went on to propose that there was yet another family of factors, or "morphogens," that were solely responsible for the reduction of adhesiveness of heterologous cells.* Curtis proposed that con-

---

*It should be noted that Curtis uses a viscometric assay to measure the probability of cells adhering to each other upon collision, i.e., a measure of their adhesiveness which is presumed to represent their ability to form aggregates (Curtis, 1969). Most other workers directly measure the time-dependent increase in size of cellular aggregates (e.g., Müller and Zahn, 1973).

centration gradients of such a morphogen, set up around a developing aggregate, would prevent the incorporation of a heterologous cell into that aggregate, and might actually cause the disintegration of nearby aggregates made up of heterologous cells. McClay (1974) isolated similar inhibitory components from each of a number of different sponges and he speculated that they might be active in substratum competition between individuals or as a defense against territorial invasion, a role described for the allelochemicals mentioned in Section 3.

More recently, Evans and Curtis (1979) have acknowledged that the morphogens, now called "interaction modulation factors," consist of both components that promote aggregation (PAF) and those that inhibit aggregation (IAF). The recognition of a promoter component presumably validates the involvement of aggregation factor in this experimental system. A great value of the work with which Curtis has been associated (Curtis and Van de Vyver, 1971; Curtis, 1979a, 1979b; Evans and Curtis, 1979) is the appreciation that non-self recognition in sponge cell aggregation is apparent not only in xenogeneic combinations, but also in allogeneic combinations. Thus, the IAF from an individual sponge of *Hymeniacidon perleve* caused a reduction of adhesiveness when applied to xenogeneic or allogeneic cells, but it had no effect when applied to syngeneic cells (Curtis, 1979b). On the other hand, a PAF seemed to be effective in enhancing aggregation when applied to both syngeneic and allogeneic cells, but it had no effect when applied to xenogeneic cells. Specific allogeneic effects may have been missed by other workers because factors were isolated from bulk sponge material, potentially derived from multiple allogeneic individuals. This problem may well require the reinterpretation of some experimental results, and should certainly be taken into consideration in future studies.

So, at the very simplest, interaction between cells from different sponge individuals depends on (1) species-specific self recognition (PAFs), (2) allospecific non-self recognition (IAFs), and (3) species-specific non-self recognition (the heteroagglutinins). There may well be a phylogenetic relationship between any two or all of these categories (e.g., MacLennan, 1974; Vaith *et al.*, 1979b), but critical evidence is currently lacking.

Antagonistic interactions between factors are not, however, restricted to cases of allo-reaggregation; they also occur in syngeneic cell reaggregation. Müller *et al.* (1979a) have recently described a second cell-membrane-bound receptor which binds aggregation factor, but in the process abolishes aggregation. This "antiaggregation receptor" is thought to operate during the third stage of reaggregation when cellular adhesions must presumably be labilized in order to allow a random cellular aggregate to be transformed into a morphologically functional sponge. Furthermore, a lectin isolated from *G. cydonium* binds homologous antiaggregation receptor and prevents it from dissociating cell clumps (Müller *et al.*, 1979b). All of these factors must obviously be integrated into any hypothesis which attempts to explain sponge cell aggregation.*

---

*It seems appropriate to point out that to date little attempt has been made to correlate the work of individual research teams. It is certainly true that each group uses different biochemical techniques to isolate their "factors," and that whereas some factors are cell surface macromolecules, others are soluble components, but it is inexcusable that there are often no references at all to the parallel work of other groups.

There is considerable interest in separating sponge cell types to determine whether different cells have different potentialities for recognition and adhesion in reaggregation assays. Separate fractions of dissociated sponge cells, obtained from sucrose step gradients (Leith, 1979), from Ficoll step gradients (Burkart and Burger, 1977), or from continuous Ficoll gradients (DeSutter and Van de Vyver, 1977, 1979), sometimes show quantitative differences in the kinetics of aggregation, but most show specific binding of homologous aggregation factor. The recombination of cell fractions in different proportions results in very different kinds of third-phase aggregates, some of which are nonviable. This has not been attributed to lack of appropriate cell-to-cell recognition, but rather to the fact that some cells are unable to dedifferentiate in order to produce the range of cell types that must be present for a sponge to be functional.

6.2. SPONGE GRAFTING

Besides the sponge cell reaggregation assay, other types of experiments, namely sponge tissue grafting, have also shown that specific recognition and subsequent specific interactions occur when sponge individuals contact each other. For the purpose of this review, these interactions are again presumed to be expressions of cellular defense mechanisms that preserve the genetic integrity of the individual. Sponge grafts may occur naturally or be artificially manipulated. Natural grafts occur when sponge larvae settle out in contact with each other or when gemmules hatch in contact. The sessile benthic nature of adult sponges often leads to eventual contact between adjacent sponges as a consequence of simple growth processes. Therefore, the investigation of artificially manipulated parabiotic or tissue transplantation reactions is highly relevant to the biology of sponges.

A number of workers have noted incompatibility in orthotopic xenogeneic sponge grafts as expressed principally in the failure of tissue fusion, for example, causing grafts to fall out unless physically held in place. In some instances, Paris (1961) described "partial fusion" of xenografts, but histological examination after 10 days showed a regenerated "neoepithelium" on both sides of the graft interface and a "mucoid barrier" separating the two. Such a mucoid layer may represent a device that effectively isolates the xenogeneic tissues from further contact, and as such it could well be present in stable epibiotic associations that involve sponges (see Section 3). Besides the failure of tissue fusion, Moscona (1968) also noted vacant skeletal areas on both sides of a graft interface between *Microciona* sp. and *Haliclona* sp. He interpreted this as the soft tissues withdrawing away from each other in a manner similar to the segregation that occurs during reaggregation of bispecific mixtures of suspended cells. McClay (1974) found dead tissue in some xenogeneic graft interfaces. This tissue death correlated with his ability to isolate "inhibitory substances" when the same xenogeneic combinations were tested in mixed aggregation assays. In the xenogeneic, but epibiotic, combination of *Hymeniacidon perleve* and *Amphilectus fucorum*, Evans and Curtis (1979) detected tissue nonfusion but no tissue necrosis. Xenogeneic tissues are obviously recognized as non-self, but the ability of sponges to recognize allogeneic tissues as non-self has until recently been controversial.

There are reports of compatible fusion of allogeneic sponge larvae, either prior to metamorphosis or following larval settling (e.g., Borojevic, 1967). Van de Vyver (1970), working with the marine sponge *Crambe crambe*, failed to observe fusion among sexually derived larvae. When gemmules of *E. fluviatilis* hatch in contact, their tissues fuse together if they are derived from parent sponges of the same strain (Van de Vyver, 1970). The separate strains in this freshwater sponge each represent a distinct allotype. When gemmules are of a different allotype, their growing tissues fail to fuse and instead a collagenlike barrier is secreted between the two. This "zone of nonconfluence" apparently serves to cut off previous allogeneic cell contact. Based on Van de Vyver's observations of the frequency of compatible fusion, Du Pasquier (1974) speculated that the degree of polymorphism at histocompatibility loci in sponges might be low. However, the opposite is probably the rule in view of more recent findings with marine sponges.

Simpson (1973) allowed small tissue explants of *Microciona prolifera* to regenerate and grow toward each other on glass slides. True syngeneic isografts always fused together to form a single functional sponge, whereas allogeneic pairs were incompatible and a distinct "line of nonfusion" separating the two explants was always apparent. These observations on natural allografts suggest allogeneic recognition but no antagonistic cytotoxic reactions were reported.

Orthotopic grafting within a single sponge species has provided conflicting data. Moscona (1968) and McClay (1974) both claimed that allogeneic tissues were fully compatible and that such grafts healed into their graft beds without any sign of subsequent alloreactivity. Moscona's "isospecific" *Microciona* grafts, which according to his experimental description were allografts, were observed for 3 weeks, and McClay's "homografts" within each of five different Bermuda sponges were observed for only 16 days. It is quite possible that these time intervals were not long enough for the appearance of alloincompatibility reactions.

In contrast, definite allogeneic incompatibility in orthotopic grafts has been independently described by several groups of investigators (Paris, 1961; Curtis, 1979a; Evans and Curtis, 1979; Hildemann *et al.*, 1979). Paris (1961) showed that allografts in two separate species of Mediterranean sponge were effectively rejected, after an initial period of tissue fusion, by the gradual physical ejection of the graft from the graft bed over a 2-month period. It is interesting to note that Paris achieved these results during the winter months when the seasonal water temperatures were lowest (~ 11°C). At other times of the year when ambient temperatures were above 16°C, the grafts and their hosts both became necrotic very rapidly. Paris attributed this tissue death to a bacterial pathogenesis, but in light of subsequent data it may well have been initiated by cytotoxic alloreactivity. Evans and Curtis (1979) described allograft rejection in *H. perleve* as: proceeding through a period of tissue fusion, followed by tissue separation, and finally, in a limited number of cases, by necrosis at the graft interface. These workers interpret alloreactivity in terms of the effects of "interaction modulation factors," principally the IAFs which are said to cause the individual cells of allogeneic tissues to lose their normal adhesion properties and essentially to fall apart. By this hypothesis, necrosis presumably results from the disruption of normal tissue architecture. It is interesting that in these experiments, allograft incompatibility was correlated with the ability to isolate from the same allogeneic combinations IAFs that caused reduction of adhesiveness in reaggregation assays. Compatible graft combinations produced IAFs that had no inhibitory effects on each other's cells in such assays.

While sampling very limited populations of *H. perleve*, e.g., those individuals living within 1 m of each other on a single rock, Curtis (1979a) noted that about 20% of the graft combinations were fully compatible. This prompted him to speculate that as few as five allotypic strains were present within these populations. It is inappropriate to assume from these results, and from those of Van de Vyver (cited in Du Pasquier, 1974), that limited numbers of allotypes are characteristic of sponges in general. The absence of information about the larger populations from which the sponges used in these experiments were small samples, leaves an inadequate understanding of the overall genetic diversity involved. Among individuals sampled from a limited geographical location, this is especially complicated by the fact that asexual reproduction, e.g., by fragmentation or gemmulation, may co-occur with various types of sexual reproduction. The potential for sexual reproduction ranges from true outbreeding to the possibility of self-fertilization, and within the Porifera there are both monoecious and dioecious species. It would therefore not be surprising to find compatible allotypes in groups of sponges sampled from a single rock, or between randomly chosen larvae derived from a single or limited number of parental sponges.

Our own investigations of allogeneic interactions in sponges (Hildemann *et al.*, 1979, 1980a,b) have used both parabiosis-type grafting of intact sponge fingers and orthotopically sutured grafts. The parabiosis technique mimics naturally occurring contact, as intact pinacoderm surfaces are immobilized in intimate contact with each other (Figure 2). Using a tropical coral reef sponge, *Callyspongia diffusa*, sampled from several different patch reefs from widely separated areas of Kaneohe Bay, Hawaii, we have yet to record a single instance of compatibility among more than 1400 intercolony pairings. Thus, there is extensive allogeneic polymorphism of histocompatibility (H) molecules in this species. We have, however, maximized the chances of incompatibility so far by avoiding grafts between individuals found growing close to each other on the same patch reef. On the other hand, intracolony isografts or isoparabionts are always fully compatible. The course of allograft reactivity begins with a period of limited tissue fusion. Contrary to the observations of Evans and Curtis (1979), after only 4 to 6 hr in contact, this fusion is morphologically quite distinct from that occurring in isografts (Hildemann *et al.*, 1980b). We have termed this transient allogeneic tissue fusion, "tissue bridging," and it may be apparent for up to several days. The cellular basis of tissue bridging is a massive influx into the graft interface of mesohyl cells, identified by Paris (1961) and by Evans and Curtis (1979) as archeocytes. We are presently involved in a more detailed description of this cellular influx, with the goal of identifying potential immu-

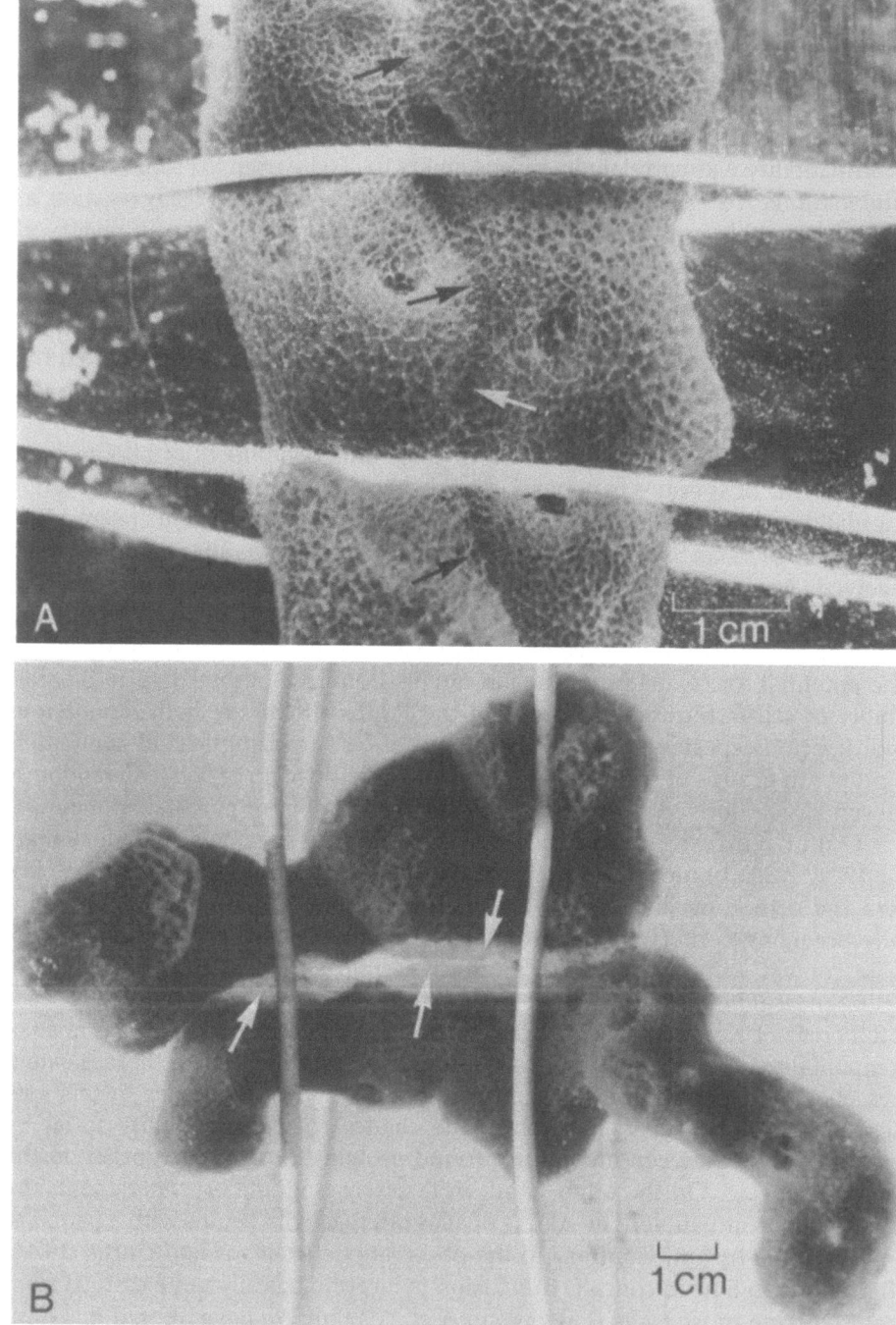

FIGURE 2. Parabiotic reactions between intact fingers of *Callyspongia diffusa* held together by vinyl-covered wire tie-downs on Plexiglas plates. (A) Compatible interfacial fusion (arrows) of syngeneic or isogeneic parabionts after 2 days in contact. (B) Incompatible bilateral cytotoxicity between allogeneic parabionts showing exposed skeletal framework (arrows) after local soft tissue necrosis at 7 days. (From Hildemann et al., 1979.)

nocytes. Whereas in isografts the skeletal frameworks from each piece of sponge are fused across a graft within 24 hr, there is no such skeletal fusion across tissue bridges of an allograft. Tissue bridging is soon followed by the appearance of a whitish hyperplasia either unilaterally or more usually bilaterally at the graft interface. This buildup or mobilization of cells is followed very rapidly by the death and disintegration of tissue in the immediate contact zone, leaving the vacant skeletal network as evidence of the extent of tissue death (Figure 2). Behind each zone of killing, a regenerated pinacoderm can be laid down, leading to a stable "stand off" between the allogeneic tissues.

In order to study the kinetics of alloreactivity, we have chosen an objectively convenient point in the sequence of events just described, that allows us to compare different genetic combinations, and to compare the same combinations under different conditions. Definitive allogeneic cytotoxicity or graft rejection is recorded when 1 mm of vacant skeleton is apparent on either side of a graft interface. This appearance of brown skeletal fibers associated with disappearance of purple soft tissue is reproducibly quantifiable by multiple observers. Such a sharply quantitative reaction endpoint becomes critical in experiments testing for specific sensitization or immune memory. However, as little as 0.1 mm of interfacial tissue death can be scored under optimal conditions of lighting with the aid of a binocular headband magnifier. Table 1 gives some representative data on reaction times (i.e., the number of days in contact before this arbitrary endpoint is reached) for first-set, second-set (repeat grafts), and naive third-party grafts. Very simply, second-set grafts react faster than first-set grafts, whereas third-party grafts yield intermediate median reaction times. This is usually taken to mean that specific alloimmune memory is operative; the fraction of third-party parabionts exhibiting accelerated reactions is attributable to sharing of polymorphic H antigens among the three sources of *Callyspongia* tested. Depending on the particular interclonal combinations of sponge individuals, alloincompatibility reactions may be strong, moderate, or weak in terms of both timing and severity, but within a single combination, replicated grafts react in a very similar manner (Table 2).

Based on these kinds of results, we would argue that sponges fulfil three minimal criteria that allow sponge alloreactivity to be classified with adaptive immunity as recognized in higher animals. These three criteria are: (1) cytotoxic reactions following a period of sensitization; (2) selective or specific reactivity; and (3) inducible memory or selectively altered reactivity on secondary contact. Both *C. diffusa* from Hawaii and *H. perleve* from Scotland (Evans *et al.*, 1980) exhibit highly discriminating alloimmunity with a specific memory component evidenced by selectively accelerated rejection of second-set grafts. We have proposed that allorecognition depends upon polymorphism of cell surface H molecules (Hildemann *et al.*, 1979). Based on the large number of incompatible allografts observed without exception, this polymorphism would indeed appear to be quite extensive in *Callyspongia*. H molecules in general are glycoproteins as are the sponge cell surface molecules that possess serological specificity (MacLennan, 1974). Because the oligosaccharides contributing specificity to these cell surface proteins depend on a family of glycosyltransferases for biosynthesis,

TABLE 1. EVIDENCE OF SPECIFICITY AND MEMORY IN ALLOGRAFT REACTION TIMES OF CALLYSPONGIA DIFFUSA[a]

Series	T (°C)	Median reaction times (MRT) and confidence limits in days[b]						
		First-sets	Pairs scored (No.)	Second-sets[c]	Pairs scored (No.)	Third-party[c]	Pairs scored (No.)	Third-party combination
COCO ⇄ MCAS[d]	24	11.3 (10.5–12.2)	47	7.2 (6.4–8.1)	26	5.6 (4.9–6.4)	11	COCO ⇄ (naive) KYC
COCO ⇄ MCAS	27	9.0 (8.3–9.8)	24	3.8 (3.3–4.4)	10	6.8 (5.9–7.8)	9	MCAS ⇄ (naive) KYC
KYC ⇄ COCO	27	8.9 (7.0–11.4)	30	4.2 (3.6–4.9)	13	5.8 (4.6–7.3)	6	KYC ⇄ (naive) MCAS
KYC ⇄ MCAS	27	7.2 (6.4–8.2)	21	4.0 (3.4–4.8)	11	5.7 (4.3–7.5)	7	COCO ⇄ (naive) MCAS
						4.6 (4.1–5.2)	12	KYC and MCAS to (naive) COCO combined

[a] Adapted from Hildemann et al. (1979).
[b] These statistics were determined by the nomographic method of Litchfield (1949); ranges in parentheses are 95% confidence limits and are equivalent to 2 standard errors of the median.
[c] Interval between first-set and second-set or third-party allografts was 15 to 16 days in the initial series at 24°C and 12 days in the remaining series at 27°C.
[d] COCO, MCAS, and KYC refer to three separate collecting sites on separate patch reefs in Kaneohe Bay; COCO ⇄ MCAS refers to parabiosis-type allografts between individual sponges obtained from these two sites.

TABLE 2. CYTOTOXIC REACTION TIMES OF FIRST-SET PARABIONT GRAFTS AMONG EIGHT RANDOM ALLOGENEIC PAIRS OF *CALLYSPONGIA DIFFUSA*[a]

Alloparabiont combination[b]	1	8	4	7	5	3	6	2
Mean reaction time (days)	4.4	4.6	5.3	5.8	8.0	9.4	9.4	9.9
Standard deviation of mean reaction time[c]	0.7	0.5	0.5	0.5	2.2	1.4	2.3	1.6

[a] Experiment performed at an ambient temperature of 25.2 ± 0.2°C.
[b] Different sets of parabionts arbitrarily numbered 1 through 8, are arranged in ascending order of mean reaction time; each mean is derived from the definitive reaction times of eight identical replicate pairings.
[c] A measure of the similarity between reaction times among replicates of the same graft pair.

some have argued that these enzymes may control cell–cell recognition and thereby serve also as a protective system against infectious agents (Parish, 1977; Rothenberg, 1978). Whatever the genetic sources of H molecule polymorphism, such molecules could serve as a general immunorecognition system underlying cell-mediated immunity as Hildemann (1977) has postulated. An identity between sponge H molecules and certain sponge "factors" remains a fascinating possibility (e.g., Vaith *et al.*, 1979b; Curtis, 1979c).

We have investigated the effect of seawater temperature on allograft reaction times and have seen a significant increase with a mere 2°C decrease in temperature (Johnston *et al.*, 1981). By extrapolation, first-set reaction times have a Q_{10} of 2.7 (over the temperature range 23–27°C). Such a temperature effect in an ectothermic animal may not be surprising, but it tends to explain why Moscona (1968), for example, working in temperate waters off Woods Hole, Massachusetts, failed to see alloincompatibility within 21 days of grafting; i.e., in these colder waters, the grafts should probably have been observed for a much longer period of time. Another possibility, however, is that at low temperatures, the cytotoxic effector phase of alloreactivity may be suppressed altogether. Such immunosuppression could explain why Paris (1961) saw correlates of alloincompatibility but failed to see cytotoxic allograft reactions within 2½ months at water temperatures of 11°C and below.

A variable, but nevertheless finite, sensitization period between allorecognition and alloreactivity would explain why those workers studying cell reaggregation, whose assays rarely lasted more than 24 hr, generally failed to see cytotoxic reactions directed against heterologous cells. The exceptions are brief reports of immediate cytolytic effects in bispecific mixtures of dissociated cells (Galtsoff, 1929; Van de Vyver, 1975). Such immediate effects are presently unexplained, as are other observations indicating that phagocytosis by archeocytes is very much increased in 6-hr aggregates from allogeneic mixtures of dissociated cells as opposed to that occurring in control aggregates derived from syngeneic cells alone (Van de Vyver and Buscema, 1977). Mechanical or chemical disruption of sponge tissues, however, might well trigger the same effector systems which operate in allograft reactions. Recent evidence (Bigger *et al.*, 1981) has shown that this effector system is indeed nonspecific; i.e., once a sponge recognizes its graft partner as "allo-non-self," the cytotoxic mechanism that is set in motion does not discriminate between the original target and any other substituted allogeneic target tissue.

Although the immunogenetic basis of alloincompatibilities and their underlying cytotoxic mechanisms are presently unknown, sponges do exhibit sharply discriminating adaptive immunity, and at this lower level of phylogeny they could reflect the origins of both cell-mediated immunity and of the histocompatibility systems of higher vertebrates.

REFERENCES

Bagby, R. M., 1970, The fine structure of pinacocytes in the marine sponge *Microciona prolifera*, *Z. Zellforsch. Mikrosk. Anat.* **105**:579.

Bakus, G. J., and Green, G., 1974, Toxicity in sponges and holothurians: A geographical pattern, *Science* **185**:951.

Bergquist, P. R., 1978, *Sponges*, University of California Press, Berkeley.

Bigger, C. H., Hildemann, W. H., Jokiel, P. L., and Johnston, I. S., 1981, Afferent sensitization and efferent cytotoxicity in allogeneic tissue responses of the marine sponge *Callyspongia diffusa*, *Transplantation* **31**:461.

Borojevic, R., 1967, La ponte et le développement de *Polystomia robusta* (Demosponges), *Cah. Biol. Mar.* **7**:1.

Boury-Esnault, N., 1977, A cell type in sponges involved in the metabolism of glycogen, *Cell Tissue Res.* **175**:523.

Bretting, H., and Kabat, E. A., 1976, Purification and characterization of the agglutinins from the sponge *Axinella polypoides* and a study of their binding sites, *Biochemistry* **15**:3228.

Bretting, H., and Konigsmann, K., 1979, Investigations on the lectin-producing cells in the sponge *Axinella polypoides* (Schmidt), *Cell Tissue Res.* **201**:487.

Brill, B., 1973, Ultrastructure of choanocytes in *Ephydatia fluviatilis*, *Z. Zellforsch. Mikrosk. Anat.* **144**:231.

Burger, M. M., Turner, R. S., Kuhns, W. J., and Weinbaum, G., 1975, A possible model for cell–cell recognition via surface macromolecules, *Philos. Trans. R. Soc. London Ser. B* **271**:379.

Burger, M. M., Burkart, W., Weinbaum, G., and Jumblatt, J., 1978, Cell–cell recognition: Molecular aspects, recognition and its relation to morphogenetic processes in general, *Symp. Soc. Exp. Biol.* **32**:1.

Burkart, W., and Burger, M. M., 1977, Studies on cell populations from *Microciona prolifera* separated by Ficoll gradients, *Biol. Bull.* **153**:417.

Burkholder, P. R., 1973, The ecology of marine antibiotics and coral reefs, in: *Biology and Geology of Coral Reefs*, Volume II, *Biology 1* (O. A. Jones and R. Endean, eds.), pp. 117–182, Academic Press, New York.

Cheng, T. C., Yee, H. W. F., and Rifkin, E., 1968a, Studies on the internal defense mechanisms of sponges, I. The cell types occurring in the mesogloea of *Terpios zeteki* (de Laubenfels) (Porifera: Demospongiae), *Pac. Sci.* **22**:395.

Cheng, T. C., Rifkin, E., and Yee, H. W. F., 1968b, Studies on the internal defense mechanisms of sponges. II. Phagocytosis and elimination of Indian ink and carmine particles by certain parenchymal cells of *Terpios zeteki*, *J. Invert. Pathol.* **11**:302.

Cheng, T. C., Yee, H. W. F., Rifkin, E., and Kramer, M. D., 1968c, Studies on the internal defense mechanisms of sponges. III. Cellular reaction in *Terpios zeteki* to implanted heterologous biological materials, *J. Invert. Pathol.* **12**:29.

Cohen, E. (ed.), 1974, *Biomedical Perspectives of Agglutinins of Invertebrate and Plant Origins*, *Ann. N.Y. Acad. Sci.* **234**.

Connes, R., 1966, Aspects morphologiques de la régénération de *Tethya lyncurium* Lamark, *Bull. Soc. Zool. Fr.* **91**:43.

Connes, R., 1967, Reactions de défense de l'éponge *Tethya lyncurium* Lamark, vis-à-vis des microorganismes et de l'amphipode *Leucothoe spinicarpa* Abildg., *Vie Milieu Ser. A* **18**:281.

Curtis, A. S. G., 1969, The measurement of cell adhesiveness by an absolute method, *J. Embryol. Exp. Morphol.* **22**:305.
Curtis, A. S. G., 1974, The specific control of cell positioning, *Arch. Biol.* **85**:105.
Curtis, A. S. G., 1978, Cell–cell recognition: Positioning and patterning systems, *Symp. Soc. Exp. Biol.* **32**:51.
Curtis, A. S. G., 1979a, Individuality and graft rejection in sponges, or, a cellular basis for individuality in sponges, *Syst. Assoc. Spec. Vol.* **11**:39.
Curtis, A. S. G., 1979b, Recognition by sponge cells, *Colloq. Int. CNRS* **291**:205.
Curtis, A. S. G., 1979c, Histocompatibility systems, recognition and cell positioning, *Dev. Comp. Immunol.* **3**:379.
Curtis, A. S. G., and Van de Vyver, G., 1971, The control of cell adhesion in a morphogenetic system, *J. Embryol. Exp. Morphol.* **26**:295.
Dayton, P. K., Robilliard, G. A., Paine, R. T., and Dayton, L. B., 1974, Biological accommodation in the benthic community at McMurdo Sound, Antarctica, *Ecol. Monogr.* **44**:105.
DeSutter, D., and Van de Vyver, G., 1977, Aggregative properties of different cell types of the fresh water sponge *Ephydatia fluviatilis* isolated on Ficoll gradients, *Wilhelm Roux Arch. Entwicklungsmech. Org.* **181**:151.
DeSutter, D., and Van de Vyver, G., 1979, Isolation and recognition properties of some definite sponge cell types, *Dev. Comp. Immunol.* **3**:389.
Donadey, C., and Vacelet, J., 1977, Les cellules`a inclusions de l'éponge *Pleraplysilla spinifera* (Schulze) (Demosponges: Dendroceratides), *Arch. Zool. Exp. Gen.* **118**:273.
Du Pasquier, L., 1974, The genetic control of histocompatibility reactions: Phylogenetic aspects, *Arch. Biol.* **85**:91.
Egami, N., and Ishii, S., 1956, Differentiation of sex cells in united heterosexual halves of the sponge *Tethya serica*, *Annot. Zool. Jpn.* **29**:199.
Evans, C. W., and Bergquist, P. R., 1974, Initial cell contact in sponge aggregates, *J. Microsc. (Paris)* **21**:185.
Evans, C. W., and Curtis, A. S. G., 1979, Graft rejection in sponges: Its relation to cell aggregation studies, *Colloq. Int. CNRS* **291**:211.
Evans, C. W., Kerr, J., and Curtis, A. S. G., 1980, Graft rejection and immune memory in marine sponges, in: *Phylogeny of Immunological Memory* (M. J. Manning, ed.), pp. 27–34, Elsevier/North Holland Biomedical Press, Amsterdam.
Faulkner, D. J., 1977, Interesting aspects of marine natural products chemistry, *Tetrahedron* **33**:1421.
Galtsoff, P. S., 1929, Heteroagglutination of dissociated sponge cells, *Biol. Bull.* **57**:250.
Galtsoff, P. S., Brown, H. H., Smith, C. L., and Walton-Smith, F. G., 1939, Sponge mortality in the Bahamas, *Nature (London)* **143**:807.
Garrone, R., and Pottu, J., 1973, Collagen biosynthesis in sponges: Elaboration of spongin by spongocytes, *J. Submicrosc. Cytol.* **5**:199.
Harrison, F. W., 1972, The nature and role of the basal pinacoderm of *Corvomeyenia carolinensis*: A histochemical and developmental study, *Hydrobiologia* **39**:495.
Hildemann, W. H., 1977, Specific immunorecognition by histocompatibility markers: The original polymorphic system of immunoreactivity characteristic of all multicellular animals, *Immunogenetics* **5**:193.
Hildemann, W. H., Johnston, I. S., and Jokiel, P. L., 1979, Immunocompetence in the lowest metazoan phylum: Transplantation immunity in sponges, *Science* **204**:420.
Hildemann, W. H., Bigger, C. H., Jokiel, P. L., and Johnston, I. S., 1980a, Characteristics of immune memory in invertebrates, in: *Phylogeny of Immunological Memory* (M. J. Manning, ed.), pp. 9–14, Elsevier/North-Holland, Amsterdam.
Hildemann, W. H., Bigger, C. H., Johnston, I. S., and Jokiel, P. L., 1980b, Characteristics of transplantation immunity in the sponge, *Callyspongia diffusa*, *Transplantation* **30**:362.
Humphreys, S., Humphreys, T., and Sano, J., 1977, Organization and polysaccharides of sponge aggregation factor, *J. Supramol. Struct.* **7**:339.
Humphreys, T., 1963, Chemical dissolution and *in vitro* reconstruction of sponge cell adhesion. I. Isolation and functional demonstration of components involved, *Dev. Biol.* **8**:27.
Hyman, L. H., 1940, *The Invertebrates*, Volume 1, *Protozoa through Ctenophora*, McGraw–Hill, New York.

Jackson, J. B. C., and Buss, L., 1975, Allelopathy and spatial competition among coral reef invertebrates, *Proc. Natl. Acad. Sci. USA* **72:**5160.

Jakowska, S., and Nigrelli, R. F., 1960, Antimicrobial substances from sponges, *Ann. N.Y. Acad. Sci.* **90:**913.

Johnston, I. S., Jokiel, P. L., Bigger, C. H., and Hildemann, W. H., 1981, The influence of temperature on the kinetics of allograft reactions in a tropical sponge and a reef coral, *Biol. Bull.* **160:**280.

Johnston, I. S., and Hildemann, W. H., 1982, Cellular organization in the marine demosponge *Callyspongia diffusa*, *Mar. Biol.* **67:**1.

Kartha, S., and Mookerjee, S., 1979, Cell contact in aggregating sponge cells: An ultrastructural study, *Mikroskopie* **35:**213.

Lauckner, G., 1980, Diseases of Porifera, in: *Diseases of Marine Animals, Vol. 1, General Aspects, Protozoa to Gastropoda* (O. Kinne, ed.), pp. 139–165, John Wiley & Sons, Chichester.

Leith, A., 1979, Role of aggregation factor and cell type in sponge cell adhesion, *Biol. Bull.* **156:**212.

Levi, C., 1970, Sponge cells, *Symp. Zool. Soc. London* **25:**353.

Litchfield, J. T., 1949, A method for rapid graphic solution of time–per cent effect curves, *J. Pharmacol. Exp. Ther.* **97:**399.

MacLennan, A. P., 1970, Polysaccharides from sponges and their possible significance in cellular aggregation, *Symp. Zool. Soc. London* **25:**299.

MacLennan, A. P., 1974, The chemical basis for taxon-specific cellular reaggregation and self–not-self recognition in sponges, *Arch. Biol.* **85:**53.

MacLennan, A. P., and Dodd, R. Y., 1967, Promoting activity of extracellular materials on sponge cell reaggregation, *J. Embryol. Exp. Morphol.* **17:**473.

McClay, D. R., 1974, Cell aggregation properties of cell surface factors from five species of sponge, *J. Exp. Zool.* **188:**89.

McIntyre, D. E., Faulkner, D. J., Van Engen, D., and Clardy, J., 1979, Renierone, an antimicrobial metabolite from a marine sponge, *Tetrahedron Lett.* **43:**4163.

Moscona, A. A., 1968, Cell aggregation: Properties of specific cell-ligands and their role in the formulation of multicellular systems, *Dev. Biol.* **18:**250.

Muller, C. H., 1970, The role of allelopathy in the evolution of vegetation, in: *Biochemical Coevolution* (K. L. Chambers, ed.), pp. 13–31, Oregon State University Press, Corvallis.

Müller, W. E. G., and Zahn, R. K., 1973, Purification and characterization of a species-specific aggregation factor in sponges, *Exp. Cell Res.* **80:**95.

Müller, W. E. G., Müller, I., Kurelec, B., and Zahn, R. K., 1976a, Species specific aggregation factor in sponges. IV. Inactivation of the aggregation factor by mucoid cells from another species, *Exp. Cell Res.* **98:**31.

Müller, W. E. G., Müller, I., Zahn, R. K., and Kurelec, B., 1976b, Species-specific aggregation factor in sponges, VI. Aggregation receptor from the cell surface, *J. Cell Sci.* **21:**227.

Müller, W. E. G., Müller, I., and Zahn, R. K., 1978a, Aggregation in sponges, *Res. Mol. Biol. (Akad. Wiss. Lit.)* **8:**1.

Müller, W. E. G., Zahn, R. K., Kurelec, B., Uhlenbruck, G., Vaith, P., and Müller, I., 1978b, Aggregation of sponge cells. XVIII. Glycosyltransferases associated with the aggregation factor, *Hoppe-Seyler's Z. Physiol. Chem.* **359:**529.

Müller, W. E. G., Zahn, R. K., Kurelec, B., Müller, I., Vaith, P., and Uhlenbruck, G., 1979a, Aggregation of sponge cells: Isolation and characterization of an inhibitor of aggregation receptor from the cell surface, *Eur. J. Biochem.* **97:**585.

Müller, W. E. G., Kurelec, B., Zahn, R. K., Müller, I., Vaith, P., and Uhlenbruck, G., 1979b, Aggregation of sponge cells: Function of a lectin in its homologous biological system, *J. Biol. Chem.* **254:**7479.

Muscatine, L., Karakashian, S. J., and Karakashian, M. W., 1967, Soluble extracellular products of algae symbiotic with a ciliate, a sponge and a mutant hydra, *Comp. Biochem. Physiol.* **20:**1.

Paris, J., 1961, Greffes et sérologie chez les éponges silicieuses, *Vie Milieu Ser. A Suppl.* **11:**1.

Parish, C. R., 1977, Simple model for self–non-self discrimination in invertebrates, *Nature (London)* **267:**711.

Pavans de Ceccatty, M., and Garrone, R., 1971, Fibrogenèse du collagène chez l'éponge *Chondrosia reniformis* Nardo (démosponge, tétractinellide): Origine et évolution des lophocytes, *C.R. Acad. Sci. Ser. D* **273:**1957.

Randall, J. E., and Hartman, W. D., 1968, Sponge-feeding fishes of the West Indies, *Mar. Biol.* **1**:216.
Reiswig, H. M., 1971a, In situ pumping activities of tropical Demospongiae, *Mar. Biol.* **9**:38.
Reiswig, H. M., 1971b, Particle feeding in natural populations of three marine demosponges, *Biol. Bull.* **141**:568.
Rothenberg, B. E., 1978, The self recognition concept: An active function for the molecules of the histocompatibility complex based on the complementary interaction of protein and carbohydrate, *Dev. Comp. Immunol.* **2**:23.
Rützler, K., 1970, Spatial competition among Porifera: Solution by epizoism, *Oecologia (Berlin)* **5**:85.
Sara, M., 1968, Bispecific cell aggregation of the sponges *Haliclona elegans* and *Tethya citrina*, *Acta Embryol. Morphol. Exp.* **10**:228.
Sara, M., 1970, Competition and cooperation in sponge populations, *Symp. Zool. Soc. London* **25**:273.
Sara, M., and Vacelet, J., 1973, Ecologie des démosponges, in: *Traité de Zoologie, Anatomie, Systematique, Biologie: Spongiares* (P. P. Grassé, ed.), pp. 462–576, Masson, Paris.
Simpson, T. L., 1973, Coloniality among the Porifera, in: *Animal Colonies* (R. S. Boardman, A. H. Cheetham, and W. A. Oliver, eds.), pp. 549–565, Dowden, Hutchinson and Ross, Stroudsburg.
Simpson, T. L., and Vaccaro, C. A., 1974, An ultrastructural study of silica deposition in the freshwater sponge *Spongilla lacustris*, *J. Ultrastruct. Res.* **47**:296.
Smith, D., Muscatine, L., and Lewis, D., 1969, Carbohydrate movement from autotrophs to heterotrophs in parasitic and mutualistic symbiosis, *Biol. Rev.* **44**:17.
Spiegel, M., 1955, The reaggregation of dissociated sponge cells, *Ann. N.Y. Acad. Sci.* **60**:1056.
Tuzet, O., and Paris, J., 1964, Réactions tissulaires de l'éponge *Suberites domuncula* (Olivi) Nardo, vis-à-vis de ses commensaux et parasites, *Vie Milieu Ser. A Suppl.* **17**:147.
Tyler, J. C., and Böhlke, J. E., 1972, Records of sponge dwelling fishes, primarily of the Caribbean, *Bull. Mar. Sci.* **22**:601.
Vacelet, J., 1975, Étude en microscopie electronique de l'association entre bacteriés et spongiaires du genre *Verongia* (Dictyoceratida), *J. Microsc. Biol. Cell.* **23**:271.
Vacelet, J., and Donadey, C. J., 1977, Electron microscope study of the association between some sponges and bacteria, *J. Exp. Mar. Biol. Ecol.* **30**:301.
Vaith, P., Müller, W. E. G., and Uhlenbruck, G., 1979a, On the role of D-glucuronic acid in the aggregation of cells from the marine sponge *Geodia cydonium*, *Dev. Comp. Immunol.* **3**:259.
Vaith, P., Uhlenbruck, G., Müller, W. E. G., and Holz, G., 1979b, Sponge aggregation factor and sponge hemagglutinin: Possible relationships between two different molecules, *Dev. Comp. Immunol.* **3**:399.
Van de Vyver, G., 1970, La non-confluence intraspécifiques chez les spongiaires et la notion d'individu, *Ann. Embryol. Morphog.* **3**:251.
Van de Vyver, G., 1975, Phenomena of cellular recognition in sponges, *Curr. Top. Dev. Biol.* **10**:123.
Van de Vyver, G., and Buscema, M., 1977, Phagocytic phenomena in different types of freshwater sponge aggregates, in: *Developmental Immunobiology* (J. B. Solomon and J. D. Horton, eds.), pp. 3–8, Elsevier/North-Holland, Amsterdam.
Walton-Smith, F. G., 1941, Sponge disease in British Honduras and its transmission by water currents, *Ecology* **22**:415.
Weinbaum, G., and Burger, M. M., 1973, A two component system for surface guided reassociation of animal cells, *Nature (London)* **244**:510.
Wilkinson, C. R., 1978, Microbial associations in sponges. III. Ultrastructure of the *in situ* associations in coral reef sponges, *Mar. Biol.* **49**:177.
Wilkinson, C. R., 1979, Bdellovibrio-like parasite of cyanobacteria symbiotic in marine sponges, *Arch. Microbiol.* **123**:101.
Wilkinson, C. R., and Fay, P., 1979, Nitrogen fixation in coral reef sponges with symbiotic cyanobacteria, *Nature (London)* **279**:527.
Wilkinson, C. R., and Vacelet, J., 1979, Transplantation of marine sponges to different condition of light and current, *J. Exp. Mar. Biol. Ecol.* **37**:91.
Wilson, H. V., 1907, On some phenomena of coalescence and regeneration in sponges, *J. Exp. Zool.* **5**:245.

Cellular Defense Systems of the Coelenterata

CHARLES H. BIGGER and WILLIAM H. HILDEMANN

1. INTRODUCTION

The phylum Coelenterata is composed of three classes: Hydrozoa (e.g., *Hydra*, hydroids), Scyphozoa (the true jellyfish), and Anthozoa (e.g., sea anemones, sea fans, and corals). Although some, such as the corals and sea whips, are truly sessile, most coelenterates are capable of some form of movement, ranging from creeping on a pedal disc and burrowing to freely swimming. Coelenterates include both marine and freshwater species. They are found from the deepest reaches of the ocean to the intertidal zone, and in some habitats they are one of the dominant animals. Representatives of the phylum have been identified in Precambrian fossils and there are an estimated 11,000 extant species (Russell-Hunter, 1969). The phylum Coelenterata must therefore be counted among the oldest and more successful of the animal groups.

These aquatic animals have in common a basic body plan consisting of some variation of a three-layered sac, the three layers being known as the endoderm (gastroderm), mesoglea, and ectoderm (epidermis). In colonial species, the interior cavity of the sac (coelenteron), and the tissue layers are continuous among the various individual polyps.

GENERAL CYTOLOGY

In studying a coelenterate cell type, it must be realized that a particular cell may serve not a single function but rather several, either concurrently or in developmental succession. Although it would be convenient to be able to

---

CHARLES H. BIGGER and WILLIAM H. HILDEMANN • Immunogenetics Group, School of Medicine and Dental Research Institute, University of California, Los Angeles, California 90024.

pigeonhole a cell and its function, i.e., as "digestive" or as a "leukocyte," at this level of tissue organization such an attitude on the part of the investigator may be counterproductive. However, for the purpose of this review, we will consider a generalized picture of coelenterate cytology and morphology except where it directly relates to our main theme. Thus, the various specialized sensory or skeleton-secreting cells will be considered as a group.

All coelenterates have an endoderm essentially one cell layer thick with the main cell type being the basically cuboidal nutritive-muscular (gastrodermal) cell. These cells are bounded by the mesoglea internally and front on the coelenteron (see Figure 1). By means of secretions from enzymatic-gland cells, extracellular digestion does occur in the coelenteron, but most or all nutritive-muscular cells are probably capable of phagocytosis. Also present are other secretory cells, neurons, sensory cells, and nematocytes (lacking in hydras).

Hydrozoan mesoglea is an acellular collagenous material. Scyphozoan and anthozoan mesoglea, while of similar matrix composition, are populated by cells variously called amebocytes, mesogleal cells, or granulocytes. Other cell types may also be capable of ameboid movement. The cell type under discussion is found in the ectoderm and endoderm (see Figure 2) as well as in the mesoglea and this cell type may sometimes lose and regenerate its granules (Minasian, personal communication). We will use the term amebocyte for this cell type, without implying a function or analogy.

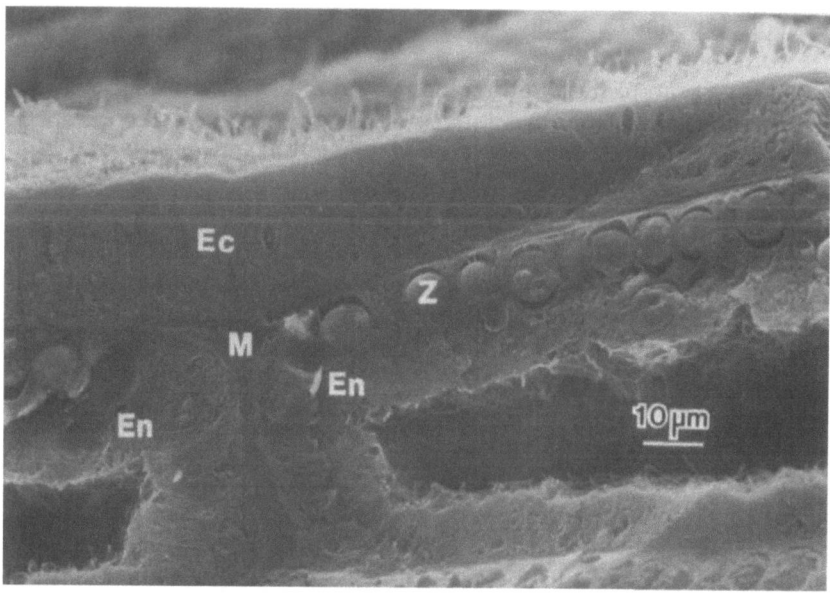

FIGURE 1. Scanning electron micrograph of ethanol cryofractured coral tissue. In this view of the coenosteal region (colonial tissue) of *Pocillipora damicornis*, the ectoderm (Ec), mesoglea (M), and endoderm (En) are readily apparent. Note the zooxanthellae (Z) in the endoderm. (Courtesy of I. Johnston.)

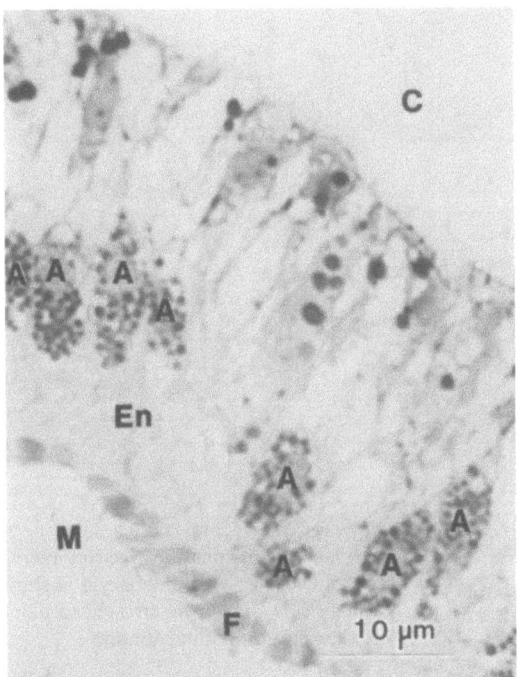

FIGURE 2. Light micrograph of a septum of the sea anemone *Haliplanella luciae*. Amebocytes (A) are commonly found in the tissue as well as in the mesoglea (M). C, coelenteron; En, endoderm; F, muscle fibers. (Courtesy of L. Minasian.)

Amebocytes are relatively small cells (e.g., 5.5- to 8.5-μm cell width in the scyphozoan *Aurelia*, D. Chapman, 1974; 4.5–12.5 μm in the anthozoan *Anthopleura krebsi*, Bigger, unpublished) whose shape varies from round to elongated with a surface topography ranging from smooth to filose projections. The cells are weakly basophilic and possess a variety of organelles and cellular inclusions (Figure 3). These include mitochondria, Golgi bodies, some rough endoplasmic reticulum, small vesicles, membrane-bound granules, and even PAS-positive 0.4-μm-wide granules as reported by D. Chapman (1974) in a scyphozoan. D. Chapman (1974) also asserts that anthozoan amebocytes contain more granules and are more filose than those found in scyphozoans. Amebocyte granules are generally reported as eosinophilic. In the anemone *Haliplanella luciae*, they range in size from 30 to 70 nm (Minasian, personal communication). D. Chapman (1974) noted a 2-μm-wide object staining for lipid in many *Aurelia* amebocytes. Patterson and Landolt (1979) have reported neutrophilic as well as eosinophilic granules in amebocytes of the sea anemone *Anthopleura elegantissima*. Consideration of amebocyte roles led Robson (1957, p. 272) to suggest:

> They [amebocytes] may form a physiological system which extends throughout the body of the sea-anemone. In this case, the intercellular fluid of the tissues, comprising not only the subepithelial layer but also the slightly hypertonic fluid of the mesoglea (see Chapman), could provide a continuous

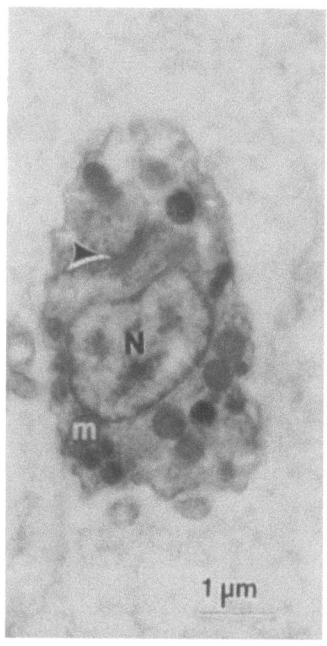

FIGURE 3. Transmission electron micrograph of a sea anemone amebocyte. This amebocyte is shown in the mesoglea of *Anthopleura krebsi*. The arrowhead points to a Golgi complex, N, nucleus; m, mitochondria.

transport medium in which the amoebocytes might function. The passage of materials such as dissolved food and excretory products between endoderm and ectoderm, for example, and the reversible changes in the mesoglea which accompany growth, or regression during starvation, could perhaps be mediated by enzymes from these cells.

There appear to be distinct categories of amebocytes as found in the sea anemone *Calliactis* (G. Chapman, 1974). These are granular and nongranular types which may or may not contain large densely staining, membrane-bound granules in the cytoplasm. Although G. Chapman (1974) suggests these granules might secrete mesogleal fibers, nongranulated cells were also observed in conjunction with such fibers. Buisson (1970) thinks this cell type can also develop into scleroblasts and other cell types. Indeed, amebocytes appear to include a stem cell population and its various ontogenetic products. Other morphologically and functionally distinct cell types which develop into scleroblasts and nematocytes, for example, often tend to be lumped together in the category of amebocytes. Metchnikoff in his classic monograph (1892) described the association of amebocytes with foreign body responses of jellyfish and others have discussed their role in wound healing (see below).

Ectoderm, although classically considered as a single cell layer, is extremely stratified in some cases. The basic cell is the epithelio-muscular cell. This columnar cell fronts on the exterior, passes through the thickness of the ectoderm, and then spreads into processes running parallel to the mesoglea. These processes contain muscle fibers. Interspersed among the epithelio-muscular cells are neurons (mostly toward the base of the tissue), sensory cells, amebocytes (in scyphozoans and anthozoans only), secretory cells, and cnidocytes (cells containing the stinging capsules or related organelles).

In the hydrozoans, a population of cells, called interstitial cells, are found at the base of the ectoderm. They are strongly basophilic and rather homogeneous in appearance, being small, round or oval, and having a high nucleus/cytoplasm ratio. The interstitial cells give rise to eggs and sperm and at one time were thought to be pleuripotent progenitors of all other cell types. Brien (1951) estimated that during a 45-day period, all cells in a hydra were replaced from the interstitial cell pool. Although the cell turnover rate for hydra may be accurate, later work (e.g., David and Campbell, 1972) has shown that other cell types can replicate. Also, mutant strains and chemically treated hydra have been developed that lack or are deficient in interstitial cells. The other cell types affected (lacking) in such hydra are the nematocytes, neurons, gametes, and endodermal gland cells (Marcum and Campbell, 1978). Nematocytes and neurons have also been shown to develop from interstitial cells (e.g., David and Gierer, 1974). Interstitial cells are motile but do not appear to be phagocytic. Some investigators (e.g., David and Challoner, 1974) have listed several different classes of interstitial cells. Scyphozoan and anthozoan amebocytes have been identified as interstitial cells, but as Table 1 points out, amebocytes and interstitial cells appear to be separate cell types although some of their functions may be similar. Cells conforming to the above morphological and functional descriptions of interstitial cells have been reported in the sea anemones *Metridium senile* (Westfall, 1966) and *Aiptasia diaphana* (Singer, 1971); alternatively, Young (1974) and Van-Praet and Doumenc (1974) failed to find that type of cell in the anemones *Calliactis parasitica* and *Actinia equina*, respectively. Much obviously remains to be learned about structure–function correlations among coelenterate cell types.

TABLE 1. CHARACTERISTICS OF INTERSTITIAL CELLS AND AMEBOCYTES

Characteristic	Interstitial cells	Amebocytes
Class occurrence	Hydrozoa, Anthozoa(?)	Scyphozoa and Anthozoa
Body location	Mostly ectoderm	Ectoderm, mesoglea, endoderm
Staining	Strongly basophilic	Generally less basophilic, but variable
Shape	Round or oval	Variable shape, some very drawn out and filose
Cellular details	Central nucleus with a prominent nucleolus and a high nucleus/cytoplasm ratio. Few small mitochondria and membrane-bound vesicles, sparse endoplasmic reticulum, inconspicuous Golgi complex, absence of granular endoplasmic reticulum, and numerous free ribosomes (D. Chapman, 1974)	Contain mitochondria, Golgi bodies, some rough endoplasmic reticulum, various numbers of small vesicles, membrane-bound eosinophilic or neutrophilic granules; some have PAS-positive granules

(continued)

TABLE 1. (Continued)

Characteristic	Interstitial cells	Amebocytes
Phagocytic	No	Yes
Motile	Yes	Yes
Proliferate	Mitosis occurs	Mitotic figures uncommon in a scyphozoan (D. Chapman, 1974). Not seen to be mitotically active in anemone *A. elegantissima* (Patterson and Landolt, 1979). Mitosis seen in anemone *Haliplanella luciae* (Minasian, personal communication).
Functions	Develop into gametes, neurons, and nematocytes. May participate in graft rejection/wound healing but are not required	Develop into nematocytes, gametes, and other cell types. Take part in wound healing and tissue reorganization. Probably secrete mesogleal fibers. Phagocytose foreign material
Can be divided into subclasses	Yes	Yes

2. CELLULAR DEFENSE

Although coelenterates have been widely utilized as experimental animals, the usual context of the research has been in the fields of developmental biology, neurophysiology/behavior, and ecology. Also, until recently, the majority of the work was with a restricted number of species, to the extent that in some cases, one received the impression that hydra biology = coelenterate biology. For these reasons, one can sometimes glean useful information from earlier work; but, with a few exceptions (e.g., Metchnikoff, 1892), experimental studies directly concerned with coelenterate cellular defense are comparatively new and are only now gaining momentum, given the impetus of modern immunology. In this treatise, we will generally consider cellular defense in terms of reactions against living material divided into categories of defense against: (1) macroscopic animals, (2) pathogens/parasites, and (3) foreign or cancerous tissue.

2.1. DEFENSE AGAINST MACROSCOPIC ANIMALS

Coelenterates utilize several interspecific and intraspecific (conspecific) responses that would be classified as behavioral in a mechanistic sense. However, some of these interactions involve specific cellular effectors. Accordingly, it has been suggested (Hildemann et al., 1975a,b) that the recognition component may often hinge on cell surface immunorecognition, i.e., the molecular specificity could result from histocompatibility molecules that function as antigen-recognizing molecules on the cell surface. It is appropriate to include these discriminating interactions in this discussion of coelenterate cellular defense.

Nematocysts, commonly known as "stinging capsules," are one of the largest (some > 100 μm long) and most complex intracellular structures secreted by a cell. These organelles are common to all coelenterates, although their distribution within an animal, both in regard to type and density, differs between species. Twenty-seven different structural types were described in the latest review (Mariscal, 1974) and other types have been discovered since (e.g., Conklin *et al.*, 1977). Each nematocyte holds a single nematocyst consisting of a thick, double-walled capsule containing a coiled thread folded complexly but in a basic triple-helix pattern. Upon receiving the proper stimulus, the nematocyst discharges. During discharge, the thread everts and, in the most studied nematocysts, releases toxin from the open tip of the thread. Toxin of the sea wasp (*Chironex flexneri*) is one of the most deadly poisons known; other jellyfish and the man-of-war cause severe pain for swimmers and fishermen in some localities. However, as there is a large body of literature dealing with nematocyst toxicology, the reader is referred to Mariscal (1974) for further information.

Nematocysts are involved in a variety of functions, aside from the widely ascribed prey capture. Among the less commonly considered functions are defense against predators and interspecific and intraspecific aggressions. There is evidence that the function of some nematocysts is specific, i.e., a nematocyst used in aggression is not used in prey capture and vice versa (Bigger, 1976, 1980; Conklin *et al.*, 1977). For example, the type II holotrich nematocyst is the cellular effector of the sea anemone acrorhagial response (Bigger, 1976, 1980). The acrorhagi are small hollow structures located at the base of the tentacles of some species of anemones (see Figure 4). After an acrorhagi-bearing animal touches

FIGURE 4. *Anthopleura krebsi* acrorhagial aggressive behavior. The anemone on the left is in the midst of an acrorhagial attack on the anemone on the right, which has contracted its tentacles. The aggressor is moving its expanded acrorhagi (Ac) toward the target anemone. Holotrich nematocysts are located in the light-colored acrorhagi tips (arrowhead).

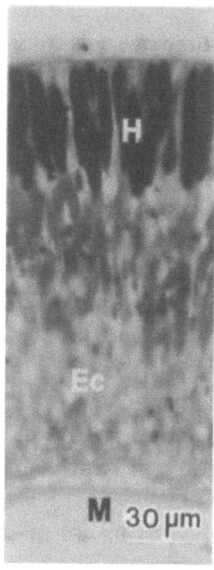

FIGURE 5. Light micrograph of a section through the tip of an *A. krebsi* acrorhagus. This region contains a dense concentration of holotrich nematocysts (H). Ec, ectoderm; M, mesoglea.

allogeneic conspecifics, or certain other coelenterates, the acrorhagi in the area of contact swell and elongate. The expanded acrorhagi, containing a dense concentration of type II holotrich-bearing nematocytes (Figure 5), are placed on the other animal, withdrawn, and then the application process is repeated. Pieces of the acrorhagial ectoderm attached by the fired nematocyst threads remain on the target animal. Within as short a time as 20 min (Bonnin, 1964), the target animal exhibits signs of necrosis in the area of the attached ectoderm. The various mechanical and physiological damages, and behavioral changes occurring in the target animal are presented by Bigger (1980). The acrorhagial response exhibits specificity, via associated receptors, with a directed nature in at least two of its three components (Bigger, 1976, 1980). In particular, the nematocysts will not discharge on contact with metal, glass, prey animals (although tentacle nematocysts will), or syngeneic clonemates, but will discharge with other conspecifics and certain other coelenterates. An intriguing similarity between the acrorhagial response and the vertebrate T-cell response is that both appear to have a stronger reaction to allogeneic cells than to xenogeneic cells (Francis, 1973; Bigger, 1976; Benacerraf and Burakoff, 1978). A similar anemone response called "catch tentacle" aggression involving a restricted type of nematocyte effector has recently been well described (Purcell, 1977). The cellular and molecular bases of these interesting aggressive responses are only now being uncovered. Although there have been suggestions concerning a relationship between these responses and other invertebrate or vertebrate immunological reactions (see Bigger, 1980), any detailed comparison of recognition mechanisms must await further study.

There are well-documented cases where animals are able to avoid discharge of a coelenterate's nematocysts, in some cases to the advantage of the coelenterate and in others to its certain detriment. An example of the former is the number of symbioses that exist between sea anemones and various shrimps, crabs (Stanton, 1977), and fish (Mariscal, 1970) where both symbionts are in intimate contact without discharge of the nematocysts. Some fish apparently require an acclimation period during which there is limited stinging by the anemone. A situation where the coelenterate suffers is nudibranch (a shell-less mollusc) predation. Some nudibranchs are well known to eat sea anemones or hydroids without the nematocysts discharging, and then selectively retain (digest some types and store others) certain unfired nematocysts in specialized structures (cerrata) on their backs for use in their own defense (e.g., Conklin and Mariscal, 1977).

Lang (1971, 1973) described a system in corals, whereby certain species extrude gastric filaments from the coelenteron and digest other species of corals in their immediate vicinity. The response is directed toward a specific target and the recipient is certainly injured. Although the attacking coral may receive some nutritive value from the externally digested target coral, this is not its usual mode of feeding; the major stimulus for such action appears to be competition for space, or xenogeneic incompatibility.

Most coelenterates are capable of secreting copious quantities of mucus. Mucus plays a role in some coral feeding (Lewis and Price, 1975), provides environmental protection, and, as will be discussed in more detail later, may function in protecting against pathogens. Other uses that should be considered are protection from aggression and as an offensive weapon. Bigger (1976) demonstrated mucus (freshly secreted or old) of the anemone *Anthopleura krebsi* would not elicit an acrorhagial attack, even though the secreting animal would. During several weeks of observation of two acrorhagially aggressive anemones *Bunodosoma cavernata* and *A. krebsi* (Bigger, unpublished), the *B. cavernata* secreted a very heavy layer of mucus. The mucus was conspicuously heavier in the area of the column in range of the *A. krebsi*. No *A. krebsi* attacks, or evidence thereof, were seen after the mucus secretion. Although such indirect evidence suggests that mucus could shield an anemone from recognition leading to an acrorhagial attack, a critical examination is required before one can definitively say that mucus could or does serve such a protective role in the animal's normal environment. Preliminary evidence (Hildemann *et al.*, 1977a,b) suggests that the mucus of the coral *Fungia scutaria*, secreted in response to contact with certain other corals or in rough human handling, contains molecules cytotoxic to other corals. Histologically, there are a variety of secretory cells in any individual coelenterate (Bigger, personal observation). The mucus of a coelenterate is probably not a homogeneous substance; its composition as well as the quantity may vary in response to external stimuli.

Our appreciation of the coelenterate behavioral repertoire has grown beyond the views of 100 years ago that coelenterates were mostly sessile animals with very little behavior except simple feeding, or swimming in jellyfish. Knowledge of selective and discriminating reactivities in coelenterates is still limited by sparse experimental evidence. Many interesting and fundamental insights will surely result from examining the mechanisms underlying selective behavioral responses.

2.2. PATHOGEN AND PARASITE DEFENSES

2.2.1. Parasites

True parasites of coelenterates appear to be uncommon. One exception is the trematode *Plagioporous* sp. whose metacercariae encyst in the polyps of the Hawaiian coral *Porites* spp. In describing this infection, Cheney (1975, p. 79) states, "Thus cellular response is usually lacking, cytology is probably normal. . . . Despite the abundance of nodules on affected colonies (up to 300–400/m^2), there does not appear to be any morbidity associated with them." On the other hand, there is certainly a response to certain algae that penetrate the Caribbean sea fan, *Gorgonia ventalina* (Morse *et al.*, 1977). They report that such infestations were accompanied by "peripheral necrosis" and gross damage to the colony. The areas of infiltration by the filamentous algae are characterized by "tumorlike growths" of the host. Microscopic examination reveals the host colonial tissue (coenenchyme) is extensively filled with the algal filaments, each surrounded by a tube of gorgonin, the same material that is secreted by the sea fan for its axial skeleton. There are also abnormally high numbers of "mesogleal amoebocytes" and "abnormal and extensive cellular production and extracellular elaboration of collagen throughout. . . ." The algal cells were not found in normal tissue or unsurrounded by gorgonin. Morse and his colleagues mention that a similar condition occurs in the gorgonians *Pseudoplexaura flagellosa* ("contain foreign cells of a different class") and *Plexaura homomalla*. Such "tumors" appear to represent not a cancerous growth, but an encapsulation or isolation response of the host gorgonian to the invading algal tissue.

2.2.2. Intracellular Symbionts

Many coelenterates contain algal intracellular symbionts called zoochlorellae or zooxanthellae depending on whether the algae are green or brown, respectively (see Figure 1). This association is extremely common and one would judge successful as evidenced by its occurrence in many species representing over 100 genera and 13 of the 21 coelenterate orders. Although the details are obscure in many instances, there is little doubt that both parties receive benefit from the association. For example, the asexual reproduction of jellyfish from the polyp stage of the scyphozoan *Cassiopea xamachana* requires the presence of zooxanthellae, and the process of calcium carbonate deposition by reef-building corals is enhanced by the presence of zooxanthellae. We restrict this discussion to two facets of this symbiosis, symbiont acquisition and specificity. For further details of this intimate association, the reader is referred to two recent reviews (Muscatine, 1974; Trench, 1979).

In some species, the coelenterate larvae carry algae derived from the maternal parent. In other instances, the offspring are devoid of symbionts and must be reinfected by the unicellular algae during each generation. Some cross-infection may occur following laboratory exposure of aposymbiotic (lacking endosymbionts) individuals of a symbiont-carrying coelenterate species to a variety of isolated or "wild" algae. However, when multiple criteria of uptake rate of symbiont, persistent reinfection, and normal morphology and physiology are applied, there is a high degree of specificity between the coelenterate host and its particular symbiotic algae.

The intial specificity of the association appears to occur during the active uptake process by the coelenterate host and involves a recognition of cell surface markers. To date, these processes have been best characterized in the *Hydra viridis–Chlorella* system (Muscatine et al., 1975; Pool, 1979; Pool and Muscatine, 1980; McNeil, 1981; McNeil et al., 1981), although more recently investigations have been expanded to include marine scyphozoans and anthozoans (e.g., Schoenberg and Trench, 1976). The zoochlorellae are taken up by the same endodermal cells that actively phagocytose food particles, heat-killed *Chlorella*, or latex beads. However, the three phagocytic processes differing in uptake kinetics and in the ultrastructural mechanism of uptake occur (Pool, 1979; McNeil, 1981; McNeil et al., 1981; and McNeil, personal communication). Following initial contact, a meshwork of microvilli or a "funnel-like" extension of the cell surface extend and surround the living symbiont. While the microvillar mode is restricted to the uptake of living symbionts, the funnel mode is also used in the uptake of latex spheres, free-living algae (*Chlorella vulgaris*), and heat-treated symbiotic algae. Uptake of 6.25 μm pieces of *Artemia* (brine shrimp) was accomplished by the third method, extension of multiple, broad surface folds of the digestive cell surface (McNeil, 1981). All material internalized is enclosed in a vacuole but, while there may be several food particles or dead symbiont cells in a vacuole, there is always a single living symbiont per vacuole. The algal symbionts are then transported to the proximal (closest to the mesoglea) portion of the endodermal cell where they remain. The food vacuoles generally remain in the distal area of the cell although McNeil (personal communication) has found evidence for proximal movement of food vacuoles also. The host selection of the symbiont appears to occur at the time of intercellular contact, with segregation and sequestration of the algae being a continuation of the process; alternatively, it could be argued that the initial ingestion is nondiscriminatory and that selection occurs intracellularly. In an attempt to resolve this question, Pool (1979) examined the uptake rate of *Hydra* algae and those of a cross-infective *Paramecium bursaria* endosymbiotic *Chlorella* (NC64A) in *H. viridis*. He found that pretreatment of these algae with an antiserum directed against "normal" *Hydra* algae reduced the uptake and sequestration by 60%. This result suggests that masking of the antigenic sites on the algal surface interfered with intercellular recognition. In that case, one could explain the 40% uptake by the possibilities that some of the bound antibodies were digested away in the coelenteron and/or that some algal cells did not have all the recognition sites bound by antibody. Possibly, phenotypic variation in the algal cultures used in infection experiments and to evoke antibodies was involved. However, McNeil, et al. (1981) suggest that the antibody binding could have neutralized the net negative charge on the algal cell surface and through a system of electrostatic forces (for details see their discussion) have inhibited the uptake of the symbionts. Of basic importance, Pool (1979) demonstrated that the algal uptake process involves cell surface recognition of some kind. In addition, he showed serologically that the *Chlorella* cell surface molecules change during residence in the host, although it is not known whether the new markers are produced by the alga or acquired from the host. In addition to the initial uptake being a discriminatory process, it appears that retention of the symbiont is also selective, in that other materials are digested or egested (Pool and Muscatine, 1980).

2.2.3. Pathogen Defense

Although there have been preliminary suggestions of antibodylike molecules in anemones (Phillips, 1960), no antibodies have actually been found in coelenterates. Hemagglutinins and bactericidins of various molecular weights have been isolated from body fluids of several invertebrates of other phyla (Acton *et al.*, 1969; Evans *et al.*, 1969; Bizot, 1971). Although no such molecules have been isolated from a coelenterate, no conclusive effort has been expended in a search. Quite possibly, the immunoglobulin antibody system may be unique to the vertebrates. Multicellular invertebrates may utilize a system of cell-mediated immunity precedent to that found among vertebrates. More to the point, coelenterates certainly live in an environment rich in potential pathogens (in the intertidal habitat of the anemone *Anthopleura elegantissima*, bacterial counts can reach levels of 10^5 bacteria/ml; Phillips, 1963) and have evolved discriminating mechanisms to deal successfully with the problem.

Many coelenterates, particularly those with which the authors have practical experience, have a plethora of secretory cells and secrete copious quantities of mucus. Phillips (1963) used carbon to demonstrate that the constant secretion of mucus physically trapped particles on the anemone *A. elegantissima*. The mucus strands containing the carbon were then cleared from the surface by a combination of ciliary action and slow rhythmic contractions of the anemone. Phillips also found that the freshly secreted mucus behaved like a weak acid. On removing overlying mucus and placing an electrode directly on the anemone surface, he recorded a pH of 5.9. Of 120 bacterial colonies Phillips isolated from seawater and the surface of anemones, 43% were unable to grow at that pH (seawater is normally pH 8). He also reported that *A. elegantissima* mucus contained "an enzyme that resembles lysozyme in its ability to lyse the bacterium, *Micrococcus lysodeikticus*" (Phillips, 1963, p. 428).

The natural antimicrobial activity of coelenterates has been most clearly demonstrated with the stony corals (madreporarians) and soft corals (alcyonarians). For example, in the studies of Burkholder (1973), pieces of coral or paper discs of coral extract were put in petri plates of common test bacteria or marine bacteria. After a 16-hr incubation, the inhibition zones (lack of bacteria) around the test materials were measured. Some species showed a marked antimicrobial activity toward only some bacteria, some against a broad spectrum, and some exhibited little or no response. Under such conditions (i.e., short incubation times and use of extracts), the investigator is measuring preexisting molecules directed toward the test microbe, not the induction of a response.

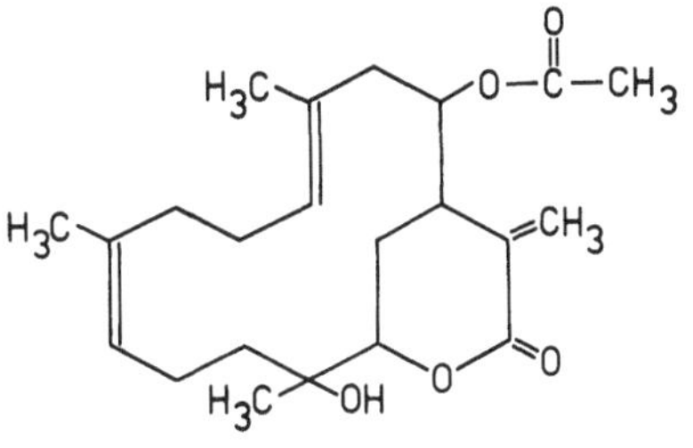

FIGURE 6. Crassin acetate. This substance has been extracted from the gorgonians *Pseudoplexaura crassa* and *P. wagenaari* and their endosymbiotic zooxanthellae (Ciereszko, 1962). Crassin acetate has antimicrobial and antiprotozoan activity and is toxic to parrot fish.

This is especially pertinent when one realizes that the corals had no prior exposure to the common "test bacteria." Particularly intriguing is Burkholder's account of a delayed type of antibiotic activity. In that instance, five species of *Parazoanthus* (a zooanthid, related to corals and sea anemones) did not display any antimicrobial activity during the first day but were allowed to remain in the culture plates. After 1 week, the zooanthids showed a delayed antibacterial activity that "appeared to involve lysis of the three cultures after their growth had taken place" (Burkholder, 1973, p. 174). At least one of the bacteria showing a susceptibility to a diffusible zooanthid substance was of marine origin but the origin of the others was not specified.

Although some stony corals have displayed activity against gram-positive bacteria in these short-term tests, the siphonophore *Porpita* (hydrozoan) showed activity against gram-negative bacteria. The gorgonians (sea whips and sea fans) have displayed some of the strongest antimicrobial activities, with more than 24 species exhibiting some degree of antimicrobial action (Burkholder and Burkholder, 1958; Burkholder, 1973). One of the better documented examples of a potentially valuable antibiotic isolated from gorgonians is crassin acetate. Ciereszko (1962) isolated crassin acetate (Figure 6) from *Pseudoplexaura crassa* and *P. wagenaari*. Crassin acetate, aside from antimicrobial activity including inhibition of *Endamoeba* at 20 µg/ml (Burkholder, 1973), is toxic to parrot fish and inhibits the development of fertilized sea urchin eggs at a concentration of 10 ppm. Crassin acetate is but one of eight crystalline terpenoid substances that have been isolated from gorgonians. Because Ciereszko was also able to extract crassin acetate from isolated *P. crassa* zooxanthellae, he believes those endosymbionts may be the source of the terpenes. "The secretion of the terpenes by the zooxanthellae may prevent the invasion of the polyps by competing microorganisms and may also help to keep the gorgonian colony clean by preventing the settlement of larvae of other animals" (Ciereszko, 1962, p. 503).

2.2.4. Wound Healing and Inflammation

The powers of coelenterate regeneration, first discovered by Trembley in 1744, are phenomenal. As noted by Metchnikoff (1892) and later by Sparks (1972), the extreme rapidity and completeness of tissue repair following a wound significantly reduces the possibility of a secondary infection and perhaps eliminates the necessity for the more complex events of the mammalian inflammatory process. The ability of a small piece of coelenterate tissue to regenerate an entire animal, among other things, has long attracted developmental biologists. Indeed, *Hydra* was the first experimental animal in the field. As a consequence, an old and extensive literature has accumulated. To further examine hydrozoan and scyphozoan regeneration, the reader is referred to Tardent (1963), the 1974 issue of the *American Zoologist* (Vol. 14, No. 2), and the proceedings of the Fourth International Coelenterate Symposium held in Interlaken, Switzerland, in 1979 (Tardent and Tardent, 1980). Of particular interest are some recent cytological studies of anthozoan wound healing/regeneration (Young, 1974; Van-Praet and Doumenc, 1974; Patterson and Landolt, 1979).

In her 1974 study, Young followed the cellular events associated with cutting away a 6-mm^2 area of the column ectoderm of the sea anemone *Calliactis parasitica*. Wound healing in *C. parasitica* lasts about 7 days at which time a new, normal-appearing cell layer has been formed. Within 1 hr, 3-μm-diameter cells have migrated to the wound surface followed by larger cells, up to 9 μm, by 12 hr postinjury. Electron micrographs show that the cells have a few mitochondria and numerous, electron-dense, membrane-bound granules about 80 nm in diameter. These "wound cells" completely cover the wound in *C. parasitica* by 24 hr, sometimes two or three cells deep. The cells at this time have cell inclusions larger than the granules and some Golgi bodies are seen. At 48 hr after wounding, the diffuse layer of mesoglea in the wound area had increased such that it sometimes exceeded 100 μm and the "wound cells" contained numerous mitochondria with large cristae and many Golgi bodies. Few membrane-bound granules were seen in the cytoplasm of these cells but more irregularly shaped granules, up to 1.2 μm long containing 26-nm-diameter fibers morphologically similar to those of mesogleal collagen fibers, were evident. By 72 hr to 14 days, the wound area had the appearance of normal ectoderm, but with fewer cells and about one-quarter the thickness. This new ectoderm is mainly formed by differentiated cells of the surrounding ectoderm migrating in from the edges of the wound. Although it also incorporates some "wound cells," other "wound cells" appear to return to the original mesoglea; additional observations of Young suggest that formation of new ectoderm requires this cell influx. Although a much smaller area is involved, the acrorhagial ectoderm of *Anthopleura krebsi* can be replaced within 6 hr of being lost, and by 48 hr, all cellular components are present and apparently capable of functioning (Bigger, unpublished observations). After inflicting larger wounds, including repeated wounding with removal of mesoglea as well as ectoderm, Young (1974) found similar events with a different time course. The results of colchicine and paradichlorobenzene treatment of animals during wound healing indicated mitosis of "mesogleal cells" was not required for simple wound healing as those cells and their migration into the wound area were unaffected by the drugs. However, ectodermal cell migration was affected, since little or no new ectoderm was found in specimens treated with colchicine. Autoradiographs of tritiated thymidine incorporation in both wounded and unwounded *Calliactis* did not reveal any increase in mitosis during the first 18 hr; very little incorporated thymidine appeared in cells on the wound surface.

Singer, in his earlier examination (1971) of tritiated thymidine uptake during large-scale regeneration in the sea anemone *Aiptasia diaphana*, also found cell proliferation was not required for wound healing and that increased cell division was not immediately stimulated but occurred later in the ectoderm, associated with such events as the outgrowth of new tentacles. Van-Praet and Doumenc (1974) were mainly concerned with the later events of tentacle regeneration in the sea anemone *Actinia equina* and gave little detail of early events. However, they confirmed the later mitotic activity in the ectoderm and observed lysosomal production by ectodermal cells including spirocytes. They also report the appearance, in the third week after tentacle amputation, of macrophages ("macrophagiques") distinct from the normal amebocytes; these cells contained numerous mitochondria, glycogen granules, osmophilic globules, and vacuoles. Such macrophages were found in the endoderm and mesoglea, but chiefly in the ectoderm where they were seen to phagocytose nematocytes and epitheliomuscle cells.

Patterson and Landolt (1979) studied cytological events following wounding of the sea anemone *Anthopleura elegantissima* with a hot probe. The damage was confined to a small area of the ectoderm and mesoglea. In normal control animals, they identified a "homogeneous population" of mitotically inactive amebocytes with eosinophilic or neutrophilic granules, and few Golgi or endoplasmic reticular structures. There were $4.9 \times 10^3/\text{mm}^2$ of these cells in the mesoglea. Twenty-four hours after wounding, the amebocyte numbers in the wound area had incresed slightly but not significantly and cells had started to appear that contained "secondary lysosomes." At 48 hr, the amebocyte numbers had increased and "secondary phagocytes" generally containing only mitochondria and granules in their cytosol were moving through the mesoglea to the wound surface whereupon they ruptured, releasing their contents. By 72 hr, the surrounding ectodermal cells were encroaching on the wound surface and the mesogleal cell population had risen to 1.2×10^4 cells/mm² within the damaged area. In addition to the above-mentioned amebocytes, "atypical cells" had appeared, seen only in the repair zone. One was a small basophilic cell with a proliferation of rough endoplasmic reticulum, an increased number of mitochondria, and a lack of specific granules. Those cells comprised less than 10% of the population and were not present by 7 days postinjury. The third cell type identified with hematoxylin and eosin was a neutrophil exhibiting an extensive Golgi and smooth endoplasmic reticulum, few mitochondria, and no specific granules. These cells were present as late as 10 days postinjury. In contrast to the situation in *Calliactis parasitica*, the amebocytes never completely covered the wound surface. As they saw no mitotic activity in the wound area and no intermediate forms progressing from normal amebocytes to the atypical cell types, Patterson and Landolt suggested that these atypical cells were produced at a remote site and then migrated to the wound. In their opinion, such a specific migration implied control, perhaps chemical in nature. They explained the role of the immediate influx of amebocytes as that of "phagocytizing necrotic tissue and eliminating it by injection into the surrounding water" (p. 196) and speculated that the other infiltrating atypical cells synthesized various substances including possible cell lysins.

Metchnikoff (1892) noted a basic difference between the hydrozoans and the other two coelenterate classes. Following a wound in the hydrozoans, regeneration processes took place, but there was no infiltration and accumulation of "phagocytes" in the wound area. Accordingly, the usual phenomenon of inflammation did not occur. However, that is not to say that there is no phagocytosis, for the endodermal cells of hydra are stationary phagocytes and are capable of putting out ameboid processes to ingest foreign bodies or even their own cells. In marine hydrozoans, the ectodermal cells are also capable of such prophylactic action. Alternatively, in the scyphozoans and anthozoans, infiltration and accumulation of inflammatory cells do take place after a lesion. As coelenterates have no circulatory system and are ectothermic, one must rely on the criterion of cellular infiltration in an attempt to define the existence of inflammation in these animals. By this criterion, one would judge the scyphozoans and anthozoans to possess an inflammatory response. Another fundamental observation made by Metchnikoff (1892) that still seems substantially valid is separation of the basic nutritive function from prophylactic activity in the scyphozoans and anthozoans. Although basic nutrition is confined to the endodermal stationary cells (there is some evidence for uptake of dissolved organics, etc., by ectodermal cells; see Mariscal and Bigger, 1977), the "mesodermic phagocytes" apparently lack a major nutritive role, but are capable of responding to foreign bodies, or their own cells, by encapsulating or phagocytosing them.

2.2.5. Nonspecific Response to Foreign Bodies

Metchnikoff (1892) was the first to examine the coelenterate response to foreign material introduced into its tissue. He observed that within 24 hr after pushing a splinter of wood or a pin into the bell of a jellyfish such as *Rhizostomum cuvieri* or *Aurelia aurita* (scyphozoans), numerous amebocytes collected and accumulated around the large foreign body. If the splinter or pin has been soaked in carmine, the assembled amebocytes phagocytosed the carmine. Now some 90 years later we wonder about the effect of microbes introduced along with the splinter, the detailed cytological events, etc. However, at present we have yet to advance much beyond Metchnikoff's original observations.

Tokin and Yericheva (1959, 1961) examined the phagocytosis of foreign material in various hydrozoans. When a whole normal *Hydra oligactis* was placed in a suspension of India ink, carmine, or dead *Bacillus subtilis*, phagocytosis rarely occurred. However, when regenerating *Hydra* tissue was exposed to the same conditions, there was increased phagocytic activity. The same type of experiments were repeated with the marine hydroids *Laomedea flexuosa, Coryne loveni,* and *Clava multicornis.* The ectodermal and endodermal cells of normal *L. flexuosa* actively phagocytosed India ink but not the dead bacteria. The cells of normal *Coryne loveni* and *Clava multicornis* phagocytosed neither the ink nor the dead bacteria. Following tissue damage/trauma, the cells of the three species actively phagocytosed both the ink particles and the dead bacteria. Tokin and Yericheva also showed that pushing a celloidin sliver or a hair into either *Hydra oligactis* or *Coryne loveni* will induce phagocytosis of ink particles. Their results suggest that such traumatic activities activate phagocytosis throughout the whole animal and not just in the immediate vicinity of the wound.

With quite different results, Prazdnikov and Mikhailova (1962; as related in Sparks, 1972) introduced carmine-saturated cotton thread into the mesoglea of the hydrozoan jellyfish *Staurophora mertensii* and the scyphozoan jellyfish *Aurelia aurita*. No phagocytosis of either carmine or damaged cells was seen during the injury response of *S. mertensii*. By 12 hr, a layer of homogenized collagen fibrils surrounded the thread and ectodermal cells had started to move onto the wound area. After $3\frac{1}{2}$ days, the thread was completely enclosed in a "ring" of homogeneous mesoglea and the wound surface was covered by large cells with a granular cytoplasm. Numerous mitotic figures were seen in the ectoderm adjacent to the wound and were considered to be the source of the new ectodermal cells.

Within 3 hr following insertion of the thread into *Aurelia* mesoglea, the injured ectoderm sloughed and amebocytes began to infiltrate. The collagen fibers around the thread became homogeneous and numerous "enlarged, vacuolated, dying amoebocytes" occurred in the vicinity of the thread. What little phagocytosis of carmine occurred was mostly restricted to ectodermal cells. After 6 to 12 hr, migration of cells into the damaged mesoglea continued, spindle-shaped ectodermal cells moved over and covered the wound area, and amebocytes and dedifferentiated ectodermal cells constituted a "phagocytic band" around the thread. Necrotic amebocytes and ectodermal cells were found among the thread fibers. After 24 hr, most of the amebocytes had left the area of the wound, and by the fourth day no live cells were seen among the thread fibers. Also on day 4, "large cells" were seen in the mesoglea and disorganized ectoderm of the wound and mitotic figures were frequently seen in both the incoming ectodermal cells and the amebocytes. From this, the authors believe that amebocytes in *Aurelia* move into and out of the wound zone, multiply, but are only weakly phagocytic. They feel that most of the phagocytosis is done by incoming ectodermal cells, which also multiply after entering the traumatized area. Thus, one may conclude that both the surrounding ectoderm and the migrating amebocytes play a role in defense but their functions are not well characterized.

Based on limited observations, there is apparently a basic difference between the foreign body response of the hydrozoans and the scyphozoans, if not also the anthozoans. However, the events are not well defined, even in the few species where they have been examined. Much more investigation is required before a clear picture of these phenomena emerges in coelenterates.

2.3. DEFENSE AGAINST CANCEROUS OR FOREIGN TISSUE

2.3.1. Cancer

Until recently, it was generally accepted that invertebrates in general, and coelenterates in particular were not, for some reason, susceptible to cancer in the vertebrate sense. Particularly as some coelenterates are among the longer-lived animals known (sea anemones are known to live 70–80 years in captivity), one wonders why they are not afflicted with cancer or whether this is really attributable to lack of investigation. In contrast to the gorgonian encapsulation of filamentous algae mentioned earlier, Cheney (1975) has described what may be the only well-documented coelenterate neoplasms. Cheney examined the coral *Acropora formosa* on Guam and Saipan in the Marianas, Palau and Ponape in the Carolinas, and Enewetak in the Marshalls. Tumorous colonies were rare, with the distribution of the tumors usually confined to one or a few colonies. Compared to normal *A. formosa* branches, the tumors have a disorganized distribution of polyps that vary widely in size and shape. The calcium carbonate skeleton of the tumor mass was very porous and reflected a relatively low calcium deposition per unit volume compared to normal branches. The coenosarc (colonial tissue) is extensive and seems to spread through the tumor mass, as opposed to the normal coenosarc that is only a few millimeters deep. In his various investigations, Cheney could find no sign of a causative agent. Invasive-type growth was also seen in some tumors. Although no distinctive cellular pathology was evident, Cheney (1975) concluded that the *A. formosa* tumors resulted from rapid, uncontrolled, and abnormal growth. Since these neoplasms were invasive and perhaps metastasized as well, cancer may occur in corals, at least in *A. formosa*. At this time, possible relationships between vulnerability to neoplasms and the histocompatibility or immunorecognition systems of corals have not been examined.

2.3.2. Transplantation

In our view (Hildemann *et al.*, 1979), only three functional criteria are necessary to define immunological competence: (1) antagonistic or cytotoxic reaction after sensitization, (2) selective or specific reactivity, and (3) inducible memory or selectively altered (positive or negative) reactivity on secondary contact. While use of defined antigens or naturally occurring microbes in controlled experiments might avoid some of the complexities inherent in tissue grafting, at the present time only tissue transplantation or parabioses have proven to be an immunological assay that is adaptable across a wide phylogenetic spectrum. In the vertebrates, tissue grafting is of course an artificial rather than naturally occurring phenomenon, although maternal/fetal interactions may be regarded as an exception. In a crowded marine habitat such as a coral reef, instances of grafting/parabiosis occur frequently. According to one viewpoint (Manning and Turner, 1976, p. 56),

> An attractive hypothesis to account for the existence of these tissue incompatibility reactions, at least in crowded stationary colonial forms (e.g. on a coral reef), is that intercolony incompatibility (repulsion) enables genetic and territorial integrity to be maintained, whilst intracolony compatibility (fusion) enhances the structural integrity of the colony. In these circumstances, specificity would be important, but speed of reaction and a memory component less so. It is thus conceivable that, at its most primitive level, the immune system evolved, not to resist infection or cancer, but simply for reasons of space.

Maintenance of the integrity of the body is surely essential for all multicellular animals. Such exquisite allodiscrimination should also preclude the survival of cells with new surface molecules arising by mutation or other genetic aberrations. Sponges, like any other animal, must cope with a variety of infectious and probable tumorogenic agents and the same discriminating immunorecognition system could serve concurrently to fend off macroscopic as well as microscopic (or macromolecular) intruders. In this larger perspective, any immune system worthy of the name probably evolved for multiple purposes.

2.3.3. Hydrozoan Histoincompatibilities

Campbell and Bibb (1970) present a good overview of tissue transplantation in hydra, and Kanaev (1952, as edited by Lenhoff, 1969) reviews the earlier literature in greater detail. Much of the hydra grafting was performed by developmental biologists and for a variety of reasons is very difficult to interpret in an immunological context. With the proper surgical techniques, hydra auto- or isografts heal within 12 hr and survive indefinitely. However, changes in the graft position (polarity) will affect graft survival because tissue position and developmental pattern are very important in hydra. It is also well established that genetic differences contribute to allograft and xenograft failure. Typically, allografts and xenografts appear to heal well but then separate after a few days or weeks.

Although a xenograft contact zone may macroscopically appear healed, Campbell and Bibb (1970) have shown that grafts between *Hydra littoralis* and *H. pseudoligactis* are different from normal tissue at the ultrastructural level. Xenogeneic cell contacts are contorted and have only "irregular and scattered bridge-like elements" rather than the long, regular septate desmosomes that characterize autogeneic cell contact. The cytoplasm of cells along the graft zone is abnormal, being highly granular and devoid of the vesicular and cytoplasmic elements found in normal ectodermal cells. Interstitial cells accumulate in the graft region approximately 4 to 5 days after transplantation between *H. pirardi* and *H. pseudoligactis* (Diehl and Burnett, 1969). However, Diehl and Burnett (1969) showed that xenograft rejection proceeded normally in animals whose interstitial cells had been removed by mustard treatment. Therefore, interstitial cells may play a role in foreign graft rejection, but they are not obligatory.

There have been reports of indefinite allo- or xenograft survival (see Campbell and Bibb, 1970, and Kanaev, 1969); however, for a variety of reasons, they must be interpreted with reservations. Most claims do not include decisive histological evidence. Tissue cell replacement constantly occurs in hydra, with cells of the apical region continually proliferating and moving downward en masse to be sloughed at the polyp base. Supposedly "permanent" foreign tissue survival could in fact represent gradual replacement of grafted tissue by host tissue to reconstitute a genetically homogeneous individual. A major fault in the experimental design of many of these experiments was failure to observe the graft for a long enough time. For example, the rapidity of coral allograft rejection has been shown to have a strong inverse relationship to temperature and at 23°C to have a median reaction time (MRT) of 22.5 days. In surveying the literature, one finds many hydra grafts are observed for several days or a few weeks at most; moreover, the temperatures in these experiments were usually below

23°C. Allograft rejection may be very slow even in vertebrates. Small hagfish skin allografts, for example, yielded an MRT of 71.9 days at 18°C, whereas similar allografts in desert iguanas exhibited an MRT of 55 days at 25°C (Hildemann and Thoenes, 1969; Cooper and Aponte, 1968). Foreign graft survival must therefore be monitored for several months before much weight can be given to graft acceptance, especially when test animals are maintained at lower temperatures. Two other concerns arise in evaluating reports of hydra allograft survival; first is the lack of information on genetic relationship among the animals used, i.e., how closely related they were, and second is the usually high technical loss rate. When large numbers of animals from a population with little genetic heterogeneity (e.g., a small pond) are grafted, one would expect to find some grafts displaying prolonged survival. An extreme example of these problems is a recent paper reporting "successful" xenogeneic hydra grafting where "dozens" of grafts were attempted, two were successful, and these two were sacrificed within 6 days after grafting. Although reports of this type must be viewed with skepticism, some reports do suggest long-term hydra xenograft survival (e.g., Issayev, 1924; Kolenkine, 1958a,b). Cytological differences between the hydrozoans and the other two classes of coelenterates could conceivably underlie fundamental differences in responses to foreign tissue; however, definitive studies of more appropriate experimental design are needed to resolve this question.

There have been several studies of histocompatibility reactions in colonial hydroids. Hauenschild (1954, 1956) examined the responses of interacting *Hydractinia echinata* colonies. He reported three types of responses to these allogeneic contacts: fusion, nonfusion, and a partial fusion followed by rejection. Interactions were tested between different strains, F_1 hybrids, F_1 and parental, F_2 and parental, and F_3 progeny. Hauenschild originally proposed a single locus with six alleles controlling *H. echinata* histocompatibility, but later realized that was too simple to explain the complexity of his accumulated data. Although, as Du Pasquier (1974) points out, the majority of Hauenschild's results fit a haplotype effect hypothesis, the data overall cannot be explained in this manner either and one must envisage more than one *H* locus or a complex, with varying degrees of polymorphism in *Hydractinia*. Other workers (Crowell, 1950; Müller, 1964; Toth, 1967) have described similar allogeneic *H. echinata* incompatibilities although Toth claimed a "temporal specificity," i.e., within a particular time period all colonies would fuse but after that period even isogeneic colonies failed to fuse.

Ivker (1972) added a different dimension to *H. echinata* responses. Strongly refuting Toth's findings (1967), she demonstrated there was no temporal specificity and in addition gave the first description of an active allogeneic response. This active response, "overgrowth," was only elicited by conspecifics, was limited to the zone of contact, and could be lethal. Overgrowth consisted of hyperplastic stolonic growth over the other colony. The hyperplastic stolons grew together to form a continuous mat of colonial tissue that completely isolated the other colony. When overgrowing tissue crossed over the opposing colony, it resumed its normal stolonic growth form. Ivker found the overgrowth response to be specific and suggests that the specificity resided in receptors and recognition molecule(s), both of which are bound to the surface of ectodermal cells. On the basis of preliminary data, Ivker suggests the *H. echinata* histoincompatibility is genetically controlled albeit there is insufficient evidence to distinguish a major immunogene complex from multiple locus sources of H polymorphism.

Kato and his colleagues studied xenogeneic interactions between marine hydroids in the context of regenerating disassociated cells (1963), normal colony contact, and stolonic grafting (1967). In their earlier work, the interspecific interaction between six hydroid species was studied by mixing disassociated stolonic cells of pairs of colonies and observing subsequent regeneration. The cells segregated to form colonies within a dominance/subordination relationship. This relationship was species specific, reflected in what Kato and colleagues call "evasion growth." In evasion growth, both colonies actively grow in contact but soon the cells of the proximal portion of the subordinate stolon disappear, leaving only the empty periderm. Concurrently, there is active growth at the distal side (away from the other colony) of the subordinate colony; i.e., the subordinate colony grows away from, and subsequently out of contact with the other colony. If the dominant colony is removed, the subordinate coenosarc (colonial tissue) will extend proximately through the empty stolon eventually forming new stolons and then a normal colony. Surface contact between the two colonies appears to play a major role in the response, although a diffusible substance may also be involved.

Kato *et al.* (1967) then examined the interactions between pairs of normal growing colonies of three hydrozoans (*Bougainvillia* sp., *Clytia volubilis*, and *Cladonema radiatum*) and side-by-side and end-to-end stolonic grafts of *Bougainvillia* and *Clytia*. Although the authors were more concerned with regeneration and developmental considerations, in the context of this discussion, two important facts emerged: (1) isografts fused (with some allowance for polarity/positional problems discussed above) and (2) not only did xenografts not fuse, but there was in at least some combinations a consistent pathology leading to cell lysis. Detailed cytological information was not given, but it appears that cell migration into the contact zone was minimal and the cytolysis was evoked by the preexisting ectodermal cells in the area. Following local contact cytolysis, the remaining cells of the damaged colony withdrew from that area of the stolon, similar to the evasion growth response.

The preliminary report of Buehrer and Tardent (1980) suggests that while allogeneic polyp stolons of the hydroid *Podocoryne carnae* will not fuse (i.e., recognition of self/not-self), medusa allografts and cell aggregate chimeras persist. However, until the long-term persistence of grafts and chimeric nature of the allogeneic aggregates is definitively established, such suggestions must be viewed with caution.

2.3.4. Scyphozoan and Anthozoan Transplantation

There has been no critical examination of tissue transplantation in the scyphozoans and the only work of note in the anthozoans centers on two groups, the gorgonians and the true corals.

Theodor's extensive studies (Theodor, 1966, 1969, 1970, 1976; Serre and

Theodor, 1967; Theodor and Carriere, 1975; Theodor and Senelar, 1975) with *Eunicella stricta* and *Lophogorgia sarmentosa* and the preliminary study of Bigger and Runyan (1979) with *Leptogorgia virgulata*, *Pseudopterogorgia elisabethae*, and *Plexaura flexuosa* have identified some of the major characteristics of histocompatibility in the gorgonians (Anthozoa: Alcyonaria), an order of the "soft corals" typified by the sea fans and sea whips. Naturally occurring isografts (within a colony) are commonly seen in the field and all experimentally manipulated isografts rapidly fuse within 24 hr. As would be expected, these colonial animals recognize a separated part of the colony as self and merge compatibly with that tissue. Two separate gorgonian responses to foreign tissue have been documented in these studies: (1) separate independent growth (nonfusion) during intraspecific contacts and (2) necrosis, acute in interspecific contacts and more chronic (the precise timing has not been reported) in intraspecific contacts. When Theodor (1976) conducted an extensive allograft study (1479 grafts) in a wild population of *Eunicella stricta*, he found that only 0.7% of the grafts remained in compatible fusion after 12 weeks.

In studying gorgonian histocompatibility, Theodor (e.g., Theodor, 1970) has focused on the rapidly occurring xenogeneic necrotic reactions. He hypothesized a system of "induced suicide" underlying graft rejection. Thus, the "killer" tissue was supposedly passive and did not directly kill foreign tissue but in some manner, perhaps involving a diffusible substance, induced the target tissue to kill itself. In more recent studies, Theodor and Senelar (1975) contend that grossly visible necrosis represents a late stage in the cytolytic sequela. Surprisingly, a 30-minute contact period is sufficient to induce 100% lysis of small xenogeneic tissue explants. When there was higher killer : target cell ratio (tissue mass), only 10–15% lysis resulted vs. the normal 70–80%; however, vital staining showed that more target cells were killed when there were more killer cells present. This cytotoxic effect is detectable in the target explant after only 1–2 min. Therefore, Theodor and Senelar (1975) suggest the toxic factor is preformed in the killer explant and only requires the proper stimulus for its release. They view the xenogeneic cytotoxic sequelae as follows: "contact induced inactivation—death of the inactivated cells—activation by chemotaxis of the surviving cells—synthesis of some lytic factor—lysis—disintegration of the explant" (Theodor and Senelar, 1975, p. 198). The delay in the cytotoxic response encountered with allografts implies fundamental differences in mechanisms operative in xenogeneic versus allogeneic cytotoxic incompatibilities. However, these mechanisms in gorgonians remain to be characterized at both cellular and molecular levels. As regards the question of immunocompetence, the gorgonians certainly fulfil the first criterion of cytotoxic alloincompatibility; however, the other requirements of memory and specificity have not yet been tested, so the issue of gorgonian immunological competence awaits further investigation.

Connell (1973, 1976) observed that reef corals are commonly interacting in some manner that causes injury to one or both colonies that touch. However, his long-term ecological studies were not designed to investigate the mechanisms of coral interaction. In 1974, Hildemann and his associates, having earlier noted compatible coral isografts and incompatible allografts and xenografts in the field, reported experimental investigations of such interactions from an immunologist's perspective. These studies, mostly with the coral *Montipora verrucosa*, have since made *Montipora* allograft rejection one of the better characterized models of invertebrate immunocompetence. Early studies (Hildemann *et al.*, 1974, 1975a,b) showed that diverse coral species were capable of a selective xenogeneic response that led to necrosis in one or both parties and that cell contact was required to initiate the reaction.

Later studies (Hildemann *et al.*, 1977a,b, 1979, 1980b; Raison *et al.*, 1976) concentrated on *Montipora* allogeneic interactions which displayed an exquisite specificity and avoided the theoretical constraints of nonimmunological, molecular incompatibilities inherent in xenogeneic grafting. As do other invertebrates, *Montipora* allografts display cytotoxicity (see Figure 7) after a lag period and yield MRTs of about 18–22 days at 25°C. Actually, strong, intermediate, and weak incompatibilities have all been observed, depending on the interclonal combination or genetic constitution of the parabionts. In all *Montipora* grafting experiments, a 1-mm-wide zone of tissue death on one or both parabionts was used for the rejection endpoint. Such a quantitative endpoint can be determined with relative ease and any person can reproducibly score the same material using this criterion.

Of major immunological importance, *M. verrucosa* allograft rejection fulfils all three criteria of an immunological response: this coral exhibits discriminating alloincompatibility with a memory component and the response also shows specificity. Second-set or repeat grafts between first-set colonies elicit an accelerated response (MRT ≃ 11 days at 25°C). Both primary and secondary allograft

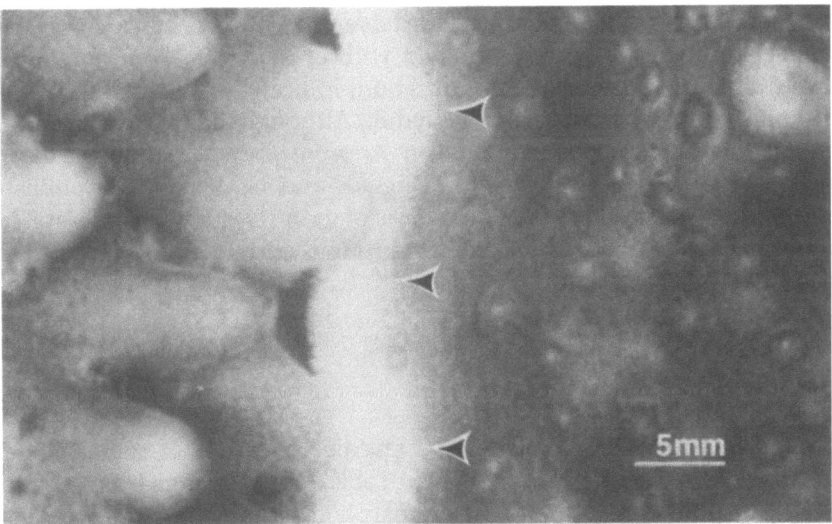

FIGURE 7. Unidirectional alloparabiotic rejection reaction in *Montipora verrucosa*. The zone of tissue death in these reactions is sharply defined by the contrast between the dark, living tissue and the bare, white skeleton where the tissue has been destroyed. Tissue death is restricted to the contact zone (arrowheads). (Courtesy of I. Johnston.)

rejections are sensitive to temperature. A four degree increase from 23° to 27°C (within the normal environmental regime) significantly accelerated allogeneic graft rejection in that coral. Unrelated third-party grafts evoked responses with a bimodal distribution in a large experimental series; about 61% showed accelerated second-set type reactions, while the remainder reacted within the primary or first-set range of 16 to 40 days. If this immune memory were nonspecific, all third-party combinations should have shown accelerated reactivity. These experiments utilized animals from three widely separated, wild populations that are maintained by both sexual (with unknown mating combinations and larval settling patterns) and asexual reproduction. Under these circumstances, cross-immunity resulting from 61% antigen sharing (the probability of any two colonies selected sharing one or more potent H antigens = 61%) is not unusual (Hildemann et al., 1980a). The more than 890 allogeneic combinations of M. verrucosa involving six widely separated populations tested thus far have all proven to be incompatible. These results suggest a multilocus array of independent and polymorphic H genes, perhaps including a major immunogene complex. Initial sensitization or induction of memory requires a minimum of 2 days of parabiotic contact; after 8 days' presensitization, all test animals show accelerated second-set rejection. This progressively increasing frequency of heightened reactivity indicates early acquisition of immune memory probably consequent upon a gradual rather than an "all-or-none" process. The alloimmune memory of M. verrucosa persists at a peak level up to 4 weeks after termination of primary stimulation, but is barely detectable at 8 weeks, and is negligible or perhaps even in a negative phase at 16 weeks (Hildemann et al., 1980b). Perhaps the short duration of potent immune memory in M. verrucosa hinges on a short-lived immunocyte or a cell population with a rapid turnover (Hildemann et al., 1980b). Hydra (Brien, 1951) seem to have a cell turnover of ~ 45 days but such investigations have not been adequately pursued with anthozoans. Some preliminary evidence (Minasian, personal communication) indicates anthozoan cell dynamics are quite different from those of hydra. Although there have been some preliminary histological observations on Acropora formosa (Hildemann et al., 1975b) and on M. verrucosa (Johnston, Bigger, and Hildemann, unpublished) graft rejections, the underlying cellular basis of graft rejection in corals still lacks definition and the UCLA immunogenetics group is currently engaged in investigations of this important topic.

3. SUMMARY

Limited investigations to date have revealed a variety of cellular defense systems in coelenterates as summarized in Table 2. These range from seemingly nonspecific responses to a finely tuned, sharply discriminating immune system characterized in corals by the beginnings of both cell-mediated immunity and a major immunogene complex. The discriminating responses presumably require specific cell surface receptors that could occur on specialized immunocytes or on diverse types of cells. Both interstitial cells and amebocytes may function as

TABLE 2. SELECTED ASPECTS OF COELENTERATE CELLULAR DEFENSE

	Antimicrobial activity described	Nonspecific cellular response to foreign materials	Allogeneic incompatibility reactions observed	Viable allografts/ xenografts persist indefinitely	Immunological responsiveness[a]	Cancer found
Hydrozoa						
Hydra	ND[b]	Yes	Yes	Yes(?), No	ND	ND
Hydroids	ND	Yes	Yes	No	ND	ND
Scyphozoa	ND	Yes	ND	ND	ND	ND
Anthozoa						
Gorgonians	Yes	Yes	Yes	No	ND	ND
Sea anemones	ND	ND	Yes	ND	ND	ND
Corals	Yes	Yes	Yes	No	Yes	Yes

[a]Based on three criteria of discriminating cytotoxic reactivity, specificity, and memory.
[b]ND = no data, not examined; Yes = described for some species of the group; No = not found in species examined.

effector cells in maintaining the integrity of individual coelenterates. We can only guess that the effector macromolecules are probably nonspecific in their actions in the same sense as the lymphokines or complement molecules of higher vertebrates. Now that the features of some coelenterate cellular defense systems have been identified and others suggested, the more demanding, but promising task of dissecting the components and mechanisms of these various responses remains.

REFERENCES

Acton, R. T., Bennett, J. C., Evans, E. E., and Schrohenloher, R. E., 1969, Physical and chemical characterization of an oyster hemagglutinin, J. Biol. Chem. 244:4128.

Benacerraf, B., and Burakoff, S. J., 1978, The biological significance of alloreactivity, in: Genetic Control of Autoimmune Disease (N. R. Rose, P. E. Bigazzi, and N. L. Warner, eds.), pp. 321–326, Elsevier/North-Holland, Amsterdam.

Bigger, C. H., 1976, The acrorhagial response in Anthopleura krebsi: Intraspecific and interspecific recognition, in: Coelenterate Ecology and Behavior (G. O. Mackie, ed.), pp. 127–136, Plenum Press, New York.

Bigger, C. H., 1980, Interspecific and intraspecific acrorhagial aggressive behavior among sea anemones, a recognition of self and not-self, Biol. Bull. 159:117.

Bigger, C. H., and Runyan, R., 1979, An in situ demonstration of self-recognition in gorgonians, Dev. Comp. Immunol. 3:591.

Bizot, M., 1971, Hemagglutinin from the snail Eobania vermiculata, Vox Sang. 21:465.

Bonnin, J. P., 1964, Recherches sur la "réaction d'agression" et sur le fonctionnement des acrorrhages d'Actinia equina L., Bull. Biol. Fr. Belg. 98:225.

Brien, P., 1951, Contribution a l'ètude des hydres d'eau douce, Bull. Soc. Zool. Fr. 76:277.

Buehrer, M. and Tardent, P., 1980, Compatibilities and incompatibilities in Podocoryne carnea (anthomedusae), in: Developmental and Cellular Biology of Coelenterates (P. Tardent and R. Tardent, eds.), pp. 477–480, Elsevier/North-Holland Biomedical Press, Amsterdam.

Buisson, B., 1970, Les supports morphologiques de l'intègration dans la colonie de Veretillum cynamorium Pall. (Cnidaria, Pennatularia), Z. Morphol. Tiere 68:1.

Burkholder, P. R., 1973, The ecology of marine antibiotics and coral reefs, in: *Biology and Geology of Coral Reefs*, Volume II, *Biology 1* (O. A. Jones and R. Endean, eds.), pp. 117–182, Academic Press, New York.

Burkholder, P. R., and Burkholder, L. M., 1958, Antimicrobial activity of horny corals, *Science* **127**:1174.

Campbell, R. D., and Bibb, C., 1970, Transplantation in coelenterates, *Transplant. Proc.* **2**:202.

Chapman, D., 1974, Cnidarian histology, in: *Coelenterate Biology: Reviews and New Perspectives* (L. Muscatine and H. M. Lenhoff, eds.), pp. 2–92, Academic Press, New York.

Chapman, G., 1974, The skeletal system, in: *Coelenterate Biology: Reviews and New Perspectives* (L. Muscatine and H. M. Lenhoff, eds.), pp. 93–128, Academic Press, New York.

Cheney, D. P., 1975, Hard tissue tumors of scleractinian corals, in: *Immunologic Phylogeny* (W. H. Hildemann and A. A. Benedict, eds.), pp. 77–87, Plenum Press, New York.

Ciereszko, L. S., 1962, Chemistry of coelenterates. III. Occurence of antimicrobial terpenoid compounds in the zooxanthellae of alcyonarians, *Trans. N.Y. Acad. Sci. Sec. II* **24**:502.

Conklin, E., and Mariscal, R. N., 1977, Feeding behavior, ceras structure, and nematocyst storage in the aeolid nudibranch, *Spurilla neapolitana* (Mollusca), *Bull. Mar. Sci.* **27**:658.

Conklin, E., Bigger, C. H., and Mariscal, R. N., 1977, The formation and taxonomic status of the microbasic q-mastigophore nematocyst of sea anemones, *Biol. Bull.* **152**:159.

Connell, J. H., 1973, Population ecology of reef-building corals, in: *Biology and Geology of Coral Reefs*, Volume II, *Biology 1* (O. A. Jones and R. Endean, eds.), pp. 205–245, Academic Press, New York.

Connell, J. H., 1976, Competitive interactions and the species diversity of corals in: *Coelenterate Ecology and Behavior* (G. O. Mackie, ed.), pp. 51–58, Plenum Press, New York.

Cooper, E. L., and Aponte, A., 1968, Chronic allograft rejection in the iguana *Ctenosaura pectinata*, *Proc. Soc. Exp. Biol. Med.* **128**:150.

Crowell, S., 1950, Individual specificity in the fusion of hydroid stolons and the relationship between stolonic growth and colony growth, *Anat. Rec.* **108**:560.

David, C. N., and Campbell, R. D., 1972, Cell cycle kinetics and development of *Hydra attenuata*. I. Epithelial cells, *J. Cell Sci.* **11**:557.

David, C. N., and Challoner, D., 1974, Distribution of interstitial cells and differentiating nematocysts in nests in *Hydra attenuata*, *Am. Zool.* **14**:537.

David, C. N., and Gierer, A., 1974, Cell cycle kinetics and development of *Hydra attenuata*. III. Nerve and nematocyte differentiation, *J. Cell Sci.* **16**:359.

Diehl, F., and Burnett, A. L., 1969, The role of interstitial cells in the maintenance of hydra. IV. Migration of interstitial cells in homografts and heterografts, *J. Exp. Zool.* **163**:125.

Du Pasquier, L., 1974, The genetic control of histocompatibility reactions: Phylogenetic aspects, *Arch. Biol.* **85**:91.

Evans, E. E., Weinheimer, P. F., Painter, B., Acton, R. T., and Evans, M. L., 1969, Secondary and tertiary responses of the induced bactericidin from the West Indian spiny lobster, *Panulirus argus*, *J. Bacteriol.* **98**:943.

Francis, L., 1973, Intraspecific aggression and its effect on the distribution of *Anthopleura elegantissima* and some related sea anemones, *Biol. Bull.* **144**:73.

Hauenschild, C., 1954, Genetische und Entwicklung-physiologische Untersuchungen über Intersexualitat und Gewebeverträglichkeit bei *Hydractinia echinata* Flemm. (Hydrox. Bougainvill.), *Wilhelm Roux Arch. Entwicklungsmech. Org.* **147**:1.

Hauenschild, C., 1956, Über die Vererbung einer Gewebeverträglichkeits—Eigenschaft bei dem Hydroidpolypen *Hydractinia echinata*, *Z. Naturforsch.* **11**:132.

Hildemann, W. H., and Thoenes, G. H., 1969, Immunological responses of Pacific hagfish. I. Skin transplantation immunity, *Transplantation* **7**:506.

Hildemann, W. H., Dix, T. G., and Collins, J. D., 1974, Tissue transplantation in diverse marine invertebrates, in: *Contemporary Topics in Immunobiology*, Vol. 4 (E. L. Cooper, ed.), pp. 141–150, Plenum Press, New York.

Hildemann, W. H., Linthicum, D. S., and Vann, D. C., 1975a, Immunoincompatibility reactions in corals (Coelenterata), in: *Immunologic Phylogeny* (W. H. Hildemann and A. A. Benedict, eds.), pp. 105–114, Plenum Press, New York.

Hildemann, W. H., Linthicum, D. S., and Vann, D. C., 1975b, Transplantation and immunoincompatibility reactions among reef-building corals, *Immunogenetics* **2:**269.

Hildemann, W. H., Raison, R. L., Hull, C. J., Akaka, L., Okamoto, J., and Cheung, G., 1977a, Tissue transplantation immunity in corals, in: *Proceedings, Third International Coral Reef Symposium*, Vol. I (D. L. Taylor, ed.), pp. 537–544, Rosenstiel School of Marine and Atmospheric Science, University of Miami.

Hildemann, W. H., Raison, R. L., Cheung, G., Hull, C. J., Akaka, L., and Okamoto, J., 1977b, Immunological specificity and memory in a scleractinian coral, *Nature (London)* **270:**219.

Hildemann, W. H., Bigger, C. H., and Johnston, I. S., 1979, Histoincompatibility reactions and allogeneic polymorphism among invertebrates, *Transplant. Proc.* **11:**1136.

Hildemann, W. H., Bigger, C. H., Jokiel, P. L., and Johnston, I. S., 1980a, Characteristics of immune memory in invertebrates, in: *Phylogeny of Immunological Memory* (M. J. Manning, ed.), pp. 9–14, Elsevier/North-Holland, Amsterdam.

Hildemann, W. H., Jokiel, P. L., Bigger, C. H., and Johnston, I. S., 1980b, Allogeneic polymorphism and alloimmune memory in the coral, *Montipora verrucosa*, *Transplantation* **30:**297.

Issayev, W., 1924, Researches on animal chimeras, *J. Genet.* **14:**273.

Ivker, F. B., 1972, A hierarchy of histo-incompatibility in *Hydractinia echinata*, *Biol. Bull.* **143:**162.

Johnston, I. S., Jokiel, P. L., Bigger, C. H., and Hildemann, W. H., 1981, The influence of temperature on the kinetics of allograft reactions in a tropical sponge and a reef coral, *Biol. Bull.* **160:**280.

Kanaev, I. I., 1969, Hydra, in: *Essays on the Biology of Fresh Water Polyps* (H. M. Lenhoff, editor of original Russian text), privately printed and circulated.

Kato, M., Hirai, E., and Kakinuma, Y., 1963, Further experiments on the interspecific relation in the colony formation among some hydrozoan species, *Sci. Rep. Tohoku Univ. Ser.* 4 **29:**317.

Kato, M., Hirai, E., and Kakinuma, Y., 1967, Experiments on the coaction among hydrozoan species in the colony formation, *Sci. Rep. Tohoku Univ. Ser.* 4 **33:**359.

Kolenkine, X., 1958a, Les modalités de l'association tissulaire après hétéro-greffe entre *Hydra attenuata* et *Pelmatohydra oligactis*, *C.R. Acad. Sci. Ser. D* **246:**1605.

Kolenkine, X., 1958b, Evolution des hydres chimères obtenues après hétéro-greff entre *Hydra attenuata* et *Pelmatohydra oligactis*, *C.R. Acad. Sci. Ser. D* **246:**1748.

Lang, J., 1971, Interspecific aggression by scleractinian corals. 1. The rediscovery of *Scolymia cubensis* (Milne Edwards & Haime), *Bull. Mar. Sci.* **21:**952.

Lang, J., 1973, Interspecific aggression by scleractinian corals. 2. Why the race is not only to the swift, *Bull. Mar. Sci.* **23:**260.

Lewis, J. B., and Price, W. S., 1975, Feeding mechanisms and feeding strategies of Atlantic reef corals, *J. Zool.* **176:**527.

Manning, M. J., and Turner, R. J., 1976, *Comparative Immunobiology*, Wiley, New York.

Marcum, B. A., and Campbell, R. D., 1978, Development of hydra lacking nerve and interstitial cells, *J. Cell Sci.* **29:**17.

Mariscal, R. N., 1970, A field and laboratory study of the symbiotic behavior of fishes and sea anemones from the tropical Indo-Pacific, *Univ. Calif. Berkeley Publ. Zool.* **91:**1.

Mariscal, R. N., 1974, Nematocysts, in: *Coelenterate Biology: Reviews and New Perspectives* (L. Muscatine and H. M. Lenhoff, eds.), pp. 129–178, Academic Press, New York.

Mariscal, R. N., and Bigger, C. H., 1977, Possible ecological significance of octocoral epithelial ultrastructure, in: *Proceedings, Third International Coral Reef Symposium*, Vol. I (D. L. Taylor, ed.), pp. 127–133, Rosenstiel School of Marine and Atmospheric Science, University of Miami.

McNeil, P. L., 1981, Mechanisms of nutritive endocytosis. I. Phagocytic versatility and cellular recognition in *Chlorohydra* digestive cells, a scanning electron microscope study, *J. Cell Sci.* **49:**311.

McNeil, P. L., Hohman, T., and Muscatine, L., 1981, Mechanisms of nutritive endocytosis. II. The effect of changed agents on phagocytic recognition by digestive cells, *J. Cell Sci.* **52:**243.

Metchnikoff, E., 1892, *Leçons sur la Pathologie Comparée de l'Inflammation*, Masson, Paris; reissued (1968) in English as *Lectures on the Comparative Pathology of Inflammation*, Dover, New York.

Morse, D. E., Morse, A. N. C., and Duncan, H., 1977, Algal "tumors" in the Caribbean sea-fan, *Gorgonia ventalina*, in: *Proceedings, Third International Coral Reef Symposium*, Vol. I (D. L. Taylor, ed.), pp. 623–629, Rosenstiel School of Marine and Atmospheric Science, University of Miami.

Müller, W., 1964, Experimentelle Untersuchungen über Stockentwicklung, Polypendifferenzierung und sexual Chimaren bei *Hydractinia echinata*, *Wilhelm Roux Arch. Entwicklungsmech. Org.* **155**:181.

Muscatine, L., 1974, Endosymbiosis of cnidarians and algae, in: *Coelenterate Biology: Reviews and New Perspectives* (L. Muscatine and H. M. Lenhoff, eds.), pp. 359–395, Academic Press, New York.

Muscatine, L., Pool, R. R., and Trench, R. R., 1975, Symbiosis of algae and invertebrates: Aspects of the symbiont surface and the host–symbiont interface, *Trans. Am. Microsc. Soc.* **94**:450.

Patterson, M. J., and Landolt, M. L., 1979, Cellular reaction to injury in the anthozoan *Anthopleura elegantissima*, *J. Invert. Pathol.* **33**:189.

Phillips, J. H., 1960, Antibodylike materials of marine invertebrates, *Ann. N.Y. Acad. Sci.* **90**:760.

Phillips, J. H., 1963, Immune mechanisms in the phylum Coelenterate, in: *The Lower Metazoa* (E. C. Dougherty, Z. N. Brown, E. D. Hanson, and W. D. Hartman, eds.), pp. 425–431, University of California Press, Berkeley.

Pool, R. R., 1979, The role of algal antigenic determinants in the recognition of potential algal symbionts by cells of *Chlorohydra*, *J. Cell Sci.* **35**:367.

Pool, R. R., and Muscatine, L., 1980, Phagocytic recognition and the establishment of the *Hydra viridis-Chlorella* symbiosis, in: *Endosymbiosis and Cell Biology*, Vol. I. (W. Schwemmler and H. E. A. Schenk, eds.), pp. 223–238, Walter de Gruyter & Co., Berlin.

Prazdnikov, E. V., and Mikhailova, I. G., 1962, A note on the problem of the character of the early inflammatory reaction in some coelenterates (*Staurophora mertensii* Brandt, 1935, *Aurelia aurita* L., *Beroe cucumis* Fabr.), *Tr. Murm. Morsk. Biol. Inst.* **4**:221.

Purcell, J. E., 1977, Aggressive function and induced development of catch tentacles in the sea anemone *Metridium senile* (Coelenterate, Actinaria), *Biol. Bull.* **153**:355.

Raison, R. L., Hull, C. J., and Hildemann, W. H., 1976, Allogeneic graft rejection in *Montipora verrucosa*, a reef-building coral, in: *Phylogeny of Thymus and Bone Marrow-Bursa Cells* (R. K. Wright and E. L. Cooper, eds.), pp. 3–8, Elsevier/North-Holland, Amsterdam.

Robson, E. A., 1957, The structure and hydromechanics of the musculoepithelium in *Metridium*, *Q. J. Microsc. Sci.* **98**:256.

Russell-Hunter, W. D., 1969, *A Biology of Higher Invertebrates*, Collier-Macmillan London.

Schoenberg, D. A., and Trench, R. K., 1976, Specificity of symbiosis between marine cnidarians and zooxanthellae, in: *Coelenterate Ecology and Behavior* (G. O. Mackie, ed.), pp. 423–432, Plenum Press, New York.

Serre, A., and Theodor, J., 1967, Mise en évidence d'unne reconnaissance immunologique de tissus chez un Invertébré, *C.R. Acad. Sci. Ser. D* **264**:513.

Singer, I., 1971, Tentacular and oral-disc regeneration in the sea anemone, *Aiptasia diaphana*. III. Autoradiographic analysis of patterns of tritiated thymidine uptake, *J. Embryol. Exp. Morphol.* **26**:253.

Sparks, A. K., 1972, *Invertebrate Pathology*, Academic Press, New York.

Stanton, G., 1977, Habitat partitioning among associated decapods with *Lebrunia danae* at Grand Bahama, in: *Proceedings, Third International Coral Reef Symposium*, Vol. I (D. L. Taylor, ed.) pp. 169–176, Rosenstiel School of Marine and Atmospheric Science, University of Miami.

Tardent, P., 1963, Regeneration in the Hydrozoa, *Biol. Rev.* **38**:293.

Tardent, P., and Tardent, R., 1980, "Developmental and Cellular Biology of Coelenterates," Elsevier/North-Holland Biomedical Press, Amsterdam.

Theodor, J., 1966, Contribution à l'étude des Gorgones (V) Les greffes chez les Gorgones: Etude d'un système de reconnaissance de tissus, *Bull. Inst. Oceanogr.* **66**(1374):1.

Theodor, J., 1969, Histotoxicité *in vivo* et *in vitro* entre tissus xénogéniques et entre tissus allogéniques chez un Invertébré, *C.R. Acad. Sci. Ser. D* **268**:2534.

Theodor, J., 1970, Distinction between "self" and "not-self" in lower invertebrates, *Nature (London)* **227**:690.

Theodor, J., 1976, Histo-incompatibility in a natural population of gorgonians, *Zool. J. Linn. Soc.* **58**:173.

Theodor, J., and Carriere, J., 1975, Direct evidence of heterolysis of gorgonian target cells, in: *Immunologic Phylogeny* (W. H. Hildemann and A. A. Benedict, eds.), pp. 101–103, Plenum Press, New York.

Theodor, J., and Senelar, R., 1975, Cytotoxic interaction between gorgonian explants: Mode of action, *Cell. Immunol.* **19**:194.

Tokin, B. P., and Yericheva, F. N., 1959, Phagocytic characteristics of cells of *Hydra oligactis* Poll., *Nauchn. Dokl. Vyssh. Shk. Biol. Nauki* **2**:43.

Tokin, B. P., and Yericheva, F. N., 1961, Phagocytal reaction in the course of regeneration and somatic embryogenesis in lower coelenterates, *Tr. Murm. Morsk. Biol. Inst.* **3**:182.

Toth, S. E., 1967, Tissue compatibility in regenerating explants from the colonial marine hydroid *Hydractinia echinata* (Flem.), *J. Cell. Physiol.* **69**:125.

Trembley, A., 1744, *Mémoires pour servir a l'histoire d'un genre de polypes d'eau douce a bras en forme de cornes*, Leyden.

Trench, R. K., 1979, The cell biology of plant–animal symbiosis, *Annu. Rev. Plant Physiol.* **30**:485.

Van-Praet, M., and Doumenc, D., 1974, Morphologie et morphogenèse expérimentale du tentacule chez *Actinia equina* L., *J. Microsc. Biol. Cell.* **23**:29.

Westfall, J. A., 1966, The differentiation of nematocysts and associated structures in the Cnidaria, *Z. Zellforsch. Mikrosk. Anat.* **75**:381.

Young, J. A. C., 1974, The nature of tissue regeneration after wounding in the sea anemone *Calliactis parasitica* (Couch), *J. Mar. Biol. Assoc. U.K.* **54**:599.

4

Cellular Defense Systems of the Platyhelminthes, Nemertea, Sipunculida, and Annelida

PIERRE VALEMBOIS, PHILIPPE ROCH, and
DOMINIQUE BOILEDIEU

1. TAXONOMY OF THE PLATYHELMINTHES, NEMERTEA, SIPUNCULIDA, AND ANNELIDA

The animals considered in this chapter belong to phyla closely related to the Platyhelminthes. We will first consider the position of Platyhelminthes in the animal kingdom (Figure 1). The Platyhelminthes, also called flatworms, consist of three main classes: (1) Turbellaria, which are free-living worms such as the well-known freshwater planarians; (2) Trematoda, which are parasitic worms (e.g., the liver flukes of ruminants); and (3) Cestoda, commonly known as tapeworms.

Before the Platyhelminthes appeared, a split of the main trunk of the phylogenetic tree occurred giving rise to the so-called protostome and deuterostome groups (Figure 1). The protostome derivation probably arose from an extinct flatwormlike stock and consists of three main branches: (1) the modern flatworms and nemerteans, (2) the Mollusca, and (3) the Annelida–Arthropoda lineage, including various small phyla such as Sipunculida. Since all these groups possess a mouth, arising from or near the blastopore, they are called Protostomia. They are also called hyponeurians because their nervous chain is placed ventrally.

The deuterostome groups are so named because their mouth develops some distance from the blastopore. The most important deuterostome phyla are: Echinodermata (sea stars or urchins), Protochordata (amphioxus and sea squirts),

---

PIERRE VALEMBOIS, PHILIPPE ROCH, and DOMINIQUE BOILEDIEU • Départment de Biologie du Développement de l'Université de Bordeaux I et Centre de Morphologie Expérimentale du CNRS, 33405 Talence Cedex, France. Supported in part by Contract 77 7 1364 with DGRST.

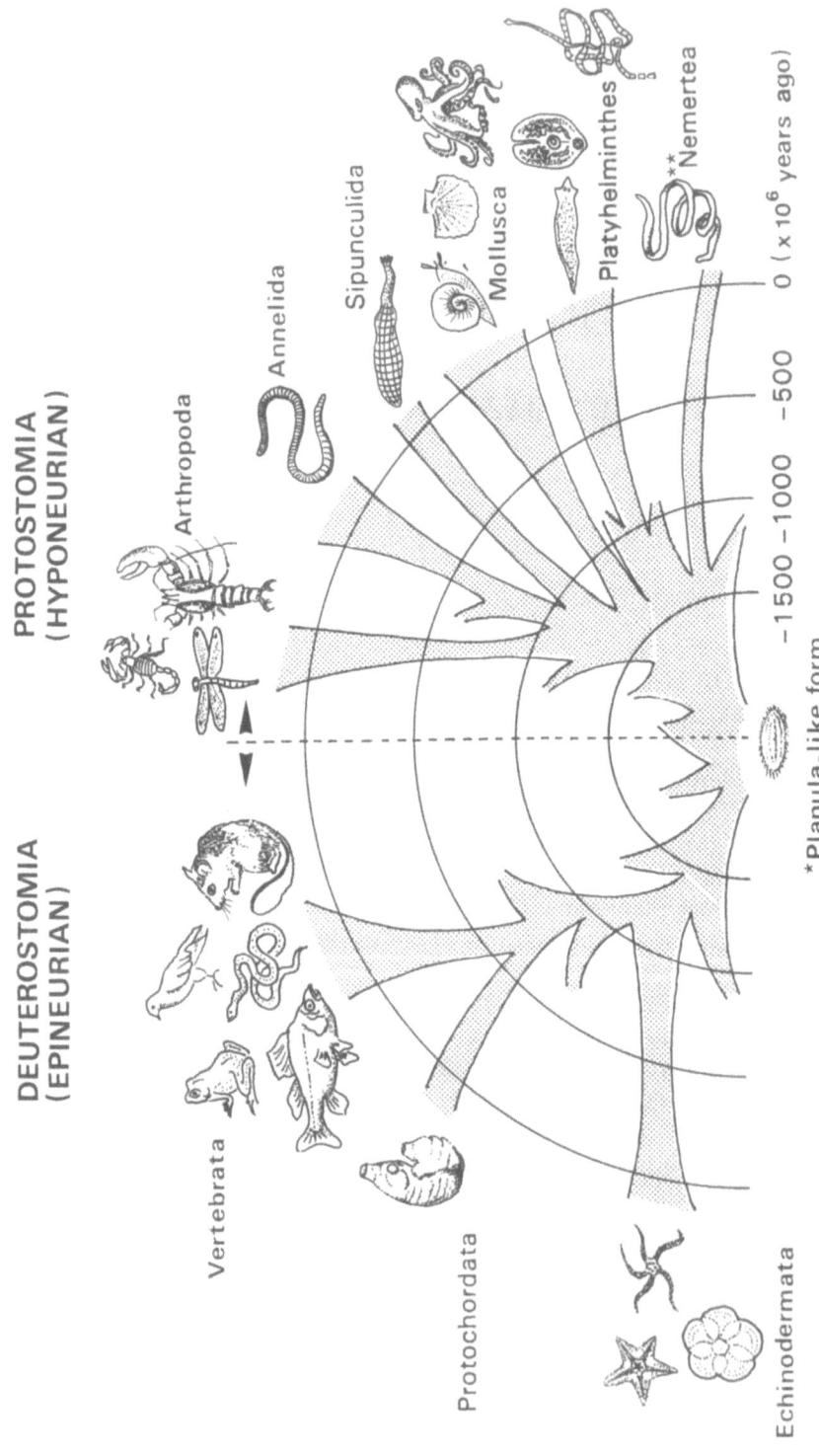

FIGURE 1. Phylogenetic relationships between principal groups of triploblastic animals and approximate time of emergence of these groups. Planula: larva of Cnidaria. Nemertea seem to have many affinities with Platyhelminthes but they have been said to be related with Deuterostomia by several authors (see text).

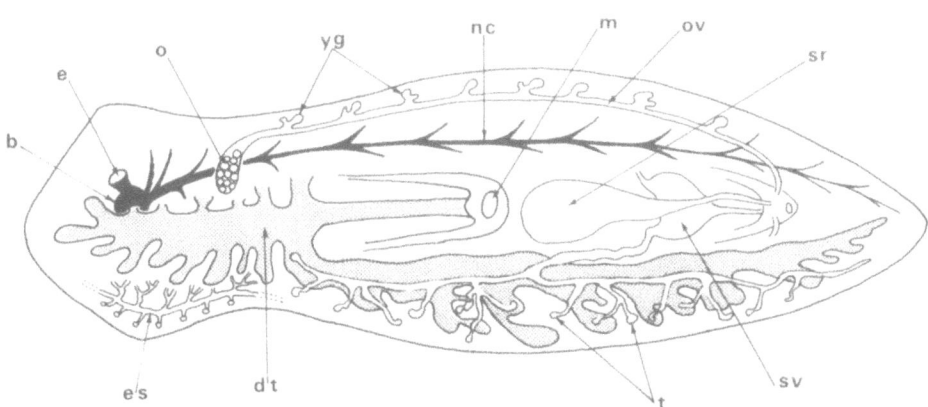

FIGURE 2. Internal structure of a planarian (diagrammatic). Only portions of the genital apparatus and the excretory system are shown. b, brain; dt, digestive tract; e, eye; es, excretory system; m, mouth; nc, nerve cord; o, ovary; ov, oviduct; sr, seminal receptacle; sv, seminal vesicle; t, testes; yg, yolk glands.

and Vertebrata. These animals are also called epineurians because their nervous tract is dorsally located.

Platyhelminths are bilaterally symmetrical multicellular organisms. The major part of their nervous system and sensory receptors are concentrated in the anterior part of the body (Figure 2). Platyhelminths are the first group to exhibit characteristics which correlate with motility. They have not yet differentiated a body cavity (or coelom) between their body wall and their digestive tract, and for this reason they are called acoelomates.

The phylum Nemertea seems to be closely allied to that of the Platyhelminthes, although certain authors (Jensen, 1960; Willmer, 1974) consider nemerteans as having affinities with deuterostomes. Platyhelminths and nemerteans do not possess a coelom. Nevertheless, nemerteans are considered more highly organized (Figure 3). Their epidermis is ciliated and their body is devoid of external segmentation. Unlike the platyhelminths, the gut of nemerteans has an anus situated at the posterior part of the body. The mouth is usually located ventrally, near the anterior end. The most characteristic organ of the phylum is the rynchostome, a muscular proboscis capable of eversion through an anterior terminal pore. These worms also possess a circulatory system consisting of a pair of lateral vessels and sometimes an unpaired dorsal vessel.

The main feature which distinguishes annelids and sipunculids from platyhelminths or nemerteans is the advantageous acquisition of a coelom. Locomotor functions have greatly improved following the development of the coelom. In acoelomates, the body wall which arises from the embryonic ectoderm, and the digestive tract which develops from the embryonic endoderm, have difficulties in moving independently. In primitive coelomates, independent movements of the body wall and digestive tract are possible because the coelomic cavity, which is filled with an incompressible fluid, acts as a hydrostatic skeleton.

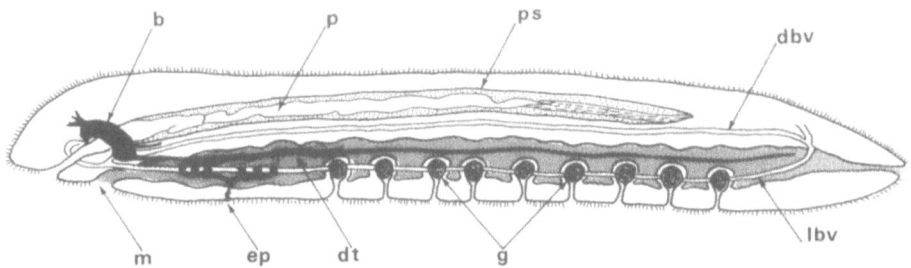

FIGURE 3. Internal structure of a nemertean (diagrammatic) with proboscis retracted into sheath (see text). b, brain; dbv, dorsal blood vessel; dt, digestive tract; ep, excretory pore; g, gonads; lbv, lateral blood vessel; m, mouth; p, proboscis; ps, proboscis sheath.

In protostomes, such as annelids or sipunculids, the coelom arises from a split inside the initial mesodermal masses (the mesoderm is the third germinal layer). Primordial mesodermal cells are originally located on both sides of the body. In metameric protostomes, such as annelids, the mesodermal masses form a linear series of segmented blocks of cells (Figure 4). In any event, a split forms within each mesodermal mass and the resulting cavity enlarges to form the coelom (Dawydoff, 1959). The portion of the mesoderm surrounding the coelomic cavity which adheres to the inner surface of the body wall is called somatopleure, and the portion which adheres to the outer surface of the gut is called splanchnopleure. As the coelom expands, the splanchnic parts of the mesoderm meet above and below the digestive tract to form the dorsal and ventral mesenteries.

One consequence of the development of the coelomic cavity is the separation of the digestive tissue from some other areas of the body by an appreciable distance. Thus, a fluid tissue or a circulatory system becomes obligatory to carry away various trophic substances. In primitive coelomates, such as annelids and sipunculids, the coelom also furnishes a place to deposit waste substances, invading organisms, or degenerative cells. It is probably for this reason that tissues surrounding the coelomic cavity have given rise to cells involved in phagocytosis and defense mechanisms.

The phylum Annelida (also called segmented worms) includes three main classes: (1) the Polychaeta, which are almost exclusively marine; (2) the Oligochaeta, which contain most of the freshwater annelids and earthworms; and (3) the Hirudinea or leeches, a uniform group of bloodsuckers or predatory worms.

One of the most striking characteristics of the annelids is the division of the body into similar parts (segments or metameres). This metameric condition is considered to be an adaptation for different kinds of locomotion: swimming in the case of certain polychaetes, or burrowing in the case of earthworms. All annelids possess a circulatory system with longitudinal vessels and lateral branches in each segment (Figure 5).

The sipunculids, often called peanut worms, form a group of approximately 300 species of marine animals. Sipunculid worms are generally considered to

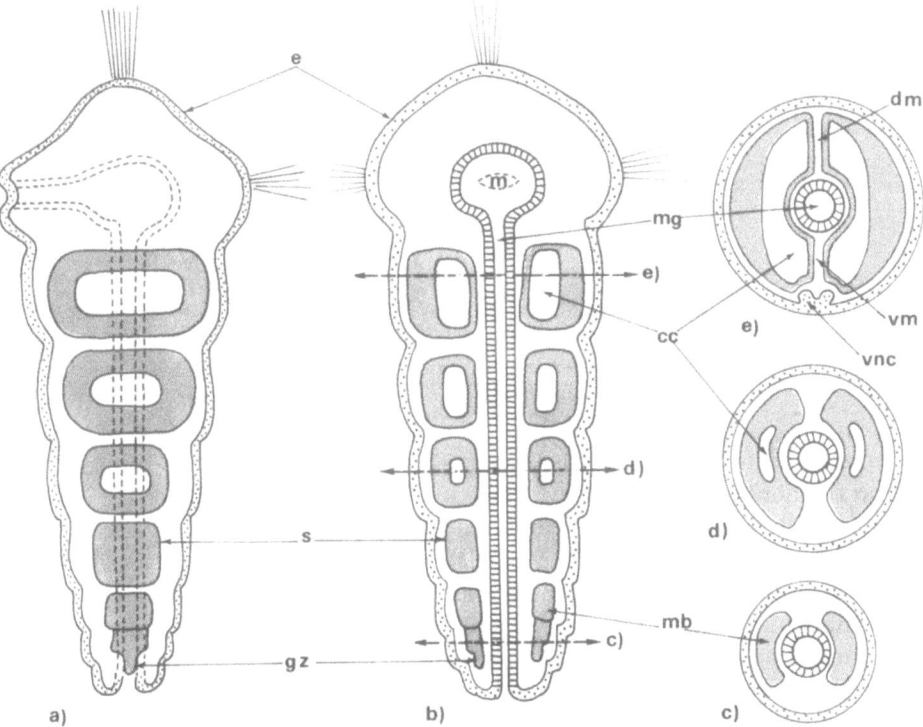

FIGURE 4. Diagrammatic sections illustrating the formation of the mesoderm and the coelomic cavity in Annelida. In young larvae, posterodorsally placed cells proliferate paired mesodermal bands. The bands segment into an anteroposterior succession of paired somites which become hollow and encircle the gut to meet in the dorsal and ventral midlines. (a) Parasagittal section; (b) frontal section; (c, d, e) transversal sections through: c, a posterior segment; d, a median segment; e, an anterior segment. cc, coelomic cavity; dm, dorsal mesentery; e, ectoderm; gz, growth zone; m, mouth; mb, mesodermal band; mg, midgut; s, somite; vm, ventral mesentery; vnc, ventral nerve cord.

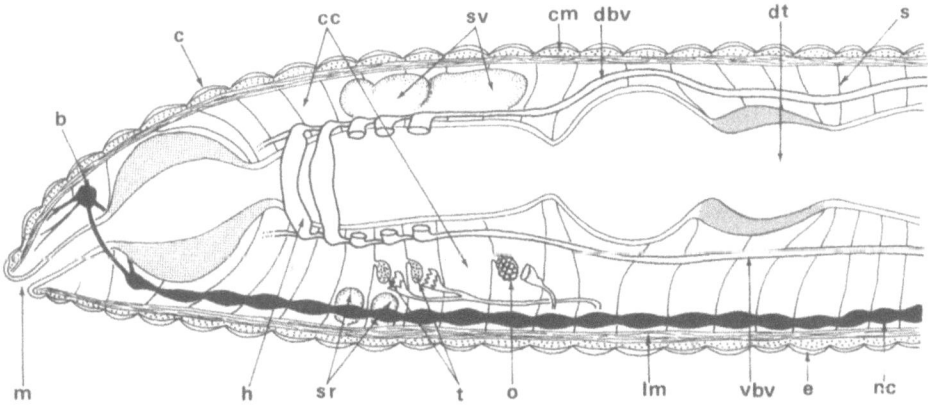

FIGURE 5. Parasagittal section through the anterior portion of an earthworm. Only two hearts (h) are shown in place. The coelomic cavity (cc) is divided into a succession of compartments by septa (s). Small pores in the septa permit the coelomic fluid to pass from segment to segment. b, brain; c, cuticle; cc, coelomic cavity; cm, circular muscles; dbv, dorsal blood vessel; dt, digestive tract; e, epiderm; h, heart; lm, longitudinal muscle; m, mouth; nc, nerve cord; o, ovary; s, septa; sr, seminal receptacles; sv, seminal vesicles. t, testes; vbv, ventral blood vessel.

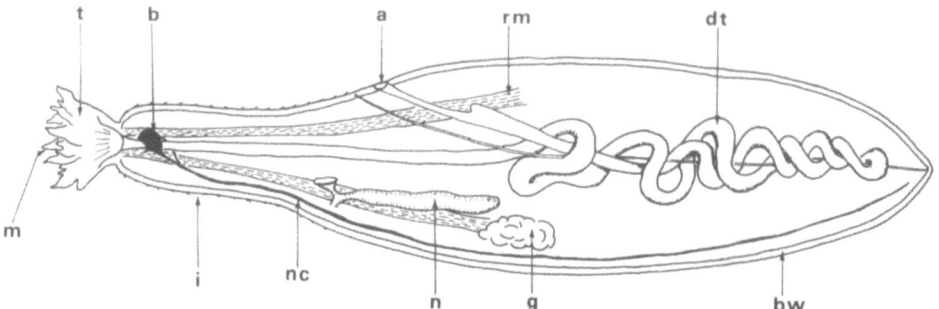

FIGURE 6. Internal structure of *Sipunculus nudus* (diagrammatic; lateral incision) showing introvert and head partially extended. a, anus; b, brain; bw, body wall; dt, digestive tract; g, gonad; i, introvert; m, mouth; n, nephridium; nc, nerve cord; rm, retractor muscles; t, tentacles.

have some affinities with annelids (Tetry, 1959), but unlike annelids, sipunculids have no segmentation of the body. Some of those that live in sand and mud are active burrowers (*Sipunculus nudus*). Some live in mucus-lined excavations, others live in coral crevices, empty snail shells (*Phascolion*) or annelid tubes. They all possess an elongated body with an anterior crown of retractile lobes or tentacles, an invertible proboscis, and a looped and twisted intestine with an anterior and mediodorsal anus (Figure 6). Their large uncompartmentalized coelomic cavity is filled with a fluid containing numerous free cells, especially erythrocytes and leukocytes.

It is possible that some invertebrate fossils which are difficult to identify and date from the Cambrian (-500 to -400×10^6 years) are closely related to sipunculids. Forms which have obvious characters of annelids existed as early as the Cambrian period. So, if modern protostomes such as annelids and contemporary deuterostomes (echinoderms or vertebrates) exhibit phylogenetic affinities, those affinities have been inherited from a common ancestor that lived one or two thousand million years ago (Figure 1).

2. THE LEUKOCYTES OF WORMS: THEIR STRUCTURE AND GENERAL PROPERTIES

Cells that somewhat resemble vertebrate leukocytes have been described for most invertebrates. The main difference between invertebrate and vertebrate leukocytes is undoubtedly the greater specialization of the latter.

In vertebrates, different types of free and fixed cells, working in concert, defend the body against foreign substances, invading organisms, etc. Polymorphonuclear leukocytes, monocytes, and eosinophilic and basophilic leukocytes are mainly concerned with inflammation and phagocytosis. Plasma cells and lymphocytes are more particularly involved in specific and anamnestic responses. It is now well known that a morphological distinction is inadequate since each of the previously mentioned categories is physiologically heterogeneous. Among the lymphocytes, for example, T cells are more particularly concerned with cellular immunity, whereas B cells are involved in humoral immunity.

In invertebrates, the different defense mechanisms are mediated by only one or two kinds of leukocytes. As might be expected, the invertebrate leukocytes, being multifunctional, are not as highly specialized or as differentiated as the vertebrate leukocytes. Some invertebrate leukocytes resemble vertebrate macrophages that are involved in numerous defense mechanisms and have kept some undifferentiated features. One of the more striking features of the macrophage is its possession of large numbers of lysosomes and its ability to synthesize lysosomal enzymes. Most lysosomal enzymes are hydrolases catalyzing reactions (schematically, type $AB + H_2O \rightarrow AH + BOH$) that occur at an optimal acidic pH. The formation of lysosomes has been well described by De Duve and Wattiaux (1966) and Novikoff (1963). It seems that most eukaryotes, including protists, possess lysosomes. In most primitive organisms, these organelles play a main role in nutrition. In primitive metazoans, the lysosomes have kept this function but have become progressively more involved in defense mechanisms.

2.1. LEUKOCYTES OF PLATYHELMINTHS AND NEMERTEANS

Planarians, which possess only a blind (without anus) rudimentary digestive tract, have a totally intracellular digestion. Phagocytosis and digestion of food particles are carried out by endodermal (gastrodermal) cells (Hay and Coward, 1975; Pedersen, 1961; Skaer, 1961). Acid phosphatase activity in these cells has been described (Jennings, 1957; Osborne and Miller, 1962; Ryder and Bowen, 1975). Cells somewhat resembling amebocytes have been described in the mesenchyme of planarians, but it seems that they do not have a defense function. Rather, they are involved in regenerative processes. Whether these cells should be considered as a stock of undifferentiated embryonic cells, or as dedifferentiated cells that arise from well-differentiated tissues, is undecided.

In trematodes, which are parasitic platyhelminths, there is a system of mesenchymal vessels which seems to represent a primitive circulatory system. Some of the hemocytes that have been found in these vessels are phagocytic (Cheng and Streisfeld, 1963).

As previously mentioned, the nemerteans, which are less primitive acoelomates than platyhelminths, have a closed circulatory system of vessels and lacunae. This system contains a cell-filled fluid, usually called blood (Hyman, 1951; Gibson, 1972). Hemoglobin-containing cells have been called red blood cells; others are called white blood cells. There have been no ultrastructural investigations of the blood cells of nemerteans. Light microscopic observations (Ohuye, 1942; Vernet and Gontcharoff, 1975, 1976), however, have revealed four or five types of leukocytes. The different leukocytes described by Vernet and Gontcharoff are: (1) basophilic granulocytes, round cells of 10-μm diameter containing basophilic granules; (2) eosinophilic granulocytes with reniform eccentric nuclei; (3) neutrophilic granulocytes very similar to eosinophilic types but with neutrophilic granules; (4) macrophages which resemble the cells previously described but exhibit a positive Schmorl's ferric ferricyanide reaction like certain vertebrate macrophages (Pearse, 1961); and (5) small-lymphocyte-like cells with a round nucleus and thin pseudopodia (Gibson, 1972; Vernet and Gontcharoff, 1976). There is a lack of experimental studies on the role and ontogenetic development of nemertean leukocytes. Thus, it is difficult to speculate about phylogenetic relationships concerning leukocytes of this and other phyla. One should note that leukocytes have exhibited a morphological polymorphism since their emergence, and have a similar appearance to either small lymphocytelike cells with a high nucleus/cytoplasm ratio, or to macrophagelike cells that often contain granules of various chromatic affinities.

2.2. LEUKOCYTES OF SIPUNCULID WORMS

Sipunculid worms generally possess a large coelomic cavity that is filled with a fluid containing different kinds of free cells. Most of these cells (about 90%) are nucleated erythrocytes. It should be noted that if numerous invertebrates possess respiratory pigments, these are generally dissolved in the coelomic fluid. Only a few invertebrate species possess erythrocytes specialized to carry pigments. The different kinds of leukocytes found in the coelomic fluid of sipunculids have been described by light microscopy (Volkonsky, 1933; Marcou and Volkonski, 1933) and by electron microscopy (Stang-Voss, 1970, 1974; Valembois and Boiledieu, 1981).

Most sipunculid leukocytes are granular cells characterized by an abundance of dense granules (Figure 7). Although the chemical nature of these granules is unknown, it seems that the granules are related to phagocytosis and probably contain lysosomal hydrolases. Electron microscopic observations have shown engulfment and phagocytosis of bacteria and cellular wastes by granulocytes and participation of granules in neutralization of foreign particles (Valembois and Boiledieu, 1981).

Other sipunculid leukocytes capable of exhibiting phagocytic activities are large hyalocytes. These cells are characterized by the presence of residual bodies and by organelles that often show degenerative aspects. Many dense granules are seen by electron microscopy. These features have led to the belief that large hyalocytes could be considered as old degranulated granulocytes.

A third category of leukocytes, small hyalocytes with a high nucleus/cytoplasm ratio, has been described. The reduced cytoplasm is characterized by a well-developed ribosomal component. Most ribosomes occur individually rather than as polyribosomes. Such features make the small hyalocytes of sipunculid worms appear very similar to vertebrate lymphocytes (Figure 8). Physiologically, similarities are also seen between small hyalocytes of the sipunculids and vertebrate lymphocytes since the small hyalocytes mediate reactions of natural killing (to be discussed).

Sipunculid worms also possess ciliated multicellular structures which swim in the coelomic fluid and which have been called urns. The urn is composed of two different parts, a ciliated structure that provides a constant swimming

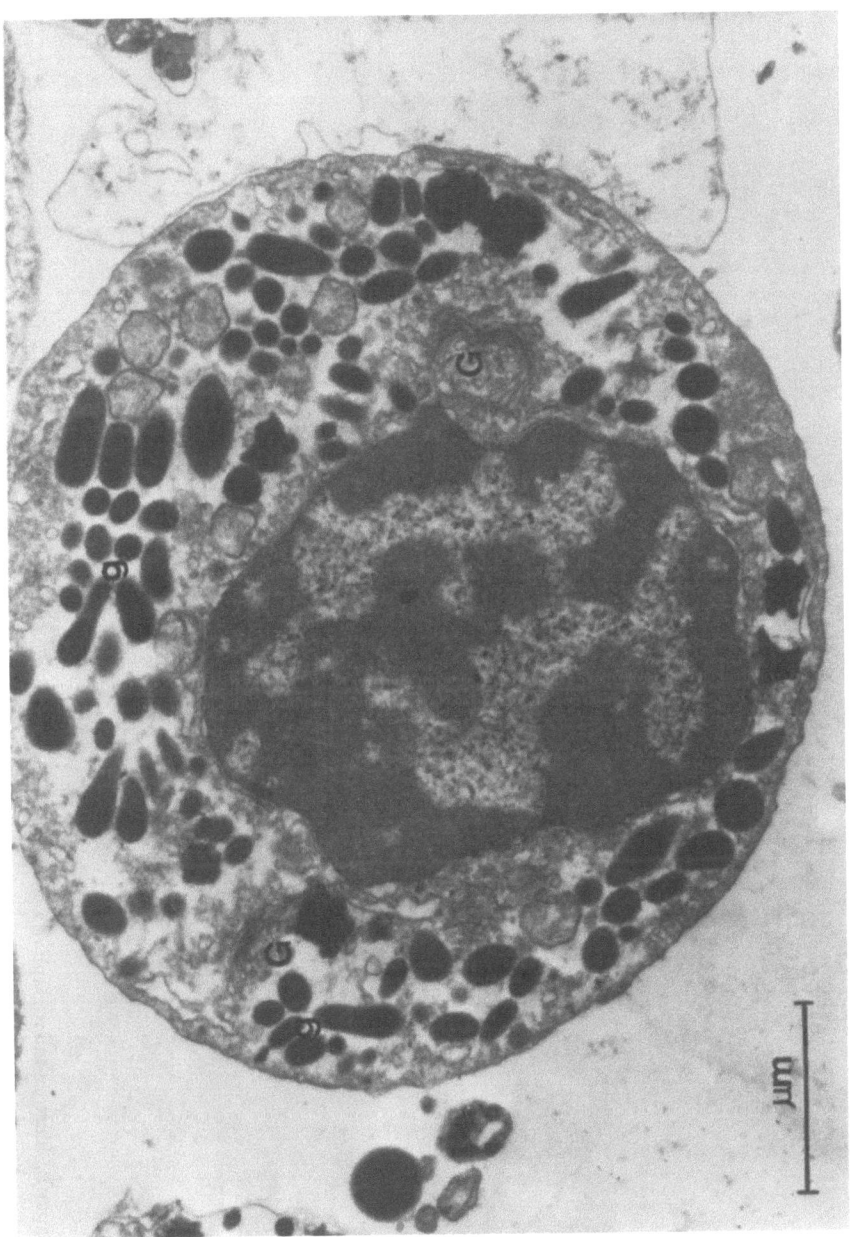

FIGURE 7. Electron micrograph of a large granulocyte of *Sipunculus nudus* showing numerous dense granules (g) and several Golgi apparatus (G). From Valembois and Boiledieu, 1981.

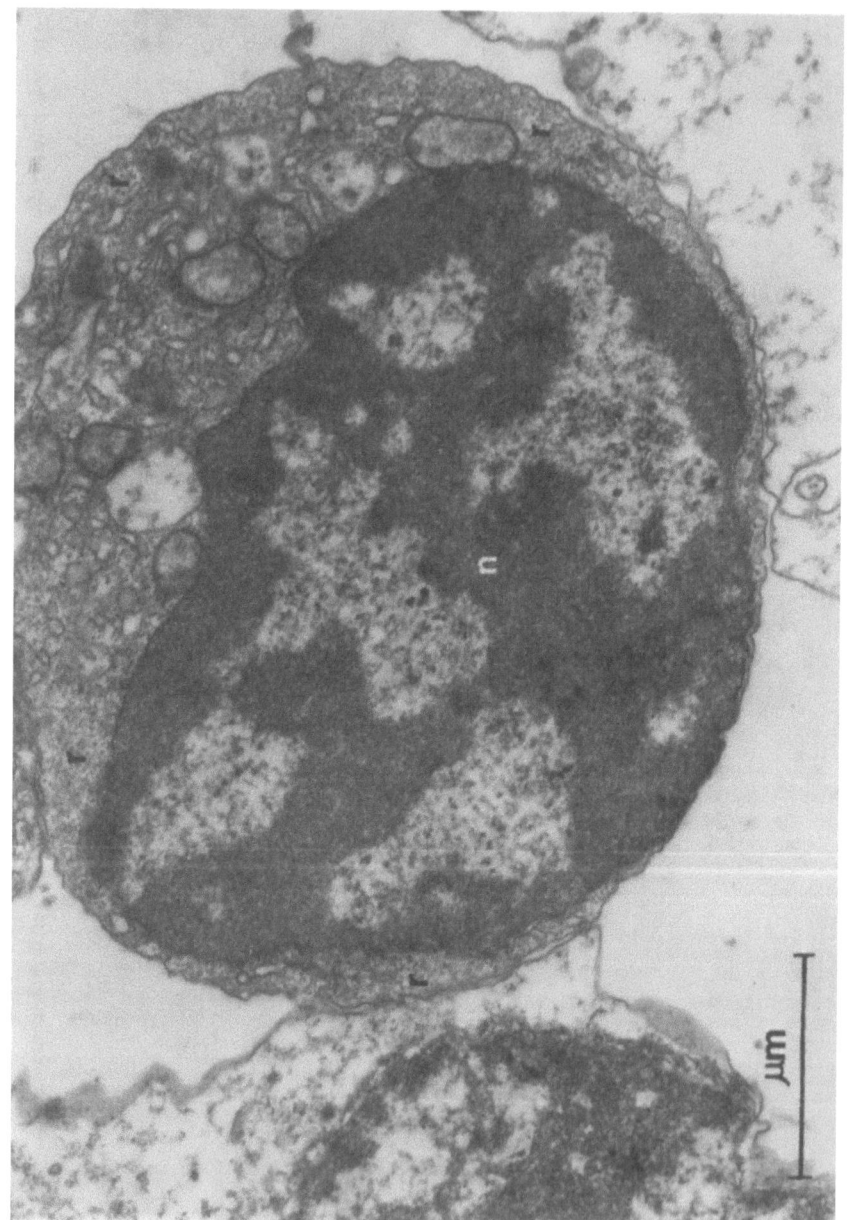

FIGURE 8. Electron micrograph of a small hyalocyte of *Sipunculus nudus*. A large nucleus (n) and numerous free ribosomes (r) make this cell very similar to the vertebrate lymphocyte. From Valembois and Boiledieu, 1981.

movement and an air vacuole that ensures floatability (Ohuye et al., 1961; Bang and Bang, 1962, 1965). Urns swim continuously, trailing a mucosidic tail behind their ciliary pole. An electron microscopic study performed by Dybas (1976) revealed the existence of three urn-cell types: (1) ciliated cells with a possible function of capturing cell debris and foreign particles; (2) cupola cells capable of phagocytosing latex particles; and (3) lobe cells capable of phagocytosing carbon particles. Both the resemblance of cupola cells and lobe cells to granulocytes (Dybas, 1976), and observations of intermediate forms between small leukocytes and ciliary cells (Valembois and Boiledieu, 1981) disprove Isern's assertion (1969) that the urn represents a parasitic organism.

2.3. LEUKOCYTES OF ANNELIDS

The defense reactions and physiology of the leukocytes have been extensively studied for a long time in annelids, particularly earthworms (Kükenthal, 1885; Rosa, 1896; Kollmann, 1908; Joseph, 1910). Authors who have investigated the leukocytes of the three classes of Annelida concur that in Annelida, the coelomocytes are implicated in two important functions, trophic and defensive.

2.3.1. Trophic Coelomocytes in Annelids

In the coelomic cavity of polychaetes, one finds cells (eleocytes) that contain lipid droplets and glycogen deposits, but no dense granules (Picton, 1898; Kollmann, 1908; Romieu, 1921, 1923). The eleocytes, which develop from the peritoneal layer, are generally considered to be degenerating coelomocytes transformed by the uptake of lipids, glycogen, and other substances. According to Dales (1961, 1964), coelomocytes take up nutrients directly in the coelomic fluid or in the blood. According to others, such a loading takes place in the peritoneal tissues (Eckelbarger, 1976), directly in the wall or in the lumen of the gut (Marsden, 1966). Many investigations (Dhainaut, 1966; Schroeder, 1971) indicate a transfer of a large amount of nutrients from coelomocytes to developing oocytes.

In oligochaetes, particularly in earthworms, the trophic function seems to be attributable to chloragogue cells that develop from the splanchnic peritoneum. These stem cells are characterized by an extensive cytoplasmic network of rough endoplasmic reticulum (Valembois, 1971c). They give rise to pedunculated cells which fall into the coelom and are called eleocytes (Liebmann, 1926; Semal-Van Gansen, 1956; Semal-Van Gansen and Van der Meersche, 1958; Duprat and Bouc-Lassalle, 1967). The cytoplasm of eleocytes (Figure 9) contains glycogen deposits, lipid droplets, and chloragosomes (yellow granules containing lipids, purines, silicates, and some other poorly elucidated components) (Semal-Van Gansen, 1957). Chloragogue cells (or chloragocytes) seem to have numerous functions which include: (1) defense (Du Pasquier and Duprat, 1968; Châteaureynaud-Duprat and Izoard, 1977a,b); (2) excretion (Bahl, 1947; Semal-Van Gansen, 1958); (3) trophic functions (Roots, 1957, 1960; Valem-

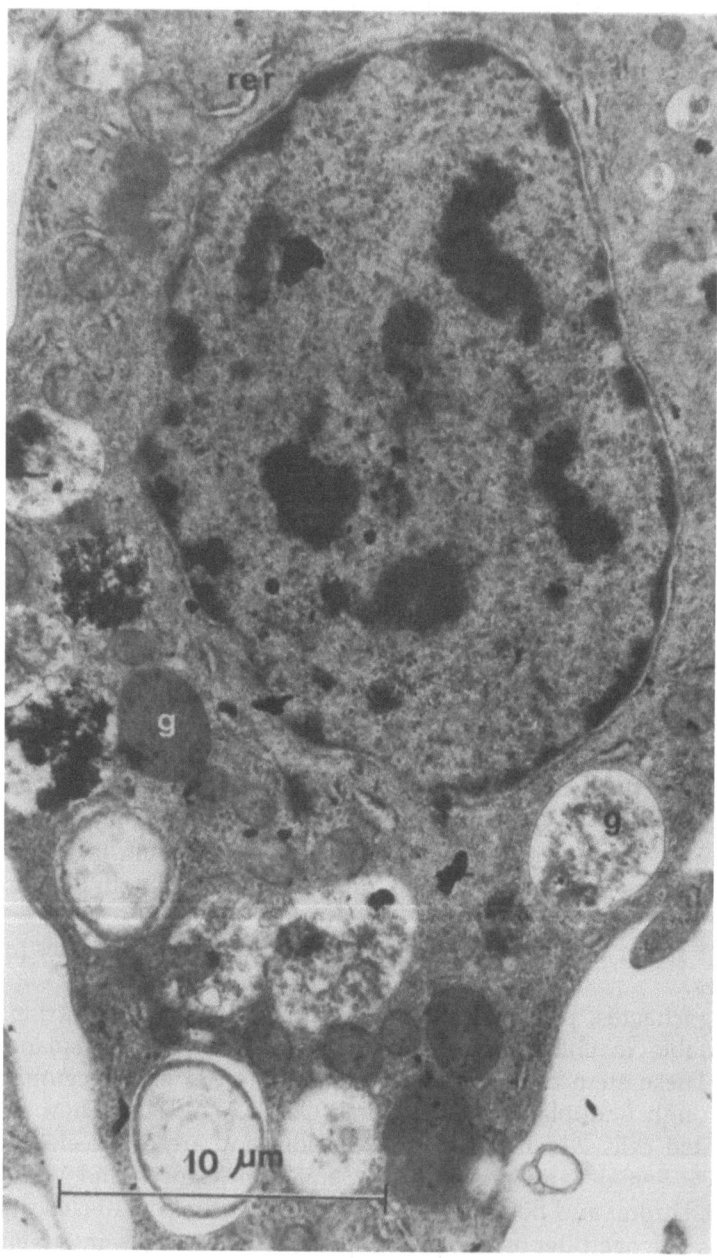

FIGURE 9. Chloragocyte of *Eisenia foetida*. Note the presence in the cytoplasm of short segments of rough endoplasmic reticulum (rer) and numerous granules (g) containing dense- or light-staining material.

bois and Cazaux, 1970); and (4) synthesis of hemoglobin (Breton-Gorius, 1963; Lindner, 1965). Leeches seem to possess chloragoguelike cells (Harant and Grassé, 1959), and cells which accumulate lipids have been described. It is unclear, however, whether these lipids are later used as nutrients.

2.3.2. Phagocytic Cells of Annelids

In various phyla of coelomates, phagocytic cells often contain electron-dense cytoplasmic granules. These granules often found associated, or in close contact, with phagocytic vacuoles, are probably lysosomelike structures. In polychaetes, investigations by electron microscopy (Sichel, 1964; Baskin, 1974; Eckelbarger, 1976) have disclosed granular leukocytes very similar to those of vertebrates. No contemporary study on the origin and function of granular cells has been made. Different authors have suggested that they could be derived from splanchnic cells.

Oligochaetes are better understood. Numerous types of cells have been described by light microscopy (Liebmann, 1942a; Duprat and Bouc-Lassalle, 1967; Stein et al., 1977) and by electron microscopy (Stang-Voss, 1971; Valembois, 1971c; Linthicum et al., 1977a). Considering the morphological aspects of the cells, it seems that two main categories of leukocytes exist: hyalocytes and granulocytes. Using electron microscopy, it appears that leukocytes arising from the somatic layer and from the splanchnic layer have different functions (Valembois, 1971c). The leukocytes arising from stem cells of the somatic layer evolve into granulocytes and hyalocytes and are involved in wound healing and the formation of scar tissue (Chapron, 1970). Leukocytes of splanchnic origin comprise granular and agranular cells which participate in immunological processes such as rejection of grafts. In earthworms, as in sipunculid worms, it is possible that the same cell first shows aspects of a small hyalocyte (Figure 11) and then evolves into a granulocyte (Figure 10) which loses its granulations with age. Perhaps a similar type of evolution exists in vertebrates where a degranulation of active granulocytes has been described (Zucker-Franklin and Hirsch, 1965). It has been suggested (Burke, 1974) that all types of nonchloragogue coelomocytes could represent different stages of the same cell (Figures 12–15). Morphological heterogeneity is a characteristic of earthworms and sophisticated oligochaetes. On the contrary, a morphological homogeneity has been noted in primitive oligochaetes (Stolte, 1955; Hess, 1970).

Few investigations have concerned leeches. In these animals, waste material is eliminated by amebocytes of the coelomic cavity and by amebocytes found in a peculiar part of the nephridia (the nephridial capsule) (Bradbury, 1959).

2.4. PHYSIOLOGICAL PROPERTIES OF WORM LEUKOCYTES

Phagocytic properties of coelomocytes have been investigated for a long time (Metchnikoff, 1892; Cameron, 1932). The finding of various enzymatic activities such as acid phosphatase activity (Valembois, 1971c) or β-glucuronidase

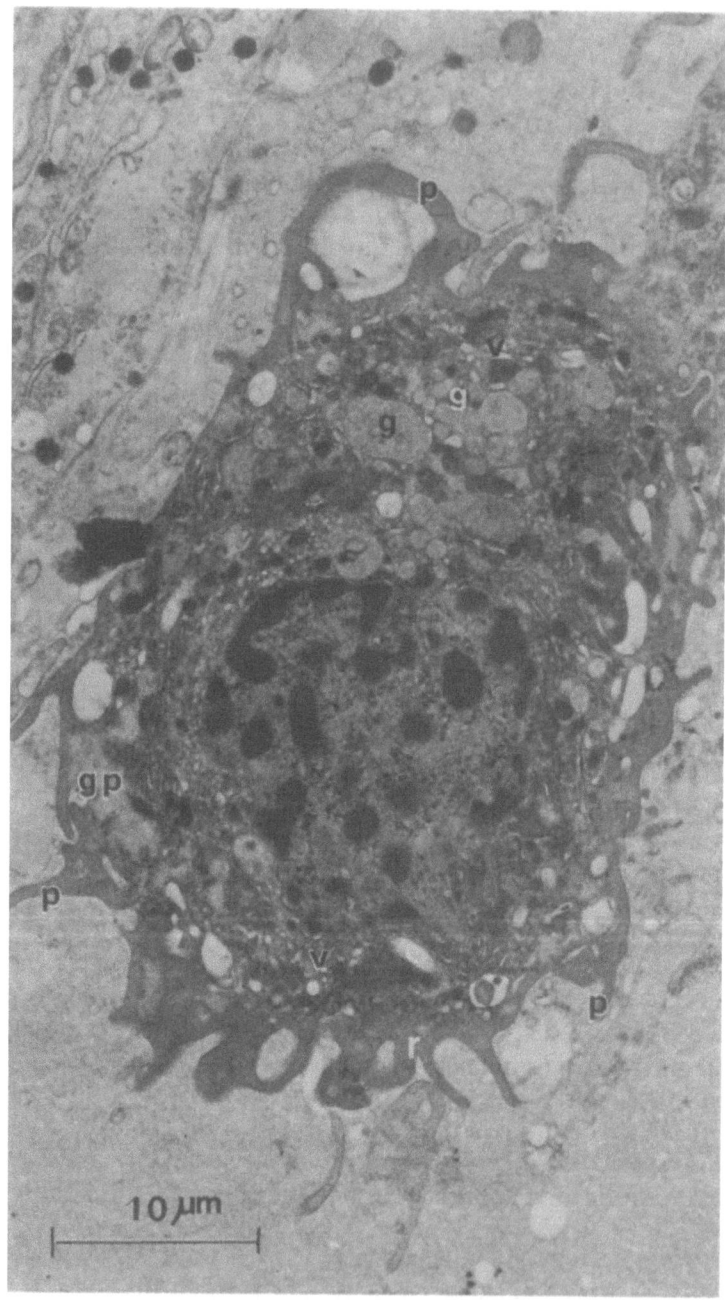

FIGURE 10. Granulocytic leukocyte of *Eisenia foetida*. Numerous granules (g) and vesicles (v) are seen in the cytoplasm. Note also the presence of glycogen particles (gp) and numerous pseudopodia (p).

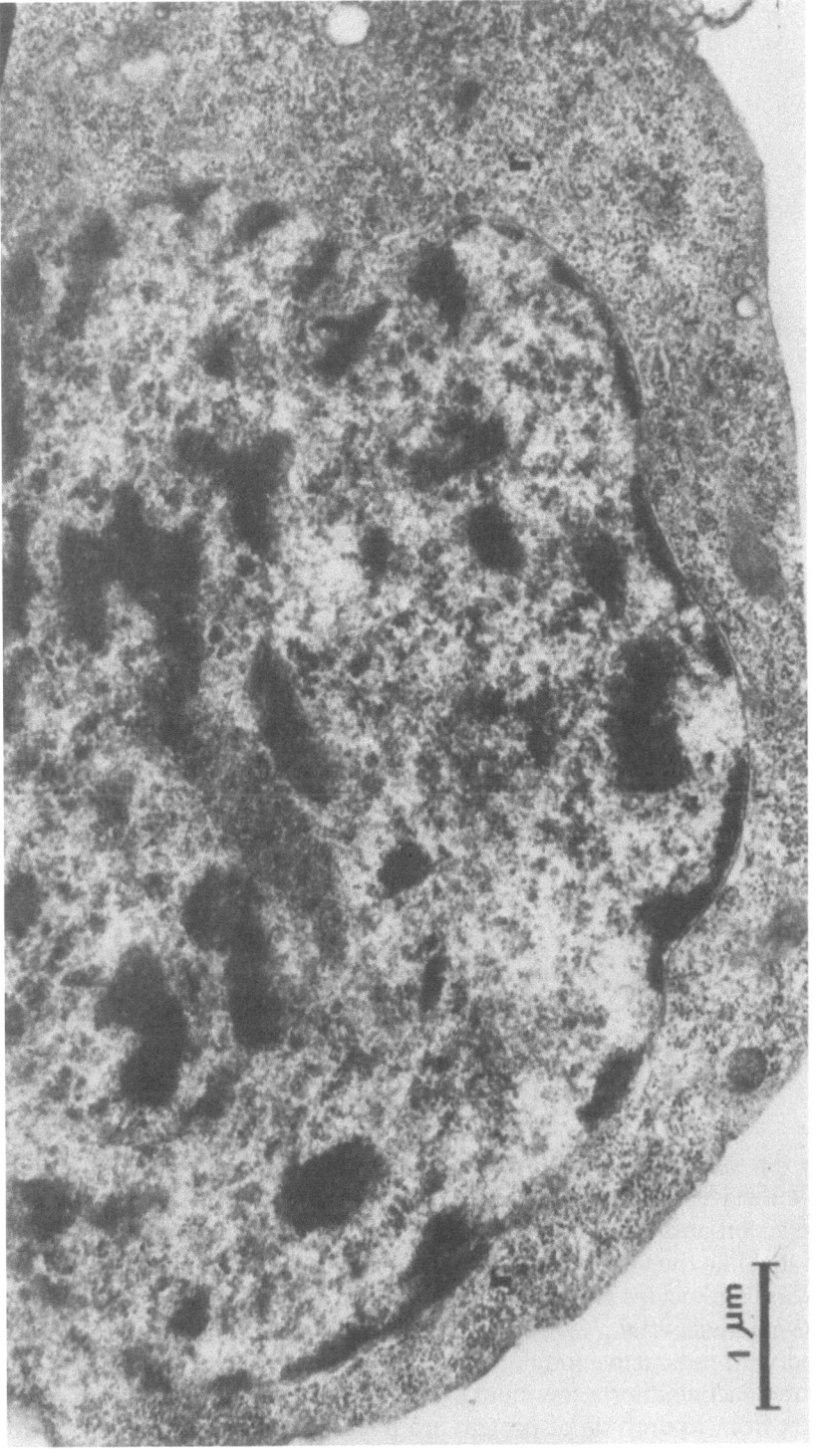

FIGURE 11. Lymphocytelike cell of *Eisenia foetida*. This cell is characterized by its very numerous free ribosomes (r) and its high nucleus/cytoplasm ratio.

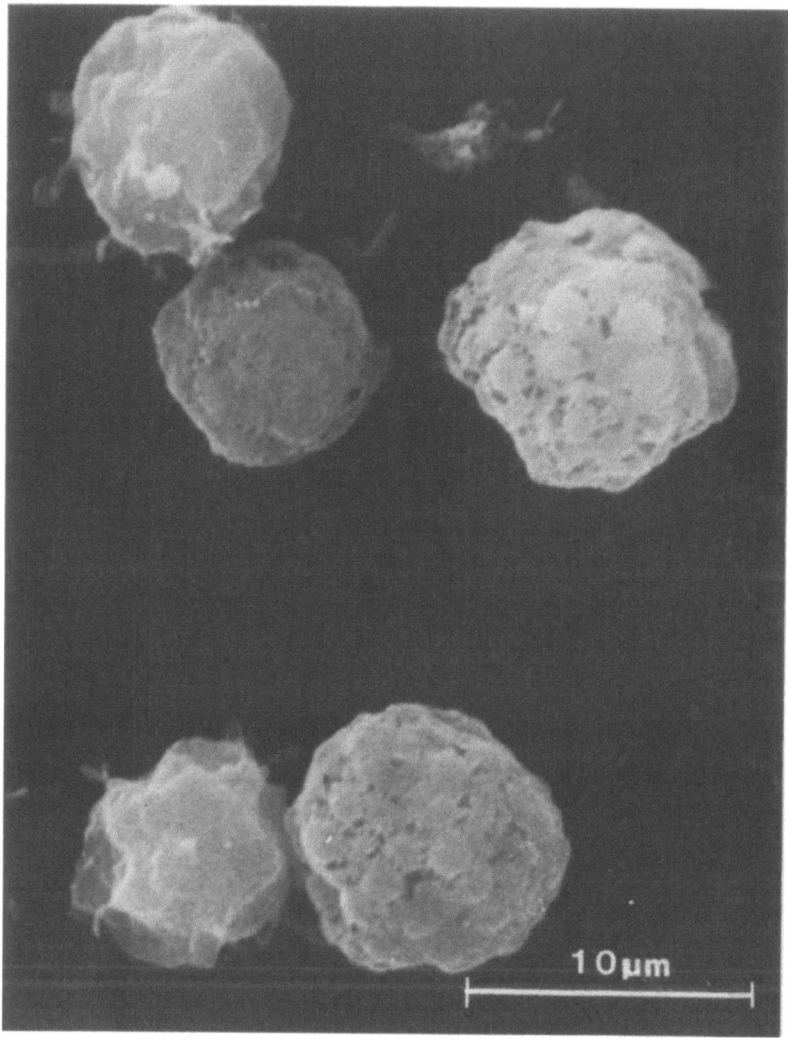

FIGURE 12. Scanning electron micrograph of small-lymphocyte-like cells in *Eisenia foetida*.

activity (Stein and Cooper, 1978) is probably related to phagocytic properties of worm leukocytes. It is now well known that leukocytes play a prominent role in graft rejection and immunological memory (details to follow). It has also been shown that like the vertebrate macrophages (Ehrenreich and Cohn, 1967), some earthworm leukocytes are even able to take up foreign proteins and metabolize them (Valembois *et al.*, 1973). For instance, when earthworm leukocytes are cultured with radioactive iodinated human serum albumin, their cytoplasm becomes radioactive after a few hours. As shown with autoradioelectrophoresis, products from hydrolysis of protein are released in the extracellular medium. Chloragocytes, known for a long time to be cells that perform trophic functions, also participate in defense mechanisms (Châteaureynaud-Duprat and Izoard, 1977b).

The role of amebocytes in wound healing and regeneration has been extensively investigated. Scar tissue is mainly constituted of leukocytes that also play a trophic role during the histogenesis which follows the formation of scar tissue (Liebmann, 1942a; Chapron, 1970; Burke, 1974).

The *in vitro* reactivity of earthworm leukocytes towards a few mitogens has been studied to see if these cells, like vertebrate lymphocytes, undergo DNA synthesis following the binding of certain nonantigenic ligands on their surface,

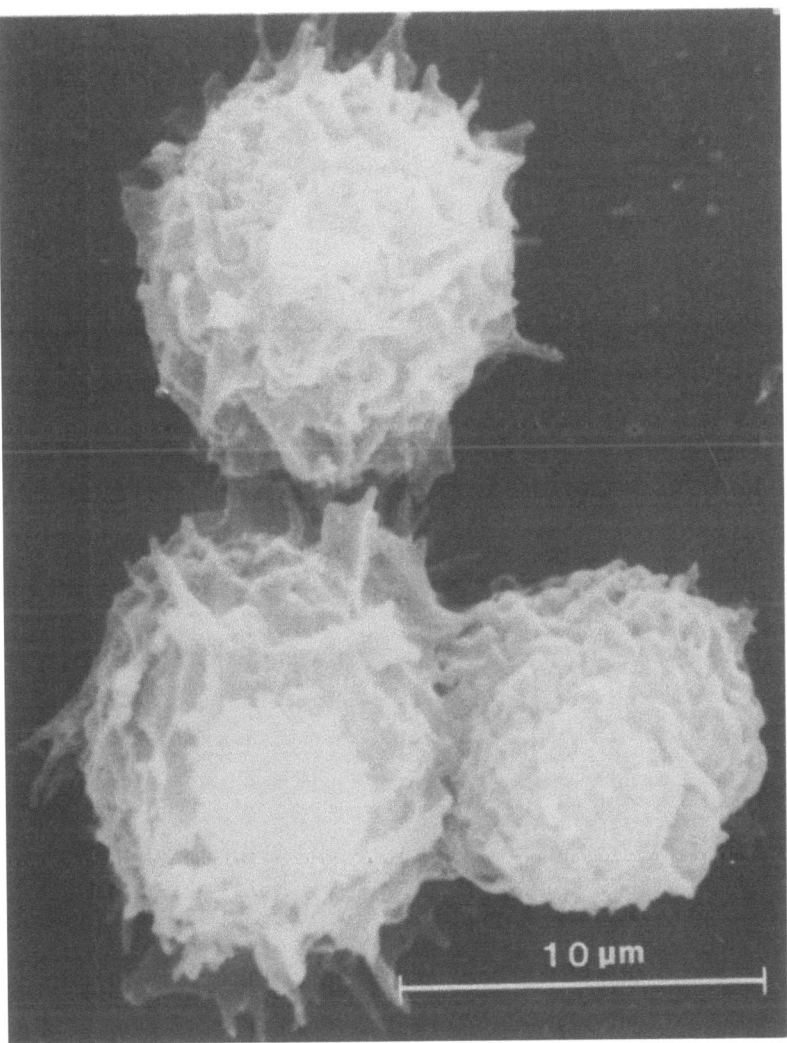

FIGURE 13. Scanning electron micrograph of lymphocytelike cells in *Eisenia foetida* after stimulation with Con A.

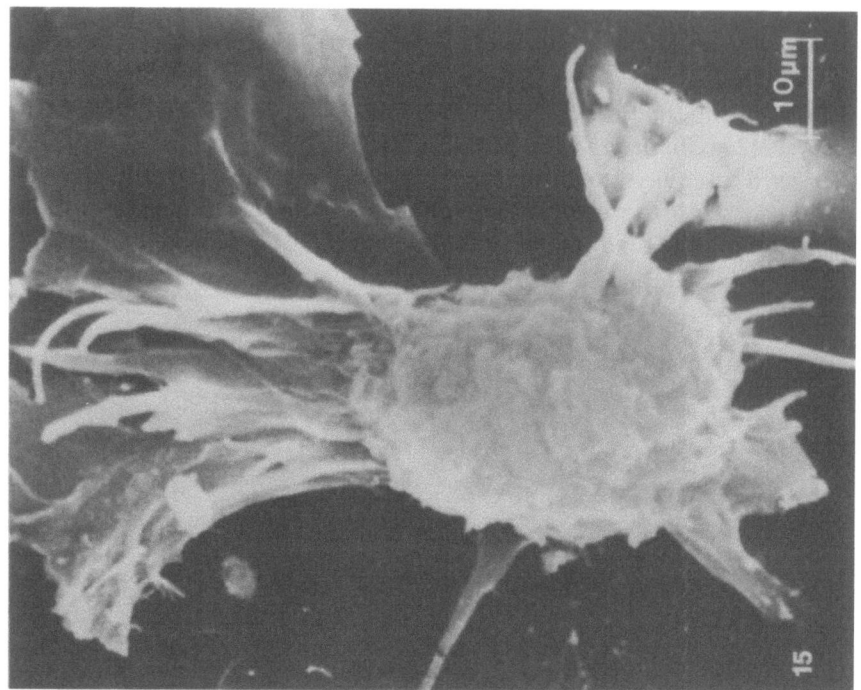

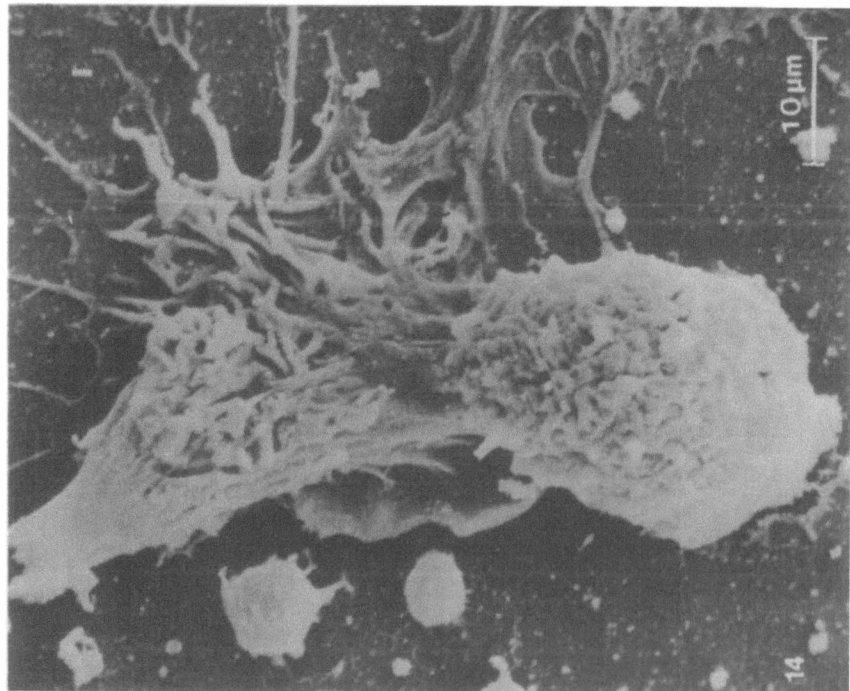

FIGURES 14 AND 15. Large macrophagelike cells in *Eisenia foetida* showing different kinds of pseudopodia, principally lamellipodia and axopodia.

and to provide an invertebrate model of differentiation adaptable to studies of cell surface dynamics.

Like vertebrate T lymphocytes, some leukocytes of *Eisenia foetida* respond to soluble Con A by increasing their DNA synthesis (Roch *et al.*, 1975; Roch, 1977). Stimulation occurs after 3–4 days of culture of the whole leukocyte population in the presence of 1–50 µg/ml soluble Con A (Figure 16b). The existence of Con A receptors on earthworm leukocyte membranes has been demonstrated with fluorescent and radioactive Con A (Roch and Valembois, 1978). Binding and

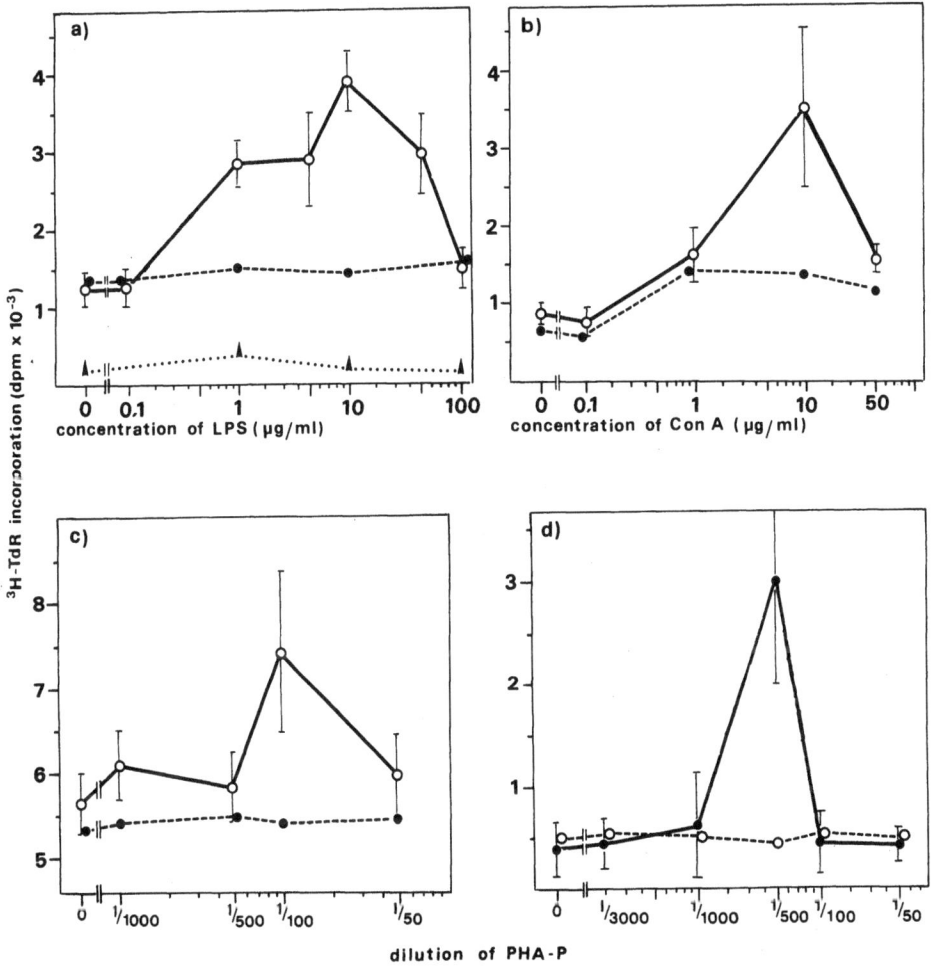

FIGURE 16. *In vitro* stimulation of DNA synthesis induced by mitogens in *Eisenia foetida andrei* leukocyte subpopulations (a) LPS stimulation after culturing for 3 days at 20°C. (▲) Chloragogue cells. (●) large adherent leukocytes; (○) small nonadherent leukocytes. (b) Con A stimulation of the whole leukocyte population. Incubation for 4 days (○) and 5 days (●) at 20°C. (c, d) PHA stimulation. (c) Small nonadherent leukocytes incubated for 3 days (○) and 4 days (●). (d) Large adherent leukocytes incubated for 5 days (●) and small nonadherent leukocytes incubated for 5 days (○). From Roch, 1977.

TABLE 1. CHARACTERISTICS OF *EISENIA FOETIDA* LEUKOCYTES INCUBATED WITH CON A[a,b]

Categories of leukocytes	Categories (%)	Cell diameter (μm)	Average length of Con A-induced villi (μm)
Nonadherent (40%)			
Spherical			
Capping	6	5 ± 1	1 (max)
Villosities		8 ± 1	3–7
Nonresponding	14	4–10	0
Irregular	80	8–15	0 ⎫ Pseudopodia and
Adherent (60%)		10–40	0 ⎭ lamellipodia

[a] From Roch and Valembois (1977).
[b] Incubation for 30 min at 4°C, in 100 μg/ml of Con A. Observations after 60–90 min redistribution at 20°C. The whole coelomic population was fracionated into "adherent" and "nonadherent" cells by passage through a short nylon wool column at room temperature.

stimulation are both inhibited by culturing the cells in the presence of methyl-α-D-mannose, the specific inhibitor of Con A-binding sites of T lymphocytes. The Con A receptors, originally scattered over the whole leukocyte membrane, become redistributed in patches and caps after 60 min incubation at 20°C. Capping occurs in 6% of the nonadherent leukocytes (Table 1). Electron microscopic observations reveal two patterns of Con A receptor redistribution in *Eisenia* (Roch and Valembois, 1977). The first pattern, associated with small undifferentiated cells (diameter 4 to 6 μm), involves classical capping (Figure 17) followed by endocytosis of receptor–ligand complexes. The second pattern concerns slightly more differentiated leukocytes (diameter 7 to 9 μm). These cells do not display true caps. Rather, they exhibit an intense membrane activation followed by a release of the microvilli carrying Con A receptors (Figure 18).

Coelomocytes from another earthworm, *Lumbricus terrestris*, were cultured with PHA and certain subpopulations were stimulated. This stimulation could not be achieved without the addition of adherent leukocytes. Among the adherent leukocytes, only trypsin-resistant cells (5 to 10%) with a high phagocytic activity, were capable of inducing the stimulation of the nonadherent leukocytes (Toupin and Lamouraux, 1976a). No separated leukocyte subpopulation could be stimulated by itself. In *E. foetida*, PHA stimulated DNA synthesis of the subpopulation of large adherent leukocytes after 5 days of culture (Figures 16c, d), and, to a lesser extent, stimulated the subpopulation of small nonadherent leukocytes after 3 days of culture (Roch, 1977). This is perhaps due to contaminations by some small macrophages. The PHA-induced stimulation of *Lumbricus* cells may indicate cooperation and/or inhibition between two cell types. Phenomenologically, these results are similar to those obtained with shark lymphocytes by Lopez et al. (1974).

Studies using LPS, a B-lymphocyte mitogen for vertebrates, suggest a physiologically heterogeneous leukocyte population of *E. foetida*. LPS is not mitogenic when it is cultured with unseparated leukocytes or with just the large adherent subpopulation (Roch, 1977). But the subpopulation of small nonadherent leukocytes (separated by nylon wool filtration) reacts with LPS concentrations ranging from 1 to 50 µg/ml after 3 days culture at 20°C (Figure 16a).

In conclusion, except for the ability to mediate specific reactions of humoral immunity, worm leukocytes seem to perform most of the reactions performed by vertebrate leukocytes. Worm leukocytes do not generally offer as large a variety of morphological aspects as vertebrate leukocytes. Consequently, invertebrate leukocytes are always less specialized and often less differentiated than vertebrate leukocytes. But, as shown by experiments with mitogens, invertebrate leukocyte populations with a uniform morphology may be physiologically heterogeneous. In invertebrates as in vertebrates, cooperation certainly occurs between cells morphologically similar but physiologically different.

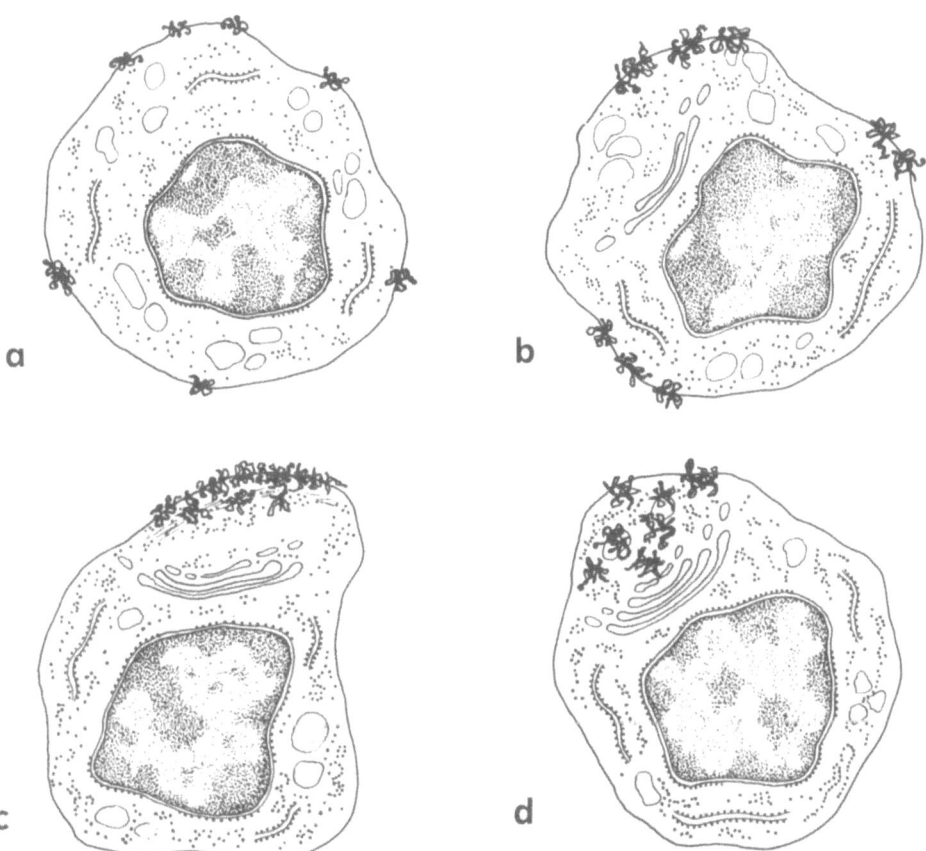

FIGURE 17. Redistribution pattern of the Con A-binding sites on undifferentiated small leukocytes of *Eisenia foetida* incubated with [^{3}H]-Con A. (a) Native distribution of the Con A-binding sites; (b) patch formations after 30 min redistribution; (c) cap formation; and (d) endocytosis of the Con A-binding sites after 2 hr redistribution. From Valembois *et al.*, 1977.

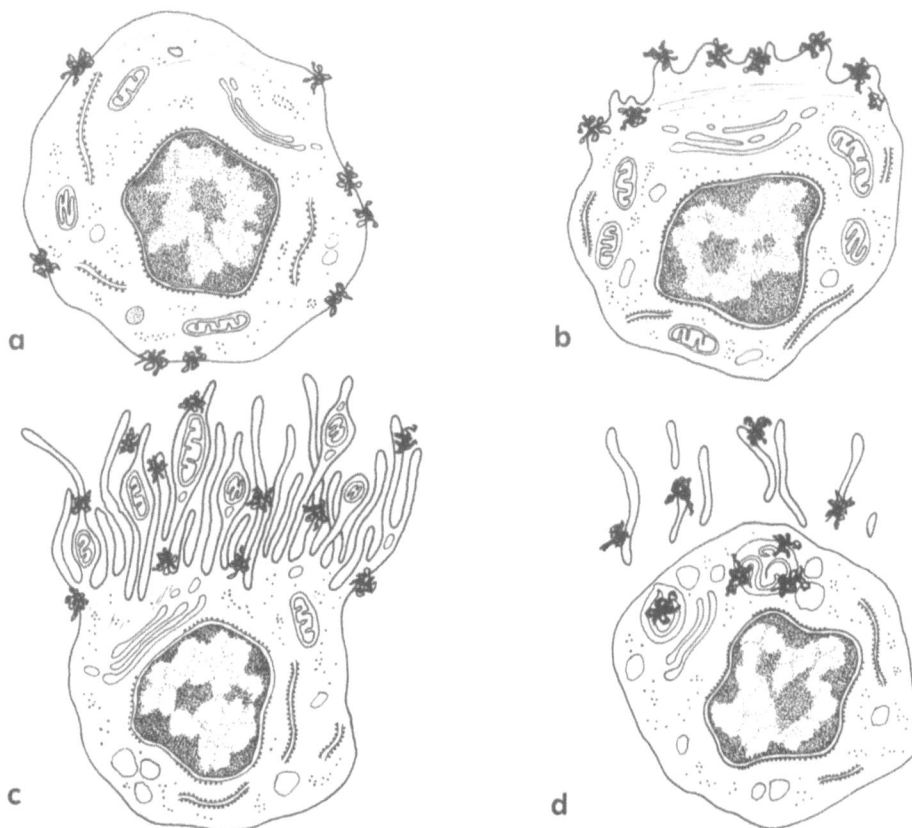

FIGURE 18. Redistribution pattern of the Con A-binding sites on slightly differentiated *Eisenia foetida* leukocytes incubated with [^{3}H]-Con A. (a) Native distribution; (b) repartition of the Con A-binding sites at one pole of the cell; (c) villi emission; (d) cut off and release of the villi containing the Con A-binding sites and some differentiated organelles. From Valembois *et al.*, 1977b.

3. THE GRAFT RESPONSE IN PLATYHELMINTHES, NEMERTEA, SIPUNCULIDA, AND ANNELIDA

For 80 to 100 years, numerous data have repeatedly shown that invertebrates are capable of rejecting grafts when a more or less strong genetic difference exists between host and donor tissues (see Loeb, 1945, for review). Nevertheless, until recently, it was thought that a rejection reaction involving specific recognition of the graft antigens and memory was the exclusive prerogative of vertebrates (Burnet, 1968; Hilgard, 1970). For the last 15 years, worms, particularly nemerteans and earthworms, have provided an ideal material with which to demonstrate specific recognition and immunological memory in some invertebrates.

We shall examine successively: (1) the behavior of an incompatible graft in worms; (2) the arguments for specificity and anamnesis of the grafted host's response; (3) the cellular aspects of the recipient's response; and (4) the degree of genetic heterogeneity required between donor and recipient to obtain rejection.

3.1. BEHAVIOR OF INCOMPATIBLE GRAFTS

In planarians (Platyhelminthes), grafting has been and is becoming a more and more commonly used tool for studying regeneration. The only experiments that constitute an attempt to understand the success or failure of grafts in planarians, however, are those of Lindh (1959). He showed that autografts, allografts, and certain intergeneric xenografts were not permanent, and the grafted tissues were progressively replaced by tissues arising from a proliferation of host neoblasts.

In nemerteans (Figure 19), which generally reject xenografts, only a few histological studies have attempted to define stages of the graft reaction. It appears that an infiltration of macrophagelike cells in grafted tissues occurs after partial necrosis of the graft (Langlet and Bierne, 1977). The rejection criteria used by these authors were the disappearance of pigmentation, a collapse of the marginal part of the graft, and an edematous swelling of the tissues (Langlet and Bierne, 1973).

Sipunculid worms possess an integument with an important connective component which makes exchanges of parts of the body wall extremely difficult. An attempt has been made to transplant different organs or cells such as tentacles (Triplett *et al.*, 1958), eggs, and sperm (Cushing *et al.*, 1965; Cushing and Boraker, 1975). Following such transplantations, both auto- and allograft tentacles were encapsulated by host leukocytes at similar rates. Homologous eggs that have not been chemically or physically altered are not recognized as nonself.

It is only in earthworms that the behavior of the grafted tissues has been detailed. Interesting data were available as early as the beginning of this century (Leypoldt, 1910, 1911; Korschelt, 1914, 1927). We shall now describe the acute rejection of a xenograft. Grafts, consisting of fragments of the dorsal body wall of an earthworm, *Allolobophora caliginosa*, were transferred onto another earthworm, *E. foetida*, and kept there with silver wire sutures. The grafted animals were maintained at 20°C. The body wall of the earthworms (Figure 5) is comprised of an epithelium formed by a single layer of cells covered by a collagenous cuticle. Beneath the epithelium lie (in order): a thin connective membrane, a layer of circular muscle fibers, a much thicker layer of longitudinal muscle fibers, and a thin layer of peritoneum. The muscle fibers are separated from each other by a connective framework of collagenous fibers. (For more details, see Stephenson, 1930; Avel, 1959; Laverack, 1963; Edwards and Lofty, 1977; Mill, 1978.)

In earthworms, the grafts are rarely eliminated in toto, but they generally degenerate progressively. As described by Valembois (1971a,b) and confirmed by Hostetter and Cooper (1972), an intense lysosomal activity occurs in the grafted tissues at about 4 to 8 days postgrafting. It appears that the grafts' muscle cells synthesize lysosomal enzymes which provoke self destruction of

FIGURE 19. Response of *Lineus* (Nemertea) to grafts. (a) A well-healed-in *L. longissimus* xenograft on *L. ruber* (17 days postgrafting). (b, c) Allografts exchanged between *L. ruber* of different pigmentation (100 days postgrafting). (d, e) First-set response of a xenografted *L. ruber* (d: 29 days postgrafting; e: 42 days postgrafting). From Langlet and Bierne, 1973.

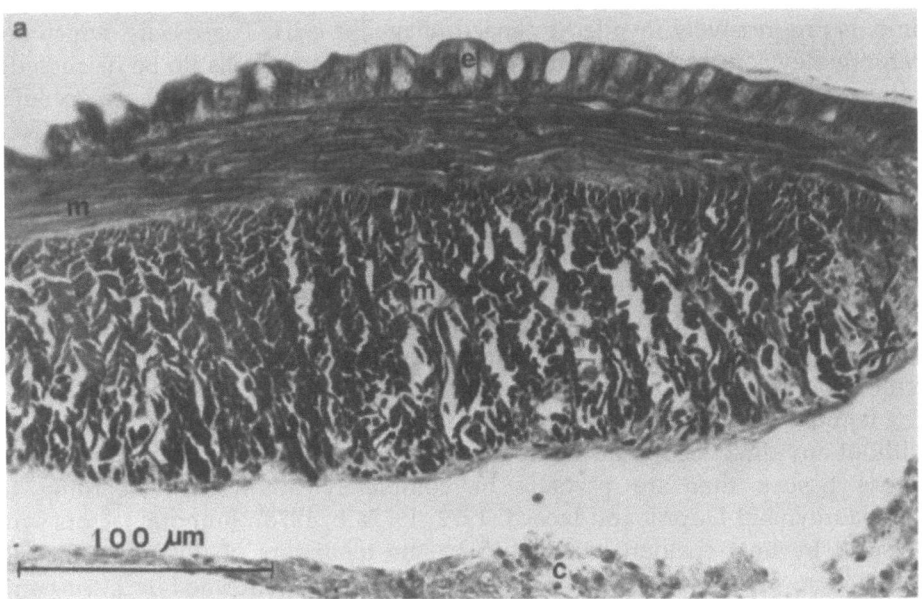
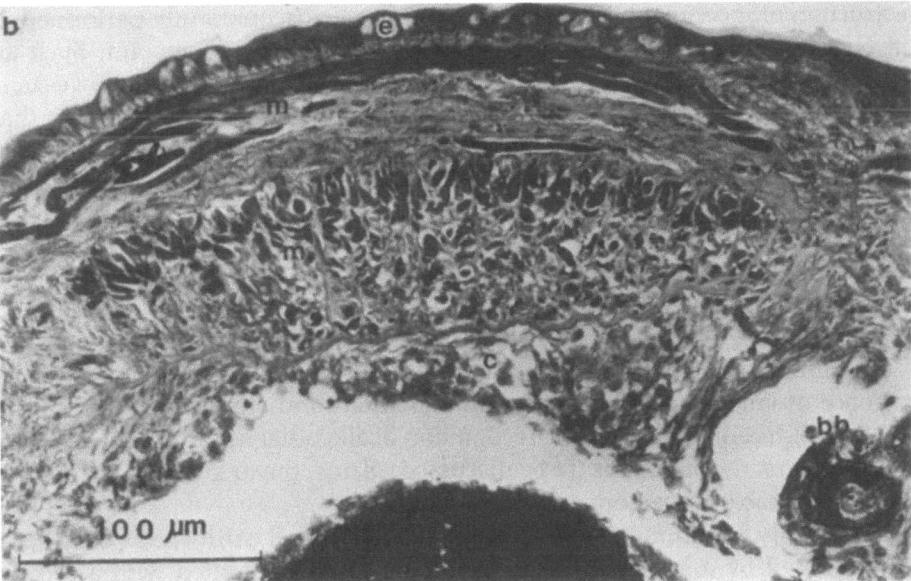

FIGURE 20. Comparison between first-set (a) and second-set response (b) in *Eisenia foetida* grafted with a fragment of dorsal body wall of *Allolobophora caliginosa*. Five days after a first graft, the epidermis (e) and the musculature (m) of the transplant are intact. At the same time, a second-set graft is partially destroyed and invaded by numerous coelomocytes (c). Brown bodies (bb) made of waste particles and coelomocytes are found in the coelomic cavity of the host.

the transplanted tissues (Valembois, 1968). By 12 days, the musculature is completely degenerated. The connective framework, which is undamaged at this time, is progressively invaded by macrophagelike cells. These cells, which apparently develop from small undifferentiated host stem cells (to be discussed), adhere to or engulf partially autolysed muscle cells. Macrophages, cellular waste, and pigmented granules are then enclosed in brown bodies (Figure 20b) which fall into the coelomic cavity (Valembois, 1963a, 1974). At the same time, myoblasts arising from the host tissues penetrate the graft connective framework and regenerate the muscle fibers. The connective framework is later destroyed and new collagenous fibers are synthesized by host fibroblasts. Finally, 2 months after grafting, a new body wall with a normal structure completely replaces the grafted tissues. Thus, because of the simultaneous occurrence of elimination of the grafted tissues and regeneration of new tissues, rejection is achieved without any breach in the integument.

It must be noted that the destruction of the graft in *Eisenia* is achieved without any direct action of the host leukocytes. These latter cells only phagocytose tissues that are partially or completely autolysed. According to Châteaureynaud-Duprat and Izoard (1972, 1977a,b, 1978), humoral factors synthesized by host coelomocytes could be the mediators of this apparent self destruction. These authors have compared rejections performed by *Eisenia* and by *Lumbricus* and have shown that *Lumbricus* leukocytes play a direct role in the destruction of incompatible grafts. An electron microscopic study performed in *Lumbricus* (Linthicum *et al.*, 1977b) seems to confirm this viewpoint. Such an acute xenograft response leading to a complete rejection within a few weeks is generally observed in earthworms. So-called chronic rejection of allografts that leads to a more or less complete graft destruction within 30 to 150 days has also been observed, but a detailed structural analysis has not been reported.

3.2. ARGUMENTS FOR SPECIFIC RECOGNITION AND IMMUNOLOGICAL MEMORY IN NEMERTEANS AND EARTHWORMS

Experiments of second-set grafting have not provided any clue as to the existence of immunological memory in planarians and sipunculid worms. The existence of receptors for specific recognition of alloantigens, however, has been shown *in vitro* in sipunculids (to be discussed). Thus, the following discussion is mainly applicable to nemerteans and annelids.

Specificity and immune memory in nemerteans and earthworms have been demonstrated by a hastened response of the recipient grafted with a second implant genetically identical with or nearly related to the first. Simultaneously with this second-set response, the sensitized recipient exhibits a first-set response against a third-party or unrelated graft. The specificity of the anamnestic component is clearly demonstrated.

A second argument that provides evidence of vertebratelike adaptive immune response in worms is the ability to adoptively transfer graft immunity from a sensitized animal to one nonsensitized by injecting the latter with leukocytes from the former.

In the following paragraphs, we shall review the characteristics of the second-set response and passive immunity.

3.2.1. Second-Set Response in Nemerteans and Earthworms

In nemerteans, Langlet and Bierne (1975, 1977) investigated the comparative features of first-set and second-set responses of *Lineus ruber* grafted with tissues from either *L. longissimus* or *L. sanguineus*. First-set xenograft of *L. sanguineus* had a median survival time (MST) of 16 days. The MST of the second-set response was about 9 days. These authors noticed that in nemerteans, the immune response is systemic, i.e., the rejection of second-set grafts transplanted to anatomical sites distant from the first grafts was still accelerated and intensified. They also noticed that the memory did not last more than 3 or 4 months in that a second-set graft performed after this delay survived as long as the first-set graft.

In earthworms, several differences between first-set and second-set responses can be noticed. Five days after transplantation of a second-set graft of *A. caliginosa* onto *E. foetida*, a large number of macrophagelike cells were seen underlying the graft; many of them were still invading the grafted tissues (Figure 20b). We previously reported that such an accumulation of leukocytes only occurred 12 to 15 days following first-set grafting. About 15 days postgrafting, the completely necrotic graft was being replaced by host tissues (Valembois, 1963b, 1971a,b). It should be recalled that a similarly complete first-set graft destruction (with identical graft size, donor origin, temperature conditions, etc.) did not occur before 1 or 2 months.

Second-set grafts of *E. foetida* onto *Lumbricus terrestris* (Cooper, 1968; Hostetter and Cooper, 1973, 1974) gave slightly different results from those obtained with the combination of *Allolobophora* onto *Eisenia*. First-set grafts of *Eisenia* onto *Lumbricus* were either slowly rejected (39% rejected in 36 ± 11 days) or quickly rejected (52% in 15 ± 3 days). Curiously, second-set grafts in slow rejectors were rejected in an accelerated fashion (17 ± 3.7 days), but worms of the fast rejector group required a longer time for second-set graft rejection (24 ± 5 days). This unusual negative memory has been described in other taxonomic groups (Hildemann and Cohen, 1967; Cooper and Aponte, 1968; Cohen and Borysenko, 1970).

The second-set response in earthworms is systemic and memory lasts for 1 to 2 months. Duprat (1964, 1967), Cooper (1969b), and Châteaureynaud-Duprat (1970), investigating the allogeneic reactivity of *E. foetida*, found that allografts between worms from different geographical locations were destroyed, but generally in a prolonged or chronic fashion. They almost always observed a faster rejection of second-set implants. Duprat (1967) exchanged allografts between animals from the same colony and generally observed an integration of the first-set graft with the host tissues. Under the same conditions, second-set grafts were always attacked, and sometimes destroyed, by the host phagocytes.

Parry (1978) denied an anamnestic component of the graft reactions of earthworms since he did not observe a heightened response following second-set grafting. His results, however, may relate to the fact that the interval between first-set and second-set grafting was only 5 days. Roch (1973) has shown that such a short interval only allows a very slightly stimulated response when grafted animals are maintained at 20°C. Parry's experiments were performed at 15°C, and this lower temperature probably caused a further delay before the immune response. Dales' results (1978), which also disagree with the existence of an enhanced second-set response, could have been due to the low temperature of the breeding medium (12°C) and to a too long interval between first-set and second-set grafting.

3.2.2. Adoptive Transfer of Transplantation Immunity

It has been known since 1953 (Mitchison) that graft immunity in vertebrates can be passively transferred from immunized to nonimmunized animals by transferring the lymphocytes. This acquired immunity is specific since a memory response is demonstrable only against grafts possessing antigens analogous to those carried by the sensitizing tissues.

A passive acquisition of adoptive immunity in earthworms was first demonstrated by Duprat (1967). She transferred blood cells or coelomic fluid cells from immune *E. foetida* to naive *Eisenia* from the same geographical origin. Adoptive transfers of xenograft immune responses were later obtained by Cooper (1970), Bailey *et al.* (1971), and Valembois (1971a,b) at 20°C. Adoptively immunized *E. foetida* exhibited a maximum reactivity only 4 days following the challenge. Roch (1973) reported that leukocytes or pieces of peritoneum carrying leukocyte stem cells from passively, but not actively, immunized animals can also confer adoptive immunity. Animals immunized by such a secondary transfer are still able to confer, by a tertiary transfer, an adoptive immunity to a normal worm. After this tertiary transfer, however, the level of sensitization is less high than after a primary or secondary transfer.

3.3. CELLULAR ASPECTS OF THE HOST RESPONSE TO GRAFTS

Little data exist concerning the role of leukocytes in graft rejection in planarians and nemerteans. In sipunculid worms, it may be hypothesized that implants are recognized and attacked by leukocytes similar to those mediating *in vitro* cytolysis of xenogeneic or allogeneic erythrocytes (to be discussed). It is only from studies in earthworms that one begins to understand the role of invertebrate leukocytes in graft rejection. Adoptive transfers of graft immunity in these animals have shown an active coelomocyte participation in graft reaction. That leukocytes are implicated in the graft destruction process is now well known from analyses of the rejection process by light microscopy (Hostetter and Cooper, 1972, 1974) and electron microscopy (Valembois, 1970, 1971b;

Linthicum et al., 1977b). The cells seen adhering to and destroying the grafted tissues have been called either macrophagelike cells, or neutrophilic or granulocytic coelomocytes by American authors. These cells are generally large (10 to 30 μm) and irregular in shape with an eccentric nucleus (Figure 11). Their cytoplasm is characterized by a granular component comprised of dense granules. In old cells, these granules are often bigger but less numerous and their content is less dense. The generally well-developed Golgi apparatus is often surrounded by different kinds of vesicles: coated vesicles, vesicles filled with moderately to extremely dense material, multivesicular bodies, etc. Gomori's method for detection of acid phosphatase in electron microscopy has shown a positive activity both in dense granules and in the vesicular component surrounding the Golgi apparatus. Thus, dense granules may be equated with lysosomes. These cells are sometimes seen to have engulfed phagocytosed fragments of grafted tissues, but they can also be seen closely adhering to the cells of the grafts, suggesting that leukocytes can also destroy the graft by a contact-mediated cytotoxic action.

It seems that the large cells which attack grafted tissues differentiate from small undifferentiated cells (8 μm) that are characterized by numerous free cytoplasmic ribosomes (Valembois, 1971c). An eventual proliferation of small undifferentiated cells during the graft response has been investigated by several authors. Lemmi (1974) measured the *in vivo* incorporation of tritiated thymidine ([^{3}H]-TdR) by leukocytes of xenografted earthworms. The greatest incorporation of [^{3}H]-TdR, found 4 days postgrafting, seemed to result from nonphagocytic leukocytes. Lemmi also reported that stimulation of [^{3}H]-TdR incorporation is not only localized in the region of the graft, but is generalized throughout the coelomic cavity. These data agree with ours (Roch et al., 1975). We harvested leukocytes from the coelom of grafted animals 4 days postgrafting, and maintained these cells *in vitro* for 16 hr in a synthetic medium containing [^{3}H]-TdR. The incorporation of [^{3}H]-TdR was followed by measuring the radioactivity both by scintillation spectrometry and by autoradiography in light and electron microscopy. Two kinds of cells incorporated [^{3}H]-TdR (Valembois and Roch, 1977). Some of them were large vesicular cells exhibiting pseudopodia (Figure 21). Since such cells were also stimulated following a single wound, it seems that they are mainly implicated in inflammatory processes. The majority of the cells that incorporated [^{3}H]-TdR were small cells (6 to 10 μm) with a high nucleus/cytoplasm ratio and a well-developed free ribosome content (Figure 22). These cells can ordinarily pass through a nylon wool column, and are therefore considered nonadherent. Roch (1979a) has studied the stimulation of DNA synthesis in nonadherent cells following xenografts and allografts. First-set xenografts in *Eisenia* induced stimulation of nonadherent cells in 23 to 40% of the combinations studied. In the case of the combination of *Lumbricus terrestris* onto *E. foetida*, positive stimulations are found in 40%; 13% of these stimulations are considered as corresponding to high responders (stimulations 10 times higher than the controls) and 27% corresponding to low responders (stimulations lower than 10 times the controls). These results may be related to those of Cooper (1968) who found that first-set graft rejections of *Lumbricus* by *Eisenia* occurred in an acute

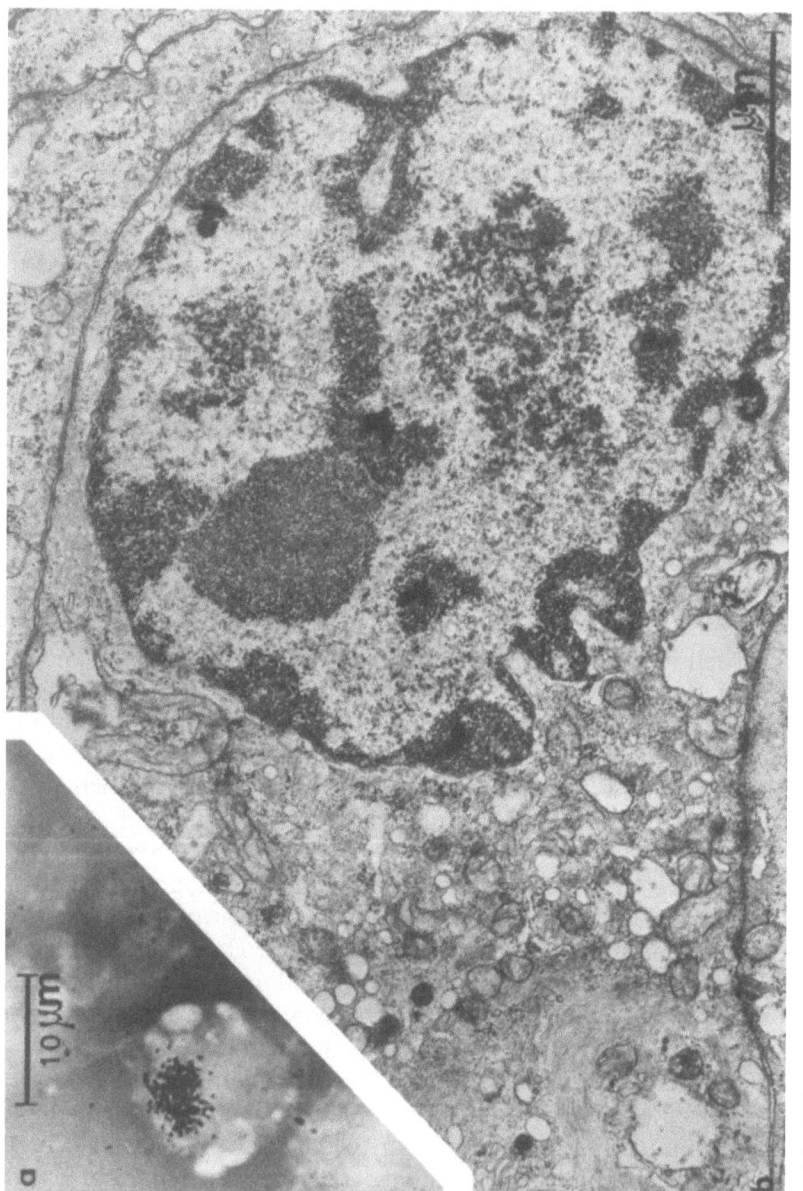

FIGURE 21. (a) Low-resolution radioautograph of a macrophagelike cell after incubation for 16 hr in [³H]-TdR (4 days postgrafting). (b) High-resolution radioautograph of the same cell. As explained in the text, the stimulation of macrophagelike cells seems to be related to inflammatory processes. From Valembois and Roch, 1977.

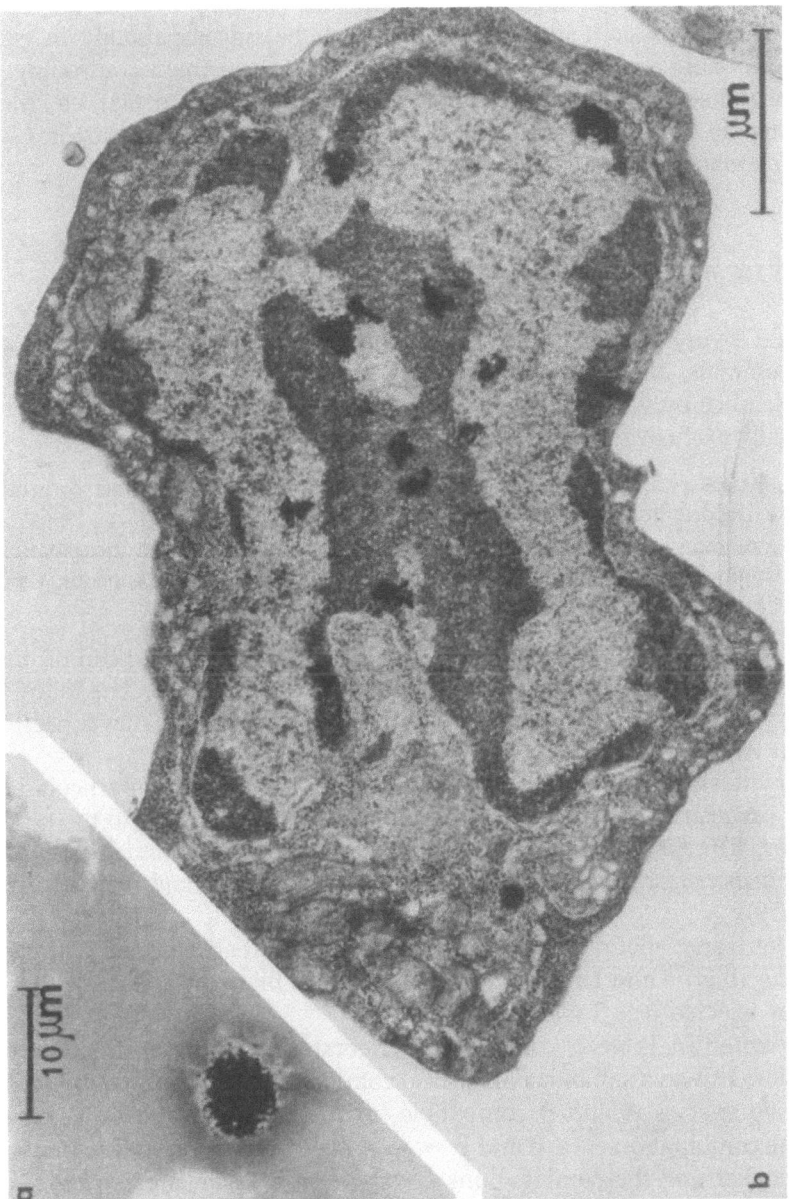

FIGURE 22. (a) Low-resolution autoradiograph of a small-lymphocyte-like cell in *Eisenia foetida* (under the same conditions as those indicated in Figure 21). (b) High-resolution autoradiograph of the same cell. The lymphocytelike cells would be implicated in processes of specific rejection (see text). From Valembois and Roch, 1977.

fashion in 15% of the experiments, a rapid chronic rejection in 25% of the animals, and a chronic rejection from 20 to more than 100 days in 58% of the combinations. Allograft combinations performed among members of the same colony or between geographically distant colonies of *E. foetida andrei* demonstrated similar percentages of positive stimulations (ranging from 23% to 30%).

From this discussion, it may be concluded that incompatibility in earthworms is probably controlled by several genes, some of them controlling the cytolytic response(s), others controlling the proliferative response(s), etc. Moreover, it must be noticed that an MLR-like reaction also seems to exist in earthworms (Valembois *et al.*, 1981).

3.4. GENETIC BASIS OF INCOMPATIBILITY

While it is very difficult to prevent the rejection of an allograft in vertebrates such as mammals, allografts are sometimes accepted in invertebrates. Such an acute intolerance by vertebrates when compared to invertebrates can have two (not mutually exclusive) meanings:

1. Vertebrates possess a greater number of polymorphic genetic loci controlling histocompatibility reactions than invertebrates.
2. In vertebrates, one haplotype difference is enough to induce an incompatibility reaction, while several haplotype differences are needed to induce incompatibility reactions in invertebrates.

Both points are probably right, but until now the genetic control of histocompatibility has been very poorly investigated in invertebrates. The following discussion will concern the different graft combinations that have been attempted in worms.

As previously stated, allografts are successful in planarians, particularly in the genus *Euplanaria*. The behavior of xenografts varies according to the type of combination. They either succeed, as in the combination *Euplanaria polychroa* → *Euplanaria lugubris*, or fail (e.g., *Euplanaria dorotocephala* → *Euplanaria maculata*) (Lindh, 1959).

In nemerteans, allografts always succeed. Xenografts performed within the genus *Lineus* (Bierne and Langlet, 1974; Langlet and Bierne, 1977) can either be accepted or rejected (see Table 2).

As indicated in Table 2, contrary to vertebrates or earthworms (to be discussed), *Lineus* always exhibits a reciprocal compatibility or incompatibility. So, between two species A and B compatible, or incompatible, the same results occur in the combination A → B and B → A. A slight exception to this rule may be noted concerning the combinations *L. ruber* → *L. lacteus* and *L. ruber* → *L. pseudolacteus*. Such a reciprocal incompatibility or compatibility suggests that rejection only occurs when donors and recipients differ by at least two haplotype differences.

In sipunculid worms, it is technically very difficult to perform grafting studies. However, an *in vitro* study of the spontaneous cytotoxic activity of leuko-

TABLE 2. COMPARATIVE FIRST-SET XENOGRAFT REJECTION TIMES IN *LINEUS* [a]

Donor → recipient combination	Graft rejection time
L. sanguineus → *L. pseudolacteus* *L. pseudolacteus* → *L. sanguineus*	No signs of rejection (> 1000 days)
L. sanguineus → *L. lacteus* *L. lacteus* → *L. sanguineus* *L. pseudolacteus* → *L. lacteus* *L. lacteus* → *L. pseudolacteus*	Delayed chronic rejection (> 200 days)
L. ruber → *L. viridis* *L. viridis* → *L. ruber* *L. ruber* → *L. lacteus* *L. ruber* → *L. pseudolacteus*	Chronic rejection (> 25 days)
L. lacteus → *L. ruber* *L. pseudolacteus* → *L. ruber*	Subacute rejection (18–25 days)
L. ruber → *L. longissimus* *L. sanguineus* → *L. ruber* *L. ruber* → *L. sanguineus*	Acute rejection (18 days or less)

cytes against allogeneic or xenogeneic erythrocytes (Boiledieu and Valembois, 1976, 1977a) has shown that in sipunculids, as in nemerteans, an incompatibility reaction only occurs when at least two haplotype differences exist (see further on).

An attempt has been made in Table 3 to briefly summarize the different combinations which have been investigated in oligochaetes (see Omodeo, 1956, and Bouché, 1972, for information on systematics).

When grafts are exchanged between worms belonging to two different families, for instance *Eudrilus* and *Lumbricus* (Cooper, 1969a) or *Hormogaster* and *Lumbricus* (Izoard, 1971), the "rejection" occurs within a few hours.

In most cases, xenografts between animals belonging to different species of lumbricids are rejected in an acute or chronic fashion. Nevertheless, exceptions have been found with regard to combinations between different species of *Lumbricus*. When a fragment of body wall of *Lumbricus casteneus* is grafted, for instance, onto a *Lumbricus terrestris* host, most of the time the graft remains permanently viable, if donor and host come from identical biotopes (Izoard, 1964, 1972). But if the biotopes are different, for eample, if grafts are exchanged between a *Lumbricus castaneous* found in decomposed plants and a *Lumbricus terrestris* coming from ordinary earth, the graft is rejected within 3 to 6 weeks after grafting. Proteins extracted from tissues of donors and recipients have been compared (Du Pasquier *et al.*, 1966; Izoard, 1973). The antigenic identity of both *Lumbricus terrestris* and *Lumbricus castaneus* is found when the animals arise from similar biotopes, whereas antigenic differences are generally noticed when the biotopes are different.

TABLE 3. SUMMARY OF THE DIFFERENT COMBINATIONS INVESTIGATED IN OLIGOCHAETA

Donors	Recipients											
	Hs	Ee	Ef	Ea	Lt	Lc	Lr	Aca	Ach	At	Dr	Dm
Hormogaster sp. (Hs)						X						
Eudrilus eugeniae (Ee)			X		X							
Eisenia foetida foetida (Ef)			X	X	X	X		X		X	X	
Eisenia foetida andrei (Ea)			X	X	X	X						
Lumbricus terrestris (Lt)			X	X	X	X	X			X		
Lumbricus castaneus (Lc)	X		X		X	X	X					
Lumbricus rubellus (Lr)			X		X	X	X					
Allolobophora caliginosa (Aca)			X	X	X			X	X			
Allolobophora chlorotica (Ach)								X	X			
Allolobophora trapezoides (At)			X		X							
Dendrobaena rubida (Dr)			X									X
Dendrobaena mammalis (Dm)											X	

Duprat (1964) showed that first-set allografts between *Eisenia* from the same geographical location healed-in and remained intact. A second-set allograft was always infiltrated by host phagocytes and was then partially or completely destroyed. Cooper and Baculi (1968) and Cooper and Rubilotta (1969), also studying *Eisenia* allografts, found that some of the first-set grafts were rejected. Rejections, when observed, occurred in a chronic fashion within 15 to 255 days. Similar results were obtained for allografts in *Lumbricus terrestris* (Cooper, 1969b).

When first-set allografts were exchanged between worms that came from different geographical locations (Duprat, 1964, 1967), two kinds of results were observed. Certain hosts attacked and partially destroyed the grafts. With other kinds of combinations, no significant attack or destruction was observed. The author suggested that the discrepancy found between allografts combining animals from the same location and from different locations could be attributed to the development of genetic homogeneity resulting from interbreeding. Duprat (1967) has studied graft exchanges between two populations of *Eisenia*: one from Talence (southwestern France), the other from Douai (northern France). First-set allografts from Douai donors onto Talence recipients provoked an acute reaction of the donors. But only a slight chronic reaction occurred when Talence worms were grafted on Douai recipients. Information concerning the genetic control of incompatibility is further demonstrated by the existence of this nonreciprocal incompatibility in *Eisenia* (Du Pasquier, 1974). The nonreciprocity of the incompatibility reaction is probably due to the fact that in worms, as in vertebrates, a one-haplotype difference is sufficient to lead to a rejection. A significant experiment performed with F_1 hybrids obtained from Bergerac and Talence (two mutually incompatible populations) reinforces the previous hypothesis. This experiment consisted of grafting F_1 hybrids onto their parents. The F_1 hybrids probably differed from each of their parents by one haplotype. In all cases, grafts of F_1 hybrids onto their parents were rejected (Duprat, 1967). Earthworms, as compared to nemerteans and sipunculids, seem to have developed a stricter system of histocompatibility and therefore reached a higher level of evolution.

4. *IN VITRO* CYTOPATHIC ACTIVITIES OF LEUKOCYTES AND BODY FLUID

The *in vitro* study of the activities of leukocytes and body fluid has greatly increased our knowledge of vertebrate and invertebrate defense mechanisms. Two essential kinds of cytopathic activities have been investigated: cytotoxic reactions that require a contact between effector cells (generally leukocytes) and target cells; and humoral cytopathic activities mediated by substances released in the extracellular compartment.

4.1. CYTOTOXICITY IN SIPUNCULIDS AND EARTHWORMS

The lysis of bacterial or animal cells or tissues can be mediated by different kinds of cytotoxic leukocytes. To our knowledge, no cytotoxicity-like mechanism has been found in platyhelminths or nemerteans. In vertebrates, four models of cytotoxicity can usually be distinguished:

- In T-dependent cytotoxicity, the cytolysis of target cells is mediated specifically by T lymphocytes from immunized donors.
- In macrophage-dependent cytotoxicity, the cytotoxic reaction is performed either specifically by macrophages from immunized donors, or nonspecifically by marcophages activated *in vitro* or *in vivo*.
- In antibody-dependent cell-mediated cytotoxicity, the complement-independent lysis of target cells coated with specific antibodies is nonspecifically achieved by killer lymphocytes obtained from nonimmunized donors.
- In natural killing, cytotoxicity is probably specifically mediated by cells arising from nonimmunized donors. The effector cells are neither T cells, B cells, nor macrophages, but are a particular lymphocytelike cell called NK cells.

4.1.1. Cytotoxicity in Sipunculids

No experimental study of the *in vitro* activity of leukocytes was undertaken before 1976, although it has long been known that homologous coelomic fluid is a good culture medium for sipunculid cells (Thomas, 1932). The fact that sipunculids generally possess a great number of leukocytes is also an advantageous factor when performing experiments requiring multiple assays and controls. Indeed, in the species *Sipunculus nudus*, each individual possesses on an average 40×10^6 leukocytes as well as several hundred million erythrocytes, which can be used as target cells.

When leukocytes of S. *nudus* are mixed *in vitro* with xenogeneic or certain allogeneic erythrocytes (20 : 1 leukocyte : erythrocyte ratio), a cytolysis of about 25 to 40% of target erythrocytes (Figure 23) occurs within 4 to 6 hr (Boiledieu and Valembois, 1976). This cytolytic activity arises spontaneously, i.e., it occurs without any previous sensitization of the leukocyte donors. This feature is characteristic of murine NK cells (Greenberg and Playfair, 1974; Herberman et al., 1975; Kiessling et al., 1975; Zarling et al., 1975). The *in vitro* cytotoxic activity of *S. nudus* is also specific (Boiledieu and Valembois, 1977a): if animal A is injected *in vivo* with erythrocytes from animal B (A and B are reciprocally incompatible), the *in vitro* cytotoxic activity of A leukocytes against B erythrocytes regularly decreases and, with time, finally disappears temporarily (Figure 24). This decreased leukocyte aggressivity (animal A) is specific in that it does not occur with erythrocytes other than those that come from animal B. The sipunculid leukocytes must be in close contact with the target cells in order to exert their cytotoxic activity (Figures 25–27) (Boiledieu and Valembois, 1977b). The effector leukocytes mediating cytotoxicity are small cells resembling vertebrate lymphocytes with many similar features in their fine structure (Valembois and Boiledieu,

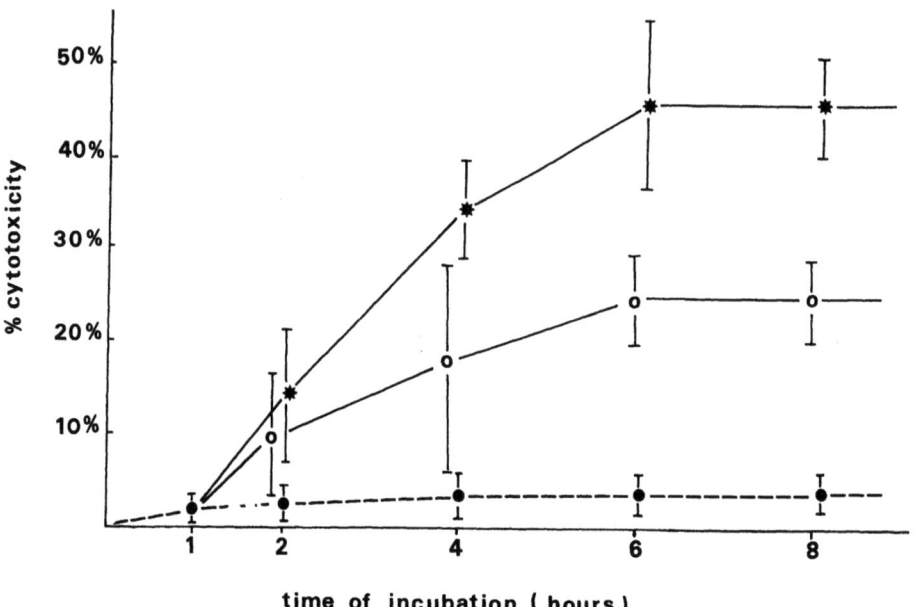

FIGURE 23. Comparison between the kinetics of xenogeneic and allogeneic cytotoxicity in sipunculids. The ratio of leukocytes to erythrocytes is 25:1. Means ± S.E., calculated from at least three independent incubation mixtures. (★) Cytotoxic effect of leukocytes from *Siphonosoma arcassonense* on target erythrocytes from *Sipunculus nudus*. (○) Allogeneic cytotoxicity between cells from incompatible *Sipunculus nudus*. (●) Percentage of spontaneous death of target cells in control wells. From Boiledieu and Valembois, 1977a.

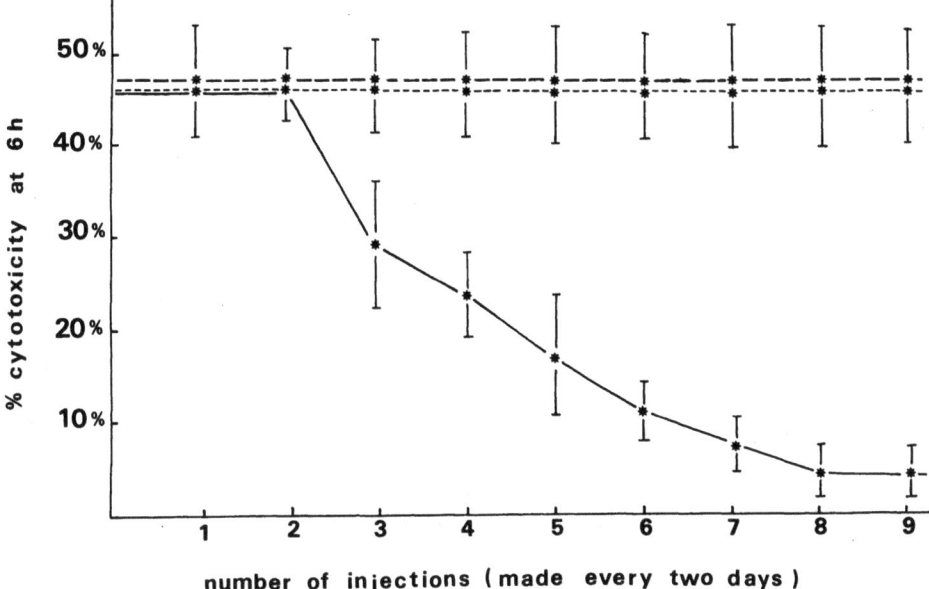

FIGURE 24. The cytotoxic effect of leucocytes from *Sipunculus nudus* injected several times with erythrocytes from *Siphonosoma arcassonense* (solid line) was abolished when it was tested against the sensitizing erythrocytes. Cytotoxicity remained significant when tested against erythrocytes from another *Siphonosoma arcassonense* (dotted line) or from *Phascolosoma vulgare* (dashed line). From Boiledieu and Valembois, 1977a.

1981). The occurrence of polypeptide synthesis during the cytotoxic reaction has been demonstrated (Boiledieu and Valembois, 1977b, 1978). It has also been shown that a leukocytic factor directly involved in cytolysis is a phospholipase which can act on the target cell membrane.

In sipunculids, a compatibility or an incompatibility shown by a positive cytotoxic reaction is always reciprocal. Thus, an incompatible reaction probably occurs only when the donor of leukocytes and the donor of erythrocytes differ by at least two haplotypes. For instance, no cytotoxicity would occur when animals MM and MN are combined, but cytotoxicity could occur between animals MN and PQ.

This *in vitro* mechanism of recognition may have a similar or identical counterpart *in vivo*. Such a mechanism could be involved, for instance, in the segregation of injected foreign eggs, observed by Cushing and Boraker (1975).

To conclude, it must be pointed out that the spontaneous cytotoxicity of sipunculids is very similar to the so-called natural killing in mammals. This kind of cytotoxicity seems to be the most primitive known, and many authors consider natural killing as the primordial system of immunosurveillance (Baldwin, 1977; Cudkowicz and Hochman, 1979).

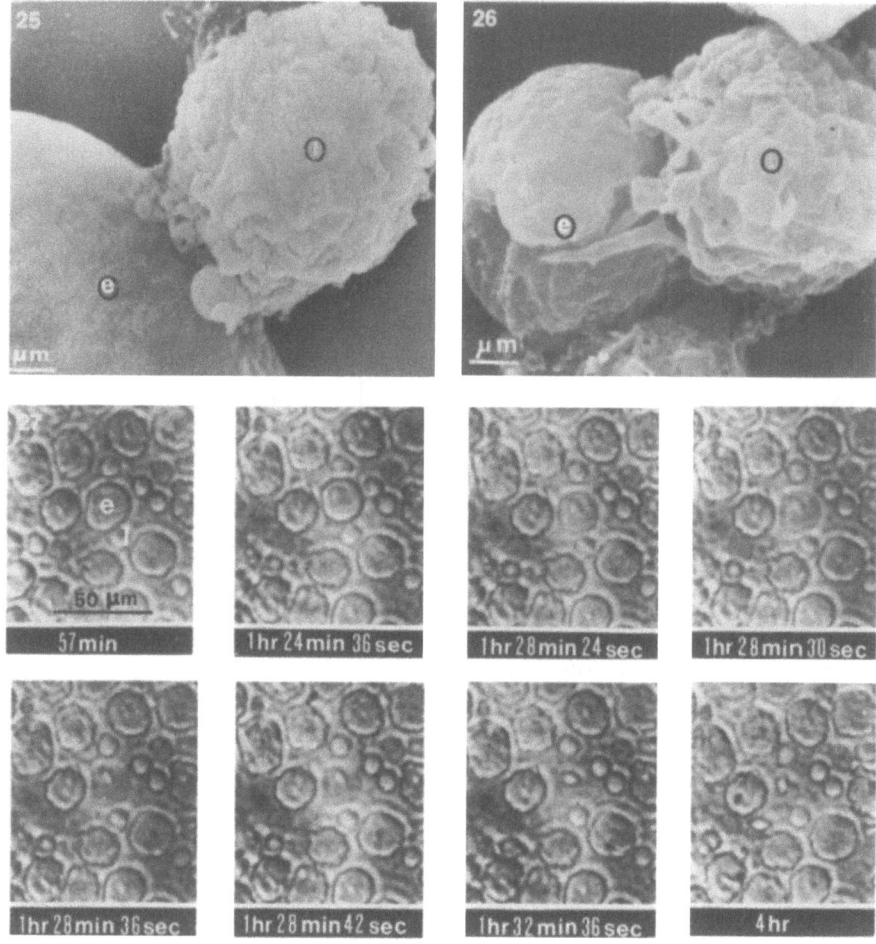

FIGURES 25–27. *In vitro* cytotoxic activity of sipunculid leucocytes against xenogeneic or allogeneic erythrocytes.
FIGURE 25. SEM of a leucocyte (1) of *Siphonosoma arcassonense* in close contact with a target erythrocyte (e) of *Sipunculus nudus*.
FIGURE 26. Close contact between a leucocyte (1) of *Sipunculus nudus* and an incompatible allogeneic target erythrocyte (e).
FIGURE 27. Sequence from time lapse film showing the killing of an erythrocyte (e) by a leucocyte (1). Each picture of the sequence is identified by the time (hr, min, sec) after the beginning of incubation. At 1hr 28 min 24 sec the erythrocyte is refringent. At 1 hr 28 min 42 sec, (18 sec later), the disintegration process of the target cell is finished: only the nucleus of the erythrocyte is visible. After 4 hours, the killer leukocyte is almost at the same place. Figures from Boiledieu and Valembois, 1977c.

4.1.2. Cytotoxicity in Earthworms

In vitro cytotoxicity in earthworms has not been as extensively studied as it has in sipunculids. Cooper (1973) found a weaker cell viability in xenogeneic and allogeneic coelomocyte mixtures than in autologous mixtures of the species

Lumbricus terrestris and *E. foetida*. His data, however, need to be confirmed in experiments that utilize more satisfactory culture conditions. Cultures of tissue fragments in the presence of allogeneic or xenogeneic tissues, coelomocytes, or cell-free coelomic fluid (Châteaureynaud-Duprat and Izoard, 1972, 1977a,b) indicate that both cellular and humoral cytopathic activities can be exerted *in vitro* by earthworm coelomocytes.

Recently, an attempt has been made to test, *in vitro*, the cytotoxicity of earthworm leukocytes against allogeneic target leukocytes by the ^{51}Cr assay. Preliminary results suggest the existence of two mechanisms of cytotoxicity: one occurs spontaneously as in sipunculids, the second is inducible by sensitizing grafts *in vivo* (Valembois *et al.*, 1980).

4.2. PROTECTIVE ACTIVITIES OF THE BODY FLUID

Extracts prepared from the tissues of several species of platyhelminths (Hyman, 1951) and nemerteans (Bacq, 1937) are highly toxic. Such products are generally considered to be protective factors that ward off predators or parasites (for review, see Halstead, 1965; Sparks, 1972). These substances probably have no relevance to leukocytic defense, in contrast to some natural toxins found in body fluids (blood, coelomic fluid, etc.) of coelomates. In the following sections, humoral protective factors of annelids and sipunculids have been classified in relation to their biological activity.

4.2.1. Hemolysins

4.2.1a. In Earthworms. A lipoprotein found in the coelomic fluid of *E. foetida* can hemolyze various vertebrate erythrocytes, including those of frogs, sheep, and humans (Du Pasquier and Duprat, 1968). It has no effect on homologous cells (Du Pasquier, 1971). This natural hemolytic factor is characterized by a low electrophoretic mobility, an ability to react at high dilutions (hemolysis of SRBC occurs at a dilution of 1 : 4000), and a heat lability at 56°C for 15 min. As determined by analytical isoelectric focusing (IEF), the hemolytic factor of *E. foetida andrei* appears to be comprised of a polymorphic system of four molecular units, each of which is characterized by a unique isoelectric point that ranges from 5.9 to 6.3 (Roch, 1979b). Each molecular unit possesses a lytic activity against SRBC and seems to have a molecular weight of about 45,000. Among all the populations tested, the different units are arranged to form six different patterns (Figure 28). Genetic studies on the transmission of the hemolytic patterns indicate that the hemolytic factor could be determined by two different genetic systems. One gene is always expressed, and a second gene possesses three alleles (Roch, 1979b). The Jerne plaque test applied to washed coelomocytes has shown synthesis and release of the hemolytic factor by chloragogue cells (Du Pasquier and Duprat, 1968). Other authors have also obtained hemolysis of a variety of vertebrate erythrocytes with a supernatant of tissue

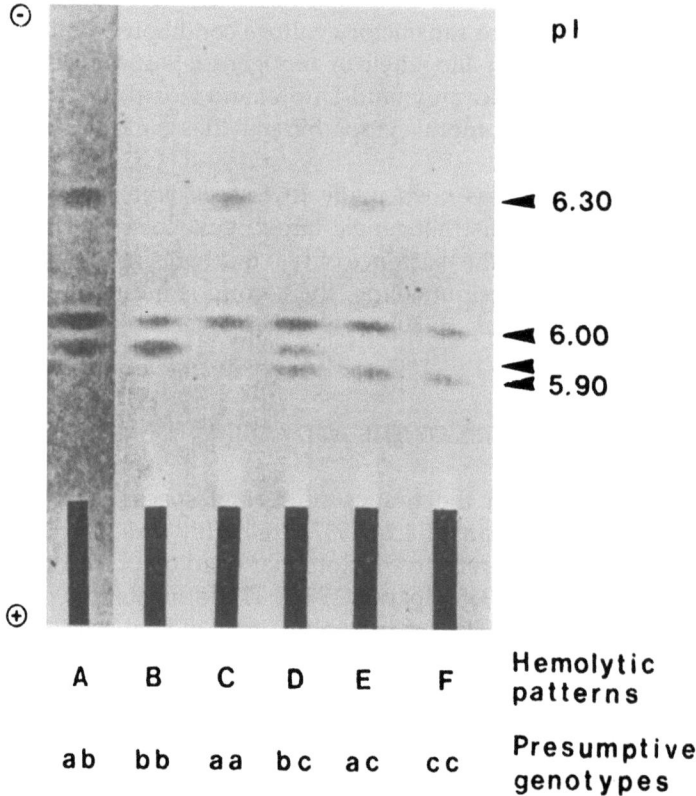

FIGURE 28. Natural polymorphism of the hemolytic factor of *Eisenia fetida andrei*. Note the presence of 4 different isoforms (arrows), only 2 or 3 of which are present simultaneously in the same animal. The isoform of pI = 6.00 is common to all the different patterns and the hemolytic pattern classification of the animals is based on the combinations of the 3 other isoforms. From Roch, 1979a.

extracts from complete worms (*E. foetida*) (Andrews and Kukulinsky, 1975). It is not impossible that such hemolysis was the result of chloragogue cells that contaminated the tissue.

The coelomic fluid of other lumbricids has been said, in certain cases, to possess a weak natural hemolytic activity (Du Pasquier, 1971; Châteaureynaud-Duprat and Izoard, 1973; Cooper et al., 1974). The titer of the hemolytic activity in worms other than *Eisenia* is always low (2 to 16 for *Lumbricus* instead of 4000 for *Eisenia*).

According to Châteaureynaud-Duprat and Izoard (1978), it is possible to enhance the hemolytic activity of *E. foetida* by sensitization with SRBC. The increased acquired activity, however, is not specific. Stimulation of the hemolytic activity of *Lumbricus terrestris* has also been found following a previous injection of erythrocytes (Cooper et al., 1974), and this acquired immunity of *Lumbricus* seems to exhibit a certain specificity and to be leukocyte-dependent (Châteaureynaud-Duprat and Izoard, 1977a).

4.2.1b. In Sipunculid Worms. The coelomic fluid of *Dendrostomum zostericolum* possesses a natural hemolysin for xenogeneic erythrocytes from several species. Its natural titer ranges from 1 : 4 for human erythrocytes (Cushing *et al.*, 1969), 1 : 16 for erythrocytes from chickens, dogs, and horses, and 1 : 32 for sheep, rabbit, calf, and turkey erythrocytes (Weinheimer *et al.*, 1970). The activity of this hemolysin is greatly reduced by heating for 20 min at 37°C; it is completely inhibited at 52°C.

As in the case of the hemolytic system of *Eisenia*, the hemolysin of *D. zostericolum* does not require divalent cations to act. Therefore, it seems to differ from the complement system of vertebrates. In another sipunculid worm, *Golfingia* sp., Day *et al.* (1970) have found that the body fluid is able to form a lytic system with cobra venom factor. Consequently, the authors conclude that the body fluid of *Golfingia* sp. probably contains a complementlike component. No other instances of complementlike activity have been seen in worms (Morgun, 1950; Gigli and Austen, 1971).

4.2.2. Bacteriolysins

One of the best known bacteriolysins is lysozyme, a mucolytic enzyme found in mammalian tissue secretions and in some microorganisms. Lysozyme is a globulin which acts by splitting the bond between N-acetylglucosamine and N-acetylmuramic acid (Salton, 1957). Lysozyme is found at high concentrations in polymorphonuclear leukocytes and alveolar macrophages (Hirsch, 1959). Reticuloendothelial cells may be the source of the enzyme in vertebrates. A lysozymelike activity has been found in a marine polychaete worm, *Nephthys hombergi* (Jolles and Zuili, 1960; Perin and Jolles, 1972), as well as in different oligochaete worms or leeches (Porchet, 1928; Schubert and Messner, 1971). These substances resemble mammalian lysozyme. In fact, it is clear that rather than there being a single lysozyme, there is a class of basic low-molecular-weight proteins that lyse certain bacteria and act as an N-acetylmuramidase (Chipman and Sharon, 1969). Very little is known about the synthesis of lysozyme by annelids, but it has been suggested that in molluscs, lysozyme is synthesized by hemocytes (Feng and Canzonier, 1970).

The hemolysins of earthworms have been tested to determine whether they exhibit a bacteriolytic activity against common bacterial genera such as *Bacillus* or *Salmonella*. No activity was found, even after a second injection (Cooper *et al.*, 1969). Such investigations should be repeated with soil bacteria, the eventual commensal of earthworms (Cameron, 1932). Despite extensive studies of several species of sipunculids (Bang, 1966; Cushing *et al.*, 1969), humoral substances with spontaneous capacity of bacteriolysis have been found only in the normal coelomic fluid of *Phascolosoma gouldii* (Bang and Krassner, 1958; Krassner, 1963; Rabin and Bang, 1964). Several strains of marine bacteria were destroyed within 24 hr after they were injected into the coelomic cavity of this species. Freezing, lipase treatment, or toluene extraction of the coelomic fluid had no effect on antibacterial properties. Since these properties were completely inhibited by trypsinization, a protein or a polypeptide is involved in this system (Krassner and Flory, 1970). Other investigations suggest that a nonprotein component is also involved. Another characteristic of the natural antibacterial factor of *P. gouldii* is its thermolability.

Synthesis of bacteriolytic factors can be induced in several species of sipunculid worms. Years ago, Cantacuzène (1922a) demonstrated that *S. nudus* developed a lytic activity for bacteria over a period of weeks, after large numbers of bacteria were injected into the body cavity. In the sipunculid worm, *D. zostericolum*, the synthesis of a nonspecific humoral factor with bactericidal activity could be induced by injections of gram-negative bacteria (Evans *et al.*, 1969). Within 60 days following the first injection, a bacteriolysin with a titer of 160 was detected. A repeated injection on the 60th day quadrupled the titer in 48 hr. This bacteriolytic factor was inactivated by heating the coelomic fluid for 20 min at 50°C, and its biological activity could not be restored by the addition of unheated coelomic fluid from unimmunized worms. Release of a similar substance from the coelomic cells of *S. nudus* occurred *in vivo*, 15 min after the injection of either the ciliated parasite *Anophrys magii*, or heat-killed bacteria (Bang, 1962, 1966, 1967). Release could also be observed *in vitro*. The substance persisted in the plasma for only a few hours, and then, 1 or 2 days later, it reappeared and persisted for 5 to 7 days. The lysin induced by *A. magii* is most active against *A. magii*, but also had toxic activity on other ciliates, and its synthesis could be induced by several other protozoans. The lysin was destroyed when heated at 45°C for 5 min and it was inactivated by ether.

4.2.3. Agglutinins

An early report on the existence of natural agglutinins for human erythrocytes in the sipunculid worm *D. zostericolum* (Triplett *et al.*, 1958) was confirmed by Cushing *et al.* (1969). They found activities ranging in titers from 32 to 128. In comparison with erythrocytes of other mammals, the titer varied from 4 (for calf erythrocytes) to 32 (for horse erythrocytes). This natural hemagglutinin of *D. zostericolum* has a molecular weight of about 10,000 and is inactivated at 40°C. No significant increase of the agglutinating activity of the body fluid of *D. zostericolum* was found following an injection of target cells or antigens (Triplett *et al.*, 1958).

The ciliary urns of certain sipunculid worms constitute a particular agglutinating system (Cuénot, 1902, 1913). Urns are generally seen *in vivo* and *in vitro* with a mucosidic tail possessing a high agglutinating capacity. Under normal conditions, the urn cells eliminate all debris, including damaged or dead cells, from the coelomic fluid (Cantacuzène, 1922b,c, 1928; Towle, 1962; Blitz, 1965). The urns seem capable of discriminating between self and non-self erythrocytes. The agglutinated foreign particles remain fixed to the tail and increase its length. After a time, tails separate from the rest of the urns and they accumulate as a "brown body" at the posterior end of the coelomic cavity. A mucous hypersecretion in *S. nudus* urns after injection of foreign material has been reported by Bang and Bang (1965, 1974, 1975).

In annelid worms, the coelomic fluid of *Lumbricus terrestris* has been said to possess a natural hemagglutinin activity (Cooper *et al.*, 1974). This activity only occurs with certain mammalian erythrocytes and it has only been demonstrated for certain worms (Châteaureynaud-Duprat and Izoard, 1977a,b). When it is detectable, the agglutinating activity has a relatively low titer, from 2 to 8. The weak natural agglutinating activity of *Lumbricus terrestris* may be increased by *in vivo* injections of mammalian erythrocytes (sheep or rabbit). One day after immunization, the agglutinin titer increased from 4 to 64 (Cooper *et al.*, 1974), but it always remained low. After treatment of erythrocytes with neuraminidase, the mean activity increased from 2 to 44 with rabbit target erythrocytes. No activity was detected at 4°C, and the factor was destroyed by heating it for 30 min at 56°C. Electrophoresis and immunoelectrophoresis of injected animals did not demonstrate new polypeptide components.

Closely related to the problem of agglutinins is that of rosette-forming cells (RFCs), whose agglutininlike cell surface receptors bind to complementary determinants of the target cell surface. Coelomocytes forming spontaneous rosettes with SRBC were shown under various experimental conditions in *Lumbricus terrestris* (Toupin and Lamoureux, 1976b). The percentage of spontaneous RFCs ranged from 7 to 17% of the unseparated coelomocytes. It varied with the pH, increased with the incubation period, and remained constant between 4 and 20°C. Twenty-four hours after injection with sheep or human erythrocytes, *Lumbricus terrestris* macrophages exhibited an increased ability to form rosettes (Châteaureynaud-Duprat and Izoard, 1973, 1977b, 1978). Binding between erythrocytes and leukocytes began within 15 to 20 min and reached a maximum within 16 hr of incubation. The increase of rosette formation seemed to be relatively specific for the injected red blood cells.

The occurrence of agglutinins has rarely been reported in other annelids. Nevertheless, body fluids of certain polychaetes have been said to contain agglutinins (Tyler, 1946). For example, Brundshuh (1966) found an "agglutinating antibody" against human erythrocytes in a homogenate of the leech, *Hirudo officinalis*.

Other toxic factors have been found in annelids (Okaichi and Hashimoto, 1962; Doeksen and Van Wingerden, 1964) and in sipunculids (Chaet, 1955, 1956), but their biological significance is not always evident. Nevertheless, to conclude, it may be said that invertebrates, particularly annelids and sipunculids, possess very efficient humoral defense mechanisms. Most of these humoral mediators occur spontaneously and do not increase after sensitization. Only a small number of species possess humoral factors which appear following a prior *in vivo* contact between effector cells and target cells. Sometimes, these induced humoral factors appear antigenically specific; other times they do not. The critical issue of specificity is far from resolved.

To recapitulate, cells that morphologically resemble and are physiologically equivalent to vertebrate leukocytes exist in all the coelomate groups and perhaps even in more primitive triploblastic animals such as nemerteans, which do not have a coelomic cavity. But the capacity to distinguish self from non-self appeared before the emergence of leukocytes (Theodor, 1970, 1971; Klein, 1977; Kolb, 1977) and is probably a general property among living creatures. When leukocytes appear in invertebrates, they are generally derived from stem cells that proliferate from the coelomic epithelium and they possess many features which characterize vertebrate leukocytes. They are capable of phagocytosis and they take up proteins *in vitro*. They can infiltrate grafts and destroy them. They can exert, at least *in vitro*, cytotoxic activities against foreign target cells. They possess receptors whose interaction with various antigenic or mitogenic ligands induces a stimulation of DNA synthesis. They have been said to synthesize different "natural antibodies" but they seem unable to secrete specific antibodies and to exhibit reactions of classical humoral immunity. (Good, 1964). Despite numerous similarities between certain invertebrate coelomocytes and vertebrate leukocytes, the existence of a phylogenetic relationship between both categories of cells remains to be proven (Wright, 1976; Warr and Marchalonis, 1978). The similarities could be the result of convergent evolution: it is common that similar demands induce similar responses from unrelated animal groups. To determine whether functional similarities between leukocytes of invertebrates and vertebrates are phylogenetically meaningful, biochemical studies on the nature of surface receptors must be undertaken.

ACKNOWLEDGMENTS. We are particularly grateful to Jacqueline Chapron and Lynda Gent for their assistance.

REFERENCES

Andrews, E. J., and Kukulinsky, N. E., 1975, Hemolysis of vertebrate erythrocytes with tissue extracts of earthworms (*Eisenia fetida*), *J. Reticuloendothelial Soc.* **17**:170.
Avel, M., 1959, Annélides Oligochètes, in: *Traité de Zoologie* (P. P. Grassé, ed.), Vol. 5, pp. 224–470, Masson, Paris.
Bacq, Z. M., 1937, L'"amphiporine" et la "nemertine" poisons des vers némertines, *Arch. Int. Physiol.* **44**:190.
Bahl, K. N., 1947, Excretion in the *Oligochaeta*, *Biol. Rev.* **22**:109.
Bailey, S., Miller, B. J., and Cooper, E. L., 1971, Transplantation immunity in annelids. II. Adoptive transfer of the xenograft reaction, *Immunobiology* **21**:81.
Baldwin, R. W., 1977, Immune surveillance revisited, *Nature (London)* **270**:557.
Bang, B. G., and Bang, F. B., 1965, Mucus hypersecretion in a normal isolated non-innervated cell, *Cah. Biol. Mar.* **6**:257.
Bang, B. G., and Bang, F. B., 1974, Invertebrate model for study of macromolecular regulation of mucus secretion, *Lancet* **30**:1292.
Bang, B. G., and Bang, F. B., 1975, Cell recognition by mucus secreted by urn cell of *Sipunculus nudus*, *Nature (London)* **253**:634.
Bang, F. B., 1962, Serological aspects of immunity in invertebrates, *Nature (London)* **196**:88.
Bang, F. B., 1966, Serological responses in marine worm *Sipunculus nudus*, *J. Immunol.* **96**:960.
Bang, F. B., 1967, Serological responses among invertebrates other than insects, *Fed. Proc.* **26**:1680.
Bang, F. B., and Bang, B. G., 1962, Studies on sipunculid blood: Immunologic properties of coelomic fluid and morphology of "urn cells," *Cah. Biol. Mar.* **3**:363.
Bang. F. B., and Krassner, S. M., 1958, Antibacterial activity of *Phascolosoma gouldii* blood, *Biol. Bull.* **115**:343.
Baskin, D. G., 1974, The coelomocytes of nereid polychaetes, *Contemp. Top. Immunobiol.* **4**:55.
Bierne, J., and Langlet, C., 1974, Recherches sur l'immunité de greffe chez les Némertines du genre *Lineus*. Etude de la réponse primaire à la transplantation hétérospécifique, *C. R. Acad. Sci. Ser. D* **278**:1445.

Blitz, R., 1965, The clearance of foreign material from the coelom of *Phascolosoma agassizii*, Ph.D. thesis, University of California, Berkeley.
Boiledieu, D., and Valembois, P., 1976, Etude *in vitro* de l'activité cytotoxique des leucocytes de Siponcles à l'encontre d'érythrocytes allogéniques et xénogéniques, *C. R. Acad. Sci. Ser. D.* **283**:247.
Boiledieu, D., and Valembois, P., 1977a, Natural cytotoxic activity of sipunculid leukocytes on allogenic and xenogenic erythrocytes, *Dev. Comp. Immunol.* **1**:207.
Boiledieu, D., and Valembois, P., 1977b, Etude d'un modèle *in vitro* de cytotoxicité naturelle chez les Siponcles, *Bull. Soc. Zool. Fr.* Suppl. 1, 57.
Boiledieu, D., and Valembois, P., 1977c, The mechanism of leukocyte cytotoxicity studied by time-lapse microcinematography and its inhibition: An example of *in vitro* specific recognition in invertebrates, in: *Developmental Immunobiology* (J. B. Solomon and J. D. Horton, eds.), pp. 51–57, Elsevier/North-Holland, Amsterdam.
Boiledieu, D., and Valembois, P., 1978, Mise en évidence d'une synthèse polypeptidique au cours de la réaction de cytotoxicité chez les Siponcles, *C. R. Soc. Biol.* **172**:98.
Bouché, M. B., 1972, Lombriciens de France: Ecologie systèmatique, *Ann. Zool. Ecol. Anim.* INRA, Paris.
Bradbury, S., 1959, The botryoidal and vaso-fibrous tissue of the leech *Hirudo medicinalis*, *Q. J. Microsc. Sci.* **100**:483.
Breton-Gorius, J., 1963, Etude au microscope électronique des cellules chloragogènes d'*Arenicola marina* L.: Leur rôle dans la synthèse de l'hémoglobine, *Ann. Sci. Nat. Zool. Biol. Anim.* **5**:211.
Brundshuh, G., 1966, Agglutinierende Antikörper gegen Humanerythrozyten bei *Hirudo officinalis*, *Z. Aerztl. Fortbild.* **60**:196.
Burke, J. M., 1974, Wound healing in *Eisenia fetida* (Oligochaeta). III. A fine structural study of the role of non epidermal tissue, *Cell Tissue Res.* **154**:83.
Burnet, F. M., 1968, Evolution of the immune process in invertebrates, *Nature (London)* **218**:426.
Cameron, G. R., 1932, Inflammation in earthworms, *J. Pathol. Bacteriol.* **35**:933.
Cantacuzène, J., 1922a, Réactions d'immunité chez *Sipunculus nudus* vacciné contre une bactérie, *C. R. Soc. Biol.* **87**:264.
Cantacuzène, J., 1922b, Sur le rôle agglutinant des urnes chez *Sipunculus nudus*, *C. R. Soc. Biol.* **87**:259.
Cantacuzène, J., 1922c, Sur le sort ultérieur des urnes chez *Sipunculus nudus* au cours de l'infection et de l'immunisation, *C. R. Soc. Biol.* **87**:283.
Cantacuzène, J., 1928, Recherches sur les réactions d'immunité chez les invertébrés: Réactions d'immunité chez le *Sipunculus nudus*, *Arch. Roum. Pathol. Exp. Microbiol.* **1**:1.
Chaet, A. B., 1955, Further studies on the toxic factor in *Phascolosoma*, *Biol. Bull.* **109**:356.
Chaet, A. B., 1956, Mechanism of toxic factor release, *Biol. Bull.* **111**:298.
Chapron, C., 1970, *Régénération céphalique chez le Lombricien Eisenia foetida unicolor*: Structure, origine et rôle du bouchon cicatriciel, *Arch. Zool. Exp. Gen.* **3**:217.
Châteaureynaud-Duprat, P., 1970, Specificity of the allograft rejection in *Eisenia foetida*, *Transplant. Proc.* **2**:222.
Châteaureynaud-Duprat, P., and Izoard, F., 1972, Etude *in vitro* de l'histocompatibilité chez les Lombriciens, *C. R. Acad. Sci. Ser. D* **275**:2795.
Châteaureynaud-Duprat, P., and Izoard, F., 1973, Etude des mécanismes de défense chez *Lumbricus terrestris*, *C. R. Acad. Sci. Ser. D* **276**:2859.
Châteaureynaud-Duprat, P., and Izoard, F., 1977a, Etude comparée *in vitro* des réactions de défense antigreffe chez deux genres de Lombriciens *Eisenia* et *Lumbricus*, *C. R. Acad. Sci. Ser. D* **284**:2581.
Châteaureynaud-Duprat, P., and Izoard, F., 1977b, Compared study of immunity between two genera of lumbriciens, *Eisenia* and *Lumbricus*, in: *Developmental Immunobiology* (J. B. Solomon and J. D. Horton, eds.), pp. 33–40, Elsevier/North-Holland, Amsterdam.
Châteaureynaud-Duprat, P., and Izoard, F., 1978, Etude comparée des mécanismes de défense de l'organisme dans deux genres de Lombriciens, *Eisenia* et *Lumbricus*, *Ann. Biol.* **17**:455.
Cheng, T. C., and Streisfeld, S. D., 1963, Innate phagocytosis in the trematodes *Megalodiscus temperatus* and *Haematoloechus* sp., *J. Morphol.* **113**:375.
Chipman, D. M., and Sharon, N., 1969, Mechanism of lysozyme action, *Science* **165**:454.

Cohen, N., and Borysenko, M., 1970, Acute and chronic graft rejection: Possible phylogeny of transplantation antigens, *Transplant. Proc.* **2**:333.

Cooper, E. L., 1968, Transplantation immunity in annelids. I. Rejection of xenografts exchanged between *Lumbricus terrestris* and *Eisenia foetida*, *Transplantation* **6**:322.

Cooper, E. L., 1969a, Specific tissue graft rejection in earthworms, *Science* **166**:1414.

Cooper, E. L., 1969b, Chronic allograft rejection in *L. terrestris*, *J. Exp. Zool.* **171**:69.

Cooper, E. L., 1970, Transplantation immunity in helminths and annelids, *Transplant. Proc.* **2**:216.

Cooper, E. L., 1973, Earthworm coelomocytes: Role in understanding the evolution of cellular immunity. I. Formation of monolayers and cytotoxicity, in: *Proceedings, III International Colloquium on Invertebrate Tissue Culture* (J. Rehácek, D. Blaskovic, and W. F. Hink, eds.), pp. 381–404, Publishing House of the Slovak Academy of Science, Bratislava.

Cooper, E. L., and Aponte, A., 1968, Chronic allograft rejection in the iguana *Clenosaura pectinata*, *Proc. Soc. Exp. Biol. Med.* **128**:150.

Cooper, E. L., and Baculi, B. S., 1968, Cell responses during xenograft rejection in annelids, *Anat. Rec.* **160**:335.

Cooper, E. L., and Rubilotta, L. M., 1969, Allograft rejection in *Eisenia foetida*, *Transplantation* **8**:220.

Cooper, E. L., Acton, R. T., Weinheimer, P., and Evans, E. E., 1969, Lack of bactericidal response in the earthworm *Lumbricus terrestris* after immunization with bacterial antigens, *J. Invert. Pathol.* **14**:402.

Cooper, E. L., Lemmi, C. A. E., and Moore, T. C., 1974, Agglutinins and cellular immunity in earthworms, *Ann. N.Y. Acad. Sci.* **234**:34.

Cudkowicz, G., and Hochman, P. S., 1979, Do natural killer cells engage in regulated reactions against self to ensure homeostasis?, *Immunol. Rev.* **44**:13.

Cuénot, L., 1902, Organes agglutinants et organes ciliophagocytaires, *Arch. Zool. Exp. Gen.* **10**:79.

Cuénot, L., 1913, Excretion et phagocytose chez les Sipunculiens, *C. R. Soc. Biol.* **74**:159.

Cushing, J. E., and Boraker, D. K., 1975, Some specific aspects of cell-surface recognition by sipunculid coelomocytes, in: *Immunologic Phylogeny* (W. H. Hildemann and A. A. Benedict, eds.), pp. 35–44, Plenum Press, New York.

Cushing, J., Boraker, D., and Keogh, E., 1965, Reactions of sipunculid worms to intracoelomic injections of homologous eggs, *Fed. Proc.* **24**:504.

Cushing, J. E., McNelly, J. L., and Tripp, M. R., 1969, Comparative immunology of sipunculid coelomic fluid, *J. Invert. Pathol.* **14**:4.

Dales, R. P., 1961, The coelomic and peritoneal cell systems of some sabellid polychaetes, *Q. J. Microsc. Sci.* **102**:327.

Dales, R. P., 1964, The coelomocytes of the terebellid polychaete *Amphitrite johnstoni*, *Q. J. Microsc. Sci.* **105**:263.

Dales, R. P., 1978, The basis of graft rejection in the earthworms *Lumbricus terrestris* and *Eisenia fetida*, *J. Invert. Pathol.* **32**:264.

Dawydoff, C., 1959, Ontogenèse des Annelides, in: *Traité de Zoologie* (P. P. Grassé, ed.), Vol. 5, pp. 594–686, Masson, Paris.

Day, N. K. B., Gewurz, H., Johannsen, R., Finstad, J., and Good, R. A., 1970, Complement and complement-like activity in lower vertebrates and invertebrates, *J. Exp. Med.* **132**:941.

De Duve, C., and Wattiaux, R., 1966, Functions of lysosomes, *Annu. Rev. Physiol.* **28**:435.

Dhainaut, A., 1966, Etude ultrastructurale de l'évolution des éléocytes chez *Nereis pelagica* L. (Annélide polychète) à l'approche de la maturité sexuelle, *C. R. Acad. Sci.* **262**:2740.

Doeksen, J., and Van Wingerden, C. G., 1964, Notes on the activity of earthworms. II. Observations on diapause in earthworm *Allolobophora caliginosa*, *Jahrb. Inst. Biol. Scheik.* **1964**:181.

Du Pasquier, L., 1971, Etude comparée d'un facteur cytolytique humoral chez une larve d'amphibien et chez un oligochète, *Arch. Zool. Exp. Gen.* **142**:81.

Du Pasquier, L., 1974, The genetic control of histocompatibility reactions: Phylogenetic aspects, *Arch. Biol.* **85**:41.

Du Pasquier, L., and Duprat, P., 1968, Aspects humoraux et cellulaires d'une immunité naturelle non spécifique chez l'Oligochète *Eisenia fetida* (Sav.), *C. R. Acad. Sci. Ser. D* **266**:538.

Du Pasquier, L., Duprat, P., and Izoard, F., 1966, Etude immunologique des hétérogreffes chez les Annélides Oligochètes, *C. R. Acad. Sci.* **262**:2389.

Duprat, P., 1964, Mise en évidence de réactions immunitaires dans les homogreffes de paroi du corps chez le Lombricien *Eisenia fetida typica, C. R. Acad. Sci.* **259:**4177.

Duprat, P., 1967, Etude de la prise et du maintien d'un greffon de paroi du corps chez le Lombricien *Eisenia fetida typica, Ann. Inst. Pasteur Paris* **118:**867.

Duprat, P., and Bouc-Lassalle, A. M., 1967, Mise au point et étude du liquide coelomique du Lombricien *Eisenia fetida* Sav., *Bull. Soc. Zool. Fr.* **92:**767.

Dybas, L., 1976, A light and electron microscopic study of the ciliated urn of *Phascolosoma agassizii* (Sipunculida), *Cell Tissue Res.* **169:**67.

Eckelbarger, K. J., 1976, Origin and development of the amoebocytes of *Nicolea zoostericola* (Polychaeta Terebellidae) with a discussion of their possible role in oogenesis, *Mar. Biol.* **36:**169.

Edwards, C. A., and Lofty, J. R., 1977, *Biology of Earthworms,* 2nd ed., Chapman & Hall, London.

Ehrenreich, B. A., and Cohn, Z. A., 1967, The uptake and digestion of iodinated human serum albumin by macrophages *in vitro, J. Exp. Med.* **126:**941.

Evans, E. E., Weinheimer, P. F., Acton, R. T., and Cushing, J. E., 1969, Induced bactericidal response in a sipunculid worm, *Nature (London)* **223:**695.

Feng, S. Y., and Canzonier, W. J., 1970, Humoral responses in the American oyster (*Crassostrea virginica*) infected with *Bucephalus* sp. and *Minchinia nelsoni,* in: *A Symposium on Diseases of Fishes and Shellfishes* (S. F. Snieszko, ed.), pp. 497–510, American Fisheries Society, Washington, D.C.

Gibson, R., 1972, *Nemerteans,* 1st ed., Hutchinson, London.

Gigli, I., and Austen, K. F., 1971, Phylogeny and function of the complement system, *Annu. Rev. Microbiol.* **25:**309.

Good, R. A., 1964, Evolution of the immune response, *J. Exp. Med.* **119:**105.

Greenberg, A. H., and Playfair, J. H. L., 1974, Spontaneously arising cytotoxicity to the P-815-Y-mastocytoma in NZB mice, *Clin. Exp. Immunol.* **16:**99.

Halstead, B. W., 1965, *Poisonous and Venomous Marine Animals of the World,* Volume I, *Invertebrates,* U.S. Government Printing Office, Washington, D.C.

Harant, J., and Grassé, P., 1959, Classe des Annélides achètes ou Hirundinées ou Sangsues, in: *Traité de Zoologie* (P. P. Grassé, ed.), Vol. 5, pp. 471–593, Masson, Paris.

Hay, E. D., and Coward, S. J., 1975, Fine structure studies on the planarian *Dugesia.* I. Nature of the "neoblast" and other cell types in non injured worms, *J. Ultrastruct. Res.* **50:**1.

Herberman, R. B., Nunn, M. E., Holden, H. T., and Lavrin, D. H., 1975, Natural cytotoxic reactivity of mouse lymphoid cells against syngeneic and allogeneic tumors. II. Characterization of effector cells, *Int. J. Cancer* **16:**230.

Hess, R. H., 1970, The fine structure of coelomocytes in the annelid *Enchytraeus fragmentosus, J. Morphol.* **132:**335.

Hildemann, W. H., and Cohen, N., 1967, Weak histocompatibilities: Emerging immunogenetic rules and generalizations, in: *Histocompatibility Testing* (E. S. Curtoni, P. L. Mattwiz, and R. M. Tosi, eds.), pp. 13–46, Munksgaard, Copenhagen.

Hilgard, H. R., 1970, Studies of protein uptake by echinoderm cells: Their possible significance in relation to the phylogeny of immune responses, *Transplant. Proc.* **2:**240.

Hirsch, J. G., 1959, Antimicrobial factors in tissues and phagocytic cells, *Bacteriol. Rev.* **24:**133.

Hostetter, R. K., and Cooper, E. L., 1972, Coelomocytes as effector cells in earthworm immunity, *Immunol. Commun.* **1:**155.

Hostetter, R. K., and Cooper, E. L., 1973, Cellular anamnesis in earthworms, *Cell. Immunol.* **9:**384.

Hostetter, R. K., and Cooper, E. L., 1974, Earthworm cellular immunity, *Contemp. Top. Immunobiol.* **4:**91.

Hyman, L. H., 1951, *The invertebrates: Platyhelminthes and Rhynchocoela* (Vol. II), McGraw–Hill, New York.

Isern, J., 1969, Sobre les urnas de *Sipunculus nudus, P. Inst. Biol. Apl.* **46:**115.

Izoard, F., 1964, Evolution des greffes hétéroplastiques de paroi du corps réalisées, chez les Lombriciens, entre animaux de même genre mais d'espèces différentes: Recherches sur le genre *Lumbricus, C. R. Acad. Sci.* **258:**5972.

Izoard, F., 1971, Contribution à l'étude des hétérogreffes de paroi du corps chez les Lombriciens: Aspects histologiques de l'évolution du greffon et relations entre le maintien et le degré de la parenté zoologique avec le portegreffe, Thèse de Sciences, Bordeaux, n° 329.

Izoard, F., 1972, Evolution de greffes hétéroplastiques de paroi du corps entre espèces différentes du genre *Lumbricus:* Influence du biotope de récolte, *C. R. Acad. Sci. Ser. D* **276:**3061.
Izoard, F., 1973, Relations antigéniques entre différentes espèces du genre *Lumbricus:* Influence du biotope de récolte, *C. R. Acad. Sci. Ser. D* **276:**673.
Jennings, J. B., 1957, Studies on feeding, digestion and food storage in free-living flatworms (Platyhelminthes: Turbellaria), *Biol. Bull.* **112:**63.
Jensen, D. D., 1960, Hoplonemertines, myxinoids and deuterostome origins, *Nature (London)* **188:**649.
Jolles, P., and Zuili, S., 1960, Purification et étude comparée de nouveaux lysozymes: Extraits du poumon de poule et de *Nephthys hombergi, Biochim. Biophys. Acta* **39:**212.
Joseph, H., 1910, Die Amöbocyten von *Lumbricus:* Ein Beitrag zur Naturgeschichte der zellulären Zentren, *Arb. Zool. Inst. Wein.* **18:**1.
Kiessling, R., Klein, E., Pross, H., and Wigzell, H., 1975, "Natural" killer cells in the mouse. II. Cytotoxic cells with specificity for mouse Moloney leukemia cells. Characteristics of the killer cells, *Eur. J. Immunol.* **5:**117.
Klein, J., 1977, Evolution and function of the major histocompatibility system: Facts and speculation, in: *The Major Histocompatibility System in Man and Animals* (D. Götze, ed.), pp. 339–378, Springer-Verlag, Berlin.
Kolb, H., 1977, On the phylogenetic origin of the immune system: A hypothesis, *Dev. Comp. Immunol.* **1:**193.
Kollmann, M., 1908, Recherches sur les leucocytes et le tissu lymphoïde des Invertébrés, *Ann. Sci. Nat. Zool. Biol. Anim Ser. 9* **8:**1.
Korschelt, E., 1914, Uber Transplantationsversuche, Ruhezustände und Lebensdauer der Lumbriciden, *Zool. Anz.* **43:**537.
Korschelt, E., 1927, *Regeneration und Transplantation*, Vol. II (in two parts), Borntraeger, Berlin.
Krassner, S. M., 1963, Further studies on an antibacterial factor in the blood of *Phascolosoma gouldii, Biol. Bull.* **125:**382.
Krassner, S. M., and Flory, B., 1970, Antibacterial factors in the sipunculid worms *Golfingia gouldii* and *Dendrostomum pyroïdes, J. Invert. Pathol.* **16:**331.
Kükenthal, W., 1885, Die lymphoiden Zellen der Anneliden, *Jena. Z. Naturwiss.* **18:**319.
Langlet, C., and Bierne, J., 1973, Recherches sur l'immunité de greffes chez les Némertines du genre *Lineus:* Evolution de transplants homospécifiques et hétérospécifiques, *C. R. Acad. Sci. Ser. D* **276:**2485.
Langlet, C., and Bierne, J., 1975, Recherches sur l'immunité de greffe chez les Némertines du genre *Lineus:* Rejet accéléré des secondes greffes hétérospécifiques incompatibles, *C. R. Acad. Sci. Ser. D* **281:**595.
Langlet, C., and Bierne, J., 1977, The immune response to xenografts in nemertines of the genus *Lineus*, in: *Developmental Immunobiology* (J. B. Solomon and J. D. Horton, eds.), pp. 17–26, Elsevier/North-Holland, Amsterdam.
Laverack, M. S., 1963, *The Physiology of Earthworms*, Pergamon Press, Elmsford, N.Y.
Lemmi, C. A. E., 1974, Tissue graft rejection mechanisms in the earthworm *Lumbricus terrestris:* Specific induction of coelomocyte proliferation, Ph.D. thesis, University of California, Los Angeles.
Leypoldt, H., 1910, Transplantationsversuche an Lumbriciden. Zur Beeinflussung der Regeneration eines kleinen Pfropfstückes durch einen grösseren Komponenten, Inaugural-Dissertation zur Erlangung der Doktor, wörde Hohen Philosophischen Fakultät der Universität Narburg, pp. 1–20.
Leypoldt, H., 1911, Transplantationsversuch an Lumbriciden. Transplantation Kleiner Hautstückchen, *Arch. Entwicklungsmech. Org.* **31:**21.
Liebmann, E., 1926, Untersuchungen über das Chloragogen der Lumbriciden, *Zool. Anz.* **69:**65.
Liebmann, E., 1942a, The role of the chloragogue in regeneration of *Eisenia foetida* Sav., *J. Morphol.* **70:**151.
Lindh, N. O., 1959, Heteroplastic transplantation of transversal body sections in flatworms, *Ark. Zool.* **12:**183.
Lindner, E., 1965, Ferritin und Hämoglobin im Chloragogen von Lumbriciden, *Z. Zellforsch. Mikrosk. Anat.* **66:**891.

Linthicum, D. S., Stein, E. A., Marks, D. H., and Cooper, E. L., 1977a, Electron microscopic observations of normal coelomocytes from the earthworm, *Lumbricus terrestris*, *Cell Tissue Res.* **185**:315.

Linthicum, D. S., Stein, E. A., Marks, D., and Cooper, E. L., 1977b, Graft rejection in earthworms: An electron microscopic study, *Eur. J. Immunol.* **7**:871.

Loeb, L., 1945, *The Biological Basis of Individuality*, Thomas, Springfield, Ill.

Lopez, D. M., Sigel, M. M., and Lee, J. C., 1974, Phylogenetic studies on T-cells. I. Lymphocytes of the shark with differential response to phytohemagglutinin and concanavalin A, *Cell. Immunol.* **10**:287.

Marcou, J., and Volkonsky, M., 1933, Les lignées leucocytaires des Sipunculidés, *Arch. Anat. Microsc. Morphol. Exp.* **29**:245.

Marsden, J. R., 1966, The coelomocytes of *Hermodice carunculata* (Polychaeta: Amphinomidae) in relation to digestion and excretion, *Can. J. Zool.* **44**:377.

Metchnikoff, E., 1892, *Leçons sur la Pathologie Comparée de l'Inflammation*, Masson, Paris.

Mill, P. J. (ed.), 1978, *Physiology of Annelids*, Academic Press, New York.

Mitchison, N. A., 1953, Passive transfer of transplantation immunity, *Nature (London)* **171**:267.

Morgun, L. I., 1950, On the question of a complement in invertebrates, *Mikorbiol. Zh.* (Akad. Nauk. Ukr. RSR) **11**:43.

Novikoff, A. B., 1963, Lysosomes in the physiology and pathology of cells: Contributions of staining methods, in: *Ciba Symposium on Lysosomes* (A.V.S. de Reuck, ed.), pp. 37–77, Churchill, London.

Ohuye, T., 1942, On the blood corpuscles and the hemopoiesis of a nemertean *Lineus fuscoviridis* and of a sipunculid *Dendrostoma minor*, *Sci. Rep. Tohoku Imp. Univ.* **17**:187.

Ohuye, T., Ochi, O., and Miyata, I., 1961, On the morphogenesis and histochemistry of the "urn" found in the coelomic fluid of a sipunculid *Phascolosoma scolops*, *Mem. Ehime Univ. Nat. Sci. Sect. II B.* **4**:145.

Okaichi, T., and Hashimoto, Y., 1962, Physiological activities of nereistoxin, *Bull. Jpn. Soc. Fish.* **28**:930.

Omodeo, P., 1956, Contribuoto alla revisione dei Lumbricidae, *Arch. Zool. Ital.* **41**:129.

Osborne, P. J., and Miller, A. T., 1962, Uptake and intracellular digestion of proteins (peroxidase) in planarians, *Biol. Bull.* **123**:589.

Parry, M. J., 1978, Survival of body wall autografts, allografts and xenografts in the earthworm *Eisenia foetida*, *J. Invert. Pathol.* **31**:383.

Pearse, A. G., 1961, *Histochemistry: Theoretical and Applied*, 2nd ed., Churchill, London.

Pedersen, K. J., 1961, Some observations on the fine structure in the planarian protonephridia and gastrodermal phagocytes, *Z. Zellforsch. Mikrosk. Anat.* **53**:609.

Perin, J. P., and Jolles, P., 1972, The lysozyme from *Nephthys hombergi* (annelid), *Biochim. Biophys. Acta* **263**:683.

Picton, L. J., 1898, On the heart-body and coelomic fluid of certain polychaetes, *Q. J. Microsc. Sci.* **41**:263.

Porchet, B., 1928, Contribution à l'étude des réactions immunitaires chez les Invertébrés, *Bull. Soc. Vaudoise Sci. Nat.* **56**:553.

Rabin, H., and Bang, F. B., 1964, *In vitro* studies of the antibacterial activity of *Golfingia gouldii* coelomic fluid, *J. Insect. Pathol.* **6**:457.

Roch, P., 1973, Contribution à l'étude du transfert de l'immunité antigreffe chez le Lombricien *Eisenia fetida* Sav., *C. R. Acad. Sci. Ser. D* **276**:1369.

Roch, P., 1977, Réactivité *in vitro* des leucocytes du Lombricien *Eisenia fetida* Sav. à quelques substances mitogéniques, *C. R. Acad. Sci. Ser. D* **284**:705.

Roch, P., 1979a, Leukocyte DNA-synthesis in grafted lumbricids: An approach to study histocompatibility in invertebrates, *Dev. Comp. Immunol.* **3**:417.

Roch, P., 1979b, Protein analysis of earthworm coelomic fluid. I. Polymorphic system of the natural hemolysin of *Eisenia fetida andrei*, *Dev. Comp. Immunol.* **3**:599.

Roch, P., and Valembois, P., 1977, Physiological heterogeneity and cellular differentiation of earthworm leukocytes studied by concanavalin A, in: *Developmental Immunobiology* (J. B. Solomon and J. D. Horton, eds.), pp. 41–49, Elsevier/North-Holland, Amsterdam.

Roch, P., and Valembois, P., 1978, Evidence for concanavalin A-receptors and their redistribution on lumbricid leukocytes, *Dev. Comp. Immunol.* **2**:51.

Roch, P., Valembois, P., and Du Pasquier, L., 1975, Response of earthworm leukocytes to concanavalin A and transplantation antigens, *Adv. Exp. Med. Biol.* **64**:45.

Romieu, M., 1921, Les inclusions cristallines des éléocytes de *Nereis* et leurs relations avec les granulations éosinophiles, *C. R. Acad. Sci.* **168**:367.

Romieu, M., 1923, Recherches histophysiologiques sur le sang et sur le corps cardiaque des annélides polychètes, *Arch. Morphol. Gen. Exp.* **15**:1.

Roots, B. I., 1957, Nature of chloragogue granules, *Nature (London)* **179**:679.

Roots, B. I., 1960, Some observations on the chloragogous tissue of earthworms, *Comp. Biochem. Physiol.* **1**:218.

Rosa, D., 1896, Les lymphocytes des Oligochètes, *Arch. Ital. Biol.* **25**:444.

Ryder, T. A., and Bowen, I. D., 1975, The fine structural localization of acid phosphatase activity in *Polycelis tenuis* (Iijima), *Protoplasma* **83**:79.

Salton, M. R. J., 1957, The properties of lysozyme and its action on microorganisms, *Bacteriol. Rev.* **27**:82.

Schroeder, P. C., 1971, Studies on oogenesis in the polychaete annelid *Nereis grubei* (Kinberg). II. Oocyte growth rates in intact and hormone-deficient animals, *Gen. Comp. Endocrinol.* **16**:312.

Schubert, V. I., and Messner, B., 1971, Unterschrengen uber das Vorkommer von Lysozym bei Anneliden, *Zool. Jahrb. Physiol.* **76**:36.

Semal-Van Gansen, P., 1956, Les cellules chloragogènes des Lombriciens, *Bull. Biol. Fr. Belg.* **90**:335.

Semal-Van Gansen, P., 1957, Le lipopigment des chloragosomes des Lombriciens, *Ann. Histochim.* **2**:41.

Semal-Van Gansen, P., 1958, Physiologie des cellules chloragogènes d'un Lombricien, *Enzymologia* **20**:98.

Semal-Van Gansen, P., and van der Meersche, G., 1958, L'ultrastructure des cellules chloragogènes, *Bull. Microsc. Appl.* **8**:7.

Sichel, G., 1964, Osservazione sull' ultrastruttura dei celomocite di *Perinereis cultrifera* (Grube), *Atti Accad. Gioenia Sci. Nat. Catania* **8**:86.

Skaer, R. J., 1961, Some aspects of the cytology of *Polycelis nigra*, *Q. J. Microsc. Sci.* **102**:295.

Sparks, A. K., 1972, *Invertebrate Pathology: Noncommunicable Diseases* Academic Press, New York.

Stang-Voss, C., 1970, Zur Ultrastruktur der Blutzellen wirbelloser Tiere. II. Über die Blutzellen von *Golfingia gouldii* (Sipunculidae), *Z. Zellforsch. Mikrosk. Anat.* **106**:200.

Stang-Voss, C., 1971, Zur Ultrastruktur der Blutzellen wirbelloser Tiere. IV. Die Hämocyten von *Eisenia fetida* L. (Sav.) (Annelidae), *Z. Zellforsch. Mikrosk. Anat.* **117**:451.

Stang-Voss, C., 1974, On the ultrastructure of invertebrate hemocytes: An interpretation of their role in comparative hematology, *Contemp. Top. Immunobiol.* **4**:65.

Stein, E. A., and Cooper, E. L., 1978, Cytochemical observations of coelomocytes from the earthworm *Lumbricus terrestris*, *Histochem. J.* **10**:657.

Stein, E., Avtalion, R. R., and Cooper, E. L., 1977, The coelomocytes of the earthworm *Lumbricus terrestris*: Morphology and phagocytic properties, *J. Morphol.* **153**:467.

Stephenson, J., 1930, *The Oligochaeta*, Oxford University Press, London.

Stolte, H. A., 1955, Bronns Klassen und Ordnungen des Tierreichs, *Oligochaeta* **4**:363.

Tetry, A., 1959, Classe des sipunculiens, in: *Traité de Zoologie* (P. P. Grassé, ed.), pp. 785–854, Masson, Paris.

Theodor, J., 1970, Distinction between "self" and "not self" in lower invertebrates, *Nature (London)* **227**:690.

Theodor, J., 1971, Reconnaissance du "self" ou reconnaissance des "not self," *Arch. Zool. Exp. Gen.* **112**:113.

Thomas, J. A., 1932, Recherches cytologiques et expérimentales sur les vésicules émigmatiques et les urnes des Siponcles, *Arch. Zool. Exp. Gen.* **73**:22.

Toupin, J., and Lamoureux, G., 1976a, Coelomocytes of earthworms: PHA responsiveness, in: *Phylogeny of Thymus and Bone Marrow-Bursa Cells* (R. K. Wright and E. L. Cooper, eds.), pp. 19–25, Elsevier/North-Holland, Amsterdam.

Toupin, J., and Lamoureux, G., 1976b, Coelomocytes of earthworms: The T-cell-like rosette, *Cell. Immunol.* **26**:127.

Towle, A., 1962, Physiological changes in *Phascolosoma agassizii* Kerferstein during the course of an annual reproductive cycle, Ph.D. thesis, Stanford University.

Triplett, E. L., Cushing, J. E., and Durall, G. L., 1958, Observations on some immune reactions of the sipunculid worm *Dendrostomum zostericolum*, *Am. Nat.* **92**:287.

Tyler, A., 1946, Natural heteroagglutinins in the body fluids and seminal fluids of various invertebrates, *Biol. Bull.* **90**:213.

Valembois, P., 1963a, Etude anatomique de l'évolution de greffons hétéroplastiques de paroi du corps chez quelques Lombriciens, *C. R. Acad. Sci.* **257**:3227.

Valembois, P., 1963b, Recherches sur la nature de la réaction antigreffe chez le Lombricien *Eisenia fetida* Sav., *C. R. Acad. Sci.* **257**:3488.

Valembois, P., 1968, Libération de phosphatase acide dans les cellules musculaires d'un greffon de paroi du corps chez un Lombricien, *J. Microsc. (Paris)* **7**:61.

Valembois, P., 1970, Etude d'une hétérogreffe de paroi du corps chez les Lombriciens: Aspects cytologiques, physiologiques et immunologiques de l'évolution du greffon et de la réaction du porte-greffe, Thèse de Doctorat ès Sciences Naturelles, Bordeaux.

Valembois, P., 1971a, Rôle des leucocytes dans l'acquisition d'une immunité antigreffe spécifique chez les Lombriciens, *Arch. Zool. Exp. Gen.* **112**:97.

Valembois, P., 1971b, Evolution de la musculature d'un xénogreffon de paroi du corps chez un Lombricien, *J. Microsc. (Paris)* **11**:339.

Valembois, P., 1971c, Etude ultrastructurale des coelomocytes du Lombricien *Eisenia fetida*, *Bull. Soc. Zool. Fr.* **96**:59.

Valembois, P., 1974, Cellular aspects of graft rejection in earthworms and some other Metazoa, *Contemp. Top. Immunobiol.* **4**:121.

Valembois, P., and Boiledieu, D., 1981, Fine structure and functions of erythrocytes and leucocytes of *Sipunculus nudus*, *J. Morphol.* **77**:163.

Valembois, P., and Cazaux, M., 1970, Etude autoradiographique du rôle trophique des cellules chloragogènes des vers de terre, *C. R. Soc. Biol.* **164**:1015.

Valembois, P., and Roch, P., 1977, Identification par autoradiographie des leucocytes stimulés à la suite de plaies ou de greffes chez un ver de terre, *Biol. Cell.* **28**:81.

Valembois, P., Roch, P., and Du Pasquier, L., 1973, Dégradation *in vitro* de protéines étrangères par les macrophages de Lombricien *Eisenia fetida* Sav., *C. R. Acad. Sci. Ser. D* **277**:5.

Valembois, P., Roch, P., and Chapron, C., 1977b, Stimulation mitogénique et différenciation cellulaire chez un Invertebré (*Eisenia fetida* Sav.), *Bull. Soc. Zool. Fr.* Suppl. **1**:51.

Valembois, P., Roch, P., and Boiledieu, D., 1980, Natural and induced cytotoxicities in sipunculid and annelid worms, in: *Phylogeny of Immunological Memory* (M. J. Manning, ed.), pp. 47–55, Elsevier/North-Holland, Amsterdam.

Valembois, P., Roch, P., and Du Pasquier, L., 1981, Evidence of a MLR-like reaction in an invertebrate, the earthworm *Eisenia fetida*, in: *Aspects of Developmental and Comparative Immunology* (J. B. Solomon, ed.), pp. 23–30, Pergamon Press, Elmsford, N.Y.

Vernet, G., and Gontcharoff, M., 1975, Etude autoradiographique de l'incorporation de l'acide Δ-aminolévulinique ^{3}H et de ^{55}Fe dans les éléments figurés du sang de *Lineus lacteus* Montagu (Hétéronémertes), *C. R. Acad. Sci. Ser. D* **280**:1413.

Vernet, G., and Gontcharoff, M., 1976, Cytological study of the blood corpuscles of *Lineus lacteus* (Rhynchocoela, Lineidae), *Cytobios* **17**:137.

Volkonsky, M., 1933, Digestion intracellulaire et accumulation des colorants acides: Etude cytologique des cellules sanguines de Sipunculides, *Bull. Biol. Fr. Belg.* **67**:135.

Warr, G. W., and Marchalonis, J. J., 1978, Specific immune recognition by lymphocytes: An evolutionary perspective, *Q. Rev. Biol.* **53**:225.

Weinheimer, P. F., Acton, R. T., Cushing, J. E., and Evans, E. E., 1970, Reactions of sipunculid coelomic fluid with erythrocytes, *Life Sci.* **9**:145.

Willmer, E. N., 1974, Nemertines as possible ancestors of the vertebrates, *Biol. Rev.* **49**:321.

Wright, R. K., 1976, Phylogenetic origin of the vertebrate lymphocyte and lymphoid tissue, in: *Phylogeny of Thymus and Bone Marrow-Bursa Cells* (R. K. Wright and E. L. Cooper, eds.), pp. 57–70, Elsevier/North-Holland, Amsterdam.

Zarling, J. M., Nowinsky, R. C., and Bach, F. H., 1975, Lysis of leukemia cells by spleen cells of normal mice, *Proc. Natl. Acad. Sci. USA* **72**:2780.

Zucker-Franklin, D., and Hirsch, M. D., 1965, Electron microscopic studies on the degranulation of rabbit peritoneal leukocytes during phagocytosis, *J. Exp. Med.* **120**:569.

Cellular Defense Systems of the Mollusca

THELMA C. FLETCHER and C. ANWYL COOPER-WILLIS

1. INTRODUCTION

By the time the oldest fossiliferous rocks were laid down, the molluscs were already represented by many different patterns (Runnegar and Pojeta, 1974). Few phyla show such wide diversity imposed on such a uniform body plan. The estimated 47,000 living molluscan species (Boss, 1971) evolved from a group related to the same stock as that from which the arthropods and annelids derive. The gastropods are the most abundant class, representing 80% of the total species, while the lamellibranchs represent 16% and the cephalopods only 1%.

The long and successful history of the molluscs is indicative of their ability to survive and that some mechanism of resistance to pathogens has conferred a selective advantage. Perhaps, as with most invertebrates, their primary method of overcoming the effects of disease on a population is by the production of large numbers of progeny. It is the longer life span and usually fewer offspring of the higher vertebrates which necessitates a sophisticated internal defense system able to respond with rapidity, specificity, and memory to assault. The extant molluscs do, however, have an internal defense system, compounded of cellular and humoral elements, which can adequately arrest and/or eliminate some invaders. Basic cellular responses include phagocytosis of small particles and encapsulation of larger bodies, but the demonstration of humoral factors having a protective role is more difficult. In this chapter they will only be discussed as an adjunct to the cellular mechanism.

2. RECOGNITION OF SELF AND NON-SELF

The basis of recognition is a central theme of biology (Marchalonis and Cohen, 1980), being a feature of all living cells which are organized into tissues

---

THELMA C. FLETCHER • NERC Institute of Marine Biochemistry, Aberdeen, Scotland.
C. ANWYL COOPER-WILLIS • Department of Microbiology and Immunology, University of Oregon Health Sciences Center, Portland, Oregon 97201.

in multicellular animals (Boyden, 1962). At the molecular level, specificity of reaction is the mark of all biological processes: to achieve specificity, macromolecules must recognize one another. Hydrophobicity is the major factor stabilizing protein–protein association, although complementarity plays a selective role in deciding which proteins may associate (Chothia and Janin, 1975). Another aspect of recognition has been observed in echinoderms. Starfish saponins lyse cells by interacting with membrane cholesterol (Δ^5-sterol), while the major Δ^7-sterol of starfish does not react with the saponins to any extent so that self tissues are protected from lysis (Mackie et al., 1977). Those echinoderms lacking saponins possess cholesterol in their membranes. The topic of recognition is discussed fully elsewhere in this volume (Ey and Jenkin; Warr and Marchalonis), but some description of the mechanisms thought to pertain to molluscs will be given here.

The unresolved problem is whether the animal recognizes self, and anything which is not self is reacted against, or whether it recognizes foreignness. Salt (1970) has been one of the chief proponents of the first view, based on his careful experiments with insects, but as Lafferty and Crichton (1973) pointed out, this does not apparently allow for the graded responses which are observed in invertebrate transplantation studies. It is possible, however, to envisage the primary recognition as being to self and the degree to which the animal responds depending on how closely the foreign material resembles self. Tauber (1976) has in fact put forward a theory that the markers of biological individuality, i.e., the histocompatibility antigens, may serve as a standard against which foreignness is directly compared. Lackie (1977), working with insect hemocytes, suggested that a threshold of difference needs to be surpassed before non-self is recognized. A plausible hypothesis for the evolution of the recognition of foreignness from earlier self-recognition structures has been put forward by Mäkelä et al. (1976). This topic lends itself to much speculation, as the experimental results available for interpretation do not yet provide a conclusive answer.

Tissue transplantation responses are probably the only indications of immune reactivity that can be applied across the whole phylogenetic spectrum (Hildemann et al., 1979). It is therefore unfortunate that in molluscs the exact level of discrimination is ill defined due to the difficulty of surgical techniques, which usually lead to inconclusive results (Cheng and Galloway, 1970). From the available evidence, however, it would appear that the majority of molluscs are unable to recognize allografts as foreign. This is based on experiments in the gastropods *Australorbis glabratus* (= *Biomphalaria glabrata*) (Tripp, 1961) and *Lymnaea stagnalis* (Sminia et al., 1974), the lamellibranch *Crassostrea gigas* (DesVoigne and Sparks, 1969), and the cephalopod *Octopus bimaculatus* (Cushing, 1962). Xenografts are consistently rejected, and recent work by Bayne et al. (1979) describes the cellular response of *Mytilus californianus* to mantle tissue transplants of *Mya arenaria*. Klein (1977) stated that a clear-cut demonstration of histoincompatibility in the Mollusca had still to be demonstrated, and this still appears to be true today. A future development must be to apply *in vitro* cellular cytotoxicity and mixed leukocyte culture techniques to molluscan cells; methods now being used with other invertebrates (Valembois et al., this volume).

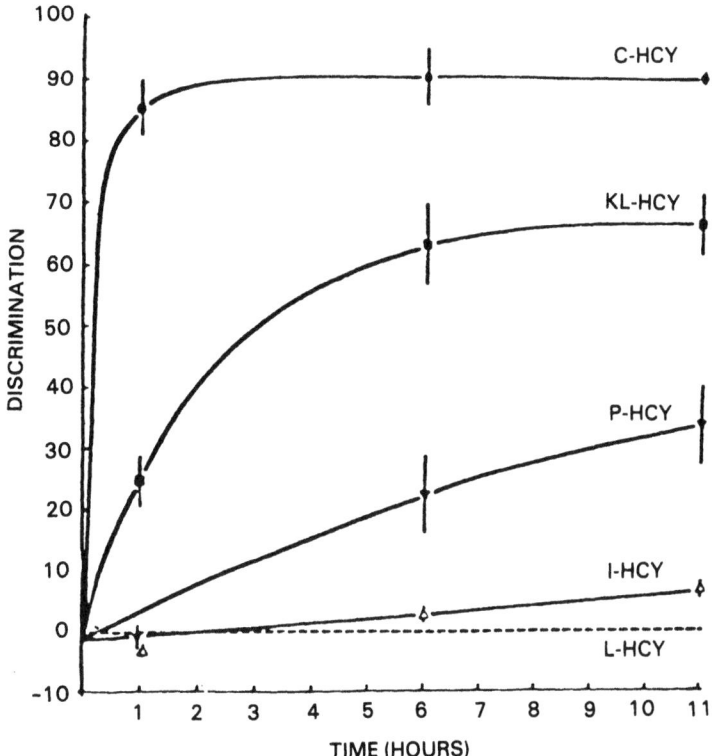

FIGURE 1. Discrimination curves showing the rate of clearance of *Ischnoradisia* (I-HCY), *Poneroplex* (P-HCY), keyhole limpet (KL-HCY), and crayfish (C-HCY) hemocyanin, relative to the clearance of *Liolophura* hemocyanin (L-HCY). Groups of animals were injected simultaneously with 0.25 mg ^{125}I-labeled L-HCY and 0.25 mg ^{131}I-labeled test hemocyanin. Each point is the mean value obtained in groups of four animals and the vertical lines show the standard error of the mean. (Reproduced from Crichton and Lafferty, 1975, with permission.)

A different approach for examining the discriminatory capacity of a mollusc to foreign proteins was described by Crichton and Lafferty (1975). They found that the chiton (*Liolophura gaimardi*) could distinguish between molecules that are structurally related and can rank these proteins in an order of foreignness that corresponds to their immunochemical relationship. They used hemocyanins, variously related to that of the chiton (Figure 1): the most foreign (crayfish) being cleared the most rapidly. Thus, the molluscs have a competent mechanism for the recognition of foreignness, and although apparently lacking the specificity of the annelids (Cooper, 1976), it contributes adequately to the well-being of this successful phylum.

3. PATHOGENS OF MOLLUSCS

Most molluscan pathogens have been identified from lamellibranchs of economic importance, such as *Crassostrea* and *Mercenaria*. Since pulmonates are often vectors for human trematode parasites, the responses of snails to larval trematodes have been extensively studied; Malek and Cheng (1974) have reviewed the literature on molluscan hosts of other human parasites. Mortalities of *Crassostrea* and other commercially fished bivalves have been associated with bacterial, protozoan, and fungal infections (Sindermann, 1970; Sprague, 1971). Chronic infections by pea crabs, copepods, and trematode and cestode larvae do not usually cause mortalities, but lower the condition, and hence the market value, of the infected animals (Sindermann, 1970). Virus particles have been recorded in over 13 species of larval (Elston, 1979) and adult molluscs (Farley, 1978). These viruses all resemble those identified from mammalian diseases, although in the molluscs they are not usually associated with overt signs of disease.

Parasitization reduces the viability of the host, although infections with the nematode, *Bucephalus cuculus*, initially stimulate the growth of *Crassostrea* (Hopkins, 1957), as do schistosome infections in *Biomphalaria* and *Lymnaea catascopium* (Pan, 1965; Loker, 1979b). Glycogen in the digestive glands of snails with established infections is reduced to a level comparable to that of starved animals (Cheng and Snyder, 1962; Christie *et al.*, 1974; Carter and Bogitsh, 1975). Glucose levels also drop (Cheng and Lee, 1971). Egg production tends to be reduced in gastropods (Pan, 1965; Loker, 1979b) or castration can occur if the parasite resides in the gonads (Hopkins, 1957; Sprague, 1965; Yevich and Berry, 1969), and life span is reduced. Death can result from massive cercarial emergence (Cheng, 1968).

Proliferative disorders and tumorous conditions are of low incidence in mollusc populations but may contain elements of cellular defense responses. At present, they are best described as neoplasticlike, until objective criteria for their classification exist. Solid tumors have been found in gastropods and lamellibranchs (Pauley, 1969). Such conditions can, however, often be ascribed to hyperplasia as a result of tissue damage or an encapsulation response. Michelson and Richards (1975) described a number of such tumors in *Biomphalaria*, where associated microorganisms could not be precluded as a cause of the pathology. The general pattern for solid tumors is of outgrowths from the foot or mantle surface, which show no cellular abnormality, in which mitotic figures are not evident and which do not appear to be invasive (Pauley, 1969; Michelson, 1972).

Intense hyperplasia of primitive-appearing cells, which resemble immature hemocytes, has been described in oysters and other lamellibranch species (Farley, 1969; Mix, 1975; Farley and Sparks, 1970; Christensen *et al.*, 1974), the whole body becoming progressively infiltrated by the abnormal cells. Ultrastructurally, although the cells have some features in common with malignant mammalian cells, such features are common to any cells undergoing rapid division (Farley, 1976; Mix *et al.*, 1979). A hyperplasia of gill epithelium in *Macoma balthica* is the only case in which any origin for the atypical cells has been found (Christensen *et al.*, 1974). Here, abnormal proliferation of the epithelium with loss of cellular polarity was observed. In more advanced cases, foci of proliferation were seen in other tissues, which resulted in general, diffuse invasion of hemal spaces and connective tissue as the disease progressed. These proliferative disorders do not always show patterns of distribution suggestive of an infectious agent, although Brown et al. (1980) found that *Mya* developed the disease, when kept in the laboratory downstream from already infected animals. In none of the reported conditions has any etiological agent been observed, although a virus may be implicated in the case of *Mya* (Brown et al., 1980).

Attempts have been made to correlate the incidence of hyperplastic disorders with pollution levels, particularly with hydrocarbons, but results so far have been inconclusive (Brown et al., 1977; Farley, 1977; Yevich and Barszcz, 1977; Lowe and Moore, 1978). Molluscs show only low abilities to metabolize these compounds (Mix et al., 1979; Bend et al., 1977) so that they may not have the same carcinogenic potential in an invertebrate species.

4. HOST RESISTANCE AND SUSCEPTIBILITY

Every animal is a potential host to all the pathogens and parasites from the environment with which it has contact, yet the parasite fauna of any given species is quite closely restricted and usually constant (Sprent, 1969). The state of the host, which will influence the outcome of the host–parasite relationship, has been described in various ways leading to some confusion in terminology, although from a practical viewpoint an animal either resists or succumbs to an infection.

Potential hosts which do not fulfill the specific physiological requirements of the parasite will remain relatively free from infection and would be described as insusceptible by Read (1958). Host insusceptibility can be of two kinds: either the potential invader cannot gain entry or, if it does, the organism cannot survive in the host (Stauber, 1961). In snails, insusceptibility can be the result of physical factors such as thick skin, the secretion of mucus which traps miracidia (Loker, 1978), or the failure of parasites to attach (Kinoti, 1971).

Susceptibility is defined by Read (1958) as "a physiological state of the host in which the parasite is supplied with its life needs" while Sprent (1969) refers to the "innate suitability" of the host for the parasite. The influence of genetics, anatomy, physiology, age, diet, and behavior on susceptibility in invertebrate, including molluscan, host–parasite relationships has been discussed by Tripp (1969).

Resistance was also defined by Read (1958) as "those alterations of the physiological state of the host which represent a response to previous or present experience with the parasite or chemically related entity." It has been customary to consider resistance as innate or acquired. The former is expressed without previous experience of the parasite and the terms "innate resistance" and "insusceptibility" are often used interchangeably, although resistance implies an active response whereas insusceptibility indicates a more passive role for the host. Snails of certain strains may show an intermediate response, since encapsulated and normally developing sporocysts can occur in the same animal (Newton, 1953; Cheng, 1978).

The evidence for the development of acquired resistance, as a result of previous exposure to the parasite, is not conclusive. Bayne et al. (1980) found no absolute confirmation of immunological memory in molluscs, although reports of elevated hemocyte counts and more rapid responses to secondary challenge would seem to indicate at least some change in nonspecific activation mechanisms (Pauley et al., 1971; Bayne, 1977; Lie and Heyneman, 1976b; Bayne et al., 1979, 1980). Acquired resistance was used to describe the results Lie and Heyneman (1979) obtained in four strains of *Biomphalaria glabrata*. These were exposed to the irradiated miracidia of three echinostome species, followed by reexposure to nonirradiated miracidia of the same species, after encapsulation of the irradiated parasites. The sensitized juvenile albino strain showed 94% resistance to a heavy challenge with *Echinostoma lindoense*, while only 8% of the nonsensitized control snails were not infected.

Resistance in the molluscs must, at present, be considered as largely innate and the result of natural selection, given their short generation times and enormous reproductive capacity (Stauber, 1961).

The reaction between host and parasite depends on factors in both which are genetically regulated (Wakelin, 1978). If the selection pressure of each species on the other is strong, this may lead to localized strain–strain specificity, as found among populations of *B. glabrata* and *Schistosoma mansoni* (Richards, 1975; Basch, 1976; Michelson and DuBois, 1978). The genetic control of molluscan defense mechanisms has practical aspects. Barr (1975) has suggested that where disease-producing organisms are transmitted by biological vectors, these could be replaced by nonvectors by the selection of refractory genotypes.

5. CELLULAR DEFENSE MECHANISMS

The internal defense mechanisms of molluscs are predominantly cellular (Michelson, 1975; Cheng, 1977) which is usually taken to be synonymous with phagocytosis. This is assumed to be protective by intracellular destruction of the pathogens or by their removal through epithelial surfaces by diapedesis, although a definitive study on how cells combat a natural disease of a mollusc is still awaited. The discovery by Bayne (1977) of a strain of *Pseudomonas aeruginosa* which kills *Helix aspersa* and *H. pomatia* has provided a model system, at least for studies on gastropod–bacterial interactions. Douglass and Haskin (1976) found no difference in the cellular responses of stocks of *Crassostrea virginica*, resistant and susceptible to the natural haplosporidian parasite *Minchinia nelsoni*. Their results in fact implied only a minor role for cellular mechanisms in defense, with humoral factors having a primary function. A great body of work has, however, accumulated on cellular responses in molluscs and, depending on the type of infection, these probably form the first line of defense (Cheng, 1975).

Work on defense functions has mainly concerned reactions in the peripheral blood, although cells from the connective tissue of molluscs have also been studied (Sminia, 1972; Sminia et al., 1979a). The involvement of tissues and organs (e.g., the digestive gland) in bacterial clearance (Bayne, 1974) indicates the importance of considering cells other than those of the blood. The morphology of blood cells from a relatively few molluscan species has been studied in detail (Cheney, 1971) but workers appear to have promoted their own descriptive terminology. It is therefore timely that Ratcliffe and Rowley (1979) have attempted to unify morphological and functional criteria of blood cells (hemocytes) of invertebrates, including molluscs. In the latter group, they differentiate two developmental series: hyaline/semihyaline cells and granular cells. The most advanced molluscs, the cephalopods, appear to have one blood cell type (Stuart, 1968; Cowden and Curtis, 1974) as does the gastropod *Lymnaea stagnalis* (Sminia, 1972).

There do not appear to be any well-defined hemopoietic organs in the molluscs, apart from the "white bodies" in the orbital pits of cephalopods (Cowden and Curtis, 1974). Lie *et al.* (1976a), however, have described an amebocyte-producing organ in *Biomphalaria* which appears to be the primary amebocyte source for encapsulation. The organ became enlarged as capsule formation proceeded around irradiated (or normal) *Echinostoma lindoense* miracidia. The posterior mantle epithelium formed the anterior border and the pericardium, the posterior limit of the organ. The careful investigations of Sminia (1974) in *L. stagnalis* revealed that the circulating amebocytes (hemocytes) were identical with those of the tissues and that there was no difference in the relative importance of the blood and connective tissue in the process of hemopoiesis. Sminia (1974) pointed out the need for caution in the interpretation of reports, based on histological studies, of the transformation of fibroblasts, endothelial and epithelial cells into amebocytes in molluscs. Thus, we can only describe the functional aspects of certain morphological cell types in relation to pathogens, without ascribing their origins.

5.1. ENCAPSULATION

Encapsulation is a response of the hemocytes to particles recognized as foreign, but too large for phagocytosis. The capsule is formed by hemocytes adhering around the particle and isolating it from the host tissues. Such a response may be engendered by invading parasites (Pan, 1965; Richards and Merritt, 1967; Cheng and Rifkin, 1970; Harris and Cheng, 1975a; Harris, 1975; Lie and Heyneman, 1975; Rachford, 1976; Krupa *et al.*, 1977), or may occur around groups of hemocytes which contain intracellular pathogens (Mackin, 1951; Michelson, 1961).

The mechanisms for hemocyte recognition are not known, but it seems probable that they are similar for both encapsulation and phagocytosis. The encapsulation response causes a concentration of hemocytes at the site of the reaction, which is probably a chemotactic response. *In vitro*, hemocytes of *Crassostrea* (and other lamellibranch species) are attracted to the encysted cercariae of *Himasthla quissetensis* (Cheng *et al.*, 1966, 1974) and to living bacteria (Cheng and Howland, 1979; Cheng and Rudo, 1976a). Hemocytes are also attracted to areas of xenogeneic tissue grafts or implants, which are then encapsulated and destroyed by the hemocytes (Tripp, 1961; Sminia et al., 1974; Bayne, et al., 1979). Generally, molluscs seem unable to distinguish between allografts and autografts, although Cheng and Galloway (1970) suggested that *Helisoma* can reject allografts. A component of damaged cells is possibly a chemotactic factor, since hemocytes are attracted to wound sites (Sminia et al., 1973; Bayne et al., 1979) and to the wounded surfaces of autologous tissue grafts, which do not otherwise elicit a response (Sminia et al., 1974; Tripp, 1961). Chemotaxis towards such actively metabolizing targets is understandable, but inert materials are also encapsulated. Polystyrene spheres, carmine particles, and pieces of yvelon sponge were encapsulated respectively by *Biomphalaria* (Tripp, 1961), *Aplysia* (Pauley and Krassner, 1972), and *Lymnaea* (Sminia et al., 1974). It is possible that the hemocytes participating in the encapsulation response release an attractant, similar to the lymphokines of mammals, which would inhibit those hemocytes entering the area from continuing in the circulation.

The process of capsule formation has been described at both the light and the electron microscopic level. The experimental infection of *Biomphalaria* with larvae of the nematode *Angiostrongylus cantonensis* has provided a model system. About 24 hr after exposure, the larvae are found in the head-foot region, with a slight infiltration of hemocytes around the parasites. This response intensified up to about 6 days, when the loose aggregation of cells reached a maximum (Harris and Cheng, 1975a). At the ultrastructural level, pseudopodial extensions from the cells, directed towards the surface of the larva, could be seen (Harris, 1975). As capsule formation proceeded, the pseudopodia became increasingly lamelliform, and came to lie more parallel with the surface of the nematode. By 8 days, the outer layer of cells forming the capsule was still rounded in outline, but the inner layer had become progressively flattened. These cells formed concentric layers of extremely thin lamellae around the parasite and were expanded in places to accommodate the nuclei and other organelles. In longstanding infections, the nematodes were surrounded by thinly drawn cells with fairly numerous mitochondria and rosettes of glycogen. Under the light microscope, these extended cells have a fibroblastlike appearance (Harris and Cheng, 1975a).

Descriptions of capsule formation in *Lymnaea* (Sminia et al., 1974) and *Bulinus* (Krupa et al., 1977) indicate a broadly similar process to that in *Biomphalaria*, although in both these species extracellular fibers are also involved. Cheng and Rifkin (1970) proposed a classification of molluscan capsules which distinguished purely cellular capsules from those which contained extracellular elements. A good example of this latter type is the capsule formed by *Crassostrea* in response to cestode larvae of *Tylocephalum* sp. (Cheng and Rifkin, 1970). Encapsulation was initiated in the loose connective tissue surrounding the gut, by the intercellular matrix becoming thickened around the larvae. This fibrous layer was then infiltrated by hemocytes, and more concentric layers of fibers added. The mature capsule was composed of a thin layer of fibroblastlike cells lining the cavity containing the parasite, surrounded by a thick fibrous coat interspersed with hemocytes. Towards the periphery of the capsule, brown cells were embedded in the matrix. These cells are thought to have some role in internal defense because of their greater abundance in diseased molluscs (Farley, 1968) but they may well have an excretory role, with their extensive tubular network communicating with the extracellular environment via membrane slits (Ruddell and Wellings, 1971). The brown cells of *Mytilus edulis* represent a cell type distinct from the hemocytes and appear to concentrate iron and lead in their cytoplasmic granules (Moore and Lowe, 1977).

The origin of the fibroblastlike cells in capsules has been in doubt. Pan (1965), studying encapsulation of *S. mansoni* cercariae by *Biomphalaria*, suggested that the fibroblasts were derived from hemocytes. India ink, injected into *Lymnaea*, is taken up by the hemocytes, and if these animals are then given an implant, the capsule that forms is black (Sminia et al., 1974). The same is true of wound repair; ink particles could be seen within vesicles in the cytoplasm of cells comprising the capsule (Sminia et al., 1973). The work of Harris (1975) confirms that the fibroblasts are in fact modified hemocytes. This process appears to be reversible; in some capsules, particularly those surrounding tissue implants, the inner layer of fibroblasts in the capsule disappears, and the implant is infiltrated by normal-looking hemocytes (Pan, 1965; Sminia et al., 1974). As the contents are destroyed and taken up by the hemocytes, the capsule regresses (Cheng and Garrabrant, 1977). Disposal of the capsule in *Biomphalaria* occasionally occurs by expulsion through ruptures in the epidermis (Lie and Heyneman, 1976a). The actual cause of death of encapsulated parasites is unknown, but in *Biomphalaria*, levels of lytic enzymes in the capsule cells are high (Harris and Cheng, 1975b; Cheng and Garrabrant, 1977). Parasite death does not invariably follow encapsulation. The larvae of *Angiostrongylus* successfully complete their metamorphosis in *Biomphalaria* despite rapid encapsulation by hemocytes which give a strong staining reaction for acid phosphatase, β-glucuronidase, and nonspecific esterases (Harris and Cheng, 1975b). Such capsules persist for long periods and viable first-stage larvae have been recovered from *Biomphalaria* up to 12 months after infection (Richards and Merritt, 1967). Hyaline cells of the prosobranch, *Cerithidea californica*, encapsulate trematode sporocysts (*Renicola buchanani*), but this does not lead to the destruction or elimination of the parasite. It does, however, show the recognition of non-self material and an active attempt to separate parasite and host tissues (Yoshino, 1976). Capsules surrounding inanimate material also persist.

In their natural molluscan hosts, most metazoan parasites elicit no cellular response and are not recognized as non-self. Similarly, the symbiotic zooxanthellae in the tridacnid clams do not stimulate a response, but moribund algal cells in various stages of digestion occur within hemocytes and digestive gland cells (Fankboner, 1971) which suggests that only senile algae are recognized as foreign by the clam. In mammals, adult schistosomes adsorb host antigens and escape detection by the immune system (Bloom, 1979). In *Biomphalaria* over a period of 2–4 weeks postinfection, schistosome antigens build up in the hemolymph and eventually saturate the hemocytes, thereby possibly blocking all the receptor sites for foreign antigens (Michelson, personal communication). Metamorphosis from miracidium to sporocyst involves extensive changes and loss of some antigenic determinants from the larval surface

(Yoshino *et al.*, 1977), and perhaps allows greater compatibility with the host. *Biomphalaria* hemocytes and schistosome sporocysts do have some surface markers in common (Heyneman *et al.*, 1971; Stanislawski *et al.*, 1976; Yoshino *et al.*, 1977; Murrell *et al.*, 1978) but these are probably not responsible for the failure of recognition, since the same determinants occur in resistant strains of snail. In susceptible snails, it is the transitional larval forms (which are probably less closely matched to the host) which tend to promote a cellular reaction (Loker, 1979a; Kinoti, 1971).

Larval trematodes may be able to paralyze the internal defense mechanisms of the host. Infection of *Biomphalaria* with normal *Echinostoma* miracidia protects subsequently introduced sporocysts from irradiated miracidia, from encapsulation. These are normally encapsulated, but prior infection with any of three species of miracidia prevents the response (Lie *et al.*, 1976b). It is likely that this is not a passive avoidance of the host's responses on the part of the parasite since that would not affect other parasites in the same snail. Paralysis of the response is also suggested by the results of Lie and Heyneman (1976b). Juvenile *Biomphalaria* which failed to destroy *Echinostoma* when first invaded, were found to be very much more susceptible to later infection, as compared with naive controls. There seemed to be a hypertrophy of the hemocytes, which just accumulated in the loose connective tissue of the digestive gland, without showing any response to the parasites. Sensitivity was lost if the initial response was not successful in destroying the parasites, and no further reaction was mounted against the existing, or later infections. *Biomphalaria* can be sensitized to *S. mansoni* (Lie and Heyneman, 1976a): encapsulation was more rapid in a second infection, the capsule larger and the migration of sporocysts within the snail's tissue slower. Sensitization, however, is not a universal molluscan response as Loker (1978) was unable to sensitize *Lymnaea catascopium* to *Schistosomatium douthitti*.

Some molluscan parasites can resist the lysosomal enzymes of their hosts. *Minchinia* is phagocytosed by the hemocytes of *Crassostrea* but is unharmed and able to multiply within the phagosomes (Farley, 1968). Hemolymph acid phosphatase is not elevated in *Biomphalaria* with encapsulated *Angiostrongylus*, and Harris and Cheng (1975b) suggested that the nematode may be able to prevent secretion of the enzymes by the capsule cells. Kassim and Richards (1978) found that lysozyme in the hemolymph of a refractory strain of *Biomphalaria* increased 10-fold on infection, although the level in the head-foot, where destruction of the parasite occurs, declined. Since a similar decline was observed in a control, susceptible strain, the authors suggest that lysozyme alone does not play a role in parasite destruction. This does not preclude the synergistic involvement of a number of hydrolytic enzymes in parasite destruction in a resistant host.

It is evident that encapsulation is not a passive "walling off" of the offending particle, but is a dynamic process with the participating cells metabolically active (Harris, 1975; Krupa *et al.*, 1977; Sminia *et al.*, 1974; Cheng and Rifkin, 1970). The cells on the inner surface of the capsule are able to resume their hemocytelike appearance when they infiltrate the foreign tissue or parasite within the capsule. The dissolution of capsules after resorption of their contents further indicates the active nature of the encapsulation response. It is interesting to speculate that similarities in the stratagems of parasites against both vertebrate (reviewed by Bloom, 1979) and invertebrate host responses could indicate that the defense systems have enough in common for similar mechanisms to be effective in circumventing them.

5.2. NACREZATION

In contrast to encapsulation, nacrezation, although a cellular response, is not a dynamic process. The foreign particle is excluded from the internal environment of the host by a wall of nacre.

Nacrezation is not a specific response, but seems to be triggered by physical pressure on the mantle. Cultured pearls are produced by implanting a piece of allogeneic tissue into the mantle, and autologous grafts have the same effect. Such grafts, in other sites, do not evoke an encapsulation response, so pearl formation is evidently triggered by different factors from encapsulation. Particles, such as sand grains or parasite larvae, trapped between the shell and the mantle will also stimulate pearl formation. This occurs within a pearl sac, made by the invagination of the mantle epithelium around the particle. Layers of nacre are then deposited around the nucleus (Cheng and Rifkin, 1970). In *Minchinia nelsoni* infections, *Crassostrea* will isolate plasmodia and cellular debris in conchiolin deposits on the nacreous surface of the shell (Farley, 1968). Pearl formation commonly occurs in lamellibranchs: *Biomphalaria* is the only gastropod species in which this response has been reported (Richards, 1970, 1972).

5.3. PHAGOCYTOSIS

Phagocytosis, the ingestion and digestion of solid particulate matter, is a fundamental attribute of many cells (Walters and Papadimitriou, 1978) and is of significance in the life of molluscs, where the process is related to nutrition and excretion, as well as to defense. These functions are facilitated by the wandering nature of the molluscan amebocytes (hemocytes). There is an appreciable variation in the number of cells in the blood between individuals of the same species and this could be related to their feeding activity, since varying numbers of amebocytes could be transporting food within the tissues (Narain, 1973). Not all circulating cells are phagocytic. *In vitro*, phagocytosis of *E. coli* by *Crassostrea virginica* hemocytes involved only 43% of the cells and these usually only contained from one to five bacteria per cell (Tripp and Kent, 1967). In *C. gigas*, only about 5% of the hemocytes appeared to be phagocytic for *E. coli* and *Vibrio anguillarum* over a 30-min period (S. W. Hardy, personal communication). In *Lymnaea stagnalis*, however, *in vivo* studies by Sminia (1972) showed 80–100% of the amebocytes contained ink particles, 1 hr after injection; similar results were obtained with bacteria. In the advanced Cephalopoda, mature leukocytes do not display a significant phagocytic capacity (Bayne, 1973a; Cowden and Curtis, 1973).

All molluscs appear to have an efficient system for clearing foreign material although there is considerable variation in the relative importance of circulating hemocytes. From evidence provided by the few species studied, circulating cells seem to be the most effective clearing mechanism in the lamellibranchs. Even here, the mechanism can be supplemented with other cells such as the epidermal cells of the mantle of *Mytilus edulis* which can phagocytose extrinsic particles in the extrapallial space (McLean, 1980). In the cephalopods, injected carbon was found localized in the gills of *Octopus dofleini* and Bayne (1973a) suggested that fixed phagocytes might be responsible. Bayne (1974) also pointed out that bacterial clearance in *Helix pomatia* was more rapid than would be expected from hemocyte phagocytosis alone and that agglutination and phagocytosis, associated within definite organs, were of importance. Sminia (1972) found no evidence for true fixed phagocytes in the connective tissue of *Lymnaea stagnalis*. Stuart (1968) concluded that in the cephalopod, *Eledone cirrosa*, the localized phagocytic cells constituted a primitive "reticuloendothelial" system.

Sminia (1972) observed that material could be phagocytosed in two ways in *L. stagnalis*. Large particles, such as bacteria, were engulfed by the fusion of the pseudopodia with each other or with the plasmalemma, forming large cytoplasmic vacuoles. Small particles caused the cell membrane to invaginate to form heterophagosomes. Moore and Eble (1977) quoted the results of time-lapse cinematography which revealed large granulocytes of *Mercenaria* enveloping yeast cells with wavelike extensions of the outer ectoplasm, while small granulocytes engulfed the yeast with rapid cell extensions around the particle. Renwrantz *et al.* (1979) described hemocytes of *C. virginica* extending a single funnellike pseudopod down which particles appeared to glide.

5.3.1. Recognition

The first stage of phagocytosis is recognition, which presumably signals a transduction mechanism, in turn evoking a number of effector steps. These include adhesion, pseudopod assembly, movement, and fusion. In many instances, recognition of foreign elements begins before the phagocyte makes contact with the particle, since, at least *in vitro*, phagocytes show directional movement towards non-self material. Chemotaxis can thus increase the frequency of encounters between phagocyte and material to be phagocytosed.

The migration of hemocytes toward non-self material was observed in encapsulation processes in molluscs (Section 5.1). There is a chemotactic attraction by live gram-positive and gram-negative bacteria *in vitro*, for hemocytes of *C. gigas*. Cheng and Howland (1979) suggested that the chemoattractant is a molecule produced by the living bacteria. They also suggested a possible correlation between absence of chemotaxis and pathogenicity, since *Vibrio parahaemolyticus*, which does not attract oyster hemocytes, is a pathogen of marine lamellibranchs. Schmid (1975) demonstrated *in vitro* chemotaxis by snail (*Viviparus malleatus*) hemocytes toward heat-killed *Staphylococcus aureus* and to N-acetylglucosamine. The presence of a bacterial agglutinin, with a specificity for glucosamine and present in the hemolymph, was necessary for the response.

Schmid postulated that the agglutinin combined with *S. aureus* and was randomly phagocytosed and that in this case, the hemocyte itself might release a chemotactic agent. It is not known whether phagocytes use the same recognition mechanisms for chemotaxis and phagocytosis.

5.3.1a. Without Opsonins. In our studies on bacterial phagocytosis by *C. gigas* hemocytes, we observed considerable *in vitro* uptake of *E. coli* and *V. anguillarum* in saline alone (Hardy *et al.*, 1977). This was usually at least 50% of the uptake found with other treatments (Figure 2) and we concluded that phagocytosis could operate via specific recognition factors on the hemocyte membrane, or by nonspecific, physicochemical mechanisms. The latter, independent of recognition of specific biological ligands, might also operate with the wide range of abiotic particles reported to be phagocytosed not only by molluscan, but also by vertebrate cells. Physicochemical mechanisms involve the surface properties of both particle and phagocyte: the former must be more hydrophobic than the phagocyte in order to be engulfed by it (van Oss, 1978). Surface charge, shape, and membrane fluidity could be involved, and Capo *et al.* (1979) have postulated the presence on phagocytes of at least one nonspecific receptor, which can bind various particles depending on their hydrophobicity. Self elements must possess physical properties which sometimes coincide with those of the non-self material. Self phagocytosis would be prevented only if self elements carry a self determinant recognized by the phagocyte.

There are no reported studies on the physical aspects of molluscan cell recognition but there is some evidence that molluscan hemocytes possess receptors capable of specific chemical recognition. Renwrantz and Cheng (1977b) found that 3–8% of an *in vitro* population of *H. pomatia* hemocytes were capable of direct attachment to erythrocytes of mice, rabbits, rats, and sheep. Heteroagglutinins are present in the hemolymph of many invertebrates (Section 5.3.1b) but it is not yet known whether these can be firmly bound to the hemocyte membrane or form a constitutive part of it. Their presence in the membrane would explain the direct rosetting of erythrocytes around the molluscan hemocyte.

5.3.1b. With Opsonins. In vertebrates, the attachment of nonself material to phagocytes can be mediated by opsonins. The onus of recognition is on humoral factors (e.g., antibody or complement) which form the link between foreign particle and phagocyte membrane receptors. Molluscan hemolymph has frequently been found to have opsonic properties although the responsible molecules have still to be identified, and the chemistry of the hemocyte membrane remains a potential field of study.

Opsonization has, in some molluscs, been related to agglutinins present in the hemolymph and usually exhibiting binding specificity for membrane carbohydrates. Pauley *et al.* (1971) observed that *Aplysia californica* rapidly cleared from the hemolymph four species of marine bacteria for which it had agglutinins, while *Serratia marcescens* was not completely cleared even after a month. Since agglutinins were lacking for the latter species, they interpreted their results as indicating that agglutinins were serving as bacterial opsonins. Agglutinins can act as opsonins, as was shown by the enhanced *in vitro* uptake by *C. gigas*

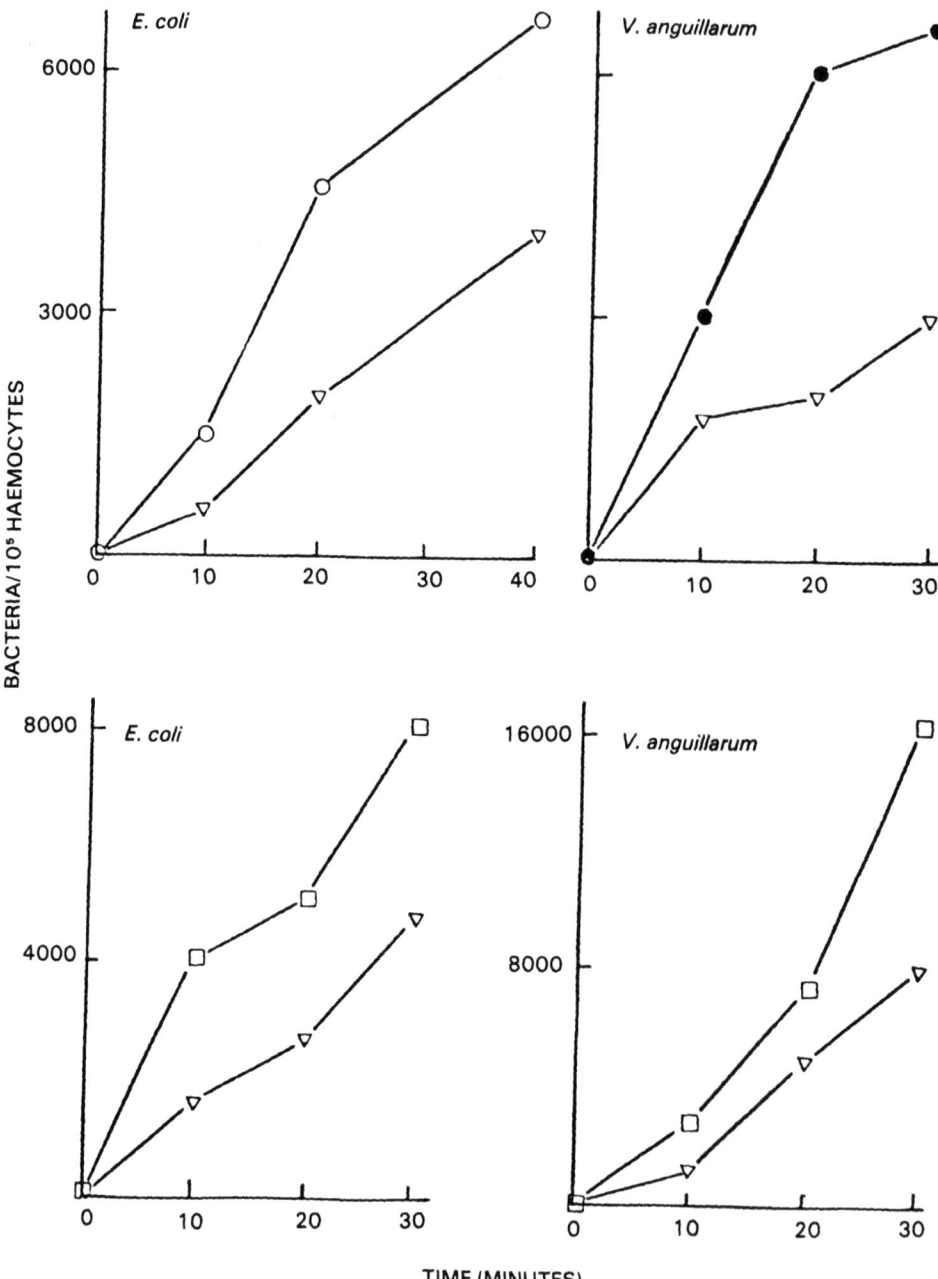

FIGURE 2. Bacterial uptake by oyster hemocytes. (▽) uptake of untreated bacteria in *Mytilus* saline; (○) uptake of bacteria in hemolymph; (●) uptake of hemolymph opsonized bacteria in *Mytilus* saline; (□) uptake of agglutinin opsonized bacteria in *Mytilus* saline. (Reproduced from Hardy et al., 1977, with permission.)

hemocytes, of bacteria previously opsonized by incubation with an agglutinin purified from the hemolymph (Hardy *et al.*, 1977). Renwrantz and Cheng (1977a,b) reported that nonnative agglutinins could link *H. pomatia* hemocytes to erythrocytes and that in this case, the opsonin (agglutinin) receptors on the hemocyte surface were carbohydrates. Parish (1977) has speculated that glycosyltransferases could be involved in the invertebrate recognition system and these have specific sugar-binding properties.

The opsonins could be molecules other than agglutinins, which are not present in all molluscs. Agglutinins might have evolved in groups such as the lamellibranchs in relation to feeding since they agglutinate the alga on which *Mercenaria mercenaria* feed (Arimoto and Tripp, 1977). Natural agglutinins were not detectable in the hemolymph of *Eledone* although human erythrocytes were only phagocytosed by the cephalopod hemocytes, after they had been opsonized with hemolymph (Stuart, 1968). Stuart believed the opsonin to have a broad spectrum of affinity for a wide range of particles. The gastropod, *Otala lactea*, also does not possess hemolymph agglutinins, and *in vitro* phagocytosis of bacteria was not influenced by the presence of hemolymph factors although they enhanced phagocytosis of formalinized yeast. This would indicate some specificity and not a generalized nonspecific stimulation (Anderson and Good, 1976). Hemolymph factors of *L. stagnalis* also promote phagocytosis (Sminia *et al.*, 1979b). In *Patella vulgata*, Cooper-Willis (1979) suggested that hemolymph enzymes might increase the apparent foreignness of invading bacteria by altering their surfaces.

Although phagocytosis can usually proceed in the absence of hemolymph opsonins, it is possible that when they are present they can modulate the cellular reactions, perhaps by increasing the potential for the recognition of non-self.

5.3.2. Fate of Phagocytosed Material

The fate of phagocytosed particles is usually either intracellular degradation, ejection from the body, or retention.

5.3.2a. Digestion. The hemocytes of molluscs have been shown to be rich in lytic enzymes (Janoff and Hawrylko, 1964; Eble, 1966; Eble and Tripp, 1968; Feng *et al.*, 1971; Sminia, 1972; Cheng, 1975, 1976a; Cheng and Rodrick, 1975; Cheng *et al.*, 1975; Cheng and Yoshino, 1976; Moore and Lowe, 1977; Yoshino and Cheng, 1976; Cooper-Willis, 1979), and are capable of degrading particles taken up by the cell (Tripp, 1958; Feng, 1966; Tripp and Kent, 1967; Sminia, 1972; Cheng and Cali, 1974; Feng *et al.*, 1977). Ultrastructurally, fusion of lysosomes with the phagocytic vacuoles containing zooxanthellae has been observed in the hemocytes and digestive gland of *Tridacna* (Fankboner, 1971). The presence of lytic enzymes was confirmed histochemically, and as digestion progressed, the phagosomes diminished in size, due probably to the loss of fluids and the products of digestion. Enzyme levels are often elevated in molluscan cells following the injection of bacteria (Cheng *et al.*, 1977) but this appears to be a nonspecific response, since the injection of sterile water has the same effect (Cheng and Butler, 1979).

Invertebrate phagocytes, unlike most mammalian cells, show little or no increase in O_2 consumption during phagocytosis (Anderson et al., 1973; Cheng, 1976b; Anderson, 1977). Metabolic energy is provided by the glycolytic pathway but without activity of the hexose monophosphate shunt; a similar condition has been reported in certain mammalian macrophages (McRipley and Sbarra, 1967). Hydrogen peroxide is generated during phagocytosis by mammalian neutrophils and in the presence of peroxidase and a halide is a highly effective bactericidal agent (Baehner, 1975). Only two molluscs have been examined for myeloperoxidase–H_2O_2 activity: a lamellibranch, *M. mercenaria*, which proved negative (Cheng, 1975), and a pulmonate, *L. stagnalis*, in which the hemocytes gave a positive staining reaction for peroxidase (Sminia et al., 1973).

It has long been suggested that molluscan hemocytes have a role in nutrition (Yonge, 1926; Zacks, 1955), but only recently has the fate of phagocytosed material been traced. Bacteria labeled with ^{14}C have been shown to be degraded, metabolized, and incorporated into the animal's tissues. In *H. pomatia*, label was rapidly cleared from the hemolymph and, over 15 days, became increasingly more evenly distributed throughout the body (Bayne, 1973b). In *C. virginica*, label first appeared in glycogen in hemocytes and body tissues 16 hr postinjection, but not in the hemolymph until 24 hr (Cheng and Rudo, 1976b). The latter authors suggested that glycogen is synthesized in the hemocytes from the phagocytosed bacteria, and then released into the hemolymph to be utilized as an energy source.

Although the peroxidase bactericidal system may not be widespread among the Mollusca, lytic enzymes within the cells evidently suffice, in most cases, to kill potential pathogens so that intracellular digestion might convert an unsuccessful parasite into nourishment for the host.

5.3.2b. Diapedesis. Particles phagocytosed by hemocytes are not always digestible and such material is usually removed from the molluscan body by migration of the laden hemocytes out of the animal (Tripp, 1958, 1960; Stauber, 1950; Feng, 1965; Reade and Reade, 1972). Diapedesis was recorded in *Ostrea* (= *Crassostrea*) *virginica* after India ink injection (Stauber, 1950). Epithelial migrations were not observed until more than 22 hr had elapsed and the peak of elimination of the ink-laden phagocytes, through the epithelia of the digestive organs, was at about 33 days. With different species of molluscs there are variations in the sites at which the cells pass to the exterior. Temperature is a factor which affects phagocytosis (Tripp, 1960) and probably also influences the rate at which the cells are eliminated (Feng and Feng, 1974; Hartland and Timoney, 1979). The externalized phagocytes are carried away in mucus and feces by the water stream set up by natural pumping action in the body of the mollusc.

Some pathogens are removed by diapedesis (Prytherch, 1940) but defense may be secondary to its primary function of nutrition and excretion. Phagocytes take up waste matter in the pericardial gland and migrate out through the gut, gills, palps, or kidney to the exterior, or possibly in some cases, return to the circulation after shedding their contents (Potts, 1967). Feng et al. (1977) do not believe that the directional movement of particle-laden hemocytes can be explained on the basis of non-self recognition. The mechanism whereby food-laden hemocytes return to the tissues while waste-laden hemocytes seek the exterior, is an unresolved question. Diapedesis has also been recorded in the gastropods (Brown, 1967; Tripp, 1961; Cheng et al., 1969).

5.3.2c. Sequestration. Not all particles taken up by phagocytes can be digested or eliminated, and in a short-lived animal with a low metabolic rate, sequestration may be a satisfactory solution. Lamellibranchs seem to rely entirely on circulating hemocytes for clearance (Stauber, 1950; Tripp, 1958; Reade and Reade, 1972), but in other molluscan groups, fixed phagocytes are also involved (Reade, 1968; Bayne, 1973a,b; Sminia et al., 1979a) so that elimination of indigestible particles by diapedesis may not be possible. Nonliving material may be sequestered either in capsules, which can persist for long periods of time (Sminia et al., 1974; Tripp, 1961; Pauley and Krassner, 1972), or in individual cells localized around the digestive gland or gill stems, or in the case of cephalopods, in the white body (Stuart, 1968; Bayne, 1973a; Pauley and Krassner, 1972).

5.4. PINOCYTOSIS

Pinocytosis is the imbibition of extracellular fluid in small membrane-bound vesicles and is a continual process. Both phagocytosis and pinocytosis are examples of endocytosis in that they involve membrane internalization (Stossel, 1977). The two mechanisms are, however, distinct: the energy of activation for pinocytosis is lower, as is its temperature minimum, and the two processes are stimulated by different ligands. While cytochalasin B and colchicine almost entirely inhibit phagocytosis, their effect on pinocytosis is slight, suggesting a greater involvement of cytoskeletal elements in the former (Silverstein et al., 1978). Pinocytosis is now recognized to include two separate processes: receptor-mediated endocytosis (RME) or adsorptive endocytosis, where specific macromolecules are bound to receptors in the cell membrane and localized in coated pits which invaginate to form intracellular coated vesicles (Goldstein et al., 1979), and fluid-phase endocytosis, where soluble unbound molecules are taken into the cell, in proportion to their concentration in the medium, via smooth vesicles (Steinman et al., 1978).

No work on RME has been done in invertebrates but coated vesicles have been observed in gill podocytes in the blue crab, *Callinectes sapidus*, in the heart and hepatopancreas of *Homarus americanus* (Johnson, 1980, and unpublished observations), and in the podocytes of the antennary gland end sac and fixed phagocytes in the hepatopancreas of the shrimp, *Palaemonetes intermedia* (Cooper-Willis, unpublished observation). These cells, termed podocytes or nephrocytes in the Crustacea (Foster and Howse, 1978), are evidently very similar in structure to the cells described in gastropods as pore cells or globular cells (Buchholz et al., 1971; Wolburg-Buchholz, 1972; Andrews, 1976, 1979; Boer and Sminia, 1976; Beltz, 1977; Curtis and Cowden, 1979). The invertebrate cell types resemble the podocytes of the mammalian glomerulus (Rodewald and Karnovsky, 1974). It seems probable that the molluscan pore cells will also be found to possess coated vesicles and be capable of RME.

Pore cells can take up very small particles, less than 20 nm in diameter in *L. stagnalis,* and smaller in *Viviparus viviparus* (Boer and Sminia, 1976). It is suggested that the particles pass through the sieve diaphragms which cover the gaps between the foot processes of the cells. These cells may be more concerned with ultrafiltration than internal defense but in *H. pomatia* and *Cepaea nemoralis* they have been shown to take up trypan blue and ferritin (Buchholz *et al.,* 1971; Wolburg-Buchholz, 1972). Curtis and Cowden (1978) found that injected, fluorescent-labeled immunoglobulins were taken up by the adventitial cells comprising the outermost layer of the blood vessels of *Limax maximus.* At the ultrastructural level, these cells have a close affinity with pore cells (Curtis and Cowden, 1979). The soluble proteins appeared to be broken down over 30 days and so the adventitial cells may, by trapping and eliminating foreign proteins, have a defense function.

Pore cells have not been described from any lamellibranch species, but Feng (1965) showed that *Crassostrea* hemocytes will take up proteins from the medium. The rate of *in vitro* uptake was a function of the ambient temperature and the protein concentration, suggesting that this is an example of fluid-phase endocytosis. This appears to be an effective clearance mechanism, since labeled hemocytes had started to migrate through the digestive tract and mantle epithelium within 15 min to 2 hr of protein injection, and *in vivo* the half-life of labeled proteins was about 80 min (Feng, 1965). In the chiton, Crichton and Lafferty (1975) showed that not only could the animal discriminate between closely related soluble proteins (Figure 1) but it was possible to competitively inhibit the removal of one protein without affecting the removal of an unrelated molecule, indicating distinct recognition units involved in the elimination of different proteins.

With increasing knowledge of cell architecture, the process of pinocytosis has started to be studied in detail, but as yet, very little is known of the mechanism and function of pinocytosis in molluscs.

6. RELATIONSHIPS OF CELLULAR DEFENSE MECHANISMS WITHIN THE MOLLUSCA AND COMPARISON WITH VERTEBRATE SYSTEMS

Most comparisons of vertebrate and invertebrate modes of internal defense have sought similarities between the vertebrate anamnestic immune response and the invertebrate defense mechanisms, particularly emphasizing parallels between thymus-dependent lymphocytes and the free hemocytes of invertebrates.

Much earlier work attempted to correlate the functions of the agglutinins and lysins found in some invertebrate hemolymphs with vertebrate antibodies. However, since it is probable that the vertebrates had their origin among the invertebrate phyla, it might be more profitable to look for primitive aspects of internal defense which have persisted in the vertebrates and which might well be found to resemble invertebrate mechanisms more closely.

Invertebrate hemocytes show functional resemblances to vertebrate leukocytes and macrophages but their morphology is obviously much less close, as evidenced by different modes of locomotion (in vitro) (Cheng, 1975; Cheng et al., 1979; Davies and Partridge, 1972) and different embryonic origin. Invertebrate hemocytes seem to have some factors in common with macrophages. Rabinovitch and De Stefano (1970) observed that in the absence of any hemolymph or serum factors, *Galleria mellonella* hemocytes and mouse macrophages were stimulated to take up SRBC, by the same treatments. They suggested that at this primitive level, the mode of recognition by the two cell types is the same. Invertebrate hemocytes, however, show neither the increased rates of respiration which occur in phagocytosing mammalian macrophages (Anderson et al., 1973) nor the peroxidase/H_2O_2 method of intracellular killing (see Section 5.3.2).

Work on the structure of soluble agglutinins of invertebrate origin indicates that these molecules have nothing in common with antibody, and their irregular distribution among molluscan species and strains would seem to indicate that they are not of universally vital importance in internal defense. Recently, lectins, built up of identical subunits, have been isolated from mammalian and other vertebrate tissues and it is possible that these could be more closely allied to the invertebrate humoral agglutinins than are antibodies (discussed by Fletcher, 1978).

The Gastropoda include at least 1640 genera and 230 families and the Lamellibranchia about 60 families (Morton, 1967). Something of the mechanisms of internal defense is known in species which are members of only about five gastropod and three lamellibranch families. Except in the case of the Helicidae, no more than one species in each family has been studied. Further work on molluscan internal defense should probably be directed at gaining a much wider knowledge of the capabilities of the species in different classes and subclasses. Experiments which have elucidated the mechanisms that exist in the species which have been studied could now well be applied to a wide range of species, selected with the intention of clarifying the evolution of internal defense within the Mollusca.

REFERENCES

Anderson, R. S., 1977, Biochemistry and physiology of invertebrate macrophages *in vitro, Comp. Pathobiol.* **3**:1.

Anderson, R. S., and Good, R. A., 1976, Opsonic involvement in phagocytosis by mollusk hemocytes, *J. Invertebr. Pathol.* **27**:57.

Anderson, R. S., Holmes, B., and Good, R. A., 1973, In vitro bactericidal capacity of *Blaberus craniifer* hemocytes, *J. Invertebr. Pathol.* **22**:127.

Andrews, E. B., 1976, The fine structure of the heart of some prosobranch and pulmonate gastropods in relation to filtration, *J. Moll. Stud.* **42**:199.

Andrews, E. B., 1979, Fine structure in relation to function in the excretory system of two species of *Viviparus*, *J. Moll. Stud.* **45**:186.

Arimoto, R., and Tripp, M. R., 1977, Characterization of a bacterial agglutinin in the hemolymph of the hard clam, *Mercenaria mercenaria*, *J. Invertebr. Pathol.* **30**:406.

Baehner, R. L., 1975, Microbe ingestion and killing by neutrophils: Normal mechanisms and abnormalities, *Clin. Haematol.* **4**:609.

Barr, A. R., 1975, Evidence for genetical control of invertebrate immunity and its field significance, in: *Invertebrate Immunity* (K. Maramorosch and R. E. Shope, eds.), pp. 129–135, Academic Press, New York.

Basch, P. F., 1976, Intermediate host specificity in *Schistosoma mansoni*, *Exp. Parasitol.* **39**:150.

Bayne, C. J., 1973a, Internal defense mechanisms of *Octopus dofleini*, *Malacol. Rev.* **6**:13.

Bayne, C. J., 1973b, Molluscan internal defense mechanism: The fate of C^{14}-labelled bacteria in the land snail *Helix pomatia* (L.), *J. Comp. Physiol.* **86**:17.

Bayne, C. J., 1974, On the immediate fate of bacteria in the land snail *Helix*, in: *Invertebrate Immunology* (E. L. Cooper, ed.), pp. 37–45, Plenum Press, New York.

Bayne, C. J., 1977, Molluscan immunobiology: The elevation of responses, in: *Developmental Immunobiology* (J. B. Solomon and J. D. Horton, eds.), pp. 67–74, Elsevier/North-Holland, Amsterdam.

Bayne, C. J., Moore, M. N., Carefoot, T. H., and Thompson, R. J., 1979, Hemolymph functions in *Mytilus californianus*: The cytochemistry of hemocytes and their responses to foreign implants and hemolymph factors in phagocytosis, *J. Invertebr. Pathol.* **34**:1.

Bayne, C. J., Sminia, T., and van der Knaap, W. P. W., 1980, Immunological memory: Status of molluscan studies, in: *Phylogeny of Immunological Memory* (M. J. Manning, ed.), pp. 57–64, Elsevier/North-Holland, Amsterdam.

Beltz, B., 1977, Transmission electron microscope study of pore cells in *Limax maximus*, Proceedings, 35th Annual Meeting of the Electron Microscopy Society of America, p. 656.

Bend, J. R., James, M. O., and Dansette, P. M., 1977, In vitro metabolism of xenobiotics in some marine animals, *Ann. N.Y. Acad. Sci.* **298**:505.

Bloom, B. R., 1979, Games parasites play: How parasites evade immune surveillance, *Nature (London)* **279**:21.

Boer, H. H., and Sminia, T., 1976, Sieve structure of slit diaphragms of podocytes and pore cells of gastropod molluscs, *Cell Tissue Res.* **170**:221.

Boss, K. J., 1971, Critical estimate of the number of recent Mollusca, *Occas. Pap. Mollusks Mus. Comp. Zool. Harv. Univ.* **3**:81.

Boyden, S. V., 1962, Cellular discrimination between indigenous and foreign matter, *J. Theor. Biol.* **3**:123.

Brown, A. C., 1967, Elimination of foreign particles by the snail *Helix aspersa*, *Nature (London)* **213**:1154.

Brown, R. S., Wolke, R. E., Saila, S. B., and Brown, C. W., 1977, Prevalence of neoplasia in 10 New England populations of the soft-shelled clam (*Mya arenaria*), *Ann. N.Y. Acad. Sci.* **298**:522.

Brown, R. S., Appledoorn, R., Brown, C. W., and Saila, S. B., 1980, The value of the multidisciplinary approach to research on marine pollution effects as evidenced in a three-year study to determine the etiology and pathogenesis of neoplasia in the soft-shelled clam, *Mya arenaria*, *Rapp. P.-V. Reun. Cons. Int. Explor. Mer* **179**:125.

Buchholz, K., Kuhlmann, D., and Nolte, A., 1971, Aufnahme von Trypanblau und Ferritin in die Blasenzellen des Bindegewebes von *Helix pomatia* und *Cepaea nemoralis* (Stylommatophora, Pulmonata). *Z. Zellforsch. Mikrosk. Anat.* **113**:203.

Capo, C., Bongrand, P., Benoliel, A.-M., and Depieds, R., 1979, Non-specific recognition in phagocytosis: Ingestion of aldehyde-treated erythrocytes by rat peritoneal macrophages, *Immunology* **36**:501.

Carter, O.S., and Bogitsh, B. J., 1975, Histologic and cytochemical observations on the effects of *Schistosoma mansoni* on *Biomphalaria glabrata*, *Ann. N.Y. Acad. Sci.* **266**:380.

Cheney, D. P., 1971, A summary of invertebrate leukocyte morphology with emphasis on blood elements of the Manila clam, *Tapes semidecussata*, *Biol. Bull.* **140**:353.

Cheng, T. C., 1968, The compatibility and incompatibility concept as related to trematodes and molluscs, *Pac. Sci.* **22**:141.

Cheng, T. C., 1975, Functional morphology and biochemistry of molluscan phagocytes, *Ann. N.Y. Acad. Sci.* **266**:343.

Cheng, T. C., 1976a, Beta-glucuronidase in the serum and hemolymph cells of *Mercenaria mercenaria* and *Crassostrea virginica* (Mollusca: Pelecypoda), *J. Invertebr. Pathol.* **27**:125.

Cheng, T. C., 1976b, Aspects of substrate utilization and energy requirement during molluscan phagocytosis, *J. Invertebr. Pathol.* **27**:263.

Cheng, T. C., 1977, Biochemical and ultrastructural evidence for the double role of phagocytosis in molluscs: Defense and nutrition, *Comp. Pathobiol.* **3**:21.

Cheng, T. C., 1978, A study of granuloma formation by molluscan cells, *Comp. Pathobiol.* **4**:97.

Cheng, T. C., and Butler, M. S., 1979, Experimentally induced elevations in acid phosphatase activity in hemolymph of *Biomphalaria glabrata* (Mollusca), *J. Invertebr. Pathol.* **34**:119.

Cheng, T. C., and Cali, A., 1974, An electron microscope study of the fate of bacteria phagocytized by granulocytes of *Crassostrea virginica*, in: *Invertebrate Immunology* (E. L. Cooper, ed.), pp. 25–35, Plenum Press, New York.

Cheng, T. C., and Galloway, P. C., 1970, Transplantation immunity in mollusks: The histoincompatibility of *Helisoma duryi normale* with allografts and xenografts, *J. Invertebr. Pathol.* **15**:177.

Cheng, T. C., and Garrabrant, T. A., 1977, Acid phosphatase in granulocytic capsules formed in strains of *Biomphalaria glabrata* totally and partially resistant to *Schistosoma mansoni*, *Int. J. Parasitol.* **7**:467.

Cheng, T. C., and Howland, K. H., 1979, Chemotactic attraction between hemocytes of the oyster, *Crassostrea virginica*, and bacteria, *J. Invertebr. Pathol.* **33**:204.

Cheng, T. C., and Lee, F. O., 1971, Glucose levels in the mollusc *Biomphalaria glabrata* infected with *Schistosoma mansoni*, *J. Invertebr. Pathol.* **18**:395.

Cheng, T. C. and Rifkin, E., 1970, Cellular reactions in marine molluscs in response to helminth parasitism, in: *A Symposium on Diseases of Fishes and Shellfishes* (S. F. Snieszko, ed.), pp. 443–496, American Fisheries Society, Washington, D.C.

Cheng, T. C., and Rodrick, G. E., 1975, Lysosomal and other enzymes in the hemolymph of *Crassostrea virginica* and *Mercenaria mercenaria*, *Comp. Biochem. Physiol. B.* **52**:443.

Cheng, T. C., and Rudo, B. M., 1976a, Chemotactic attraction of *Crassostrea virginica* hemolymph cells to *Staphylococcus lactus*, *J. Invertebr. Pathol.* **27**:137.

Cheng, T. C., and Rudo, B. M., 1976b, Distribution of glycogen resulting from degradation of ^{14}C-labelled bacteria in the American oyster, *Crassostrea virginica*, *J. Invertebr. Pathol.* **27**:259.

Cheng, T. C., and Snyder, R. W., 1962, Studies on host–parasite relationships between larval trematodes and their hosts. I. A review. II. The utilisation of the host's glycogen by the intramolluscan larvae of *Glypthelmins pennsylvaniensis* Cheng, and associated phenomena, *Trans. Am. Microsc. Soc.* **81**:209.

Cheng, T. C. and Yoshino, T. P., 1976, Lipase activity in the serum and hemolymph cells of the soft-shelled clam, *Mya arenaria*, during phagocytosis, *J. Invertebr. Pathol.* **27**:243.

Cheng, T. C., Shuster, C. N., and Anderson, A. H., 1966, Effects of plasma and tissue extracts of marine pelecypods on the cercaria of *Himasthla quissetensis*, *Exp. Parasitol.* **19**:9.

Cheng, T. C., Thakur, A. S., and Rifkin, E., 1969, Phagocytosis as an internal defense mechanism in the mollusca: With an experimental study of the role of leucocytes in the removal of ink particles in *Littorina scabra* Linn., *Proc. Symp. Mollusca* **II**:546.

Cheng, T. C., Cali, A., and Foley, D. A., 1974, Cellular reactions in marine pelecypods as a factor influencing endosymbiosis, in: *Symbiosis in the Sea* (W. A. Vernberg, ed.), pp. 61–91, University of South Carolina Press, Columbia.

Cheng, T. C., Rodrick, G. E., Foley, D. A., and Koehler, S. A., 1975, Release of lysozyme from hemolymph cells of *Mercenaria mercenaria* during phagocytosis, *J. Invertebr. Pathol.* **25**:261.

Cheng, T. C., Chorney, M. J., and Yoshino, T. P., 1977, Lysozyme-like activity in the hemolymph of *Biomphalaria glabrata* challenged with bacteria, *J. Invertebr. Pathol.* **29**:170.

Cheng, T. C., Butler, M. S., Guida, V. G., and Gerhart, P. L., 1979, A scanning electron microscope study of the pseudopodia of *Biomphalaria glabrata* granulocytes, *J. Invertebr. Pathol.* **33**:118.

Chothia, C., and Janin, J., 1975, Principles of protein–protein recognition, *Nature (London)* **256**:705.

Christensen, D. J., Farley, C. A., and Kern, F. G., 1974, Epizootic neoplasms in the clam *Macoma balthica* (L.) from Chesapeake Bay, *J. Natl. Cancer Inst.* **52**:1739.

Christie, J., Foster, W. B., and Stauber, L. A., 1974, The effect of parasitism and starvation on carbohydrate reserves of *Biomphalaria glabrata*, *J. Invertebr. Pathol.* **23**:55.

Cooper, E. L., 1976, Cellular recognition of allografts and xenografts in invertebrates, in: *Comparative Immunology* (J. J. Marchalonis, ed.), pp. 36–79, Blackwell, Oxford.

Cooper-Willis, C. A., 1979, Changes in the acid phosphatase levels in the haemocytes and haemolymph of *Patella vulgata* after challenge with bacteria, *Comp. Biochem. Physiol. A* **63**:627.

Cowden, R. R., and Curtis, S. K., 1973, Observations on living cells dissociated from the leukopoietic organ of *Octopus briareus*, *Exp. Mol. Pathol.* **19**:178.

Cowden, R. R., and Curtis, S. K., 1974, The octopus white body: An ultrastructural survey, in: *Invertebrate Immunology* (E. L. Cooper, ed.), pp. 77–90, Plenum Press, New York.

Crichton, R., and Lafferty, K. J., 1975, The discriminatory capacity of phagocytic cells in the chiton (*Liolophura gaimardi*), *Adv. Exp. Med. Biol.* **64**:89.

Curtis, S. K., and Cowden, R. R., 1978, Responsiveness of the slug (*Limax maximus*) to injections of fluorescein- and rhodamine-conjugated immunogens, *Dev. Comp. Immunol.* **2**:727.

Curtis, S. K., and Cowden, R. R., 1979, Histochemical and ultrastructural features of the aorta of the slug (*Limax maximus*), *J. Morphol.* **161**:1.

Cushing, J. E., 1962, Blood groups in marine animals and immune mechanisms of lower vertebrates and invertebrates (comparative immunology), *Proceedings, Conference on Immunoreproduction, La Jolla, California*, pp. 205–207, The Population Council, New York.

Davies, P. S., and Partridge, T., 1972, Limpet haemocytes. I. Studies on aggregation and spike formation, *J. Cell Sci.* **11**:757.

DesVoigne, D. M., and Sparks, A. K., 1969, The reaction of the Pacific oyster, *Crassostrea gigas*, to homologous tissue implants, *J. Invertebr. Pathol.* **14**:293.

Douglass, W. R., and Haskin, H. H., 1976, Oyster–MSX interactions: Alterations in hemolymph enzyme activities in *Crassostrea virginica* during the course of *Minchinia nelsoni* disease development, *J. Invertebr. Pathol.* **27**:317.

Eble, A. F., 1966, Some observations on the seasonal distribution of selected enzymes in the American oyster as revealed by enzyme histochemistry, *Proc. Natl. Shellfish. Assoc.* **56**:37.

Eble, A. F., and Tripp, M. R., 1968, Enzyme histochemistry of phagosomes in oyster leucocytes, *Bull. N.J. Acad. Sci. B* **13**:93.

Elston, R., 1979, Viruslike particles associated with lesions in larval Pacific oysters (*Crassostrea gigas*), *J. Invertebr. Pathol.* **33**:71.

Fankboner, P. V., 1971, Intracellular digestion of symbiotic zooxanthellae by host amoebocytes in giant clams (Bivalvia: Tridacnidae), with a note on the nutritional role of the hypertrophied siphonal epidermis, *Biol. Bull.* **141**:222.

Farley, C. A., 1968, *Minchinia nelsoni* (Haplosporida) disease syndrome in the American oyster *Crassostrea virginica*, *J. Protozool.* **15**:585.

Farley, C. A., 1969, Probable neoplastic disease of the hematopoietic system in oysters, *Crassostrea virginica* and *Crassostrea gigas*, *Natl. Cancer Inst. Monogr.* **31**:541.

Farley, C. A., 1976, Ultrastructural observations on epizootic neoplasia and lytic virus infection in bivalve molluscs, *Prog. Exp. Tumor Res.* **20**:283.

Farley, C. A., 1977, Neoplasms in estuarine mollusks and approaches to ascertain causes, *Ann. N.Y. Acad. Sci.* **298**:225.

Farley, C. A., 1978, Viruses and virus-like particles in marine molluscs, *Mar. Fish. Rev.* **40**:18.

Farley, C. A., and Sparks, A. K., 1970, Proliferative diseases of hemocytes, endothelial cells and connective tissue cells in molluscs, *Bibl. Haematol. (Basel)* **36**:610.

Feng, S. Y., 1965, Pinocytosis of proteins by oyster leucocytes, *Biol. Bull.* **129**:95.

Feng, S. Y., 1966, Experimental bacterial infections in the oyster, *Crassostrea virginica*, *J. Invertebr. Pathol.* **8**:505.

Feng, S. Y., and Feng, J. S., 1974, The effect of temperature on cellular reactions of *Crassostrea virginica* to the injection of avian erythrocytes, *J. Invertebr. Pathol.* **23**:22.

Feng, S. Y., Feng, J. S., Burke, C. N., and Khairallah, L. H., 1971, Light and electron microscopy of the leucocytes of *Crassostrea virginica* (Mollusca, Pelecypoda), *Z. Zellforsch. Mikrosk. Anat.* **120**:222.

Feng, S. Y., Feng, J. S., and Yamasu, T., 1977, Roles of *Mytilus coruscus* and *Crassostrea gigas* blood cells in defense and nutrition, *Comp. Pathobiol.* **3:**31.

Fletcher, T. C., 1978, Defence mechanisms in fish, *Biochem. Biophys. Perspect. Mar. Biol.* **4:**189.

Foster, C. A., and Howse, H. D., 1978, A morphological study on the gills of the brown shrimp, *Penaeus aztecus, Tissue Cell* **10:**77.

Goldstein, J. L., Anderson, R. G. W., and Brown, M. S., 1979, Coated pits, coated vesicles and receptor-mediated endocytosis, *Nature (London)* **279:**679.

Hardy, S. W., Fletcher, T. C., and Olafsen, J. A., 1977, Aspects of cellular and humoral defence mechanisms in the Pacific oyster, *Crassostrea gigas*, in: *Developmental Immunobiology* (J. B. Solomon and J. D. Horton, eds.), pp. 59–66, Elsevier/North-Holland, Amsterdam.

Harris, K. R., 1975, The fine structure of encapsulation in *Biomphalaria glabrata*, *Ann. N.Y. Acad. Sci.* **266:**446.

Harris, K. R., and Cheng, T. C., 1975a, The encapsulation process in *Biomphalaria glabrata* experimentally infected with the metastrongylid *Angiostrongylus cantonensis:* Light microscopy, *Int. J. Parasitol.* **5:**521.

Harris, K. R., and Cheng, T. C., 1975b, The encapsulation process in *Biomphalaria glabrata* experimentally infected with the metastrongylid *Angiostrongylus cantonensis:* Enzyme histochemistry, *J. Invertebr. Pathol.* **26:**367.

Hartland, B. J., and Timoney, J. F., 1979, In vivo clearance of enteric bacteria from the hemolymph of the hard clam and the American oyster, *Appl. Environ. Microbiol.* **37:**517.

Heyneman, D., Faulk, W. P., and Fudenberg, H. H., 1971, *Echinostoma lindoense:* Larval antigens from the snail intermediate host, *Biomphalaria glabrata*, *Exp. Parasitol.* **29:**480.

Hildemann, W. H., Bigger, C. H., and Johnston, I. S., 1979, Histoincompatibility reactions and allogeneic polymorphism among invertebrates, *Transplant. Proc.* **11:**1136.

Hopkins, S. H., 1957, Our present knowledge of the oyster parasite *Bucephalus*, *Proc. Natl. Shellfish. Assoc.* **47:**58.

Janoff, A., and Hawrylko, E., 1964, Lysosomal enzymes in invertebrate leucocytes, *J. Cell. Comp. Physiol.* **63:**267.

Johnson, P. T., 1980, *Histology of the Blue Crab, Callinectes sapidus (Decapoda: Portunidae): A Model for the Decapoda*, Praeger, New York.

Kassim, O. O., and Richards, C. S., 1978, *Schistosoma mansoni:* Lysozyme activity in *Biomphalaria glabrata* during infection with two strains, *Exp. Parasitol.* **46:**213.

Kinoti, G. K., 1971, Observations on the infection of bulinid snails with *Schistosoma mattheei:* The mechanism of resistance to infection, *Parasitology* **62:**161.

Klein, J., 1977, Evolution and function of the major histocompatibility system: Facts and speculation, in: *The Major Histocompatibility System in Man and Animals* (D. Götze, ed.), pp. 339–378, Springer-Verlag, Berlin.

Krupa, P. L., Lewis, L. M., and Del Vecchio, P., 1977, *Schistosoma haematobium* in *Bulinus guernei:* Electron microscopy of hemocyte–sporocyst interactions, *J. Invertebr. Pathol.* **30:**35.

Lackie, A. M., 1977, Cellular recognition of 'not-self' in insects, in: *Developmental Immunobiology* (J. B. Solomon and J. D. Horton, eds.), pp. 75–81, Elsevier/North-Holland, Amsterdam.

Lafferty, K. J., and Crichton, R., 1973, Immune response of invertebrates, in: *Viruses and Invertebrates* (A. J. Gibbs, ed.), pp. 300–320, North-Holland, Amsterdam.

Lie, K. J., and Heyneman, D., 1975, Studies on resistance in snails: A specific tissue reaction to *Echinostoma lindoense* in *Biomphalaria glabrata* snails, *Int. J. Parasitol.* **5:**621.

Lie, K. J., and Heyneman, D., 1976a, Studies on resistance in snails. 3. Tissue reactions to *Echinostoma lindoense* sporocysts in sensitized and resensitized *Biomphalaria glabrata*, *J. Parasitol.* **62:**51.

Lie, K. J., and Heyneman, D., 1976b, Studies on resistance in snails. 6. Escape of *Echinostoma lindoense* sporocysts from encapsulation in the snail heart and subsequent loss of the host's ability to resist infection by the same parasite, *J. Parasitol.* **62:**298.

Lie, K. J., and Heyneman, D., 1979, Acquired resistance to echinostomes in four *Biomphalaria glabrata* strains, *Int. J. Parasitol.* **9:**533.

Lie, K. J., Heyneman, D., and Jeong, K. H., 1976a, Studies on resistance in snails. 4. Induction of ventricular capsules and changes in the amebocyte-producing organ during sensitization of *Biomphalaria glabrata* snails, *J. Parasitol.* **62:**286.

Lie, K. J., Heyneman, D., and Jeong, K. H., 1976b, Studies on resistance in snails. 7. Evidence of interference with the defense reaction in *Biomphalaria glabrata* by trematode larvae, *J. Parasitol.* **62**:608.

Loker, E. S., 1978, *Schistosomatium douthitti*: Effects of *Lymnaea catascopium* age on susceptibility to infection, *Exp. Parasitol.* **45**:65.

Loker, E. S., 1979a, Pathology and host responses induced by *Schistosomatium douthitti* in the fresh water snail, *Lymnaea catascopium*, *J. Invertebr. Pathol.* **33**:265.

Loker, E. S., 1979b, Effects of *Schistosomatium douthitti* infection on the growth, survival and reproduction of *Lymnaea catascopium*, *J. Invertebr. Pathol.* **34**:138.

Lowe, D. M., and Moore, M. N., 1978, Cytology and quantitative cytochemistry of a proliferative atypical hemocytic condition in *Mytilus edulis* (Bivalvia, Mollusca), *J. Natl. Cancer Inst.* **60**:1455.

Mackie, A. M., Singh, H. T., and Owen, J. M., 1977, Studies on the distribution, biosynthesis and function of steroidal saponins in echinoderms, *Comp. Biochem. Physiol. B* **56**:9.

Mackin, J. G., 1951, Histopathology of infection of *Crassostrea virginica* (Gmelin) by *Dermocystidium marinum* (Mackin, Owen, and Collier), *Bull. Mar. Sci. Gulf Caribbean* **1**:72.

Mäkelä, O., Koskimies, S., and Karjalainen, K., 1976, Possible evolution of acquired immunity from self-recognition structures, *Scand. J. Immunol.* **5**:305.

Malek, E. A., and Cheng, T. C., 1974, *Medical and Economic Malacology*, Academic Press, New York.

Marchalonis, J. J., and Cohen, N. (eds.), 1980, *Self/Non-self Discrimination*, Plenum Press, New York.

McLean, N., 1980, Phagocytosis by epidermal cells of the mantle in *Mytilus edulis* L. (Mollusca: Bivalvia), *Comp. Biochem. Physiol. A* **66**:367.

McRipley, R. J., and Sbarra, A. J., 1967, Role of the phagocyte in host–parasite interactions. XII. Hydrogen peroxide–myeloperoxidase bactericidal system in the phagocyte, *J. Bacteriol.* **94**:1425.

Michelson, E. H., 1961, An acid-fast pathogen of freshwater snails, *Am. J. Trop. Med. Hyg.* **10**:423.

Michelson, E. H., 1972, A neoplasm in the giant African snail, *Achatina fulica*, *J. Invertebr. Pathol.* **20**:264.

Michelson, E. H., 1975, Cellular defense mechanisms and tissue alterations in gastropod molluscs, in: *Invertebrate Immunity* (K. Maramorosch and R. E. Shope, eds.), pp. 181–195, Academic Press, New York.

Michelson, E. H., and DuBois, L., 1978, Susceptibility of Bahian populations of *Biomphalaria glabrata* to an allopatric strain of *Schistosoma mansoni*, *Am. J. Trop. Med. Hyg.* **27**:782.

Michelson, E. H., and Richards, C. S., 1975, Neoplasms and tumor-like growths in the aquatic pulmonate snail *Biomphalaria glabrata*, *Ann. N.Y. Acad. Sci.* **266**:411.

Mix, M. C., 1975, Proliferative characteristics of atypical cells in native oysters (*Ostrea lurida*) from Yaquina Bay, Oregon, *J. Invertebr. Pathol.* **26**:289.

Mix, M. C., Hawkes, J. W., and Sparks, A. K., 1979, Observations on the ultrastructure of large cells associated with putative neoplastic disorders of mussels, *Mytilus edulis*, from Yaquina Bay, Oregon, *J. Invertebr. Pathol.* **34**:41,

Moore, C. A., and Eble, A. F., 1977, Cytochemical aspects of *Mercenaria mercenaria* hemocytes, *Biol. Bull.* **152**:105.

Moore, M. N., and Lowe, D. M., 1977, The cytology and cytochemistry of the hemocytes of *Mytilus edulis* and their responses to experimentally injected carbon particles, *J. Invertebr. Pathol.* **29**:18.

Morton, J. E., 1967, *Molluscs*, pp. 171–225, Hutchinson University Library.

Murrell, K. D., Taylor, D. W., Vannier, W. E., and Dean, D. A., 1978, *Schistosoma mansoni*: Analysis of surface membrane carbohydrates using lectins, *Exp. Parasitol.* **46**:247.

Narain, A. S., 1973, The amoebocytes of lamellibranch molluscs, with special reference to the circulating amoebocytes, *Malacol. Rev.* **6**:1.

Newton, W. L., 1953, The inheritance of susceptibility to infection with *Schistosoma mansoni* in *Australorbis glabratus*, *Exp. Parasitol.* **2**:242.

Pan, C.-T., 1965, Studies on the host–parasite relationship between *Schistosoma mansoni* and the snail *Australorbis glabratus*, *Am. J. Trop. Med. Hyg.* **14**:931.

Parish, C. R., 1977, Simple model for self-non-self-discrimination in invertebrates, *Nature (London)* **267**:711.

Pauley, G. B., 1969, A critical review of neoplasms and tumor-like lesions in molluscs, *Natl. Cancer Inst. Monogr.* **31**:509.

Pauley, G. B., and Krassner, S. M., 1972, Cellular defence reactions to particulate materials in the California sea hare, *Aplysia californica, J. Invertebr. Pathol.* **19**:18.

Pauley, G. B., Krassner, S. M., and Chapman, F. A., 1971, Bacterial clearance in the California sea hare, *Aplysia californica, J. Invertebr. Pathol.* **18**:227.

Potts, W. T. W., 1967, Excretion in the molluscs, *Biol. Rev.* **42**:1.

Prytherch, H. F., 1940, The life cycle and morphology of *Nematopsis ostrearum* sp. nov., a gregarine parasite of the mud crab and oyster, *J. Morphol.* **66**:39.

Rabinovitch, M., and De Stefano, M. J., 1970, Interactions of red cells with phagocytes of the waxmoth (*Galleria mellonella* L.) and mouse, *Exp. Cell Res.* **59**:272.

Rachford, F. W., 1976, Host–parasite relationship of *Angiostrongylus cantonensis* in *Lymnaea palustris*. II. Histopathology, *Exp. Parasitol.* **39**:382.

Ratcliffe, N. A., and Rowley, A. F., 1979, A comparative synopsis of the structure and function of the blood cells of insects and other invertebrates, *Dev. Comp. Immunol.* **3**:189.

Read, C. P., 1958, Status of behavioral and physiological "resistance," *Rice Inst. Pam.* **45**:36.

Reade, P. C., 1968, Phagocytosis in invertebrates, *Aust. J. Exp. Biol. Med. Sci.* **46**:219.

Reade, P., and Reade, E., 1972, Phagocytosis in invertebrates. II. The clearance of carbon particles by the clam, *Tridacna maxima, J. Reticuloendothelial Soc.* **12**:349.

Renwrantz, L. R., and Cheng, T. C., 1977a, Identification of agglutinin receptors on hemocytes of *Helix pomatia, J. Invertebr. Pathol.* **29**:88.

Renwrantz, L. R., and Cheng, T. C., 1977b, Agglutinin-mediated attachment of erythrocytes to hemocytes of *Helix pomatia, J. Invertebr. Pathol.* **29**:97.

Renwrantz, L., Yoshino, T., Cheng, T., and Auld, K., 1979, Size determination of hemocytes from the American oyster, *Crassostrea virginica*, and the description of a phagocytosis mechanism, *Zool. Jahrb. Abt. Allg. Zool. Physiol. Tiere* **83**:1.

Richards, C. S., 1970, Pearl formation in *Biomphalaria glabrata, J. Invertebr. Pathol.* **15**:459.

Richards, C. S., 1972, *Biomphalaria glabrata* genetics: Pearl formation, *J. Invertebr. Pathol.* **20**:37.

Richards, C. S., 1975, Genetic factors in susceptibility of *Biomphalaria glabrata* for different strains of *Schistosoma mansoni, Parasitology* **70**:231.

Richards, C. S., and Merritt, J. W., 1967, Studies on *Angiostrongylus cantonensis* in molluscan intermediate hosts, *J. Parasitol.* **53**:382.

Rodewald, R., and Karnovsky, M. J., 1974, Porous substructure of the glomerular slit diaphragm in the rat and mouse, *J. Cell Sci.* **60**:423.

Ruddell, C. L., and Wellings, S. R., 1971, The ultrastructure of the oyster brown cell, a cell with a fenestrated plasma membrane, *Z. Zellforsch. Mikrosk. Anat.* **120**:17.

Runnegar, B., and Pojeta, J., 1974, Molluscan phylogeny: The paleontological viewpoint, *Science* **186**:311.

Salt, G., 1970, Experimental studies in insect parasitism. XV, *Proc. R. Soc. London Ser. B* **176**:105.

Schmid, L. S., 1975, Chemotaxis of hemocytes from the snail *Viviparus malleatus, J. Invertebr. Pathol.* **25**:125.

Silverstein, S. C., Michl, J., and Sung, S. S. J., 1978, Phagocytosis, in: *Transport of Macromolecules in Cellular Systems* (S. C. Silverstein, ed.), pp. 245–264, Dahlem Konferenzen, Berlin.

Sindermann, C. J., 1970, *Principal Diseases of Marine Fish and Shellfish*, Academic Press, New York.

Sminia, T., 1972, Structure and function of blood and connective tissue cells of the fresh water pulmonate *Lymnaea stagnalis* studied by electron microscopy and enzyme histochemistry, *Z. Zellforsch. Mikrosk, Anat.* **130**:497.

Sminia, T., 1974, Haematopoiesis in the fresh water snail *Lymnaea stagnalis* studied by electron microscopy and autoradiography, *Cell Tissue Res.* **150**:443.

Sminia, T., Pietersma, K., and Scheerboom, J. E. M., 1973, Histological and ultrastructural observations on wound healing in the fresh water pulmonate *Lymnaea stagnalis, Z. Zellforsch. Mikrosk. Anat.* **141**:561.

Sminia, T., Borghart-Reinders, E., and van de Linde, A. W., 1974, Encapsulation of foreign materials experimentally introduced into the fresh water snail *Lymnaea stagnalis, Cell Tissue Res.* **153**:307.

Sminia, T., van der Knaap, W. P. W., and Kroese, F. G. M., 1979a, Fixed phagocytes in the fresh water snail *Lymnaea stagnalis, Cell Tissue Res.* **196**:545.

Sminia, T., van der Knaap, W. P. W., and Edelenbosch, P., 1979b, The role of serum factors in phagocytosis of foreign particles by blood cells of the fresh water snail *Lymnaea stagnalis, Dev. Comp. Immunol.* **3**:37.

Sprague, V., 1965, Observations on *Chytridiopsis mytilovum* (Microsporidia), *J. Protozool.* **12**:385.

Sprague, V., 1971, Diseases of oysters, *Annu. Rev. Microbiol.* **25**:211.

Sprent, J. F. A., 1969, Evolutionary aspects of immunity in zooparasitic infections, in: *Immunity to Parasitic Animals*, Vol. 1 (G. J. Jackson, R. Herman, and I. Singer, eds.), pp. 3–62, North-Holland, Amsterdam.

Stanislawski, E., Renwrantz, L., and Becker, W., 1976, Soluble blood group reactive substances in the hemolymph of *Biomphalaria glabrata* (Mollusca), *J. Invertebr. Pathol.* **28**:301.

Stauber, L. A., 1950, The fate of India ink injected intracardially into the oyster, *Ostrea virginica* (Gmelin), *Biol. Bull.* **98**:227.

Stauber, L. A., 1961, Immunity in invertebrates, with special reference to the oyster, *Proc. Natl. Shellfish. Assoc.* **50**:7.

Steinman, R. M., Silver, J. M., and Cohn, Z. A., 1978, Fluid phase pinocytosis, in: *Transport of Macromolecules in Cellular Systems* (S. C. Silverstein, ed.), pp. 167–179, Dahlem Konferenzen, Berlin.

Stossel, T. P., 1977, Endocytosis, in: *Receptors and Recognition*, Series A, Vol. 4 (P. Cuatrecasas and M. F. Greaves, eds.), pp. 104–141, Chapman & Hall, London.

Stuart, A. E., 1968, The reticulo-endothelial apparatus of the lesser octopus, *Eledone cirrosa, J. Pathol. Bacteriol.* **96**:401.

Tauber, J. W., 1976, "Self": Standard of comparison for immunological recognition of foreignness, *Lancet* **2**:291.

Tripp, M. R., 1958, Disposal by the oyster of intracardially injected red blood cells of vertebrates, *Proc. Natl. Shellfish. Assoc.* **48**:143.

Tripp, M. R., 1960, Mechanisms of removal of injected microorganisms from the American oyster, *Crassostrea virginica* (Gmelin), *Biol. Bull.* **119**:273.

Tripp, M. R., 1961, The fate of foreign materials experimentally introduced into the snail, *Australorbis glabratus, J. Parasitol.* **47**:745.

Tripp, M. R., 1969, General mechanisms and principles of invertebrate immunity, in: *Immunity to Parasitic Animals*, Vol. 1 (G. J. Jackson, R. Herman, and I. Singer, eds.), pp. 111–128, North-Holland, Amsterdam.

Tripp, M. R., and Kent, V. E., 1967, Studies on oyster cellular immunity, *In Vitro* **3**:129.

van Oss, C. J., 1978, Phagocytosis as a surface phenomenon, *Annu. Rev. Microbiol.* **32**:19.

Wakelin, D., 1978, Genetic control of susceptibility and resistance to parasitic infection, *Adv. Parasitol.* **16**:219.

Walters, M. N.-I., and Papadimitriou, J. M., 1978, Phagocytosis: A review, *Crit. Rev. Toxicol.* **5**:377.

Wolburg-Buchholz, K., 1972, Blasenzellen im Bindegewebe des Schlundrings von *Cepaea nemoralis* L. (Gastropoda, Stylommatophora). II, *Z. Zellforsch. Mikrosk. Anat.* **130**:262.

Yevich, P. P., and Barszcz, C. A., 1977, Neoplasia in soft-shelled clams (*Mya arenaria*) collected from oil-impacted sites, *Ann. N.Y. Acad. Sci.* **298**:409.

Yevich, P. P., and Berry, M. M., 1969, Ovarian tumors in the quahog, *Mercenaria mercenaria, J. Invertebr. Pathol.* **14**:266.

Yonge, C. M., 1926, Structure and physiology of the organs of feeding and digestion in *Ostrea edulis, J. Mar. Biol. Assoc. U.K.* **14**:295.

Yoshino, T. P., 1976, Encapsulation response of the marine prosobranch *Cerithidea californica* to natural infections of *Renicola buchanani* sporocysts (Trematoda: Renicolidae), *Int. J. Parasitol.* **6**:423.

Yoshino, T. P., and Cheng, T. C., 1976, Experimentally induced elevation of aminopeptidase activity in hemolymph cells of the American oyster, *Crassostrea virginica, J. Invertebr. Pathol.* **27**:367.

Yoshino, T. P., Cheng, T. C., and Renwrantz, L. R., 1977, Lectin and human blood group determinants of *Schistosoma mansoni*: Alteration following *in vitro* transformation of miracidium to mother sporocyst, *J. parasitol.* **63**:818.

Zacks, S. I., 1955, The cytochemistry of the amoebocytes and intestinal epithelium of *Venus mercenaria* (Lamellibranchiata) with remarks on a pigment resembling ceroid, *Q. J. Microsc. Sci.* **96**:57.

6

Cellular Defense Systems of the Arthropoda

NORMAN A. RATCLIFFE, KEITH N. WHITE,
ANDREW F. ROWLEY, and JULIA B. WALTERS

1. GENERAL INTRODUCTION

The arthropods are a vast assemblage of both aquatic and terrestrial animals, which together form a considerable proportion of the total number of species in the animal kingdom. As a group, they include such diverse organisms as insects, crustaceans, spiders, scorpions, millipedes, and centipedes which share in common a number of features such as a segmented body with paired appendages including jaws, a well-developed nervous system, chitinous cuticle, separate sexes, an open blood vascular system, and a lack of cilia in any part of the body (except *Peripatus*). Classically, zoologists believed that the arthropods represented a monophyletic group and classified these animals as the phylum Arthropoda. More recently, however, mounting evidence from fossil, embryological, and anatomical studies has shown that the arthropods are probably polyphyletic, originating from several types of primitive polychaete annelids (see Manton, 1977, and Barnes, 1980). This should be borne in mind by comparative immunologists when investigating the phylogeny of the immune system in invertebrates.

Direct equivalents to the vertebrate reticuloendothelial system (RES) in invertebrates are generally difficult to find. However, in the arthropods, the free and fixed blood cells, pericardial cells (nephrocytes), and hemopoietic tissue may be analogous.

In this review, we have divided the arthropods into four main groups: (1)

NORMAN A. RATCLIFFE, KEITH N. WHITE, ANDREW F. ROWLEY, and JULIA B. WALTERS • Department of Zoology, University College of Swansea, Wales, United Kingdom. Supported by grants from the Royal Society, Natural Environmental Research Council (GR3/3399), and Science and Engineering Research Council (B.RG.5924.3, GR.A.2286.0, and GR.B60958).

TABLE 1. SIMPLIFIED CLASSIFICATION SCHEME FOR THE ARTHROPODS[a]

1. Phylum Chelicerata
   Subphylum Merostomata (horseshoe crabs)
   Subphylum Arachnida (scorpions, spiders, ticks, mites)
2. Phylum Crustacea (crabs, lobsters, etc.)
3. Phylum Uniramia
   Subphylum Onychophora
   Subphylum Myriapoda (centipedes, millipedes)
   Subphylum Hexapoda (insects)

[a] From Manton (1977).

the chelicerates, (2) the crustaceans, (3) the myriapods and onychophorans, and (4) the insects (Table 1), and the structure and functions of the RES are dealt with separately in each group.

2. CHELICERATES

2.1. INTRODUCTION

The phylum Chelicerata is a large group of animals which includes the subphylum Merostomata (horseshoe crabs), subphylum Arachnida (scorpions, spiders, ticks, and mites), and the class Pycnogonida (sea spiders) (Manton, 1977). All of these animals have a body cavity divided into two regions, the cephalothorax (prosoma) and the abdomen (opisthosoma), feeding structures called chelicerae, and the lack of distinct antennae. The host defense systems of the arachnids and pycnogonids have scarcely been studied. Recently, however, in the horseshoe crabs there has been great interest in such systems, following the discovery of a potent coagulation reaction after challenge with bacterial endotoxin (Levin and Bang, 1964a,b).

2.2. MEROSTOMATES

The merostomates are subdivided into three classes: the Aglaspida, the Eurypterida, and the Xiphosura (horseshoe crabs), but only the latter include living representatives. The Xiphosura are a small group of animals consisting of three genera and four species with a fossil record stretching back to the Cambrian period (Barnes, 1980).

2.2.1. Circulatory System

The circulatory system of horseshoe crabs is hemocoelic but unusual within the arthropods in that it is partially closed (Clarke, 1979). It consists of a well-developed dorsal tubular heart which pumps the blood (hemolymph) through a complex series of arteries to the tissues and sinuses. From some of these sinuses, the blood flows into the gills and returns back to the heart via the pericardium. The gills not only oxygenate the blood but also aid in its circulation. The respiratory pigment in horseshoe crabs is hemocyanin which gives the plasma its characteristic blue color. The structure of the circulatory system is described in full by Shuster (1978) and Clarke (1979).

2.2.2. Cells of the Defense System

At present, the only cells known to be involved in the defenses of horseshoe crabs are the blood cells (hemocytes) and the cells of the hypodermal glands (Stagner and Redmond, 1975).

2.2.2a. Hemocytes. There is some confusion as to the number of different hemocyte types present in horseshoe crab blood. For example, many authors including Loeb (1902, 1920), Levin and Bang (1964a,b), Dumont et al. (1966), Holme and Solum (1973), and Armstrong (1977, 1979) described only a single cell type, the granular amebocyte, while Fahrenbach (1970) and Sherman (1981) identified, in *Limulus polyphemus*, both granular amebocytes and cells termed cyanoblasts/cyanocytes, which synthesize and release the respiratory pigment hemocyanin.

The cyanocytes are probably formed in the hepatopancreas and are most readily found in the blood sinuses in the neural plexus of the compound eye, where they account for between 1 and 8% of the circulating blood cells (Fahrenbach, 1970). There are no reports of these cells having any role in the host defense systems.

The structure of the granular amebocytes has been extensively studied in *L. polyphemus* in both light (Loeb, 1920; Armstrong, 1977, 1979) and electron microscopes (Dumont et al., 1966; Solum, 1970a; Ornberg and Reese, 1979; Nemhauser et al., 1980). *In vivo*, when observed with phase-contrast microscopy, these cells are ovoid and appear highly refractile, while *in vitro*, in the absence of endotoxin, they flatten out, become ameboid, and eventually degranulate (Armstrong, 1979) (Figure 1). Ultrastructural studies have shown that the cytoplasm characteristically contains a number of electron-dense amorphous granules, prominent Golgi complexes, pinocytotic vesicles, peripheral microtubules, free ribosomes, and mitochondria (Dumont et al., 1966; Ornberg and Reese, 1979). The site of formation of these cells is unknown.

In the Japanese horseshoe crab, *Tachypleus tridentatus*, the granular amebocytes have been subdivided by Shishikura and Sekiguchi (1979) into two morphologically distinct types, based on the structure of the granular inclusions. The first type, "Type A," accounts for approximately 10% of the cell population and contains bacilliform granules, while "Type B" accounts for 90% of the cells and encloses large, faintly staining and smaller, eosinophilic granules (Shishikura and Sekiguchi, 1979). Whether these are simply two distinct cell types of different stages in a maturation or degeneration series is unclear, and needs further investigation.

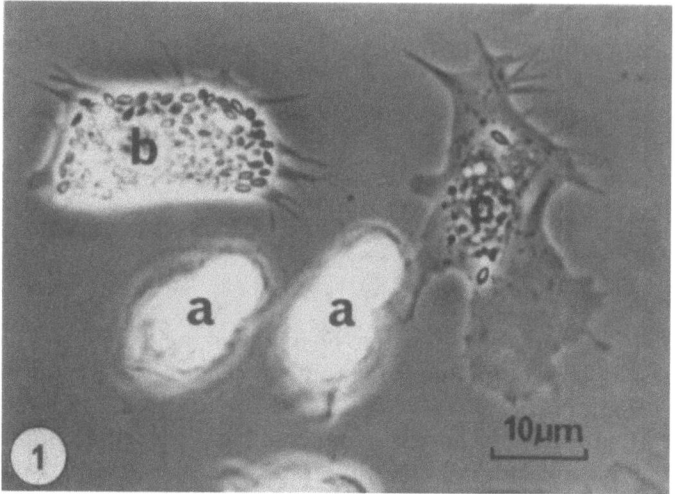

FIGURE 1. Phase-contrast micrograph of *Limulus polyphemus* amebocytes in various stages of spreading and degranulation *in vitro*, from rounded and contracted (a), through flattened and granular (b), to flattened and partially degranulated (c) forms. (From Armstrong, 1979.)

2.2.2b. Hypodermal Glands. The hypodermal glands lie below the external protective carapace and open to the outer environment via ducts. Stagner and Redmond (1975) showed that these glands produce a glycoprotein secretion which is released through the ducts in response to bacterial endotoxin or microbial-fouled seawater.

2.2.3. Cellular Defense Reactions

2.2.3a. Hemocytes. Unlike the insects and crustaceans, the role of the hemocytes of horseshoe crabs in nodule formation, encapsulation, and phagocytosis has hardly been studied. The major part of the work has been concerned with clotting and microbial killing by these cells.

Hemolymph Coagulation. The clotting process in the horseshoe crab consists of two distinct phases: (1) the agglutination of the granular amebocytes, and (2) the coagulation, or gelation, of the surrounding plasma. Levin and Bang (1964a,b) first demonstrated that the production of the gel is dependent upon factors released from the amebocytes (the coagulogen or pre-gel), and that in the absence of endotoxin, conversion of coagulogen to gel does not occur. As expected, plasma alone will not clot when mixed with endotoxin, but the addition of intact or lysed amebocytes restores this capacity (Levin, 1967). Later studies showed that the conversion of coagulogen to insoluble gel (coagulin) is mediated by a proteolytic enzyme found in the amebocytes which is activated from the zymogen form by endotoxin (Levin and Bang, 1968; Young *et al.*, 1972; Tai and Liu, 1977) (Figure 2). This clotting system provides an extremely reliable method for the clinical analysis of endotoxin levels (see reviews in Cohen, 1979).

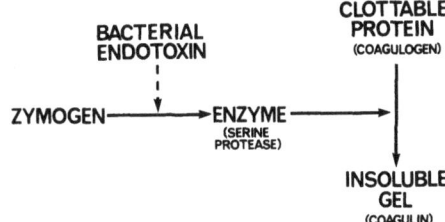

FIGURE 2. Flow diagram showing the stages in the coagulation reaction in the horseshoe crab, *Limulus polyphemus*. (Based on Levin, 1979.)

Dumont *et al.* (1966), using electron microscopy, studied the changes that occur in the amebocytes of *L. polyphemus* during clot formation. They found that clotting is associated with the discharge of granules from the amebocytes. During this process, the amorphous electron-dense granules become less electron-dense and the amorphous matrix is replaced by one containing numerous microtubules. Ultimately, the granular material is released from the cells by exocytosis (Ornberg and Reese, 1979). Similar events have been seen in both crustaceans (Bauchau, 1981) and insects (Rowley and Ratcliffe, 1976a; Rowley, 1977a) but do not seem to be reliant on the presence of endotoxin.

Murer *et al.* (1975) managed to isolate intact the granules from *L. polyphemus* amebocytes and showed that gelation occurs when the granules are mixed with endotoxin, thus demonstrating that these inclusions contain all the factors required for clotting.

Biochemical analyses of both the coagulogen and the insoluble gel have been carried out in both *L. polyphemus* (Solum, 1970a,b, 1973; Holme and Solum, 1973; Murer *et al.*, 1975; Tai and Liu, 1977; Mosesson *et al.*, 1979) and *T. tridentatus* (Nakamura *et al.*, 1976a,b,c; Shishikura *et al.*, 1977). In *L. polyphemus*, the coagulogen has a molecular weight of between 19,000 and 27,000 (Solum, 1970a,b, 1973; Gaffin, 1976; Tai *et al.*, 1977; Mosesson *et al.*, 1979), and recently has been shown to consist of two linked chains (Mosesson *et al.*, 1979). Electron microscopic examination of the insoluble gel has demonstrated a helical structure with a periodicity of 4.5–5 nm (Holme and Solum, 1973) which is different from vertebrate fibrinogen/fibrin.

In *T. tridentatus*, the coagulogen has a molecular weight of between 16,000 and 19,000 (depending on the analysis method used; Nakamura *et al.*, 1976b), which is somewhat less than that of *L. polyphemus*. During gel formation, the coagulogen is cleaved at two sites by the active proteolytic enzyme (activated from the zymogen by endotoxin), to form a peptide termed "peptide C," leaving two fragments F-A and F-B which are linked together and correspond to the gel (Nakamura *et al.*, 1976a,b,c) (Figure 3). These studies also show that the horseshoe crab clotting system may be similar to that of vertebrates, in that peptide C and the A chain (F-A) are structurally similar to primate fibrinopeptide B. Furthermore, the enzyme(s) which cleaves the horseshoe crab coagulogen is also like thrombin in its action. These similarities may well indicate that both the horseshoe crab and vertebrate clotting systems have developed from a common ancestor.

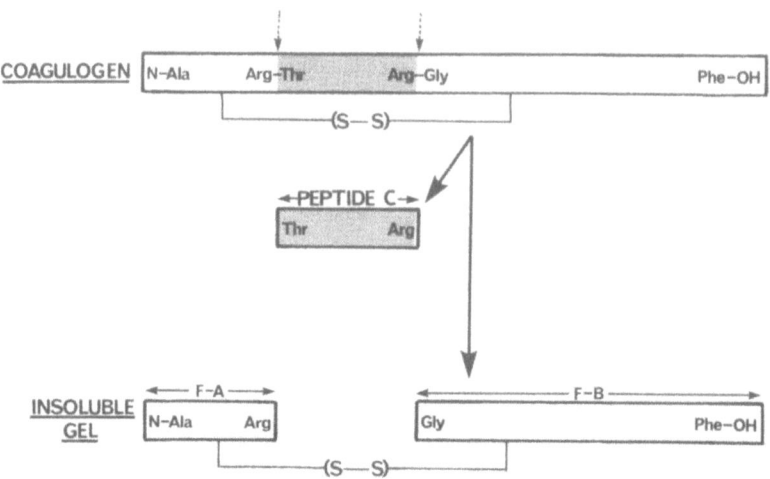

FIGURE 3. Structural changes of *Tachypleus tridentatus* coagulogen during gel formation. The arrows indicate the sites cleaved by the clotting enzyme(s). (Redrawn from Nakamura et al., 1976a.)

The coagulation reaction following the injection of either bacteria or endotoxin into the vascular system is probably of utmost importance in the animal's defense as it causes the immobilization of the microorganisms, reducing the risk of their spread throughout the hemolymph (Levin, 1967). This reaction, however, may be deleterious, as is the case during infection with *Vibrio* sp. in which hemolymph clotting spreads throughout the animal and ultimately leads to its death (Bang, 1956).

Serum Bactericidal Activity. Several authors, including Johannsen et al. (1973), Furman and Pistole (1976), and Pistole and Furman (1976), have demonstrated that the serum of *L. polyphemus* has a broad-spectrum antibacterial activity, primarily against gram-negative species. It has been suggested that because the plasma shows no significant bacterial killing, this activity originates from the amebocytes (Furman and Pistole, 1976). Amebocyte lysates have shown similar results (Pistole and Britko, 1978; Nachum et al., 1979) although these antimicrobial factors are primarily heat stable (60°C for 10 min) (Nachum et al., 1979), unlike that of serum, which is heat labile (56°C for 30 min) (Johannsen et al., 1973; Furman and Pistole, 1976). These differences may indicate the existence of two distinct killing mechanisms.

Phagocytosis. The role of phagocytosis in the defense systems of horseshoe crabs is at present unclear, although, in the absence of endotoxin, Armstrong and Levin (1979) demonstrated that carbonyl iron particles are ingested by the granular amebocytes of *L. polyphemus in vitro* (Figure 4). They put forward the attractive proposal that gram-negative bacteria are dealt with by coagulation and the subsequent release of antimicrobial factors, while other microorganisms (including gram-positive bacteria?) are eliminated primarily by phagocytosis. They

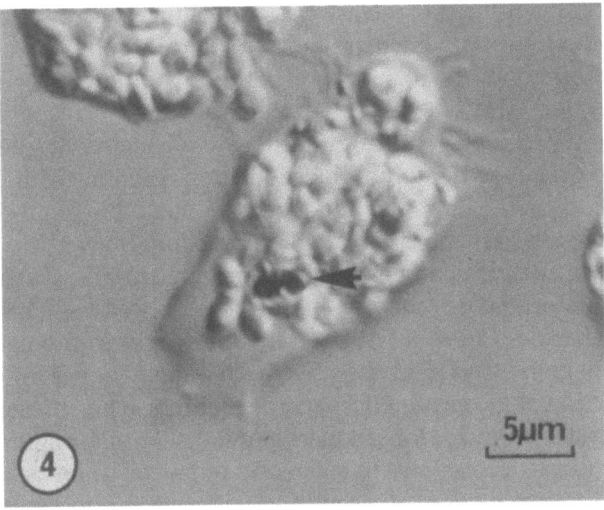

FIGURE 4. *Limulus polyphemus* amebocyte containing ingested carbonyl iron (arrow) particles. (From Armstrong and Levin, 1979.)

did not, however, test this idea by challenging the amebocytes with gram-positive bacteria. This elegant differential response to gram-positive and gram-negative bacteria, however, may not necessarily be quite so specific, since Stagner and Redmond (1975) found that gram-negative bacteria are apparently phagocytosed by the amebocytes of *L. polyphemus in vitro*. This observation disagrees with the results of other workers (e.g., Shirodkar *et al.*, 1960; Levin and Bang, 1968; Armstrong and Levin, 1979) who found that phagocytosis of gram-negative bacteria does not occur either *in vivo* or *in vitro*.

2.2.3b. Hypodermal Glands. The cells of the hypodermal glands secrete a glycoprotein onto the surface of the carapace in response to the presence of bacterial endotoxin and this has two main functions. First, due to its viscosity, it provides a mechanical barrier to invading microbes, and second, it agglutinates erythrocytes, bacteria, and algae and has potent antibacterial activity (Stagner and Redmond, 1975). Preliminary work by these authors suggests that the agglutinin has a rather low molecular weight of below 10,000 and is therefore unlike other invertebrate agglutinins, most of which are larger molecules (Acton and Weinheimer, 1974). A similar defense system is present in the cephalopod mollusc, *Loligo vulgaris*, in which the body surface is covered by a hemagglutinin of unknown origin (Marthy, 1974).

2.2.4. Recognition of Foreignness

Recognition of foreignness in *L. polyphemus* appears to occur in two distinct stages. First, upon the detection of bacteria in surrounding seawater, an agglutinin is released from the hypodermal glands that entraps the microbes. The triggering mechanism for this release has been suggested to be associated with the sensilla in the carapace which are morphologically similar to chemoreceptors (Stagner and Redmond, 1975). Second, if the invading microorganism avoids this defense system and enters the hemolymph, the next line of defense, the hemocytes, comes into play. With gram-negative bacteria, the recognition process appears to cause the hemolymph coagulation reaction, with the characteristic degranulation of the granular amebocytes, but with gram-positive forms, phagocytosis ensues. The stimuli for these reactions, however, are unknown, but undoubtedly involve changes at the amebocyte surface which may be mediated by receptors on the cell membrane for bacterial cell wall components such as lipopolysaccharide (Liu et al., 1979). The differences in the nature of the cell walls of gram-negative and gram-positive bacteria may hence influence the type of cellular reaction that occurs. If specific receptors were present for the two types of bacteria, receptors for gram-negative cell walls might cause the degranulation reaction to occur, while the receptors for gram-positive cell walls might initiate the production of pseudopodia, resulting in bacterial ingestion.

L. polyphemus hemolymph has been shown to contain both erythrocyte agglutinins (Cohen et al., 1965; Cohen, 1968; Marchalonis and Edelman, 1968; Oppenheim et al., 1974), usually termed "limulin," and bacterial agglutinins (Pistole, 1976, 1978, 1979; Gilbride and Pistole, 1979; Rostam-Abadi and Pistole, 1979). It has yet to be established if these substances have the capacity to act as opsonins, as occurs in other invertebrates (e.g., Tripp and Kent, 1967; Prowse and Tait, 1969; Hardy et al., 1977).

Limulin has a molecular weight of approximately 330,000 (Marchalonis and Edelman, 1968; Roche and Monsigny, 1974), and it is also a lymphocyte mitogen (Roche et al., 1977). Recently, further information on the physicochemical properties of limulin has become available with its partial purification by Roche and Monsigny (1979) and analysis of the amino acid sequence (Kehoe et al., 1979), which indicates that no homologies exist between limulin and vertebrate immunoglobulins.

Pistole (1978) found by simple adsorption experiments that there are probably two bacterial agglutinins present in *L. polyphemus*, one for gram-positive and one for gram-negative bacteria. Later studies (Gilbride and Pistole, 1979; Rostam-Abadi and Pistole, 1979; Pistole, 1979) extended this initial work to show that the gram-positive agglutinin has antigalactan activity while the gram-negative agglutinin has an affinity for 2-keto-3-deoxyoctonate in bacterial lipopolysaccharide. This differential response of serum agglutinins towards the two types of bacteria, as mentioned above, is reflected by the hemocyte reactions during coagulation and phagocytosis.

2.3. ARACHNIDS

The arachnids include spiders, scorpions, pseudoscorpions, mites, and ticks. The mites and ticks, in particular, are of economic importance, as many are parasitic on man or his domesticated animals and are responsible for the destruction of crops and stored products. Although these animals are of such significance, little information is available concerning their host defense systems. This has probably resulted from several factors including the difficulty in rearing them in the laboratory and their small size which makes them unsuitable subjects for such experimentation.

2.3.1. Circulatory System

In spiders and scorpions, the circulatory system is well developed with a dorsal heart and arteries, some of which lead to the book lungs. After aeration, blood collects in the ventral sinus and reenters the heart (Savory, 1964). In mites, the circulatory system is very much reduced and a distinct heart is usually absent, its role being replaced by contractions of the body musculature (Barnes, 1980).

2.3.2. Cells of the Defense System

Like other arthropods, the major cells of the defense system are the hemocytes. No widely accepted classification scheme for arachnid hemocytes exists. For example, some authors including Deevey (1941) used terms such as hyaline leucocyte, basophil, chromophobe, and eosinophil, while Brinton and Burgdorfer (1971), Sherman (1973, 1981), Ravindranath (1974a, 1977a), and Midttun and Jensen (1978) attempted to adopt the Jones (1962) classification scheme developed for insect hemocytes. Ravindranath (1974a) examined the blood cells of the scorpion, *Palamnaeus swammerdami*, and identified prohemocytes, plasmatocytes, granular hemocytes, cystocytes, spherule cells, and adipohemocytes, which morphologically, if not functionally, correspond to their namesakes in insects. The variation in morphology and classification schemes is fully reviewed by Sherman (1981) in greater depth than possible in the present account.

The blood cells in both spiders and scorpions are thought to be formed in distinct hemopoietic organs (often called lymphoid organs). In spiders, these are localized in the wall of the heart (Franz, 1904; Seitz, 1972), while in scorpions the hemopoietic organs are scattered throughout the body cavity (Cuénot, 1897; Kollmann, 1910; Blanc, 1967).

2.3.3. Cellular Defense Reactions

2.3.3a. Hemolymph Coagulation. The literature on the role of the hemocytes of scorpions and spiders in coagulation is often confusing. Grégoire (1952, 1970) showed that "fragile hemocytes" are involved in this process, but from the micrographs presented, it is unclear whether these cells correspond to either the plasmatocyte, granular hemocyte, or cystocyte types described by Ravindranath (1974a).

2.3.3b. Phagocytosis. Deevey (1941) found that the hyaline leukocytes (equivalent to plasmatocytes?) and eosinophils (granular cells?) of the Haitian tarantula, *Phormictopus cancerides*, ingest a variety of dyes, but whether these observations were made from either *in vivo* or *in vitro* experiments was not clarified. Similarly, bacteria or various dyes injected into scorpions have been found to be phagocytosed by cells in the hemopoietic organs (glande lymphatique, Kollmann, 1910).

Although these observations show that phagocytosis does occur, they do not indicate its importance as a method of clearing the hemocoel of potentially harmful microorganisms. Indeed, Millot (1926) believed that phagocytosis was only of secondary importance in spiders during infection, with the foremost defense mechanism being, as in horseshoe crabs, cell agglutination and coagulation.

2.3.4. Recognition of Foreignness

There is little information on how the hemocytes of arachnids recognize foreign particles. Brahmi and Cooper (1974) and Cohen *et al.* (1979) found a naturally occurring hemagglutinin in the blood of the scorpion, *Androctonus australis*, the levels of which can be increased by erythrocyte injections, but whether this has opsonic activity is yet to be investigated.

2.4. PYCNOGONIDS

The pycnogonids or sea spiders are a small group of marine animals. Characteristically, they have a narrow body with four to six pairs of walking legs and one pair of ovigerous legs, palps, and chelifores (King, 1973; Barnes, 1980). At the anterior end of the head is a pointed proboscis which is used in feeding.

The circulatory system comprises a dorsal heart which pumps blood anteriorly into the hemocoel. The hemocoel is divided by a membrane which is perforated by slits into upper and lower compartments.

Cells of the Defense System

Study of the morphology of the hemocytes is, to our knowledge, limited to Sanchez (1959), who noted two types of blood cells, leukocytes and granulocytes. Some of the mature leukocytes apparently become swollen with spherulelike inclusions and appear to be morphologically similar to the spherule cells of insects. The role(s) of these hemocytes in defense against invading microorganisms is at present unknown.

3. CRUSTACEANS

3.1. INTRODUCTION

The Crustacea, now elevated to the status of a separate phylum (Manton, 1977), include some of the most familiar and important arthropods, such as lobsters, crabs, crayfish, shrimps, and barnacles. They are primarily aquatic and include commercially important species of great relevance to the economy of certain areas. This, coupled with the present and future potential for crustacean mariculture, renders a knowledge of their epidemiology and disease resistance of more than academic interest. The deleterious effect of certain Crustacea, as parasites (Sindermann, 1970), predators (Warner, 1977), and fouling organisms (Pyefinch, 1947), also emphasizes their importance in the human economy.

The Crustacea are characterized by a chitinous exoskeleton, biramous jointed appendages, gills, and a free-swimming larva. They may be divided into six classes (Manton, 1977). The Cephalocarida contains small, primitive, shrimplike forms, while the Branchiopoda includes mainly small freshwater types, such as *Daphnia*. Members of the Copepoda, such as *Calanus*, are of great relevance in aquatic food chains, but the Ostracoda are of little apparent importance, although over 2000 species are known. The class Cirrepedia encompasses the various barnacle species. The sixth class, the Malacostraca, contains almost three-quarters of known crustaceans and includes decapods, such as crabs, lobsters, crayfish, shrimps, and the smaller mysids, isopods and amphipods.

3.2. CIRCULATORY SYSTEM

The crustacean circulatory system varies greatly in form from species to species. The differences center on the heart structure and the extent of the arterial system, both of which may be absent in some copepods, ostracods, and in the Cirrepedia. Primitively, the heart is a long tubular structure with numerous paired ostia. In more advanced forms, however, it is reduced to a small vessel lying within the dorsal part of the thorax and suspended in a pericardial cavity by a series of fibrous ligaments and muscles, with the number of ostia reduced, often to a single pair. A neural pacemaker is apparently present in most higher crustaceans.

In some small crustaceans, the arterial system is restricted to an anterior vessel from the heart, but in addition there may also be a posterior abdominal artery, lateral arteries, and a ventral artery, all arising from the heart. The principal arteries may branch extensively, sometimes forming capillary networks, particularly around the nerve cord and brain. Hemolymph is eventually delivered to numerous tissue spaces (sinuses). These are bounded by membranes derived from adjacent connective tissue and it is in the sinuses that gaseous exchange with the tissues occurs. In those crustaceans lacking a heart, valves may be present in the sinuses. Return of the hemolymph via the gills to the pericardium is generally mediated through distinct drainage channels, which in larger forms may be membrane-bound, thereby assuming the character of vessels. Membranous partitions may subdivide the sinuses, resulting in a more definite pathway for hemolymph flow by separating the arterial and deoxygenated blood.

Crustacean internal defenses center around cellular and humoral factors, but the latter, except where they mediate cellular responses, are beyond the scope of this survey which is concerned with the activities of the blood cells, properly termed hemocytes, and associated cell-mediated defense systems.

3.3. CRUSTACEAN DISEASES

It is not proposed to provide a comprehensive review of crustacean diseases, only to indicate the range of pathogenic organisms infecting members of this phylum. For a more extensive review see Sindermann and Rosenfield (1967), Johnson (1970), Sprague (1970), and Sindermann (1970, 1971). A major bacterial disease is gaffkemia, a fatal septicemia of lobsters (Stewart and Rabin, 1970; Sindermann, 1971; Stewart, 1975) caused by *Aerococcus viridans*. It is only mildly toxic to other crustaceans. A less serious, although sometimes fatal, disease of decapods is produced by a chitin-destroying bacterium resulting in so-called shell or spot disease (Rosen, 1967, 1970). A major fungal parasite (*Aphanomyces astaci*) occurs in the crayfish, *Astacus astacus* (Unestam and Weiss, 1970). *Pythium afertile* infection has resulted in heavy mortalities of the prawn, *Palaemon serratus* (Anderson and Conroy, 1968), and black gill disease caused by a *Fusarium* sp. can result in significant mortalities in *Penaeus japonicus* (Egusa and Veda, 1972). Of the protozoan classes, the Telosporea, Haplosporea, Microsporidea, Rhizopodea, and Ciliatea all contain types parasitic in Crustacea (see review by Sprague, 1970). Important protozoan parasites include *Nosema*, which infects the body musculature of the crabs, *Callinectes sapidus* and *Carcinus maenas* (Sprague, 1965), and *Paramoeba perniciosa*, which causes gray crab disease of *Callinectes sapidus* (Sawyer, 1969). Finally, helminth parasites are also found in the Crustacea. The trematodes, *Microphallus nicolli* and *M. carcini*, are common in many species of crab, while several cestodes are ubiquitous parasites of shrimps (Sindermann and Rosenfield, 1967; Sindermann, 1970). However, the epidemiology of such potential pathogens is often poorly understood.

3.4. CELLS OF THE DEFENSE SYSTEM

A fairly extensive, although far from comprehensive, literature exists on aspects of crustacean hemocyte morphology. This includes a series of studies on the lobster, *Homarus americanus* (Toney, 1958; Stewart *et al.*, 1967; Hearing and Vernick, 1967; Cornick and Stewart, 1978); the crayfish *Astacus fluviatilis* (Hardy, 1892; Tait and Gunn, 1918), *A. astacus* (Stang-Voss, 1971), *Orconectes virilis* (Wood and Visentin, 1967), *Orconectes* spp. (Sternshein and Burton, 1980), *Cambarus bartoni* (George and Nichols, 1948; Toney, 1958), *Procambarus* spp. (Sternshein and Burton, 1980); the crabs *Carcinus maenas* (Sewell, 1955; Johnston *et al.*, 1973; Williams and Lutz, 1975a; Smith and Ratcliffe, 1978, 1980a), *C. mediterraneus*, *Macropipus depurator* (Durand, 1973), *Emerita asiatica* (Ravindranath, 1975a), *E. talpoida* (George and Nichols, 1948), *Cancer magister* (Mix and Sparks, 1980), *Callinectes sapidus* (George and Nichols, 1948; Toney, 1958; Bodammer, 1978), *Eriocheir sinensis* (Bauchau and De Brouwer, 1972), *Pachygrapsus marmoratus* (Arvy, 1952), *Uca pugnax*, *Ocypode albicans*, *Panopeus herbsti*, and *Libinia dubia* (George and Nichols, 1948); the prawn *Metapenaeus masterii* (Dall, 1964); the amphipod *Niphargus virei* (Gibert, 1972); the isopods *Helleria brevicornis* (Hoarau, 1976) and *Ligia exotica* (Ravindranath, 1974b); and the anostracans *Ar-*

temia salina (Lochhead and Lochhead, 1941) and *Daphnia pulex* (Hardy, 1892). These accounts usually describe between two and four hemocyte types but correlations of cell types from different species, and even between descriptions of the same species, are often difficult. There are a number of reasons for this, and although some variation in morphology and composition between hemocytes of such a diverse group as the Crustacea is likely to occur, the problem is compounded since the various cell types may form a continuous differentiation series, with many intermediates present (see Section 3.4.5). Furthermore, the hemocytes may change form according to the physiological condition of the animal. For example, starvation results in a decrease in carbohydrate reserves in *C. maenas* hemocytes (Williams and Lutz, 1975a,b) and this and other changes in hemocyte metabolic processes may be reflected in concomitant morphological changes. The structural integrity of some of the hemocytes is also lost during and after their removal from the circulation, due to the initiation of the clotting reaction, and as a result of temperature and pH changes (Bauchau and De Brouwer, 1974; Ravindranath, 1975a,b).

Ravindranath (1974b) and Gupta (1979a) have extended the insect classification scheme proposed by Jones (1962) to the Crustacea, six or seven cell types being identified. However, because of the uncertainties discussed above, we have distinguished just two cell "types" in the Crustacea (see also Rabin, 1970; Smith and Ratcliffe, 1978, 1980a; Lackie, 1980), although this does not exclude the presence of intermediate forms. This is in broad agreement with Bauchau (1981) who, however, places the intermediates in a separate category.

3.4.1. Hyaline Cells (Figures 5,6)

Hyaline cells are characterized by a large, centrally placed nucleus, bounded by a relatively thin layer of cytoplasm. They are round, ovoid, or spindle-shaped, and vary in size from 4.5 to 25 μm except in the crayfish, *Astacus astacus*, in which they may exceed 30 μm in diameter (Stang-Voss, 1971). In the crab, *Carcinus maenas*, hyaline cells form 80% of the hemocyte population (Smith and Ratcliffe, 1978). As their name implies, few cytoplasmic inclusions can be discerned in fresh preparations, particularly prior to their attachment to the substratum by pseudopodia, after which vacuolation may occur and the granules appear as small dark bodies under phase contrast optics (Figure 5). In *C. maenas* and *H. americanus*, the presence of acid phosphatase has been demonstrated within some of the hyaline cell granules (Hearing, 1969; White and Ratcliffe, 1981). Leucine aminopeptidase was localized in similar structures in *Orconectes virilis* (Wood and Visentin, 1967) and there is little doubt that some, at least, are primary lysosomes (see Dean, 1977). Hyaline cells have also been shown to contain esterase and protease (Wood and Visentin, 1967; Bauchau, 1981), and amylase activity in the serum is probably of hemocyte, possibly hyaline cell, origin (Horn and Kerr, 1969). Peroxidase is not present in the hyaline cells of *C. maenas* (White, unpublished), nor is there phenoloxidase activity in this cell type in *Cancer pagurus* (Decleir and Vercauteren, 1965).

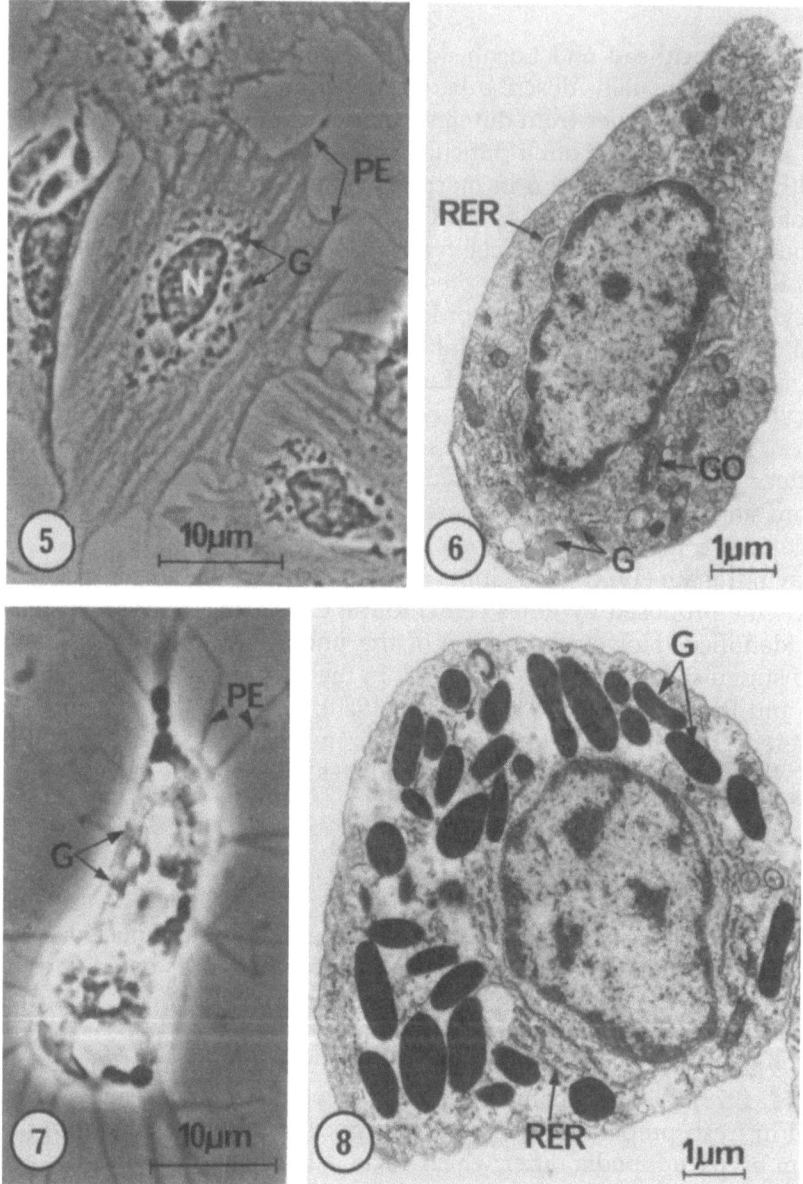

FIGURE 5. Phase-contrast micrograph of hyaline cells from the shore crab, *Carcinus maenas*. Note the central nucleus (N), small phase-dark granules (G), and peripheral cytoplasm with protoplasmic extensions (PE). (Courtesy of Dr. V. J. Smith.)

FIGURE 6. Electron micrograph of a hyaline cell of *C. maenas* with characteristic poorly developed Golgi complex (Go), rough endoplasmic reticulum (RER), and small granules (G). (Courtesy of Dr. V. J. Smith.)

FIGURE 7. Phase-contrast micrograph of a granulocyte from *C. maenas* with characteristic refractile cytoplasm containing granules (G). Spikelike pseudopodia (PE). (From Smith and Ratcliffe, 1978.)

FIGURE 8. Electron micrograph of a granulocyte from *C. maenas* with prominent, membrane-bound, electron-dense granules (G) and profiles of rough endoplasmic reticulum (RER) surrounding the nucleus. (Courtesy of Dr. V. J. Smith.)

Other histochemical observations on crustacean hemocytes have centered on the localization of carbohydrate-containing substances. Mucopolysaccharide and possibly some glycogen were detected in the hyaline cells of *O. virilis* and *C. maenas* (Wood and Visentin, 1967; Johnston *et al.*, 1973). However, Williams and Lutz (1975a) and Smith (1978) failed to detect PAS-positive material in *C. maenas* hyaline hemocytes, although there remains the possibility that seasonal or other factors may influence carbohydrate levels in these cells, as is known to be the case with lipid (Sewell, 1955).

Ultrastructural observations have confirmed the presence of a variable number of small membrane-bound granules. The often poorly developed Golgi and the small amount of endoplasmic reticulum characteristically present (Figure 6) are not indicative of an active system of sequestration and synthesis of material within hyaline cells.

3.4.2. Granulocytes (Figures 7,8)

As their name implies, this class of hemocytes is characterized by a variable, but often large number of inclusions. Under phase contrast, these granules are frequently highly refractile (Figure 7). Granulocytes are commonly ovoid or spherical in shape and vary in size from 7 to 35 µm (50µm in *A. astacus*). They are larger than the hyaline cells, and in *C. maenas* comprise some 20% of the hemocyte population (Smith and Ratcliffe, 1978). Unlike hyaline cells, granulocytes do not spread extensively on contact with the substratum but may attach by fine pseudopodial strands. *In vitro*, these cells are frequently fragile, having a tendency to lyse and release their granules.

Ultrastructural observations have revealed the granules to be membrane-bound, usually homogeneous, electron-dense bodies (Figure 8). Other characteristic cytoplasmic components include one or more well-developed Golgi and profiles of endoplasmic reticulum adjacent to the nucleus (Figure 8). Ribosomes are often scattered throughout the cytoplasm. Such components are indicative of an active process of synthesis and sequestration of material, much of it probably packaged within the granules.

Protein is present in the granules of *Homarus americanus*, *Carcinus maenas*, and *Eriocheir sinensis* (Hearing and Vernick, 1967; Bauchau *et al.*, 1975), and a polysaccharide component has been detected in *C. maenas*, *E. sinensis*, and *Helleria brevicornis* (Johnston *et al.*, 1973; Bauchau *et al.*, 1975; Hoarau, 1976) but not in *Orconectes virilis* (Wood and Visentin, 1967). The polysaccharides are principally neutral mucosubstances (Bauchau *et al.*, 1975), but may include an acid mucopolysaccharide (Dall, 1965). That there exist variations in granule composition, both between species and possibly within the same cell (Mengeot *et al.*, 1976, 1977), no doubt reflects both differences in granulocyte functions and the dynamics of granule metabolism. The granules also possess certain lysosomal characteristics, with leucine aminopeptidase activity in *O. virilis* (Wood and Visentin, 1967) and acid phosphatase in some granules of *E. sinensis* (Bauchau, 1981), *H. americanus* (Hearing, 1969), and *C. maenas* (White, unpublished). Phenoloxidase activity has also been recorded in *Cancer pagurus* (Decleir and Vercauteren, 1965) and *A. astacus* (Unestam and Nylund, 1972).

Only Williams and Lutz (1975b) have recorded glycogen in the granules, although it is ubiquitous as a cytoplasmic component in the granulocyte (Lochhead and Lochhead, 1941; Arvy, 1952; Wood and Visentin, 1967; Bauchau *et al.*, 1975), where it is present in larger amounts than in hyaline cells (Johnston *et al.*, 1973; Bauchau *et al.*, 1975; Bodammer, 1978). There is increasing evidence that hemocytes may play an important role in carbohydrate metabolism (Johnston *et al.*, 1973; Williams and Lutz, 1975a,b; Bauchau, 1981), and other nondefensive functions of granulocytes may include the synthesis and storage of hemocyanin (Stang-Voss, 1971), and the accumulation of chitin or a precursor for use during cuticle deposition (Dall, 1965; Johnston *et al.*, 1973).

3.4.3. Other Circulating Hemocytes

In addition to the hyaline and granular cells, other circulating cells have been identified in several Crustacea. Explosive cells (cystocytes/coagulocytes) have been described in *Astacus fluviatilis* (Hardy, 1892; Tait and Gunn, 1918), *Homarus americanus*, *Cambarus bartoni*, *Callinectes sapidus* (Toney, 1958), *Emerita asiatica* (Ravindranath, 1975a), *Carcinus maenas* (Johnston *et al.*, 1973), *Procambarus* spp. and *Orconectes* spp. (Sternshein and Burton, 1980). However, this may not be a separate cell type but a form of granulocyte, perhaps the final stage in the latter's development. The observation that granulocytes can transform into cystocytelike cells in a alkaline medium (Ravindranath, 1975b) supports this contention.

Another cell type which has been described is the lipoprotein cell of *C. maenas* and *E. sinensis* (Bauchau and De Brouwer, 1972), while the adipohemocytes reported in *Ligia exotica* and *Emerita asiatica* by Ravindranath (1974b, 1975a) may also be lipoprotein cells. These cells are probably transformed hyaline cells, which appear in large numbers prior to molting and are involved in cuticle secretion (Sewell, 1955).

3.4.4. Sessile Hemocytes (Figure 9)

Circulating hemocytes retain some affinity for host tissue, such as the parenchyma (connective) cells, and the hyaline cells may attach themselves to surrounding tissues, particularly during cell clumping in response to infection. However, there exist certain cells that appear permanently associated with the body tissues. Fontaine and Lightner (1973, 1974) described fixed phagocytic cells lining the sinuses of the gills, heart, and abdomen of the shrimp, *Penaeus setiferus*, while Johnson (1976) mentioned the presence of fixed rosettes of phagocytic cells in the intertubular spaces of the crab, *Callinectes sapidus* (Figure 9). Cells lining the interacinar sinuses of the hepatopancreas of the crayfish, *Parachaeraps bicarinatus*, have also been shown to be important in the removal of carbon particles from the circulation (Reade, 1968). Tyson and Jenkin (1973) are of the

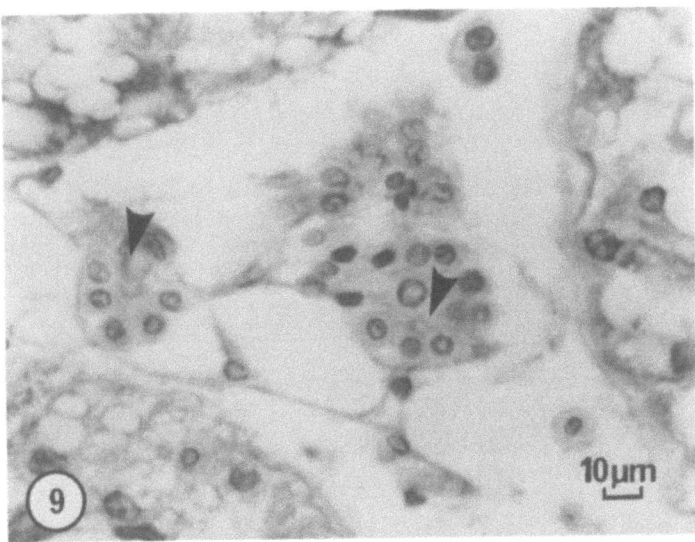

FIGURE 9. Mass of fixed hemocytes (arrowheads) between the hepatopancreatic acini of *Callinectes sapidus*. (Courtesy of Dr. P. T. Johnson.)

opinion that such cells may also be of major importance in the removal of bacteria from the hemolymph of crayfish, although confirmatory visual localization is required. Although the role of fixed phagocytes vis-à-vis cell clumping (see Section 3.5.2) has not been elucidated, comparisons with the fixed macrophage system of vertebrates are compelling.

The origin of the fixed phagocyte, however, is unclear, although some appear similar to circulating hemocytes and may represent "trapped" hyaline cells. The sessile cells described in *P. bicarinatus* by Reade (1968) have more of the characteristics of epithelial cells than of attached hemocytes. Clearly, more information is required in order to make a realistic assessment of the origin and importance of fixed phagocytic cells in the crustacean defense system.

A further type of attached hemocyte is the cyanocyte (reviewed in Bauchau, 1981) which is characterized by the synthesis, accumulation, and secretion of hemocyanin. They appear to have no immunological significance and therefore need concern us no further.

3.4.5. Hemopoietic Tissue (Figures 10, 11)

The site of production of the circulating hemocytes and cyanocytes has been localized to a series of lymphocytogenic nodules located either adjacent to the ophthalmic artery or within the wall of the foregut (Cuénot, 1897; Bruntz, 1907; Kollmann, 1908; Marrec, 1944; Gibert, 1972; Ghiretti-Magaldi *et al.*, 1977; Bazin, 1979). The hemocytogenic cells are small with large nuclei and many mitoses are observed in the nodules, resulting in the production of hyaline protohemocytes (hemoblasts) (Figure 11). The relationship of the various hemocyte types to one

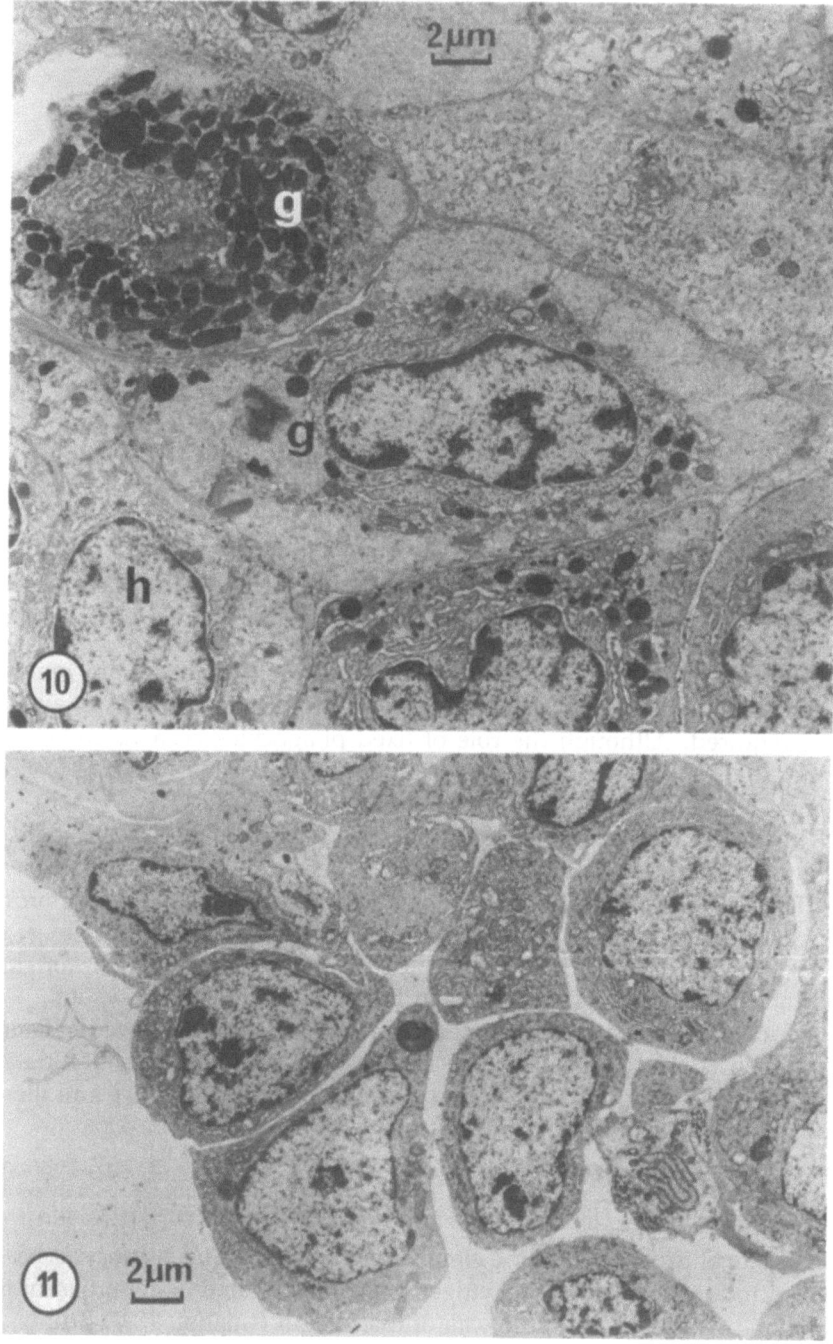

FIGURE 10. Part of a lymphocytogenic nodule of *Carcinus maenas* containing cells at various stages of maturation from hyaline cells (h) to granulocytes (g). (From Ghiretti-Magaldi et al., 1977.)

FIGURE 11. Peripheral area of a lymphocytogenic nodule of *C. maenas* showing the release of undifferentiated hemoblasts. (From Ghiretti-Magaldi et al., 1977.)

another is unclear, the general consensus being that they form a continuous developmental series (Cuénot, 1897; Arvy, 1952; Wood and Visentin, 1967; Stang-Voss, 1971; Bauchau and De Brouwer, 1972; Gibert, 1972; Hoarau, 1976; Bodammer, 1978; Mix and Sparks, 1980), circulating hyaline cells differentiating into granulocytes. However, the situation is not as clear-cut as may at first appear, for in *C. maenas* and *E. sinensis*, hemocytes at various stages of differentiation have been found in the hemopoietic organs (Ghiretti-Magaldi *et al.*, 1977; Bauchau, 1981) (Figures 10, 11), so that cell development in the circulation is not the only, and perhaps not the most important, stage in granulocyte maturation. If not, then intriguing possibilities arise as to the regulation of hemocyte types and numbers, particularly as hemopoietic activity may be under hormonal control (Matsumoto and Tongu, 1966; Charmantier-Daures, 1973). Molting and other hormone-influenced phenomena such as diurnal rhythms result in a change in hemocyte numbers (Drach, 1939; Bauchau and Plaquet, 1973; Ravindranath, 1977b). One can speculate that a similar correlation may exist between stress-induced changes in hemocyte numbers, such as that resulting from handling (Hamann, 1975), starvation (Stewart *et al.*, 1967), temperature (Schutz, 1925; Dean and Vernberg, 1966; Ravindranath, 1977b) and variations in the rate of hemocyte production. Further elucidation of the role of the hemopoietic organs, versus the circulation, in hemocyte maturation awaits the successful application of cell separation and culture methods, plus the utilization of pulse-labeling techniques.

Unlike in insects (see Section 5.6.1b), there is no evidence that the crustacean hemopoietic organ has phagocytic properties, nor is there any indication that it is involved in the production of humoral defense substances.

3.4.6. Nephrocytes (Figures 12, 13)

Nephrocytes occur in the gills of certain crustaceans (Figure 12) and are capable of removing material from the hemolymph (Drach, 1930; Lison, 1942; Balss, 1944; Ali, 1966). However, neither Johnson (1976) nor Smith and Ratcliffe (1981) could detect bacteria in these cells following natural or artificial infection, and the ameba *Paramoeba perniciosa*, found within the nephrocytes of *Callinectes sapidus* by Johnson (1977), probably actively invaded the cells. Nevertheless, ultrastructural, histochemical, and tracer studies of nephrocytes from infected *Carcinus maenas* do suggest an indirect involvement in the host response to infection (see Section 3.5.4).

Light and electron microscopic studies (e.g., Drach, 1930; Flemister, 1959; Wright, 1964; Ali, 1966; Strangways-Dixon and Smith, 1970; Foster and Howse, 1978; Smith and Ratcliffe, 1981) have revealed that these cells contain numerous vesicles, peripheral inclusions, and often a large central vacuole (Figures 12, 13). The cell boundary bears a certain resemblance to the filter slits (pedicels) of the podocyte cells of the vertebrate glomerulus, and to the end sac of the crustacean excretory organ (Kümmel, 1973), in having a series of cytoplasmic processes resting on a basement membrane (Figure 13, inset). Crustacean nephrocytes closely resemble the pericardial cells of insects and may have functional similarities (see Section 5.5.4).

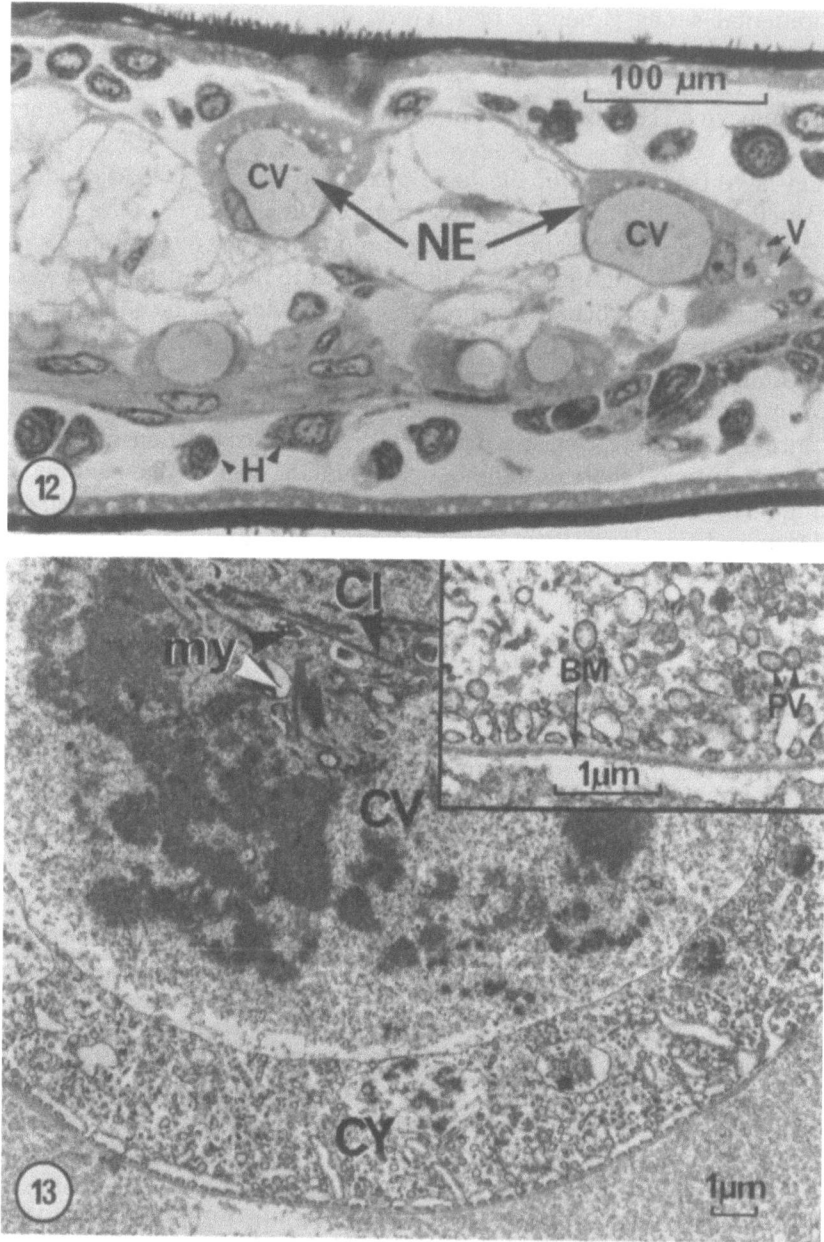

FIGURE 12. Nephrocytes (NE) in the gill lamella of the crab, *Carcinus maenas*. Note the large central vacuoles (CV) and peripheral vacuoles (V) in the nephrocyte cytoplasm. Hemocytes (H). (From Smith and Ratcliffe, 1981.)

FIGURE 13. Electron micrograph of part of a *C. maenas* nephrocyte showing the central vacuole (CV) containing small amounts of myelin configurations (my) and crystalline inclusions (CI) surrounded by a rim of cytoplasm (CY). Inset shows enlargement of peripheral area with the basement membrane (BM) and pinocytotic vesicles (PV). (From Smith and Ratcliffe, 1981.)

3.5. CELLULAR DEFENSE REACTIONS

Notwithstanding their involvement in such diverse activities as the synthesis and transport of carbohydrates (Johnston et al., 1973; Williams and Lutz, 1975b; Bauchau et al., 1975; Bodammer, 1978), the production of hemocyanin (Stang-Voss, 1971), and ecdysis (Bauchau and De Brouwer, 1974; Durliat and Vranckx, 1978; Herberts et al., 1978), in common with many other invertebrates, the role of crustacean hemocytes in the isolation and sequestration of foreign substances is probably central to the continued survival of the group. Crustacean defense reactions may be conveniently divided into phagocytosis, cell clumping/nodule formation, encapsulation, hemolymph coagulation, and wound repair.

3.5.1. Phagocytosis

Since the pioneering work of Metchnikoff (1884) with *Daphnia*, the involvement of crustacean hemocytes in the phagocytosis of material has been documented in the lobster *Homarus americanus* (Cornick and Stewart, 1968, 1978; Rabin, 1970; Paterson and Stewart, 1974; Paterson et al., 1976), the crabs *Carcinus maenas* and *Callinectes sapidus* (Johnson, 1976, 1977; Smith and Ratcliffe, 1978), the crayfish *Parachaeraps bicarinatus* (Reade, 1968; McKay and Jenkin, 1970a,b; Tyson and Jenkin, 1974), *Procambarus* spp. and *Orconectes* spp. (Sternshein and Burton, 1980), and the shrimp *Penaeus setiferus* (Fontaine and Lightner, 1974). Invariably, the hyaline cells are involved in particle uptake although the granulocytes may also have a phagocytic capacity (Paterson and Stewart, 1974; Johnson, 1976; Sternshein and Burton, 1980).

Although something is known of events initiating phagocytosis and antigenic recognition (see Section 3.6), there is a dearth of information on the process itself and the subsequent intracellular events. That phagocytosis is an energy-utilizing process in the Crustacea would account for its temperature dependence in *P. bicarinatus* and *H. americanus* (McKay and Jenkin, 1970a; Paterson and Stewart, 1974). The inhibition of phagocytosis after microfilament disruption by cytochalasin B is indicative of a role in the phagocytic event (Bauchau and Mars, unpublished, reported in Bauchau, 1981). Following sequestration within membrane-bound vesicles, the ingested material is, no doubt, subjected to the action of various degradative enzymes of lysosomal and cytosol origin. In an *in vitro* study of *H. americanus* hemocytes, Paterson and Stewart (1974) noted vacuolation around phagocytosed material and degranulation of the surrounding cytoplasm, while an ultrastructural study of phagocytosis by crayfish blood cells seems to show the fusion of the cell's granules with the phagosome and the release of their contents into the interior (Sternshein and Burton, 1980). These events may be connected with the release of enzymes from primary lysosomes into the phagosome. Lysosomal enzymes are known to be present in crustacean hemocytes (see Sections 3.4.1 and 3.4.2), and in *C. maenas* there is sometimes an increase in acid phosphatase activity after bacterial infection (White and Ratcliffe, unpublished). Peroxidase, a potent microbicide in certain vertebrate

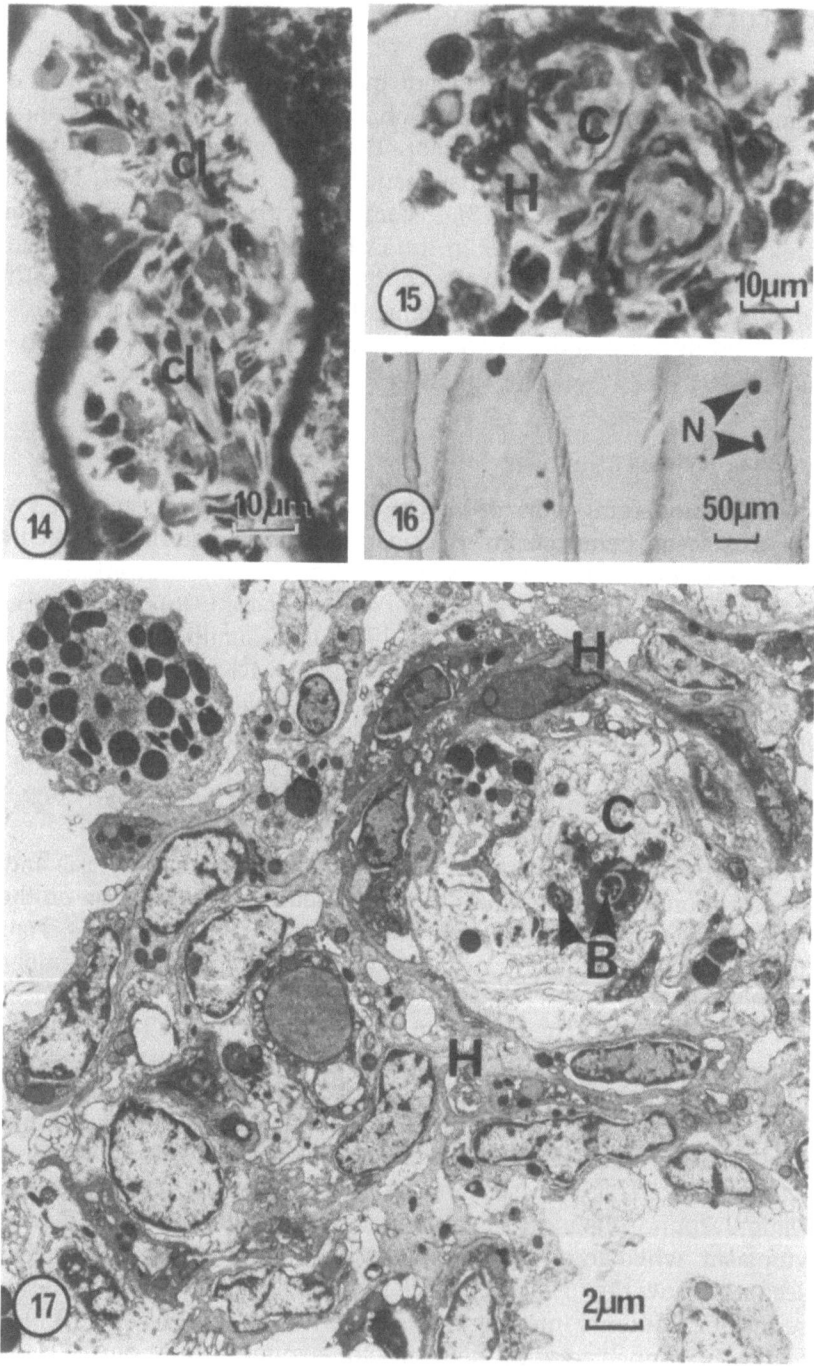

FIGURE 14. Gill lamella of *Carcinus maenas*, 1 hr after injection of *Moraxella* sp., showing an extensive cell clump (cl) formed by a loose network of hemocytes.

FIGURE 15. Gill lamella of *C. maenas*, 1 day after injection of *Moraxella* sp., showing a cell clump composed of concentric layers of hemocytes (H) surrounding a necrotic core (C).

blood cells, does not appear to be present in crustacean hemocytes (Section 3.4.1), although it has been localized in the hepatopancreas of the crayfish *Cambarus bartoni* (in the fixed phagocytic cells?), but appears to have no microbicidal properties (Merril and Glenister, 1980). Direct evidence of bacterial killing after *in vitro* phagocytosis by crayfish hemocytes suggests that, as in the vertebrates, the process is rapid, 90% of *Salmonella abortus equi* being destroyed during the first 60 min (Tyson and Jenkin, 1974). *In vivo*, killing of *Pseudomonas* by immunized crayfish hepatopancreatic tissue is similarly rapid (McKay and Jenkin, 1970c), but the importance of phagocytosis in comparison to cell clumping (see Sections 3.5.2 and 3.5.3) cannot be ascertained. However, destruction of microbes is not always as efficient as this in Crustacea and the pathogen may be able to neutralize or degrade the killing agents. For example, *Aerococcus viridans* is pathogenic at very low doses (10 or less viable bacilli can result in death in *H. americanus*; Stewart, 1975) due to its ability to overcome the effects of phagocytosis and to multiply rapidly in the hemolymph. Similarly, not all the spirochaete parasites observed in the hemocytes of the brine shrimp *Artemia salina* were within digestive vacuoles. Some lay directly in the cytoplasm and showed no signs of degeneration (Tyson, 1975). They may have represented individuals that had migrated from the phagosomes into the surrounding cytosol.

Pathogen breakdown, however, probably occurs within the phagosomes, and Tyson (1975) did observe spirochaetes within such bodies of *A. salina* hemocytes, undergoing ultrastructural changes consistent with cell breakdown. She also reported other vacuoles containing material that may have represented completely degraded parasites. Degradation of bacteria within phagosomes of the fixed phagocytes of *P. bicarinatus* was reported by Reade (1968), and myelin figures, seen commonly in the experimental crayfish but not in phagocytes of control animals, are also indicative of bacterial breakdown.

The elimination of phagocytosed material may not always necessitate its physical destruction. Fontaine and Lightner (1974) observed the elimination of phagocytosed carmine from the shrimp *Penaeus setiferus* by hemocyte migration through the gills, gut, hepatopancreas, pereiopods, and pleopods. The elimination of bacteria may occur by a similar route.

3.5.2. Hemocyte Clumping/Nodule Formation (Figures 14–17)

In addition to phagocytosis, crustacean hemocytes may respond to challenge from bacteria or other small particles by forming cell aggregates (Smith and Ratcliffe, 1980a). The foreign particle is associated with the hemocytes, often in large numbers, but not necessarily as a central encapsulated mass. This dif-

---

FIGURE 16. Gill lamellae of *C. maenas*, 1 day after injection of *Moraxella* sp., stained for the presence of melanin. Note the heavily melanized cell clumps/nodules (N) within the lamellae.
FIGURE 17. Electron micrograph of a nodule formed 12 hr after the injection of *Moraxella* sp. into *C. maenas* showing several layers of flattened hemocytes (H) surrounding a necrotic core (C) containing bacteria (B). (From Smith and Ratcliffe, 1980a.)

ference has led us to distinguish between bacteria-associated cell clumps, which may be composed of a loose network of hemocytes (Figure 14), and nodules. Characteristically, nodules consist of concentric, flattened, sometimes degenerated, layers of hemocytes surrounding a necrotic core (Johnson, 1976; Smith and Ratcliffe, 1980a) (Figures 15, 17). Peripherally, more normal hemocytes are present and the whole structure may be similar to the nodules described in insects (see Section 5.6.2). Between 24 hr and 4 days postinjection, the cores of this type of clump become melanized (Cornick and Stewart, 1968; Fontaine and Lightner, 1974; Fontaine et al., 1975; Smith and Ratcliffe, 1980a) (Figure 16). As melanization also accompanies encapsulation, consideration of the possible significance of this substance is dealt with below (see Section 3.5.5). Both hyaline cells and granulocytes are involved in clump/nodule formation, and the role of the latter cells may be crucial as they may mediate hemocyte aggregation by releasing granular material resulting in localized clot formation and the entrapping of bacteria. Both intra- and extracellular bacteria are found within such hemocyte clumps and nodules of C. maenas (Smith and Ratcliffe, 1980a), but only intracellular particles were recorded in P. setiferus aggregates (Fontaine and Lightner, 1974).

Clumps and nodules are commonly found in the gills, but other regions permeated by the hemolymph, such as the hepatopancreas, heart, excretory organ, and the appendages, may also contain similar hemocyte aggregates. Localization of clumps in these regions may simply be a passive process, but in addition, the highly vascularized gills would be expected to be a major repository of hemocyte aggregates due to the drop in hemolymph pressure in the sinuses (Maynard, 1960; Smith and Ratcliffe, 1980a). The pillar cells within the gill lamellae may also provide a physical barrier to the movement of the larger clumps through this region.

Cell aggregation appears effective in the removal of bacteria from the circulation (see Section 3.5.3) but there is, as yet, no conclusive demonstration of bacterial killing in such clumps and nodules. However, ultrastructural evidence is suggestive of bacterial breakdown (Smith and Ratcliffe, 1980a) as is the increase in acid phosphatase activity in C. maenas clumps (White and Ratcliffe, 1981). Although many of the hemocyte aggregates, in particular the loose clumps, may be cleared within the first day after injection (Smith and Ratcliffe, 1980a; see also Section 3.5.3), some of the clumps and nodules remain within the host for longer periods (Cornick and Stewart, 1968; Fontaine and Lightner, 1974; Johnson, 1976; Smith and Ratcliffe, 1980a). Melanized nodules have been observed in natural populations of C. maenas (White and Ratcliffe, unpublished), and may also occur in response to the introduction of bacteria during wounding (Fontaine and Lightner, 1973); thus, this phenomenon is not simply a laboratory-induced cellular defense reaction. As to the eventual fate of the nodules, in insects they may be lost at molt, but a similar fate has yet to be demonstrated in the Crustacea.

3.5.3. Overview of Bacterial Clearance in *Carcinus maenas*

Before considering the functions of nephrocytes and the cellular response of crustaceans to larger foreign objects, it is pertinent at this point to widen the discussion and consider, in detail, the clearance of bacteria in *C. maenas in vivo* in relation to the previously mentioned cellular defense reactions. A series of experiments carried out in our laboratory by Smith and Ratcliffe (1978, 1980a,b) and White and Ratcliffe (unpublished and 1981) will be discussed in this context.

Clearance of bacteria from the circulation is an extremely rapid process in the Crustacea. Smith and Ratcliffe (1980b) reported that of 1×10^8 bacteria injected into *C. maenas*, over 97% were removed from the circulation within 6 hr. Similar rapid clearance of foreign material has been reported in *Orconectes virilis* (Merril *et al.*, 1979), *Callinectes sapidus* (McCumber and Clem, 1977), *Homarus americanus* (Cornick and Stewart, 1968), and *Parachaeraps bicarinatus* (McKay *et al.*, 1969). This process is accompanied by hemocytopenia (Cornick and Stewart, 1968; McKay *et al.*, 1969; Tyson and Jenkin, 1973), which in *C. maenas* results in a fall in hemocyte numbers of some 90% within the first 30 min (Smith and Ratcliffe, 1980b). Evidence for a cause–effect relationship between changes in bacteria and hemocyte numbers is provided by observations on the *in vivo* clearance of live, radiolabeled *Moraxella* sp. from the circulation of *C. maenas* (White and Ratcliffe, 1981). When challenged with high doses of this bacterium (4×10^7/

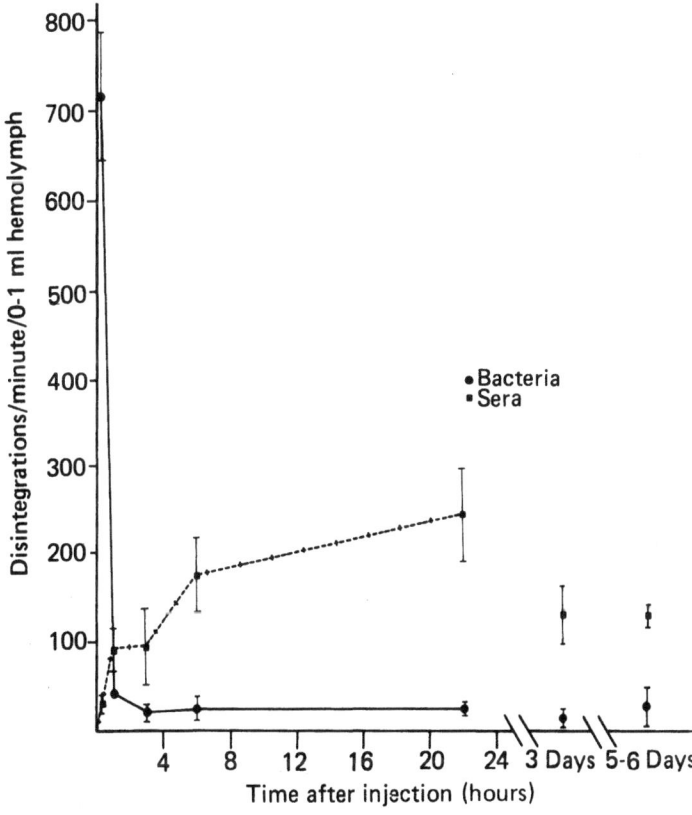

FIGURE 18. Level of radioisotope in the hemolymph of the crab *Carcinus maenas* following the injection of 4×10^7 radiolabeled *Moraxella* sp. The fraction of tracer in the sera is shown separately from that remaining in the bacteria. Vertical bars indicate the standard error of the mean.

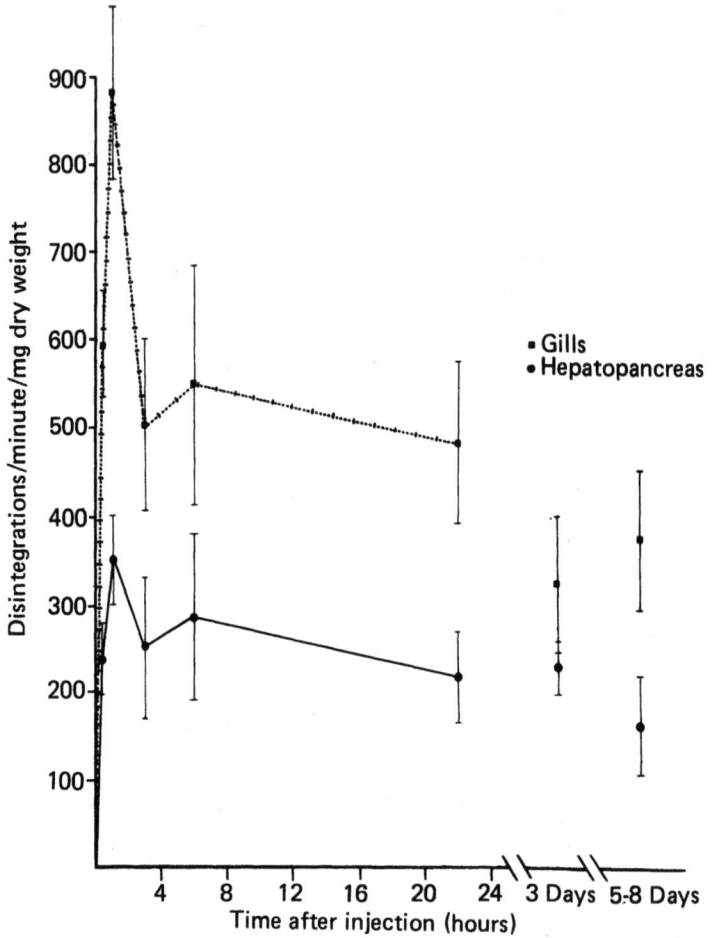

FIGURE 19. Level of radioisotope in the gills and hepatopancreas of *Carcinus maenas* following the injection of 4×10^7 radiolabeled *Moraxella* sp.

crab), rapid removal from the circulation (Figure 18) is correlated with the concentration of radioisotope in the tissues, particularly the gills, hepatopancreas, heart, subcuticular tissue, and gut within the first hour (Figures 19, 20). Autoradiographs show that the majority of the radiolabel (and hence bacteria) is localized within hemocyte clumps in these tissues, and also possibly in fixed phagocytes in the hepatopancreas. The absence of appreciable quantities of label in the serum during the first 4 hr indicates that bacterial breakdown products are not released into the hemolymph and that bacteriolytic factors are not involved. Some clearance of radiotracer occurs during the first 6 hr (Figures 19, 20) particularly from the gills where around 40% of the label present after 1 hr is lost by this time. This may reflect the breakdown of hemocytes and associated bacteria and particularly susceptible may be the loose cell clumps. Subsequent loss from the tissues occurs only slowly (Figures 19, 20); much of the label remains in the clumps and nodules for at least 18 days. Expressed in terms of amount of accumulated isotope per unit weight, the gills contain the most radiolabel, and also the highest number of aggregates. However, due to the large size of the hepatopancreas in *C. maenas*, the overall importance of this organ in bacterial sequestration is equal to that of the gills (Figure 21).

Given the importance of clump formation in the elimination of bacteria from the circulation in *C. maenas*, a logical question to ask is, to what extent is bacterial sequestration within these aggregates a result of phagocytosis? *In vitro*, the percentage of hyaline hemocytes ingesting bacteria is very low, only 5.3% for *Bacillus cereus* and 14.7% for *Moraxella* sp. even after 3 hr (Smith and Ratcliffe, 1978). Due to the ingestion of more than one bacterium, the number of *Moraxella* phagocytosed per 100 hemocytes is in excess of 23. However, given a hyaline cell population of 1.2×10^7/crab, the clearance of 95% (i.e., 9.5×10^7) of an injected load of 1×10^8 *Moraxella* (Smith and Ratcliffe, 1980b) is unlikely to be due to phagocytosis alone, unless the *in vivo* response is greatly enhanced. The importance of phagocytosis in the *in vivo* clearance of large numbers of bacteria is further diminished when it is borne in mind that over 70% (i.e., 7×10^7 *Moraxella*/ crab) are cleared from the circulation within the first 5 min postinjection (Smith and Ratcliffe, 1980b). *In vitro*, phagocytosis after 5 min is negligible and, *in vivo*, its contribution is unlikely to be so elevated as to account for the level of clearance recorded. Thus, association of bacteria with the hemocyte surface, followed by sequestration within cell clumps, may represent the major route of removal from the circulation in *C. maenas* [see Nodule Formation (insects), Section 5.6.2].

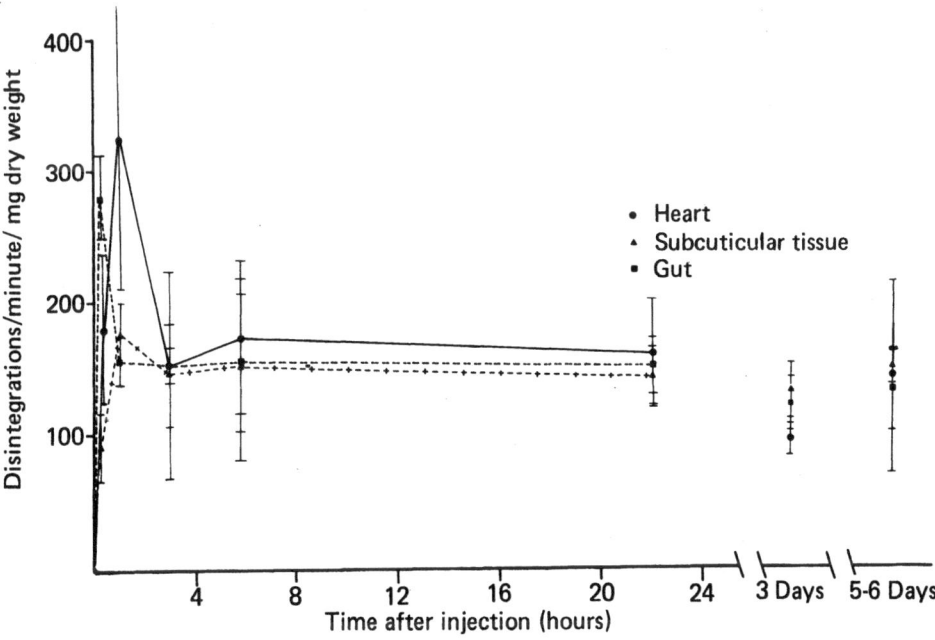

FIGURE 20. Level of radioisotope in the heart, subcuticular tissue, and gut of *Carcinus maenas* following the injection of 4×10^7 radiolabeled *Moraxella* sp.

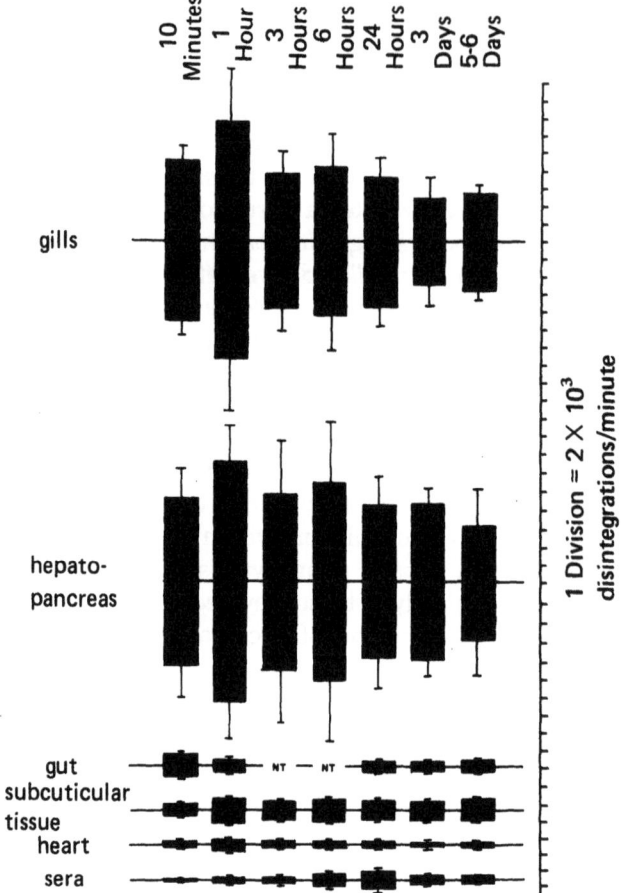

FIGURE 21. Total amount of radioisotope in each body fraction of *Carcinus maenas* at various time intervals following injection of 4×10^7 *Moraxella* sp., calculated as disintegrations per minute per total tissue weight. NT, not tested.

In conclusion, phagocytosis by hyaline cells, perhaps in conjunction with the sessile phagocytes, probably provides an adequate first line of defense against low numbers of bacteria, particularly as the host is unlikely to be challenged by the large numbers used in laboratory experiments. Extracellular association and cell clumping may provide a second line of defense if bacterial division occurs and the infection becomes more severe.

The role of humoral factors in combating infection is outside the scope of this review but merits a brief mention in the context of their relative importance vis-à-vis cellular defenses. Bactericidins are apparently absent from *C. maenas* (Smith and Ratcliffe, 1978) which emphasizes the importance of the cellular response to bacterial infection in this crab. Bactericidins, however, have been found in the sera of *Panulirus argus*, *P. interruptus*, and *Homarus americanus*

(Evans *et al.*, 1968, 1969a,b; Acton *et al.*, 1969). They are rapidly produced in response to infection but are relatively short-lived.

3.5.4. Nephrocytes

There is little doubt that in *C. maenas* these cells are actively involved in the removal of material resulting from hemocyte–bacteria interactions. Morphological changes induced by bacterial infection include an increase in the number of coated vesicles and peripheral electron-dense bodies, followed by a rise in the amount of material, probably originating from these peripheral organelles, in the central vacuole (Smith and Ratcliffe, 1981) (Figure 22). This is accompanied by the appearance of radioisotope, of bacterial origin, in these cells, particularly in the central vacuole (White and Ratcliffe, 1981). This no doubt reflects their general function in keeping the gills free of debris and so preventing the impairment of respiration.

Histochemical studies on *C. maenas* nephrocytes indicate that some of the peripheral bodies are lysosomal in nature with acid phosphatase and β-glucuronidase activity, while the central vacuole only contains β-glucuronidase. Following bacterial challenge, there is a marked increase in the β-glucuronidase activity of the central vacuole, which reaches a peak 1 day postinjection (White

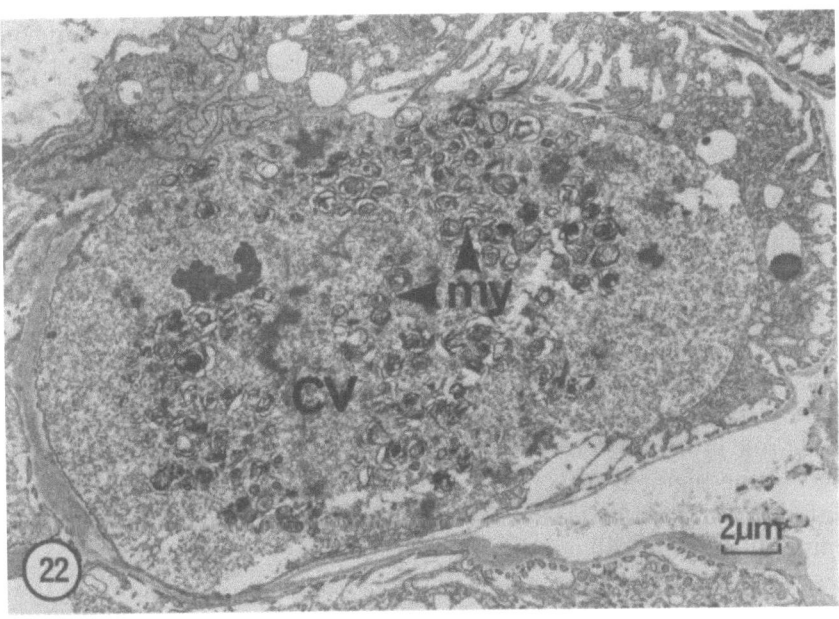

FIGURE 22. Nephrocyte, 12 hr after the injection of *Moraxella sp.*, containing many myelin configurations (my) in the central vacuole (CV). These myelin configurations are thought to be produced from the breakdown products of the bacteria/hemocytes in the gill lamellae nodules. (From Smith and Ratcliffe, 1980b.)

and Ratcliffe, 1981). This probably reflects the active catabolic processes resulting from the removal of material from the hemolymph and its sequestration within the central vacuole. There also remains the intriguing possibility that these cells are involved in the production of bactericidal enzymes for release into the hemolymph. Crossley (1972) demonstrated an increase in the production of lysozyme by the nephrocytes (pericardial cells) of the insect, *Calliphora erythrocephala*, following bacterial infection. Possibly, β-glucuronidase is released into the hemolymph by the crab nephrocytes.

3.5.5. Encapsulation

Foreign objects larger than ca. 10 μm in diameter are encapsulated rather than phagocytosed (Lackie, 1980). Cellular capsules are ubiquitous in the Crustacea (Unestam and Weiss, 1970; Unestam and Nylund, 1972; Sparks and Fontaine, 1973; Unestam, 1975; Hubert et al., 1976; Poinar and Hess, 1977; Bauchau and De Brouwer, in Bauchau, 1981). Commonly the capsule consists of multiple layers of flattened hemocytes, the innermost eventually becoming necrotic. Both hemocyte types are involved and, in addition, the presence of fibroblastlike cells has been reported in capsules formed in the shrimp, *Penaeus setiferus*, in response to the cestode, *Prochristianella penaei* (Sparks and Fontaine, 1973), although the latter may be transformed hemocytes (Bazin, 1979). The granulocytes may release substances that enhance cell–cell adhesion (see Section 5.6.6) and cell attachment may also be assisted by the secretion of a collagenlike material (Sparks and Fontaine, 1973; Hubert et al., 1976; Bauchau and De Brouwer, in Bauchau, 1981). The production of melanin around the foreign particle is a common feature, and deposition may also occur within the capsule cells themselves (Bauchau and De Brouwer, in Bauchau, 1981). The enzyme involved in melanin production, a phenoloxidase, and its substrate are both present in the granules of the granulocytes (Decleir and Vercauteren, 1965; Unestam and Nyhlén, 1974). Melanization may be correlated with the degree of resistance to fungal infection in various crayfish (Unestam and Weiss, 1970; Unestam, 1975), and Solangi and Lightner (1976) have shown that a *Fusarium* sp. isolated from a susceptible shrimp, *Penaeus californiensis*, induces little encapsulation and melanization, in contrast to the resistant shrimps, *P. setiferus* and *P. aztecus*, in which both are initiated. Despite the common occurrence of melanization in arthropods, its role in host defense is still in dispute. The action of phenoloxidase in producing the pigment may have a cytotoxic function (Taylor, 1969), and the melanin itself may act to "seal off" the parasite, preventing its access to oxygen and nutrients (Salt, 1970).

Encapsulation may eventually kill the parasite, and breakdown of the cestode, *Prochristianella penaei*, after encystment has been reported in the hepatopancreas of *P. setiferus* (Sparks and Fontaine, 1973), leaving a fibrous capsule of mainly host material. Capsules are sometimes brittle and further breakdown may be as a result of mechanical action by adjacent muscles or organs (Bauchau and De Brouwer, in Bauchau, 1981). Encapsulation, however, may not necessarily result in the death of the invading parasite (Hubert et al., 1976).

3.5.6. Hemolymph Coagulation and Wound Repair

In the Crustacea, hemolymph coagulation may occur in response to wounding and the introduction of foreign material (Bang, 1967; Rabin, 1970). Coagulation appears to be a two-stage process involving cell aggregation, followed by the formation of an extensive cellular mesh as a result of pseudopodial attachment to adjacent cells (Bang, 1967; Bauchau and De Brouwer, 1974). These events may or may not be accompanied or followed by plasma gelation (Tait, 1911). Both cell types are involved in coagulation (Tait, 1910, 1911; Wood et al., 1971; Bauchau and De Brouwer, 1974; Ravindranath, 1975a). Material released from the granulocytes is thought to induce both cell aggregation and plasma gelation (Hardy, 1892; Tait and Gunn, 1918; Bang, 1970, 1971; Ravindranath, 1975a) and may include a muco- or glycoprotein (Wood et al., 1971). In the lobster, in contrast to the vertebrates, a single enzyme appears responsible for initiating plasma gelation (Glavind, 1948; Duchâteau and Florkin, 1954). This is an intracellular transamidase (transglutaminase) (Lorand et al., 1966; Myhrman and Bruner-Lorand, 1970), and is released during hemocyte clumping or muscle damage, and catalyzes the formation of peptide links between a proto-clotting molecule (Lorand, 1972; Laki, 1972). The process of cell clump formation in response to bacterial infection bears certain resemblances to the initial stages of clotting in that cell necrosis and some granule release have been observed (Smith and Ratcliffe, 1980a). However, these processes do not proceed as far as in coagulation, where the cytoplasm is quickly reduced to a thin layer surrounding a large and degenerating nucleus (Bauchau and De Brouwer, 1974; Sternshein and Burton, 1980). That similarities exist between hemocyte–hemocyte and foreign body–hemocyte attachment is implied by the fact that the sticking and clumping of hemocytes on fungal hyphae (and hence encapsulation) and cell clumping during clotting are both inhibited by N-ethylmaleimide (Bang, 1967; Unestam and Nyhlén, 1974).

Injury to the external surface results in the initiation of clotting, which functions to seal the wound and prevent infection. Melanization of the clot occurs probably resulting in a stronger barrier (Bauchau, 1981), but possibly a bactericidal effect may also result from the action of the melanin precursors (see Section 3.5.5). Epidermal regeneration and subsequent cuticle formation are followed by the sloughing off of the melanized scar (Bazin and Demeusy, 1972; Fontaine and Lightner, 1973).

3.6. RECOGNITION OF FOREIGNNESS

Central to the survival and development of multicellular animals is the existence of a discriminatory system capable of distinguishing self from non-self and of recognizing damaged or effete self components (Boyden, 1963). In common with other invertebrates, it is apparent that in many crustaceans this recognition mechanism is partially mediated by a system of serum and cell-membrane-associated components with certain similarities of function (but not structure) to the immunoglobulin/complement molecules of vertebrates. In addition, other recognition factors may be released from the hemocytes following foreign body challenge.

3.6.1. Chemotaxis

Cell–foreign body contact mediated by a chemotactic response has been strongly implicated in the insects (see Section 5.6.6a), but to date this important aspect of the crustacean immune response has received scant attention. In this context, the role of the granulocytes could be important in that cell breakdown and degranulation may result in the release of chemotactic factors.

3.6.2. Humoral Factors and Receptors

Upon hemocyte–foreign body contact, phagocytosis may occur without the mediation of hemolymph factors (McKay and Jenkin, 1970a,b; Tyson and Jenkin, 1973, 1974; Paterson and Stewart, 1974; Paterson et al., 1976; Smith and Ratcliffe, 1978), although often only to a minimal degree. However, enhancement of phagocytosis by hemolymph factors (opsonins) may occur, although it is apparently not a universal phenomenon in the Crustacea. Increased phagocytosis by H. americanus hemocytes occurs in response to serum-incubated bacteria and erythrocytes (Paterson and Stewart, 1974; Schapiro et al., 1977). Serum-treated erythrocytes elicit a similar response in P. bicarinatus (McKay and Jenkin, 1970a,b), but despite an increase in the rate of clearance from the hemolymph after opsonization (Tyson and Jenkin, 1973), elevated phagocytosis of sensitized bacteria was not demonstrated in vitro (Tyson and Jenkin, 1974). There is evidence that the serum recognition factors are cytophilic (Tyson and Jenkin, 1974) and the differential response in P. bicarinatus may be a function of the size difference between the two test particles. The surface-bound opsonins might be sufficient for the mediation of bacterial uptake, but for the phagocytosis of the larger erythrocyte, additional recognition factors in the serum may be required (Tyson and Jenkin, 1974).

The lack of a phagocytic response to erythrocytes recorded in vitro in C. maenas (Smith and Ratcliffe, 1978) may be correlated with the apparent absence of serum opsonins in this animal. However, Schapiro et al. (1977) have shown that enhanced phagocytosis occurs after bacterial incubation in cell-free plasma, as compared with clot-free serum. They deduce from this that some of the low rates of phagocytosis reported in other papers may be due to incomplete opsonization. However, the lack of agglutinins, which could function as opsonins in C. maenas (Smith and Ratcliffe, 1978), suggests that the in vitro phagocytic rates are probably correct, as does the observation that P. bicarinatus plasma has the same opsonic titer as serum (Tyson and Jenkin, 1974).

The nature of the opsonic factors is unclear; there may be more than one class of compound capable of enhancing attachment and phagocytosis. Early work by Cantacuzène (1923) demonstrated the presence of hemagglutinins in the sera of various invertebrates, and it has been suggested that agglutinins may function as opsonins in invertebrates (McKay et al., 1969). Data from crustacean studies lend support to this thesis. Hemagglutinins have been found to enhance both the adhesion of erythrocytes to *P. bicarinatus* hemocytes (McKay *et al.*, 1969; McKay and Jenkin, 1970a) and phagocytosis (McKay and Jenkin, 1970a). Moreover, hemagglutinins are produced by *H. americanus* hemocytes and may be released into the serum (Cornick and Stewart, 1973, 1978). Attempts at achieving an opsonization effect using agglutinins from other invertebrates suggest that these factors may be phylum specific (McKay and Jenkin, 1970a). Significantly, the removal of hemagglutinin activity also results in the removal of opsonizing activity (Tyson and Jenkin, in Jenkin, 1976).

The ubiquitous nature of melanization in arthropods, in response to injury and infection, has led some workers to suggest an opsonizing role (Brewer and Vinson, 1971; Messner, 1972). Melanization can occur as a result of the release of phenoloxidase and substrate from the hemocytes into the serum (Unestam, 1975), and in the melanization response to fungal infection in crayfish, enzyme activity probably occurs by a combination of nonspecific attachment of the phenoloxidase to the hyphae and specific activation by the fungal cell wall glucans (Unestam and Söderhäll, 1977; Söderhäll and Unestam, 1979; Söderhäll *et al.*, 1979). Such enzyme activation by specific cell wall components would result in enhancement of the encapsulation response if hemocyte recognition is assisted by the process of melanization.

There is some evidence to suggest that hemocyte–foreign body contact may result in changes in the hemocyte recognition response to other self-tissues. Thus, McKay *et al.* (1969) observed that crayfish hemocytes having contacted particles have a tendency to attach to surrounding cells. Such alterations in cell–cell interactions may aid cell clumping and hemolymph coagulation.

Elucidation of the mode of action of opsonins and other recognition factors must await the characterization of hemocyte membrane receptors, of which little is known in the Crustacea. Trypsin treatment removes the membrane-associated opsonizing factors in the crayfish but does not destroy the membrane receptors (Tyson and Jenkin, 1974), although similar treatment did reduce hemocyte–foreign body interaction in the lobster (Paterson *et al.*, 1976). The observation of a differential rate of association and phagocytosis by *C. maenas* hemocytes challenged by unopsonized gram-negative and gram-positive bacteria (Smith and Ratcliffe, 1978) may be a result of the presence of more than one type of receptor. There also remains the intriguing possibility that, as in the lugworm, *Arenicola marina* (Fitzgerald and Ratcliffe, 1982), there may exist several hyaline cell subpopulations, specifically associating with certain types of bacteria. The observation that different types of foreign material are cleared to different body regions in *Callinectes sapidus* (McCumber and Clem, 1977) could also imply the presence of various cell subpopulations, and points to a relative subtle and sophisticated recognition system in the Crustacea.

3.6.3. Transplantation Reaction and Memory

We know of no transplantation studies carried out to determine the level of self–non-self recognition in the Crustacea. The tough exoskeleton and aquatic mode of life present practical difficulties, although preliminary experiments have shown that autografts are accepted by the sea slater, *Ligia oceanica* (Ratcliffe and P. M. Smith, unpublished).

Furthermore, no specific cell-mediated immunity has been demonstrated in the Crustacea, although in the crayfish and lobster a memory component induced by prior injection of live or dead bacteria, endotoxin, or antibiotics does result in an increased resistance to infection (McKay and Jenkin, 1969, 1970b,c; McKay *et al.*, 1973; Schapiro *et al.*, 1974; Stewart and Arie, 1974; Stewart and Zwicker, 1974a,b). The reaction is completely nonspecific and appears to act by inducing an increase in hemocyte phagocytosis (McKay and Jenkin, 1969, 1970b,c; Paterson *et al.*, 1976). This is reflected in the close relationship between phagocytic activity and the rate of survival of immunized *P. bicarinatus*, where an elevation in hemocyte numbers may also contribute to an increased killing response (McKay and Jenkin, 1970b,c).

The lack of a specific memory component in the Crustacea may be due to the absence of agglutinin induction by immunization (McKay and Jenkin, 1970c; Cornick and Stewart, 1973; Paterson *et al.*, 1976). If the hemagglutinins are an opsonic component, such a deficiency would deprive this group of a major potential specific defense system, and would account for the lack of inducible serum-opsonins in immunized crayfish (McKay and Jenkin, 1970b,c).

4. ONYCHOPHORANS AND MYRIAPODS

4.1. INTRODUCTION

Both numerically and in terms of their general importance, the Hexapoda (insects) overshadow the Onychophora and Myriapoda, and this is reflected in the paucity of information on the host defenses of the two latter groups. However, in view of the possible phylogenetic relationship of the onychophorans and myriapods to the other arthropods (Gupta, 1979a), further work in this direction may be of more than purely intrinsic interest.

The inclusion of the onychophorans within the phylum Uniramia (Table 1) is based largely on possible phylogenetic relationships (Manton, 1977). In many respects, i.e., the body wall structure, the nonjointed appendages, and the presence of nephridia, they betray annelid characteristics. There are only about 70 existing species of onychophorans, and these are restricted to tropical and southern temperate regions.

The myriapods are far more numerous, with some 10,500 known species, which are widely distributed in both tropical and temperate regions. In common with onychophorans, most myriapods require a humid environment such as beneath stones and wood, and in the soil. Of the four myriapod classes, the most numerous and well known are the Chilopoda, the centipedes, and the Diplopoda, the millipedes. When present in large numbers, certain millipedes can cause serious damage to cultivated plants (Baker, 1974). The Pauropoda are similar to the millipedes in appearance. They are small grublike animals and are mainly found in leaf mold and soil. The fourth class, the Symphyla, are interesting in having certain insectlike characteristics. One such symphylid, *Scutigerella immaculata*, can be a serious pest to crops.

4.2. CIRCULATORY SYSTEMS

In common with the Hexapoda, the onychophoran and myriapod circulatory system is open, with a dorsal tubular heart and, in the latter, an anterior aorta. Hemolymph is passed anteriorly along the heart to supply the hemocoelic cavities, from which it returns to the heart via a series of fissures, the ostia. An arterial system is absent in the onychophorans, nor is it well developed in the myriapods, although a cephalic arterial system is generally present and a posterior vessel may assist in the supply of hemolymph to the trunk. In the chilopods, hemolymph flow may be augmented by accessory pulsatory structures, comparable with those in certain insects. Gaseous exchange, either directly from the tissues or via the hemocoelic spaces, is by means of a tracheal system.

4.3. DISEASES

Little is known concerning the type of infectious diseases occurring in the Onychophora or Myriapoda. Their preference for humid environments and decaying vegetation is conducive to infection, but as far as we are aware, no disease of a bacterial nature has been reported in these groups, although both protozoan and fungal parasites have been recorded in the myriapods (Manier, 1954; Tuzet and Manier, 1955). The diplopod, *Narceus americanus*, can also serve as a natural intermediate host for the acanthocephalan parasite, *Macracanthorhynchus ingens*, whose definitive host is the raccoon (Crites, 1964). The myriapods are widespread and additional investigation may reveal many instances of this group harboring parasitic organisms, both as the primary and secondary host.

4.4. CELLS OF THE DEFENSE SYSTEM

There is a lack of information on cell-mediated responses to foreign materials in the Onychophora and Myriapoda, although there exist a number of studies on the hemocytes and nephrocytes (pericardial cells) of this group. For this reason, hemocyte structure will only be briefly reviewed; for a fuller account the reader is referred to Ravindranath (1981).

4.4.1. Hemocytes

Jones' (1962) system of insect hemocyte classification has been extended to the other uniramians by various authors (Gupta, 1968, 1979a; Vostal and Pir-

TABLE 2. A COMPARISON OF CERTAIN CHARACTERISTICS OF THE COMMON HEMOCYTE TYPES OF THE ONYCHOPHORA, MYRIAPODA, AND HEXAPODA (INSECTS) FROM PUBLISHED LIGHT MICROSCOPIC OBSERVATIONS

	Onychophora (*Peripatus acacioi*) (Lavallard and Campiglia, 1975)	Myriapoda–Diplopoda (*Thyropygus poseidon*) (Ravindranath, 1973)	Myriapoda–Chilopoda (*Scolopendra morsitans*) (Sundara-Rajulu, 1971b)	Hexapoda (various) (Jones, 1962; Price and Ratcliffe, 1974)
Prohemocytes				
Size (μm)	8.0–12.0	3.5–8.0	7.0–10.0	6.0–13.0
Shape	Round	Round	Round	Round/ovoid
Nucleus	Central	Central	Central	Central
Cytoplasm	Basophilic	Basophilic	—	Basophilic
Cytoplasmic inclusions	Absent/rare	Rare	Absent	Absent/few
Behavior *in vitro*	—[a]	Stable	Stable	Stable
Plasmatocytes				
Size (μm)	10.0–16.0	10.0–16.5	9.0–15.0	10.0–15.0
Shape	Round/ovoid	Ovoid/sometimes polymorphic	Round/sometimes polymorphic	Round/ovoid/spindle-shaped
Nucleus	Eccentric	Central/eccentric	Central	Central
Cytoplasm	Acidophilic	Basophilic	—	Basophilic
Cytoplasmic inclusions	Absent/rare	Rare	Variable	Variable, sometimes numerous
Behavior *in vitro*	—	Produce pseudopodia; vacuolation	Produce pseudopodia; some unstable	Produce pseudopodia; vacuolation
Granular hemocytes				
Size (μm)	10.0–12.0	15.0–19.0	11.0–14.0	10.0–17.0
Shape	Polymorphic	Polymorphic	Round/ovoid/polymorphic	Round/ovoid

Nucleus	Central	Central	Central	Central
Cytoplasm	Acidophilic	Basophilic	—	Acidophilic
Cytoplasmic inclusions	Numerous, dense acidophilic granules	Numerous dense, acidophilic or basophilic granules	Numerous dense, acidophilic granules	Numerous dense, acidophilic granules
Behavior in vitro	—	Produce pseudopodia; stable but vacuolation degranulation occurs	Produce pseudopodia; some stable, others lyse with granule release	Produce pseudopodia; degranulation occurs, with cell lysis
Spherule cells				
Size (µm)	8.0–30.0	10.0–35.0	15.0–30.0	8.0–16.0
Shape	Round/oval	Round	Round/ovoid	Ovoid/spindle-shaped
Nucleus	Central	Eccentric	Eccentric	Eccentric
Cytoplasm	Basophilic	Basophilic	—	—
Cytoplasmic inclusions	Numerous basophilic spherules	Numerous basophilic spherules	Variable no. of spherules	Numerous
Behavior in vitro	—	Stable	Degranulation occurs, some cells lyse	Stable
Cystocytes				
Size (µm)	—	12.0–17.0	—	9.0–14.0
Shape	—	Round	—	Round
Nucleus	—	Eccentric	—	Central
Cytoplasm	—	Weakly basophilic	—	Weakly basophilic
Cytoplasmic inclusions	—	Some acidophilic or basophilic granules	—	Some acidophilic granules
Behavior in vitro	—	Alteration in shape; cells lyse, with release of granules	—	Alteration in shape; cells lyse, with release of granules

[a] —, information not available.

còva, 1968; Sundara-Rajulu, 1970, 1971a,b; Sundara-Rajulu et al., 1970; Ravindranath, 1973, 1981). Notwithstanding the identification of as few as two and as many as five hemocyte types in the Onychophora (Manton and Heatley, 1937; Arvy, 1954; Grégoire, 1955; Tuzet and Manier, 1958; Sundara-Rajulu et al., 1970; Lavallard and Campiglia, 1975), four cell types appear ubiquitous in this group, namely prohemocytes, plasmatocytes, granular hemocytes, and spherule cells (Table 2). The fragile hemocytes described by Grégoire (1955) in various onychophorans could be cystocytes, although these cells may not be present in all onychophorans.

Of the myriapods, the hemocytes of the Pauropoda have not been studied, but work on the other classes categorizes between two and seven cell types (Duboscq, 1899; Bruntz, 1906a,b; Kollmann, 1908; Valeri, 1934; Bucherl, 1939; Palm, 1954; Tuzet and Manier, 1954; Grégoire, 1955; Grégoire and Jolivet, 1957; Shukla, 1964; Gupta, 1968; Kirshnan, 1968; Ravindranath, 1970, 1973, 1977a; Sundara-Rajulu, 1970, 1971a,b; Vostal, 1970; Krishnan and Ravindranath, 1973). Employing the insect terminology of Jones (1962), the following five hemocyte types are commonly found in the myriapods: prohemocytes, plasmatocytes, granular cells, spherule cells, and adipohemocytes (Ravindranath, 1981), although the latter may well be fat body cells. Cells similar to insect cystocytes may also occur throughout all three myriapod classes, although their presence in the Chilopoda requires substantiation. A seventh cell type, the oenocytoid, is present in the chilopods and diplopods but appears to be absent from the single symphylid examined to date (Gupta, 1968). Notwithstanding the many problems that accompany hemocyte classification, the extension of Jones' (1962) system to the Onychophora and Myriapoda has more validity than to the Crustacea. In the latter group, there is a dearth of information to substantiate such a scheme, but in the onychophorans and myriapods the various cell types are readily identifiable, and any criticisms leveled at such a classification could equally well apply to the insects. Thus, a close morphological similarity exists between onychophoran and myriapod hemocytes and those of insects. Table 2 compares various characteristics of the common hemocyte types in the Onychophora, Diplopoda, and Chilopoda with the description of similar insect cells, given in Jones (1962) and Price and Ratcliffe (1974). One difference lies with the degree of granulation of the plasmatocytes, although this is reflected in similar variations within the insects themselves, as is the difficulty in distinguishing plasmatocytes from granular cells.

The limited electron microscopic studies (Lavallard and Campiglia, 1975; Seifert; Seifert and Rosenberg; in Ravindranath, 1981) also reveal structural similarities between insect hemocytes and those of other arthropods, although the ultrastructure of the spherules in the spherule cells of the onychophoran, *Peripatus acacioi* (Lavallard and Campiglia, 1975), appears different from that of insects. Further ultrastructural studies on both onychophoran and myriapod hemocytes are required, particularly after challenge with non-self material.

The morphological relationship between the hemocytes of the Onychophora, Myriapoda, and Insecta may be of taxonomic significance; the similarity of hemocyte types between myriapods and insects is used by Gupta

(1979a) as evidence of a monophyletic relationship between these groups. Conversely, the absence of oenocytoids, adipohemocytes (circulating fat body cells?), and possibly also cystocytes in the Onychophora indicates to Gupta (1979a) a polyphyletic relationship to the myriapods and insects.

4.4.2. Hemopoietic Tissue

Hemopoietic tissue has been localized in the onychophoran, *Peripatopsis sp.* (Arvy, 1954), the diplopod, *Glomeris marginata* (Tuzet and Manier, 1954), and the chilopod, *Scolopendra* sp. (Duboscq, 1899), and lies adjacent to the heart or arterial vessels. Lack of information precludes a meaningful consideration of the development of the various hemocyte types, but it appears that the hemocytes may form a developmental series similar to that proposed for insects (Shrivastava and Richards, 1965). In *G. marginata*, differentiation of hemoblasts into other hemocyte types was observed by Tuzet and Manier (1954), but the problem is compounded by the fact that some of the hemocytes are capable of division in the circulation (Kollmann, 1908; Lavallard and Campiglia, 1975). Clearly, further ultrastructural investigations and tracer studies are required.

4.4.3. Nephrocytes

Pericardial cells (nephrocytes) are present in both onychophorans and myriapods. Studies by Campiglia and Lavallard (1975), Seifert and Rosenberg (1976, 1977), and Rosenberg (1978) reveal structural similarities of these cells both between these groups and within the Uniramia as a whole. They are attached, individually or in groups, to various body regions, including the fat body, muscle, connective tissue, and tracheae. Nephrocytes may also occur free in the hemolymph. In addition, the ecdysial glands of the myriapods are probably specialized nephrocytes (Seifert and Rosenberg, 1976). As in other arthropods, the presence of numerous coated vesicles, vacuoles, and lysosomes points to the active sequestration and modification of hemolymph components. The accumulation of various dyes in the nephrocytes (Bruntz, 1906b; Palm, 1954) is indicative of such uptake, although more refined confirmatory tracer studies are called for.

4.5. CELLULAR DEFENSE REACTIONS

Information pertaining to onychophoran and myriapod defense reactions is severely restricted, although there is some evidence for a phagocytic response, and hemocyte involvement in encapsulation and wound repair is documented.

4.5.1. Phagocytosis

To date, save for the early study of Kowalewsky (1894), no literature exists on the endocytosis of bacteria by the blood cells of the Onychophora or Myriapoda. Hemocytes have, however, been shown to accumulate various colloidal substances such as dyes and carbon (Kowalewsky, 1894, 1895; Duboscq, 1896; Bruntz, 1906a,b; Seifert, 1932; Palm, 1954; Ravindranath, 1973). In the diplopods, granular cells appear more active in the uptake of material than do plasmatocytes (Bruntz, 1906a,b; Seifert, 1932). However, there is often no distinction made between phagocytic and pinocytotic uptake.

4.5.2. Encapsulation

Encapsulation of the acanthocephalan parasite, *Macracanthorhynchus ingens*, by the diplopods, *Narceus americanus* and *Floridobolus penneri*, has been described by Bowen (1967). Concentric layers of hemocytes formed around the parasite but its growth was not impaired. Melanization was not observed.

4.5.3. Hemolymph Coagulation and Wound Repair

At the risk of repetition, it may be stated that information relating to wound healing is wanting in the Onychophora and Myriapoda. Hemocyte aggregation may help to prevent hemolymph flow, although in various onychophorans Grégoire (1955) noted that hemolymph was still issuing from a wound made 24 hr earlier. Cell clumping can be the only response to wounding in the onychophorans, chilopods, and diplopods, for no hemolymph coagulation occurs in these groups (Grégoire, 1955, 1970). From this, it is deduced that the clot-initiating substances are absent, despite the presence of granular hemocytes and possibly cystocytes. However, the release of material from cystocytes does induce clotting in the symphylid, *Scutigerella immaculata* (Gupta, 1968). The coagulation pattern was similar to the type II described by Grégoire (1957), in which the cystocyte forms numerous strands of cytoplasmic material.

5. INSECTS

5.1. INTRODUCTION

The insects have been classified most recently by Manton (1977) as arthropods belonging to the subphylum Hexapoda within the phylum Uniramia (Table 1). They are six-legged animals with three distinct body regions: the head, thorax, and abdomen, protected by a tough, chitinous exoskeleton, and are ramified by fine tubes, the tracheae, which transport the respiratory gases. Blood circulation is within a hemocoel in which the internal organs, bathed in hemolymph, lie. The insects usually possess hemopoietic tissue (Jones, 1970) and both free and fixed blood cells (hemocytes).

By virtue of their great versatility, the insects have colonized a large number of ecological niches which they occupy in huge numbers. They are relatively short-lived, but have incredible reproductive capabilities, thereby ensuring the survival of the whole species rather than of the individual. Such animals are unlikely to have developed sophisticated adaptive immune responses, but since many live in highly infected habitats, such as stagnant water and dung heaps, they must possess some sensitive and effective method of defending themselves against both macrobial and microbial invaders.

5.2. CIRCULATORY SYSTEM

The structure of the circulatory system has been described by Wigglesworth (1965), Jones (1977), and Clarke (1979), and consists of a hemocoel, which is partially divided by dorsal and ventral diaphragms, and a tubular, dorsal vessel, which maintains blood flow. The dorsal vessel can be separated into an abdominal portion, termed the heart, and the thoracic and cephalic aortae (Jones, 1977). The heart is composed of a variable number of chambers which open into the surrounding hemolymph via valved openings called ostia. Blood flows into the heart through the incurrent ostia and is then forced forward into the aorta by the rhythmic contractions of the muscles in the heart wall. In some cockroaches, lateral segmental vessels carry blood from the heart to the hemocoel (Nutting, 1951); however, in general, unlike the chelicerates and arachnids, no distinct arteries or veins are found associated with the heart. This simplification of the insect circulatory system is probably due to the development of the tracheal system to transport oxygen directly to the organs, so that a relatively low-pressure, sluggish circulation suffices.

In the peripheral regions, blood circulation may be aided by pulsatile organs called accessory hearts, which are found in many sites including near the bases of the legs, wings, and antennae. The undulatory movements of the ventral diaphragm also assist blood circulation.

5.3. INSECT DISEASES

In the natural environment of the insect, the cellular defenses are likely to be called upon to deal with a number of types of infective agents. Bacterial, fungal, protozoan, and viral infections have been responsible for complete annihilation of whole insect populations, and it is certain that some of these agents are permanently inherent in these animals. Insect defense reactions against all types of microbial infections have been researched extensively in recent years (Ratcliffe and Rowley, 1979a). The cuticle and gut barrier are extremely effective in preventing entry of pathogens, but once these are breached the cellular defenses are mobilized and combat infection by phagocytosis, nodule formation, and encapsulation. Insects also possess humoral factors such as agglutinins, lysozyme (Anderson, 1975), and melanin (Whitcomb et al., 1974; Nappi, 1975), which may act synergistically with the cellular defenses and oppose invaders, but these substances are beyond the scope of this review.

Bacteria frequently gain entry into the hemocoel by gut rupture following feeding of the insect on infected food, and small numbers are probably phagocytosed while larger numbers are sequestered in nodules. These cellular defense reactions against bacteria have been recorded both in laboratory-reared and natural insect populations (Bucher, 1959; Ratcliffe and Rowley, 1979a). Fungal infections of insects are also extremely common (e.g., Anderson and Magnarelli, 1980; Prior and Perry, 1980), with the pathogens entering the host mainly through the integument, either directly via wounds or by means of chitinase digestion of the cuticle. These fungal elements are combated not only by phagocytosis and nodule formation, but also by the production of cellular and melanized capsules (Götz and Vey, 1974). Protozoan and viral epizootics are also common in insects (Teakle, 1973; Brooks *et al.*, 1980), with the pathogens being ingested with the food and sequestered by phagocytosis and nodule formation. Finally, mention should also be made of the many parasitic worms and parasitoids which attack insects (Salt, 1970; Poinar, 1971; Cawthorn, 1980) and usually penetrate into the hemocoel actively (teeth, ovipositors), directly through the integument. These parasites have sophisticated means of avoiding detection in their habitual hosts, but in alien insects they rapidly become encapsulated and so may be deprived of essential metabolites and die.

5.4. INSECT HOST DEFENSES

These include the physico/chemical barriers, and the humoral and cellular mechanisms associated with the hemolymph.

The physico/chemical defenses consist of the tough, chemical-resistant and usually water-resistant exoskeleton, particularly seen in those species which spend relatively long periods as adults, and the so-called "gut barrier" of enzymes, bactericidal secretions, and cuticular and peritrophic linings, which thwart potentially pathogenic microorganisms taken in with the food.

The humoral defenses, as mentioned above, are mainly beyond the scope of this review (see Boman *et al.*, 1978), although they have been described where they influence the cellular defenses in some way.

The cellular defenses in insects have classically been defined as phagocytosis, nodule formation, encapsulation, coagulation, and wound healing, and are generally mediated by unpigmented, freely circulating blood corpuscles, the hemocytes. These originate from the hemopoietic tissue and/or by division of free cells and can give rise to humoral factors such as hemagglutinins (Amirante, 1976), and possibly also bactericidal substances. Pericardial cells (nephrocytes) are also present in insects and these together with the fixed hemocytes and phagocytic organs may be equivalent to the reticuloendothelial system of higher animals.

5.5. CELLS OF THE DEFENSE SYSTEM

5.5.1. Free Hemocytes (Figures 23–28)

The classification of insect hemocyte types has been the subject of a great deal of controversy, due to disagreement as to the nomenclature and origins of the cells. The latest review of hemocyte types by Gupta (1979b) provides a comprehensive appraisal of previous classification schemes, and Gupta (1979c) includes an identification key for the various hemocyte types. The problems encountered in hemocyte classification are also extensively discussed by Arnold (1979) and Jones (1979). Most authors, including Price and Ratcliffe (1974), who examined insects from 15 orders, have identified at least six categories of hemocytes. These are prohemocytes, plasmatocytes, granular cells, cystocytes or coagulocytes, spherule cells, and oenocytoids. Gupta (1979c) includes adipohemocytes in his classification scheme as a seventh type.

A detailed description of these cells and their functions can be found in Price and Ratcliffe (1974), Whitcomb et al. (1974), Gupta (1979a,b), Ratcliffe and Rowley (1979a,b), and Rowley and Ratcliffe (1981). Briefly, plasmatocytes, granular cells, and cystocytes (Figures 23–28) are the only cells to have well-known, specific roles in the cellular defenses. The plasmatocytes (Figures 23, 24), which

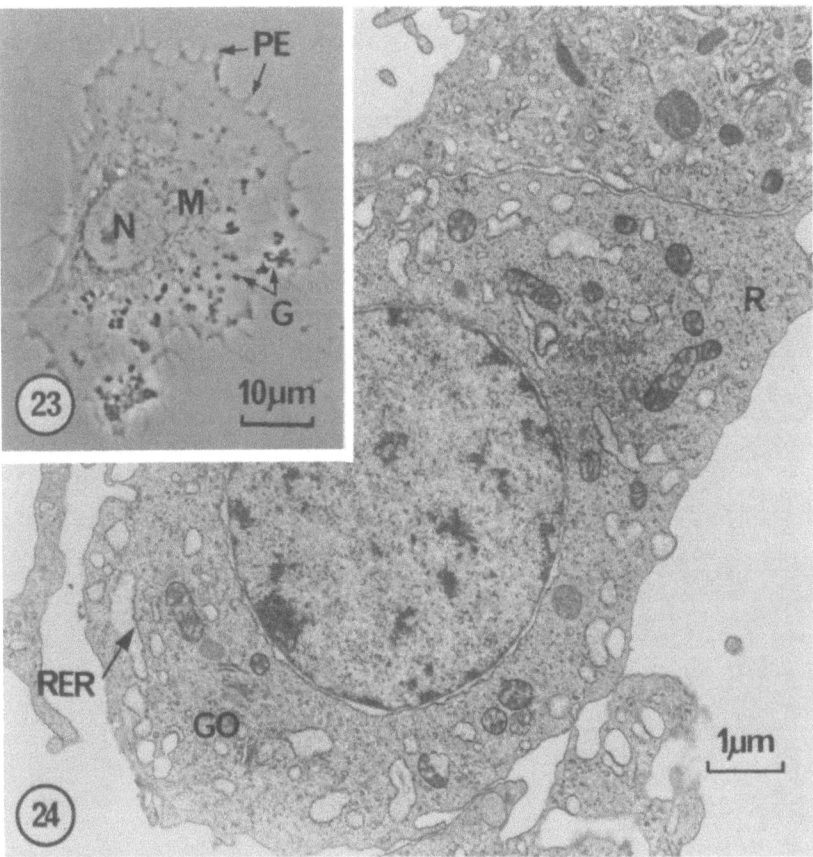

FIGURE 23. Phase-contrast micrograph of a plasmatocyte from *Periplaneta americana*. The nucleus (N) is surrounded by a cytoplasm containing granules (G) and mitochondria (M). Protoplasmic extensions (PE). (From Rowley, 1977b.)

FIGURE 24. Electron micrograph of a plasmatocyte from *Galleria mellonella*. The cytoplasm contains a poorly developed Golgi complex (GO), free ribosomes (R), and rough endoplasmic reticulum (RER).

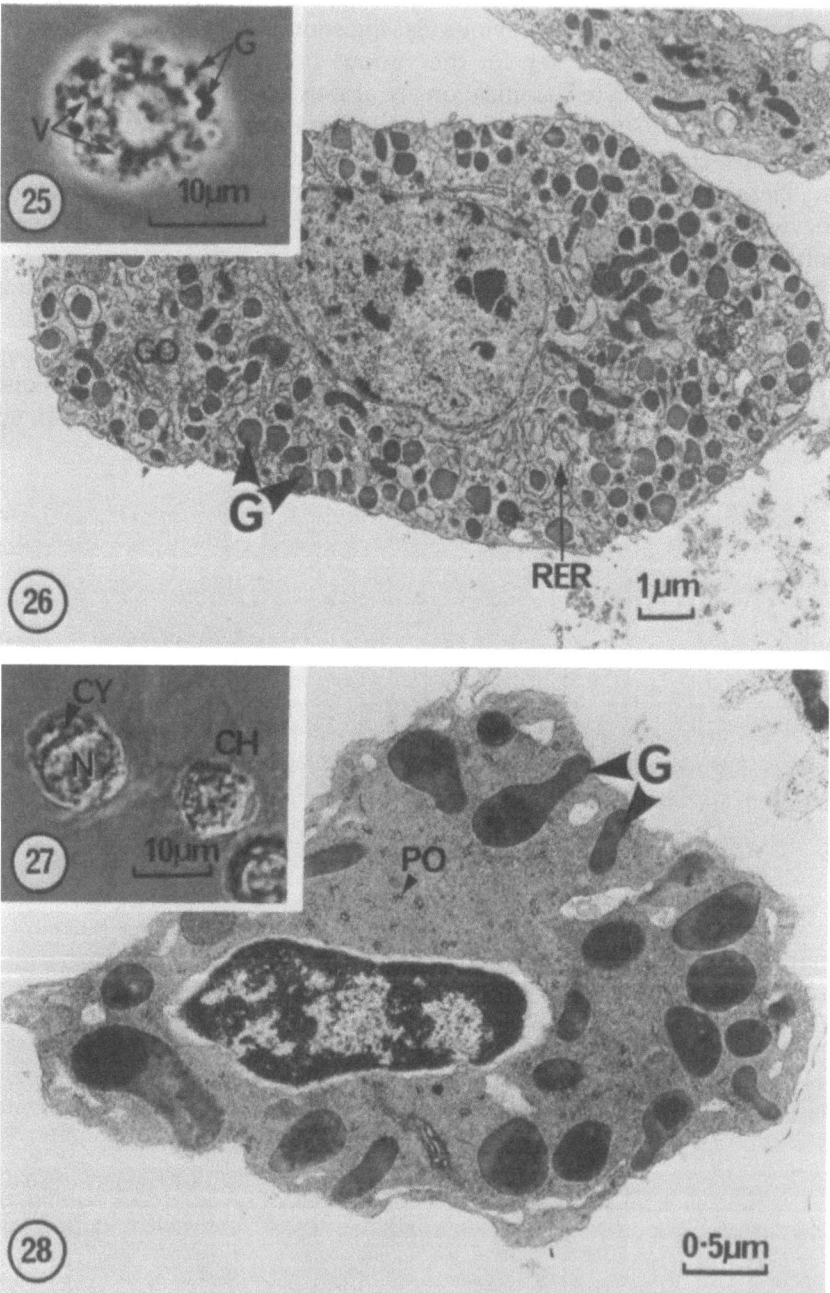

FIGURE 25. Phase-contrast micrograph of a granular cell from *Galleria mellonella*. Characteristically the cytoplasm contains many granules (G) some of which are released *in vitro* leaving vacuoles (V).
FIGURE 26. Electron micrograph of a granular cell from *G. mellonella*. Note the many electron-dense granules (G), a well-developed Golgi complex (GO), and swollen rough endoplasmic reticulum (RER) present in the cytoplasm.

are sometimes termed "insect macrophages," are usually spindle-shaped *in vivo*, 10–20 μm long, and *in vitro* rapidly form protoplasmic extensions and attach to the substrate (Figure 23). Depending on the species, the cytoplasm may contain a number of phase-dark granules which become more evident as the cells spread out *in vitro*. The granular cells are round or oval, 8–20 μm in diameter, and, like some plasmatocytes, contain cytoplasmic granules (Figures 25, 26). Indeed, the presence of these inclusions in both plasmatocytes and granular cells has led to some confusion in distinguishing these cell types; however, the granular cells, unlike the plasmatocytes, do not spread extensively *in vitro*. The cystocytes (Figures 27, 28) are 8–15 μm in diameter, round to oval, and are extremely fragile *in vitro*, so that the cytoplasmic granules are rapidly released; the cell then becomes surrounded by an "islet of coagulation" (Grégoire, 1970) and the nucleus becomes bright and prominent (Figure 27). The prohemocytes are thought to be the progenitor cells, from which some, at least, of the other cell types are derived (Shrivastava and Richards, 1965).

Both the spherule cells and oenocytoids are structurally, and probably functionally, distinct from the above cell types, and as yet have not been shown to play any major role in the cellular defenses. As well as the basic cell types, mentioned above, there are also a number of unusual hemocytes which have only been described in a few species. For example, Rizki (1962) found cells termed "crystal cells" in *Drosophila melanogaster*, and Zachary and Hoffmann (1973) discovered some unusual hemocytes, the thrombocytoids, in *Calliphora erythrocephala* which readily fragment *in vitro*. The thrombocytoids are probably equivalent to the podocytes described by Jones (1956), and in our opinion are best considered as a specialized type of plasmatocyte as they have similar phagocytic and encapsulation functions (Zachary *et al.*, 1975; Rowley and Ratcliffe, 1976b).

5.5.2. Sessile Hemocytes

In insects, not all the hemocytes making up the total cell population are freely circulating and a number of sessile hemocytes exist. Fixed hemocytes have been recorded at many sites in the body, including the corpora allata and cardiaca, Malpighian tubules, fat body, and muscle fibers (Wigglesworth, 1965; Scharrer, 1972). In some species, such as *Rhodnius prolixus*, they may account for the majority of the blood cells (Wigglesworth, 1965). These hemocyte reserves may be released under various physiological and stress conditions, such as wounding or infection, and so account for the great variations often reported in hemocyte numbers (Jones, 1962, 1979; Feir, 1979; Shapiro, 1979). To date, there

---

FIGURE 27. Nomarski interference micrograph of *Periplaneta americana* cystocytes which have discharged their granules causing the formation of a coagulum (CH) around these cells. During this process, the nucleus (N) becomes more prominent and the cytoplasm (CY) is reduced to a thin disorganized zone.

FIGURE 28. Electron micrograph of a cystocyte from *Clitumnus extradentatus* with characteristic granules (G) and polyribosomes (PO). (From Rowley, 1977a.)

is no reason to suppose that fixed hemocytes are structurally or functionally distinct from the freely circulating blood cells. Indeed, in the wax moth, *Galleria mellonella*, a two- to sixfold increase in circulating hemocytes can be induced by exposure of the larvae to acetic acid vapor, with most of the additional cells appearing identical in structure to free plasmatocytes (Walters, unpublished).

5.5.3. Hemopoietic or Fixed Phagocytic Organs

The structure and function of phagocytic or hemopoietic organs have been described in several insect orders and are reviewed fully by Jones (1970), Feir (1979), and Hoffmann et al. (1979). They have been recorded in various areas of the body of a large number of species (Jones, 1970), and also play an important part in the cellular defenses of many insects. Recent studies by Hoffmann and colleagues (unpublished) suggested that the hemopoietic tissue is of extreme importance in the attempts of *Locusta migratoria* to defend itself against invading pathogens (see Section 5.6, Cellular Defense Reactions).

Hoffmann (1970) categorized insect hemopoietic tissue into three main types: (1) groups of dissociated hemocytes, as described in *Musca domestica* (Nappi, 1974), some of which may be undergoing mitosis; (2) loosely packed blood cells, with little sign of zonation, but forming a distinct tissue in definite regions of the body, as reported in *L. migratoria* (Hoffmann, 1970), *C. erythrocephala* (Zachary and Hoffmann, 1973), and *Melolontha melolontha* (Brehélin, 1973). These tissues contain reticular cells which ingest foreign materials (Figures 29, 30) and may be the insect equivalent of the phagocytic reticular cells of the vertebrate lymphatic system (Hoffmann et al., 1974). (3) The third type of hemopoietic tissue is more highly differentiated, being divided into definite regions, and, in contrast to the two previous types, delimited by a membrane. Such well-developed hemopoietic organs are found in *Bombyx mori* (Akai and Sato, 1971), *Aglia tau*, *Antheraea pernyi*, *Manduca sexta* (Monpeyssin and Beaulaton, 1978; Beaulaton, 1980), and *Gryllus bimaculatus* (Hoffmann, 1970), and most closely resemble those found in the vertebrates (Hoffmann et al., 1979). The hemopoietic organs of *G. bimaculatus* are encapsulated by fibroblastlike cells and collagenlike fibrils. Inside the capsule is a cortical zone of reticular cells which undergo frequent mitoses to form islets of stem cells called hemocytoblasts. These cells then differentiate into recognizable hemocyte types, but only one cell type is formed per islet. The mature hemocytes congregate in the lumen of the organ from whence they are discharged into the dorsal vessel (Hoffmann et al., 1979). Monpeyssin and Beaulaton (1978) and Akai and Sato (1971) have described similar events in the formation and maturation of hemocytes in *A. pernyi* and *B. mori*, respectively.

Evidence for the role of the hemopoietic tissue in maintaining cell numbers has been obtained by irradiation (Hoffmann, 1972; Zachary and Hoffmann, 1973), ligaturing (Hinks and Arnold, 1977), and extirpation experiments (Beaulaton, 1978, 1980), all of which produced a large drop in numbers of circulating hemocytes. The maintenance of cell numbers is also achieved to a greater or lesser extent, depending on the species, by the mitosis of freely circulating

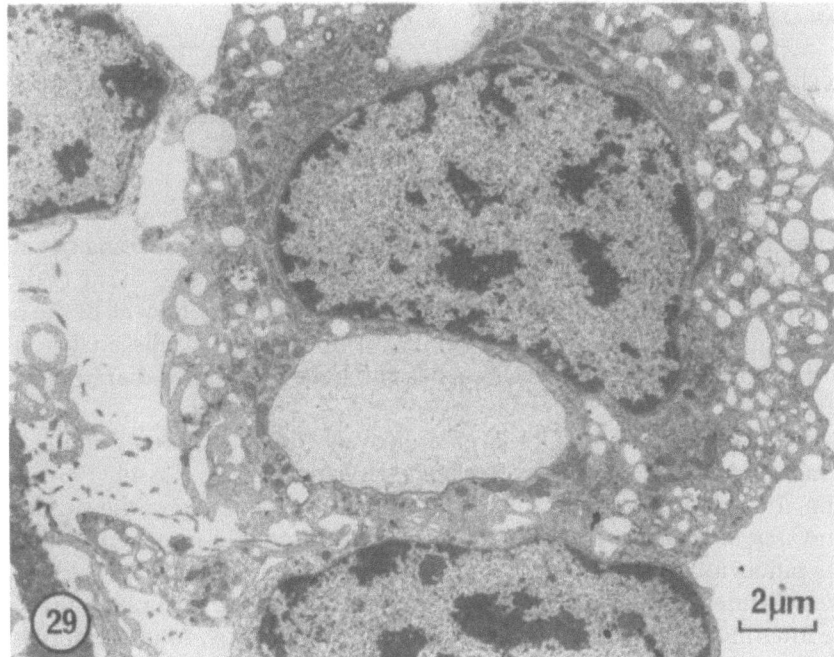

FIGURE 29. Reticular cell in the hemopoietic tissue of adult *Locusta migratoria*. (Courtesy of Dr. D. Hoffmann.)

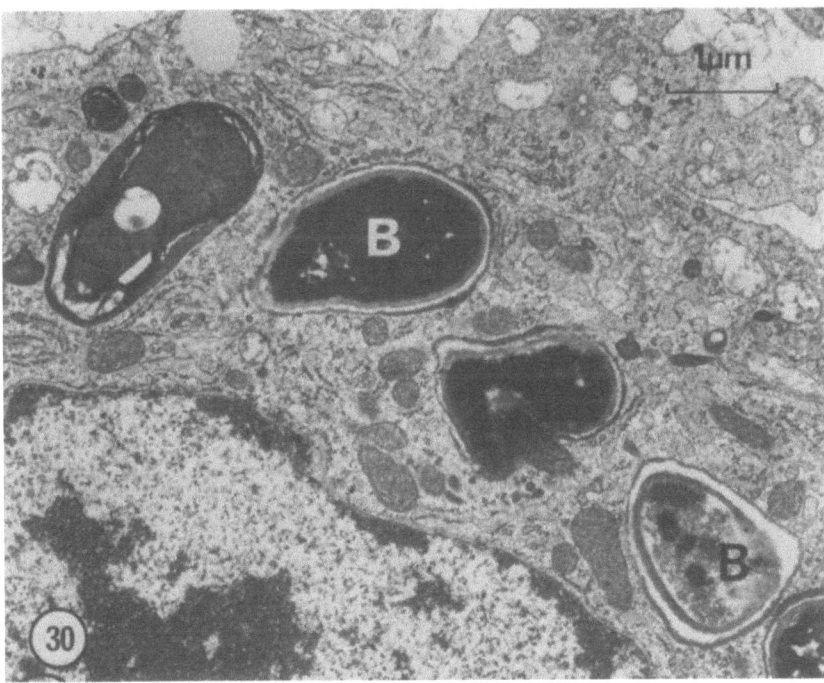

FIGURE 30. Intracellular *Bacillus thuringiensis* (B) in the cytoplasm of a reticular cell from *Locusta migratoria*, 6 hr after the injection of 5×10^3 bacteria. (Courtesy of Dr. D. Hoffmann.)

hemocytes (Jones and Liu, 1968, 1969). The functions of the hemopoietic/phagocytic organs in the cellular defenses will be discussed in detail below.

5.5.4. Pericardial Cells (Nephrocytes) (Figure 31)

These have been described by Wigglesworth (1970) as comprising the insect equivalent of the vertebrate RES. They are found in various sites in the body but are commonly present in the dorsal sinus, attached to the lateral wall of the heart between the alary muscles (Hollande, 1922; Jones, 1977). Cells analogous to pericardial cells are present in cockroaches and have been called diaphragm cells (Edwards and Challice, 1960).

Pericardial cells are usually large and oval, contain various granules and pigments, and are sometimes vacuolated (Hollande, 1922; Kessel, 1961, 1962; Brehélin, 1977). Crossley (1972) made an ultrastructural study of the pericardial cells in *C. erythrocephala*, and described an outer, infolded layer, enclosing numerous pinocytotic vesicles, surrounding an area of mitochondria, vesicles, and vacuoles. Inside this was a zone containing electron-dense vacuoles, surrounding the innermost core, which enclosed the Golgi apparatus, rough endoplasmic reticulum, and glycogen.

Pericardial cells are highly pinocytotic (Figure 31), and readily absorb small amounts of tracer materials, such as ammonium carmine particles, trypan blue,

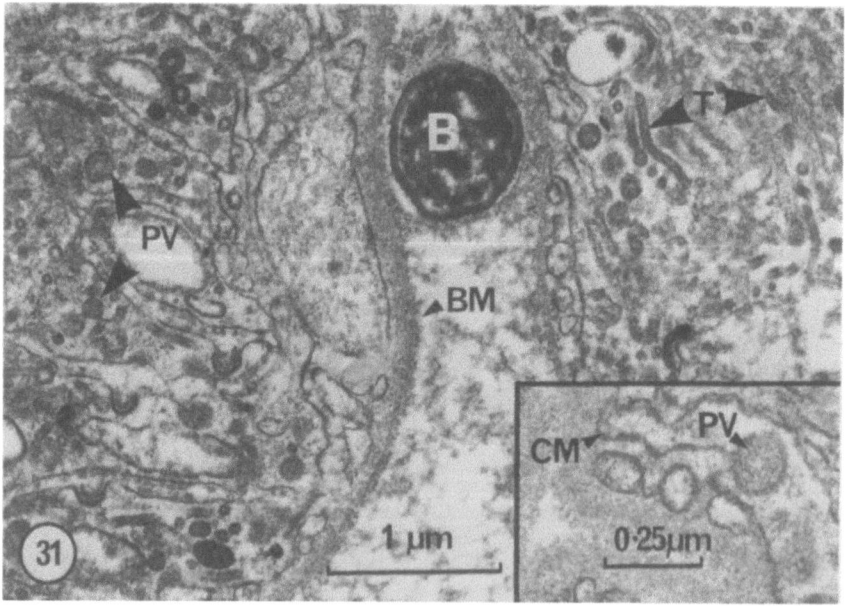

FIGURE 31. Part of the cytoplasm of two adjacent pericardial cells from *Tenebrio molitor* 10 min after the injection of *Bacillus cereus* (B). These cells are characterized by numerous pinocytotic vesicles (PV), tubules (T), and an outer basement membrane (BM). Note that the bacteria are not ingested by these cells. Inset shows a high-power electron micrograph of a portion of a *T. molitor* pericardial cell demonstrating the formation of a pinocytotic vesicle (PV) by infolding of the cell membrane (CM).

and egg white (Hollande, 1922). Cameron (1934) reported that the pericardial cells of *G. mellonella* were capable of ingesting bacteria; however, Crossley (1972) maintains that they are nonphagocytic but that a division of labor exists and that phagocytosis is carried out only by the circulating hemocytes.

The functions of the pericardial cells are discussed below, together with a comparison with the RES of vertebrates.

5.5.5. Oenocytes

These cells should not be confused with the true hemocyte type, the oenocytoid. Oenocytes are ectodermal in origin (Gupta, 1979b) and therefore not hemocytes since these are mesodermally derived. They are included in this review since, like the pericardial cells, they may take part in detoxification processes (Clark and Dahm, 1973), and form a component of the insect equivalent of the vertebrate RES.

Oenocytes are large cells usually found associated with the fat body or spiracles (Wigglesworth, 1965) or close to the epidermis (Locke, 1969). Their ultrastructure has been studied in *Calpodes ethlius* (Locke, 1969), *Culex pipiens* (Gnatzy, 1970), *Gryllus bimaculatus* (Romer, 1974), and *Oncopeltus fasciatus* (Dorn and Romer, 1976). Oenocytes are usually packed with smooth, tubular endoplasmic reticulum, contain two types of Golgi complexes and membranous and crystalline inclusions. These cells may take part in the detoxification process in *Musca domestica* (Clark and Dahm, 1973), in the manufacture of ecdysone (Locke, 1969; Dorn and Romer, 1976) which stimulates molting, and in the synthesis of hydrocarbons (Diehl, 1975) for the cuticle, during which they may interact with the oenocytoids (Wigglesworth, 1979).

5.6. CELLULAR DEFENSE REACTIONS

The cellular defense reactions of insects have been reviewed in detail most recently by Crossley (1975), Ratcliffe and Rowley (1979a), and Wigglesworth (1979). These include the processes of phagocytosis, nodule formation, and encapsulation, as well as hemolymph coagulation and wound healing. They are categorized merely for convenience in description and should not be thought of as mutually exclusive or totally separated, as they probably interact to a greater or lesser extent.

Coagulation and wound healing may be considered more as preventative measures, in that they check the influx of infective organisms following wounding, while phagocytosis, nodule formation, and encapsulation are the methods used by the insect to cope with invading organisms once they have entered the hemocoel.

5.6.1. Phagocytosis

5.6.1a. Phagocytosis by Circulating Hemocytes. Many papers have shown that insect hemocytes have the ability to phagocytose a range of inert particles and biological agents *in vivo* and *in vitro*, and these studies have been reviewed by Jones (1962), Salt (1970), Arnold (1974), Whitcomb *et al.* (1974), Ratcliffe and Rowley (1979a), and Rowley and Ratcliffe (1981).

Although *in vitro* systems are invaluable for dissecting out and studying the important components of phagocytosis such as the opsonins, it is clearly also desirable, if possible, to study this defense reaction *in vivo*. With insects, however, studies on phagocytosis *in vivo* have been hindered by rapid hemolymph coagulation, nodule formation, melanization reactions, etc., so that recent efforts have been directed towards establishing *in vitro* systems, and these have provided much valuable information about this process (Rabinovitch and De Stefano, 1970; Landureau *et al.*, 1972; Anderson *et al.*, 1973a,b; Anderson, 1974, 1975, 1976a,b, 1977; Ratcliffe and Rowley, 1975).

The main cell type involved in phagocytosis in most insects studied, both *in vitro* and *in vivo*, is the plasmatocyte (Figure 32) (Rabinovitch and De Stefano, 1970; Ryan and Nicholas, 1972; Anderson *et al.*, 1973a; Ratcliffe and Rowley, 1974, 1975). Neuwirth (1974), however, observed that the granular cells in *Calpodes ethlius* were phagocytic and capable of taking up thorium dioxide (Figure 33), India ink, and latex particles, whereas the plasmatocytes appeared to be nonphagocytic. Brehélin *et al.* (1978) and Brehélin and Hoffmann (1980) showed unequivocally that the coagulocytes of *Locusta migratoria* are also phagocytic *in vivo*. These authors maintain that discrepancies between their results and those of other workers were caused by differences between *in vivo* and *in vitro* techniques and between the nature of the test particles used. The reticular cells of the hemopoietic/phagocytic tissue have also been reported to be phagocytic (Brehélin and Hoffmann, 1980) (see Section 5.6.1b).

Various stages can be seen in phagocytosis and though these are not, of course, separated rigidly, they provide a convenient format for discussion.

Since "chemotaxis," "opsonins," and "attachment by receptors" are covered elsewhere in detail (see Section 5.6.6, Recognition of Foreignness), they will not be discussed in detail here. Suffice it to say that they form the essential first stages in the uptake and binding of foreign particles and that in insects there is

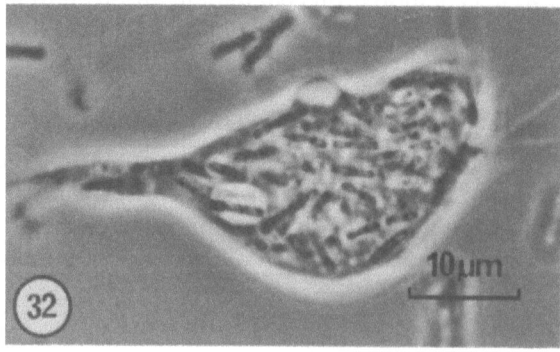

FIGURE 32. Phase-contrast micrograph showing a plasmatocyte of *Calliphora erythrocephala* containing numerous intracellular *Bacillus cereus*.

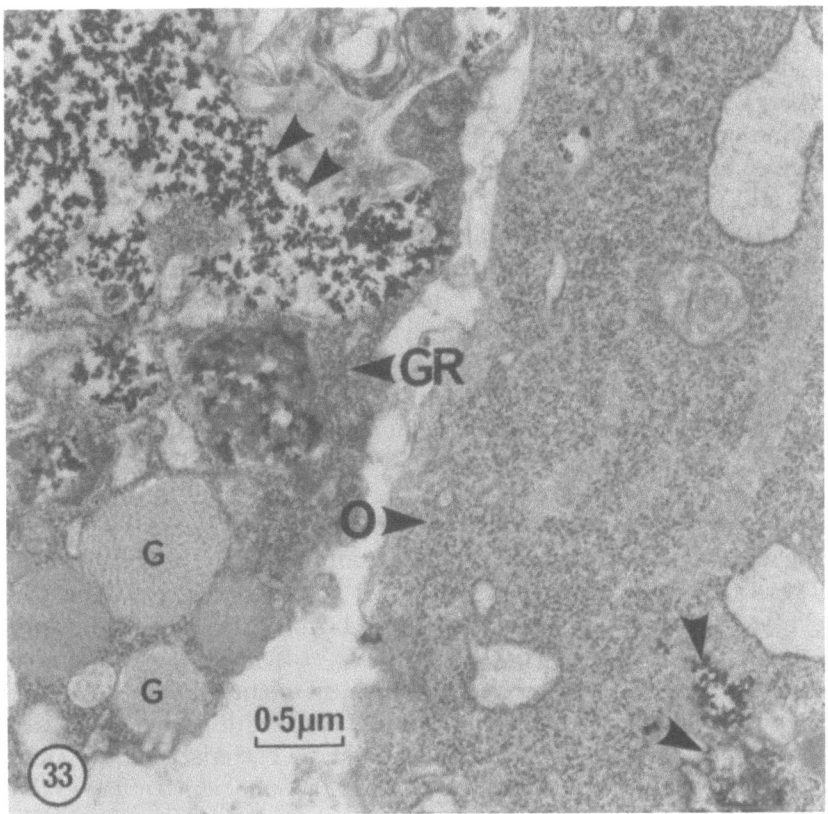

FIGURE 33. Granular cell (GR) and oenocytoid (O) containing phagocytosed thorium dioxide (arrowheads), 17 hr after injection of particles into *Calpodes ethlius*. Note the granules (G) in the granular cell. (From Neuwirth, 1974, reproduced by permission of The National Research Council of Canada from the *Canadian Journal of Zoology* Vol. 52, pp. 783–784, 1974.)

good evidence for the presence of chemotaxis but little for enhanced uptake of foreign particles presensitized in serum (Ratcliffe and Rowley, 1979a, 1982; Rowley and Ratcliffe, 1980).

As in vertebrates (Berlin and Oliver, 1978), microfilaments, and hence surface microvilli, appear to be necessary for the efficient binding of foreign particles to insect phagocytes (Anderson, 1976b). However, the addition of vinblastine and colchicine in culture has no effect on the hemocyte–erythrocyte rosetting reaction, so that microtubules can play no part in the initial attachment phase (Anderson, 1976a,b).

Ingestion follows binding to the phagocyte, and recent light microscopic and ultrastructural studies have clarified the details of this process in insects (Crossley, 1968, 1975; Vago and Vey, 1970; Ratcliffe and Rowley, 1974; Rowley and Ratcliffe, 1976b,c), and shown that it is similar to that seen in vertebrate leukocytes. During ingestion, the microorganism can be seen to be bound to the cell membrane, to become surrounded by fine cytoplasmic extensions which fuse at their tips to enclose the bacteria in phagocytic vacuoles. These phagosomes then become drawn into the cytoplasm of the cell and interact with other organelles.

Experiments by Anderson et al. (1973b) and Anderson (1974) have provided information on the biochemistry of ingestion by insect blood cells. These findings were compared with the results of experiments on vertebrate phagocytosis (Stähelin et al., 1956, 1957; Sbarra and Karnovsky, 1959; Iyer et al., 1961; Sbarra et al., 1971) during which oxygen consumption, glycolysis, and the activity of the hexose monophosphate pathway are all stimulated. Anderson et al. (1973b) and Anderson (1974) showed that phagocytosis of heat-killed *Staphylococcus aureus* or latex particles by *Blaberus craniifer* hemocytes was, as in mammalian leukocytes, accompanied by enhanced glycogen breakdown, lactate formation, and glucose flow through the glycolytic pathway. However, in contrast to these vertebrate cells, no increased activity of the hexose monophosphate pathway was recorded in this insect species.

Killing and digestion are probably the usual outcome of phagocytosis in insects, although, as in vertebrates, the phagosome may provide a protective envelope in which the pathogen can divide unhindered by antibacterial factors in the serum. From a number of reports, it is apparent that insects can kill intracellular microorganisms (Metalnikov, 1924; Hollande, 1930; Anderson et al., 1973a), but only the classic work of Anderson et al. (1973a) has attempted to quantify the process utilizing modern biochemical methods. They demonstrated the bactericidal activity of insect hemocytes *in vitro*, with *B. craniifer* hemocytes against a number of both gram+ve and gram−ve bacterial species. The plasmatocytes were the cells responsible and killed 25% of *S. aureus* 502A and *Streptococcus faecalis* after 10 min, and over 50% of *S. aureus* 502A and 45% of *S. faecalis* after 1 hr. Furthermore, *Serratia marcescens*, *Staphylococcus albus*, and *Proteus mirabilis* were also killed, but to a lesser extent, while *Pseudomonas aeruginosa*, *Escherichia coli*, *Salmonella typhosa*, and *Diplococcus pneumoniae* remained viable. Thus, *B. craniifer* cells have a bactericidal capacity but, since not all bacteria infecting them are killed, the effectiveness of *Blaberus* as a disease vector is increased (Anderson et al., 1973a).

Cytochemical and electron microscopic studies on vertebrate phagocytosis have been carried out and have shown that both primary and secondary lysosomes are involved in bacterial breakdown (Nelson et al., 1962; Leake and Myrvik, 1970; Armstrong and Hart, 1975). Similar investigations with insect phagocytes (Rowley and Ratcliffe, 1976b,c, 1979; Rowley, 1977b) have helped to elucidate the actual mechanism of bacterial breakdown following their internalization by the hemocytes of *C. erythrocephala*, *G. mellonella*, and *P. americana*. In *C. erythrocephala* (Rowley and Ratcliffe, 1976b), bacterial degradation occurs immediately following the fusion of primary lysosomelike organelles with the phagosomes containing the *E. coli*, and the whole process takes place within approximately 60 min. In *G. mellonella*, however, prolonged incubation (12–24 hr) of the plasmatocytes with *E. coli* resulted in only limited bacterial breakdown. Further experiments (Rowley, unpublished) involved overlaying *G. mellonella* hemocyte cultures with either *Staphylococcus aureus* or *Serratia marcescens*. Both of these bacterial species rapidly divided within the plasmatocytes and eventually caused the lysis of the cells. There was no indication of any interaction of the lysosomes with the bacteria. It has also been shown, in preliminary experiments with *G. mellonella*, that injection of the pathogen, *B. cereus*, enhances the level of acid phosphatase in the phagocytes at 10 min–6 hr postinjection and that this stimulatory effect is higher than that observed following the injection of the nonpathogen, *E. coli*. Thus, the *G. mellonella* hemocytes are capable of responding to the presence of bacteria but possibly this reactivity may only manifest itself in the *in vivo* state. Indeed, Rabinovitch and De Stefano (1970) have shown that *G. mellonella* plasmatocytes can ingest a greater range of bacteria *in vivo* than *in vitro*, and Möhrig and Schittek (1979) demonstrated the importance of lymphokinelike factors in the ingestion of *Bacillus thuringiensis in vivo*.

Finally, preliminary experiments involving the incubation of latex particles with *P. americana* plasmatocytes demonstrated the fusion of both nonlysosomal granules and smaller primary lysosomes with phagosomes containing the latex spheres (Rowley, unpublished). It is therefore possible that the granules may contain nonlysosomal killing factors such as cationic proteins or lysozyme. The granules of *P. americana* plasmatocytes have already been shown to contain chitinase (Landureau and Grellet, 1975), an enzyme not unlike lysozyme.

Not surprisingly, these studies show a remarkable similarity to ultrastructural observations of vertebrate phagocytosis, and close parallels can be drawn as regards lysosomal fusion and degranulation between the insect and the mammal.

Regarding the biochemistry of killing in phagocytic hemocytes, practically the only information available is from Anderson *et al.* (1973b). In vertebrate leukocytes, bacterial killing is due, at least partly, to oxidative metabolism since killing is decreased in anaerobic conditions (Holmes *et al.*, 1968). Baehner *et al.* (1968) and Klebanoff (1968) demonstrated the antimicrobial myeloperoxidase–H_2O_2–halide system in the phagolysosomes of vertebrate polymorphs. Klebanoff (1975) has also identified a number of other antimicrobial factors in polymorphs, including lysozyme, acid pH, superoxide anion, cationic proteins, and lactoferrin.

Anderson *et al.* (1973b) showed that the myeloperoxidase–H_2O_2–halide system is absent in the plasmatocytes of the cockroach, *B. craniifer*. Furthermore, they showed that the reduction of nitroblue tetrazolium (NBT) was also absent in *B. craniifer* hemocytes. In mammalian polymorphs, NBT is reduced to formazan by the H_2O_2 produced from the increased respiration associated with phagocytosis. Thus, the mode of bacterial killing by insect hemocytes is yet to be determined, although Anderson and Cook (1979) propose that bactericidal activity may be associated with lysozymelike enzymes, as in vertebrates (Biggar and Sturgess, 1977).

5.6.1b. Phagocytosis by Cells in the Hemopoietic Organs. The cellular defense role played by hemopoietic tissue was first recognised by Cuénot (1895, 1897) and has been confirmed by Hoffmann *et al.* (1974) and Brehélin and Hoffmann (1980). In recent studies by Hoffmann and colleagues (unpublished), lethal and nonlethal doses of *Bacillus thuringiensis*, a bacterial species not usually taken up by freely circulating hemocytes when injected into the hemocoel of *L. migratoria*, were cleared from the blood within 4 hr after the prior injection of a nonlethal dose of 5×10^3 bacteria per insect. No bacteria were seen in the circulating hemocytes, but intracellular bacteria were found in the reticular cells of the hemopoietic organs (Figure 30), and some of these cells subsequently

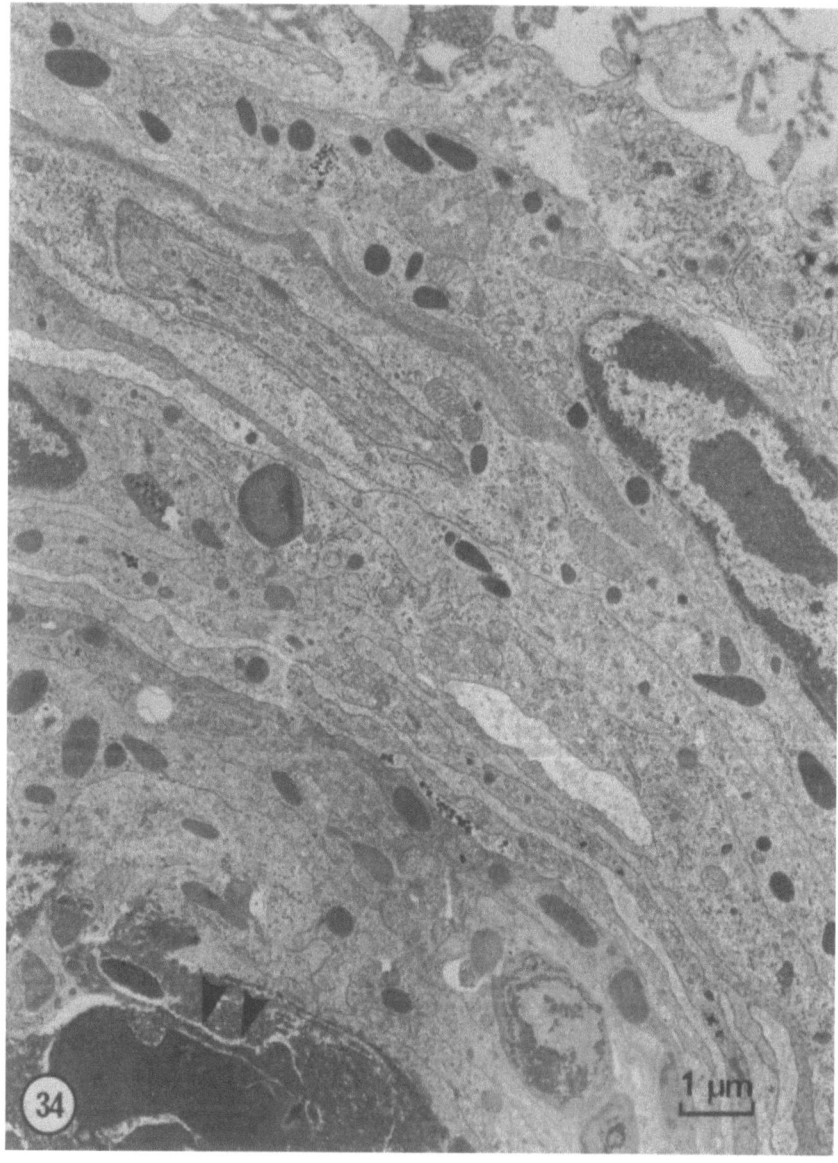

FIGURE 34. Capsule formation by granular plasmatocytes around necrotic reticular cells (arrowheads) and bacteria, 20 hr after the injection of 5×10^3 *Bacillus thuringiensis* into adult *Locusta migratoria*. (Courtesy of Dr. D. Hoffmann.)

became necrotic and enclosed in melanized capsules (Hoffmann et al., 1974, 1979) (Figure 34). Hoffmann et al. (1979) explain the immunization reaction in L. migratoria in two ways: first, the hemopoietic and phagocytic functions of the reticular cells in the hemopoietic tissue are stimulated, the former leading to the proliferation of circulating hemocytes more of which are now available for phagocytosis, and second, the hemopoietic tissue of Locusta may be producing a humoral antibacterial factor which is being secreted into the hemolymph (Hoffmann et al., 1979).

Hoffmann and colleagues (unpublished) in an ultrastructural study have examined the role of the reticulohemopoietic tissue of L. migratoria in this immunization process, and found, first, that phagocytic reticular cells become necrotic and are encapsulated by granular cells (granular plasmatocytes?) (figure 34), next, that hemopoietic differentiation is stimulated, and last, that the reticular cells show an increased protein secretion (Figure 35), with cisternae of rough endoplasmic reticulum filled with crystalloid inculsions of dense material (Hoffman et al., 1979; Brehélin and Hoffmann, 1980). It was therefore suggested that the reticular cells involved in increased protein synthesis could be producing proteinaceous antibacterial humoral factors, which render the insect immune after the first nonlethal dose of bacteria. This is further substantiated by irradiating the hemopoietic tissue, since after such treatment a normally nonlethal dose of B. thuringiensis will kill the insect, thus indicating the vital role of this tissue in the host defenses of L. migratoria.

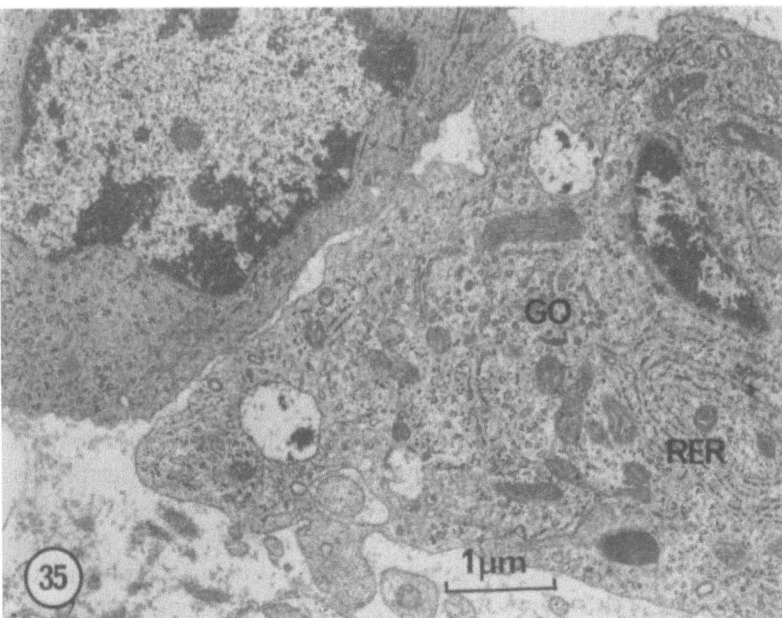

FIGURE 35. Reticular cell from a Bacillus thuringiensis-immunized Locusta migratoria containing large amounts of rough endoplasmic reticulum (RER) and an active Golgi complex (GO). (Courtesy of Dr. D. Hoffmann.)

Most studies on insect hemopoietic organs, however, report on their hemopoietic function (Jones, 1970; Feir, 1979). It is therefore suggested by Hoffmann *et al.* (1979) that the hemopoietic tissue has two main functions: namely, the production of mature and active hemocytes for release into the circulation (Feir, 1979; Hoffmann *et al.*, 1979), and the elimination of debris, unwanted hemocytes, and intruders into the hemocoel, thus maintaining a defensive role during the life of the insect. This latter role may include both phagocytosis and the production of antibacterial factors (Hoffmann and colleagues, unpublished). These workers also suggest that the hemopoietic tissues of certain insects and those of vertebrates are more similar than had hitherto been released, since immunization, as well as leading to increased protein synthesis by vertebrate reticular cells, also stimulates the transitional cells between reticular and plasma cells, so that following infection some reticular cells may become "immunologically" activated (Hoffmann *et al.*, 1979).

5.6.2. Nodule Formation

Nodule formation is another type of cellular defense mechanism recorded in a large range of insect species (Ratcliffe and Rowley, 1979a). This method of ridding the hemocoel of invading microorganisms is used when the amount of foreign material is too large to be cleared by phagocytosis alone.

The invading microorganisms form a clump with some of the blood cells and this whole structure becomes enclosed in a multicellular sheath of hemocytes. Salt (1970) described nodule formation as "clearly a mixture, in variable measure, of phagocytosis and encapsulation" during which plasmatocytes, having phagocytosed the invading organisms, form a clump which is encapsulated by other plasmatocytes, in the same way as insects react against large parasites (see Section 5.6.3. Encapsulation). Subsequent work by Gagen and Ratcliffe (1976), Ratcliffe and Gagen (1976, 1977), and Ratcliffe *et al.* (1976a,b) formation in *Galleria mellonella* and *Pieris brassicae*, however, has shown that the initial step in nodule formation is a clumping reaction of the granular cells together with the bacteria, within the first 5 min after injection of the microorganisms. This mass then becomes surrounded by a sheath of plasmatocytes and the central core melanizes (Figure 36).

The initial clumping reaction results from discharge of the granules from the granular cells so that a localized hemolymph coagulation occurs which entraps the microorganisms in a thin layer of extracellular coagulum. After only 5 min, the clumps compact and melanization begins; 2 hr later, plasmatocytes begin attaching to form a characteristic multicellular sheath. This latter process is so specific in *G. mellonella* and *P. brassicae*, i.e., only plasmatocytes are involved, that the lysing granular cells may release chemotactic factors to which only the plasmatocytes respond. This hypothesis was tested using excised nodules embedded in agarose and overlaid with plasmatocytes, and these cells were shown to move in a unidirectional manner towards the nodules (see Gagen and Ratcliffe, unpublished, reported in Ratcliffe and Rowley, 1979a, and also Section 5.6.6, Recognition of Foreignness).

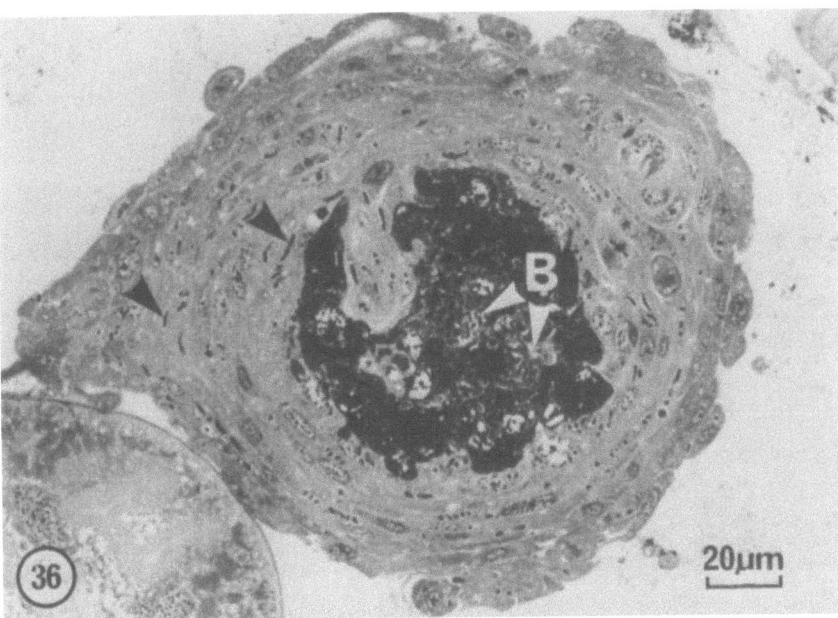

FIGURE 36. *Pieris brassicae* nodule 24 hr after the injection of *Bacillus cereus*. Note the typical structure, with central and melanized core containing bacteria (B) and sheath of plasmatocytes enclosing ingested bacteria (arrowheads). (From Ratcliffe and Gagen, 1976.)

Similar events in nodule formation have been reported for other insect species including *Tenebrio molitor*, *Schistocerca gregaria*, and *Clitumnus extradentatus* in laboratory experiments (Ratcliffe, Rowley, and Gagen, unpublished), and its importance as a natural defense reaction has been confirmed in field-collected insects by Bucher (1959), Huger (1960), and Henry (1967).

Recent work by Walters and Ratcliffe (1981) and Lacey (unpublished) has provided additional information about nodule formation. These workers have shown that nodules appear in insects not only in response to large doses of injected dead bacteria (Gagen and Ratcliffe, 1976), but also under more natural experimental conditions with much smaller doses of live bacteria (Walters and Ratcliffe, 1981) given to the insects *per os* Lacey (unpublished). These studies also indicate that nodule formation is not a nonspecific cellular reaction in insects since speed, size, and degree of clumping are very intimately related to the nature of the test bacteria.

Little is known about the fate of bacteria segregated in nodules although reports exist of microorganisms breaking free of these structures (Vey and Vago, 1969; Vey and Fargues, 1977). Experimental studies by Walters and Ratcliffe (1981) on the histochemistry of cells involved in nodule formation in *G. mellonella* have shown that acid phosphatase is present in the granular cell–plasmatocyte mass of nodules at 3–6 hr following injection of *B. cereus* and *E. coli*, but is later masked by melanin formation in the core. β-glucuronidase and β-glucosaminidase are also associated with these structures but at lower levels. The latter

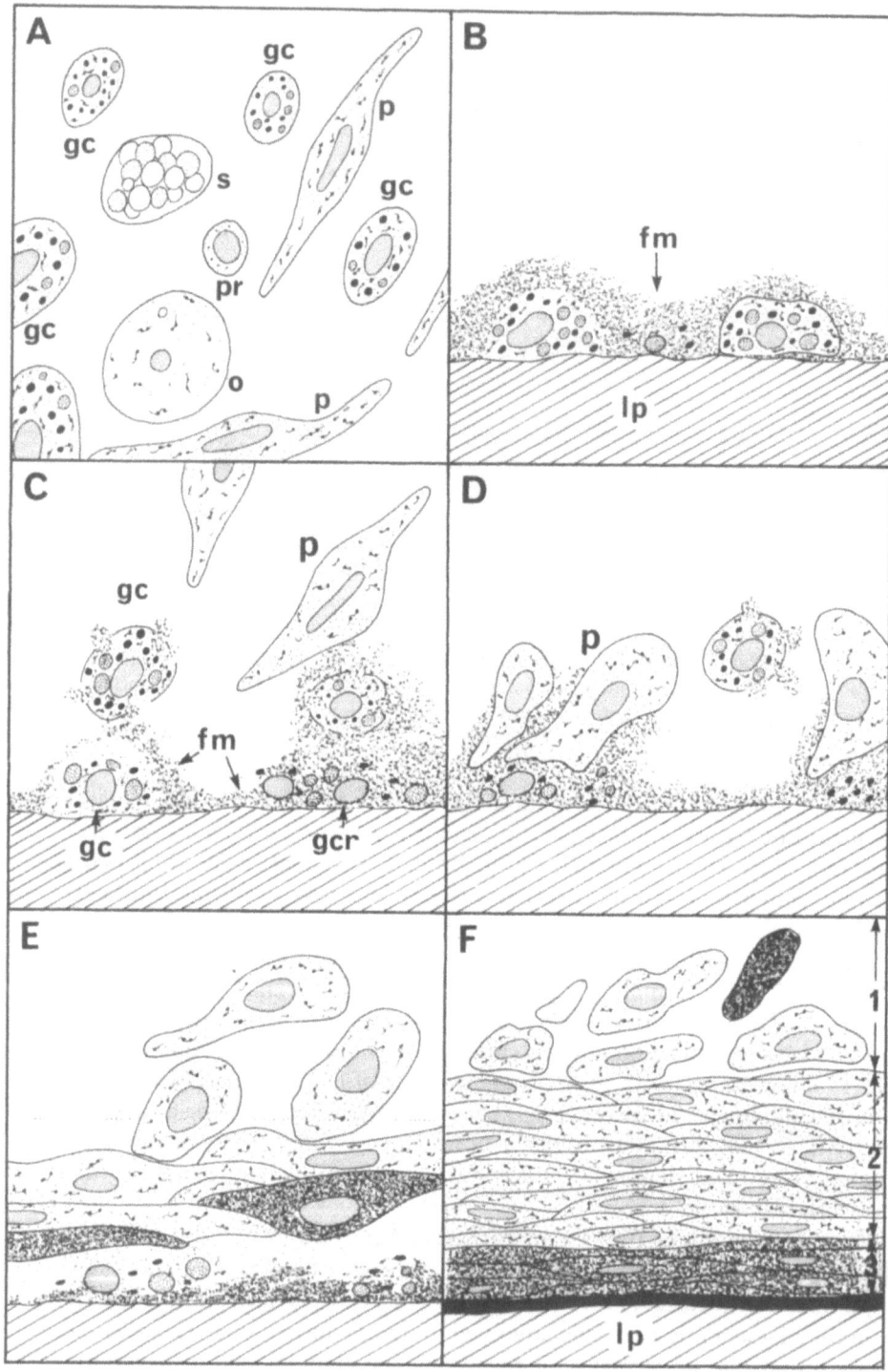

FIGURE 37. Diagrammatic representation of the main events during encapsulation in *Galleria mellonella*. (A) Hemocyte types before encapsulation commences: prohemocyte (pr), plasmatocyte (p), granular cell (gc), spherule cell (s), oenocytoid (o). (B) Initial stage of encapsulation approximately 5 min after the implantation of a piece of Araldite (Ip) into the hemocoel. Granular cells degranulate on contact with the implant and release flocculent material (fm) which coats the Araldite. (C–D) (30–60 min) More granular cell (gc) degranulation occurs with some cells breaking down

two enzymes are bacteriolytic, as well may be the melanin or its precursors (Taylor, 1969), so that some bacterial breakdown is possible within the clumps.

Initial experiments with ^{3}H-labeled *B. cereus* and *E. coli* injected into *G. mellonella* larvae (Walters and Ratcliffe, 1981) have attempted to determine the fate of sequestered bacteria and have shown that some killing does indeed occur within nodules. With *B. cereus* (a pathogen of *Galleria*), the surviving bacteria multiply after a period of 8–12 hr, and eventually swamp the hemolymph and kill the insect.

The low incidence of nodules reported from field-collected species (e.g., only 8% of grasshoppers contained these structures; Bucher, 1960) could be due to the fact that a slightly higher dose than that which stimulates nodule formation would kill the insect, so that these bodies represent a final attempt to contain overwhelming infections. Nodules could also, once formed, be lost by reabsorption or by sloughing off with the cuticle at molt. Preliminary work (Walters and Ratcliffe, 1981) has shown that after *B. cereus* injection into final-instar *Tenebrio molitor* larvae, nodules are formed but that following pupation they are absent from the adults. This is similar to observations made by Metalnikov (1924), who reported that nodules attached to the body wall and were discharged by rupture of the cuticle. Ermin (1939), however, recovered nodules from adult *G. mellonella* after their formation in the larva.

Another area which requires further study is the mechanism(s) by which some microorganisms resist segregation in nodules. Kurstak *et al.* (1969) and Gagen and Ratcliffe (1976) have stated that the pathogenic bacterial strains in their experiments blocked this cellular defense reaction. Kurstak *et al.* (1969) suggested that the avoidance of such sequestration may be due to the production of proteolytic enzymes or lipopolysaccharides. However, in other studies, Walters (unpublished) has shown that *B. cereus*, a pathogenic species in *Galleria*, stimulates faster and greater nodule formation compared with the reaction elicited by a similar dose of *E. coli* K12, a nonpathogen.

5.6.3. Encapsulation

Encapsulation is a form of cellular defense similar in many ways to nodule formation, but in this case, the multicellular sheath is formed around an invading organism, too large to be phagocytosed by a single plasmatocyte (Figure 37).

There are various agents of both biological and nonbiological origin which will become encapsulated when introduced into the hemocoel (Rowley and Ratcliffe, 1981); biological ones include protozoa, fungi, cestodes, nematodes, and insect parasitoids (Salt, 1970; Götz and Vey, 1974; Kitano, 1974; Poinar, 1974; Nappi, 1974, 1975; Lackie, 1976; Vinson, 1977; Ratcliffe and Rowley, 1979a). Encapsulation is also brought about during gut invasion by pathogens (Ka-

---

(gcr); plasmatocytes (p) begin to attach specifically to the sites where cell lysis has occurred. (E) Early stages (approx. 2–10 hr postimplantation) in the attachment and flattening of plasmatocytes over the now degenerated granular cells. (F) Mature capsule some 72 hr postimplantation with the characteristic sheath composed of three distinct regions: (1) outer region of newly attaching cells; (2) middle region of flattened cells; and (3) inner region of partially flattened cells, some melanized and necrotic with a layer of melanin on the innermost surface. (Based on Schmit and Ratcliffe, 1977.)

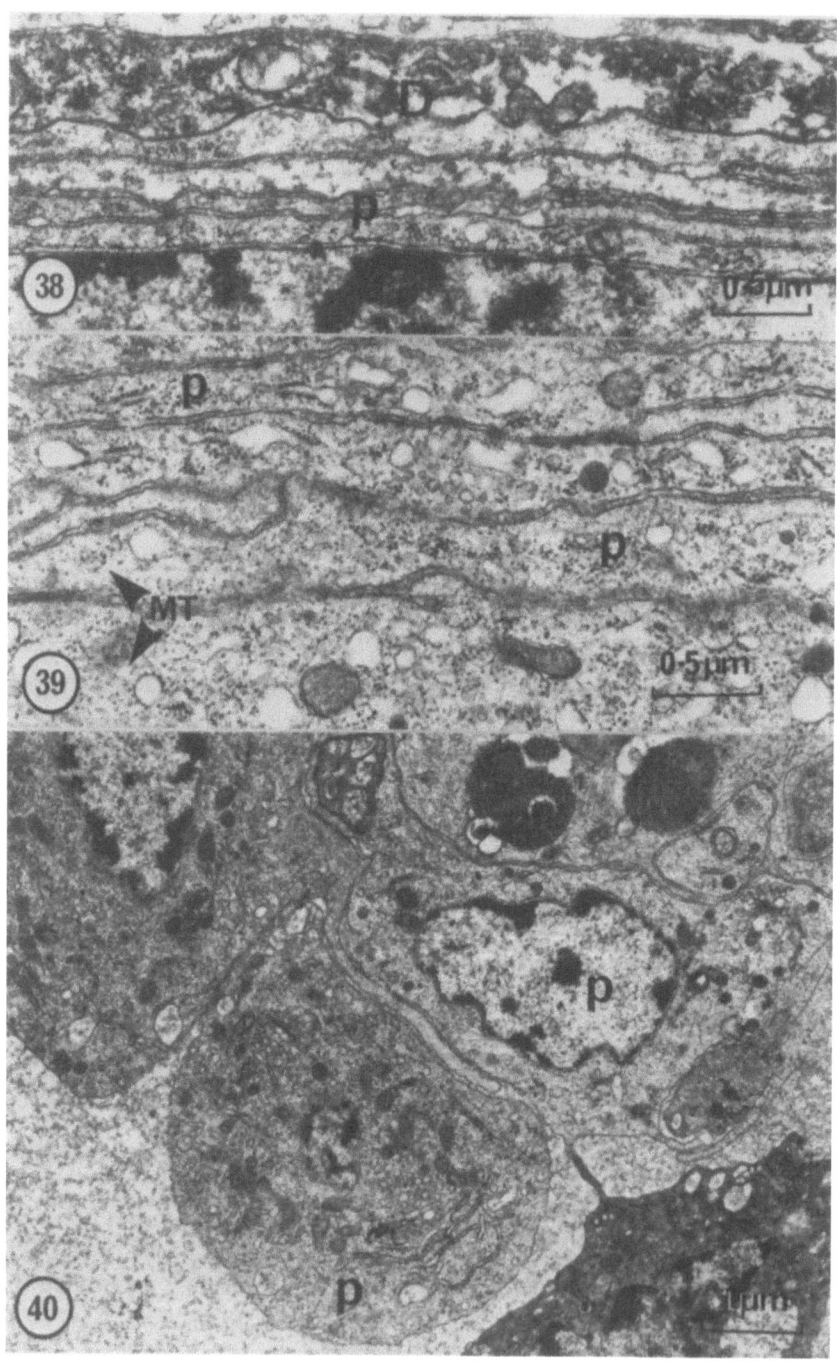

wanishi *et al.*, 1978; Splittstoesser *et al.*, 1978) and during tumorigenesis (Rizki and Rizki, 1976). Nonbiological agents encapsulated include pieces of Araldite, glass, nylon threads, and latex spheres (Salt, 1956, 1957; Lackie, 1976; Sato *et al.*, 1976; Schmit and Ratcliffe, 1978; Schmit, 1979).

The cell types which have been reported to take part in cellular encapsulation are reviewed by Ratcliffe and Rowley (1979a) and they also provide a more detailed appraisal of the process. Many workers (e.g., Grimstone *et al.*, 1967; Poinar *et al.*, 1968; Bréhélin *et al.*, 1975; Schmit and Ratcliffe, 1977, 1978) have studied the ultrastructure of capsules and have shown that there is often a common and distinct form, regardless of the type of implant or species of insect. There are often three regions: (1) an inner one of necrotic plasmatocytes and melanin adjacent to the implant; (2) a middle layer of flattened plasmatocytes showing few signs of degradation; and (3) an outer region of loosely bound plasmatocytes consisting of cells morphologically identical to the circulating hemocytes (Figures 38–40) (Grimstone *et al.*, 1967). Not all capsules, however, have a cellular, multilayered structure. Sheath capsules (Walker, 1959; Nappi and Streams, 1969; Nappi, 1970; Salt, 1970) in *Drosophila*, for example, consist of a thick layer of melanin and a few attached hemocytes, and a completely noncellular type of capsule derived from the blood cells of *P. americana*, after parasitization by *Moniliformis dubius*, has been described by Mercer and Nicholas (1967).

Schmit and Ratcliffe (1977) studied the structure of capsules formed around pieces of *Schistocerca gregaria* nerve cord implanted into the hemocoel of *G. mellonella* larvae (Figures 37–40), and for the first time described in detail the all-important early stages in the process (see Section 5.6.6, Recognition of Foreignness). The mechanism of encapsulation has also been described by Sato *et al.* (1976) and Schmit and Ratcliffe (1978). It was shown that encapsulation is divided, like nodule formation, into two distinct phases. First, contact and lysis of the granular cells and/or cystocytes take place rapidly on the surface of the foreign implant, and result in localized "clot formation" on certain sites of the implant surface, which become covered with cell debris and coagulum. Subsequently, after 20–30 min, plasmatocytes attach specifically to these areas of hemocyte degradation and gradually an outer sheath of plasmatocytes is built up (Figure 37) (Schmit and Ratcliffe, 1977, 1978). These observations suggest that products of granular cell breakdown control the attachment of the plasmatocytes. Bréhélin *et al.* (1975) have also described coagulation around cellophane im-

---

FIGURES 38–40. Characteristic cells in three regions of a 72-hr capsule of *Galleria mellonella*. (From Schmit and Ratcliffe, 1977.)

FIGURE 38. Inner necrotic region with characteristic granular cell remnants (D) close to implant (just off the top of the micrograph) and flattened plasmatocytes (p).

FIGURE 39. Middle region of capsule with extremely flattened plasmatocytes (p) with prominent microtubules (MT).

FIGURE 40. Outer region of capsule consisting of unflattened newly attaching plasmatocytes (p).

plants in *L. migratoria* and *Melolontha melolontha,* and Crossley (1975) has postulated that the release of a "recognition signal," by "disrupted labile hemocytes," during parasitization of *Calliphora* sp. could lead to the subsequent accumulation of more hemocytes.

Control of termination of encapsulation is not yet understood, but one hypothesis suggested by Ratcliffe and Rowley (1979a) and Rowley and Ratcliffe (1981) is that the concentration of the products of cell lysis gradually decreases as the wall of enveloping plasmatocytes increases in thickness, until eventually free hemocytes are no longer drawn into the chemotactic gradient around the capsule.

Studies have also been made of the role of the endocrine system in the control of encapsulation (reviewed by Nappi, 1974, 1975). It was noted that encapsulation of parasites by *D. melanogaster, Orthellia caesarion,* and *Musca domestica* leads to similar changes in hemocyte morphology and number as seen during the pupation process of control insects. Since the latter process is known to be under the control of juvenile hormone and ecdysone (molting hormone), it was postulated that encapsulation could also be. Nappi (1975) also studied the effect of ligaturing the endocrine center of *Drosophila algonquin* from the rest of the body during parasitization with *Pseudeucoila bochei,* and reported a decrease in encapsulation. Lynn and Vinson (1977), however, did not find that juvenile hormone or molting hormone analogs influenced encapsuation in *Heliothis virescens* or *H. zea.* Furthermore, Condon and Gordon (1977) studied the larvae of the blackfly, *Prosimulium mixtum fuscum,* and found no evidence for stimulation of endocrine activity during mermithid parasitization.

Whether capsules actively kill or merely prevent parasites from multiplying and then invading the rest of the body is unknown. If killing does occur, then, as in nodules, melanin and its toxic by-products may be involved (Taylor, 1969; Brewer and Vinson, 1971; Nappi, 1973; Crossley, 1975; Cawthorn and Anderson, 1977; Schmit *et al.,* 1977; Ratcliffe and Rowley, 1979a) or the parasite may simply be asphyxiated and starved (Salt, 1959, 1970; Wigglesworth, 1959).

Finally, it is well known that habitual parasites develop successfully in insects and are somehow able to block the encapsulation process. This may be due to the production of encapsulation-inhibiting factors either free in the hemolymph (Kitano, 1969a,b; Nappi and Streams, 1969; Kitano and Nakatsuji, 1978; Osman, 1978) or attached to the outside of the parasite (Salt, 1965; Rotheram, 1967, 1973a,b; Brennan and Cheng, 1975; Bedwin, 1979a,b), which suppress or prevent a host response. Alternatively, the surface of the parasite may mimic the host tissues in some way so that it is not recognized as non-self (Lackie, 1976). Lackie (1980) has recently reviewed the whole subject of recognition of non-self in insects and discusses the evasion of encapsulation by parasites.

5.6.4. Hemolymph Coagulation and Wound Repair

Hemolymph coagulation has recently been extensively reviewed by Grégoire (1970), Crossley (1975, 1979), and Rowley and Ratcliffe (1981). This defense reaction is found throughout the Arthropoda, but varies in importance from one group to another. Hemolymph coagulation serves to seal off the wound site following injury, thus preventing both the influx of bacteria into the hemocoel and excessive bleeding. The involvement of hemocytes in hemolymph coagulation has been reviewed (Grégoire, 1970, 1974; Grégoire and Goffinet, 1979; Rowley and Ratcliffe, 1981). There appear to be two methods by which blood cells instigate coagulation; in the first, a gelation process occurs involving discharge from cystocytes and granular cells (Rowley and Ratcliffe, 1976a; Rowley, 1977a), while in the second the blood fails to gelate but modified plasmatocytes, such as thrombocytoids or podocytes, agglutinate and interconnect to seal the wound (Crossley, 1975).

Wound repair has been studied in detail in *T. molitor* and *Ephestia kühniella* (Ries, 1932), *Rhodnius prolixus* (Wigglesworth, 1937, 1973; Locke, 1966; Lai-Fook, 1968, 1970), and *G. mellonella* (Rowley and Ratcliffe, 1978). In this latter study, it was shown that the process of wound healing begins as soon as 10 min after wounding. Degeneration and granule release by granular cells result in hemolymph coagulation, which, together with fat body and other extruded tissues, seals the wound. After 1–2 hr, a similar reaction to nodule formation is seen so that the wound compacts and melanizes, then, after 6 hr, a large number of plasmatocytes attach to the wound area and flatten out (Figure 41), so that by 12 hr postwounding, the structure resembles a capsule. It is suggested that the plasmatocytes are under the influence of a "wound factor" produced by the

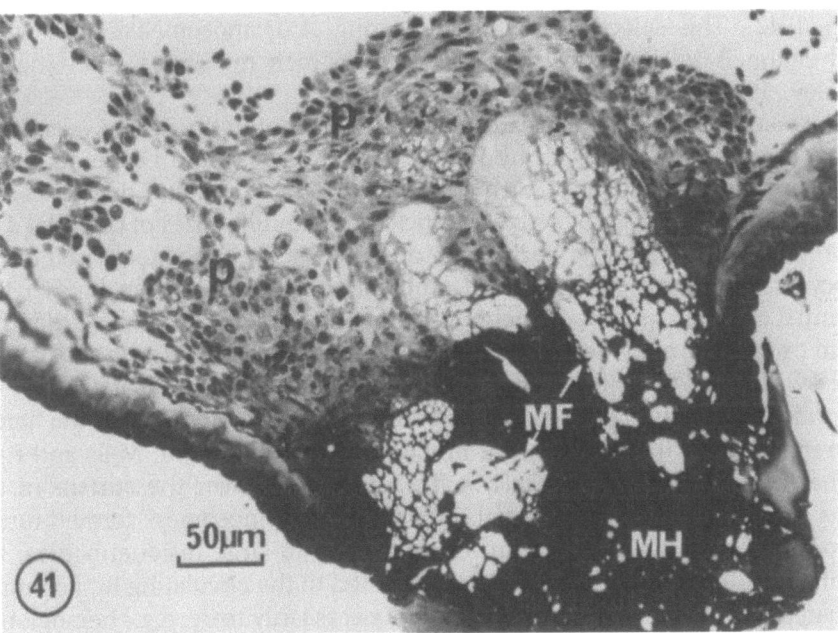

FIGURE 41. Section through a cuticular wound site in *Galleria mellonella* 6 hr after wounding, showing the massive influx of plasmatocytes (p) into the area, to form a sheath similar to that seen during nodule formation and encapsulation. The melanized fat body (MF) and hemolymph (MH) form an amorphous scab which protects against microbial invasion. (From Rowley and Ratcliffe, 1978.)

degranulating hemocytes or epidermal cells (Harvey and Williams, 1961; Cherbas, 1973; Bohn, 1975). Another aspect of wound repair which is not fully understood is whether the hemocytes have a role in repair of the new endocuticle and basement membrane (Lazarenko, 1925; Wigglesworth, 1973) or whether this process is undertaken solely by the epidermal cells (Wigglesworth, 1937; Lai-Fook, 1968; Ashhurst, 1979). Bohn (1977a,b,c) has postulated that epidermal cell migration is under the control of a "conditioning factor," produced mainly by the plasmatocytes which come to line the wound area during healing. This factor may result from the release of plasmatocyte granules as observed *in vitro*.

5.6.5. Detoxification and Secretion by Pericardial Cells

The ultrastructural study of Crossley (1972) described the organelles present in the pericardial cells of *C. erythrocephala*, and reviewed the previous work (Kessel, 1961, 1962; Bowers, 1964; Porter *et al.*, 1967).

The infoldings of the plasma membrane, the system of channels, and the many pinocytotic vesicles in the cytoplasm reflect the main function of these cells, which is to absorb various colloidal substances and detoxify the blood (Hollande, 1922). Crossley (1972) also showed that small tracer molecules, for example, horseradish peroxidase, introduced into the hemocoel are hydrolyzed within lysosomal vacuoles, but this uptake of material from the hemolymph is not random and it appears that there is some selection of molecules for pinocytosis. Pericardial cells mainly take up particles within a size range of 16–20Å (Lison, 1942). This selection process, occurring at desmosomelike sites which guard channels leading into the cells, limits which molecules enter the cell (Crossley, 1972).

The comparison of the pericardial cells of insects with the RES of vertebrates was hypothesized in the 1930s by Poll (1934) and Grassé and Lesperon (1935), and is supported by Wigglesworth (1943, 1970). It appears that the pericardial cells of insects take up smaller-sized particles and limited doses of colloidal dyes and proteins, but that larger particles or doses are actively phagocytosed by the hemocytes. Since the early work of Balbiani (1886), it has been known that pericardial cells detoxify the hemolymph. Hollande (1922) also showed that they contain proteolytic enzymes with which to solubilize absorbed toxic substances. Wigglesworth (1970) believes that the pericardial cells of *R. prolixus* have an active role in detoxifying the insect's hemolymph after a blood meal, the hemoglobin ingested by these cells being broken down to biliverdin. Mills and King (1965) also observed in *Drosophila*, active pinocytosis over the surface of the plasma membrane of the pericardial cells, with the "pinosomes" formed fusing with primary lysosomes, which in turn gave rise to large residual bodies, the detoxified contents of which were then returned to the circulating hemolymph. The materials pinocytosed, however, are not necessarily toxic, e.g., hemoglobin, but may yield useful substances after breakdown.

More recent studies by Brehélin and Hoffmann (1980) described the localization of iron saccharate in the pericardial cells of *L. migratoria* for up to 3 weeks postinjection. This substance was present both in cytoplasmic vacuoles and in the cisternae of the rough endoplasmic reticulum. These authors suggest that the existence of iron in the rough endoplasmic reticulum might be related to the detoxification function of the pericardial cells, although its presence could also be a normal feature of cells engaged in the synthesis of iron-containing proteins.

Another function assigned to pericardial cells is lysozyme synthesis (Crossley, 1979). The maximum lysozyme concentration of uninfected *C. erythrocephala* occurs simultaneously with degranulation of pericardial cells (Crossley, 1972). Stimulation of lysozyme production also follows the infection of *C. erythrocephala* with *E. coli* and *B. cereus*. It is, however, unlikely that the synthesis of this enzyme is activated by the ingestion of intact invading bacteria since pericardial cells do not usually sequester such large particles.

5.6.6. Recognition of Foreignness

In all the types of cellular defense reaction discussed in this review, the initial requirement has been for some interaction between the hemocytes and the foreign object or injured self material. The specific attachment of plasmatocytes during nodule and capsule formation and the unidirectional movements of hemocytes during wound healing (see above and Rowley and Ratcliffe, 1978; Ratcliffe and Rowley, 1979a) indicate that this interaction is unlikely to be solely a random process but is probably mediated by factors present in the hemolymph which may or may not be blood cell derived. Such substances exist in vertebrates and include the immunoglobulins, complement, and the lymphokines which greatly facilitate chemotaxis, phagocytosis, and the movement of polymorphs and macrophages, thus mediating leukocyte–antigen interactions. These factors exert their influence via the membrane-bound receptors on the cell surface which receive and amplify the various signals. In the remainder of this review we shall consider the evidence for the presence of chemotaxis, opsonins, and cell surface receptors in insects and then briefly discuss the transplantation studies which have been undertaken to determine the level of non-self recognition in these animals.

5.6.6a. Chemotaxis. Insect hemolymph contains neither immunoglobulins nor classical complement, both of which facilitate chemotaxis and opsonization in vertebrates. Furthermore, Salt (1970) believes that no attraction process occurs between hemocytes and foreign material. He pointed out that even inert objects, such as pieces of latex, polyfluoro-carbon, and glass, are encapsulated (Salt, 1970). He argued that this would not be expected to occur if chemotactic gradients were involved since all these implants are biologically inert. However, it is possible that the attaching hemocytes respond to material produced by the interaction between the very first set of randomly attaching blood cells and the foreign objects (Crossley, 1975; Gagen and Ratcliffe, 1976). Hemocytes also often only attach to cut ends of nerve cord and tracheal implants following intraspecific organ transplantation (Salt, 1970), and the production of a specialized "wound factor" from injured tissue has been postulated to occur to explain this phenomenon (Clarke and Harvey, 1965; Crossley, 1975). Such injury factors have also been described as responsible for the stimulation of hemocyte movement into damaged areas of the insect (Wigglesworth, 1937; Harvey and Williams, 1961; Lea and Gilbert, 1961; Wyatt and Linzen, 1965; Cherbas, 1973; Bohn, 1975).

Additional evidence favoring the existence of chemotactic responses by insect hemocytes is provided by reports that G. *mellonella* hemocytes are attracted to *Aspergillus flavus* conidia *in vitro* (Vey et al., 1968), and by the premature release of hemocytes from posterior clumps in *Musca* spp. in order to encapsulate invading parasitic nematodes, the suggestion being that the cells were attracted to the parasite by a chemotactic factor (Nappi and Stoffolano, 1972; see also Section 5.6.3).

Ratcliffe and Gagen (unpublished, reported in Ratcliffe and Rowley, 1979a) (see also section 5.6.2) have also described the unidirectional movement of G. *mellonella* plasmatocytes towards nodules formed in *Galleria* against *B. cereus*, after 30 min *in vitro*. The specific attachment of plasmatocytes to form the multicellular sheaths around nodules and capsules (Ratcliffe and Gagen, 1977; Schmit and Ratcliffe, 1977) also provides evidence for chemotaxis.

In G. *mellonella*, the initial reaction to contact with injected *B. cereus* or *Schistocerca* nerve cord implants is degranulation of the granular cells to form localized clots which either entrap the bacteria, in the case of nodule formation, or adhere to certain areas of the implant surface, in the case of capsule formation. After 20 min to 6 hr, plasmatocytes begin to attach to the bacteria/granular cell clumps and implant/granular cell clots, and eventually form the typical multicellular sheaths. The important points to emphasize are that the initial reaction between foreign antigen and hemocytes is probably a random process since it occurs within 5 min of introduction of the bacteria or nerve cord into the hemocoel, while the subsequent attachment of the plasmatocytes is highly specific and only involves tbe plasmatocytes, despite the presence of the other hemocyte types in the near vicinity (Schmit, 1979). Thus, granular cell lysis following contact with non-self appears to release some chemotactic agent ("injury factor"?) to which the plasmatocytes, and the plasmatocytes alone, respond. Why or how the granular cells degranulate and lyse in contact with foreign material is unknown, but in some way foreign antigens effect the permeability of the cell membrane of this cell type alone. Granular cells are therefore responsible for the recognition of foreignness prior to nodule and capsule formation in G. *mellonella*, and they or their equivalents (cystocytes) may operate in a similar manner in other insects.

5.6.6b. Humoral Factors, Opsonins, and Receptors. In vertebrates, specific antibody is formed which identifies the foreign antigen and assists in its removal, often by phagocytosis. The antibody frequently acts as an opsonin, coating the foreign particle prior to ingestion. Complement also has an important role to play in this recognition process and as such the vertebrate phagocytes have surface receptors for both IgG and C3b (Weir, 1973; Roitt, 1974). In insects, neither immunoglobulins nor the complete complement system have been found (Bernheimer et al., 1952; Anderson et al., 1972). Furthermore, although insect hemocytes have trypsin-labile receptors (Scott, 1971a; Anderson, 1976a,b), there is no evidence that these surface structures are able to bind the Fc or C3b components of immunoglobulin and complement, respectively (Rabinovitch and De Stefano, 1970; Scott, 1971a; Anderson, 1976a,b). In other invertebrates, however, there is considerable evidence for the presence of serum-dependent phagocytosis, with the hemagglutinins, which are present in the body fluids of many invertebrates, acting as opsonins (e.g. Tripp, 1966; Stuart, 1968; Prowse and Tait, 1969; Paterson and Stewart, 1974; Anderson and Good, 1976; Sminia et al., 1979). These substances are proteinaceous, but with respect to many of their other characteristics, such as molecular weight, specificity, and electrophoretic mobilities, they are unlike vertebrate immunoglobulins (Acton and Weinheimer, 1974; Brahmi and Cooper, 1974; Donlon and Wemyss, 1976). Renwrantz and Uhlenbruck (1974) suggested that the sugar residues to which these agglutinins bind are similar to those found on bacteria, viruses, and other infective organisms and that they may represent a primitive recognition system of invertebrates.

In insects, however, although hemagglutinins are present (Bernheimer, 1952; Ratcliffe and Rowley, 1982), there is no evidence that they act as opsonins (Scott, 1971a; Anderson et al., 1973a; Rowley and Ratcliffe, 1980). Even so, recent work does indicate that recognition factors are present in insect hemolymph and that these substances stimulate recognition and phagocytosis. In G. mellonella, Rabinovitch and De Stefano (1970) showed that a greater variety of foreign particles were recognized by the plasmatocytes in vivo than in vitro so that additional recognition factors may be present in the intact insects. Furthermore, Ratcliffe and Rowley (1982) showed that when P. americana hemocyte monolayers were overlaid with a B. cereus isolate which had been preincubated in serum, there was a significant enhancement in bacterial ingestion, which indicates that an opsoninlike factor may be present in the hemolymph towards this bacterium. Homologous adsorption experiments undertaken with the same bacterium showed that prior incubation of B. cereus in insect hemolymph not only removed the opsonic activity but also imparted an inhibitory effect upon phagocytosis (Ratcliffe and Rowley, 1982). Further experiments (Rowley and Ratcliffe, unpublished) have demonstrated that this opsonic activity cannot be enhanced by immunization with B. cereus. It is suggested that this opsonin may be similar to the C3 complementlike factor reported in the cockroach, B. craniifer, by Anderson et al. (1972). Finally, Möhrig and Schittek (1979) and Möhrig et al. (1979a,b) have reported in G. mellonella larvae that immediately following the injection of latex beads (but not bacteria), normally nonphagocytosable Bacillus thuringiensis subtoxicus are ingested, and they demonstrated that this serum factor could be transfused into other larvae and also induce phagocytosis of the bacteria in these. They refer to this activated "principle" as a lymphokinelike factor and showed by fractionation that it consists of two fractions. Thus, in insects, humoral factors do indeed exist which mediate the recognition of foreignness; much, however, has yet to be learnt about the occurrence, specificity, and nature of these molecules.

5.6.6c. **Transplantation Studies.** In investigating the evidence for memory and the level of recognition in insect defense systems, two experimental methods have been adopted: implantation and grafting.

Implantation experiments have been carried out with abiotic materials such as cotton, nylon thread, and artificial sponge, and biological material such as parasites, gonads, nerve cord, and hemocyte capsules (Salt, 1970; Ratcliffe and Rowley, 1979a). Salt (1970) discusses the specificity of the insect's host response when challenged with implanted parasites and reviews the evidence for enhanced encapsulation reaction on secondary parasitization (Salt, 1960; Fisher, 1961), but most of these results were in fact negative. Other work (Salt, 1960; Scott, 1971b) has been carried out on the recognition of foreignness to tissue implants. Lackie's (1979) results from implantation of pieces of ventral nerve cord, testis, or ovariole have provided information on recognition of non-self in insects. She implanted allografts and xenografts from six donor species into host *Periplaneta americana* and *Schistocerca gregaria* and found that the only donor tissues not encapsulated by *P. americana* were the allogeneic implants and the grafts of the closely related *Blatta orientalis*, whereas *Gryllus domesticus* was the only donor species which was encapsulated by *Schistocerca*. From her experiments, Lackie (1979) suggests that there is a relationship between the taxonomic nearness of the host and donor tissue and the ease with which the recognition response is facilitated and encapsulation carried out. The ability of *Schistocerca* to tolerate xenografts from insects of other orders is not reciprocated when host and donor species are reversed. Lackie (1979) proposes that "hemocytes require a threshold of difference to be surpassed before 'non-self' is recognised" and that hemocytes of *Periplaneta* could have a lower threshold, that is, are able to recognize more insect tissue as non-self than those of *Schistocerca*.

This differential response of hemocytes to xenografts can be compared with the response of some insects to parasites. It has already been mentioned that some habitual parasites of insects avoid encapsulation, but if these are transplanted from their normal host to a new insect species they may be encapsulated. Lewis and Vinson (1968) worked on the braconid, *Cardiochiles nigriceps*, which is not usually encapsulated by its normal host *Heliothis virescens*, but is encapsulated by *H. zea*, and found that when eggs which had not yet been encapsulated by *H. zea* were transferred to *H. virescens*, they were then encapsulated by this their normal host. They suggested that this could be due to a change in the parasite surface by a "substance" produced by *H. zea*. They also made the analogy between this "substance" and mammalian heterophile antigens (Jenkin, 1963).

The effect of the genetic relationship of the implanted material and the host on the degree of cellular response has been examined by Carton (1976). Genetically pure strains of *Pimpla instigator* eggs, eggs of *Drosophila melanogaster*, and those of *Apechthis pecthis compuctor* were used as implants into stock *Pimpla* larvae. From the results, Carton showed that there was an increasing encapsulation reaction with increasing genetic difference between host and donor. *Pimpla* larvae could even distinguish between strains of itself, i.e., showed allograft rejection as well as having the ability to tell self from non-self.

An alternative method of surgical investigation of the immune response is by integumental grafting experiments. The grafting of pieces of the donor's outer surface onto or into the surface of the host animal has been carried out in a

TABLE 3. FATE OF FIRST-SET GRAFTS ON TWO INSECT SPECIES[a]

Graft recipient	Type of graft	Original No. grafted	No. surviving with grafts on	Grafts accepted	Grafts rejected	Rejection times (days)	
						Range	Mean
Extatosoma tiaratum	Autografts	11	10	10	0	—	—
	Allografts	16	9	9	0	—	—
	Clitumnus	27	11	6	5	18–35	28.2 ± 6.6
	Blaberus	34	32	0	32	10–31	22.4 ± 5.8
	Homocoryphus	16	15	0	15	14–24	18.9 ± 3.2
	Schistocerca	10	8	0	8	10–15	12.5 ± 1.4
	Galleria	10	8	0	8	10–15	13.0 ± 2.0
	Tenebrio	10	7	0	7	6–11	8.6 ± 2.0
Blaberus craniifer	Autografts	33	16	14	2	—	—
	Allografts	46	17	15	2	—	—
	Periplaneta	93	23	0	23	10–27	18.2 ± 4.6
	Extatosoma	34	13	0	13	17–29	24.1 ± 4.0
	Homocoryphus	28	10	0	10	12–26	17.1 ± 4.1
	Schistocerca	75	31	0	31	7–17	13.0 ± 2.7
	Galleria	37	13	0	13	7–18	11.5 ± 3.6
	Tenebrio	35	19	0	19	5–13	8.0 ± 1.8

[a] From Thomas and Ratcliffe (1981).

large number of invertebrates (Cooper, 1976; Lackie, 1980), but such studies are wanting in insects except for recent work by Thomas and Ratcliffe (1981, 1982) and Jones and Bell (1982). Since in such experiments the host animal does not have to be killed in order to examine the graft, secondary responses to subsequent transplants can be investigated and the possibility of enhanced secondary rejection examined. Three types of grafts—autografts, allografts, and xenografts—were made by Thomas and Ratcliffe (1981, 1982) using the giant stick insect, *Extatosoma tiaratum* and the cockroach, *Blaberus craniifer*, as hosts. The criteria for graft rejection were the presence of new host integument under the graft, and discoloration, melanization, and sloughing off of graft tissue. Rejection time was noted and from Table 3 it can be seen that: (1) as had been shown previously with the implantation work on insects (Lackie, 1979), all the autografts and allografts were accepted; (2) of the xenografts, all were rejected by *Blaberus* and *Extatosoma* except for 2 of the 10 *Clitumnus* grafts on the latter species; (3) generally, with the xenografts, the more closely related the donor was to the host, the longer the rejection process took.

In some preliminary experiments, second-set grafting was also carried out in an effort to demonstrate enhanced responses and memory (Thomas and Ratcliffe, unpublished). The hosts were again *Extatosoma* and *Blaberus*, and although initial results appeared to show some enhanced rejection, subsequent experiments have failed to confirm these findings. However, these results have to be verified, and great care taken to ensure that the first-set grafts are fully rejected before second-set grafting (Lackie, 1980).

ACKNOWLEDGMENTS. We are very grateful to the following scientists: Dr. J. Armstrong, Mr. S. J. Gagen, Dr. A. Ghiretti-Magaldi, Dr. D. Hoffmann, Professor J. A. Hoffmann, Dr. P. Johnson, Dr. M. Neuwirth, Dr. A. R. Schmit, and Dr. V. J. Smith who have provided both published and unpublished material for this chapter.

Thanks also go to Professor J. S. Ryland in whose department this work was carried out and Mrs. D. Bowditch, Miss L. Millett, and Mrs. M. Thomas for secretarial assistance.

REFERENCES

Acton, R. T., and Weinheimer, P. F., 1974, Hemagglutinins: Primitive receptor molecules operative in invertebrate defense mechanisms, in: *Contemporary Topics in Immunobiology* Volume IV, *Invertebrate Immunology* (E. L. Cooper, ed.), pp. 271–282, Plenum Press, New York.

Acton, R. T., Weinheimer, P. F., and Evans, E. E., 1969, A bactericidal system in the lobster *Homarus americanus*, *J. Invertebr. Pathol.* **13**:463.

Akai, H., and Sato, S., 1971, An ultrastructural study of the haemopoietic organs of the silkworm, *Bombyx mori*, *J. Insect Physiol.* **17**:1665.

Ali, M., 1966, The histology of the gills of *Carcinus maenas* (L.) and other decapod Crustacea, Ph.D. thesis, University of Newcastle-upon-Tyne.

Amirante, G. A., 1976, Production of heteragglutinins in hemocytes of *Leucophaea maderae* L., *Experientia* **32**:526.
Anderson, J. F., and Magnarelli, L. A., 1980, *Entomophthora tabanivora*, a new pathogen in horseflies (Diptera: Tabanidae), *J. Invertebr. Pathol.* **34**:263.
Anderson, J. W., and Conroy, D. A., 1968, The significance of disease in preliminary attempts to raise Crustacea in seawater, Proc. 3rd Symp. Mond. Comm. Off. Int. Epizool. Etud. Mal. Poissons **3**:8.
Anderson, R. S., 1974, Metabolism of insect hemocytes during phagocytosis, in: *Contemporary Topics in Immunobiology*, Volume 4, *Invertebrate Immunology* (E. L. Cooper, ed.), pp. 47–54, Plenum Press, New York.
Anderson, R. S., 1975, Phagocytosis by invertebrate cells *in vitro*: Biochemical events and other characteristics compared with vertebrate phagocytic systems, in: *Invertebrate Immunity* (K. Maramorosch and R. E. Shope, ed.), pp. 153–180, Academic Press, New York.
Anderson, R. S., 1976a, Expression of receptors by insect macrophages, in: *Phylogeny of Thymus and Bone Marrow—Bursa Cells* (R. K. Wright and E. L. Cooper, eds.), pp. 27–34, Elsevier/North-Holland, Amsterdam.
Anderson, R. S., 1976b, Macrophage function in insects, in: *Proceedings, First International Colloquium on Invertebrate Pathology* (T. A. Angus, P. Faulkner, and A. Rosenfield, eds.), pp. 215–19, Queens University Printing Dept., Canada.
Anderson, R. S., 1977, Rosette formation by insect macrophages inhibited by cytochalasin B, *Cell. Immunol.* **29**:331.
Anderson, R. S., and Cook, M. L., 1979, Induction of lysozyme-like activity in the hemolymph and hemocytes of an insect, *Spodoptera eridania*, *J. Invertebr. Pathol.* **33**:197.
Anderson, R. S., and Good, R. A., 1976, Opsonic involvement in phagocytosis by mollusk hemocytes, *J. Invertebr. Pathol.* **27**:57.
Anderson, R. S., Day, N. K. B., and Good, R. A., 1972, Specific hemagglutinin and a modulator of complement in cockroach hemolymph, *Infect. Immun.* **5**:55.
Anderson, R. S., Holmes, B., and Good, R. A., 1973a, In vitro bactericidal capacity of *Blaberus craniifer* hemocytes, *J. Invertebr. Pathol.* **22**:127.
Anderson, R. S., Holmes, B., and Good, R. A., 1973b, Comparative biochemistry of phagocytozing insect hemocytes, *Comp. Biochem. Physiol.* B **46**:595.
Armstrong, J. A., and Hart, P. D., 1975, Phagosome–lysosome interactions in cultured macrophages infected with virulent tubercle bacilli: Reversal of the usual non-fusion pattern and observations on bacterial survival, *J. Exp. Med.* **142**:1.
Armstrong, P. B., 1977, Interaction of the motile blood cells of the horseshoe crab, *Limulus*: Studies on contact paralysis of pseudopodial activity and cellular overlapping in vitro, *Exp. Cell Res.* **107**:127.
Armstrong, P. B., 1979, Motility of the *Limulus* blood cell, *J. Cell Sci.* **37**:169.
Armstrong, P. B., and Levin, J., 1979, In vitro phagocytosis by *Limulus* blood cells, *J. Invertebr. Pathol.* **34**:145.
Arnold, J. W., 1974, The hemocytes of insects in: *The Physiology of Insecta*, Vol. 5, 2nd ed. (M. Rockstein, ed.), pp. 201–254, Academic Press, New York.
Arnold, J. W., 1979, Controversies about hemocyte types in insects, in: *Insect Hemocytes: Development, Forms, Functions and Techniques* (A. P. Gupta, ed.), pp. 231–259, Cambridge University Press, London.
Arvy, L., 1952, Contribution a l'étude du sang et de la leucopoiése chez *Pachygrapsus marmoratus* Fabr., *Ann. Sci. Nat. Zool.* (Ser. 11) **14**:1.
Arvy, L., 1954, Présentation de documents sur la leucopoiése chez *Peripatopsis capensis* Grube, *Bull. Soc. Zool. Fr.* **79**:13.
Ashhurst, D. E., 1979, Hemocytes and connective tissue: A critical assessment, in: *Insect Hemocytes: Development, Forms, Functions and Techniques* (A. P. Gupta, ed.), pp. 319–331, Cambridge University Press, London.
Baehner, R. L., Karnovsky, M. J., and Karnovsky, M. L., 1968, Degranulation of leukocytes in chronic granulomatous disease, *J. Clin. Invest.* **48**:187.

Baker, A. N., 1974, Some aspects of the economic importance of millipedes, *Symp. Zool. Soc. London* **32**:621.

Balbiani, C. R., 1886, Études bacteriologiques sur les arthropods, *Acad. Sci. Paris* **103**:952.

Balss, H., 1944, Die blukorperchen in Klassen und ordnungen des tierreichs (H. G. Bronn, ed.): 1(5) Decapoda, pp. 512–516, Akademische Vertagsges, Leipzig.

Bang, F. B., 1956, A bacterial disease of *Limulus polyphemus*, *Bull. Johns Hopkins Hosp.* **98**:325.

Bang, F. B., 1967, Blood clot formation in the antenna of the hermit crab *Pagurus longicarpus*, *Biol. Bull.* **133**:456.

Bang, F. B., 1970, Cellular aspects of blood clotting in the seastar and hermit crab, *J. Reticuloendothelial Soc.* **7**:167.

Bang, F. B., 1971, A factor in crab amoebocytes which stimulates in vitro clotting of crab blood, *J. Invertebr. Pathol.* **18**:280.

Barnes, R. D., 1980, *Invertebrate Zoology*, 4th ed., Saunders College/Holt, Rinehart & Winston, Philadelphia.

Bauchau, A. G., 1981, Crustaceans, in: *Invertebrate Blood Cells Vol. 2* (N. A. Ratcliffe and A. F. Rowley, eds.), pp. 385–420, Academic Press, London.

Bauchau, A. G., and De Brouwer, M. B., 1972, Ultrastructure des hémocytes d'*Eriocheir sinensis*, Crustacé Decapodé Brachyoure, *J. Microsc. (Paris)* **15**:171.

Bauchau, A. G., and De Brouwer, M. B., 1974, Étude ultrastructurale de la coagulation de l'hémolymphe chez les crustacés, *J. Microsc. (Paris)* **19**:37.

Bauchau, A. G., and Plaquet, J. C., 1973, Variations du nombre des hémocytes chez les crustacés brachyoures, *Crustaceana (Leiden)* **24**:215.

Bauchau, A. G., De Brouwer, M. B., Passelecq-Gérin, E., and Mengeot, J. C., 1975, Étude cytochemique des hémocytes chez Crustacés Decapodés Brachyoures, *Histochemistry* **45**:101.

Bazin, F., 1979, Ultrastructure de l'organe hématopoïétique chez le crabe *Carcinus maenas* (L.) (Crustacea, Decapoda), *Arch. Anat. Microsc. Morphol. Exp.* **68**:12.

Bazin, F., and Demeusy, N., 1972, Processus de cicatrisation consecutif á l'autotomie d'un péréipode chez le crabe *Carcinus maenas* (L.), *C.R. Acad. Sci. Ser. D* **274**:2603.

Beaulaton, J., 1978, L'hemocytoïèse chez le ver à soie du Chêne *Antheraea pernyi* (Lépidoptère): Effets de l'ablation bilatérale des organes méso-et métathoraciques juxta-alaires sur l'évolution de l'hemogramme au cours du cinquième âge larvaire, *C.R. Acad. Sci. Ser. D* **287**:713.

Beaulaton, J., 1980, Physiologie des invertébrés—Effets de l'ablation des organes hémocytopoïetiques et de ligatures sur l'évolution de l'hémocytogramme différentiel chez *Antheraea pernyi* (Lépidoptère) au dernier âge larvaire, *C.R. Acad. Sci. Ser. D* **290**:247.

Bedwin, O.R., 1979a, The particulate basis of the resistance of a parasitoid to the defense reactions of its insect host, *Proc. R. Soc. London Ser. B* **205**:267.

Bedwin, O. R., 1979b, An insect glycoprotein: A study of the particles responsible for the resistance of a parasitoid's egg to the defence reactions of its insect host, *Proc. R. Soc. London Ser. B* **205**:271.

Berlin, R. D., and Oliver, J. M., 1978, Analogous ultrastructure and surface properties during capping and phagocytosis in leukocytes, *J. Cell Biol.* **77**:789.

Bernheimer, A. W., 1952, Haemagglutinins in caterpillar blood, *Science* **155**:150.

Bernheimer, A. W., Caspari, E., and Kaiser, A. D., 1952, Studies on antibody formation in caterpillars, *J. Exp. Zool.* **119**:23.

Biggar, W. D., and Sturgess, J. M., 1977, Role of lysozyme in the microbicidal activity of rat alveolar macrophages, *Infect. Immun.* **16**:974.

Blanc, P. L., 1967, Étude des elements figures de l'hemolymphe de quelques scorpions africains, *Biol. Eco. Rapport particulier* **14**:1.

Bodammer, J. E., 1978, Cytological observations on the blood and hemopoietic tissue in the crab *Callinectes sapidus*. I. The fine structure of hemocytes from intermolt animals, *Cell Tissue Res.* **187**:79.

Bohn, H., 1975, Growth promoting effect of haemocytes on insect epidermis in vitro, *J. Insect Physiol.* **21**:1283.

Bohn, H., 1977a, Differential adhesion of the haemocytes of *Leucophaea maderae* (Blattaria) to a glass surface, *J. Insect Physiol.* **23**:185.

Bohn, H., 1977b, Conditioning of a glass surface for the outgrowth of insect epidermis (*Leucophaea maderae*, Blattaria), *In Vitro* **13**:100.

Bohn, H., 1977c, Enzymatic and immunological characterization of the conditioning factor for epidermal outgrowth in the cockroach *Leucophaea maderae*, *J. Insect Physiol.* **23**:1063

Boman, H. G., Faye, I., Pye, A., and Rasmusan, T., 1978, The inducible immunity system of giant silk moths, in:*Comparative Pathobiology*, Vol. 4 (L. A. Bulla, Jr., and T. C. Cheng, eds.), pp. 145–163, Plenum Press, New York.

Bowen, R. C., 1967, Defense reactions of certain spirobolid millipedes to larval *Macracanthorhynchus ingens*, *J. Parasitol.* **53**:1092.

Bowers, B., 1964, Coated vesicles in the pericardial cells of the aphid (*Myzus persicae*, Sulz.), *Protoplasma* **59**:351.

Boyden, S., 1963, Cellular recognition of foreign matter, *Int. Rev. Exp. Pathol.* **2**:311.

Brahmi, Z., and Cooper, E. L. 1974, Characteristics of the agglutinin in the scorpion, *Androctonus australis*, in: *Contemporary Topics in Immunobiology*, Volume IV, *Invertebrate Immunology* (E. L. Cooper, ed.), pp. 261–270, Plenum Press, New York.

Brehélin, M., 1973, Presence d'un tissu hématopoiétique chez le Coleoptère *Melolontha melolontha*, *Experientia* **29**:1539.

Brehélin, M., 1977, Etude Morphologique et Fonctionnelle des Hémocytes d'Insectes, Ph.D. thesis, University of Strasbourg, France.

Brehélin, M., and Hoffmann, J. A., 1980, Phagocytosis of inert particles in *Locusta migratoria* and *Galleria mellonella*: Study of ultrastructure and clearance, *J. Insect Physiol.* **26**:103.

Brehélin, M., Hoffmann, J. A., Matz, G., and Porte, A., 1975, Encapsulation of implanted foreign bodies by hemocytes in *Locusta migratoria* and *Melolontha melolontha*, *Cell Tissue Res.* **160**:283.

Brehélin, M., Zachary, D., and Hoffmann, J. A., 1978, A comparative ultrastructural study of blood cells from nine insect orders, *Cell Tissue Res.* **195**:45.

Brennan, B. B., and Cheng, T. C., 1975, Resistance of *Moniliformis dubius* to the defense reactions of the American cockroach *Periplaneta americana*, *J. Invertebr. Pathol.* **26**:65.

Brewer, F. D., and Vinson, S. B., 1971, Chemicals affecting the encapsulation of foreign material in an insect, *J. Invertebr. Pathol.* **18**:287.

Brinton, L. P., and Burgdorfer, W., 1971, Fine structure of normal hemocytes in *Dermacentor andersoni* Stiles (Acari: Ixodidae), *J. Parasitol.* **57**:1110.

Brooks, W. M., Montrass, D. B., Sprenkel, R. K., and Carner, G., 1980, Microsporidioses of coleopterous pests of soybeans, *J. Invertebr. Pathol.* **35**:93.

Bruntz, L., 1906a, L'organe phagocytaire des *Polydesmus*, *C.R. Soc. Biol.* **61**:1.

Bruntz, L., 1906b, La phagocytose chez les Diplopodes, *Arch. Zool. Exp. Gén.* **5**:491.

Bruntz, L., 1907, Étude sur les organes lymphoides phagocytaires et excreteurs des crustacés superieurs, *Arch. Zool. Exp. Gén.* **7**:1.

Bucher, G. E., 1959, Bacteria of grasshoppers of western Canada. III. Frequency of occurrence, pathogenicity, *J. Insect Pathol.* **1**:391.

Bucher, G. E., 1960, Potential bacterial pathogens of insects and their characteristics, *J. Insect Pathol.* **2**:172.

Bucherl, W., 1939, Os quilopodes do Brasil, *Mem. Inst. Butantan Sao Paulo* **13**:49.

Cameron, G. R., 1934, Inflammation in the caterpillars of Lepidoptera, *J. Pathol. Bacteriol.* **38**:441.

Campiglia, P. S., and Lavallard, R., 1975, Contributions a l'hématologie de *Peripatus acacioi* Marcus et Marcus (Onychorphore). II. Structure et ultrastructure de globularé pericardial, *Ann. Sci. Nat. Zool. Biol. Anim.* **17**:93.

Cantacuzène, J., 1923, Le probleme de l'immunité chez les invertébrés, *C.R. Soc. Biol.* **88**(Suppl.):48.

Carton, Y., 1976, Isogenic, allogenic and xenogenic transplants in an insect species, *Transplantation* **21**:17.

Cawthorn, R. J., 1980, The cellular responses of migratory grasshoppers (*Melanoplus sanguinipes* F.) and African desert locusts (*Schistocerca gregaria* L.) to *Diptotriaena tricuspis* (Nematoda: Diplotriaenoidea), *Can. J. Zool.* **58**:109.

Cawthorn, R. J., and Anderson, R. C., 1977, Cellular reactions of field crickets (*Acheta pennsylvanicus* Burmeister) and German cockroaches (*Blatella germanica* L.) to *Physaloptera maxillaris* Molin (Nematoda: Physalopteroidea), *Can. J. Zool.* **55**:368.

Charmantier-Daures, M., 1973, Activité de l'organe léucopoiétique de *Pachygrapsus marmoratus* au counts du cycle d'internue: Influence possible chez hormones pedonaulaires, *Bull. Soc. Zool. Fr.* **98**:221.

Cherbas, L., 1973, The induction of an injury reaction in cultured haemocytes from saturniid pupae, *J. Insect Physiol.* **19**:2011.

Clark, M. K., and Dahm, P. A., 1973, Phenobarbital-induced, membrane-like scrolls in the oenocytes of *Musca domestica* Linnaeus, *J. Cell Biol.* **56**:870.

Clarke, K. U., 1979, Visceral anatomy and arthropod phylogeny, in: *Arthropod Phylogeny* (A. P. Gupta, ed.), pp. 467–550, Van Nostrand–Reinhold, Princeton, N.J.

Clarke, R. M., and Harvey, W. R., 1965, Cellular membrane formation by plasmatocytes of diapausing cercropia pupae, *J. Insect Physiol.* **11**:161.

Cohen, E., 1968, Immunologic observations on the agglutinins of the hemolymph of *Limulus polyphemus* and *Birgus latro*, *Trans. N.Y. Acad. Sci.* **30**:427.

Cohen, E. (ed.), 1979, *Biomedical Applications of the Horseshoe Crab (Limulidae)*, Liss, New York.

Cohen, E., Rowe, A. W., and Wissler, F. C., 1965, Heteroagglutinins of the horseshoe crab *Limulus polyphemus*, *Life Sci.* **4**:2009.

Cohen, E., Ilodi, G. H. U., Brahmi, Z., and Minowada, J., 1979, The nature of cellular agglutinins of *Androctonus australis* (Saharan scorpion) serum, *Dev. Comp. Immunol.* **3**:429.

Condon, W. J., and Gordon, R., 1977, Some effects of mermithid parasitism on the larval blackflies *Prosimulium mixtum fuscum* and *Simulium venestum*, *J. Invertebr. Pathol.* **29**:56.

Cooper, E. L., 1976, Cellular recognition of allografts and xenografts in invertebrates, in: *Comparative Immunology* (J. J. Marchalonis, ed.), pp. 36–79, Prentice–Hall, Englewood Cliffs, N.J.

Cornick, J. W., and Stewart, J. E., 1968, Interaction of the pathogen *Gaffkya homari* with natural defense mechanisms of *Homarus americanus*, *J. Fish. Res. Board Can.* **25**:695.

Cornick, J. W., and Stewart, J. E., 1973, Partial characterization of a natural agglutinin in the hemolymph of the lobster, *Homarus americanus*, *J. Invertebr. Pathol.* **21**:255.

Cornick, J. W., and Stewart, J. E., 1978, Lobster (*Homarus americanus*) hemocytes: Classification, differential counts, and associated agglutinin activity, *J. Invertebr. Pathol.* **31**:194.

Crites, J., 1964, A milliped, *Narceus americanus*, as a natural intermediate host of an acanthocephalan, *J. Parasitol.* **50**:293.

Crossley, A. C. S., 1968, The fine structure and mechanism of breakdown of larval intersegmental muscles in the blowfly, *Calliphora erythrocephala* (Meig.), *J. Insect Physiol.* **14**:1389.

Crossley, A. C. S., 1972, The ultrastructure and function of pericardial cells and other nephrocytes in an insect: *Calliphora erythrocephala*, *Tissue Cell* **4**:529.

Crossley, A. C. S., 1975, The cytophysiology of insect blood, *Adv. Insect Physiol.* **11**:117.

Crossley, A. C. S., 1979, Biochemical and ultrastructural aspects of synthesis, storage and secretion in hemocytes, in: *Insect Hemocytes: Development, Forms, Functions and Techniques* (A. P. Gupta, ed.), pp. 423–475, Cambridge University Press, London.

Cuénot, L., 1895, Études physiologiques sur les Orthoptères, *Arch. Biol.* **14**:293.

Cuénot, L., 1897, Les globules sanquins et les organes lymphoides des invertébrés (Revue critique et nouvelles recherches), *Arch. Anat. Microsc. Morphol. Exp.* **1**:153.

Cuénot, L., 1898, Études physiologiques sur les Oligochètes, *Arch. Biol.* **15**:114.

Dall, W., 1964, Studies on the physiology of a shrimp *Metapenaeus mastersii* (Hashwell) (Crustacea: Decapoda: Penaeidae). I. Blood constituents, *Aust. J. Mar. Freshwater Res.* **15**:145.

Dall, W., 1965, Studies on the physiology of a shrimp, *Metapenaeus* sp. (Crustacea: Decapoda: Penaeidae). III. Composition and structure of the integument, *Aust. J. Mar. Freshwater Res.* **16**:13.

Dean, J. M., and Vernberg, F. J., 1966, Hypothermia of blood of crabs, *Comp. Biochem. Physiol. B* **17**:19.

Dean, R. T., 1977, *Lysosomes*, Arnold, London.

Decleir, W., and Vercauteren, R., 1965, Activite phenoloxydasique dans les leucocytes de crabe au cours du cycle d'intermue, *Cah. Biol. Mar.* **6**:163.

Deevey, G. B., 1941, The blood cells of the Haitian tarantula and their relation to the moulting cycle, *J. Morphol.* **68**:457.

Diehl, P. A., 1975, Synthesis and release of hydrocarbons by the oenocytes of the desert locust, *Schistocerca gregaria*, *J. Insect Physiol.* **21**:1237.

Donlon, W. C., and Wemyss, C. T., 1976, Analysis of the hemagglutinin and general protein element of the hemolymph of the West Indian leaf cockroach, *Blaberus craniifer, J. Invertebr. Pathol.* **28**:191.

Dorn, A., and Romer, F., 1976, Structure and function of prothoracic glands and oenocytes in embryos and last larval instars of *Oncopeltus fasciatus* Dallas (Insecta, Heteroptera), *Cell Tissue Res.* **171**:331.

Drach, P., 1930, Étude sur le systéme branchial des crustacés décapodes, *Arch. Anat. Microsc. Morphol. Exp.* **26**:83.

Drach, P., 1939, Mue et cycle d'intermue chez les crustacés décapodes, *Ann. Inst. Oceanogr.* **19**:103.

Duboscq, D., 1896, La termination des vaisscaux et les corpuscules de Kowalevsky chez les Scolopendrides, *Zool. Anz.* **19**:1.

Duboscq, D., 1899, Recherches sur les Chilopodes, Thesis, University of Paris.

Duchâteau, G., and Florkin, M., 1954, La coagulation du sang des arthropodes. IV. Sur le fibrinogène et sur la coaguline musculaire du homard, *Bull. Soc. Chim. Biol.* **36**:295.

Dumont, J. N., Anderson, E., and Winner, G., 1966, Some cytologic characteristics of the hemocytes of *Limulus* during clotting, *J. Morphol.* **119**:181.

Durand, J., 1973, Ph.D. thesis, University of Montpellier.

Durliat, M., and Vranckx, R., 1978, Changes in water-soluble proteins from integument of *Astacus leptodactylus* during the molt cycle, *Comp. Biochem. Physiol. B* **59**:123.

Edwards, G. A., and Challice, C. E., 1960, The ultrastructure of the heart of the cockroach *Blatella germanica, Ann. Entomol. Soc. Am.* **53**:369.

Egusa, S., and Veda, T., 1972, A *Fusarium* sp. associated with black gill disease of the Kuruma prawn *Penaeus japonicus* (Bate), *Bull. Jpn. Soc. Sci. Fish* **38**:1253.

Ermin, R., 1939, Über Bau & Funktion der Lymphocyten bei Insekten (*Periplaneta americana* L.), *Z. Zellforsch. Mikrosk. Anat.* **29**:613.

Evans, E. E., Painter, B., Evans, M. L., Weinheimer, P., and Acton, R. T., 1968, An induced bactericidin in the spiny lobster, *Panulirus argus, Proc. Soc. Exp. Biol. Med.* **128**:394.

Evans, E. E., Weinheimer, P. F., Painter, B., Acton, R. T., and Evans, M. L., 1969a, Secondary and tertiary response of the induced bactericidin from the West Indian spiny lobster *Panulirus argus, J. Bacteriol.* **98**:943.

Evans, E. E., Cushing, J. E., Sawyer, S., Weinheimer, P. F., Acton, R. T., and McNeely, J. L., 1969b, Induced bactericidal response in the spiny lobster, *Panulirus interruptus, Proc. Soc. Exp. Biol. Med.* **132**:111.

Fahrenbach, W. F., 1970, The cyanoblast: Hemocyanin formation in *Limulus polyphemus, J. Cell Biol.* **44**:445.

Feir, D., 1979, Multiplication of hemocytes, in: *Insect Hemocytes: Development, Forms, Functions and Techniques* (A. P. Gupta, ed.), pp. 67–83, Cambridge University Press, London.

Fisher, R. C., 1961, A study in insect multiparasitism. II. The mechanism and control of competition for possession of the host, *J. Exp. Biol.* **38**:605.

Fitzgerald, S. W., and Ratcliffe, N. A., 1982, Evidence for the presence of subpopulations of *Arenicola marina* (L.) coelomocytes identified by their selective response towards gram +ve and gram −ve bacteria, *Dev. Comp. Immunol.*, **6**:23.

Flemister, S. G., 1959, Histophysiology of gill and kidney of the crab *Ocypode albicans, Biol. Bull.* **116**:37.

Fontaine, C. T., and Lightner, D. V., 1973, Observations on the process of wound repair in penaid shrimp, *J. Invertebr. Pathol.* **22**:23.

Fontaine, C. T., and Lightner, D. V., 1974, Observations on the phagocytosis and elimination of carmine particles injected into the abdominal musculature of the white shrimp *Penaeus setiferus, J. Invertebr. Pathol.* **24**:141.

Fontaine, C. T., Bruss, R. G., Sanderson, I. A., and Lightner, D. V., 1975, Histopathological response to turpentine in the white shrimp *Penaeus setiferus, J. Invertebr. Pathol.* **25**:321.

Foster, C. A., and Howse, H. D., 1978, A morphological study on gills of the brown shrimp *Penaeus aztecus, Tissue Cell* **10**:77.

Franz, V., 1904, Ueber die Struktur des Herzens und die Entstehung von Blutzellen bei Spinnen, *Zool. Anz.* **27**:192.

Furman, R. M., and Pistole, T. G., 1976, Bactericidal activity of hemolymph from the horseshoe crab, *Limulus polyphemus, J. Invertebr. Pathol.* **28**:245.

Gaffin, S. L., 1976, The clotting of the lysed white cells of *Limulus* induced by endotoxin-I: Preparation and characterization of clot-forming proteins, *Biorheology* **13**:273.

Gagen, S. J., and Ratcliffe, N. A., 1976, Studies on the *in vivo* cellular reactions and fate of injected bacteria in *Galleria mellonella* and *Pieris brassicae* larvae, *J. Invertebr. Pathol.* **28**:17.

George, W. C., and Nichols, J., 1948, A study of the blood of some Crustacea, *J. Morphol.* **83**:425.

Ghiretti-Magaldi, A., Milanesi, C., and Tognon, G., 1977, Haemopoiesis in Crustacea Decapoda: Origin and production of haemocytes and cyanocytes of *Carcinus maenas, Cell Differ.* **6**:167.

Gibert, J., 1972, Contribution a l'étude de l'hémolymphe de *Niphargus virei* Chevreux (Amphipode hypogé): ses constituants leur origine et leurs fonctions, *Crustaceana Suppl. (Leiden)* **3**:342.

Gilbride, K. J., and Pistole, T. G., 1979, Isolation and characterization of a bacterial agglutinin in the serum of *Limulus polyphemus*, in: *Biomedical Applications of the Horseshoe Crab (Limulidae)* (E. Cohen, ed.), pp. 525–535, Liss, New York.

Glavind, D., 1948, *Studies on the Coagulation of Crustacean Blood*, pp. 7–137, Nyt. Nordish, Forlag. Arnold Busck, Copenhagen.

Gnatzy, W., 1970, Struktur und Entwicklung des Integuments und der Oenocyten von *Culex pipiens* L. (Dipt.), *Z. Zellforsch. Mikrosk, Anat.* **110**:401.

Götz, P., and Vey, A., 1974, Humoral encapsulation in Diptera (Insecta): Defense reactions of *Chironomus* larvae against fungi, *Parasitology* **68**:1.

Grassé, P. P., and Lesperon, L., 1935, Accumulation de colorants acides chez le ver à soie par des tissus différents selon la voie d'accès, *C.R. Acad. Sci.* **201**:618.

Grégoire, Ch., 1952, Sur la coagulation du sang des araignées, *Arch. Int. Physiol.* **60**:100.

Grégoire, Ch., 1955, Blood coagulation in arthropods. VI. A study of phase contrast microscopy of blood reactions *in vitro* in Onychophora and in various groups of arthropods, *Arch. Biol.* **66**:489.

Grégoire, Ch., 1957, Studies by phase contrast microscopy on distribution of patterns of hemolymph coagulation in insects, *Smithson. Misc. Collect.* **134**(6):1.

Grégoire, Ch., 1970, Haemolymph coagulation in arthropods, in: *The Haemostatic Mechanism in Man and Other Animals* (R. G. Macfarlane, ed.), *Symp. Zool. Soc. London* **27**:45.

Grégoire, Ch., 1974, Hemolymph coagulation, in: *The Physiology of Insecta*, Vol. 5, 2nd ed. (M. Rockstein, ed.), pp. 309–360, Academic Press, New York.

Grégoire, Ch., and Goffinet, G., 1979, Controversies about the coagulocyte, in: *Insect Hemocytes: Development, Forms, Functions and Techniques* (A. P. Gupta, ed.), pp. 189–231, Cambridge University Press, London.

Grégoire, Ch., and Jolivet, P., 1957, Coagulation du sang chez les Arthropodes (VIII). Réactions du sang et de l'hémolymphe *in vitro*, étudiées au microscope à contraste de phase, chez 210 especes d'Arthropodes africains, *Explor. Parc Nat. Albert deux Sér. 2* **4**:1.

Grimstone, A. V., Rotheram, S., and Salt, G., 1967, An electron-microscope study of capsule formation by insect blood cells, *J. Cell Sci.* **2**:281.

Gupta, A. P., 1968, Hemocytes of *Scutigerella immaculata* and the ancestry of Insecta, *Ann. Entomol. Soc. Am.* **61**:1028.

Gupta, A. P., 1979a, Arthropod hemocytes and phylogeny, in: *Arthropod Phylogeny*, pp. 669–735, Van Nostrand–Reinhold, Princeton, N.J.

Gupta, A. P., 1979b, Hemocyte types: Their structures, synonymies, interrelationships and taxonomic significance, in: *Insect Hemocytes: Development, Forms, Functions and Techniques*, pp. 85–127, Cambridge University Press, London.

Gupta, A. P., 1979c, Identification key for hemocyte types in hanging-drop preparations, in: *Insect Hemocytes: Development, Forms, Functions and Techniques*, pp. 527–530, Cambridge University Press, London.

Hamann, A., 1975, Stress induced changes in cell-titre of crayfish haemolymph, *Z. Naturforsch.* **30**:850.

Hardy, S. W., Fletcher, T. C., and Olafsen, J. A., 1977, Aspects of cellular and humoral defence mechanisms in the Pacific oyster, *Crassostrea gigas*, in: *Developmental Immunobiology* (J. B. Solomon and J. D. Horton, eds.), pp. 59–66, Elsevier/North-Holland, Amsterdam.

Hardy, W. B., 1892, Blood corpuscles of the Crustacea, together with a suggestion as to the origin of crustacean fibrin-ferment, *J. Physiol. (London)* **13**:165.

Harvey, W. R., and Williams, C. M., 1961, The injury metabolism of the cecropia silkworm. I. Biological amplication of the effects of localized injury, *J. Insect Physiol.* **7**:81.

Hearing, V. J., 1969, Demonstration of acid phosphatase activity in the granules of the blood cells of the lobster, *Homarus americanus*, *Chesapeake Sci.* **10**:24.

Hearing, V. J., and Vernick, S. H., 1967, Fine structure of the blood cells of the lobster, *Homarus americanus*, *Chesapeake Sci.* **8**:170.

Henry, J. E., 1967, *Nosema acridophagous* sp. n., a microsporidian isolated from grasshoppers, *J. Invertebr. Pathol.* **9**:331.

Herberts, C., Andrieux, N., and De Frescheville, J., 1978, Variations des protéines de l'hemolymphe et de l'hypoderme au cours du cycle de mue chez *Carcinus mediterraneus* Czerniavsky: Analyses électrophoretique et immunochimique, *Can. J. Zool.* **56**:1735.

Hinks, C. F., and Arnold, J. W., 1977, Haemopoiesis in Lepidoptera. II. The role of the haemopoietic organs, *Can. J. Zool.* **55**:1740.

Hoarau, F., 1976, Ultrastructure des hémocytes de l'oniscoide, *Helleria brevicornis* Ebner (Crustacés Isopode), *J. Microsc. Biol. Cell.* **27**:47.

Hoffmann, D., Brehélin, M., and Hoffmann, J. A., 1974, Modifications of the hemogram and of the hemocytopoietic tissue of male adults of *Locusta migratoria* (Orthoptera) after injection of *Bacillus thuringiensis*, *J. Invertebr. Pathol.* **24**:238.

Hoffmann, J. A., 1970, Les organes hématopoiétique de deux Insectes Orthoptères: *Locusta migratoria* et *Gryllus bimaculatus*, *Z. Zellforsch. Mikrosk. Anat.* **106**:451.

Hoffmann, J. A., 1972, Modifications of the haemogramme of larval and adult *Locusta migratoria* after selective X-irradiations of the haemocytopoietic tissue, *J. Insect Physiol.* **18**:1639.

Hoffmann, J. A., Zachary, D., Hoffmann, D., Brehélin, M., and Porte, A., 1979, Postembryonic development and differentiation: Hemopoietic tissues and their functions in some insects, in: *Insect Hemocytes: Development, Forms, Functions and Techniques* (A. P. Gupta, ed.), pp. 29–67, Cambridge University Press, London.

Hollande, A. C., 1922, La cellule pericardiale des insectes, *Arch. Anat. Microsc.* **18**:85.

Hollande, A. C., 1930, La digestion des bacilles tuberculeux par les leucocytes du sang des chenilles, *Arch. Zool. Exp. Gén.* **70**:231.

Holme, R., and Solum, N. O., 1973, Electron microscopy of the gel protein formed by clotting of *Limulus polyphemus* hemocyte extracts, *J. Ultrastruct. Res.* **44**:329.

Holmes, B., Page, A. R., Winhorst, D. B., Willie, P. G., White, J. G., and Good, R. A., 1968, The metabolic pattern and phagocytic function of leukocytes from children with chronic granulomatous disease, *Ann. N.Y. Acad. Sci.* **155**:888.

Horn, E. C., and Kerr, M. S., 1969, The hemolymph proteins of the blue crab *Callinectes sapidus*. I. Hemocyanins and certain other major protein constituents, *Comp. Biochem. Physiol.* **29**:493.

Hubert, M., Chassard-Bouchard, C., and Bocquet-Vedrine, J., 1976, Aspects ultrastructuraux des hémocytes de *Carcinus maenas* L. (Crustacé Décapode), parasité par *Sacculina carcini* Thompson (Crustacé Cirripède): Activité réactionnelle, genése de collagéne, *C.R. Acad. Sci. Ser. D* **283**:789.

Huger, A., 1960, Untersuchungen zur Pathologie iner Mikrosporodiose von *Agrotis segetum* (Schiff.) (Lepidopt, Noctuidae), verursacht durch *Nosema perizoides* nov. spec., *Z. Pflanzenker. (Pflanzenpathol.) Pflanzenschutz* **67**:65.

Iyer, G. Y. N., Islam, M. F., and Quastel, J. M., 1961, Biochemical aspects of phagocytosis, *Nature (London)* **192**:535.

Jenkin, C. R., 1963, Heterophile antigens and their significance in the host–parasite relationship, *Adv. Immunol.* **3**:351.

Jenkin, C. R., 1976, Factors involved in the recognition of foreign material by phagocytic cells from invertebrates, in: *Comparative Immunology* (J. J. Marchalonis, ed.), pp. 80–97, Blackwell, Oxford.

Johannsen, R., Anderson, R. S., Good, R. A., and Day, N. K., 1973, A comparative study of the bactericidal activity of horseshoe crab (*Limulus polyphemus*) hemolymph and vertebrate serum, *J. Invertebr. Pathol.* **22**:372.

Johnson, P. T., 1976, Bacterial infection in the blue crab *Callinectes sapidus*: Course of infection and histopathology, *J. Invertebr. Pathol.* **28**:25.

Johnson, P. T., 1977, Paramoebiasis in the blue crab *Callinectes sapidus*, *J. Invertebr. Pathol.* **29**:308.

Johnson, T. W., 1970, Fungi in marine crustaceans, in: *Symposium on Diseases of Fishes and Shellfishes* (S. F. Snieszko, ed.), pp. 405–408, American Fisheries Society, Washington, D.C.

Johnston, M. A., Elder, H. Y., and Davies, P. S., 1973, Cytology of *Carcinus* haemocytes and their function in carbohydrate metabolism, *Comp. Biochem. Physiol. A* **46**:569.

Jones, J. C., 1956, The hemocytes of *Sarcophaga bullata* Parker, *J. Morphol.* **99**:233.

Jones, J. C., 1962, Current concepts concerning insect hemocytes, *Am. Zool.* **2**:209.

Jones, J. C., 1970, Hemocytopoiesis in insects, in: *Regulation of Hematopoiesis* (A. S. Gordon, ed.), pp. 7–65, Appleton, New York.

Jones, J. C., 1977, *The Circulatory System of Insects*, Thomas, Springfield, Ill.

Jones, J. C., 1979, Pathways and pitfalls in the classification and study of insect hemocytes, in: *Insect Hemocytes: Development, Forms, Functions and Techinques* (A. P. Gupta, ed.), pp. 279–301, Cambridge University Press, London.

Jones, J. C., and Liu, D. P., 1968, A quantitative study of mitotic divisions of haemocytes of *Galleria mellonella* larvae, *J. Insect Physiol.* **14**:1053.

Jones, J. C., and Liu, D. P., 1969, The effects of ligaturing *Galleria mellonella* larvae on total haemocyte counts and on mitotic indices among haemocytes, *J. Insect Physiol.* **15**:1703.

Jones, S. E., and Bell, W. J., 1982, Cell-mediated immune-type response of the American cockroach, *Dev. Comp. Immunol.* **6**:35.

Kawanishi, C. Y., Splittstoesser, C. M., and Tashiro, H., 1978, Infection of the European chaffer, *Amphimallon majalis*, by *Bacillus popillae*. II. Ultrastructure, *J. Invertebr. Pathol.* **31**:91.

Kehoe, J. M., Kaplan, R., and Steven, S.-L. Li., 1979, Functional implications of the covalent structure of limulin: An overview, in: *Biomedical Applications of the Horseshoe Crab (Limulidae)* (E. Cohen, ed.), pp. 617–623, Liss, New York.

Kessel, R. G., 1961, Electron microscope observations on the sub-microscopic vesicular component of the suboesophageal body and pericardial cells of the grasshopper *Melanoplus differentialis differentialis*, Thomas., *Exp. Cell Res.* **22**:108.

Kessel, R. G., 1962, Light and electron microscope studies on the pericardial cells of nymphal and adult grasshoppers, *Melanoplus differentialis differentialis*, Thomas., *J. Morphol.* **110**:79.

King, P. E., 1973, *Pycnogonids*, Hutchinson, London.

Kitano, H., 1969a, Defensive ability of *Apanteles glomeratus* L. (Hymenoptera: Braconidae) to the hemocytic reaction of *Pieris rapae crucivora* Boisduval (Lepidoptera: Pieridae), *Appl. Entomol. Zool.* **4**(1):51.

Kitano, H., 1969b, Experimental studies on the parasitism of *Apanteles glomeratus* L. with special reference to its encapsulation-inhibiting capacity, *Bull. Tokyo Gekugei Univ.* **21**(4):95.

Kitano, H., 1974, Effects of the parasitization of a braconid, *Apanteles*, on the blood of the host, *Pieris*, *J. Insect Physiol.* **20**:315.

Kitano, H., and Nakatsuji, N., 1978, Resistance of *Apanteles* eggs to the haemocytic encapsulation by their habitual host, *Pieris*, *J. Insect Physiol.* **24**:261.

Klebanoff, S. J., 1968, Myeloperoxidase–halide–hydrogen peroxide antibacterial system, *J. Bacteriol.* **95**:2131.

Klebanoff, S. J., 1975, Antimicrobial systems of the polymorpho-nuclear leukocyte, in: *The Phagocytic Cell in Host Resistance* (J. Bellanti and D. M. Dayton, eds.), pp. 45–56, Raven Press, New York.

Kollmann, M., 1908, Recherches sur les leucocytes et le tissue lymphoïde des Invertébrés, *Ann. Sci. Nat. Zool. Biol. Anim.* Ser. 9 **8**:1.

Kollmann, M., 1910, Notes sur les functions de la glande lymphatique des scorpionides, *Bull. Soc. Zool. Fr.* **35**:25.

Kowalewsky, A. O., 1894, Investigations of the lymphatic systems of insects and myriapods, [in Russian], *Izv. Imp. Akad. Nuak* **4**:1.

Kowalewsky, A., 1895, Étude des glandes lymphatiques de quelques Myriapodes, *Arch. Zool. Exp. Gén.* **3**:3.

Krishnan, G., 1968, *The Millipede Thyropygus*, C.S.I.R. Zod. Mem. I, Publications and Information Directorate, New Delhi.

Krishnan, G., and Ravindranath, M. H., 1973, Blood cell phenoloxidase of millipedes, *J. Insect Physiol.* **19**:647.

Kümmel, G., 1973, Filtration structures in excretory systems, a comparison, in: *Comparative Physiology* (L. Bolls, K. Schmidt—Nielsen, and S. H. P. Madrell, eds.), pp. 221–240, North-Holland, Amsterdam.

Kurstak, E., Goring, I., and Vago, C., 1969, Cellular defense in an arthropod in response to infection with a *Salmonella typhimurium* strain, *Antonie van Leeuwenhoek J. Microbiol. Serol.* **35**:45.

Lackie, A. M., 1976, Evasion of the haemocytic defense reaction of certain insects by larvae of *Hymenolepis diminuta* (Cestoda), *Parasitology* **73**:97.

Lackie, A. M., 1979, Cellular recognition of foreignness in two insect species, the American cockroach and the desert locust, *Immunology* **36**:909.

Lackie, A. M., 1980, Invertebrate immunity, *Parasitology* **80**:393.

Lai-Fook, J., 1968, The fine structure of wound repair in an insect (*Rhodnius prolixus*), *J. Morphol.* **124**:37.

Lai-Fook, J., 1970, Haemocytes in the repaid of wounds in an insect (*Rhodnius prolixus*), *J. Morphol.* **139**:79.

Laki, K., 1972, Our ancient heritage in blood clotting and some of its consequences, *Ann. N.Y. Acad. Sci.* **202**:297.

Landureau, J. C., and Grellet, P., 1975, Obtention de lignées permanentes d'hémocytes de Blatte: Caractéristiques physiologiques et ultrastructurales, *J. Insect Physiol.* **21**:137.

Landureau, J. C., Grellet, P., and Bernier, I., 1972, Caractérisation, en culture *in vitro*, d'un rôle inconnu des hémocytes des Insectes: Sa signification physiologique, *C.R. Acad. Sci. Ser. D* **274**:2200.

Lavallard, R., and Campiglia, S., 1975, Contributions à l'hématologie de *Peripatus acacioi* Marcus et Marcus (Onychophore). I. Structure et ultrastructure de hémocytes, *Ann. Sci. Nat. Zool.* (Ser. 12) **17**:67.

Lazarenko, T., 1925, Beiträge zure vergleichenden Histologie des Blutes und des Bindehewebes. II. Die morphologische. Bedeutung der Bhitund Bindegewebe-elemente der Insekten, *Z. Zellforsch. Mikrosk. Anat.* **3**:409.

Lea, M. S., and Gilbert, L. E., 1961, Cell division in diapausing silkworm pupae, *Am. Zool.* **1**:368.

Leake, E. S., and Myrvik, Q. N., 1970, Interaction of lysosome-like structures and phagosomes in normal and granulomatous alveolar macrophages, *J. Reticuloendothelial Soc.* **8**:407.

Levin, J., 1967, Blood coagulation and endotoxin in invertebrates, *Fed. Proc.* **26**:1707.

Levin, J., 1979, The reaction between bacterial endotoxin and amebocyte lysate, in: *Biomedical Applications of the Horseshoe Crab (Limulidae)* (E. Cohen, ed.), pp. 131–146, Liss, New York.

Levin, J., and Bang, F. B., 1964a, The role of endotoxin in the extracellular coagulation of *Limulus* blood, *Bull. Johns Hopkins Hosp.* **115**:265.

Levin, J., and Bang, F. B., 1964b, A description of cellular coagulation in the *Limulus*, *Bull. Johns Hopkins Hosp.* **115**:337.

Levin, J., and Bang, F. B., 1968, Clottable protein in *Limulus*: Its localization and kinetics of its coagulation by endotoxin, *Thromb. Diath. Haemorrh.* **19**:186.

Lewis, W. J., and Vinson, S. B., 1968, Immunological relationships between the parasite *Cardiochiles nigriceps* Viereck and certain *Heliothis* species, *J. Insect Physiol.* **14**:613.

Lison, L., 1942, Recherches sur l'histophysiologie comparée de l'excretion chez les arthropodes, *Acad. R. Belg. Mem. Cl. Sci. Collect. 8* **19**:1.

Liu, T.-Y., Seid, R. C., Tai, J. Y., Liang, S.-M., Sakmar, T. P., and Robbins, J. B., 1979, Studies on *Limulus* lysate coagulating system, in: *Biomedical Applications of the Horseshoe Crab (Limulidae)* (E. Cohen, ed.), pp. 147–158, Liss, New York.

Lochhead, J. H., and Lochhead, M. S., 1941, Studies on the blood cells and related tissues in *Artemia*, *J. Morphol.* **68**:593.

Locke, M., 1966, Cell interactions in the repair of wounds in an insect (*Rhodnius prolixus*), *J. Insect Physiol.* **12**:389.

Locke, M., 1969, The ultrastructure of the oenocytes in the molt/intermolt cycle of an insect, *Tissue Cell* **1**:103.

Loeb, L., 1902, On the blood lymph cells and inflammatory processes of *Limulus*, *J. Med. Res.* **7**:145.

Loeb, L., 1920, The movements of the amoebocytes and the experimental production of amoebocyte (cell fibrin) tissue, *Wash. Univ. Stud.* **8**:3.

Lorand, J., Urayama, T., and Lorand, L., 1966, Transglutaminase as a blood clotting enzyme, *Biochem. Biophys. Res. Commun.* **23**:828.

Lorand, L., 1972, Fibrinoligase: The fibrin-stabilizing factor system of blood plasma, *Ann. N.Y. Acad. Sci.* **202**:6.

Lynn, D. C., and Vinson, S. B., 1977, Effects of temperature, host age, and hormones upon the encapsulation of *Cardiochiles nigriceps* eggs by *Heliothis* spp., *J. Invertebr. Pathol.* **29**:50.

Manier, J. F., 1954, Essnis de culture des *Eccirina fexilis* Leger et Duboscq trichomycetic endocommensaux des *Glomeris marinata* Villers, *Ann. Parasitol. Hum. Comp.* **29**:265.

Manton, S. M., 1977, *The Arthropoda: Habits, Functional Morphology, and Evolution*, Oxford University Press, London.

Manton, S. M., and Heatley, N. G., 1937, VI. Studies on the Onychophora. II. The feeding, digestion, excretion and food storage of *Peripatopsis*, *Philos. Trans. R. Soc. London Ser. B* **227**:411.

Marchalonis, J. J., and Edelman, G. M., 1968, Isolation ad characterization of a natural hemagglutinin from *Limulus polyphemus*, *J. Mol. Biol.* **32**:453.

Marrec, M., 1944, L'organ lymphocytogéne des crustacés décapods: Son activité cyclique, *Bull. Inst. Océanogr.* **41**:1.

Marthy, H. J., 1974, Evidence and significance of a haemagglutinin from the skin of cephalopods, *Z. Immunitaetsforsch.* **148**:225.

Matsumoto, K., and Tongu, Y., 1966, Effects of eyestalkectomy on the leucopoietic organ of a crab, *Zool. Mag. Tokyo* **75**:203.

Maynard, D. M., 1960, Circulation and heart function, in: *The Physiology of Crustacea*, Vol. I (T. H. Waterman, ed.), pp. 161–226, Academic Press, New York.

McCumber, L. J., and Clem, L. W., 1977, Recognition of viruses and xenogeneic proteins by the blue crab, *Callinectes sapidus*. I. Clearance and organ concentration, *Dev. Comp. Immunol.* **1**:5.

McKay, D., and Jenkin, C. R., 1969, Immunity in the invertebrates. II. Adaptive immunity in the crayfish (*Parachaeraps bicarinatus*), *Immunology* **17**:127.

McKay, D., and Jenkin, C. R., 1970a, Immunity in the invertebrates: The role of serum factors in the phagocytosis of erythrocytes by haemocytes of the freshwater crayfish (*Parachaeraps bicarinatus*), *Aust. J. Exp. Biol. Med. Sci.* **48**:139.

McKay, D., and Jenkin, C. R., 1970b, Immunity in the invertebrates: Correlation of the phagocytic activity of haemocytes with the resistance to infection in the crayfish (*Parachaeraps bicarinatus*), *Aust. J. Exp. Biol. Med. Sci.* **48**:609.

McKay, D., and Jenkin, C. R., 1970c, Immunity in the invertebrates: The fate and distribution of bacteria in normal and immunised crayfish (*Parachaeraps bicarinatus*), *Aust. J. Exp. Biol. Med. Sci.* **48**:599.

McKay, D., Jenkin, C. R., and Rowley, D., 1969, Immunity in the invertebrates. I. Studies on the natural occurring haemagglutinins in the fluid from invertebrates, *Aust. J. Exp. Biol. Med. Sci.* **47**:125.

McKay, D., Jenkin, C. R., and Tyson, C. J., 1973, Effect of endotoxin on resistance of the freshwater crayfish (*Parachaeraps bicarinatus*) to infection, *J. Infect. Dis.* **128**(Suppl. 1):165.

Mengeot, J. C., Bauchau, A. G., De Brouwer, M. B., and Passelecq-Gérin, E., 1976, Separation des granules presents dans les hémocytes des crustacés par exocytose provoque, *Comp. Biochem. Physiol. A* **54**:145.

Mengeot, J. C., Bauchau, A. G., De Brouwer, M. B., and Passelecq-Gérin, E., 1977, Isolement des granules des hémocytes de *Homarus vulgaris*: Examens electrophoretiques du contenu proteique des granules, *Comp. Biochem. Physiol. A* **58**:393.

Mercer, E. M., and Nicholas, W. L., 1967, The ultrastructure of the capsule of the larval stages of *Moniliformis dubius* (Acanthocephala) in the cockroach *Periplaneta americana*, *Parasitology* **57**:169.

Merril, D. P., and Glenister, R. F., 1980, Crayfish hepatopancreas peroxidase: Absence of a microbiocidal function, *J. Invertebr. Pathol.* **35**:214.

Merril, D. P., Mongeon, S. A., and Fisher, G., 1979, Distribution of fluorescent latex particles following clearance from the haemolymph of the freshwater crayfish *Orconectes virilis* (Hagen), *J. Comp. Physiol.* **132**:363.

Messner, B., 1972, Die rolle des Tyrosinase-Systems in der Immunologischen Abwehrreaktion bei Wirbellosen. I. Insecten, *Zool. Jahrb. Physiol.* **76**:368.

Metalnikov, S., 1924, Phagocytose et réactions des cellules dans l'immunité, *Ann. Inst. Pasteur Paris* **38**:787.

Metchnikoff, E., 1884, Über eine sprosspilzkrankheit de Daphnien, *Arch. Pathol. Anat.* **96**:177.

Midttun, B., and Jensen, H., 1978, Ultrastructure of oenocytoids from two spiders, *Pisaura mirabilis* and *Trochosa terricola* (Araneae), *Acta Zool. (Stockholm)* **59**:157.

Millot, J., 1926, Contribution à l'histophysiologie des aranéides, *Bull. Biol. Fr. Belg.* **8**:1.
Mills, R. P., and King, R., 1965, The pericardial cells of *Drosophila melanogaster*, *Q. J. Microsc. Sci.* **106**:261.
Mix, M. C., and Sparks, A. K., 1980, Hemocyte classification and differential counts in the Dungeness crab, *Cancer magister*, *J. Invertebr. Pathol.* **35**:134.
Möhrig, W., and Schittek, D., 1979, Phagocytosis-stimulating mediators in insects, *Acta Biol. Med. Ger.* **38**:953.
Möhrig, W., Schittek, D., and Hanschke, R., 1979a, Investigations on cellular defense reactions with *Galleria mellonella* against *B. thuringiensis*, *J. Invertebr. Pathol.* **34**:207.
Möhrig, W., Schittek, D., and Hanschke, R., 1979b, Immunological activation of phagocytic cells in *Galleria mellonella*, *J. Invertebr. Pathol.* **34**:84.
Monpeyssin, M., and Beaulaton, J. C., 1978, Hemocytopoiesis in the oak silkworm, *Antheraea pernyi* and some other Lepidoptera, *J. Ultrastruct. Res.* **64**:35.
Mosesson, M. W., Wolfenstein-Todel, C., Levin, J., and Bertrand, O., 1979, Characterization of amebocyte coagulogen from the horseshoe crab (*Limulus polyphemus*), *Thromb. Res.* **14**:765.
Murer, E. H., Levin, J., and Holme, R., 1975, Isolation and studies of the granules of the amebocytes of *Limulus polyphemus*, the horseshoe crab, *J. Cell. Physiol.* **86**:533.
Myhrman, R., and Bruner-Lorand, J., 1970, Lobster muscle transpeptidase, *Methods Enzymol.* **19**:765.
Nachum, R., Watson, S. W., Sullivan, J. D., Jr., and Siegel, S. E., 1979, Antimicrobial defense mechanisms in the horseshoe crab, *Limulus polyphemus*: Preliminary observations with heat-derived extracts of *Limulus* amoebocyte lysate, *J. Invertebr. Pathol.* **33**:290.
Nakamura, S., Takagi, T., Iwanaga, S., Niwa, M., and Takahashi, K., 1976a, A clottable protein (coagulogen) of horseshoe crab hemocytes: Structural changes of its polypeptide chain during gel formation, *J. Biochem.* **80**:649.
Nakamura, S., Iwanaga, S., Harada, T., and Niwa, M., 1976b, A clottable protein (coagulogen) from amoebocyte lysate of the Japanese horseshoe crab (*Tachypleus tridentatus*): Its isolation and biochemical properties, *J. Biochem.* **80**:1011.
Nakamura, S., Takagi, T., Iwanaga, S., Niwa, M., and Takahashi, K., 1976c, Amino acid sequence studies on the fragments produced from horseshoe crab coagulogen during gel formation: Homologies with primate fibrinopeptide B, *Biochem. Biophys. Res. Commun.* **72**:902.
Nappi, A. J., 1970, Defense reactions of *Drosophila euronotus* larvae against the hymenopterous parasite *Pseudeucoila bochei*, *Parasitology* **66**:23.
Nappi, A. J., 1973, The role of melanization in the immune reaction of larvae of *Drosophila algonquin* against *Pseudeucoila bochei*, *Parasitology* **66**:23.
Nappi, A. J., 1974, Insect hemocytes and the problems of host recognition of foreignness, in: *Contemporary Topics in Immunobiology*, Volume 4, *Invertebrate Immunology* (E. L. Cooper, ed.), pp. 207–224, Plenum Press, New York.
Nappi, A. J., 1975, Parasite encapsulation in insects, in: *Invertebrate Immunity* (K. Maramorosch and R. E. Shope, eds.), pp. 293–326, Academic Press, New York.
Nappi, A. J., and Stoffolano, J. G., Jr., 1972, Distribution of haemocytes in larvae of *Musca domestica* and *Musca autumnalis* and possible chemotaxis during parasitization, J. Insect. Physiol. **18**:169.
Nappi, A. J., and Streams, F. A., 1969, Hemocytic reactions of *Drosophila melanogaster* to the parasites *Pseudeucoila mellipes* and *P. bochei*, *J. Insect Physiol.* **15**:1551.
Nelson, E., Blinzinger, K., and Hager, H., 1962, Ultrastructural observations on phagocytosis of bacteria in experimental (*E. coli*) meningitis, *J. Neuropathol. Exp. Neurol.* **21**:155.
Nemhauser, I., Ornberg, R., and Cohen, W. D., 1980, Marginal bands in blood cells of invertebrates, *J. Ultrastruct. Res.* **70**:308.
Neuwirth, M., 1974, Granular hemocytes, the main phagocytic blood cells in *Calpodes ethlius* (Lepidoptera, Hesperiidae), *Can. J. Zool.* **52**:783.
Nutting, W. L., 1951, A comparative anatomical study of the heart and accessory structures of the orthopteriod insects, *J. Morphol.* **89**:501.
Oppenheim, J. D., Nachbar, M. S., Salton, M. R. J., and Aull, F., 1974, Purification of a hemagglutinin from *Limulus polyphemus* by affinity chromatography, *Biochem. Biophys. Res. Commun.* **58**:1127.
Ornberg, R. L., and Reese, T. S., 1979, Secretion in *Limulus* amebocytes is by exocytosis, in: *Biomedical Applications of the Horseshoe Crab (Limulidae)* (E. Cohen, ed.), pp. 125–130, Liss, New York.

Osman, S. E., 1978, Die Wirkung der sekrete der weiblichen Genital anhangsdrüsen von *Pimpla turionellae* L. (Hym., Ichneumonidae) auf die Hämocyten und die Einkapselungsreaktion von Wirtspuppen, *Z. Parasitenkd.* **57**:89.

Palm, N. B., 1954, The elmination of injected vital dyes from the blood in myriapodes, *Ark. Zool.* **6**:219.

Paterson, W. D., and Stewart, J. E., 1974, In vitro phagocytosis by hemocytes of the American lobster (*Homarus americanus*), *J. Fish. Res. Board Can.* **31**:1051.

Paterson, W. D., Stewart, J. E., and Zwicker, B. M., 1976, Phagocytosis as a cellular immune response mechanism in the American lobster *Homarus americanus*, *J. Invertebr. Pathol.* **27**:95.

Pauley, G. B., Granger, G. A., and Krassner, S. M., 1971, Characterization of a natural agglutinin present in the hemolymph of the California sea hare, *Aplysia californica*, *J. Invertebr. Pathol.* **18**:207.

Pistole, T. G., 1976, Naturally occurring bacterial agglutinin in the serum of the horseshoe crab, *Limulus polyphemus*, *J. Invertebr. Pathol.* **28**:153.

Pistole, T. G., 1978, Broad-spectrum bacterial agglutinating activity in the serum of the horseshoe crab, *Limulus polyphemus*, *Dev. Comp. Immunol.* **2**:65.

Pistole, T. G., 1979, Bacterial agglutinins from *Limulus polyphemus*—An overview in: *Biomedical Applications of the Horseshoe Crab (Limulidae)* (E. Cohen, ed.), pp. 547–553, Liss, New York.

Pistole, T. G., and Britko, J. L., 1978, Bactericidal activity of amebocytes from the horseshoe crab, *Limulus polyphemus*, *J. Invertebr. Pathol.* **31**:376.

Pistole, T. G., and Furman, R. M., 1976, Serum bactericidal activity in the horseshoe crab, *Limulus polyphemus*, *Infect. Immun.* **14**:888.

Poinar, G. O., Jr., 1971, Use of nematodes for microbial control of insects, in: *Microbial Control of Insects and Mites* (H. D. Burges and N. W. Hussey, eds.), pp. 181–203, Academic Press, New York.

Poinar, G. O., Jr., 1974, Insect immunity to parasitic nematodes, in: *Contemporary Topics in Immunobiology*, Volume 4, *Invertebrate Immunology* (E. L. Cooper, ed.), pp. 167–178, Plenum Press, New York.

Poinar, G. O., Jr., and Hess, R. T., 1977, Cellular responses in decapod crustaceans to *Ascarophis* sp. (Spirurida: Nematoda), in: *Comparative Pathobiology*, Volume 3, *Invertebrate Immune Responses* (L. A. Bulla, Jr., and T. C. Cheng, eds.), pp. 135–154, Plenum Press, New York.

Poinar, G. O., Jr., Leutenegger, R., and Götz, P., 1968, Ultrastructure of the formation of a melanotic capsule in *Diabrotica* (Coleoptera) in response to a parasitic nematode (Mermithidae), *J. Ultrastruct. Res.* **25**:293.

Poll, M., 1934, Recherches histophysiologiques sur les tubes de malpighi du *Tenebrio molitor* L., *Rec. Inst. Zool. Torley-Rousseau* **5**:73.

Porter, K., Kenyon, K., and Badenhausen, S., 1967, Specialisations of the unit membrane, *Protoplasma* **63**:262.

Price, C. D., and Ratcliffe, N. A., 1974, A reappraisal of insect haemocyte classification by the examination of blood from fifteen insect orders, *Z. Zellforsch. Mikrosk. Anat.* **147**:537.

Prior, C., and Perry, C. M., 1980, Infection of *Promecotheca papuana* with *Synnematium jonesii*, *J. Invertebr. Pathol.* **35**:14.

Prowse, R. H., and Tait, N. N., 1969, In vitro phagocytosis by amoebocytes from the haemolymph of *Helix aspersa* (Müller). I. Evidence for opsonic factor(s) in serum, *Immunology* **17**:437.

Pyefinch, K. A., 1947, Biology of ship fouling, *New Biology (London)* **3**:128.

Rabin, H., 1970, Hemocytes, hemolymph and defense reactions in crustaceans, *J. Reticuloendothelial Soc.* **7**:195.

Rabinovitch, M., and De Stefano, M. J., 1970, Interactions of red cells with phagocytes of the wax moth (*Galleria mellonella* L.) and mouse, *Exp. Cell Res.* **59**:272.

Ratcliffe, N. A., and Gagen, S. J., 1976, Cellular defense reactions of insect hemocytes in vivo: Nodule formation and development in *Galleria mellonella* and *Pieris brassicae* larvae, *J. Invertebr. Pathol.* **28**:373.

Ratcliffe, N. A., and Gagen, S. J., 1977, Studies on the in vivo cellular reactions of insects: An ultrastructural analysis of nodule formation in *Galleria mellonella*, *Tissue Cell* **9**:73.

Ratcliffe, N. A., and Rowley, A. F., 1974, In vitro phagocytosis of bacteria by insect blood cells, *Nature (London)* **252**:391.
Ratcliffe, N. A., and Rowley, A. F., 1975, Cellular defense reactions of insect hemocytes *in vitro*. Phagocytosis in a new suspension culture system, *J. Invertebr. Pathol.* **26**:225.
Ratcliffe, N. A., and Rowley, A. F., 1979a, Role of hemocytes in defense against biological agents, in: *Insect Hemocytes: Development, Forms, Functions and Techniques* (A. P. Gupta, ed.), pp. 331–415, Cambridge University Press, London.
Ratcliffe, N. A., and Rowley, A. F., 1979b, A comparative synopsis of the structure and function of the blood cells of insects and other invertebrates, *Dev. Comp. Immunol.* **3**:189.
Ratcliffe, N. A., and Rowley, A. F., 1982, Opsonic activity of insect hemolymph, in: *Comparative Pathobiology* (L. A. Bulla, Jr., and T. C. Cheng, eds.), in press.
Ratcliffe, N. A., Gagen, S. J., Rowley, A. F., and Schmit, A. R., 1976a, Studies on insect cellular defense mechanisms and aspects of the recognition of foreignness, in: *Proceedings, First International Colloquium on Invertebrate Pathology* (T. A. Angus, P. Faulkner, and A. Rosenfield, eds.), Queen's University Printing Department, Kingston, Ontario.
Ratcliffe, N. A., Gagen, S. J., Rowley, A. F., and Schmit, A. R., 1976b, The role of granular hemocytes in the cellular defense reactions of the wax moth *Galleria mellonella*, in: *Proceedings, Sixth European Congress on Electron Microscopy* (Y. Ben-Shaul, ed.), pp. 295–297, Tal International, Israel.
Ravindranath, M. H., 1970, Comparative studies on the blood of chilopods and diplopods in relation to cuticle formation, Ph.D. thesis, University of Madras.
Ravindranath, M. H., 1973, The hemocytes of a millipede, *Thyropygus poseidon*, *J. Morphol.* **141**:257.
Ravindranath, M. H., 1974a, The hemocytes of a scorpion *Palamnaeus swammerdami*, *J. Morphol.* **144**:1.
Ravindranath, M. H., 1974b, The hemocytes of an isopod *Ligia exotica* Roux, *J. Morphol.* **144**:11.
Ravindranath, M. H., 1975a, Effects of temperature on the morphology of hemocytes and coagulation process in the mole crab *Emerita* (=*Hippa*) *asiatica*, *Biol. Bull.* **148**:286.
Ravindranath, M. H., 1975b, Effects of hydrogen ion concentration on the morphology of hemocytes of the mole-crab *Emerita asiatica*, *Biol. Bull.* **149**:226.
Ravindranath, M. H., 1977a, A comparative study of the morphology and behaviour of granular haemocytes of arthropods, *Cytologia* **42**:743.
Ravindranath, M. H., 1977b, The circulating haemocyte population of the mole crab *Emerita* (=*Hippa*) *asiatica* (Milne-Edwards), *Biol. Bull.* **152**:415.
Ravindranath, M. H., 1981, Onychophorans and myriapods, in: *Invertebrate Blood Cells* Vol. 2 (N. A. Ratcliffe and A. F. Rowley, eds.), pp. 327–354, Academic Press, New York.
Reade, P. C., 1968, Phagocytosis in invertebrates, *Aust. J. Exp. Biol. Med. Sci.* **46**:219.
Renwrantz, L. R., and Uhlenbruck, G., 1974, Blood-group-like substances in some marine invertebrates. III. Glycoproteins with blood-group A specificity in the cephalopods *Sepia offinalis* L. and *Loligo vulgaris* Lam, *J. Exp. Zool.* **188**:65.
Ries, E., 1932, Experimentelle Symbiosestudien. II. Mycetomtransplantationen, *Z. Morphol. Oekol. Tiere* **25**:184.
Rizki, M. T. M., 1962, Experimental analysis of hemocyte morphology in insects, *Am. Zool.* **2**:247.
Rizki, M. T. M., and Rizki, R. M., 1976, Cell interactions in hereditory melanotic tumor formation in *Drosophila*, in: *Proceedings, First International Colloquium on Invertebrate Pathology* (T. A. Angus, P. Faulkner, and A. Rosenfield, eds.), pp. 137–141, Queen's University Printing Department, Kingston, Ontario.
Roche, A. C., and Monsigny, M., 1974, Purification and properties of limulin: A lectin (agglutinin) from hemolymph of *Limulus polyphemus*, *Biochim. Biophys. Acta* **371**:242.
Roche, A. C., and Monsigny, M., 1979, Limulin (*Limulus polyphemus* lectin): Isolation, physiochemical properties, sugar specificity and mitogenic activity, in: *Biomedical Applications of the Horseshoe Crab (Limulidae)* (E. Cohen, ed.), pp. 603–616, Liss, New York.
Roche, A. C., Perrodon, Y., Halpern, B., and Monsigny, M., 1977, Limulin (*Limulus polyphemus* lectin): Mitogenic effect on human peripheral lymphocytes, *Eur. J. Immunol.* **7**:263.
Roitt, I., 1974, *Essential Immunology*, 2nd ed., Blackwell, Oxford.

Romer, F., 1974, Ultrastructural changes of the oenocytes of *Gryllus bimaculatus* DEG (Saltatoria, Insecta) during the moulting cycle, *Cell Tissue Res.* **151**:27.

Rosen, B., 1967, Shell diseases of the blue crab, *Callinectes sapidus*, *J. Invertebr. Pathol.* **9**:348.

Rosen, B., 1970, Shell disease of aquatic crustaceans, in: *Symposium on Diseases of Fishes and Shellfishes* (S. F. Snieszko, ed.), pp. 405–408, American Fisheries Society, Washington, D.C.

Rosenberg, J., 1978, Zur ultrastruktur der nephrozyten von erdlaufern (Chilopoda: Pleurostigmorphora: Geophilomorpha), *Entomol. Germ.* **4**:24.

Rostam-Abadi, H., and Pistole, T. G., 1979, Sites on the lipopolysaccharide molecule reactive with *Limulus* agglutinins, in: *Biomedical Applications of the Horseshoe Crab (Limulidae)* (E. Cohen, ed.), pp. 537–545, Liss, New York.

Rotheram, S. M., 1967, Immune surface of eggs of a parasitic insect, *Nature (London)* **214**:700.

Rotheram, S. M., 1973a, The surface of the egg and first-instar larva of *Nemeritis*, *Proc. R. Soc. London Ser. B* **183**:179.

Rotheram, S. M., 1973b, The surface of the egg of a parasitic insect. II. The ultrastructure of the particulate coat on the egg of *Nemeritis*, *Proc. R. Soc. London Ser. B* **183**:195.

Rowley, A. F., 1977a, The role of the haemocytes of *Clitumnus extradentatus* in haemolymph coagulation, *Cell Tissue Res.* **182**:513.

Rowley, A. F., 1977b, Studies on insect cellular defences in vitro, Ph.D. thesis, University of Wales.

Rowley, A. F., and Ratcliffe, N. A., 1976a, The granular cells of *Galleria mellonella* during clotting and phagocytic reactions in vitro, *Tissue Cell* **8**:437.

Rowley, A. F., and Ratcliffe, N. A., 1976b, An ultrastructural study of the in vitro phagocytosis of *Escherichia coli* by the hemocytes of *Calliphora erythrocephala*, *J. Ultrastruct. Res.* **55**:193.

Rowley, A. F., and Ratcliffe, N. A., 1976c, The intracellular fate of bacteria and latex particles in insect blood cells, in: *Proceedings, Sixth European Congress on Electron Microscopy* (Y. Ben-Shaul, ed.), pp. 301–303, Tal International, Israel.

Rowley, A. F., and Ratcliffe, N. A., 1978, A histological study of wound healing and hemocyte function in the wax moth, *Galleria mellonella*, *J. Morphol.* **157**:181.

Rowley, A. F., and Ratcliffe, N. A., 1979, An ultrastructural and cytochemical study of the interaction between latex particles and the haemocytes of the wax moth *Galleria mellonella* in vitro, *Cell Tissue Res.* **199**:127.

Rowley, A. F., and Ratcliffe, N. A., 1980, Insect erythrocyte agglutinins. In vitro opsonization experiments with *Clitumnus extradentatus* and *Periplaneta americana* haemocytes, *Immunology* **40**:483.

Rowley, A. F., and Ratcliffe, N. A., 1981, Insects, in: *Invertebrate Blood Cells* Vol. 2 (N. A. Ratcliffe and A. F. Rowley, eds.), pp. 421–488. Academic Press, London.

Ryan, M., and Nicholas, W. L., 1972, The reaction of the cockroach, *Periplaneta americana*, to the injection of foreign particulate material, *J. Invertebr. Pathol.* **19**:299.

Salt, G., 1956, Experimental studies in insect parasitism. IX. The reactions of a stick insect to an alien parasite, *Proc. R. Soc. London Ser. B* **146**:93.

Salt, G., 1957, Experimental studies in insect parasitism. X. The reactions of some endopterygote insects to an alien parasite, *Proc. R. Soc. London Ser. B* **147**:167.

Salt, G., 1959, The fate of a braconid parasite, *Rogas testaceus*, in four species of hosts, *Biologia (Lahore)* **5**(1):84.

Salt, G., 1960, Experimental studies in insect parasitism. XI. The haemocytic reaction of a caterpillar under varied conditions, *Proc. R. Soc. London Ser. B* **151**:446.

Salt, G., 1965, Experimental studies in insect parasitism. XIII. The haemocytic reaction of a caterpillar to eggs of its habitual parasite, *Proc. R. Soc. London Ser. B* **162**:303.

Salt, G., 1970, *The Cellular Defence Reactions of Insects*, Cambridge Monographs in Experimental Biology, No. 16, Cambridge University Press, London.

Sanchez, S., 1959, Le développement des pycnogonides et leurs affinités avec les arachnides, *Arch. Zool. Exp. Gén.* **98**:1.

Sato, S., Akai, H., and Sawada, H., 1976, An ultrastructural study of capsule formation by *Bombyx* hemocytes, *Annot. Zool. Jpn.* **49**:177.

Savory, T., 1964, *Arachnida*, Academic Press, New York.

Sawyer, T. K., 1969, Preliminary study on the epizoology and host–parasite relationship of *Paramoeba* sp. in the blue crab *Callinectes sapidus*, *Proc. Natl. Shellfish. Assoc.* **59**:60.
Sbarra, A. J., and Karnovsky, M. L., 1959, The biochemical basis of phagocytosis. I. Metabolic changes during the ingestion of particles by polymorphonuclear leukocytes, *J. Biol. Chem.* **234**:1355.
Sbarra, A. J., Jacobs, A. A., Strauss, R. R., Paul, B., and Mitchell, G. E., 1971, The biochemical and antimicrobial activities of phagocytizing cells, *Am. J. Clin. Nutr.* **24**:272.
Schapiro, H. C., Mathewson, J. H., Steenbergen, J. F., Kellogg, S., Ingram, C., Nievengarten, G., and Rabin, H., 1974, Gaffkemia in the California spiny lobster *Panulirus interruptus*: Infection and immunization, *Aquaculture* **3**:403.
Schapiro, H. C., Steenbergen, J. F., and Fitzgerald, Z. A., 1977, Hemocytes and phagocytosis in the American Lobster, *Homarus americanus*, in: *Comparative Pathobiology*, Volume 3, *Invertebrate Immune Responses* (L. A. Bulla, Jr., and T. C. Cheng, eds.), pp. 126–134, Plenum Press, New York.
Scharrer, B., 1972, Cytophysiological features of hemocytes in cockroaches, *Z. Zellforsch. Mikrosk. Anat.* **129**:301.
Schmit, A. R., 1979, Studies on encapsulation in insects, Ph.D. thesis, University of Wales.
Schmit, A. R., and Ratcliffe, N. A., 1977, The encapsulation of foreign tissue implants in *Galleria mellonella* larvae, *J. Insect Physiol.* **23**:175.
Schmit, A. R., and Ratcliffe, N. A., 1978, The encapsulation of Araldite implants and recognition of foreignness in *Clitumnus extradentatus*, *J. Insect Physiol.* **24**:511.
Schmit, A. R., Rowley, A. F., and Ratcliffe, N. A., 1977, The role of *Galleria mellonella* hemocytes in melanin formation, *J. Invertebr. Pathol.* **29**:232.
Schutz, F. N., 1925, Physiologie de körpersäfte: Crustaceen, in: *Handbuch der Vergleichenden Physiologie* (H. Winterstein, ed.), pp. 669–746, Fischer, Jena.
Scott, M. T., 1971a, Recognition of foreignness in invertebrates. II. *In vitro* studies of cockroach phagocytic hemocytes, *Immunology* **21**:817.
Scott, M. T., 1971b, Recognition of foreignness in invertebrates. I. Transplantation studies using the American cockroach (*Periplaneta americana*), *Transplantation* **11**:78.
Seifert, B., 1932, Anatomie ünd biologie des Diplopoden *Strongylosoma pallipes* (Oliv.), *Z. Morph. Oekol. Tiere* **25**:362.
Seifert, G., and Rosenberg, J., 1976, Die ultrastruktur der nephrozyten von *Orthomorpha gracilis* (C. L. Koch, 1847) (Diplopoda, Strongylosomidae), *Zoomorphologie* **85**:23.
Seifert, G., and Rosenberg, J., 1977, Die ultrastruktur der nephrozyten von *Peripatoides kuckarti* (Saenger, 1869) (Onychophora, Peripatopsidae), *Zoomorphologie* **86**:169.
Seitz, K.-A., 1972, Zur histologie and feinstruktur des Herzens und der Hämocyten von *Cupiennius salei* Keys (Araneae Ctenidae). II. Zur Funktionsmorphologie der Phagocyten, *Zool. Jahrb. Anat.* **89**:385.
Sewell, M. T., 1955, Lipoprotein cells in the blood of *Carcinus maenas*, and their cycle of activity correlated with the moult, *Q. J. Microsc. Sci.* **96**:73.
Shapiro, M., 1979, Changes in hemocyte populations, in: *Insect Hemocytes: Development, Forms, Functions and Techniques* (A. P. Gupta, ed.), pp. 475–525, Cambridge University Press, London.
Sherman, R. G., 1973, Ultrastructurally different hemocytes in a spider, *Can. J. Zool.* **51**:1155.
Sherman, R. G., 1981, Chelicerates, in: *Invertebrate Blood Cells* Vol. 2 (N. A. Ratcliffe and A. F. Rowley, eds.), pp. 355–384, Academic Press, London.
Shirodkar, M. V., Warwick, A., and Bang, F. B., 1960, The *in vitro* reaction of *Limulus* amebocytes to bacteria, *Biol. Bull.* **118**:324.
Shishikura, F., and Sekiguchi, K., 1979, Comparative studies on hemocytes and coagulogens of the Asian and the American horseshoe crabs, in: *Biomedical Applications of the Horseshoe Crab (Limulidae)* (E. Cohen, ed.), pp. 185–201, Liss, New York.
Shishikura, F., Chiba, J., and Sekiguchi, K., 1977, Two types of hemocytes in localization of clottable protein in Japanese horseshoe crab, *Tachypleus tridentatus*, *J. Exp. Zool.* **201**:303.
Shrivastava, S. C., and Richards, A. G., 1965, An autoradiographic study of the relation between hemocytes and connective tissue in the wax moth *Galleria mellonella* L., *Biol. Bull.* **128**:337.
Shukla, G. S., 1964, Studies of *Scolopendra morsitans* (Linn). Part IV. Blood vascular system and associated structures, *Agra Univ. J. Res. Sci.* **13**:227.

Shuster, C. N., Jr., 1978, The circulatory system of blood of the horseshoe crab, *Limulus polyphemus* L.: A review, U.S. Department of Energy, Federal Energy Regulatory Commission, DOE/FERC/ 0014.

Sindermann, C. J., 1970, *Principal Diseases of Marine Fish and Shellfish* Academic Press, New York.

Sindermann, C. J., 1971, Internal defences of Crustacea: A review, *Fishery Bull. Fish Wildl. Serv.* **69**:455.

Sindermann, C. J., and Rosenfield, A., 1967, Principal diseases of commercially important marine bivalve Mollusca and Crustacea, *Fishery Bull. Fish Wildl. Serv.* **66**:335.

Sminia, T., van der Knaap, W. P. W., and Edelenbosch, P., 1979, The role of serum factors in phagocytosis of foreign particles by blood cells of the freshwater snail *Lymnaea stagnalis*, *Dev. Comp. Immunol.* **3**:37.

Smith, V. J., 1978, Cellular defence reactions of *Carcinus maenas* (Crustacea), Ph.D thesis, University of Wales.

Smith, V. J., and Ratcliffe, N. A., 1978, Host defence reactions of the shore crab *Carcinus maenas* (L.) *in vitro*, *J. Mar. Biol. Assoc. U.K.* **58**:367.

Smith, V. J., and Ratcliffe, N. A., 1980a, Cellular reactions of the shore crab *Carcinus maenas*: In vivo hemocytic and histopathological responses to injected bacteria, *J. Invertebr. Pathol.* **35**:65.

Smith, V. J., and Ratcliffe, N. A., 1980b, Host defence reactions of the shore crab, *Carcinus maenas* (L.): Clearance and distribution of injected test particles, *J. Mar. Biol. Assoc. U.K.* **60**:89.

Smith, V. J., and Ratcliffe, N. A., 1981, Pathological changes in the nephrocytes of the shore crab, *Carcinus maenas*, following injection of bacteria, *J. Invertebr. Pathol.*, **38**:113.

Söderhäll, K., and Unestam, T., 1979, Activation of serum prophenoloxidase in arthropod immunity. The specificity of cell wall glucan activation and activation by purified fungal glyco-proteins of crayfish phenoloxidase, *Can. J. Microbiol.* **25**:406.

Söderhäll, K., Häll, L., Unestam, T., and Nyhlén, L., 1979, Attachment of phenoloxidase to fungal cell walls in arthropod immunity, *J. Invertebr. Pathol.* **34**:285.

Solangi, M. A., and Lightner, D. V., 1976, Cellular inflammatory response of *Penaeus aztecus* and *P. setiferus* to the pathogenic fungus *Fusarium* sp. isolated from the Californian brown shrimp *P. californiensis*, *J. Invertebr. Pathol.* **27**:77.

Solum, N. O., 1970a, Coagulation in *Limulus*—some properties of the clottable protein of *Limulus polyphemus* blood cells, in: *The Haemostatic Mechanism in Man and Other Animals* (R. G. Macfarlane, ed.), *Symp. Zool. Soc. London* **27**:207.

Solum, N. O., 1970b, Some characteristics of the clottable protein of *Limulus polyphemus* blood cells, *Thromb. Diath. Haemorrh.* **23**:170.

Solum, N. O., 1973, The coagulogen of *Limulus polyphemus* hemocytes: A comparison of the clotted and non-clotted forms of the molecule, *Thromb. Res.* **2**:55.

Sparks, A. K., and Fontaine, C. T., 1973, Host response in the white shrimp *Penaeus setiferus* to infection by the larval trypanorhynchid cestode *Prochristianella penaei*, *J. Invertebr. Pathol.* **22**:213.

Splittstoesser, C. M., Kawanishi, C. Y., and Tashiro, H., 1978, Infection of the European chafer, *Amphimallon majalis*, by *Bacillus popilliae*. I. Light and electron microscope observations, *J. Invertebr. Pathol.* **31**:84.

Sprague, V., 1965, *Nosema* sp. (Microsporidia: Nosematidae) in the musculature of the crab, *Callinectes sapidus*, *J. Protozool.* **12**:66.

Sprague, V., 1970, Some protozoan parasites and hyper-parasites in marine decapod Crustacea, in: *Symposium on Diseases of Fishes and Shellfishes* (S. F. Snieszko, ed.), pp. 416–430, American Fisheries Society, Washington, D.C.

Stagner, J. I., and Redmond, J. R., 1975, The immunological mechanisms of the horseshoe crab, *Limulus polyphemus*, *Mar. Fish. Rev.* **37**:11.

Stähelin, H., Suter, E., and Karnovsky, M. L., 1956, Studies on the interaction between phagocytes and tubercle bacilli. I. Observations on the metabolism of guinea pig leucocytes and the influence of phagocytosis, *J. Exp. Med.* **104**:121.

Stähelin, H., Karnovsky, M. L., Farnham, A. E., and Suter, E., 1957, Studies on the interaction between phagocytes and tubercle bacilli. III. Some metabolic effects in guinea pigs associated with infection with tubercle bacilli, *J. Exp. Med.* **105**:265.

Stang-Voss, C., 1971, Zur ultrastructur der blutzellen wir-belloser Tiere. V. Über die hämocyten von *Astacus astacus* (L.) (Crustacea), *Z. Zellforsch. Mikrosk. Anat.* **122**:68.

Sternshein, D. J., and Burton, P. B., 1980, Light and electron microscopic studies of crayfish hemocytes, *J. Morphol.* **165**:67.

Stewart, J. E., 1975, Gaffkemia, the fatal infection of lobsters (genus *Homarus*) caused by *Aerococcus viridans* (var) homari: A review, *Mar. Fish Rev.* **37**:20.

Stewart, J. E., and Arie, B., 1974, Effectiveness of vancomycin against gaffkemia, the bacterial disease of lobsters (genus *Homarus*), *J. Fish. Res. Board Can.* **31**:1873.

Stewart, J. E., and Rabin, H., 1970, Gaffkemia, a bacterial disease of lobsters, in: *Symposium on Diseases of Fishes and Shellfishes* (S. F. Snieszko, ed.), pp. 405–408, American Fisheries Society, Washington, D.C.

Stewart, J. E., and Zwicker, B. M., 1974a, Comparison of various vaccines for inducing resistance in the lobster *Homarus americanus* to the bacterial infection gaffkemia, *J. Fish. Res. Board Can.* **31**:1887.

Stewart, J. E., and Zwicker, B. M., 1974b, Induction of internal defence mechanisms in the lobster *Homarus americanus*, in: *Contemporary Topics in Immunology*, Volume 4, *Invertebrate Immunology* (E. L. Cooper, ed.), pp. 233–239, Plenum Press, New York.

Stewart, J. E., Cornick, J. W., and Dingle, J. R., 1967, An electronic method for counting lobster (*Homarus americanus*) Milne-Edwards hemocytes and the influence of diet on hemocyte numbers and hemolymph proteins, *Can. J. Zool.* **45**:291.

Strangways-Dixon, J., and Smith, D. S., 1970, The fine structure of gill podocytes in *Panulirus argus* (Crustacea), *Tissue Cell* **2**:611.

Stuart, A. E., 1968, The reticuloendothelial apparatus of the lesser octopus (*Eledone cirrosa*), *J. Pathol. Bacteriol.* **96**:401.

Sundara-Rajulu, G., 1970, A study of haemocytes of a centipede *Ethmostigmus spinosus* (Chilopoda: Myriapoda), *Curr. Sci.* **20**:324.

Sundara-Rajulu, G., 1971a, A study of haemocytes of millipede *Cingalobulus bugnioni* Carl. (Diplopoda: Myriapoda), *Indian J. Zool.* **2**:73.

Sundara-Rajulu, G., 1971b, A study of haemocytes in a centipede *Scolopendra morsitans* (Chilopoda: Myriapoda), *Cytologia* **36**:515.

Sundara-Rajulu, G., Krishnan, N., and Singh, M., 1970, The haemocytes of *Eoperipatus weldoni* (Onychophora: Arthropoda), *Zool. Anz.* **184**:220.

Tai, J. Y., and Liu, T.-Y., 1977, Studies on *Limulus* amoebocyte lysate. Isolation of pro-clotting enzyme, *J. Biol. Chem.* **252**:2178.

Tai, J. Y., Seid, R. C., Jr., Huhn, R. D., and Liu, T.-Y., 1977, Studies on *Limulus* amoebocyte lysate. II. Purification of the coagulogen and the mechanism of clotting, *J. Biol. Chem.* **252**:4773.

Tait, J., 1910, Crustacean blood coagulation as studied in the Arthropoda, *Q. J. Exp. Physiol.* **3**:1.

Tait, J., 1911, Types of crustacean blood coagulation, *J. Mar. Biol. Assoc. U.K.* **9**:191.

Tait, J., and Gunn, J. D., 1918, The blood of *Astacus fluviatilis*: Study in crustacean blood with special reference to coagulation and phagocytosis, *Q. J. Exp. Physiol. Cogn. Med. Sci.* **12**:35.

Taylor, R. L., 1969, A suggested role for the polyphenol-phenoloxidase system in invertebrate immunity, *J. Invertebr. Pathol.* **14**:427.

Teakle, R. E., 1973, Records of virus diseases in insects in Queensland, *Queensl. J. Agric. Anim. Sci.* **30**:191.

Thomas, I. G., and Ratcliffe, N. A., 1981, Studies on recognition of foreignness in insects utilizing integumental transplants, in: *Developmental and Comparative Immunology*, Vol. 1 (J. B. Solomon, ed.), pp. 105–110, Pergamon Press, Elmsford, N.Y.

Thomas, I. G., and Ratcliffe, N. A., 1982, Integumental grafting and immunorecognition in insects. *Dev. Comp. Immunol.* **6**: (in press).

Toney, M. E., 1958, Morphology of the blood cells of some Crustacea, *Growth* **22**:35.

Tripp, M. R., 1966, Hemagglutinin in the blood of the oyster (*Crassostrea virginica*), *J. Invertebr. Pathol.* **8**:478.

Tripp, M. R., and Kent, V. E., 1967, Studies on oyster cellular immunity *In Vitro* **3**:129.

Tuzet, O., and Manier, J. F., 1954, Les Organes hématopoiétiques et le sang des Myriapodes Diplopodes (Etude par le microscope à contraste de phase), *Bull. Biol. Fr. Belg.* **88**:90.

Tuzet, O., and Manier, J. F., 1955, Sporazoaires et cilies parasite de myriapodes diplopodes recoltes dans la forét de la Mandraka (Madagascar), *Mem. Inst. Sci. Madagascar Ser. A* **9**:15.

Tuzet, O., and Manier, J. F., 1958, Recherches sur *Peripatopsis moseleyi* Wood-Mason. Peripate du Natal I. Etude sur le sang II. La spermatogenese, *Bull. Biol. Fr. Belg.* **91**:7.

Tyson, C. J., and Jenkin, C. R., 1973, The importance of opsonic factors in the removal of bacteria from the circulation of the crayfish (*Parachaeraps bicarinatus*), *Aust. J. Exp. Biol. Med. Sci.* **51**:609.

Tyson, C. J., and Jenkin, C. R., 1974, Phagocytosis of bacteria in vitro by haemocytes from the crayfish (*Parachaeraps bicarinatus*), *Aust. J. Exp. Biol. Med. Sci.* **32**:341.

Tyson, G. E., 1975, Phagocytosis and digestion of spirochaetes by amoebocytes of infected brine shrimp, *J. Invertebr. Pathol.* **26**:105.

Unestam, T., 1975, Defence reactions in and susceptibility of Australian and New Guinean freshwater crayfish to European crayfish plague fungus, *Aust. J. Exp. Biol. Med. Sci.* **53**:349.

Unestam, T., and Nyhlén, L., 1974, Cellular and noncellular recognition of and reactions to fungi in crayfish, in: *Contemporary Topics in Immunobiology*, Volume IV, *Invertebrate Immunology* (E. L. Cooper, ed.), pp. 189–206, Plenum Press, New York.

Unestam, T., and Nylund, J. E., 1972, Blood reactions in vitro in crayfish against a fungal parasite *Aphanomyces astaci*, *J. Invertebr. Pathol.* **19**:94.

Unestam, T., and Söderhäll, K., 1977, Soluble fragments from fungal cell walls elicit defence reactions in crayfish, *Nature (London)* **267**:45.

Unestam, T., and Weiss, D. W., 1970, Host–parasite relationship between crayfish and crayfish disease fungus *Aphanomyces astaci*: Responses to infection by susceptible and resistant species, *J. Gen. Microbiol.* **60**:77.

Vago, C., and Vey, A., 1970, *Mycoses d'invertebres*, Service du film Recherche Scientifique, Paris.

Valeri, O. M., 1934, Osservazioni sulla morfologia degli elementi del sangue di *Pachyiulus communis* (Savi), *Memories Soc. tosc. Sci. Nat.* **43**:1.

Vey, A., and Fargues, J., 1977, Histological and ultrastructural studies of *Beauveria bassiana* infection in *Leptinotarsa decemlineata* larvae during ecdysis, *J. Invertebr. Pathol.* **30**:207.

Vey, A., and Vago, C., 1969, Recherches sur la guérison dans les infections cryptogamiques d'insectes: Infections a *Aspergillus niger* V. Tiegh. chez *Galleria mellonella* L., *Ann. Zool. Ecol. Anim.* **1**:121.

Vey, A., Quiot, J. M., and Vago, C., 1968, Formation in vitro de réactions d'immunité cellulaire chez les insects, in: *Proceedings, Second International Colloquium on Invertebrate Tissue Culture* (C. Barigozzi, ed.), pp. 254–263, Instituto Lombardo di Scienze e Lettere, Milan.

Vinson, S. B., 1977, *Microplitis croceipes*: Inhibitions of the *Heliothis zea* defense reaction to *Cardiochiles nigriceps*, *Exp. Parasitol.* **41**:112.

Vostal, A., and Pirčová, E., 1968, Zur kenntnis der hämocyten der vielfüsser (Diplopoda), *Biologia (Bratislava)* **23**:161.

Vostal, Z., 1970, On typification of tracheate hemocytes, [in Czech], *Biologica (Bratislava)* **25**:811.

Walker, I., 1959, Die Abwehrreaktion des Wirtes *Drosophila melanogaster* gegen die zoophage Cynipidae *Pseudeucoila bochei* Weld., *Rev. Suisse Zool.* **68**:569.

Walters, J. B., and Ratcliffe, N. A., 1981. A comparison of the immune response of the wax moth *Galleria mellonella* to pathogenic and non-pathogenic bacteria, in: *Developmental and Comparative Immunology*, Vol. I (J. B. Solomon, ed.), pp. 147–152, Pergamon Press, Elmsford, N.Y.

Warner, G. F., 1977, *The Biology of Crabs*, Elek Science, London.

Weir, D. M., 1973, *Immunology for Undergraduates*, 3rd ed., Churchill Livingstone, Edinburgh.

Whitcomb, R. F., Shapiro, M., and Granados, R. R., 1974, Insect defence mechanisms against microorganisms and parasitoids, in: *The Physiology of Insecta*, Vol. 5, 2nd ed. (M. Rockstein, ed.), pp. 447–536, Academic Press, New York.

White, K. N., and Ratcliffe, N. A., 1981, Crustacean internal defence mechanisms: Clearance and distribution of injected bacteria by the shore crab *Carcinus maenas* (L.), in: *Developmental and Comparative Immunology*, Vol. I (J. B. Solomon, ed.), pp. 153–158, Pergamon Press, Elmsford, N.Y.

Wigglesworth, V. B., 1937, Wound healing in an insect, *Rhodnius prolixus* (Hemiptera), *J. Exp. Biol.* **14**:364.

Wigglesworth, V. B., 1943, The fate of haemoglobin in *Rhodnius prolixus* (Hemiptera) and other blood sucking arthropods, *Proc. Soc. London Ser. B* **131**:313.

Wigglesworth, V. B., 1959, Insect blood cells, *Annu. Rev. Entomol.* **4**:1.
Wigglesworth, V. B., 1965, *The Principles of Insect Physiology*, 6th ed., Methuen, London.
Wigglesworth, V. B., 1970, The pericardial cells of insects: Analogue of the reticuloendothelial system, *J. Reticuloendothelial Soc.* **7**:208.
Wigglesworth, V. B., 1973, Haemocytes and basement membrane formation in *Rhodnius*, *J. Insect Physiol.* **19**:831.
Wigglesworth, V. B., 1979, Hemocytes and growth in insects, in: *Insect Hemocytes: Development, Forms, Functions and Techniques* (A. P. Gupta, ed.), pp. 303–318, Cambridge University Press, London.
Williams, A. J., and Lutz, P. L., 1975a, Blood cell types in *Carcinus maenas* and their physiological role, *J. Mar. Biol. Assoc. U.K.* **55**:671.
Williams, A. J., and Lutz, P. L., 1975b, The role of the haemolymph in the carbohydrate metabolism of *Carcinus maenas*, *J. Mar. Biol. Assoc. U.K.* **55**:667.
Wood, P. J., and Visentin, L. P., 1967, Histological and histochemical observations on haemolymph cells in the crayfish, *Orconectes virilis*, *J. Morphol.* **123**:559.
Wood, P. J., Podlewski, J., and Shenk, T. E., 1971, Cytochemical observations of hemolymph cells during coagulation in the crayfish *Orconectes virilis*, *J. Morphol.* **134**:479.
Wright, K. A., 1964, The fine structure of the nephrocytes of the gills of two marine decapods, *J. Ultrastruct. Res.* **10**:1.
Wyatt, G. R., and Linzen, B., 1965, The metabolism of ribonucleic acid in *Cecropia* silkmoth pupae in diapause, during development and after injury, *Biochim. Biophys. Acta* **103**:588.
Young, N. S., Levin, J., and Prendergast, R. A., 1972, An invertebrate coagulation system activated by endotoxin: Evidence for enzymatic mediation, *J. Clin. Invest.* **51**:1790.
Zachary, D., and Hoffmann, J. A., 1973, The haemocytes of *Calliphora erythrocephala* (Meig.) (Diptera), *Z. Zellforsch. Mikrosk. Anat.* **141**:55.
Zachary, D., Brehélin, M., and Hoffmann, J. A., 1975, Role of the "thrombocytoids" in capsule formation in the dipteran *Calliphora erythrocephala*, *Cell Tissue Res.* **162**:343.

Cellular Defense Systems of the Echinodermata

RICHARD D. KARP and KATHERINE A. COFFARO

1. INTRODUCTION

The echinoderms are of particular interest in phylogenetic studies because of their relatedness to the vertebrate subphylum. During the early Precambrian period, the animal kingdom is believed to have diverged into two major groups, known as the deuterostomes and the protostomes. Both the echinoderms and the vertebrates are deuterostomes, while all other invertebrates are classified as protostomes. The two groups are distinguished on the basis of major differences in early development, which may suggest that they were derived from different ancestors. "Protostome" is Greek for "mouth first," and in this group, the mouth is derived from the embryonic blastopore. The blastopore gives rise to the anus in deuterostomes ("mouth later"), while the mouth develops as a separate invagination. The pattern of cell cleavage also varies in the two groups. Cells are oriented in a spiral pattern in protostomes, whereas the blastomeres are positioned directly upon one another in the deuterostomes, resulting in a radial cleavage pattern. Early separation of the blastomeres in deuterostomes gives rise to complete individuals; this type of development is known as indeterminate cleavage. It does not occur in protostomes, where removal of blastomeres results in deformed embryos. Cleavage is said to be determinate in the latter case. During later development, the two groups also differ in formation of mesoderm and the coelomic cavity.

The echinoderms are thought to have diverged from their common vertebrate ancestor before the beginning of the Cambrian period, over 600 million years ago. Due to their abundance and calcareous shells, they have been well preserved as fossils. Over 20,000 extinct species have been described, and the phylum is presently represented by 6000 species (Hyman, 1955). A hard shell

---

RICHARD D. KARP • Department of Biological Sciences, University of Cincinnati, Cincinnati, Ohio 45221. KATHERINE A. COFFARO • Division of Natural Sciences, Thimann Laboratories, University of California, Santa Cruz, California 95064.

(test) is the most characteristic feature of the group, and is present in all classes except for the holothurians (sea cucumbers), where it takes the form of microscopic ossicles embedded in the dermis. Most echinoderms are of pentameric radial symmetry, although the larvae are bilaterally symmetrical. They lack a well-defined brain; the nervous system consists of a rather primitive nerve net. The epidermis is covered by various external appendages such as spines in several groups, which aid in locomotion, grooming, and defense. Many of the pedicellaria contain poison that is able to immobilize predators. Present-day echinoderms are represented by five major classes. The Asteroidea, commonly known as starfish or sea stars, usually have five arms, each of which contains gonads and part of the digestive tract. The Ophiuroidea, or brittle stars, also have arms that radiate from a central disk, which is more sharply demarcated than in sea stars, and does not contain lobes of the alimentary tract. The echinoidea (sea urchins and sand dollars) are generally spherical and may be either rounded or flat. The Holothuroidea, commonly known as sea cucumbers, are elongated and lack arms. The Crinoidea, or sea lilies and feather stars, are characterized by elaborately branched arms.

Several coelomocyte types have been identified in each of the five classes of echinoderms. These cells are found to circulate freely in the coelomic fluid and to also pervade body tissues. They have been studied by several investigators beginning in the late 1800s. These early studies have been reviewed by Hyman (1955) and Endean (1966). Most are limited to gross morphological descriptions, and to speculation as to the functions of coelomocytes. They have been mentioned in reference to defense mechanisms only in recent studies (Johnson, 1969a,b,c). Nearly every conceivable function, such as hormonal, respiratory, and nutritive, has been attributed to coelomocytes. These roles have not been substantiated by later studies. However, two major early observations, which have been corroborated by recent studies, are the participation of coelomocytes in phagocytosis and clotting.

2. COELOMIC CELL TYPES

2.1. ASTEROIDEA

Sea stars in general appear to have the fewest types of coelomocytes of all the echinoderms. In fact, cells other than the amebocyte are found very infrequently in the coelomic fluid of these animals (Boolootian and Giese, 1958; Endean, 1966; Kaneshiro and Karp, 1980). The coelomic amebocyte is a relatively large, colorless cell that may assume two morphological forms (Boolootian and Giese, 1958; Johnson and Beeson, 1966; Kaneshiro and Karp, 1980). The first form is the so-called bladder amebocyte (Figure 1) in which the cell sends out distinctive pseudopods that take on a petaloid arrangement, making the cell appear to be surrounded by bags or "bladders." The other type is the filiform amebocyte (Figure 2), which displays long extended pseudopods that may or may not be branched. It would appear that these two morphological forms have

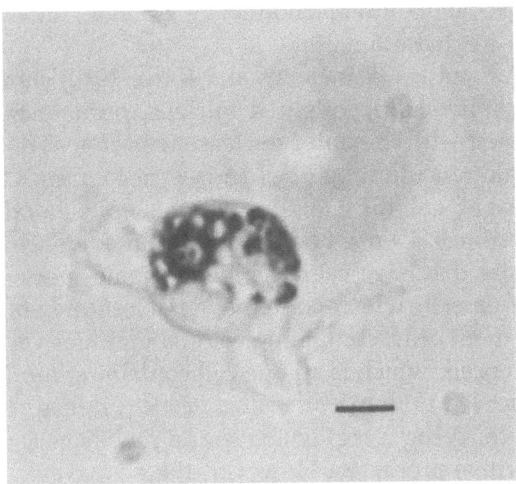

FIGURE 1. Bladder amebocyte from *Dermasterias imbricata* (hanging-drop suspension). Stained with 1% neutral red. Bar = 2 μm.

separate functions. The bladder amebocytes carry out phagocytosis of foreign effete material, whereas the filiform amebocyte primarily participates in the clotting reaction (Johnson and Beeson, 1966). Early evidence indicates that the two amebocytes are actually the same cell, and when conditions favoring coagulation arise, the bladder type converts to the filiform type. Boolootian and Giese (1958), studying both asteroid and holothurian amebocytes, found that if

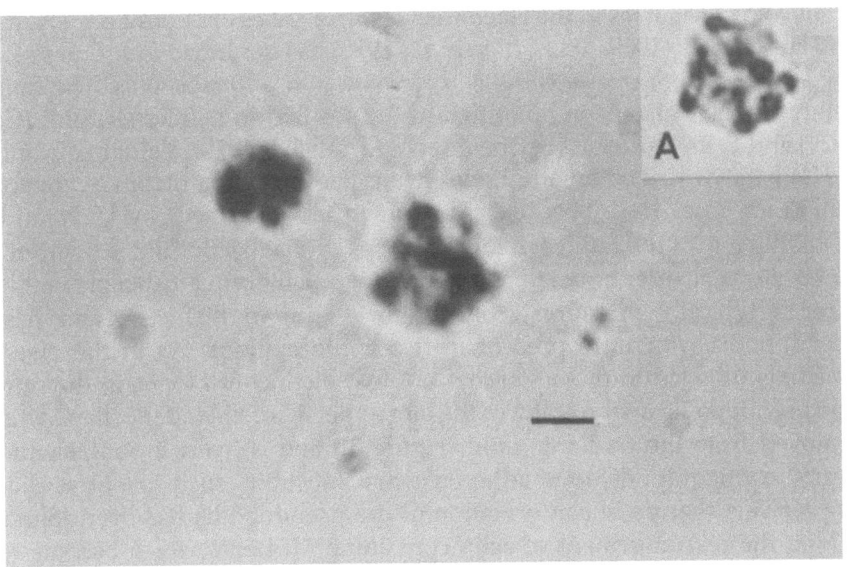

FIGURE 2. Filiform amebocyte from *Dermasterias imbricata* emphasizing filipodia (smear preparation). Inset (a) depicts the cell proper in focus for comparison. Stained with 1% neutral red. Bar = 2 μm.

they inhibited the coagulation reaction using cysteine, only bladder amebocytes were observed in treated animals.

Ultrastructural studies (Kaneshiro and Karp, 1980) show that the coelomic amebocyte has a spherical or indented nucleus, numerous nuclear pores, and ribosomes associated with the outer nuclear membrane. The cytoplasm contains rough endoplasmic reticulum, as well as free ribosomes. Other organelles observed to be present are the Golgi apparatus, mitochondria, and a striking frequency of lysosomes. Dense bodies are also in evidence.

Other cell types that have been ascribed to some species of sea stars are: (1) the colorless morula cell, which contains variable numbers of refractile spherules; (2) small pigment cells, containing colored granules varying in size; and (3) the hyaline plasma cell, which is a large cell containing numerous granules and vacuoles, but observed only in *Poraniopsis inflata* (Endean, 1966).

The amebocyte, being a highly active and mobile cell, was first thought to participate in nutritive processes. However, there is no evidence at this point to suggest the participation of these coelomocytes in the digestion, carriage, or storage of food materials (Endean, 1966). It appears that the primary functions of the asteroid amebocyte are to carry out phagocytosis and participate in the clotting reaction.

2.2. ECHINOIDEA

There has been considerable disagreement over the years as to the types of coelomocytes present in sea urchins. Chien *et al.* (1970) credit this confusion to the interchangeable conformation of some of the cell types, and to the presence of mucoid substances in the coelomocytes. In a series of elegant papers, Johnson and co-workers (1969a,b,c; Chien *et al.*, 1970) has described four types of cells in the sea urchins *Strongylocentrotus purpuratus* and *S. franciscanus*. The same four cell types have also been identified in the sea urchin *Lytechinus pictus* (Coffaro, 1979) and *S. drobachiensis* (Bertheussen and Seljelid, 1978). Vethamany and Fung (1971) have described an additional two types in the sea urchin *S. drobachiensis*. Still other types have been described in *Arbacia punctulata* by Liebman (1950).

Only one of the four coelomocyte types is phagocytic, and is referred to as a leukocyte, a bladder or petaloid amebocyte, or a filiform amebocyte in the literature. It typically measures from 20 to 30 μm in diameter, and has large pseudopodia which are present in two conformations. As in the case of the asteroids, this led to the assignment of two distinct cell types to the same cell. The pseudopodia are typically in the large "petal" or "bladder" shape when first removed from the coelomic fluid (Figure 3), and convert to long slender "filiform" configurations upon adherence to a substrate, such as a glass slide (Figure 4). This change in conformation of the pseudopodia has been found to be due to the rearrangement of actin-containing filaments which become radially oriented, beginning at the periphery of the cell (Edds, 1977). This is accompanied by the withdrawal of cytoplasm between the filaments, which is reversible. The bladder type is often referred to as the active stage, as the other is

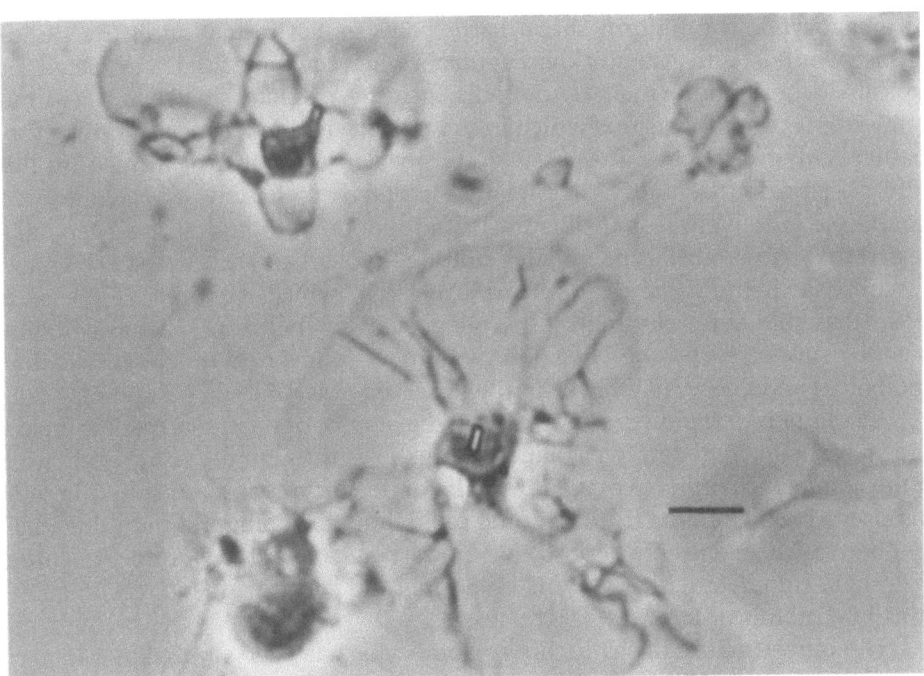

FIGURE 3. Bladder amebocyte from *Lytechinus pictus* (smear preparation). Bar = 9 μm.

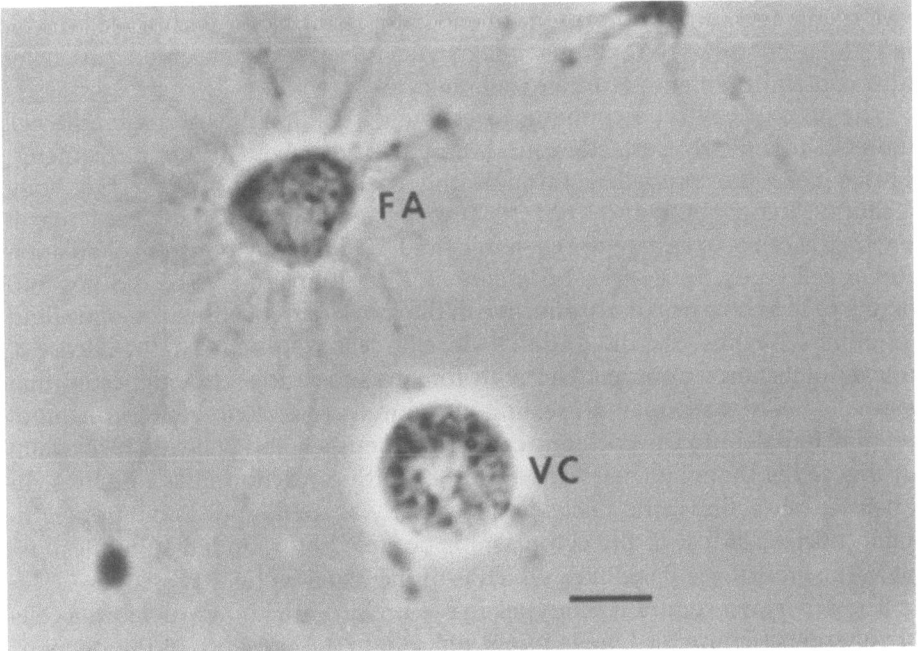

FIGURE 4. Filiform ameboycte (FA) and vibratile cell (VC) from *Lytechinus pictus* (smear preparation). Bar = 5 μm.

nonphagocytic. The filiform shape participates in clot formation at the site of surface wounds, and also quickly coagulates when coelomic fluid is extracted. In typical clot formation, the pseudopodia lose the petaloid shape, and assume long ectoplasmic protrusions which become enmeshed in a syncytium-type formation. Other cell types are passively trapped in the network formed by the filiform pseudopodia. It is generally accepted that only the phagocytic amebocyte is actively responsible for clot formation, although it has been suggested that another cell type initiates clotting. Boolootian and Giese (1959) have described something they call an explosive cell, which ruptures to release a factor that stimulates clotting by the phagocytic amebocytes. This observation was not supported by the later observations of Johnson, who suggested that Boolootian and Giese may have simply observed a despherulated granular cell.

In addition to rapidly forming clots at the surface of wound areas, echinoids have remarkable regenerative abilities. Large holes through the entire body wall of the animals are rapidly covered by a film of phagocytes and red spherule cells (Kindred, 1924; Coffaro, 1979). Dermal cells, thought to migrate from the adjacent body wall, eventually deposit calcium carbonate over the wound. An entire new shell, complete with overlying epidermis and surface appendages, regenerates in a month or two.

The ultrastructure of sea urchin coelomocytes has been studied by Chien et al. (1970). The nuclei of phagocytic amebocytes are found to vary in shape from oval or round to a lobed conformation. The chromosomes are situated around the periphery of the nuclei, and nucleoli are observed. The nuclear membrane contains pores and has ribosomes on the outer surface. Both rough and smooth endoplasmic reticulum are present in the cytoplasm, along with a well-developed Golgi complex. Numerous phagocytic vesicles are observed, many of which contain fragments of other coelomocyte.

The second cell type common to echinoids is known as the vibratile cell (Figure 4). It is nearly perfectly round, and averages about 10 μm in diameter. Vibratile cells are propelled through the coelomic fluid by a single long flagellum. Cuénot (1891) and Kuhl (1937) suggested that the physiological role of the vibratile cell is to agitate the coelomic fluid, and thereby prevent coagulation of other cell types. In a series of in vitro studies, Johnson (1969a) did not find vibratile cells to be particularly effective in the movement of either coelomic fluid or of other cell types. She did find that vibratile cells were active in the release of a mucoid substance upon contact with foreign substances and suggested that these cells might participate in sealing off injured areas to prevent intrusion of foreign materials into the coelomic cavity. Bertheussen and Seljelid (1978) claim that this cell is the major participant in the clotting reaction in sea urchins. In ultrastructure sections, the nuclei are irregularly shaped, and do not appear to contain nucleoli. Most of the cytoplasm is occupied by rounded spherules, except in the area of the flagellum, which is of the usual 9 plus 2 type.

The other two common cell types in sea urchins are known as morula cells and spherules (Figure 5). One is bright red due to the presence of the pigment echinochrome. Kuhn and Wallenfels (1940) found the pigment to be a naphthaquinone of the formula $C_{12}H_{8-10}O_{7-8}$. The other spherule cell is colorless. Both

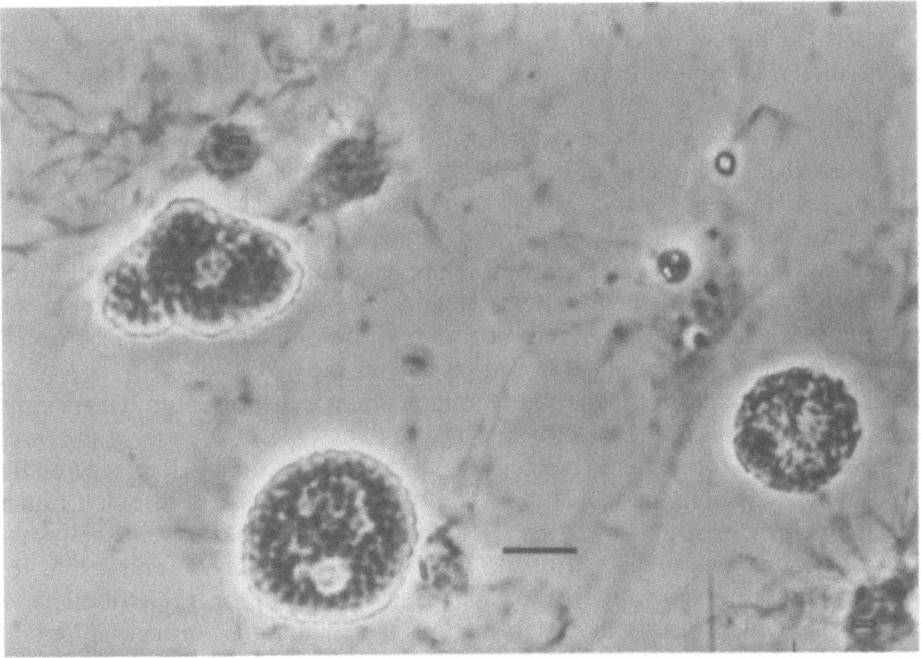

FIGURE 5. Red morula cell from *Lytechinus pictus* (smear preparation). Bar = 4 μm.

cell types move by extension of short, blunt pseudopodia, and have numerous round spherules in the cytoplasm. They generally measure 10–15 μm in diameter, with spherical inclusions of 1–1.5 μm across. *Chien et al.* (1970) found the nuclei to be of irregular shape, and to rarely contain nucleoli. Nuclear pores are occasionally observed. Most of the cytoplasm of the cells contains spherules.

2.3. HOLOTHUROIDEA

There have been several excellent studies detailing the morphology and function of holothurian coelomocytes. Several reviews (Boolootian, 1962; Hetzel, 1963; Endean, 1966; Fontaine and Lambert, 1977) have described either eight or nine cell types circulating in the coelom of sea cucumbers. One cell type that has been well defined is the hemocyte. This is a medium-sized cell (dimensions vary between 12–16 × 13–23 μm) that may exist in a variety of forms depending upon the species being studied. These cells contain hemoglobin, and display a definite similarity to the morphology of both nucleated erythrocytes of lower vertebrates, and early erythroblast stages of mammals (Fontaine and Lambert, 1973). In contrast to erythrocytes of vertebrates, holothurian hemocytes maintain an impressive number of active organelles, such as mitochondria and ribosomes (Fontaine and Lambert, 1973). Although the cells usually appear to have a smooth cytoplasmic membrane, they may have an occasional filiform pseudopod that is no greater than 5 μm in length (Hetzel, 1963). Fontaine and Lambert (1973) also described the formation of blebs on the cytoplasmic membrane that are formed during the exocytosis process resulting from autophagy.

Ultrastructurally (Fontaine and Lambert, 1973), the hemocyte has an ovoid nucleus. The observation of a nucleolus is rather rare. Nuclear pores are present in the nuclear membrane, presumably acting as transfer sites for hemoglobin. Ribosomes are also found attached to the outer nuclear membrane. The cytoplasm will typically contain a few mitochondria, a Golgi apparatus, with attending centrioles, ribosomes, numerous lysosomes, a marginal band of microtubules, and a canicular system which may be analogous to endoplasmic reticulum. Close inspection of the entire cytoplasmic membrane of the hemocyte reveals that it is moderately active in pinocytotic activity.

Sea cucumbers also have the ubiquitous amebocyte present in the coelomic fluid. In fact, this is usually the most abundant type of leukocyte observed (Hetzel, 1963). Once again, this cell is found as either the bladder type or the filiform type. Fontaine and Lambert (1977) also distinguish a so-called transitional amebocyte, which is characterized by collapsed bladders and the formation of short lamellipodlike protrusions. Hetzel (1963) has observed that the bladder pseudopodia appear to collapse, instigating a redistribution of cytoplasm, which results in the formation of branching filiform pseudopodia that may reach impressive lengths. As an example, a typical bladder pseudopod may be 5–15 μm long, whereas filiform pseudopods may reach up to 50 μm in length.

Ultrastructural studies (Fontaine and Lambert, 1977) reveal that the amebocyte usually has a bean-shaped nucleus with a single nucleolus. The nuclear membrane has numerous nuclear pores and irregularly spaced ribosomes attached to the outer membranes. The cytoplasm contains rough endoplasmic reticulum, some free ribosomes, and smooth vesicles, some of which may represent smooth endoplasmic reticulum. The Golgi apparatus is well developed, with centrioles lying nearby. There are abundant mitochondria which tend to cluster near RER. Frequent lysosomes appear in the cytoplasm and are particularly abundant in bladder amebocytes. Other vacuoles associated with pinocytosis and exocytosis are also in evidence.

Fontaine and Lambert (1977) studied the sequence of morphological changes in the amebocytes using time-lapse photography. They found that the bladderlike pseudopods of the first type are in constant motion, resulting in shape changes, collapse, and reformation of new bladders. It was also determined that these bladders are not necessarily closed vesicles, since an unattached leading edge close to the cell body was often observed. Bladders could therefore be interpreted as lamellipods which reflect back upon themselves forming petaloid protrusions. The bladder phase was observed to progress to the transitional phase by the collapse of the bladders, followed by the formation of short lamellipods containing an inner core of microfilaments. The filiform phase comes about presumably by the extension of filipods by means of extensive microfilament and possibly microtubule systems.

The holothurian amebocyte carries out the same functions of phagocytosis and clot formation as other echinoderms. However, unlike asteroid amebocytes, the sea cucumber counterpart apparently has the ability to digest and transport food materials (Hetzel, 1963; Endean, 1966). Evidence indicates that amebocytes transport spherules containing digestive enzymes from the coelomic epithelium to the digestive tract. Amebocytes may also transport the products of digestion from the digestive tract to other parts of the sea cucumber body.

The third cell type found in holothurian coelomic fluid is the lymphocyte. This is the second most abundant leukocyte, and has been well described by several investigators (Hetzel, 1963; Endean, 1966; Fontaine and Lambert, 1977). The lymphocyte is a relatively small cell, 6–8 μm in diameter, with a spherical nucleus that almost fills the cell, surrounded by clear cytoplasm. These cells may bear one or more filipods, but they rarely exceed 15 μ in length. Although these pseudopodia may be branched, they never appear as complex as those of amebocytes. The lymphocytes do not appear to have any means of locomotion, nor do the filipods show movement.

The ultrastructural studies of Fontaine and Lambert (1977) reveal that the lymphocyte is the least complicated of the coelomocytes, and in fact, its lack of cytoplasmic sophistication is taken as being distinctive for this cell. A deeply indented nucleus with a single central nucleolus dominates the cell's contents. Evenly spaced ribosomes are found associated with the nuclear membrane. Another distinctive feature of the lymphocyte is the abundance of both free and RER-bound ribosomes in the cytoplasm. There does not appear to be any evidence of either pinocytotic or phagocytic activity in the cell. Other organelles in evidence are a single Golgi apparatus, small mitochondria, and abundant microtubules. Microfilaments are not observed.

Hetzel (1965) states that the lymphocyte may be the basic stem cell from which hemocytes, amebocytes, and morula cells arise. This is based on the observation of intermediate cell types between lymphocyte–hemocyte and lymphocyte–amebocyte. Fontaine and Lambert (1977) observed intermediates between lymphocytes and amebocytes, and between lymphocytes and morula cells. Unlike Hetzel, they did not find intermediates between lymphocytes and hemocytes. Lymphocytes themselves may differentiate from mesenchymal connective tissue cells residing in the walls of the hemal vessels.

The least numerous of the leukocytes are the morula cells. These medium-sized cells (8–20 μm) appear to be either spheroid or ovoid, and are characterized by the presence of numerous colorless spherules in the cytoplasm (Hetzel, 1963, 1965; Endean, 1966; Fontaine and Lambert, 1977). The spherules consist of a core of protein probably complexed to carbohydrate, and surrounded by an outer shell of sulfated acid mucopolysaccharide (Hetzel, 1965). Morula cells also appear to display a form of true ameboid locomotion which involves the extrusion of a pseudopod from the leading edge of the cell membrane, and the rolling of the spherules into the bleb, followed by the retraction of the trailing edge (Fontaine and Lambert, 1977).

Ultrastructural studies (Fontaine and Lambert, 1977) reveal that the morula cell has an irregularly shaped nucleus, which has no definite location since it rolls about the cell in response to the locomotory activity. The nucleus contains a single nucleolus. The spherules unique to these cells dominate the cytoplasm, and are 5 μm in diameter. It appears that morula cells will frequently degranulate when present in connective tissue. The spherules do not leave much room for cytoplasm. The small cytoplasmic area in conjunction with the nucleus contains a Golgi apparatus, free ribosomes, some RER, and mitochondria. The remaining cytoplasm elsewhere in the cell contains mitochondria, free ribosomes, RER, and some smooth vacuoles, but lysosomes are quite infrequent. On the other hand, microtubules are very abundant.

Morula cells do not appear to be active phagocytes, since they fail to endocytose injected carbon particles (Hetzel, 1965). These cells may play a role in food storage (Hetzel, 1963). Most investigators have theorized that the major contribution of morula cells is in aiding wound healing (Endean, 1958; Menton and Eisen, 1973; Fontaine and Lambert, 1977). This fiber-producing cell has been observed infiltrating wound sites, and may be responsible in part for the formation of connective tissue and the reconstruction of the superficial dermis and deeper dermal layers. However, their specific function(s) has not been totally defined.

Several other "cells" have been described in holothurians (Hyman, 1955; Hetzel, 1963; Endean, 1966); however, not all of these appear to be legitimate cells. The so-called crystal cell contains highly refractile bodies that appear to be crystalline in nature. These cells are extremely fragile and rapidly disintegrate when subjected to any stress. Thus, there has been no identification of the crystalline substance, nor of the function of these cells. Minute corpuscles have also been described, which are only 1–6 μm in diameter. These appear to be enucleated fragments of other coelomocytes that have been cytolyzed. Fusiform cells are structureless columns of cytoplasm that may represent torn tissue fragments. Brown bodies are aggregations of granular material (yellow to brown in color) that accumulate in the coelomic cavities, hemal systems, and tissues. These forms may represent degenerating cells or collections of cell debris including concentrations of hemoglobin. Vibratile cells are spherical in shape (about 8 μm in diameter) and may have one or two flagella that may be up to 23 μm in length. There has been some question as to whether or not these cells are contaminants or bona fide coelomocytes. Most would agree that hemocytes, amebocytes, lymphocytes, and morula cells are the cell types that should be considered to be the normal cellular constituents of holothurian coelomic fluid.

2.4. CRINOIDEA

Cuénot (1891) described the presence of both colored and colorless morula cells in the coelomic fluid of *Antedon rosacea*. Large spherical corpuscles have also been observed (Boolootian and Giese, 1958). Crinoids also have an amebocyte which is quite granular in appearance, with short pseudopodia that are unlike those typical of phagocytic cells in the asteroids, echinoids, or holothurians (Cuénot, 1891). However, it has been demonstrated that these cells are indeed phagocytic (Reichensperger, 1912). Another cell type that is quite elongated, and contains numerous rods and granules, has no known function (Reichensperger, 1912).

2.5. OPHIUROIDEA

This class of animal has probably received less attention than any of the echinoderms. The ophiuroids appear to have a cell analogous to the holothurian hemocyte since it contains hemoglobin and can assume a variety of shapes (Foettinger, 1880). Granular cells demonstrating short pseudopodia and phagocytic activity have also been reported (Cuénot, 1888; Kindred, 1924), as well as colorless morula cells and vibratile cells (Kindred, 1924). Unfortunately, the limited nature of these studies did not allow for the elucidation of the physiological roles of the ophiuroid coelomocytes (nor of crinoids for that matter). However, there is enough information to ensure that phagocytic cells are phylogenetically distributed among the various classes of echinoderms.

3. ORIGIN OF THE COELOMOCYTES

There is a great deal of interest in determining where coelomocytes originate, and if echinoderms possess any organs analogous to that of vertebrate lymphoid tissue.

Sea stars possess some anatomical structures that could possibly play the role of lymphoid-type tissue. One such organ is the blood axial sinus, which is in close conjunction with the stone canal of the water vascular system and is an integral part of the blood vascular system. This organ is in constant contact with the hemolymph, which passes through it, and thus might be an ideal site for stimulation by antigen. Leclerc (1973) has reported that repeated injections of horseradish peroxidase in or about the axial organ induces the appearance of a substance which specifically reacts with the enzyme. Other possible candidates are the Tiedemann bodies which are organized paired structures attached to the basal circular water ring. These organs appear to be highly specialized for absorption and phagocytosis. Ultrastructural studies reveal that these organs contain a great deal of coelomocytes, as well as ciliated cells, and are divided into segments by canallike structures that radiate out to the periphery (Kaneshiro and Karp, 1980). Furthermore, Tiedemann bodies are in communication with all major body spaces of the sea star, and thus are capable of circulating the animal's fluids around their cells (Ferguson, 1966). These organs are thus possible candidates for being a primitive system of lymph node analogs in the sea star. Ferguson (1966) carried out studies in *Asterias forbesi*, looking for increased cellular activity in these organs. Using autoradiography, he reported only modest increases in the number of new cells in these structures. However, these animals may have been too immature, and also had not been antigenically stressed. So the question of whether or not these organs can generate expanding cell populations remains open. Another possibility has been suggested by van den Bossche and Jangoux (1976) who reported data claiming that the sole origin of coelomocytes in the sea star *Asterias rubens* is the coelomic epithelium.

The extensive studies by Hetzel (1965) on holothurians revealed that there were no indications of free coelomocytes in the coelomic fluid undergoing any stage of cell division in experimentally stressed or control animals. In sea stars, it has similarly been observed that animals recovering from intermittent bleedings do not have any free coelomocytes displaying mitotic figures (Kaneshiro and Karp, 1980). Holland *et al.* (1965), based on tritiated thymidine labeling experiments, have claimed that the sea urchin may be an exception. However, the mere labeling of cells does not necessarily mean they will go on to divide. Moreover, only one of the three animals followed for a prolonged period of time (2 months) showed a significant increase in labeled amebocytes. The question of whether these cells were recruited or were products of cell division in the coelom could not be answered.

If coelomocytes are not dividing in the free coelomic fluid, then one would assume that such division is occurring in the tissues. The studies on sea cucumbers (Hetzel, 1965) could find no suggestion of coelomocyte origin in the walls of the respiratory trees, the body wall, the wall of the digestive tract, the polian vesicles, the water vascular system in general, nor from any part of the reproductive system. Hetzel believes that it is the hemal system of holothurians that may be regarded as lymphoid tissue and may serve as the source of coelomocytes. As was discussed in the preceding section, Hetzel's hypothesis is that the holothurian lymphocyte acts as the basic stem cell from which amebocytes, hemocytes, and morula cells arise. The lymphocytes themselves are thought to possibly differentiate from mesenchymal connective tissue cells in the walls of the hemal vessels. The axial gland immediately comes to mind as an organ that could serve a primary function in this way. Although the axial complex is poorly developed in holothurians, the axial gland may be represented by some lacunar tissue filled with amebocytes found at the base of the stone canal (Hyman, 1955). The work of Smith (1978) on the polian vesicles of the sea cucumber suggests that they may be immunologically reactive tissues. The vesicles from sea cucumbers receiving repeated injection of milkfish (*Chanos chanos*) serum revealed marked enlargement of the walls of these organs as compared to controls. These reactive vesicles often contained large clumps of coelomocytes, which appeared to be mostly lymphocytes.

Although the actual source of coelomocytes has yet to be definitively ascertained, there are obviously many promising avenues of investigation to explore which may lead to the solution of this puzzle.

4. DEFENSE REACTIONS

4.1. PHAGOCYTOSIS

Echinoderm amebocytes are very efficient at clearing the coelomic fluid of foreign material by phagocytosis. The studies of Bang and Lemma (1962) revealed that the coelomic fluid of untraumatized sea stars was free of bacteria.

When these animals were injected with either various concentrations of *Vibrio* sp., or India ink, it was found that it took only 3 days for phagocytes to totally clear the coelomic fluid of these particles. This phenomenon has been confirmed by Kaneshiro and Karp (1980). Reinisch and Bang (1971) injected sea urchin cells (*Arbacia punctulata*) into the sea star *Asterias vulgaris*. They found an abrupt drop in the number of circulating amebocytes in the host. The reason for this was that the injected sea urchin cells adhered to and were phagocytosed by host amebocytes, which then clumped within the dermal papulae and later ruptured to the exterior. Several studies (Bang and Lemma, 1962; Endean, 1966; Reinisch and Bang, 1971) have reported that the papulae provide the main sites of evacuation of amebocytes filled with ingested particles. Other sites that may provide passage to the exterior for coelomocytes are the madreporite and alimentary canal (Endean, 1966). In sea cucumbers, it appears that amebocytes carrying phagocytosed material can be eliminated by migrating through the walls of the respiratory trees, and then pass to the exterior by the expiratory current (Endean, 1966). Those species of holothurians lacking respiratory trees accumulate amebocytes carrying ingested material in the ciliary urns, and then pass these cells to the exterior via the body wall (Endean, 1966).

Sea urchins have also been demonstrated to clear bacteria *in vivo* (Unkles and Wardlaw, 1976; Wardlaw and Unkles, 1978). In the initial study, 10^7 particles of unidentified marine bacteria were injected into the coelomic cavity of the sea urchin, *Echinus esculentus*. The coelomic fluid was found to be clear of bacteria in about 24 hr. These studies were extended (Wardlaw and Unkles, 1978) in a series of *in vitro* tests, using the same species of sea urchin and a marine bacterium, *Pseudomonas* strain No. 111. This strain was chosen because it leaves easily identifiable black colonies when plated out on agar. One thousand colony-forming units were incubated with 1.8 ml of coelomic fluid over a temperature range of 4–22°C. At 10°C, 90% of the bacteria were dead by 24 hr, and by 48 hr, most of the animals had either sterile coelomic fluid or only a few colonies present. There was no seasonal effect in the bactericidal activity. Although coelomic fluid rapidly clots upon withdrawal from the animals, Wardlaw and Unkles found no reduction in the bactericidal activity of coelomic fluid that had been allowed to clot for 2 hr before the addition of bacteria. The most effective temperature was 4°C, although the activity was still present at 22°C.

Coelomocytes were found to be necessary for the elimination of bacteria, as the coelomic fluid alone had no effect. However, intact cells were not required, as disrupted ones effectively eliminated bacteria. The bactericidal activity was heat stable, and was entirely lost upon dialysis against control fluid, suggesting that the responsible factor was of low molecular weight.

Johnson (1969a) has found that coelomocytes can be maintained in hanging-drop cultures for extended periods of time, and she has used this method for observing the behavior of coelomocytes inoculated with bacteria (Johnson, 1969c). The sea urchins *S. purpuratus* and *S. franciscanus* were used in these studies. Both were injected with five gram-negative marine bacterial strains, seven gram-positive strains, and one terrestrial form, which was known to be an insect pathogen. Coelomic extracts from each of the two species of sea urchins reacted in a similar manner to the bacteria. Variable reactions to the different strains of bacteria were observed. Typically, the gram-negative forms were walled off from the rest of the drop by a layer of phagocytic amebocytes, following an initial retreat of the amebocytes from the bacteria. The rapidity of the response varied with the different strains, and wall formation was not observed with the gram-positive forms. When used in heavy doses, one gram-positive strain, *Gaffkya homari*, resulted in temporary wall formation.

Two of the gram-negative strains caused the attraction of red spherule cells to the site of bacterial inoculation. The red cells formed a wall facing the bacteria, and sometimes contacted the foreign material. Generally, the red cells were observed to migrate through a layer of phagocytic amebocytes. Only the gram-positive forms were phagocytosed as whole bacteria. The gram-negative strains were apparently phagocytosed after their disintegration. Johnson (1969c) has speculated that the gram-positive strains were more readily phagocytosed because they are relatively rare in marine environments, and hence were more foreign to the animals. The red cells were found to release echinochrome from their cytoplasmic spherules, which had the effect of lysing the bacteria. When the bacterial strains were incubated in predominately red cell cultures or phagocytic amebocyte drops, the red cells were more effective in suppressing bacterial activity.

In addition to eliminating bacteria, sea urchins have also been found to remove the virus T4 from the coelomic fluid (Coffaro, 1978). Sea urchins belonging to the species *Lytechinus pictus* were injected with 2×10^9 phage particles. Following a 4-day lag period, phage concentration was reduced in logarithmic intervals to a level of 10^2 particles/ml in 10 days. The animals were given a second phage injection of the same dose 1 month after the first one. Both the overall kinetics and the time of clearance were the same as those observed in the primary response. No neutralizing action was detected in the coelomic fluid of animals that had cleared the phage.

There have been few reports of diseased sea urchins. Johnson and Chapman (1970b) have described some specimens of *S. franciscanus* collected with infected spines. Many of the damaged spines were in the process of regeneration. Several microorganisms were observed in histological sections of the spines. Among these were blue-green algae, green algae, diatoms, ciliates, unidentified wormlike organisms, and fungal growth. Red spherule cells were common at the infected sites, and many had released echinochrome from their spherules. Phagocytic amebocytes were also present. Johnson and Chapman have concluded that echinochrome acts as a general disinfectant, and is of considerable importance in walling off and immobilizing foreign materials. Vevers (1963) found that echinochrome acts as an inhibitor of algal growth, and noted that pale specimens were far more apt to be infested with blue-green algae than were normally pigmented animals.

In an earlier study, Johnson and Chapman (1970a) described some specimens of *S. franciscanus* with abnormal tissue swelling on their spines. Large quantities of red spherule cells were present in the growths, giving them a red color. Phagocytes were also observed. Histological sections indicated the presence of organic material which resembled fungal ascospores, although most of the material could not be identified. It seemed likely that the echinochrome had effectively dissolved the foreign materials. Several of the swellings were observed to detach from the epidermis during culture, suggesting a simple means of ridding the sea urchins of infecting organisms.

The phagocytic amebocyte could be considered to be the main line of defense for echinoderms. These cells are extremely efficient in recognizing and clearing foreign particulate matter from the coelomic fluid. This coupled to the ability of the echinoderm to externalize the unwanted agent once it has been phagocytosed, allows these animals to deal with such problems as bacterial or viral infection on a very simple, but effective, basis.

4.2. CLOTTING REACTION

Clot formation in echinoderm coelomic fluid is basically brought about by the aggregation of coelomic amebocytes, but unlike vertebrates, there is no gelation of the fluid proper (Endean, 1966). The clotting reaction of asteroids and holothurians is relatively weaker than that found in echinoids (Binyon, 1972). One of the primary purposes for clot formation is to prevent the loss of coelomic fluid that may occur following an injury. Sea stars and sea cucumbers can effectively prevent such losses by muscular contraction around the afflicted area, whereas echinoids do not usually demonstrate such flexibility due to the nature of their test. It has therefore been theorized that because of this physical difference, there has been more selective pressure on echinoids to develop a quick and effective clotting reaction (Boolootian and Giese, 1959; Binyon, 1972).

As already outlined in Section 2, studies in all three classes indicate that it is the filiform phase of the coelomic amebocyte which participates in clot formation. [Bertheussen and Seljelid (1978) hold that the vibratile cell is more important in this regard in echinoids.] The sequence of events, which has been well described by Fontaine and Lambert (1977), begins with the bladder amebocyte. The bladder-type pseudopods collapse, and the formation of short lamellipod protusions on the membrane surface takes place. This step can be inhibited by the addition of EDTA (Noble, 1970), and is thought to be Ca^{2+} dependent (Boolootian and Giese, 1959). This cell form is now known as the transitional amebocyte, which has the property of developing an increasingly sticky cell membrane that aids in cell-to-cell adherence. The transitional amebocyte acts as the nucleus for clot formation. The change to the filiform amebocyte is thought to be due to the extension of filipods and lamellipods brought about by an extensive system of microfilaments and possible microtubules.

The studies of Boolootian and Giese (1959) and Fontaine and Lambert (1977), in both asteroids and holothurians, reveal that there may be two distinct types of clots formed in echinoderms. The first type of aggregate is relatively loose, and the cells maintain their individual identity. This form can be separated either by mechanical or by chemical means. Fontaine and Lambert believe that aggregates occurring naturally *in vivo* are in this category. The second form results in intimate cell contacts that lead to the formation of a plasmodium. The extensive filipods of the amebocytes involved are retracted and appear to fold back upon themselves in a rather complex fashion (Fontaine and Lambert, 1977). A soluble tissue factor may also be involved in the formation of both of these types of clots (Boolootian and Giese, 1959).

In actuality, the different forms of clots may just be different stages of the same process. If the inducing stimulus is not very intense, the participating cells can disengage and continue to function as individual coelomocytes. If the inducing stimulus is intense, the primary clot continues on to form a growing syncytial mass that can aid in alleviating the problem. Presumably, clot reactions are very important in walling off other internal threats (i.e., infectious agents) besides injuries.

4.3. CELL-MEDIATED IMMUNE RESPONSIVENESS

The ability to reject allografts has now been demonstrated in three classes of echinoderms. Hildemann and Dix (1972) first reported that the tropical sea cucumber *Cucumaria tricolor* could reject integumentary allografts. They also presented preliminary evidence that the sea star *Protoreaster nodosus* might also have this capability. In 1976, Karp and Hildemann reported that *Dermasterias imbricata*, a sea star living in colder water temperatures, specifically rejected allografts, and displayed short-term immunological memory. Coffaro and Hinegardner (1977) have observed that the sea urchin *Lytechinus pictus* can also specifically recognize and reject allografts. Thus, it would appear that echinoderms in general have a well-established form of adaptive cell-mediated immunity.

The work of Hildemann and Dix (1972) revealed that the tropical sea cucumber, *Cucumaria tricolor*, was able to generate a cell-mediated allograft rejection reaction. Full-thickness 8- to 10-mm integumentary allografts were exchanged between partner animals, and sutured in place. The onset of chronic rejection began 40–60 days postgrafting. This was reflected by the appearance of slight discoloration and hyperplasia in the peripheral contact zone, some edema, and the first signs of fading pigmentation. By 70 days postgrafting, rejection was well under way. The median survival time (MST) for the 15 first-set grafts was 165 days (Table 1). Graft rejection was characterized by loss of pigmentation, graft contraction, and the gradual ingrowth of recipient tissue. Histologically, the organization of the tissue elements was completely disrupted, with conspicuous infiltration of the cell layers by what the authors described as phagocytic and small lymphocytelike cells. Three second-set grafts were placed on animals still undergoing first-set rejection. The reaction to these grafts was accelerated with an MST of 43 days (Table 1), and characterized by the early development of inflammation and discoloration, as well as invasive resorption of graft tissue.

Hildemann and Dix (1972) also presented preliminary data indicating that the sea star *Protoreaster nodosus* may also possess the ability to reject allografts. Three 5- to 6-mm full-thickness first-set integumentary allografts were successfully placed on animals by literally tying them down with monofilament nylon.

TABLE 1. ALLOGRAFT REJECTION IN VARIOUS ECHINODERMS

Species	Reference	Allografts	Number of grafts scored	Median survival time (days)
Cucumaria tricolor (Holothuroidea)	Hildemann and Dix (1972)	First-set	15	165
		Second-set	3	43
Protoreaster nodosus (Asteroidea)	Hildemann and Dix (1972)	First-set	3	163
Dermasterias imbricata (Asteroidea)	Karp and Hildemann (1976)	First-set	17	213
		Second-set	5	44
		Third-set	4	8
Lytechinus pictus[a] (Echinoidea)	Coffaro and Hinegardner (1977)	First-set	14	35
		Second-set	14	12

[a] Data for wild-type animals.

These grafts were rejected with an approximate MST of 163 days (Table 1). The gross characteristics of rejection were similar to those of *Cucumaria*, with loss of pigmentation, hyperplasia, and eventual invasion of the graft by recipient tissue. Histologically, the reaction appeared to be mediated by small lymphocytelike cells and phagocytes. The authors also reported finding an eosinophilic granulocyte involved, but this was more likely a small pigment cell or possibly a phagocyte which had engulfed some of the highly pigmented tissue. No repeat grafting was carried out in this study.

Karp and Hildemann (1976) completed a more extensive study on the sea star *Dermasterias imbricata*. Animals were randomly paired, and each partner had a reciprocal 5-mm full-thickness integumentary allograft sutured into place. Once again, the criteria for rejection were loss of normal pigmentation, edema, and necrosis. Seventeen animals rejected their first-set allografts with a MST of 213 days (Table 1). Rejected grafts lost their normal pigmentation, had a fibrous disorganized appearance, and many became contracted. The five second-set allografts completed were rejected in an accelerated fashion with an MST of 44 days (Table 1). The reaction was more intense as reflected by the development of severe edema and large necrotic patches, as well as eventual sloughing of the graft. Four third-set allografts were attempted in this study, with the result that all were rejected with an MST of only 8 days (Table 1). This acute reactivity was characterized by edematous swelling, necrosis, and sloughing in quick succession. Four third-party allografts were placed at the same time as three of the third-set grafts and one of the second-set grafts. None of these allografts were rejected in an accelerated fashion, providing further evidence of the specificity of the response. Furthermore, none of the control autografts performed throughout the study were rejected or lost due to technical reasons. Histologically, rejected allografts, whether primary or repeat, showed complete disruption of the normal tissue cytoarchitecture. This was mediated by a heavy cellular infiltration of the tissue with what appeared to be large phagocyticlike cells, and a second smaller mononuclear cell type. Autografts, as well as tissue adjacent to allografts, showed no evidence of infiltration or tissue disturbance.

It is quite obvious in both of the studies on sea cucumbers and sea stars that there is definitely cellular participation in the graft rejection process. However, there is no indication as yet as to whether the phagocytic cells involved are recruited from the circulating coelomocyte pool or elsewhere. Also, it is still not clear whether two cell types are involved, or two phases of the same cell type. In any case, the histological profile of echinoderm graft reactivity is much the same as what we might expect for a vertebrate type of chronic response.

As part of a broad study to characterize the nature of sea star effector cells, the coelomocytes of *Dermasterias* have been subjected to incubation with various mitogenic substances that are known stimulators of vertebrate immunocytes (Karp and Johns, 1978; Karp, Beckham, and Johns, unpublished results). Assays for mitogenic activity were carried out by incubating 2×10^6 pooled coelomocytes per culture (all cultures performed in triplicate) with various concentrations of the mitogen being tested. A significant increase in the uptake of [^{3}H]thymidine over that of controls receiving no mitogen, was taken as being a positive result. The results from extensive experimentation indicate that phytohemagglutinin and pokeweed mitogen have no stimulatory effect on coelomocytes in culture. However, coelomocytes incubated 7 days with 10 μg of bacterial lipopolysaccharide (LPS) showed a two- to threefold increase over that of controls in the incorporation of [^{3}H]thymidine. In addition, *Dermasterias* coelomocytes incubated with 100 μg or more of concanavalin A (Con A) for 3 days also demonstrated a significant two- to threefold increase in the uptake of [^{3}H]thymidine. The response to Con A could be totally inhibited by the addition of 0.1 M α-methyl-D-mannoside. Since other experiments (Karp and Beckham, unpublished results) have shown that there is no mixed lymphocyte reactivity among cells from conventional animals, the stimulation of the pooled cells was due to the mitogens and not to mixed lymphocyte culture activity.

The reactivity to LPS does not necessarily indicate that sea stars have analogs to vertebrate B lymphocytes. Rather, it may be that because these animals are constantly exposed to marine gram-negative bacteria in nature, they have developed receptors to these substances without regard to cellular subpopulations.

Although it took a relatively large amount of Con A to achieve stimulation, the reaction only took 3 days to manifest itself. It would seem that the receptors for Con A may either be very sparse on the cell surface, or be of extremely low affinity. However, once the concentration is high enough, the reaction occurs quite quickly. It is apparent from these data that *Dermasterias* will consistently react to a known T-lymphocyte mitogen.

The cells that are stimulated in culture appear to be amebocytes, since these cells dominate the coelomic fluid of sea stars. This taken with the fact that these cells are actively involved in graft rejection, makes them a strong candidate for the role of echinoderm immunocyte.

Extensive grafting studies have also been carried out on the sea urchin *Lytechinus pictus* (Coffaro and Hinegardner, 1977; Coffaro, 1979). Reciprocal 3-mm full-thickness integumentary allografts were exchanged between pairs of animals (including groups of laboratory inbred animals, as well as wild-type animals). The first gross response occurred with the appearance of red spherule cells on the grafts. The brightly colored cells were frequently first observed in the border between the host and the graft, and appeared to migrate onto the graft from the underlying tissue. These cells were present in slight-to-moderate numbers on most of the autografts, and decreased between the first and third weeks after surgery until no longer present. The pigmented cells were present in considerably greater density on the allografts, and were observed throughout the rejection period.

For all rejected grafts, the gross morphology of the rejection process was similar. The graft epidermis receded from its borders, sometimes forming large blisters. Concurrent with this event, the graft surface structures, such as spines and pedicellaria, came to lie flat against the graft test. In a period of about 3 to 6 weeks, the receding epidermis, including the external appendages, formed a clump in the center of the graft, which continued to deteriorate over a period of several more weeks, leaving what finally appeared to be a cell-free piece of test on the host. Rejection time has been defined as the point of appearance of this clump of degenerating tissue. Grafts between unrelated donors were rejected in about 1 month, whereas grafts exchanged between related animals survived for over 2 months in some cases. First-set allografts in wild-type animals were rejected with an MST of 35 days (Table 1).

Another set of experiments (Coffaro, 1979) examined second-set grafts. The specificity of second-set rejection was tested by giving some animals a graft from an animal from a different geographical location (a third-party graft). Second-set and/or third-party grafts were placed on animals either immediately after first-set graft rejection, 1 month later, or 2 months later. The average rate of both second-set and third-party rejection was about half that of primary graft rejection (Table 1).

There was no significant difference among the rates of rejection for the second-set grafts, indicating that memory does not change appreciably between the time of primary graft rejection and 2 months later. There was also no significant difference among the rates of third-party graft rejection. For each of the experimental groups, the rate of second-set graft rejection was compared to the rate of third-party rejection. Again, no significant difference was found between the rates of rejection.

Although the second-set and third-party grafts did not test to be different with the Mann–Whitney or Kurskle–Wallis tests (which allow for data that do not have a normal distribution), two tendencies suggested dividing the data for second-set and third-party rejectors into groups of those which rejected at an accelerated rate and those which failed to do so. The means for the second-set groups tended to be lower than the means for the third-party groups, and the third-party groups also tended to have more grafts which failed to show accelerated rejection. Twenty-four days was chosen as the cutoff value for accelerated rejection. This number was selected because 90% of the 400 primary grafts that were exchanged during the most recent series of experiments rejected at 24 or more days. Therefore, all second-set and third-party grafts were divided according to the 24-day value and analyzed by the χ^2 contingency method. A difference between the rate of second-set and third-party rejection was found to be present at the 0.01 level. The same significant difference was present using cutoff values of all days from 20 to 30 in the χ^2 analysis.

Although the evidence for specificity was not striking, it does suggest that some degree of specificity is present. These results corroborate findings in the tissue incompatibility systems of the sponges and corals, where antigen sharing between second-set and third-party donors has been suggested as the reason for accelerated rejection of some third-party grafts (Hildemann et al., 1977, 1979).

The position of the allograft was found to affect the rejection rate. Grafts were placed in four different areas of the oral, aboral, and central surfaces of the hosts. The area in front of the madreporite was designated as 0°. Moving counterclockwise, grafts were positioned at 0, 90, 180, and 270°. Twenty grafts were placed in each of the four positions on the oral, aboral, and central surfaces of the animals. The 270° oral grafts tested to be significantly different in their rejection rates from the other positions. A delayed rejection of 42.1 ± 8.8 days was found.

The sex of the animals did not affect the rate of rejection. However, a dosage effect was present. Grafts measuring 4 mm were rejected with a mean of 34.4 ± 8.6 days; grafts of 5 mm were rejected in 42.5 ± 8.1 days; and grafts of 6 mm were rejected in 56.7 ± 13.0 days.

Allografts have also been exchanged between individuals belonging to other species of sea urchins (Coffaro, 1979). *S. purpuratus* rejected all allografts in the same time period and with an almost identical gross morphology as did *L. pictus*. The limited data obtained with a species other than *L. pictus* do suggest that the ability to reject allografts is not limited to this group, and may be present in all echinoids.

Sea urchins rapidly reject xenografts (Coffaro, 1979). Xenografts have been exchanged between the species *S. purpuratus* and *L. pictus*. Both species responded in the same manner to xenogeneic tissue. About one-half of the grafts for each group were entirely covered with a dense filmlike layer of red spherule cells by the second day after surgery. By the fifth day, these grafts were considered to be rejected, as the film of red cells had resulted in complete inactivity of the grafts' external appendages. The spines and pedicellaria were collapsed on the graft test. Many of these structures deteriorated around the base, and fell off the graft. The other half of the grafts in each group were rapidly infiltrated by red cells, as opposed to being covered by a film layer. By the fourth day, the graft epidermis began to recede from the periphery of the grafts, and by the eighth day, all of the grafts which responded in this manner were scored as rejected. The overall morphology resembled an accelerated version of typical allograft rejection in *S. purpuratus* and *L. pictus*. The response to xenografts in sea urchins appears to follow the same pattern observed in vertebrates for such grafts. That is, they are rejected far more rapidly than are allografts. In sea urchins, as in higher vertebrates, the more foreign a tissue is, the more rapidly it is rejected. Inbred sea urchins show delayed rejection and the acceptance of some allografts. Allografts between unrelated donors are rejected in about 1 month, whereas xenografts did not survive longer than 8 days.

Histological sections through rejected allografts indicated the presence of two infiltrating cell types. One was easily identified as the red spherule cell due to its dark color and cytoplasmic inculsions. The other was not positively identified. However, the association of phagocytes with red cells during contact with foreign material makes it likely that phagocytic amebocytes were present in the rejected tissue.

In order to determine if the quantity of all or some of the coelomocytes varied during graft rejection, coelomic fluid was withdrawn from the animals every 3 or 4 days and the cells counted during and after graft rejection. Cell numbers were also monitored in animals that had received an autograft only. During the 70-day observation period, the total number of coelomocytes more than doubled in both groups. At the starting time, when animals received either an allograft or an autograft, slightly under 15,000 cells/ml coelomic fluid were counted in each of 20 animals. Seventy days later, each group had an average of 35,000 cells/ml. One variation was noted in the kinetics of the two groups. The experimental group doubled in about 12 days, whereas the control group lagged behind, and did not reach the maximum level until about day 30. After rapidly doubling, the experimental group fell to the level of the control group by day 20. From this time until the end of the observation period at day 70, the two groups fluctuated in a similar pattern. Both groups regained near-maximum levels by the end of the observation period.

The levels of the four individual cell types were also monitored. The phagocytes, colorless spherule cells, and red spherule cells all followed the same pattern that was observed in the total cell population for both the experimental and control groups. A notable exception was the vibratile cell, which fluctuated slightly from day to day in each group, but did not increase significantly in either. During the observation period, the body cavities of the animals were exposed to tank water until the grafts healed in well. It is possible that some of the increased cell counts in both groups were due to exposure to foreign material in the seawater, or to perturbations caused by removal of the cells. However, the experimental group reached maximum cell levels before the control group did, suggesting that the presence of the allograft may have caused the more rapid increase in cells.

In another series of experiments (Coffaro, 1979), the levels of coelomocytes were manipulated during the rejection period. These experiments were initiated following an observation which suggested a positive correlation between the level of circulating coelomocytes and the rate of graft rejection. An animal was observed to reject a primary graft in just 4 days. A count of the circulating coelomocytes revealed that the animal had 60,000 cells/ml, which is about four times the normal level. As in the case of cell counts during graft rejection, the phagocytic, colorless spherule, and red spherule cells had increased proportionally, while the vibratile cells were not much above their typical level. Following this observation, grafts were placed on animals known to have a high initial level of circulating coelomocytes. The seven animals with highly elevated counts rejected allografts at an accelerated rate of 11.6 ± 3.3 days. All were able to heal-in autografts.

Since there were data suggesting that the level of circulating coelomocytes was correlated to the rate of graft rejection, grafts were placed on animals, and the circulating cell level was kept low. Coelomic fluid was withdrawn every 3–4 days from animals, as required, to maintain the initial cell level. This manipulation had no effect on the rate of graft rejection. All allografts were rejected in the usual time of about 1 month.

In a similar experiment, 10 animals were selected which had recently rejected a primary graft. All but one had elevated coelomocyte levels. Coelomic fluid was withdrawn, and the cell counts were reduced and maintained at the average level. Grafts were rejected at the typical accelerated rate found for second-set grafts. When the level of circulating coelomocytes was not allowed to increase during both primary and secondary graft rejection, typical rejection rates were noted. These results might be explained by the observations of Holland *et al.* (1965) who suggest that cell division of coelomocytes is coordinated for the phagocytes and spherule cells, and that cells in both the circulating and body compartments divide synchronously. Perhaps graft rejection is mediated by body tissue coelomocytes and not those in the circulating coelomic fluid.

Bertheussen (1979) has carried out *in vitro* experiments with sea urchin coelomocytes searching for mixed lymphocyte-type reactivity. It appears that enchinoid phagocytes are able to recognize and react to allogeneic and xenogeneic cells in culture. The results of these experiments demonstrated that approximately 70% of the cell combinations within a species resulted in cytotoxic reactions. This activity increased to 90% when cells of different species were placed in contact. This remains one of the few reports of an *in vitro* demonstration of a cell-mediated immune phenomenon in echinoderms.

5. CONCLUSIONS

The echinoderms represent a highly successful biological group that has stood the test of time almost unchanged. Until fairly recently, it was thought that these animals defended themselves against environmental insults solely by effective nonspecific phagocytic mechanisms. Studies on sea cucumbers, sea stars, and sea urchins reveal that these phagocytic cells may also be capable of becoming immunocytes that will specifically react to foreign tissue. We have only begun to scratch the surface, and other specific defense mechanisms and activities may yet be discovered.

Stang-Voss (1974) has referred to the echinoderm amebocyte as being analogous to the RES cells of vertebrates. The coelomic amebocyte is a highly diverse cell that can serve as a phagocyte, a clotting cell, and in some cases may carry oxygen and nutrients to other tissues. Its importance is enhanced by the fact that of all the cell types discovered in the echinoderms, it appears to be the only one common to all classes of echinoderms.

Current knowledge would indicate that coelomocytes do not perpetuate themselves in the coelomic fluid (with the possible exception of the echinoids). Experimental evidence shows that depleting sea cucumbers (Hetzel, 1965) or sea stars (Kaneshiro and Karp, 1980) of coelomocytes by repeated bleedings does not give rise to dividing cells in the recovery population. This implies that coelomocytes originate elsewhere, most likely from organized tissues. At present, there is no direct evidence indicating which organ(s) may serve as a coelomocytogenic structure.

Echinoderms are very efficient at removing particulate matter from the circulating coelomic fluid and externalizing it. In the case of sea cucumbers, phagocytic cells laden with engulfed material pass through the respiratory trees to the exterior, whereas in sea stars, the cells clump in the dermal papulae, and then are ruptured to the exterior. Echinoderms may also possess organized tissues that are analogous to lymphoid tissue. Smith (1978) has demonstrated that the polian vesicles of sea cucumbers are immunologically reactive, since delayed-type hypersensitivity reactions to milkish serum can be experimentally induced in these organs. Phagocytic cells in the Tiedemann bodies of the sea star *Dermasterias imbricata* have been observed to be full of engulfed bacteria (Kaneshiro and Karp, 1980), and thus may act as biological filters similar to the vertebrate spleen or lymph node. The blood axial gland complex is a structure common to many echinoderms that has been implicated as a source of coelomocytes, and may also prove to play a role in defense.

Burnet (1968) has suggested that vertebrate immunocompetent cells are equivalent to, and descended from, invertebrate coelomocytes. This hypothesis is strongly supported by the data accumulated from the allograft studies on three classes of echinoderms, which have established that a specific cell-mediated immune response is characteristic of these animals. The histological evidence would indicate that either the coelomic amebocyte, or a tissue counterpart, is playing a major role in these reactions. The origin of the cell participating in the rejection process remains to be determined.

There appears to be a much quicker rejection of allografts among the echinoids as compared to asteroids or holothurians. As in the case of clotting efficiency, this may be a result of differential selective pressures due to the physiological and structural variations between the classes of echinoderms. Although echinoids react much faster to alloantigenic differences, the reaction does not appear to be as specific as that of asteroids. One possible reason for this discrepancy in specificity between these classes may be due to the fact that the spherule cell (a major participant in the echinoid nonspecific wound-healing reaction) is involved in the sea urchin graft response, and not in the sea star or sea cucumber responses.

The echinoderms provide many interesting models for studying the evolution of both nonspecific and specific effector cells and their functions, as well as organ systems that may have evolved to become the framework of vertebrate RES defense mechanisms. This phylum of animals has already provided us with several enlightening discoveries, and many more appear to be waiting.

ACKNOWLEDGMENTS. Portions of the work concerning *Dermasterias imbricata* were supported by NIAID Research Grant AI15601 awarded to R.D.K. Portions of the work concerning *Lytechinus pictus* were supported by Research Grant PHS 443150-23664 awarded to K.A.C.

This chapter is dedicated to the memory of Mr. Joseph Lubov whose support and inspiration never wavered.

REFERENCES

Bang, F. B., and Lemma, A., 1962, Bacterial infection and reaction to injury in some echinoderms, *J. Insect. Pathol.* **4**:401.

Bertheussen, K., 1979, The cytotoxic reaction in allogeneic mixtures of echinoid phagocytes, *Exp. Cell Res.* **120**:373.

Bertheussen, K., and Seljelid, R., 1978, Echinoid phagocytes in vitro, *Exp. Cell Res.* **111**:401.

Binyon, J., 1972, *Physiology of Echinoderms*, Pergamon Press, Elmsford, N.Y.

Boolootian, R. A., 1962, The perivisceral elements of echinoderm body fluids, *Am. Zool.* **2**:275.
Boolootian, R. A., and Giese, A. C., 1958, Coelomic corpuscles of echinoderms, *Biol. Bull.* **115**:53.
Boolootian, R. A., and Giese, A. C., 1959, Clotting of echinoderm coelomic fluid, *J. Exp. Zool.* **140**:207.
Burnet, F. M., 1968, Evolution of the immune process in vertebrates, *Nature (London)* **218**:426.
Chien, P. K., Johnson, P. T., Holland, N. D., and Chapman, F. A., 1970, The coelomic elements of sea urchins (*Strongylocentrotus*). IV. Ultrastructure of the coelomocytes, *Protoplasma* **71**:419.
Coffaro, K. A., 1978, Clearance of bacteriophage T4 in the sea urchin *Lytechinus pictus*, *J. Invertebr. Pathol.* **32**:304.
Coffaro, K. A., 1979, Ph.D. dissertation, University of California, Santa Cruz.
Coffaro, K. A., and Hinegardner, R. T., 1977, Immune response in the sea urchin *Lytechinus pictus*, *Science* **197**:1389.
Cuénot, L., 1888, Études anatomiques et morphologique sur les ophiures, *Arch. Zool. Exp. Gén. Sér. 2* **6**:3.
Cuénot, L., 1891, Études sur le sang et les glandes lymphatiques dans la série animale (2° parties: Invertébrés), *Arch. Zool. Exp. Gén. Sér. 2* **9**:593.
Edds, K. T., 1977, Dynamic aspects of filopodial formation by reorganization of microfilaments, *J. Cell Biol.* **73**:179.
Endean, R., 1958, The coelomocytes of *Holothuria leucospilota*, *Q. J. Microsc. Sci.* **99**:47.
Endean, R., 1966, The coelomocytes and coelomic fluids, in: *Physiology of Echinodermata* (R. A. Boolootian, ed.), pp. 301–328, Interscience, New York.
Ferguson, J. C., 1966, Cell production in the Tiedemann bodies and haemal organs of the starfish, *Asterias forbesi*, *Trans. Am. Microsc. Soc.* **85**:200.
Foettinger, A., 1880, Sur l'existence de l'hémoglobine chez les échinodermes, *Arch. Biol.* **1**:405.
Fontaine, A. R., and Lambert, P., 1973, The fine structure of the haemocyte of the holothurian, *Cucumaria miniata* (Brandt), *Can. J. Zool.* **51**:323.
Fontaine, A. R., and Lambert, P., 1977, The fine structure of the leucocytes of the holothurian, *Cucumaria miniata*, *Can. J. Zool.* **55**:1530.
Hetzel, H. R., 1963, Studies on holothurian coelomocytes. I. A survey of coelomocyte types, *Biol. Bull.* **125**:289.
Hetzel, H. R., 1965, Studies on holothurian coelomocytes. II. The origin of coelomocytes and the formation of brown bodies, *Biol. Bull.* **128**:102.
Hildemann, W. H., and Dix, T. G., 1972, Transplantation reactions of tropical Australian echinoderms, *Transplantation* **15**:624.
Hildemann, W. H., Raison, R. L., Cheung, G., Hull, C. J., Akata, L., and Okamoto, J., 1977, Immunological specificity and memory in a scleractinian coral, *Nature (London)* **270**:219.
Hildemann, W. H., Johnston, I. S., and Jokiel, P. L., 1979, Immunocompetence in the lowest metazoan phylum: Transplantation immunity in sponges, *Science* **204**:420.
Holland, N. D., Philips, J. H., and Giese, A. C., 1965, An autoradiographic investigation of coelomocyte production in the purple sea urchin (*Strongylocentrotus purpuratus*), *Biol. Bull.* **128**:259.
Hyman, L., 1955, *The Invertebrates: Echinodermata, the Coelomate Bilateria*, McGraw–Hill, New York.
Johnson, P. T., 1969a, The coelomic elements of sea urchins (*Strongylocentrotus*). I. The normal coelomocytes, their morphology and dynamics in hanging drops, *J. Invertebr. Pathol.* **13**:25.
Johnson, P. T., 1969b, The coelomic elements of sea urchins (*Strongylocentrotus*). II. Cytochemistry of the coelomocytes, *Histochemie* **17**:213.
Johnson, P. T., 1969c, The coelomic elements of sea urchins (*Strongylocentrotus*). III. In vitro reaction to bacteria, *J. Invertebr. Pathol.* **13**:42.
Johnson, P. T., and Beeson, R. J., 1966, In vitro studies on *Patria miniata* (Brandt) coelomocytes, with remarks on revolving cysts, *Life Sci.* **5**:1641.
Johnson, P. T., and Chapman, F. A., 1970a, Abnormal epithelial growth in sea urchin spines (*Strongylocentrotus franciscanus*), *J. Invertebr. Pathol.* **16**:116.
Johnson, P. T., and Chapman, F. A., 1970b, Infection with diatoms and other microorganisms in sea urchin spines (*Strongylocentrotus franciscanus*), *J. Invertebr. Pathol.* **16**:268.
Kaneshiro, E. S., And Karp, R. D., 1980, The ultrastructure of coelomocytes of the sea star *Dermasterias imbricata*, *Biol. Bull.* **159**:295.

Karp, R. D., and Hildemann, W. H., 1976, Specific allograft reactivity in the sea star *Dermasterias imbricata*, *Transplantation* **22**:434.

Karp, R. D., and Johns, J. D., 1978, Evolution of immune reactivity: Mitogenic responsiveness in the sea star *Dermasterias imbricata*, Abstracts, 78th Annual Meeting of the American Society for Microbiology, p. 48.

Kindred, J. E., 1924, The cellular elements in the perivisceral fluid of echinoderms, *Biol. Bull.* **46**:228.

Kuhl, W., 1937, Die zellelemente in der Liebeshohlenflüssigheit des seeigels *Psammechinus miliaris* und ihr Bewegungsphysiologisches Verhalten, *Z. Zellforsch. Mikrosk. Anat.* **27**:1.

Kuhn, R., and Wallenfels, K., 1940, Echinochrome als prosthetiche Gruppen hochmolekular symplex in den Eiern von *Arbacia pustulosa*, *Ber. Dtsch. Chem. Ges.* **73**:458.

Leclerc, M., 1973, Étude ultrastructurale des réactions D'*Asterina gibbosa* (Échinoderme, Astéride) au niveau de l'organe axial aprés injection de protéines, *Ann. Immunol. (Inst. Pasteur)* **124C**:363.

Liebman, E., 1950, The leucocytes of *Arbacia punctulata*, *Biol. Bull.* **98**:46.

Menton, D. N., and Eisen, A. Z., 1973, Cutaneous wound healing in the sea cucumber, *Thyone briareus*, *J. Morphol.* **141**:185.

Noble, P. B., 1970, Coelomocyte aggregation in *Cucumaria frondosa*. Effect of ethylenediaminetetraacetate, adenosine, and adenosine nucleotides, *Biol. Bull.* **139**:549.

Reichensperger, A., 1912, Beiträge zur histologie und zum verlauf der regeneration bei crinoiden, *Z. Wiss. Zool.* **101**:1.

Reinisch, C. L., and Bang, F. B., 1971, Cell recognition reactions of the sea star (*Asterias vulgaris*) to the injection of amebocytes of the sea urchin (*Arbacia punctulata*), *Cell. Immunol.* **2**:496.

Smith, A. C., 1978, A proposed phylogenetic relationship between sea cucumber polian vesicles and the vertebrate lymphoreticular system, *J. Invertebr. Pathol.* **31**:353.

Stang-Voss, C., 1974, On the ultrastructure of invertebrate hemocytes: An interpretation of their role in comparative hematology, *Contemp. Top. Immunobiol.* **4**:65.

Unkles, S. E., and Wardlaw, A. C., 1976, Antibacterial activity in the sea urchin *Echinus esculentus*, *Soc. Gen. Microbiol. Proc.* **3**:182.

van den Bossche, J. P., and Jangoux, M., 1976, Epithelial origin of starfish coelomocytes, *Nature (London)* **261**:227.

Vethamany, V. G., and Fung, M., 1971, The fine structure of coelomocytes of the sea urchin *Strongylocentrotus drobachiensis* (Muller O. F.), *Can. J. Zool.* **50**:77.

Vevers, H. G., 1963, Pigmentation of the echinoderms, *Proc. XIV Int. Congr. Zool. Washington D.C.* **3**:120.

Wardlaw, A. C., and Unkles, S. E., 1978, Bactericidal activity of coelomic fluid from the sea urchin *Echinus esculentus*, *J. Invertebr. Pathol.* **32**:25.

8

Cellular Defense Systems of the Protochordata

RICHARD K. WRIGHT and THOMAS H. ERMAK

1. INTRODUCTION

Chordates are the largest phylum of deuterostomes and are divided into three subphyla: the Vertebrata, Cephalochordata, and Urochordata. Although cephalochordates and urochordates lack a vertebral column, they do possess the three distinguishing characteristics of chordates at some time in their life cycle, i.e., a notochord, a dorsal tubular nerve cord, and pharyngeal clefts or gill slits. They are frequently referred to as the Protochordata, the subject of this chapter.

Cephalochordates are small, translucent, free-living, filter-feeding marine animals found in many parts of the world buried in sand on the sea bottom. The only members are various species of *Branchiostoma* (often called *Amphioxus*) and *Asymmetron*. The means by which cephalochordates protect themselves from infections and other antigenic challenges are unknown since no studies exist on their internal defense system. They have no circulating cellular elements within their blood (Franz, 1927; Moller and Philpott, 1973) and the presence of analogous counterparts within the coelom (coelomocytes) is uncertain. Natural hemagglutinins, similar to those found in most invertebrates, are present (Bretting and Renwrantz, 1973; Renwrantz and Uhlenbruck, 1974), but whether they function as part of an internal defense system has not been examined. Thus, the main focus of this chapter will be on the Urochordata.

Urochordates, commonly referred to as tunicates, are marine animals with a worldwide distribution. Their larval stages are free-living whereas adults are generally sessile. There are, however, some pelagic species. As a group, urochordates are divided into three classes: Ascidiacea, Thaliacea, and Larvacea. The Ascidiacea contain the majority of species and are the most common and

---

RICHARD K. WRIGHT • Department of Anatomy, School of Medicine, Center for the Health Sciences, University of California, Los Angeles, California 90024. THOMAS H. ERMAK • Department of Physiology, University of California School of Medicine, San Francisco, California 94143. Supported in part by USPHS Grant HD 09333-05 (R.K.W.).

typical tunicates. They may be solitary or colony forming. The other two classes are pelagic and specialized for a planktonic existence. Of these three classes, only the Ascidiacea will be considered since information on the internal defense system of Thaliacea and Larvacea is lacking.

The Ascidiacea are the most generalized class of tunicates and are believed to be the stock from which the other tunicate classes and vertebrates evolved (Berrill, 1955). Except for the larval stage, they are sessile animals attached to various substrates in littoral marine waters. Most large ascidians, e.g., *Ciona*, *Styela* and *Ascidia*, are solitary, whereas others are colonial and composed of discrete individuals (zooids) united by stolons as in *Perophora* or, like *Botryllus*, composed of many individuals embedded in a common matrix. For ease of reference, the principal anatomical structures mentioned in the text are illustrated in Figure 1. Additional details can be found in invertebrate zoology texts (Meglitsch, 1967; Barnes, 1968) or in the excellent review by Goodbody (1974).

The external surface of ascidians is a special mantle called the tunic or test. It is usually thick, varies in consistency from gelatinous to fibrous, and is composed of a type of cellulose, protein, and inorganic compounds. Within the tunic, there are two distinct body regions, an anterior pharyngeal region containing the pharynx (branchial basket) and an abdominal region containing primarily the digestive tract, heart, and gonads. The heart, a single layer of myoepithelial cells, pumps blood through the circulatory system by means of peristaltic contractions. From each end of the heart, a single large blood vessel exits and branches to all parts of the body and tunic. The blood vessels are not lined by an endothelium but are open channels and lacunae which run through

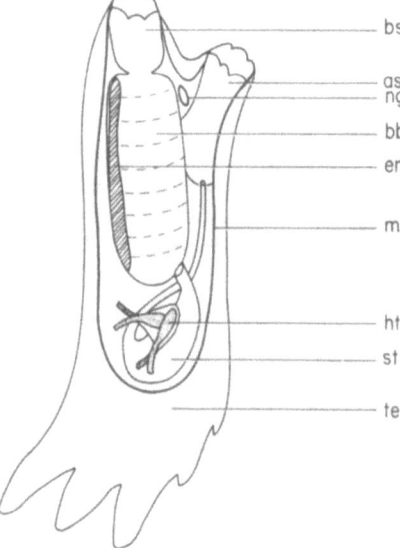

FIGURE 1. The principal anatomical structures of an adult solitary ascidian, *Ciona intestinalis*. Abbreviations: as, atrial siphon; bb, branchial basket; bs, buccal siphon; en, endostyle; ht, heart; mn, mantle (atrial) epithelium; ng, neural gland; st, stomach; te, test or tunic. (Redrawn from Millar, 1953.)

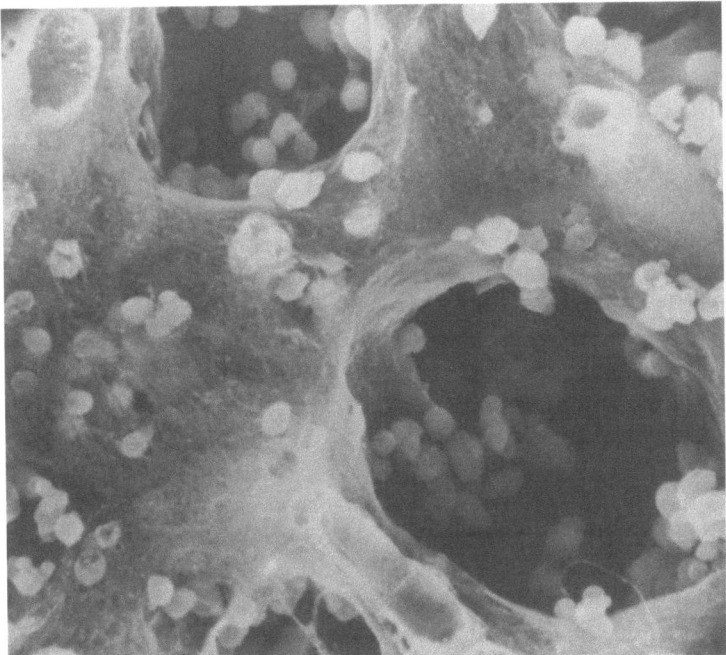

FIGURE 2. Scanning electron micrograph of blood cells in blood channels beneath the stomach epithelium of *Styela clava*. The blood channels are lined only by connective tissue and not an endothelium. × 1000. (From Ermak, 1975a.)

connective tissue (Figure 2). Circulating through this network of channels is a colorless blood plasma and various types of blood cells. These cells are actively ameboid and pass from the blood spaces into the connective tissues and tunic where they can be found randomly distributed or in large concentrations. It is within the circulatory system, connective tissues, and tunic that they perform a variety of functions (reviewed by Wright, 1981). These functions include coagulation, excretion, nutrition, budding, germ cell formation, tunic formation, pigmentation, and immunological surveillance.

Immunological surveillance rests in the ability of an animal to recognize and react to foreign, non-self materials. This recognition and reaction system is a function of the blood cells and plasma components. The circulation of blood cells and their ability to migrate throughout the connective tissues and tunic ensures that an effective immunosurveillance system is present at all times. In this chapter, we will examine the types of ascidian blood cells and their different functional roles in immunosurveillance. Because of the phylogenetic link between tunicates and vertebrates, we will attempt to identify those components which might have led to the evolution of the vertebrate reticuloendothelial system.

2. ASCIDIAN INTERNAL DEFENSE SYSTEM: COMPONENTS, ORIGIN, AND ONTOGENY

2.1. ASCIDIAN BLOOD CELLS

Ascidian blood is colorless, iso-osmotic with seawater, slightly alkaline, acid, or neutral in pH, and contains a low concentration of sulfate ions and proteins (Henze, 1911, 1912; Fulton, 1920; Webb, 1939, 1956; Robertson, 1954; Endean, 1955; Weiss et al., 1976). The percentage by volume of blood occupied by blood cells is on the order of 1–2% (Endean, 1955; Vallee, 1967). Considerable differences are present in blood cell numbers, not only between species but also between individuals of the same species (Table 1).

Blood cell morphologies have been described for over 70 species from live and stained preparations (Cúenot, 1891; Kollmann, 1908; Hecht, 1918; Fulton, 1920; George, 1926, 1930, 1939; Ohuye, 1936; Azéma, 1937; Pérès, 1943; Millar, 1953; Endean, 1955, 1960; Andrew, 1961; Freeman, 1964; Vallee, 1967; Smith, 1970a; Fuke, 1979) and by electron microscopy (Kalk, 1963; Gansler et al., 1963; Overton, 1966; Ermak, 1976; Botte and Scippa, 1977; Milanesi and Burighel, 1978). Although the terminology used by these investigators is not always the same, it is possible to classify the various cell types into several categories, as follows: (1) undifferentiated cells (hemoblasts or lymphocytes), (2) leukocytes (hyaline or granular), (3) vacuolated cells (signet ring, compartment, or morula), (4) various pigmented cells, and (5) nephrocytes (Figure 3). In addition, transitional blood cells, representing immature stages of the various differentiated cell types, can also be distinguished. Of these cell types, lymphocytes (hemoblasts?), leukocytes, and vacuolated cells are involved in immunosurveillance and will be described in greater detail. Their proportions within the blood cell population show considerable differences among species (Table 2). In general,

TABLE 1. Total Blood Cell Counts in Adult Ascidian Blood

	Mean No./ml ($\times 10^7$)	Range ($\times 10^7$)	Reference
Ascidia ceratodes	7.7[a]	4.7–14.5[a]	Biggs and Swinehart (1979)
Ascidia nigra	5.3[a]	3.2–7.9[a]	Vallee (1967)
Ciona intestinalis	1.0[a]	0.8–1.4[a]	Brown and Davies (1971)
Halocynthia aurantium	1.7[a]	0.7–2.7[a]	Smith (1970a)
	1.2	1.0–1.4	Fuke (1979)
Halocynthia roretzi	1.4	1.1–1.7	Fuke (1979)
Molgula manhattensis	2.1[a]	1.7–2.5[a]	Anderson (1971)
Phallusia mammillata	6.8[a]	—	Endean (1960)
Pyura mirabilis	1.2	1.0–1.4	Fuke (1979)
Pyura stolonifera	3.7[a]	1.8–6.8[a]	Endean (1955)
Styela clava	0.4	0.3–0.5	Fuke (1979)
Styela plicata	0.8	0.6–1.0	Fuke (1979)

[a]Original values reported as mean number of cells/mm^3.

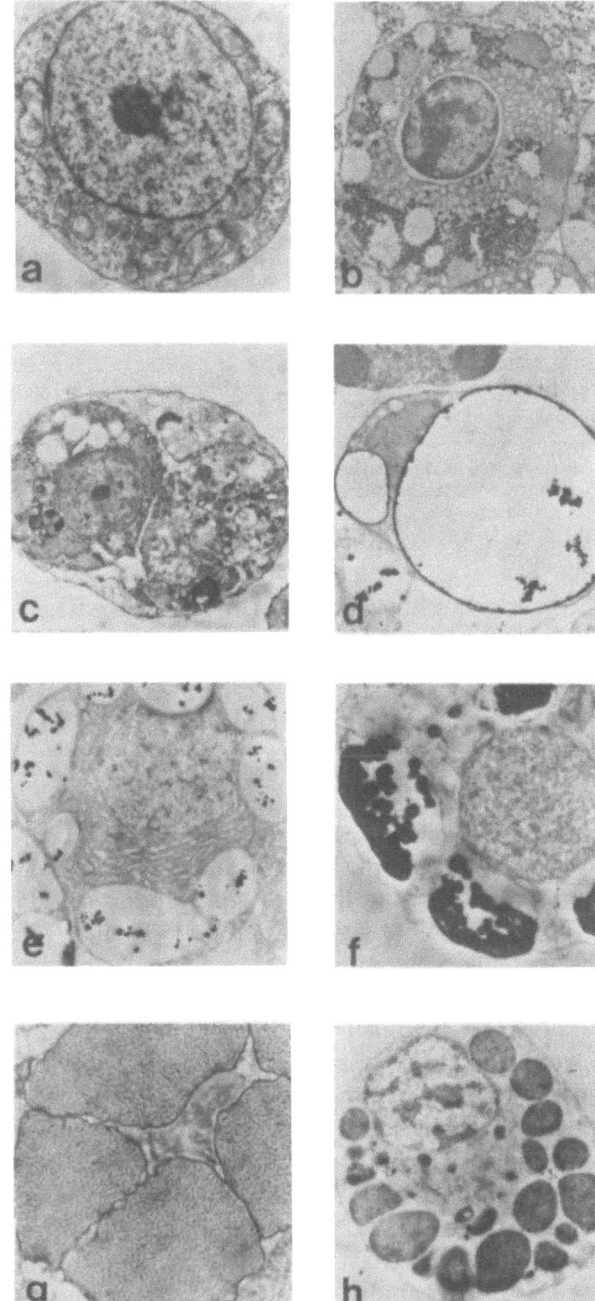

FIGURE 3. Fine structure of blood cells in *Perophora viridis*. (a) Lymphocyte (hemoblast?), × 5700; (b) granular leukocyte, × 4800; (c) granular leukocyte (macrophage) containing phagocytosed material, × 5000; (d) signet ring cell, × 6000; (e) compartment cell, × 5000; (f) morula cell, × 7000; (g) morula cell, × 5000; and (h) orange pigment cell, × 5700. (From Overton, 1966, with permission.)

TABLE 2. PROPORTIONS OF BLOOD CELL TYPES IN ADULT ASCIDIAN BLOOD[a]

	Lymphocytes	Leukocytes	Vacuolated cells	Others[b]	Reference
Ascidia atra	< 1	15	81	3–5	Fulton (1920)
Ascidia nigra	< 1	4	94	1	Vallee (1967)
Ciona intestinalis	17	27	43	13	Millar (1953)
Perophora viridis	< 1	6–17	76–87	—	Freeman (1964)
Phallusia mammillata	< 1	7	92	< 1	Endean (1960)
Halocynthia aurantium	5	60	34	—	Smith (1970a)
	< 1	42–52	40–70	2–4	Fuke (1979)
Halocynthia roretzi	< 1	17–22	54–77	2	Fuke (1979)
Molgula manhattensis	3	6	88	3	Anderson (1971)
Pyura mirabilis	< 1	12–17	75–85	< 1	Fuke (1979)
Pyura stolonifera	5	3	90	2	Endean (1955)
Styela clava	< 1	42–52	40–50	< 1	Fuke (1979)
Styela plicata	ND[c]	80	15	—	Fuke (1979)

[a] Values reported as percentages.
[b] Pigment cells and nephrocytes.
[c] ND, not determined.

however, lymphocytes occur in low percentages, leukocytes at intermediate levels, and vacuolated cells predominate.

2.1.1. Lymphocyte or Hemoblast?

Small, undifferentiated ascidian blood cells 4–8 μm in diameter containing no granules or vacuoles (Figure 3a) have traditionally been called lymphocytes (George, 1939; Pérès, 1943; Millar, 1953; Endean, 1955, 1960; Freeman, 1964; Overton, 1966; Smith, 1970a; Ermak, 1975a,b; Fuke, 1979), even though it has been recognized that this cell type produced the other blood cells, i.e., it was involved in hemopoiesis. Other studies which have recognized the role of undifferentiated cells as stem cells have, more appropriately, designated them hemoblasts (Pérès, 1943; Mukai and Watanabe, 1976; Ermak, 1976; Milanesi and Burighel, 1978). Small, undifferentiated cells have also been found to be involved in immune responses (Reddy *et al.*, 1975). These cells, too, have been called lymphocytes, and an effort has been made, based upon morphological criteria established years ago (Pérès, 1943), to distinguish hemoblasts, involved in hemopoiesis, from lymphocytes, involved in immune-type responses (Ermak, 1976; Wright, 1976, 1981). Lymphocytes have been distinguished from hemoblasts on the basis of their size and nuclear morphology. Hemoblasts are larger than lymphocytes, have a distinctive large nucleolus, and little chromatin, characteristics of a rapidly dividing cell. Lymphocytes, however, like most other blood cells, have no nucleolus and the chromatin occurs in patches along the inner surface of the nuclear membrane and in the interior of the nucleus. Both cell types have a narrow ring of finely granular, basophilic cytoplasm containing several small, round mitochondria, a Golgi apparatus, a few cisternae of rough endoplasmic reticulum, and abundant free ribosomes often arranged in polysomes (Endean, 1960; Freeman, 1964; Overton, 1966; Ermak, 1976; Milanesi and Burighel, 1978; Fuke, 1979). Conventional histochemical staining indicates that the cytoplasm contains proteins, carbohydrates, but no lipids or glycogen (Endean, 1955, 1960; Fuke, 1979). They also exhibit ameboid activity, producing pseudopodia, but they are not phagocytic (Endean, 1960; Freeman, 1964; Fuke, 1979).

Based upon these criteria or cell kinetic behavior, it can be seen that many descriptions or illustrations of lymphocytes (Figure 3a) are unquestionably hemoblasts (Freeman, 1964; Overton, 1966; Ermak, 1975a,b). It is, thus, possible that these two cell types are identical, i.e., the undifferentiated cells involved in both hemopoiesis and immune responses are hemoblasts. If so, it is possible that so-called lymphocytes are actually immature blood cells (transitional cells) whose differentiating organelles are too small to be seen at the light microscope level, or out of the plane of section at the electron microscopic level. Lymphocytelike cells with small granules in the cytoplasm (Warr *et al.*, 1977) are probably transitional cells, not lymphocytes (see Section 2.2.1). Naturally, it is an easy task to find lymphocytelike cells in electron microscopic sections of blood cells or blood-forming tissues. However, this does not functionally define these cells as lymphocytes as we know them in the vertebrates. Milanesi and Burighel (1978) recognized undifferentiated cells with a variety of nuclear forms; however, no studies have correlated any of the diverse morphological forms with specific functions. On the other hand, it is also possible that hemopoietic and immunologically active undifferentiated cells are different cell types or represent subpopulations within the same cell type. This remains to be demonstrated experimentally. In the meantime, the evidence suggests that there are small, nongranular cells with a high nucleus/cytoplasm ratio involved in hemopoiesis and in immune responses which may or may not be the same cell type or subpopulation.

2.1.2. Leukocytes

Leukocytes (Figures 3b,c, and 4) consist of two cell types: hyaline and granular. They are frequently referred to as amebocytes (George, 1930; Ohuye, 1936; Endean, 1960; Overton, 1966; Smith, 1970a; Fuke, 1979), granulocytes (Milanesi and Burighel, 1978), macrophages (George, 1939; Endean, 1955; Milanesi and Burighel, 1978), and phagoyctes (Overton, 1966; Fuke, 1979). In the living state, they assume a variety of shapes due to their ameboid activity and pseudopod formation. In fixed preparations, they are spherical, varying in size from 6 to 12 μm. Both appear to be phagocytic.

Hyaline leukocytes (George, 1930, 1939; Ohuye, 1936; Smith, 1970a; Fuke, 1979) have a centrally located, round nucleus that may or may not contain a nucleolus. Within the nucleus, chromatin is condensed along the inner surface of the nuclear membrane. The cytoplasm is homogeneous and much more abundant than in its hemoblast precursor. It is filled with granules of uniform size and contains a well-developed Golgi complex, rough endoplasmic reticulum,

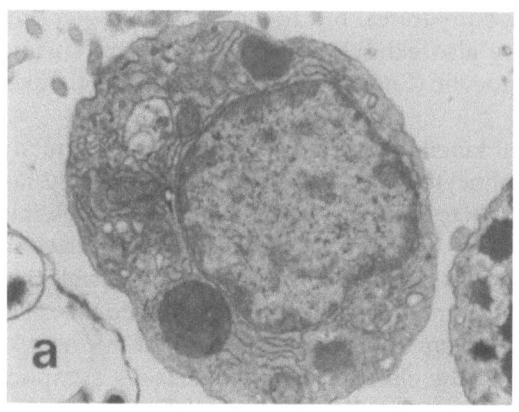
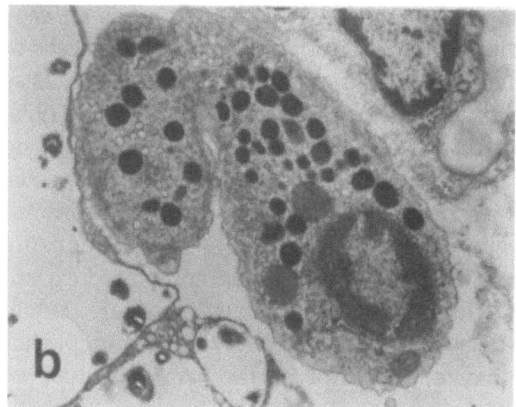
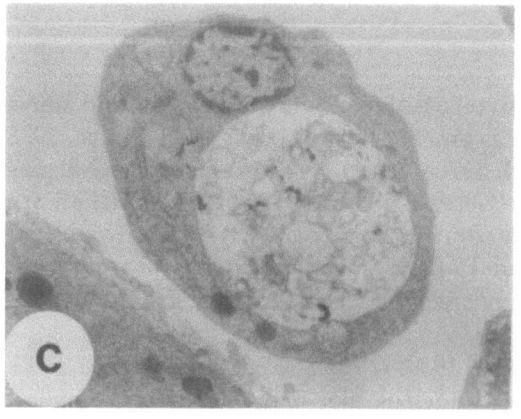

FIGURE 4. Electron micrographs of amebocytes (leukocytes) of *Botryllus schlosseri*. (a) Macrogranular amebocyte (hyaline leukocyte), × 16,000; (b) microgranular amebocyte (granular leukocyte), × 19,500; and (c) macrophage (hyaline leukocyte?), × 10,700. (From Milanesi and Burighel, 1978, with permission.)

and mitochondria. Under phase illumination, the cytoplasm appears reticulated or clumped. A few vacuoles containing granular inclusions exhibiting Brownian motion may also be present. Histochemical staining reveals the presence of cytoplasmic protein, diffuse PAS-positive material, carbohydrate (Endean, 1955; Smith, 1970b; Fuke, 1979), and a positive oxidase material (Ohuye, 1936). Lipids and acid mucopolysaccharides appear to be absent (Endean, 1955; Smith, 1970b).

Granular leukocytes (George, 1930, 1939; Ohuye, 1936; Overton, 1966; Smith, 1970a; Milanesi and Burighel, 1978) have a small, centrally located, round nucleus with no nucleolus. Chromatin is condensed and clumped around the inner margin of the nuclear membrane. The cytoplasm has a thin clear ectoplasm and a coarsely granular endoplasm containing distinct oval or spherical granules measuring 0.2 to 3.5 μm in diameter. Under bright-field illumination, the granules have refractive properties and appear to be associated with a reticular network when viewed under phase illumination. The granules are homogeneous, strongly electron dense, and with conventional blood stains, may be acidophilic, basophilic, or neutrophilic. They give positive oxidase and PAS reactions and contain protein (Ohuye, 1936; Endean, 1960; Smith, 1970b; Fuke, 1979). In addition to the granules, lipid droplets and masses of dense 200- to 300-Å particles characteristic of glycogen (Overton, 1966; Milanesi and Burighel, 1978) also occur in the cytoplasm. Surrounding the nucleus is a peripheral layer of rough endoplasmic reticulum.

2.1.3. Vacuolated Cells

Vacuolated cells consist of three cell types: signet ring cells, compartment cells, and morula cells (Figures 3d,e,f, and g). They have a considerable size range and characteristically contain one or more large vacuoles. Vacuolated cells appear to be interrelated, passing through a progressive cycle from signet ring to compartment to morula cell (Endean, 1955, 1960; Kalk, 1963). During the course of their development, they accumulate and concentrate heavy metals, forming a heavy metal–protein complex within the vacuoles (reviewed by Goodbody, 1974). Morula cells are also unique, containing large quantities of sulfuric acid, varying from 0.39 N in *Pyura stolonifera* (Endean, 1955) to 1.83 N in *Ascidia mammillata* (Webb, 1939). Sulfuric acid appears to be bound to the heavy metal–protein complex in the vacuoles (Bielig *et al.*, 1954; Endean, 1955).

Signet ring cells resemble signet rings and measure 6–12 μm in diameter (George, 1926, 1930; Ohuye, 1936; Endean, 1960; Kalk, 1963; Freeman, 1964; Overton, 1966; Smith, 1970a). They have a peripheral rim of clear, structureless cytoplasm enclosing a single large vacuole that occupies over 70% of the cell. The peripheral cytoplasmic ring is thickened at one place, forming a polar cap that contains a small nucleolus. The inner surface of the vacuole is electron dense with irregular masses of electron-dense granular material adhering to it. Within the vacuole is a clear, colorless fluid that frequently contains granules exhibiting Brownian motion or a mass of granular material that is electron dense.

These intravacuolar granules and masses, considered to be heavy metal complexes, have an affinity for methylene blue and neutral red; in the presence of osmic acid, they stain gray or black (George, 1926, 1930; Ohuye, 1936; Endean, 1955, 1960; Fuke, 1979). Histochemically, the vacuoles and granules are positive for protein in some species, carbohydrates in others, but negative for lipids (Endean, 1960, 1961; Smith, 1970b).

Compartment cells are spherical, 8–20 µm diameter in the fixed state (George, 1926; Ohuye, 1936; Endean, 1960; Overton, 1966) but assume an elongated shape with blunt and long slender pseudopodia in living preparations (George, 1926, 1930; Freeman, 1964). The nucleus may be eccentrically or centrally located and has no nucleolus. It is surrounded on one side by mitochondria and numerous profiles of dense endoplasmic reticulum and ribosomes (Kalk, 1963). Each cell contains a variable number of large round or angular boxlike vacuoles at the periphery of the cell, giving it a compartmental appearance. The vacuoles are separated from each other by nongranular cytoplasmic partitions continuous with the ectoplasm and cytoplasm around the nucleus. Each vacuole is filled with a clear, colorless fluid containing granules exhibiting Brownian motion. These granules stain with methylene blue and neutral red and blacken with osmic acid (George, 1926, 1930; Ohuye, 1936; Endean, 1960). Histochemically, the vacuoles stain positive for protein, give a faint PAS-positive reaction, but are negative for lipid and neutral or acidic polysaccharides (Endean, 1960; Smith, 1970b).

Morula cells are spherical, 8–16 µm in diameter (George, 1926; Ohuye, 1936; Freeman, 1964), and have a berrylike or morula appearance. They have been called green cells (George, 1926; Ohuye, 1936; Endean, 1955; Freeman, 1964; Overton, 1966), colorless morula cells (George, 1926; Ohuye, 1936), vanadocytes (Webb, 1939), and ferrocytes (Endean, 1955; Smith, 1970a), the latter two names derived from the fact that in some species, they contain vanadium, or, in others, iron. These cells contain a variable number (up to 20) of tightly packed, symmetrically arranged membrane-bound vacuoles or globules 1–3.5 µm in diameter that appear wedge-shaped in some cells. Nongranular cytoplasmic extensions pass from the perinuclear cytoplasm between and around the vacuoles. In living cells, the vacuoles appear colorless to yellow-green and are highly refractive.

Morula cell vacuoles contain electron-dense material which may be homogeneous, granular, or filamentous (Gansler et al., 1963; Overton, 1966; Botte and Scippa, 1977; Milanesi and Burighel, 1978). This material is PAS positive and may be protein or polysaccharide sulfates complexed to vanadium or iron (Endean, 1955, 1960; Smith, 1970b; Milanesi and Leban, 1974). The nucleus is usually obscured by the vacuoles and may be eccentrically or centrally located. It contains scattered patches of chromatin and has no nucleolus. To one side of the nucleus is a well-developed Golgi complex containing dense irregularly shaped masses and an endoplasmic reticulum whose cisternae enclose dense granules. These masses and granules are similar to those observed in the vacuoles (Overton, 1966; Milanesi and Burighel, 1978).

2.2. BLOOD CELL ORIGINS

2.2.1. Hemopoietic Tissues and Hemopoiesis

Ascidian blood cells are a renewing cell population (Messier and Leblond, 1960) composed of stem, transitional, and mature cells. They have a rapid proliferation rate which, at the steady state, is balanced by the rate of cell loss. Cell renewal occurs in the circulating blood and in hemopoietic tissues (commonly called "lymph nodules") located in various parts of the body. Three principal areas have been identified: (1) in the pharynx, (2) around the digestive tract, and (3) in the body wall of advanced species (Millar, 1953; Ermak, 1976). Each ascidian family exhibits a characteristic relationship between pharynx, postpharyngeal gut, and gonads, and the distribution of blood-forming tissues follows accordingly.

Hemopoietic tissue is organized as either diffuse clusters of cells or discrete nodules. In the pharynx, a greatly enlarged organ perforated by a multitude of ciliated gill slits (stigmata), these cells are located in the transverse and longitudinal bars (Figure 5a) and along the endostyle. Around the gut, they are associated with connective tissue or blood channels. In the body wall, they lie in the connective tissue next to the atrial epithelium or blood channels (Figure 5b).

A hemopoietic nodule is composed of a cluster of hemoblasts in the center surrounded by maturing blood cells in various stages of differentiation (Figure 6). A single nodule may contain more than one cluster of hemoblasts. As stem cells, they divide and reproduce themselves as well as produce all other blood cell types, either directly or indirectly. Hemoblasts, which also occur in the circulating blood, have a high nuclear/cytoplasmic ratio, a prominent nucleolus, sparse chromatin, numerous polyribosomes, and a few cytoplasmic organelles (Figure 7). They are specialized for cell division as shown by the incorporation of tritiated thymidine ($[^3H]$-TdR) into newly synthesized DNA (Figure 5b) and they are also sensitive to X-rays (Freeman, 1964, 1970b; Ermak, 1975b, 1976).

Differentiating blood cells (transitional cells) around the hemoblasts lose their prominent nucleolus and gain chromatin within the nucleus. As one procedes away from the center of a nodule, these blood cells increase in size. Adjacent to the hemoblasts, small, electron-dense granules about 0.5 μm in diameter appear in the cytoplasm of differentiating transitional cells (Figures 6 and 7). These granules become larger and more numerous as the cells mature (Figure 6). Thus, many transitional cells are characterized by having granules smaller and less numerous than those of mature blood cells (Ermak, 1976; Milanesi and Burighel, 1978) and should not be confused with lymphocytelike cells. Cell differentiation is also marked by the loss of polyribosomes and the development of long cisternae of rough endoplasmic reticulum, elongate mitochondria, and a larger Golgi apparatus.

Blood cell proliferation rates have been determined in these nodules by using $[^3H]$-TdR and autoradiography (Ermak, 1975b). After a 1-hr $[^3H]$-TdR exposure, blood cells engaged in premitotic DNA synthesis were labeled in the

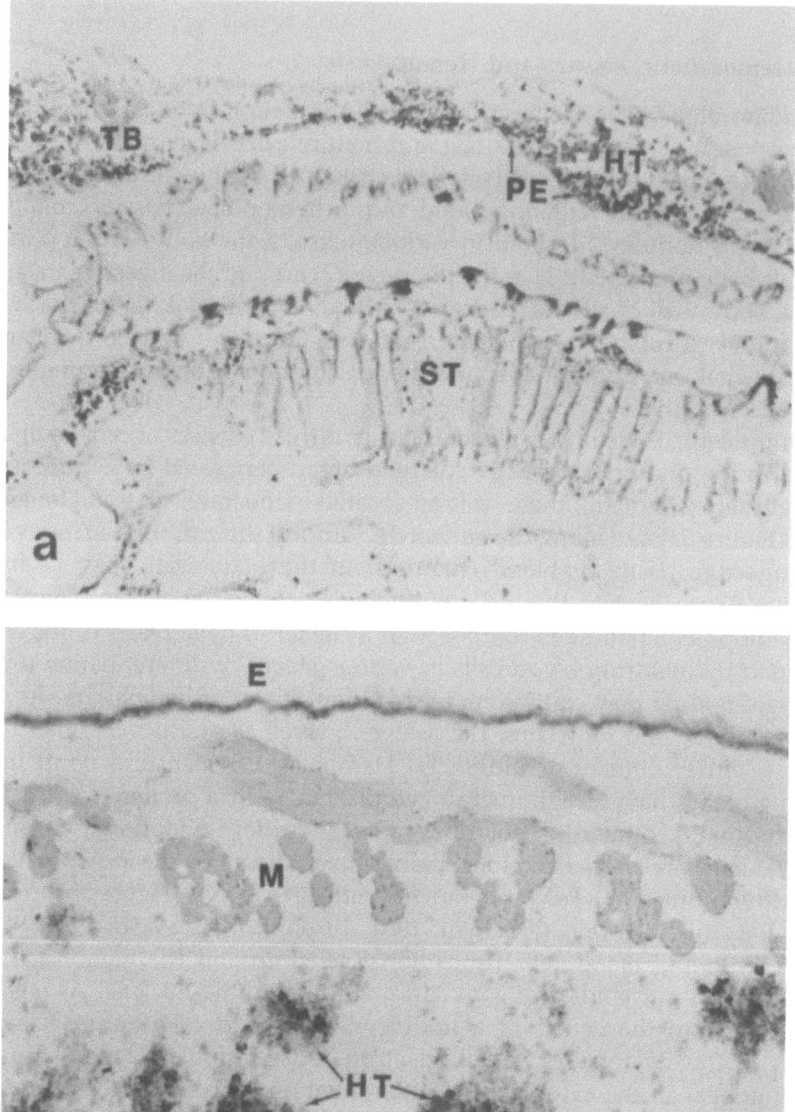

FIGURE 5. Histological locations of adult ascidian hemopoietic tissues as shown in autoradiograms 1 hr after the administration of tritiated thymidine. (a) Longitudinal section through transverse bar (TB) and rows of stigmata (ST) in pharynx of *Ciona intestinalis*. Hemopoietic tissue (HT) occurs as diffuse clusters adjacent to the pharyngeal epithelium (PE). × 100. (b) Cross-section through body wall of *Styela clava*. Hemopoietic tissue (HT) occurs in connective tissue next to the atrial epithelium (AE). Blood cells engaged in premitotic DNA synthesis are labeled in the center of the nodules. × 250. E, epidermis; M, muscle. (From Ermak, 1976, with permission.)

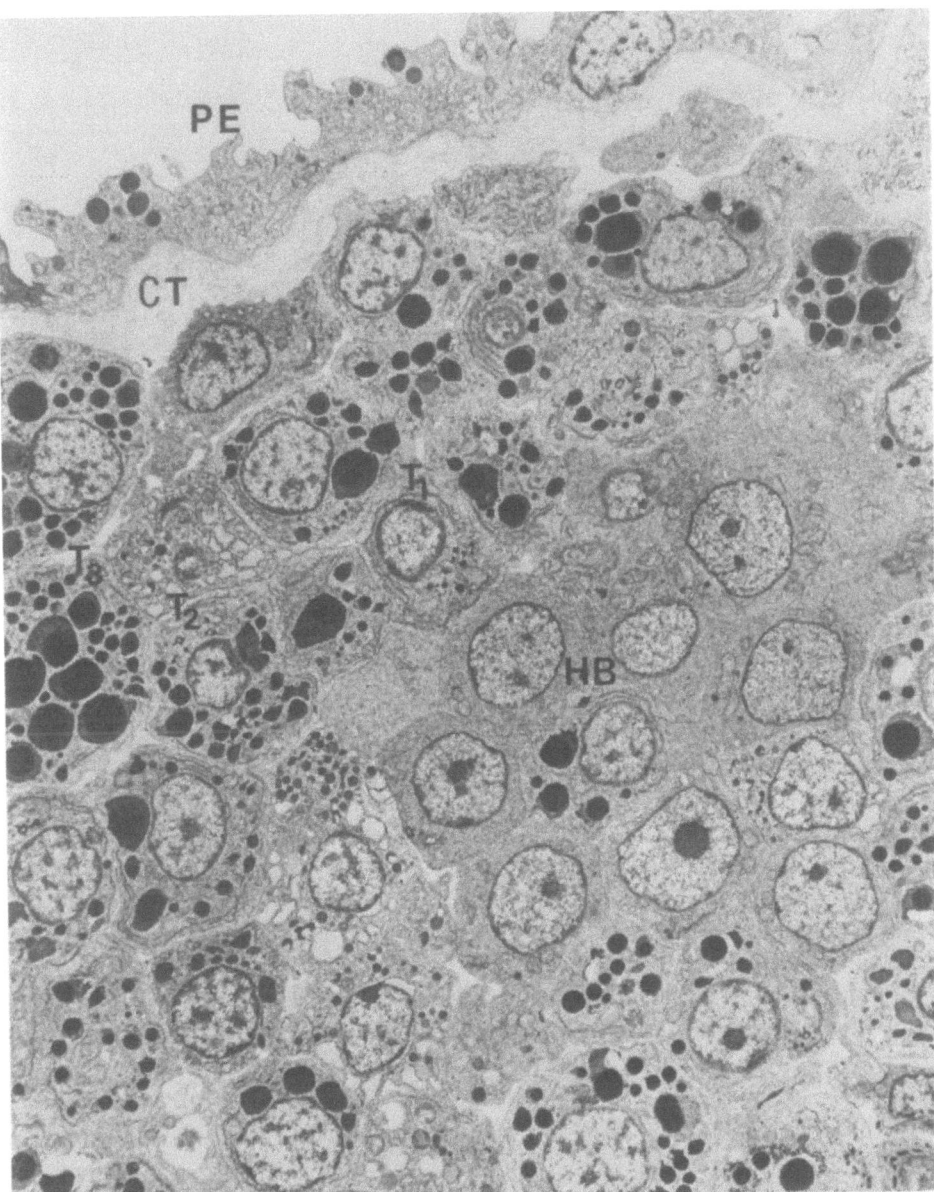

FIGURE 6. Electron micrograph of hemopoietic (lymph) nodule in transverse bar from pharynx of *Styela clava*. HB, hemoblasts surrounded by differentiating blood cells. CT, connective tissue fibers; PE, pharyngeal epithelium; T_1, T_2, T_3, transitional cells (immature leukocytes) containing increasingly larger dense granules. × 12,000. (From Ermak, 1976, with permission.)

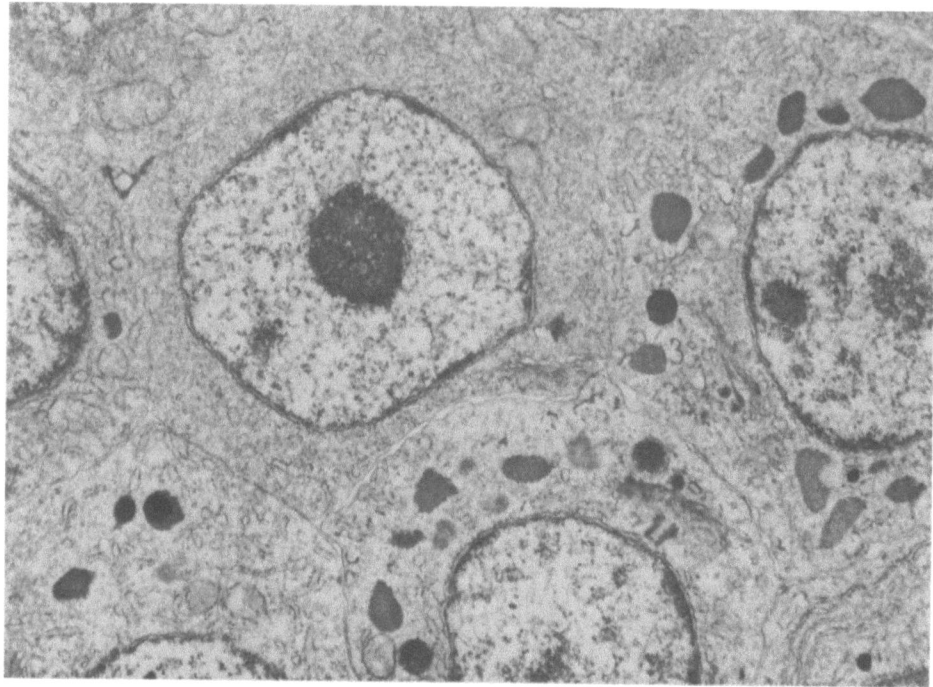

FIGURE 7. Electron micrograph of a hemoblast and several differentiating leukocytes from a hemopoietic nodule from the pharynx of *Styela clava*. Transitional cells lose their conspicuous nucleolus, acquire greater nuclear chromatin, and develop small electron-dense granules in the cytoplasm which increase in size with continued differentiation. × 12,000. (Unpublished micrograph by T. H. Ermak.)

center of nodules (Figure 5b). These are probably hemoblasts. Labeled blood cells could also be observed in the circulating blood, suggesting that cell proliferation outside the nodule can also occur. By day 20, most of the labeled cells were in the peripheral parts of nodules; cells in the interior were no longer labeled. Many cells in the circulating blood were also labeled. Sixty days after [^{3}H]-TdR administration, most blood cells in the nodules were unlabeled and only a few labeled cells occurred in the blood channels. Ascidian blood cells are thus renewed on the order of several weeks.

2.2.2. Ontogeny

Within developing embryos, prospective blood-forming mesodermal cells arise from a cell mass formed by cells of the archenteron (Cowden, 1968). Morphologically, these cells resemble the hemoblasts found in adult hemopoietic tissues. In early tail bud stages, these cells extend around the coelom, lying adjacent to the ectoderm. During the progressive differentiation of tail bud stages, some of these cells rapidly differentiate into various blood cell types while others retain their hemoblast character.

Upon hatching, the free-swimming larvae have a functional blood vascular system and circulating blood cells (Andrew, 1961). Lymphocytes (hemoblasts?), granular leukocytes, vacuolated cells, and pigment cells can be found in the circulating blood (Andrew, 1961; Cowden, 1968). Morula cells are abundant and their appearance corresponds to the time when significant vanadium concentration occurs (Bielig et al., 1963). After a brief larval period, metamorphosis occurs, transforming the larvae into a sessile juvenile adult. Following metamorphosis, numerous hemoblasts cluster in the pharyngeal wall (Ermak, 1976). With continued division and migration, the pharyngeal hemoblasts give rise to the hemopoietic tissue of the pharynx, gut, and body wall.

3. ASCIDIAN INTERNAL DEFENSE SYSTEM: REACTIONS

The ascidian internal defense system is composed of two components, one humoral, the other cellular, that recognize and react to the presence of foreign materials. Humoral components consist of natural agglutinins and bactericidins present in the blood plasma and various antimicrobial and antitumor substances present in body tissues. Some of these factors also appear to function in facilitating the cellular components which consist of lymphocytes (hemoblasts?), hyaline and granular leukocytes, and vacuolated cells. The types of reactions usually associated with these cells include inflammatory responses with phagocytosis or granuloma (encapsulation) formation and various histoincompatibility responses toward allogeneic tissues that result in their rejection. Together, these components account for the maintenance and integrity of the animal in an aseptic condition.

3.1. SERUM AND TISSUE FLUID FACTORS

Bactericidins for several species of gram-negative obligate marine bacteria isolated from marine invertebrate tissues, are present in the blood and serum of *Ciona intestinalis* (Johnson and Chapman, 1970). Both fluids killed two of the species within 72 hr, depressed bacterial numbers in two other species, but had no effect on a gram-positive facultative marine micrococcus or a gram-negative terrestrial rod when assayed *in vitro*. Nothing is known, however, about the specificity, source, mode of activity, or chemical nature of these bactericidins or whether they can be induced.

Tissue extracts from *Polyandrocarpa* sp. (Cheng and Rinehart, 1978) and *Aplidium* sp. (Carter and Rinehart, 1978) possess inhibitory activity toward fungi, gram-positive and gram-negative bacteria, are cytotoxic to tumor cells (L1210 and KB), and show slight activity against herpesvirus, type I. The bioactive component from *Polyandrocarpa* extracts, called polyandrocarpidine, contains a highly strained cyclopropene ring and has a molecular formula of $C_{18}H_{30}N_4O$. In the *Aplidium* extract, the bioactive component, called aplidiasphingosine, is a terpenoid (2-amino-5,9,13,17-tetramethyl-8,16-octadeca-

TABLE 3. CHARACTERISTICS OF TUNICATE BLOOD AGGLUTININS

	Heat lability (°C)	pH activity range	Dialyzable	Precipitable[a]	Divalent ions for activity[b]	Proteinase sensitive[c]	2-ME sensitive[d]	Periodate sensitive	Reference
Ascidia malaca	75	2–10	No	Yes	No	Yes	Yes	No	Parrinello and Patricolo (1975)
Ciona intestinalis	70	—	No	—	Yes	—	—	—	Wright (1973, 1974)
Halocynthia pyriformis	50	6–10	No	Yes	Yes	Yes	—	—	Anderson and Good (1975)
Phallusia mammillata	75	2–10	No	Yes	No	Yes	Yes	No	Parrinello and Patricolo (1975)
Styela plicata	>140	2–10	No	Yes	No	No	—	Yes	Fuke and Sugai (1972)

[a] Ammonium sulfate or distilled water precipitation.
[b] Calcium or magnesium ions required for activity.
[c] Activity destroyed by trypsin or pepsin digestion.
[d] 2-ME, 2-mercaptoethanol.

diene-1,3,14-triol) that can be regarded as a derivative of sphingosine. Tissue extracts of *Ecteinascidia turbinata* have also been demonstrated to exert *in vivo* and *in vitro* antitumor cell activity (Lichter *et al.*, 1975). *Ecteinascidia* extracts can also suppress antibody production to sheep erythrocytes in mice, inhibit graft-vs.-host and host-vs.-graft reactions, and inhibit human and mouse lymphocyte blastogenic transformation in response to mitogenic and antigenic stimulation. There are no unequivocal reports of neoplasia in tunicates and whether these substances function in the prevention of neoplasia is unknown. Furthermore, their relationship to the ascidian internal defense system and to blood bactericidins has yet to be examined.

Ascidian serum also contains naturally occurring substances that agglutinate a wide range of vertebrate erythrocytes. Chemical and biophysical characterization of these agglutinins (Table 3) suggests they are glycoproteins except in *Styela plicata*, where they appear to be polysaccharides. These agglutinins appear to be part or all of the plasma proteins since their levels (titers) are proportional to protein concentrations (Wright, 1973; Wright and Cooper, 1975; Anderson and Good, 1975) and can be reduced or abolished by removal of plasma proteins (Anderson and Good, 1975). Adsorption studies indicate a considerable degree of cross-reactivity and lack of specificity since adsorption with erythrocytes from one species not only decreases activity toward the adsorbing cells but also decreases activity toward erythrocytes from other species. Agglutinin inhibition studies with monosaccharides indicate that the receptor site(s) of *Halocynthia pyriformis* agglutinin is N-acetylneuraminic acid (Anderson and Good, 1975). It is likely that the agglutinins from the other species in Table 3 will also have binding specificities for carbohydrates.

H. pyriformis agglutinin has also been isolated by affinity chromatography on insolubilized bovine submaxillary mucin (BSM) (Form *et al.*, 1979). Sucrose density gradient ultracentrifugation and polyacrylamide gel electrophoresis (PAGE) analysis indicate that the agglutinin is a polymer with an intact mass of approximately 800,000 daltons and is composed of 20,000-dalton subunits. Its binding specificity appears to be complex since BSM but none of its monosaccharide components will bind to the agglutinin. Serum agglutinins from *Pyura stolonifera* not only bind erythrocytes but also bind the hapten 2,4-dinitrobenzene sulfonic acid (DNP) (Marchalonis and Warr, 1978). Characterization of the DNP-binding protein by PAGE analysis indicated a population of molecules consisting of single subunits of approximately 65,000–70,000 daltons with a pronounced charge heterogeneity and a resemblance to vertebrate immunoglobulin μ heavy chain. Two-dimensional electrophoretic separations also suggested that this population of molecules resembled vertebrate immunoglobulin in heterogeneity.

The source and physiological significance of these agglutinins remain to be defined. It has been suggested that they function as opsonins (Wright, 1974) by enhancing phagocytosis as has been observed in other invertebrates (Dales, 1979). Alternatively, they may be necessary for cell–cell adherence (Fuke and Sugai, 1972) or humoral recognition factors of foreignness (Anderson and Good, 1975) or some as yet undefined function. Agglutinin similarity in subunit structure to μ chain suggests they may be related to vertebrate immunoglobulin, possibly a primitive μ-like chain ancestral to all subsequent vertebrate immunoglobulin polypeptide chains (Marchalonis and Warr, 1978). It is not unlikely that further studies will reveal not only their origins but also their biological role(s) within the ascidian immune system.

3.2. INFLAMMATORY RESPONSES

Inflammation is a process whereby fluid and blood cells accumulate at a site of injury regardless of the nature of the injurious agent. Inflammatory responses accomplish two things, the removal of foreign substances and the disposal of damaged tissue, allowing healing to occur. All kinds of infection (e.g., bacteria, fungi) and tissue damage (e.g., wounds) elicit inflammatory responses. The response may be acute, lasting for only a few hours or days, or chronic, lasting much longer. Tunic injuries and a variety of foreign materials evoke inflammatory reactions to ascidians (Table 4). Sequestering and elimination of these foreign substances is a function of their nature and size. Small particulate material such as bacteria or carbon are effectively removed by phagocytosis. When the offending substance is too large to be degraded by phagocytosis (e.g., parasites) or the irritant persists for a long time period (e.g., large accumulations of bacteria, glass fragments), granulation tissue forms around the material, encapsulating it.

3.2.1. Phagocytosis

Phagocytosis involves recognition of non-self. After recognition, the foreign material attaches to the cell surface, is ingested and internalized in a vacuole. There it is degraded and destroyed. Ascidian hyaline and granular leukocytes are actively phagocytic cells. How they recognize non-self materials is unknown. Opsonization by a factor(s) may be necessary and the natural agglutinins present in the blood are likely candidates (Dales, 1979). *In vitro* studies using *Styela plicata* agglutinin, however, suggest that it has no apparent opsonic effect on phagocytosis of rabbit erythrocytes (Fuke and Sugai, 1972). The effects of other tunicate agglutinins on phagocytosis remain to be examined. Once ingested and internalized into vacuoles, the elmination of the engulfed particles depends on their chemical nature. Organic materials such as bacteria, erythrocytes, etc. are probably degraded by lysosomal enzymes. Histochemical examination of hyaline and granular leukocytes to confirm the presence of lysosomal enzymes, however, is lacking. Inert substances that cannot be degraded such as carbon or carmine particles appear to be transported and released to the external environment.

Phagocytic reactions in the tunic, vascular system, and peritoneal spaces toward organic and inert particles have been observed in a number of ascidian species. Small quantities of human or duck erythrocytes injected into the tunic of *Ciona intestinalis* are phagocytosed and cleared from the tunic matrix within 24–48 hr (Wright, 1974). Their mode of destruction was not studied but is probably the result of intracellular digestion. Hyaline leukocytes actively engulf car-

TABLE 4. ACUTE AND CHRONIC INFLAMMATORY REACTIONS OBSERVED IN ASCIDIANS

	Agent inducing reaction	Reference
Ascidia mentula	Bacteria	Cantacuzène (1919), Thomas (1931a, 1932)
	Chemical oncogens	Thomas (1931b)
Ciona intestinalis	Tunic injury	Pérès (1948)
	Vertebrate erythrocytes	Wright (1974), Parrinello et al. (1977)
	Tunic transplants	Reddy et al. (1975)
Cynthia papillosa	Tunic injury	St. Hilaire (1931)
Distaplia unigermis	Carmine gruel	Ivanova-Kazas (1966)
Halocynthia aurantium	Tunic injury; carbon particles	Smith (1970a)
Halocynthia pyriformis	Marine hydrozoans; lesions of unknown etiology	Anderson et al. (1977)
Molgula manhattensis	Glass fragments; tissue transplants	Anderson (1971)
Phallusia mammillata	Tunic injury	Endean (1961)
Unknown	Wood splinters	Metchnikoff (1892)

bon particles introduced into the tunic of *Halocynthia aurantium* (Smith, 1970a). After injection, there is an increase in the number of carbon-containing cells that reaches a maximum at 18 hr and is maintained over a 48-hr period. The eventual fate of the phagocytosed carbon was not determined.

Carmine particles injected intracardially in *Molgula manhattensis* are phagocytosed by leukocytes and cleared from the vascular system within 24 hr (Anderson, 1971). Carmine-bearing leukocytes could be identified for 7–10 days after the injection but not after 14 days. No carmine was observed to be deposited within any tissues other than leukocytes. Trypan blue aggregates were also phagocytosed by the leukocytes. Unlike carmine particles, trypan blue appeared to be released to the branchial sac epithelial cells from which it was voided to the environment. Perivisceral injections of colloidal thorium dioxide into *C. intestinalis* followed by subsequent radiographs revealed that the particles were phagocytosed by granular leukocytes (Brown and Davies, 1971). Elimination occurred via the intestine and vas deferens and was complete 9 days after injection.

Cantacuzène (1919) studied bacterial infections in *Ascidia mentula* by injecting cultures of bacteria isolated from *Aplysia punctata* intestine into the endostylar sinus. No free bacteria were observed within the circulation during the first 24 hr. Thereafter, bacteria became numerous and by day 4, septicemia had developed; however, all injected animals recovered. Beginning with day 2, hyaline leukocytes were observed phagocytosing and digesting the bacteria. Masses of bacteria were also observed sticking to leukocyte surfaces in such a way that they were bound together in a transparent gel that surrounded the leukocytes. By day 9, no free bacteria could be observed in the vascular fluid. More recently, Bang (1975) has investigated the course of bacterial infections in *Ascidia mentula*. Active bacterial infections were established by endostylar sinus injections of *Pseudomonas*, *Vibrio*, and a natural bacterial infection. *Pseudomonas* was able to initiate an infection. None of the injected animals died and in only one was infection still apparent by day 3. Infection with *Vibrio* was often lethal, and overwhelming bacterial infection could be demonstrated just before death. A spontaneous natural bacterial infection that was readily passed from animal to animal killed two-thirds of the injected *Ascidia* in 2 to 3 days.

While it is clear from these studies that ascidian blood leukocytes recognize and respond to foreign material in the tunic and vascular system, the parameters governing phagocytosis and clearance of the foreign material are still unknown. Many questions such as the phagocytic recognition mechanism(s), clearance mechanism(s), and why some bacterial infections become established and others do not, need to be resolved. Answers to these questions will provide a better understanding of the role phagocytosis plays in the ascidian internal defense system.

3.2.2. Encapsulation

Foreign materials too large to be phagocytosed, such as parasites, inert objects, or large accumulations of organic materials, are encapsulated. Naturally occurring encapsulation responses have been observed toward copepods which live as parasites in ascidians. These copepods live in the epithelial lining of the peribranchial cavity (Bresciana and Lützen, 1960), the branchial circulatory sinuses (Monniot, 1963), or the subendostylar blood vessel (Dudley, 1968) where they are encapsulated by host tissue. These capsules, however, do not appear to be formed by the blood cells, and the mode of their formation, once the copepod has gained entrance, is unknown.

Gonophysema gullmarensis occurs throughout the peribranchial cavity exterior wall of *Ascidiella aspersa* (Bresciana and Lützen, 1960). They are situated behind the epithelial lining of the cavity and are enclosed on all sides by the vascular tissue of the ascidian. If the parasite is dissected out from the tissue, it is found to be surrounded by a thin mesenchymatous membrane. A continuous space filled with the blood of the ascidian is left between this membrane and the wall of the parasite. *Kystodelphys drachi* parasitizes the branchial wall vessels of *Microcosmus savignyi* (Monniot, 1963). The capsules (cysts) formed consist of three cell layers of host tissue. Blood cells can be observed migrating through the cyst walls into the cavity containing the copepod. Occasionally, large blood cell accumulations can be found within the cyst. The significance of their presence, however, is unknown. *Scolecodes huntsmani* occurs in the subendostylar blood vessel of several ascidian species, but primarily *Styela gibbsii*, where it is encapsulated in a cellular cyst composed of a single layer of epithelium with some connective tissue holding it in place (Dudley, 1968). Blood cells can pass by diapedesis into the cyst. In cases where only the copepod exuviae are found, many phagocytic cells are present. It is not known whether the tissue of the copepod has been degraded by these cells, leaving only the cuticle, or whether the copepod has departed the cyst, leaving a true exuvium. As yet, no blood cell immunological reactivity has been observed toward any of these parasites and the ascidian hosts do not appear to be harmed by their presence.

Encapsulation responses within the tunic, however, are a type of chronic inflammation characterized by the accumulation of blood cells around the object followed by a gradual fibrotic alteration of the cells and the formation of a capsule. Ascidian blood cells responsible for these reactions are those of the vacuolated category, predominantly the morula cell. Experimentally, encapsulation responses have been elicited by introducing inert objects such as splinters (Metchnikoff, 1892) and glass fragments (Anderson, 1971) or by injecting bacteria (Thomas, 1931b, 1932) and vertebrate erythrocytes (Wright, 1973, 1974; Parrinello et al., 1977) into the tunic.

Injection of Bacterium tumefaciens into the tunic of Ascidia mentula produces an intense inflammatory response with concomitant phagocytosis and necrosis at the injection site (Thomas, 1931b, 1932). Concentric rings of blood cells surround the necrotic region, walling off the inflamed area and producing a nodule which regresses after several weeks. Cellular responses to glass fragments inserted into the tunic of Molgula manhattensis are immediate, with many blood cells covering the glass within 30 min (Anderson, 1971). One day after insertion, morula cells and signet ring cells were present in equal numbers, constituting more than 95% of the blood cells coating the glass. Two days postinsertion, 90% of the adhering cells were morula cells. The most commonly encountered capsules surrounding glass fragments were composed of multilayered, 8–20 cell layers thick, morula cell aggregates (Figure 8a). In some areas, morula cell aggregates did not expand and thicken. Instead, individual cells appear to separate and radiate from the periphery of the aggregate. These cells became oriented so that their long axes were parallel and formed monolayers of cuboidal or columnar cells (Figure 8b). Morula cells did not aggregate on these cellular monolayers. Morula cells adhering to the glass surface were also active in producing tunic matrix material (Figure 8c). Production of tunic material (tunicin) proceeded slowly and 1 week was required for the development of tunicin sheets upon the glass.

Large concentrations of human, duck, rabbit, or sheep erythrocytes injected into the tunic of C. intestinalis (Wright, 1973, 1974; Parrinello et al., 1977) become agglutinated into a mass (Figure 9a). Injections that are not cleared within 24 hr are encapsulated by the infiltrating blood cells. These capsules may assume a light tan color, increase in size, and acquire a firm consistency (Wright, 1974). Alternatively, a whitish halo may appear around the mass (Figure 9b), which also increases in size with time (Parrinello et al., 1977). Within 2–3 weeks after capsule formation, a large blister filled with gelatinous or liquid material forms (Figure 9c). The overlying tunic becomes thinner and after a few hours ruptures, producing a wound in the tunic (Figure 9d). Encapsulated material is then released to the environment and if the wound is not too severe, the animal survives and the wound heals.

Histological observations (Parrinello et al., 1976, 1977) of a 1- to 2-day-old halo reveal this zone to be filled with granular material and vacuolated cells

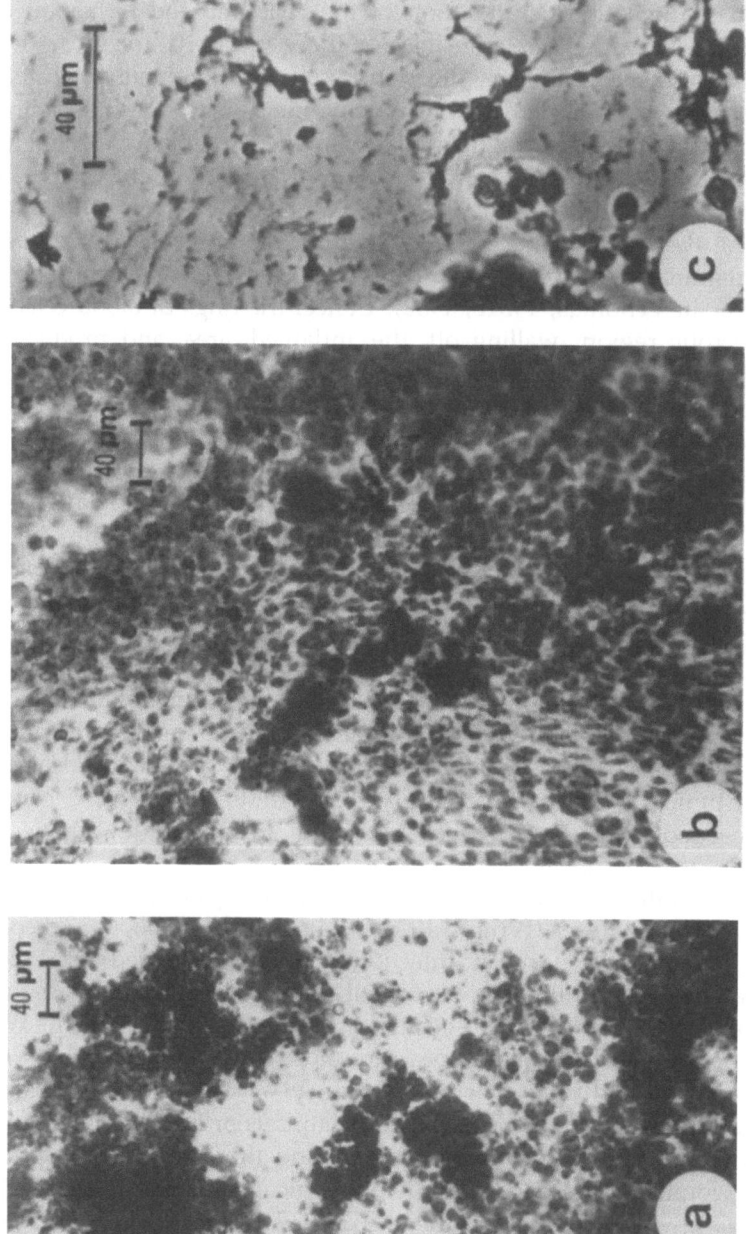

FIGURE 8. Encapsulation of a glass fragment inserted into *Molgula manhattensis* branchial sac tissue. (a) Two days after insertion, formation of multilayered vanadocyte (morula cell) aggregates can be seen in several areas. (b) Four days after insertion, the formation of a cellular monolayer (left) from a vanadocyte (morula cell) aggregate (upper right) can be seen. (c) Phase-contrast microscopy of vanadocytes (morula cells) connected to each other by strands of tunic material on a glass fragment after 3 days. (From Anderson, 1971, with permission.)

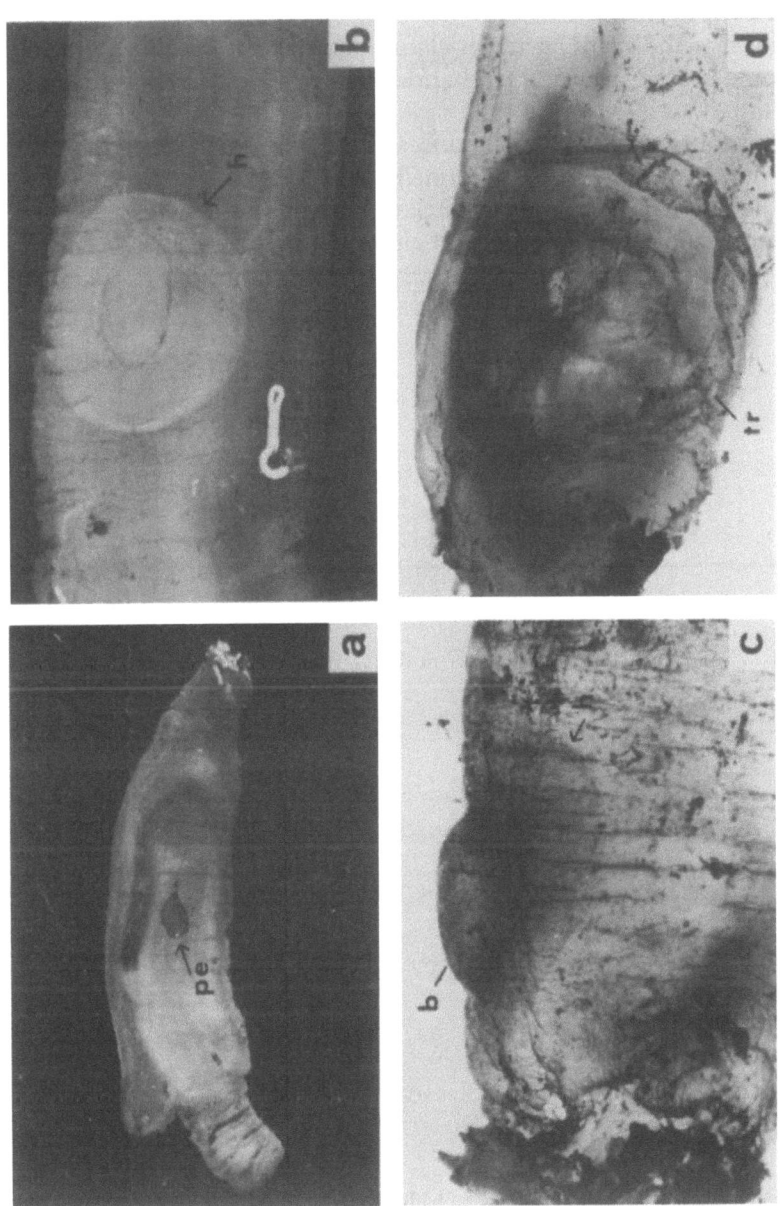

FIGURE 9. Sequential phases of the tunic reaction in *Ciona intestinalis* injected with 4×10^7 sheep red blood cells. (a) Packed erythrocytes (pe) visible through the tunic. (b) Halo (h) formation. (c) Blister (b) formation. (d) Tunic rupture (tr). (From Parrinello et al., 1977, with permission.)

(Figure 10). Both lie over the mantle and can be observed around the erythrocyte mass. The granules spread throughout the tunic surrounding the erythrocytes, form granular streams from the mantle epithelium or collect in distinct clusters. Near the erythrocytes, the material is more conspicuous and scattered, whereas toward the periphery of the halo, it gradually decreases. Histochemical studies demonstrate that the granular material and some of the cells in the halo zone have the same protein and polysaccharide staining reactions as the mantle epithelial cells.

Erythrocyte encapsulation responses appear to be nonspecific (Wright, 1973; Wright and Cooper, 1975; Parrinello et al., 1977). Primary injections of nonencapsulating concentrations of duck or human erythrocytes, followed 6 or 35 days later by secondary injections of either the same or opposite erythrocyte type at the same concentration given in the primary injection, result in capsule

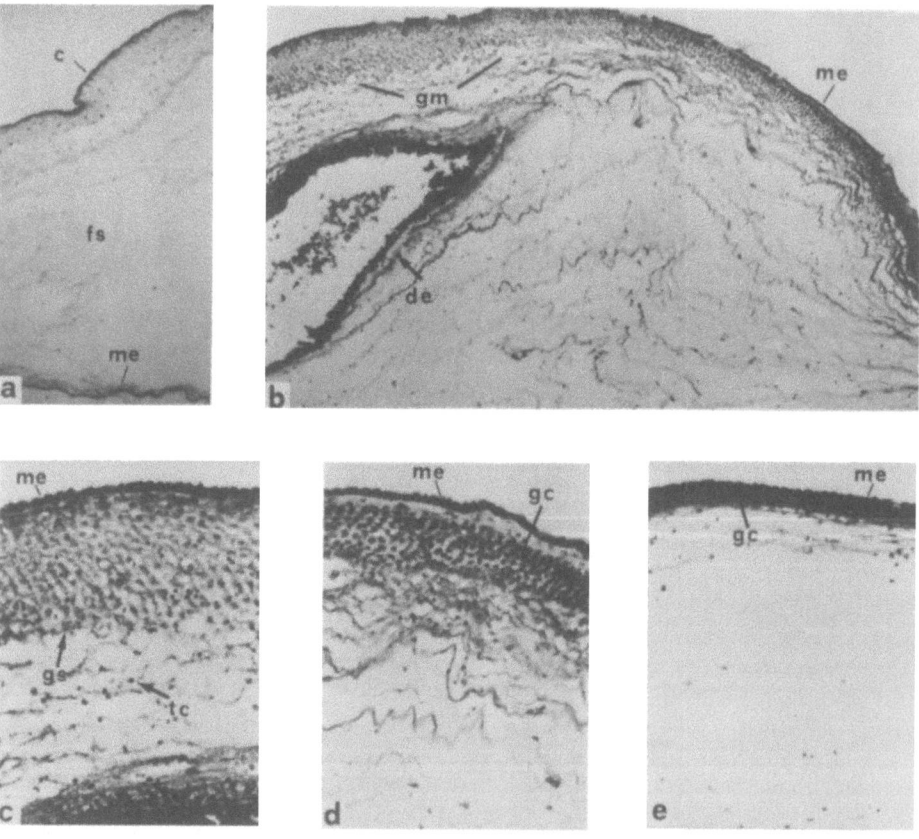

FIGURE 10. Histological sections of tunic injected with sheep red blood cells. (a) Transverse section of tunic from control (uninjected) animals. c, cuticle; me, mantle epithelium; fs, fundamental substance. × 60. (b) Transverse section of tunic with halo around degenerated erythrocytes (de). Monolayered epithelium (me) overlies halo components which contains granular material (gm). × 90. (c) Higher magnification (× 225) of zone near erythrocytes in (b) showing granular streams (gs) and tunic cells (tc). (d) Higher magnification (× 225) of lateral zone in (b) showing granules collected in clusters (gc). (e) Peripheral zone of halo in which newly synthesized material is packed over the mantle epithelium. × 200. (From Parrinello et al., 1977, with permission.)

formation. Similar results have been obtained with sheep and rabbit erythrocytes (Parrinello et al., 1977). Since capsules are formed against the secondary erythrocyte type injected, the response is nonspecific. Secondary erythrocyte encapsulation reactions are usually stronger (heightened) than primary reactions. They occur more rapidly and in larger numbers of animals. Differential responses with respect to animal age are also apparent (Wright and Cooper, 1975). Sexually immature animals exhibit fewer erythrocyte encapsulation responses than sexually mature animals. Reasons for these differential responses are unknown but may reflect differences in blood cell populations of young and old animals. The heightened responses toward secondary exposures may be due to a more rapid blood cell infiltration or to the fact that cells are already in the neighboring area as a result of the primary injection. Alternatively, primary injections may have shifted blood cell type homeostasis in favor of increased vacuolated cell production.

Encapsulation responses have not been observed to occur in the vascular system or perivisceral cavity (Anderson, 1971; Wright, 1973, 1974). They appear to be confined to the tunic. Capsules, formed in response to glass fragments, are composed of multilayered structures made up of morula cells, monolayers of cells derived from morula cell aggregates and tunicin produced by morula cells. Erythrocyte encapsulation also involves vacuolated cells and granular material produced by the mantle epithelium. Since tunic formation and tunic integrity involves vacuolated blood cells and the secretion of mantle epithelial carbohydrates (see review by Goodbody, 1974), encapsulation responses appear to involve a similar process. They are, therefore, an extension of tunic formation and wound healing.

3.3. HISTOINCOMPATIBILITY RESPONSES

Each individual is unique with respect to its cell surface antigens. These antigens are polymorphic, encoded by genes mapping to the individual's histocompatibility gene complex. Histocompatible differences existing between individuals of the same species and different species can be recognized and reacted against by various blood leukocytes. Two sensitive indicators of histoincompatibility are allogeneic leukocyte interactions (mixed leukocyte cultures, MLC) and tissue graft rejection. Positive MLC reactions between allogeneic leukocytes are reflected in DNA synthesis and mitosis of the responding cells. It requires that the cells be alive and antigenically disparate. Graft rejection is also a manifestation of cell-mediated immunity and occurs when transplanted cells or tissues are antigenically disparate from the graft recipient. The mechanism(s) leading to cell and tissue destruction, however, is poorly understood.

3.3.1. Mixed Leukocyte Reactions

Allogeneic interactions between blood leukocytes have been studied only in *Halocynthia hirgendorfi* (Tanaka, 1975) and *Pyura stolonifera* (Warr et al., 1977). In preliminary experiments with cultured blood cells from *H. hirgendorfi*, two out of nine combinations among six animals gave a 25–35% higher increase in ^{14}C-labeled amino acid incorporation in mixed cultures than in control cultures. In the other seven combinations, incorporation was at the same level in both mixed and control cultures. Although these results do not demonstrate DNA synthesis or cell proliferation, they do suggest that the interaction between the mixed cells in some combinations is inducing a response in their biochemical machinery. Whether these changes result in the manifestation of some as yet undefined response other than proliferation remains to be investigated.

Mixtures of blood leukocytes from *P. stolonifera* collected from the same or distant locality do not show enhanced DNA synthesis when assayed at time intervals ranging from 3 to 10 days (Warr et al., 1977). Although the leukocytes remained viable during the culture period and incorporated [^{125}I]dUrd, there was no significant increase in [^{125}I]dUrd incorporation over control levels. While these negative results suggest that ascidian blood leukocytes do not proliferate in response to allogeneic differences, it should be pointed out that only two combinations among three animals were tested. Definitive conclusions must await the investigation of many other combinations among natural populations.

Positive MLC reactions depend upon the ability of responding cells to proliferate. Circulating ascidian blood cells apparently can divide *in vivo* (Ermak, 1975b); however, stimulation of blood cells by plant lectins *in vitro* has produced conflicting results. Plant lectins such as concanavalin A (Con A) and phytohemagglutinin (PHA) are mitogenic for vertebrate lymphocytes. These lectins bind to carbohydrate moieties of glycoproteins on the cell surface, stimulating the cell to divide. Tam et al. (1976) have reported that *C. intestinalis* blood cells synthesize new DNA and proliferate in response to PHA. To obtain positive blastogenic responses, high PHA concentrations and a long culture period were required. *In vitro* cultivation of *P. stolonifera* blood leukocytes with Con A, wheat germ agglutinin, or soybean lectin at several concentrations for a period of 4 to 13 days, however, produced no significant proliferative response over cultures containing no lectin(Warr et al., 1977). These results were obtained from only a small number of animals. Experiments with ^{125}I-labeled lectins clearly showed that the membranes of tunicate blood cells could bind these lectins and that the appropriate sugar residues to allow binding were present.

Absence of a positive MLC reaction and the questionable ability of tunicate blood cells to proliferate in response to plant lectins do not imply that allogeneic differences cannot be recognized and reacted against. In mammals, naturally occurring rapidly cytolytic cells exist (Kiessling and Wigzell, 1979). No prior sensitization or proliferation is required for these natural killer cells to recognize and lyse allogeneic cells. Thus, it is possible that tunicate blood leukocytes may respond in a similar fashion as a result of allogeneic interactions (see Section 3.3.3). Whether or not ascidian leukocytes possess cytotoxic potential remains to be tested.

3.3.2. Tissue Transplantation

Orthotopic tissue transplantation is another sensitive indicator of histoincompatibility. Available experimental evidence on graft rejection in tunicates is limited to two solitary species, *M. manhattensis* (Anderson, 1971) and *C. intes-*

tinalis (Reddy *et al.*, 1975). While it is clear from these studies that blood cells from these two species can recognize and react to tissue grafts, technical problems have not allowed definitive rejection times to be obtained. In addition, tests for immunological memory and specificity toward second-set and third-party grafts remain to be conducted.

Autogeneic and allogeneic branchial sac tissue surgically implanted within the branchial wall does not fuse with host tissues in *M. manhattensis*. The presence of a mucus-filled space between graft and host tissue apparently precludes an intimate contact between them, a necessary event for initial healing. This space, however, did not present a barrier to blood cell migration into the implanted autogeneic and allogeneic grafts. For the first 2 days, morphology and blood cell composition of graft and host branchial sac tissues were normal. On day 3, signet ring and morula cells in the graft and surrounding tissues in-

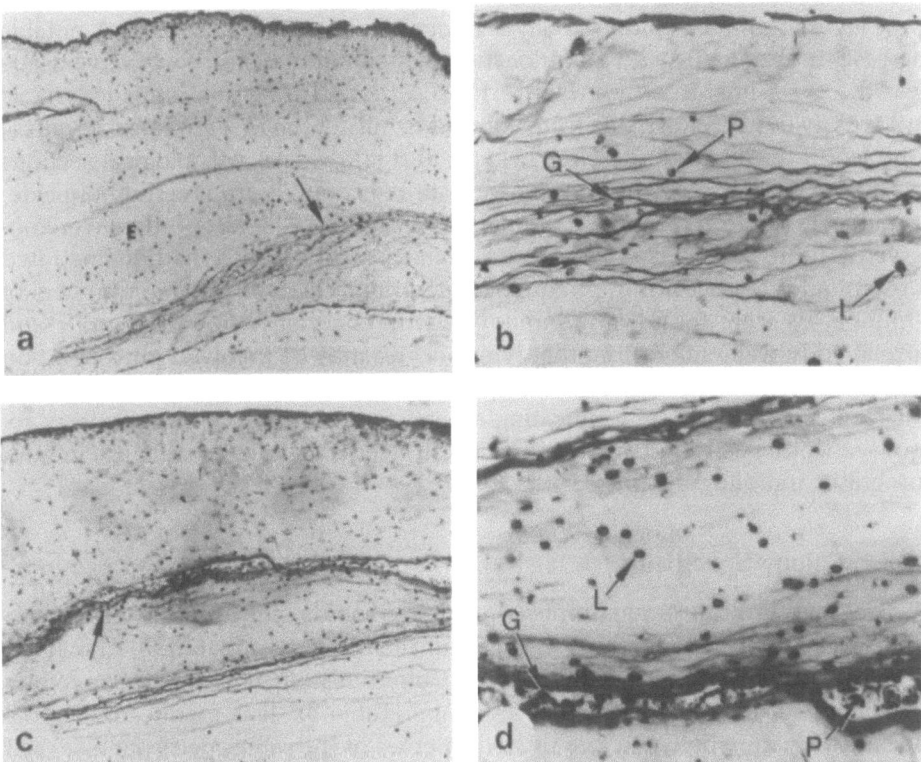

FIGURE 11. Tunic autograft and allograft reactions in *Ciona intestinalis*. (a) Transverse section of autograft 8 weeks after grafting showing complete fusion of graft (arrow) and normal appearance of outer tunic layer (T) and inner epidermal layer (E). × 100. (b) Higher magnification (× 400) of (a) showing an increase in granulocytes (G) and phagocytes (P) associated with inflammatory responses observed along the zones of graft–host interface with no noticeable increase in lymphocytes (L). (c) Transverse section of allograft 8 weeks after grafting showing clear demarcation of graft from host tissue (arrow) and infiltration by different blood cell types. × 100. (d) Higher magnification (× 400) of (c) showing graft–host interface zone invaded with large numbers of lymphocytes (L), granulocytes (G), and phagocytes (P) associated with progressive graft rejection. (From Reddy *et al.*, 1975, with permission.)

creased. After day 4, grafts contained a morula cell concentration (cells per unit of section) at least three times that present at the time of implantation. Grafts became completely infiltrated with morula cells after 7 days, the cells totally masking graft morphology. Within the graft, blood cells lost their typical morphology and eventually ruptured. Twelve to fourteen days after implantation, the grafts became necrotic which may be a reflection of morula cell rupturing and release of their acid contents into the graft (see Section 2.1.3). Although there are resemblances in timing and cell types involved between graft infiltration and encapsulation of glass fragments (see Section 3.2.2), blood cell encapsulation of the graft was not observed.

Tunic transplants in *C. intestinalis* behave differently. Autografts and allografts effectively healed to the graft bed after 5–7 days. Eight weeks after the initial transplantation, however, the study had to be discontinued due to an increased mortality in all experimental animals. In histological sections, autografts were indistinguishable from normal host tissue (Figure 11a). The outer tunic layer and inner epidermal layer retained a normal appearance. A notable increase in phagocytes and granulocytes was seen in the loose connective tissue near the graft–host interface (Figure 11b). Although allografts remained viable and grossly indistinguishable from autografts during the first 4 weeks, a gradual contraction from host tissue was evident at 6 to 8 weeks. Histological sections revealed different stages of progressive graft rejection(Figure 11c). Lymphocyte, granulocyte, and phagocyte accumulations along the allograft borders were observed (Figure 11d). Total cell counts from histological sections demonstrated substantial differences in the frequency of various cell types. Large numbers of lymphocytes were found only in allografts, in contrast to both autografts and normal body wall, suggesting that lymphocytes may be involved in recognition and rejection of allografts. In both autografts and allografts, granulocytes and phagocytes also increased, indicating a persistent nonspecific inflammatory response. Morula cells and signet ring cells showed a significant decrease in both autografts and allografts.

3.3.3. Colony Specificity

Colonial tunicates are composed of many zooids embedded in a common tunic. A vascular network runs throughout the tunic, terminating in numerous ampullae at the periphery of the colony. Colony histoincompatibility or colony specificity exists in many colonial ascidians (Table 5). It is manifested by the fusibility in nature between two related colonies of the same species to fuse (self recognition), forming a single colony. During fusion, there is a complete union of blood vessels from one colony with those of the other, thus establishing a common vascular system between the two. Unrelated colonies do not fuse (nonself recognition). Sister and brother colonies developed from larvae released by one parental colony sometimes fuse and sometimes do not. Two pieces taken from the same colony or a related colony, when brought in contact, will also fuse, whereas pieces from unrelated colonies will not. Nonfusion reactions entail the formation of tissue necrosis in contact areas. Thus, colony histoincom-

TABLE 5. OCCURRENCE OF COLONY SPECIFICITY IN COLONIAL TUNICATES

	Reference
Botrylloides violaceus	Mukai and Watanabe (1974, 1975b)
Botryllus primigenus	Watanabe (1953), Oka and Watanabe (1957, 1960, 1967), Mukai (1967), Tanaka (1973), Tanaka and Watanabe (1973), Mukai and Watanabe (1974, 1975a)
Botryllus schlosseri	Bancroft (1903), Sabbadin (1962), Karakashian and Milkman (1967)
Didemnum moseleyi	Mukai and Watanabe (1974)
Ecteinascidia tortugenesis	Freeman (1970a)
Perophora bermudensis	Freeman (1970a)
Perophora viridis	Freeman (1970a)
Symplegma reptans	Mukai and Watanabe (1974)

patibility is analogous to tissue transplantation and graft rejection and can be considered as a type of allogeneic recognition.

Colony specificity was first described by Bancroft (1903) in *Botryllus schlosseri* and has been extensively studied in this species and in *B. primigenus* (Table 5). Fusibility is genetically controlled by a series of multiple alleles at a single locus (Oka and Watanabe, 1960). Each colony in nature is heterozygotic with respect to the gene governing fusibility. Colonies containing at least one gene in common are fusible with each other. Colonies can be expressed at AB, CD, EF, etc. and since they have no common allele, do not fuse. If two colonies, for example AB and CD, are crossed, the offspring will be composed of four groups, AC, AD, BC, and BD. Since all have a common allele with both parents, they will fuse with them. Estimations of the effective number of alleles in natural populations range from approximately 39 in *B. primigenus* (Mukai and Watanabe, 1975a) to 80 in *B. schlosseri* (Karakashian and Milkman, 1967).

The process of nonfusion in *B. primigenus* is temperature dependent and can be divided into six stages (Tanaka and Watanabe, 1973). When two incompatible colonies are placed in contact with each other, their ampullar tips actively extend toward each other and the margins of the test matrices make contact (stage 1). One to two hours after contact, the ampullar tips push against each other, penetrating into the test matrices of the opposite colony, forming a tip-to-side contact between the ampullae of both colonies (stage 2). Two to three hours later, the first sign of nonfusion is apparent when the test cells surrounding the tips of the penetrating ampullae become dark and opaque (stage 3). Seven hours after contact, the ampullae of both colonies have ceased further penetration into the test matrices of the opposite colony. Test cells and brown cells released from the ampullae aggregate at the ampullar tips (stage 4). Areas surrounding the penetrating ampullae then become dark and opaque, the ampullae begin to contact and become thinner at their proximal parts, and blood flow decreases (stage 5). Two hours after reaching stage 5, there is further ampullar contraction and blood flow in the ampullae has completely ceased. Ampullae of the contact area become separated from the healthy parts of the proximal vascular system and new walls are formed, separating the degenerated zone from the healthy parts of both colonies facing each other (stage 6).

Histological and electron microscopic examinations reveal that at stage 3, test cells were being destroyed and numerous filaments were appearing around the disintegrated test cells. Brown cells released through the ampullar walls into the test matrix were also destroyed. No noticeable change, however, was observed in the blood cells retained in the vascular system, even at stage 6.

Nonfusion reactions are irreversible (Tanaka, 1973). Results from nonfusible combinations show that when removal was made after the fusion of tunics, the nonfusion reaction continues irreversibly at the same speed and in the same manner as in unseparated cases. No evidence of the nonfusion reaction is observed if removal occurs before the tunics fuse. Fusion can also be altered (Oka and Watanabe, 1957; Mukai, 1967). If an AB colony is allowed to fuse with an AC colony, a nonfusion reaction occurs when the AB colony is separated and brought into contact with a BD colony. Since AB and BD share a common allele (B), one would not expect a nonfusion reaction. Similar results were obtained when the AC colony was brought into contact with a CD colony. These results suggest that some blood element is playing a role in the incompatibility reaction.

To determine what role(s) blood cells and humoral factor(s) may play in such combinations, Tanaka (1973) examined the interaction between three colonies, AC, BC, and BD, placed side by side in the order AC–BC–BD. Colonies AC and BD are not fusible since they lack a common allele and their blood cannot mix. Placement of the BC colony between them, however, allows the blood of AC and BD to mix within the vessels of the BC colony after the establishment of connections on both sides (Figure 12a). In both contact areas, blood could distinctly be seen passing from one colony to the other. Afterwards, blood flow stopped and blood cell clusters along with ampullae contraction were observed (Figures 12b,c). After a complete fusion of AC and BC (Figure 12d), a nonfusion reaction was observed between BC and BD (Figure 12e). Some blood vessels in the middle region of the central colony were also observed to disintegrate with formation of blood cell clusters and blood vessel constriction as the result of blood cell inflows from both side colonies.

Histological and electron microscopic findings showed that granular leukocytes and brown cells were dominant in the blood cell clusters. Test cells of the BC colony neighboring the necrotized vessels were destroyed and thin filaments were formed around them. Reactions occurring in the blood vessels consisted of granular leukocyte breakdown (cytolysis?) and filament formation in the intercellular spaces of the blood cell clusters. Granular leukocytes were also observed to contain similar filaments in their cytoplasm.

From these findings, it appears that the fusion characteristics of the BC colony can be modified by blood from a compatible AB colony. While BC would normally be compatible with CD, it is no longer compatible after an initial fusion with AB. According to Tanaka (1973, 1975), the blood contains colony specificity factors that destroy granular leukocytes and test cells when two incompatible colonies make contact. Upon their destruction, they release a factor(s) that causes further cell destruction and contraction of the ampullae and blood ves-

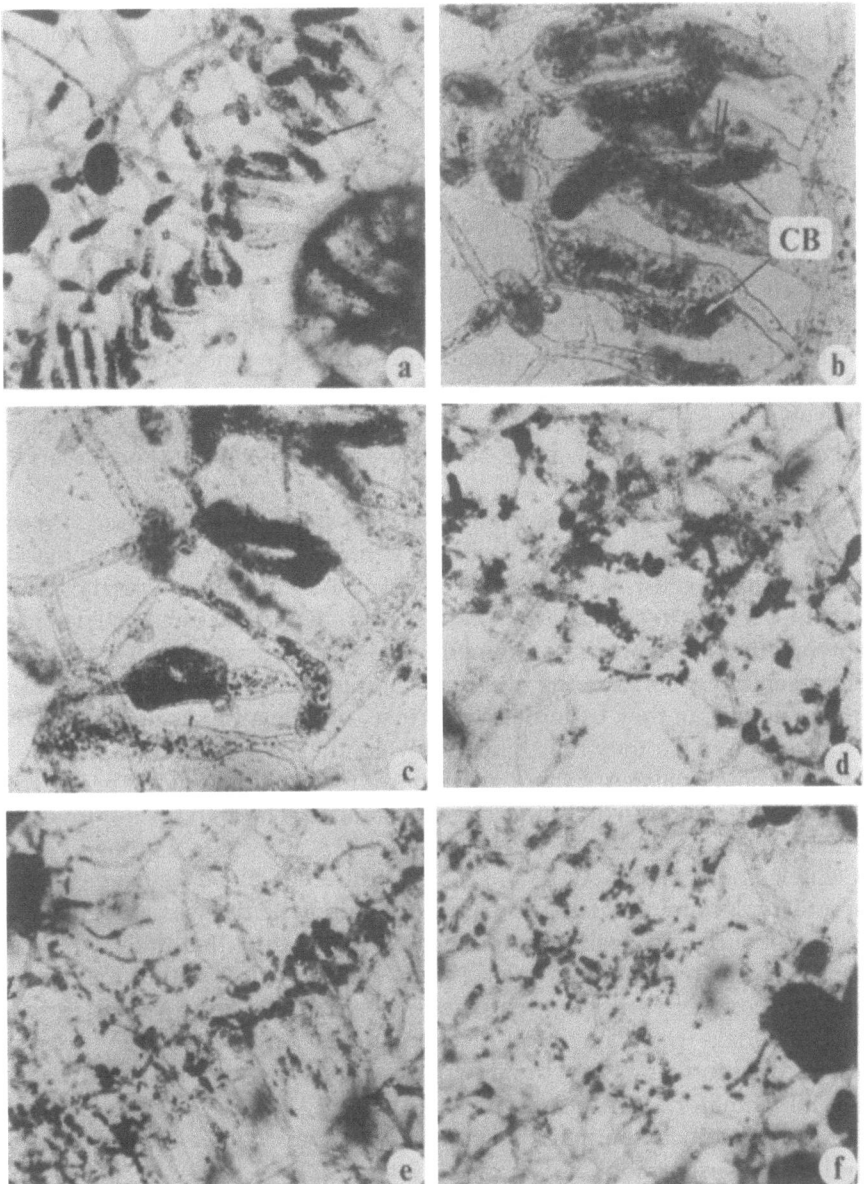

FIGURE 12. Fusion/nonfusion reaction features in the vascular system of *Botryllus primigenus* among three colonies (AC–BC–BD) where fusion occurred at a 5-min lag between BC and two side colonies. (a) Thirty minutes after establishing tip-to-side connection of ampullae (arrow) between AC and BC colonies. × 15. (b) Ninety minutes after fusion between AC and BC. × 45. Ampullar contraction (double arrows) and clusters of blood cells (CB) are observed. (c) Four hours and 30 min after fusion of AC and BC. × 45. Blood flow through the connections between AC and BC is at a standstill. (d) Twenty hours after fusion of AC and BC. × 15. A complete fusion by newly established connections between AC and BC. (e) Nonfusion reaction between BC and BD colonies is evidenced by formation of a necrotic zone. × 15. (f) Partial destruction of vascular system in central BC colony. × 15. (From Tanaka, 1973, with permission.)

sels. Nonfusion reactions, therefore, appear to involve both cellular and humoral factors, suggesting that an immunological mechanism may be responsible for the incompatibility reaction.

4. CONCLUSIONS AND SUMMARY

The ascidian immunosurveillance system has two major components, one cellular, the other humoral. Together, they account for the maintenance and integrity of the animal in an aseptic condition. Humoral components consist of various antimicrobial, antitumor, and antifungal factors, bactericidal factors, colony specificity factors and agglutinins present in the blood plasma and body tissues. Presumably, these factors are involved in the neutralization of non-self material, functioning alone or in conjunction with cell-mediated responses. However, little is known about their precise function(s) and role(s) in immunosurveillance.

Many questions concerning humoral factors need to be answered. Do they function as recognition factors, opsonins to enhance phagocytosis, or in mobilizing inflammatory cells? What is their origin, mode of action, specificity, and relationship, if any, to each other? Although the chemical nature of some factors is known, others remain to be defined. Are they components of a hemolytic system resembling perhaps the alternate complement pathway, lysosomal, or other enzymes? Can they be induced and if so, are they the result of cell activation or cell injury? Do they represent newly synthesized material or the release of preformed materials? Are there quantitative and qualitative changes in them after infection or introduction of foreign substances? Through further investigation, many of these questions will be answered.

Cellular components consist of several different blood cell types whose origin and replacement occurs in discrete hemopoietic tissues. These cells are actively ameboid, diapedesing from the blood spaces into the connective tissues where they occur randomly or in large concentrations. It is within the vascular system and connective tissues that they recognize and react toward foreign, non-self materials. How they recognize foreignness and the nature of their cell surface recognition units are unknown. Reactions toward non-self materials include inflammatory responses with concomitant phagocytosis and/or encapsulation and histoincompatibility responses toward allogeneic tissues leading to cell and tissue destruction.

Three cell types are involved in ascidian immune responses: undifferentiated cells commonly referred to as lymphocytes, hyaline and granular leukocytes, and vacuolated cells. The identity of a true lymphocytic cell in ascidians is not clear, and there is much confusion regarding the role of this cell, or undifferentiated cells involved in hemopoiesis (hemoblasts), in the immune response. Lymphocytelike cells have been observed to infiltrate and accumulate in allografts undergoing rejection. Whether they are the effector cells responsible for rejection remains to be determined. Hyaline and granular leukocytes are phagocytic cells, actively removing and sequestering foreign materials. They accumulate at injury sites regardless of the nature of the injurious agent. Whether chemotactic factors are responsible for their accumulation is unknown. Organic materials (e.g., bacteria) are probably degraded by lysosomal enzymes. Histochemical studies confirming the presence of these or other enzymes, however, are lacking. Inert particles such as carbon appear to be transported and released to the external environment.

Granular leukocytes also appear to be involved in allogeneic histoincompatibility reactions. They have been observed to infiltrate and accumulate in allografts undergoing rejection. Additionally, they appear to be destroyed (cytolysis?) when two incompatible colonies are brought together. Whether this destruction is the result of cell–cell interactions analogous to vertebrate natural killer cell cytolysis of allogeneic cells or due to some humoral factor(s) is unknown. Substances too large to be sequestered by phagocytosis are encapsulated. Encapsulation responses are a function of the vacuolated cells which accumulate around the object, forming multilayered structures composed of tunic material and cells. Their primary function, however, is tunic formation and maintenance of tunic integrity. Encapsulation responses are therefore an extension of this function.

Ascidians are generally considered to be the phylogenetic group that gave rise to the vertebrates. Although the fossil record does not preserve an account of the phylogenetic origin and development of the vertebrate reticuloendothelial system and its immune functions, the structural basis leading to its evolution appears to be present in the ascidians (Freeman, 1970b; Burnet, 1971; Wright, 1976). The location and microanatomy of adult tunicate hemopoietic tissues are similar to those of agnathan fish. In both groups, these tissues are located in the pharyngeal (a "primitive thymus") and gut (a "primitive spleen") regions. With further evolutionary development, the definitive vertebrate thymus and spleen developed in these areas (Wright, 1976).

Another similarity between the ascidian and the vertebrate reticuloendothelial system is the proliferative characteristics and kinetics of blood cell development. In both groups, blood cells are a renewing population with finite life spans. Hematological studies on over 70 ascidian species confirm the presence of blood cells called lymphocytes or hemoblasts. Morphologically, these cells resemble vertebrate lymphocytes and hemoblasts. In ascidians, these cells are apparently involved in graft rejection and can form rosettes with sheep erythrocytes (Hildemann and Reddy, 1973), characteristics they have in common with vertebrate T lymphocytes. However, the complete immunological function of these cells remains to be defined. Recent findings that tunicate plasma contains a DNP-binding protein subunit resembling μ chain of vertebrate immunoglobulin and its similarity to the immunoglobulin heavy chain of the vertebrate T-lymphocyte receptor, suggest that ascidian lymphocytelike cells might express a primitive μ-like chain (Marchalonis and Warr, 1978). Additional functional studies and characterization of cell surface components are needed, however, to determine whether tunicate hemoblasts or lymphocytes have homologous functions to vertebrate lymphocytes.

Finally, the genetics of recognition in colonial tunicates bears similarities to the histocompatibility antigen systems present in vertebrates (Freeman, 1970b; Burnet, 1971). Each individual is unique with respect to its cell surface components. These components are encoded at the genetic level by multiple alleles mapping to the individual's histocompatibility complex. In vertebrates, differences existing between individuals at these loci can be recognized and reacted against by blood cell populations. The scheme in tunicates differs slightly since individuals sharing one allele but differing at the other, are compatible. Histoincompatible reactions occur only when they differ at both loci. These reactions, however, are mediated by some blood element that appears to be associated with a histocompatibility system for distinguishing between self and non-self. Clearly, this system could serve as a foundation leading, with further refinement through evolution, to the development of the vertebrate histocompatibility system.

The past decade has seen significant progress and insight toward our understanding of the ascidian internal defense system, its components and immune reactions. The relationship of this system to that of other invertebrates and vertebrates is becoming apparent. Future investigations will increase our understanding of ascidian blood cells, their role in the ascidian reticuloendothelial system, and their relationship to invertebrate and vertebrate immune systems.

ACKNOWLEDGMENTS. We wish to thank Drs. Robert S. Anderson, Paolo Burighel, Carla Milanesi, Jane Overton, Nicolò Parrinello, A. Lakshma Reddy, and Kunio Tanaka who provided the figures used in this chapter. We thank Academic Press, Acta Zoologica (Stockholm), Elsevier/North-Holland, Springer-Verlag, Unione Zoologica Italiana presso l'Instituto di Biologica Animale, Wistar Institute Press, and Woods Hole Marine Biological Laboratory for permission to publish these figures.

REFERENCES

Anderson, R. S., 1971, Cellular responses to foreign bodies in the tunicate *Molgula manhattensis* (DeKay), *Biol. Bull.* **141**:91.

Anderson, R. S., and Good, R. A., 1975, Naturally-occurring hemagglutinin in a tunicate *Halocynthia pyriformis*, *Biol. Bull.* **148**:357.

Anderson, R. S., Jordan, L. A., and Harshbarger, J. C., 1977, Tunic abnormalities of the urochordate *Halocynthia pyriformis*, *J. Invertebr. Pathol.* **30**:160.

Andrew, W., 1961, Phase microscope studies of living blood cells of the tunicates under normal and experimental conditions, with a description of a new type of motile cell appendage, *Q. J. Microsc. Sci.* **102**:89.

Azéma, M., 1937, Recherches sur le sang et l'excrétion chez les Ascidies, *Ann. Inst. Oceanogr. Monaco* **17**:1.

Bancroft, F. W., 1903, Variation and fusion of colonies in compound ascidians, *Proc. Calif. Acad. Sci. 3rd Ser.* **3**:137.

Bang, F. B., 1975, A search in *Asterias* and *Ascidia* for the beginnings of vertebrate immune responses, *Ann. N.Y. Acad. Sci.* **266**:334.

Barnes, R. D., 1968, *Invertebrate Zoology*, 2nd ed., Saunders, Philadelphia.

Berrill, N. J., 1955, *The Origin of Vertebrates*, Oxford University Press, London.
Bielig, H. J., Bayer, E., Califano, L., and Wirth, L., 1954, Vanadium-containing blood pigments. II. Hemovanadium, a sulfate complex of trivalent vanadium, *Pubbl. Stn. Zool. Napoli* **25**:26.
Bielig, H. J., Pfleger, K., Rummel, W., and de Vicentis, M., 1963, Beginning of the accumulation of vanadium during the early development of the ascidian *Phallusia mamillata* Cuvier, *Nature* **197**:1223.
Biggs, W. R., and Swinehart, J. H., 1979, Studies of the blood of *Ascidia ceratodes:* Total blood cell counts, differential blood cell counts, hematocrit values, seasonal variations, and fluorescent characteristics of blood cells, *Experientia* **35**:1047.
Botte, L., and Scippa, S., 1977, Ultrastructural study of vanadocytes in *Ascidia malaca*, *Experientia* **33**:80.
Bresciana, J., and Lützen, J., 1960, *Gonophysema gullmarensis* (Copepoda Parasitica). An anatomical and biological study of an endoparasite living in the ascidian *Ascidiella aspersa*. I. Anatomy, *Cah. Biol. Mar.* **1**:157.
Bretting, H., and Renwrantz, L., 1973, Untersuchungen von Invertebraten des Mittelmeeres auf ihren Gehalt an hämagglutininierenden Substanzen, *Z. Immunitaetsforsch. Bd.* **145**:242.
Brown, A. C., and Davies, A. B., 1971, The fate of thorium dioxide introduced into the body cavity of *Ciona intestinalis* (Tunicata), *J. Invertebr. Pathol.* **18**:276.
Burnet, F. M., 1971, 'Self-recognition' in colonial marine forms and flowering plants in relation to the evolution of immunity, *Nature (London)* **232**:230.
Cantacuzène, J., 1919, Étude d'une infection expérimentale chez *Ascidia mentula*, *C.R. Soc. Biol.* **82**:1019.
Carter, G. T., and Rinehart, K. L., Jr., 1978, Aplidiasphingosine, an antimicrobial and antitumor terpenoid from an *Aplidium* sp. (marine tunicate), *J. Am. Chem. Soc.* **100**:7441.
Cheng, M. T., and Rinehart, K. L., Jr., 1978, Polyandrocarpidines: Antimicrobial and cytotoxic agents from a marine tunicate (*Polyandrocarpa* sp.) from the Gulf of California, *J. Am. Chem. Soc.* **100**:7409.
Cowden, R. R., 1968, The embryonic origin of blood cells in the tunicate *Clavelina picta*, *Trans. Am. Microsc. Soc.* **87**:521.
Cuénot, L., 1891, Études sur le sang et les glandes lymphatiques dans le série animale (2° parties: Invertébrés), *Arch. Zool. Exp. Gen. Ser. 2* **9**:13.
Dales, R. P., 1979, Defence of invertebrates against bacterial infection, *J. R. Soc. Med.* **72**:688.
Dudley, P. L., 1968, A light and electron microscopic study of tissue interactions between a parasitic copepod, *Scolecodes huntsmani* (Henderson), and its host ascidian, *Styela gibbsii* (Stimpson), *J. Morphol.* **124**:263.
Endean, R., 1955, Studies of the blood and tests of some Australian ascidians. I. The blood of *Pyura stolonifera* (Heller), *Aust. J. Mar. Freshwater Res.* **6**:35.
Endean, R., 1960, The blood cells of the ascidian *Phallusia mammillata*, *Q. J. Microsc. Sci.* **101**:177.
Endean, R., 1961, The test of the ascidian *Phallusia mammillata*, *Q. J. Microsc. Sci.* **102**:107.
Ermak, T. H., 1975a, Cell proliferation in the ascidian *Styela clava*: An autoradiographic and electron microscopic investigation emphasizing cell renewal in the digestive tract of this and fourteen other species of ascidians, Ph.D. dissertation, University of California, San Diego.
Ermak, T. H., 1975b, An autoradiographic demonstration of blood cell renewal in *Styela clava* (Urochordata: Ascidiacea), *Experientia* **31**:837.
Ermak, T. H., 1976, The hematogenic tissues of tunicates, in: *Phylogeny of Thymus and Bone Marrow-Bursa Cells* (R. K. Wright and E. L. Cooper, eds.), pp. 45–56, Elsevier/North-Holland, Amsterdam.
Form, D. M., Warr, G. W., and Marchalonis, J. J., 1979, Isolation and characterization of a lectin from the hemolymph of a tunicate, *Halocynthia pyriformis*, *Fed. Proc.* **38**:934.
Franz, V., 1927, Morphologie der Akranier, *Ergeb. Anat. Entwicklungsgesch.* **27**:464.
Freeman, G., 1964, The role of blood cells in the process of asexual reproduction in the tunicate *Perophora viridis*, *J. Exp. Zool.* **156**:157.
Freeman, G., 1970a, Transplantation specificity in echinoderms and lower-chordates, *Transplant. Proc.* **2**:236.
Freeman, G., 1970b, The reticuloendothelial system of tunicates, *J. Reticuloendothelial Soc.* **7**:183.

Fuke, M. T., 1979, Studies on the coelomic cells of some Japanese ascidians, *Bull. Mar. Biol. Stn. Asamushi Tohoku Univ.* **16**:143.

Fuke, M. T., and Sugai, T., 1972, Studies on the naturally occurring hemagglutinin in the coelomic fluid of an ascidian, *Biol. Bull.* **143**:140.

Fulton, J. F., 1920, The blood of *Ascidia atra* Lesueur; with special reference to pigmentation and phagocytosis, *Acta Zool.* (Stockholm) **1**:381.

Gansler, H., Pfleger, K., Seifen, E., and Bielig, H. J., 1963, Submikroskopische Struktur von Vanadocyten. Ein Beitrag zur Vanadin-Anhaufung bei Tunicaten, *Experientia* **19**:232.

George, W. C., 1926, The histology of the blood of *Perophora viridis* (Ascidian), *J. Morphol. Physiol.* **41**:311.

George, W. C., 1930, The histology of the blood of some Bermuda ascidians, *J. Morphol. Physiol.* **49**:385.

George, W. C., 1939, A comparative study of the blood of the tunicates, *Q. J. Microsc. Sci.* **81**:391.

Goodbody, I., 1974, The physiology of ascidians, *Adv. Mar. Biol.* **12**:1.

Hecht, S., 1918, The physiology of *Ascidia atra* Lesueur. III. The blood system, *Am. J. Physiol.* **45**:157.

Henze, M., 1911, Untersuchungen über das Blut der Ascidien. I. Die Vanadiumbindung der Blutkörporchen, *Hoppe-Seyler's Z. Physiol. Chem.* **72**:494.

Henze, M., 1912, Untersuchungen über das Blut der Ascidien. II. Mitteilung, *Hoppe-Seyler's Z. Physiol. Chem.* **79**:215.

Hildemann, W. H., and Reddy, A. L., 1973, Phylogeny of immune responsiveness: Marine invertebrates, *Fed. Proc.* **32**:2188.

Ivanova-Kazas, O. M., 1966, Phagocytic reaction in the ascidian *Distaplia unigermis*, *Ark. Anat. Gistol. Embriol.* **51**:48.

Johnson, P. T., and Chapman, F. A., 1970, Comparative studies on the *in vitro* response of bacteria to invertebrate body fluids. II. *Aplysia californica* (sea hare) and *Ciona intestinalis* (tunicate), *J. Invertebr. Pathol.* **16**:259.

Kalk, M., 1963, Intracellular sites of activity in the histogenesis of tunicate vanadocytes, *Q. J. Microsc. Sci.* **104**:483.

Karakashian, S., and Milkman, R., 1967, Colony fusion compatibility types in *Botryllus schlosseri*, *Biol. Bull.* **133**:473.

Kiessling, R., and Wigzell, H., 1979, An analysis of the murine NK cell as to structure, function and biological relevance, *Immunol. Rev.* **44**:165.

Kollmann, M., 1908, Recherches sur les leucocytes et le tissue lymphoïde des Invertébrés, *Ann. Sci. Nat. Zool. Biol. Anim.* Ser. 9 **8**:1.

Lichter, W., Lopez, D. M., Wellham, L. L., and Sigel, M. M., 1975, *Ecteinascidia turbinata* extracts inhibit DNA synthesis in lymphocytes after mitogenic stimulation by lectins, *Proc. Soc. Exp. Biol. Med.* **150**:475.

Marchalonis, J. J., and Warr, G. M., 1978, Phylogenetic origins of immune recognition: Naturally occurring DNP-binding molecules in chordate sera and hemolymph, *Dev. Comp. Immunol.* **2**:443.

Meglitsch, P. A., 1967, *Invertebrate Zoology*, Oxford University Press, London.

Messier, B., and Leblond, C. P., 1960, Cell proliferation and migration as revealed by radioautography after injection of thymidine-H^3 into male rats and mice, *Am. J. Anat.* **106**:247.

Metchnikoff, E., 1892, *Leçons sur la Pathologie Comparée de l'Inflammation*, Masson, Paris; reissued (1968) in English as *Lectures on the Comparative Pathology of Inflammation*, Dover, New York.

Milanesi, C., and Burighel, P., 1978, Blood cell ultrastructure of the ascidian *Botryllus schlosseri*. I. Hemoblast, granulocytes, macrophage, morula cell and nephrocyte, *Acta zool.* (Stockholm) **59**:135.

Milanesi, C., and Leban, E., 1974, Iron occurrence in the blood cells of *Botryllus schlosseri* (Ascidiacea), *J. Submicrosc. Cytol.* **6**:123.

Millar, R. H., 1953, *L.M.B.C. Memoirs on Typical British Marine Plants and Animals*. XXXV. Ciona, University of Liverpool Press, Liverpool.

Moller, P. C., and Philpott, C. W., 1973, The circulatory system of *Amphioxus* (*Branchiostoma floridae*). I. Morphology of the major vessels of the pharyngeal area, *J. Morphol.* **139**:389.

Monniot, C., 1963, *Kystodelphys drachi* n.g.n.sp., copepode enkyste dans une branchie d'Ascidie, *Vie et Milieu* **14**:263.

Mukai, H., 1967, Experimental alteration of fusibility in compound ascidians, *Sci. Rep. Tokyo Kyoiku Daigaku Sect. B* **13**:51.

Mukai, H., and Watanabe, H., 1974, On the occurrence of colony specificity in some compound ascidians, *Biol. Bull.* **147**:411.

Mukai, H., and Watanabe, H., 1975a, Distribution of fusion incompatibility types in natural populations of the compound ascidian, *Botryllus primigenus*, *Proc. Jpn. Acad.* **51**:44.

Mukai, H., and Watanabe, H., 1975b, Fusibility of colonies in natural populations of the compound ascidian *Botrylloides ciolaceus*, *Proc. Jpn. Acad.* **51**:48.

Mukai, H., and Watanabe, H., 1976, Studies on the formation of germ cells in a compound ascidian *Botryllus primigenus* Oka, *J. Morph.* **148**:337.

Ohuye, T., 1936, On the coelomic corpuscles in the body of some invertebrates. III. The histology of the blood of some Japanese ascidians, *Sci. Rep. Tohoku Univ.* Ser. 4 **11**:191.

Oka, H., and Watanabe, H., 1957, Colony-specificity in compound ascidians as tested by fusion experiments (a preliminary report), *Proc. Jpn. Acad.* **33**:657.

Oka, H., and Watanabe, H., 1960, Problems of colony-specificity in compound ascidians, *Bull. Biol. Stn. Asamushi* **10**:153.

Oka, H., and Watanabe, H., 1967, Problems of colony specificity, with special reference to the fusibility of ascidians, *Kagaku (Tokyo)* **37**:307.

Overton, J., 1966, The fine structure of blood cells in the ascidian *Perophora viridis*, *J. Morphol.* **119**:305.

Parrinello, N., and Patricolo, E., 1975, Erythrocyte agglutinins in the blood of certain ascidians, *Experientia* **31**:1092.

Parrinello, N., DeLeo, G., and Patricolo, E., 1976, Evolution of the immune response. Tunic reaction of *Ciona intestinalis* L. to erythrocyte injection. Some ultrastructural aspects, *Boll. Zool.* **43**:390.

Parrinello, N., Patricolo, E., and Canicattì, C., 1977, Tunicate immunobiology. I. Tunic reaction of *Ciona intestinalis* L. to erythrocyte injection, *Boll. Zool.* **44**:373.

Pérès, J. M., 1943, Recherches sur le sang et les organes neuraux des tuniciers, *Ann. Inst. Oceanog. Monaco* **21**:229.

Pérès, J. M., 1948, Recherches sur la genèse et la régénération de la tunique chez *Ciona intestinalis* L., *Bull. Inst. Oceanogr.* **936**:923.

Reddy, A. L., Bryan, B., and Hildemann, W. H., 1975, Integumentary allograft versus autograft reactions in *Ciona intestinalis:* A protochordate species of solitary tunicate, *Immunogenetics* **1**:584.

Renwrantz, L., and Uhlenbruck, G., 1974, Blood-group-like substances in some marine invertebrates. I. Blood-group A reactive substances in the ascidian *Phallusia mammillata* (Cuvier) and in the lancelet *Amphioxus (Branchiostoma) lanceolatus* (Pallas), *Vox Sang.* **26**:385.

Robertson, J. D., 1954, The chemical composition of the blood of some aquatic chordates, including members of the Tunicata, Cyclostomata and Osteichthyes, *J. Exp. Biol.* **31**:424.

Sabbadin, A., 1962, Le basi genetiche della capacitá di fusione fra colonie in *Botryllus schlosseri* (Ascidiacea), *Rend. Accad. Naz. Lincei Ser.* 8 **32**:1031.

Smith, M. J., 1970a, The blood cells and tunic of the ascidian *Halocynthia aurantium* (Pallas). I. Hematology, tunic morphology and partition of cells between blood and tunic, *Biol. Bull.* **138**:354.

Smith, M. J., 1970b, The blood cells and tunic of the ascidian *Halocynthia aurantium* (Pallas). II. The histochemistry of the blood cells and tunic, *Biol. Bull.* **138**:379.

St. Hilaire, K., 1931, Morphogenetische untersuchungen des Ascidienmantels, *Zool. Jahrbuch. Abt. Anat. Ontog. Tiere* **54**:455.

Tam, M. R., Reddy, A. L., Karp, R. D., and Hildemann, W. H., 1976, Phylogeny of cellular immunity among vertebrates, in: *Comparative Immunology* (J. J. Marchalonis, ed.), pp. 98–119, Blackwell, Oxford.

Tanaka, K., 1973, Allogeneic inhibition in a compound ascidian, *Botryllus primigenus* Oka. II. Cellular and humoral responses in "nonfusion" reaction, *Cell. Immunol.* **7**:427.

Tanaka, K., 1975, Allogeneic distinction in *Botryllus primigenus* and in other colonial ascidians, *Adv. Exp. Med. Biol.* **64**:115.

Tanaka, K., and Watanabe, H., 1973, Allogeneic inhibition in a compound ascidian, *Botryllus primigenus* Oka. I. Processes and features of "nonfusion" reaction, *Cell. Immunol.* **7**:410.

Thomas, J. A., 1931a, Sur les réactions de la tunique d'*Ascidia mentula* Müll., à l'inoculation de *Bacterium tumefaciens* Sm., *C.R. Soc. Biol.* **108**:694.

Thomas, J. A., 1931b, Réactions de deux invertébrés: *Ascidia mentula* Müll. et *Nereis diversicolor* O.F.M., à l'inoculation de substances à propriétés cancérigènes, *C.R. Soc. Biol.* **108**:667.

Thomas, J. A., 1932, Contribution à l'étude des réactions de quelques invertébrés à l'inoculation de substances à propriétés cancérigènes et du *Bacterium tumefaciens* Sm. et Town, *Ann. Inst. Pasteur Paris* **49**:234.

Vallee, J. A., 1967, Studies of the blood of *Ascidia nigra* (Savigny). I. Total blood cell counts, differential blood cell counts, and hematocrit values, *Bull South. Calif. Acad. Sci.* **66**:23.

Warr, G. W., Decker, J. M., Mandel, T. E., DeLuca, D., Hudson, R., and Marchalonis, J. J., 1977, Lymphocyte-like cells of the tunicate *Pyura stolonifera*: Binding of lectins, morphological and functional studies, *Aust. J. Exp. Biol. Med. Sci.* **55**:151.

Watanabe, H., 1953, Studies on the regulation in fused colonies in *Botryllus primigenus* (Ascidiae Compositae), *Sci. Rep. Tokyo Kyoiku Daigaku Sect. B.* **7**:183.

Webb, D. A., 1939, Observations on the blood of certain ascidians with special reference to the biology of vanadium, *J. Exp. Biol.* **16**:499.

Webb, D. A., 1956, The blood of tunicates and the biochemistry of vanadium, *Pubbl. Stn. Zool. Napoli* **28**:273.

Weiss, J., Goldman, Y., and Morad, M., 1976, Electromechanical properties of the single cell-layered heart of the tunicate *Boltenia ovifera* (Sea potato), *J. Gen. Physiol.* **68**:503.

Wright, R. K., 1973, Immunobiological studies of the ascidian urochordate *Ciona intestinalis* Linneaus, Ph.D. dissertation, University of California, Santa Barbara.

Wright, R. K., 1974, Protochordate immunity. I. Primary response of the tunicate *Ciona intestinalis* to vertebrate erythrocytes, *J. Invertebr. Pathol.* **24**:29.

Wright, R. K., 1976, Phylogenetic origin of the vertebrate lymphocyte and lymphoid tissue in: *Phylogeny of Thymus and Bone Marrow-Bursa Cells* (R. K. Wright and E. L. Cooper, eds.), pp. 57–70, Elsevier/North-Holland, Amsterdam.

Wright, R. K., 1981, Urochordates, in: *Invertebrate Blood Cells 2* (N. A. Ratcliffe and A. F. Rowley, eds.), pp. 565–626. Academic Press, New York.

Wright, R. K., and Cooper, E. L., 1975, Immunological maturation in the tunicate *Ciona intestinalis*, *Am. Zool.* **15**:21.

9

Molecular Basis of Self/Non-self Discrimination in the Invertebrata

PETER L. EY and CHARLES R. JENKIN

1. INTRODUCTION

In 1907 Reudiger and Davis claimed that hemolymph from several different species of invertebrates promoted the ingestion of bacteria by human polymorphonuclear leukocytes and suggested on this basis that invertebrates may possess factors (opsonins) which play an important role in enchancing phagocytosis. This particular field lay relatively dormant for almost 60 years until Tripp (1966) showed *in vitro* that hemolymph from the oyster *Crassostrea virginica* increased the rate of phagocytosis of erythrocytes by amebocytes from this animal. Two years later Stuart (1968) found that the blood cells of the octopus *Eledone cirrosa* would phagocytose erythrocytes only if they had been pretreated with hemolymph. Unfortunately, in neither investigation was the specificity of these factors investigated.

The phagocytic cells of animals display a high degree of selectivity in what they will or will not ingest. In the vertebrates, overwhelming evidence points to the importance of immunoglobulin molecules (antibodies) in providing this selectivity. Two decades ago it was not uncommon for immunologists to consider that invertebrates had only a crude ability to respond to foreign material and that the observed responses to foreignness lacked the fine specificity observed in the vertebrates. However, many recent observations have shown that the phagocytic cells of invertebrates do react efficiently to a variety of foreign particles and that the recognition frequently involves the prior interaction of the foreign material with factors in the hemolymph (review: Jenkin, 1976). Thus, the data accumulated to date indicate that self/non-self recognition by invertebrate phagocytes may involve a system functionally analogous to that found in the vertebrates. In this present review an attempt will be made to explain this phenomenon at a molecular level.

PETER L. EY and CHARLES R. JENKIN • Department of Microbiology and Immunology, The University of Adelaide, Adelaide 5001, South Australia.

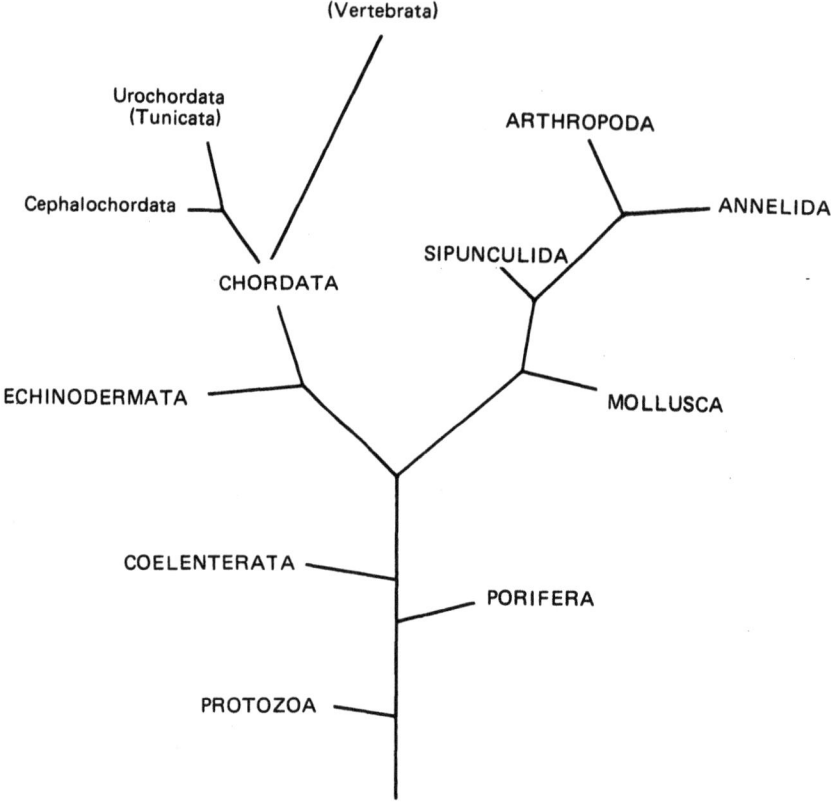

FIGURE 1. Taxonomic tree, showing probable relationships between selected invertebrate phyla and subphyla.

There are two serious problems which face one in reviewing this particular area. First, unlike the vertebrates, the invertebrates are a polyphyletic group of animals and thus subject to a vast diversity (Figure 1), so that observations and conclusions regarding recognition drawn from one group of animals may not necessarily be applicable to others. Second, very few groups of investigators have attempted to study the factors involved in the recognition of foreignness by phagocytic cells from invertebrates at both the biological and the chemical level. That is to say, after showing that substances in the hemolymph of the animal promote phagocytosis, few have attempted to characterize these factors physicochemically or to examine the nature and specificity of the interaction between the foreign particles, the opsonic factors, and the phagocytic cells.

Extensive studies by Cantacuzène (1919a) showed that the hemolymph from different species of invertebrates from various phyla such as the Annelida, Arthropoda, and Mollusca contained agglutinins for vertebrate erythrocytes. Numerous other investigators have amply confirmed these observations (Tyler and Metz, 1945; Tyler, 1946; Bernheimer, 1952; Tripp, 1966; Gilbertson and Etges, 1967; Stuart, 1968; Marchalonis and Edelman, 1968; Brown et al., 1968;

McKay *et al.*, 1969; Cooper *et al.*, 1974). Tripp (1966) considered that the opsonic activity of oyster hemolymph for erythrocytes was due to the naturally occurring hemagglutinin, although this point was not clearly established. The hemolymph of the octopus (*E. cirrosa*) also agglutinated erythrocytes that Stuart (1968) used in his studies, and as mentioned previously, human erythrocytes had to be pretreated with hemolymph in order to be ingested by octopus hemocytes.

The suggestion that invertebrate hemagglutinins may have antibodylike properties is not new (Boyden, 1966), but until the experiments of Tripp (1966), Stuart (1968), and Prowse and Tait (1969), there was no good experimental evidence to indicate that factors in the hemolymph of various invertebrates promoted the phagocytosis of foreign material by their own phagocytes.

In view of the possibility that the hemagglutinins may represent a family of molecules which function in recognition, it seems pertinent to consider the available information regarding the physicochemical properties of these molecules. It should be emphasised, however, that the proposed opsonic role of these molecules is with two exceptions based on purely circumstantial evidence. [See sections on Mollusca (3.1) and Protochordata (4.2).] This evidence is founded on experiments which show that following the adsorption of hemolymph with erythrocytes, both the hemagglutinating and the opsonic activities are lost. Nevertheless, it is not unreasonable to expect that the two properties may be associated with the same molecule. Hemagglutinins occur in the sera of vertebrates and these with few exceptions are immunoglobulin molecules which, depending on the isotype, possess opsonic properties.

2. PRIMITIVE METAZOA (PORIFERA)

The reaggregation of sponges which involves essentially a self/self and self/non-self recognition system has received attention for almost 75 years. While we know very little about the mechanism(s) sponges use to combat potential pathogens, it is possible that the factors involved in tissue recognition, i.e., self from non-self, are utilized in the recognition of foreign particles, as has been recently suggested by Vaith *et al.* (1979).

The occurrence of hemagglutinins in marine sponges was first reported by Dodd *et al.* (1968). These agglutinins have since been studied in some detail. Khalap *et al.* (1970, 1971) examined extracts from various *Axinella* species and from *Cliona celata* and found that they agglutinated a variety of erythrocytes from different species of vertebrates. Pretreating the erythrocytes with various hydrolytic enzymes invariably increased the hemagglutinating titer of the extracts, sometimes as much as 50-fold. The hemagglutinins present in *Axinella* extracts showed specificity for Gal and for oligosaccharides and glycoproteins containing this sugar in a terminal position. Extracts of *Cliona celata* were more strongly inhibited by Lac than by Gal. The *Axinella* sp. agglutinin was purified by Gold *et al.* (1974) who showed it to be a metalloprotein (molecular weight 15,000–18,000) containing 15 moles of Ca^{2+} and 12 moles of Fe^{3+} per mole. The isoelectric point of the agglutinin was very low (pH 3.9). Bretting (1973) and Bretting and

Renwrantz (1974) studied the agglutinins present in extracts of *Axinella polypoides* and *Aaptos papillata*. *Axinella polypoides* contained two agglutinins, one with an apparent molecular weight of 17,000 and the other, 21,000. The hemagglutinin from *Aaptos papillata* had a molecular weight of 65,000. Agar diffusion studies showed that the *Axinella* agglutinins precipitated galactogens, while that from *Aaptos* reacted with blood-group A substances.

The agglutinins from these two species of sponges have been characterized by Bretting and Kabat (1976) and Bretting *et al.* (1976). Those from *Axinella polypoides* were purified by affinity chromatography on Sepharose 4B, which contains 1,3-linked β-D-Gal residues, using Gal as the eluant. Chromatography of the eluted material on DEAE-cellulose yielded two fractions, agglutinin I (eluting at pH 3.5) and agglutinin II (pH 2.6). The fractions were assayed against human type A, B, and O erythrocytes. Polyacrylamide gel electrophoresis showed that the agglutinin II fraction was a mixture of at least two agglutinins. Using preparative polyacrylamide gel electrophoresis, sufficient quantities of agglutinin II were obtained to warrant further study. The apparent molecular weight of agglutinin I and agglutinin II was approximately 21,000 and 15,000, respectively, as estimated by SDS-PAGE. Ultracentrifuge studies showed the agglutinins to have sedimentation coefficients of 2.6 S and 2.8 S, respectively. Amino acid analyses revealed both to be rich in Glu and Asp and low in Cys and Met (see Table 11). Agglutinin I contained no Tyr or His. Sugar-inhibition studies revealed minor differences in specificity between the two hemagglutinins, although both reacted best with oligosaccharides containing β-linked D-Gal. Similar results have been found by Baldo *et al.* (1977), who also reported that structures containing α-linked D-Gal residues gave marked inhibition of agglutination.

Affinity chromatography using the immunoadsorbent, polyleucyl hog A + H substance, was utilized for the initial preparation of the hemagglutinin fractions from *Aaptos papillata* (Bretting *et al.*, 1976). Chromatography of the eluate from this column of DEAE-cellulose yielded three hemagglutinating fractions which were eluted with 3 M $MgCl_2$ at different pH, *Aaptos* I eluting at pH 6.2 and two other fractions eluting at pH 3.7 and pH 3.0, respectively. The fraction eluting at pH 3.0 was separated further by preparative polyacrylamide electrophoresis into two active components termed *Aaptos* II and *Aaptos* III. Immunodiffusion studies using a rabbit antiserum against a crude extract of the sponge, revealed that *Aaptos* II and III were antigenically related whereas *Aaptos* I was quite distinct. The amino acid composition of *Aaptos* II and III was similar (Table 11). *Aaptos* I contained no Cys but had an unusually high content of Tyr. All three hemagglutinins were rich in Asp and Glu, a common feature of the hemagglutinins isolated from invertebrates. The sedimentation coefficients of the agglutinins were 3.5 S (*Aaptos* I), 6.0 S (*Aaptos* II), and 5.5 S (*Aaptos* III). Analysis of each agglutinin by SDS-PAGE revealed a single polypeptide subunit (molecular weight 16,000) for *Aaptos* II and III (reduced and unreduced), whereas *Aaptos* I yielded two bands corresponding to molecular weights of 21,000 and 12,000. These data indicate that *Aaptos* II and III are probably composed of identical subunits and that *Aaptos* I consists of two different subunits.

The binding specificity of the *Aaptos* agglutinins appeared to be directed against blood-group substances with terminal GlcNAc residues. *Aaptos* II was precipitated completely by blood-group substances and glycoproteins containing terminal GalNAc, GlcNAc, or AcNeu residues. *Aaptos* III had a similar specificity to *Aaptos* II.

Recently Bretting *et al.* (1978) have described another agglutinin from the sponge *Axinella polypoides* which agglutinates a strain of bacteria (*Nitrobacter winogradskyi*). This agglutinin appears to be unrelated to those previously described by Bretting and Kabat (1976). It did not agglutinate erythrocytes, nor was the agglutination of the bacteria inhibited by Gal.

An agglutinin for human and bovine erythrocytes from the sponge *Geodia cydonium* has been purified by affinity chromatography on Sepharose 4B (Vaith *et al.*, 1979). This molecule, a glycoprotein, appeared to consist of subunits with a molecular weight of about 12,000. Agglutination of erythrocytes was independent of Ca^{2+} ions. Experimental evidence indicated that the specificity of this hemagglutinin was directed against terminal, nonreducing β-linked D-Gal residues.

3. PROTOSTOMIA

3.1. MOLLUSCA

The presence of natural agglutinins in the body fluids of molluscs was first established by the pioneer study of Tyler (1946). It was not until the early 1960s, however, that the occurrence of blood-group active substances in various species was examined more closely (Cheng and Sanders, 1962; Cushing *et al.*, 1963). Soon afterwards, a blood-group A_1 specific agglutinin was reported to be present in saline extracts of the marine bivalve *Saxidomus giganteus* (Johnson, 1964). Agglutinins predominantly specific for human type A erythrocytes were also detected in extracts of the German land snails *Helix hortensis* (Prokop *et al.*, 1965a; Rackwitz *et al.*, 1965) and *H. pomatia* (Prokop *et al.*, 1965b; Kim *et al.*, 1966; Uhlenbruck and Prokop, 1966) and in the closely related Mediterranean land snail *Otala lactea* (Boyd and Brown, 1965; Boyd *et al.*, 1966).

These discoveries prompted studies of other molluscs, especially pulmonate gastropods, in the hope of discovering additional blood-group reagents. As a result, the molluscs have probably been investigated for the presence of agglutinins more than any other group of invertebrates. The phylogenetic relationships between many of the species studied for this purpose were depicted in Figure 2. The majority of these species belong to the Gastropoda and comprise different pulmonate species, particularly the land snails (order Stylommatophora). Several comprehensive reviews concerning the prevalence and specificity of blood-group specific agglutinins among the gastropods have been published and the reader is referred to these for more detailed information (Prokop *et al.*, 1968; Uhlenbruck *et al.*, 1972; Pemberton, 1974; Gold and Balding, 1975).

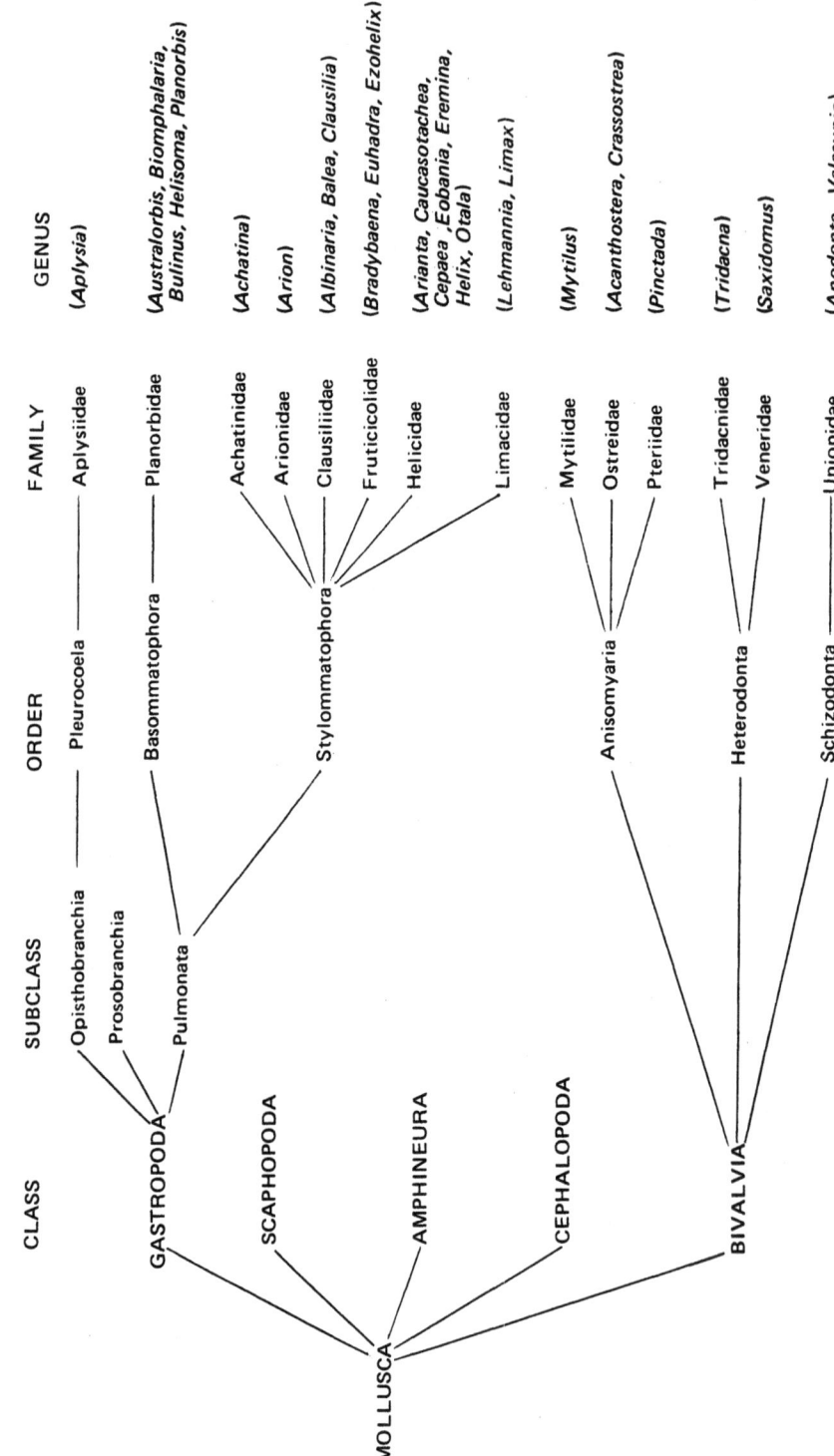

FIGURE 2. Phylogenetic relationships between selected members of the Mollusca. Most of the genera for which data are available are shown within parentheses.

In most of these studies, tissue or whole-body homogenates have been screened for agglutinins (and occasionally, hemolysins) against native and enzyme-treated human erythrocytes. Hemolymph has been largely ignored, since initial studies had shown that it contained only small amounts of hemagglutinin(s) compared with extracts. Considering the attention which has been accorded to this group of animals, it is disappointing from the standpoint of this review that virtually nothing is known about the activity, nature, or function of the agglutinins which are present in low titer in their hemolymph and that few of the many agglutinins detected in tissue extracts have been isolated and characterized in any detail. It seems reasonable to expect that if humoral factors such as opsonins or agglutinins are involved in the phagocytic defense system of these animals, they would be present in the hemolymph. However, if these substances are secreted in limited amounts, they may be more abundant intracellularly in certain tissues than free in the hemolymph and it is quite possible that some of the agglutinins detected in tissue extracts have a natural opsonic role.

Relatively little attention has been directed to the question of whether or not molluscs possess opsonins as part of their defense system. There is a small amount of data which indicates that such factors do occur in the hemolymph of several species. However, confirmation of the existence of these opsonins and information about their properties, their relationships with the agglutinin(s) present in the hemolymph, and their importance for recognition and phagocytosis *in vivo* await further study.

The requirement for serum opsonins by phagocytes present in the blood of the lesser octopus *E. cirrosa* (class Cephalopoda) was examined by Stuart (1968) after he had observed phagocytosis of erythrocytes following their injection into a tentacle vein. Phagocytosis of human erythrocytes was observed *in vitro* after the cells had been preincubated with *E. cirrosa* serum or when the serum was included in the suspending medium. Cells incubated with a rabbit anti-human erythrocyte serum were not ingested. It was concluded that phagocytosis required an opsonin which was present in the serum.

The hemolymph of the snail *Helix aspersa* was shown by Prowse and Tait (1969) to promote the phagocytosis of yeast cells and sheep erythrocytes by *H. aspersa* amebocytes *in vitro*. The opsonic effect was due to the adsorption of hemolymph components onto the foreign particles. Neither yeast nor erythrocytes were phagocytosed significantly in a medium supplemented with hemolymph adsorbed with the homologous particle, whereas phagocytosis of each occurred using hemolymph adsorbed with the heterologous particle. No significant difference was observed between the extent of phagocytosis of both yeast and erythrocytes in normal hemolymph and in hemolymph adsorbed with the heterologous particle. Thus, it appeared that two or more distinct opsonins were present in the hemolymph. As far as we are aware, neither the biological importance nor the physicochemical nature of these opsonic factors has been investigated.

More recently, Renwrantz and his colleagues (Renwrantz and Mohr, 1978; Renwrantz, 1979; Harm and Renwrantz, 1980; Renwrantz *et al.*, 1981) investigated the role of opsonins in the clearance of human type A and B erythrocytes and yeast cells from the circulation of *H. pomatia*. The clearance of these particles, like that of carbon particles and bacteria (Reade, 1968; Bayne and Kime, 1970; Bayne, 1974), followed an exponential curve. The elimination of erythrocytes was significantly faster ($p < 0.05$) if the cells were incubated in hemolymph prior to injection, as evidenced by the data summarised in Table 1. Unsensitized erythrocytes injected several hours after an identical primary (blockading) dose were removed very slowly, but the clearance rate returned to normal if the cells comprising the second dose were preincubated with hemolymph. These and other results (Harm and Renwrantz, 1980; Renwrantz et al., 1981) showed that the elimination of human and rabbit erythrocytes, as well as radiolabelled human serum albumin, by the phagocytic system of *H. pomatia* is strongly influenced by opsonic substances present in the hemolymph of this snail. The rate of removal of both type B erythrocytes and serum albumin, like that of type A erythrocytes, was significantly reduced in snails injected 3 hr beforehand with a blockading dose of type A erythrocytes, indicating that a single opsonin may mediate the clearance of all these particles.

The highly efficient elimination of living yeast cells seemed, in contrast, to be independent of the opsonin level of the hemolymph since the clearance index in snails given a prior blockading dose of yeast or type A erythrocytes was not significantly different from that observed in untreated snails (Renwrantz et al., 1981). It is pertinent that the elimination of yeast cells coated *in vitro* with bovine serum albumin were cleared at the same rate as uncoated cells whereas cells coated with *Helix* hemocyanin showed a significantly reduced rate of clearance, a result consistent with the hypothesis that the removal of yeast cells from the circulation may involve the direct binding of the latter to specific recognition molecules on fixed cells inside *Helix* organs. This conclusion is supported by the finding that the elimination of yeast cells as well as type A and B erythrocytes was impaired if the cells were injected in a 0.2 M solution of GalNAc or GlcNAc (Harm and Renwrantz, 1980; Renwrantz et al., 1981). Fucose was inactive in this respect, indicating that the effect was specific. The hemagglutinin(s) in *H. pomatia* hemolymph, whose detection requires the use of pronase-treated type A erythrocytes as indicator cells, were similarly inhibited by GalNAc, GlcNAc and the blood-group A substance from peptone but not by fucose (Renwrantz, 1979). It seems likely, on the basis of these results, that the opsonin and hemagglutinin in the hemolymph of *H. pomatia* may be one and the same molecule.

Renwrantz and Cheng (1977a,b) had previously demonstrated that a number of plant and animal lectins could mediate the attachment of human erythrocytes to *H. pomatia* hemocytes *in vitro*, thereby showing that carbohydrate moieties to which these lectins bind are present on the hemocyte surface. Subsequently, several known hemagglutinins were tested for their capacity to function as opsonins in *H. pomatia* by measuring the rate of clearance of second-dose erythrocytes which had been pre-incubated with a sub-agglutinating dose of extract or agglutinin prior to injection into blockaded snails. These results, summarized in Table 1, showed that albumin gland extracts of *H. pomatia* and a closely related snail, *Cepaea nemoralis*, could reverse the blockade effect whereas

TABLE 1. CLEARANCE OF HUMAN ERYTHROCYTES FROM THE CIRCULATION OF THE VINEYARD SNAIL, HELIX POMATIA[a,b]

Pretreatment of erythrocytes used for injection[c]	Type A erythrocytes		Type B erythrocytes	
	Primary dose	Secondary dose[d]	Primary dose	Secondary dose[e]
None	1.64 ± 0.29 (14)	0.78 ± 0.13 (8)	0.85 ± 0.12 (13)	0.40 ± 0.09 (17)
H. pomatia hemolymph	1.87 ± 0.23 (10)	1.49 ± 0.19 (8)	1.01 ± 0.16 (7)	—
H. pomatia albumin gland (extract)	—	3.46 ± 0.67 (9)	—	—
(purified agglutinin)	—	1.7 ± 0.4 (8)*	—	—
C. nemoralis albumin gland (extract)	—	2.49 ± 0.13 (5)	—	—
Aaptos papillata (extract)	—	0.82 ± 0.04 (5)	—	—
Axinella polypoides (purified agglutinin)	—	0.6 ± 0.3 (7)*	—	—
Concanavalin A	—	0.6 ± 0.2 (7)*	—	—

[a] Data from Renwrantz and Mohr, 1978; Harm and Renwrantz, 1980; Renwrantz et al., 1981.
[b] Phagocytic index ± one standard deviation. For comparative purposes, all values have been multiplied by 100. The number of snails used in each experiment is shown within parentheses. The rate of clearance is proportional to the magnitude of the index.
[c] Extracts and purified hemagglutinins were used at about one-quarter of an agglutinating dose.
[d] 1.7×10^8 erythrocytes given either 12-19 hr or 3 hr (*) after a primary (blockading) injection of 1.7×10^8 cells.
[e] 1×10^9 erythrocytes given 3 hr after a similar primary dose of type A erythrocytes.

two sponge agglutinins and the plant lectin concanavalin A were unable to do so. In particular, the purified blood group A-specific *H. pomatia* albumin gland agglutinin (anti-A_{HP}, p. 338) was found also to reverse the blockade effect. This hemagglutinin can be inhibited by GalNAc and less effectively by GlcNAc, i.e., the same sugars which inhibit both yeast and erythrocyte clearance in *H. pomatia* and the hemagglutinin in *H. pomatia* hemolymph. The situation has therefore been reached where it seems not unreasonable to suggest that the anti-A agglutinin in the albumin gland may be slowly released into the open circulatory system of the snail (see Section 3.1.1), where it may have a natural role as an opsonin and perhaps as a cell-bound receptor. The test of this hypothesis clearly lies in the enrichment, purification, and characterization of both the opsonic and hemagglutinating factors from the hemolymph of *H. pomatia* (Harm and Renwrantz, 1980).

The recognition and phagocytosis of normal and formalin-treated foreign cells by glass-adherent hemocytes from the snail *Otala lactea* have been studied by Anderson and Good (1976). The phagocytes were capable of ingesting *Escherichia coli* B, *Staphylococcus aureus* 502A, and a snail-associated *Pseudomonas* species, but uptake was variable and was not influenced by the presence of *O. lactea* hemolymph. However, hemolymph was found to enhance the phagocytosis of formalinized yeast cells, although a low baseline level of phagocytosis of both untreated and formalinized yeast was consistently observed. The effect of hemolymph on the ingestion of formalinized yeast cells was significant at a level of 0.0005, based on data from eight experiments. In contrast, the hemolymph showed no capacity to promote the phagocytosis of untreated yeast cells, or untreated or formalinized sheep erythrocytes. Although no adsorption experiments were carried out, the data clearly suggest that opsonins specific for formalinized yeast cells were present in the hemolymph. It is noteworthy that the addition of *O. lactea* albumin gland extracts at dilutions of 1 : 8 to 1 : 32 caused a significant increase in the extent of phagocytosis of formalinized sheep erythrocytes ($p = 0.05$). In view of the agglutinins which are known to be present in the albumin gland (see Gastropoda, Section 3.1.1), it would be most interesting to know whether the opsonic effect of the extract can be prevented by GalNAc, L-Fuc, or other sugars. The possibility cannot yet be excluded that agglutinins possessing opsonic activity and secreted by the albumin gland or other tissues may be present in the hemolymph at low concentrations and on the surface of both fixed and circulating hemocytes.

Attempts to demonstrate adherence of human A_1 erythrocytes to hemocytes from the water snail *Biomphalaria glabrata* were unsuccessful (Stein and Basch, 1979). In view of the above-mentioned results with *H. pomatia*, it is noteworthy that erythrocytes treated with the anti-A agglutinin purified from extracts of *B. glabrata* albumin glands also failed to adhere to hemocytes. Crude albumin gland extracts and hemolymph were not tested. The results of the two studies are not strictly comparable, however, since one used *in vivo* and the other *in vitro* techniques. Moreover, the hemocytes of another freshwater snail, *Lymnaea stagnalis*, have been shown *in vitro* by another group to bind and phagocytose sheep erythrocytes and yeast cells, particularly if the latter are first preincubated with snail hemolymph (Sminia et al., 1979).

The sea hare *Aplysia californica*, an opisthobranch gastropod, has been shown by Pauley et al., (1971b) to be capable of rapidly clearing at least four marine bacteria (*Micrococcus aquivivus*, *Pseudomonas* sp., *Gaffkya homari*, and a gram-negative rod, Ap5Y) from its hemolymph *in vivo*. It is probably significant that hemolymph from normal animals contained agglutinins for all these bacteria, but not for the terrestrial bacterium *Serratia marcescens* which was not cleared *in vivo* (Pauley et al., 1971a,b). Large numbers of *S. marcescens* were found to persist in the hemolymph for more than 1 month postinjection, whereas *M. aquivivus* were completely cleared within 24 hr. Bacterial clearance was in all cases faster at 20°C than at 14°C and appeared to be accelerated by previous exposures of the animal to the bacterium. The hemolymph agglutination titers and the number of circulating hemocytes decreased markedly during the early rapid clearance phase, both later returning to normal levels as the bacteria were cleared. The hemolymph was not bacteriolytic (Johnson and Chapman, 1970). The phagocytosis of chicken erythrocytes by glass-adherent *Aplysia californica* hemocytes was enhanced when the erythrocytes were presensitized with normal hemolymph. This appeared to be due to heat-labile opsonins, since heated hemolymph was much less effective.

The *in vitro* phagocytosis of rabbit erythrocytes by hemocytes from the oysters *Crassostrea virginica* and *C. gigas* (class Bivalvia) was found to be enhanced if the erythrocytes were pretreated with oyster hemolymph, but the nature and specificity of the opsonic factor(s) were not investigated (Tripp, 1966; Tripp and Kent, 1968). Moreover, although clearance studies have shown phagocytosis to be a major defensive reaction in the oyster (Stauber, 1950; Tripp, 1960), the importance of opsonins in this process has still to be demonstrated *in vivo*.

Arimoto and Tripp (1977) reported that the *in vitro* phagocytosis of a marine gram-negative bacterium (RS-005) by hemocytes from the clam *Mercenaria mercenaria* was enhanced by the presence of *M. mercenaria* hemolymph. Hemolymph which had been adsorbed with the bacteria was unable to enhance phagocytosis, and unlike unadsorbed hemolymph, it could not agglutinate RS-005 or other bacteria. *Salmonella schottmülleri* and *Vibrio marinus* were also phagocytosed, but to the same extent in the presence and absence of hemolymph. The results are consistent with the hypothesis that the agglutinin and opsonin for RS-005 are identical, but whether this is in fact so remains to be shown. It is possible that the RS-005 opsonin also occurs on the surface of the hemocytes and that these cell-bound molecules are responsible for the recognition of the bacteria in the absence of hemolymph.

Studies on the clearance of various proteins by the chiton *Liolophura gaimardi* (class Amphineura) have demonstrated that the phagocytic systems of at least some invertebrates are capable of distinguishing degrees of foreignness (Crichton et al., 1973; Crichton and Lafferty, 1975). By measuring the rate of clearance of rhodamine-labeled proteins, bovine serum albumin was shown to be removed from the circulation and taken up predominantly by fixed phagocytic cells at a much faster rate than homologous proteins isolated from *Liolophura* hemolymph. The discrimination was such that 80–90% of the labeled albumin but only 10% of the labeled homologous protein was eliminated from the hemolymph during the initial 1 to 12 hr postinjection (Crichton *et al.*, 1973).

In order to exclude the possibility that this difference in clearance was due to major differences in the physical properties of these molecules, the authors used double-isotope techniques to investigate the elimination of hemocyanins purified from the hemolymph of two chitons belonging to the closely related genera *Ischnoradsia* and *Poneroplax*, the more distantly related keyhole limpet (*Megathura crenulata*), and a crustacean, the crayfish *Jasus lalandii* (Crichton and Lafferty, 1975). Significant differences were observed in the rates of clearance of the various hemocyanins. These differences could be correlated with the relatedness of each protein to *Liolophura* hemocyanin. Thus, 85% of the crayfish hemocyanin (which exhibited no antigenic cross-reaction with *Liolophura* hemocyanin) and 25% of the slightly cross-reactive keyhole limpet hemocyanin was cleared within 1 hr postinjection. The more closely related hemocyanins were cleared much more slowly, the order (in decreasing rate) being *Poneroplax* >> *Ischnoradsia* > *Liolophura*. Further experiments demonstrated that it was possible to competitively inhibit the elimination of one protein without affecting the removal of an unrelated molecule. Thus, chitons injected with a large dose (25 mg) of unlabeled human serum albumin 1 hr before a test dose (0.25 mg) of labeled protein eliminated *Poneroplax* hemocyanin at the same rate as uninjected control animals but removed human and bovine serum albumin at a significantly reduced rate. It therefore seems that distinct recognition factors were involved in the clearance of *Poneroplax* hemocyanin and mammalian serum albumins. There is as yet no evidence to indicate whether these factors are humoral or cell-bound.

In view of the possible identity of hemagglutinins and opsonins in some invertebrates, we shall consider briefly what is known regarding the prevalence of hemagglutinins in the Mollusca and then discuss in more detail the results of studies in which the properties of these molecules have been examined. It is clear that any phylogenetic relationships which may exist between the agglutinins and opsonins of various invertebrate species will only be recognized by comparing the physicochemical properties of the molecules involved.

3.1.1. Class Gastropoda

Pemberton (1974) reviewed the data collected from 134 species of Gastropoda and concluded that hemagglutinins occur in a large proportion. The occurrence of agglutinins for human erythrocytes among these species is summarized in Table 2. These substances seem to be especially prevalent in the Stylommatophora (land snails), but it should be noted that this group has been studied most intensively. Pemberton mentions that in many cases only one sample of each species and in some instances only a single specimen had been examined. Moreover, few of the extracts were tested against a full panel of erythrocyte types, although in some studies enzyme-treated erythrocytes were

TABLE 2. THE PREVALENCE OF HEMAGGLUTININS AMONG THE GASTROPODA[a]

Taxonomic group	Incidence of hemagglutinins in different species (No. positive/No. tested)		No. of species containing agglutinins specific for native human erythrocytes of type		
			A	B	Other
Class Gastropoda					
Subclass					
Opisthobranchia	0/3	(0%)	0	0	0
Prosobranchia	17/31	(55%)	2	1	14
Pulmonata	62/100	(62%)	30	4	28
Subclass Pulmonata					
Order					
Basommatophora (water snails)	6/21	(29%)	2	0	4
Stylommatophora (land snails)	56/79	(71%)	28	4	24
Order Stylommatophora					
Family					
Clausiliidae	15/15	(100%)	7	3	5
Fruticicolidae	6/6	(100%)	6	0	0
Helicidae	16/24	(67%)	12	0	4
Limacidae	4/8	(50%)	0	0	4

[a] Adapted from Pemberton (1974, Tables 1 and 2).

also used to detect agglutinins which may not be active against untreated cells. The data should be viewed with these considerations in mind.

Of the 134 species surveyed, 59% yielded hemagglutinins and 24% contained agglutinins primarily of anti-blood-group A specificity. Hemagglutinins were detected in 55% and 62% of the prosobranch and pulmonate species, respectively. The three species representing the Opisthobranchia were negative, but at least one opisthobranch, the sea hare *Aplysia californica* which was not included in Pemberton's data, possesses a macromolecular agglutinin in its hemolymph (Pauley et al., 1971a; Pauley, 1974a). It is noteworthy that although 62% of the pulmonate species contained hemagglutinins, the frequency of positive species in the Basommatrophora (water snails) and Stylommatophora (land snails) was markedly different, being 29% and 71%, respectively (Table 2).

Of the 17 families of Stylommatophora for which data were available (Pemberton, 1974), only four (Clausiliidae, Fruticicolidae, Helicidae, Limacidae) were represented by more than five species (Table 2). Hemagglutinins were detected in most of the species comprising these families. Of the 15 species of the Clausiliidae which yielded hemagglutinins, seven contained predominantly bloodgroup A specific agglutinins, compared with all six species of the Fruticicolidae and 12 of the 16 positive species of the Helicidae. The agglutinating activity of the 24 species of Helicidae was found to correlate with taxonomic groups. Thus,

11 out of the 12 species belonging to the subfamily Helicinae (which includes *Helix* and *Cepaea* spp.) consistently yielded blood-group A specific agglutinins in high titer (1 : 500 to 1 : 50,000 against native type A erythrocytes), whereas species belonging to the other subfamilies either contained weak agglutinins of varying specificities or were negative.

Investigations on the source of agglutinins within the snail have been largely restricted to the strong blood-group A specific agglutinin (anti-A_{HP}) of *H. pomatia*. This agglutinin is apparently present, at least in high titer, only in sexually mature, adult snails where it occurs predominantly in the albumin gland (part of the sexual apparatus) and also in the egg-mass (Prokop *et al.*, 1965b; Kilias *et al.*, 1972; Kothbauer, 1970, 1972; Kothbauer *et al.*, 1972). The relative size of the albumin glands show distinct annual variations, being maximal during the breeding season, but no significant changes in the agglutination titer have been observed. The occurrence of hemagglutinins in male and female individuals of several nonhermaphroditic snail species has also been investigated (Kothbauer and Schenkel-Brunner, 1971). The content of agglutinin(s) in individuals of both sexes showed considerable variation, such that some males contained more agglutinin than some females and vice versa. Moreover, hemagglutinins were detected in both whole-body and egg extracts of some individuals but only in the body extracts of others. The reasons for these variations remain unestablished.

It has yet to be shown whether the agglutinins present in the hemolymph and albumin glands of snails are in any way related and whether either has a role in self-protection. Prokop *et al.* (1968) believed that the agglutinin(s) present in the albumin gland and egg-mass of the snail was important in the protection of the eggs and therefore suggested that they be called "protectins." The agglutination of bacteria by snail agglutinins such as anti-A_{HP} (see Stylommatophora, Section 3.1.1b) is used as supporting evidence for the protection of snail eggs against infection by soil bacteria. However, Gold and Balding (1975, p. 289) comment that "unless these agglutinins have some further destructive effect against bacteria, one would expect that agglutination would have an undesirable effect by concentrating bacteria around the egg." Efforts to demonstrate such properties, for example lysozyme activity, have been unsuccessful (Kühnemund *et al.*, 1972). Kothbauer (1970) observed that young snails (*H. pomatia*) will feed on dead eggs (which do not seem to rot) and suggested that anti-A_{HP} may prevent the decomposition of the eggs. Since this agglutinin apparently disappears between the 9th and 16th day of life, reappearing only with the full development of the reproductive system, one cannot rule out such possibilities. However, these hypotheses must be considered speculative until some firm experimental evidence becomes available to either support or reject them.

3.1.1a. Subclass Pulmonata, Order Basommatophora. Studies conducted on members of this group have revealed hemagglutinins in relatively few species (Table 2). For example, Pemberton (1971b) examined whole-body extracts of *Ancylus fluviatilis*, two species of *Lymnaea*, and five of *Planorbis* and found only one sample (of *L. stagnalis*) to possess weak hemagglutinating activity. Extracts of *A. fluviatilis*, *P. corneus*, and *P. albus* contained a weak hemolysin. Brown *et al.* (1968) reported that whole-body extracts of *Bulinus truncatus* and *Biomphalaria alexandrinus* contained agglutinins for papain-treated human A erythrocytes (titers 1 : 16 and 1 : 256, respectively) and that the latter extract also contained anti-AB and anti-H-like agglutinins. Pemberton (1971b, 1974) concluded from his own and others' data that the presence of strong agglutinins such as those found in *Helix* and *Cepaea* spp. is unusual and that the presence of agglutinins in different populations of a particular species is variable and that the specificity of such agglutinins is not constant.

Investigations which are perhaps more likely to lead to the identification of humoral factors involved in immunity in these animals began with the work of Gilbertson and Etges (1967). These investigators detected agglutinins in the hemolymph of individuals from three Latin American populations of the planorbid snail *Australorbis glabratus* and from an African population of *Biomphalaria sudanica* (titers 1 : 2 to 1 : 8). The *A. glabratus* agglutinins could be distinguished by their reactivity for human and rabbit erythrocytes, whereas *B. sudanica* hemolymph agglutinated human, rabbit, and hamster erythrocytes (titer 1 : 8). The agglutinins were inactivated by repeated freeze-thawing, heat (56°C, 30 min), or EDTA. The hemolymph of some *Bulinus* species also appears capable of agglutinating mammalian erythrocytes (Rudolph, 1973).

Michelson and Dubois (1977) examined the occurrence of hemagglutinins and hemolysins for normal human erythrocytes in the hemolymph and in extracts of the egg-masses and albumin glands of 31 strains (representing 8 species and 4 genera) of planorbid snails. Hemagglutinins and hemolysins were not uniformly present in all the members of Planorbidae which were tested. Their detection depended in part on the source of the material to be assayed, and differences were observed between genera and species as well as between populations of a single species. Some of these results are summarized in Table 3. Agglutinins were detected in the egg and albumin gland extracts of all 17 populations of *B. glabrata*. However, only six populations exhibited agglutinins in their hemolymph. Particular samples clearly contained more than one type of agglutinin, since erythrocytes of type A_1 only, O only, A_1 and B, or A_1, B, and O were agglutinated by different samples. Moreover, there was considerable variation in the relative titer for A_1 and B erythrocytes among those samples which agglutinated both types of cell, some exhibiting higher titers against A_1 than against B cells and others the reverse. Heating the albumin gland or egg-mass extracts at 56°C for 1 hr apparently had no effect on their agglutinating activity.

The authors suggested that agglutinins may be synthesized by the albumin gland and secreted into the sperm–oviduct, from whence they disseminate into the hemolymph; the failure to identify agglutinins in the hemolymph of some populations could then be explained on the basis that too little agglutinin is entering the hemolymph. Interestingly, the origin of the agglutinin(s) in *Helisoma* sp. may be different from those of *Biomphalaria*, since in the former activity was not detected in any albumin gland extract but was restricted almost entirely to the hemolymph (Table 3). Furthermore, the agglutinins from the two species seemed quite distinct, as GalNAc inhibited the O-specific agglutinin from *Helisoma* hemolymph, whereas the *Biomphalaria* albumin gland agglutinins were

TABLE 3. OCCURRENCE OF HEMAGGLUTININS IN PLANORBID SNAILS (ORDER BASOMMATOPHORA, FAMILY PLANORBIDAE)[a]

Species of snail	Source of sample	Number of strains containing agglutinins for human erythrocytes of type			Proportion of strains in which agglutinins were detected	Maximum titer	Binding specificity Inhibition by 0.1 M sugar	
		A_1	A_1 and B	A_1, B, and O			GalNAc	Others[b]
Biomphalaria glabrata (17 strains)	Albumin gland	3	12	1	16/16 (100%)	1:128	−[c]	+
	Egg-mass	5	10	2	17/17 (100%)	1:256	nt	nt
	Hemolymph	3	0	3	6/17 (35%)	1:32	nt	nt
Helisoma sp. (8 strains)	Albumin gland	0	0	0	0/3 (0%)	—	nt	nt
	Egg-mass	1	0	0	1/8 (12%)	1:1	nt	nt
	Hemolymph	0	0	6	6/8 (75%)	1:2048	+	−

[a] Data from Michelson and Dubois (1977).
[b] D- and L-fucose, glucose, mannose, maltose, sorbose, sucrose, fructose.
[c] −, no inhibition; +, inhibition; nt, not tested.

inhibited not by GalNAc but by a number of other sugars which had no effect on the *Helisoma* activity. These results indicate that the types and site(s) of synthesis of agglutinins may differ considerably between different snail species. Nonspecific hemolysins were detected primarily in samples of hemolymph, rarely in albumin gland extracts and never in egg-mass extracts of *B. glabrata* (Michelson and Dubois, 1977). The titers were extremely low, rarely exceeding 1 : 2.

An alternative explanation to account for the variable hemagglutinating titers of different samples is suggested from reports that soluble blood-group active substances are present in the tissues of snails and other invertebrates (e.g., Prokop *et al.*, 1965a; Gold and Thompson, 1969b; Pemberton, 1970; Fischer *et al.*, 1972; Renwrantz and Berliner, 1978). The hemolymph of *B. glabrata* has also been shown to contain such substances in addition to hemagglutinins which seemed to be ABO-nonspecific (Stanislawski *et al.*, 1976). Obviously, the apparent agglutinating titer may not be indicative of the actual level of agglutinin(s) if inhibitors are present. It is conceivable that the opsonins (agglutinins?) produced by these animals react weakly with self and have therefore to be maintained in a "nonreactive" state by inhibitors which become displaced in the presence of foreign substances to which the opsonins bind with higher affinity. In the absence of data on the natural function of either agglutinins or blood-group active substances, however, we can only speculate on their relevance to the animal.

The concept that the hemolymph contains hemagglutinins derived from the albumin gland is supported by data published by Stein and Basch (1979), who purified hemagglutinins from the hemolymph and egg-mass and albumin-gland extracts of *B. glabrata* by affinity chromatography. When each sample was mixed at 4°C and pH 7.2 with an equal volume of packed Sephadex G-150, all of the agglutinating activity for human A_1 and A_2 erythrocytes became adsorbed to the gel. The agglutinins were subsequently recovered by elution with 0.1 M glycine·HCl (pH 3.0). The purification, based on activity per milligram of protein, was in each case approximately 100-fold. The agglutinins purified from each source exhibited similar serological specificity for A_1 and A_2 erythrocytes, being 10–20 times more active against the former. They were inactivated by EDTA which had no effect on the cells and therefore seem to require divalent cations for functional integrity. The agglutinins appeared to be identical on the basis of immunodiffusion and immunoelectrophoretic analyses. A single cathodal precipitin line was observed upon immunoelectrophoresis against antisera from rabbits immunized with either crude extracts or homologous erythrocytes sensitized with agglutinin. The anti-A from *H. aspersa* (see Stylommatophora) showed no reactivity with any of these antisera. On the basis of its electrophoretic mobility, the *B. glabrata* agglutinin appears to have a pI $\geq$ 8. A molecular weight of about 55,000 was estimated from SDS-PAGE and ultrafiltration data.

Stein and Basch (1979) also reported that the agglutinin became bound to different stages of *Schistosoma mansoni*, for which *B. glabrata* is an intermediate host. However, attempts to demonstrate adherence of hemagglutinin-sensitized A_1 erythrocytes to hemocytes from *B. glabrata* were unsuccessful. Erythrocytes treated with subagglutinating doses of the purified agglutinin were prone to lysis upon the addition of fresh human serum. This phenomenon appeared to involve the complement system, since it was not observed if heated (56°C, 30 min) serum was used. EDTA was also inhibitory, but this would be expected to inactivate the agglutinin as well as the complement pathway.

3.1.1b. Subclass Pulmonata, Order Stylommatophora. The blood-group A specific agglutinin of *H. pomatia* (anti-A_{HP}) is the only snail agglutinin for which detailed physicochemical and structural data are available (reviews: Uhlenbruck *et al.*, 1972; Hammarström, 1972, 1974; Gold and Balding, 1975; Goldstein and Hayes, 1978). Anti-A_{HP} is confined almost exclusively to the albumin gland of the snail (Prokop *et al.*, 1965b; Kilias *et al.*, 1972) and comprises approximately 0.5% of the wet weight and up to 8% of the soluble (extracted) protein of this gland (Hammarström, 1972, 1974). It has been purified by chromatography on DEAE- and CM-cellulose (Knobloch *et al.*, 1970) and by affinity chromatography on Sephadex (Kühnemund and Köhler, 1969; Ishiyama and Uhlenbruck, 1972; Schnitzler *et al.*, 1972) and insolubilized polyleucyl blood-group A substance (Hammarström and Kabat, 1969; Svensson *et al.*, 1970). Agglutinins which appear to be similar to anti-A_{HP} have also been isolated from several closely related species of snail (see Table 4). There are indications that small amounts of other agglutinins occur in the tissues and hemolymph of *H. pomatia* and other snails (Haferland *et al.*, 1967; Oehme *et al.*, 1968; Uhlenbruck and Reifenberg, 1971; Schnitzler *et al.*, 1971a; Kilias *et al.*, 1972; Ishiyama *et al.*, 1973).

Anti-A_{HP} is a glycoprotein containing 7–13% carbohydrate, the predominant sugar being Gal with smaller amounts of Man and GalNAc (Vogt *et al.*, 1969; Hammarström and Kabat, 1969; Salfner *et al.*, 1971). Although it is electrophoretically homogeneous under acidic as well as alkaline conditions, isoelectric focusing has revealed up to nine components having pI values between pH 6 and pH 8 (Kühnemund and Köhler, 1975; Vretblad *et al.*, 1979). It is soluble in 20% but not in 40% saturated ammonium sulfate (Vogt *et al.*, 1969; Knobloch *et al.*, 1970) and migrates, upon electrophoresis, in the region of mammalian γ-globulins (Uhlenbruck *et al.*, 1967). The molecule appears to be resistant to proteolysis, since its hemagglutinating activity is unaffected by most proteolytic enzymes (Vogt *et al.*, 1969; Uhlenbruck *et al.*, 1971). It is active over a wide range of pH values (5–12) and is also relatively heat-stable, temperatures > 80°C being required to reduce its agglutinating activity during a 30-min incubation period (Vogt *et al.*, 1969). Divalent cations do not seem to be required for activity, although we have not seen any rigorous examination of this aspect.

The anti-A_{HP} purified by affinity chromatography was found to be homogeneous by gel filtration, ultracentrifugation, and immunoelectrophoresis (Hammarström and Kabat, 1969). Its molecular weight was estimated from sedimentation equilibrium data to be 79,000 ± 4000 (Hammarström *et al.*, 1972b) and, more recently, 72,400 ± 2500 (Kühnemund *et al.*, 1974). The latter authors reported the sedimentation and diffusion coefficients of the molecule to be 5.07 S and 6.2 × 10^{-7} cm²/sec, respectively, and its partial specific volume to be 0.730 g/ml. From these data, a molecular weight of 73,500 ± 4400 and a molar frictional ratio (f/f_0) of 1.23 were calculated. Optical rotatory dispersion and circular dichroism spec-

TABLE 4. PROPERTIES OF MOLLUSC HEMAGGLUTININS

Species	Agglutinin	Source	Molecular weight ($\times 10^{-3}$)	Sensitivity to Heat	Sensitivity to Reduction	Sensitivity to Protease(s)	Divalent cation requirement	Binding site specificity
Helix pomatia	Anti-A_{HP}	Albumin gland	73–79	L^a	L	R	None	GalNAc > GlcNAc
Helix aspersa	Anti-A_{HA}	Albumin gland	~80	—	(L)	—	—	GalNAc
Helix hortensis	Anti-A_{HH}	Albumin gland	—	—	—	—	—	GalNAc
Otala lactea	Anti-A_{OL}	Albumin gland	42	—	L	—	—	GalNAc
Cepaea nemoralis	Anti-A_{CN}	Albumin gland	73	—	L	—	—	—
Caucasotachea atrolabiata	Anti-A_{CA}	Albumin gland	77	—	L	—	—	α-(1-6)-Gal-Gal-Gal
Arianta arbustorum	Anti-A_{AA}	Albumin gland	—	L	L	—	—	GalNAc
Bradybaena fruticum	Anti-A_{BF}	Albumin gland	~100	L	—	—	—	GalNAc = GlcNAc
Euhadra callizona amaliae	Anti-A_{EC}	Albumin gland	89	—	—	R	—	GalNAc = GlcNAc
Euhadra periomphala	Anti-A_{EP}	Albumin gland	—	—	—	—	—	AcNeu
Achatina granulata		Body extract	~200	—	—	—	—	Complex
Arion empiricorum		Mucus	~43	—	—	—	—	GalNAc?
Biomphalaria glabrata		Albumin gland	~55	—	—	—	Ca^{2+}	
Aplysia californica		Hemolymph	>400	L	L	R	None	see Table 5
Crassostrea virginica		Hemolymph	—	L	—	—	Ca^{2+}	GalNAc, GlcNAc
Mercenaria mercenaria		Hemolymph	—	L	—	L	Ca^{2+}	GalNAc, GlcNAc
Saxidomus giganteus		Extract	—	L	—	—	—	GalNAc ≫ GalN
Tridacna maxima		Hemolymph	470	—	—	—	Ca^{2+}	—
Velesunio ambiguus		Hemolymph	—	L	R	L	—	—
Octopus vulgaris		Hemolymph	—	L	L	R	—	Lac

^{a}L, labile; R, resistant; —, not known.

tra (Hammarström, 1974) indicated that the agglutinin contained a significant percentage of β-structure. The molecules appeared as elliptical structures having dimensions of 2.6 × 3.2 nm in the electron microscope (Schnitzler et al., 1971c).

Using [^{14}C]iodoacetate to alkylate free sulfhydryl groups, the native molecule was shown to contain 18 moles of half-cystine, all as disulfide bonds, in agreement with amino acid composition data (Table 11) (Hammarström et al., 1972b). Under denaturing conditions, the agglutinin dissociated into a subunit of molecular weight 26,000–30,000 while complete reduction in 7 M guanidine·HCl yielded a single component with a molecular weight of 13,000 (Hammarström et al., 1972b; Uhlenbruck et al., 1972). A single polypeptide subunit (molecular weight 12,500–16,700) was also obtained when the agglutinin was reduced in 7 M guanidine·HCl plus 0.1 M GalNAc although in this case only 3–4 disulfide bonds were cleaved (Hammarström et al., 1972b). Peptide mapping data of trypsin digests were consistent with the presence of a single type of subunit. Equilibrium dialysis against a tritium-labeled, blood-group A active pentasaccharide showed that the agglutinin possessed six homogeneous, noninteracting carbohydrate-binding sites with $K°_{assoc} = 5 \times 10^3$ liter/mole at 25°C (Hammarström and Kabat, 1971).

On the basis of these results, it is believed that the anti-A_{HP} agglutinin consists of six identical polypeptide subunits (molecular weight 13,000), each containing one intrachain disulfide bond and a single carbohydrate-binding site. The subunits are present as disulfide-bonded dimers (molecular weight 26,000), the third half-cystine residue present in each subunit contributing to a single interchain disulfide bond. The native hexameric agglutinin therefore comprises three dimers held together by noncovalent interactions and contains altogether nine disulfide bonds. An aggregated, "superagglutinating" form of the molecule has been detected by some investigators (Uhlenbruck et al., 1972). It is possible that this "superagglutinin" is an artifact of some purification procedures, particularly those involving an acidic eluant (which may cause the molecular to dissociate into dimers) for the recovery of the agglutinin from an adsorbent such as Sephadex. Many preparations also seem to have been lyophilized, which could cause some aggregation.

Treatment of anti-A_{HP} with 2-mercaptoethanol in the absence of dissociating agents converts it reversibly into an "incomplete" agglutinin against A_1 erythrocytes (detecting using anti-A_{HP} specific rabbit antibodies) with a loss of reactivity for A_2 neuraminidase-treated O erythrocytes and Sephadex (Uhlenbruck et al., 1967; Ishiyama and Uhlenbruck, 1972; Ishiyama et al., 1971, 1973). This change is irreversible if the reduced agglutinin is alkylated with iodoacetate. The "incomplete" form of anti-A_{HP} does not seem to differ antigenically from the parent molecule and retains its specific sugar-binding sites. However, it is smaller, having a molecular weight similar to a subunit dimer (30,000) (Uhlenbruck et al., 1972), and apparently results from dissociation of the native agglutinin. Similar fragments are formed by partial reduction of the agglutinin in 6 M urea plus 0.1 M GalNAc (Hammarström et al., 1972b; Hammarström, 1973). These findings indicate that the interchain (but not the intrachain) disulfide bonds are reduced by mercaptoethanol in the absence of denaturing agents, allowing the agglutinin to dissociate into noncovalently linked dimers. Noncovalent bonding between the disulfide-linked subunits must therefore be extremely weak or nonexistent.

The specificity of anti-A_{HP} has been demonstrated by its capacity to bind to and agglutinate human A (but not B or O) erythrocytes (Prokop et al., 1965b; Uhlenbruck and Prokop, 1966; Prokop et al., 1968; Hammarström and Kabat, 1969; Hammarström, 1973) and to precipitate polysaccharides and glycoproteins such as human blood-group A substance, desialyzed ovine submaxillary mucin, group C streptococcal polysaccharide, and hog group A + H substance (Uhlenbruck and Prokop, 1966; Prokop et al., 1968; Hammarström and Kabat, 1969, 1971; Hammarström et al., 1972a; Hammarström, 1972, 1974; Ishiyama et al., 1974). The reactivity of all these substances is due to their content of terminal, nonreducing α-D-linked GalNAc groups. A large number of strains of different bacteria have also been examined for agglutination by anti-A_{HP} (Uhlenbruck et al., 1966; Köhler and Prokop, 1967; Prokop and Köhler, 1967; Prokop et al., 1968). Most of the reactive strains belonged to the genera *Corynebacterium, Staphylococcus, Sarcina,* and *Bacillus.*

Using a quantitative precipitation system, the binding properties of the agglutinin have been studied in detail (Hammarström and Kabat, 1969, 1971; Hammarström et al., 1972a; Hammarström, 1972, 1974). Polysaccharides and glycoproteins containing nonreducing (terminal) α-D-linked GalNAc were all precipitated. A galactomannan (containing nonreducing terminal α-D-Gal groups) and *Staphylococcus aureus* teichoic acid (containing reducing, terminal α-D-GlcNAc groups) were also precipitated, but polymers containing β-linked GalNAc or GlcNAc were nonreactive. The agglutinin precipitated lipopolysaccharides isolated from *Salmonella typhimurium* rough mutants of chemotype Ra and Rb but not those from the smooth parent strain or from Rc, Rd, and Re mutants. Nonreducing 6-O-α-D-galactosyl-D-glucosyl groups present in the lipopolysaccharides may have provided the binding sites for the agglutinin. The binding specificity of the agglutinin was further investigated by testing a variety of sugars for their ability to inhibit the precipitation of human blood-group A substance or *S. typhimurium* SH 180 lipopolysaccharide. The most effective monosaccharide inhibitor was 1-O-methyl-α-D-GalNAc, which was twice as active as 1-O-phenyl-α-D-GalNAc or GalNAc and > 12 times more active than GlcNAc and GalN. Since the molar inhibitory capacity of the methyl glycoside was similar to that of blood-group type A di-, tri-, and pentasaccharides, it was concluded that the combining site accommodates a single α-D-linked glycosyl unit (Hammarström and Kabat, 1969). Carbohydrate-binding studies involving hemagglutination-inhibition and displacement of anti-A_{HP} from Sephadex or insolubilized blood-group A substance gave essentially the same results (Uhlenbruck and Prokop, 1966; Prokop et al., 1968; Ishiyama and Uhlenbruck, 1972).

The presence of hemagglutinins in extracts of *H. aspersa* has been reported from several laboratories and in all instances the agglutination was predominantly A-specific (Gold and Thompson, 1969a; Grace, 1969; Khalap et al., 1971). The bulk of the agglutinin (anti-A_{HA}) is apparently confined like anti-A_{HP} to the albumin gland and egg-mass of the snail. Extracts of these two tissues have also (in the case of *H. aspersa* only) been found to be hemolytic for human A and rabbit but not horse and guinea pig erythrocytes in the presence of fresh human serum (Gold and Thompson, 1969b). Because no lysis occurred using serum containing EDTA or inactivated by heat (56°C, 30 min) or with normal serum at a dilution greater than 1 : 10, the activation of complement seemed to be necessary. In the presence of undiluted serum, albumin gland extracts were hemolytic up to a dilution of 1 : 4. The hemolysin could be separated from anti-A_{HA} by adsorption onto human O erythrocytes or by chromatography on DEAE-cellulose (Khalap *et al.*, 1972). The agglutinin, partially purified by chromatography through Sephadex G-200, was active against human A_1, sheep, and guinea pig erythrocytes and showed slight activity towards human O and B cells. The agglutination of A_1 cells was inhibited by GalNAc and more weakly by GlcNAc. Eluates from A_1 cells agglutinated sheep erythrocytes. Ishiyama *et al.* (1972) purified anti-A_{HA} from albumin gland extracts by affinity chromatography on Sephadex G-200. The agglutinin was eluted with 2 mM GlcNAc and gave a single precipitin line upon immunoelectrophoresis. It possessed no hemolytic activity in the presence of complement.

Uhlenbruck and Weis (1973) reported serological differences between the agglutinins derived from the albumin glands of *H. aspersa* collected in France and Spain. The agglutinins purified from snails belonging to these two and a third (South African) population were compared with anti-A_{HP} from *H. pomatia* (Ishiyama *et al.*, 1973). There were marked differences in both the serological and the immunoelectrophoretic properties of the *H. aspersa* agglutinins and it was concluded that those from the snails collected in France and South Africa consisted predominantly of nonprecipitating agglutinins, analogous to the "incomplete" agglutinin obtained by partial reduction of anti-A_{HP}. Reduction and alkylation had little effect on these two preparations, whereas the activity of anti-A_{HP} and the anti-$A_{HA(Spain)}$ preparation was significantly reduced. The former preparations were almost insoluble and it is possible that the agglutinins from these two snail populations are particularly prone to denaturation during purification. All the agglutinins were inhibited by GalNAc and GlcNAc and more weakly by raffinose (cf. Ishiyama and Uhlenbruck, 1971). A strong antigenic cross-reaction was observed between anti-A_{HP} and the anti-A_{HA} derived from the Spanish snails.

The amino acid composition of the *H. aspersa* agglutinin was reported by Salfner *et al.* (1972) and is shown in Table 11. The agglutinin differed from anti-A_{HP} in its content of Thr, Gly, Asp, Phe, and Trp. Both agglutinins were also shown to contain covalently bound carbohydrate (anti-A_{HA}: 4.65%; anti-A_{HP}: 1.35%). The molecular weight of anti-A_{HA} is reported to be approximately 90,000 (Uhlenbruck *et al.*, 1972). These findings indicate that anti-A_{HA} and anti-A_{HP} are similar but not identical molecules and that the genes coding for these two proteins may be derived from a common ancestral gene.

Schnitzler *et al.* (1971b) studied the anti-A agglutinins extracted from the albumin glands of *Cepaea nemoralis* (syn. *Helix nemoralis*) and *Caucasotachea atrolabiata*. Each agglutinin was purified by adsorption to Sephadex G-100 at pH 7.2

and elution of pH 3.4. The *C. nemoralis* (CN) extract contained two hemagglutinins; one, specific for human O erythrocytes, was recovered in the pH 7.2 effluent and the other, specific for A_1 erythrocytes, in the pH 3.4 eluate. These findings confirmed previous data (Gold and Thompson, 1969a; Schnitzler and Kilias, 1970) which had indicated the presence of at least two agglutinins distinguishable by their susceptibility to inhibition by GalNAc or AcNeu. The *C. atrolabiata* (CA) extract also exhibited multiple specificities, agglutinating type A_1 and Pronase-treated type O erythrocytes (Schnitzler et al., 1971b), but both activities were retained on the Sephadex and recovered together in the acid eluate. These activities are due to distinct agglutinins, which can be separated by adsorption to various types of erythrocytes (Sprenger and Uhlenbruck, 1971). Inhibition tests have shown that the A_1-specific agglutinin recognizes α-D-GalNAc residues like anti-A_{CN} and anti-A_{HP}, whereas the agglutinin for Pronase-treated O cells recognizes subterminal 1-6-α(or β)-D-GalNAc 1→ plus terminal α- or β-D-GalNAc structures (Sprenger and Uhlenbruck, 1971).

The *C. nemoralis* and *C. atrolabiata* agglutinins obtained by elution from Sephadex at pH 3.4 were predominantly A-specific and were studied without further purification (Schnitzler et al., 1971b). Like anti-A_{HP} (from *H. pomatia*), both were reversibly inactivated by 2-mercaptoethanol. Each migrated towards the cathode and appeared pure by immunoelectrophoresis. However, anti-A_{CA} was found to be heterogeneous in electrophoresis on cellulose acetate. The authors believed this to be due to the presence of aggregates or polymers. The molecular weight of the agglutinins was determined by sedimentation equilibrium to be 77,000 (anti-A_{CA}) and 73,000 (anti-A_{CN}), with a maximal error of 5%. Examination of the preparations in the electron microscope revealed elongated, oval particles with flat surfaces, similar in size to those of anti-A_{HP}.

Whole-body and albumin gland extracts of *Otala lactea* also contain a strong A-specific hemagglutinin (inhibited by GalNAc) as well as a second agglutinin, which agglutinates papain- or ficin-treated type O erythrocytes and is inhibited by L-Fuc (Boyd and Brown, 1965; Boyd et al., 1966; Bhatia et al., 1967). The A-specific agglutinin (anti-A_{OL}), a euglobulin, was purified by fractional precipitation with ethanol followed by chromatography on Biogel P-200, when it migrated as a single component. From chromatography and ultracentrifugation data, anti-A_{OL} was calculated to have a molecular weight of approximately 42,000 and a sedimentation coefficient of 4.3 S (Bhatia et al., 1967, 1968). The purified protein migrated toward the cathode during immunoelectrophoresis and formed a single precipitin line with A substance. Approximately 22% of the protein present in the purified preparation was precipitated with A substance, indicating a reasonable degree of purity. The agglutinating activity of anti-A_{OL} was diminished by at least 94% upon reduction with 2-mercaptoethanol.

Eggs and albumin glands of *Arianta arbustorum* have been shown to contain a powerful agglutinin which reacts specifically with rabbit erythrocytes (Renwrantz and Berliner, 1978). Agglutinating titers of 1 : 4000 for rabbit erythrocytes but only 1 : 4 for human (A, B, and O) and guinea pig cells were reported for extracts from either tissue. The agglutination could be inhibited completely by di-, tri-, and oligosaccharides possessing α-glycosidically (1→6) bound Gal residues (e.g., melibiose, raffinose, and stachyose, melibiose being slightly less efficient on a molar basis). Gal, GlcNAc, and Lac were inactive. The specificity of this agglutinin is therefore quite different from that of the anti-A and other agglutinins detected in other snails. The activity of the agglutinin was not altered by dialysis against Ca^{2+}-free saline or by the addition of $CaCl_2$, but the titer decreased from 1 : 5096 to 1 : 8 after treatment with 2-mercaptoethanol, showing that the molecule contains disulfide bonds. The agglutinin was relatively heat-stable, retaining full activity after a 1-hr incubation at 60°C but becoming inactivated at temperatures $\geq$ 70°C. It has not been purified.

Less work has been done on *H. hortensis*, which seems to contain a second (AcNeu-specific) hemagglutinin in addition to the anti-A (GalNAc) specific agglutinin (Prokop *et al.*, 1965a; Rackwitz *et al.*, 1965; Uhlenbruck and Pardoe, 1969; Pardoe *et al.*, 1970; Khalap *et al.*, 1971). Fractionation of albumin gland extracts on Sephadex G-75 and G-200 resolved these two activities (Oehme *et al.*, 1968) but the properties of the agglutinins were not investigated.

The Fruticicolidae have received less attention than the Helicinae, although several species (listed by Gold and Balding, 1975) have been shown to contain strong A-specific agglutinins. The agglutinin extracted from the albumin gland of *Bradybaena fruticum* (anti-A_{BF}) reacts preferentially with human A_1 erythrocytes and is inhibited by GalNAc (Schnitzler and Kilias, 1970; Schnitzler and Geserick, 1971). In contrast with the anti-A agglutinins from snails of the Helicinae group, anti-A_{BF} was recovered in the effluent when extracts were chromatographed on Sephadex G-100 at pH 7.2. From its elution position, the agglutinin was estimated to have a molecular weight of < 100,000. However, since the agglutinating peak exhibited a pronounced trailing edge, it seems likely that anti-A_{BF} has a weak affinity for Sephadex and its chromatographic behavior may therefore not indicate its true size. The agglutinin did not appear to be labile at 60 or 80°C. In contrast to the agglutinins from *C. nemoralis*, *C. atrolabiata*, and *H. lucorum*, anti-A_{BF} gave no reaction with antisera specific for anti-A_{HP} from *H. pomatia*.

The albumin glands of Japanese snails belonging to the genus *Euhadra* (Fruticicolidae) also contain agglutinins which react predominantly with A_1 substances, although their activities are weak in comparison with those from *Helix* spp. (Ishiyama *et al.*, 1974). The *Euhadra* agglutinins apparently react with the same terminal GalNAc residue, but seem to have binding site specificities different from that of anti-A_{HP} since they can be inhibited as effectively by GlcNAc as by GalNAc. The agglutinin from *E. periomphala* can be adsorbed by various Salmonellae, particularly by *S. riogrande* which was the only strain to be agglutinated (Ishiyama and Takatsu, 1970). The activity of the *E. periomphala* agglutinin is reduced at pH values $\leq$ 5 and $\geq$ 10, but not by proteolytic enzymes.

The anti-A agglutinin from *E. callizona amaliae* (anti-A_{EC}) has been purified by affinity chromatography on (meconium A)-aminoethyl cellulose since like anti-A_{BF} it exhibits no remarkable affinity for Sephadex (Mukaida *et al.*, 1974). The agglutinin was eluted in 5 mM GalNAc at pH 7.2 and then chromatographed on Bio-Gel P-300 from which it emerged in a single, symmetrical peak. It accounted for about 2% of the protein in the albumin gland extract. The agglutinin was shown to be pure by immunoelectrophoresis. Ultracentrifugational analysis revealed a single component having a sedimentation coefficient of 5.3 S. The molecular weight of the agglutinin was calculated to be 89,000, using a partial specific volume of 0.72 g/ml which was calculated from the amino acid composition of the molecule (Table 11). The low isoelectric point of anti-A_{EC} (pI = 3.6) was consistent with the amino acid composition, which showed a predominance of acidic residues. Like anti-A_{HP} and anti-A_{HA}, anti-A_{EC} is a glycoprotein and has a hexose content of approximately 5.1%. Anti-A_{EC} could precipitate various blood-group A substances, but inhibition experiments indicated that its affinity for these substances was lower than that of anti-A_{HP} (Mukaida et al., 1974; Ishiyama et al., 1974).

An agglutinin with an affinity for AcNeu has been purified to homogeneity from extracts of *Achatina granulata* (Khalap, 1972, cited by Gold and Balding, 1975). The molecule, a glycoprotein, had an apparent molecular weight of 200,000 and an isoelectric point of pH 4.9. It was present in the extracts in only small amounts.

The only member of the Arionidae from which an agglutinin has been isolated and partially characterized is *Arion empiricorum* (Habets et al., 1979). The molecule was purified by ion-exchange chromatography from snail mucus and was homogeneous as determined by immunoelectrophoresis and polyacrylamide gel electrophoresis. It had an apparent molecular weight of about 43,000. The agglutinin precipitated at least two proteins from mammalian sera, showed no blood-group specificity, and was inhibited only partially by Gal, Glc, GalNAc, and GlcNAc at high concentration (0.2 M). It appeared to possess proteinase-inhibitor activity.

The hemolymph of the sea hare *Aplysia californica* (subclass Prosobranchia) was found to agglutinate a number of bacteria and vertebrate erythrocytes (Pauley et al., 1971a; Pauley, 1974a). The results of adsorption experiments indicated that a single agglutinin might have been responsible. The agglutinin did not require divalent cations, since it was fully active in the presence of sodium citrate. Although resistant to trypsin, Pronase, and several other hydrolytic enzymes, it was inactivated by heat (> 70°C) and partially inactivated by 2-mercaptoethanol. These properties suggest that the agglutinin is a stable protein containing disulfide bonds. Its molecular weight seems to be > 400,000 as it was completely excluded by Sephadex G-200 and had a sedimentation coefficient of approximately 18 S as estimated from sucrose density gradient ultracentrifugation data.

The hemolymph of an *Aplysia* sp. has also been reported by McKay et al. (1969) to contain agglutinins for a number of vertebrate erythrocytes. No attempt was made to characterize the molecules involved.

3.1.2. Class Bivalvia

Agglutinins have been detected in a limited number of species representing each of the three orders of Bivalvia (see Figure 2) and several have been studied in some detail.

Tyler (1946) tested the body and seminal fluids of *Mytilus californianus* (order Anisomyaria) and detected agglutinins in both for various invertebrate spermatozoa and several mammalian erythrocytes. Other Anisomyaria species shown to possess erythrocyte agglutinins include *M. edulis* (Brown et al., 1968; Uhlenbruck et al., 1970), *Pinctada nigromarginata* (McKay et al., 1969), and the oysters *Crassostrea virginica* (Tripp, 1966; McDade and Tripp, 1967; Li and Flemming, 1967; Acton et al., 1969) and *Acanthostera gemmata* (McKay et al., 1969).

C. virginica has been studied in some detail (Tripp, 1974). Although the hemolymph opsonic factor(s) has not yet been identified, several hemagglutinins appear to exist in the hemolymph and these are clearly potential candidates (Tripp, 1966). The agglutinins were initially distinguished by titrating unadsorbed and adsorbed hemolymph against different vertebrate indicator erythrocytes. Low titers (1 : 4 to 1 : 16) were observed using mouse, chicken, guinea pig, and bovine erythrocytes and the agglutinins for all of these cells were removed by adsorption with any single type of erythrocyte (chicken, guinea pig, human, sheep, rabbit, and horse). Higher titers (1 : 64–1 : 512) were apparent with human, sheep, rabbit, and horse erythrocytes, no difference being observed between human A, B, and O cells. When hemolymph was adsorbed with each type of erythrocyte, however, differential adsorption of the activity for the various cells occurred, indicating that a number of different agglutinins might have been present. Adsorption with human erythrocytes caused a reduction in hemolymph protein concentration from 19 to 0.4 mg/ml. Several strains of bacteria were also tested for agglutination and for their capacity to adsorb the hemagglutinins. The results were uniformly negative.

McDade and Tripp (1967) reported further evidence for the existence of multiple agglutinins in *C. virginica* hemolymph. The agglutination of human A, B, and O erythrocytes was found to be Ca^{2+} dependent, activity being abolished by the addition of citrate or by dialyzing hemolymph against Ca^{2+}-free saline. This inactivation was reversible, since activity was fully recovered upon the addition of Ca^{2+} ions. In contrast, the activity of the agglutinins for sheep and rabbit erythrocytes was unaffected by dialysis or by addition of citrate or Ca^{2+} ions. The agglutinins were stabilized against heat-denaturation in the presence of Ca^{2+} ions, but unfortunately the type(s) of erythrocyte used to measure activity in this experiment was not indicated. The latter point is important, since it would be particularly interesting to know whether the sheep and rabbit erythrocyte agglutinins were also stabilized by Ca^{2+} ions. In this regard, it is noteworthy that the sheep erythrocyte agglutinin purified by Acton et al. (1969) was inactivated by citrate (see below).

The binding specificity of the agglutinins was also examined by testing 13 monosaccharides for their capacity to inhibit the agglutination of various types of erythrocytes. The agglutinins for each type of cell exhibited distinct sugar-inhibition patterns (see Table 5) and must therefore have differed in binding specificity. On the basis of these and other results, McDade and Tripp (1967) suggested that oyster hemolymph contains a family of protein molecules, each having a slightly different specificity and capable of reacting with one or a few structurally related sugars. They suggested that the hemagglutinin molecules

TABLE 5. THE CAPACITY OF VARIOUS MONOSACCHARIDES TO INHIBIT HEMAGGLUTINATION BY OYSTER (CRASSOSTREA VIRGINICA) HEMOLYMPH[a]

Sugar (0.2–0.3 M)	Reduction in hemagglutination titer (%)					
	Type of erythrocyte tested					
	Human	Monkey	Rabbit	Horse	Mouse	Sheep
D-Rib	0	0	≥ 97	75	0	0
D-Gal	0	0	0	75	0	75
D-GalN	≥ 97	0	75	75	50	0
D-GalNAc	≥ 97	0	0	0	0	0
D-GlcN	88	50	0	0	0	0
D-GlcNAc	≥ 97	0	0	0	0	0

[a] Calculated from McDade and Tripp (1967, Table 1).

might be polymers comprising a number of these protein subunits. Attempts to purify the agglutinins were unsuccessful.

In another study, Li and Flemming (1967) observed that *C. virginica* hemolymph agglutinated human (type O), rabbit, and trout erythrocytes. The agglutinins, diluted in 0.05 M phosphate (pH 8.0), were inactivated at temperatures above 56°C. Upon ultracentrifugation of hemolymph at 98,000g for 2–4 hr, all of the agglutinating activity but only 10% of the protein was recovered in the supernatant. This result appears inconsistent with the data of Tripp (1966) and Acton *et al.* (1969), who found that more than 90% of the hemolymph protein could be adsorbed to human and sheep erythrocytes. It may be that some of the protein(s) becomes adsorbed nonspecifically to erythrocytes. Li and Flemming separated the agglutinating activity into two distinct fractions by chromatography on Sephadex G-75, which excludes globular proteins having molecular weights greater than about 70,000. The first peak, presumably containing excluded macromolecules, accounted for approximately half of the recovered activity. In view of the sugar specificity of the hemagglutinins reported by McDade and Tripp (1967; Table 5), the agglutinins comprising the second peak may have been retarded by interaction with the Sephadex. If so, their actual molecular size may be considerably larger than indicated by their elution position.

In 1969, Acton and colleagues reported the purification of a hemagglutinin from *C. virginica* hemolymph by chromatography on Sepharose 4B (Acton *et al.*, 1969). Adsorption of hemolymph with sheep erythrocytes removed 98% of the soluble protein, suggesting that the hemagglutinin(s) constituted a large proportion of the protein present in the hemolymph. A similar result was reported by Tripp (1966) using human erythrocytes (see above). The purity of the agglutinin was assessed by SDS-PAGE and N-terminal amino acid analysis, which revealed a single polypeptide component (apparent molecular weight 20,000) and Thr as the only N-terminal residue. The molecules contained 13% carbohydrate and had an unusual amino acid composition (Table 11), being relatively rich in His and Cys but devoid of Lys. Subsequent studies yielded a lower carbohydrate content (8.8%) but confirmed Man (3.7%), GlcN (2.7%), and Gal (1.5%) as the predominant sugars (Acton *et al.*, 1973). Divalent cations were apparently required for agglutinating activity (measured with sheep erythrocytes), since citrate caused complete inactivation. The native agglutinin was analyzed by analytical ultracentrifugation and was shown to consist of a heterogeneous group of polymeric molecules, the predominant species having a sedimentation coefficient of approximately 30 S. The subunits were held together by noncovalent bonds. Some of the properties of this agglutinin are summarized in Tables 4 and 12. Neither the binding specificity nor the effect of divalent cations on polymerization was investigated.

Several species of clam (order Heterodonta) have been reported to contain agglutinins. Johnson (1964) found that saline extracts of the butter clam *Saxidomus giganteus* contained an agglutinin specific for type A_1 human erythrocytes. The agglutinin was adsorbed completely by A_1 and A_1B cells, partially by A_2 cells, but not by B and O cells. It was heat-labile and could be inhibited by GalNAc and GlcNAc. McKay *et al.* (1969) reported the presence of *Tridacna fossor* hemolymph of agglutinins for erythrocytes from a number of different vertebrate species, the highest titer (1 : 128) being observed against guinea pig cells. Adsorption tests with various types of erythrocytes indicated that hemagglutinins with distinct specificities were present. Galactan-reactive hemagglutinins have been detected in the hemolymph of two closely related species, one in *T. maxima* (Baldo and Uhlenbruck, 1975a,b) and two in *T. crocea* (Uhlenbruck *et al.*, 1977, 1979). Similar agglutinins which exhibit distinct specificities have been found in the hemolymph of other *Tridacna* species (Uhlenbruck *et al.*, 1979).

The *T. maxima* agglutinin has been purified by affinity chromatography and extensively characterized (Baldo *et al.*, 1978). It is a homogeneous, β-galactosyl-binding glycoprotein (molecular weight 470,000; $s_{20,w}$ = 15.0) consisting of multiple subunits and contains approximately 7% carbohydrate (see Table 12). The amino acid composition (Table 11) revealed a significant amount of Cys and a lack of Met residues. The molecule required Ca^{2+} ions for its hemagglutinating and precipitating activities. Precipitates were formed with polysaccharides and glycoproteins having terminal nonreducing β-D-galactosyl residues. GalNAc was the best inhibitor of precipitation, being 20-fold more active than Gal and nearly 10-fold more active than GalN on a molar basis. Circular dichroism studies showed a relatively high proportion of β-structure.

The hemolymph from another clam, *Mercenaria mercenaria*, was found to agglutinate a marine alga and several strains of bacteria (Arimoto and Tripp, 1977). A single agglutinin appeared to be responsible, since adsorption with one type of bacterium also removed the agglutinins for the alga and the other bacteria. Calcium ions contributed to the heat stability of the agglutinin molecules and were essential for activity. The agglutinin was inactivated by protease treatment and the authors suggested, on the basis of results from SDS-PAGE of adsorbed and unadsorbed hemolymph, that it might be composed of polypeptide subunits having a molecular weight of approximately 21,000. Several sugars, particularly GalNAc, GlcNAc, GlcN, and D-Fuc, partially inhibited bacterial agglutination, indicating that the agglutinin recognized oligosaccharides containing these sugars. It was suggested, on the basis of adsorption and *in vitro* phagocytosis experiments, that the clam agglutinin possessed opsonic properties and that it may be utilized for feeding and/or defense purposes.

Agglutinins have also been detected in several mussel species (order Schizodonta). Gibb *et al.* (1967) examined homogenates of pond mussels (*Anodonta* spp.). They demonstrated the presence of blood-group H- and P_1-like substances, but could not detect agglutinins. An agglutinin for human erythrocytes was reported by Pemberton (1969) to occur in extracts of the swan mussel *A. cygnea* and to be inhibited specifically by lactose. An extract of a specimen of *A. anatina* also agglutinated bromelain-treated human erythrocytes. Further investigations indicated that whole-body extracts of *A. anatina* contained an agglutinin which was strongly inhibited by Lac and weakly by terminal Gal-containing oligosaccharides (Pemberton, 1971a). Extracts of *A. complanata* agglutinated bromelain-treated human A and B erythrocytes (Pemberton, 1971b).

The hemolymph of the Murray mussel *Velesunio ambiguus* contained agglutinins for erythrocytes from several vertebrate species, particularly those from the rat, rabbit, mouse, and guinea pig (titers 1 : 64 to 1 : 512; McKay *et al.*, 1969; Jenkin and Rowley, 1970). Adsorption experiments revealed the presence of agglutinins specific for one or another type of cell. The molecules sedimented in sucrose density gradients as a single, homogeneous band, the material from which exhibited a sedimentation coefficient of 28 S in the analytical ultracentrifuge (Jenkin and Rowley, 1970). The agglutinin(s) for rabbit and rat erythrocytes was purified by adsorption to rabbit erythrocyte stroma. Treatment of the stroma with glycine·HCl (pH 3.0) released almost 1 mg of protein which was considered to represent the pure agglutinin(s). The recovery of agglutinating activity was only about 6%, but the agglutinins were shown separately to be inactivated by the acid conditions. The purified material was used for amino acid analyses (Table 11). The data showed a remarkably high content of Phe and Ser and a low content of Ile, Tyr, Cys, and Met. The subunit structure of these agglutinins has not been determined.

3.1.3. Class Cephalopoda

Using a rabbit antiserum raised against hemolymph from the octopus *Eledone cirrosa*, Stuart (1968) studied the capacity of the latter to sensitize various types of cells. When human and rabbit erythrocytes, yeast cells, and *Salmonella typhi* were incubated with *E. cirrosa* hemolymph, they became sensitized for agglutination by the rabbit antiserum but not by normal rabbit serum. Incubation with either hemolymph or rabbit antiserum caused no agglutination. It therefore appeared that one or more antigenic substances present in the hemolymph became adsorbed to the cells. To confirm this result, rabbit erythrocytes were incubated with octopus hemolymph, washed, and then injected back into the donor rabbit. The serum obtained after five injections gave a strong precipitin line upon immunoelectrophoresis against the hemolymph, indicating that the rabbit cells had indeed been coated with some hemolymph component(s). On the basis of these results and the observation that the octopus hemolymph promoted the *in vitro* phagocytosis of human erythrocytes by octopus hemocytes, it was suggested that a single polyvalent opsonin with broad-spectrum affinity for a wide range of particles might occur in *E. cirrosa* hemolymph. The nature of the molecules responsible for these effects has not been determined, although ultracentrifugational analysis of the hemolymph apparently revealed one major (hemocyanin?) and two minor components.

To our knowledge, *Octopus vulgaris* is the only cephalopod in which hemagglutinins have been directly detected (Bretting and Renwrantz, 1973; Renwrantz and Uhlenbruck, 1974a). The hemolymph of this animal agglutinated a variety of native and enzyme-treated vertebrate erythrocytes. The agglutinins, inhibited by Lac, were heat-labile, inactivated by reduction, and resistant to Pronase. A blood-group A-like substance was also detected (Renwrantz and Uhlenbruck, 1974a). A similar substance was found in the hemolymph of *O. bimaculatus* by Cushing *et al.* (1963).

The properties of the hemagglutinins detected in various molluscan species are summarized in Table 4.

3.2. ARTHROPODA

3.2.1. Crustacea

Tyler and Metz (1945) found that the hemolymph from the spiny lobster *Panulirus interruptus* agglutinated erythrocytes from various species of vertebrates and spermatozoa from some other invertebrates. Cross-adsorption experiments indicated the presence of 8–10 different agglutinins.

Attempts to purify the hemagglutinins by isoelectric precipitation at pH 4.8–5.0 separated them from the hemocyanin (Tyler and Scheer, 1945). By electrophoresis, the precipitated material was resolved into two components, the more slowly moving one possessing the hemagglutinating activity. This material reacted with an antiserum raised against purified *P. interruptus* agglutinin and Tyler and Scheer (1945) suggested that the hemagglutinins were antigenically similar to the hemocyanin molecules. It is unlikely, however, that the material obtained by these procedures was homogeneous and one cannot therefore rule out the possibility of cross contamination.

Another member of the Crustacea which has been investigated in some detail is the American lobster *Homarus americanus*. In 1973, Cornick and Stewart described some of the properties of the hemagglutinin(s) present in the hemolymph of this animal. They found that the agglutinin to rabbit erythrocytes was partially inactivated when incubated for 30 min at temperatures above 45°C and completely inactivated at 65°C. Dialysis of the hemolymph against Tris-buffered saline to remove Ca^{2+} ions increased the hemagglutinating titer. This is a surprising finding in view of the reports by other investigators that the activity of the hemagglutinins from this species is Ca^{2+} dependent. However, it is uncertain to what extent bound Ca^{2+} would have been lost, since this would depend on dissociation rates.

In testing the capacity of various sugars to inhibit the agglutinating activity for monkey, horse, sheep, and rabbit erythrocytes, it was found that GlcN inhibited the agglutination of monkey, sheep, and rabbit (but not horse) erythrocytes. Furthermore, Man inhibited partially the agglutination of monkey cells, but not that of the others. Although only a limited number of sugars were tested (all at the same concentration), the results suggested that the hemolymph of *Homarus americanus* contained hemagglutinins of differing specificity. Cornick and Stewart (1973) were able to demonstrate also that the hemocytes from these animals were a rich source of the hemagglutinin(s), although it is not clear from their studies whether the agglutinin(s) was synthesized by these circulating cells or associated only with their surface membranes via specific receptors. Similar findings have been reported by Tyson and Jenkin (1974) for the freshwater crayfish, *Cherax destructor* (formerly known as *Parachaeraps bicarinatus*), and for the protochordate, *Botrylloides leachii* (Coombe, Schluter, Ey, and Jenkin, unpublished observations). A later study investigating the phagocytosis of sheep erythrocytes by lobster hemocytes *in vitro* showed that the uptake of erythrocytes was dependent on a factor (opsonin) in the hemolymph (Paterson and Stewart, 1974). These authors also made a very interesting observation which is germane to data considered elsewhere in this review on the specificity of some of these hemagglutinins. They state "there is a factor that sensitizes SRBC [sheep red blood cells] and it seems to be loosely attached to the erythrocytes as adsorption rates dropped markedly when the sensitized SRBC were washed with and then resuspended in LIII [lobster haemolymph medium] prior to use." Since the lobster hemolymph medium contained 0.5 mg/ml of dextrose, it is possible that the specificity of the opsonin was directed against sugar moieties similar in structure to Glc. If this were the case, one might expect that in the presence of Glc, the association/dissociation equilibrium for the erythrocyte–opsonin interaction would be favored towards dissociation, thus affecting the adsorption of the erythrocytes to the phagocytes (see Section 4.2.1).

The physicochemical characteristics of the hemagglutinins in the hemolymph of *Homarus americanus* were first investigated in detail by Hall and Rowlands (1974a). They found that the highest yield of agglutinating activity against human and mouse erythrocytes could be obtained from both hemolymph and hemocyte lysates, the latter being the richer source on a total protein basis. The hemagglutinin(s) partially purified by precipitation with 35% saturated ammonium sulfate in the presence of 0.02 M $CaCl_2$ was dissolved in 0.1 M barbital buffer (pH 8.6) and subjected to chromatography on Sepharose 4B. This resulted in the resolution of two peaks of hemagglutinating activity. The material which eluted first (designated L Ag-1) agglutinated human erythrocytes, while the second peak (L Ag-2) was active against mouse erythrocytes. Although the activities were clearly separated, there was considerable trailing of each which may have been the result of weak interaction between the agglutinins and the Sepharose.

The L Ag-1 preparation reacted with human, mouse, horse, hamster, chicken, and sheep erythrocytes, while L Ag-2 reacted with only mouse, horse, and hamster. The existence of at least two distinct lobster hemagglutinins was confirmed by Pevikon block electrophoresis and by velocity sedimentation studies of the material precipitated by ammonium sulfate. Two distinct bands of agglutinating activity were obtained by electrophoresis, one specific for mouse erythrocytes only and the other for both mouse and human.

Sucrose density gradient ultracentrifugation partially resolved the agglutinins such that the maximal activity against horse, mouse, and human erythrocytes was present in different parts of the gradient. Further purification of L Ag-1 and L Ag-2 was achieved by Pevikon block electrophoresis followed by chromatography on either Sepharose 6B (L Ag-2) or Sepharose 4B (L Ag-1). The activity of the hemagglutinins was labile to heat (56°C, 15 min) and to the effects of EDTA. The latter characteristic appears to be common to many invertebrate hemagglutinins.

Rate zonal ultracentrifugation of the purified hemagglutinins through sucrose density gradients showed that the L Ag-2 had a sedimentation constant of approximately 11.5 S. Analytical ultracentrifugation of the same material gave values of 10.8 S and 11.3 S. The L Ag-1 was much larger and an approximate S value was not determined. Chromatography of both L Ag-1 and L Ag-2 on Sephadex G-200 in the presence of 6 M urea suggested that the hemagglutinin(s) in both fractions was composed of noncovalently linked subunits possessing a molecular weight of 55,000.

Similar studies on the spiny lobster *Panulirus argus* by Weinheimer (1971) (quoted by Hall and Rowlands, 1974a) indicated that the hemagglutinin(s) of this species contained subunits with a molecular weight of 68,000 (see also Acton and Weinheimer, 1974). The hemagglutinating activity was Ca^{2+} dependent. The heat lability of the agglutinins confirms the results of Paterson and Stewart (1974). However, the Ca^{2+} dependency of the hemagglutinating activity was at variance with the observations of the latter group, possibly because different treatments were used by the two groups to remove Ca^{2+} ions. The hemagglutinin molecules, as shown by Hall and Rowlands (1974a), are composed of subunits linked noncovalently. These contain approximately 4.6% covalently bound carbohydrate, consisting predominantly of Man residues (Acton et al., 1973). The integrity of the native molecule may, in part, depend on bound Ca^{2+} ions, and incomplete removal of these ions by dialysis might lead to partial dissociation of the agglutinin(s) without substantially affecting their capacity to cross-link erythrocytes. The net result would be to increase the hemagglutinating activity. On the other hand, EDTA (a powerful chelating agent) probably removes all Ca^{2+} ions and may lead to dissociation of the molecules accompanied by a loss in binding capacity.

Highly purified preparations of L Ag-1 and L Ag-2 were compared for possible antigenic cross-reactivity in Ouchterlony plates. The results of this analysis indicated that the two fractions were antigenically distinct.

Hall and Rowlands (1974b) extended their investigations to an examination of the binding specificities of the hemagglutinins present in the L Ag-1 and L Ag-2 fractions. A wide variety of cells (erythrocytes, lymphocytes, and a species of *Achromobacter*) were agglutinated by these preparations. In preliminary experiments using hemolymph, it was found that the agglutination of human erythrocytes could be inhibited by several simple sugars, the most potent being AcNeu. Using the purified hemagglutinins to agglutinate mouse erythrocytes, Hall and Rowlands showed that the agglutination by L Ag-2 could be inhibited by GalNAc and to some extent by either ManNAc or GlcNAc. In contrast, the agglutination by L Ag-1 was not inhibited by a variety of simple sugars nor by a combination of GalNAc, AcNeu, and GlcNAc, each present at a concentration of 0.1 M.

The question then arose as to whether a given agglutinin possessed heterogeneous binding sites. An answer was sought in a rather novel manner, which involved observing microagglutinates of two types of erythrocytes in the presence of either L Ag-1 or L Ag-2. The experiments involved the use of human and mouse erythrocytes, the human erythrocytes having first been conjugated with fluorescein to allow them to be distinguished from mouse cells. Equal proportions of the two cell types were added to one or the other purified hemagglutinin and the macroagglutinates were observed under the microscope. In the presence of L Ag-1, mixed agglutinates were formed whereas L Ag-2 agglutinated only mouse erythrocytes. When similar experiments were carried out with L Ag-1 in the presence of AcNeu, the agglutinates contained only mouse erythrocytes. This was suggestive evidence that binding sites with different affinities were present on the same molecule. If further experimental evidence of this nature were available, then it would lend some support to the hypothesis put forward by Jenkin and Hardy (1975) on the generation of diversity of recognition factors (opsonins) in the invertebrates. However, one must add a note of caution which the investigators themselves acknowledge. The results are equally consistent with the possibility that L Ag-1 molecules have homologous binding sites which react, with different affinity, with structurally related moieties on the two different cells. An examination of the cell type agglutinated by L Ag-1 showed, for instance, that 78% of the cells making up the microagglutinates present in the mixed suspension were mouse erythrocytes, while the cells remaining unagglutinated were mostly human. These results detract from the idea that single L Ag-1 molecules have binding sites with more than one specificity.

Human erythrocytes sensitized with L Ag-1 bound to lobster hemocytes and there was some indication of phagocytosis though this did not appear to be striking (Hall and Rowlands, 1974b).

Further work on the hemagglutinins from *Homarus americanus* has been carried out by Hartman *et al.* (1978). Hemolymph was first made 35% saturated with ammonium sulfate and the precipitate was collected and dissolved in 0.1 M Tris·HCl/0.1 M $CaCl_2$ (pH 7.2). The hemagglutinins were adsorbed to an affinity column consisting of Sepharose to which bovine submaxillary mucin had been coupled. When the column was washed with a solution of GalNAc, agglutinins for mouse but not human erythrocytes (L Ag-2) were eluted. Subsequent elution with ManNAc yielded a peak of hemagglutinating activity for mouse and human erythrocytes (L Ag-1). These data confirm the findings of Hall and Rowlands (1974a,b), that lobster hemolymph contains at least two distinct hemagglutinins.

The recognition of foreignness by phagocytic cells from the crayfish *Cherax destructor* (previously *Parachaeraps bicarinatus*) has been examined by McKay and Jenkin (1970). They found that the uptake and ingestion of vertebrate erythrocytes was dependent on certain factors in the hemolymph. These recognition factors appeared to be phylum specific, insofar as the hemolymph from molluscs such as *H. aspersa*, although possessing agglutinins for sheep erythrocytes, would not promote the phagocytosis of these cells by phagocytes from the crayfish. A rabbit anti-sheep erythrocyte serum was also inactive, a finding similar to that described by Stuart (1968) using phagocytic cells from the octopus, *E. cirrosa*. McKay and Jenkin (1970) observed some degree of specificity in this reaction, since hemolymph adsorbed with erythrocytes from one species of vertebrate remained opsonic for erythrocytes from other species.

Further investigations concerned with the phagocytosis of various strains of bacteria indicated that these particles, unlike erythrocytes, could be recognized and ingested by hemocytes in the absence of hemolymph (Tyson and Jenkin, 1974). These *in vitro* findings were somewhat contradictory to those previously published (Tyson and Jenkin, 1973), where it was found that *in vivo* the elimination of bacteria from the circulation of the crayfish was dependent on recognition factors circulating in the hemolymph. The data indicating the importance of circulating opsonic factors in promoting the removal of bacteria were derived from "blockading"-type experiments which have been used successfully in the vertebrates. For example, if vertebrates that have just eliminated a primary dose of bacteria from the circulation are given a second similar dose, the latter is removed much more slowly than the primary dose. However, if the bacteria in the second dose are first pretreated (opsonized) with specific antibody, their rate of clearance may be returned to normal, indicating that the "blockade" effected by the primary dose is due to a depletion of circulating antibody. Similar experiments to these were carried out in the crayfish and it was found that blockade in these animals could be reversed by pretreating the second dose of bacteria with hemolymph. Prior adsorption of the hemolymph used for opsonization of the second dose with the specific strain of bacteria prevented the reversal of the blockade. Further, if the primary (blockading) dose of bacteria was pretreated with hemolymph prior to injection, thus minimizing the removal of circulating recognition factors, the second dose was cleared at normal rates.

The apparent ability of hemocytes from *Cherax destructor* to recognize bacteria but not erythrocytes *in vitro* in the absence of hemolymph may be accounted for by the concentration of molecules of the specific factor(s) associated with the surface of the phagocytes. There is a great difference between the surface area of an erythrocyte and that of a bacterium and it is possible that the number of recognition molecules present on a hemocyte is insufficient to bind erythrocytes. Experiments in the vertebrates would support the above suggestion. Miescher *et al.* (1963) found that the amount of antibody required to facilitate the uptake of erythrocytes by phagocytes is greater than that required for bacteria.

Further *in vitro* studies showed that treating the hemocytes with trypsin destroyed their ability to recognize bacteria unless the bacteria had been previously opsonized with hemolymph. These data suggested that recognition factors for bacteria were both free in the hemolymph and associated with the plasma membrane of the phagocytic cell, raising the possibility that a single factor may be involved. Preliminary experiments indicated that the hemolymph and hemocyte factors were similar. Monolayers of hemocytes were treated with trypsin and divided into two groups. One group was incubated with a tissue culture medium supplemented with crayfish hemolymph, while the other group was incubated in medium alone. The cells were then washed *in situ*, following which unopsonized bacteria were added to the monolayers. Only those cells preincubated with hemolymph were able to ingest the bacteria. It could be argued that incubating the hemocytes in the presence of hemolymph provided a nutritionally richer environment which enabled the cells to resynthesize their membrane receptors. However, this seems unlikely since hemocytes incubated in hemolymph which had previously been adsorbed with bacteria did not regain their ability to recognize unopsonized bacteria (Tyson and Jenkin, 1974).

Tyson and Jenkin (unpublished observations) partially characterized the hemagglutinins and recognition factors for bacteria in the hemolymph from *C. destructor*. Rabbits were hyperimmunized with sheep erythrocytes that had been pretreated with hemolymph and subsequently washed. The immunoglobulin fraction of the rabbit antiserum was first passed down a Bio-Gel A50M column containing covalently linked hemocyanin to remove antihemocyanin antibodies and finally coupled to Bio-Gel A50M for use as an affinity column. The latter immunoadsorbent removed not only all the hemagglutinating activity from hemolymph but also the opsonic activity for both bacteria and erythrocytes. Earlier studies had indicated that there were opsonins displaying some specificty for various strains of bacteria as well as for erythrocytes from different species of vertebrates (McKay *et al.*, 1969; Tyson and Jenkin, 1974). The adsorption of all these factors to the immunoadsorbent suggested that they were antigenically related. Both the hemagglutinating and the opsonic activity required Ca^{2+} ions. Fractionation of the material eluted in 3 M sodium thiocyanate from the affinity column on a calibrated Bio-Gel P-150 column indicated that both the opsonins and the hemagglutinins were molecules with a molecular weight of 81,000. SDS-PAGE analysis showed that the molecules could be dissociated into subunits with a molecular weight of 13,500.

The ability of the crayfish (*Procambarus clarkii*) to recognize various proteins as foreign and to clear them from its circulation was investigated by Sloan *et al.* (1975). Their experiments were based on the premise that if the animal possessed recognition molecules, whether free or cell-bound, they must be of finite number (and thus saturable) and have some discernable degree of specificity. By following the kinetics of clearance of 5 µg amounts of ^{125}I-labelled proteins injected intracardially, this crayfish was found to be capable of discriminating between two of its own proteins (hemocyanin and another uncharacterized hemolymph protein, neither of which was eliminated to any significant extent from the circulation in the 24 hr period after injection) and several foreign proteins (bovine serum albumin [BSA] and human gamma globulin [HGG]) which were cleared with considerable rapidity. BSA was removed faster than

HGG, clearance of 75% of the radioactivity requiring 4 and 24 hr respectively. The radioactivity became concentrated mainly in the gills during this period.

The specificity of this discriminatory system was investigated by injecting animals with 5 μg of ^{125}I-labelled protein mixed with 5 mg of an unlabelled protein. It was found that the clearance of labelled BSA was retarded by excess BSA and also by other serum albumins (human, rabbit, and horse) but not by HGG, bovine gamma globulin (BGG), or keyhole limpet hemocyanin (KLH). Both HGG and BGG slowed down the initial elimination of HGG, whereas BSA and KLH had no effect. The clearance of labelled KLH was similarly specifically inhibited by KLH but not by BSA or BGG. These results clearly supported the conclusion that this crayfish possesses at least three groups of naturally occurring recognition molecules which are specific for albumins, gamma globulins, and hemocyanins, respectively.

Earlier studies by Miller and his colleagues (1972) had demonstrated that the hemolymph of *P. clarkii* contained agglutinins in low titer to both erythrocytes and bacteria. These molecules displayed some degree of specificity since adsorption of the hemolymph with chicken erythrocytes did not remove any of the activity against rabbit erythrocytes. Heating at 60°C for 30 min partially inactivated the hemagglutinins. Fractionation of the hemolymph on Sephadex G-200 indicated that the chicken erythrocyte agglutinin was a macromolecule.

The discriminatory capability of the blue crab *Callinectes sapidus* has been studied by McCumber and Clem (1977), who investigated the clearance and subsequent organ distribution of viruses and xenogeneic proteins following injection into the infrabranchial sinus. Blue crab hemocyanin, injected as a control (self) protein, showed no significant clearance except for an initial period of equilibrium and hemocyanins from closely related crustaceans exhibited only slight to negligible clearance. In contrast, hemocyanins from more distantly related species (*Limulus* and the keyhole limpet) and other foreign proteins (BSA, BGG) were, with some differences in rates, rapidly removed from the circulation. Interestingly, poliovirus and bacteriophages T2 and T4 were rapidly cleared whereas phages T3, T7 and ØII remained at or near initial levels for much longer time periods. Analysis of various organs 1 hr after injection revealed that cleared proteins became concentrated in the gills (as was reported for the crayfish *P. clarkii* by Sloan *et al.*, 1975) where they were apparently degraded, whereas T2 and T4 phages were concentrated in the hepato-pancreas. Attempts to enhance the clearance rates of proteins and viruses by injecting crabs with 0.1–20 mg amounts of BSA and BGG or 10^8 plaque-forming units of T2 phage 2 weeks prior to challenge were unsuccessful.

In later work, McCumber *et al.* (1979) found that *C. sapidus* possessed a naturally occurring humoral factor which neutralises T2 bacteriophage in relatively specific fashion. The molecule responsible for this neutralization was isolated from crab plasma and shown to be distinct from hemocyanin and the agglutinin for mouse erythrocytes. The activity of the molecule, which appeared to be a 6–13 S polymer of noncovalently linked subunits each with a molecular weight of about 80,000, was lost upon exposure to denaturing solvents but was resistant to mild heat or EDTA. It remains to be shown whether it is this factor

TABLE 6. PROPERTIES OF AGGLUTININS DETECTED IN THE HEMOLYMPH OF VARIOUS ARTHOPODS

	Sensitivity to			Divalent cation requirement	Binding specificity
Species	Heat	Reduction	Protease(s)		
Homarus americanus					
Ag-1	L[a]	L	L	Ca^{2+}	AcNeu
Ag-2	L	L	L	Ca^{2+}	GalNAc
Panuliris argus	L	—	—	Ca^{2+}	—
Procambarus clarkii	L	—	R	—	—
Cherax destructor	L	L	L	Ca^{2+}	—
Callinectes sapidus	L	—	R	None	—
Androctonus australis	L	—	—	—	N-acetylamino-sugars
Limulus polyphemus	L	L	L	Ca^{2+}	AcNeu
Tachypleus tridentatus	L	—	—	Ca^{2+}	N-acetylamino-sugars

[a] L, labile; R, resistant; —, not known.

which mediates the clearance of T2 phage *in vivo*. Earlier work by Pauley (1974a,b) had shown that the hemolymph of this animal contained hemagglutinins which were labile to heat (60°C, 30 min) and appeared by chromatography on Sephadex G-200 to be macromolecules.

The properties of the hemagglutinins from these arthropods are summarized in Table 6.

3.2.2. Insecta

A considerable amount of work has been done on insect defense mechanisms which has been the subject of an excellent review by Whitcomb *et al.* (1974). However, as these reviewers state, the importance of opsonins in phagocytosis by insect hemocytes has yet to be demonstrated. Nevertheless, it is clear that these hemocytes play an important role in self/non-self recognition.

Scott (1971) demonstrated that the adherence of chicken and sheep erythrocytes to the surface of hemocytes of the cockroach (*Periplaneta americana*) is independent apparently of humoral factors. Adherence of the erythrocytes to the hemocytes was reduced considerably by pretreatment of the monolayer with 0.2% trypsin. Incubation of the hemocytes with hemolymph failed to restore their activity unlike the situation reported for the crayfish *Cherax destructor* (*Parachaeraps bicarinatus*, Tyson and Jenkin, 1974). This suggests that recognition of non-self is associated with receptors on the membrane of the phagocytic cells. Furthermore, these receptors, if free and of a cytophilic nature, were not present in the hemolymph to any significant level. Anderson *et al.* (1973) found that hemocytes from the cockroach *Blaberus craniifer* are capable of phagocytosing *in vitro* a number of different strains of bacteria independent of the presence of hemolymph.

Hemagglutinins have been reported in the hemolymph of the large milkweed bug *Oncopeltus fasciatus* and the small milkweed bug *Lygaues kalmii* (Feir and Waltz, 1964), the American cockroach *P. americana* (Wemyss, 1951; Scott, 1971), and the West Indian leaf cockroach *B. craniifer* (Anderson *et al.*, 1972). None of these agglutinins have been well characterized. Some preliminary work by Donlon and Wemyss (1976) indicates that the hemagglutinin is of a high molecular weight with a sedimentation coefficient > 21.4 S.

The possible role of these hemagglutinins and bacterial agglutinins in host defense is at present unknown. However, Amirante (1976) has shown that the hemocytes from *Leucophaca maderae* have hemagglutinins associated either with the cytoplasm or on the cell membrane.

3.2.3. Arachnida

The horseshoe crab *Limulus polyphemus*, a member of the Arachnida, has been in existence and has remained (at least morphologically) relatively unchanged over a period of 600 million years. A great deal of attention has been centered on this animal both from the point of view of its defense mechanisms against bacteria and also from the fact that the hemolymph contains hemagglutinins (Cohen, 1968). A decade or more ago, it was hoped that structural analysis of the hemagglutinins from this and other invertebrates, might throw some light on the evolution of the immunoglobulin molecule, since their agglutinating activity appeared to mimic one function of vertebrate antibody.

Early studies by Bang (1956) had shown that *Limulus* amebocytes degranulated in the presence of gram-negative bacteria. More recently, Pistole and Britko (1978) have investigated the bactericidal activity of these amebocytes and produced some, but not very convincing, evidence that the inactivation of a strain of *Escherichia coli* by amebocytes relied on a factor present in the hemolymph. Previously, Pistole (1976) had shown that the hemolymph contained agglutinating activity against *Salmonella minnesota* and more recently against other members of the Enterobacteriaceae and certain gram-positive strains of bacteria (Pistole, 1978). Unfortunately, however, as far as we are aware there is no convincing evidence which indicates that these agglutinins function as opsonins. Nevertheless, because evidence from other species suggests that they might function as such, it would be worth considering what is known about the structure of a hemagglutinin from *Limulus*.

This particular agglutinin, which agglutinates horse erythrocytes, was first purified and characterized by Marchalonis and Edelman in 1968. Initially, hemocyanin was separated from the hemolymph by ultracentrifugation and the clear supernatant was subjected to zone electrophoresis on starch gel. The material present in the fractions containing the hemagglutinating activity was then chromatographed on Sephadex G-200. Analysis of the purified agglutinin showed it to have a molecular weight of 400,000 and a sedimentation constant of 13.4 S. The molecule could be dissociated into polypeptide subunits of 67,000 molecular weight at pH 2.0 or pH 9.0. Dissociation into smaller subunits (molecular weight 22,500) occurred in either 20% acetic acid or 8 M urea. Thus, the hemagglutinin appears to be made up of six basic units each composed of three subunits.

Electron micrographs of the purified agglutinin showed molecules with a very uniform hexagonal ring shape structure (Fernández-Morán et al., 1968).

Finstad et al. (1972) also purified this agglutinin by ultracentrifugation at 100,000g and then by zone electrophoresis in agar. The agglutinating activity was Ca^{2+} dependent. The molecule had a sedimentation constant of 12.6 S and was composed of subunits with a molecular weight of 25,300 as determined by SDS-PAGE. Further work has revealed some minor discrepancies between various investigators regarding the molecular weight of this molecule and its subunits and its gross amino acid composition (Table 11). A summary of these data including that for hemagglutinins from other species of invertebrates is given in Table 12.

Roche and Monsigny (1974) purified the *Limulus* hemagglutinin (now termed limulin) by chromatography on Sephadex G-50, DEAE-Sephadex, and finally on Sepharose 6B columns. The molecular weight of limulin (335,000) calculated on the basis of diffusion and sedimentation coefficients was lower than that reported by Marchalonis and Edelman (1968), as was the molecular weight of the reduced subunits (19,000). The agglutination of horse erythrocytes by limulin was inhibited by AcNeu. Amino acid sequence analysis of limulin and of various cyanogen bromide fragments by Kaplan et al. (1977) revealed an absence of His and Ala from the N-terminal 50 residues. Five of the eight Pro residues present in the molecule are located between positions 13 and 30, and a possible carbohydrate attachment site consisting only of the amino acids Pro and Ser between residues 13 and 19. Roche and Monsigny (1974) found that the total sugar content of limulin was 3.6% (1.8% neutral sugars, 1.8% GlcN). Kaplan et al. (1977) were unable to find any sequence homologies with vertebrate immunoglobulins, which supports the conclusion of Marchalonis and Edelman (1968) that the structure of limulin is unrelated to that of vertebrate immunoglobulins.

Another arachnid that has been studied is the Japanese horseshoe crab *Tachypleus tridentatus* (Schimizu et al., 1977). The results of this study are of interest since they appear to indicate that this animal has at least four different hemagglutinins. The lectins were purified by affinity chromatography using bovine submaxillary mucin coupled to Sepharose. Initially the hemolymph was stirred with the conjugated Sepharose, resulting in almost complete adsorption of the agglutinating activity. The Sepharose was then washed before being poured into a column and eluted with 0.1 M NaCl in 0.1 M sodium borate (pH 8.5), then with 50 mM GlcNAc, and finally with 1.0 NaCl in 50 mM Tris·HCl at pH 7.0. Four peaks of hemagglutinating activity were separated by this procedure and designated M_1, M_2, M_3, and M_4. The M_3 peak was purified further on a GalNAc-Sepharose affinity column.

Electrophoresis of the lectins under nonreducing conditions indicated that M_1, M_3, M_4 were heterogeneous but M_2 gave only one sharp band. In the presence of SDS, M_2 and M_3 each gave one broad band with an estimated molecular weight of 20,000 and 23,000, respectively, whereas M_1 and M_4 gave several bands. All four fractions appeared to be antigenically unrelated by immunodiffusion. The titers of the four agglutinins against a panel of erythrocytes varied. For example, M_1 agglutinated pig erythrocytes at a dilution of 1 : 256

whereas M_3 was inactive. All the hemagglutinins appeared to be inactivated to varying degrees by glycoproteins containing sialic acid in their carbohydrate chains.

A further species of Arachnida that has been investigated, but not in great detail in relationship to the topic under discussion, is the scorpion, *Androctonus australis* (Brahmi and Cooper, 1974; Cohen *et al.*, 1979). The hemagglutinin(s) in the hemolymph of this animal to human erythrocytes is apparently inhibited by GalNAc, GlcNAc, and most markedly by AcNeu (Cohen *et al.*, 1979).

4. DEUTEROSTOMIA

4.1. ECHINODERMATA

The Echinodermata are represented by two subphyla, the Pelmatozoa (one class: Crinoidea) and the Eleutherozoa (four classes: Asteroidea, Echinoidea, Holothuroidea, Ophiuroidea). Information concerning the existence of humoral agglutinins in species of Eleutherozoa is rather limited, although representatives of all four classes have been examined. The data are summarized in Table 7. No similar work seems to have been done on the Pelmatozoa.

4.1.1. Class Asteroidea (Sea Stars)

The body fluids of two asteroid species, *Patiria miniata* and *Pisaster ochraceus*, were reported by Tyler (1946) to possess agglutinins for spermatozoa from a variety of invertebrate species. Adsorption tests revealed the presence of at least four distinct heteroagglutinins in *Patiria* body fluid. Much later, Brown *et al.* (1968), screening a variety of invertebrates for human erythrocyte agglutinins, noticed that the body fluid of both *Asterias forbesi* and *Oreaster reticulatus* contained "nonspecific" agglutinins for cells of all ABO-types. In a similar survey, Bretting and Renwrantz (1973) observed lysins but not agglutinins for A, B, and O erythrocytes in homogenates of *Asterina gibbosa* and *Marthasterias glacialis*. Two other species were also examined. Extracts of the gonads of *Astropecten aurantiacus* contained hemagglutinins, whereas that of the digestive tract exhibited both agglutinating as well as lytic activity. *Echinaster sepositus* mucus contained both agglutinins and lysins, whereas gonad extracts contained only agglutinins and digestive tract extracts only lysins. The properties of these factors were not examined in any of these studies.

Agglutinin(s) from the coelomic fluid of *Asterias forbesi* was studied in some detail by Finstad *et al.* (1972). These investigators purified the agglutinin(s) by electrophoresis in agar and by ultracentrifugation in sucrose density gradients. The purified protein(s) agglutinated erythrocytes from a number of different mammalian species, although cross-adsorption tests using horse, sheep, and rabbit erythrocytes indicated that distinct agglutinins for each cell type might have been present. Activity was lost when both purified and unpurified samples were heated (65–70°C). The addition of EDTA also caused inactivation and the

TABLE 7. PROPERTIES OF HEMAGGLUTININS AND HEMOLYSINS DETECTED IN THE BODY FLUIDS OF ECHINODERMS

Species	Activity	Sensitivity to						Divalent cation requirement	Binding site specificity
		Heat	Reduction	Periocate	Pepsin	Trypsin	Bromelain		
Asterias forbesi	Agglutinin	L[a]	—	—	—	—	—	Ca^{2+}	—
Anthocidaris crassispina	Lysin	L	L	—	—	L	—	Ca^{2+}	—[b]
	Agglutinin	L	L	L	—	R	L	Ca^{2+}	—
Pseudocentrotus depressus	Lysin	L	—	—	—	—	—	Ca^{2+}	—
	Agglutinin	L	L	L	—	R	L	Ca^{2+}	Gal, Lac, etc.
Hemicentrotus pulcherrimus	Lysin	L	—	—	—	—	—	Ca^{2+}	—
	Agglutinin	R	(L)	R	—	R	R	Ca^{2+}	—[b]
Holothuria polii	Agglutinin	L	L	—	L	—	—	None	—[c]
Holothuria tubulosa	Agglutinin	—	—	—	—	—	—	None	—[c]

[a] L, labile; R, resistant; — not known.
[b] Not inhibited by 0.05 M Gal, Glc, or other sugars tested.
[c] Not inhibited by 0.2 M Gal(NAc), Glc(NAc), AcNeu, Man, Fuc, Ara, Xyl, Fru, or sorbose.

subsequent addition of Ca^{2+} ions proved ineffective in restoring activity. The purified molecules appeared homogeneous in the analytical ultracentrifuge, having an apparent sedimentation coefficient of 6.5 S which would correspond to a molecular weight of 120,000–150,000, and formed a single precipitin line upon immunoelectrophoresis using antisera from rabbits immunized with whole coelomic fluid. However, N-terminal amino acid analysis revealed multiple residues (Asp, Glu, Ser, and Thr), leading the authors to suggest the existence of either distinct molecules, each with a different N-terminal residue, or a single agglutinin exhibiting N-terminal heterogeneity. Since the agglutinin(s) had not been purified by affinity chromatography, however, it is also possible that contaminants were present. The native molecules appeared to be comprised of noncovalently linked subunits, since SDS-PAGE analyses revealed two polypeptide bands, a major subunit with a molecular weight of approximately 30,000 and a minor component of approximately 13,000. The amino acid composition (Table 11) showed a preponderance of Glu, Asp, Thr, Gly, Ser, and Leu and a low content of Phe, His, Met, and Tyr. Intrachain disulfide bonds were probably present, as 5 moles of half-cystine were detected per 30,000 g. Unfortunately, the specificity of the agglutinin(s) was not investigated.

4.1.2. Class Echinoidea (Sea Urchins)

Two echinoid species, *Lytechinus pictus* and *Strongylocentrotus purpuratus*, were reported by Tyler (1946) to possess spermagglutinins in their body fluids. More recently, *L. pictus* was shown to clear T4 bacteriophage from its coelomic fluid, although the mechanism(s) responsible is not known (Coffaro, 1978). Agglutinins specific for human type O and B erythrocytes were detected in the body fluid of 5 out of 15 specimens of *S. drobachiensis*, whereas "non-ABO specific" hemagglutinins were present in the body fluid of *Arbacia punctulata*, *Echinometra lucunter*, and three out of six *Echinarachnius parma* (Brown *et al.*, 1968). Extracts of *L. variegatus* were found to be hemolytic. The presence of human erythrocyte agglutinins in *Echinus esculentus* and *Psammechinus miliaris* has also been reported (Uhlenbruck *et al.*, 1970).

A detailed study of echinoid humoral substances has been carried out by Ryoyama (1973, 1974). The dialized coelomic fluid of two species, *Anthocidaris crassispina* and *Pseudocentrotus depressus*, was hemolytic for rabbit, mouse, and human erythrocytes (titer 1 : 512, 1 : 4, and 1 : 16, respectively) but not for erythrocytes from other species such as the rat, guinea pig, sheep, and chicken (titer (1 : < 2). Coelomic fluid from a third echinoid, *Hemicentrotus pulcherrimus*, was lytic for rat and guinea pig erythrocytes (titer 1 : 16 and 1 : 32, respectively) in addition to those of rabbit, mouse, and man (titer 1 : 1024, 1 : 64, and 1 : 256, respectively). The hemolytic activity of each species was enhanced in the presence of Ca^{2+} but not Mg^{2+} ions and was reduced by EDTA, by heat (56°C, 30 min), and by treatment with trypsin or 2-mercaptoethanol (Table 7). These results, together with the finding that the hemolytic activity was associated with a single polypeptide band after electrophoresis of coelomic fluid on polyacrylamide gels, indicate that the active factor(s) was probably protein.

The coelomic fluid from these three echinoids also contained hemagglutinins (Ryoyama, 1974). All three species agglutinated rabbit and human erythrocytes (titer 1 : 64 to 1 : 512 and 1 : 2 to 1 : 32, respectively). However, differences in the agglutinating titer for other types of erythrocytes were evident. For example, *A. crassispina* contained agglutinins for dog erythrocytes (titer 1 : 64) but not mouse or guinea pig erythrocytes (titer 1 : < 2); *H. pulcherrimus* fluid agglutinated dog (titer 1 : 64) and guinea pig (1 : 32) but not mouse (1 : < 2) erythrocytes; *P. depressus* fluid agglutinated guinea pig (1 : 8) but not dog or mouse erythrocytes (1 : < 2). The results of cross-adsorption experiments indicated that the activity of *A. crassispina* fluid was probably due to a single agglutinin, that of *P. depressus* to at least two and *H. pulcherrimus* to three or more agglutinins. The molecular weight of all these molecules was apparently > 200,000 as estimated by chromatography on Sephadex G-200. The presence of Ca^{2+} ions was essential for the agglutination of rabbit erythrocytes by fluid from each species. EDTA caused reversible inactivation of the agglutinins. The *A. crassispina* and *P. depressus* agglutinins were inactivated by heat (70–85°C, 30 min), by exposure to 2-mercaptoethanol (0.4 M), and by incubation with bromelain. They were resistant to trypsin. Interestingly, the activity of each species was completely destroyed by 25 mM periodate. In contrast, the activity of the *H. pulcherrimus* fluid was affected neither by heating at 100°C for 30 min nor by incubation with protease (trypsin and bromelain). Furthermore, although mercaptoethanol effected a 50% reduction in activity, periodaae ($\leq$ 100 mM) caused no change. These results are summarized in Table 7.

The specificity of the rabbit erythrocyte agglutinins was investigated by testing the inhibitory capacity of about 40 mono- and oligosaccharides (Ryoyama, 1974). The activity of the coelomic fluid from *A. crassispina* and *H. pulcherrimus* was not affected by any of the tested sugars. However, nine sugars including Gal, L-Glc, Lac, melibiose, and raffinose inhibited the agglutination by *P. depressus*. Complete inhibition was effected by these sugars in the concentration range 10–50 mM. Human saliva was also examined for its ability to inhibit the agglutination of both rabbit and human erythrocytes. The agglutination of rabbit erythrocytes by *A. crassispina* fluid was completely inhibited by saliva, irrespective of ABO blood group, but the activity of *P. depressus* and *H. pulcherrimus* was unaffected. When human erythrocytes were used, however, saliva completely inhibited the agglutinating activity of the fluid from the three different species of sea urchin. The agglutination of dog erythrocytes by *A. crassispina* and of guinea pig erythrocytes by *P. depressus* was also reduced by human saliva, whereas the activity of *H. pulcherrimus* for both cell types was unaffected. These results provide tentative support for the hypothesis that a number of distinct agglutinins were present in these fluids.

Coelomic fluid of *Echinus esculentus* has been shown to kill a marine *Pseudomonas* sp. *in vitro* (Wardlaw and Unkles, 1978). The bactericidal activity was localized mainly in the coelomocytes, since removal of these cells substantially reduced the activity of the fluid and cell lysates exhibited a considerable activity. However, the data do not distinguish between the possibilities that the cells secreted the active substance(s) or that they phagocytosed the bacteria and killed them intracellularly. Johnson (1969) has in fact observed immobilization and phagocytosis of bacteria *in vitro* by coelomocytes from *Strongylocentrotus purpuratus* and *S. franciscanus*.

Previous studies by Hilgard and co-workers had shown that *S. purpuratus* coelomocytes could take up foreign proteins, labeled with diazotized [^{14}C]benzoate, both *in vivo* and *in vitro* (Hilgard *et al.*, 1967). Animals which had been injected 14 days beforehand with 100 mg of unlabeled bovine serum albumin (BSA) or 750 mg of bovine γ-globulin (BGG) exhibited a substantially reduced capacity to clear a small challenge dose (approx. 1 mg) of labeled protein. The inhibitory effect was greater for the homologous than for the heterologous protein. The number of coelomocytes in the coelomic fluid of seawater-pretreated and BSA-pretreated animals at the time of challenge was not significantly different. However, the protein concentration of cell-free coelomic fluid from the latter group was four times that of the controls and this fluid was shown by immunodiffusion to contain residual BSA. Since coelomocytes from seawater-, BSA-, and BGG-pretreated animals showed no difference in their ability to take up labeled BSA and BGG *in vitro*, the reduced *in vivo* clearance in pretreated animals was attributed to isotope dilution and to the differential "blockade" of receptors by the unlabeled proteins. Similar studies on the clearance and uptake of human and chicken serum albumins and of ^{14}C-labeled *S. purpuratus* coelomic fluid proteins (Hilgard and Phillips, 1968; Hilgard *et al.*, 1974) indicated that the uptake process has some specificity with respect to these different proteins and that foreign molecules are cleared from the coelomic fluid more rapidly than are native molecules. The nature of the presumptive receptors and whether they occur free in the coelomic fluid as well as on the cells have yet to be determined. The authors failed to elicit accelerated coelomocyte uptake or clearance of BSA by immunizing animals with various doses (0.01–25 mg) of BSA for different periods of time before testing.

4.1.3. Class Holothuroidea (Sea Cucumbers)

Agglutinins have been detected in the body fluids of a number of *Holothuria* and *Stichopus* species (families Holothuriidae and Stichopodidae, respectively). Spermagglutinins and hemagglutinins were found in extracts of *S. californicus* (Tyler, 1946) and bacterial agglutinins in *S. tremulus* (Johnson and Chapman, 1971). Brown *et al.* (1968) reported that extracts of *S. badionotus* and four *Holothuria* species were hemolytic. A number of different vertebrate erythrocytes were agglutinated by the fluid from *H. atra* and *H. leucospilata* (McKay *et al.*, 1969) and *H. polii* and *H. tubulosa* (Parrinello *et al.*, 1976). The latter authors also observed hemolytic activity against some erythrocyte types.

The only species to have been studied in any detail are *H. polii* and *H. tubulosa* (Parrinello *et al.*, 1976). Adsorption of *H. polii* coelomic fluid with rabbit erythrocytes completely removed the agglutinins for rabbit, calf, chicken, pig, and human erythrocytes but only slightly reduced the titer for horse, fish, and sheep cells. Similar adsorption of *H. tubulosa* coelomic fluid left the agglutinins for horse, rat, fish, and human B erythrocytes. On the other hand, adsorption of the fluid from both species with horse erythrocytes removed the agglutinins for horse, calf, and human erythrocytes but did not remove those for rabbit cells. These results demonstrate clearly that a number of agglutinins with different specificities occur in the coelomic fluid of both species. The specificity of the agglutinins from both species for rabbit, sheep, and human A and B erythrocytes was examined by testing the capacity of 0.2 M solutions of Glc, Gal, Man, L-Fuc, L-Ara, D-Xyl, GlcNAc, GalNAc, AcNeu, Fru, and L-sorbose to inhibit agglutination. None was inhibitory.

Some of the properties of the *H. polii* hemagglutinins were investigated by Parrinello *et al.* (1976). The agglutinins for rabbit, sheep, horse, calf, pig, and human A and B erythrocytes exhibited no requirement for divalent cations. Those for rabbit erythrocytes were stable at 45°C but became inactivated at temperatures > 75°C. The rabbit erythrocyte agglutinins were soluble in 25% but not in 50% saturated solutions of ammonium sulfate and were inactivated by pepsin at pH 2 and by reduction with 0.1 M 2-mercaptoethanol followed by alkylation with iodoacetamide.

Smith (1977) studied the reactions of *H. cinerascens* to serum from the milkfish (*Chanos chanos*). There was no evidence of an induced circulating precipitin, but a naturally occurring precipitin as well as a transferable, tissue-sensitizing substance were apparently detected in the coelomic fluid. No data were reported.

4.1.4. Class Ophiuroidea (Serpent/Brittle Stars)

As far as we are aware, the only study available on members of this group is that of Brown *et al.* (1968) who reported that extracts of *Ophiocoma echinata* were hemolytic.

4.2. PROTOCHORDATA

The protochordates are generally considered to be the survivors of the stem line of evolution giving rise directly to the vertebrates (Garstang, 1928; Berrill, 1955). As such, these animals may yield information on the evolutionary origins of the various components of vertebrate immunity. While it is clear that tunicates lack the ability shown by the vertebrates to produce inducible, highly specific antibodies upon challenge, they possess an efficient phagocytic system and can recognize and respond to a variety of foreign matter (Freeman, 1970; Anderson, 1971; Hildemann and Reddy, 1973; Tanaka and Watanabe, 1973; Wright and Cooper, 1975). Nothing is known about the molecular aspects of the recognition process in these animals. The blood of a number of tunicates has been shown to contain agglutinins for vertebrate erythrocytes, but the exact nature of these substances and their significance in the discrimination of self from non-self are at present obscure.

The protochordates are grouped within three subphyla (Hemichordata, Urochordata, and Cephalochordata), although the relationship of the Hemichor-

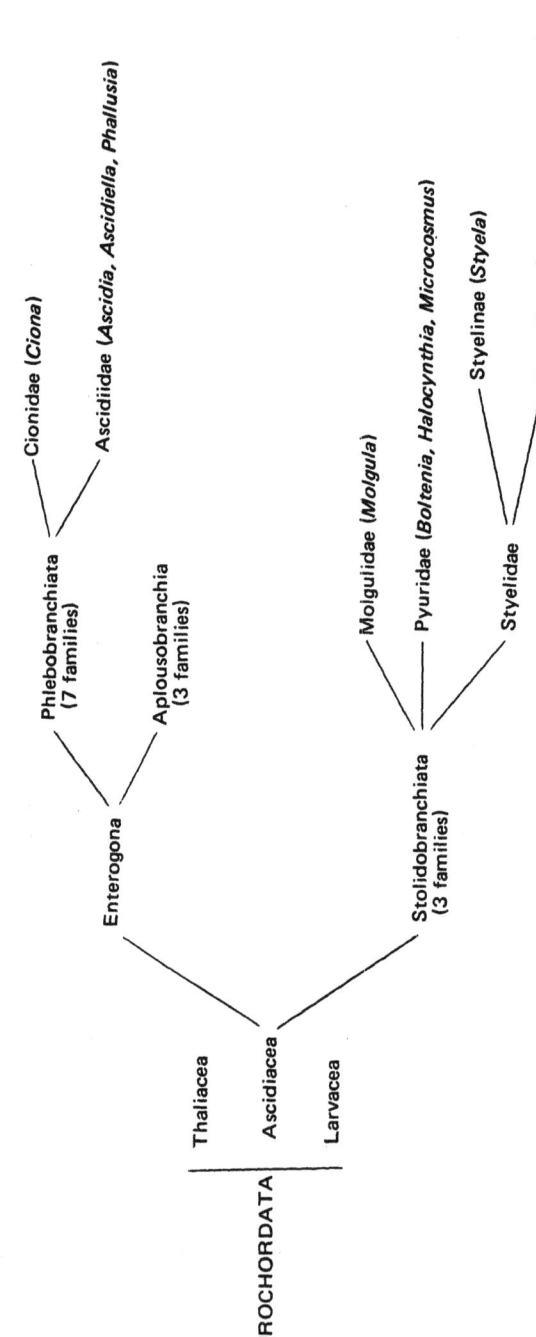

FIGURE 3. Proposed phylogenetic relationships between selected members of the Urochordata (after Berrill, 1950). Genera for which data are available are shown within parentheses.

data is controversial (Berrill, 1955). As far as we are aware, the natural defense systems of the hemichordates and cephalochordates have not been examined. However, extracts of the lancelot *Amphioxus* (*Branchiostoma*) *lanceolatus* have been reported to contain human blood-group A-like substances and hemagglutinins for human type A, B, and O erythrocytes (Bretting and Renwrantz, 1973; Renwrantz and Uhlenbruck, 1974b). Among the Urochordata (or Tunicata), only a few species of ascidian (class Ascidiacea) have been studied. The relationships between these organisms are shown in Figure 3.

4.2.1. Urochordata (Tunicata), Class Ascidiacae

The existence of agglutinins in the hemolymph of tunicates was first reported by Cantacuzène (1919a). One of these substances, present in extracts of *Phallusia mammillata* and active against sheep erythrocytes, was destroyed by heating to 55–60°C (Cantacuzène, 1923). Later, Tyler (1946) observed that body and seminal fluids of both *Ciona intestinalis* and *Styela barnharti* agglutinated erythrocytes from the amphibian *Bufo halophilus* and spermatozoa from a variety of invertebrates. These findings have been confirmed and extended by recent work.

4.2.1a. Family Cionidae. The hemolymph of the solitary ascidian *C. intestinalis* contains low titers (1 : < 8) of agglutinins for different avian and mammalian erythrocytes (Uhlenbruck *et al.*, 1970; Wright, 1974; Wright and Cooper, 1975; Anderson and Good, 1975; Parrinello and Patricolo, 1975). Wright (1974) and Wright and Cooper (1975) studied the tissue and humoral responses of *C. intestinalis* to human and duck erythrocytes injected into the tunic tissue or perivascular cavity. Following injection, large clumps of agglutinated erythrocytes were observed in the tunic tissue and small aggregates within the perivascular cavity. In each instance, the aggregates attracted and became infiltrated by phagocytic amebocytes. The rate of clearance of the erythrocytes was shown to be dose dependent, human cells being cleared more rapidly than equivalent numbers of duck cells. Erythrocytes were observed in many of the amebocytes present in samples of blood collected during the clearance phase (e.g., 12 hr after injection). The serum hemagglutinin titers were also measured. These were reduced in a dose-dependent fashion during the first 12–18 hr after injection, but returned to normal levels as the injected cells were cleared or encapsulated. No increase in titer was observed. These results show that the *C. intestinalis* hemagglutinin(s), like the bacterial agglutinin(s) in the gastropod *Aplysia californica* (Pauley *et al.*, 1971b), can be replaced within a short period. It is possible that the hemagglutinin(s) is opsonic and promotes the phagocytosis of the erythrocytes, but although the data are compatible with such a hypothesis, more evidence is required before such a function could be attributed with any confidence to these factors. The coelomic fluid of *C. intestinalis* has also been reported to contain bactericidins for several gram-negative marine bacteria (Johnson and Chapman, 1970).

There is virtually no information available concerning the properties of the hemagglutinin(s) of *Ciona* species. Wright and Cooper (1975) mentioned that the

TABLE 8. PROPERTIES OF HEMAGGLUTININS FROM VARIOUS ASCIDIANS

Species	Sensitivity to					Divalent cation requirement	Binding specificity
	Heat	Reduction	Periodate	Trypsin	Pepsin		
Enterogona							
Ciona intestinalis[1,2]	L[a]	L	R	—	—	$Ca^{2+} \pm Mg^{2+}$ (Ref. 1) None (Ref. 2)	—
Phallusia mammillata[2]	L	L	R	—	L	None	—
Ascidia malaca[2]	L	L	R	—	L	None	—
Stolidobranchiata							
Halocynthia pyriformis[3]	L	—	—	—	—	Ca^{2+}	AcNeu
Styela plicata[4]	R	—	L	R	—	None	—
Botrylloides leachii[5]							
HA-1	L	L	—	L	L	Ca^{2+}	Lac ≡ Gal
HA-2	L	R	—	R	—	None	Lac >> Gal

References: 1, Wright and Cooper (1975); 2, Parrinello and Patricolo (1975); 3, Anderson and Good (1975); 4, Fuke and Sugai (1972); 5, Schluter *et al.* (1981, and unpublished data).
[a]L, labile; R, resistant; —, not known.

responsible molecules are nondialyzable, heat labile (70°C), and require Ca^{2+} and Mg^{2+} ions for activity, but no data were provided. Similarly, Parrinello and Patricolo (1975) mentioned (without presenting data) that the agglutinating activity for rat and rabbit erythrocytes (titer 1 : ≤ 4) was destroyed when hemolymph was heated (75–100°C) or reduced (0.1 M 2-mercaptoethanol) and alkylated. The hemagglutinating activity was apparently not altered by periodate (which oxidizes 1,2-dihydroxy sugars), nor after incubation for 12 hr at room temperature in the pH range 2–10. In contrast with the results of Wright and Cooper (1975), however, Parrinello and Patricolo (1975) found that the activity was not altered when hemolymph was dialyzed against saline containing EDTA or various concentrations of Ca^{2+} or Mg^{2+} ions. From these observations, which are summarized in Table 8 together with information on agglutinins from other ascidians, it appears that the *C. intestinalis* agglutinins are proteins which may require Ca^{2+} ions for maximal activity. The homogeneity, binding site specificity, structure, and function of these molecules are unknown.

4.2.1b. Family Ascidiidae. Two members of this family, closely related to *Ciona* (Fig. 3), have been shown to possess agglutinins. Cantacuzène (1923) first reported the presence of heat-labile hemagglutinins in the hemolymph of *Phallusia mammillata*. This was later confirmed by Bretting and Renwrantz (1973), who found that hemolymph and homogenates of the whole animal agglutinated human A, B, and O erythrocytes. Blood-group A reactive substances are also present in these extracts (Renwrantz and Uhlenbruck, 1974b).

Parrinello and Patricolo (1975) investigated the *P. mammillata* hemagglutinins in some detail. These authors used hemolymph to agglutinate a variety of vertebrate erythrocytes, including those from amphibians, fish, mammals, and birds. Rabbit, sheep, human, rat, and horse cells gave the highest titers (range 1 : 8 to 1 : 256). When hemolymph was incubated with an equal volume of packed rabbit, sheep, or rat erythrocytes, the agglutinins for all three cell types were removed, suggesting that the various types of cells may have been agglutinated by a single agglutinin. The agglutinin is probably a protein, as it was said to be inactivated by heat (75–100°C), reduction (0.1 M 2-mercaptoethanol) and alkylation, and by incubation with pepsin at pH 2, but resistant to pH extremes (2–10) and periodate (Table 8). It did not require Ca^{2+} or Mg^{2+} ions for activity.

In the same study, Parrinello and Patricolo (1975) examined the hemolymph of *Ascidia malaca* and detected agglutinins for rabbit and rat erythrocytes (titers 1 : 8 to 1 : 32) as well as for human, calf, pig, and dolphin erythrocytes (titers 1 : < 4). Sheep, horse, and chicken erythrocytes were not agglutinated. The activity for rabbit and rat cells could be adsorbed by either cell type, indicating that a single agglutinin may have been involved. The proterties of this agglutinin (Table 8) suggest that it also is a protein.

With regard to the possible importance of these agglutinins in recognition and opsonization, it is noteworthy that Cantacuzène (1919b), who established infections in *A. mentula* by tunic injections of a gram-negative bacterium isolated from an opisthobranch gastropod, observed an increase in the number of bacteria adhering to amebocytes taken from the ascidian as the infection progressed. All the animals survived the infection and it is relevant to consider whether humoral or cell-associated agglutinins mediated the attachment of the bacteria to the amebocytes.

4.2.1c. Family Molgulidae. To our knowledge, there is only one published study which deals with immunity in members of Molgulidae. Anderson (1971) investigated the cellular reactions of *Molgula manhattensis* to foreign particles. The discriminatory capacity of these animals was evidenced by their ability to rapidly clear injected dyes by phagocytosis, to encapsulate glass fragments, and to infiltrate and reject grafted tissue. However, recognition mechanisms or agglutinins were not discussed.

4.2.1d. Family Pyuridae. Hemagglutinins have been detected in three species of *Halocynthia*, *H. hilgendorfi* (Fuke and Sugai, 1972), *H. papillosa* (Bretting and Renwrantz, 1973), and *H. pyriformis* (Anderson and Good, 1975; Form *et al.*, 1979). The coelomic fluid of *H. hilgendorfi* did not agglutinate sheep, guinea pig, or mouse erythrocytes, although very high titers were observed with rabbit and rat cells (1 : 4096 and 1 : 512, respectively). Fish, frog, and snake erythrocytes were not agglutinated. The specificity of the agglutinin(s) was investigated by adsorbing coelomic fluid with erythrocytes. The agglutinins for both rabbit and rat erythrocytes were depleted by either cell type, but not by mouse erythrocytes. It appeared therefore that the fluid contained a single agglutinin. In another study, Bretting and Renwrantz (1973), screening a variety of invertebrates for human erythrocyte agglutinins, observed agglutinins for type A, B, and O cells in the body fluid of *H. papillosa*. No characterization was attempted.

Anderson and Good (1975) found that *H. pyriformis* hemolymph agglutinated a wide range of avian and mammalian erythrocytes. Low titers (1 : 2 to 1 : 32) were observed using human, pigeon, rabbit, sheep, pig, goat, and calf erythrocytes, whereas higher titers (1 : 64 to 1 : 512) were recorded for duck, chicken, goose, turkey, guinea pig, and horse cells. The agglutinins were Ca^{2+} dependent, since EDTA abolished the activity for human type A and B and horse erythrocytes while the addition of Ca^{2+}, but not Mg^{2+}, increased the titers for all the cell types. The results of adsorption experiments indicated that a single agglutinin might have been responsible for the activity against all the cells. The agglutinin(s) became inactivated at temperatures > 48°C, could be precipitated in 30–50% saturated ammonium sulfate, and were insoluble in deionized water. They are therefore most probably proteins. No attempt was made to purify or characterize the molecules. However, the agglutination of both horse and human type A erythrocytes by *H. pyriformis* hemolymph was inhibited by AcNeu at a concentration of 0.25 mM. A number of other sugars (including Gal, Glc, Man, L-Fuc, L-Ara, D-Xyl, GalNAc, and GlcNAc) were noninhibitory at concentrations up to 10^{-4} M. The agglutinating titer for certain erythrocyte types was considerably enhanced if the cells were first treated with trypsin or Pronase, an effect also evident with human erythrocytes and snail agglutinins (e.g., Uhlenbruck *et al.*, 1972).

The purification of a hemagglutinating lectin from the hemolymph of *H. pyriformis* by affinity chromatography on insolubilized bovine submaxillary mucin has recently been reported by Form *et al.* (1979). Data from sucrose density gradient ultracentrifugation and SDS-PAGE indicate that the lectin has a native molecular weight of approximately 800,000 and is composed of subunits having a molecular weight of about 20,000. The binding site specificity was described as complex.

Agglutinins for sheep, goose, chicken, pigeon, turkey, horse, ox, and human erythrocytes were detected in very low titers (1 : 2 to 1 : 4) in the hemolymph of *Boltenia ovifera* (Anderson and Good, 1975). However, the hemolymph lacked a powerful agglutinin for either avian or mammalian erythrocytes (such as was found in *H. pyriformis*), and duck, goat, calf, and guinea pig erythrocytes were not agglutinated. Bretting and Renwrantz (1973) detected agglutinins for human erythrocytes in the hemolymph of *Microcosmus sulcatus*, but the titers were not indicated.

4.2.1e. Family Styelidae. This group is split into two subfamilies, the Styelinae (six genera; all solitary) and the Botryllinae (eight genera; colony-forming).

Two members of the Styelinae have been studied for agglutinins. The body fluid of *Styela barnharti* was reported by Tyler (1946) to contain agglutinins for erythrocytes from the amphibian *Bufo halophilus* and for spermatoxoa from various invertebrates. The titers were not indicated. The coelomic fluid of *S. plicata* has been shown to agglutinate rabbit erythrocytes to a titer of about 1 : 8000 (Fuke and Sugai, 1972). Mouse erythrocytes were also agglutinated, but at a much lower titer (1 : 16). Sheep, guinea pig, and rat erythrocytes were not agglutinated. A single agglutinin was believed to be responsible for the activity against rabbit and mouse cells, since the activity for both was adsorbed by either cell type but not by rat erythrocytes. The agglutinin was found to be a large macromolecule (nondialyzable; excluded by Sephadex G-100) which could be precipitated in 50% saturated ammonium sulfate. The agglutinating activity was resistant to heat (140°C), trypsin, and extreme pH (2–10) and was not altered by EDTA or Ca^{2+} or Mg^{2+} ions. However, the activity was destroyed by periodate oxidation. Because of its resistance to heat and inactivation by periodate, the authors concluded that the agglutinin was a polysaccharide or mucopolysaccharide. If this is true, the S. plicata agglutinin is most unusual. Considering the properties of the other ascidian hemagglutinins (Table 8) and of invertebrate and plant agglutinins and lectins in general, it is probable that the S. plicata agglutinin is in fact a protein or glycoprotein. Without further data, however, the character of this agglutinin must remain in doubt.

The ability of S. plicata body fluid to promote the phagocytosis of fixed rabbit erythrocytes by coelomic cells was also investigated by Fuke and Sugai (1972). Although no effect was observed, the results are inconclusive since a considerable proportion (50%) of the cells ingested erythrocytes in the absence of added coelomic fluid, indicating that any soluble factor(s) which might be involved in adherence and phagocytosis may have been associated with the phagocytes even after washing.

The Botryllinae have been of particular interest to investigators concerned with self/non-self recognition since, together with several other groups of colonial tunicates, they exhibit colony specificity. This phenomenon, first noted in Botryllus species by Bancroft in 1903, is characterized by the ability of particular colonies to fuse and form a common vasculatory system with some but not other colonies of the same species (Freeman, 1970; Tanaka and Watanabe, 1973). The nonfusion reactions of incompatible colonies may be analogous to transplantation reactions (Scofield et al., 1982).

Until quite recently, no information was available regarding the existence of humoral factors in members of the Botryllinae. Bretting and Renwrantz (1973) tested homogenates of colonies of Botryllus schlosseri for agglutinins to human erythrocytes, but observed no reaction. However, the hemolymph of a closely related ascidian, Botrylloides leachii, contains agglutinins for various avian and mammalian erythrocytes (Schluter et al., 1981). The titer for guinea pig cells (1 : 10,000 to 1 : 20,000) was considerably greater than that for other cells (1 : 16 to 1 : 128). A number of sugars were found to inhibit the agglutination, Lac being the most efficient inhibitor irrespective of cell type. Chromatography of hemolymph on Sephadex G-200 equilibrated with 0.1 M Gal and 0.05 M Lac resolved two peaks of activity. The agglutinins present in the first (HA-1) peak were specific for guinea pig erythrocytes and could be adsorbed by these but not by sheep or mouse cells. Ca^{2+} ions were essential for activity. In contrast, the second (HA-2) peak contained agglutinins for all the different types of erythrocytes, including those from the guinea pig. Divalent cations were not required for activity. The susceptibility of the HA-1 and HA-2 agglutinins to

TABLE 9. CAPACITY OF SUGARS TO INHIBIT THE ACTIVITY OF THE HEMAGGLUTININS FROM *BOTRYLLOIDES LEACHII* HEMOLYMPH[a]

	Concentration of sugar (mM) required to halve the hemagglutination titer					
	HA-1 agglutinin		HA-2 agglutinin			
Sugar	Guinea pig RBC	Pigeon RBC	Guinea pig RBC	Sheep RBC	Human RBC	Mouse RBC
Lactose	0.5	0.1	4	8	8	8
D-Galactose	1	2	> 67	> 67	> 67	> 67
Melibiose	0.5	4	> 67	> 67	> 67	> 67
L-Arabinose	4	> 67	> 67	> 67	> 67	> 67
D-Glucose	8	> 67	> 67	> 67	> 67	> 67
D-Xylose	16	> 67	> 67	> 67	> 67	> 67
L-Fucose	16	> 67	> 67	> 67	> 67	> 67
D-Arabinose	32	> 67	> 67	> 67	> 67	> 67
Maltose	67	> 67	> 67	> 67	> 67	> 67

[a]Data from Schluter *et al.* (1981).

inhibition with various sugars (Table 9) indicated that they had distinct binding site specificities. It was concluded on the basis of exhaustive adsorption tests that a single agglutinin accounted for the HA-2 activity.

The HA-1 and HA-2 agglutinins have been purified by affinity chromatography using Sepharose 4B to which Lac had been covalently attached (Schluter *et al.*, 1983). They were eluted with solutions of Gal and Lac, respectively. Each agglutinin was homogeneous with respect to size, as estimated by gel filtration and by velocity sedimentation in sucrose density gradients. It was calculated from the results of these experiments that the two molecules had a molecular weight of approximately 140,000 (HA-1) and 64,000 (HA-2).

Analysis of the HA-1 agglutinin by SDS-PAGE under nonreducing conditions revealed a single band (apparent molecular weight 138,000) which was detected in gels stained for protein (Coomassie blue) and carbohydrate (periodic acid–Schiff). The agglutinin is therefore a glycoprotein. Only one polypeptide band, having an apparent molecular weight of about 27,000, was detected under reducing conditions. The arrangement of the subunits in the asymmetrical penta- or hexameric native molecule (Stokes radius = 6.5 nm; molar frictional ratio $(f/f_0) = 1.73$) is currently under study.

In addition to the HA-1 agglutinin, an antigenically cross-reactive protein, which we have called Lactose Binding Protein 3 (LBP-3), has been purified from *B. leachii* hemolymph. LBP-3 is very similar to HA-1 in several respects—it has a similar native molecular weight (approximately 140,000), apparently identical binding site specificity (binding to guinea pig erythrocytes only and exhibiting similar sugar-inhibition characteristics) and a dependence on Ca^{2+} ions for its binding activity. Unlike HA-1, however, LBP-3 exhibits no capacity to agglutinate guinea pig erythrocytes. The two proteins copurify during affinity chromatography and chromatography on DEAE-cellulose at pH 8.2, but can be separated by chromatography on Sephacryl S-300 in the presence of a chelating agent such as EDTA. LBP-3 contains two different types of polypeptide chain subunits, S_1 and S_2 (molecular weights approximately 27,000 and 22,000, respectively), which are covalently linked within the native molecule. The S_1 subunit appears to be identical to the HA-1 subunit, a finding which explains the antigenic relationship and apparently identical binding site characteristics of the two proteins. Each LBP-3 molecule contains no more than two S_1 subunits interlinked with about four S_2 subunits. The inability of LBP-3 to act as an agglutinin may therefore be explained in terms of reduced valency (compared with HA-1) together with possible restrictions of conformational flexibility imposed by the S_2 subunits.

The HA-2 agglutinin is quite different from the HA-1 molecule. It is a globular protein (Stokes radius = 3.6 nm; f/f_0 = 1.35) and consists of two probably identical polypeptide chains (apparent molecular weight 32,000–33,000) which are held together by noncovalent bonds. The subunit is not stained by the periodic acid-Schiff reagent and therefore contains little or no covalently linked carbohydrate. These results, together with the differences in binding specificity and divalent cation requirement, and the finding that rabbit antisera raised against each agglutinin exhibit complete specifity for the immunising antigen, show conclusively that the two agglutinins are unrelated. The properties of the agglutinins are summarized in Tables 8 and 12 and their amino acid compositions are given in Table 11.

Studies aimed at elucidating the natural functions of these molecules have also been carried out. Using specific rabbit antisera, both agglutinins have been detected on the surface of many of the cells present in *B. leachii* hemolymph (Coombe, Ey, Schluter, and Jenkin, unpublished data), although it has yet to be determined whether cells expressing HA-1 determinants bear HA-1 and/or LBP-3. The molecules seem to be tightly bound to the cells, since extensive washing did not remove them.

When hemocytes were incubated in suspension with sheep erythrocytes, they were observed to bind and ingest significant numbers of erythrocytes. Lac had no obvious effect on the attachment and phagocytosis of these cells. However, binding was completely abolished when a polysaccharide fraction, obtained by phenol extraction of the Cohn fraction V of bovine serum, was included in the incubation medium. This polysaccharide preparation also inhibited the HA-2 but not the HA-1 agglutinin. These results are consistent with the hypothesis that the HA-2 molecules present on the hemocyte surface act as receptors for the attachment of erythrocytes and perhaps other particles, since affinity differences may have accounted for the lack of inhibition by Lac.

It has also been found that the HA-2 agglutinin mediates the attachment of sheep and mouse erythrocytes to mouse peritoneal macrophages *in vitro* (Coombe *et al.*, 1981, 1982). The binding, effected by presensitizing the erythrocytes or the macrophage monolayers with hemolymph or the purified agglutinin, seems to involve a simple heteroagglutination reaction since mac-

TABLE 10. AGGLUTINATION TITERS OF THE HEMOLYMPH COLLECTED FROM VARIOUS ASCIDIANS[a,b]

		Type of erythrocyte					
		Sheep		Mouse		Guinea pig	
Classification	Species	$Ca^{2+} + Mg^{2+}$	Citrate	$Ca^{2+} + Mg^{2+}$	Citrate	$Ca^{2+} + Mg^{2+}$	Citrate
Aplousobranchia							
Didemnidae	Didemnum patulum	16 (16)[c]	2	16	32 (32)	0	0
	Diplosoma sp.	8	8 (8)	0	64	2	0
Polyclinidae	Aplidium australiensis	32	32 (32)	128	64 (64)	128 (L=32)	128 (L=32)
Clavelinidae	Sycozoa tenuicaulis	0	0	0	0	0	0
	Atapozoa fantasiana	4	2	4	2	4	2
	Podoclavella cylindrica	0	0–2	0	0	0	0
Phlebobranchiata							
Ascidiidae	Phallusia depressiuscula	2	2	4	4	2	4
	Ascidia thompsoni	0–2	0	0	0	0	0
Cionidae	Ciona intestinalis	2	2	0	0	0	0
Stolidobranchiata							
Pyuridae	Microcosmus nichollsi	4–8	8 (L,M<4)	4	0	0	0
	Pyura praeputiculis	4	8	2	2	0	0
	Pyura irregularis	0	0	32 (all<4)	0	2	0
	Halocynthia hispida	16 (16)	4	8	8	512 (L,S=64) (G=128)	2

Species						
Styelidae (Botryllinae)						
Herdmania momus	2	2	2	2	0	0
Botrylloides leachii	64	64 (L=2)	64	64	12,800 (G,L,M<200) (A=1600; S=6400)	128 (L<4)
Botrylloides magnicoecus	0	2	0	0	4	4
Botryllus schlosseri	8	8 (G,L,M<4)	8	8 (G,L,M<4)	8	8 (G,L,M<4)
(Styelinae)						
Cnemidocarpa etheridgii	4	4	64 (64)	4	128 (128)	8 (8)
Polycarpa pedunculata	0	0	0	0	0	0
Polycarpa obtecta	0–2	0–2	2	2	0	0
Polycarpa papillata	2–4	2–4	256	128 (128)	32	32 (32)
(Polyzoinae)						
Stolonica australis	2	2	2	2	2	2

[a] Unpublished data of D. R. Coombe, P. L. Ey, and C. R. Jenkin.
[b] Twofold serial dilutions of each sample were mixed with an equal volume of a 0.5% erythrocyte suspension. The diluent was 0.13 M trisodium citrate, or Tris-buffered saline (pH 8) containing 5 mM CaCl$_2$ plus 5 mM MgCl$_2$. Reciprocal titers are shown.
[c] Titers using diluent containing D-arabinose (A), D-galactose (G), lactose (L), melibiose (M), or sucrose (S). Each sample was tested against every sugar, the final sugar concentration being in each case 67 mM. Inhibitory sugar(s) and titer(s) are indicated.

rophages could be agglutinated by the HA-2 agglutinin and the erythrocyte–macrophage binding was prevented by including Lac in the medium during presensitization of either cell type. The HA-1 agglutinin neither agglutinated macrophages nor promoted the attachment of guinea pig erythrocytes to macrophage monolayers. This phenomenon is artifactual, but it serves to indicate that HA-2 molecules have the potential to agglutinate cells or particles which carry the appropriate carbohydrate structures. If these or similar structures (cf. Renwrantz and Cheng, 1977a,b) occurred on the surface of ascidian blood cells, they would normally be saturated with HA-2 molecules since there is an excess of these molecules in the hemolymph. Because many of the HA-2 molecules would probably be bound monovalently, these could act as receptors for the cells and compete with free HA-2 molecules for carbohydrate structures which may enter the animal, e.g., on a bacterium. In other words, the agglutinin may act as a natural heteroagglutinin and enable phagocytes to bind and ingest particles which are recognized by the HA-2 molecule.

More recent work from our laboratory has given attention to the hemagglutinin levels in hemolymph from B. leachii following injections with sheep and chicken erythrocytes (Coombe, Ey, and Jenkin, 1982). The rationale for these experiments was to determine whether the hemagglutinating titer could be induced to rise, in a manner analagous to that observed in vertebrates, by injecting colonies with erythrocytes bearing the relevant agglutinin receptors. Hemolymph was collected from colonies of B. leachii killed at various times after injection with sheep or chicken erythrocytes and the agglutination titer for sheep, chicken and guinea pig erythrocytes was determined. Controls included groups of uninjected colonies and others injected with balanced salt solution. Animals given a single injection of erythrocytes exhibited no change in titer until the second week when a marginal (two- to four-fold) but statistically significant rise in titer was detected using each of the three types of indicator erythrocytes. A return to control values was observed by six weeks. The response to a second injection of erythrocytes, given six weeks after the first, appeared to be identical in magnitude and time of onset to the primary response, i.e., there was no evidence of an anamnestic response. At present we have no information about the duration of the response after the second injection, but the minor change in titer is markedly different from that observed in the secondary antibody response of higher vertebrates.

The effect of injecting sheep or chicken erythrocytes appears to be restricted to the HA-2 agglutinin, since the HA-1 titer exhibited no change. This finding, together with the fact that neither sheep nor chicken erythrocytes possess binding sites for the HA-1 agglutinin, suggests that the erythrocytes stimulated production of HA-2 specifically, perhaps to enhance elimination of the erythrocytes from the circulation, rather than evoking some nonspecific physiological change. Further experimentation will be required to determine whether the HA-2 agglutinin does indeed function as an opsonin in this species.

If the HA-2 (and/or HA-1) agglutinin does fulfill a receptor function for phagocytes in B. leachii, it seems appropriate to ask whether such agglutinins are common among members of the Ascidiacea. Because of the lack of detailed

TABLE 11. AMINO ACID COMPOSITION OF PURIFIED INVERTEBRATE HEMAGGLUTININS

Amino acid	Porifera						Mollusca						Arthropoda	Echinodermata	Protochordata	
	Axinella polypoides[1] Agglutinin		Aeptos papillata[2] Agglutinin			Gastropoda			Bivalvia						Botrylloides leachii[13]	
	I	II	I	II	III	Helix pomatia[3-5] anti-A_{HP}	Helix aspersa[4] anti-A_{HA}	Euhadra callizona amaliae[5] anti-A_{EC}	Crassostrea virginica[6]	Velesunio ambiguus[7]	Tridacna maxima[8]	Limulus polyphemus[9-12]	Asterias forbesi[10]	HA-1	HA-2
									moles per 100 moles of recovered amino acids							
Asp	16.2	18.1	10.7	12.4	11.1	12.6 ± 1.3[a]	10.9 ± 1.2[a]	15.7	17.6	14.6	12.2	10.3 ± 0.7[a]	12.4	11.8	12.5
Thr	7.0	6.1	6.3	7.3	7.6	5.8 ± 0.6	7.0 ± 0.4	10.1	6.2	6.5	7.0	6.2 ± 0.3	8.0	6.1	8.9
Ser	6.9	3.7	8.8	6.5	6.6	10.4 ± 0.7	7.9 ± 0.6	8.2	3.7	15.0	2.6	7.2 ± 0.6	7.3	7.8	5.9
Glu	12.8	13.1	9.5	12.3	12.4	9.3 ± 1.8	9.9 ± 1.1	9.3	9.8	9.2	12.3	12.7 ± 0.5	12.7	14.1	9.0
Pro	7.1	5.3	5.0	4.4	5.5	8.0 ± 2.7	9.8 ± 1.2	6.5	3.6	2.2	1.1	4.4 ± 0.5	5.5	1.8	5.2
Gly	8.4	8.0	9.3	7.4	6.6	5.2 ± 0.7	8.0 ± 0.6	6.1	9.2	9.9	12.1	8.6 ± 0.9	7.7	6.8	9.9
Ala	7.0	9.2	7.6	10.6	10.9	5.1 ± 1.1	5.6 ± 0.8	5.6	7.8	6.4	9.1	4.9 ± 0.3	5.5	6.7	5.9
Val	5.9	4.6	4.8	7.2	7.9	7.9 ± 0.8	8.0 ± 0.4	6.5	2.9	6.5	4.5	6.5 ± 1.1	6.8	3.5	8.3
Cys	1.1	0.8	0.0	8.5	7.6	2.0 ± 0.9	2.9 ± 0.3	0.0	5.1	tr	5.0	2.4 ± 0.5	4.8	3.9	0.7
Met	0.9	0.4	0.4	0.6	0.6	1.0 ± 0.5	0.7 ± 0.1	1.1	1.2	tr	<0.1	1.2 ± 0.1	1.8	1.9	1.6
Ile	3.6	2.4	4.6	1.5	1.6	6.0 ± 0.4	5.7 ± 0.8	7.3	4.1	1.1	2.4	4.9 ± 0.3	4.5	4.7	8.2
Leu	11.7	11.9	8.3	6.7	6.8	6.9 ± 1.1	6.3 ± 0.8	4.0	7.6	5.0	6.3	9.6 ± 0.5	6.8	9.7	4.0
Tyr	0.0	4.9	8.4	3.0	3.0	4.6 ± 0.1	3.2 ± 1.2	5.0	2.1	tr	3.3	2.3 ± 1.1	1.7	4.4	4.4
Phe	5.8	4.3	4.1	3.1	3.5	2.3 ± 0.2	4.2 ± 0.1	3.2	2.8	13.0	6.0	4.1 ± 0.3	2.6	4.6	6.9
Lys	3.2	3.1	2.4	3.0	2.7	4.6 ± 1.6	3.3 ± 1.0	5.9	0.0	4.3	8.3	6.4 ± 0.6	5.6	7.0	4.3
His	0.0	0.6	2.4	1.0	1.0	1.2 ± 0.1	tr[b]	0.8	10.9	2.5	5.8	5.1 ± 0.6	2.0	0.0	0.3
Arg	2.2	1.5	4.2	4.2	4.3	5.7 ± 0.9	5.5 ± 0.5	4.6	4.4	3.8	2.1	2.3 ± 0.2	4.3	4.9	4.1
Trp	ND	1.9	3.2	0.3	0.2	2.5 ± 2.8	1.3 ± 0.1	ND	1.2	ND	5.0	2.3	ND	ND	ND

References: 1, Bretting and Kabat (1976); 2, Bretting et al. (1976); 3, Hammarström and Kabat (1969); 4, Salfner et al. (1972); 5, Mukaida et al. (1974); 6, Acton et al. (1969); 7, Jenkin and Rowley (1970); 8, Baldo et al. (1978); 9, Marchalonis and Edelman (1968); 10, Finstad et al. (1972); 11, Roche and Monsigny (1974); 12, Kaplan et al. (1977); 13, Schluter et al. (1983).
[a] Mean values ± standard deviations, calculated using data from references 3–5 (H. pomatia), reference 4 (H. aspersa), and references 9–12 (L. polyphemus).
[b] tr, trace; ND, not determined.

TABLE 12. PROPERTIES OF AGGLUTININS PURIFIED FROM VARIOUS INVERTEBRATES

Species	Agglutinin	Source	Sedimentation coefficient (S) ($s_{20,w}$)	Molecular weight ($\times 10^{-3}$) Native molecule	Molecular weight ($\times 10^{-3}$) Subunit(s)	N-terminal amino acid residue	Isoelectric point (pI)	Carbohydrate content (%)	Structure	Agglutinating activity Divalent cation requirement	Agglutinating activity Binding site specificity
Axinella polypoides[1]	I	Extract	2.6		15 (& 67)		3.9	~0.5		None	β1 → 6-D-Gal
	II	Extract	2.8		15		3.9	~0.5			β1 → 6-D-Gal
Aaptos papillata[2]	I	Extract	3.5		12 & 21		4.7–7.6	0			GlcNAc[a] >> GalNAc
	II	Extract	6.0		16		3.8–5.3	0			GlcNAc[b] = AcNeu > GalNAc
	III	Extract	5.5		16		3.8–5.3	0			GlcNAc[b] = AcNeu > GalNAc
Helix pomatia[3,4]	Anti-A$_{HP}$	Albumin gland	5.3	73–79	13		6–8	~7	Hexamer	None	GalNAc > GlcNAc
Euhadra callizona[5]	Anti-A$_{EC}$	Albumin gland	5.3	89			3.6	5.1		None?	GalNAc = GlcNAc
Velesunio ambiguus[6]	—	Hemolymph	28								
Crassostrea virginica[7]	—	Hemolymph	33.4		20	Thr		9–13		$Ca^{2+} \pm Mg^{2+}$	GalNAc, GlcNAc
Tridacna maxima[8]	—	Hemolymph	15.0	470	10, 20, & 40		4.0–4.2	≥ 7		Ca^{2+}	β-D-GalNAc
Limulus polyphemus[9–12]	Limulin	Hemolymph	13.9	335–400	~20	Leu	5.1	3.6	Polymer	Ca^{2+}	AcNeu
Homarus americanus[13]	L Ag-1	Hemolymph	>19		55					Ca^{2+}	AcNeu
	L Ag-2	Hemolymph	~11.0		55					Ca^{2+}	GalNAc > GlcNAc
Panulirus argus[14]	—	Hemolymph	10.3	400	68.5	Phe		4.6	Hexamer	Ca^{2+}	
Cherax destructor[15]	—	Hemolymph		81	13.5	Asp, Glu, Ser, Thr			Hexamer	Ca^{2+}	
Asterias forbesi[10]	—	Hemolymph	6.5	140	30					Ca^{2+}	
Botrylloides leachii[16]	HA-1	Hemolymph	5.7	140	27		4–5	+	Pentamer	Ca^{2+}	Lac=Gal
	HA-2	Hemolymph	4.3	64	32		6–8	−	Dimer	None	Lac >> Gal
Halocynthia pyriformis[17]		Hemolymph		~800	~20						

References: 1, Bretting and Kabat (1976); 2, Bretting et al. (1976); 3, Hammarström (1972, 1974); 4, Kühnemund et al. (1974); 5, Mukaida et al. (1974); 6, Jenkin and Rowley (1970); 7, Acton et al. (1969, 1973); 8, Baldo et al. (1978); 9, Marchalonis and Edelman (1968); 10, Finstad et al. (1972, 1974); 11, Roche and Monsigny (1974); 12, Kaplan et al. (1977); 13, Hall and Rowlands (1974a,b); 14, Acton et al. (1973), Acton and Weinheimer (1974); 15, Tyson and Jenkin (unpublished data); 16, Schluter et al. (1981, 1983); 17, Form et al. (1979).

[a] N,N',N''-Tetraacetylchitotetraose was 2000 times more inhibitory than GlcNAc.
[b] N,N',N''-Triacetylchitotriose was 13 times more inhibitory than GlcNAc.

comparative data, we have tested the hemolymph of a variety of species (representing most of the taxonomic groups included with the Ascidiacea) for agglutinins against sheep, mouse, and guinea pig erythrocytes using diluents both supplemented with or devoid of Ca^{2+} and Mg^{2+} ions (Coombe, Ey, and Jenkin, unpublished data). Some of the results of this work are presented in Table 10.

Agglutinins were detected in most but not all of the species examined and considerable variations in titer were observed among the positive samples. For example, *Didemnum patulum* hemolymph agglutinated sheep and mouse but not guinea pig erythrocytes. Divalent cations were required by the sheep erythrocyte agglutinin(s) but not by those for mouse erythrocytes. In contrast, hemolymph from *Aplidium australiensis* agglutinated all three cell types, the highest titer being against guinea pig erythrocytes. None of the *Aplidium* agglutinins seemed to require divalent cations. The specificity of the hemagglutinins in some of the more active samples was also examined by testing the inhibitory effect of a number of sugars (Table 10). Marked differences in sensitivity were observed, some samples showing no inhibition with any sugar and others exhibiting partial or complete inhibition by one or more sugars. The high agglutinating activity of *B. leachii* hemolymph for guinea pig erythrocytes, which was due to the HA-1 agglutinin, did not appear in general to be shared by the other species tested. These results indicate that a variety of agglutinins, whose functions have still to be appreciated, occur in the hemolymph of many different ascidians.

5. DISCUSSION

By now the reader will be aware of the paucity of information relating to the molecular basis for the recognition of foreign material by phagocytic cells from invertebrates. A small number of papers dealing mainly with a few species from two major phyla, the Arthropoda and Mollusca, indicate that the ability of the phagocytes to discriminate between self and non-self is determined by factors which may be free in the hemolymph and/or associated with the membrane of the cell. These molecules could be synthesized by the phagocyte for use as integral, membrane-bound receptors. Alternatively, they could occur as opsonins, secreted by the phagocytes or other cells and becoming attached to phagocytes in a manner functionally analogous to the binding of vertebrate immunoglobulins to macrophage F_c receptors.

Opsonins have been demonstrated in the hemolymph of several species of mollusc (Tripp and Kent, 1968; Stuart, 1968; Prowse and Tait, 1969; Pauley *et al.*, 1971b; Anderson and Good, 1976; Renwrantz and Mohr, 1978; Sminia *et al.*, 1979) and the freshwater crayfish (McKay and Jenkin, 1970; Tyson and Jenkin, 1974). It is still an open question as to whether or not the cell–associated recognition factors are similar to those which may be free in the hemolymph, but two studies, one with the arthropod *Cherax destructor* and another with the protochordate *Botrylloides leachii*, suggest that they are similar (Tyson and Jenkin, 1974; Keough, Schluter, Ey, and Jenkin, unpublished observations). Additionally, phagocytes from the sea urchin *Strongylocentrotus purpuratus*, the chiton *Liolophura gaimardi*, and the crayfish *Procambarus clarkii*, have been shown to be capable of distinguishing various foreign proteins, presumably through the agency of specific recognition factors (Hilgard et al., 1974; Crichton and Lafferty, 1975; Sloan et al., 1975). Although few in number, these papers indicate that at least some invertebrates possess recognition factors with functional properties comparable to those of vertebrate immunoglobulins. Adsorption studies using a variety of particles (e.g., erythrocytes and bacteria) have shown that both the opsonins and hemagglutinins of these animals display a degree of specificity. However, with the exception of the crayfish, none of these opsonins or cell-bound recognition factors have been isolated or characterized to any extent.

Considerably more effort has been spent in the detection, isolation, and characterization of hemagglutinins from invertebrates. These molecules, at least in their ability to agglutinate erythrocytes, appear to behave like the naturally occurring hemagglutinins (specific antibodies) from vertebrates and in reviewing this field we have consequently felt it necessary to concentrate on the structure and physicochemical properties of the hemagglutinins. However, it should be emphasised that the evidence for the view that the hemagglutinins of invertebrates function as opsonins is at present purely circumstantial and is based on the observation that adsorption of the agglutinins from the hemolymph with erythrocytes also removes the opsonic properties of the hemolymph for that particular cell. To our knowledge, there is only one instance (*Helix pomatia*, c.f., Harm and Renwrantz, 1980; Renwrantz et al., 1981) in which a highly purified invertebrate hemagglutinin or bacteragglutinin has been tested for opsonic properties in the homologous species and it remains to be shown unambiguously whether this or any other invertebrate agglutinin has a natural role in phagocyte or self/non-self recognition. Indeed, Uhlenbruck and Steinhausen (1977) have suggested that the lectinlike system of invertebrates plays a more important role in the nutrition of the animal than in its defense against potential pathogens. Nevertheless, in the absence of detailed data concerning the function(s) of hemagglutinins in invertebrates, we believe that these molecules remain likely candidates for opsonins and that there is sufficiently strong justification for the detailed review of their occurrence, structure and physiochemical properties presented in the preceding pages. The excellent recent work of Renwrantz and his colleagues (Renwrantz et al., 1981) on the opsonic properties of the anti-A_{HP} agglutinin in *Helix pomatia* can only support this point of view.

For comparative purposes, the properties of all the invertebrate hemagglutinins which have been purified are summarized in Table 12. It is evident and to be expected, in view of the variety of sources, that there is a considerable diversity of structure between these different molecules. However most, if not all, are carbohydrate-specific and a surprising number require Ca^{2+} ions for maintenance of their activity. The amino acid compositions of some of these proteins are shown in Table 11.

An important point that has to be borne in mind is that the invertebrates are a polyphyletic group of animals and may in the course of evolution have solved the problems of recognition of foreignness at a molecular level in different ways, as they have evolved different respiratory pigments such as hemerythrins, hemocyanins, and chlorocruorins. There is clearly insufficient information presently available for any valid conclusions to be expressed on the similarity or otherwise of the phagocyte recognition systems of different invertebrate groups or on the nature of the recognition system within even a single group.

In contrast to the vertebrates, invertebrates appear to maintain a rather constant level of opsonins which is not significantly altered by exposure to foreign material. Although the range and degree of specificity of invertebrate recognition factors is not known, a number of opsonins, each with a particular specificity, have been shown by adsorption studies to exist in hemolymph from the snail *Helix aspersa* and the crayfish *Cherax destructor* (Prowse and Tait, 1969; Tyson and Jenkin, 1974). The hemagglutinins in the hemolymph of the oyster *Crassostrea virginica* appear to be equally heterogeneous (McDade and Tripp, 1967). Based on the apparent hexameric structure of the crayfish opsonins, Jenkin and Hardy (1975) proposed an hypothesis to account for the existence of opsonins differing in specificity. They suggested that six different subunits (A, B, C, D, E, F) may be synthesized by the same cell and combine in a random fashion to form a variety of hexameric molecules. Assuming that the combination of A with B engenders a different specificity than A with C, the random association of the six subunits could yield several thousand specificities. Since all the cells would be producing the same variable specificities, it would not be possible to select and expand a clone of cells producing a specific factor as occurs in the vertebrates. This hypothesis could explain why the titer of specific recognition factors does not vary greatly between individuals of the same species and is not altered by prior exposure to the specific "antigen." There is some evidence favoring this hypothesis (Jenkin and Hardy, 1975).

An ingenious model for the structure of invertebrate recognition factors which guarantees strict self/non-self discrimination has also been proposed by Parish (1977). The recognition factors are envisaged as being composed of some of the glycosyltransferases which the invertebrate uses to synthesize its own carbohydrate side chains. The random polymerization of these enzymes into a recognition factor would be initiated extracellularly by an additional protein having an acceptor site on the phagocyte surface. This protein subunit would impart cytophilic and opsonic properties to the recognition factors, but in order to prevent interaction between phagocytes only one subunit could be included in each polymer. This model is well suited to experimental testing and appears most attractive, particularly in view of the possible widespread importance of cell-bound glycosyltransferases in cellular recognition (Roth, 1973) and the prevalence of carbohydrate-specific agglutinins in invertebrate body fluids. Carbohydrate determinants occur on the surface of most organisms, especially bacteria and some viruses, and recognition factors possessing such specificities could be expected to react with many potential pathogens.

Both of the above-mentioned models postulate the random association of different subunits into polymeric recognition factors. An excellent example of such a system is provided by the hemocyanin of the crayfish *C. destructor* (Murray and Jeffrey, 1974; Jeffrey *et al.*, 1978). This protein is composed of three distinct subunits, M_1 and M_2 having similar molecular weights of about 75,000 and M'_3, a disulfide-bonded dimeric subunit having a molecular weight of 144,000. The hemocyanin exists predominantly as a 17 S hexamer and a 25 S "dodecamer" both of which occur in a number of electrophoretically distinct forms stabilized by Ca^{2+} ions. The hexamers and "dodecamers" are formed by the random association of the different subunits, the hexamers consisting of M_1 and M_2 subunits only. The 25 S component consists of a population of 11 compositional isomers which each contain one dimeric (M'_3) subunit and 10 monomeric subunits, the latter being present in all the combinations of M_1 and M_2. The dimer is essential for the formation of the 25 S molecules. There is therefore a precedent for systems such as those proposed by Jenkin and Hardy (1975) and Parish (1977).

ACKNOWLEDGMENTS. We would like to express our gratitude to the numerous authors who kindly forwarded reprints and preprints of their work for our perusal. We thank Deirdre Coombe and Sam Schluter for their willingness in providing unpublished data and discussing various aspects of this review. The early part of the work was done under the support of a grant from the Australian Research Grants Committee. P.L.E. is supported by a Research Fellowship from the National Health and Medical Research Council of Australia.

REFERENCES

Acton, R. T., and Weinheimer, P. F., 1974, Hemagglutinins: Primitive receptor molecules operative in invertebrate defense mechanisms, *Contemp. Top. Immunobiol.* **4**:271.

Acton, R. T., Bennett, J. C., Evans, E. E., and Schrohenloher, R. E., 1969, Physical and chemical characterization of an oyster hemagglutinin, *J. Biol. Chem.* **15**:4128.

Acton, R. T., Weinheimer, P. F., and Niedermeier, W., 1973, The carbohydrate composition of invertebrate hemagglutinin subunits isolated from the lobster *Panulirus argus* and the oyster *Crassostrea virginica, Comp. Biochem. Physiol. B* **44**:185.

Amirante, G. A., 1976, Production of heteroagglutinins in haemocytes of *Leucophaea maderae* L., *Experientia* **32**:526.

Anderson, R. S., 1971, Cellular responses to foreign bodies in the tunicate *Molgula manhattensis* (DeKay), *Biol. Bull.* **141**:91.

Anderson, R. S., and Good, R. A., 1975, Naturally-occurring hemagglutinin in a tunicate *Halocynthia pyriformis, Biol. Bull.* **148**:357.

Anderson, R. S., and Good, R. A., 1976, Opsonic involvement in phagocytosis by mollusk hemocytes, *J. Invertebr. Pathol.* **27**:57.

Anderson, R. S., Day, N. K. B., and Good, R. A., 1972, Specific haemagglutinin and a modulator of complement in cockroach haemolymph, *Infect. Immun.* **5**:55.

Anderson, R. S., Holmes, B., and Good, R. A., 1973, In vitro bactericidal capacity of *Blaberus craniifer* haemocytes, *J. Invertebr. Pathol.* **22**:127.

Arimoto, R., and Tripp, M. R., 1977, Characterization of a bacterial agglutinin in the hemolymph of the hard clam, *Mercenaria mercenaria, J. Invertebr. Pathol.* **30**:406.

Baldo, B. A., and Uhlenbruck, G., 1975a, Purification of tridacnin, a novel anti-β-(1,6)-digalactobiose precipitin from the haemolymph of *Tridacna maxima* (Röding), *FEBS Lett.* **55**:25.

Baldo, B. A., and Uhlenbruck, G., 1975b, Tridacnin, a potent anti-galactan precipitin from the haemolymph of *Tridacna maxima* (Röding), *Adv. Exp. Med. Biol.* **64**:3.

Baldo, B. A., Uhlenbruck, G., and Steinhausen, G., 1977, Anti-galactan agglutinins from the marine sponge *Axinella polypoides* (Schmidt), *Biol. Zentralbl.* **96**:723.

Baldo, B. A., Sawyer, W. H., Stick, R. V., and Uhlenbruck, G., 1978, Purification and characterization of a galactan-reactive agglutinin from the clam *Tridacna maxima* (Röding) and a study of its combining site, *Biochem. J.* **175**:467.

Bancroft, F. W., 1903, Variation and fusion of colonies in compound ascidians, *Proc. Calif. Acad. Sci. 3rd Ser.* **3**:137.

Bang, F. B., 1956, A bacterial disease of *Limulus polyphemus*, *Bull. Johns Hopkins Hosp.* **98**:325.

Bayne, C. J., 1974, On the immediate fate of bacteria in the land snail *Helix*, *Contemp. Top. Immunobiol.* **4**:37.

Bayne, C. J., and Kime, J. B., 1970, In vivo removal of bacteria from the hemolymph of the land snail *Helix pomatia* (Pulmonata: Stylommatophora), *Malacol. Rev.* **3**:103.

Bernheimer, A. W., 1952, Haemagglutinins in caterpillar blood, *Science* **115**:150.

Berrill, N. J., 1950, *The Tunicata*, The Ray Society, Bernard Quaritch Ltd., London.

Berrill, N. J., 1955, *The Origin of Vertebrates*, Oxford University Press, London.

Bhatia, H. M., Boyd, W. C., and Brown, R., 1967, Serological and immunochemical studies of snail (*Otala lactea*) anti-A: A simple purification method, *Transfusion* **7**:53.

Bhatia, H. M., Kim, Y. C., and Boyd, W. C., 1968, Serological and immunochemical studies on the snail (*Otala lactea*), *Vox Sang.* **14**:170.

Boyd, W. C., and Brown, R., 1965, A specific agglutinin in the snail *Otala lactea* (*Helix*), *Nature (London)* **208**:593.

Boyd, W. C., Brown, R., and Boyd, L. C., 1966, Agglutinins for human erythrocytes in molluscs, *J. Immunol.* **96**:301.

Boyden, S. V., 1966, Natural antibodies and the immune response, *Adv. Immunol.* **5**:1.

Brahmi, Z., and Cooper, E. L., 1974, Characterization of the agglutinin in the scorpion, *Androctonus australis*, *Contemp. Top. Immunobiol.* **4**:261.

Bretting, H., 1973, Serologische und immunelektrophoretische Untersuchungen über zwei Agglutinine aufgefunden in den Schwämmen *Aaptos papillata* (Keller) und *Axinella polypoides* (Schmidt), *Z. Immunitaetsforsch.* **146**:239.

Bretting, H., and Kabat, E. A., 1976, Purification and characterisation of the agglutinins from the sponge *Axinella polypoides* and a study of their combining sites, *Biochemistry* **15**:3228.

Bretting, H., and Renwrantz, L., 1973, Investigations of invertebrates of the Mediterranean Sea with regard to their contents of haemagglutinins, *Z. Immunitaetsforsch.* **145**:242.

Bretting, H., and Renwrantz, L., 1974, Further investigations of the sponge hemagglutinins from *Aaptos papillata* and *Axinella polypoides*, *Z. Immunitaetsforsch.* **147**:250.

Bretting, H., Kabat, E. A., Liao, J., and Pereira, M. E. A., 1976, Purification and characterisation of the agglutinins from the sponge *Aaptos papillata* and a study of their combining sites, *Biochemistry* **15**:5029.

Bretting, H., Kalthoff, H., and Fehr, S., 1978, Studies on the relationship between lectins from *Axinella polypoides* agglutinating bacteria and human erythrocytes, *J. Invertebr. Pathol.* **32**:151.

Brown, R., Almodovar, L. R., Bhatia, H. M., and Boyd, W. C., 1968, Blood group specific agglutinins in invertebrates, *J. Immunol.* **100**:214.

Cantacuzène, J., 1919a, Anticorps normaux et expérimentaux chez qualques invertebrés marins, *C.R. Soc. Biol.* **82**:1087.

Cantacuzène, J., 1919b, Etude d'une infection expérimentale chez *Ascidia mentula*, *C.R. Soc. Biol.* **82**:1019.

Cantacuzène J., 1923, in: *Celébration du 75ème anniversaire de la Fondation de la Société de Biologie*, p. 48, Masson, Paris.

Cheng, T. C., and Sanders, B. G., 1962, Internal defense mechanisms in molluscs and an electrophoretic analysis of a naturally occurring serum haemagglutinin in *Viviparus malleatus* Reeve, *Proc. Penn. Acad. Sci.* **36**:72.

Coffaro, K., 1978, Clearance of bacteriophage T4 in the sea urchin *Lytechinus pictus*, *J. Invertebr. Pathol.* **32**:384.

Cohen, E., 1968, Immunologic observations of the agglutinins of the haemolymph of *Limulus polyphemus* and *Birgus latro*, *Trans. N.Y. Acad. Sci.* **20**:427.

Cohen, E., Ilodi, G. H. U., Brahmi, Z., and Minowada, J., 1979, The nature of cellular agglutinins of *Androctonus australis* (Saharan scorpion) serum, *Dev. Comp. Immunol.* **3**:429.

Coombe, D. R., Ey, P. L., Schluter, S. F., and Jenkin, C. R., 1981, An agglutinin in the haemolymph of an ascidian promoting adhesion of sheep erythrocytes to mouse macrophages, *Immunology* **42**:661.

Coombe, D. R., Schluter, S. F., Ey, P. L., and Jenkin, C. R., 1982, Identification of the HA-2 agglutinin in the haemolymph of the ascidian *Botrylloides leachii* as the factor promoting adhesion of sheep erythrocytes to mouse macrophages, *Dev. Comp. Immunol.*, **6**:65.

Coombe, D. R., Ey, P. L., and Jenkin, C. R., 1982, Haemagglutinin levels in haemolymph from the colonial ascidian *Botrylloides leachii* following injection with sheep or chicken erythrocytes, *Aust. J. Exp. Med. Sci.* **60**: (in press).

Cooper, E. L., Lemni, C. A. E., and Moore, T. C., 1974, Agglutinins and cellular immunity in earthworms, *Ann. N.Y. Acad. Sci.* **234**:34.

Cornick, J. W., and Stewart, J. E., 1973, Partial characterisation of a natural agglutinin in the haemolymph of the lobster (*Homarus americanus*), *J. Invertebr. Pathol.* **21**:255.

Crichton, R., and Lafferty, K. J., 1975, The discriminatory capacity of phagocytic cells in the chiton (*Liolophura gaimardi*), *Adv. Exp. Med. Biol.* **64**:89.

Crichton, R., Killby, V. A. A., and Lafferty, K. J., 1973, The distribution and morphology of phagocytic cells in the chiton *Liolophura gaimardi*, *Aust. J. Exp. Biol. Med. Sci.* **51**:357.

Cushing, J. E., Calaprice, N. L., and Trump, G., 1963, Blood group reactive substances in some marine invertebrates, *Biol. Bull* **125**:69.

Dodd, R. Y., Maclennan, A. P., and Hawkins, D. C., 1968, Haemagglutinins from marine sponges, *Vox Sang.* **15**:386.

Donlon, W. C., and Wemyss, C. T., 1976, Analysis of the haemagglutinin and general protein element of the haemolymph of the West Indian leaf cockroach, *Blaberus craniifer*, *J. Invertebr. Pathol.* **28**:191.

Feir, D., and Waltz, M. A., 1964, An agglutinating factor in insect haemolymph, *Ann. Entomol. Soc. Am.* **57**:388.

Fernández-Morán, H., Marchalonis, J. J., and Edelman, G. M., 1968, Electron microscopy of a hemagglutinin from *Limulus polyphemus*, *J. Mol. Biol.* **32**:467.

Finstad, C. L., Litman, G. W., Finstad, J., and Good, R. A., 1972, The evolution of the immune response. XIII. The characterization of purified erythrocyte agglutinins from two invertebrate species, *J. Immunol.* **108**:1704.

Finstad, C. L., Good, R. A, and Litman, G. W., 1974, The erythrocyte agglutinin from *Limulus polyphemus* hemolymph: Molecular structure and biological function, *Ann. N.Y. Acad. Sci.* **234**:170.

Fischer, K., Poschmann, A., Reuther, K., and Prokop, O., 1972. Über das gleichzeitige Vorkommen von "Antigen" und "Antikörper" bei Schnecken: Autoantikörper oder Transportmechanismus?, *Immun-Information* **2**:20.

Form, D. M., Warr, G. W., and Marchalonis, J. J., 1979, Isolation and characterization of a lectin from the hemolymph of a tunicate, *Halocynthia pyriformis*, *Fed. Proc.* **38**:934.

Freeman, G., 1970, Transplantation specificity in echinoderms and lower chordates, *Transplant. Proc.* **2**:236.

Fuke, M. T., and Sugai, T., 1972, Studies on the naturally occurring hemagglutinin in the coelomic fluid of an ascidian, *Biol. Bull.* **143**:140.

Garstang, W., 1928, The morphology of the Tunicata, and its bearings on the phylogeny of the Chordata, *Q. J. Microsc. Sci.* **72**:51.

Gibb, B., Zahn, I., and Scheibe, E., 1967, Immunhämatologische Untersuchungen an Teichmuscheln (Anodonta). I. Blutgruppenaktive Substanzen und Agglutinine, *Z. Immunitaetsforsch.* **133**:385.

Gilbertson, D. E., and Etges, F. J., 1967, Haemagglutinins in the haemolymph of planorbid snails, *Ann. Trop. Med. Parasitol.* **61**:144.

Gold, E. R., and Balding, P., 1975, *Receptor-Specific Proteins: Plant and Animal Lectins*, Excerpta Medica, Amsterdam.

Gold, E. R., and Thompson, T. E., 1969a, Serological differences between related species of snails. I. Revealed by reverse passive agglutination tests, *Vox Sang.* **16**:63.

Gold, E. R., and Thompson, T. E., 1969b, Serological differences between related species of snails. II. Revealed by haemolysis tests, agglutination tests with tumour cells and content of B-like substances, *Vox Sang.* **16**:119.

Gold, E., Phelps, C. F., Khalap, S., and Balding, P., 1974, Observations on *Axinella* sp. haemagglutinins, *Ann. N.Y. Acad. Sci.* **234**:122.

Goldstein, I. J., and Hayes, C. E., 1978, The lectins: Carbohydrate-binding proteins of plants and animals, *Adv. Carbohyd. Chem. Biochem.* **35**:127.

Grace, H. J., 1969, A potent snail haemagglutinin with anti-A specificity, *J. Forensic Med.* **16**:100.

Habets, L., Vieth, U. C., and Hermann, G., 1979, Isolation and new biological properties of *Arion empiricorum* lectin, *Biochim. Biophys. Acta* **582**:154.

Haferland, W., Kim, Z., Uhlenbruck, G., and Nelson, D. S., 1967, Zur Frage der Einheitlichkeit von anti-A_{hel}. *Z. Immunitaetsforsch.* **132**:93.

Hall, J. L., and Rowlands, D. T., 1974a, Heterogeneity of lobster agglutinins. I. Purification and physiochemical characterisation, *Biochemistry* **13**:821.

Hall, J. L., and Rowlands, D. T., 1974b, Heterogeneity of lobster agglutinins. II. Specificity of agglutinin-erythrocyte binding, *Biochemistry* **13**:828.

Hammarström, S., 1972, Purification and properties of *Helix pomatia* A hemagglutinin, *Methods Enzymol.* **28B**:368.

Hammarström, S., 1973, Binding of *Helix pomatia* A hemagglutinin to human erythrocytes and their cells: Influence of multivalent interaction on affinity, *Scand. J. Immunol.* **2**:53.

Hammarström, S., 1974, Structure, specificity, binding properties, and some biological activities of a blood group A-reactive hemagglutinin from the snail *Helix pomatia*, *Ann. N.Y. Acad. Sci.* **234**:183.

Hammarström, S., and Kabat, E. A., 1969, Purification and characterization of a blood-group A reactive hemagglutinin from the snail *Helix pomatia* and a study of its combining site, *Biochemistry* **8**:2696.

Hammarström, S., and Kabat, E. A., 1971, Studies on specificity and binding properties of the blood group A reactive hemagglutinin from *Helix pomatia*, *Biochemistry* **9**:1684.

Hammarström, S., Lindberg, A. A., and Robertsson, E. S., 1972a, Precipitation of lipopolysaccharides from rough mutants of *Salmonella typhimurium* by an A hemagglutinin from *Helix pomatia*, *Eur. J. Biochem.* **25**:274.

Hammarström, S., Westöö, A., and Björk, I., 1972b, Subunit structure of *Helix pomatia* A hemagglutinin, *Scand. J. Immunol.* **1**:295.

Harm, H., and Renwrantz, L., 1980, The inhibition of serum opsonins by a carbohydrate and the opsonizing effect of purified agglutinin on the clearance of nonself particles from the circulation of *Helix pomatia*, *J. Invertebr. Pathol.* **36**:64.

Hartman, A. L., Campbell, P. A., and Abel, C. A., 1978, An improved method for the isolation of lobster lectins, *Dev. Comp. Immunol.* **2**:617.

Hildemann, W. H., and Reddy, A. L., 1973, Phylogeny of immune responsiveness: Marine invertebrates, *Fed. Proc.* **12**:2188.

Hilgard, H. R., and Phillips, J. H., 1968, Sea urchin response to foreign substances, *Science* **161**:1243.

Hilgard, H. R., Hinds, W. E., and Phillips, J. H., 1967, The specificity of uptake of foreign proteins by coelomocytes of the purple sea urchin, *Comp. Biochem. Physiol.* **23**:815.

Hilgard, H. R., Wander, R. H., and Hinds, W. E., 1974, Specific receptors in relation to the evolution of immunity, *Contemp. Top. Immunobiol.* **4**:151.

Ishiyama, I., and Takatsu, A., 1970, Anti-A haemagglutinin from the garden snail *Euhadra periomphala*: Inhibition by N-acetyl-D-galactosamine and N-acetyl-D-glucosamine, *Vox Sang.* **19**:522.

Ishiyama, I., and Uhlenbruck, G., 1971, On the nature of the anti-dextran activity of the *Helix pomatia* "anti-A" agglutinin, *Z. Naturforsch. Teil B* **26**:1198.

Ishiyama, I., and Uhlenbruck, G., 1972, Some problems concerning the adsorption mechanism of anti-A agglutinins from *Helix pomatia* onto Sephadex G-200, *Z. Immunitaetsforsch.* **143**:147.

Ishiyama, I., Takatsu, A., Uhlenbruck, G., Reifenberg, U., Schnitzler, S., and Prokop, O., 1971, Serological behaviour of an "incomplete and superagglutinating" anti-A from the snail *Helix pomatia*, *Z. Naturforsch. Teil B* **26**:171.

Ishiyama, I., Uhlenbruck, G., and Hermann, G., 1972, Isolation of an anti-A agglutinin from *Helix aspera*, *Blut* **24**:178.

Ishiyama, I., Dietz, W., and Uhlenbruck, G., 1973, Comparative studies of anti-A agglutinins from various snails of the genus *Helix* (*Helix pomatia* and *Helix aspersa*), *Comp. Biochem. Physiol. B* **44**:529.

Ishiyama, I., Mukaida, M., and Takatsu, A., 1974, Hemagglutinins and enzyme inhibitions: Comparative studies on the reactivity of anti-A agglutinins of *Helix pomatia* and *Euhadra callizona amaliae*, *Ann. N.Y. Acad. Sci.* **234**:75.

Jeffrey, P. D., Shaw, D. C., and Treacy, G. B., 1978, Hemocyanin from the Australian freshwater crayfish *Cherax destructor*: Characterization of a dimeric subunit and its involvement in the formation of the 25 S component, *Biochemistry* **17**:3078.

Jenkin, C. R., 1976, Factors involved in the recognition of foreign material by phagocytic cells from invertebrates, in: *Comparative Immunology* (J. J. Marchalonis, ed.), pp. 80–94, Blackwell, Oxford.

Jenkin, C. R., and Hardy, D., 1975, Recognition factors of the crayfish and the generation of diversity. *Adv. Exp. Med. Biol.* **64**:55.

Jenkin, C. R., and Rowley, D., 1970, Immunity in invertebrates. The purification of a haemagglutinin to rat and rabbit erythrocytes from the haemolymph of the Murray mussel (*Velesunio ambiguus*), *Aust. J. Exp. Biol. Med. Sci.* **48**:129.

Johnson, H. M., 1964, Human blood group A_1 specific agglutinin of the butter clam *Saxidomus giganteus*, *Science* **146**:548.

Johnson, P. T., 1969, The coelomic elements of sea urchins (*Strongylocentrotus*). III. *In vitro* reaction to bacteria, *J. Invertebr. Pathol.* **13**:42.

Johnson, P. T., and Chapman, F. A., 1970, Comparative studies on the *in vitro* response of bacteria to invertebrate body fluids. II. *Aplysia californica* (sea hare) and *Ciona intestinalis* (Tunicate), *J. Invertebr. Pathol.* **16**:259.

Johnson, P. T., and Chapman, F. A., 1971, Comparative studies on the *in vitro* response of bacteria to invertebrate body fluids. III. *Stichopus tremulus* (sea cucumber) and *Dendraster excentricus* (sand dollar), *J. Invertebr. Pathol.* **17**:94.

Kaplan, R., Li, S. S. L., and Kehoe, J. H., 1977, Molecular characterisation of limulin, a sialic acid binding lectin from the haemolymph of the horseshoe crab, *Limulus polyphemus*, *Biochemistry* **16**:4297.

Khalap, S., Thompson, T. E., and Gold, E. R., 1970, Haemagglutination and haemagglutination inhibition reactions of extracts from snails and sponges. I, *Vox Sang.* **18**:501.

Khalap, S., Thompson, T. E., and Gold, E. R., 1971, Haemagglutination and haemagglutination inhibition reactions of extracts from snails and sponges. II. *Vox Sang.* **20**:1150.

Khalap, S., Phelps, C. F., Fudenberg, H. H., and Gold, E. R., 1972, Separation of haemagglutinins from haemolysins in extracts of the albumin gland of *Helix aspersa*, *Vox Sang.* **23**:218.

Kilias, R., Schnitzler, S., Kothbauer, H., Stober, D., and Prokop, O., 1972, Further investigations on haemagglutinins in pulmonate snails (haemagglutinin tests on four more species, comparative biotope studies with previously investigated species, and tests on organs of *Helix pomatia* Linnaeus), *Z. Immunitaetsforsch.* **144**:157.

Kim, S., Uhlenbruck, G., Prokop, O., and Schlesinger, D., 1966, Über die B Substanz und das anti-A von *Helix pomatia*, *Z. Immunitaetsforsch.* **130**:290.

Knobloch, W., Knobloch, I., Vogt, W. E., Schnitzler, S., and Böttger, M., 1970, Untersuchungen über die Isolierung und Reindarstellung eines Agglutinins aus der Eiweissdrüse der *Helix pomatia*, *Z. Immunitaetsforsch.* **139**:119.

Köhler, W., and Prokop, O., 1967, Agglutination von Streptokokken der Gruppe C durch ein Agglutinin aus *Helix pomatia*, *Z. Immunitaetsforsch.* **133**:50.

Kothbauer, H., 1970, Die Bedeutung von anti-A_{HP}, einem Agglutinin aus der Eiweissdrüse der Weinbergschnecke (*Helix pomatia*), *Oecologie (Berlin)* **6**:48.

Kothbauer, H., 1972, Grösse und Anti-A_{HP}-Gehalt der Eiweissdrusen von Weinbergschnecken (*Helix pomatia* L.) zu verschiedenen Jahreszeiten, *Acta Biol. Med. Ger.* **28**:845.

Kothbauer, H., and Schenkel-Brunner, H., 1971, Hämagglutinine aus Schnecken: Zur Frage ihrer biologischen Funktion, *Naturwissenschaften* **26b**:1082.

Kothbauer, H., Nopp, H., and Schenkel-Brunner, H., 1972, Haemagglutinine aus Schnecken: Auswirkung der Amputation der Augententakel: Einfluss des Entwicklungszustandes des Genitaltraktes, *Immun-Information* **6**:2.

Kühnemund, O., and Köhler, W., 1969, Untersuchungen über die Reinigung des Protektins anti-A_{hel} (Anti-A_{HP}) aus *Helix pomatia, Experientia* **25**:1137.

Kühnemund, O., and Köhler, W., 1975, Gelelectrofocusing of protectin anti-A_{HP} from the albumin gland of *Helix pomatia, Z. Immunitaetsforsch.* **149**:94.

Kühnemund, O., Köhler, W., and Prokop, O., 1972, Investigations about an alleged lysozyme-like behaviour of the protectin anti-A_{HP} from the albumin gland of the edible snail (*Helix pomatia*), *Z. Immunitaetsforsch.* **144**:344.

Kühnemund, O., Strassburger, J., and Triebel, H., 1974, Molecular weight determination on the protectin anti-A_{HP} from the albumin gland of *Helix pomatia, Z. Immunitaetsforsch.* **147**:127.

Li, M. F., and Flemming, C., 1967, Hemagglutinins from oyster hemolymph, *Can. J. Zool.* **45**:1225.

Marchalonis, J. J., and Edelman, G. M., 1968, Isolation and characterisation of a haemagglutinin from *Limulus polyphemus, J. Mol. Biol.* **32**:453.

McCumber, L. J., and Clem, L. W., 1977, Recognition of viruses and xenogeneic proteins by the blue crab, *Callinectes sapidus.* I. Clearance and organ distribution, *Develop. Compar. Immunol.* **1**:5.

McCumber, L. J., Hoffmann, E. M., and Clem, L. W., 1979, Recognition of viruses and xenogeneic proteins by the blue crab *Callinectes sapidus:* A humoral receptor for T2 bacteriophage, *J. Invertebr. Pathol.* **33**:1.

McDade, J. E., and Tripp, M. R., 1967, Mechanism of agglutination of red blood cells by oyster hemolymph, *J. Invertebr. Pathol.* **9**:523.

McKay, D., and Jenkin, C. R., 1970, Immunity in the invertebrates: The role of serum factors in the phagocytosis of erythrocytes by haemocytes of the fresh water crayfish (*Parachaeraps bicarinatus*), *Aust. J. Exp. Biol. Med. Sci.* **48**:139.

McKay, D., Jenkin, C. R., and Rowley, D., 1969, Immunity in the invertebrates. I. Studies on the naturally occurring haemagglutinins in the fluid from invertebrates, *Aust. J. Exp. Biol. Med. Sci.* **47**:125.

Michelson, E. H., and Dubois, L., 1977, Agglutinins and lysins in the molluscan family Planorbidae: A survey of hemolymph, egg-masses, and albumen-gland extracts, *Biol. Bull.* **153**:219.

Miescher, P. A., Spiegelberg, H., and Benacerraf, B., 1963, Studies on the mechanism of immune phagocytosis of sensitised bacteria and red cells by the reticulo-endothelial system in mice, in: *Role du Système Réticulo-Endothélial dans l'Immunité Antibactérienne et Antitumorale* (B. N. Halpern, ed.), pp. 463–475, CNRS, Paris.

Miller, V. H., Ballback, R. S., Pauley, G. B., and Krassner, S. M., 1972, A preliminary physicochemical characterisation of an agglutinin found in the haemolymph of the crayfish, *Procambarus clarkii, J. Invertebr. Pathol.* **19**:83.

Mukaida, M., Takatsu, A., and Ishiyama, I., 1974, Purification and characterization of anti-A agglutinin from *Euhadra callizona amaliae, Vox Sang.* **27**:347.

Murray, A. C., and Jeffrey, P. D., 1974, Hemocyanin from the Australian freshwater crayfish *Cherax destructor:* Subunit heterogeneity, *Biochemistry* **13**:3667.

Oehme, P., Schnitzler, S., and Vogt, W. E., 1968, Untersuchungen zur Antikörpercharakteristik der Helixagglutinine, *Z. Immunitaetsforsch.* **136**:421.

Pardoe, G. I., Uhlenbruck, G., and Bird, G. W. G., 1970, Studies on some heterophile receptors of the Burkitt EB2 lymphoma cell, *Immunology* **18**:73.

Parish, C. R., 1977, Simple model for self–non-self discrimination, *Nature (London)* **267**:711.

Parrinello, N., and Patricolo, E., 1975, Erythrocyte agglutinins in the blood of certain ascidians, *Experientia* **31**:1092.

Parrinello, N., Canicattì, C., and Rindone, D., 1976, Naturally-occurring haemagglutinins in the coelomic fluid of the echinoderms, *Holothuria polii* Delle Chiaje and *Holothuria tubulosa* Gmelin, *Boll. Zool.* **43**:259.

Paterson, W. D., and Stewart, J. E., 1974, *In vitro* phagocytosis by haemocytes of the American lobster, *Homarus americanus, J. Fish. Res. Board Can.* **31**:1051.

Pauley, G. B., 1974a, Physicochemical properties of the natural agglutinins of some mollusks and crustaceans, *Ann. N.Y. Acad. Sci.* **234**:145.

Pauley, G. B., 1974b, Comparison of a natural agglutinin in the haemolymph of the blue crab, *Callinectes sapidus,* with agglutinins of other invertebrates, *Contemp. Top. Immunobiol.* **4**:241.

Pauley, G. B., Granger, G. A., and Krassner, S. M., 1971a, Characterization of a natural agglutinin present in the hemolymph of the California sea hare, *Aplysia californica*, *J. Invertebr. Pathol.* **18**:207.

Pauley, G. B., Krassner, S. M., and Chapman, F. A., 1971b, Bacterial clearance in the California sea hare, *Aplysia californica*, *J. Invertebr. Pathol.* **18**:227.

Pemberton, R. T., 1969, Studies on the human red cell agglutinins of the swan mussel (*Anodonta cygnea*), *Vox Sang.* **16**:457.

Pemberton, R. T., 1970, Blood group A reactive substance in the common limpet (*Patella vulgata*), *Vox Sang.* **18**:71.

Pemberton, R. T., 1971a, Observations on a haemagglutinin from the freshwater mussel *Anodonta anatina*, *Vox Sang.* **21**:159.

Pemberton, R. T., 1971b, Haemagglutinins from some British non-marine Mollusca, *Vox Sang.* **21**:509.

Pemberton, R. T., 1974, Anti-A and anti-B of gastropod origin, *Ann. N.Y. Acad. Sci.* **234**:95.

Pistole, T. G., 1976, Naturally occurring bacterial agglutinin in the serum of the horseshoe crab, *Limulus polyphemus*, *J. Invertebr. Pathol.* **28**:153.

Pistole, T. G., 1978, Broad-spectrum bacterial agglutinating activity in the serum of the horseshoe crab *Limulus polyphemus*, *Dev. Comp. Immunol.* **2**:65.

Pistole, T. G., and Britko, J. L., 1978, Bactericidal activity of amebocytes from the horseshoe crab, *Limulus polyphemus*, *J. Invertebr. Pathol.* **31**:376.

Prokop, O., and Köhler, W., 1967, Agglutinationsreaktionen von Mikroorganismen mit *Helix pomatia* Eiweissdrusenextrakt (Anti-A_{hel} Agglutinin), *Z. Immunitaetsforsch.* **133**:176.

Prokop, O., Rackwitz, A., and Schlesinger, D., 1965a, A "new" human blood group receptor A_{hel} tested with saline extracts from *Helix hortensis* (garden snail), *J. Forensic Med.* **12**:108.

Prokop, O., Schlesinger, D., and Rackwitz, A., 1965b, Über eine thermostabile "antibody-like substance" (Anti-A_{hel}) bei *Helix pomatia* und deren Herkunft, *Z. Immunitaetsforsch.* **129**:402.

Prokop, O., Uhlenbruck, G., and Köhler, W., 1968, A new source of antibody-like substances having anti-blood group specificity: A discussion of the specificity of *Helix* agglutinins, *Vox Sang.* **14**:321.

Prowse, R. H., and Tait, N. N., 1969, *In vitro* phagocytosis by amoebocytes from the haemolymph of *Helix aspersa* (Müller). I. Evidence for opsonic factor(s) in serum, *Immunology* **17**:437.

Rackwitz, A., Schlesinger, D., and Prokop, O., 1965, Über ein Blutgruppenprinzip B (anti-A) bei *Helix hortensis*: Ein neuer menschlicher A-Rezeptor A_{hel}, *Aca Biol. Med. Ger.* **15**:187.

Reade, P. C., 1968, Phagocytosis in invertebrates, *Aust. J. Exp. Biol. Med. Sci.* **46**:219.

Renwrantz, L., 1979, An investigation of molecules and cells in the hemolymph of *Helix pomatia* with special reference to immunobiologically active components, *Zool. Jahrb. Physiol.* **83**:283.

Renwrantz, L., and Berliner, U., 1978, A galactose specific agglutinin, a blood-group H active polysaccharide, and a trypsin inhibitor in albumin glands and eggs of *Arianta arbustorum* (Helicidae), *J. Invertebr. Pathol.* **31**:171.

Renwrantz, L., and Cheng, T. C., 1977a, Identification of agglutinin receptors on hemocytes of *Helix pomatia*, *J. Invertebr. Pathol.* **29**:88.

Renwrantz, L., and Cheng, T. C., 1977b, Agglutinin-mediated attachment of erythrocytes to hemocytes of *Helix pomatia*, *J. Invertebr. Pathol.* **29**:97.

Renwrantz, L., and Mohr, W., 1978, Opsonizing effect of serum and albumin gland extracts on the elimination of human erythrocytes from the circulation of *Hel:x pomatia*, *J. Invertebr. Pathol.* **31**:164.

Renwrantz, L., and Uhlenbuck, G., 1974a, Blood-group-like substances in some marine invertebrates. II. An agglutinin, which can be inhibited by lactose besides a blood-group A like glycoprotein in the hemolymph of *Octopus vulgaris* (Lam.), *Z. Immunitaetsforsch.* **148**:16.

Renwrantz, L., and Uhlenbuck, G., 1974b, Blood-group-like substances in some marine invertebrates. I. Blood-group A reactive substances in the ascidian *Phallusia mammillata* (Cuvier) and in the lancelet *Amphioxus* (*Branchiostoma*) *lanceolatus* (Pallas), *Vox Sang.* **26**:385.

Renwrantz, L., Schäncke, W., Harm, H., Erl, H., Liebsch, H., and Gercken, J., 1981, Discriminative ability and function of the immunobiological recognition system of the snail *Helix pomatia*, *J. Compar. Physiol.* **141**:477.

Reudiger, G., and Davis, D. J., 1907, Phagocytosis and opsonins in the lower animals, *J. Infect. Dis.* **4**:333.

Roche, A. C., and Monsigny, M., 1974, Purification and properties of limulin: A lectin (agglutinin) from haemolymph of *Limulus polyphemus*, *Biochim. Biophys. Acta* **371**:242.

Roth, S., 1973, A molecular model for cell interactions, *Q. Rev. Biol.* **48**:541.

Rudolph, P. H., 1973, The occurrence of hemagglutinins in some Basommatophora and Stylommatophora, *Malacol. Rev.* **6**:48.

Ryoyama, K., 1973, Studies on the biological properties of coelomic fluid of sea urchin. I. Naturally occurring hemolysin in sea urchin, *Biochim. Biophys. Acta* **320**:157.

Ryoyama, K., 1974, Studies on the biological properties of coelomic fluid of sea urchin. II. Naturally occurring haemagglutinin in sea urchin, *Biol. Bull.* **146**:404.

Salfner, B., Ishiyama, I., and Uhlenbruck, G., 1971, Determination of the carbohydrate moiety of purified anti-A from *Helix pomatia* by gas-liquid-chromatography, *Z. Klin. Chem. Klin. Biochem.* **9**:460.

Salfner, B., Ishiyama, I., and Uhlenbruck, G., 1972, Über die Aminosäurezusammensetzung antikörperähnlicher Agglutinine aus Schnecken (Eiweissdrüsen), *Hoppe-Seyler's Z. Physiol. Chem.* **353**:1977.

Schluter, S. F., Ey, P. L., Keough, D. R., and Jenkin, C. R., 1981, Identification of two carbohydrate-specific erythrocyte agglutinins in the haemolymph of the protochordate *Botrylloides leachii*, *Immunology* **42**:241.

Schluter, S. F., Ey, P. L., Coombe, D. R., and Jenkin, C. R., 1983, The purification and properties of two lectins from the hemolymph of the ascidian *Botrylloides leachii*, *Biochemistry*, in preparation.

Schnitzler, S., and Geserick, G., 1971, Anti-A_1 from *Bradybaena fruticum*, *Z. Immunitaetsforsch.* **141**:317.

Schnitzler, S., and Kilias, R., 1970, Über das Vorkommen von Hämagglutininen bei Landlungen Schnecken, *Blut* **20**:221.

Schnitzler, S., Geserick, G., Krüger, W., Gogochia, S. D., Mirvis, A. B., and Annenkow, H. A., 1971a, Ungewöhnliche Blutgruppenreaktionen durch Protektine, *Aerztl. Lab.* **17**:236.

Schnitzler, S., Krüger, W., Felix, D., David, H., Uerlings, I., Böttger, M., and Kuhn, W., 1971b, Reinigung und Eigenschaften der Hämagglutinine anti-A_{CN} und anti-A_{CA}, *Z. Klin. Chem. Klin. Biochem.* **9**:304.

Schnitzler, S., Uerlings, I., and David, H., 1971c, Elektronenmikroskopische Darstellung der Antikörper aus *Helix pomatia*, *Acta Biol. Med. Ger.* **26**:193.

Schnitzler, S., Oehme, P., Krüger, W., and Pardoe, G. I., 1972, Zur Adsorption von Hämagglutininen an Sephadex: Eine einfache Reinigungsmethode für das anti-A Agglutinin aus der Schnecke *Helix pomatia*, *Acta Biol. Med. Ger.* **29**:889.

Scofield, V. L., Schlumpberger, J. M., West, L. A., and Weissman, I. L., 1982, Protochordate allorecognition is controlled by a MHC-like gene system, *Nature (London)* **295**:499.

Scott, H. T., 1971, Recognition of foreignness in invertebrates. II. *In vitro* studies of cockroach phagocytic haemocytes, *Immunology* **21**:817.

Shimizu, S., Ito, M., and Niwa, M., 1977, Lectins in the haemolymph of Japanese horseshoe crab *Tachypleus tridentatus*, *Biochim. Biophys. Acta* **500**:71.

Sloan, B., Yocum, C. H., and Clem, L. W., 1975, Recognition of self from nonself in crustaceans, *Nature (London)* **258**:521.

Sminia, T., van der Knaap, W. P. W., and Edelenbosch, P., 1979, The role of serum factors in phagocytosis of foreign particles by blood cells of the freshwater snail *Lymnaea stagnalis*, *Dev. Comp. Immunol.* **3**:37.

Smith, A. C., 1977, Immunologic reactions of the sea cucumber, *Holothuria cinerascens*, to serum from the milkfish, *Chanos chanos*, *J. Invertebr. Pathol.* **29**:326.

Sprenger, I., and Uhlenbruck, G., 1971, On the specificity of broad spectrum agglutinins. XI. The reaction of the agglutinin from the snail *Caucasotachea atrolabiata*, *Z. Immunitaetsforsch.* **142**:254.

Stanislawski, E., Renwrantz, L., and Becker, W., 1976, Soluble blood group reactive substances in the hemolymph of *Biomphalaria glabrata* (Mullusca), *J. Invertebr. Pathol.* **28**:301.

Stauber, L. A., 1950, The fate of India ink injected intracardially into the oyster, *Ostrea virginica* Gmelin, *Biol. Bull.* **98**:227.

Stein, P. C., and Basch, P. F., 1979, Purification and binding properties of hemagglutinin from *Biomphalaria glabrata*, *J. Invertebr. Pathol.* **33**:10.

Stuart, A. E., 1968, The reticulo-endothelial apparatus of the lesser octopus, *Eledone cirrosa*, *J. Pathol. Bacteriol.* **96**:401.

Svensson, S., Hammarström, S., and Kabat, E. A., 1970, The effect of borate on polysaccharide–protein and antigen–antibody reactions and its use for the purification and fractionation of cross-reacting antibodies, *Immunochemistry* **7**:413.

Tanaka, K., and Watanabe, H., 1973, Allogeneic inhibition in a compound ascidian *Botryllus primagenus* Oka. I. Processes and features of "nonfusion" reaction, *Cell. Immunol.* **7**:410.

Tripp, M. R., 1960, Mechanisms of removal of injected microorganisms from the American oyster *Crassostrea virginica* (Gmelin), *Biol. Bull.* **119**:210.

Tripp, M. R., 1966, Haemagglutinin in the blood of the oyster *Crassostrea virginica*, *J. Invertebr. Pathol.* **8**:478.

Tripp, M. R., 1974, Oyster hemolymph proteins, *Ann. N.Y. Acad. Sci.* **234**:18.

Tripp, M. R., and Kent, V. E., 1968, Studies on oyster cellular immunity, *In Vitro* **3**:129.

Tyler, A., 1946, Natural heteroagglutinins in the body fluids and seminal fluids of various invertebrates, *Biol. Bull.* **90**:213.

Tyler, A., and Metz, C., 1945, Natural heteroagglutinins in the serum of the spiny lobster, *Panulirus argus*. I. Taxonomic range of activities, electrophoretic, and immunising properties, *J. Exp. Zool.* **100**:387.

Tyler, A., and Scheer, B. T., 1945, Natural haemagglutinins in the serum of the spiny lobster (*Panulirus interruptus*). II. Chemical and antigenic relation to blood proteins, *Biol. Bull.* **89**:93.

Tyson, C. J., and Jenkin, C. R., 1973, The importance of opsonic factors in the removal of bacteria from the circulation of the crayfish (*Parachaeraps bicarinatus*), *Aust. J. Exp. Biol. Med. Sci.* **51**:609.

Tyson, C. J., and Jenkin, C. R., 1974, Phagocytosis of bacteria *in vitro* by haemocytes from the crayfish (*Parachaeraps bicarinatus*), *Aust. J. Exp. Biol. Med. Sci.* **52**:341.

Uhlenbruck, G., and Pardoe, G. I., 1969, Serologische Besonderheiten eines heterophilen Rezeptors aus Entenblutkörperchen, *Z. Naturforsch.* **24**:142.

Uhlenbruck, G., and Prokop, O., 1966, An agglutinin from *Helix pomatia* which reacts with terminal N-acetyl-galactosamine, *Vox Sang.* **11**:519.

Uhlenbruck, G., and Reifenberg, U., 1971, Über ein Agglutinin in der Hämolymphe von *Helix pomatia*, *Immun-Information* **1**:14.

Uhlenbruck, G., and Steinhausen, G., 1977, Tridacnins: Symbiosis-profit or defense-purpose?, *Dev. Comp. Immunol.* **1**:183.

Uhlenbruck, G., and Weis, A., 1973, Studies on broad-spectrum agglutinins. XIV. Heterogeneity of *Helix aspersa* agglutinins, *Z. Immunitaetsforsch.* **145**:356.

Uhlenbruck, G., Prokop, O., and Haferland, W., 1966, Agglutination von *E. coli* durch ein Agglutinin aus *Helix pomatia*, *Zentralbl. Bakteriol. Parasitenkd. Infektionskr. Hyg. Abt. Orig.* **199**:271.

Uhlenbruck, G., Kim, Z., and Prokop, O., 1967, Reversible inactivation of *Helix* (*pomatia*) agglutinin by 2-mercapto-ethanol, *Nature* (*London*) **213**:76.

Uhlenbruck, G., Reifenberg, U., and Heggen, M., 1970, On the specificity of broad spectrum agglutinins. IV. Invertebrate agglutinins: Current status, conceptions and further observations on the variation of the Hel receptor in pigs, *Z. Immunitaetsforsch.* **139**:486.

Uhlenbruck, G., Reifenberg, U., and Prokop, O., 1971, Resistance to proteases of *Helix pomatia* anti-A: Consequences for tumour cell A-like antigen, *Acta Biol. Med. Ger.* **27**:455.

Uhlenbruck, G., Pardoe, G. I., Prokop, O., and Ishiyama, I., 1972, The serological specificity of snail agglutinins (protectins), *Anim. Blood Groups Biochem. Genet.* **3**:125.

Uhlenbruck, G., Steinhausen, G., and Baldo, B. A., 1977, Different anti-galactans in the haemolymph of *Tridacna maxima* and *Tridacna gigas*, *Comp. Biochem. Physiol. B* **56**:329.

Uhlenbruck, G., Karduck, D., and Pearson, R., 1979, Different tridacnins in different tridacnid clams: A comparative study, *Comp. Biochem. Physiol. B* **63**:125.

Vaith, P., Uhlenbruck, G., Müller, W. E. G., and Holz, G., 1979, Sponge aggregation factor and sponge haemagglutinin: Possible relationships between two different molecules, *Dev. Comp. Immunol.* **3**:399.

Vogt, W. E., Oehme, P., Knobloch, W., and Schnitzler, S., 1969, Zur Antikörpernatur von Helixagglutininen, *Z. Immunitaetsforsch.* **138**:62.

Vretblad, P., Hjorth, R., and Låås, T., 1979, The isolectins of *Helix pomatia*. Separation by isoelectric focusing and preliminary characterization, *Biochim. Biophys. Acta* **579**:52.

Wardlaw, A. C., and Unkles, S. E., 1978, Bactericidal activity of coelomic fluid from the sea urchin *Echinus esculentus*, *J. Invertebr. Pathol.* **32**:25.

Wemyss, C. T., 1951, Reponses, of the American cockroach *Periplaneta americana* to certain infected materials, Thesis, Rutgers University, New Brunswick, N.J.

Whitcomb, R. F., Shapiro, H., and Granados, R. R., 1974, Insect defense mechanisms against microorganisms and parasitoids, in: *The Physiology of the Insecta*, Vol. 5, 2nd ed. (M. Rockstein, ed.), pp. 447–536, Academic Press, New York.

Wright, R. K., 1974, Protochordate immunity, I. Primary immune response of the tunicate *Ciona intestinalis* to vertebrate erythrocytes, *J. Invertebr. Pathol.* **24**:29.

Wright, R. K., and Cooper, E. L., 1975, Immunological maturation in the tunicate *Ciona intestinalis*, *Am. Zool.* **15**:21.

10

RES Structure and Function of the Fishes

LARRY J. MC CUMBER, M. MICHAEL SIGEL,
RICHARD J. TRAUGER, and MARVIN A. CUCHENS

1. INTRODUCTION

Fish are the earliest vertebrates which have a well-developed immune system characterized by both cellular and humoral branches endowed with specificity and memory. More specifically, since invertebrates apparently lack a molecule resembling the vertebrate antibody molecule (Bang, 1973; Shapiro, 1975), fish are the first group of animals which clearly possess an immunoglobulin component. In addition, although there is evidence indicating cellular function in recognition of non-self in invertebrates (reviewed in Cooper, 1976), again fish are the first group of animals where T-like and B-like lymphocytes may actually attain a level of specialization.

Although fish appear to have developed elements of the immune system common to the mammalian immune system, such as immunoglobulins and lymphocyte heterogeneity, they have also retained more primitive molecules analogous to those of invertebrates, such as precipitins and agglutinins (Harisdangkul et al., 1972a,b; Alexander, 1980). Even though the precise structural relationship between the carbohydrate-binding molecules of fish and the carbohydrate-specific hemagglutinins of invertebrates is unclear, it would appear that fish may utilize a spectrum of defense mechanisms ranging from "nonspecific" lectinlike molecules and "natural killer" cells (Hinuma et al., 1980) to the highly specific antibody and lymphocyte systems.

We will first review the morphology of the lymphoreticular system of fish and consider its constituent cells, and subsequently present a discussion of the

---

LARRY J. MC CUMBER, M. MICHAEL SIGEL, and RICHARD J. TRAUGER • Department of Microbiology and Immunology, University of South Carolina School of Medicine, Columbia, South Carolina 29208. MARVIN A. CUCHENS • Department of Microbiology, University of Mississippi Medical Center, Jackson, Mississippi 39216.

functions of this system. We will address the question of lymphocyte heterogeneity in this group, with particular emphasis on an important aspect of fish, the effect of temperature on functions of certain cells.

2. MORPHOLOGY OF THE LYMPHORETICULAR SYSTEM OF FISH

2.1. ORGANS

Although the different classes of fish show considerable intraclass variation in development of the lymphoreticular system (LRS), as a general rule the more advanced Osteichthyes (bony fish) show greater development of the LRS than the Chondrichthyes (cartilaginous fish), while the latter are clearly more advanced than the primitive Agnatha (jawless fish). However, within any class one may find a marked degree of variation between different orders and species. As an example, the bony fish show development of the LRS somewhat analogous to the amphibian situation wherein the anurans and urodeles show considerable dichotomy for several LRS and immune parameters. It would appear that in some ways there is more similarity between certain fish and urodeles than between some fish or between urodeles and anurans (Ruben and Edwards, 1980). Nevertheless, what characterizes all fish *as a group* is that they lack distinctive bone marrow and discrete regional lymph nodes, though some species possess what appears to be an anlage for bone marrow and many species manifest local aggregates of lymphoid cells approaching the form of nodes. Some species possess, while others lack, *mammalianlike* thymuses. What is also of great interest is that unique lymphomyeloid structures have appeared in certain groups: the organ of Leydig and the epigonal organ. Moreover, in some species, at least one tissue, the pronephros, serves as the site of erythro-, myelo-, and lymphopoiesis.

There is very little *new* information about the lymphomyeloid system at the tissue level and much of what we have reviewed comes from old reports. The cellular elements of the system have not been clearly defined but are being scrutinized with the aid of modern instruments. The only real progress has been seen in the functional studies and molecular determinations which are experimental approaches borrowed from mammalian immunology. In this section we will discuss the three extant classes of fish from the standpoint of the organ development of the LRS, contrasting the variations between species and classes.

2.1.1. Agnatha

The jawless fish lack a definite thymus (Harboe, 1963), although in the sea lamprey small groups of lymphoid cells in the pharyngeal region are considered primitive thymus tissue (Good *et al.*, 1966). In the sea lamprey, a primitive spleen is present as an invagination of anterior gut tissue. This organ, along with primitive "bone marrow-like" tissue located in the fibrocartilaginous protovertebral arch, appears to function in hemopoiesis (Good *et al.*, 1966). The Atlantic hagfish (*Myxine glutinosa*) has hemopoietic foci which are present in the lamina propria of the gut (Good *et al.*, 1966). This intestinal myeloid tissue, along with a kidneylike organ called the pronephros, or head kidney, would appear to be the two major tissues functioning in erythropoiesis and lymphopoiesis in the hagfish. The origin of the various peripheral blood cells in this species is not totally clear. Holmgren (1950) reports that the pronephros not only contains erythrocytes, granulocytes, lymphocytes, spindle cells, and phagocytes in various developmental stages but functions as the organ of their origin. In contrast, Mattisson and Fange (1977) suggest that all lines of hagfish blood cells originate from agranulate lymphocytelike stem cells, most of which are produced by the intestinal myeloid tissue.

2.1.2. Chondrichthyes

According to some authors, some of the cartilaginous fish are characterized by the presence of a distinct thymus. Even the primitive holocephalan, *Chimaera monstrosa*, has a thymus consisting of cortical and medullary regions (Fange and Sundell, 1969). Claims have been made for the existence of similarly differentiated thymuses in elasmobranches (Good *et al.*, 1966). A more recent comparative evaluation by Fange (1977) indicates that there is a good deal of variation in the development of this tissue in different species of sharks and rays. There is also controversy as to whether there is involution of the thymus with age. It would appear that, at least in some species, the thymus is not a constant feature. There is no marked involution, according to descriptions by Fange and Sundell (1969) for the *Chimaera* and Good *et al.* (1966) for the elasmobranches. We want to again emphasize that evaluations are complicated by the fact that they are being done with wild-caught specimens of unknown age and with relatively few specimens. Often, the findings are based on old literature and on pooled specimens representing multiple species. Although Good *et al.* (1966) spoke of the thymus of the nurse shark, Fange and Mattisson (1981) comment on the lack of thymic tissue in this species based on macroscopic examination of "not-fully-adult" specimens. In our own work, we were able to observe the thymus only in a few instances while dissecting young sharks. Switching to the rays, a well-developed thymus, especially in young specimens, was found in *Raja* and *Torpedo* (Fange, personal communication).

The spleen, which is well developed in the cartilaginous fish, contains red and white pulp and appears to be important in both erythropoiesis and lymphopoiesis (Scottizzi, 1932; Good *et al.*, 1966; Fange and Sundell, 1969). The kidney seems to play a considerably less important role in hemopoiesis in the cartilaginous fish when compared to the head kidney, or pronephros, of bony fish (Jordan and Speidel, 1924; Good *et al.*, 1966), although aggregates of lymphomyeloid cells are often found in this organ. The cartilaginous fish do, however, have other hemopoietic organs such as the cranial lymphomyeloid tissue in *C. monstrosa* (Fange and Sundell, 1969) and the organ of Leydig of the esophagus and epigonal organ associated with the gonads in elasmobranches. In addition, aggregates of lymphomyeloid cells are also often found in the spiral valve of the intestine (Fange, 1977).

Several interesting aspects of lymphomyeloid tissue development and distribution in cartilaginous fish have been presented by Fange (1977). In comparing this system with the mammalian system, Fange calls attention to the fact that in mammals, the total weight of lymphomyeloid tissues is on the order of 1% of body weight and that of the red bone marrow corresponds to about 2% for a total of 3% of body weight. In contrast, the total weight of lymphomyeloid structures of sharks and rays varies between 0.6% and 1% of body weight. In *Chimaera*, the corresponding value is 1.7%. The white myeloid system is apparently granulopoietic and also contains lymphocytes, plasma cells, and macrophages. Although the spleen and thymus are considered part of this system, in certain species of elasmobranchs the system is primarily represented by Leydig's organ and the epigonal organ. These organs vary considerably in size; Leydig's organ may weight 0.5 kg and the epigonal organ up to 6.6 kg, but in many instances the organs may be rudimentary or absent. Fange presents evidence for a reciprocal relationship in the development of these organs. If the epigonal organ is large, the spleen may be relatively small and Leydig's organ may be absent. If, on the other hand, Leydig's organ is the dominant lymphomyeloid tissue, the spleen is small and the epigonal organ may be absent. While granulopoiesis and lymphopoiesis occur in the white myeloid tissues and the spleen, erythropoiesis may occur mainly in the circulating blood.

2.1.3. Osteichthyes

There is a good deal of variability in the reports on the morphology of the thymus in bony fish. Thus, a chondrostean, the paddlefish, and one holostean, the bowfin, were alleged to have a rather well-developed thymus (Good *et al.*, 1966), whereas another holostean, the gar, apparently lacks a readily identifiable thymus (McKinney, 1974). This same disparity seems to apply to reports on the thymuses of teleosts.

The thymus originates from the wall of one or more pharyngeal pouches, and in *most animals* the early rudiments become fused and separated from the epithelium. In teleosts, however, the thymus remains an integral part of the pharyngeal epithelium. Studies by Ellis (1977) and Grace and Manning (1980) in salmon and rainbow trout showed that the thymus precedes the spleen in its development. In the salmon, lymphocytes appeared to differentiate in the thymus on day 22 before hatching while lymphoid cells of the kidney were first seen on day 14 before hatching and the spleen did not fully develop until day 12 after hatching. The differences were not quite as exaggerated in the rainbow trout, but even here the thymus became fully differentiated by day 6 posthatch while other LRS structures lagged behind and the spleen did not fully mature by day 25 posthatch. In fact, the spleen remains primarily erythroid rather than lymphoid throughout life.

Despite the early appearance and function, the thymuses of the salmon and rainbow trout lack the structure and organization of distinct cortical and medullary zones that are present in mammalian thymuses. This is somewhat surprising since these fish are considered to be among the phylogenetically most recent, and as will be pointed out later, some of these fish are alleged to have separate populations of T and B cells. In contrast, there are reports on other teleosts describing differentiated thymuses. For example, a thymus with cortical and medullary areas and Hassall's corpuscles was described for *Tilapia mossambica* (the mouth breeder) (Sailendri and Muthukkaruppan, 1975). Earlier studies on the eel by von Hagan (1936) reported the absence of Hassall's corpuscles in *Anguilla vulgaris*. Fange reports (personal communication) that he found thymus glands in 15 different species of teleosts which he has examined, though histologically the tissue seemed less differentiated into cortex and medulla than in mammals and lacked Hassall's corpuscles. He points out that in the some cases the thymus was distinctly developed only in *young* individuals.

It would appear that, for bony fish in general, the spleen plays a subsidiary role to the pronephros. This latter tissue has hemopoietic rather than excretory functions in adult fish with primarily myeloid and lymphoid cells present (Jordan and Speidel, 1924; Ellis and deSousa, 1974). While there are lymphoid aggregates in the intestinal tissues of teleosts, the pronephros appears to serve as a stem cell compartment and primary lymphoid organ, the bony fish lacking bone marrow, bursa, or lymph node tissue. In addition, the mesonephros is an organ of teleosts which also appears to play some role in immunity (Rijkers, 1980).

2.2. CELLS

We should emphasize that the nomenclature of fish leukocytes is a problematic issue, much more complicated than is mammalian leukocyte taxonomy, because markers are missing, information is sketchy, and classification is most incomplete. In spite of this, we will attempt to discuss fish leukocytes by comparison to mammalian classification schemes. We should also point out that there is considerable variation in relative numbers and ratios of different cell types for different species within a class, as well as in different classes of fish. Studies of the rosy barb (*Barbus conchonius*) and carp (*Cyprinus carpio*) by Davina *et al.* (1980) indicated that the leukocytes in peripheral blood of these two species consist mainly of lymphocytes and, to a lesser degree, of heterophilic granulocytes. Morphologically, these lymphocytes resembled those of other vertebrates. PAS-positive granulocytes and monocytes were rarely found. When one adopts for teleosts the morphological and histochemical criteria used to identify mammalian leukocytes, neither eosinophilic nor basophilic granulocytes nor monocytes have been observed in blood of these fish species. Switching to an elasmobranch, the nurse shark (in our work) shows not only a much greater number of peripheral blood leukocytes, but a rather dramatic increase in the percentage of granulocytic cells (although lymphocytic cells are still present in significant numbers). As we shall see, the presence or absence of different cell types and the relative ratios of each appear to be quite species specific, although this may, in some cases, be a reflection of different classification techniques and investigators' interpretations using methods insufficient to differentiate certain cell types.

In this section we will be looking at monocytes and macrophages, as well as granulocytes and thrombocytes. Lymphocytes have been studied in many species of fish and the question of specialization has become a matter of controversy. While it is certainly clear that fish possess lymphocytes, based on both morphological and functional analyses, the question of lymphocyte heterogeneity is still undergoing considerable investigation. The concept of T and B cells and lymphocyte heterogeneity in fish will be discussed in depth in Section 3.5.

2.2.1. Monocytes and Macrophages

For convenience, circulating monocytes and tissue macrophages will be considered together. According to Ellis *et al.* (1976), monocytes represented 0.1 to 0.2% of the total white blood cells of the plaice (*Pleuronectes platessa*). They became more numerous in the blood 24 hr after injection of carbon particles. The cells varied in size from 7 to 14 μm. The nucleus occupied less than one-half of the cell and often was notched or horseshoe-shaped with fairly loosely arranged chromatin. Nucleoli were not visible. Histochemical tests revealed a small number of rather coarse and scattered granules which stained positively for acid phosphatase. With PAS, the cytoplasm sometimes stained a faint pink. The other histochemical tests were largely inconclusive.

Although fish thrombocytes have been reported to engage in phagocytic activity, and Ferguson (1976) has demonstrated ingested carbon particles in vacuoles of plaice thrombocytes, Ellis makes a point that in his studies *the only cells* shown to contain ingested colloidal carbon within the circulation were *monocytes, thus apparently excluding thrombocytes as phagocytic cells.*

Ellis describes three morpological types of macrophages:

1. Free rounded cells, 10–20 μm in diameter, found in smears of kidney, spleen, thymus, mesentery, and peritoneal fluid. The nucleus was eccentric and usually round with fairly loosely packed chromatin. The abundant cytoplasm was only slightly basophilic with Romanowsky dyes. It gave a positive PAS reaction and was shown to contain coarse granules which manifested acid phosphatase activity.
2. Fixed macrophages forming incomplete linings to blood sinuses in the kidney, spleen, and heart were termed by Ellis *et al.* (1976) as reticuloendothelial (RE) cells. "Macrophages which might be considered analogous to Kupffer cells were not observed in the plaice liver. Fixed macrophages or RE cells form an interlacing network of elongated and bifurcating cells. In the atrium of the heart, they line the blood sinuses which permeate the muscle bundles. These cells appeared to break free of their fixation points on taking up small amounts of carbon and become free macrophages, which presumably enter the general circulation for a short time and probably settle in the kidney parenchyma. In the spleen, RE cells are found attached to the reticulin fibers which surround the endothelium of the arterial capillaries where they form a sheath of highly phagocytic cells constituting the ellipsoid sheaths."
3. The term "melanomacrophage," introduced by Roberts (1974), has been applied

to macrophagelike cells which contain melanin pigment. It is unclear whether these cells, seen easily in smears of kidney and spleen, are melanogenic or merely phagocytic for melanin granules which are released by other cells. In some instances, the melanomacrophages did not contain many black granules but merely refractile vesicles of yellow-brown pigment. Roberts (1974) reported that these cells stained with Schmorl's reaction, indicating that this pigment is probably lipofuscin. Melanomacrophages occasionally stained positively with Perl's method for iron, indicating they are phagocytic for hemoglobin breakdown products.

Plaice immunoglobulin (Ig) was detected, by immunofluorescence, on the cell surface of about 10% of macrophages and such cells did not manifest capping.

2.2.2. Granulocytes

According to Ellis *et al.* (1976), only one type of granular leukocyte has been observed in the plaice (Figure 1). It has tentatively been termed a neutrophil because of similarities to mammalian neutrophils in its histochemical staining characteristics and certain ultrastructural features as reported by Ferguson (1976). Approximately 5–10% of the blood leukocytes were represented by granulocytes. These neutrophillike granulocytes were present in peritoneal fluid, kidney, and, to a lesser extent, spleen. Using Romanowsky stain, the cells (8–10 µm in diameter) had a grayish granular cytoplasm. The nucleus was eccentric and occupied about one-third or less of the cell. It was usually round or oval and only very rarely dumbbell-shaped. Hence, the alternative mammalian term "polymorphonuclear leukocyte" cannot be applied meaningfully to the plaice neutrophil. The plaice neutrophils gave a strong PAS reaction which was, at least in part, due to the presence of glycogen. Alkaline phosphatase activity was located in the nucleus and faint traces could be detected in the cytoplasm. Acid phosphatase stains demonstrated positive cytoplasmic granules which were finer than those observed in monocytes.

According to Ellis, little is known of the function of the plaice neutrophil. *Apparently, this cell does not ingest carbon particles,* yet the number of neutrophils in peritoneal fluid increased markedly after intraperitoneal injection of India ink.

Ultrastructural studies of the plaice neutrophil revealed a hazy, rather irregular outline, a nuclear chromatin that was denser and more patchy in distribution than seen in the monocyte. The cytoplasm contained numerous large, round, or elongated granules. These granules showed varying degrees of organization but they were nearly all fibrillar, with fibrils running either parallel or irregularly throughout the granule. Some had a banded appearance. The Golgi apparatus was not as well developed as in the monocyte.

Some authors (Hines and Spira, 1973) refer to granulocytes as neutrophils because they do not stain with Romanowsky dyes; nevertheless, the granules contain crystalloid inclusions which are characteristic of mammalian eosinophils. For this reason, Davina *et al.* (1980) prefer to term them heterophilic granulocytes. The PAS-positive granulocyte has been designated by others as eosinophil (Smith *et al.*, 1970; Davies and Haynes, 1975), fine reticular cell (Hines

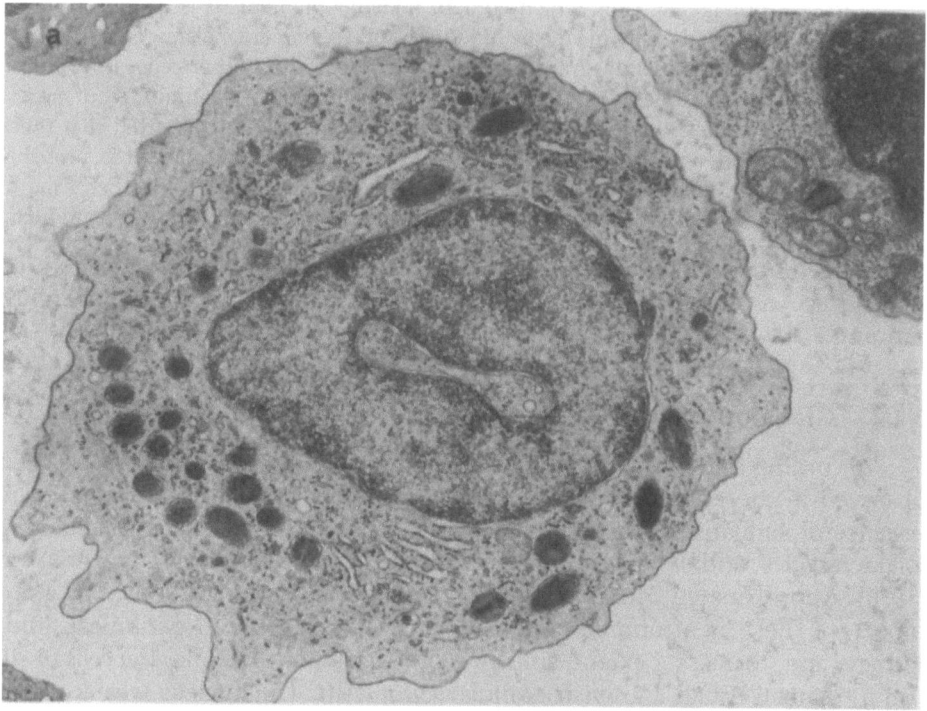

FIGURE 1. Neutrophils. The nucleus was often more irregular than that of the monocyte, but was never seen to be multilobed (Ferguson, 1976). The cytoplasm contains numerous ovoid granules. (a) × 16,000; (b) × 10,800.

and Spira, 1973), basophil (Watson *et al.*, 1963; Weinreb, 1963), and mast cell (Huebner and Chee, 1978).

2.2.3. Thrombocytes

One of the most interesting cells in the circulation of fish is the thrombocyte. As first described in phase-contrast microscopic studies by Wardle (1971), the thrombocyte is an elongated fusiform cell. Using the Romanowsky staining technique, Ellis *et al.* (1976) described four morphological forms of thrombocytes in the plaice. The *spiked form* was the commonest when the blood samples were collected with minimal stress to the fish. It was an elongated cell, rounded at one pole and extended into a tapering spike at the other pole. The cytoplasm was pale or very light blue-gray. The oval nucleus often appeared striped. The second most common form was the *spindle cell* in which both poles appeared to be rounded. The third form was a shortened version of the second form, referred to as the *ovoid form*, and appeared to be fairly rare in blood smears. The fourth form was the naked nucleus, or *lone-nucleus* cell, which appeared as small, naked, and densely staining nuclei although occasionally surrounded by a small amount of

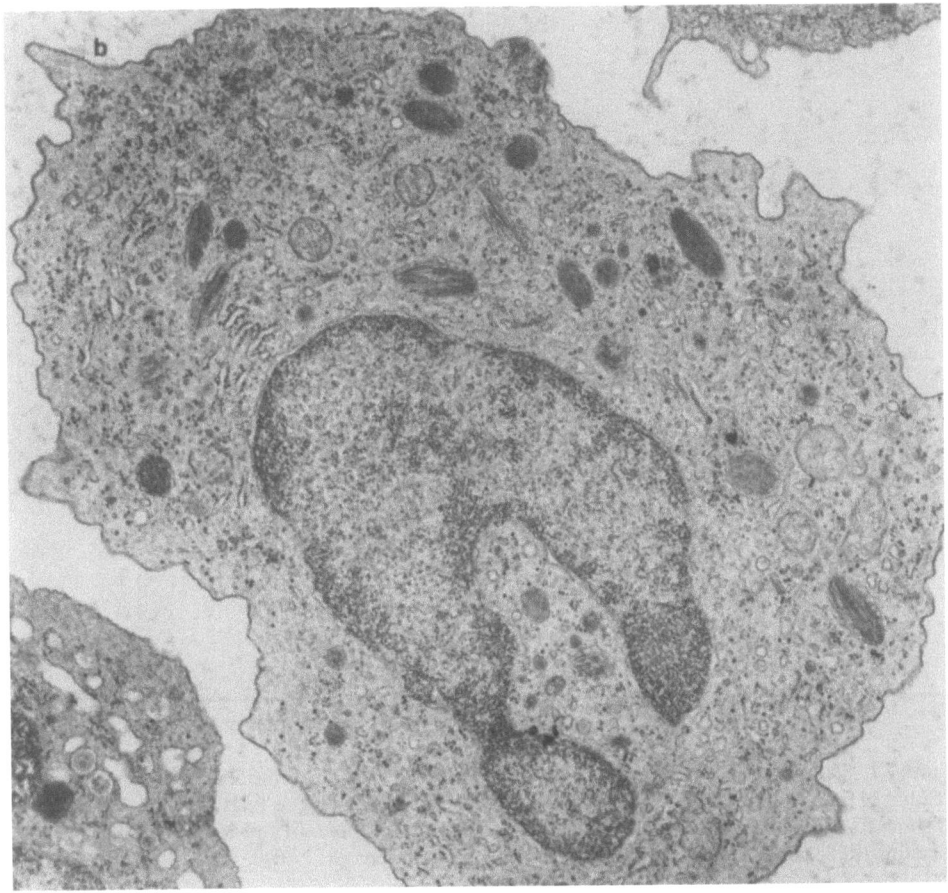

FIGURE 1. (*Continued*)

ragged, gray-blue cytoplasm. The last form was seen primarily in preparations from fish that were stressed prior to collection of blood or when the blood was thick and clotted rapidly. Histochemical examinations have not been very revealing. Thrombocytes gave negative stains with Sudan black B and the benzidine–peroxidase and alkaline phosphatase tests. Acid phosphatase stains demonstrated single granules at the spiked pole of the cell. Small numbers of coarse PAS positive granules were found throughout the cytoplasm. The number of thrombocytes (at least in the plaice) approximated the number of lymphocytes (the ratio was approximately 1 : 1.4). In this connection, it should be mentioned that thrombocytes do not carry surface Ig and this can be used as one of the markers for distinguishing thrombocytes from lymphocytes. The combined number of lymphocytes and thrombocytes in the plaice ranges from 65,000 to 120,000 per mm^3 but most fish sampled were in the range of 100,000 to 110,000 per mm^3.

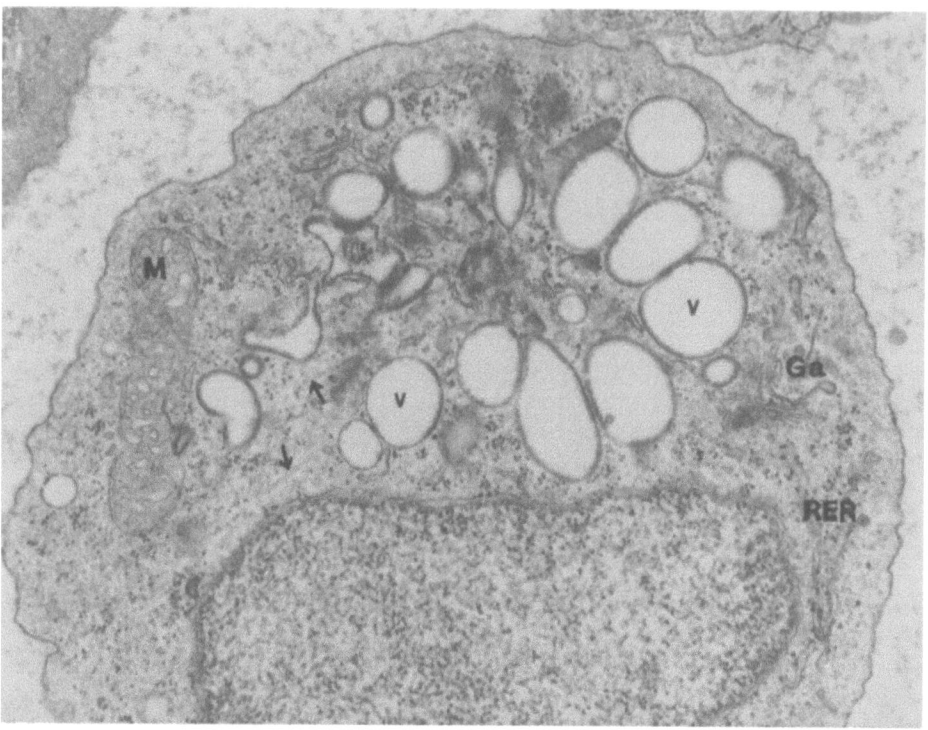

FIGURE 2. Plaice thrombocyte. Microtubules (arrows) radiate from the centriole. The most prominent feature of the thrombocyte cytoplasm is the large electronlucent vesicles (v), which are lined with a distinct fuzzy coat (Ferguson, 1976). Mitochondria (M), small amounts of rough endoplasmic reticulum (RER), and a Golgi apparatus (Ga) are also present. × 27,000.

An excellent description of the fine structure of thrombocytes was given by Ferguson (1976). There was usually an indentation in the plasma membrane above the nucleus. The cytoplasm displayed a large number of vesicles lined with a "fuzzy" coat (Figure 2). Moreover, fish thrombocytes were highly phagocytic and injection with colloidal carbon suspension led to accumulation of carbon particles within some of these vesicles. This function should distinguish thrombocytes from lymphocytes with which they share morphological features. Other characteristics that should help distinguish thrombocytes from lymphocytes include a well-developed microtubular system in thrombocytes in addition to the abundant network of vacuoles. The most prominent distinguishing features of lymphocytes are the large and often elongated mitochondria.

There is still a good deal of uncertainty regarding the various functions performed by thrombocytes in fish and the exact distribution of this type of cell. Little is known about their homogeneity and origin. Cotton (1951) could recognize large numbers of thrombocytes in brown trout but could not identify thrombocytes in the perch. This author considered these cells primarily hemoglobin carriers, whereas other investigators were more concerned with their phagocytic properties (Yokoyama, 1960; Fange, 1968). Fish thrombocytes are, of course, very much involved in blood clotting (Gardener and Yevich, 1969; Wardle, 1971).

3. FUNCTIONS OF THE LYMPHORETICULAR SYSTEM OF FISH

3.1. PHAGOCYTOSIS

The primary function of the LRS is the defense of the animal against pathogenic organisms. In mammals, phagocytosis of foreign substances by mononuclear phagocytes and polymorphonuclear leukocytes is the first line of defense, leading to elimination of the intruding material as well as functioning as an initial step in the onset of specific immune response. The phagocytic system of lower vertebrates may be of even more crucial importance to this group of animals than to higher vertebrates. We particularly have in mind poikilothermic vertebrates whose body temperature is at the mercy of the environment. Thus, as will be brought out in Section 3.5.2, when the temperature of the water drops, fish are less able to respond to infectious agents with antibody production because the lowered environmental temperature causes a decrease in the temperature of the body and a corresponding diminution in antibody response. It has been suggested, therefore, that under such circumstances the animal may be dependent on phagocytosis for protection and survival. Such speculations appear to be justified by the observation that, although the rate of phagocytosis may decrease at lower temperatures, those temperatures which prevent antibody synthesis in some species may still allow phagocytosis to occur, albeit at a slower rate (Avtalion et al., 1973; Wojdani et al., 1979). In fact, there appears to be an inverse relationship between antibody production and phagocytic cells under the influence of chemical as well as physical factors. Rijkers et al. (1980) observed that doses of the antibiotic oxytetracycline sufficient to suppress the humoral and cellular immune responses of the carp, led to increased numbers of granulocytic cells in the spleen. They speculated that in those cases where specific lymphoid defense mechanisms were blocked, the phagocytic defense system became more active. Whether this is a true compensatory mechanism or simply a drug toxicity reaction unrelated to defense against other foreign substances is still open to speculation. It is clear, however, that the phagocytic system of fish is critical to recognition of some foreign substances, although we must avoid oversimplification. Certainly, one may safely conclude that temperature is a factor in phagocytosis *in vivo* (particle clearance) in light of the data discussed below.

The clearance of viruses and carbon particles from the blood has been used as an indicator of phagocytic activity. Studies by Sigel et al. (1968) and O'Neill (1980) indicate that cartilaginous and bony fish can completely clear bacteriophage particles from the circulation in 4–7 days when kept at optimal temperatures. Studies on carp and rainbow trout (O'Neill, 1980) as well as nurse shark and lane snapper (Russell et al., 1976) indicate that the rate of bacteriophage clearance decreases with lowering temperatures. No viable phages were found in the blood by 9 days at 30.7°C, whereas 14–20 days was required for their elimination at 21.6°C (Russell *et al.*, 1976). Species such as the icefish take 42–56 days for clearance of MS2 bacteriophage when maintained at 2°C (O'Neill, 1980), in spite of their presumed adaptation to this "optimal" environment. It would appear that below certain temperatures, even long-term adaptation of a species, as in the case of the icefish, does not allow the phagocytic system to compensate, an observation which is not surprising in light of the fact that phagocytosis (the presumed effector of clearance) requires energy expenditure and contractile activity which are dependent upon hydrolysis of ATP. Species less "adapted" to cold would probably suffer greater consequences from low temperatures interfering with their phagocytic as well as humoral responses.

Although mammals utilize both mononuclear and granulocytic phagocytes, the phagocytic component of fish shows species variation with regard to the functioning of various cell types. Studies on the holostean gar showed that, while monocytes and macrophages were capable of phagocytosing bacteria, yeast, and SRBC, the granulocytes appeared to be unable to take up these particles (McKinney *et al.*, 1977). The plaice was found to have phagocytic activity in the monocytes (Ellis *et al.*, 1976; Ferguson, 1976) and thrombocytes (Ferguson, 1976) of peripheral blood, but not the granulocytes. Thus, there is agreement that plaice *granulocytes* are not involved in phagocytosis, whereas *monocytes* clearly are involved. The question is raised as to whether *thrombocytes* actually play any role in phagocytosis. Even in a single species like the plaice, some investigators have found phagocytosis by thrombocytes while others have not. One of the problems lies in a clear-cut definition of what is a thrombocyte under all circumstances. As already stated, the neutrophillike granulocytes are not phagocytic in the plaice. In contrast, in addition to monocytes, the granulocytes of rainbow trout, goldfish, carp, cunner, and guppy were reported to possess phagocytic activity (Mackmull and Michels, 1932; Pliszka, 1939; Jakowski and Nigrelli, 1953; Watson *et al.*, 1963; Weinreb and Weinreb, 1969; Finn and Nielsen, 1971).

3.2. GRAFT REJECTION

The extent to which a graft is rejected is dependent upon the host's ability to recognize specific allogeneic histocompatibility antigens, for example, those encoded for by the *K* and *D* regions of the mouse major histocompatibility complex (MHC). This implies that the host must be able to distinguish self from non-self with a high degree of specificity and maintain the ability to mount an attack against foreign tissue. In fish, this reaction characteristically follows a pattern of inflammatory response, lymphocyte infiltration, capillary hemorrhage, and pigment cell destruction (Sailendri, 1973; Rijkers, 1980). In mammals, the T cell has generally been regarded as the effector cell responsible for graft rejection. Clearly, the rejection of grafts is an artificial manipulation in both mammals and fish, but the same factors or cells may be critical in cell-mediated immunity (CMI) and surveillance against tumors.

Nothing is known about the MHC antigens of fish and in fact it has not been determined whether fish possess a functional MHC. Nonetheless, transplantation studies have been carried out in various classes and orders of fish.

Allograft rejection in jawless and cartilaginous fish is chronic, that is, a median survival time (MST) of greater than 30 days for the transplanted tissue. These fish represent the most primitive vertebrates. Although they are capable of antibody production and will respond to some antigens with a proliferation of mononuclear cells, they appear to manifest a low level of immunocompetence. Complete destruction of allografts may take as long as 40 days and is characterized by mononuclear cell infiltration with some lymphocytes and heterophils present (Finstad and Good, 1966).

The Osteichthyes appear to be the first major class of vertebrates capable of acute graft injection, that is, an MST of less than 14 days. Hildemann (1972) reported that rejection of first-set allografts had already begun by day 2 in two species of fish, *Chromis caeruleus* and *Dascyllus arvanas*. The rejection was complete by day 11, with the MST for each fish being 5.2 and 6.6 days, respectively. Control autografts were healed by day 3. The sequence of events leading to rejection, i.e., hyperplasia and capillary hemorrhage, seems to be characteristic of cutaneous allograft rejection in many bony fish. The effect of temperature on graft rejection and the role of lymphocyte classes or subsets will be discussed in Section 3.5.

Acute rejection was also found in the work by McKinney *et al.* (1981) on the gar, *Lepisosteus platyrhincus*, indicating that by day 8 posttransfer a significant response was in progress as evidenced by the marked inflammatory infiltrate involving the connective tissue between the skeletal muscle bundles and the scale. This resulted in an extensive necrosis of the connective tissue, edema, degeneration of the muscle fibers, and, finally, erosion of the scale plate. By day 18 the rejection was complete. Control autographs initially showed numerous inflammatory cells as well, but this reaction had subsided by day 6. The putative effector cells responsible for rejecting the graft were mostly lymphocytes and monocytes, with some granulocytes present.

3.3. INTERFERON

Of the many soluble effectors of the immune system of vertebrates, interferon is one of the most studied and least understood. Originally described as an antiviral agent by Isaacs and Lindenmann in 1957, this glycoprotein appears to function as a message from an infected cell to the surrounding cells that enables them to produce an antiviral protein which inhibits virus replication.

The first evidence for the *in vitro* production of interferon in fish cells comes from the work of Gravell and Malsberger (1965) and Beasley *et al.* (1966). While establishing a marine fish cell line, Beasley *et al.* (1966) followed the fate of grunt fin (GF) cells that survived a viruslike infection and, upon subsequent passage, they found the cells to be chronically infected with a viruslike agent called GFA. A carrier line (designated G1A) was established and the cells appeared to have survived, despite the coexistence of GFA, by virtue of protection from interferon. Interferon activity could actually be transferred from these carrier cultures using supernatants as well as intact cells. Lysis of these carrier cells diminished protection, presumably due to loss of interferon production (Beasley et al., 1966). This interferon protection was species specific, stable at pH 2.0, nonsedimentable, nondialyzable, moderately heat stable, and able to provide protection against both GFA and IPN (infectious pancreatic necrosis) viruses.

Studies on the rainbow trout, *Salmo gairdneri*, by de Kinkelin and Dorson (1973) indicate that interferon production may be an important line of defense against virally induced diseases such as viral hemorrhagic septicemia (VHS) and infectious hematopoietic necrosis (IHN). This protection appears to be highly temperature sensitive. At 6–12°C, VHS infection is extremely widespread as indicated by the significant losses experienced by hatcheries operating in this temperature range. Elevating the temperature to 15°C appears to spontaneously remedy the situation, even though this temperature exhibits no deleterious effect on viral growth *in vitro*. Although able to produce circulating antibodies at 15°C, these animals possess no demonstrable titer of neutralizing antibody to VHS postinfection, but serum from these animals could protect trout cells from infection with IPN virus *in vitro*. The antiviral factor was characterized as a nondialyzable serum component that was stable at 56°C and pH 2.0. It was also found to be trypsin labile and RNAse resistant. These qualities, along with the rapid increase and decrease of the protective effect in serum (maximum 3 days postinfection), strongly implicate the presence of interferon. These authors raise the possibility that the apparent relationship between the water temperature and the appearance of VHS and IHN diseases may be a result of the temperature dependence of interferon production by the trout. Below 15°C the fish made neither antibody nor interferon; at 15°C they produced interferon; and at higher water temperature they could also form antibodies.

We thus again see an example of the recurring theme that fish may possess "tiered" defense systems which are capable of compensating for a depression of antibody production, whether this depression is due to a lowering of the body temperature or is due to an effect of a chemical substance, as in the case of antibiotics and phagocytosis, as mentioned in Section 3.1.

3.4. CYTOTOXICITY

3.4.1. Specific T-Cell Cytotoxicity

In Section 3.5 we discuss the controversy of heterogeneity of fish lymphocytes, i.e., multipotential lymphocytes, vs. specialized classes of T-like and B-like cells. Be that as it may, to our knowledge no reports of cytotoxicity mediated by specifically activated T cells have been published. Rather, most of the cytotoxicity reactions appear to be of the nonspecific or "natural" kind.

3.4.2. Nonspecific or Natural Cytotoxicity

Among the very interesting nonspecific cytotoxic reactions are those exerted by fish cells against xenogeneic target cells. One such reaction, described by Hinuma et al. (1980), manifested two intriguing aspects: the killer cells were detectable in freshwater fish but not in marine fish and these effectors were lytic for established cell lines but not for primary cultures. Although the effector cells were located principally in the kidney, they were also at times demonstrable in peripheral blood. The cytotoxicity occurred immediately after mixing of effector with target cells and the optimal temperature was 25°C. The effect was substantially decreased at 37°C. The cytotoxicity increased exponentially until 6 hr after incubation, after which time it reached a plateau. Table 1 lists a number of

TABLE 1. DISTRIBUTION OF NATURAL CYTOTOXICITY OF LEUKOCYTES OF VARIOUS VERTEBRATES ON FISH LYMPHOCYTES[a]

	Organ	Cytotoxicity[b] (% ^{51}Cr release)
Human	Peripheral blood	13.8
Rabbit	Peripheral blood	5.9
Mouse	Bone marrow	3.8
	Peripheral blood	3.6
	Spleen	4.3
Chicken	Peripheral blood	1.9
Turtle	Peripheral blood	0.8
	Spleen	0.0
Frog	Peripheral blood	8.0
	Spleen	0.3
Fish		
Freshwater		
Cyprinus carpio	Kidney	66.9
	Peripheral blood	65.9
	Spleen	35.8
Carassius cuvieri	Kidney	54.9
Ctenopharyngodon idellus	Kidney	57.1
Misgurnus anguillicaudatus	Kidney	27.8
Channa argus	Kidney	24.4
Lepomis macrochirus	Kidney	2.5
Salmo gairdneri	Kidney	0.0
Anguilla japonica	Kidney	0.5
Saltwater		
Amanses modestus	Kidney	0.3
Lateolabrax japonicus	Kidney	0.0
Hexagrammos otakii	Kidney	0.5
Hexagrammos stelleri	Kidney	0.5
Mylio macrocephalus	Kidney	0.6
Platichthys stellatus	Kidney	0.4

[a]From Hinuma et al. (1980).
[b]10^5 ^{51}Cr-labeled fish lymphocytes were incubated with 2 + 10^6 leukocytes of various vertebrates for 6 hr. The data are given as the mean values of the ^{51}Cr release in duplicate assays.

species and indicates which ones display activity in the cytotoxicity assays. The broad spectrum of activity against the variety of unrelated cell types indicated that the reaction was not being effected by specific immune T cells, especially since the fish were not immunized prior to the performance of cytotoxicity assays. The fact that the presence of antibody was not demonstrated and antibody was not required for the reaction would imply that the effector cells were not killer cells involved in antibody-dependent cellular cytotoxicity (ADCC) reactions. The authors, therefore, assumed that the effector cells may be similar to mammalian natural killer (NK) cells (Herbermann et al., 1975; West et al., 1977). However, it should be pointed out that, under the conditions of the present testing, NK cells of birds and mammals failed to react against the target cells which were being lysed by the fish effector cells. The NK cells of fish were analogous to activated mammalian macrophages in that they manifested greater cytotoxicity toward transformed, compared to untransformed, fibroblasts (Hibbs et al., 1972; Fauve, 1978). In spite of this analogy, it is important to keep in mind that, unlike mammalian macrophages, the cytotoxic kidney cells of fish do not seem to require overt activation.

The reaction was not like SINEC (specifically induced, nonspecifically expressed cytotoxicity). In SINEC, lymphocytes recognize the antigen to which they are immune and, in reacting to it, produce a product which can arm additional effector cells toward nonspecific cytotoxicity against irrelevant target cells. In the cytotoxic reaction of fish, there was no detectable lymphokine or monokine. Intact living cells had to be used to cause cytotoxicity (Hinuma et al., 1980). The authors do not identify the effector cell but intimate, at the end of the article, that the responsible cell appears to be a phagocytic cell.

A paradigm of different forms of cytotoxicity was demonstrated in our laboratory for peripheral blood leukocytes of the nurse shark, *Ginglymostoma cirratum* (McKinney et al., 1977). The cytotoxic reactions were nonspecific but demonstrated degrees of selectivity which depended to a large extent on the conditions under which the effector cells were isolated. Thus, when peripheral blood leukocytes were separated gently, with relatively little trauma, they manifested no cytotoxic activity against any target cells. These cells could be rendered cytotoxic only against xenogeneic red blood cells (not autologous or allogeneic) by exposure to phytohemagglutinin (PHA). (For the sake of osmotic compatibility, marine fish red blood cells were used as targets.) PHA appeared to act as a "glue" facilitating the attraction of leukocytes toward red blood cells and did not seem to activate the cells. When the leukocytes were separated by a more drastic procedure, entirely different results were obtained. The cells from the interface were not cytotoxic to autologous or allogeneic red blood cells but were cytotoxic toward xenogeneic cells *in the absence* of PHA and were not apparently potentiated in their activity by the addition of PHA. Moreover, cells from the pellet were indiscriminantly cytotoxic toward all target cells. It is of further interest that antibody against shark Ig could inhibit the selective xenogeneic cytotoxicity but failed to diminish the nonselective killing by the pellet. This suggested that the latter effect may have been due to the production of a soluble cytotoxic mediator such as a lymphokine or a proteolytic enzyme. This interpretation would be in line with the supposition that the drastic treatment caused damage to the effector cells, a supposition which was, in part, supported by trypan blue staining demonstrating loss of cell viability.

None of the shark leukocyte cytotoxicities were of the ADCC type. Shark blood is known to possess antibody to a large variety of antigens. In the present work, it was found that antibody did *not* contribute to the cytotoxicity but, in fact, had an inhibitory effect on the selective cytotoxicity of leukocytes. Absorption of shark plasma with target cells prior to its use in the cytotoxicity assays caused enhancement of cytotoxic activity, presumably due to removal of natural antibody.

In subsequent studies, Pettey and McKinney (1981) demonstrated that increased selective cytotoxicity could be achieved by the addition of stimuli other than PHA, e.g., Con A or lipopolysaccharide (LPS). All three agents (PHA, Con A, LPS) rendered shark leukocytes cytotoxic for xenogeneic target cells but not for autologous or allogeneic target cells. This is in contrast to mitogen-induced cytotoxic reactions in higher vertebrates where PHA-stimulated leukocytes are found to lyse autologous target cells as readily as foreign target cells. As we have intimated previously, these agents may not actually activate the system but bring about a better juxtaposition or interaction between the effector cell and the target. Pettey and McKinney (1981) considered the possibility that the shark leukocyte activity may be analogous to that of NK cells in higher vertebrates. The authors point out, however, that the effector cells in the shark system were glass-adherent, and thus the effector cells may not be lymphocytes. The dominant cell population among the adherent cells is the monocyte, with only a small percentage of lymphocytes. The nonadherent cells which are not involved in cytotoxicity are predominantly composed of lymphocytes. Thus, the implication at the moment is that the cell in question may be a monocyte.

Natural cytotoxicity can apparently occur spontaneously but still in a selective manner when fish are stressed by environmental conditions. It is as if, under these circumstances, the reaction could represent a form of defense against infection or disease. Pettey and McKinney (1980) discovered this type of activity in leukocytes of the nurse shark. It was found to be present only during the months of January through March when the water temperature ranged from 21 to 23°C, as contrasted with the 25–30°C temperatures during the warmer months. This spontaneous cytotoxicity resembles the cytotoxicity described previously in that it was selective for xenogeneic target cells and was not manifested against allogeneic or autologous targets.

All these forms of cytotoxicity appear to involve the same type of effector cell, a cell which is glass-adherent. There may be additional heterogeneity in that cells which are plastic-adherent may display even more effective cytotoxicity. Cells which are nonadherent are not involved in positive cytotoxicity but instead may play a regulatory role, for example, as suppressor cells.

3.5. CELL COOPERATION AND ANTIBODY PRODUCTION: LYMPHOCYTE HETEROGENEITY

It is generally agreed that the earliest manifestation of antibody function occurred in fish. The ability to make antibodies intimates that fish possess cells equivalent in at least one function to the B lymphocytes of mammals. What about T cells? At issue are two questions. When in phylogeny did T cells evolve? Do phylogenetically more primitive species such as fish possess distinctive populations of lymphocytes, endowed with specialized functions, or multipotential lymphocytes capable of carrying out a variety of different functions? In this section we will attempt to review the data supporting the multipotential function of lymphocytes and then contrast this with the evidence supporting the possibility that fish may, in fact, have specialized subsets of lymphocytes which carry out T-like and B-like functions.

3.5.1. Evidence Supporting the Concept of Multipotential Lymphocytes

There is little doubt that fish possess cells with a B-cell function, antibody production. The antibody is restricted to the IgM class and in some fish there is an absence of immunological memory (reviews: Cooper, 1976; Rijkers, 1980). Such limitations in immunological responses of mammals occur only with thymus-independent (TI) antigens, e.g., pneumococcal polysaccharide. The usual antibody production in mammals is thymus dependent (TD) and involves: (1) interaction of B and T cells; (2) production of IgM followed by a switch to IgG; and (3) immunological memory. It would appear that the response of B cells to TD antigens requires two signals (Watson et al., 1973), one by the antigen and the other by a product of T cells. Since in some fish the immune mechanism handles all antigens tested as if they were TI, it follows that their B cells may not require the participation of T cells. It is as if their lymphocytes become triggered and fully differentiated by one signal delivered by the antigen.

The following observations also support the hypothesis that in some fish the antibody response may be independent of T cells. The thymus of some fish species is often primitive and devoid of the differentiated characteristics of mammalian thymuses (in some fish no discernible thymus can be detected). In some species the thymus appears to be *the* source of all lymphocytes, not just T cells as in mammals. The fish thymus, in sharp contrast to the mammalian thymus, participates in antibody production (Ortiz-Muniz and Sigel, 1968, 1971; Sailendri, 1973) and possesses lymphocytes with surface Ig as well as plasma cells (Ellis and Parkhouse, 1975; Fiebig et al., 1977; Warr et al., 1977; Yamaga et al., 1977; DeLuca et al., 1978). As reported by McKinney et al. (1976), allogeneic reactions (MLR) with lymphoid cells of sharks and snappers gave negative results. In xenogeneic combinations, leukocytes of these fish failed to respond (T function) but caused stimulation (B-cell or monocyte function) of human leukocytes. Blastogenic transformation reactions with mitogens gave weak or variable results with lymphocytes from sharks, gars, and catfish (McKinney et al., 1976). In addition, allograft rejection in several classes of fish is of the chronic type.

It should be noted that allograft rejection is not always the function of T cells as this type of immunity can be effected by macrophages, killer cells, etc. We should also emphasize at this point, however, that conclusions which are reached on the basis of weak or poorly developed responses must be tempered by the possibility that the optimal conditions for the experiment may not have been found. For example, results of *in vitro* mitogen studies with catfish lymphocytes can be *dramatically* affected by the choice of medium supplements and incubation temperature (Clem, personal communication).

Studies using mammals have demonstrated that immunization with carrier protein followed by immunization with hapten–carrier leads to enhanced primary response to the hapten (Rajewsky *et al.*, 1969). This has prompted development of the concept of cell cooperation wherein T cells respond to the carrier and B cells respond to the hapten. This hapten–carrier effect has been observed in fish in several laboratories. The results of the experiments on carrier function in fish would seem to argue in favor of the presence of cellular elements analogous to the mammalian T and B cells. However, this would require an assumption that the carrier is recognized by T cells. Yet, pending formal proof of such an assumption, one can challenge this argument on the basis that the carrier may be recognized by non-T cells. In fact, a hapten–carrier effect has been discerned in snappers, a species which appears to be deficient in some T-cell functions. It is possible that the carrier (or helper) effect may be mediated by B-cell subsets. There is precedent for this among mammalian TI responses where one subset of B-cells appears to recognize a carrier (Ficoll) while another B-cell subset may respond to the hapten DNP (Schott and Merchant, 1979). On the other hand, the carrier helper effect may be delivered by a multipotential (bipotential) cell performing both T and B functions. It may be pertinent at this point to cite the work of Ruben and Selker (1975) who demonstrated in a primitive amphibian, the newt, the production of lymphocytes binding both carrier and hapten following immunization to demonstrate hapten–carrier interaction (a multipotential cell?).

3.5.2. Evidence Supporting the Concept of Specialized Subsets of Lymphocytes

3.5.2a. Mitogen Studies. Perhaps the best evidence for delineating subsets of lymphocytes in fish has come from *in vitro* studies of mitogen stimulation of fish lymphocytes. In mammals, LPS is considered a B-cell mitogen whereas Con A and PHA are considered T-cell mitogens. Mitogenic responses in Agnatha and Chondrichthyes suggest, by analogy to mammals, the presence of T-like cells (Lopez *et al.*, 1974; Tam *et al.*, 1977; Sigel *et al.*, 1978), although the requirement for relatively high concentrations of mitogen and the rather low responses suggest that the functions are less well developed in these primitive fish. Again, we should point out, however, that optimal conditions may not have been tested.

Results of mitogenic studies in teleosts, although variable for different species, have given more definitive support to the concept of specialized subsets of fish lymphocytes. Thymocytes from the rainbow trout respond to the T-cell mitogen Con A but not to the B-cell mitogen LPS, while lymphocytes from the pronephros respond to LPS but not to Con A (Etlinger *et al.*, 1976). The observed organ distribution of mitogenic responsiveness suggests an analogy with mammalian mitogenic responses, thereby implying existence of T-like and B-like cellular classes in the rainbow trout.

Support for the concept of specialized lymphocyte classes in fish is further derived from studies combining mitogenic responsiveness and depletion of T-like (or B-like) subpopulations. Thy-1 is a cell surface antigen used as a marker for mouse T cells (Raff, 1970). Immunizing rabbits with mouse brain cells, which have Thy-1-related antigen on their cell surface, and absorbing the antiserum with mouse erythrocytes and liver cells, yields an antiserum specific for mouse T cells, i.e., anti-Thy-1. A similar approach was used by Cuchens and Clem (1977) to prepare a rabbit antiserum against bluegill brain tissue. Treating lymphocytes with this antiserum and with complement specifically depleted the PHA (mammalian T-cell mitogen)-responsive cells whereas the LPS (mammalian B-cell mitogen)-responsive cells were unaffected. These results imply that the PHA-responsive lymphocytes from the bluegill share a common antigenic determinant with brain tissue, which may be similar to the murine Thy-1 antigen, suggesting a T-like population in the bluegill. Additional support for this concept of T-like and B-like cells in fish was provided by these same authors (Cuchens and Clem, 1977) who observed that while the subpopulation that was PHA- and Con A-reactive and shared antigenicity with brain could not form spontaneous rosettes with rabbit erythrocytes, the LPS-reactive population did form rosettes.

3.5.2b. Temperature Effects on Several *in Vitro* Responses. Recent results obtained by Cuchens and Clem (unpublished observations) utilizing differential temperature effects support the concept of a temperature-sensitive T-cell helper stage in the fish immune response. As shown in Table 2, the primary *in vitro* response of bluegill lymphocytes to SRBC, a T-dependent antigen in mammals, is quite good at 32°C but is totally inhibited at 22°C. This is consistent with their earlier observation (Cuchens and Clem, 1977) that the optimal temperature for *in vitro* LPS stimulation of bluegill pronephros lymphocytes was 22°C whereas responses to PHA and Con A were optimal at 32°C. More recent results with catfish peripheral blood leukocytes (Figure 3; Faulman, Cuchens, and Clem, unpublished) are also in agreement with the concept of a temperature-sensitive T-cell stage. As shown in Figure 3, the *in vitro* LPS response of catfish lymphocytes is unaffected in the 17–27°C range, whereas the Con A response is dramatically decreased at lower temperatures. Likewise, the bluegill MLR responses (presumed T-cell function) were found to show a temperature dependence. There was a positive MLR at 32°C but no response at 22°C (Cuchens and Clem, 1977). More will be said in this context below.

3.5.2c. Hapten–Carrier Studies. Studies of the hapten–carrier phenomenon have shown a wide disparity in results obtained with different classes of fish. In the brook lamprey, no hapten–carrier effect was observed (Fujii *et al.*, 1979a,b). The authors are unaware of published studies of this aspect in Chondrichthyes, although Clem obtained negative results with the nurse shark (personal communication). The hapten–carrier effect, however, has been demonstrated in several teleost species (Yocum *et al.*, 1975; McKinney *et al.*, 1976; Ruben *et al.*, 1977; Warr *et al.*, 1977; Weiss and Avtalion, 1977). As discussed previously, one conclusion which might be reached from these findings is that fish possess T-like and B-like cells capable of cooperating in producing an im-

TABLE 2. PRIMARY IN VITRO IMMUNIZATION OF BLUEGILL LYMPHOID ORGAN CELL SUSPENSIONS WITH SHEEP RED BLOOD CELLS[a]

Experiment	Culture	Days in culture	PFC/culture		% recovered[c]		% viable[d]	
			22°C[b]	32°C	22°C	32°C	22°C	32°C
1	Control	5	ND[e]	57	ND	95	ND	89
	Immunized	5	ND	660	ND	115	ND	93
	Control	7	0	50	72	49	88	96
	Immunized	7	0	1045	65	177	94	97
	Control	10	ND	38	ND	62	ND	86
	Immunized	10	ND	810	ND	101	ND	90
2	Control	7	0	0	83	56	89	91
	Immunized	7	0	147	77	109	95	93
	Control	10	0	0	36	40	73	94
	Immunized	10	0	82	36	68	69	90

[a]From Cuchens (1977).
[b]Cultures were maintained at the indicated temperatures.
[c]Cell recoveries are expressed as a percent of the initial number of cells (day 0).
[d]Viability was determined by trypan blue exclusion and is expressed as a percent of the total number of cells recovered from cultures.
[e]ND, not done.
[f]Immunized cultures received SRBC. Cultures were Mischell–Dutton with Bass serum as supplement. Cuchens, 1977, Lymphocyte heterogeneity in teleosts and reptiles, Ph.D. dissertation, University of Florida.

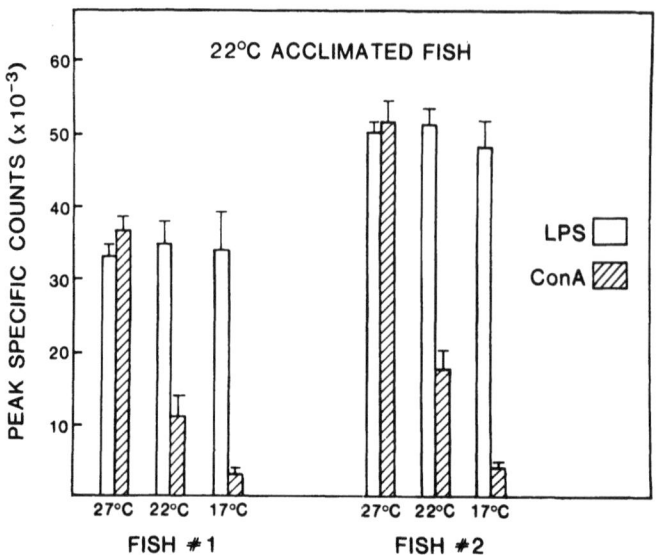

FIGURE 3. Mitogen responses of catfish peripheral blood leukocytes (about 70% of which appeared to be lymphocytes) at various temperatures *in vitro*. These data represent maximal responses at the various temperatures. The background (unstimulated) counts were ≤ 1000 cpm incorporation of [^{3}H]thymidine. (Faulman, Cuchens, and Clem, unpublished observations.)

mune response, presupposing, however, that carrier recognition in fish is a T-cell function. More conclusive evidence to support T helper function is provided by the observations of Ruben *et al.* (1977) and Warr *et al.* (1977). They found that hapten-reactive and carrier-reactive lymphocytes could actually be delineated as separate populations using the nylon wool adherence method (Julius *et al.*, 1973). Conservatively, results of studies of the hapten–carrier effect in fish indicate that there is, in fact, *heterogeneity* of lymphocytes at the functional level. The designation of T-like and B-like subsets by analogy to mammals may be premature, however, based on the above data alone, particularly in light of the observation that DNP–Ficoll, a presumed TI antigen, apparently acts on two subsets of B cells, one recognizing the DNP and the other the Ficoll (Schott and Merchant, 1979).

3.5.2d. **Temperature Effects on *in Vivo* Antibody Responses.** As early as 1948, Bisset proposed that the regulatory effect of temperature on immunity in cold-blooded animals was the result of the inhibition of antibody production and release, not the inhibition of the acquisition of the potential for antibody production (Bisset, 1946, 1947a,b,c, 1948). However, later work by Avtalion *et al.* (1976) demonstrated the synthesis and release of antibody at low temperature when the fish were exposed to the antigen at a higher temperature. These authors suggested that, while priming of immunocompetent cells could occur at low temperatures, the temperature-sensitive event was related to cellular interactions and/or cellular differentiation. This possibility is supported by their observation that carp, when immunized at high temperature with rabbit γ-globulin as a carrier and then transferred to a lower temperature, were able to produce antihapten antibodies following immunization with the hapten–carrier conjugate in the cold. Additional evidence is provided by the use of acetylated BSA (AcBSA) which does not induce antibody production in carp. If AcBSA-primed animals are given BSA at 40 days after priming, the anti-BSA response is of the secondary type. Most important, priming with AcBSA at high temperature leads to a secondary-type response to BSA in animals moved to a lower temperature after priming (Avtalion *et al.*, 1980).

These observations are consistent with the concept of the presence of T-like carrier-dependent helper cells in fish. The results imply that the helper cells are involved in the regulatory effect of temperature. Presumably, at high temperatures the T-like helper cells would function normally, whereas at lower temperatures the primed T-like cells would be unable to become active or useful helper cells.

It is clear that the role of temperature in regulation of the immune response in cold-blooded animals is an area of great potential for dissecting the intricacies of the immune system. This is emphasized by the recent report of Avtalion *et al.* (1980) who demonstrated that not only is the putative T-like "helper" function in fish apparently temperature sensitive, a T-like "suppressor" function may also be subject to regulation by temperature. Comparisons of the effects of temperature on the results obtained with *in vivo* immune responses and *in vitro* assays such as mitogen responsiveness, MLR, and response to SRBC (as discussed above) suggest that the temperature-sensitive component in all of these assays may be a T-like cell which is unable to function at lower temperatures.

3.5.2e. Temperature Effects on Allograft Rejection. The effect of temperature on allograft rejection, a T-cell function in mammals, has also been used in attempting to delineate the presence of subpopulations of lymphocytes in fish. Several laboratories have demonstrated that allograft rejection in fish is delayed at lower temperatures (Hildemann, 1957; Goss, 1961; Stutzman, 1967; Sailendri, 1973; Botham *et al.*, 1980), although a secondary response can still be observed and the degree of the effect varies with the species. Based on observations in amphibians (Marcela and Romanovsky, 1969) and reptiles (Borysenko and Hildemann, 1970), wherein elevation of temperature *during* allograft rejection leads to an accelerated rejection, the hypothesis could be made that, while antigen processing and recognition may take place at lower temperatures, a temperature-dependent event such as cell proliferation or helper cell function may be retarded, similar to the hapten–carrier studies with temperature. The above experiment and the reverse of this experiment, temperature depression during rejection, may be useful approaches to dissecting the effects of temperature on lymphocyte subsets in fish, particularly in light of the effects of temperature on the hapten carrier effect, as discussed above.

3.5.2f. Cell Surface Immunoglobulin Studies. The presence or absence of surface Ig (sIg) has also been used to define T and B cells in higher vertebrates and this technique has been applied to fish. The presence of sIg is usually considered a B-cell characteristic in mammals (Raff, 1970). The observation by several laboratories that fish thymocytes, as well as peripheral blood, splenic, and pronephric lymphocytes, react with antiserum raised against fish serum Ig may seem to suggest that only B-like cells are present in fish. However, several factors must be taken into consideration.

1. First, the quantity of sIg has been reported to be lower on most species of fish thymocytes than on splenic, peripheral blood, and pronephric lymphocytes (Ellis and Parkhouse, 1975; Fiebig et al., 1977; Warr and Marchalonis, 1977; Yamaga et al., 1977; DeLuca et al., 1978), although the bluegill shows similar levels of sIg on lymphocytes from peripheral blood, spleen, pronephros, and thymus (Clem et al., 1977). One might conclude that differences in quantity of sIg in most fish species may reflect differences between subsets of lymphocytes.

2. Fish thymocyte sIg also shows molecular weight and solubility properties different from pronephric and splenic lymphocyte sIg (Fiebig and Ambrosius, 1976; Warr et al., 1976; Ruben et al., 1977).

3. *Some* authors report sIg on mammalian thymocytes. These observations may relate to the possibility that the sIg on both fish and mammalian thymocytes might not be of a recognized isotype but may be comprised of a variable region common to serum Ig but a constant region unique to T cells (IgT?).

4. Cross-reactivity of some anti-fish Ig antisera for carbohydrate moieties may cause artifactually high reactivity against thymocytes (Yamaga et al., 1977). This possibility has now been supported by the observations of Lobb and Clem (personal communication) who found that with 17 different monoclonal antibodies (hybridomas) to catfish Ig, there are only about 20% of the peripheral blood leukocytes which are Ig positive, as opposed to close to 100% with conventional polyclonal antisera to catfish Ig. Thus, since mixtures of different monoclonal antisera never gave over about 20% Ig-positive leukocytes, the implication is that polyclonal antisera recognize antigenic determinants that are present on structures other than Ig, e.g., carbohydrates on cell membranes.

3.5.3. Brief Recapitulation and Overview

With regard to antibody production, while the Agnatha appear to have a rather poorly developed immune response (Good and Papermaster, 1964; Finstad and Good, 1966; Thoenes and Hildemann, 1969), the Chondrichthyes and Osteichthyes clearly are capable of producing specific antibody to a wide variety of antigens (Clem and Sigel, 1963; Finstad and Good, 1966; Clem and Small, 1967; Corbel, 1975; Rijkers, 1980). As discussed earlier, the antibodies produced, however, are restricted to one major class (IgM) and there is a lack of immunological memory in some fish species (Sigel and Clem, 1966), suggesting that some fish respond to immunization as though the antigens are T cell independent. It should be pointed out, however, that other teleosts respond to antigens with a good secondary response (see reviews of Cushing, 1970; Carton, 1973; Corbel, 1975), and a report by Fuller et al. (1978) suggests that sharks may possess an IgG-like molecule. The point should be made, and emphasized, that conclusions reached with regard to lymphocyte heterogeneity in one class of fish may possibly be invalid for another class, especially considering the extremely wide spectrum of different species which have been lumped under the title "fish."

The ability of all extant fish which have been studied to produce antigen-specific antibody clearly indicates the presence of B cells. It is unclear, however, whether the different collaborative interactions observed in fish represent true *thymus-derived T cells* or *subsets of B cells*. Functions normally ascribed to T cells in the mouse, for example, graft rejection and MLR, are relatively poorly developed in fish. Likewise, mitogenic responses are less pronounced in this group (given the previous condition that optimal conditions may not have been found for these assays). Even when one considers that within each ectothermic class (amphibian as well as fish) there is often a heightened efficiency of T-cell functions which correlates with increased structural differentiation of the thymus gland, one must address specific examples, such as the rainbow trout which shows a relatively well-developed MLR, mitogen response, and helper function but a rather poorly differentiated thymus. Additionally, in spite of the thymus differentiation of the different classes of fish, there is no apparent antibody class switch, a function attributed to T cells in mice. Likewise, there are not reports of T-cell-mediated cytotoxicity in fish.

The presence in different vertebrate classes of clear-cut B-cell function, combined with an apparent heightening of those functions attributable to T cells in mammals during evolution of a particular class, suggests that T-cell functions may have evolved several times, by convergent evolution, in each of the vertebrate classes (see Cohen, 1977, and Ruben and Edwards, 1980, for excellent reviews of this area). This concept would essentially involve the fine-tuning and amplification of a primordial B-cell function by the evolution of T cells from the primitive B cells. The specialized T cells could thus function in cellular collaboration to amplify the B-cell function, as well as regulate in a negative fashion. The common origin of both specialized classes of cells would also allow a sharing of the genetic library of combining site specificities, thus allowing the simplest genetic mechanism for explaining recent observations which suggest that T and B cells share common idiotypes and therefore, presumably, common variable regions. Extrapolation of this concept to the invertebrates, *which may or may not be valid*, might suggest that some invertebrates may possess a primitive Ig-like cell surface receptor on primordial "immunocytes." The apparent absence of antibody in the invertebrates could thus be the result of a lack of evolution of this cell population to secretion of the receptor molecule.

We are thus left with the conclusion that while specialized classes of lymphocytes appear to exist in some species of fish, the designation of these subsets as T cells and B cells may be premature. Clearly, by analogy to mammals there is sufficient evidence, particularly for the more advanced fish, to tentatively assign this designation to these subsets, but whether one of these cellular subsets is strictly a thymus-derived T cell remains to be established.

REFERENCES

Alexander, J. B., 1980, Precipitins in the serum of the Atlantic salmon, *Dev. Comp. Immunol.* 4:641.
Avtalion, R. R., Wojdani, A., Malik, Z., Shahrabani, R., and Duczyminer, M., 1973, Influence of environmental temperature on the immune response in fish, *Curr. Top. Microbiol. Immunol.* 61:1.

Avtalion, R. R., Weiss, E., Moalem, T., and Milgram, L., 1976, Regulatory effects of temperature upon immunity in ectothermic vertebrates, in: *Comparative Immunology* (J. J. Marchalonis, ed.), pp. 227–238, Blackwell, Oxford.

Avtalion, R. R., Wishkovsky, A., and Katz, D., 1980, Regulatory effect of temperature on specific suppression and enhancement of the humoral response in fish, in: *Phylogeny of Immunological Memory* (M. J. Manning, ed.), pp. 113–121, Elsevier/North-Holland, Amsterdam.

Bang, F., 1973, Immune reactions among marine and other invertebrates, *BioScience* **23**:584.

Beasley, A. R., Sigel, M. M., and Clem, L. W., 1966, Latent infection in marine fish cell tissue cultures, *Proc. Soc. Exp. Biol. Med.* **121**:1169.

Bisset, K. A., 1946, The effect of temperature on non-specific infections of fish, *J. Pathol. Bacteriol.* **58**:251.

Bisset, K. A., 1947a, Bacterial infection and immunity in lower vertebrates and invertebrates, *J. Hyg.* **45**:128.

Bisset, K. A., 1947b, The effect of temperature on immunity in Amphibia, *J. Pathol. Bacteriol.* **59**:301.

Bisset, K. A., 1947c, Natural and acquired immunity in frogs and fish, *J. Pathol. Bacteriol.* **59**:79.

Bisset, K. A., 1948, The effect of temperature upon antibody protection in cold-blooded vertebrates, *J. Pathol. Bacteriol.* **60**:87.

Borysenko, M., and Hildermann, W. H., 1970, Reactions to skin allografts in the horn shark, *Heterodontis francisci*, *Transplantation* **10**:545.

Botham, J. W., Grace, M. F., and Manning, M. J., 1980, Ontogeny of first set and second set alloimmune reactivity in fishes, in: *Phylogeny of Immunological Memory* (M. J. Manning, ed.), pp. 83–92, Elsevier/North-Holland, Amsterdam.

Carton, Y., 1973, La response immunitaire chez les agnathes et les poissons. Structure des immunoglobulines, *Ann. Biol.* **12**:139.

Clem, L. W., and Sigel, M. M., 1963, Comparative immunochemical and immunological reactions in marine fishes with soluble viral and bacterial antigens, *Fed. Proc.* **23**:1138.

Clem, L. W., and Small, P. A., 1967, Phylogeny of immunoglobulin structure and function. I. Immunoglobulins of the lemon shark, *J. Exp. Med.* **125**:893.

Clem, L. W., McLean, W. E., Shankey, V. T., and Cuchens, M. A., 1977, Phylogeny of lymphocyte heterogeneity. I. Membrane immunoglobulins of teleost lymphocytes, *Dev. Comp. Immunol.* **1**:105.

Cohen, N., 1977, Phylogenetic emergence of lymphoid cells and tissues, in: *The Lymphocyte* (J. J. Marchalonis, ed.), pp. 149–202, Dekker, New York.

Cooper, E. L., 1976, *Comparative Immunology*, p. 88, Prentice–Hall, Englewood Cliffs, N.J.

Corbel, M. J., 1975, The immune response in fish: A review, *J. Fish Biol.* **7**:539.

Cotton, W. T., 1951, Blood cell formation in certain teleost fishes, *Blood* **6**:39.

Cuchens, M. A., and Clem, L. W., 1977, Phylogeny of lymphocyte heterogeneity. II. Differential effects of temperature in fish T-like and B-like cells, *Cell. Immunol.* **34**:219.

Cushing, J. E., 1970, Immunology of fish, in: *Fish Physiology IV* (W. S. Hoar and D. F. Randall, eds.), pp. 465–500, Academic Press, New York.

Davies, H. G., and Haynes, M. E., 1975, Light and electron microscope observations on certain leukocytes in Teleost fish, *J. Cell Sci.* **17**:263.

Davina, J., Hans, M., Rijkers, G. T., Rombout, J. H., Timmermans, L., and van Muiswinkel, W. B., 1980, Lymphoid and non-lymphoid cells in the intestine of cyprinid fish, in: *Development and Differentiation of Vertebrate Lymphocytes* (J. D. Horton, ed.), pp. 129–140, Elsevier/North-Holland, Amsterdam.

de Kinkelin, P., and Dorson, M., 1973, Interferon production in rainbow trout (*Salmo gairdneri*) experimentally infected with Egted virus, *J. Gen. Virol.* **19**:125.

DeLuca, D., Warr, G. W., and Marchalonis, J. J., 1978, Phylogenetic origins of immune recognition: Lymphocyte surface immunoglobulins and antigen binding in the genus *Carassius* (Teleostei), *Eur. J. Immunol.* **8**:25.

Ellis, A. E., 1977, Ontogeny of the immune response in *Salmo salar*: Histogenesis of the lymphoid organs and appearance of membrane immunoglobulin and mixed leucocyte reactivity, in: *Developmental Immunobiology* (J. B. Solomon and J. D. Horton, eds.), pp. 225–231, Elsevier/North-Holland, Amsterdam.

Ellis, A. E., and deSousa, M., 1974, Phylogeny of the lymphoid system. I. A study of the fate of circulating lymphocytes in plaice, *Eur. J. Immunol.* **4**:338.

Ellis, A. E., and Parkhouse, R. M. E., 1975, Surface immunoglobulins on the lymphocytes of the skate, *Raja naevus, Eur. J. Immunol.* **5**:726.

Ellis, A. E., Munroe, A. L., and Roberts, R. J., 1976, Defense mechanisms in fish. I. A study of the phagocytic system and the fate of intraperitoneally injected particulate material in the plaice (*Pleuronectes platessa*), *J. Fish Biol.* **8**:67.

Etlinger, H. M., Hodgins, H. O., and Chiller, J. M., 1976, Evolution of the lymphoid system. I. Evidence for lymphocyte heterogeneity in rainbow trout revealed by the organ distribution of mitogenic responses, *Immunology* **116**:1547.

Fange, R., 1968, The formation of eosinophilic granulocytes in the eosophageal lymphomyeloid tissue in the elasmobranchs, *Acta Zool. (Stockholm)* **58**:125.

Fange, R., and Mattisson, A., 1981, The lymphomyeloid system of the Atlantic nurse shark, *Biol. Bull.* **160**:240.

Fange, R., and Sundell, G., 1969, Lymphomyeloid tissues, blood cells and plasma proteins in *Chimaera monstrosa* (Pisces, Holocephali), *Acta Zool. (Stockholm)* **50**:155.

Fauve, R. M., 1978, Phagocytes, in: *Immunology* (J. F. Bach, ed.), p. 92, Wiley, New York.

Ferguson, H. W., 1976, The ultrastructure of plaice (*Pleuronectes platessa*) leucocytes, *J. Fish Biol.* **8**:132.

Fiebig, H., and Ambrosius, H., 1976, Cell surface immunoglobulin of lymphocytes in lower vertebrates, in: *Phylogeny of Thymus and Bone Marrow-Bursa Cells* (R. K. Wright and E. L. Cooper, eds.), pp. 195–203, Elsevier/North-Holland, Amsterdam.

Fiebig, H., Scherbaum, I., and Ambrosius, H., 1977, Zelloberflachen-immunoglobulin von Lymphozyten des Karpfens (*Cyprinus carpio*), *Acta Biol. Med. Ger.* **36**:1167.

Finn, J. P., and Nielsen, N. O., 1971, The inflammatory response in rainbow trout, *J. Fish Biol.* **3**:463.

Finstad, J., and Good, R. A., 1966, Phylogenetic studies of adaptive immune response in the lower vertebrates, in: *Phylogeny of Immunity* (R. T. Smith, P. A. Miescher, and R. A. Good, eds.), pp. 173–189, University of Florida Press, Gainesville.

Fujii, T., Nakagawa, H., and Murakawa, S., 1979a, Immunity in lamprey. I. Production of haemolytic and haemagglutinating antibody to sheep red blood cells in Japanese lampreys, *Dev. Comp. Immunol.* **3**:441.

Fujii, T., Nakagawa, H., and Murakawa, S., 1979b, Immunity in lamprey. II. Antigen-binding response to sheep erythrocytes and hapten in the ammocoete, *Dev. Comp. Immunol.* **3**:609.

Fuller, L., Murray, J., and Jensen, J., 1978, Isolation from nurse shark serum of immune 7S antibodies with two different molecular weight H-chains, *Immunochemistry* **15**:251.

Gardener, G. R., and Yevich, P. O., 1969, Studies on the blood morphology of three estuarine cyprinodontiform fishes, *J. Fish. Res. Board Can.* **26**:433.

Good, R. A., and Papermaster, B. W., 1964, Ontogeny and phylogeny of adaptive immunity, *Adv. Immunol.* **4**:1.

Good, R. A., Finstad, J., Pollara, B., and Gabrielsen, A. E., 1966, Morphological studies on the evolution of the lymphoid tissue among the lower vertebrates, in: *Phylogeny of Immunity* (R. T. Smith, P. A. Miescher, and R. A. Good, eds.), pp. 149–168, University of Florida Press, Gainesville.

Goss, R. J., 1961, Metabolic antagonists and prolonged survival of scale homografts in *Fundulus heteroclitis, Biol. Bull.* **121**:162.

Grace, M. F., and Manning, M. J., 1980, Histogenesis of the lymphoid organs in rainbow trout, *Salmo gairdneri, Dev. Comp. Immunol.* **4**:255.

Gravell, M., and Malsberger, R. G., 1965, *Ann. N.Y. Acad. Sci.* **126**:55.

Harboe, M., 1963, A note on the absence of immune reactions in myxinoids, in: *The Biology of Myxine* (A. Brodal and R. Fange, eds.), pp. 456–458, Universitetforlaget, Oslo.

Harisdangkul, V., Kabat, E. A., McDonough, R. J., and Sigel, M. M., 1972a, A protein in normal nurse shark serum which reacts specifically with fructosans. I. Purification and immunochemical characterization, *J. Immunol.* **108**:1244.

Harisdangkul, V., Kabat, E. A., McDonough, R. J., and Sigel, M. M., 1972b, A protein in normal nurse shark serum which reacts specifically with fructosans. II. Physiochemical studies, *J. Immunol.* **108**:1259.

Herbermann, R. B., Nunn, M. E., Holden, H. T., and Lavrin, D. H., 1975, Natural cytotoxic reactivity of mouse lymphoid cells against syngeneic and allogeneic tumors. II. Characterization of effector cells, *Int. J. Cancer* **16**:230.

Hibbs, J. B., Lambert, L. H., Jr., and Remington, J., 1972, Control of carcinogenesis: A possible role for the activated macrophage, *Science* **117**:998.

Hildemann, W. H., 1957, Scale homotransplantation in goldfish (*Carassius auratus*), *Ann. N.Y. Acad. Sci.* **64**:775.

Hildemann, W. H., 1972, Phylogeny of transplantation reactivity, in: *Transplantation Antigens: Markers of Biological Individuality* (B. D. Kahan and R. A. Reisfeld, eds.), pp. 3–73, Academic Press, New York.

Hines, R. A., and Spira, D. T., 1973, Ichthyophthiriasis in the mirror corp II: Leukocyte response, *J. Fish Biol.* **5**:527.

Hinuma, S., Abo, T., Kumagai, K., and Hata, M., 1980, The potent activity of fresh water fish kidney cells in cell-killing: Characterization and species-distribution of cytotoxicity, *Dev. Comp. Immunol.* **4**:653.

Holmgren, N., 1950, On the pronephros and the blood in *Myxine glutinosa*, *Acta Zool.* (*Stockholm*) **31**:234.

Huebner, E., and Chee, G., 1978, Histological and ultrastructural specialization of the digestive tract of *Hoplosternum thoracatum*, *J. Morphol.* **157**:301.

Isaacs, A., and Lindenmann, J., 1957, Virus interference. I. The interferon, *Proc. R. Soc. London Ser. B* **147**:1258.

Jakowski, S., and Nigrelli, R. F., 1953, Localized response in fish to experimental inflammation caused by pathogenic bacteria, *Anat. Rec.* **117**:526.

Jordan, H., and Speidel, C., 1924, Studies on lymphocytes. II. The origin, function and fate of the lymphocytes in fishes, *J. Morphol.* **38**:529.

Julius, M. H., Simpson, E., and Herzenberg, L. A., 1973, A rapid method for the isolation of functional thymus-derived murine lymphocytes, *Eur. J. Immunol.* **3**:645.

Lopez, D. M., Sigel, M. M., and Lee, J. C., 1974, Phylogenetic studies on T cells. I. Lymphocytes of the shark with differentiated response to PHA and Con A, *Cell. Immunol.* **10**:287.

Marcela, A., and Romanovsky, A., 1969, The role of temperature in separate stages of the immune response in anurans, *Folia Biol.* **15**:157.

Mattisson, A. G. M., and Fange, R., 1977, Light and electron microscopic observations on the blood cells of the Atlantic hagfish, *Myxine glutinosa*, *Acta Zool.* (*Stockholm*) **58**:205.

McKinney, E. C., 1974, The cellular immune response of the gar, Ph.D. dissertation, University of Miami.

McKinney, E. C., Ortiz, G., Lee, J. C., Sigel, M. M., Lopez, D. M., Epstein, R. S., and McLeod, T. F., 1976, Lymphocytes of fish: Multipotential or specialized?, in: *Phylogeny of Thymus and Bone Marrow-Bursa Cells* (R. K. Wright and E. L. Cooper, eds.), pp. 73–82, Elsevier/North-Holland, Amsterdam.

McKinney, E. C., Smith, S. B., Haines, H. G., and Sigel, M. M., 1977, Phagocytosis by fish cells, *J. Reticuloendothelial Soc.* **21**:89.

McKinney, E. C., McLeod, T. F., and Sigel, M. M., 1981, Allograft rejection in a holostean fish, *Lepisosteus platyrhincus*, *Dev. Comp. Immunol.* **5**:65.

Mackmull, G., and Michels, N. A., 1932, Absorption of colloidal carbon from the peritoneal cavity in the teleost, *Tautogolabrus adspersus*, *Am. J. Anat.* **51**:5.

O'Neill, J. G., 1980, Temperature and the primary and secondary immune responses of three teleosts, *Salmo trutta*, *Cyprinus corpis* and *Notothenia rossii*, to MS2 bacteriophage, in: *Phylogeny of Immunological Memory* (J. J. Manning, ed.), pp. 123–130, Elsevier/North-Holland, Amsterdam.

Ortiz-Muniz, G., and Sigel, M. M., 1968, In vitro synthesis of anti-BSA antibodies by fish lymphoid organs, *Bacteriol. Proc.* **118**:66.

Ortiz-Muniz, G., and Sigel, M. M., 1971, Antibody synthesis in lymphoid organs of two marine teleosts, *J. Reticuloendothelial Soc.* **9**:42.

Pettey, C. L., and McKinney, E. C., 1980, Effect of decreased environmental temperature on spontaneous cytotoxicity in the nurse shark, *Fed. Proc. Abstr.* **39**(Part II):934 (abstr. 3508).

Pettey, C. L., and McKinney, E. C., 1981, Mitogen-induced cytotoxicity in the nurse shark, *Dev. Comp. Immunol.* **5**:53.

Pliszka, F., 1939, Weitere Untersuchungen uber Immunitatsreacktionen und uber Phagozytose bei Karpken, *Zentralbl. Bakteriol. Parasitenkd. Infektionskr. Hyg. Abt. Orig.* **143**:451.

Raff, M. C., 1970, Two distinct populations of peripheral lymphocytes in mice distinguishable by immunofluorescence, *Immunology* **19**:637.

Rajewsky, K., Schirrmacher, V., Nase, S., and Jerne, N. K., 1969, The requirement for more than one antigenic determinant for immunogenicity, *J. Exp. Med.* **129**:637.

Rijkers, G. T., 1980, The immune system of cyprinid fish, Ph.D. thesis, Agricultural University, Wageningen, The Netherlands.

Rijkers, G. T., Teunissen, A. G., Oosterom, R., and van Muiswinkel, W., 1980, The immune system of cyprinid fish. Immunosuppressive effect of the antibiotic oxytetracycline in carp, *Aquaculture* **19**:177.

Roberts, R. T. J., 1974, Melanin-containing cells of teleost fish and their relation to disease, in: *Anatomic Pathology of Teleost Fish* (W. R. Ribelin and G. Migaki, eds.), University of Wisconsin Press, Madison.

Ruben, L. N., and Edwards, B. F., 1980, Phylogeny of the emergence of T–B collaboration in humoral immunity, in: *Contemporary Topics in Immunobiology*, Vol. 9 (J. J. Marchalonis and N. Cohen, eds.), pp. 55–84, Plenum Press, New York.

Ruben, L. N., and Selker, E. U., 1975, Polyfunctional antigen-binding specificity in hapten/carrier responses of the newt, *Triturus viridescens*, *Adv. Exp. Biol. Med.* **64**:387.

Ruben, L. N., Warr, G. W., Decker, J. M., and Marchalonis, J. J., 1977, Phylogenetic origins of immune recognition: Lymphoid heterogeneity and the hapten/carrier effect in goldfish, *Carassius auratus*, *Cell. Immunol.* **31**:266.

Russell, W. J., Taylor, S. A., and Sigel, M. M., 1976, Clearance of bacteriophage in poikilothermic vertebrates and the effect of temperature, *J. Reticuloendothelial Soc.* **9**:91.

Sailendri, K., 1973, Studies on the development of lymphoid organs and immune responses in the teleost, *Tilapia mossambica*, Ph.D. thesis, Madurai University.

Sailendri, K., and Muthukkaruppan, VR., 1975, The immune response of the teleost, *Tilapia mossambica*, to soluble and cellular antigens, *J. Exp. Zool.* **191**:79.

Schott, C. F., and Merchant, B., 1979, Carrier-specific immune memory to a thymus-dependent antigen in congenitally athymicmice, *J. Immunol.* **122**:1710.

Scottizzi, I., 1932, La milza di *Chimaera monstrosa*, *Arch. Ital. Anat.* **29**:560.

Shapiro, H. C., 1975, Immunity in decapod crustaceans, *Am. Zool.* **15**:13.

Sigel, M. M., and Clem, L. W., 1966, Immunologic anamnesis in elasmobranchs, in: *Phylogeny of Immunity* (R. T. Smith, P. A. Miescher, and R. A. Good, eds.), pp. 190–197, University of Florida Press, Gainesville.

Sigel, M. M., Acton, R. T., Evan, E. E., Russell, W. J., Wells, T. G., Painter, B., and Lucas, A., 1968, T2 bacteriophage clearance in the lemon shark, *Proc. Soc. Exp. Biol. Med.* **128**:977.

Sigel, M. M., Lee, J. C., McKinney, E. C., and Lopez, D. M., 1978, Cellular immunity in fish as measured by lymphocyte stimulation, *Mar. Fish Rev.* **40**:6.

Smith, A. M., Wivel, N. A., and Potter, M., 1970, Plasmacytopoiesis in the pronephros of the carps, *Anat. Rec.* **167**:351.

Stutzman, W. J., 1967, Combined effects of temperature and immuno-suppresive drug therapy on allograft rejection in goldfish, *Transplantation* **5**:1344.

Tam, M. R., Reddy, A. L., Karp, R. D., and Hildemann, W. H., 1977, Phylogeny of cellular immunity among vertebrates, in: *Comparative Immunology* (J. J. Marchalonis, ed.), pp. 98–119, Blackwell, Oxford.

Thoenes, G. H., and Hildemann, W. H., 1969, Immunological responses of Pacific hagfish. II. Serum antibody production to soluble antigen, in: *Developmental Aspects of Antibody Formation and Structure* (J. Sterzl and J. Rina, eds.), pp. 711–722, Czechoslovakia Academy of Sciences, Prague.

von Hagen, F., 1936, Die Wichtigsten Endokrinen des Flussaals, Thyreoidea, Thymus, und Hypophyse im Lebenszyklus des Flussaals (*Anguilla vulgaris*) nebst einiger Untersuchunger uber das chromophile und chromophobe Kolloid der Thyreoidea, *Zool. Jahrb.* **61**:467.

Wardle, C. S., 1971, New observations on the lymph system of the plaice, *Pleuronectes platessa*, and other teleosts, *J. Mar. Biol. Assoc. U.K.* **51**:977.

Warr, G. W., and Marchalonis, J. J., 1977, Lymphocyte surface immunoglobulin of the goldfish differs from its serum counterpart, *Dev. Comp. Immunol.* **1**:15.

Warr, G. W., DeLuca, D., and Marchalonis, J. J., 1976, Phylogenetic origins of immune recognition: Lymphocyte surface immunoglobulins in the goldfish, *Carassius auratus*, *Proc. Natl. Acad. Sci. USA* **73**:2476.

Warr, G. W., DeLuca, D., Decker, J. M., Marchalonis, J. J., and Ruben, L. N., 1977, Lymphoid heterogeneity in teleost fish: Studies on the genus *Carassius*, in: *Developmental Immunobiology* (J. B. Solomon and J. D. Horton, eds.), pp. 241–248, Elsevier/North-Holland, Amsterdam.

Watson, J., Trenkner, E., and Cohn, M., 1973, The use of bacterial LPS to show that two signals are required for the induction of antibody synthesis, *J. Exp. Med.* **138**:699.

Watson, L. J., Schechmeister, I. L., and Jackson, L. L., 1963, The haematology of goldfish (*Carassius auratus*), *Cytologia* **28**:118.

Weinreb, E. L., 1963, Studies on the fine structure of Teleost blood cells. I. Peripheral blood, *Anat. Rec.* **147**:219.

Weinreb, E. L., and Weinreb, S., 1969, A study of experimentally induced endocytosis in a teleost. I. Light microscopy of peripheral blood cell response, *Zoologica (N.Y.)* **54**:25.

Weiss, E., and Avtalion, R. R., 1977, Regulatory effect of temperature and antigen upon immunity in ectothermic vertebrates. II. Primary enhancement of anti-hapten antibody response at high and low temperatures, *Dev. Comp. Immunol.* **1**:93.

West, W. H., Cannon, G. B., Kay, H. D., Bonnard, G. D., and Herberman, R. B., 1977, Natural cytotoxic reactivity of human lymphocytes against a myeloid cell line: Characterization of effector cells, *J. Immunol.* **118**:355.

Wojdani, A., Katz, E., Shahraboni, R., and Avtalion, R. R., 1979, Influence of environmental temperature on the fate of soluble and bacterial antigens in carp. *Am. Zool.* **19**:932.

Yamaga, K., Etlinger, H. M., and Kubo, R. T., 1977, Partial characterization of membrane immunoglobulins on rainbow trout lymphocytes, in: *Immune System: Genetics and Regulation* (E. E. Sercarz, L. A. Herzenberg, and C. F. Fox, eds.), pp. 297–304, Academic Press, New York.

Yocum, D., Cuchens, M., and Clem, L. W., 1975, The hapten/carrier effect in teleost fish, *J. Immunol.* **114**:925.

Yokoyama, H. O., 1960, Studies on the origin, development and seasonal variations in the blood cells of the perch, *Perca flavescens*, *J. Wildl. Dis.* **6**:1.

RES Structure and Function of the Amphibia

MARGARET J. MANNING and JOHN D. HORTON

1. INTRODUCTION

The reticuloendothelial system of Aschoff comprises those phagocytic cells which take up carbon and dyes after these have been introduced into the animal (Clark, 1965). Both morphologically and functionally, this system is so intimately related to lymphoid cells and to the immune response that it would be unrealistic to consider one without the other. Furthermore, in amphibians, myeloid and lymphoid tissue occur along with macrophages and other reticuloendothelial cells, in close relationship within the same organs, the spatial separation between myelopoiesis and lymphopoiesis being less distinct in poikilotherms than in mammals (Yoffey, 1960). The organs concerned with these functions in the Amphibia are listed in Table 1. In this chapter we shall consider the structure and function of these organs.

2. THYMUS

2.1. STRUCTURE

Studies on the anatomy, morphology, and embryology of the amphibian thymus extend back to the last century (see, for example, Cooper, 1967a, 1973, 1976, for reviews). As in all jawed vertebrates, the thymus is associated embryologically with the pharyngeal epithelium. In the order Gymnophiona (= caecilians or apodans, an aberrant group of legless amphibians), the thymic anlagen retain the primitive branchiomeric arrangement, epithelial buds developing dorsally in relation to each pair of pharyngeal pouches including those of

---

MARGARET J. MANNING • Department of Biological Sciences, Plymouth Polytechnic, Plymouth, England.   JOHN D. HORTON • Department of Zoology, University of Durham, England.

TABLE 1. ORGAN DISTRIBUTION OF LYMPHOID TISSUE IN THE AMPHIBIA[a]

			Anura			
			Primitive anurans (e.g., *Xenopus*)		Higher anurans (Ranidae and Bufonidae)	
	Gymnophiona	Urodela	Larva	Adult	Larva	Adult
Thymus	+	+	+	+	+	+
Spleen	+	+	+	+	+	+
Kidney	?	±	+	+	+	+
Lymphomyeloid nodes (e.g., jugular body, lymph gland)	−	−	−	−	+	+
GALT[b]						
Lymphoepithelial (e.g., tonsils, ventral cavity bodies)	?	+	+	−	+	+
Nodular gut tissue	−	−	−	+	−	+
Bone marrow	−	+[c]	−	+	−	+
Liver	+	+	+	+	+	±

[a] +, present; −, absent; ±, some lymphoid cells present, but not a major lymphoid organ.
[b] Gut-associated lymphoid tissue.
[c] Bone marrow present in Plethodontidae, but absent in other families of urodeles.

the spiracular region. Of these, the first and last atrophy and the remaining four pairs undergo thymic histogenesis to produce the four-lobed definitive organ (Figure 1a) (Wiedersheim, 1879).

In the order Urodela, the embryo may initially produce five pairs of thymic buds of which the first two disappear, leaving those from visceral pouches III, IV, and V to differentiate and form the thymus (Baldwin, 1918, in *Ambystoma punctatum*). In this case, the thymus isolates as a paired three-lobed organ lying in connective tissue behind the mandible (Figure 1b). In some newts, however, the paired thymus appears as a single lobe, as in *Triturus alpestris* (Tournefier, 1973) and *Pleurodeles waltlii* (Charlemagne, 1977).

In the order Anura, the thymus forms as a single pair of buds from the dorsal epithelium of the second visceral pouches (Figures 2 and 3). The thymus later detaches from the pharyngeal epithelium and undergoes histogenesis to become a lymphoid organ: in the adult, it comes to lie immediately posterior to the tympanum (Figure 1c). As in all vertebrates, the thymus is the first organ to become lymphocytic (Fabrizio and Charipper, 1941, in *Rana sylvatica*; Cooper, 1967a, in *R. catesbeiana*; Manning and Horton, 1969, in *Xenopus laevis*; Curtis and Volpe, 1971, in *R. pipiens*; Horton, 1971a, in *R. pipiens*). In anurans, thymic lymphopoiesis begins within 1 or 2 weeks postfertilization, but in urodeles the lymphoid system is slower to develop and the thymus does not become lympho-

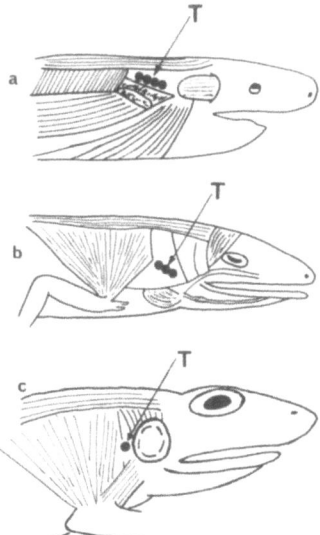

FIGURE 1. Diagram showing the position of the thymus (T) in the head region of members of three amphibian orders. (a) Four-lobed thymus of *Coecilia lumbricoides* (Gymnophiona). (After Wiedersheim, 1879, with modifications.) (b) Three-lobed thymus of *Ambystoma punctatum* (Urodela). (From Baldwin, 1918, with modifications.) (c) Single-lobed thymus of *Xenopus laevis* (Anura).

cytic until a relatively late age—approximately 9 to 10 weeks postfertilization in *T. alpestris* (Tournefier, 1973).

The architecture of the amphibian thymus is similar to that of other vertebrates. The organ is enclosed by a thin connective tissue capsule from which trabeculae may penetrate inwards, carrying capillaries with them deep into the gland. The thymus often lies embedded in fatty tissue. For details of its blood supply and innervation in various amphibian species, see the reviews of James (1939) and Sterba (1950). The epithelial component of the thymus forms a three-dimensional meshwork of cells, with lymphocytes in the interstices. The epithelial cells are connected to each other via their irregularly branched cytoplasmic processes by desmosomal junctions (Nagata, 1976, in *X. laevis*). In urodeles, there is no distinction between thymic cortex and medulla (Webster, 1934, in *Necturus maculosus*; James, 1939, in *N. maculosus*; Klug, 1967, in *Ambystoma mexicanum*; Hightower and St. Pierre, 1971, in *Notophthalmus viridescens*; Curtis et al., 1979b, in *Plethodon glutinosus*). In anurans, on the other hand, there is an obvious distinction between the cortex, which is densely packed with lymphocytes, and the medulla, in which the epithelial component is more conspicuous (Figure 4). Derivatives of the epithelial component include myoid cells, multicellular and unicellular Hassall's corpuscles, cysts, and cells with eosinophilic inclusions.

Myoid cells are recognized by their large size and by their concentrically oriented bundles of striated myofibrils (Figure 5). They have been described in the thymus of a number of different vertebrate groups (see Rimmer, 1977). The function of these cells is not understood but it has been suggested that they are involved in promoting circulation of tissue fluids within the thymus (Törö et al., 1969) or that they may provide a source of self antigen, which might play a role

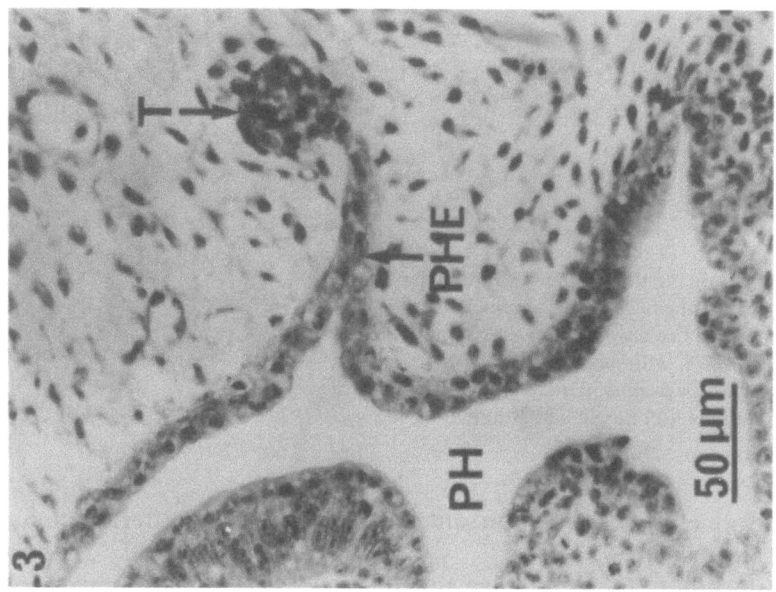

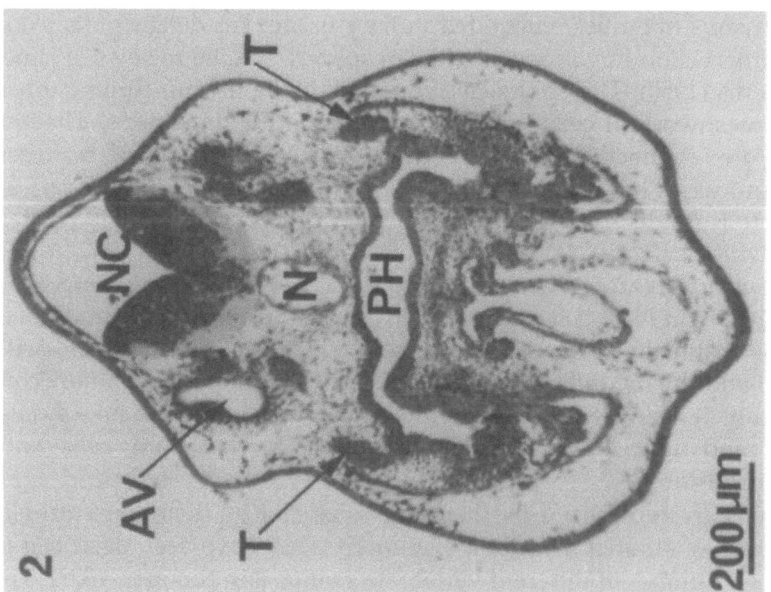

FIGURES 2 AND 3. Sections passing through the pharyngeal region of anuran larvae, showing early formation of the thymic buds. Stain: hematoxylin and eosin. In Figure 2, the two thymic anlagen (T) of an 8-day *Rana pipiens* larva are seen budding dorsally from the second visceral pouches. Figure 3 shows one epithelial thymic bud of a 3-day *Xenopus laevis* larva and its attachment to the pharyngeal epithelium (PHE). AV, auditory vesicle; PH, pharynx; N, notochord; NC, nerve cord.

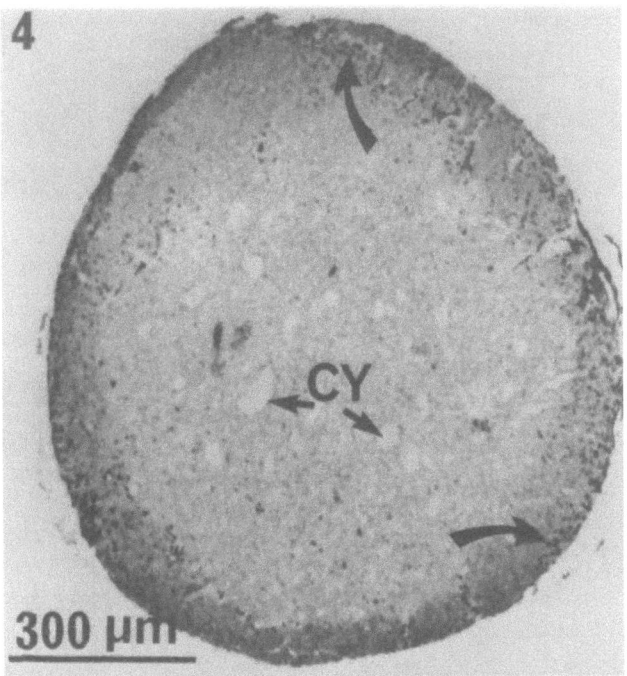

FIGURE 4. Autoradiograph of thymus from *Xenopus laevis*. The toadlet had been injected 4 hr prior to killing with 1 μCi/g body wt of tritiated thymidine (specific activity 22.4 Ci/mmole). Stain: methyl green–pyronine. DNA-synthetic cells (arrows), indicating a rapidly proliferating population of lymphocytes, predominate in the densely staining cortex. Numerous cysts (CY) are apparent in the thymic medulla.

in the development of natural self tolerance (Mackay and Goldstein, 1967). In amphibians, myoid cells exist in both urodeles (see Curtis *et al.*, 1979b) and larval and adult anurans (Sterba, 1950; Kapa *et al.*, 1968; Manning and Horton, 1969; Curtis *et al.*, 1972; Nagata, 1976; Rimmer, 1977; Hanzlikova, 1979). They seem to be lacking from the thymus of some urodeles, however (Hightower and St. Pierre, 1971).

Unicellular or multicellular concentric bodies containing degenerating cells and resembling the Hassall's corpuscles of mammals have been described in all three extant orders of amphibians. Garcia-Herrera and Cooper (1968) found structures similar to Hassall's corpuscles in the thymus of the apodan, tentatively identified as *Typhlonectes compressicaudata*, but later found to be *Nectrocaecilia cooperi* (Cooper, 1976). Hassall's corpuscles occur in urodeles (Curtis *et al.*, 1979b), although they may not always be very conspicuous (James, 1939), and are lacking in *N. viridescens* (Hightower and St. Pierre, 1971) and in *A. mexicanum* (Klug, 1967). They are common in the thymus of anuran amphibians (see Kapa, 1963) where they can usually be detected from a fairly early stage in larval development (Sterba, 1950, in *X. laevis*; Curtis and Volpe, 1971, in *R. pipiens*), although in *R. sylvatica*, Fabrizio and Charipper (1941) reported the first appearance of Hassall's corpuscles around the time of metamorphosis.

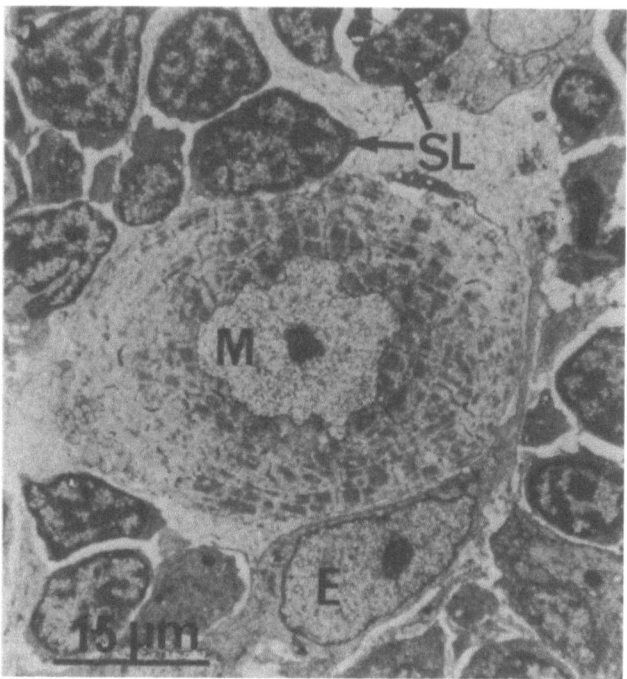

FIGURE 5. Electron micrograph showing various cell types in the thymic medulla of a *Xenopus laevis* larva. M, myoid cell, with its characteristic concentrically arranged myofibrils. SL, small lymphocytes with typical chromatin pattern and thin rim of cytoplasm. E, epithelial cell with cytoplasmic extension around myoid cell. (Courtesy of Dr. J. J. Rimmer, University of Aston.)

There has been considerable recent interest in the various cysts and secretory cells which occur in the amphibian thymus (Curtis et al., 1979b), especially in view of the possible occurrence of thymic hormonal secretions in these poikilotherms (Dardenne et al., 1973). Inter- and intracellular cysts have been described in both urodeles (Dustin, 1911; James, 1939) and anurans (Sterba, 1950; Canaday, 1968; Nagata, 1976; Rimmer, 1977). Granular epithelial cells may be associated with the cysts, as in the slimy salamander, *P. glutinosus*, a species which possesses unusually conspicuous populations of thymic secretory cells (Curtis et al., 1979b). From their studies on the leopard frog thymus, Curtis et al. (1972) suggested that the substance in the cysts might be a candidate for the origin of thymic humoral activity.

The thymic lymphocytes of amphibians show essentially the same morphological features as those of mammals (Figure 5). This is indicated from the ultrastructural studies of a number of workers including Klug (1967), Kapa et al. (1968), Curtis et al. (1972), Nagata (1976), and Rimmer (1977). Studies using tritium-labeled thymidine reveal a high level of proliferation in both the urodele (Hightower, 1975) and the anuran thymus (Horton and Horton, 1975). In the anuran thymus, most of the labeling occurs in the cortex (Figure 4). In the adult newt, *N. viridescens*, Hightower's evidence suggests that large and medium-sized lymphocytes present in the peripheral parenchyma of the thymus give rise to smaller lymphocytes which move centrally and emigrate from the organ.

Plasma cells and their precursors have been described in the thymus of a number of amphibians, including *P. glutinosus* (Curtis *et al.*, 1979b), *R. esculenta* (Kapa *et al.*, 1968), *R. catesbeiana* (Minagawa *et al.*, 1975; Moticka *et al.*, 1973), and *Bufo marinus* (Evans *et al.*, 1966). The significance of B cells within the amphibian thymus is discussed below.

Basophils may occur in the connective tissue thymic capsule and these cells have also been described within the thymic parenchyma in some amphibians (Klug, 1967; Kapa *et al.*, 1968; Csaba *et al.*, 1970; Kapa and Csaba, 1972).

Age-related involution of the thymus is, perhaps, less marked in the Amphibia than in some other vertebrate classes. However, there is a decrease in thymic lymphocyte numbers in older animals (Du Pasquier and Weiss, 1973), and a progressive increase in the amount of connective tissue and collagen is associated with the adult gland (James, 1939; Cooper and Hildemann, 1965). An infiltration of adipose tissue and a decrease in ratio of cortex to medulla also occur and basophils and erythrocytes become more numerous (Cooper and Hildemann, 1975). In *X. laevis*, nests of cells which stain yellow with azan dyes (termed "Gelbzellen" by Sterba, 1950, 1952) form within the adult thymus. Seasonal changes in the anuran thymus have been described in frogs by Sklower (1925), Holzapfel (1937), and Töró *et al.* (1969). During hibernation, the thymus (and bone marrow) lose their lymphoid characteristics, lymphopoiesis ceases, and lymphocyte numbers decrease. Ultrastructural studies of *R. esculenta* show that in the hibernating frog, large numbers of thymocytes leave the thymus, the remaining tissue being rich in myoid cells (Töró *et al.*, 1969). In the spring, the thymus regains its lymphoid character. Seasonal changes in the frog thymus are described in detail by Cohen (1977).

In anurans, changes in the thymus occur during metamorphosis itself; these may well be of significance in relation to the appearance of new self antigens at this critical period. Alteration in lymphocyte numbers associated with metamorphosis in *X. laevis* has been reported by Du Pasquier and Weiss (1973), who described a reduction in the number of thymocytes, amounting to a 5- to 10-fold depletion by the end of metamorphosis when compared with the maximum number of lymphocytes found in the larval thymus. After metamorphosis is complete, the thymus starts to grow again and lymphocyte numbers rise to reach maximum levels at 4 to 6 months of age. The beginning of this second period of thymic growth is accompanied by a diminished expression of thymocyte membrane-bound immunoglobulin as measured by the immunofluorescence "spot" technique (Du Pasquier and Weiss, 1973). These workers suggest that this phenomenon reflects the emergence of certain functionally mature T-cell populations, such as those involved in production by the adult of low-molecular-weight, high-affinity (IgY or IgG-like) antibody.

The relative ease with which membrane-bound immunoglobulin can be detected on thymic lymphocytes of both urodele and anuran amphibians (Du Pasquier *et al.*, 1972; Charlemagne and Tournefier, 1975; Jurd and Stevenson, 1976; Nagata and Katagiri, 1978) and other poikilotherms is of considerable interest. Thus, there is still controversy as to the nature of the antigen receptor on mammalian T cells (see Warr *et al.*, 1980). In fact, some reappraisal of the poikilotherm experiments may be necessary in view of the finding of cross-reactive anticarbohydrate activity in anti-trout and anti-frog immunoglobulin antisera (Mattes and Steiner, 1978; Yamaga *et al.*, 1978). Indeed, *R. pipiens* thymocytes incubated with F(ab)^2 anti-IgM or anti-Ig"G" antisera lack the membrane-associated immunoglobulin determinants detected on B cells (Zettergren *et al.*, 1980). Similarly, a monoclonal anti-*Xenopus* IgM that detects LPS-reactive B cells by immunofluorescence, fails to stain splenic lymphocytes that respond to PHA, Con A, or alloantigens. This reagent also fails to stain larval or adult *Xenopus* thymocytes (Bleicher and Cohen, 1981).

2.2. FUNCTION

The thymus is not involved in antigen trapping or carbon clearance (Turner, 1969). Like the mammalian thymus, it is a primary lymphoid organ concerned with the differentiation of T cells and their export to the periphery. For example, the role of the thymus in establishing a fully competent lymphocytic population capable of effecting normal allograft rejection has been demonstrated both in urodeles (Charlemagne and Houillon, 1968; Cohen, 1969, 1970; Tournefier, 1972, 1973; Charlemagne, 1974; Fache and Charlemagne, 1975) and in anurans (Cooper and Hildemann, 1965; Du Pasquier, 1965; Manning, 1971; Curtis and Volpe, 1971; Horton and Manning, 1972; Tochinai and Katagiri, 1975; Rimmer and Horton, 1977).

In the Anura, there is a clear-cut functional dichotomy of the immune system into thymus-dependent and thymus-independent components which closely resemble their avian and mammalian counterparts. Thus, following early thymectomy, the peripheral lymphocytes are unable to respond *in vitro* in mixed lymphocyte reactions or to T-cell mitogens, while still retaining their normal reactivity to B-cell mitogens (Du Pasquier and Horton, 1976; Manning *et al.*, 1976; Donnelly *et al.*, 1976; Green and Cohen, 1979). Furthermore, the early thymectomized *Xenopus* toad can produce normal levels of antibody in response to "classical" thymus-independent antigens (Collie *et al.*, 1975; Tochinai, 1976a) but lacks the capacity for antibody production to antigens known to be thymus-dependent in mammals (Turner and Manning, 1974; Horton and Manning, 1974b; Tochinai and Katagiri, 1975; Horton *et al.*, 1977a).

There is strong evidence that the thymus of anurans generates a population of helper T cells with a collaborative role in antibody production. Thus, experiments with the hapten trinitrophenyl (TNP) conjugated to a foreign erythrocyte carrier have revealed that good antihapten responses require low-dose carrier-specific priming (see review of Ruben and Edwards, 1979). The morphology and kinetics of the rosette-forming cell (RFC) response in the spleen and thymus suggest the possibility that helper (carrier-reactive) cells emanate from the thymus (Ruben, 1975). Recently, thymectomy experiments have strengthened this suggestion, since TNP–sheep red blood cell reactivity is abolished following thymic removal in larval and adult life (Horton *et al.*, 1979; Gruenwald and

Ruben, 1979). In contrast, anti-TNP RFC reactivity following injection of the same hapten conjugated to a thymus-independent carrier (LPS from *Escherichia coli*) is unaffected by thymus removal. Furthermore, it has been shown that *in vitro* immunoglobulin production to the hapten DNP requires collaboration of primed B- and T-like lymphocytes (Blomberg et al., 1980).

Cellular cooperation in antihapten responses also occurs in urodeles (Ruben et al., 1973; Ruben, 1975) but it is far from certain whether the thymus plays the same role as in anurans. Thus, in *P. waltlii*, the response to *Salmonella* H antigen (which is thymus-dependent in mammals) is only minimally impaired by thymectomy (Tournefier and Charlemagne, 1975), while in the axolotl thymectomy actually increases the antibody levels elicited by horse erythrocytes (Charlemagne and Tournefier, 1977). These authors suggest that thymic regulation in urodeles may be effected by means of immunosuppressive mechanisms rather than through helper cells. The existence of suppressor cell subsets in anurans has also recently been indicated (Du Pasquier and Bernard, 1980).

One fascinating aspect of thymic function in amphibians is the fact that the thymus needs to be present for only a very short period in order to establish a full complement of peripheral cells capable of effecting normal alloimmune responses and mixed lymphocyte culture reactivity, whereas it is required for a considerably longer larval period to establish normal antibody production and T mitogen reactivity (Horton et al., 1977a; Horton and Sherif, 1977; Manning and Collie, 1977). This suggests the possibility of a sequential generation of different subsets of T cells, a suggestion which is backed up by preliminary thymus transplantation studies using isogeneic *Xenopus* (Du Pasquier and Cohen, quoted in Cohen and Turpen, 1978).

In anurans, it is possible to demonstrate mixed lymphocyte responses and T-cell mitogen reactivity of thymus cells *in vitro* (Du Pasquier and Weiss, 1973; Du Pasquier and Miggiano, 1973). In urodeles, on the other hand, this *in vitro* reactivity appears to be acquired only after lymphocytes have left the thymus for the periphery (Collins and Cohen, 1976). Anuran thymic cells differ from those of mammals, however, in their ability to respond to the B-cell mitogen LPS (Horton et al., 1980). This finding of intrathymic B cells is consistent with the presence of antibody-producing cells in the thymus of antigen-stimulated anurans (Evans et al., 1966; Moticka et al., 1973; Minagawa et al., 1975). It is inconsistent, however, with recent reports that *Xenopus* thymocytes are negative with respect to membrane-associated IgM (Bleicher and Cohen, 1981). This inconsistency could be resolved if LPS were mitogenic for subpopulations of T cells or for membrane-Ig-negative B cells. Indeed, Bleicher and Cohen (1981) have recently demonstrated that in the presence of fetal calf serum, membrane-Ig-negative thymocytes do proliferate when they are cultured with LPS. The failure of thymectomized *Xenopus* to produce antihapten responses following TNP–Ficoll injection (this hapten–carrier complex is thymus-independent in mammals) suggests the possibility that the anuran thymus may actually spawn some B-cell subsets (Horton et al., 1979). However, reconstitution experiments on thymectomized *Xenopus*, both *in vitro* (Ruben et al., 1977) and *in vivo* (Katagiri et al., 1980; Kawahara et al., 1980), indicate, at least in terms of hemolytic response to foreign erythrocytes, that the thymus provides helper lymphocytes rather than cells which actually produce antibody.

3. SPLEEN

3.1. STRUCTURE

The spleen is an important lymphoid organ in all jawed vertebrates. Like other secondary lymphoid organs, its structure is such as to provide the anatomical framework for the meeting of antigen with antigen-reactive cells, to accommodate the proliferating lymphoid populations which result from antigenic stimulation, to permit cellular cooperation, to house developing effector cells, and to provide for the appropriate release of these cells and their products (see Manning and Turner, 1976).

The spleen begins as a mesenchymal thickening in the dorsal mesogastrium. In apodans, the adult spleen of *Nectrocaecilia cooperi* has been described as an elongated organ which resembles the spleen of other vertebrates both in appearance and in its content of lymphopoietic and myelopoietic cells (Garcia-Herrera and Cooper, 1968). In the apodans *Siphonopus indistinctus* and *Hypogeophis rostratus*, Weilacher (1933) noted that the spleen produces "round cells," presumably lymphocytes. In urodeles, also, the spleen is an elongated organ, whereas in anurans it is ovoid. Furthermore, the urodele spleen differs from that of anurans in being vascularized by multiple intermediate-sized branches of the major splenic arteries and veins (Cowden and Dyer, 1971). As in all vertebrates, lymphoid tissue accompanies the arterial vasculature. In urodeles, however, the spleen is not demarcated into well-defined white pulp and red pulp areas (Hightower and St. Pierre, 1971; Cowden and Dyer, 1971). Distinction between red and white pulp is also absent in the primitive anuran *Ascaphus truei* (Cooper and Wright, 1976). Cooper and Wright describe a progression of splenic architecture from the primitive Ascaphidae to the more advanced Ranidae and Bufonidae. In *Scaphiopus couchii* of the family Pelobatidae, lymphocytes occur in relatively discrete units suggestive of white pulp areas, while in *Rana* spp. and *Bufo* spp. the red and white pulp regions are well defined. The most distinctive white pulp organization, however, occurs not in these advanced families but in the relatively primitive aglossan *X. laevis*. This may perhaps be accounted for by the fact that *Xenopus* differs from ranid and bufid frogs and toads in its lack of lymphomyeloid nodes, the spleen being the only lymphoid organ with any complexity of organization (Turner, 1969).

The spleen of *X. laevis* is described in detail by Sterba (1951). Each white pulp region (Milzfollikel) is clearly delineated by a boundary layer (Grenschichtmembran). Outside this, there is a cuff of scattered lymphocytes which extends irregularly into the red pulp as a perifollicular zone (Figure 6) (Turner and Manning, 1973). The central arteriole provides capillaries which pass outwards to the periphery of the follicle: the capillary walls contribute to the boundary layer and they then terminate just beyond this position (Figure 7) to empty their contents into the red pulp perifollicular area. It is perhaps not surprising therefore that it is this red pulp region that first receives injected carbon or antigen (Turner, 1969; Collie, 1974). It is also of interest that the perifollicular zone is a thymus-dependent area in *Xenopus*, since it is a region which shows severe lymphocytic depletion after early thymectomy (Manning, 1971; Horton and Manning 1974a; Tochinai, 1976b). Moreover, it has now been shown that histocompatible thymus (Tochinai *et al.*, 1976) or injected splenic or peripheral blood lymphocytes (Nagata and Tochinai, 1978) have the capacity to restore the lymphocytic nature of the perifollicular region in reconstitution experiments on thymectomized *Xenopus*. Furthermore, this region is depleted following carcinogen (*N*-methyl-*N*-nitrosurea) treatment and, together with the thymic cortex, it remains sparsely populated with lymphocytes, even when other areas have recovered (Balls *et al.*, 1980).

3.2. FUNCTION

The reticuloendothelial function of the spleen has been studied in detail by Diener and Nossal (1966) and by Turner (1969) following the injection of colloidal carbon by various routes. Carbon particles can be detected in the red pulp shortly after injection (within 1 to 3 hr). In *B. marinus*, the particles remain in the red pulp and do not enter the white pulp (Diener and Nossal, 1966). Similarly, in *A. truei*, carbon-laden phagocytic cells circumscribe the white pulp area (Cooper and Wright, 1976). In *X. laevis*, however, the white pulp, although clear of carbon at day 1 after injection, later begins to accumulate carbon and by 3 to 4 weeks contains dense aggregates of the injected material in its central region (Figure 6).

During larval development in *X. laevis*, the free macrophages of the body cavities are the first line of defense against intraperitoneally injected particles: these macrophages function at an early age (about 7 days at 23°C). Once the lymphoid organs have developed, carbon particles borne by blood and lymph are removed mainly by cells of the spleen and liver in both larvae and adults (Turner, 1969). In studies using colloidal thorium dioxide, Coleman and Phillips (1972) reported that particles taken into macrophages of the spleen or into Kupffer cells of the liver were not eliminated from the body, but could still be detected 3 years after injection.

When the injected material is antigenic in nature, it may be located initially in the same areas where carbon first appears. Subsequently, however, antigens are handled in a different manner from nonbiotic substances and they may be subject to long-term retention extracellularly on the surfaces of dendritic cells. This type of antigen localization, which is in the form of immune complexes in birds and mammals, is probably the only way in which antigens are held in the body for a considerable period of time (Nossal and Ada, 1971). Its functional significance in terms of B-cell proliferation and immunological memory has recently been reviewed by van Rooijen (1980). In amphibians, Diener and Marchalonis (1970) showed extracellular trapping of ^{125}I-labeled *Salmonella ade-*

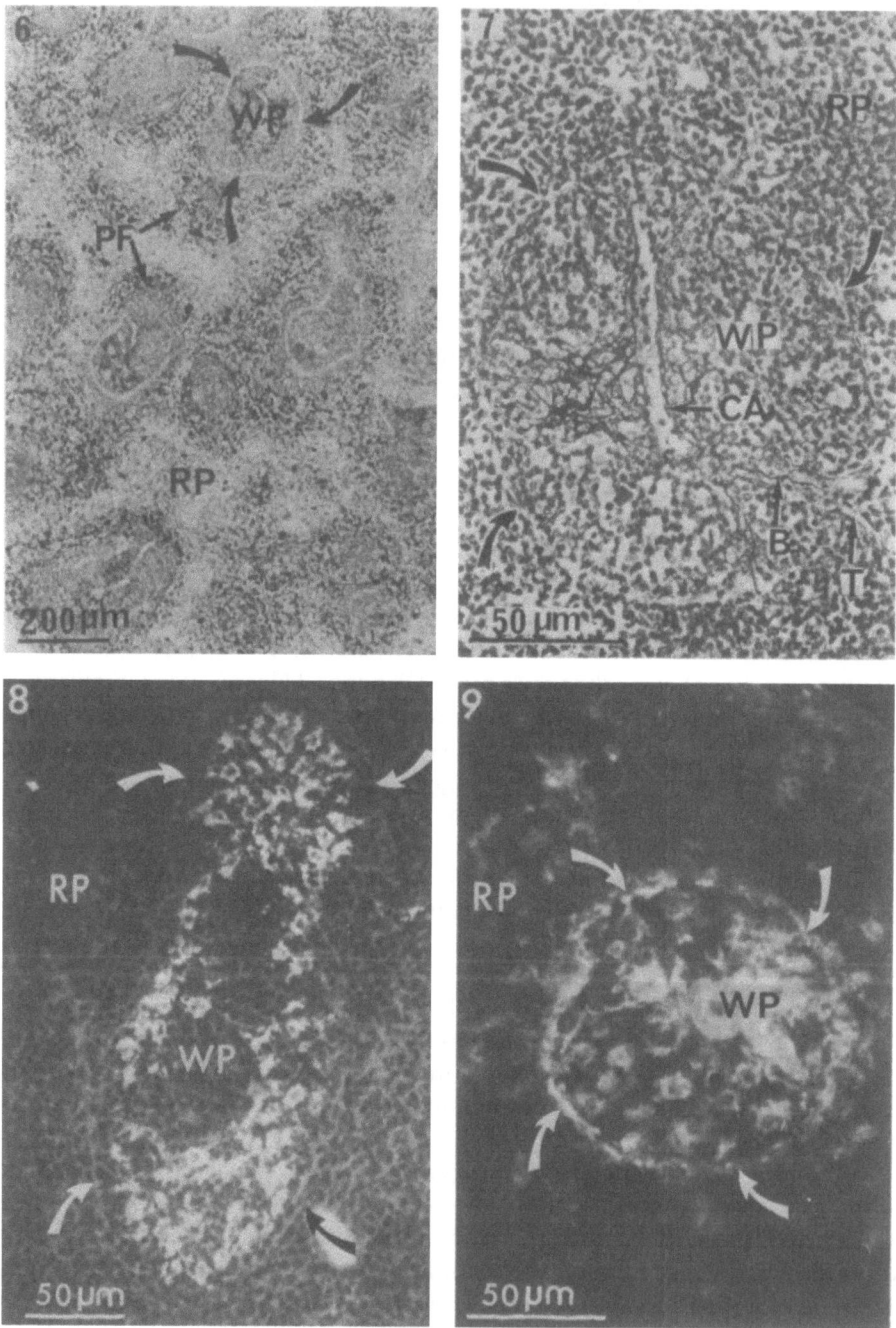

FIGURE 6. Spleen section from *Xenopus laevis* toadlet that had been injected 3 weeks prior to killing with India ink via the dorsal lymph sac. Stain: methyl green–pyronine. At this time the ink is concentrated in the perifollicular (PF) red pulp zone and in the central region of the white pulp follicles (WP). Arrows point to boundary layer around one white pulp follicle. RP, red pulp.

laide flagellae in *B. marinus*. Antigen localization displayed random distribution on dendritic cells in the jugular bodies and in the splenic red pulp. The antigen, like carbon, failed to enter the white pulp. In contrast, in *X. laevis* (which has no jugular bodies), the spleen is the only organ to retain antigen and it does so in a distinct zone near the periphery of the white pulp (Figure 8). Following injection of a soluble protein antigen (human γ-globulin, HGG) in adjuvant and subsequent assay of spleen sections with fluorescein-labeled antibody, the bright fluorescence in this region was attributed to the presence of a mass of tiny fluorescent spots. The latter often follow the contour of just one cell or may surround a group of smaller cells (Horton and Manning, 1974b). The failure of this form of antigen retention to occur in early thymectomized *Xenopus*, and the general similarity to the picture seen in the germinal centers of birds and mammals, has led to the suggestion that the antigen is being trapped extracellularly in the form of antigen–antibody complexes. Particulate (cellular) antigens such as formalin-killed *Aeromonas salmonicida* bacteria are also retained by the spleen of *Xenopus*, but in a different pattern (Figure 9), the fluorescence being more randomly distributed in the white pulp, together with conspicuous staining of the boundary layer (Secombes and Manning, 1980). In contrast to these experiments on *Xenopus*, unpublished studies (M. J. Manning) on *R. pipiens* show no distinct fluorescent areas in the spleen at any time after injection of HGG in adjuvant: instead, antigen is seen retained within the jugular bodies as discussed below. It seems that the exact site at which antigen is trapped may vary from one amphibian group to another, but the dendritic pattern of antigen localization remains similar.

The spleen usually responds to antigen by cellular proliferation (Figure 10) and by an increase in the number of large pyroninophilic cells (Figure 11). This has been described in *X. laevis* following injection of soluble or particulate antigens (Turner and Manning, 1973) and in response to the application of skin allografts (Horton *et al.*, 1977b). In *R. catesbeiana*, on the other hand, pyroninophilia was more conspicuous in other lymphoid organs rather than in the spleen (Minagawa *et al.*, 1975). In *Xenopus* injected with *Salmonella tennessee*, pyroninophilic cells failed to appear in any lymphoid organ despite antibody production (Mitsuhashi *et al.*, 1971). In *B. marinus*, the pyroninophilic cells appeared at the periphery of the white pulp areas, also around blood vessels, after

---

FIGURE 7. Silver-stained preparation of spleen from *Xenopus laevis* toadlet showing reticulin network in a white pulp (WP) follicle. CA, central arteriole with capillary branches (B) passing to the periphery of the white pulp follicle. T, termination of one of these capillaries at the boundary layer. Arrows point to boundary layer. RP, red pulp.

FIGURES 8 AND 9. Immunofluorescent preparations showing antigen localization in spleen of *Xenopus laevis* toadlets that had been injected 3 weeks prior to killing with either HGG in adjuvant (Figure 8) or formalin-killed *Aeromonas salmonicida* (Figure 9). For details of injection schedule, see Secombes and Manning (1980). The spleen sections were incubated with the relevant fluorescein-labeled antiserum and the bright fluorescence indicates the presence of antigen. In Figure 8, HGG is localized within the white pulp (WP), in a peripheral zone near its border. In Figure 9, *Aeromonas* is more randomly distributed in the white pulp and there is conspicuous staining of the boundary layer. Arrows point to boundary around the white pulp. RP, red pulp.

the injection of bovine serum albumin (BSA): furthermore, immunofluorescent studies demonstrated the presence of plasma cells containing anti-BSA antibody (Evans et al., 1966). Booster injections of antigen do not appear to enhance splenic pyroninophilia (Diener and Nossal, 1966; Manning and Turner, 1972).

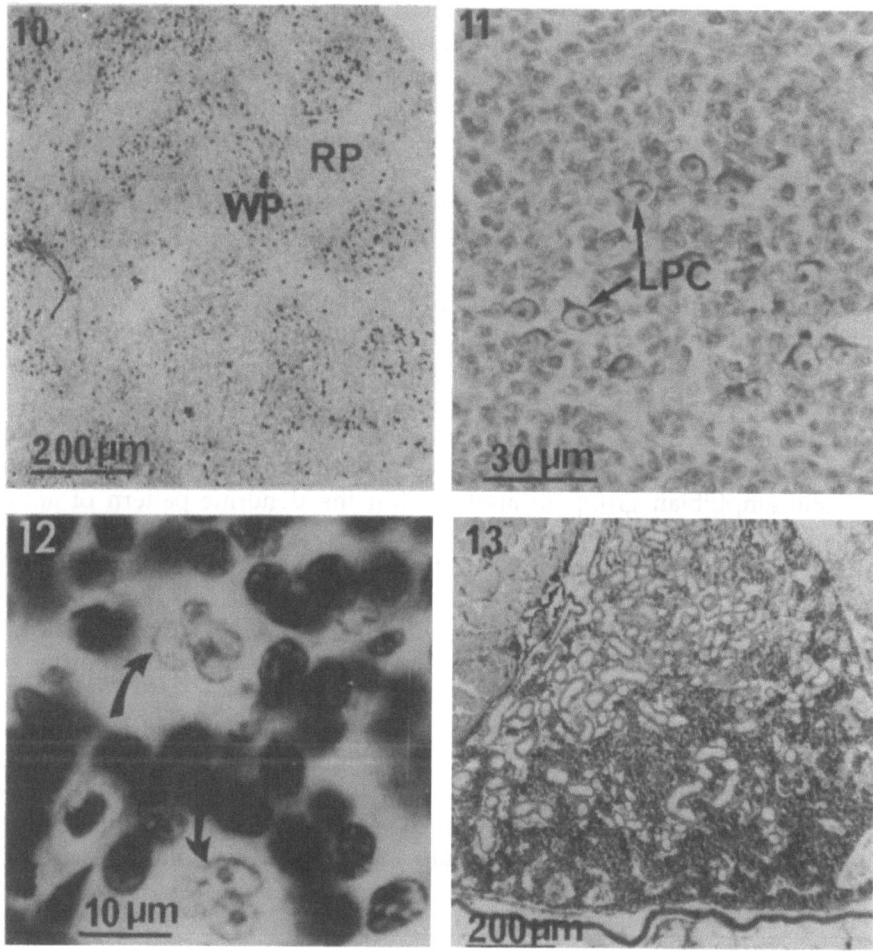

FIGURES 10 AND 11. Enhanced cellular activity in spleens of *Xenopus laevis* following antigen administration. Stain: methyl green pyronine. Figure 10 is an autoradiograph (for technical details see Figure 4) of spleen taken from a toadlet that was in the process of rejecting a skin allograft. High levels of [^{3}H]thymidine-labeled cells [particularly noticeable in the white pulp (WP regions] are seen in allografted but not autografted animals (see Horton et al., 1977b, for details). RP, red pulp. Figure 11 shows large pyroninophilic cells (LPC) in the red pulp of a toadlet injected with foreign red cells (see Turner and Manning, 1973, for details).

FIGURE 12. Two "degenerating macrolymphocytes" (arrows) are seen among other lymphoid cells in the splenic white pulp of *Xenopus laevis*. These cells become conspicuous in hyperimmunized animals and when adjuvants are injected. Stain: hematoxylin and eosin.

FIGURE 13. Extensive intertubular lymphomyeloid tissue located in the mesonephros of a *Rana pipiens* larva. Stain: hematoxylin and eosin.

Cells with the appearance by light microscopy of plasma cells and large pyroninophilic cells are often found in considerable numbers in the dorsal mesogastrium and connective tissue adjacent to the spleen of immunized animals: they are also seen in the mesentery between the kidneys (M. J. Manning, unpublished observations). This histological picture is not inconsistent with circumstantial evidence from other studies, such as those on *N. viridescens* (see Ruben and Edwards, 1979), where it seems that sensitized cells may be leaving the spleen after secondary challenge, to continue antibody synthesis elsewhere in the body. Kraft and Shortman (1972), using *B. marinus*, also noted that the spleen is not a closed system with respect to the production of antibody-forming cells.

Proliferative responses and the increase in pyroninophilia reach a peak in the spleen of *Xenopus* at approximately 3 weeks after antigen administration, the peak serum antibody response occurring at week 8 (Turner and Manning, 1973). Although the pyroninophilic cells resemble those of avian and mammalian germinal centers, they are not arranged into any closely packed accumulations to which the term germinal center could be applied. Initially, the pyroninophilia occurs mainly in places where antigen is retained, but later it becomes more widespread and pyroninophilic cells extend into the perifollicular zone and deeper into the red pulp. Vacuole formation is often very marked in the spleens of immunized amphibians, as noted in *Xenopus* injected with HGG in adjuvant (Manning and Turner, 1972) and in *R. catesbeiana* larvae injected with foreign erythrocytes (Moticka *et al.*, 1973). The vacuolated areas appear to correspond with regions of phagocytic activity. In hyperimmunized animals and when adjuvants are used, the reactive zones of the spleen eventually become filled with pale-staining histiocytic tissue (Turner *et al.*, 1974) and "degenerating macrolymphocytes" (Sterba, 1950) become conspicuous.

The "degenerating macrolymphocytes" of *X. laevis* (Sterba, 1950, 1951) have recently been studied by Baldwin and Cohen (1981) who present evidence that these cells are neither lymphoid nor the effete by-products of lymphocyte proliferation. Rather, they are large mitotically active cells with abundant electron-lucent cytoplasm, large hyperlobated nuclei, and prominent nucleoli (Figure 12). Like mammalian dendritic cells, these cells (which the authors provisionally name XL cells) are located in B-lymphocyte-rich splenic follicles and have long cytoplasmic processes that are in intimate contact with adjacent lymphocytes. Some of the cytoplasmic processes extend through the T-lymphocyte-rich marginal zone into red pulp and appear to trap and transport foreign material (colloidal carbon and human IgG) from its initial site of entry in the splenic red pulp into the white pulp. In distinction to macrophages, XL cells do not phagocytose large quantities of carbon. They also fail to stain for the nonspecific esterase, either in a diffuse pattern characteristic for macrophages or in the punctate pattern found in most lymphoid cells in this species. In contrast to adjacent B lymphocytes in the white pulp follicle, XL cells do not stain for cytoplasmic Ig. Unlike T lymphocytes, early larval thymectomy does not appear to interfere with their development (Manning, 1971). It is for these reasons that Baldwin and Cohen (1981) propose that this cell may be a primitive follicular dendritic cell.

When the antigen is a live organism, as in the experiments of Clothier and Balls (1973) using *Mycobacterium marinus* bacilli in *Xenopus*, the initial splenic response follows the pattern of proliferation and pyroninophilia described above, but if large doses are administered, this may be superceded by granulomata formation. The latter occurs readily in thymectomized animals, presumably as the result of failure of an effective immune response (Clothier, 1972).

The ability of amphibian spleen cells to secrete antibody has been demonstrated following both *in vitro* (Auerbach and Ruben, 1970; Azzolina, 1975) and *in vivo* immunization (see Ambrosius and Hanstein, 1971). Antibody has been detected by immunofluorescence (Kent *et al.*, 1964; Evans *et al.*, 1966) and by plaque-forming cell (PFC) production (Moticka *et al.*, 1973). Specific PFC production is even seen in spleens of larval *Alytes obstetricans* when less than one million lymphocytes are found in total in the body (Du Pasquier, 1970). Lymphocytes that bind multilayers of foreign erythrocytes (secretory RFCs) are also thought to represent antibody-releasing cells and these can be found in several anuran and urodele species after immunization (see Ruben and Edwards, 1979). The splenic RFC response seems slow to develop and is of low magnitude in the urodele *P. waltlii* (Debons and Deparis, 1973). The immunocyto-adherence assay, combined with hapten–carrier immunization, has proved a powerful tool for analysis of cellular cooperation between helper and antibody-producing cells and for investigating immunoregulation in amphibians (Ruben, 1976; Ruben and Edwards, 1979). It has also been used for examining the specificity of antibody production (Amirante, 1968) and the effect of temperature on the kinetics of the immune response (Cone and Marchalonis, 1972). Antibody-forming cells appear to be the produce of a rapidly proliferating population of stimulated lymphocytes and mature in a sequential fashion as in mammals (Diener and Marchalonis, 1970; Kraft and Shortman, 1972). The morphology of amphibian plasma cells has been described by a number of workers, both in urodeles (Tooze and Davies, 1968) and in anurans (Cowden *et al.*, 1968a,b). Plasma cells may be more conspicuous in the kidney and other lymphoid tissues than in the spleen. Electron microscopic studies of *B. marinus* reveal that these amphibian antibody-producing cells have abundant endoplasmic reticulum and a complex Golgi zone: in some cases, the mature plasma cells show considerable intracisternal enlargement, which has led Cowden and Dyer (1971) to suggest that anuran plasma cells store a larger amount of immunoglobulin than their mammalian counterparts.

Functional heterogeneity of amphibian lymphocytes (from both spleen and other peripheral lymphoid tissues) can be demonstrated *in vitro* following mitogen stimulation or mixed lymphocyte reactivity. Such *in vitro* responses were first described in *B. marinus* by Goldshein and Cohen (1972). From the many subsequent studies performed on anurans (see review by Wright *et al.*, 1978), important points to emerge are: (1) that different culture conditions are required for achieving optimal stimulation with T- or B-cell mitogens (Goldstine *et al.*, 1976), and (2) that early thymectomy abrogates T mitogen reactivity (Manning *et al.*, 1976) and mixed lymphocyte responses (Du Pasquier and Horton, 1976) of cells in the periphery, whereas B mitogen reactivity is left intact. The percentage of spleen lymphocytes bearing surface immunoglobulins is significantly elevated in thymectomized *Xenopus* (Weiss et al., 1973). The level of response of urodele spleen cells to B-cell mitogens is comparable to that seen in anurans (Collins et al., 1975). In contrast, urodele lymphocytes display only a poor and delayed response to T-cell mitogens: mixed lymphocyte culture reactivity is also significantly lower in urodeles than in anurans (see Cohen and Collins, 1977). These findings could possibly indicate an earlier evolution of B-cell reactivity, which afforded protection against naturally occurring bacterial products.

The ability of both urodele and anuran spleen cells to manifest delayed hypersensitivity reactions has been demonstrated *in vitro* in macrophage migration-inhibition tests (Ambrosius and Drössler, 1972; Tahan and Jurd, 1979; Rimmer and Gearing, 1980). Tahan and Jurd were able to transfer the hypersensitivity passively in *Ambystoma* by means of splenocytes. There is also evidence in *A. mexicanum* that nonsensitized spleen populations include cells with K-cell-like activity, which can effect cytotoxic killing of antibody-coated target cells (Jurd and Doritis, 1977).

A number of experiments have demonstrated the alloimmune competence of amphibian spleen cells. Thus, in Japanese newts (*Cynops pyrrhogaster*), autologous spleen cells transferred to irradiated hosts can restore alloimmunity (Murakawa, 1968). Alloreactive spleen cells in *Xenopus* have also been demonstrated in spleen transfer and autoradiographic studies (Horton et al., 1977b). In contrast, in *R. catesbeiana*, although antibody synthesis could be restored by autologous spleen cell transfer to irradiated larvae, allograft rejection could not (Brown and Cooper, 1976). Graft-versus-host reactive spleen cells have also been demonstrated in amphibians (Clark and Newth, 1972). Graft-versus-host reactivity is strong in anurans (Brown et al., 1975) but weak in urodeles (see Cohen and Collins, 1977). These findings, along with allograft rejection and mixed lymphocyte culture studies, are consistent with the thesis that anurans possess a major histocompatibility complex, whereas urodeles do not.

Splenectomy has no marked effect upon the general health and growth of amphibians, presumably because sufficient numbers of functionally equivalent cells remain elsewhere in the body. Thus, splenectomy does not affect allograft rejection in anurans (Brown and Cooper, 1976) or in urodeles (Cohen, 1971; Charlemagne, 1972a; Deparis and Flavin, 1973) either in first-set or in memory responses (Cohen and Horan, 1977). Antibody production usually remains efficient (Turner, 1973) except to threshold doses of antigen in *X. laevis* (Collie and Turner, 1975), although in *R. catesbeiana* larvae a reduction in humoral antibody production was noticed (Brown and Cooper, 1976).

Besides its role as a major lymphoid organ, the spleen is intimately involved in myelopoiesis. Thus, it is a major site of erythropoiesis and thrombocytopoiesis in the majority of urodeles (Dawson, 1932; Tooze and Davies, 1968; Charlemagne, 1972b). Basophil precursors are also produced in the spleen of urodeles (Cowden, 1965). The spleen is also a major erythropoietic organ in adult anurans (Jordan and Spiedel, 1923a,b; Jordan, 1938; Fey, 1962; Maclean and Jurd, 1972; Carver and Meints, 1977). In the tadpole, on the other hand, the spleen only destroys erythrocytes: it is after metamorphosis that it becomes a major organ for production of erythrocytes, thrombocytes, and granulocytes (Le Douarin, 1966). Seasonal changes in level of hemopoiesis in spleen and liver have been described by Garavini (1970). There is considerable output of immature elements into the circulation in spring, while in autumn and winter the circulating elements decrease and immature cells are retained in the lymphoid organs.

4. KIDNEY

In both anuran and urodele amphibians, as in fish, the kidney is a hemopoietic organ (Cowden and Dyer, 1971; see also the review by Cohen, 1977). In urodeles, the kidney plays a role in immune responses (see Cohen, 1977). Thus, RFCs have been described in the kidney of the newt, *N. viridescens*, following immunization with foreign erythrocytes, although RFCs were less numerous in the kidney than in the spleen. This immunologic activity of the newt kidney may well be generated *in situ*, since RFC numbers were unaffected by splenectomy (Ruben et al., 1973). The kidney is possibly also involved in immune reactivity in apodans (Cooper, 1976).

In anurans, lymphocytic accumulations are numerous in the intertubular tissue of the kidney, often near the blood sinuses in areas where the blood flow is probably sufficiently slow to permit easy passage of lymphocytes out of the circulation to cluster in response to the presence of antigens. In *X. laevis*, lymphocytic accumulations exist along the entire length of the mesonephric kidney (Figure 13) occupying positions between the renal tubules, occasionally in association with glomeruli. Antigen may reach the kidney via the circulation, or in mobile macrophages (Turner, 1973). Carbon and antigenic materials have been traced to the intertubular areas (Diener and Nossal, 1966; Turner, 1969, 1970) and, following intraperitoneal injection, are phagocytosed by the tubule cells themselves (Turner, 1969). Antigen in the intertubular tissue is associated with scattered cells or cells in small groups. Pyroninophilic cells and plasma cells, rarely found in the kidney of nonimmunized animals, become conspicuous after immunization (Kent et al., 1964; Diener and Nossal, 1966; Evans et al., 1966; Minagawa et al., 1975).

Cowden and Dyer (1971) have reviewed the respective roles of kidney and spleen as secondary lymphoid organs in anurans. In *B. marinus*, following injection of the soluble antigen BSA, most antibody-forming cells were either in granulomata at the site of injection or in the kidney, rather than in the spleen (Cowden et al., 1968b). On the other hand, in *R. pipiens*, the response to *Aeromonas hydrophilia* (the bacterium which causes the disease "red leg") was chiefly in the spleen.

Myeloid tissue predominates over lymphoid tissue in the kidney of urodeles in both adults (Cowden and Dyer, 1971) and neotenic and larval forms (Dawson, 1932). In anurans, the kidney has repeatedly been considered the major center of erythropoiesis in the larva (Jordan, 1933; Hollyfield, 1966). However, recent studies suggest that the liver, rather than the kidney, is the major site of erythropoiesis prior to metamorphosis (Turpen *et al.*, 1979; see Chapter 15). Experimental evidence is now at hand to suggest that the mesonephros is the major source of pluripotential hemopoietic stem cells (Turpen, 1980). The pronephros appears to be an initial site of granulopoiesis (Karpenter and Turpen, 1979) and lymphopoiesis (Horton, 1971a) in *R. pipiens*. The adult anuran kidney is also a major granulopoietic and lymphopoietic organ (Le Douarin, 1966).

5. LYMPHOMYELOID NODES

In the higher anurans, a number of lymphoid structures occur in the neck and upper thoracic region. These structures were described by a number of earlier workers, including von Braunmühl (1926) who called them "lymphomyeloid organs," based on their function of lymphopoiesis and granulopoiesis. These and similar organs have been tabulated, classified, and further described by Cooper (1967a,b, 1968, 1976), Baculi and Cooper (1967, 1968) Baculi *et al.* (1970), and Horton (1971b). These organs phagocytose carbon (see Myers, 1928). Their involvement in the immune system was first demonstrated by Kent *et al.* (1964) and Evans *et al.* (1966), who discovered in *B. marinus* some 4 to 10 pairs of nodular structures, 0.5 to 0.8 mm in maximum diameter, situated along the great vessels at the base of the neck and their branches in the neck and axilla. These nodes were shown to be sites of antibody production (see below). A variety of lymphomyeloid nodes have been described in a number of adult anurans (Baculi *et al.*, 1970; Horton, 1971b), but they are lacking in the more primitive *X. laevis* (Sterba, 1950; Turner, 1969). They are absent in urodeles and apodans.

In adult Ranidae, the lymphomyeloid nodes include a propericardial body (on either side of the larynx), a paired procoracoid body (anterior to the truncus arteriosus and near the coracoid bone), a pair of small epithelial bodies (in close proximity to the parathyroids), and a pair of jugular bodies (dorsal and lateral to the sternohyoideus muscle) (Cooper, 1967a,b, in *R. catesbeiana*; Horton, 1971b, in *R. pipiens*; see Figure 14). The pear-shaped jugular body is the largest of these nodes. It is composed of numerous lobules of lymphocytic (and some granulocytic) tissue separated by blood sinusoids lined with reticuloendothelial cells (Figure 15). Macrophages which take up India ink particles are also found in the stroma that supports the sinusoid walls and lymphoid parenchyma (Baculi and Cooper, 1968).

The jugular bodies show a marked increase in pyroninophilia following immunization (Diener and Nossal, 1966) with the appearance of antibody-producing plasma cells to soluble antigens (Kent *et al.*, 1964; Evans *et al.*, 1966) and plaque-forming cells to sheep erythrocytes (Minagawa *et al.*, 1975; Wright and Cooper, 1980). They are organs in which antigen is retained extracellularly on the processes of dendritic cells, as described above for the spleen (Diener and Marchalonis, 1970). This antigen trapping occurs randomly throughout the lymphocytic nodules of the organ (Figure 16) and pyroninophilia is similarly widespread. There is no suggestion of germinal center formation as occurs in the mammalian lymph node (Diener and Nossal, 1966). Furthermore, these anuran

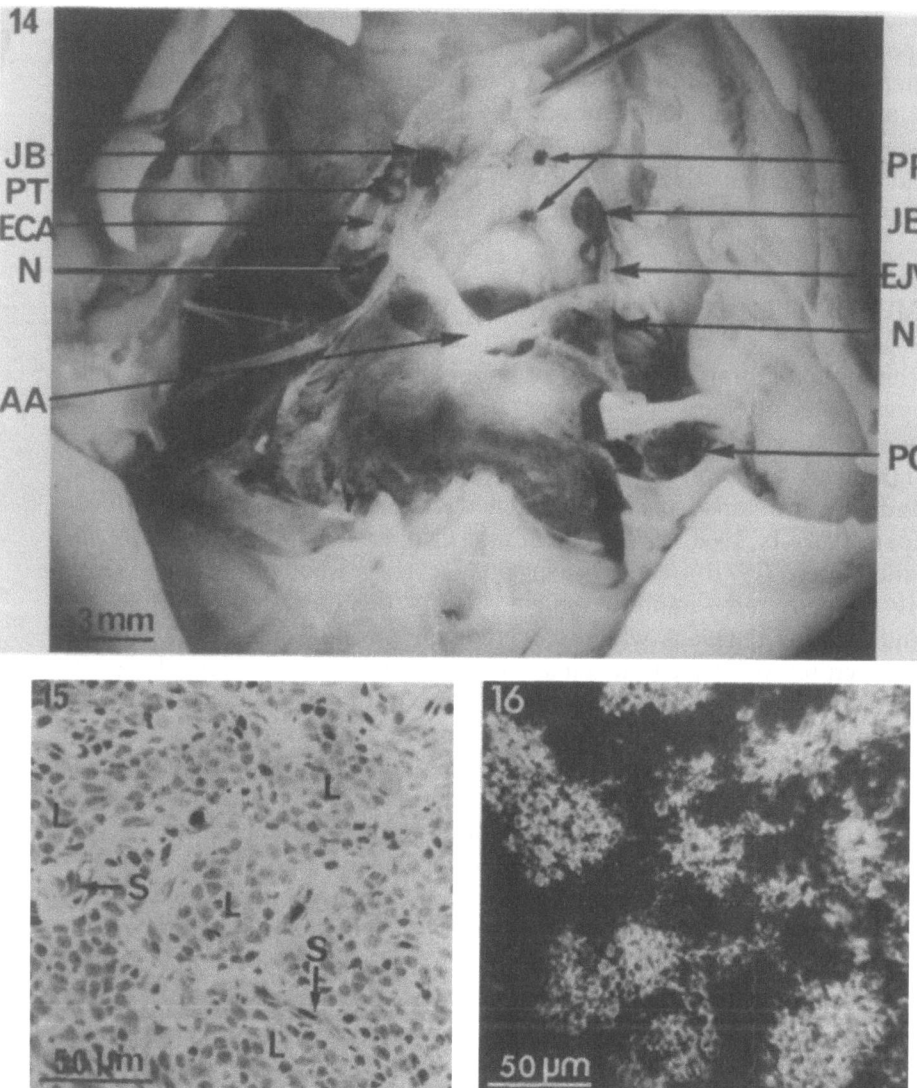

FIGURE 14. Ventral dissection of adult *Rana pipiens*, displaying the lymphomyeloid nodes of the throat and axillary regions. Twenty-four hours prior to killing, the frog had been injected with India ink, which blackens these phagocytic structures. The lymphomyeloid nodes seen here are the large jugular bodies (JB), the prepericardial bodies (PP), the procoracoid bodies (PC), and another paired node (N) ventral to the systemic and pulmocutaneous arteries. AA, aortic arches; EJV, external jugular vein; ECA, external carotid artery; PT, parathyroid glands.

FIGURE 15. Histology of jugular body from *Rana pipiens* adult. One-micrometer section, cut from glutaraldehyde-fixed and epon-embedded material. Stain: methylene blue + azure II + basic fuchsin. Lobules of lymphoid tissue (L) are seen with intervening sinusoids (S).

FIGURE 16. Immunofluorescent preparation showing antigen localization in the jugular body of adult *Rana pipiens* that had been injected 3 weeks prior to killing with HGG in adjuvant. For details of injection schedule, see Secombes and Manning (1980). The section has been incubated with anti-HGG (fluorescein-labeled) and the bright fluorescence indicates presence of antigen, randomly distributed in a dendritic pattern within the lymphocyte lobules.

nodes filter mainly from blood sinusoids (Baculi and Cooper, 1967) and it is doubtful whether at this stage of evolution they represent anything more than additional organs with sinusoidal blood flow which, like the kidney, provide sites where lymphoid cells accumulate and respond to antigenic stimulation. It should be noted, however, that in *R. pipiens* it is the jugular bodies, rather than the spleen, in which long-term antigen retention takes place (M. J. Manning, unpublished observations): thus, HGG appeared in the jugular bodies in the first week after injection (via either intramuscular, intraperitoneal, or dorsal lymph sac route), and did not begin to disappear until about week 9.

The tadpoles of many higher anurans possess a lymphomyeloid node different from that found in adults. This is the larval lymph gland, a bilateral structure in the branchial chamber near the developing anterior limb (Cooper, 1967a,b; see Figures 17 and 19). The lymph gland is found in several ranid species, but is absent in *R. temporaria* (H. Vickers and J. Horton, unpublished observations). The lymph gland comprises lobules of predominantly lymphoid cells which surround blood sinusoids lined with reticuloendothelial cells. The parenchyma of the lymph gland is supported by stromal cells and also contains macrophages, granulocytes, and plasma cells (Baculi and Cooper, 1968). Some of the macrophages and the reticuloendothelial cells phagocytose injected carbon particles of India ink from the blood sinusoids (Baculi and Cooper, 1968; see Figure 20). However, it should be noted that the larval lymph gland lies in the anterior lymphatic sinus (Horton, 1971a), and unpublished morphologic observations (J. Horton) suggest that cells on the outer surface of the gland, next to this lymphatic sinus, are capable of taking up ink directly from the body fluids. The lymph gland may thus be both a blood *and* lymph filtering organ.

The larval lymph gland plays a role in humoral immunity. Blast cell transformation and plaque-forming cells are seen in lymph glands of larvae injected with foreign erythrocytes (Moticka *et al.*, 1973; Cooper *et al.*, 1975). Tadpoles with lymph glands removed fail to synthesize antibody but are able to reject allografts (Cooper, 1968; Cooper *et al.*, 1971). However, the gland can restore not only antibody synthesis to foreign erythrocytes, but also partial restoration of transplantation immunity is seen in larvae rendered immunologically incompetent by total-body irradiation (Cooper *et al.*, 1975). These authors therefore postulate that the lymph gland is a source of both T and B stem cells.

Other viscera, such as the lungs, may also be involved in hemopoiesis. For example, the extremities of the lungs, together with visceral mesenteries, contain large numbers of lymphocytes and granulocytes in *A. obstetricans* larvae (Du Pasquier, 1968) and in *R. temporaria* larvae (H. Vickers and J. Horton, unpublished).

6. GUT-ASSOCIATED LYMPHOID TISSUE

6.1. LYMPHOEPITHELIAL TISSUE

In addition to the lymph gland, ranid tadpoles also possess ventral cavity bodies in the branchial region. These are lymphoepithelial organs lying as paired

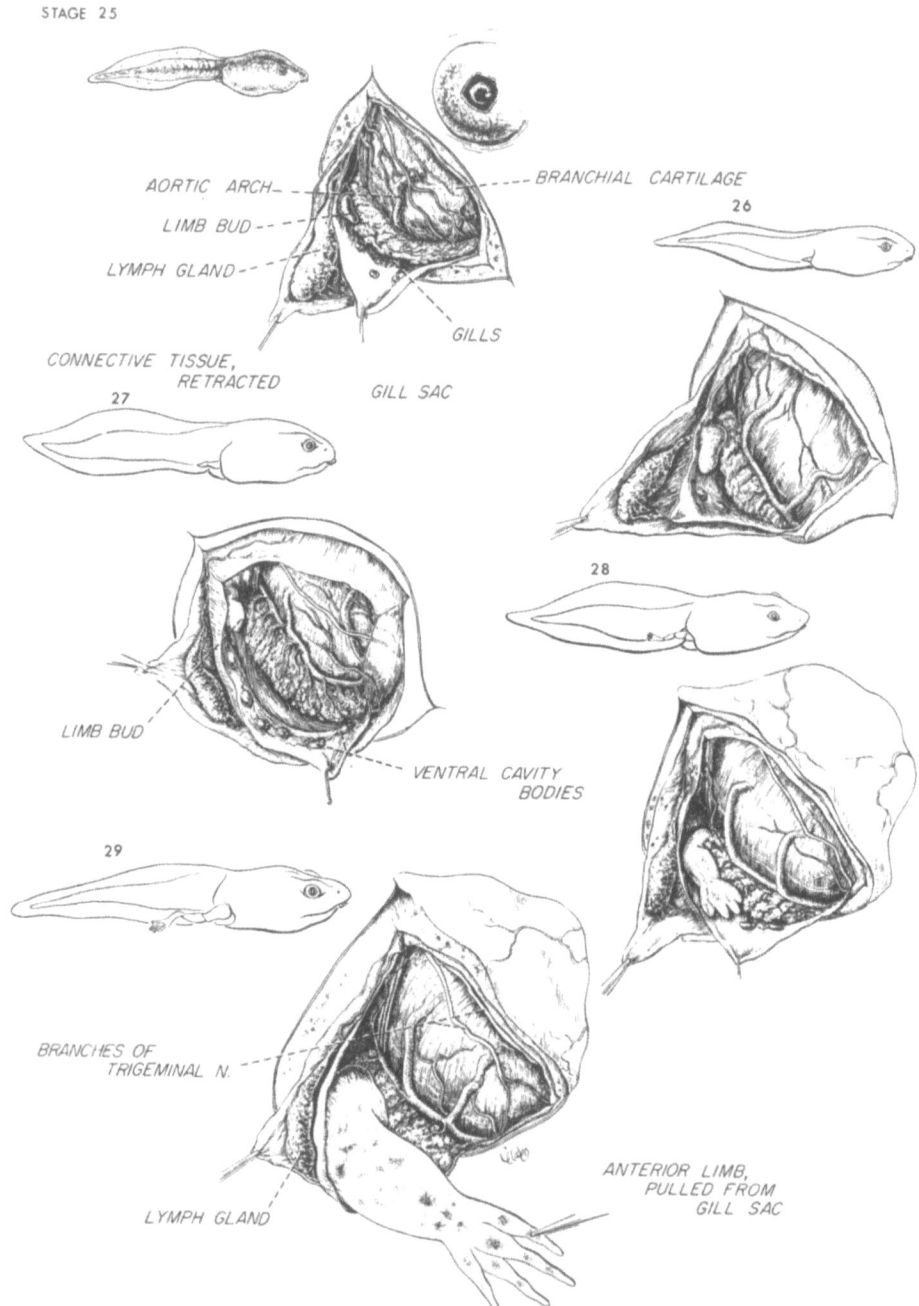

FIGURE 17. Diagrams of *Rana catesbeiana* larvae at different stages of development (stages according to Witschi, 1956) showing the location of the lymph gland and ventral cavity bodies in relation to other branchial-associated structures. (Courtesy of Dr. E. L. Cooper; taken from Cooper, 1967b.)

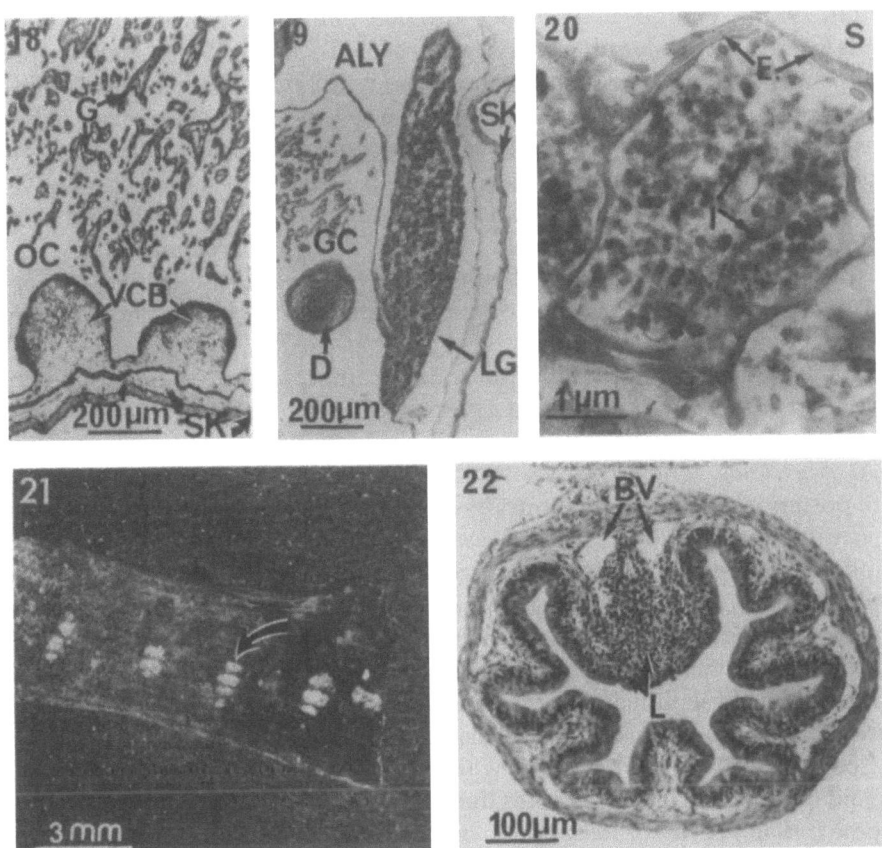

FIGURE 18. Ventral cavity bodies (VCB) are located ventrolaterally in the branchial region of anuran larvae: two VCB are shown in this section that passes through the opercular chamber (OC) of a *Rana pipiens* larva. Lymphocytes are found within the epithelium and connective tissue region of the CVB. G, gills; SK, skin of ventrolateral surface. Stain: hematoxylin and eosin.

FIGURE 19. The lymph gland, found in several *Rana* species, is located dorsolaterally, close to the developing anterior limb, in the branchial region. This section shows the lymph gland (LG) of a *Rana pipiens* larva: the gland consists of an extensive lymphoid parenchyma (stains darkly) and a network of intervening sinusoids (pale-staining regions). The organ is dorsoventrally elongated, is attached (ventrally) to the epithelium lining the gill chamber (GC), and projects into the anterior lymphatic (ALY). D, digit of anterior limb lying in gill chamber; SK, lateral skin. Stain: hematoxylin and eosin.

FIGURE 20. Electron micrograph of lymph gland from *Rana pipiens* injected 24 hr previously with India ink. Carbon particles of the ink (I) are seen being engulfed by pseudopodial extensions (E) of the reticuloendothelial cell cytoplasm. The reticuloendothelial cells (i.e., the littoral cells of Baculi and Cooper, 1968) line the sinusoids of the lymph gland. S, sinusoid lumen.

FIGURE 21. Piece of small intestine from *Xenopus laevis* toadlet after treatment with acetic acid to facilitate macroscopic visualization of nodular gut-associated lymphoid tissue. An orderly arrangement of GALT nodules (arrow) is seen in the dorsal intestinal wall.

FIGURE 22. Section through small intestine of *Xenopus laevis* toadlet, revealing dorsally situated nodular GALT. Lymphocytes (L) are seen both in the lamina propria and in the overlying gut epithelium: the epithelial basement membrane becomes difficult to detect in areas where infiltration of epithelium is high. BV, blood vessels. Stain: hematoxylin and eosin.

bodies in the ventral part of the pharynx (Figures 17 and 18) (Cooper, 1967a,b; Horton, 1971a). Ventral cavity bodies also occur in *X. laevis*, which lacks lymph glands (Sterba, 1950; Manning and Horton, 1969; Tochinai, 1975). They comprise mainly small lymphocytes and appear to be lymphocytic rather than lymphopoietic organs. Ventral cavity bodies occupy a sentinel position in relation to the water current passing over the gills and the opercular openings from the pharyngeal cavity (Manning and Horton, 1969). In addition to several pairs of ventral cavity bodies, similar lymphocytic accumulations may occur in the dorsal part of the larval pharynx (Horton, 1971a, in *R. pipiens*; Tochinai, 1975, in *X. laevis*). Ventral and dorsal cavity bodies in larval *Xenopus* are depleted of lymphocytes following early thymectomy (Manning, 1971; Horton and Manning, 1974a; Tochinai, 1975).

The ventral cavity bodies disappear at metamorphosis. Gut-associated lymphoid tissue (GALT) is then seen at a more regular frequency, although lymphocyte accumulations have been described in association with the larval esophagus [in *A. obstetricans*, where they are thymus-dependent (Du Pasquier, 1968)] and more posterior gastrointestinal tract of larval *A. obstetricans* (Du Pasquier, 1968), *R. pipiens* (Horton, 1971a), and *X. laevis* (Tochinai, 1975). In addition to the lymphoid nodules of the postpharyngeal region of the adult (see below), the oropharyngeal cavity may acquire tonsillike structures. These "tonsils" are found in both adult anurans and urodeles (Kingsbury, 1912; Jolly, 1919). These are classified as lymphoepithelial tissues (see Jolly, 1923) since they take the form of tonsillike epithelial crypts infiltrated with lymphocytes. Some of the anuran tonsillike structures phagocytose carbon (Myers, 1928) and are classified by Cooper (1967a,b, 1976) along with other lymphoepithelial and lymphomyeloid organs. Nigam (1977) notes that the lymphopoietic activity of the tonsils in some Indian species of *Rana* occurs only during the spring and summer; from October to February, adipose cells infiltrate the organ.

All vertebrates show occasional infiltration of the gut epithelium and lamina propria by scattered lymphocytes and granulocytes. Indeed, Hightower and St. Pierre (1971) suggest that the lamina propria of the adult newt (*N. viridescens*) is a site of lymphopoiesis. In the urodele *Necturus maculosus*, Good *et al.* (1966) observed that cells with the morphology of plasma cells were abundant in the lamina propria. Cowden and Dyer (1971), however, believe this to be unusual among the urodeles, since they could not find many plasma cells in this situation in *Ambystoma, Desmognathus, Notophthalmus, Amphiuma,* and *Siren*.

6.2. NODULAR GUT-ASSOCIATED LYMPHOID TISSUE

Nodules of GALT are regularly seen in adults of diverse anuran species (see review by Goldstine *et al.*, 1975), but are apparently lacking in urodeles. Anuran GALT nodules are numerous (Figure 21) and are similar in their histology (Figure 22) to a single nodule of a mammalian Peyer's patch but, in common with other amphibian lymphoid tissues, they lack germinal centers. Although they have been observed in the intestine of all anurans so far examined, they are less numerous in *Xenopus* than in the more "advanced" ranid and bufid anurans. Lymphoid nodules have also been described in the adult anuran esophagus (Wong, 1972) and in the cloaca and urogenital system (Goldstine et al., 1975). Amphibia, however, lack any cloacal structure comparable with the avian bursa of Fabricius.

The anuran GALT nodules contain phagocytic cells which take up carbon (Chin and Wong, 1977, in *Bufo melanostictus*) and they respond to antigenic stimulation with the appearance of pyroninophilic and plasma cells (Goldstine et al., 1975; Minagawa et al., 1975). The proportion of lymphocytes in the GALT nodules of *Xenopus* expressing surface IgM (51–68%) or low-molecular-weight immunoglobulin (4%) is comparable with that seen in peripheral blood lymphocytes. These two Ig classes are also found in gut secretions Jurd (1977) and GALT lymphocytes have been demonstrated to be cytoplasmic Ig-positive cells by indirect immunofluorescence (Michea-Hamzehpour, 1977). Interestingly, there appears to be no gut-associated IgA production in amphibians (Jurd, 1977).

7. BONE MARROW

Bone marrow has not yet been detected in the Gymnophiona (Garcia-Herrera and Cooper, 1968). It is also absent in urodeles with the exception of the lungless salamanders of the family Plethodontidae. In the newt *N. viridescens*, for example, Hightower and St. Pierre (1971) describe the bone marrow cavity as being filled with fat cells and fibroblasts with no hemopoietic activity. A trend towards bone marrow hemopoiesis is seen in certain Proteidae (for example, the giant Japanese salamander) which possess lymphopoietic tissue in the meninges (see Cowden and Dyer, 1971). Within the Plethodontidae, on the other hand, most genera produce both lymphocytes and granulocytes in the marrow of all bones large enough to possess marrow cavities (Barrett, 1936). Curtis et al. (1979a) describe the bone marrow of *Plethodon glutinosus* as containing large numbers of developing neutrophils and eosinophils, lymphocytes of various sizes, plasmablasts, and plasma cells. The bone marrow is also part of the phagocytic system: macrophages which take up particulate matter are present in extravascular locations. Unlike higher vertebrates, however, the bone marrow of *Plethodon* does not produce erythrocytes or basophils.

In anurans also, the bone marrow is mainly lymphogranulopoietic. Erythropoiesis occurs in the bone marrow (Maniatis and Ingram, 1971) but is seasonal and takes place mainly during the spring, the spleen being the more important erythropoietic organ in adults (Jordan and Spiedel, 1923a,b; Le Douarin, 1966). Macrophages are found in the extravascular tissue and free in the sinuses of the marrow: many lymphocytes are present and plasma cells have been identified by electron microscopy (Campbell, 1970; Cooper et al., 1980). In *X. laevis*, bone marrow develops in the postmetamorphic animal where it is the site of considerable DNA-synthetic activity (Green and Cohen, 1979): lymphoid cells are found both in the extravascular tissue and in the blood capillaries of the marrow (Horton and Manning, 1974a). The lymphocytes present in the bone marrow of adult frogs *R. pipiens* and *R. catesbeiana* are equivalent in number to those present in the spleen. In young *R. catesbeiana*, the ratio of marrow to spleen leukocytes is considerably greater than in the adult (Eipert et al., 1979).

Plasma cells and large pyroninophilic cells appear in the marrow of ranid frogs following immunization with BSA (Cowden and Dyer, 1971) or with sheep erythrocytes (Minagawa et al., 1975). Bone marrow of ranid frogs in fact gives a higher primary PFC response against both the hapten TNP (conjugated to LPS) and the thymus-dependent antigen sheep erythrocytes, than does the spleen (Eipert et al., 1979; Cooper et al., 1980). Furthermore, this heightened marrow reactivity is also true for *in vitro* proliferative responses to both B and T mitogens (LPS and Con A, respectively) (Eipert et al., 1977). Functionally mature T and B lymphocytes have, in contrast, not been demonstrated in *Xenopus* marrow. Thus, Green and Cohen (1979) found no positive reactivity of 1- to 2-year-old toadlet marrow to Con A, PHA, and the B-cell mitogen purified protein derivative of tuberculin (PPD). Their mitogen experiments with dextran sulfate also failed to identify precursor B cells in toadlet marrow and liver, although such cells were present in the spleen. Interestingly, N. H. Williams (personal communication) has preliminary evidence suggesting proliferative responses of marrow lymphocytes to the B-cell mitogen LPS in early thymectomized *Xenopus*.

In an immune system damaged by irradiation, the bone marrow can restore immune reactivity, bone shielding of an otherwise totally irradiated leopard frog being sufficient to maintain alloimmunity (Cooper and Schaefer, 1970) and to restore the PFC response to sheep erythrocytes (Cooper et al., 1980). In evolutionary terms, the bone marrow of tetrapods probably represents a rehousing of cell populations into the available space of the marrow cavity, where blood flow is suitably slow in the sinusoids (see Manning and Turner, 1976). A further advantage of this location is that the radiosensitive hemopoietic stem cells are protected from irradiation by bone which, as animals became more terrestrial, may have substituted for water as a radiopaque shield (Cooper et al., 1980).

8. LIVER

The liver of amphibians is an important phagocytic organ; it also plays a role in hemopoiesis. In apodans, the liver and spleen probably are the major sites of myelopoiesis and lymphopoiesis. Welsch and Storch (1972) have described a cortical layer of blood-forming tissue containing two types of granulocyte, together with lymphocytelike cells in the liver of the apodan *Ichthyophis kohtaoensis*. The liver is particularly important in the production of granulocytes and is probably the only source of many granulocyte types in some urodeles (Le Douarin, 1966). Hepatic granulopoiesis involves eosinophils and neutrophils, basophils being produced in the spleen. In urodeles, the hemopoietic cells are found in a perihepatic subcapsular layer (Maximow, 1927; Hightower and Haar, 1975) which entirely covers the urodele liver. This layer extends in various degrees of thickness inwards towards the liver parenchyma. In *P. waltlii*, the subcapsular layer has been shown to contain lymphocytes as well as granulocytes: it also produces plasmacytes and plasma cells. The latter are scattered throughout the perihepatic layer and are also found in the liver sinusoids. Their number is small compared with the number of granulocytes present, but large when the whole liver surface is considered. The perihepatic layer of urodeles may therefore be a B-cell source (Henry and Charlemagne, 1977). These authors also note that the peritoneal epithelium which limits the perihepatic layer comprises cells whose free surfaces contain numerous micropinocytotic vesicles. These cells retain *Salmonella* bacteria injected intraperitoneally. Ruben *et al.* (1973) have shown a good RFC response in the liver of *N. viridescens* following erythrocyte injection, thereby indicating an important role of this tissue in immune reactivity in urodeles.

During development, the liver is an important source of hemopoietic cells both in urodeles (Deparis and Jaylet, 1975) and in anurans (Turpen *et al.*, 1979). It appears to be the main site of maturation of erythrocytes in the anuran tadpole (Maniatis and Ingram, 1971; Turpen *et al.*, 1979). B-lymphocyte differentiation takes place in the larval liver in anurans: thus, Zettergren *et al.* (1977) have demonstrated cytoplasmic IgM-positive cells in the subcapsular hemopoietic areas of the liver in mid-larval *X. laevis* larvae. In *R. pipiens*, cytoplasmic Ig-positive cells are also seen in the larval liver: however, in this species, the kidney may be the initial source of B cells (Zettergren *et al.*, 1980). In addition to lymphocytes, granulocytes are also found in the liver of larval and adult *Xenopus* (Manning and Horton, 1969).

The liver contains Kupffer cells and, by virtue of its large size, is the main site of trapping circulating particulate matter (Turner, 1970). Melanin deposits occur in the amphibian liver and may be associated with sites of phagocytosis (Zylbersac, 1936; Manning and Turner, 1972). The liver of *Xenopus* showed no areas of lymphocytic stimulation after administration of human serum in adjuvant (Manning and Turner, 1972). However, *Xenopus* liver cells do proliferate *in vitro* following stimulation with PHA and PPD, but not with Con A (Green and Cohen, 1979). Macrophages containing pigment were seen beneath the subcapsular layer of the liver in *A. mexicanum*. They were apparently derived from Kupffer cells and were rare in young animals, but abundant at an age of 5 to 8 years (Andrew, 1969).

9. BLOOD

Finally, it should be remembered that the circulating blood itself contains phagocytic cells (e.g., monocytes) that take up India ink (Jordan, 1938). The buffy coat cells of the blood are frequently used in studies on mitogen responses (see Wright *et al.*, 1978) and mixed leukocyte culture reactivity (see Du Pasquier and Horton, 1976). Furthermore, in urodeles, the circulating blood is involved in erythropoiesis (Dawson, 1932). Also, the heart itself shows hemopoietic activity in the trabecular tissue of the ventricle in urodeles (Jolly and Lievre, 1931), while in *R. temporaria*, epicardiac hemopoiesis takes place in the larva, prior to the appearance of the adult bone marrow (see Le Douarin, 1966).

10. DISCUSSION: PHYLOGENETIC CONSIDERATIONS

The immune system of amphibians displays structural and functional diversity. Thus, anurans show incipient development of at least some of the "advanced" characteristics of the immune system of higher vertebrates, including well-developed antigen retention mechanisms, clear-cut T cell/B cell dichotomy, a major histocompatibility complex (MHC), the production of antibody of a new low-molecular-weight class, nodular GALT, the regular occurrence of hemopoietic bone marrow, and in the higher anurans, lymphomyeloid nodes such as the jugular bodies. The immune system of urodeles, on the other hand, seems more "primitive" in several respects. For example, they possess only one class of immunoglobulin (Marchalonis and Cohen, 1973) and probably lack a MHC (Cohen and Collins, 1977). Furthermore, in the majority of urodeles, the major lymphoid organs are limited to the thymus, spleen, and the perihepatic layer of the liver (although the bone marrow of the Plethodontidae and the kidney of newts may also be implicated in the immune system).

"Primitive" features common to both anurans and urodeles include the lack of germinal centers in the peripheral lymphoid tissues and the fact that the latter possess lymphocytes admixed with myeloid and other hemopoietic cells. Other "primitive" characteristics include the retention of an extensive renal portal system with sinusoidal blood flow, and important role of the kidney as a lymphoid organ in anurans. Amphibians possess a lymphatic system with large lymph sacs and pulsatile lymph hearts, but these differ from the valve-bearing lymphatic system of mammals, whose vessels are interrupted by the intercalation of lymph nodes. These mammalian lymph nodes allow regional responses to local antigenic stimulation. The lymph nodes, however, are not isolated but are linked and finely integrated with the entire immune system, which presents a considerable advance on the amphibian condition.

It now seems that the fundamental cell types of the immune system—from hemopoietic stem cells, through macrophages, dendritic antigen-trapping cells, functional T and B lymphocytes, to large pyroninophilic and plasma cells—are essentially similar from fish to mammals. Although more diverse immunoglobulin classes and more sophisticated immunoregulatory mechanisms exist in mammals compared with lower vertebrates, in terms of the cellular basis of the immune system, it is the organ distribution and architecture of the lymphoid tissues which show considerable advance from lower to higher vertebrates.

The intermediate condition of the amphibian immune system reflects the transitional status of apodans, urodeles, and anurans in terms of their overall anatomical and physiological adaptations to life on land. Thus, compared with fish, amphibians show changes in their heart and blood vascular system which result from the need for more efficient circulation of body fluids to cater for active life in the terrestrial environment (Manning, 1978). These adaptations influence the circulation of body fluids, and as a consequence affect lymphoid tissue structure, the movements of immunocompetent cells, the distribution of antibody molecules, and the dispersal and sequestration of antigens.

ACKNOWLEDGMENTS. The photographic assistance of David Hutchinson is gratefully acknowledged. Our special thanks go to Mrs. Trudy L. Horton for help with preparation of the manuscript.

REFERENCES

Ambrosius, H., and Drössler, K., 1972, Spezifische zellvermittelte Immunität bei Froschlurchen. I. Quantitative Nachweistechnik mit dem Makrophagen—Migrations—Hemmtest für Peritoneal exsudat zellen und Milzstücken, *Acta Biol. Med. Ger.* **29**:437.

Ambrosius, H., and Hanstein, R., 1971, Beiträge zur Immunbiologie poikilothermer Wirbeltiere. VI. Die Dynamik Antikörper produzierender Zellen in den lymphoiden Organen des Wasser frosches *Rana esculenta* L., *Acta Biol. Med. Ger.* **27**:771.

Amirante, G. A., 1968, Induction of antibodies against pure proteins in *Xenopus laevis* Daud., *Experientia* **24**:171.

Andrew, W., 1969, The nature of pigment cells in the liver of *Ambystoma mexicanum* and their changes with age, *J. Cell Biol.* **43**:7.

Auerbach, R., and Ruben, L. N., 1970, Studies of antibody formation in *Xenopus laevis*, *J. Immunol.* **104**:1242.

Azzolina, L. S., 1975, A primary immune response of *Bufo marinus* spleen cells *in vitro*. *Eur. J. Immunol.* **5**:795.

Baculi, B. S., and Cooper, E. L., 1967, Lymphomyeloid organs of Amphibia. II. Vasculature in larval and adult *Rana catesbeiana*, *J. Morphol.* **123**:463.

Baculi, B. S., and Cooper, E. L., 1968, Lymphomyeloid organs of Amphibia. IV. Normal histology in larval and adult *Rana catesbeiana*, *J. Morphol.* **126**:463.

Baculi, B. S., Cooper, E. L., and Brown, B. A., 1970, Lymphomyeloid organs of Amphibia. V. Comparative histology in diverse anuran species, *J. Morphol.* **131**:315.

Baldwin, T. M., 1918, Pharyngeal derivatives of *Ambystoma*, *J. Morphol.* **30**:605.

Baldwin, W. M., III, and Cohen, N., 1981, A primitive dendritic splenocyte in *Xenopus laevis* with morphological similarities to Reed-Sternberg cells, in: *Aspects of Developmental and Comparative Immunology I* (J. B. Solomon, ed.), pp. 179–182, Pergamon Press, Oxford.

Balls, M., Clothier, R., Hodgson, R., and Berridge, D., 1980, Effects of N-methyl-N-nitrosurea and cyclosporin A on the lymphocytes and immune responses of *Xenopus laevis* and other amphibians, in: *Development and Differentiation of Vertebrate Lymphocytes* (J. D. Horton, ed.), pp. 183–184, Elsevier/North-Holland, Amsterdam.

Barrett, W. C., Jr., 1936, A comparative survey of the hemopoietic foci in the urodele Amphibia, with especial reference to the bone marrow of the Plethodontidae, *Folia Haematol. (Leipzig)* **54**:165.

Bleicher, P. A., and Cohen, N., 1981, Monoclonal anti-IgM can separate T cell from B cell proliferative responses in the frog, *Xenopus laevis*, *J. Immunol.* **127**:1549.

Blomberg, B., Bernard, C. C. A., and Du Pasquier, L., 1980, *In vitro* evidence for T–B lymphocyte collaboration in the clawed toad, *Xenopus laevis*, *Eur. J. Immunol.* **10**:869.

Brown, B. A., and Cooper, E. L., 1976, Immunological dichotomy in the larval bullfrog spleen, *Immunology* **30**:299.

Brown, B. A., Wright, R. K., and Cooper, E. L., 1975, Lymphoid organs and amphibian immunity, in: *Immunologic Phylogeny* (W. H. Hildemann and A. A. Benedict, eds.), pp. 267–275, Plenum Press, New York.

Campbell, F. R., 1970, Ultrastructure of the bone marrow of the frog, *Am. J. Anat.* **129**:329.

Canaday, S. D., 1968, Light and electron microscopy of the thymus in adult *Rana pipiens*, *Anat. Rec.* **160**:326.

Carver, F. J., and Meints, R. H., 1977, Studies of development of frog haemopoietic tissue *in vitro*. 1. Spleen culture assay of an erythropoietic factor in anaemic frog blood, *J. Exp. Zool.* **201**:37.

Charlemagne, J., 1972a, Les réactions immunitaires chez les amphibiens urodèles. I. Resultats acquis et possibilités expérimentales, in: *Phylogenic and Ontogenic Study of the Immune Response and Its*

Contribution to the Immunological Theory, Colloque Inserm, pp. 89–95, Ministry of Public Health, Paris.

Charlemagne, J., 1972b, Aspects morphologiques de la différenciation des éléments sanguins chez l'Axoltol, *Ambystoma mexicanum* Shaw, *Z. Zellforsch. Mikrosk. Anat.* **123:**224.

Charlemagne, J., 1974, Larval thymectomy and transplantation immunity in the urodele *Pleurodeles waltlii* Michah (Salamandridae), *Eur. J. Immunol.* **4:**390.

Charlemagne, J., 1977, Thymus development in amphibians: Colonization of thymic endodermal rudiments by lymphoid stem-cells of mesenchymal origin in the urodele *Pleurodeles waltlii* Michah, *Ann. Immunol. (Inst. Pasteur)* **128c:**897.

Charlemagne, J., and Houillon, C., 1968, Effets de la thymectomie larvaire chez l'amphibien urodèle, *Pleurodeles waltlii* Michah. Production a l'état adulte d'une tolérance aux homogreffes cutanées, *C.R. Acad. Sci.* **267:**253.

Charlemagne, J., and Tournefier, A., 1975, Cell surface immunoglobulins of thymus and spleen lymphocytes in urodele amphibian *Pleurodeles waltlii* (Salamandridae), in: *Immunologic Phylogeny* (W. H. Hildemann and A. A. Benedict, eds.), pp. 251–256, Plenum Press, New York.

Charlemagne, J., and Tournefier, A., 1977, Anti-horse red blood cells antibody synthesis in the Mexican axolotl *Ambystoma mexicanum*: Effect of thymectomy, in: *Developmental Immunobiology* (J. B. Solomon and J. D. Horton, eds.), pp. 267–275, Elsevier/North-Holland, Amsterdam.

Chin, K. N., and Wong, W. C., 1977, Some ultrastructural observations on the intestinal mucosa of the toad (*Bufo melanostictus*), *J. Anat.* **123:**331.

Clark, J. C., and Newth, D. R., 1972, Immunological activity of transplanted spleens in *Xenopus laevis*, *Experientia* **28:**951.

Clark, W. E., Le Gros, 1965, *The Tissues of the Body*, 5th ed., p. 56, Oxford University Press, London.

Clothier, R. H., 1972, The histopathology of a lymphoreticular disease in *Xenopus laevis*, Ph.D. thesis, University of East Anglia.

Clothier, R. H., and Balls, M., 1973, Mycobacteria and lymphoreticular tumours in *Xenopus laevis*, the South African clawed toad. I. Isolation, characterization and pathogenicity for *Xenopus* of *M. marinum* isolated from lymphoreticular tumour cells, *Oncology* **28:**445.

Cohen, N., 1969, Immunogenetic and developmental aspects of tissue transplantation immunity in urodele amphibians, in: *Biology of Amphibian Tumours* (M. M. Mizell, ed.), pp. 153–168, Springer-Verlag, Berlin.

Cohen, N., 1970, Tissue transplantation immunity and immunological memory in Urodela and Apoda, *Transplant. Proc.* **2:**275.

Cohen, N., 1971, Amphibian transplantation reactions: A review, *Am. Zool.* **11:**193.

Cohen, N., 1977, Phylogenetic emergence of lymphoid tissues and cells, in: *The Lymphocyte: Structure and Function* (J. J. Marchalonis, ed.), pp. 149–202, Dekker, New York.

Cohen, N., and Collins, N. H., 1977, Major and minor histocompatibility systems of ectothermic vertebrates, in: *The Major Histocompatibility System in Man and Animals* (D. Götze, ed.), pp. 313–337, Springer-Verlag, Berlin.

Cohen, N., and Horan, M., 1977, Lack of correlation between the rapidity of newt allograft rejection and the frequency and magnitude of stimulation in the mixed lymphocyte culture reaction, in: *Developmental Immunobiology* (J. B. Solomon and J. D. Horton, eds.), pp. 259–266, Elsevier/North-Holland, Amsterdam.

Cohen, N., and Turpen, J. B., 1978, Early ontogeny of heterogeneous populations of lymphocytes in anuran amphibians, in: *Animal Models of Comparative and Developmental Aspects of Immunity and Disease* (M. E. Gershwin and E. L. Cooper, eds.), pp. 37–47, Pergamon Press, Elmsford, N.Y.

Coleman, R., and Phillips, A. D., 1972, Longterm retention of colloidal thorium dioxide in the liver and spleen of *Xenopus laevis* Daudin, *Experientia* **28:**1326.

Collie, M. H., 1974, The location of soluble antigen in the spleen of *Xenopus laevis*, *Experientia* **30:**1205.

Collie, M. H., and Turner, R. J., 1975, Influence of antigen dose on antibody production of intact and splenectomised *Xenopus laevis*, *J. Exp. Zool.* **192:**173.

Collie, M. H., Turner, R. J., and Manning, M. J., 1975, Antibody production to lipopolysaccharide in thymectomized *Xenopus*, *Eur. J. Immunol.* **5:**426.

Collins, N. H., and Cohen, N., 1976, Phylogeny of immunocompetent cells. II. *In vitro* behaviour of lymphocytes from the spleen, blood and thymus of the urodele *Ambystoma mexicanum*, in:

Phylogeny of Thymus and Bone Marrow—Bursa Cells (R. K. Wright and E. L. Cooper, eds.), pp. 143–151, Elsevier/North-Holland, Amsterdam.

Collins, N. H., Manickavel, V., and Cohen, N., 1975, In vitro responses of urodele lymphoid cells: Mitogenic and mixed lymphocyte culture reactivities, in: *Immunologic Phylogeny* (W. H. Hildemann and A. A. Benedict, eds.), pp. 305–314, Plenum Press, New York.

Cone, R. E., and Marchalonis, J. J., 1972, Cellular and humoral aspects of the influence of environmental temperature on the immune response of poikilothermic vertebrates, *J. Immunol.* **108:** 952

Cooper, E. L., 1967a, Some aspects of the histogenesis of the amphibian lymphomyeloid system and its role in immunity, in: *Ontogeny of Immunity* (R. T. Smith, R. A. Good, and P. A. Miescher, eds.), pp. 87–101, University of Florida Press, Gainesville.

Cooper, E. L., 1967b, Lymphomyeloid organs of Amphibia. I. Appearance during larval and adult stages of *Rana catesbeiana*, *J. Morphol.* **122:**381.

Cooper, E. L., 1968, Lymphomyeloid organs of Amphibia, III. Antibody synthesis and lymph glands in larval bullfrogs, *Anat. Rec.* **162:**453.

Cooper, E. L., 1973, The thymus and lymphomyeloid system in poikilothermic vertebrates, in: *Contemporary Topics in Immunobiology*, Volume 2, *Thymus Dependency* (A. J. S. Davies and R. L. Carter, eds.), pp. 13–38, Plenum Press, New York.

Cooper, E. L., 1976, Immunity mechanisms, in: *Physiology of the Amphibia*, Vol. III (B. Lofts, ed.), pp. 163–272, Academic Press, New York.

Cooper, E. L., and Hildemann, W. H., 1965, Allograft reactions in bullfrog larvae in relation to thymectomy, *Transplantation* **3:**446.

Cooper, E. L., and Schaefer, D. A., 1970, Bone marrow restoration of transplantation immunity in the leopard frog, *Rana pipiens*, *Proc. Soc. Exp. Biol. Med.* **135:**406.

Cooper, E. L., and Wright, R. K., 1976, The anuran amphibian spleen. An evolutionary model for terrestrial vertebrates, in: *Immuno-Aspects of the Spleen* (J. R. Battiste and J. W. Streilen, eds.), pp. 47–58, Elsevier/North-Holland, Amsterdam.

Cooper, E. L., Brown, B. A., and Baculi, B. S., 1971, New observations on lymph gland (LMI) and thymus activity in larval bullfrogs (*Rana catesbeiana*), in: *Morphological and Functional Aspects of Immunity* (K. Lindahl-Kessling, G. Alm, and M. G. Hanna, eds.), pp. 1–10, Plenum Press, New York.

Cooper, E. L., Brown, B. A., and Wright, R. K., 1975, New ideas on amphibian immunity. The lymph gland: A generator of both T and B cells, *Am. Zool.* **15:**85.

Cooper, E. L., Klempau, A. E., Ramirez, J. A., and Zapata, A. G., 1980, Source of stem cells in evolution. in: *Development and Differentiation of Vertebrate Lymphocytes* (J. D. Horton, ed.), pp. 3–14, Elsevier/North-Holland, Amsterdam.

Cowden, R. R., 1965, Quantitative and qualitative cytochemical studies on the *Amphiuma* basophil leucocyte, *Z. Zellforsch. Mikrosk, Anat.* **67:**219.

Cowden, R. R., and Dyer, R. F., 1971, Lymphopoietic tissue and plasma cells in amphibians, *Am. Zool.* **11:**183.

Cowden, R. R., Dyer, R. F., Gebhardt, B. M., and Volpe, E. P., 1968a, Amphibian plasma cells, *J. Immunol.* **100:**1293.

Cowden, R. R., Gebhardt, B. M., and Volpe, E. P., 1968b, The histophysiology of antibody-forming sites in the marine toad, *Z. Zellforsch. Mikrosk. Anat.* **85:**196.

Csaba, G., Oläh, I., and Kapa, E., 1970, Phylogenesis of the mast cells. II. Ultrastructure of the mast cells in the frog, *Acta Biol. Acad. Sci. Hung.* **21:**255.

Curtis, S. K., and Volpe, E. P., 1971, Modification of responsiveness to allografts in larvae of the leopard frog by thymectomy, *Dev. Biol.* **25:**177.

Curtis, S. K., Volpe, E. P., and Cowden, R. R., 1972, Ultrastructure of the developing thymus of the leopard frog (*Rana pipiens*), *Z. Zellforsch. Mikrosk. Anat.* **127:**323.

Curtis, S. K., Cowden, R. R., and Nagel, J. W., 1979a, Ultrastructure of the bone marrow of the salamander *Plethodon glutinosus* (Caudata: Plethodontidae), *J. Morphol.* **159:**151.

Curtis, S. K., Cowden, R. R., and Nagel, J. W., 1979b, Ultrastructural and histochemical features of the thymus glands of the adult lungless salamander, *Plethodon glutinosus* (Caudata: Plethodontidae), *J. Morphol.* **160:**241.

Dardenne, M., Tournefier, A., Charlemagne, J., and Bach, J. F., 1973, Studies on thymus products. VII. Presence of thymic hormone in urodele serum, *Ann. Immunol. (Inst. Pasteur)* **124c:**465.

Dawson, A. B., 1932, Haemopoietic loci in *Necturus maculosus*, *Anat. Rec.* **52:**367.

Debons, M. C., and Deparis, P., 1973, Mise en évidence des immunocytes par immunocytoadhérence chez l'Amphibien Urodèle *Pleurodeles waltlii* Michah après immunisation par les globules rouges de mouton, *C.R. Soc. Biol.* **167:**568.

Deparis, P., and Flavin, M., 1973, Les effects de la splénectomie précoce chez l'Amphibien Urodèle *Pleurodeles waltlii* Michah, *J. Physiol. (Paris)* **66:**19.

Deparis, P. and Jaylet, A., 1975, Recherches sur l'origine des différentes lignées de cellules sanguines chez l'Amphibien Urodèle *Pleurodeles waltlii*, *J. Embryol. Exp. Morphol.* **33:**665.

Diener, E., and Marchalonis, J., 1970, Cellular and humoral aspects of the primary immune response of the toad, *Bufo marinus*, *Immunology* **18:**279.

Diener, E., and Nossal, G. J. V., 1966, Phylogenetic studies on the immune response. I. Localization of antigens and immune response in the toad, *Bufo marinus*, *Immunology* **10:**535.

Donnelly, N., Manning, M. J., and Cohen, N., 1976, Thymus dependence of lymphocyte subpopulations in *Xenopus laevis*, in: *Phylogeny of Thymus and Bone Marrow-Bursa Cells* (R. K. Wright and E. L. Cooper, eds.), pp. 133–141, Elsevier/North-Holland, Amsterdam.

Du Pasquier, L., 1965, Aspects cellulaires et humoraux d l'intolérance aux homogreffes de tissu musculaire chez le têtard d'*Alytes obstetricans*; rôle du thymus, *C.R. Acad. Sci.* **261:**1144.

Du Pasquier, L., 1968, Les protéines sériques et le complexe lympho-myéloide chez le têtard d'*Alytes obstetricans* normal et thymectomisé, *Ann. Inst. Pasteur Paris* **114:**490.

Du Pasquier, L., 1970, Ontogeny of the immune response in animals having less than one million lymphocytes: The larvae of the toad *Alytes obstetricans*, *Immunology* **19:**353.

Du Pasquier, L., and Bernard, C. C. A., 1980, Active suppression of the allogeneic histocompatibility reactions during the metamorphosis of the clawed toad *Xenopus*, *Differentiation* **16:**1.

Du Pasquier, L., and Horton, J. D., 1976, The effect of thymectomy on the mixed leucocyte reaction and phytohaemagglutinin responsiveness in the clawed toad, *Xenopus laevis*, *Immunogenetics* **3:**105.

Du Pasquier, L., and Miggiano, V. C., 1973, The mixed leucocyte reaction in the toad *Xenopus laevis*: A family study, *Transplant. Proc.* **5:**1457.

Du Pasquier, L., and Weiss, N., 1973, The thymus during the ontogeny of the toad *Xenopus laevis*: Growth, membrane-bound immunoglobulins and mixed lymphocyte reaction, *Eur. J. Immunol.* **3:**773.

Du Pasquier, L., Weiss, N., and Loor, F., 1972, Direct evidence for immunoglobulins on the surface of the thymus lymphocytes of amphibian larvae, *Eur. J. Immunol.* **2:**366.

Dustin, A. P., 1911, Le thymus de l'Axolotl, *Arch. Biol.* **26:**557.

Eipert, E. F., Wright, R. K., and Cooper, E. L., 1977, Comparison of spleen and bone marrow mitogen responses in *Rana catesbeiana*, *Am. Zool.* **17:**892.

Eipert, E. F., Klempau, A. E., Lallone, R. L., and Cooper, E. L., 1979, Bone marrow and antibody synthesis in *Rana*, *Cell. Immunol.* **46:**275.

Evans, E. E., Kent, S. P., Bryant, R. E., and Moyer, M., 1966, Antibody formation and immunological memory in the marine toad, in: *Phylogeny and Immunity* (R. T. Smith, R. A. Good, and P. A. Miescher, eds.), pp. 218–226, University of Florida Press, Gainesville.

Fabrizio, M., and Charipper, H. A., 1941, The morphogenesis of the thymus gland of *Rana sylvatica* as correlated with certain stages of metamorphosis, *J. Morphol.* **68:**179.

Fache, B., and Charlemagne, J., 1975, Influence on allograft rejection of thymectomy at different stages of larval development in urodele amphibian *Pleurodeles waltlii* Michah (Salamandridae), *Eur. J. Immunol.* **5:**155.

Fey, F., 1962, Haematologische Untersuchungen an *Xenopus laevis* Daudin. I. Die Morphologie des Blutes mit einigen vergleichenden Betrachtungen bei *Rana esculenta* und *Rana temporaria*, *Morphol. Jahrb.* **103:**9.

Garavini, C., 1970, Seasonal variations in the hematic picture in *Triturus cristatus*, *Riv. Biol.* **63:**459.

Garcia-Herrera, F., and Cooper, E. L., 1968, Organos lifoides del anfibio apoda *Typhlonectes compressicaudata*, *Sombretiro de Acta Medica IV* **1968:**157.

Goldshein, S. J., and Cohen, N., 1972, Phylogeny of immunocompetent cells. I. *In vitro* blastogenesis and mitosis of toad (*Bufo marinus*) splenic lymphocytes in response to phytohemagglutinin and in mixed lymphocyte cultures. *J. Immunol.* **108**:1025.

Goldstine, S. N., Manickavel, V., and Cohen, N., 1975, Phylogeny of gut-associated lymphoid tissue, *Am. Zool.* **15**:107.

Goldstine, S. N., Collins, N. H., and Cohen, N., 1976, Mitogens as probes of lymphocyte heterogeneity in anuran amphibians, in: *Immunologic Phylogeny* (W. H. Hildemann and A. A. Benedict, eds.), pp. 343–352, Plenum Press, New York.

Good, R. A., Finstad, J., Pollara, B., and Gabrielsen, A. E., 1966, Morphologic studies on the evolution of the lymphoid tissues among the lower vertebrates, in: *Phylogeny of Immunity* (R. T. Smith, R. A. Good, and P. A. Miescher, eds.), pp. 149–170, University of Florida Press, Gainesville.

Green, N., and Cohen, N., 1979, Phylogeny of immunocompetent cells. III. Mitogen response characteristics of lymphocyte subpopulations from normal and thymectomized frogs (*Xenopus laevis*), *Cell. Immunol.* **48**:59.

Gruenwald, D. A., and Ruben, L. N., 1979, The effect of adult thymectomy upon helper function in *Xenopus laevis*, the South African clawed toad, *Immunology* **38**:191.

Hanzlikova, V., 1979, Histochemical and ultrastructural properties of myoid cells in the thymus of the frog, *Cell Tissue Res.* **197**:105.

Henry, M., and Charlemagne, J., 1977, Plasmocytic series in the perihepatic layer of the urodele amphibian *Pleurodeles waltlii* Michah (Salamandridae), *Dev. Comp. Immunol.* **1**:23.

Hightower, J. A., 1975, DNA synthesis in the thymus of the adult newt *Notophthalmus viridescens*, *Acta Anat.* **92**:454.

Hightower, J. A., and Haar, J. L., 1975, A light and electron microscopic study of the myelopoietic cells in the perihepatic and subcapsular region of the liver in the adult aquatic newt *Notophthalmus viridescens*, *Cell Tissue Res.* **159**:63.

Hightower, J. A., and St. Pierre, R. L., 1971, Haemopoietic tissue in the adult newt, *Notophthalmus viridescens*, *J. Morphol.* **135**:299.

Hollyfield, J. G., 1966, The origin of erythroblasts in *Rana pipiens* tadpoles, *Dev. Biol.* **14**:461.

Holzapfel, R. A., 1937, The cyclic character of hibernation in frogs, *Q. Rev. Biol.* **12**:65.

Horton, J. D., 1971a, Histogenesis of the lymphomyeloid complex in the larval leopard frog, *Rana pipiens*, *J. Morphol.* **134**:1.

Horton, J. D., 1971b, Ontogeny of the immune system in amphibians, *Am. Zool.* **11**:219.

Horton, J. D., and Horton, T. L., 1975, Development of transplantation immunity and restoration experiments in the thymectomized amphibian, *Am. Zool.* **15**:73.

Horton, J. D., and Manning, M. J., 1972, Response to skin allografts in *Xenopus laevis* following thymectomy at early stages of lymphoid organ maturation, *Transplantation* **14**:141.

Horton, J. D., and Manning, M. J., 1974a, Lymphoid organ development in *Xenopus* thymectomized at eight days of age, *J. Morphol.* **143**:385.

Horton, J. D., and Manning, M. J., 1974b, Effect of early thymectomy on the cellular changes occurring in the spleen of the clawed toad following administration of soluble antigen, *Immunology* **26**:797.

Horton, J. D., and Sherif, N. E. H. S., 1977, Sequential thymectomy in the clawed toad: Effect on mixed leucocyte reactivity and phytohaemagglutinin responsiveness, in: *Developmental Immunobiology* (J. B. Solomon and J. D. Horton, eds.), pp. 283–290, Elsevier/North-Holland, Amsterdam.

Horton, J. D., Rimmer, J. J., and Horton, T. L., 1977a, Critical role of the thymus in establishing humoral immunity in amphibians: Studies on *Xenopus* thymectomized in larval and adult life, *Dev. Comp. Immunol.* **1**:119.

Horton, J. D., Horton, T. L., and Rimmer, J. J., 1977b, Splenic involvement in amphibian transplantation immunity, *Transplantation* **24**:247.

Horton, J. D., Edwards, B. F., Ruben, L. N., and Mette, S., 1979, Use of different carriers to demonstrate thymic-dependent and thymic-independent anti-trinitrophenyl reactivity in the amphibian *Xenopus laevis*, *Dev. Comp. Immunol.* **3**:621.

Horton, J. D., Smith, A. R., Williams, N. H., Smith, A., and Sherif, N. E. H. S., 1980, Lymphocyte reactivity to 'T' and 'B' cell mitogens in *Xenopus laevis:* Studies on thymus and spleen, *Dev. Comp. Immunol.* **4**:75.

James, E. S., 1939, The morphology of the thymus and its changes with age in the neotenous amphibian *Necturus maculosus, J. Morphol.* **64**:445.

Jolly, J., 1919, Sur les organes lymphoïdes céphaliques des betraciens, *C.R. Soc. Biol.* **82**:200.

Jolly, J., 1923, *Traité Technique d'Hématologie,* Maloine, Paris.

Jolly, J., and Lievre, C., 1931, Hematopoïese intra-cardiaque chez les Urodèles, *C.R. Soc. Biol.* **106**:74.

Jordan, H. E., 1933, The evolution of blood-forming tissues, *Q. Rev. Biol.* **8**:58.

Jordan, H. E., 1938, Comparative hematology, in: *Handbook of Hematology* (H. Downey, ed.), pp. 704–862, Harper (Hoeber), New York.

Jordan, H. E., and Spiedel, C. C., 1923a, An experimental study of the spleen of the frog. *Rana pipiens, Anat. Rec.* **25**:136.

Jordan, H. E., and Spiedel, C. C., 1923b, Blood cell formation and destruction in relation to the mechanism of thyroid accelerated metamorphosis in the larval frog. *J. Exp. Med.* **38**:529.

Jurd, R. D., 1977, Secretory immunoglobulins and gut-associated lymphoid tissue in *Xenopus laevis,* in: *Developmental Immunobiology* (J. B. Solomon and J. D. Horton, eds.), pp. 307–314, Elsevier/North-Holland, Amsterdam.

Jurd, R. D., and Doritis, A., 1977, Antibody-dependent cellular cytotoxicity in poikilotherms, *Dev. Comp. Immunol.* **1**:341.

Jurd, R. D., and Stevenson, G. T., 1976, Surface immunoglobulins on *Xenopus laevis* lymphocytes, *Comp. Biochem. Physiol. A* **53**:381.

Kapa, E., 1963, Histological and histochemical analysis of the thymus in tailless amphibians, *Acta Morphol. Acad. Sci. Hung.* **12**:1.

Kapa, E., and Csaba, G., 1972, Phylogenesis of mast cells. III. Effect of hormonal induction on the maturation of mast cells in the frog, *Acta Biol. Acad. Sci. Hung.* **23**:47.

Kapa, E., Oláh, I., and Törö, I., 1968, Electron-microscopic investigation of the thymus of adult frog (*Rana esculenta*), *Acta Biol. Acad. Sci. Hung.* **19**:203.

Karpenter, K. L., and Turpen, J. B., 1979, Experimental studies on hemopoiesis in the pronephros of *Rana pipiens, Differentiation* **14**:167.

Katagiri, C., Kawahara, H., Nagata, S., and Tochinai, S., 1980, The mode of participation of T-cells in immune reactions as studied by transfer of triploid lymphocytes into early-thymectomized diploid *Xenopus,* in: *Development and Differentiation of Vertebrate Lymphocytes* (J. D. Horton, ed.), pp. 163–171, Elsevier/North-Holland, Amsterdam.

Kawahara, H., Nagata, S., and Katagiri, C., 1980, Role of injected thymocytes in reconstituting cellular and humoral immune responses in early thymectomized *Xenopus:* Use of triploid markers, *Dev. Comp. Immunol.* **4**:679.

Kent, S. P., Evans, E. E., and Attleberger, M. H., 1964, Comparative immunology. Lymph nodes in the amphibian, *Bufo marinus, Proc. Soc. Exp. Biol. Med.* **116**:456.

Kingsbury, B. F., 1912, Amphibian tonsils, *Anat. Anz.* **42**:593.

Klug, H., 1967, Submikroskipische zytologie des Thymus von *Ambystoma mexicanum, Z. Zellforsch. Mikrosk, Anat.* **78**:388.

Kraft, N., and Shortman, K., 1972, Differentiation of antibody-forming cells in toad spleen. A study using density and sedimentation velocity cell separation, *J. Cell Biol.* **52**:438.

Le Douarin, N., 1966, L'hématopoïese dans les formes embryonnaires et jeunes des vertébrés, *Annee Biol.* **5**(ser. 4):105.

Mackay, I. R., and Goldstein, G., 1967, Thymus and muscle, *Clin. Exp. Immunol.* **2**:139.

Maclean, N., and Jurd, R. D., 1972, The control of haemoglobin synthesis, *Biol. Rev.* **47**:393.

Maniatis, G. M., and Ingram, U. M., 1971, Erythropoiesis during amphibian metamorphosis. I. Site of maturation of erythrocytes in *Rana catesbeiana, J. Cell Biol.* **49**:372.

Manning, M. J., 1971, The effect of early thymectomy on histogenesis of the lymphoid organs in *Xenopus laevis, J. Embryol. Exp. Morphol.* **26**:219.

Manning, M. J., 1978, The amphibian immune system and emerging adaptations to life on land, in: *Proceedings of the Zodiac Symposium on Adaptation,* pp. 88–91, Pudoc, Centre for Agricultural Publishing and Documentation, Wageningen, The Netherlands.

Manning, M. J., and Collie, M. H., 1977, The ontogeny of thymic dependence in the amphibian *Xenopus laevis*, in: *Developmental Immunobiology* (J. B. Solomon and J. D. Horton, eds.), pp. 291–298, Elsevier/North-Holland, Amsterdam.

Manning, M. H., and Horton, J. D., 1969, Histogenesis of lymphoid organs in larvae of the South African clawed toad, *Xenopus laevis* (Daudin), *J. Embryol. Exp. Morphol.* **22**:265.

Manning, M. J., and Turner, R. J., 1972, Some responses of the clawed toad, *Xenopus laevis*, to soluble antigens administered in adjuvant, *Comp. Biochem. Physiol. A* **42**:735.

Manning, M. J., and Turner, R. J., 1976, *Comparative Immunobiology*, Blackie, Glasgow.

Manning, M. J., Donnelly, N., and Cohen, N., 1976, Thymus-dependent and thymus-independent components of the amphibian immune system, in: *Phylogeny of Thymus and Bone Marrow-Bursa Cells* (R. K. Wright and E. L. Cooper, eds.), pp. 123–132, Elsevier/North-Holland, Amsterdam.

Marchalonis, J. J., and Cohen, N., 1973, Isolation and partial characterization of immunoglobulin from a urodele amphibian (*Necturus maculosus*), *Immunology* **24**:395.

Mattes, M. J., and Steiner, L. A., 1978, Anti-sera to frog immunoglobulins cross-react with a periodate-sensitive cell surface determinant, *Nature (London)* **273**:761.

Maximow, A., 1927, cited by Weidenreich, R., 1933, in: *Handbuch der vergleichenden Anatomie der Wirbeltiere* (L. Bolk, E. Göppert, and E. Kallius, eds.), Vol. VI, pp. 375–447, Urban & Schwarzenberg, Munich.

Michea-Hamzehpour, M., 1977, Indirect immunofluorescent identification of 19S immunoglobulin-containing cells in the intestinal mucosa of *Xenopus laevis*, *J. Exp. Zool.* **201**:109.

Minagawa, Y., Ohnishi, K., and Murakawa, S., 1975, Structure and immunological function of lymphomyeloid organs in the bullfrog *Rana catesbeiana*, in: *Immunologic Phylogeny* (W. H. Hildemann and A. A. Benedict, eds.), pp. 257–266, Plenum Press, New York.

Mitsuhashi, S., Kurashige, S., Mishima, S., Yamaguchi, N., and Fukai, K., 1971, Antibody production without reactive proliferation of pyroninophilic cells in the rainbow trout and African clawed toad, *Tohoku J. Exp. Med.* **103**:7.

Moticka, E. J., Brown, B. A., and Cooper, E. L., 1973, Immunoglobulin synthesis in bullfrog larvae, *J. Immunol.* **110**:855.

Murakawa, S., 1968, Studies on the transplantation immunity in the Japanese newt, *Cynops pyrrhogaster*, *SABCO J.* **4**:17.

Myers, M. A., 1928, A study of the tonsillar developments in the lingual region of anurans, *J. Morphol.* **45**:399.

Nagata, S., 1976, An electron microscopic study on the thymus of larval and metamorphosed toads, *Xenopus laevis* Daudin, *J. Fac. Sci. Hokkaido Univ. Ser. 6* **20**:263.

Nagata, S., and Katagiri, C., 1978, Lymphocyte surface immunoglobulin in *Xenopus laevis*. Light and electron microscopic demonstration by immunoperoxidase method, *Dev. Comp. Immunol.* **2**:277.

Nagata, S., and Tochinai, S., 1978, Isolated lymphocytes can restore allograft rejection capacity of early-thymectomized *Xenopus*, *Dev. Comp. Immunol.* **2**:637.

Nigam, H. C., 1977, Tonsils and other leucocytopoietic centres in the Indian water-skipping frog, *Rana cyanophlyctis* (Boulanger), *Curr. Sci.* **46**:435.

Nossal, G. J. V., and Ada, G. L., 1971, *Antigens, Lymphoid Cells, and the Immune Response*, Academic Press, New York.

Rimmer, J. J., 1977, Electron microscopic studies of developing amphibian thymus, *Dev. Comp. Immunol.* **1**:321.

Rimmer, J. J., and Gearing, A. J. H., 1980, Antigen specific migration inhibition of peritoneal exudate cells in an anuran (*Rana temporaria*), in: *Development and Differentiation of Vertebrate Lymphocytes* (J. D. Horton, ed.), pp. 195–200, Elsevier/North-Holland, Amsterdam.

Rimmer, J. J., and Horton, J. D., 1977, Allograft rejection in larval and adult *Xenopus* following early thymectomy, *Transplantation* **23**:142.

Ruben, L. N., 1975, Ontogeny, phylogeny and cellular cooperation, *Am. Zool.* **15**:93.

Ruben, L. N., 1976, Phylogeny of cell–cell cooperation in immunity, in: *Comparative Immunology* (J. J. Marchalonis, ed.), pp. 120–166, Blackwell, Oxford.

Ruben, L. N., and Edwards, B. F., 1979, The phylogeny of the emergence of "T–B" collaboration in humoral immunity, in: *Contemporary Topics in Immunobiology* (N. Cohen and J. J. Marchalonis, eds.), Vol. 9, pp. 55–89, Plenum Press, New York.

Ruben, L. N., van der Hoven, A., and Dutton, R. W., 1973, Cellular cooperation in hapten-carrier responses in the newt, *Triturus viridescens*, *Cell. Immunol.* **6**:300.

Ruben, L. N., Clothier, R., Hodgson, R., and Balls, M., 1977, The *in vitro* reconstitution of thymus cell dependent humoral immune response of thymectomized *Xenopus laevis* with allogeneic thymocytes, in: *Developmental Immunobiology* (J. B. Solomon and J. D. Horton, eds.), pp. 277–282, Elsevier/North-Holland, Amsterdam.

Secombes, C. J., and Manning, M. J., 1980, Comparative studies on the immune system of fishes and amphibians: Antigen localization in the carp *Cyprinus carpio* L., *J. Fish Dis.* **3**:399.

Sklower, A., 1925, Das inkretorische System im Lebenszyklus der Frösche *Rana temporaria* L. I. Schildrüse, Hypophyse, Thymus und Keimdrüsen, *Z. Vgl. Physiol.* **2**:474.

Sterba, G., 1950, Über die morphologischen und histogenetischen Thymus-probleme bei *Xenopus laevis* Daudin nebst einigen Bermerkungen über die Morphologie der Kaulquappen, *Abh. Sachs. Akad. Wiss. Leipzig Math. Naturwiss. Kl.* **44**:1.

Sterba, G., 1951, Untersuchungen an der Milz des Krallenfrosches (*Xenopus laevis* Daudin), *Morphol. Jahrb.* **90**:221.

Sterba, G., 1952, Mitteilungen über die Altersinvolution des Amphibien thymus. I. Volumetrische Bestimmungen am Thymus des Krallenfrosches *Xenopus laevis* Daud., *Anat. Anz.* **99**:106.

Tahan, A. M., and Jurd, R. D., 1979, Delayed hypersensitivity in *Ambystoma mexicanum*, *Dev. Comp. Immunol.* **3**:299.

Tochinai, S., 1975, Distribution of lympho-epithelial tissues in the larval African clawed toad, *Xenopus laevis* (Daudin), *J. Fac. Sci. Hokkaido Univ. Ser. 6* **19**:803.

Tochinai, S., 1976a, Demonstration of thymus-independent immune system in *Xenopus laevis*. Response to polyvinylpyrrolidone, *Immunology* **31**:125.

Tochinai, S., 1976b, Lymphoid changes in *Xenopus laevis* following thymectomy at the initial stage of its histogenesis, *J. Fac. Sci. Hokkaido Univ. Ser. 6* **20**:175.

Tochinai, S., and Katagiri, C., 1975, Complete abrogation of immune response to skin allografts and rabbit erythrocytes in the early thymectomized *Xenopus*, *Dev. Growth Differ.* **17**:383.

Tochinai, S., Nagata, S., and Katagiri, C., 1976, Restoration of immune responsiveness in early thymectomized *Xenopus* by implantation of histocompatible adult thymus, *Eur. J. Immunol.* **6**:711.

Tooze, J., and Davies, H. G., 1968, Light and electron microscopic observations on the spleen and the splenic leucocytes of the newt *Triturus cristatus*, *Am. J. Anat.* **123**:521.

Törö, I., Oláh, I., Röhlich, P., and Virágh, S., 1969, Electron microscopic observations on myoid cells of the frog's thymus, *Anat. Rec.* **165**:329.

Tournefier, A., 1972, Les reactions immunitaires chez les Amphibiens Urodèles. III. Rôle du thymus dans l'immunité de transplantation. Capacité d'immunisation aux antigènes particulaires chez le Pleurodèle et le Triton alpestre adultes, in: *Phylogenic and Ontogenic Study of the Immune Response and Its Contribution to the Immunological Theory*, Colloque Inserm, pp. 105–112, Ministry of Public Health, Paris.

Tournefier, A., 1973, Développement des organes lymphoïdes chez l'Amphibien Urodèle *Triturus alpestris* Laur.; tolérance des allogreffes après la thymectomie larvaire, *J. Embryol. Exp. Morphol.* **29**:383.

Tournefier, A., and Charlemagne, J., 1975, Antibodies against salmonella and SRBC in urodele amphibians: Synthesis and characterizations, in: *Immunologic Phylogeny* (W. H. Hildemann and A. A. Benedict, eds.), pp. 161–172, Plenum Press, New York.

Turner, R. J., 1969, The functional development of the reticuloendothelial system in the toad, *Xenopus laevis* (Daudin), *J. Exp. Zool.* **170**:467.

Turner, R. J., 1970, The influence of colloidal carbon on hemagglutinin production in the toad, *Xenopus laevis*, *J. Reticuloendothelial Soc.* **8**:434.

Turner, R. J., 1973, Response of the toad, *Xenopus laevis*, to circulating antigens. II. Responses after splenectomy, *J. Exp. Zool.* **183**:35.

Turner, R. J., and Manning, M. J., 1973, Response of the toad, *Xenopus laevis*, to circulating antigens. I. Cellular changes in the spleen, *J. Exp. Zool.* **183**:21.

Turner, R. J., and Manning, M. J., 1974, Thymic dependence of amphibian antibody responses, *Eur. J. Immunol.* **4**:343.

Turner, R. J., Tâm, N. D., and Manning, M. J., 1974, Effects of *Corynebacterium parvum* and Freund's adjuvants on amphibian antibody responses, *J. Reticuloendothelial Soc.* **16**:232.

Turpen, J. B., 1980, Early embryogenesis of hemopoietic cells in *Rana pipiens*, in: *Development and Differentiation of Vertebrate Lymphocytes* (J. D. Horton, ed.), pp. 15–24, Elsevier/North-Holland, Amsterdam.

Turpen, J. B., Turpen, C. J., and Flajnik, M., 1979, Experimental analysis of hematopoietic cell development in the liver of larval *Rana pipiens*, *Dev. Biol.* **69**:466.

van Rooijen, N., 1980, Immune complex trapping in lymphoid follicles: A discussion on possible functional implications, in: *Phylogeny of Immunological Memory* (M. J. Manning, ed.), pp. 281–290, Elsevier/North-Holland, Amsterdam.

von Braunmühl, A., 1926, Über einige myelolymphoide und lymphoepitheliale Organe der Anuren, *Z. Mikrosk, Anat. Forsch.* **4**:635.

Warr, G. W., Deluca, D., and Marchalonis, J. J., 1980, Phylogeny and ontogeny of antigen-specific T cell receptors, in: *Development and Differentiation of Vertebrate Lymphocytes* (J. D. Horton, ed.), pp. 99–110, Elsevier/North-Holland, Amsterdam.

Webster, W. D., 1934, The development of the thymus bodies in *Necturus maculosus*, *J. Morphol.* **56**:295.

Weilacher, S., 1933, Die Milz der Gymnophionen Beitrag z. Kenntnis der Gymnophionen, *Morphol. Jahrb.* **72**:469.

Weiss, N., Horton, J. D., and Du Pasquier, L., 1973, The effect of thymectomy on cell surface associated and serum immunoglobulin in the toad, *Xenopus laevis*, Daudin: A possible inhibitory role of the thymus on the expression of immunoglobulins, in: *Phylogenetic and Ontogenetic Study of the Immune Response and Its Contribution to the Immunological Theory*, Colloque Inserm, pp. 165–174, Ministry of Public Health, Paris.

Welsch, U., and Storch, V., 1972, Elektronenmikroskopische Untersuchungen an der Leber von *Ichthyophis kohtaoensis* (Gymnophiona), *Zool. Jahrb, Anat. Abt. Ontog. Tiere* **89**:621.

Wiedersheim, R., 1879, cited by Pischinger, A., 1933, in: *Handbuch der vergleichenden Anatomie der Wirbeltiere* (L. Bolk, E. Göppert, E. Kallius, and W. Lubosch, eds.), Vol. III, pp. 279–348, Urban & Schwarzenberg, Munich.

Witschi, E., 1956, *Development of Vertebrates*, Saunders, Philadelphia.

Wong, W. C., 1972, Lymphoid aggregations in the oesophagus of the toad (*Bufo melanostictus*), *Acta Anat.* **83**:461.

Wright, R. K., and Cooper, E. L., 1980, Temperature and immunological memory in anuran amphibians, in: *Phylogeny of Immunological Memory* (M. J. Manning, ed.), pp. 155–160, Elsevier/North-Holland, Amsterdam.

Wright, R. K., Eipert, E. F., and Cooper, E. L., 1978, Regulatory role of temperature on the development of ectothermic vertebrate lymphocyte populations, in: *Animal Models of Comparative and Developmental Aspects of Immunity and Disease* (M. E. Gershwin and E. L. Cooper, eds.), pp. 80–92, Pergamon, Press, Elmsford, N.Y.

Yamaga, K. M., Kubo, R. T., and Etlinger, H. M., 1978, Studies on the question of conventional immunoglobulin on thymocytes from primitive vertebrates. II. Delineation between Ig-specific and cross-reactive membrane components, *J. Immunol.* **120**:2074.

Yoffey, J. M., 1960, The lymphomyeloid complex, in: *Haemopoiesis* (G. E. W. Wolstenholme and M. O'Conner, eds.), Ciba Foundation Symposium, pp. 1–36, Churchill, London.

Zettergren, L. D., Lydyard, P. M., and Parkhouse, R. M. E., 1977, Liver as a site of B cell generation in *Xenopus laevis*, *Fed. Proc.* **36**:1239.

Zettergren, L. D., Kubagawa, H., and Cooper, M. D., 1980, Development of B cells in *Rana pipiens*, in: *Phylogeny of Immunological Memory* (M. J. Manning, ed.), pp. 177–185, Elsevier/North-Holland, Amsterdam.

Zylbersac, S., 1936, Sur la nature des cellules pigmentaires dans le foie des Amphibiens, *Arch. Int. Med. Exp.* **11**:545.

12

RES Structure and Function of the Reptilia

VR. MUTHUKKARUPPAN, MYRIN BORYSENKO, and RASHIKA EL RIDI

1. INTRODUCTION

A phylogenetic approach to understanding the development and mechanisms of immunity has gained attention only in recent years. On the evolutionary scale, reptiles are pivotal since they are the progenitors of both avian and mammalian classes. It is now agreed that the basic pattern of adaptive immune responsivity was established when the first vertebrates arose. However, this basic pattern has been constantly altered by evolutionary forces, and thus, the existing reptilian species possess a particular level of organization of their reticuloendothelial systems, much like other organ systems, such as the vascular systems, urogenital systems, nervous systems, etc. In this light, it is important to understand the structure and function of the reptilian immune systems. In particular, these ectothermic amniotes provide an organizational level which could help us to delineate the nature of the immune systems of endotherms. For example, the avian bursa of Fabricius, the generator of B lymphocytes, probably evolved from reptilian ancestors and its equivalent may still exist among living representatives. With the advent of immunohistochemical and *in vitro* techniques, structural and functional markers of reptilian lymphoid cells can now be identified and characterized. This review attempts to synthesize the existing literature concerning the structure and function of the reticuloendothelial systems of reptiles. Naturally, a major portion of this information has to do with the immune systems proper. For lack of space, certain aspects such as naturally occurring antibodies, fever in relation to immune response, and the development and morphology of lymphatics will not be considered.

VR. MUTHUKKARUPPAN • Department of Immunology, Madurai-Kamaraj University, Madurai 625 021, India.   MYRIN BORYSENKO • Department of Anatomy and Cellular Biology, Tufts University, School of Medicine, Boston, Massachusetts 02111.   RASHIKA EL RIDI • Zoology Department, Faculty of Science, Cairo University, Cairo, Egypt.

2. LYMPHOID ORGANS AND CELLS

2.1. SURVEY OF REPTILIAN LYMPHOID TISSUES AND ORGANS

Reptilian lymphoid cells and tissues are widely distributed throughout the organ systems. Lymphocytes and monocytes are common blood elements in all reptiles. Although diffuse lymphoid aggregates are commonly found in members of all reptilian classes, they are not as extensive or well developed as those of mammals and birds. For example, gut-associated lymphoid aggregates of turtles and lizards are commonly found throughout the lamina propria of the alimentary tract. They appear as solitary units, however, and never as extensive or complex as mammalian tonsils and Peyer's patches or the avian bursa of Fabricius (Borysenko and Cooper, 1972; Hussein et al., 1978b). Thus, while the mere presence of reptilian gut-associated lymphoid tissues (GALT) suggests a possible evolutionary precursor of the secretory IgA system of antibody production and/or a bursal equivalent, there is no compelling functional evidence to support this. Neither serum nor secretory IgA nor their equivalent has ever been demonstrated in reptiles. Likewise, histological evidence does not suggest the presence of a specialized organization for B-lymphocyte generation in the cloacal region of reptiles although definitive lymphoid accumulations are observed in this region in turtles and lizards (Sidky and Auerbach, 1975; Muthukkaruppan et al., 1976b; Hameed, 1980). Such simple aggregates are also found in the lungs, kidneys, and urinary bladder in some reptiles and may well be transient rather than permanent structures (Borysenko, 1978). Recent studies on a number of lizard and snake species show that the number and size of the gut-associated lymphoid aggregates vary tremendously with the season of year. They are particularly sparse in the winter months in those reptiles which hibernate and in one lizard, *Agama stellio*, they remain poorly developed (Hussein et al., 1978a,b, 1979a,d).

There have been several reports of lymph node-like structures in reptiles. Small encapsulated lymphoid organs occur in the axillary regions of the turtle, *Chelydra serpentina* (Borysenko and Cooper, 1972), and the lizard, *Tiliqua rugosa* (Wetherall and Turner, 1972). Their anatomical location and histological composition suggest that they may be lymph node precursors. More definitive evidence of lymph node-like structures has been presented for the snake, *Elaphe quadrivirgata* (Kotani, 1959), and the lizard, *Gehyra variegata* (Johnston, 1973). These structures consist of diffuse lymphoid tissue in the walls of blood vessels and perivascular lymphatics. Their relationship to the lymphatic vessel reflects at least a superficial similarity to the lymph nodes of birds (Biggs, 1957) and anuran amphibians (Horton, 1971). However, there is not enough functional information to determine the degree of homology between them. It may be appropriate to mention that the reptiles apparently do not possess lymphopoietic anterior kidneys like those of fish and anuran amphibians (Diener and Nossal, 1966; Sailendri and Muthukkaruppan, 1975). Likewise, in adult reptiles, the liver is not an important lymphoid organ as it is in urodele amphibians (Henry and Charlemagne, 1977).

The most prominent and well-developed reptilian lymphoid organs are the thymus and spleen. In the four orders of reptiles, the thymus varies somewhat in terms of location and number of lobes, reflecting some variation in embryonic origin. For example, in the tuatara and in most lizards and snakes, there are two lobes on each side of the neck. Each lobe consists of a peripheral cortex and central medulla, but is not subdivided into lobules. The thymus of crocodilians consists of an elongated, chainlike structure, beginning at the base of the skull, traversing the length of the neck, and extending nearly to the heart. Thus, the thymus of this reptilian group most closely resembles that of birds. Turtles are unique in that the thymus usually consists of a distinct single lobe on each side of the neck at the bifurcation of the common carotid arteries and each lobe is subdivided into a number of partial lobules (Bockman, 1970). The turtle thymus is also in close association with the parathyroid glands.

Morphological and immunological criteria indicate that the reptilian spleen is the best developed and functionally most important of the peripheral lymphoid organs, subserving both cell-mediated and humoral immune functions. Although there is much variation from species to species with regard to splenic size, shape, anatomical location, and histological organization, in all cases the spleen is a large, heavily encapsulated lymphoid organ in close association with the systemic blood circulation.

2.2. HISTOLOGY OF THYMUS, SPLEEN, AND GALT

In all reptiles studied, the thymus is clearly separated into cortical and medullary regions (Figure 1). Arteries and veins are prominent in the connective tissue septae and in the medulla, whereas only capillaries are encountered in the cortex proper. The reptilian thymic cortex consists primarily of densely packed small lymphocytes (thymocytes) in a delicate "framework" of stellate epithelial cells. There are fewer lymphocytes in the medulla. The predominant cells in this region are the lighter-staining epithelial cells, which account for its relative paleness. Large myoid cells are consistent cellular elements of the medulla, but occur in small numbers in the cortex. In addition, small numbers of macrophages and eosinophils occur in the medulla as well (Borysenko and Cooper, 1972). Thus, in these respects, the reptilian thymus is similar to that of mammals (Clark, 1963).

A detailed ultrastructural study of several reptilian species describes three distinct epithelial cell types in the thymus, based strictly on morphological criteria. Of particular interest are those which were described as having a similar appearance to the beta cells of pancreatic islands, in that their nuclei are euchromatic and their cytoplasm contains a well-developed Golgi complex and a number of dense granules (Bockman and Winborn, 1967). Similar secretorylike cells have been described in mammals (Clark, 1966), and one is tempted to consider that this cell may be the producer of thymic hormone(s) which induces thymocyte maturation. Such notions remain speculative since direct histochemical evidence is still lacking in both mammals and reptiles.

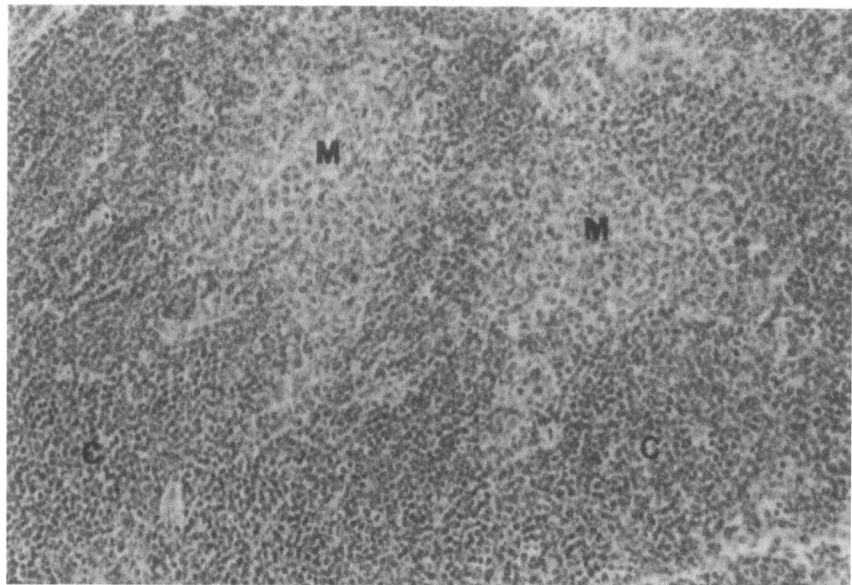

FIGURE 1. Thymic lobule of a young snapping turtle. The cortex (C) and medulla (M) are well differentiated. Hematoxylin and eosin.

Ultrastructural examination of reptilian myoid cells reveals the presence of both mature and immature cells, containing varying amounts of myofibrils. The presence of immature myoid cells in the adult reptilian thymus suggests that myogenesis may occur there (Raviola and Raviola, 1967). Myoid cells are very large and well developed in reptiles compared to those of mammals.

Although thymic (Hassall's) corpuscles are not present in the reptilian thymus, a variety of epithelially lined cysts occur and appear to be associated with degenerative processes. Thymic involution occurs with age in reptiles as in other vertebrates. The distinction between cortex and medulla becomes reduced and much of the lymphoid tissue is replaced by connective tissue elements. Factors other than aging may cause thymic involution in reptiles. Seasonal fluctuation (decreased thymocyte numbers in the winter) is transitory and reversible, while long-term starvation and severe disease may lead to permanent involution, as with aging (Bockman, 1970).

The spleens of turtles (Borysenko, 1976a), snakes (Kanakambika and Muthukkaruppan, 1973), and the tuatara (Marchalonis et al., 1969) have well-defined red and white pulp regions. On the other hand, in several species of lizards, the splenic pulp is not as well demarcated. Here, the red pulp consists of narrow strands of blood sinuses between confluent areas of white pulp (Kanakambika and Muthukkaruppan, 1973). The spleen of the lizard, *Calotes versicolor*, is even more unique in that the splenic pulp is divided into a number of lobules by connective tissue septae. Each lobule consists of a central white pulp region associated with an arteriole, surrounded by a thin red pulp zone (Pitchappan and Muthukkaruppan, 1977c). The more typical spleens, such as those of turtles, contain well-defined periarteriolar lymphocyte sheaths separated from the red pulp cords and sinuses by a marginal zone (Figure 2). In lizards there is also a prominent and continuous layer of reticular cells interposed between the arteriole and the lymphocyte sheath. The functional significance of these cells is not known. Reticular cells, attached to reticular fibers, are also found among the cells of the lymphocyte sheath and the marginal zone. Reticular cells and macrophages are prominent in the red pulp as well. Intracardiac injection of colloidal carbon results in entrapment of carbon particles throughout the red pulp and along the marginal zone (Figure 3).

Although it is clear that the spleen houses lymphocytes which elicit both cell-mediated and humoral immune responses, T- and B-cell zones have not been defined by immunohistochemical means. However, some histological observations suggest that T and B cells may be segregated into such regions. In the lizard, *Calotes versicolor*, adult thymectomy or treatment with rabbit anti-*Calotes* thymocyte serum results in severe depletion of the white pulp. During lymphoid regeneration, the juxta-arteriolar (reticular) region is repopulated by medium, then small lymphocytes (Pitchappan and Muthukkaruppan, 1977c). In the turtle, *Chelydra serpentina*, adult specimens (post-thymic involution) exhibit some lymphoid cell depletion in the juxta-anteriolar region of the white pulp sheaths (Borysenko and Cooper, 1972). Reticular cells are seen as the predominant cell type in this zone. On the other hand, in juvenile turtles, there is a continuous sheath of lymphocytes from the arteriole to the marginal zone. At the start of a humoral response (antigen trapping), blastogenesis and plasma cell development occur primarily in the outer lymphocyte sheath and extend into the red

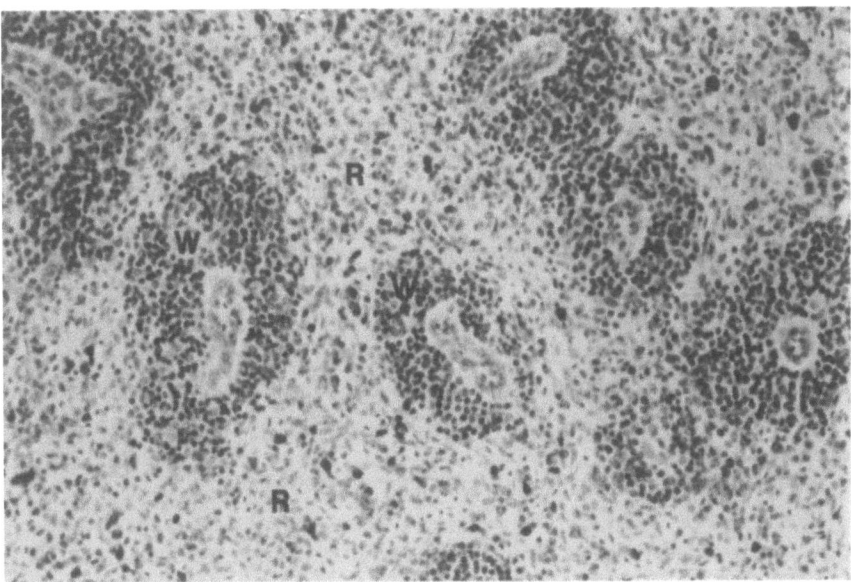

FIGURE 2. Spleen of a young snapping turtle. The white (W) and red (R) pulp regions are clearly defined. Darkly stained cells in the red pulp are plasma cells. Methyl green–pyronine Y.

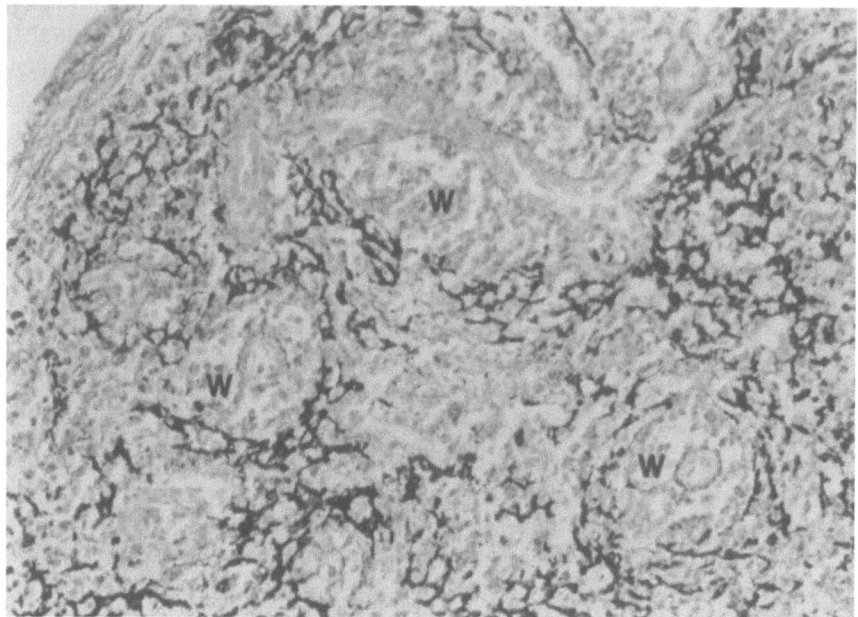

FIGURE 3. Spleen of a young snapping turtle. Colloidal carbon was injected into the heart and the spleen was removed 1 hr later. Note the carbon trapping pattern in the red pulp and marginal zones. The white pulp regions (W) are relatively free of carbon. Gridley's reticulum stain.

pulp (Borysenko, 1976a). These preliminary observations suggest that the inner lymphocyte sheath consists predominantly of T cells, while the outer zone of the lymphocyte sheath consists of predominantly of B cells. Studies on the rabbit spleen indicate this kind of arrangement (Langevoort, 1963).

Although reptiles do not possess complex GALT such as tonsils, Peyer's patches, and the appendix, diffuse aggregates of lymphoid cells occur in the lamina propria and epithelia of the gut in most reptiles. A recent study on a variety of reptilian species described a number of cell types in the connective tissues and epithelia of the alimentary tract. Lymphocytes, plasma cells, macrophages, basophils, and "globular" leukocytes (of unknown geneology) have been identified in these regions (Solas and Zapata, 1980). The origin and function of "globular" cells have not been elucidated, although it is reasonable to assume that they play some role in the immune system. For the most part, a similar variety of cell types is found in the gut of mammals and birds.

2.3. DEVELOPMENT OF THYMUS, SPLEEN, AND GALT

Among reptiles there is much variability in thymic development. This is reflected in the diversity of anatomical locations of the mature thymus and its lobar pattern. Like the thymuses of fish, amphibians, and birds, reptilian thymic primordia are dorsal outgrowths of the pharyngeal pouches. The definitive thymuses of lizards, snakes, and turtles develop from different pairs of pharyngeal pouches. The lizard thymus originates from the second and third, the snake thymus from the fourth and fifth, and the turtle thymus from the third and fourth. Transitory or rudimentary thymic buds may also arise from other pharyngeal pouches. Thus, small aberrant foci of thymic tissue may occur, separated from the major lobes of the definitive thymus (review: Bockman, 1970).

The thymus and parathyroid originate in close proximity in the pharyngeal pouches. In many reptilian species, parathyroid tissue is frequently embedded within the parenchyma of the thymus (Bockman, 1970). In the snapping turtle, *Chelydra serpentina*, the thymus not only surrounds the parathyroid but frequently envelopes the common carotid artery, thereby making surgical thymectomy in this species virtually impossible (Borysenko, 1978). In contrast, the thymus of the lizard, *Calotes versicolor*, is distinct and is accessible to surgical removal (Pitchappan and Muthukkaruppan, 1976).

As in the case in all vertebrates possessing lymphoid organs, the thymus in reptiles is the first lymphoid organ to develop. A detailed study of the lizard, *Calotes versicolor*, describes the early embryonic development of the thymus (Pitchappan and Muthukkaruppan, 1977d). Stem cells, first observed in the yolk sac islets, are later seen infiltrating the epithelial thymic rudiments. Subsequently, large and small lymphocytes appear in the thymus, followed by differentiation into the cortical and medullary zones. These sequential events suggest that the lymphocytes of the lizard thymus are derived from blood-borne stem cells, although the origin of these stem cells remains to be determined. Similarly, there is no direct evidence which points to the origin of immunoglobulin-producing cells (bursal equivalent). The yolk sac islets as well as embryonic liver and bone marrow are likely candidates in this regard, but definitive studies have not yet conducted in any reptilian species.

The spleen in reptiles is functionally the most important secondary lymphoid organ, containing vast numbers of immunocompetent cells of diverse origin. As one might expect, the spleen develops into a lymphopoietic organ much later than the thymus. In the lizard (Pitchappan and Muthukkaruppan, 1977d) and turtle (Borysenko, 1978), the thymus is fully differentiated before any lymphocytes appear in the splenic rudiment. At about the time of hatching, the spleen is fairly well differentiated and continues to grow in size.

In the early phases of splenic development, before the organ becomes lymphopoietic, it contains a large number of granulocytes. In the turtle, the eosinophils are restricted to the subcapsular region while basophils are scattered throughout the parenchyma. It appears as though the eosinophils proliferate in the subcapsular spleen during this stage of development, while basophils enter the spleen as blood elements. In later development, the subcapsular eosinophils disappear and the spleen becomes primarily lymphopoietic. Plasma cells first appear in the spleen around the time of hatching (Borysenko, 1978). Similar observations were reported in the lizard embryo (Kanakambika and Muthukkaruppan, 1973).

In reptiles, the GALT develops somewhat later than the lymphopoietic spleen. The lymphoid aggregates of the cloacal complex of the snapping turtle,

Chelydra serpentina, are first evident at the time of hatching when the spleen is already well formed (Sidky and Auerbach, 1968). In the lizard, *Calotes versicolor*, the lymphoid aggregates of the cloacal region do not appear until 2 months posthatching (Pitchappan and Muthukkaruppan, 1977d). These studies suggest that the GALT is a secondary lymphoid tissue, acquiring cells from the thymus and possible other stem-cell sources which develop earlier. In this sense, the GALT is unlike the avian bursa of Fabricius, which develops in parallel with the thymus and primarily serves as a generator of B cells.

2.4. CHANGES IN SPLEEN HISTOLOGY FOLLOWING ANTIGENIC STIMULATION

Upon antigenic challenge with sheep erythrocytes, the lizard spleen undergoes a proliferative response, but shows no further changes after secondary challenge (Kanakambika and Muthukkaruppan, 1972). Similarly, turtles stimulated with keyhole limpet hemocyanin (KLH) show a strong and prompt proliferative response in the white pulp sheaths (Figure 4), followed by migration and differentiation of lymphoblasts into plasma cells in the red pulp (Figure 5). Upon further antigenic stimulation, no new proliferative response was observed in the spleen (Borysenko, 1976a). The kinetics of the primary lymphoblast response and lack of responsiveness following secondary challenge correlate well with observations in amphibians (Diener and Nossal, 1966; Moticka *et al.*, 1973).

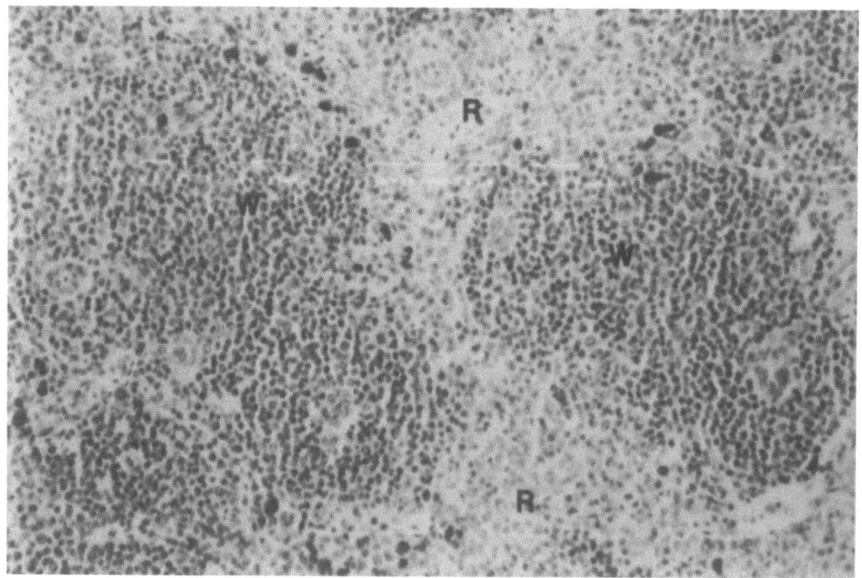

FIGURE 4. Spleen of a snapping turtle 8 days after immunization with KLH. Note the expansion of the white pulp (W) regions and diminution of red pulp (R) due to lymphocyte proliferation. Compare to Figure 2. Methyl green–pyronine Y.

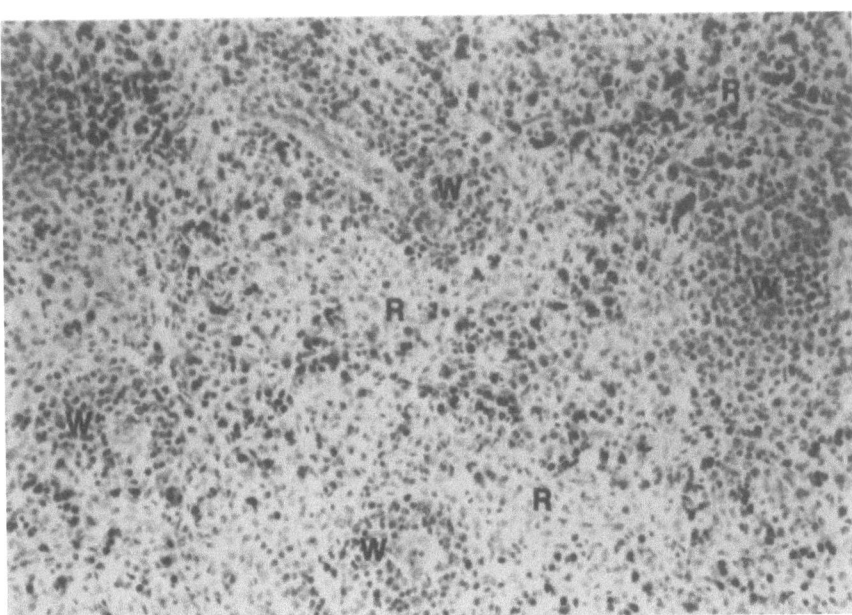

FIGURE 5. Spleen of a snapping turtle 18 days after immunization with KLH. The white pulp (W) appears somewhat depleted, whereas the darkly stained plasma cells are increased in number in the red pulp (R). Methyl green–pyronine Y.

Although detailed histological studies have not been performed in other reptilian species, the absence of nodules with germinal centers has also been reported in immunized tuatara (Marchalonis et al., 1969) and the lizard, Tiliqua rugosa (Wetherall and Turner, 1972). In spite of this apparent deficiency, it is evident that most reptiles, when challenged with potent antigens and maintained at a warm temperature, produce good secondary responses (Lerch et al., 1967; Ambrosius et al., 1970; Wetherall and Turner, 1972; Wright and Shapiro, 1973). Although nodules with germinal centers are associated with secondary humoral immune responses in mammals (Thorbecke et al., 1974), such responses in reptiles and other ectotherms apparently occur without such structures. It has been suggested that in the evolutionary scheme, the complex structure of peripheral lymphoid organs of endotherms probably reflects the "centralization" of immunocompetent cells and immune functions resulting in greater efficiency and diversity in immune responsiveness, rather than the capacity to mount secondary responses per se. In ectothermic vertebrates, immune functions are less centralized and occur to a greater extent in connective tissue microenvironments, such as the lamina propria of the gut (Borysenko, 1975). Such microenvironments may be of primary importance in the hagfish, which is immunocompetent without any organized peripheral lymphoid organs (Hildemann and Thoenes, 1969; Linthicum and Hildemann, 1970), and in the fetal lamb, which is capable of an array of immune responses prior to the development of peripheral lymphoid organs (Fahey, 1974).

2.5. ULTRASTRUCTURE OF IMMUNOCOMPETENT CELLS

Ultrastructural studies of reptilian lymphoid organs and cells derived from these organs reveal a variety of cell types which correspond closely to immunocompetent cells in other vertebrate classes, birds and mammals on one side of the phylogenetic tree and fish and amphibians on the other. A detailed study has been conducted on the snapping turtle (Borysenko, 1976b).

Monocytes and macrophages are seen in substantial numbers in turtle spleens, particularly following intravenous antigenic challenge. They occur most frequently along the sinus walls and in the red pulp cords. The nucleus of this cell type is usually U-shaped, and moderate amounts of heterochromatin are distributed along the nuclear envelope and nucleolus. In the cytoplasm are found a scattering of mitochondria, short segments of rough endoplasmic reticulum, and lysosomes of various sizes and densities (Figure 6). Lysosomal

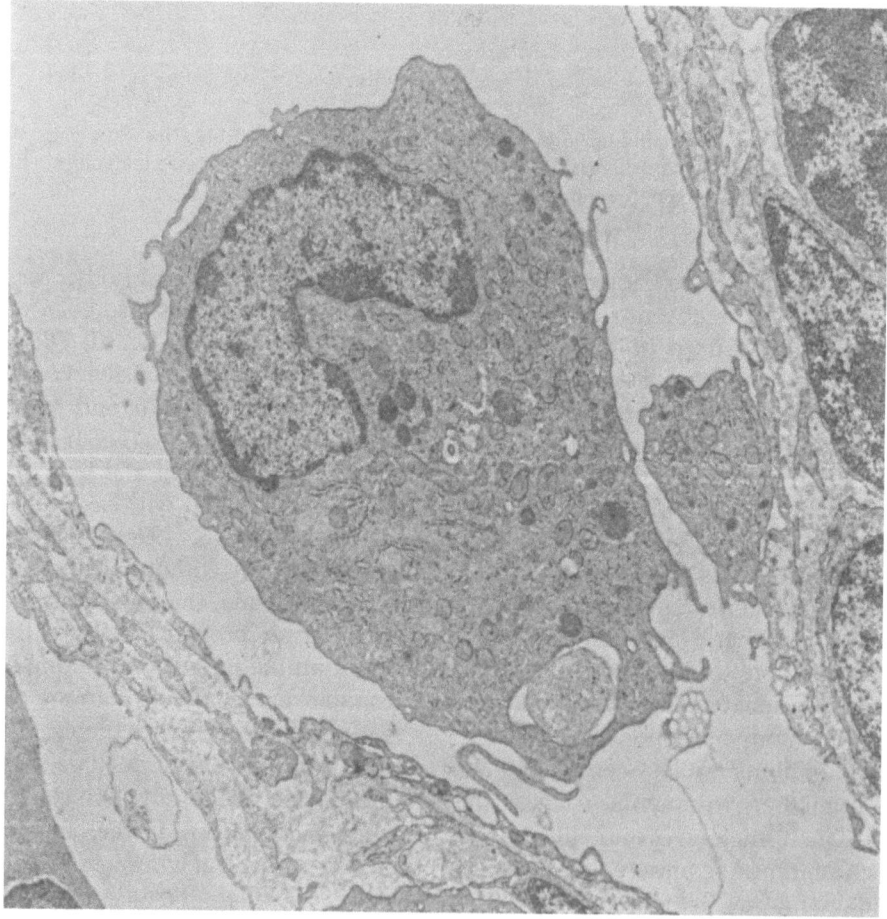

FIGURE 6. Electron micrograph of a monocyte or immature macrophage within a red pulp sinus of the snapping turtle spleen.

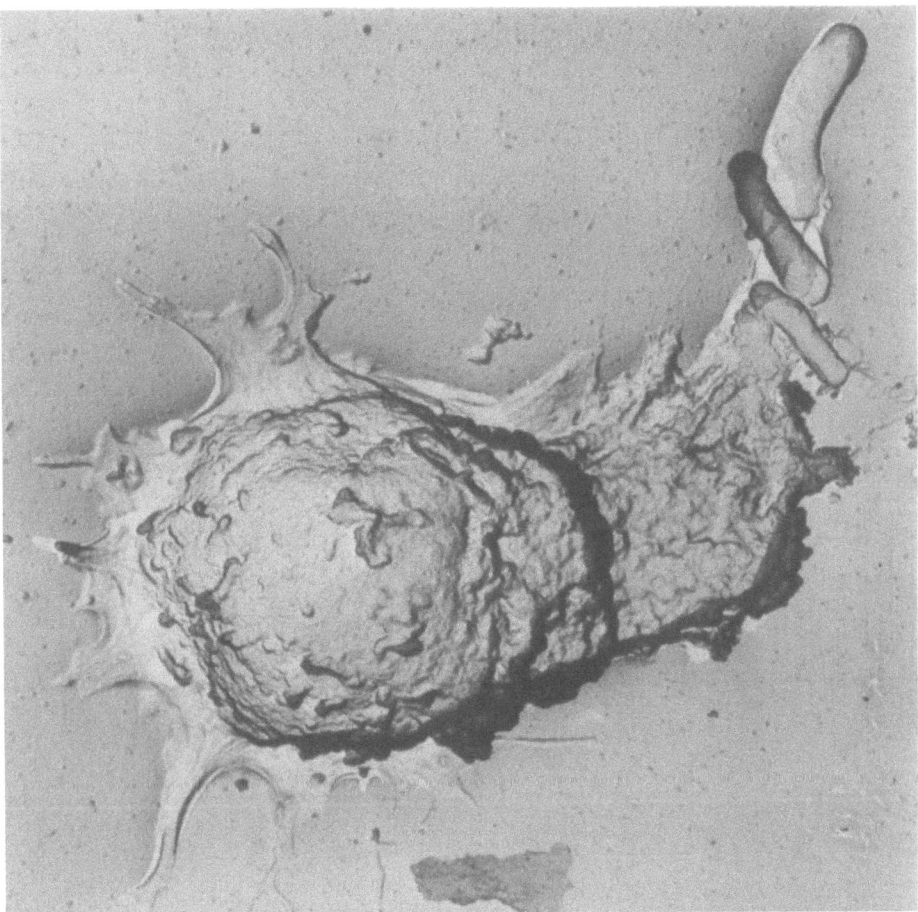

FIGURE 7. Electron micrograph of a surface replica of a macrophage derived from a turtle spleen. The cylindrical profiles in the upper right are bacteria being phagocytosed.

content varies considerably. Presumably the less mature macrophages contain fewer lysosomes. Even in immature macrophages, numerous pseudopodia can be seen extending from the cell surface. Surface replicas of turtle macrophages frequently show primary pseudopods extending from the cells, as is seen in Figure 7.

Turtle lymphocytes are similar to those of other vertebrates in their ultrastructure. The nuclei are nearly round with a thin rim of cytoplasm around them. Heterochromatin and a prominent nucleolus comprise most of the nuclear material. Cytoplasmic organelles consist of free ribosomes and a number of mitochondria, usually located at one pole of the cell. Much larger cells, identified as lymphoblasts, occur in small numbers in the normal turtle spleen as well (Figure 8). Here the nucleus is larger, has a prominent nucleolus and very little dense chromatin material. The euchromatic nucleus reflects its mitotic potential. Lymphoblasts contain proprotionally more cytoplasm than lymphocytes. Ribo-

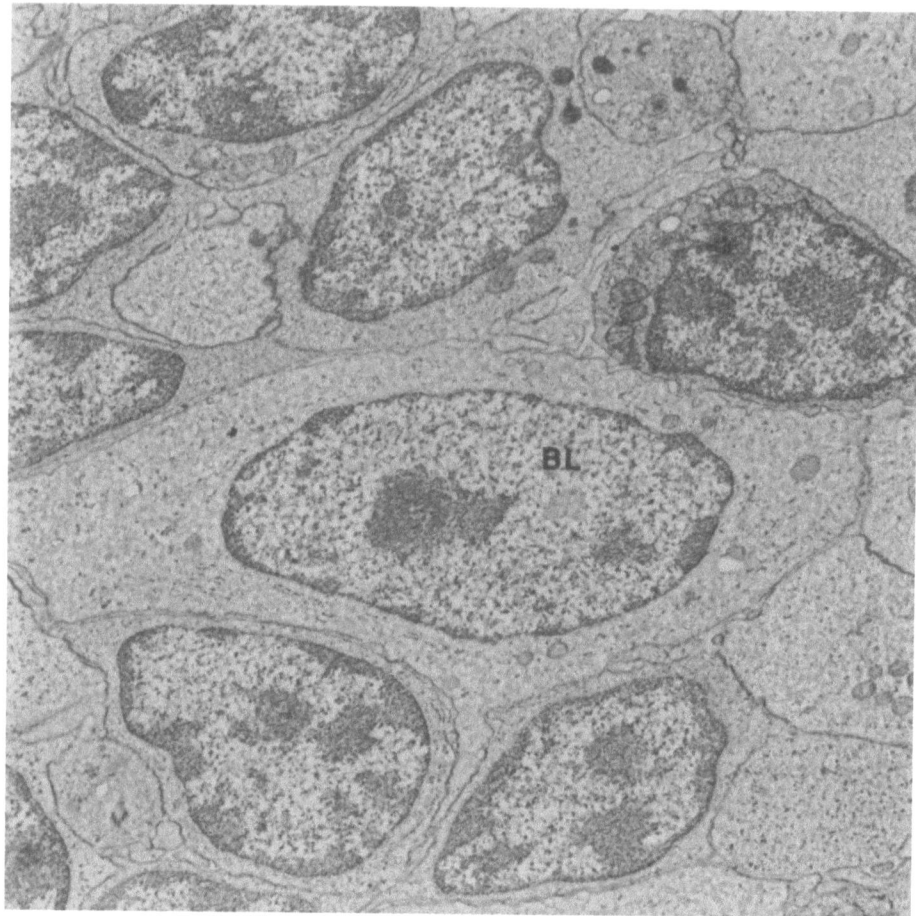

FIGURE 8. Electron micrograph of turtle splenic white pulp showing a lymphoblast (BL) surrounded by numerous small lymphocytes.

somes are generally found in small clusters (polyribosomes), scattered evenly throughout the cytoplasm, and mitochondria are usually located at one pole of the cell. A few short segments of rough endoplasmic reticulum are seen in some of these cells.

Reticular cells are seen among the lymphoid elements of the white pulp, but are more evident at the periphery of these areas where their circumferential arrangement forms a marginal zone between the white and red pulp. These cells are rather undistinguished in their ultrastructural features except for their long cell processes. Reticular cells are thought to be involved in antigen trapping (see Section 2.2). It is not known whether they have phagocytic properties in reptiles.

Electron microscopic descriptions of antibody-producing cells have been made in the tortoise, *Agrionemys horsfieldi* (Ambrosius and Hoheisel, 1973), and the turtle, *Chelydra serpentina* (Borysenko, 1976b). In the tortoise, plaque-forming cells (PFC) were comprised of a heterogeneous population of lymphoid cells, including transitional types, from lymphoblasts to plasma cells. However, these investigators found no cells which possessed a well-developed lamellar pattern of rough endoplasmic reticulum, a dense chromatic pattern, or significant numbers of Golgi vesicles, as are characteristic of mammalian plasma cells. Tortoise PFC were also reported to have phagocytic properties. For these reasons, it was suggested that tortoise plasma cells were "primitive," being significantly different in appearance and function from those of mammals (Ambrosius and Hoheisel, 1973). More recently, electron microscopic analysis of immunized turtle spleen over a broad time course has revealed the presence of classical plasma cells as well as less-mature transitional types. Mature plasma cells possess a dense chromatic pattern, but the pattern of the rough endoplasmic reticulum can be either lamellar or saccular. Golgi vesicles are also plentiful in active cells

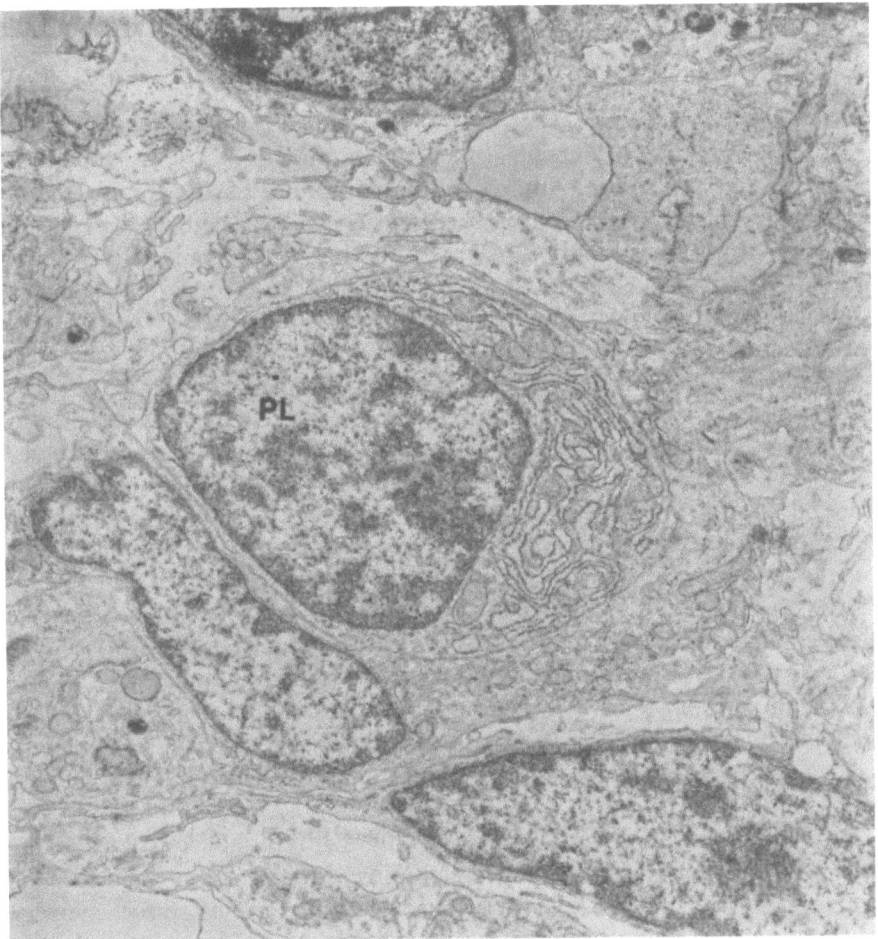

FIGURE 9. Electron micrograph of a typical mature plasma cell (PL) of the turtle spleen located in a red pulp cord. Note the eccentric nucleus and extensive rough endoplasmic reticulum.

(Borysenko, 1976b). Furthermore, in this study there was no evidence of phagocytosis in lymphocytes or plasma cells. If one allows sufficient time to elapse following immunization with a potent antigen, mature plasma cells of the classical variety are abundant and do not differ significantly from those of birds and mammals (Figure 9).

2.6. LYMPHOCYTE HETEROGENEITY

Lymphocytes of birds and mammals are heterogeneous in terms of their membrane-associated antigens and receptors. The major classes of lymphocytes, T and B cells and their subsets, are known to be involved in a diversity of immune functions. However, while the functional heterogeneity of lymphocytes has been demonstrated in reptiles, specific membrane markers to distinguish different populations have not yet been well characterized.

Attempts have been made to evaluate major categories of lymphocytes by means of heterologous antisera raised in rabbits against thymocytes (ATS) and immunoglobulins of a few reptilian species. The cytotoxic activity of such an antiserum against lizard (*Calotes versicolor*) thymocytes was several orders of magnitude greater for thymocytes than for lymphocytes of spleen, bone marrow, and peripheral blood. Among the nonlymphoid tissues, brain was more efficient than kidney and liver in absorbing the cytotoxic activity of ATS (Pitchappan and Muthukkaruppan, 1977a). In another study, carried out with the lizard, *Chalcides ocellatus*, ATS, absorbed with lizard erythrocytes, kidney and liver cells, in the presence of guinea pig complement, selectively killed 100% of thymocytes, about 85% of peripheral blood (PB) lymphocytes, and 56% of splenic cells (plateau of cytotoxicity). Further, ATS lysed a negligible proportion of PB and splenic lymphocytes collected from lizards thymectomized 4 months previously, thus indicating that the fraction of PB and splenic lymphocytes recognized by ATS are predominantly T cells (El Ridi and Kandil, 1981). Similar results have been reported in the snake, *Spalerosophis diadema* (Mansour et al., 1980). More recently, two distinct lymphocyte surface antigens were detected in another reptile, the snapping turtle, *Chelydra serpentina*. Rabbit anti-turtle thymocyte sera, absorbed repeatedly with nonlymphoid tissues, were tested by indirect immunofluorescence. These sera labeled splenic lymphocytes as well as thymocytes, but not other turtle cells, such as kidney, demonstrating common lymphocyte markers. However, when the antisera were absorbed extensively with splenic cells and retested, only the thymocytes were labeled, demonstrating the exclusive presence of thymocyte markers (Mead and Borysenko, unpublished).

Attempts have also been made to distinguish lymphocyte subpopulations on the basis of their surface immunoglobulins. An antiserum directed against lizard serum Ig and absorbed with erythrocytes stained, in indirect membrane immunofluorescence, no thymocytes but reacted with a proportion of PB and splenic lymphocytes (El Ridi and Kandil, 1981). A similar study in the lizard, *Calotes versicolor*, showed that 40 to 50% of splenic cells and less than 2% of thymocytes are Ig positive (Kannan, unpublished). Similarly, rabbit anti-snake IgM (*Elaphe quadrivirgata*) stained 10% of the thymocytes and 35% of the splenic cells in young adult snakes. Surface positive cells were not detected in the thymuses of newly hatched snakes. Cytoplasmic IgM-bearing cells were also found in small numbers in the lymphoid organs of adult snakes. The IgM-positive cells were considered to be of the B-cell lineage (Kawaguchi *et al.*, 1980). The above studies clearly demonstrate lymphocyte heterogeneity in reptiles, at least at the level of T- and B-cell diversity.

Despite the fact that the anti-lizard Ig serum was raised in rabbits which are phylogenetically distant enough to recognize putative cross-reacting determinants shared by the T-cell surface and serum Ig (Hämmerling *et al.*, 1976; Marchalonis *et al.*, 1979), lizard thymocytes were not labeled and thus appear to lack readily detectable membrane Ig. In this sense they are similar to thymocytes of mammals (Warner, 1974), birds (Kincade *et al.*, 1971), frogs (Mattes and Steiner, 1978a), and some fish (Yamaga *et al.*, 1978a). On the other hand, rabbit anti-Ig sera reacted in immunofluorescence with the majority of thymocytes of several fish species (Marchalonis *et al.*, 1978). However, it has been argued recently that thymocyte reactivity appears to be due primarily, if not solely, to cross-reactive, carbohydrate-containing cell surface structures, rather than to Ig determinants (Mattes and Steiner, 1978a,b; Yamaga *et al.*, 1978a,b).

The nature of lymphocyte heterogeneity has also been studied in the Florida alligator, *Alligator mississippiensis* (Cuchens *et al.*, 1976; Cuchens and Clem, 1979). PB lymphocytes were isolated by Ficoll–Hypaque centrifugation, fractionated in glass-wool column, and subjected to the classical T- and B-cell mitogens *in vitro*. The studies showed that a nonadherent population of lymphocytes responded to phytohemagglutinin (PHA), but not to lipopolysaccharide (LPS). On the other hand, the adherent population was stimulated by LPS. Thus, on the basis of mitogenic responsiveness and in analogy to mammalian lymphocytes, it appears that there are at least two distinct populations of lymphocytes in reptiles.

3. HUMORAL IMMUNE RESPONSES

3.1. CHARACTERISTICS OF ANTIBODY RESPONSE TO VARIOUS ANTIGENS

Several studies have revealed that all extant groups of reptiles—turtles, tortoises, snakes, lizards, alligators, and tuataras—are capable of responding to a variety of antigens: proteins (Evans, 1963; Ambrosius and Lehmann, 1965; Grey, 1966; Lerch *et al.*, 1967; Lykakis, 1968), bacteria (Metchnikoff, 1901; Evans, 1963; Maung, 1963), and heterologous erythrocytes (Rothe and Ambrosius, 1968). These earlier studies have been reviewed by Evans (1963), Grey (1966), Ambrosius *et al.* (1970), and Cohen (1971). Although naturally occurring antibodies have been detected in a variety of reptilian species (review: Kawaguchi *et al.*, 1978), this area is beyond the scope of our discussion, which we will limit to adaptive immune responses.

In recent years, detailed analysis has been made on the kinetics of the antibody response in several species of lizards and snakes to soluble and particulate antigens (Wetherall and Turner, 1972; Kanakambika and Muthukkaruppan, 1972; Coe et al., 1976; El Kes, 1978; El Rouby, 1978; Hussein et al., 1979b,d; El Deeb et al., 1980; El Ridi et al., 1981). A more comprehensive study on the development, morphology, and the diverse functions of the immune system has been made in the garden lizard, *Calotes versicolor* (Muthukkaruppan et al., 1970; Kanakambika, 1971; Manickavel, 1972; Pitchappan, 1975; Jayaraman, 1976; Subramonia Pillai, 1977; Ramila, 1978; Baskar, 1978). Using this species, the sequence of early cellular events and the kinetics of the serum antibody response to sheep red blood cells (SRBC) have also been investigated.

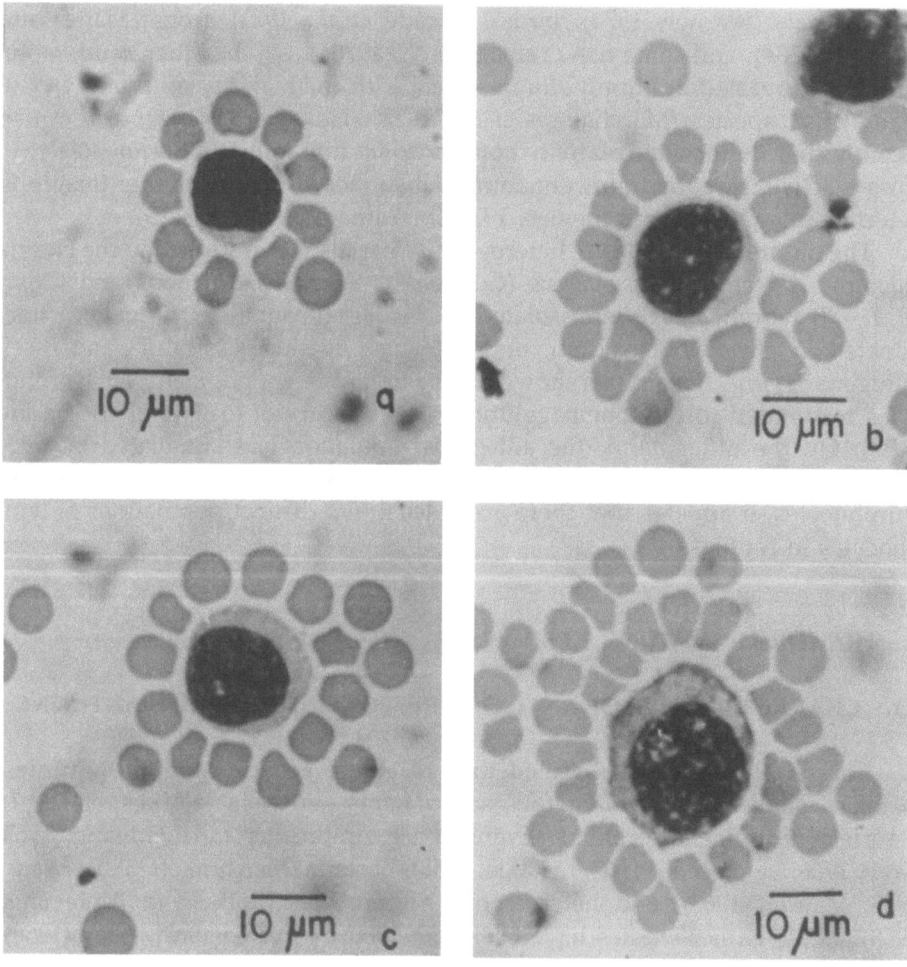

FIGURE 10. Morphology of antigen-binding cells in the garden lizard. Splenic cells were subjected to rosette formation with SRBC following the technique of Subramonia Pillai and Muthukkaruppan (1977); smeared using a cytocentrifuge and stained with May–Grunwald–Giemsa. (a) Small lymphocyte RFC; (b) medium lymphocyte RFC; (c) medium lymphocyte RFC; (d) lymphoblast RFC.

Antigen-specific rosette-forming cells (RFC) and antibody-producing cells (Figures 10 and 11) were enumerated in the spleen after immunization with 0.1 ml of 25% SRBC. As shown in Figure 12, within 3 days after immunization there is a significant increase in the number of specific antigen-binding cells in the spleen over the background, reaching the peak level by the 7th day (Subramonia Pillai and Muthukkaruppan, 1977). This is followed by the appearance of plaque-forming cells (PFC) (Figure 12), attaining the peak level by the 14th day. The serum antibody response peaks by the 21st day (Kanakambika and Muthukkaruppan, 1972). This illustrates the sequence of early cellular events leading to antibody formation. However, survey of the literature on reptilian immune response reveals that there is a wide range of variability in the kinetics, level and type of antibody generated, depending on the nature of antigen, dose and route of antigen injected, temperature at which the animals are held, the type of assay used, species-specific characteristics, as well as reproductive and seasonal rhythms (Cohen, 1971).

Specificity and memory, the two important characteristics of immunity, have been demonstrated in some turtles, lizards, and alligators, with early appearance of antibody, higher peak titers, and a change in molecular species (Lerch et al., 1967; Rothe and Ambrosius, 1968; Ambrosius et al., 1970; Wetherall and Turner, 1972). However, this pattern of anamnestic response was not observed in a number of other reptilian species studied (Maung, 1963; Grey, 1966; Rothe and Ambrosius, 1968). For instance, after a second injection of the same dose of SRBC into lizards, the rise in PFC and serum antibody was more rapid

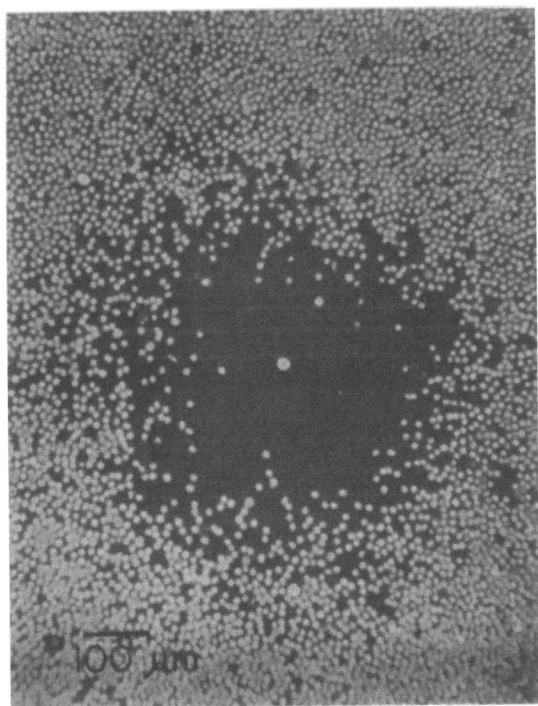

FIGURE 11. Enlarged view of a single plaque with the anti-SRBC antibody-producing cell in the center. SRBC-immunized lizard splenic cells, mixed with 10% SRBC and lizard complement, were plated between the two microslides as a monolayer. The plaques were developed by incubation at 37°C for 30 min.

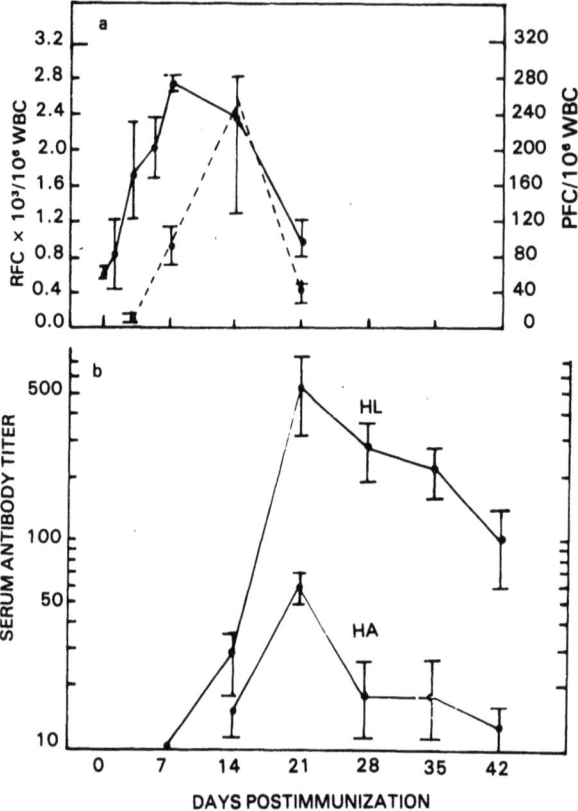

FIGURE 12. Immune response of the lizard, *Calotes versicolor*, to 0.1 ml of 25% SRBC, injected intraperitoneally. (a) The kinetics of RFC (———) and PFC (- - - -) responses in the spleen. (b) The kinetics of serum antibody titers, hemolysin (HL), and hemagglutinin (HA).

Each point represents the mean ± S.E. of 6–10 animals. Animals were maintained at ambient temperature of 26 ± 4°C. Significant difference in RFC number between day 0 and 7, $p < 0.001$; between days 7 and 21, $p < 0.001$.

with a shorter latent period, but without a concomitant increase in antibody level (Kanakambika and Muthukkaruppan, 1972).

3.2. IMMUNOGLOBULIN CLASSES

Reptiles have been shown to synthesize at least two immunoglobulin classes (Grey, 1963, 1966; Ambrosius, 1966; Ambrosius *et al.*, 1970; Lykakis, 1968; Coe, 1972). Salient findings are that during the primary response, "heavy" 18 S and "light" 7 S antibodies are synthesized, both of which are sensitive to mild treatment with 2-mercaptoethanol (2ME). With multiple immunizations 7 S and subsequently 4.5 S molecular species, both resistant to 2ME, appear in the serum. The existence of three different sizes of immunoglobulins has been confirmed by Benedict and Pollard (1972) and Leslie and Clem (1972) from their studies on different species of turtles. The sedimentation coefficients of these immunoglobulins are ~ 17 S, 7.5 S, and 5.7 S. Antigenic analysis suggests that the 5.7 S molecule may be a fragment of the 7.5 S molecule, and that the H chain of 7.5 S is distinct from that of the 17 S molecule.

With reference to the molecular species of antibodies, again there is considerable variation among reptiles. For example, Wetherall and Turner (1972) have shown that both 19 S and 7 S antibodies are produced against rat erythrocytes and bovine serum albumin in the lizard, *Tiliqua rugosa*, but *Salmonella typhimurium* could induce the formation of only 19 S antibodies which are susceptible to mild reduction by 2ME. An interesting finding among reptiles is that the 19 S response is quite prolonged even in the presence of high titers of 7 S antibody. This, in fact, is a common feature in several other reptilian species studied (Maung, 1963; Ambrosius and Lehmann, 1965; Grey, 1966; Lerch *et al.*, 1967; Lykakis, 1968; Marchalonis *et al.*, 1969). Furthermore, the humoral response to certain antigens (e.g., *S. typhimurium*) is characterized by the formation of only 19 S antibody even after repeated immunization (Wetherall and Turner, 1972), as already reported for other bacterial antigens in tortoises (Maung, 1963) and tuataras (Marchalonis *et al.*, 1969). Thus, bacterial antigens appear to induce only IgM antibodies in reptiles.

A more detailed analysis has been made to characterize the antibodies produced by tuatara, *Sphenodon punctatum* (Marchalonis *et al.*, 1969). The serum of immunized animals consists of 18 S and 7 S immunoglobulins. Antiflagellin antibody activity is confined to the high-molecular-weight (18 S) immunoglobulin type, which resembles the IgM of higher vertebrates in size and polypeptide chain structure. The 7 S molecule possesses light chain resembling those of the 18 S protein, but the heavy chains of the 7 S molecules differ from those of the 18 S, thereby indicating the presence of distinct immunoglobulin classes. Further, immunoelectrophoretic analysis revealed the presence of slow-moving proteins corresponding to IgM and IgG of mammals. There is very little information with reference to other immunoglobulin classes in reptiles. The presence of reaginic antibody (IgE) may be inferred from the demonstration of the hypersensitivity reaction type I, anaphylaxis in the sand turtle (Downs, 1928) and in the garden lizard (Jayaraman and Muthukkaruppan, 1975) to heterologous serum proteins.

3.3. ANAPHYLAXIS

The earliest study on the hypersensitivity reaction type I (anaphylaxis) was that of Downs (1928) who induced specific anaphylaxis in the sand turtle by sensitizing with mammalian serum. More recently, Jayaraman and Muthukkaruppan (1975) have demonstrated the typical anaphylactic response in the lizard, *Calotes versicolor*, to egg albumin (Table 1). The symptoms of anaphylaxis observed in *Calotes* are remarkably similar to those found in birds and mammals. They are produced within 2 min after the injection of the shocking dose of antigen, resulting in the death of the animal 3 to 10 min later. The maximum mortality was observed 14 days after sensitization (Table 1).

3.4. ROSETTE-FORMING CELLS AND PLAQUE-FORMING CELLS

The types of cells involved in antigen recognition and binding have been identified in the lizard, *Calotes versicolor*. The morphology of each RFC can be clearly identified in the smears prepared using a cytocentrifuge (Figure 10).

TABLE 1. ANAPHYLACTIC RESPONSE TO EGG ALBUMIN IN THE LIZARD, *CALOTES VERSICOLOR*[a,b]

Postsensitization period (days)	Group I				Group II				Group III			
	N	−	+	D	N	−	+	D	N	−	+	D
5	11	5	6	5	3	3	0	0	2	2	0	0
12	7	3	4	4	3	3	0	0	2	2	0	0
14	15	2	13	12	5	5	0	0	3	3	0	0
20	7	3	4	3	3	3	0	0	2	2	0	0
27	4	2	2	1	3	3	0	0	2	2	0	0

[a]Animals were sensitized by a single intracardiac injection of 100 mg/kg body wt of egg albumin (The British Drug Houses Ltd., Poole, England) in 0.85% normal saline. Shocking was performed by a single injection of 2 g/kg body wt of egg albumin by the same route. Animals were shocked on days 5, 12, 14, 20, and 27 after sensitization.
[b]Group I: Sensitized lizards shocked with egg albumin. Group II: Sensitized lizards shocked with BSA. Group III: Saline-injected lizards shocked with egg albumin. N, No. of animals; −, no symptoms; +, mild to severe symptoms; D, death due to anaphylaxis.

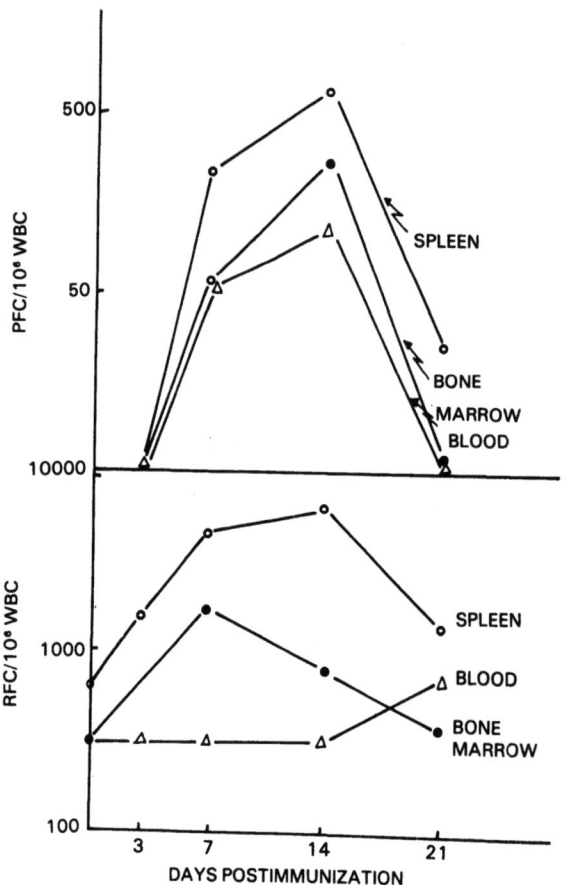

FIGURE 13. Kinetics of RFC and PFC responses in different lymphoid organs in the garden lizard, after immunization intraperitoneally with 0.1 ml of 25% SRBC. Each point represents mean value of pooled cells from 6 to 10 animals.

Small, medium, and large lymphocytes exhibit the ability of specific antigen binding, as shown by the formation of rosettes with SRBC. The study shows that RFC in normal spleen are predominantly of small and medium lymphocytes (76%), whereas 7 days after immunization with 0.1 ml of 25% SRBC, more than 55% rosettes are formed by blast cells (Subramonia Pillai, 1977).

Utilizing the Jerne plaque assay, antibody-forming cells have been identified in the spleen but not kidney, thymus, or intestine of chelonians (Rothe and Ambrosius, 1968; Sidky and Auerbach, 1968; Kassin and Pevenitskii, 1969). More detailed analysis in the lizard shows the presence of PFC in the spleen, blood, and bone marrow, but not in the thymus, lung, liver, kidney, or cloacal complex, after immunization with SRBC (Kanakambika and Muthukkaruppan, 1972; Subramonia Pillai, 1977).

The distribution and kinetics of RFC and PFC in various lymphoid tissues following immunization with SRBC have also been studied in the lizard, *Calotes versicolor* (Subramonia Pillai, 1977). A rapid increase in the number of specific antigen-binding cells was observed both in the spleen and bone marrow, attaining the peak level on the 7th day. However, in peripheral blood the RFC remained at the base level except for a slight increase on the 21st day. The PFC peak response occurred on day 14 in the spleen, bone marrow, and peripheral blood (Figure 13). In general, the spleen contains more RFC and PFC than bone marrow and peripheral blood. Peritoneal exudate cells contain very few PFC and no immune RFC. These results suggest that the cellular events leading to the differentiation of antibody-producing cells occur in the spleen of intact lizards and that only a proportion of PFC is recirculated. This is further evidenced by the finding that splenectomy in the lizard completely suppresses the appearance of RFC and PFC in bone marrow and blood (Subramonia Pillai, 1977).

3.5. ROUTE AND DOSE–RESPONSE KINETICS

The quality, strength, and kinetics of the immune response depend on the dose and route of antigen administration in any animal species. This basic information is necessary before one can elucidate the mechanism of immunoregulation as well as the nature of functional heterogeneity among the subpopulations of lymphocytes. For this reason, a detailed study was undertaken to determine the effects of varying doses and routes of immunization in the lizard, *Calotes versicolor* (Muthukkaruppan *et al.*, 1976b; Subramonia Pillai, 1977; Baskar, 1978; Jayaraman and Muthukkaruppan, 1978b).

Immunization with an optimal dose of antigen, 6.0×10^8 SRBC (0.1 ml of 25%), induces a peak PFC response on day 7 after intramuscular injection and on day 14 after intraperitoneal injection. The response to intracardiac immunization is consistently low at all dose levels tested (Baskar, 1978). Four different antigen doses were determined with regard to ability to induce a PFC response; nonimmunogenic (10^4), subimmunogenic (10^7), optimal (6.0×10^8), and supraoptimal (2.0×10^9) (Baskar and Muthukkaruppan, 1980). Intramuscular immunization was found to induce the development of a faster and higher level of PFC re-

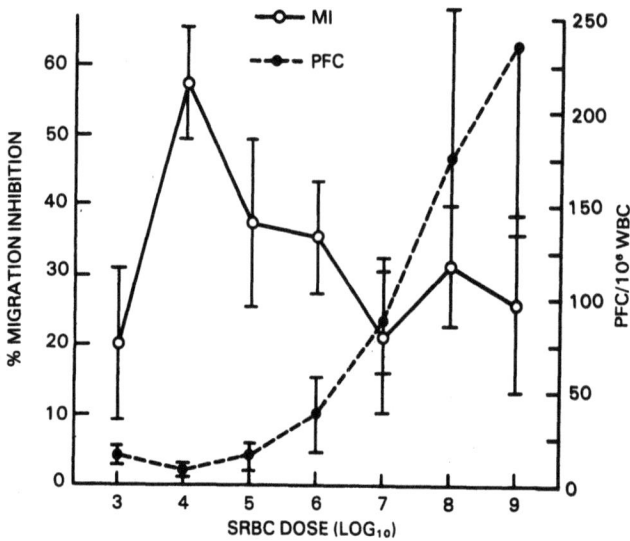

FIGURE 14. The PFC and *in vitro* capillary MI response of the garden lizard to different doses of SRBC, injected intramuscularly. Assays were performed with splenic cells, 14 days after immunization. Each point represents the mean ± S.E. of 8–14 animals.

sponse than the other routes tested. Another point of interest is that there was an inverse relationship between PFC and migration inhibition (MI) responses (Figure 14). Low dose of antigen induces a strong and rapid MI response with concomitant delay in the development of humoral immune response (Jayaraman and Muthukkaruppan, 1978b). If a supraoptimal dose is used for immunization, the MI level produced by splenic cells is very low at a time when the PFC response is at its maximum and the degree of MI rises after the decline in PFC number. Thus, the MI response to SRBC can be induced in lizards with the minimal antigen dose. However, the humoral immune response requires a higher threshold level of antigen. These results indicate that the type of immune response is profoundly influenced by the route and dose of antigen administration in the garden lizard as it is in mice (Lagrange *et al.*, 1974; Tamura and Egashira, 1975).

3.6. HELPER FUNCTION

The procedure of low-dose priming with heterologous erythrocytes is known to stimulate the helper activity maximally in mice (Playfair, 1971; Grantham and Fitch, 1975). Pretreatment with a suboptimal dose of SRBC a few days before challenge primed the mice to produce an enhanced antibody response to subsequent injection of the same antigen. This is due to selective triggering of T helper cells by low antigen dose (Mitchison *et al.*, 1970) or by chemically modified antigen (Dennert and Tucker, 1972). This model, known to selectively prime T cells, has been employed to understand the nature of helper

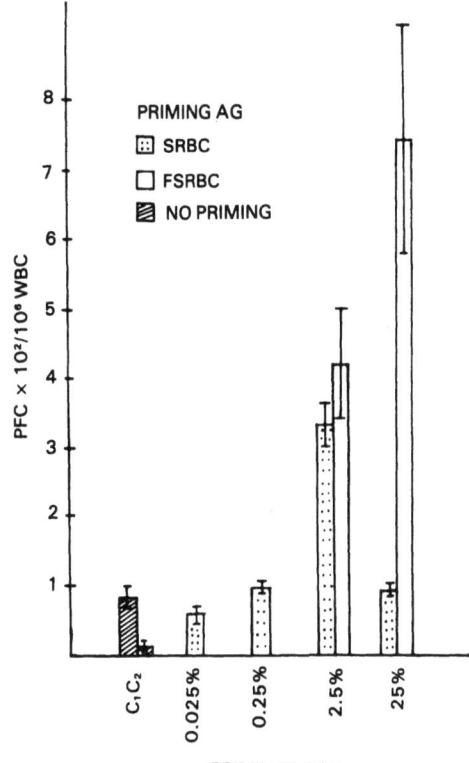

FIGURE 15. Influence of dose of priming by SRBC or F-SRBC on the accelerated PFC response. Lizards were primed intraperitoneally with 0.1 ml of indicated dose of SRBC or F-SRBC and 7 days later challenged intraperitoneally with 0.1 ml of 2.5% SRBC. The direct PFC response to SRBC was determined in splenic cells 14 days after challenge. Controls were not primed, but immunized intraperitoneally with 2.5% SRBC (C_1) or 25% F-SRBC (C_2). Each bar represents the mean (± S.E.) PFC/10^6 viable white blood cells.

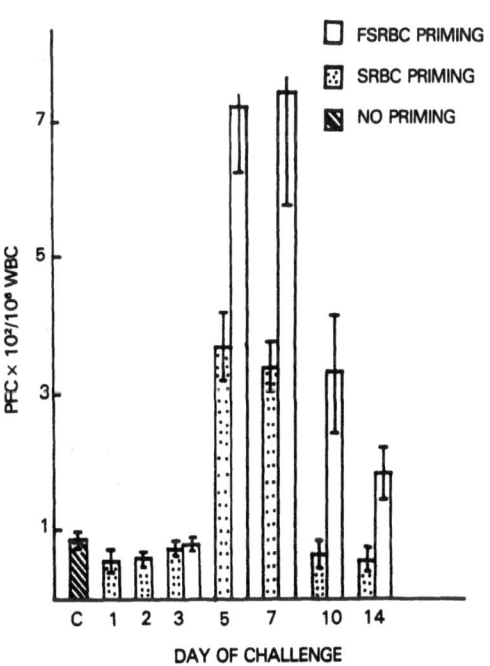

FIGURE 16. Temporal relationship between priming and challenge on the enhanced PFC response to SRBC in the garden lizard. Groups of lizards were primed with 0.1 ml of 25% F-SRBC or 2.5% SRBC intraperitoneally and 1 to 15 days later challenged with 0.1 ml of 2.5% SRBC intraperitoneally. PFC assay was performed with splenic cells 14 days after challenge. Each bar represents the mean (± S.E.) PFC/10^6 viable white blood cells of 8–10 animals.

function in the lizard, *Calotes versicolor*, with reference to the priming dose, and the duration and specificity of helper activity (Muthukkaruppan et al., 1976b; Ramila, 1978).

A significant increase in the number of PFC is generated in the spleen, when lizards are primed with 2.5% SRBC or formaldehyde-treated SRBC (F-SRBC) before challenge (Figure 15). The enhancement is as much as four times that of controls which were immunized with a single dose of 2.5% SRBC. Further, 25% F-SRBC is more effective in evoking an accelerated anti-SRBC response than the other doses tested, even though F-SRBC by itself is unable to induce the anti-SRBC PFC response in lizards (Muthukkaruppan et al., 1976b) as in mice (Dennert and Tucker, 1972). The study also shows the specificity of the helper activity generated by SRBC and F-SRBC priming (Ramila, 1978). Furthermore, the helper function is maximal from 5 to 7 days after priming with 2.5% SRBC and from 5 to 10 days after priming with F-SRBC, thus demonstrating the involvement of short-lived (helper) memory cells in the accelerated humoral immune response in reptiles (Figure 16).

3.7. IMMUNE RESPONSE TO HAPTENS

It is well known that the cooperative interaction between carrier-specific and hapten-specific lymphoid cells is essential for the development of anti-hapten antibody response (Katz and Benacerraf, 1972). Hapten–carrier conjugates are widely used in determining the function of the various cells in the immune response and to understand helper and effector cell interactions in antibody formation. This model has been successfully used to study the mechanism of cellular interaction in antibody formation in fish (Stolen and Makela, 1975; Yocum et al., 1975; Ruben et al., 1977), amphibians (Ruben, 1975), and tortoises (Ambrosius and Frenzel, 1972).

Recently, the mechanism and types of lymphocytes involved in cell collaboration have been studied in the garden lizard using TNP as hapten and SRBC and ovalbumin as carriers (Ramila, 1978). The peak anti-TNP PFC response occurs 10 days after immunization with TNP–SRBC (0.1 ml 25%) or TNP_{30} ovalbumin. Lower concentrations of SRBC (2.5%) or F-SRBC effectively prime the lizard for an accelerated anti-TNP PFC response (Figure 17). Maximum carrier effect was observed when the hapten–carrier complex was injected 10 days after priming for both TNP–SRBC or TNP–protein conjugates (Figure 18). An accelerated anti-TNP response is elicited only when the same carrier is used for both priming and challenge, thereby indicating the carrier specificity in anti-hapten response in the lizard. These results are in general agreement with the findings in mice, with reference to effective priming by F-SRBC (Dennert and Tucker, 1972), requirement for critical time interval between priming and challenge (Mitchison, 1971), and carrier specificity (Cheers and Breitner, 1971). Thus, in analogy it may be suggested that the carrier-primed enhancement in the anti-TNP antibody response is the result of cell collaboration between carrier-specific helper T cells and hapten-specific antibody-producing precursor B

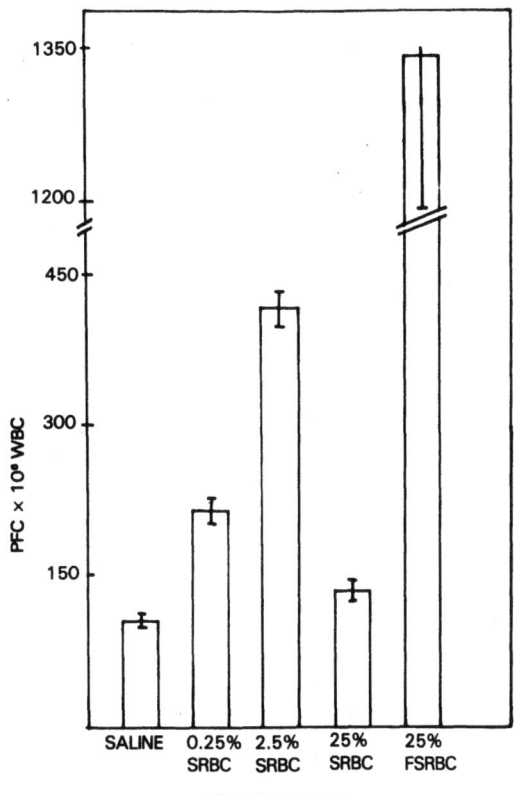

FIGURE 17. Influence of various doses of carrier pretreatment on the anti-TNP PFC response. 0.1 ml of each of the carriers was injected intraperitoneally 10 days before immunization with 0.1 ml of 25% TNP–SRBC intraperitoneally. Controls were injected with saline followed by TNP–SRBC. PFC assay was performed 10 days after challenge with splenic cells using TNP–RRBC. Each bar represents the mean ($\pm$ S.E.) PFC/10^6 viable white blood cells of 6–10 animals.

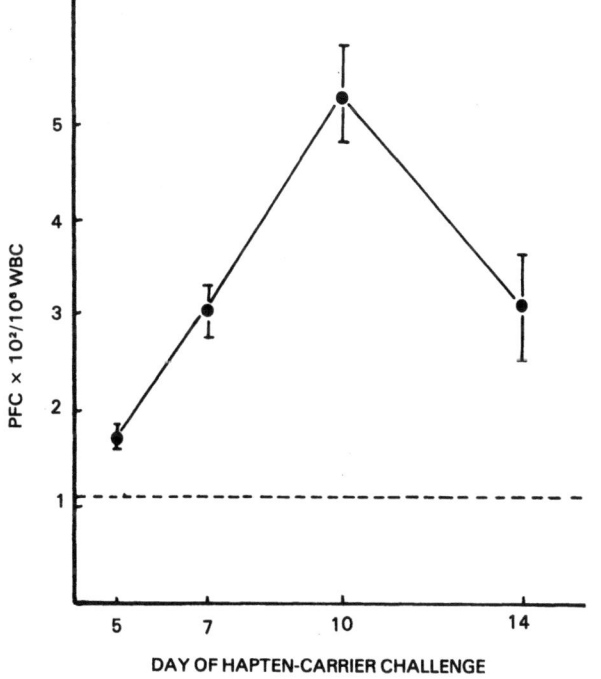

FIGURE 18. Temporal relationship between the carrier priming and hapten–carrier challenge: 100 μg of ovalbumin was injected intraperitoneally on various days before challenge with 500 μg of TNP_{30} ovalbumin. PFC assay was performed with splenic cells using TNP–SRBC, 10 days after challenge. Each point represents the mean ($\pm$ S.E.) PFC/10^6 white blood cells of 8–10 animals. The baseline represents the level of normal response to a single injection of TNP_{30} ovalbumin without carrier priming.

cells in the lizard. This contention is confirmed by adult thymectomy experiments in lizards (Ramila, 1978).

4. CELL-MEDIATED IMMUNE RESPONSES

4.1. TRANSPLANTATION REACTIONS

The first account of allograft rejection in reptiles is the classic study by May (1923), using the chameleon, *Anolis carolinensis*. He reported that while autografts survived permanently, allografts appeared healthy for a time, but were totally destroyed by 60–90 days posttransplantation at 23.5°C. He provided a vivid and highly accurate description of the inflammatory reaction, round cell infiltration, and graft destruction, long before these events were established as hallmarks of cell-mediated immunity. With the renewed interest in transplantation immunity in the past two decades, a number of other reptilian species were examined. Although one sees much variation from species to species, the general pattern is one of chronic rejection of first-set skin allografts and accelerated rejection of the second-set, at moderate temperatures (reviews: Borysenko, 1970; Worley and Jurd, 1979). The relative slowness of allograft rejection has been attributed to the absence of a major histocompatibility locus, rather than to a primitive or deficient immune system (Cohen and Borysenko, 1970).

Studies on several species of reptiles, the snapping turtle, *Chelydra serpentina* (Borysenko, 1970), the teiid lizard, *Cnemidophorus sexlineatus* (Maslin, 1967), the iguana, *Ctenosaura pectinata* (Cooper and Aponte, 1968), the garden lizard, *Calotes versicolor* (Manickavel and Muthukkaruppan, 1969; Manickavel, 1972), the whiptail lizard, *Cnemidophorus tigris* (Cuellar and Smart, 1977), the European green lizard, *Lacerta viridis* (Worley and Jurd, 1979), the garter snake, *Thamnophis sirtalis* (Terebey, 1972), have documented a number of characteristic features of transplantation immunity in Reptilia. (1) Skin xenografts are rejected faster than allografts. (2) Second-set allografts and xenografts are rejected more rapidly than the first-set. (3) The rate of graft rejection is markedly affected by environmental temperature. With regard to the last point, in turtles the relationship between allograft survival time and temperature is not linear. A break occurs at about 25°C, implying that there are at least two temperature-sensitive phases in allograft rejection. Although allograft rejection is completely suppressed at 10°C, antigen recognition apparently is not inhibited, since transfer of turtles to higher temperatures resulted in accelerated rejection (day of transfer considered day 1). Alternatively, transfer of turtles to low temperatures after graft rejection had initiated curtails the rejection process. These findings led to the hypothesis that at low temperatures an early, but postrecognition event, such as proliferation, was primarily affected, while at higher temperature ranges, effector functions such as antibody and lymphokine production were primarily affected (Borysenko, 1979).

4.2. *IN VITRO* CORRELATE OF CELL-MEDIATED IMMUNITY

The *in vitro* capillary leukocyte migration inhibition (MI) technique has been successfully adapted to assess the cell-mediated immune (CMI) response to skin

TABLE 2. MIGRATION INHIBITION OF SPLENIC CELLS FROM SKIN-ALLOGRAFTED LIZARDS IN THE PRESENCE OF SPECIFIC DONOR ANTIGEN[a]

Day of assay	Morphological status of melanophores[b]	Number of animals tested	% MI (mean ± S.E.)
—	Controls[c]	10[d]	3.8 ± 2.0
4	Normal	7	28.8 ± 13.7
7	Normal	7	49.4 ± 13.5
14	Beginning of fragmentation	16	37.2 ± 6.3
30	Beginning of granulation	5	34.2 ± 9.9
35	Complete granulation	5	61.9 ± 8.8

[a]Specific donor histocompatibility antigen was prepared from the spleen of the respective donor, following the method of Green and Clunie (1974).
[b]The level of allograft reaction was determined by the morphological status of melanophores as described earlier (Manickavel and Muthukkaruppan, 1969).
[c]Unsensitized lizards were tested with general antigen (prepared by mixing four or five specific donor antigens) at different time intervals along with experimental lizards.
[d]Numbers in parentheses indicate number of animals tested.

allografts in the garden lizard, *Calotes versicolor* (Jayaraman and Muthukkaruppan, 1977). After grafting allogeneic skin tissue in a unidirectional fashion (Manickavel and Muthukkaruppan, 1969), sensitized splenic cells were cultured in capillary tubes in the presence or absence of the antigen, prepared from the donor spleen. For the sequential study of the onset of CMI to skin allografts, splenic cells were tested at specified intervals using the MI assay and compared with morphological criteria of rejection. It is evident from the data presented in Table 2 that the lizards respond to skin allografts, as indicated by melanophore degranulation, and that a significant inhibition of migration of sensitized splenic cells occurs by the seventh day, in the presence of the respective donor antigen. This response is maintained as long as 1 month after grafting, the time at which the clinical manifestation of allograft rejection is observed. The MI of allograft-sensitized splenic cells is an antigen-specific event (Jayaraman and Muthukkaruppan, 1977). Thus, the *in vitro* MI technique is a sensitive and specific method of measuring the CMI response to skin allografts in reptiles. This is a highly useful assay mainly because the assessment of allograft reaction in reptiles is problematic due to its chronic nature and due to the use of multiple morphological criteria as the endpoint. Recently, the ability of peripheral blood lymphocytes to participate in two-way mixed lymphocyte reactions has been demonstrated in alligators (Cuchens and Clem, 1979).

4.3. GRAFT-VS.-HOST REACTIONS

The graft-vs.-host (GVH) reaction has been studied in only one reptilian species, *Chelydra serpentina*. *In vitro* experiments have shown that splenomegaly

TABLE 3. INCIDENCE OF MORTALITY (IN 120 DAYS) DUE TO GVH DISEASE AT DIFFERENT ENVIRONMENTAL TEMPERATURES[a]

Temperature (°C)	No. of animals dead/total No. (%)	
	Experimental	Control
30	19/26 (73%)	0/9
20	8/23 (35%)	0/10
10	0/20 (0%)	0/9

[a]Experimental animals were injected with 5×10^6 allogenic splenic cells from an adult donor from a distant geographical origin. Control animals were injected with 0.1 ml saline or 5×10^6 allogeneic kidney cells.

can be induced in immature splenic explants upon exposure to adult allogeneic splenic cells (Sidky and Auerbach, 1968). Another study, using the same turtle species, showed a wide spectrum of *in vivo* GVH reactions, from acute to chronic, which again suggests the presence of multiple weak histocompatibility differences in outbred populations of turtles. The incidence of acute reactivity, as evidenced by splenomegaly and early mortality, was highest among those turtles that received splenic cells from donors of distant geographical locations. The incidence of lethal GVH reactions and the rapidity of these reactions were also greater at elevated temperatures (Borysenko and Tulipan, 1973) and completely suppressed at 10°C (Borysenko, 1972) (Table 3). The parallels between GVH reactions and graft rejection are quite striking, with the single exception that xenogeneic splenic cells were ineffective in inducing GVH reactivity, while skin xenografts were rejected more rapidly than allografts.

4.4. DELAYED HYPERSENSITIVITY REACTIONS

Lizards (*Tarentola annularis*) stimulated in summer with 1 mg Bacillus Calmette Guérin displayed cutaneous hypersensitivity following intradermal challenge with purified protein derivative (PPD) at 21 to 34 days postsensitization. Lesions involving desquamation, erythema, induration, and dryness began to appear 3 to 5 days postchallenge, persisted 5 to 6 days, and then subsided. Histologically, degeneration of horny layer, degranulation of melanophores, and infiltration of lymphoid cells were observed. Thus, the lizards' cutaneous response to tuberculin has many of the features characteristic of the mammalian delayed hypersensitivity reaction (Badir *et al.*, 1981).

4.5. CMI RESPONSE TO SHEEP ERYTHROCYTES AND HAPTEN–CARRIER COMPLEX

The capillary leukocyte MI technique has also been applied to detect CMI directed against SRBC in the lizard, *Calotes versicolor* (Jayaraman and Muthukkaruppan, 1978a). The migration of splenic cells obtained from SRBC-injected lizards is inhibited in the presence of specific antigen. The MI response is mediated by sensitized lymphocytes and is an antigen-specific phenomenon. Sensitization of splenic cells against SRBC is evident as early as 4 days postimmunization by the inhibition of their migration *in vitro* in the presence of specific antigen. In this system, MI reaches a maximum on day 7 and is maintained steadily at least up to 30 days. Administration of a very low dose of SRBC (10^4) results in a high degree of MI, without the production of PFC. Furthermore, F-SRBC, which unable to stimulate the generation of PFC response in mice (Dennert and Tucker, 1972) and in lizards (Ramila, 1978), is shown to induce a good MI response (Table 4). These results and the finding that MI is a thymus-dependent phenomenon (Jayaraman, 1976) indicate that the MI response of splenic cells is an *in vitro* manifestation of CMI function in the lizard.

The nature of CMI response to hapten–carrier complex has been investigated in the garden lizard, using the *in vitro* capillary MI assay (Ramila, 1978). A significant percentage of MI was observed in splenic cells obtained from lizards immunized with 500 µg of TNP_{30} ovalbumin. It was also shown that the migration of cells is not inhibited in the presence of TNP–bovine γ-globulin, thereby indicating the specificity of MI response to hapten–carrier complex as in mice (Snippe *et al.*, 1975; David *et al.*, 1964).

4.6. NORMAL LYMPHOCYTE TRANSFER REACTION

Chronic graft destruction in mammals results when skin and other tissue grafts are transplanted across minor histocompatibility (H) differences. By analogy, it was suggested that chronic rejection in reptiles reflects the activities of antigenic products of minor H loci acting in the absence of a major H system

TABLE 4. THE IMMUNE RESPONSE TO SRBC IN THE LIZARD, *CALOTES VERSICOLOR*

Dose of SRBC[a]	$PFC/10^6$ WBC (mean ± S.E.)	% MI (mean ± S.E.)
PBS	5.2 ± 1.6 (5)[b]	7.0 ± 3.5 (22)
10^4	0.7 ± 0.5 (4)	72.9 ± 3.9 (4)
10^6	22.5 ± 6.6 (4)	52.5 ± 7.3 (4)
6×10^8	286.2 ± 77.3 (5)	22.0 ± 7.2[c] (5)
6×10^8 F-SRBC	19.9 ± 6.9 (8)	21.1 ± 21.1 (4)

[a]Fourteen days after intramuscular immunization with varying doses of SRBC or F-SRBC, PFC and MI assays were performed. The area of migration was traced at 24 hr of incubation at 37°C.
[b]Number of animals investigated is given in parentheses.
[c]Significantly different from controls ($p < 0.001$).

(Cohen and Borysenko, 1970; Cohen, 1971; Cohen and Collins, 1977). This problem has been approached by utilizing the normal lymphocyte transfer (NLT) reaction. In mammals, NLT reaction develops at the injection site, following intradermal inoculation of immunocompetent lymphocytes into unimmunized, allogeneic hosts differing from the donor at the major H locus. The reaction is a manifestation of a local GVH reaction, as well as a host cellular immune response against strong H antigens on the inoculated cells (Streilein and Billingham, 1970; Zakarian and Billingham, 1972; Sidky and Auerbach, 1975). The appearance and tempo of NLT reactions in irradiated or nonirradiated lizards, *Tarentola annularis*, following intradermal injection of 2–3 × 10^6 viable allogeneic splenic lymphocytes are similar to those observed in guinea pigs, hamsters, dogs, rats, and mice (Badir *et al.*, 1981). The data thus suggest that lizards are capable of eliciting vigorous CMI responses to alloantigens. This implies that strong antigenic disparity occurs in *T. annularis*. Furthermore, recent studies clearly demonstrate the ability of the sensitized lymphocytes of the lizard against alloantigens to manifest the specific CMI response *in vitro*, as early as 7 days after grafting (Jayaraman and Muthukkaruppan, 1977).

Thus, in light of the findings that lizards show a strong delayed hypersensitivity reaction to tuberculin, a strong lymphocyte transfer reaction (Badir *et al.*, 1981), a specific MI response *in vitro* to alloantigens, and a pronounced and rapid MI response to SRBC (Jayaraman and Muthukkaruppan, 1977, 1978a,b), chronic allograft rejection in these reptiles cannot be ascribed to a deficiency in CMI. The positive NLT reaction also raises questions about the postulated lack of a major H complex (Cohen and Collins, 1977) and the allelic polymorphism of its constituent loci (Cohen, 1980). It is quite possible that the chronicity of allograft rejection might be due to the characteristic organization of reptilian integument rather than to the sluggish nature of the immune response induced by transplantation antigens. For this reason it would be of interest to investigate the nature of the allograft reaction in reptiles grafting other tissues in appropriate sites.

5. ROLE OF THYMUS, SPLEEN, AND GALT

5.1. ROLE OF THYMUS IN IMMUNOGENESIS

Ever since the discovery of the role of thymus in immunogenesis by neonatal thymectomy in mammals (Miller, 1974), attempts have been made to elucidate the phylogeny of thymic development and function in several species of lower vertebrates (Sailendri, 1973; Charlemagne, 1974; Manning, 1975; Tochinai and Katagiri, 1975; Pitchappan and Muthukkaruppan, 1977b,d). The ideal animal model is one in which ablation can be performed at the earliest stage of life history, when the thymus is just a rudiment. The anuran amphibian, *Xenopus laevis* (Tochinai and Katagiri, 1975), as well as the teleost, *Tilapia mossambica* (Sailendri, 1973), are two such models. Such an early intervention has not been possible in reptilian forms. Nevertheless, attempts have been made to study the effect of adult thymectomy on a variety of immune functions in two species of lizards, *Calotes versicolor* (Pitchappan, 1975; Muthukkaruppan, *et al.*, 1976a) and *Chalcides ocellatus* (El Masri, 1979). In the absence of inbred lines of reptiles, experimental protocols involving irradiation and bone marrow transfusion cannot be carried out. Nevertheless, it is possible to identify immune deficiencies brought about by adult thymectomy without irradiation if immunization is performed long after thymectomy. In fact, this is true even in mice in which the duration may be as long as 6 months postthymectomy for effective elimination of T cells. Studies on the short-term effects of adult thymectomy in mice indicate an important role for the adult thymus in the secondary immune response and immunosuppression (Jacobs and Byrd, 1975; Simpson and Cantor, 1975). Pronounced deficiencies in both CMI and HI responses are evident 6–9 months postthymectomy in mice (Miller, 1961; Metcalf, 1965).

Another approach is to use specific antithymocyte or anti-T-cell serum to augment the effects of adult thymectomy (Pitchappan, 1975; Pitchappan and Muthukkaruppan, 1977a,b). Treatment of adult thymectomized lizards (*Calotes versicolor*) with ATS results in the abrogation of PFC response to SRBC. A definite recovery of anti-SRBC response after ATS treatment occurs in the presence of the thymus (Pitchappan and Muthukkaruppan, 1976, 1977b; Muthukkaruppan *et al.*, 1976a). This functional depletion correlates with the depletion of lymphocytes in the periarteriolar region of the splenic white pulp. The repopulation of this area requires the presence of an intact thymus (Pitchappan and Muthukkaruppan, 1977c; Pitchappan, 1980).

Bilateral thymectomy performed 1 month before immunization significantly reduces the PFC response to optimal dose of SRBC (0.1 ml of 25%, i.e., 6×10^8 cells), but the response to supraoptimal dose is not altered (Jayaraman, 1976) (Fig. 19). Furthermore, the periarteriolar region becomes depleted by lymphocytes 1 month after adult thymectomy (Pitchappan and Muthukkaruppan, 1977c). This effect is also confirmed by the profile of the PB lymphocytes in adult thymectomized lizards (Table 5). The total number of white blood cells as well as the percentage of small and medium lymphocytes is significantly reduced in thymectomized lizards (Murugan, 1980). These results indicate the importance of thymus-derived lymphocytes in the anti-SRBC PFC response to optimal antigenic stimulation. High doses of erythrocyte antigen seem to obviate, at least partially, the requirement for thymus-derived lymphocytes by stimulating the B cells directly (Taylor and Wortis, 1968). However, it is possible that the residual T cells could have been stimulated by the large dose of antigen. Thus, xenogeneic erythrocytes may be considered T-cell-dependent antigens in the lizard. On the other hand, the antibody response to a known T-cell-independent antigen, namely polyvinyl pyrrolidone (PVP), is not affected by adult thymectomy performed 1 month before immunization (Table 6) (Jayaraman and Baskar, unpublished).

A more sensitive probe to study the effect of adult thymectomy is to measure the level of impairment of the helper-carrier-induced humoral immune (HI) response. Recently, Ramila (1978) has demonstrated that the helper-induced anamnestic response to SRBC is suppressed 1 month after adult thymectomy (Table 7). The results clearly demonstrate the thymus-dependent nature of help-

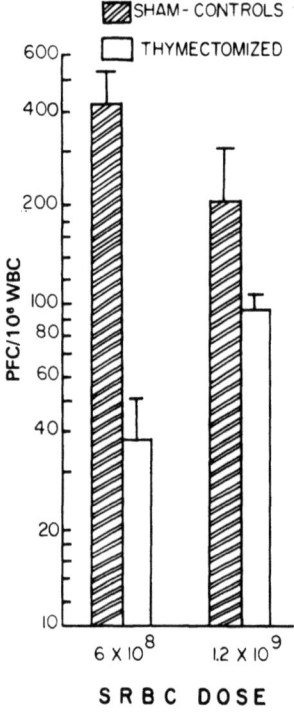

FIGURE 19. Effect of adult thymectomy on the PFC response to SRBC in the garden lizard. Sham-operated and thymectomized lizards were immunized intramuscularly with 6×10^8 or 1.2×10^9 SRBC 1 month after surgery and assays were performed with splenic cells 7 days later. Mean ± S.E. of 5–8 animals per group.

TABLE 5. Effect of Adult Thymectomy on Peripheral Blood Leukocytes[a]

	Total count per mm³	Differential count (%)				
		Small and medium lymphocytes	Large lymphocytes	Lymphocytes (all types)	Monocytes	Granulocytes
Normal lizards	6704 ± 387[b]	—	—	59 ± 4	2 ± 2	39 ± 3
ATx—15 days	6087 ± 141	40 ± 10	15 ± 9	55 ± 10	10 ± 2	36 ± 8
SATx—15 days	6853 ± 226	59 ± 5	6 ± 2	65 ± 7	7 ± 2	35 ± 6
ATx—30 days	5605 ± 412[c]	26 ± 3[d]	16 ± 3	42 ± 5[e]	10 ± 3	48 ± 5
SATx—30 days	6707 ± 99[c]	55 ± 4[d]	7 ± 1	62 ± 3[e]	5 ± 3	38 ± 4

[a]For total WBC count, fresh blood was diluted with PBS (pH 7.2) in diluting pipette and counted using a hemocytometer under phase. For differential count, blood smear was prepared and stained with May–Grunwald–Giemsa. At least 100 leukocytes were counted to determine the percentage of each cell type.
[b]Mean ± S.D. of five lizards in each group.
[c-e]Significance between ATx (adult thymectomized) and SATx (sham-operated) groups: [c]$p < 0.001$; [d]$p < 0.001$; [e]$p < 0.001$.

TABLE 6. INFLUENCE OF ADULT THYMECTOMY ON THE ANTIBODY RESPONSE TO PVP

Treatment	Antigen dose[a]	Antibody titer (mean ± S.E.)[b]	
		Day 14	Day 21
ATx	10 mg	300.0 ± 57.7 (7)[c]	500.0 ± 100.0 (7)
SATx	10 mg	350.0 ± 50.0 (5)	300.0 ± 67.7 (5)
ATx	20 mg	8960.0 ± 1280.0 (8)	3840.0 ± 548.0 (8)
SATx	20 mg	7680.0 ± 1478.0 (5)	2304.0 ± 236.6 (5)

[a]Both ATx (adult thymectomized) and SATx (sham-operated) lizards were immunized with a single injection of desired amounts of PVP, 30 days after surgery.
[b]Lizards were bled at different time intervals after immunization and the sera were titrated for anti-PVP antibodies.
[c]Number of animals investigated is given in parentheses.

er function induced by F-SRBC in the lizard. Similarly, the carrier-primed specific enhancement of an antihapten response is dependent on the lymphocytes derived from adult thymus (Figure 20), thereby confirming the earlier finding in *X. laevis* (Gruenwald and Ruben, 1979). Thus, the helper-carrier function is impaired in adult lizards 1 month after the ablation of the thymus. Such a pronounced depletion is not surprising in light of the absence of major secondary lymphoid centers other than the spleen in this species of lizard. It belongs to the family Agamidae in which the GALT is poorly developed except for a few cloacal lymphoid aggregates (Pitchappan, 1980).

Yet another model to study the role of adult thymus in immunogenesis is the use of a reptilian species which exhibits pronounced lymphoid depletion as a consequence of seasonal changes. This approach has been initiated with the lizard, *Chalcides ocellatus*, in which the thymus is involuted and devoid of viable lymphocytes in winter. Therefore, removal of the thymic remnant during this

TABLE 7. EFFECT OF ADULT THYMECTOMY ON THE HELPER FUNCTION IN THE LIZARD, *CALOTES VERSICOLOR*

Treatment	Priming[a]	Challenge[b]	PFC/10⁶ WBC[c]	
SATx	—	SRBC	109 ± 14	
SATx	F-SRBC	SRBC	592 ± 103	
ATx	F-SRBC	SRBC	72 ± 9	$p < 0.001$
ATx	—	SRBC	62 ± 8	

[a]One month after adult thymectomy (ATx) or sham operation (SATx), lizards were primed with 0.1 ml of 25% F-SRBC intraperitoneally.
[b]Ten days after priming, they were challenged with 0.1 ml of 2.5% SRBC intraperitoneally.
[c]PFC assay was performed with splenic cells 14 days after challenge. Mean ± S.E. PFC/10⁶ viable WBC of 8–10 animals.

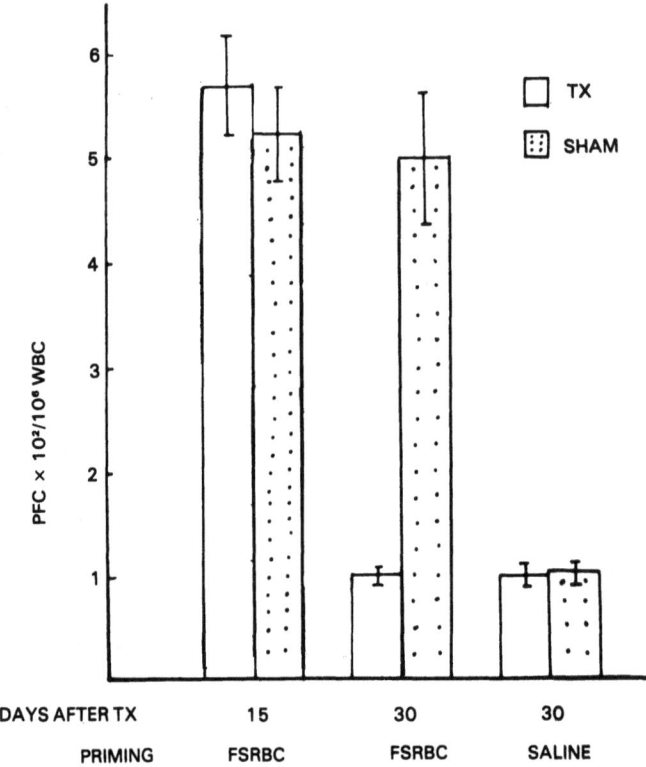

FIGURE 20. Effect of adult thymectomy on carrier-primed anti-TNP response in the garden lizard. Fifteen or thirty days after adult thymectomy (Tx) or sham operation, lizards were primed intraperitoneally with 0.1 ml of 25% F-SRBC and 10 days later challenged with 0.1 ml of 25% TNP-SRBC intraperitoneally. PFC assay was performed with splenic cells using TNP–RRBC, 10 days after challenge injection. Each bar represents the mean (± S.E.) PFC/10^6 viable white blood cells of 8–10 animals.

period of the year would result in a total depletion of T cells without resorting to additional irradiation, corticosteroid, or antilymphocyte serum treatment. El Masri (1979) examined the lizards 3–5 months after adult thymectomy for peripheral lymphoid organ development, skin allograft rejection, and humoral response to rat red blood cells (RRBC), human serum albumin (HSA), and PVP. In such lizards, the periarteriolar region of the spleen was depleted of lymphocytes and skin allograft survival was slightly prolonged. On the other hand, antibody responses to HSA and PVP were not affected and anti-RRBC antibody response was enhanced (Figure 21).

Thus, several approaches are now available and further studies would contribute to a fuller understanding of reptilian thymus in immunity.

5.2. ROLE OF SPLEEN AND GALT IN IMMUNE RESPONSES

Splenectomy in the agamid lizard, *Calotes versicolor*, did not impair the skin allograft rejection or the MI response of peritoneal exudate cells to varying doses

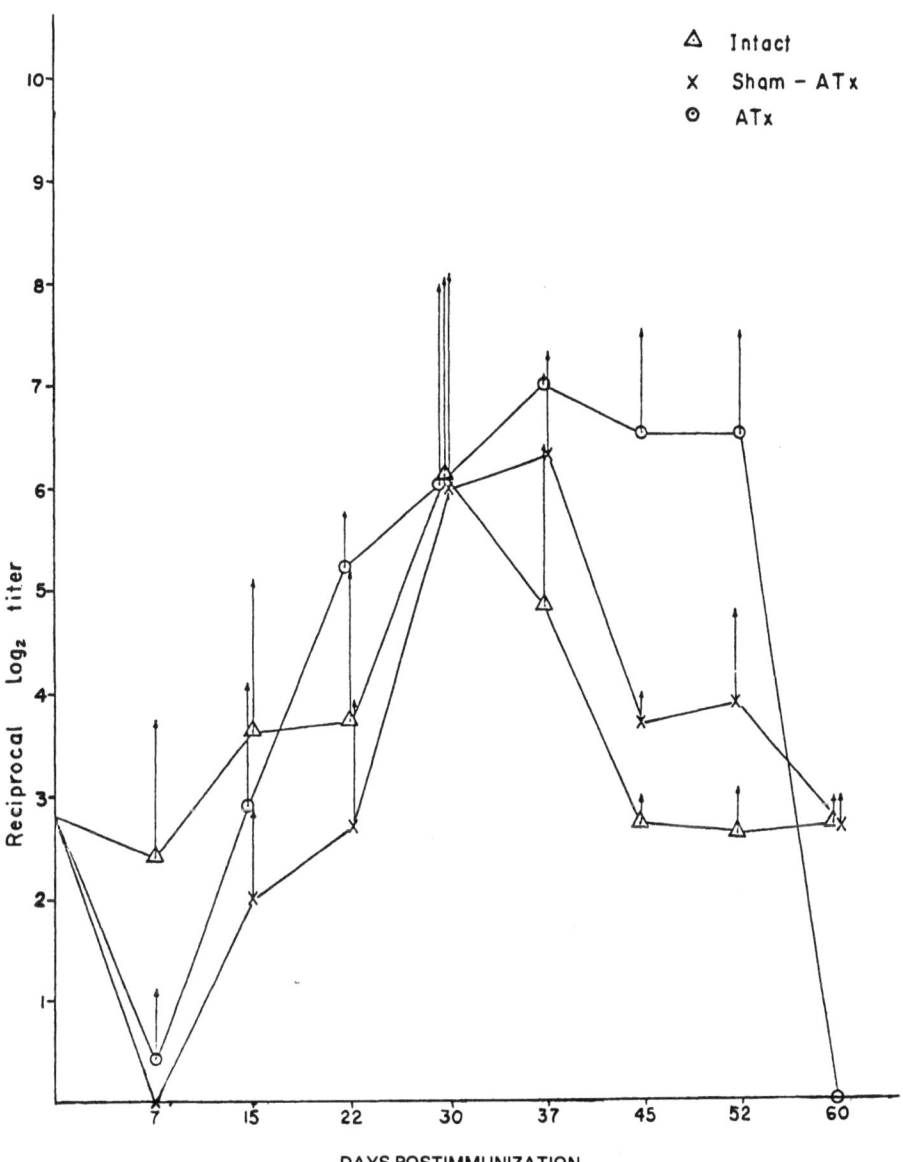

FIGURE 21. Mean serum hemagglutinin titers in intact, sham-thymectomized (Sham-ATx), or adult thymectomized (ATx) *Chalcides ocellatus* following intraperitoneal immunization with 0.1 ml of 10% RRBC suspension. Vertical bars indicate standard errors.

of SRBC (10^4 to 10^9 cells) (Muthukkaruppan *et al.*, 1976b; Jayaraman and Muthukkaruppan, 1977, 1978a,b). In contrast, in the lizard, *Chalcides ocellatus*, splenectomy of adult animals led to a significant prolongation of allograft survival time compared to intact and sham-operated controls (El Masri, 1979). With regard to MI, splenectomy of *Calotes versicolor* completely abrogated antibody production against 0.1 ml of 25% SRBC suspension or 10 mg BSA, administered 5–7 days postsurgery, thus suggesting that in this species, the spleen is the major site for the generation of antibody-producing cells (Kanakambika and Muthukkaruppan, 1972; Muthukkaruppan et al., 1976b). Again in contrast, antigenic stimulation of splenectomized lizards (*Chalcides ocellatus*) with 0.1 ml of 10% RRBC suspension or 10 mg HSA resulted in a strong antibody response with no significant difference in serum antibody titers between splenectomized, sham-operated, and nonoperated controls (El Masri, 1979). Similar results were obtained in the lizard, *Scincus scincus*, in that splenectomy did not reduce the HI response to an injection of 0.3 ml of a 20% RRBC suspension until 30 days postimmunization, but depressed subsequent antibody production (Hussein et al., 1979c). Apparently, immunological tasks are shared between the spleen and other lymphoid tissues such as the GALT, at least in some species of lizards. In this regard, it is noteworthy that the GALT is very well developed in *Chalcides ocellatus* (Hussein et al., 1978a) and *Scincus scincus* (Hussein et al., 1979d). However, it remains to be determined whether the antibody-producing cells are generated in the GALT after immunization in splenectomized lizards. It would also be of interest to look into the functional nature of lymph node-like structures in these forms (Johnston, 1973).

In the lizards, *Agama stellio* and *Chamaeleon chamaeleon*, the GALT is not well developed. *A. stellio* rejects skin allografts in a manner and tempo comparable to those recorded in the GALT-rich lizards, *Chalcides ocellatus* and *T. annularis* (Table 8). However, *A. stellio* and *Chamaeleon chamaeleon* are able to produce only minimal amounts of antibody when challenged with various mammalian erythrocytes or HSA over a wide dose range, by the intraperitoneal, intramuscular, or subcutaneous routes, regardless of whether the antigen is in Freund's adjuvant or not (El Deeb, 1978; El Kes, 1978; Badir et al., 1981). Nevertheless, *A. stellio* possesses small amounts of natural heterohemagglutinins and does mount a good HI response against *S. typhimurium* (Charmy, 1979; Hussein et al., 1979b). In another species of agamid lizard, *Calotes versicolor*, the GALT is poorly developed except for a few definitive lymphoid accumulations in the cloacal region (Hameed, 1980). This species is capable of producing a good antibody response to a variety of antigens (Muthukkaruppan et al., 1976a). Thus, further work is required to determine the role of GALT in the HI response, particularly in the generation of IgA-producing cells in reptiles.

6. THE EFFECTS OF NUTRITION AND ENVIRONMENT ON THE IMMUNE SYSTEM

6.1. THE INFLUENCES OF MALNUTRITION AND HOUSING MODE ON IMMUNITY

The thymuses of reptiles involute gradually with age and also undergo seasonal fluctuations, since thymocyte numbers diminish in winter and increase in spring. This observation was reported over 70 years ago (Dustin, 1909). Another long-term fact is that starvation also causes reversible thymic involution in reptiles (Salkind, 1915). This observation has recently been confirmed and ex-

TABLE 8. ALLOGRAFT REJECTION IN LIZARDS

Species	GALT	Onset of rejection day (mean ± S.E.)	Completion of rejection day (mean ± S.E.)
Chalcides ocellatus	rich	18.6 ± 0.1	32.3 ± 0.4
Tarentola annularis	rich	27.5 ± 0.7	53.0 ± 1.6
Agama stellio	poor	14.7 ± 0.1	43.5 ± 0.4

tended. It was found that nutritional deficiency impairs both HI and CMI in young snapping turtles. The thymus undergoes acute involution during which the thymocyte count is decreased more than 10-fold to less than 1 million cells. The splenic white pulp is also depleted of lymphocytes, but to a lesser extent. In advanced stages of malnutrition, susceptibility to bacterial infection is greatly increased. However, with improved nutrition, the histopathology and immunodeficiency are readily reversible. Interestingly, long-term tolerance to skin allografts can be induced in some malnourished turtles which are later renourished (Borysenko and Lewis, 1979).

Seasonal and nutritional factors which affect immunity can be explained in terms of the known temperature and nutritional requirements of reptiles. If temperature is low and nutrition inadequate, the immune system is depressed. However, there are more subtle and as yet poorly defined factors which affect immunity. In a recent study on the lizard, *Lacerta viridis*, large differences in skin allograft survival times were observed between those maintained in the laboratory compared to those maintained in an outdoor environment. Even though the diet, mean temperatures, and light–dark cycles were nearly identical, allograft rejection was very slow indoors (220–320 days) but much faster (32–64 days) in a more natural outdoor environment. The environmental factors which account for this difference have not been elucidated. Besides the temperature range, which was wider for the outdoor group, differences in humidity and exposure to sunlight may have influenced the immune responses (Worley and Jurd, 1979).

6.2. LYMPHOID TISSUES AND CELLS IN THE DIFFERENT SEASONS

Extensive studies have been conducted in several Egyptian lizards and snakes on the effect of seasonal variation on the structure and function of lymphoid tissues. By the end of spring through summer, the thymuses of lizards are differentiated into a peripheral cortex and central medulla. Lymphocytes are very numerous, while myoid cells and epithelial aggregates are few and restricted to the medulla and corticomedullary area. Thymic involution starts in early autumn and proceeds steadily through winter. It is characterized by progressive reduction in size, depletion of lymphoid elements, and increase in number and distribution of myoid cells and cysts. The thymus is eventually transformed into a mass of cell remnants, cysts, and fibrous tissue. By the onset of spring, the gradual regeneration of thymus begins (El Deeb, 1978; El Kes,

1978; Hussein et al., 1978a,b, 1979d). In snakes, the thymus involutes both in winter and in summer, while during spring and autumn it acquires a rich lymphoepithelial organization (Hussein et al., 1979a).

The lymphoid organization of the spleen of lizards and snakes is also influenced by seasonal variations. Lymphoid depletion occurs during both summer and winter in snakes and during winter only in lizards (El Deeb, 1978; El Kes, 1978; Hussein et al., 1978a,b 1979a,d).

In lizards, the gut-associated lymphoid aggregates are generally more numerous and larger in size in spring through autumn than in winter, but in contrast to thymus and spleen, GALT remains well developed in winter, thus suggesting that lizard lymphoid tissues display distinct sensitivity to seasonal changes (Hussein et al., 1978a,b, 1979d). In snakes, the number of lymphoid aggregates in the esophagus, small and large intestines is usually rather constant in the four seasons, while patches of the small intestine are numerous in winter and autumn but are sharply reduced in number and size in spring and disappear in summer (Hussein et al., 1979a).

The seasonal effects on PB lymphocytes have been reported in the Florida alligator, A. mississippiensis. The number of lymphocytes was found to be reduced to about 10 million/ml blood during the winter from the maximum of 20 to 40 million/ml during the warm months (Cuchens et al., 1976). Thus, reptilian lymphoid tissues atrophy and regenerate in a cyclic, seasonal fashion (Mansour et al., 1980).

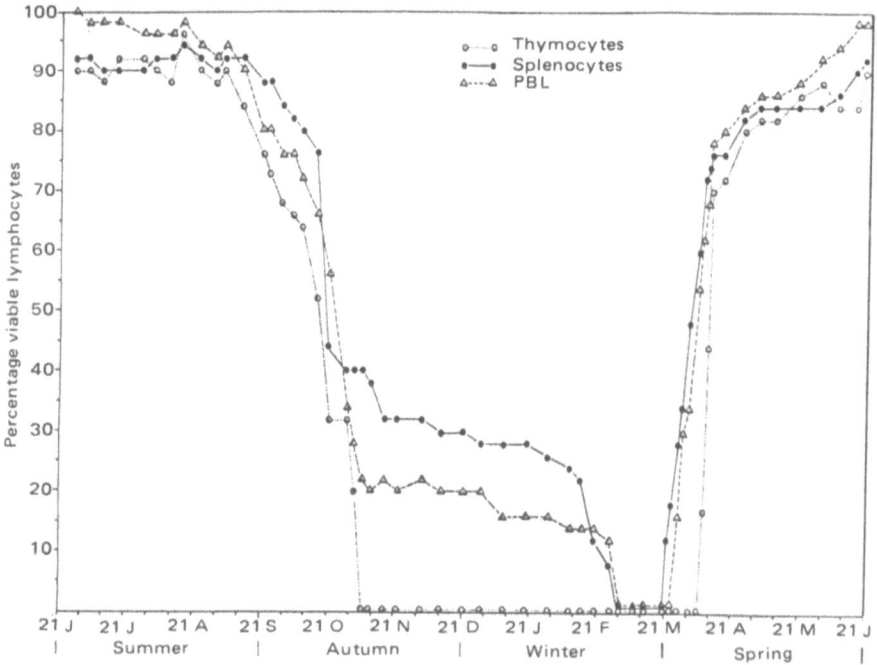

FIGURE 22. Percentage lymphocyte viability of the lizard, Chalcides ocellatus, in the different seasons. Trypan blue dye exclusion test.

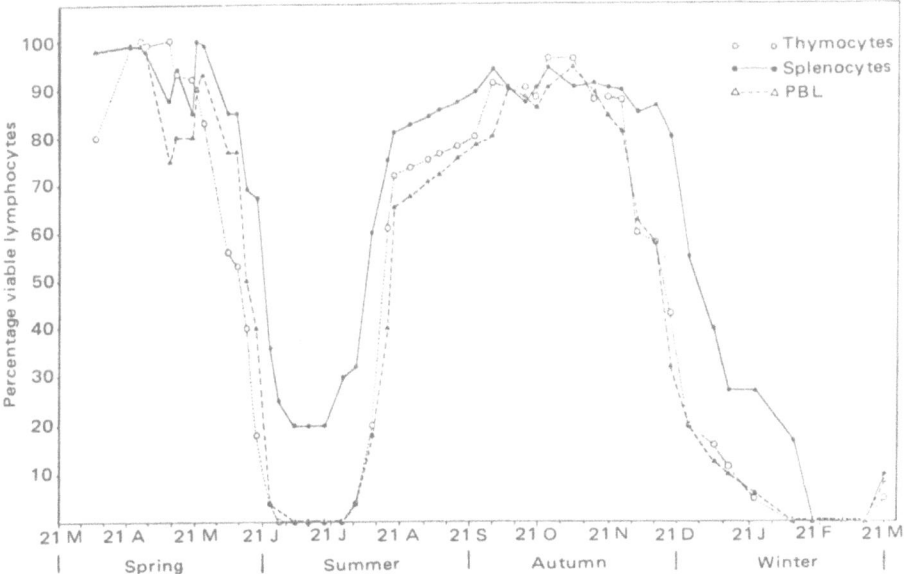

FIGURE 23. Percentage lymphocyte viability of the snake, *Spalerosophis diadema*, in the different seasons. Trypan blue dye exclusion test.

The seasonal involution of reptilian thymus differs from the thymic age-related involution recorded in all vertebrate classes in that it is reversible and is accompanied by a drastic decrease in number and viability of lymphocytes in the thymus, spleen, and peripheral blood (Figures 22, 23). In fact, the increase in the proportion of dead cells in the lymphoid organs of winter animals may be due to the depressed efficiency of phagocytic system in winter, as evidenced by the colloidal carbon studies (El Ridi *et al.*, 1981).

The percentages of lymphocytes carrying thymocyte-specific antigen and readily detectable surface Ig were estimated in the four consecutive seasons in the lizard, *Chalcides ocellatus*, and the snake, *Spalerosophis diadema*. In spring lizards, the percentages of thymocytes, PB and splenic lymphocytes labeled with anti-Ig serum was 0, 14, and 43%, respectively (Figures 24, 25). Concurrently, 100% thymocytes, 85% of PB lymphocytes, 56% of splenocytes were lysed by ATS (El Ridi and Kandil, 1980). Similarly, in spring snakes, the percentage of lymphocytes bearing surface Ig complemented the percentage of cells recognized by ATS (Mansour *et al.*, 1981). In no season did thymocytes react with anti-Ig serum. In winter lizards and snakes, thymocytes and thymus-derived cells in blood and spleen are totally absent. During this season, most of the viable lymphocytes in peripheral lymphoid organs are surface Ig positive (Mansour *et al.*, 1981; El Ridi and Kandil, 1981). However, Cuchens *et al.* (1976) have reported that LPS responsiveness of alligator PB lymphocytes was significantly reduced during winter months. In general, the complete absence of Ig-positive cells in thymus throughout the year suggests that the Ig-bearing lymphocyte population is totally independent of the thymus for its development and is equivalent to the avian and mammalian B cells. Furthermore, these results indicate that lizard and

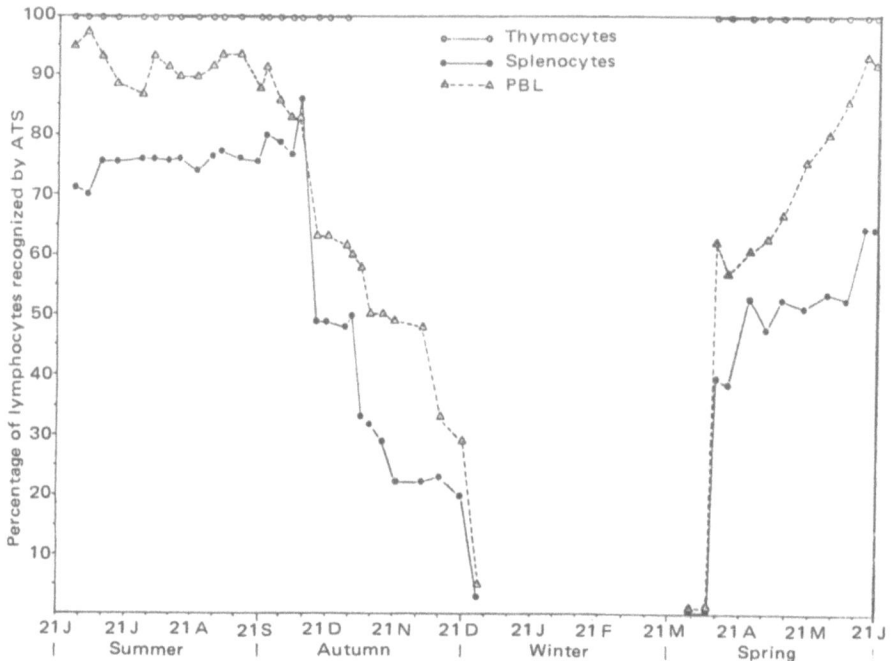

FIGURE 24. Percentage of lizard lymphocytes recognized by rabbit anti-lizard thymocyte serum (ATS) in the different seasons in complement-dependent cytotoxic assay (used when lymphocyte viability was > 50%) or indirect membrane immunofluoresence. Each point on the curves represents value obtained with lymphocytes pooled from 5 to 10 lizards.

snake B cells differ from thymocytes in their sensitivity to various environmental factors during different seasons.

6.3. HUMORAL IMMUNE RESPONSES IN THE DIFFERENT SEASONS

Systematic studies, in which variables such as dosages and routes of administration of antigens (RRBC, HSA, PVP) and environmental conditions at which animals were kept were carefully controlled, showed that the kinetics of serum antibody response is strikingly similar in several species of Egyptian lizards and snakes tested, but is severely affected by seasonal variations (Cohen, 1971; El Kes, 1978; El Rouby, 1978; Hussein et al., 1979d; El Deeb et al., 1980; El Ridi et al., 1981). In winter, lizards fail to produce appreciable amounts of antibody to intraperitoneal injections of 0.3–0.4 ml of a 20% RRBC suspension. In the other seasons, the response is vigorous, although slower in spring than in summer or autumn. The kinetics and magnitude of lizards' humoral response correlate well with the state of lymphoid tissue development during the different seasons. Further studies showed that the kinetics of splenic RFC response was identical in the four seasons, implying that antigen recognition and binding are unaffected by seasonal changes (Abdel Khader, unpublished). It is quite possible that the proliferative and differentiative stages of B-lymphocyte response follow-

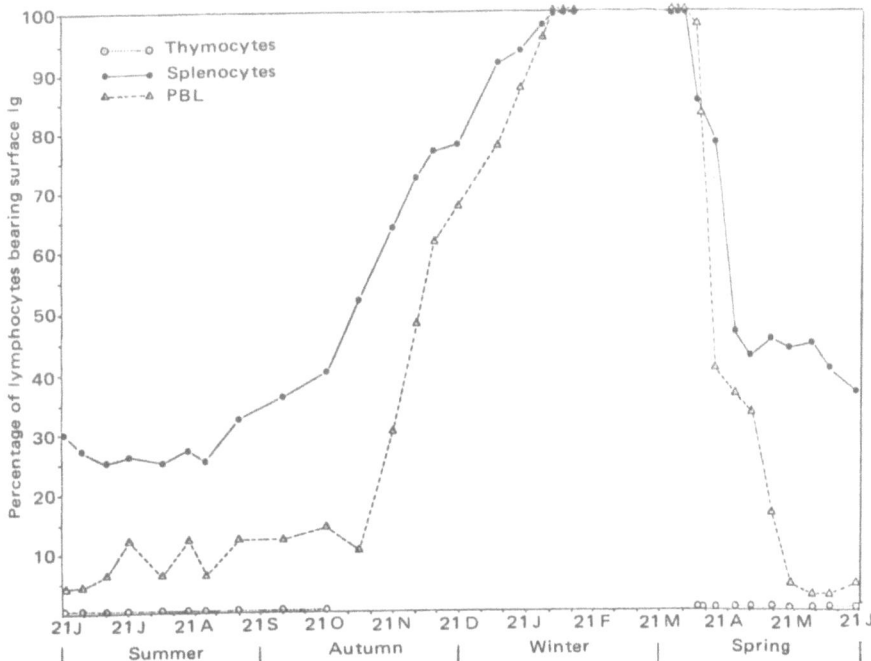

FIGURE 25. Percentage of lizard lymphocytes stained, in indirect membrane immunofluorescence, by a specific rabbit anti-lizard immunoglobulin serum in the four seasons. Each point represents value of 5–10 pooled animals.

ing antigen binding are inhibited during winter in reptiles, as in other ectotherms (Wright et al., 1978).

The effect of seasonal variation on the immune system of the snake, *Psammophis schokari*, has also been studied. Serum antibody responses to a single intraperitoneal injection of 0.4 ml of 20% RRBC suspension, 15 mg HSA, or 10 µg PVP were assayed in the different seasons. In spring and autumn, snakes respond to these antigens by promptly synthesizing a considerable amount of 2ME-sensitive and 2ME-resistant antibodies. In summer and winter, a negligible level of immune antibodies is produced with the exception that PVP elicits a strong, 2ME-sensitive antibody response in summer snakes (El Ridi et al., 1981).

When these data are correlated with findings acquired by histological, viability, and cell surface marker studies (see Section 6.2), it becomes evident that vigorous humoral reactivity in spring and autumn occurs with full development of lymphoid tissues and the presence of large numbers of T and B lymphocytes. The strong humoral responses of summer snakes to PVP correspond to the considerable quantity of B cells in spleen during this season. Concomitant feeble response to RRBC and HSA is probably due to the depletion of T lymphocytes necessary for cell cooperation. Finally, despite the fact that B lymphocytes and some T lymphocytes persist in spleen until February, winter humoral response to RRBC, HSA, and PVP is conspicuously poor, suggesting that other mechanisms of immunosuppression may be involved (El Ridi, unpublished).

Further studies are required to shed light on the evolutionary significance of these phenomena.

7. CONCLUDING REMARKS

It is evident from this review that all reptiles are capable of mounting highly specific immune responses mediated by heterogeneous populations of immunocompetent cells. These cells produce several immunoglobulin types and in some instances different subclasses of lymphocytes cooperate to amplify or suppress immune responses. All of the above demonstrate that the highly diverse orders of reptiles are in no way primitive or deficient with respect to their immune systems. Indeed, among the reptiles, one finds some variation in the structure of lymphoid organs and the distribution of lymphoid cells. Likewise, the diversity of reptilian orders is reflected in the ontogeny of their immune systems. Furthermore, because reptiles are distributed so widely geographically and occupy so many ecological niches, environmental influences on immune functions are quite complex and not simply related to environmental temperature and nutritional factors alone. The immune responses of the endotherms, both birds and mammals, are much more predictable in this sense. For this reason, comparative immunologists must take special precautions to simulate natural environments when studying the immune systems of reptiles and not to overlook either environmental or internal factors which may enhance or hinder their ability to resist disease.

The future prospects for the use of reptiles as immunological models are many. Several species have been successfully maintained in laboratory environments for years and an abundance of baseline data is presently available. The current applications of *in vitro* techniques to reptilian immunocompetent cells greatly enhance the prospects of acquiring more detailed knowledge of cellular interactions and functions.

Two major evolutionary problems have been with us for a time and still remain unanswered. The first concerns the evolutionary origins of the avian bursa or its equivalent. Although B cells and plasma cells have been identified in reptiles, their organ or site(s) of origin is still a mystery. The second concerns the problem of immunological memory. Although anamnestic responses against some antigens have been demonstrated in a number of reptilian species, the changes associated with such responses in the secondary lymphoid organs (i.e., formation of germinal centers) apparently do not occur in reptiles or in any other endothermic classes of animals. Thus, the mediating mechanisms of immunological memory in ectotherms have not been characterized. As evolutionary precursors of both birds and mammals, reptiles represent fertile ground for exploring these questions. Modern fluorescent and isotope labeling techniques, in conjunction with highly specific antibodies, can be applied in both *in vivo* and *in vitro* studies to solve these riddles of evolution and the many other problems which still remain.

REFERENCES

Ambrosius, H., 1966, Comparative investigation of immune globulins of various vertebrate classes, *Nature (London)* **209**:524.

Ambrosius, H., and Frenzel, E. M., 1972, Anti-DNP antibodies in carps and tortoises, *Immunochemistry* **9**:65.

Ambrosius, H., and Hoheisel, G., 1973, Ultrastructure of antibody producing cells of reptiles. I. Spleen cells of the tortoise, *Agrionemys horsfieldi*, *Acta Biol. Med. Ger.* **31**:733.

Ambrosius, H., and Lehmann, R., 1965, Beitraege zur Immuno-biologie poikilothermer Wirbeltiere. II. Immunologische untersuchungen an Schilokroeten (*Testudo hermanni* Gmelin), *Z. Immun. Allergieforsch.* **128**:81.

Ambrosius, H. J., Hammerling, R., Richter, R., and Schimke, R., 1970, Immunoglobulins and the dynamics of antibody formation in poikilothermic vertebrates (Pisces, Urodela, Reptilia), in: *Developmental Aspects of Antibody Formation and Structure*, Vol. II (J. Sterzl and I. Riha, eds.), pp. 727–744, Academic Press, New York.

Badir, N., Afifi, A., and El Ridi, R., 1981, Cell-mediated immunity in the gecko, *Tarentola annularis*, *Folia Biol. (Praha)* **27**:28.

Baskar, S., 1978, Mechanism of low-zone tolerance to sheep erythrocytes in the lizard, *Calotes versicolor*, Ph.D. thesis, Madurai University, Tamilnadu, India.

Baskar, S., and Muthukkaruppan, VR., 1980, Induction of low zone tolerance to sheep erythrocytes in the lizard, *Calotes versicolor*, *Dev. Comp. Immunol.* **4**:491.

Benedict, A. A., and Pollard, L. W., 1972, Three classes of immunoglobulins found in the sea turtle, *Chelonia mydas*, *Folia Microbiol. (Prague)* **17**:75.

Biggs, P. M., 1957, The association of lymphoid tissue with the lymph vessels in the domestic chicken, *Gallus domesticus*, *Acta Anat.* **29**:36.

Bockman, D. E., 1970, The thymus, in: *Biology of the Reptilia* (C. Gans and T. S. Parson, eds.), Vol. 3, pp. 111–133, Academic Press, New York.

Bockman, D. E., and Winborn, W. B., 1967, Electron microscopy of the thymus in two species of snakes, *Crotalus atrox* and *Lampropeltis getulus*, *J. Morphol.* **121**:277.

Borysenko, M., 1970, Transplantation immunity in Reptilia, *Transplant. Proc.* **2**:299.

Borysenko, M., 1972, Immunosuppressive effects of low temperature on the graft-vs-host reaction in the snapping turtle, *Anat. Rec.* **175**:275.

Borysenko, M., 1975, Cellular aspects of humoral immune responsiveness in *Chelydra*, in: *Immunologic Phylogeny* (W. H. Hildemann and A. A. Benedict, eds.), *Adv. Exp. Med. Biol.* **64**:277.

Borysenko, M., 1976a, Changes in spleen histology in response to antigenic stimulation in the snapping turtle, *Chelydra serpentina*, *J. Morphol.* **149**:223.

Borysenko, M., 1976b, Ultrastructural analysis of normal and immunized spleen of the snapping turtle, *Chelydra serpentina*, *J. Morphol.* **149**:243.

Borysenko, M., 1978, Lymphoid tissues and cellular components of the reptilian immune system, in: *Animal Models of Comparative and Developmental Aspects of Immunity and Disease* (M. E. Gershwin and E. L. Cooper, eds.), pp. 63–79, Pergamon Press, Elmsford, N.Y.

Borysenko, M., 1979, Evolution of lymphocytes and vertebrate alloimmune reactivity, *Transplant. Proc.* **11**:1123.

Borysenko, M., and Cooper, E. L., 1972, Lymphoid tissue in the snapping turtle, *Chelydra serpentina*, *J. Morphol.* **138**:487.

Borysenko, M., and Lewis, S., 1979, The effect of malnutrition on immunocompetence and whole body resistance to infection in *Chelydra serpentina*, *Dev. Comp. Immunol.* **3**:89.

Borysenko, M., and Tulipan, P., 1973, The graft-vs-host reaction in snapping turtle, *Chelydra serpentina*, *Transplantation* **16**:496.

Charlemagne, J., 1974, Larval thymectomy and transplantation immunity in the urodele, *Pleurodeles waltlii* Michah (Salamandridae), *Eur. J. Immunol.* **4**:390.

Charmy, R., 1979, Studies of natural and acquired immunity in the lizard, *Agama stellio*, M.Sc. thesis, Faculty of Science, Cairo University.

Cheers, C., and Breitner, J. C. S., 1971, Co-operation between carrier reactive and hapten sensitive cells *in vitro*, *Nature New Biol.* **232**:180.

Clark, S. L., 1963, The thymus in mice studied with the electron microscope, *Am. J. Anat.* **112**:1.
Clark, S. L., 1966, Cytological evidence of secretion in the thymus, in: *The Thymus* (G. E. W. Wolstenholme and R. Porter, eds.), pp. 1–30, Little, Brown, Boston.
Coe, J. E., 1972, Immune response in the turtle (*Chrysemys picta*), *Immunology* **23**:45.
Coe, J. E., Leong, D., Portis, J. L., and Thomas, L. A., 1976, Immune response in the garter snake (*Thamnophis ordinoides*), *Immunology* **31**:417.
Cohen, N., 1971, Reptiles as models for the study of immunity and its phylogenesis, *J. Am. Vet. Med. Assoc.* **159**:1662.
Cohen, N., 1980, Salamanders and the evolution of the major histocompatibility complex, in: *Contemporary Topics in Immunobiology*, Vol. 9 (J. J. Marchalonis and N. Cohen, eds.), pp. 109–139, Plenum Press, New York.
Cohen, N., and Borysenko, M., 1970, Acute and chronic graft rejection: Possible phylogeny of transplantation antigens, *Transplant. Proc.* **2**:333.
Cohen, N., and Collins, N. H., 1977, Major and minor histocompatibility systems of ectothermic vertebrates, in: *The Major Histocompatibility System in Man and Animals* (D. Götze, ed.), pp. 313–337, Springer-Verlag, Berlin.
Cooper, E. L., and Aponte, A., 1968, Chronic allograft rejection in the iguana, *Ctenosaura pectinata*, *Proc. Soc. Exp. Biol. Med.* **128**:150.
Cuchens, M. A., and Clem, L. W., 1979, Phylogeny of lymphocyte heterogeneity. III. Mitogenic responses of reptilian lymphocytes, *Dev. Comp. Immunol.* **3**:287.
Cuchens, M., McLean, E., and Clem, L. W., 1976, Lymphocyte heterogeneity in fish and reptiles, in: *Phylogeny of Thymus and Bone Marrow-Bursa Cells* (R. K. Wright and E. L. Cooper, eds.), pp. 205–213, Elsevier/North-Holland, Amsterdam.
Cuellar, O., and Smart, C., 1977, Analysis of histoincompatibility in a natural population of the bisexual whiptail lizard, *Cnemidophorus tigris*, *Transplantation* **24**:127.
David, J. R., Lawrence, H. S., and Thomas, L., 1964, Delayed hypersensitivity *in vitro*. III. The specificity of the hapten protein conjugates in the inhibition of cell migration, *J. Immunol.* **93**:279.
Dennert, G., and Tucker, D. F., 1972, Selective priming of T-cells by chemically altered cell antigens, *J. Exp. Med.* **136**:656.
Diener, E., and Nossal, G. J. V., 1966, Phylogenetic studies on the immune response. I. Localization of antigens and the immune response in the toad, *Bufo marinus*, *Immunology* **10**:535.
Downs, C. M., 1928, Anaphylaxis. VII. Active anaphylaxis in turtles, *J. Immunol.* **15**:77.
Dustin, A. P., 1909, Contribution à l' étude du thymus des reptiles. Cellules épithéloides, cellules myoides et corps de Hassal, *Arch. Zool. Exp. Gen.* **2**:43.
El Deeb, S., 1978, Study of the immune system of some Egyptian lizards, M.Sc. thesis, Faculty of Science, Cairo University.
El Deeb, S., El Ridi, R., and Badir, N., 1980, Effect of seasonal and temperature changes on humoral response of *Eumeces schneideri*, *Dev. Comp. Immunol.* **4**:753.
El Kes, N., 1978, Experimental studies on immunologic performance of some Egyptian reptiles, M.Sc. thesis, Faculty of Science, Cairo University.
El Masri, M., 1979, Role of thymus and spleen in immune response of the lizard, *Chalcides ocellatus*, M.Sc. thesis, Faculty of Science, Cairo University.
El Ridi, R., and Kandil, O., 1981, Membrane markers of reptilian lymphocytes, *Dev. Comp. Immunol. Suppl. 1* **5**:143.
El Ridi, R., Badir, N., El Rouby, S., 1981, Effect of seasonal variation on the immune system of the snake, *Psammophis schokari*, *J. Exp. Zool.* **216**:357.
Evans, E. E., 1963, Antibody response in Amphibia and Reptilia, *Fed. Proc.* **22**:1132.
Fahey, K. J., 1974, Immunological reactivity in the fetus and the structure of fetal lymphoid structure, in: *Progress in Immunology II* (B. Amos, ed.), pp. 49–60, Academic Press, New York.
Grantham, W. G., and Fitch, F. W., 1975, The role of antibody feedback inhibition in the regulation of the secondary antibody response after high and low dose priming, *J. Immunol.* **114**:394.
Green, M. K., and Clunie, G. J. A., 1974, *In vitro* leukocyte migration in renal transplantation, *Med. J. Aust.* **1**:740.
Grey, H., 1963, Phylogeny of the immune response: Studies on some physical, chemical, and serologic characteristics of antibody produced in the turtle, *J. Immunol.* **91**:819.

Grey, H. M., 1966, Structure and kinetics of formation of antibody in the turtle, in: *Phylogeny of Immunity* (R. T. Smith, P. S. A. Miescher, and R. A. Good, eds.), pp. 227–235, University of Florida Press, Gainesville.

Gruenwald, D. A., and Ruben, L. N., 1979, The effect of adult thymectomy upon helper function in *Xenopus laevis*, the South African clawed toad, *Immunology* **38**:191.

Hameed, N. S. S., 1980, Studies on the distribution of the gut-associated lymphoid tissues (GALT) in vertebrates, M. Phil. thesis, Madurai University, Tamilnadu, India.

Hämmerling, U., Mack, C., and Pickel, H. S., 1976, Immunofluorescence analysis of Ig determinants of mouse thymocytes and T cells, *Immunochemistry* **13**:525.

Henry, M., and Charlemagne, J., 1977, Plasmocytic series in the perihepatic layer of the urodele amphibian, *Pleurodeles waltlii*, *Dev. Comp. Immunol.* **1**:23.

Hildemann, W. H., and Thoenes, G., 1969, Immunologic response of Pacific hagfish. I. Skin transplantation immunity, *Transplantation* **7**:506.

Horton, J. D., 1971, Histogenesis of the lymphomyeloid complex in the larval leopard frog, *Rana pipiens*, *J. Morphol.* **134**:1.

Hussein, M. F., Badir, N., El Ridi, R., and Akef, M., 1978a, Differential effect of seasonal variation on the lymphoid tissue of the lizard, *Chalcides ocellatus*, *Dev. Comp. Immunol.* **2**:297.

Hussein, M. F., Badir, N., El Ridi, R., and Akef, M., 1978b, Effect of seasonal variation on lymphoid tissues of the lizards, *Mabuya quinquetaeniata* and *Uromastyx aegyptia*, *Dev. Comp. Immunol.* **2**:496.

Hussein, M. F., Badir, N., El Ridi, R., and Akef, M., 1979a, Lymphoid tissues of the snake, *Spalerosophis diadema*, in the different seasons, *Dev. Comp. Immunol.* **3**:77.

Hussein, M. F., Badir, N., El Ridi, R., and Charmy, R., 1979b, Natural heterohaemagglutinins in the serum of the lizard, *Agama stellio*, *Dev. Comp. Immunol.* **3**:643.

Hussein, M. F., Badir, N., El Ridi, R., and El Deeb, S., 1979c, Effect of splenectomy on the humoral immune response in the lizard, *Scincus scincus*, *Experientia* **35**:869.

Hussein, M. F., Badir, N., E. Ridi, R., and El Deeb, S., 1979d, Effect of seasonal variation on immune system of the lizard, *Scincus scincus*, *J. Exp. Zool.* **209**:91.

Jacobs, D. M., and Byrd, W., 1975, Adult thymectomy results in loss of T-dependent mitogen response in mouse spleen cells, *Nature (London)* **255**:153.

Jayaraman, S., 1976, Modulation of humoral and cell-mediated immune response to sheep erythrocytes in the lizard, *Calotes versicolor*, Ph.D. thesis, Madurai University, Tamilnadu, India.

Jayaraman, S., and Muthukkaruppan, VR., 1975, Manifestation of anaphylaxis to egg albumin in the lizard, *Calotes versicolor*, *Experientia* **31**:1468.

Jayaraman, S., and Muthukkaruppan, VR., 1977, In vitro correlate of transplantation immunity: Spleen cell migration inhibition in the lizard, *Calotes versicolor*, *Dev. Comp. Immunol.* **1**:133.

Jayaraman, S., and Muthukkaruppan, VR., 1978a, The detection of cell-mediated immunity to sheep erythrocytes by the capillary migration inhibition technique in the lizard, *Calotes versicolor*, *Immunology* **34**:231.

Jayaraman, S., and Muthukkaruppan, VR., 1978b, Influence of route and dose of antigen on the migration inhibition and plaque forming cell responses to sheep erythrocytes in the lizard, *Calotes versicolor*, *Immunology* **34**:241.

Johnston, M. R. L., 1973, Perivascular lymphoid tissue associated with the axillary lymph sinus and lateral vein of the *Gehyra variegata*, *J. Morphol.* **139**:431.

Kanakambika, P., 1971, Studies on the morphology, development and immunological function of the spleen in the lizard, *Calotes versicolor*, Ph.D. thesis, Madurai University, Tamilnadu, India.

Kanakambika, P., and Muthukkaruppan, VR., 1972, Immune response to sheep erythrocytes in the lizard, *Calotes versicolor*, *J. Immunol.* **109**:415.

Kanakambika, P., and Muthukkaruppan, VR., 1973, Lymphoid differentiation and organization of the spleen in the lizard, *Calotes versicolor*, *Proc. Indian Acad. Sci. Sect. B* **78**:37.

Kassin, L. F., and Pevenitskii, L. A., 1969, Detection of antibody-forming cells in turtles' spleens by means of a modified method of local hemolysis in gel, *Bull. Exp. Biol. Med.* **67**:287.

Katz, D. H., and Benacerraf, B., 1972, Regulatory influence of activated T-cells on B-cell response to antigens, *Adv. Immunol.* **15**:1.

Kawaguchi, S., Muramatsu, F., and Mitsuhashi, S., 1978, Natural hemolytic activity of snake serum. I. Natural antibody and complement, *Dev. Comp. Immunol.* **2**:287.

Kawaguchi, S., Hiruki, T., Harada, T., and Morikawa, S., 1980, Frequencies of cell-surface or cytoplasmic IgM-bearing cells in the spleen, thymus and peripheral blood of the snake, *Elaphe quadrivirgata*, *Dev. Comp. Immunol.* **4**:559.

Kincade, P. W., Lawton, A. R., and Cooper, M. D., 1971, Restriction of surface immunoglobulin determinants to lymphocytes of the plasma cell line, *J. Immunol.* **106**:1421.

Kotani, M., 1959, Lymphgefasse, lymphatische apparate und extravasculare saftbahnen der Schlange (*Elaphe quadrivirgata*), *Acta Sch. Med. Univ. Imp. Kioto* **36**:121.

Lagrange, P. H., Mackaness, G. B., and Miller, T. E., 1974, Influence of dose and route of antigen injection on the immunological induction of T cells, *J. Exp. Med.* **139**:528.

Langevoort, H. L., 1963, The histophysiology of the antibody respcnse. I. Histogenesis of the plasma cell reaction in rabbit spleen, *Lab. Invest.* **12**:106.

Lerch, E. G., Huggins, S. E., and Bartel, A. H., 1967, Comparative immunology. Active immunization of young alligators with haemocyanin, *Proc. Soc. Exp. Biol. Med.* **124**:448.

Leslie, G. A., and Clem, L. W., 1972, Phylogeny of immunoglobulin structure and function. VI. 17S, 7.5S, and 5.7S anti-DNP of the turtle, *Pseudemys scripta*, *J. Immunol.* **108**:1656.

Linthicum, D. S., and Hildemann, W. H., 1970, Immunologic responses of Pacific hagfish. III. Serum antibodies to cellular antigens. *J. Immunol.* **105**:912.

Lykakis, J. J., 1968, Immunoglobulin production in the European pond tortoise, *Emys orbicularis*, immunized with serum protein antigens, *Immunology* **14**:799.

Manickavel, V., 1972, Studies on skin transplantation immunity in the lizard, *Calotes versicolor*, Ph.D. thesis, Madurai University, Tamilnadu, India.

Manickavel, V., and Muthukkaruppan, VR., 1969, Allograft rejection in the lizard, *Calotes versicolor*, *Transplantation* **8**:307.

Manning, M. J., 1975, The phylogeny of thymic dependence, *Am. Zool.* **15**:63.

Mansour, M. H., El Ridi, R., and Badir, N., 1980, Surface markers of lymphocytes in the snake, *Spalerosophis diadema*. I. Investigation of lymphocytes surface markers, *Immunology* **40**: 605

Marchalonis, J. J., Ealey, E. H. M., and Diener, E., 1969, Immune response of the tuatara, *Sphenodon punctatum*, *Aust. J. Exp. Biol. Med. Sci.* **47**:367.

Marchalonis, J. J., Warr, G. W., and Ruben, L. N., 1978, Evolutionary immunobiology and the problem of the T-cell receptor. *Dev. Comp. Immunol.* **2**:203.

Marchalonis, J. J., Warr, G. W., Bucana, C., and Hoyer, L. C., 1979, The immunoglobulin-like T-cell receptor. I. *In situ* demonstration of immunoglobulin Fab-region determinants of rodent T- and B-lymphocytes using chicken antibodies, *J. Immunogenet.* **6**:289.

Maslin, T. P., 1967, Skin grafting in the bisexual teiid lizard, *Cnemidophorus sexlineatus*, and in the unisexual, *C. tesselatus*, *J. Exp. Zool.* **165**:137.

Mattes, M. J., and Steiner, L. A., 1978a, Surface immunoglobulin on frog lymphocytes: Identification of two lymphocyte populations, *J. Immunol.* **121**:1116.

Mattes, M. J., and Steiner, L. A., 1978b, Antisera to frog immunoglobulins cross-react with a periodate-sensitive cell surface determinant, *Nature (London)* **273**:761.

Maung, R. T., 1963, Immunity in the tortoise, *Testudo ibera*, *J. Pathol. Bacteriol.* **85**:51.

May, R. M., 1923, Skin grafts in the lizard, *Anolis carolinensis*, *Br. J. Exp. Biol.* **1**:529.

Metcalf, D., 1965, Delayed effect of thymectomy in adult life on immunological competence, *Nature (London)* **208**:1336.

Metchnikoff, E. L., 1901, *L'immunité dans les maladies infectieuses*, p. 349, Masson, Paris.

Miller, J. F. A. P., 1961, Immunological function of the thymus, *Lancet* **2**:748.

Miller, J. F. A. P., 1974, Role of the cells which originate from the thymus and bone marrow, *Ann. Immunol. (Inst. Pasteur)* **125C**:213.

Mitchison, N. A., 1971, The carrier effect in the secondary response to hapten–protein conjugates. II. Cellular co-operation, *Eur. J. Immunol.* **1**:18.

Mitchison, N. A., Rajewsky, K., and Taylor, R. S., 1970, Co-operation of antigenic determinants and of cells in the induction of antibodies, in: *Developmental Aspects of Antibody Formation and Structure*, Vol. II (J. Sterzl and I. Riha, eds.), pp. 547–564, Academic Press, New York.

Moticka, E. J., Brown, B. A., and Cooper, E. L., 1973, Immunoglobulin synthesis in bullfrog larvae, *J. Immunol.* **110**:855.

Murugan, T. S., 1980, The effect of adult thymectomy and splenectomy on the peripheral lymphocyte population in the lizard, *Calotes versicolor*, M. Phil. thesis, Madurai University, Tamilnadu, India.

Muthukkaruppan, VR., Kanakambika, P., Manickavel, V., and Veeraraghavan, K., 1970, Analysis of the development of the lizard, *Calotes versicolor*. I. A series of normal stages in the embryonic development, *J. Morphol.* **130**:479.

Muthukkaruppan, VR., Pitchappan, RM., and Ramila, G., 1976a, Thymic dependence and regulation of the immune response to sheep erythrocytes in the lizard, in: *Phylogeny of Thymus and Bone Marrow-Bursa Cells* (R. K. Wright and E. L. Cooper, eds.), pp. 185–194, Elsevier/North-Holland, Amsterdam.

Muthukkaruppan, VR., Subramonia Pillai, P., and Jayaraman, S., 1976b, Immune functions of the spleen in the lizard, in: *Immuno-Aspects of the Spleen* (J. R. Battisto and J. W. Streilein, eds.), pp. 61–73, Elsevier/North-Holland, Amsterdam.

Pitchappan, RM., 1975, Studies on the development and immune functions of the thymus in the lizard, *Calotes versicolor*, Ph.D. thesis, Madurai University, Tamilnadu, India.

Pitchappan, RM., 1980, Review: On the phylogeny of the splenic structure and function, *Dev. Comp. Immunol.* **4**:395.

Pitchappan, RM., and Muthukkaruppan, VR., 1976, Procedure for thymectomy in the lizard, *Calotes versicolor*, *Proc. Indian Acad. Sci. Sect. B* **84**:42.

Pitchappan, RM., and Muthukkaruppan, VR., 1977a, In vitro properties of heterologous anti-lizard thymocyte serum, *Proc. Indian Acad. Sci. Sect. B* **85**:1.

Pitchappan, RM., and Muthukkaruppan, VR., 1977b, Role of the thymus in the immune response to sheep erythrocytes in the lizard, *Calotes versicolor*, *Proc. Indian Acad. Sci. Sect. B* **85**:25.

Pitchappan, RM., and Muthukkaruppan, VR., 1977c, Thymus-dependent lymphoid regions in the spleen of the lizard, *Calotes versicolor*, *J. Exp. Zool.* **199**:177.

Pitchappan, RM., and Muthukkaruppan, VR., 1977d, Analysis of the development of the lizard, *Calotes versicolor*. II. Histogenesis of the thymus, *Dev. Comp. Immunol.* **1**:217.

Playfair, J. H. L., 1971, Cell co-operation in the immune response. *Clin. Exp. Immunol.* **8**:839.

Ramila, G., 1978, Studies on cellular interactions in the immune response and antigenic competition in the lizard, *Calotes versicolor*, Ph.D. thesis, Madurai University, Tamilnadu, India.

Raviola, E., and Raviola, G., 1967, Striated muscle cells in the thymus of reptiles and birds: An electron microscopic study, *Am. J. Anat.* **121**:623.

Rothe, F., and Ambrosius, H., 1968, Beitraege zur Immunologie poikilothermer Wirbeltiere. V. Die Proliferation Antikoerper bildender Zellen bei Schildkrieten, *Acta Biol. Med. Ger.* **21**:525.

Ruben, L. N., 1975, Ontogeny and phylogeny of cellular collaboration, *Am. Zool.* **15**:93.

Ruben, L. N., Warr, G. W., Decker, J. M., and Marchalonis, J. J., 1977, Phylogenic origin of immune recognition: Lymphoid heterogeneity and hapten/carrier effect in the gold fish, *Carassius auratus*, *Cell Immunol.* **31**:266.

Sailendri, K., 1973, Studies on the development of lymphoid organs and immune responses in the teleost, *Tilapia mossambica* (Peters), Ph.D. thesis, Madurai University, Tamilnadu, India.

Sailendri, K., and Muthukkaruppan, VR., 1975, Morphology of lymphoid organs in a cichlid teleost, *Tilapia mossambica* (Peters), *J. Morphol.* **147**:109.

Salkind, J., 1915, Contributions histologiques à la biologie comparée du thymus, *Arch. Zool. Exp. Gen.* **55**:81.

Sidky, Y. A., and Auerbach, R., 1968, Tissue culture analysis of immunological capacity of snapping turtles, *J. Exp. Zool.* **167**:187.

Sidky, Y. A., and Auerbach, R., 1975, Lymphocyte-induced angiogenesis: A quantitative and sensitive assay of the graft-vs-host reaction, *J. Exp. Med.* **141**:1084.

Simpson, E., and Cantor, H., 1975, Regulation of the immune response by sub-class of T lymphocytes. II. The effect of adult thymectomy upon humoral and cellular responses in mice, *Eur. J. Immunol.* **2**:114.

Snippe, H., Williams, P. J., Graven, W. G., and Kamp, C., 1975, Delayed hypersensitivity in the mouse induced by hapten–carrier complexes, *Immunology* **28**:897.

Solas, M. T., and Zapata, A., 1980, Gut-associated lymphoid tissues (GALT) in reptiles: Intraepithelial cells, *Dev. Comp. Immunol.* **4**:87.

Stolen, J. S., and Makela, G., 1975, Carrier preimmunization in the anti-hapten response of a marine fish, *Nature (London)* **254**:718.

Streilein, J. W., and Billingham, R. E., 1970, An analysis of the genetic requirements for delayed cutaneous hypersensitivity reactions to transplantation antigens in mice, *J. Exp. Med.* **131**:409.

Subramonia Pillai, P., 1977, Studies on the role of antigen-binding cells in the immune response to sheep erythrocytes in the lizard, *Calotes versicolor*, Ph.D. thesis, Madurai University, Tamilnadu, India.

Subramonia Pillai, P., and Muthukkaruppan, VR., 1977, The kinetics of rosette-forming cell response against sheep erythrocytes in the lizard, *J. Exp. Zool.* **199**:97.

Tamura, S. I., and Egashira, Y., 1975, Cellular and humoral immune responses in mice. II. Effect of intraperitoneal or subcutaneous injection of carrier on anti-hapten antibody and delayed hypersensitivity responses, *Immunology* **28**:909.

Taylor, R. B., and Wortis, H. H., 1968, Thymus dependence of antibody response: Variation with dose of antigen and class of antibody, *Nature (London)* **220**:927.

Terebey, N., 1972, A light microscopic study of the mononuclear cells infiltrating skin homografts in the garden snake, *Thamnophis sirtalis* (Reptilia: Colubridae), *J. Morphol.* **137**:149.

Thorbecke, G. J., Romano, T. J., and Lerman, S. P., 1974, Regulatory mechanisms in proliferation and differentiation of lymphoid tissue, with particular reference to germinal center development, in: *Progress in Immunology II* (B. Amos, ed.), pp. 25–34, Academic Press, New York.

Tochinai, S., and Katagiri, C., 1975, Complete abrogation of immune response to skin allografts and rabbit erythrocytes in the early-thymectomized *Xenopus*, *Dev. Growth Differ.* **17**:383.

Warner, N. L., 1974, Membrane immunoglobulins and antigen receptors on B and T lymphocytes, *Adv. Immunol.* **19**:67.

Wetherall, J. D., and Turner, K. J., 1972, Immune response of the lizard, *Tiliqua rugosa*, *Aust. J. Exp. Biol. Med. Sci.* **50**:79.

Worley, R., and Jurd, R. D., 1979, The effect of a laboratory environment on graft rejection in *Lacerta viridis*, the European green lizard, *Dev. Comp. Immunol.* **3**:653.

Wright, R. K., Eipert, E. F., and Cooper, E. L., 1978, Regulating role of temperature on the development of ectothermic vertebrate lymphocyte populations, in: *Animals Models of Comparative and Developmental Aspects of Immunity and Disease* (M. E. Gershwin and E. L. Cooper, eds.), pp. 80–92, Pergamon Press, Elmsford, N.Y.

Wright, R. K., and Shapiro, H. C., 1973, Primary and secondary immune responses of the desert iguana, *Dipsosaurus dorsalis*, *Herpetologica* **29**:275.

Yamaga, K. M., Kubo, R. T., and Etlinger, H. M., 1978a, Studies on the question of conventional immunoglobulin on thymocytes from primitive vertebrates. II. Delineation between Ig-specific and cross-reactive membrane components, *J. Immunol.* **120**:2074.

Yamaga, K. M., Kubo, R. T., and Etlinger, H. M., 1978b, Studies on the question of conventional immunoglobulin on thymocytes from primitive vertebrates. I. Presence of anti-carbohydrate antibodies in rabbit anti-trout Ig sera, *J. Immunol.* **120**:2068.

Yocum, D., Cuchens, M., and Clem, L. W., 1975, The hapten–carrier effect in teleost fish, *J. Immunol.* **114**:925.

Zakarian, S., and Billingham, R. E., 1972, Studies on normal immune lymphocyte transfer reactions in guinea pigs, with special reference to the cellular contribution of the host, *J. Exp. Med.* **138**:1545.

13

RES Structure and Function of the Aves

BRUCE GLICK

1. INTRODUCTION

In ancient times, prior to the first sundial, the transition of night into day was marked by the crowing of the rooster (Lind, 1963). Perhaps this was only a prelude to the major role that the chicken was to play in the 20th century in awakening the scientific community to fundamental concepts that would help formulate immunobiology and the T and B cell concept (Glick *et al.*, 1956; Warner and Burnet, 1961; Cooper *et al.*, 1966; Weber, 1975; Glick, 1977a). While this chapter will include historical milestones in avian immunology, the major thrust will be to emphasize structural and functional characteristics of the lymphomyeloid complex and the interrelationships of these representatives of the chicken's RES.

2. MORPHOLOGICAL AND STRUCTURAL STUDIES

2.1. BURSA OF FABRICIUS

The bursa of Fabricius is a dorsal diverticulum of the proctadael region of the cloaca (Jolly, 1915) (Figure 1). In the chicken, it is round or oval (Figure 2), whereas in the duck, it is elongated (Jolly, 1915; Glick, 1963). A thin layer of connective tissue and smooth muscle form the outer serosal layer of the bursa (Retterer, 1885; Jolly, 1915, Boyden, 1922; Calhoun, 1933). The anlage of the bursa appears about day 5 of embryonic (E) development (Romanoff, 1960), and by day 13, the epithelial cells that line the numerous plicae thicken and extend into the tunica propria as epithelial buds where subsequently lymphopoiesis occurs (Ackerman and Knouff, 1959; Ackerman, 1962). The ontogeny of the

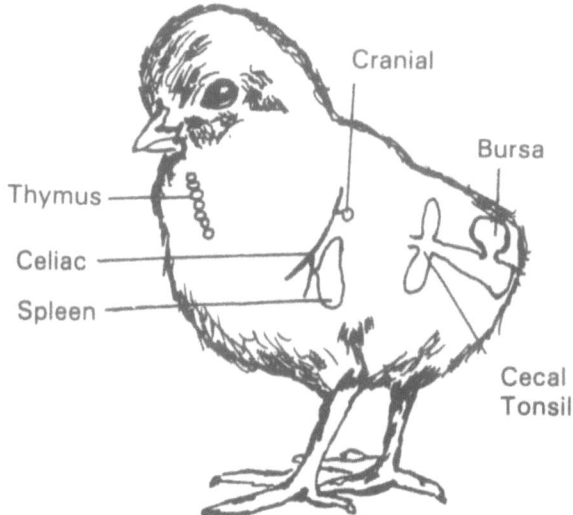

FIGURE 1. Immunobiological tissue of the chick. Cranial identifies an accessory spleen along the celiac artery. (From Glick, 1970, *BioScience* **20**:602; reproduced with permission from the American Institute of Biological Sciences.)

avian bursal and thymic lymphocytes will not be discussed because it has been exquisitely described in this volume by Le Douarin and co-workers (Chapter 16). The bursal follicle, the repository of bursal lymphocytes, appears after day 15 E. At this time, the lumen of the bursa is lined with 11 to 14 primary folds or plicae and 6 or 7 secondary plicae (Figure 3) (Jolly, 1915; Ackerman and Knouff, 1959;

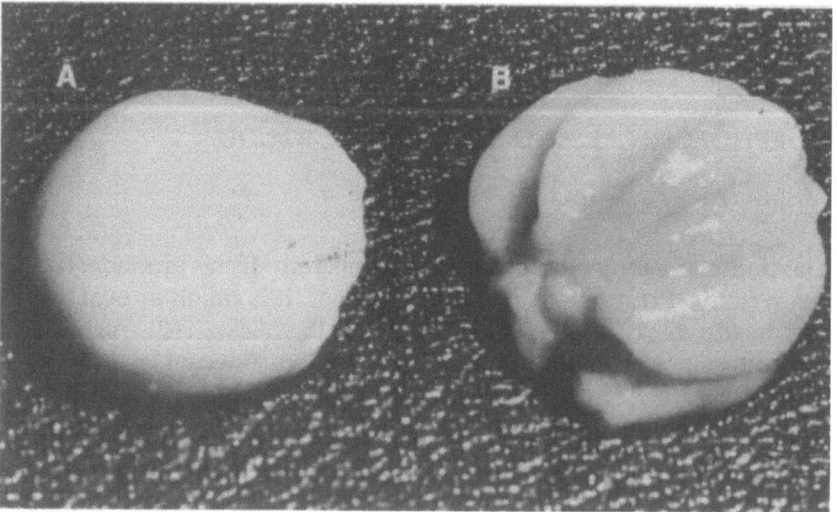

FIGURE 2. Bursa of Fabricius from a 2-week-old chicken. (A) A freshly removed bursa; (B) the opened bursa, revealing numerous plicae. [From Glick, 1964, in: *The Thymus in Immunobiology* (R. A. Good and A. Gabrielson, eds.), pp. 343–358; reproduced with permission from Harper & Row.]

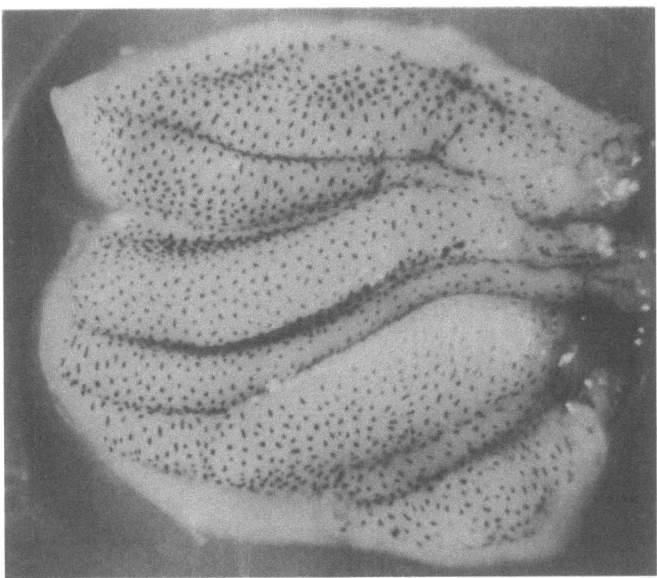

FIGURE 3. A stereo micrograph of the bursal plicae subsequent to a vent application of carbon. The follicular epithelia appear as black dots. × 4.8. (From Oláh and Glick, 1978a; reproduced with permission from the Poultry Science Association.)

Oláh and Glick, 1978a). Each bursal follicle is divided into a cortex and medulla by epithelial cells whose basal lamina is continuous with that of the epithelial cells lining the plicae (Jolly, 1915; Boyden, 1922; Ackerman and Knouff, 1959). Transmission electron microscopy (TEM) revealed two kinds of surface epithelium (Bockman and Cooper, 1973). One type immediately above the follicle, follicle-associated epithelium, had micropinocytotic activity. Employing scanning electron microscopy (SEM), we demonstrated the presence of two distinct types of surface epithelium (Holbrook et al., 1974). The epithelium which covered the follicle, follicular epithelium (FE), was polygonal, sparsely covered with microvilli, and possessed a well-delineated border (Figure 4). On the other hand, the epithelium present between follicles, interfollicular epithelium (IFE), was irregularly shaped, revealed pitlike depressions, contained a florid cover of microvilli, and lacked distinct cellular borders (Figure 5). The embryonic development of Fe and IFE has not been studied extensively. Naukkarinen et al. (1978) reported differentiation of both types of epithelium between day 13 and 15 E. Our data agree that FE appears at this time, but we suggest a somewhat later development for the IFE (Glick et al., 1977a). Apical vacuoles appeared on day 15 E in the FE (Naukkarinen et al., 1978). The vacuoles of the IFE, unlike those of the FE, contained mucin. Microvilli appeared on the surface of the IFE, and invaginations of the apical plasma membrane occurred in the FE on day 17 E. The latter suggested that the endocytotic potential for FE begins approximately at day 17 E. However, active endocytosis of carbon particles was not observed at this time and it was only rarely seen at day 19 E. The greater activity of succinate,

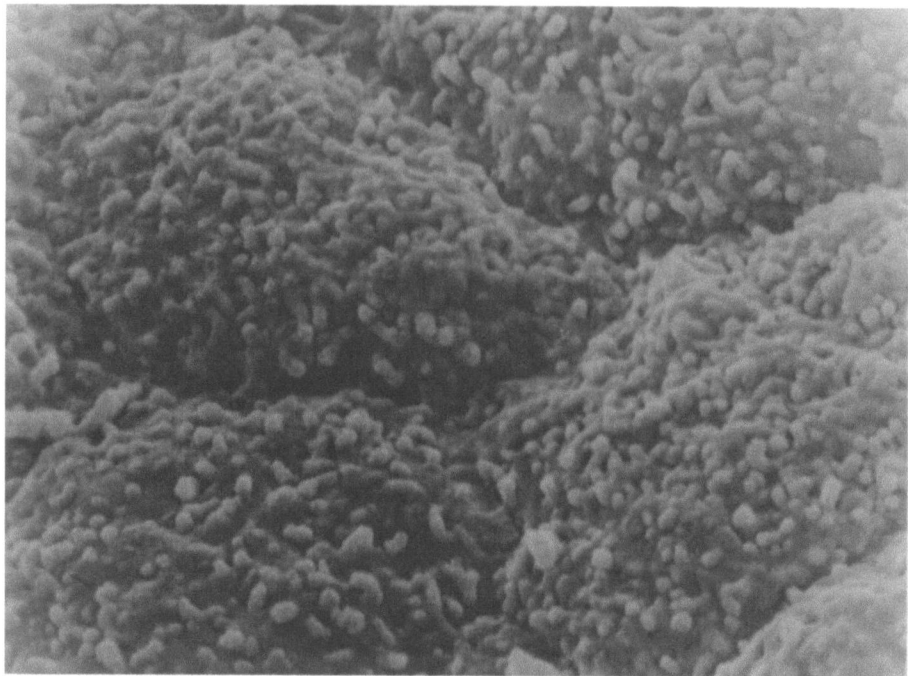

FIGURE 4. The follicular epithelium which covers bursal follicles is polygonal, sparsely covered with microvilli, and possesses a well-delineated border. × 8500.

lactate, and glucose-6-phosphate dehydrogenase and ATPase in the IFE than in the FE may allow enzymatic differentiation of these two sites (Ruuskanen et al., 1977). Normal differentiation of FE is not dependent on contact with the gastrointestinal tract since morphological and functional development of FE occurred in 10- to 12-day E bursa transplanted to the chorioallantoic membrane (Bockman, 1979).

We have revealed that the surface of the bursal plicae possesses two types of bursal follicles (Holbrook et al., 1974). The buttonlike follicles (BLF) are covered by a continuous layer of epithelium and separated from the IFE by a cryptlike depression. The projecting follicles (PF) possess a disrupted surface epithelium and lack crypts. The existence of both types of follicles in posthatched chicks led us to speculate concerning the ontogeny of each follicle. SEM of the embryonic bursa revealed the earlier appearance of PF, with BLF predominating as the bird aged (Glick et al., 1977a). We were unable to determine if there was a transition from PF to BLF. The events occuring prior to lymphoid follicle development are clear. For example, the culturing of 10-, 11-, and 12-day E bursal explants for 4 to 7 days demonstrated that epithelial thickening must occur before the appearance of basophilic follicles which preceded the appearance of lymphoid follicles (Ritter and Lebacq, 1977).

Estimates of the number of bursal follicles per fold range from 40 to 60 (Wenckebach, 1896; Boyden, 1922). We have addressed the question of the number and geometry of bursal follicles by employing the original observation of Schaffner *et al.* (1974) that carbon applied to the vent of a chicken will be internalized, and it is the FE and not IFE that exhibits pinocytosis (Bockman and Cooper, 1973). Anal sucking can be induced in 15-day embryos, but is spontaneous in chicks at hatching (Sorvari *et al.*, 1977). A carbon solution (64 mg/ml) was applied to the vent (0.1 ml) of 4-week-old chickens, and 2 hr later the bursa was removed (Oláh and Glick, 1978a). The number of FE areas was counted on six folds (Figure 3) and found to average 800 per fold. Each bursa contains 10 to 15 folds and therefore possesses 8000 to 12,000 follicular areas per bursa. A trapezoid was outlined on a given fold and its area determined as 6.6 mm^2. Within this defined geometry, the area of each FE was measured, and the mean was 0.6 mm^2. Thus, about 10% of the fold is FE and 90% IFE. Viewing the folds as a prism of trapezoid base and a pyramid of rectangular base, bursal fold dimensions and bursal surface areas were calculated. The bursal surface area was about 10 cm^2, and since 10% of the area is occupied by FE, then one might conclude that approximately 1 cm^2 of the bursal surface areas is immunologically oriented. Damage to the immunologically oriented surface, FE, by cloacal applications of ^{239}PU (Schaffner *et al.*, 1976) or cyclophosphamide injections into newly hatched chicks (Sachs *et al.*, 1979) was translated functionally by a failure of the bursa-treated birds to concentrate carbon previously applied to the cloaca. Further emphasis on the immunological role of the FE comes from the research

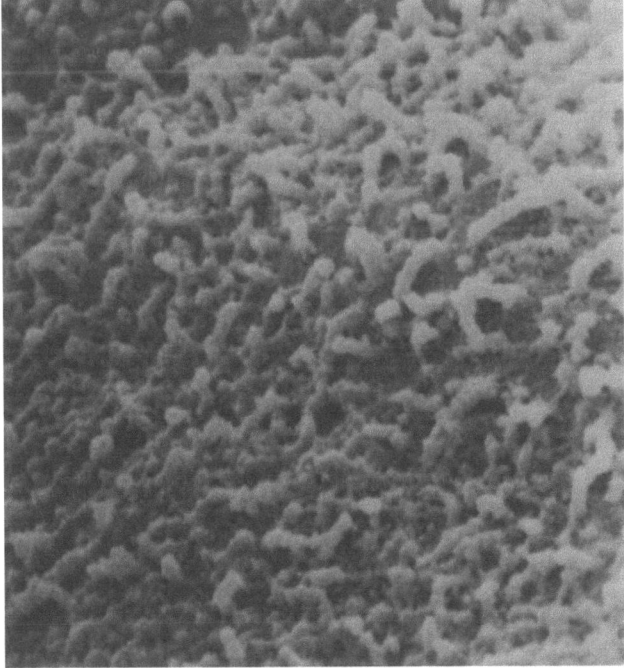

FIGURE 5. The interfollicular epithelium of the bursal folds reveals pitlike depressions, contains a florid cover of microvilli, and lacks distinct cellular borders. × 8500.

of Bockman and Stevens (1976). Horseradish peroxidase (HRP) administered intraluminally (bursa) and intravenously was concentrated in FE and, at least, after intraluminal administration the HRP was shed into the lumen. Similar results were observed with rabbit appendix. Perhaps the movement of HRP out of the FE may be similar to immunocompetent units of the bursa. Stimulation by way of FE would elicit a positive response from the cellular elements of the bursal follicle, namely, antibody or immunoglobulin production, and their release into the lumen. Our observations of lymphocytes on the surface of the plicae suggests a luminal migration of cells (Holbrook et al., 1974). There are numerous reports of an antibody response by bursal lymphocytes following the direct introduction of antigen into the bursal lumen or by anal application (Van Alten and Meuwissen, 1972; Sorvari et al., 1975; Sorvari and Sorvari, 1977). Anal application of sheep red blood cells (SRBC) or *Brucella abortus* (BA) will stimulate the production of antibody to each antigen (Sorvari and Sorvari, 1977). Surgical bursectomy at 10 weeks followed by anal lip application of either SRBC or BA inhibited the response only to BA. This suggests that the bursa was directly involved in BA immunity and that some other site (e.g., cecal tonsil) might be utilized in the response to SRBC. It is noteworthy that in birds lacking a bursa, the anal sucking is retained, but more of the imbibed material is directed to the cecum (Sorvari et al., 1977). While BA could only be found in the bursal lumen, its failure to appear in the bursa medulla emphasizes a possible role for lymphocytes which we have observed on the bursal fold. There is evidence from intrabursal priming that bursal cells mature earlier for BA (day 4 posthatch) than for *E. coli* and that intrabursal priming could not potentiate an antibody response to *S. pullorum* (Matsuda et al., 1976). The importance and plausibility of intrafollicular involvement in the immune response is emphasized by the presence of contact sites (characterized by subplasmalemmal densities and coated pits) between lymphocytes and between lymphocytes and macrophages within the cortex of the bursa (Holbrook et al., 1977). Numerous experimenters have timed the rate of passage from the anal lip to the bursa (Schaffner et al., 1974; Sorvari and Sorvari, 1977; Sorvari et al., 1977; Glick, 1977b). A single application of colloidal carbon to the anal lip was followed by its appearance in the lumen in 2 min, FE in 15 min, medulla in 6 hr, and cortex and interfollicular stroma in 72 hr (Sorvari and Sorvari, 1977). Application of 0.1 ml tritiated thymidine ([^{3}H]-TdR, 100 µCi) to the anal lip of 3-week-old chickens revealed a progressive increase in the uptake by the bursa for 6 hr (Glick, 1977b). No other site with the exception of the large intestine showed an equivalent concentration of [^{3}H]-TdR. The [^{3}H]-TdR was utilized by the cells of the bursal follicle since 24.4% of the bursal small lymphocytes were labeled within 6 hr.

We have shown by SEM that lymphocytes often reside on the surface of the folds and that these lymphocytes are smooth (Holbrook et al., 1974). These original observations have been confirmed and extended by our SEM of single-cell suspensions of thymic, bursal, and splenic lymphocytes (Stinson and Glick, 1977). More than 90% of the lymphocytes from the thymus and bursa were smooth with only an occasional cell showing intermediate or numerous microvilli. Marked differences between thymic and bursal cells were not noted by

Nazerian *et al.* (1976). These data suggest that distinct surface differences cannot be revealed by SEM between lymphocytes residing in the thymus and bursa. Of course, the intermingling of cells in peripheral lymphoid tissue could lead to surface alterations.

Internal differences may exist between T and B cells. Tubular structures, anastomosing tubules in endoplasmic reticulum, were observed in bursal lymphocytes from a 12-day-old chicken (Matos and Sousa, 1978). They were not associated with viral particles. The significance of the tubular structures is not clear unless its occurrence is characteristic of B lymphocytes since it was induced in a B-cell line. It would be interesting to determine if tubular structures exist in a diffusely infiltrated area that is dorsal to the bursal duct since it has been suggested that this area contains T cells (Odend'hal and Breazile, 1979).

The bursa of Fabricius receives its innervation from sympathetic fibers, the pelvic nerve, and intestinal nerves that enter the first bursacloacal ganglion at the anterior pole of the bursa (Pintea *et al.*, 1967; Cordier, 1969). The major blood vessels are the pudendal arteries and veins and posterior mesenteric vein (Pintea *et al.*, 1967). Lymphatics have been observed draining the bursal area (Dransfield, 1945). We have observed lymphatics in bursal sections (unpublished). An early observation of carbon appearing in the cloaca and bursa 24 hr after an intratesticular injection of carbon might be explained by lymphatics (Atwal and McFarland, 1971).

2.2. THYMUS

The thymus develops from the third and fourth pharyngeal pouches (Hammond, 1954; Ruth *et al.*, 1964). Two sets of seven lobes each make up the thymic tissue. They are located on either side of the neck extending from the lower jaw to the thorax where the last may be partially embedded in the thyroid (Payne, 1971). Cortical and medullary components are evident in the thymus. On approximately day 11 E, lymphocytes appear in the thymus (Lucas and Jamroz, 1961). The predominantly large lymphocytes (11 µm) of day 11 thymus are replaced by 8-µm cells between days 11 and 13 (Sugimoto *et al.*, 1977a). The lymphocytes became progressively smaller between days 13 and 15 (8–7.5 µm), and by day 16 E, lymphocytes of 5.5 µm predominate. These data on decreasing thymic size during embryonic development agree with previous reports (Sherman and Auerbach, 1966; Peterson and Good, 1965). Embryonic thymus-specific antigen (T-Ag) appeared at day 12 or 13 E (Sugimoto *et al.*, 1971a). The presence of T-Ag correlated with morphological changes in thymic lymphocytes. Cytoplasmic processes were evident at day 11 or 12 E, and by day 12 the presence of heterochromatin became more distinct. On day 13 E, the cells rounded, lost cytoplasmic processes, and showed distinct heterochromatin. The loss of cytoplasm, similar to the occurrence in spermiogenesis, was suggested as a possible prefunctional step in the life of the thymic lymphocyte. At day 12 or 13 E, reticuloepithelial cells of the thymus showed the presence of electronlucent vacuoles and dense bodies (Sugimoto *et al.*, 1977b). A fusion of some of the latter bodies also occurred. Because these morphological changes paralleled the period of embryonic T-Ag, it was theorized that material in the inclusion vacuoles is released and triggered the events leading to T-Ag differentiation by the thymic lymphocytes. The authors postulated that the material in the inclusion vacuoles might have systemic or local ramifications. Therefore, the action of the humoral substance would either be on prethymic cells which require conditioning in the microenvironment of the thymus before responding to thymic humors or on postthymic cells which have experienced the microenvironment of the thymus and are capable of responding to humoral thymic factors (Stutman et al., 1970a,b). Terminal deoxynucleotidyl transferase (Tdt) directs the synthesis of DNA sequences without benefit of template (Bollum, 1974). Tdt appears in the embryonic thymus (Penit and Chaperille, 1977) as early as day 12 E (Sugimoto and Bollum, 1979). It is rare in bursal cells. The appearance of Tdt in the thymus at a time when lymphocytes are being produced suggests that it may be important in cellular differentiation of the thymus (Sugimoto and Bollum, 1979).

Although thymic and bursal development are independent events, there is now evidence for some reciprocal involvement between the thymus and the bursa in the developmental process. Jankovic and Isakovic (1964) demonstrated that thymectomy at hatching significantly reduced the small-lymphocyte population in the bursa of Fabricius and spleen. The latter might be expected since it receives a high percentage of its cells from the thymus, but the bursa has been considered somewhat autonomous and possesses primarily surface immunoglobulin-positive (sIg^+) cells (Glick, 1977a). Bursa from 10-day-old normal chicks reconstituted antibody potential and bursal morphology of cyclophosphamide-treated chicks (Hirota and Bito, 1978). However, thymectomized, cyclophosphamide-treated, irradiated chicks did not respond to the bursal transfers, i.e., their bursal remnants remained atrophic. These data suggested that either T cells or the thymus are necessary for the regeneration of the bursa. Since Toivanen et al. (1972) found 1- or 3-day-old bursal cells to be the most effective in adoptive transfer reconstitution studies, Hirota and Bito's observations may need further evaluation. Yet, their data do confirm that thymectomy reduced the medullary cell development in the bursa (Jankovic and Isakovic, 1964). That the bursa may control thymic development has been suggested by Medina and Pederneva (1976), who report that *in ovo* bursectomy (BSX) at 62-hr E leads to reduction in allograft response. Further assessment of the bursal control of thymic development comes from Fitzimmons et al. (1977). *In ovo* BSX at 72 hr reduced thymic size by 70% and the amount of cortical mass of 18-day embryos. Hormonal BSX between days 2 and 5 and *in ovo* BSX at 72 hr significantly reduced thymocyte count of 18- and 19-day embryos. Thymocyte count was not influenced by hormonal BSX after day 5 E. Thus, the bursa's influence on thymic development is exerted early in its ontogeny. *In ovo* BSX after day 3 should be performed to verify the latter statement and eliminate the onus of the peripheral effects of the hormone. The allograft response to *in ovo* BSX was deficient. Loss of thymocytes by hormonal BSX on day 2 to 5 E, but not later, would tend to agree with our observation that occasionally embryos treated with testosterone propionate prior to day 5, but not later than day 11, exhibited increased thymocyte activity (Mitlin and Glick, 1980). The increased activity may be an attempt to compensate for the reduced cell numbers.

2.3. SPLEEN

The spleen is a reddish-brown oval or round structure located dorsal to the proventriculus (Figure 6). In our strain of New Hampshire chicken, one-third of the birds possessed accessory spleens that were located cranial, adjacent, or caudal to the spleen (Glick and Sato, 1964). Following splenectomy, the cranial accessory spleen hypertrophied, often reaching a mean weight of 203 mg (Glick, 1970). The spleen appears to grow most rapidly during the first 6 weeks after hatching, with the maximum spleen-to-body weight ratio attained by 10 weeks of age (Norton and Wolfe, 1949; Wolfe et al., 1962).

The splenic white pulp is associated with arterioles and appears to be less clearly differentiated from the red pulp than in mammals (Jankovic and Isakovic, 1964; Payne, 1971). The central arterioles are surrounded by periarteriolar lymphatic sheaths (PAL) (Taliaffero and Taliaffero, 1955; Hoffman-Fezer et al., 1977). The PAL is populated by thymus-derived lymphocytes with occasional B cells located between the numerous T cells (Hoffman-Fezer et al., 1977). The penicillate arteries, radiating from the central artery, lead into arterial capillaries which are encapsulated by the Schweiger–Seidel sheaths (SSS) (Taliaffero and Taliaffero, 1955; Hoffman-Fezer et al., 1977). These arterial capillaries terminate in the red pulp. According to Hoffman-Fezer et al. (1977), the periellipsoid lymphatic (PE) tissue surrounding the SSS is populated by B lymphocytes. Germinal centers, with their predominant B-cell population, are adjacent to central arterioles. The white pulp includes the PAL, PE, germinal centers, and the cords of Billroth which extend into the red pulp and contain lymphocytes, macrophages, blast cells, granulocytes, reticulum cells, and plasma cells (Taliaffero and Taliaffero, 1955). Plasma cells may also be seen in the vicinity of germinal centers and arteries. On the basis of a cytotoxic test 60 to 70% of the splenic lymphocytes are T cells and 40 to 50% are B cells (Hoffman-Fezer et al., 1977). In general, these values agree with other investigators who found more T than B cells in the spleen (Hudson and Roitt, 1973; Potworowski, 1972). Central T and B cells may differ antigenically from the peripheral T and B cells of the spleen and peripheral blood (Schauenstein, 1979). The T and B cells of the latter sites contain a new determinant, peripheral lymphocyte-specific antigen. The presence of this new antigen on peripheral T and B cells may suggest a previous suppression by the microenvironment of the bursa and thymus or stimulation by humoral or cellular interactions in the periphery leading to biochemical changes on the surface membranes of lymphocytes.

2.4. CECAL TONSIL

Another major structure possessing RE cells is the cecal tonsil (Muthmann, 1913; Looper and Looper, 1929). The cecal tonsil is located in the proximal 6 to 8

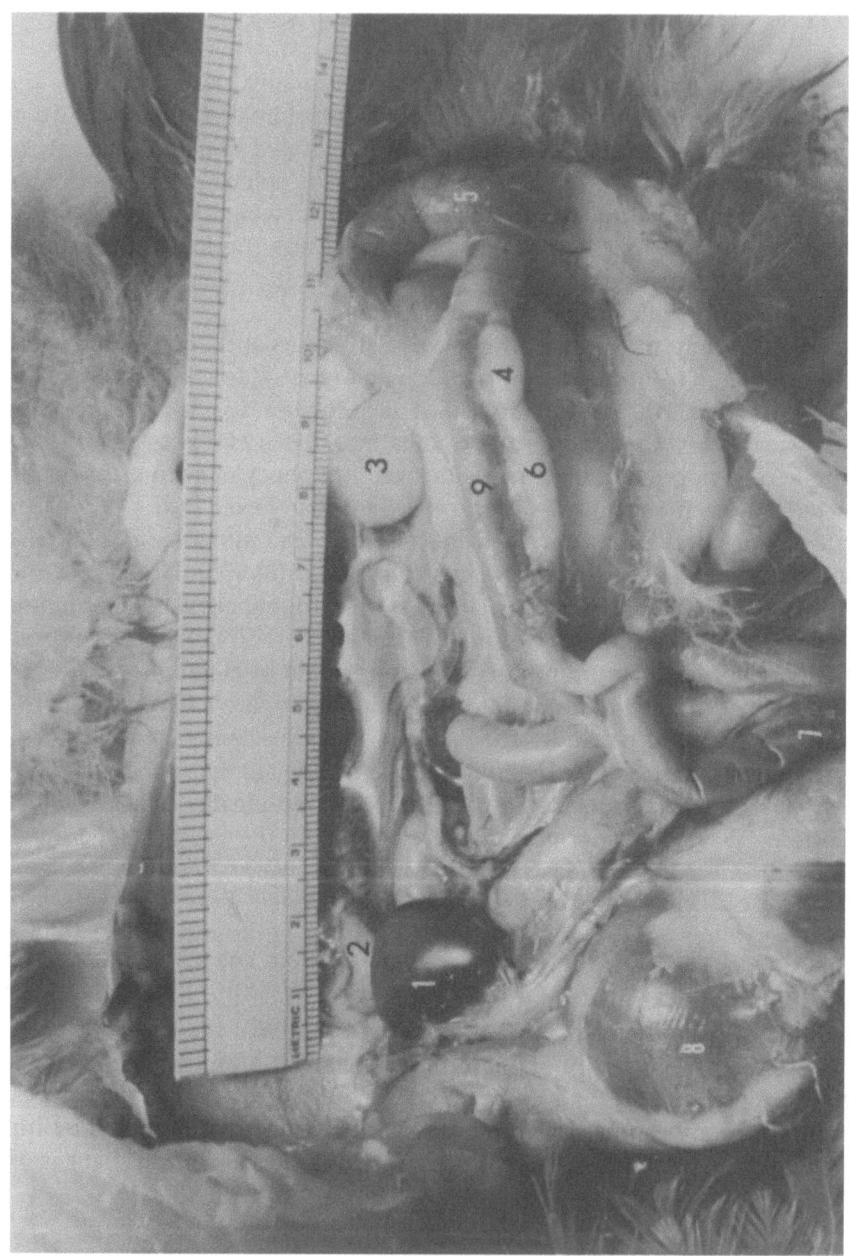

FIGURE 6. The exposed soft tissues in the abdominal cavity of the chicken. (1) Spleen; (2) testes; (3) bursa of Fabricius; (4) cecal tonsil; (5) large intestine; (6) cecal pouch; (7) cecal pouch's distal portion; (8) gizzard; (9) ileum.

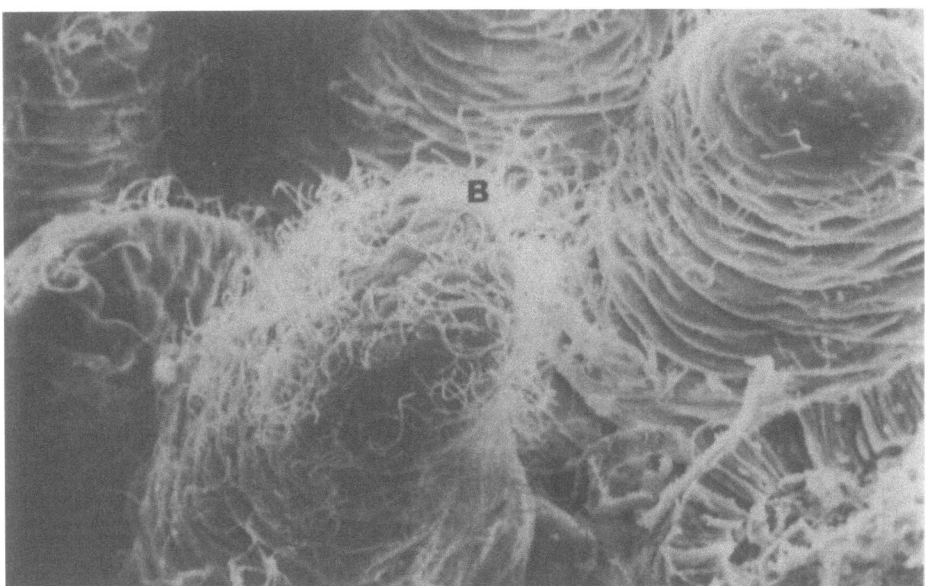

FIGURE 7. Bacteria (B) are embedded in the epithelium of villi projecting from the cecal tonsil. Scanning electron micrograph. × 266.

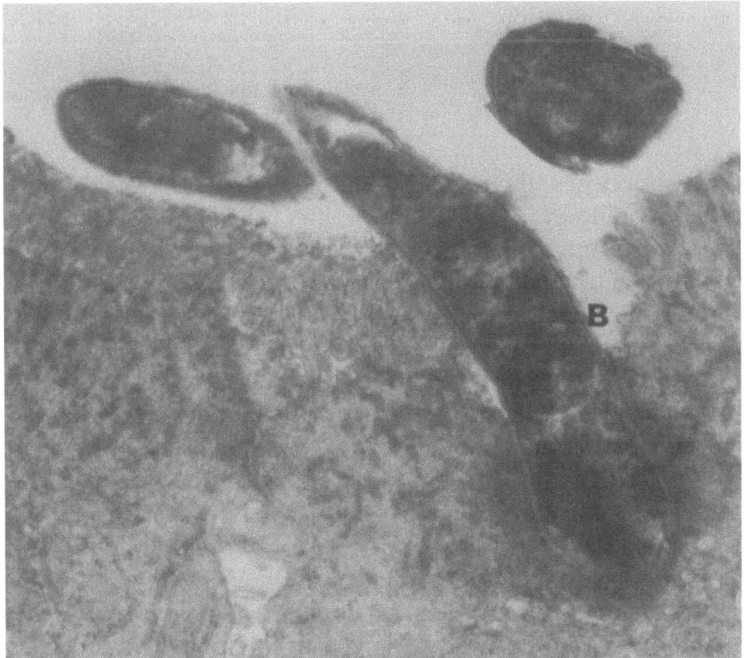

FIGURE 8. Bacteria (B) are attached to the epithelium of the cecal tonsil. × 31,500. (From Glick *et al.*, 1978; reproduced with permission from the Poultry Science Association.)

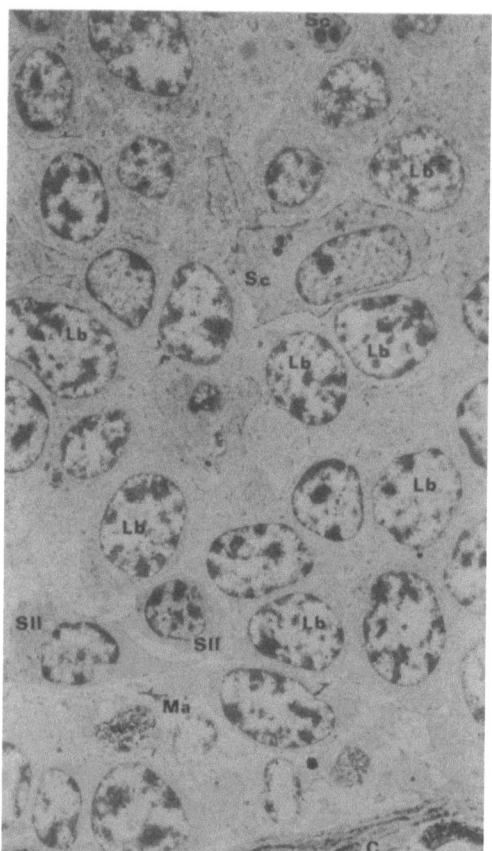

FIGURE 9. The majority of cells in the cortical zone of a germinal center in the cecal tonsil are lymphoblasts (Lb). Macrophage (Ma), small-lymphocyte-like cells (SIL), and secretory cells (Sc) with dense granules are present. The Sc are well defined by the intercellular substance. C, capsule of the germinal center. × 4200. (From Oláh and Glick, 1979; reproduced with permission from the Poultry Science Association.)

mm of the cecal pouches (Figure 6). The luminal view of the cecal tonsil reveals numerous villi (Calhoun, 1933; Whitlock et al., 1975; Glick et al., 1978) with bacteria attached (Figures 7 and 8). Between the villi are mounts filled with lymphatic tissue. Encapsulated and partially encapsulated germinal centers have been identified in the cecal tonsil (Oláh and Glick, 1979). The cortical portion of the germinal centers is rich in small lymphoblasts with occasional macrophages, small-lymphocyte-like cells, and secretory cells (Figure 9). The small lymphoblasts exhibited a greater mitotic activity than the large lymphoblasts of the germinal center medulla, and 30 min subsequent to a topical application of [^{3}H]-TdR, only the small lymphoblasts of the cortex were heavily labeled. Therefore, our data suggest a transition from small to large lymphoblasts within the germinal centers of the cecal tonsil (Oláh and Glick, 1979). The decline in mitotic activity of germinal centers from 2.2 to 0.4% from 4 weeks to 1 year suggests an aging factor for the cecal tonsil. The secretory cells of the cecal tonsil are similar to secretory cells observed in the medullary area close to the corticomedullary border of bursal follicles in normal and cyclophosphamide-treated birds (Oláh and Glick, 1978b; Oláh et al., 1979). There appear to be bursa-

and thymus-dependent areas in the cecal tonsil 6 days after hatching (Hoshi and Mori, 1973), and these are more numerous in immunized birds (Jankovic *et al.*, 1966). The cecal tonsil is capable of producing antibody to several soluble antigens (Jankovic, 1968; Orlans and Rose, 1970).

2.5. HARDERIAN GLAND

The RES extends to the paraocular and paranasal tissue (Bang and Bang, 1968; Aitken and Survashe, 1977). The Harderian gland is the most prominent of these structures (Kennedy, 1970). In the chicken, it is located ventral and posterior-medial to the eyeball and is a compound tubuloacinar structure (Wight *et al.*, 1971a; Rothwell *et al.*, 1972). Three types of Harderian gland have been identified based on their lobular structure: Type I, compound tubuloacinar; Type II, compound tubular; and Type III, mixed compound tubuloacinar and compound tubular (Figure 10) (Burns, 1975; Aitken and Survashe, 1977). While chickens belong to Type I, Burns (1975) reports that Type III is characteristic of the more evolutionarily advanced birds. The peripheral portion of the lobules exhibits a collection of acini which are connected to secondary tubules by tertiary tubules (Wight *et al.*, 1971a). The secondary tubules, like the acini, are lined by columnar epithelium and lead to a single collecting tubule which extends the length of the "straplike" Harderian gland. Light and electron microscopy have revealed the epithelial cells lining the Harderian gland ducts to be similar to ducks, fowl, and turkeys (Burns and Maxwell, 1979; Maxwell and Burns, 1979). The type of secretion is merocrine (Burns, 1975; Kuhnel and Beier, 1973). The mucosecretion of the duck Harderian gland has been characterized as sialomucins and sulfomucins (Wight *et al.*, 1971b; Wight and MacKenzie, 1974).

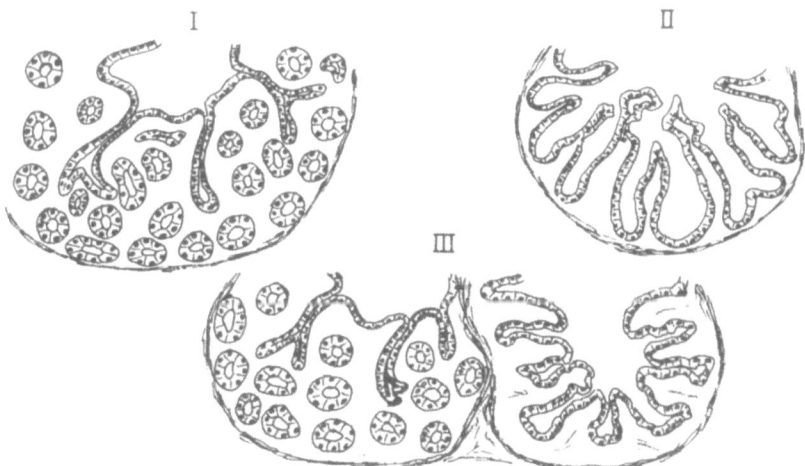

FIGURE 10. Three morphological types of Harderian glands have been described by Burns (1975): Type I, compound tubuloacinar; Type II, compound tubular; and Type III, a mixture of Types I and II. (From Aitken and Survashe, 1977; reproduced with permission from Pergamon Press.)

With the exception of the duck, plasma cells are prominent in the Harderian gland of most birds (Bang and Bang, 1968; Kuhnel and Beier, 1973; Burns, 1975; Glick et al., 1977b). The presence of Russell bodies in plasma cells increases with age (Wight et al., 1971a). Chicken plasma cells, unlike murine plasma cells, have a high density of Ig receptors (Glick et al., 1977b). The percentage of sIg$^+$ plasma cells exceeds 90% in 8-month-old chickens, while the sIg$^+$ small lymphocyte in the Harderian gland declines from approximately 12% at 14 weeks to less than 2.0% at 8 months. Whereas vaccination with infectious bronchitis virus by intranasal and conjunctival routes will increase the number of Harderian gland plasma cells (Davelaar and Kouwenhoven, 1976), birds fed a diet deficient in vitamin A and infected with Newcastle disease virus have a deficit of plasma cells of the Harderian gland (Bang et al., 1975). The lacrimal and Harderian glands of the duck have the potential to convert corticosterone to 11-dehydrocorticosterone (Butler et al., 1978). The significance of this is unclear, unless the Harderian gland may become a salt-secreting gland since the latter glands bind corticosterone and convert it to 11-dehydrocorticosterone.

2.6. LYMPH NODULES

For a detailed discussion of the avian lymphatic system, the reader is referred to Kampmeier (1969) and Dransfield (1945). Lymph nodes appear to be present in water, marsh, and shore birds but not chickens or higher species of birds (Payne, 1971). On the other hand, mural lymphoid nodules associated with lymph vessels have been described in numerous species of birds including the chicken (Kondo, 1937; Biggs, 1957). The nodules appeared to be formed by the condensation of lymphocytes along the wall of the lymphatic vessel. The majority of the mural nodules ranged between 0.3 and 0.5 mm in diameter (Biggs, 1957). The observation of Good and Finstad (1967) that footpad injection of Freund's adjuvant with antigen would enlarge the nodules has been confirmed in our laboratory (McCorkle et al., 1979). Also, our data revealed an increase in germinal centers and plasma cells, more B (53%) than T (38%) cells, and the production of plaque-forming cells in femoral lymph nodules following a foodpad injection of SRBC (McCorkle et al., 1979).

One should not be surprised to find aggregations of lymphocytes residing comfortably in diverse locations of the chicken's anatomy. Our interest has been to detect a role for such lymphocytes. Cogburn has systematically studied the ontogeny and function of pineal lymphocytes (Cogburn and Glick, 1979a,b). Lymphocytes were not observed in the pineal parenchyma at 2 days of age, but lymphoblasts and lymphocytes were present in the vessels of the pineal. Germinal centers were evident by 2 weeks and maximal lymphocyte accumulation in the parenchyma by 4 weeks. By 3–4 months of age, the lymphocytes had disappeared from the pineal parenchyma. Bursectomy–thymectomy and irradiation reduced the number of lymphocytes. While plaque-forming cells were not produced to SRBC, antibody to bovine serum albumin (BSA) was demonstrated in the plasma cells and plasmacytoid cells of the pineal. Failure of the pineal to respond to SRBC was not a reflection of a deficiency of T cells since the pineal possessed 51% T cells and 42% B cells (Cogburn and Glick, 1979b).

3. FUNCTIONAL STUDIES

3.1. Fc RECEPTORS

It is well known that the bursa contains the largest pool of sIg$^+$ cells (Glick, 1977a). Recent evidence has revealed that B cells (Ewald et al., 1976a; Ewald and Sanders, 1977; de Kruyff et al., 1977) and other mononuclear cells (Chiao et al., 1978; Duncan and McArthur, 1978) of the chicken possess Fc receptors. Sensitized SRBC were bound as early as day 12 E by splenic and bone marrow cells and at day 13 E by bursal cells (Ewald et al., 1976b). These rosette-forming cells (RFC) increased markedly only in the embryonic bursa so that by day 17 there were 36.6% RFC in the bursa. At hatching, the RFC had declined to 3.2% in the bursa. These data demonstrated the presence of Fc receptors on chicken lymphocytes. The presence of RFC in spleen and bone marrow prior to their appearance in the bursa might suggest a lack of dependency of the Fc receptor on the bursal environment. A high percentage of RFC is present in the spleen of agammaglubulinemic chickens (Ewald and Sanders, 1977). A portion of the RFC lacked B-specific antigen. The RFC of agammaglobulinemic chickens may be B cells that escaped surgical bursectomy (SBSX) and cyclophosphamide treatment or, alternatively, enhance the concept of an extrabursal site (see Section 3.3).

Conflicting data concerning the nature of RFC have been reported. Duncan and McArthur (1978) suggested that the majority of the RFC are monocytes or macrophages since more than 80% take a nonspecific esterase stain. In normal birds, 10 to 15% of RFC were esterase negative and failed to bind anti-T serum. Since these cells were missing from SBSX birds, one could conclude that a small population of the RFC is B$^+$. Previously cited work that the Fc$^+$ splenic cells are derived from the bursa may be questioned because in this study of Duncan and McArthur (1978), the majority of the Fc$^+$ cells were esterase positive and were still present in agammaglobulinemic birds. Evidence for a T-RFC has been reported (Chiao et al., 1978). Mononuclear cells of agammaglobulinemic chickens formed rosettes with sensitized red cells. Since the cells did not possess membrane Ig and did not take up latex particles, they were considered to be T cells. The number of RFC was increased in the spleen of agammaglobulinemic birds following hyperimmunization. Inhibition of rosettes was obtained by incubating the lymphocytes with IgG but not Fab. Thus, the activated T cell may possess IgG-Fc receptors (T_G^+). An assay for helper activity by the adoptive transfer technique revealed that T_G^+ cells are T_H cells while T_G^- are not. Glick (1977a) has reviewed earlier papers on RFC.

The primary in vitro antibody response by splenic and peripheral blood (PB) lymphocytes appears to be dependent on the generation of RFC (Ivanyi and Evans, 1979). Seven days after SRBC immunization, the RFC were five times more frequent in PB than in spleen, whereas PB lymphocytes showed no plaque-forming cells (PFC) and the spleen possessed IgG and IgM PFC. Four days in culture yielded IgG PFC in PB lymphocytes. The removal of RFC at the time of harvest eliminated PFC in 4-day cultures within the PB population and reduced the PFC in spleen. SBSX at 3 days significantly depressed RFC and IgG PFC in spleen harvested 8 days after immunization and cultured for 4 days. On the other hand, harvesting 20 to 30 days after primary immunization and *in vitro* for 4 days led to significantly high levels of IgG PFC. Therefore, these authors suggested that memory responses were recovered in the SBSX birds, and RFC were not necessary for memory B cells.

3.2. KINETICS OF LYMPHOCYTES IN RETICULOENDOTHELIAL STRUCTURES

The kinetics of bursal lymphocytes has been reviewed (Glick, 1977a). Since that review, we have identified long-lived and rapidly turning over lymphocyte populations in members of the RES (Glick, 1977b) and others have reported on the kinetics of cytoplasmic and surface immunoglobulin (Grossi *et al.*, 1977). Single or multiple application of [^{3}H]-TdR to the air cell at day 12 to 18 E will label 100% of bursal small lymphocytes (SL), whereas labeling of 100% of thymic, splenic, or bone marrow SL required multiple applications of [^{3}H]-TdR (Glick, 1976). These data indicated that the SL turned over more rapidly in the bursa than in the other compartments of the RES. After 100% of bursal and thymic SL were labeled, it required 37 and 16 days, respectively, until all labeled SL disappeared from these two tissues. The longest-lived population of SL reside in the spleen and bone marrow where 1.5 and 3.5%, respectively, of labeled SL remained 58 days after the last application of [^{3}H]-TdR. The absence of labeled SL in the thymus and bursa at this time reinforced the evidence of Weber (1975) that precursor cells from bone marrow do not home to the bursa in growing chickens. Only SL of the bursa and bone marrow approached 100% labeling after multiple injections of [^{3}H]-TdR in posthatched chickens (Figure 11; Glick, 1977b). The low percentage of labeled thymic SL could be explained, in part, by the paucity of large and medium lymphocytes in the thymus of posthatched chicks. These experiments also confirmed that bone marrow appeared to be the repository of a long-lived population of SL. We have not characterized the T and B nature of the long- and short-lived populations. Our preliminary data suggested that each cell type contributed to the long-lived population.

Two week after hatching, none of the spleen, bone marrow, cecal tonsil, thymus, or PB SL had detectable cytoplasmic IgM (cIgM), while bursal cells at this age showed a Type 3 (single or multiple droplets of Ig) cIgM (Grossi *et al.*, 1977). The only cells in the spleen and bone marrow to show both cIgM and sIgM were plasmablasts and plasma cells. On day 12 E, bursal cells stained for both cIgM and sIgM. At this time, the cIgM was diffuse (Type I). Low levels of anti-μ removed sIgM at day 15 E, whereas higher levels were required at posthatch ages. This correlated with the type of IgM pattern. Type 3 was the

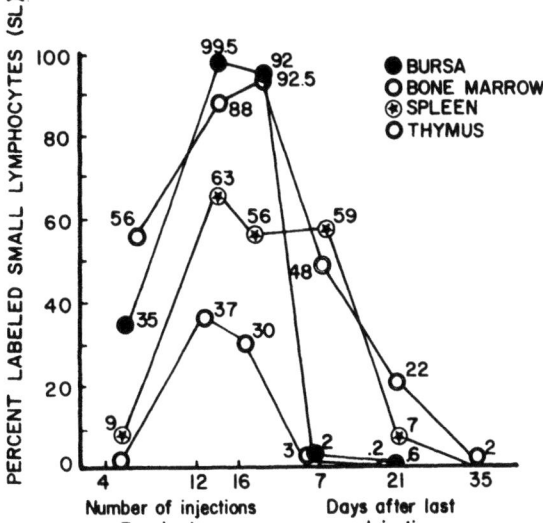

FIGURE 11. The percentage of labeled small lymphocytes (SL) in the bursa, bone marrow, spleen, and thymus subsequent to multiple injections of [^{3}H]-TdR (1 μCi/g body wt) into 14-day-old chickens. (From Glick, 1977b; reproduced with permission from Elsevier/North-Holland.)

most difficult to modulate. The authors suggested, on the basis of their data, that B-cell tolerance is best achieved *in ovo*.

Age did not appear to alter the B-cell capacity for Ig synthesis (Lifter and Choi, 1977). There appear to be more IgM$^+$ cells in the medulla than cortex, and the higher miotic rate of cortical cells suggested this as a site of clonal expansion (Grossi *et al.*, 1976). The greater turnover rate for secretory Ig suggests that the biosynthetic pathways leading to membrane Ig may be different from those leading to secretory Ig (Choi, 1976).

The identification of subpopulations of B cells was reported by Lifter *et al.* (1976). A larger percentage of bursal cells than splenic cells sedimented at a rate greater than 3.5 mm/hr but the majority of bursal cells sedimented at 2.3 mm/hr. Eighty-five percent of all bursal cells showed sIg and no cIg, whereas the rapidly sedimenting splenic cells possessed cIg. Small amounts of Ig were released both by bursal populations and by the slowly sedimenting cells of the spleen. The rapidly sedimenting population from the spleen proved to be plasma cells and released 3700-fold more Ig than the other population. Thus, on the basis of Ig biosynthesis, these authors were able to distinguish between bursal B cells and splenic B cells.

3.3. IMMUNOGLOBULIN SYNTHESIS

A major role of the bursa of Fabricius appears to be the regulation of sequential synthesis of specific Ig isotypes (Glick, 1977a). Early experiments revealed the presence of an IgM-like protein subsequent to the culturing of a bursa from a day 16 embryo (Marinkovich and Baluda, 1966). IgM was identified in acrylamide gel electrophoresis of bursal extracts from 1-week-old chickens (Glick and Whatley, 1967). The concentration of IgM at 1 week was more intense in the bursa than in the spleen, but at 5 weeks the reverse was true. The identification of both IgM and IgG in the bursa was made by Thorbecke et al. (1968). These investigators reported (1) IgM in the bursa of day 18 embryos, (2) IgM in the bursa at hatching, (3) the presence of Ig in the bursal medulla, (4) a more active synthesis of Ig in the bursa during the first week after hatching and less active by 3 months, and (5) synthesis of IgM was not antigen dependent. A sequential development of Ig synthesis within the bursa was suggested by Cooper and colleagues, on the basis of (1) a reduction in antibody and Ig following SBSX at hatch and (2) a reduction or loss of IgG before IgM following *in ovo* BSX (Long and Pierce, 1963; Carey and Warner, 1964; Ortega and Der, 1964; Pierce et al., 1966; Van Alten et al., 1968; Cooper et al., 1969; Moticka and Van Alten, 1972a,b). Kincade and Cooper (1971) observed IgM synthesis in the bursa by day 14 E, and by days 17 and 19, IgM was present in the cecal tonsil and spleen, respectively. IgG appeared in the bursa at hatch and in the spleen 4 days later. Also, bursal medullary cells were labeled with a fluoresceinated anti-μ serum and were occasionally observed to be positive for both μ and γ chains (Kincade and Cooper, 1971). These authors suggested that a developmental switch from IgM to IgG occurred within the bursa because: (1) the bursa was the initial site of IgM synthesis; (2) IgG appeared in only those bursal medullas which possessed IgM; and (3) cells positive for both μ and γ chains were found only in the bursa. In an experimental test of this hypothesis, anti-μ serum injected on day 13 E eliminated the synthesis of IgM by the bursa and when followed by SBSX at hatch suppressed both IgM and IgG (Kincade et al., 1970; Cooper et al., 1972; Kincade and Cooper, 1973). Reinforcement of the concept of sequential Ig development and apparent verification of the dependency of Ig synthesis on the bursa comes from the observations that birds BSX at hatch respond to a $1°$, $2°$, $3°$, or $4°$ immunization by producing antibody predominantly of the IgM isotype (Jankovic and Isakovic, 1966, 1967; Rose and Orlans, 1968; Eyckmans et al., 1968; Marvanova and Hayek, 1969; Alm, 1970). However, our thesis that "other sites in the chicken were capable of conditioning or supplying immunocompetent cells" (Glick, 1968a) was based on equally valid experiments in which birds, hormonally BSX *in ovo* and lacking a visible bursa at hatch, were capable of antibody synthesis (Glick and Sadler, 1961; Sadler and Glick, 1961; Claflin et al., 1966; Glick, 1968a,b). The concept of a bursa-independent site for Ig synthesis is further supported by the observation that normal or elevated amounts of IgM are found in birds hatched from eggs dipped on day 3 of incubation in testosterone propionate (TPD) (Merkenschlager et al., 1966; Glick, 1968b; Morgan and Glick, 1972; Hoffman-Fezer and Losch, 1973) and the failure to produce agammaglobulinemia by irradiation of BSX birds (Ivanyi et al., 1969; Van Meter et al., 1969). In support of our thesis, we have reported that TPD birds were: (1) devoid of a bursa embryonically or, at least, failed to develop bursal follicles (Glick and McDuffie, 1974, 1975); (2) agamma- or hypogammaglobulinemic for IgG; (3) hypergammaglobulinemic for IgM; (4) capable of an enhanced response to SRBC, with significantly high titers of IgM antibody; and (5) deficient in splenic and cecal tonsil germinal centers and plasma cells (Lerner et al., 1971). Because it is possible that elevated levels of IgM is BSX birds may represent a response to environmental antigens or viruses, we designed experiments in which TPD birds would be raised in pathogen-free (PF) environments (Subba Rao et al., 1978). Significantly lower levels of IgM were observed in TPD-PF birds than in TPD-conventional birds. However, the TPD-PF birds still maintained a significantly higher level of IgM than did normal-conventional birds. The IgM levels of TPD-PF birds returned to the IgM levels of TPD-conventional birds within 5 weeks after they were returned to the conventional environment. Therefore, exclusion of viral particles did reduce but did not eliminate the excessively high IgM. Our results suggest that elevated levels of IgM in TPD birds are, in part, a response to environmental antigens in the absence of modulation of IgG and provided support for the development of IgM capability in the absence of the bursa.

Our thesis for a bursa-independent IgM pathway finds supportive evidence from *in ovo* SBSX performed prior to 72 hr of incubation (Fitzsimmons et al., 1973; Jankovic et al., 1975, 1976, 1977). At hatching, the *in ovo* BSX birds lacked a bursa and portion of their large intestine. Yet, they possessed PFC, antibody, Ig, and B cells. The nonbursal microenvironment may function only in the absence of the bursal environment. If so, one might dissect the bursal environment and determine which factors are present in the peripheral lymphoid tissue in the presence and absence of the bursa. Perhaps certain intrinsic factors are necessary for differentiation. Extrinsic factors (e.g., neural and humoral) may influence the nonbursal microenvironment differently, depending on the presence or absence of the bursa (Jankovic et al., 1977). A negative feedback system by the bursa may dampen the response, whereas removal of the bursa would create a change in the internal environment conducive to allowing the development of a nonbursal site.

3.4. PHAGOCYTOSIS

Organs of the RES that possess blood space–sinusoidal pathways, e.g., liver and spleen, experience enhanced phagocytosis during the developmental period (Mizejewski and Ramm, 1969). Our observations have been similar. India ink (Pelikan, CII/1431a, Gunther Wagner) injected intravenously into growing chickens was concentrated within the liver, spleen, and bone marrow, sparsely present in the kidney and lung, and absent from the bursa and thymus (Glick et al., 1964). The heterophiles, monocytes, and thrombocytes from circulating blood of both normal and SBSX at hatch chicks were efficient phagocytic cells. These data demonstrated the phagocytic ability of thrombocytes and suggested that the bursa did not influence the phagocytic ability of circulating cells. To test the possible influence of the bursa on the postengulfment phase of phagocytosis, thrombocytes were collected from normal and BSX birds, and O_2 uptake was measured by the cells in the presence and absence of carbon (Yarborough et al., 1971). While the addition of carbon to the thrombocyte culture increased the uptake of O_2, there were no significant differences between cells from normal or

SBSX birds. Therefore, the postengulfment phase did not appear to be influenced by bursa removal. That the cells of the RES are not influenced by loss of the bursa has been reported by Waltenbaugh et al. (1976). These investigators did observe SBSX birds to be deficient in the clearance rate of a lipid emulsion. However, the deficiency was attributed to a missing serum factor in SBSX birds and not the cells of the RES.

3.5. SUBSETS OF T CELLS

3.5.1. T Helper Cells

T helper cells in chickens were first demonstrated in the laboratories of Thorbecke (McArthur et al., 1972; Weinbaum et al., 1973) and recently confirmed by Choi (1977). Normal birds were immunized with trinitrophenyl– keyhole limpet hemocyanin (TNP–KLH) and birds BSX with cyclophosphamide at hatch were immunized with BSA. In adoptive transfer experiments, spleens from normal birds supplied a source of B cells primed to make antibody to TNP while spleens from cyclophosphamide-treated birds supplied T cells sensitized to KLH. The adoptive transfer into cyclophosphamide-treated recipients of either cell accompanied by TNP–BSA failed to induce anti-TNP plaques, but a mixture of the two was successful (Weinbaum et al., 1973).

3.5.2. T Suppressor Cells

Antibody to KLH and Ig was suppressed in irradiated 4-month-old chickens following the adoptive cell transfer of bone marrow cells from agammaglobulinemic birds (Blaese et al., 1974). Transfer of bone marrow cells from normal birds to irradiated birds did not influence the immune response. It is now clear that a T suppressor (Ts) cell is responsible for these results (Palladino et al., 1976; Grebenau et al., 1976; Moticka, 1977; Kermani-Arab and Leslie, 1977). The embryonic thymic cells (day 16 to 20 E) from syngeneic and allogeneic sources suppressed the rejection of skin grafts but had no graft-versus-host (GVH) activity (Droege, 1976). On the other hand, adult thymic cells showed good GVH ability and poor suppressive activity. The cells responsible for Ts and GVH appear to be distinct. Three subpopulations of SL reside in the thymus (Droege et al., 1973, 1974). Cell type I is found only in embryonic and early hatched chicks and therefore must be the Ts cell. Only cell type III is present by 8 months and since these thymuses lack Ts and contain GVH ability, cell type III is equivalent to the GVH cell. The lack of cortex at this age would suggest that Ts cells (type I) reside in the cortex. The embryonic predisposition of Ts cells may explain tolerance and nonreactivity to antigens by the embryo (Droege, 1976). Grebenau et al. (1979) tested the role of Ts cells in the tolerogenic response of normal and agammaglobulinemic birds to human γ-globulin (HGG). Splenic cells from tolerized donors did not interfere with T helper cell activity nor did adoptive transfer of such splenic cells to normal or agammaglobulinemic birds influence delayed hypersensitivity (T_{DH} cells) to HGG plus Freund's complete adjuvant. Therefore, these authors concluded that the Ts cells are apparently not involved in HGG tolerance in the chick.

The Ts cells appear to be dependent on signals from the bursa since neonatal BSX markedly reduced the suppressor ability of thymic cells from 7- to 11-week-old donors (Droege, 1976). The bursal feedback to the thymus observed in the data from Droege's laboratory concerning the partial control of Ts differentiation or function does not stand as an isolated phenomenon. SBSX at 62 hr of incubation resulted in a reduced allograft response (Medina and Pedernera, 1976; Fitzsimmons *et al.*, 1977). Also, Fitzsimmons *et al.* (1977) reported that *in ovo* BSX at 72 hr reduced thymic size by 70%, and hormonal BSX between 2 and 5 days and SBSX at 72 hr significantly reduced thymocyte counts and gland size. Because thymocyte count was not affected by hormonal BSX performed after 5 days, the bursal influence on thymus is most likely humoral in nature. Mention has been made of the reduced cellularity of bursal follicles in thymectomized birds (Jankovic and Isakovic, 1964). Confirmation of a possible control by the thymus on bursal cellularity comes from the report of Hirota and Bito (1978). Bursal cells from 10-day-old normal chicks reconstituted antibody and bursal cellularity of 7-day-old cyclophosphamide-treated chicks. However, thymectomized, cyclophosphamide-treated, irradiated chicks did not respond to the adoptive transfer of bursal cells. Therefore, one might conclude that T cells or thymic hormone interact with transferred B cells in the reconstitution of bursal follicles.

3.5.3. T Cytotoxic Cells

Cytotoxic effector lymphocytes (CEL) are stimulated in the bone marrow, spleen, and thymus but not the bursa by phytohemagglutinin (PHA), pokeweed mitogen (PWM), and concanavalin A (Con A) (Kirchner and Blaese, 1973). Bone marrow lymphocytes of agammaglobulinemic birds were altered to CEL by PHA and, unlike splenic lymphocytes, did not proliferate in the presence of mitogen. The T-cell nature of the CEL was emphasized in experiments in which cytotoxicity was not modified following embryonic or hormonal BSX (Radzichovaskaja, 1967a,b; Yamanouchi *et al.*, 1971; Hayami *et al.*, 1972; Calder *et al.*, 1974; Granlund *et al.*, 1974; Granlund and Loan, 1974). Perhaps, unlike Ts cells, T-cytotoxic cells are not influenced by the bursa.

Natural killer cells. Chickens, like mammals, do not seem wholly dependent on T-cell subpopulations for surveillance (Lam and Linna, 1979; Sharma and Coulson, 1979). Spleens from pathogen-free chickens expressed natural killer cell activity against Marek's disease lymphoma line MSB-1 (Sharma and Coulson, 1979). The natural cytotoxicity of spleen cells was not T cell related and was not found in the thymic, bursal, or PB lymphocytes.

3.6. LYMPHOKINES

A mitogenic factor was first revealed by Oates *et al.* (1972) when buffy coats from Rhode Island Red chickens treated 4 weeks previously with *Mycobacterium tuberculosis* were exposed to purified protein derivative (PPD) and found to release a substance that increased the uptake of [^{3}H]-TdR. The heterogeneity of the mitogenic factor was suggested when they noted a reduced release of the factor by cells of BSX–irradiated and thymectomized–irradiated birds. Spleens from 6- to 10-week-old agammaglobulinemic birds were cultured in the presence of Con A or PHA and were found to produce a mononuclear chemotactic factor (Altman and Kirchner, 1972; Kirchner *et al.*, 1974). They concluded that the factor was not produced by a T cell. The extraction (over a 24-hr period) of bursal cells removed a substance that inhibited the mitogenic response of splenic cells to PHA (Van Alten *et al.*, 1976; Danielson and Van Alten, 1974). This bursal inhibitory factor was resistant to heat (50°C) but lost its activity on dialysis, suggesting that it was a low-molecular-weight compound. Another inhibitory factor which might be similar to that studied by Van Alten's group has been characterized in our laboratory. Thymic and bursal lymphocytes exhibited active migration from capillary tubes within 4 and 18 hr, respectively (Subba Rao and Glick, 1977). Bursal or thymic cells previously sensitized to PPD will fail to migrate in the presence of PPD, whereas naive cells are not inhibited by PPD. If the supernatant produced by sensitized bursal or thymic cells in the presence of PPD is transferred to naive bursal or thymic cells, migration will cease. Therefore, a lymphocyte inhibitory factor was released by the sensitized bursal and thymic cells. The release of this factor is age dependent: peak release by the bursal cells occurred between 5 and 10 weeks and then declined rapidly, whereas activity appeared in thymic cells at 9 weeks and remained high until 27 weeks. T and B cells from PPD-sensitized birds in the presence of PPD will release a lymphokine (thrombocyte inhibitory factor) that inhibits the migration of thrombocytes (Stinson *et al.*, 1979). The specificity of lymphocyte inhibitory factor from T and B cells, the relation of lymphocyte inhibitory factor to the thrombocyte inhibitory factor and the mitogenic factor, and the influence of lymphokines on other components of the RES are under investigation.

3.7. HARDERIAN GLAND

Prior to the revelation that the Harderian gland possessed plasma cells (Bang and Bang, 1968), the suggestion that it was a site of antibody and Ig synthesis would have been an abstruse thought. However, once plasma cells had been demonstrated there (Bang and Bang, 1968), one might, a priori, include the Harderian gland as a site of antibody production. In our laboratory, 4- to 6-week-old birds received SRBC either intravenously or by inuncting the eye orbit (Mueller *et al.*, 1971). The Harderian gland and spleen were assayed for PFC 3 to 5 days later (Sato and Glick, 1970; Mueller *et al.*, 1971). PFC were present in the spleen but not the Harderian gland after an intravenous injection. On the other hand, eye-orbit application was an efficient way of stimulating PFC in the Harderian gland but not the spleen. While these observations have been confirmed (Neumann and Kaleta, 1977; Survashe *et al.*, 1979), Burns (1976) reported antibody to BSA in the Harderian gland subsequent to ocular or intravenous injections. The ability of the Harderian gland to respond immunologically to intravenously administered BSA but not SRBC might reflect solubility differences between the two antigens or that the multiple intravenous injections employed by Burns mobilized cells from other sites to the Harderian gland. The ability of the Harderian gland to mount an antibody response to thymus-dependent antigens (e.g., SRBC) would suggest the presence of T cells. Cells from the Harderian gland were poor mobilizers of the GVH response (Sundick et al., 1973). However, the Harderian gland possessed a small percentage of T cells up to 5 weeks of age and 10% T cells by 9 weeks (Albini and Wick, 1974). The predominant cell appeared to be a B cell with some null cells present. Cells positive for surface Ig determinants (sIg$^+$) peaked at 6 weeks of age (90%) and were more numerous in the Harderian gland than in the bone marrow or cecal tonsil (Albini and Wick, 1973). These investigators suggested that the sIg$^+$ cells were plasmacytoid. Autoradiographic studies have revealed the sIg$^+$ cell to be a plasma cell (Glick et al., 1977b). A paucity of lymphocytes are present at all ages in the Harderian gland. Only 7–16% of the SL were sIg$^+$ at 14 weeks and fewer than 2% were sIg$^+$ between 21 and 32 weeks. The percentage of sIg$^+$ plasma cells did not fluctuate with age. Vaccination with infectious bronchitis virus, via intranasal or conjunctival route, of day-old chicks stimulated the formation of lymphoid follicles and increased the number of plasma cells in the Harderian gland (Davelaar and Kouwenhoven, 1976).

During the first weeks after hatching, sIgM$^+$ cells are numerous in the Harderian gland (Albini et al., 1974). IgG and IgA appear as surface determinants between 5 and 9 weeks with IgA isotype becoming dominant by 11 weeks. While sIgG$^+$ cells in the Harderian gland may be low in some older birds (Bienenstock et al., 1973), the majority of chickens seem to possess a predominance of this cell type (Parry and Aitken, 1973; Albini et al., 1974). It is difficult to explain the concentration of sIg$^+$ plasma cells in the Harderian gland since our data and those of others have not revealed an age-dependent increase in sIg$^+$ SL. The emigration of preformed plasma cells or an explosive transformation of SL to plasma cells might explain the rich population of plasma cells in the Harderian gland (Wight et al., 1971a; Glick et al., 1977a; Glick, 1978). The ultimate source of the plasma cells appears dependent on the bursa of Fabricius since the number of plasma cells in the Harderian gland was significantly reduced following hormonal BSX or SBSX at hatch (Mueller et al., 1971). Splenectomy at 6 weeks of age did not influence the Harderian gland plasma cell populations of 7-week-old birds (Survashe et al., 1979). It is difficult to understand the rationale for the splenectomies or the expectations that a lapse of 1 week would be sufficient time to note an effect. Birds fed a diet deficient in vitamin A and injected with Newcastle disease virus had a deficit of Harderian gland plasma cells and basophilic bone marrow cells (Bang et al., 1975). Neumann (1975) surgically removed the Harderian gland and reported lower IgM production in these birds (Neumann and Kaleta, 1977). After bilateral removal of the chicken Harderian gland, the lacrimal gland appeared to compensate by an increased secretory activity and proliferation of plasma cells (Burns, 1979). This approach to studies of the Harderian gland should further our understanding of this important reticuloendothelial structure.

ACKNOWLEDGMENTS. Sincere appreciation is extended to Terry Bragg who typed and retyped the manuscript, Beverly Beck for assisting with my reference file, Hieronymus Fabricius for being a marvelous observer, and the National Institutes of Health for past and present research support (Grant R01CA20169-04, awarded by the National Cancer Institute). This is journal article No. 5095 from the Mississippi Agricultural and Forestry Experiment Station.

REFERENCES

Ackerman, G. A., 1962, Electron microscopy of the bursa of Fabricius of the embryonic chick with particular reference to the lymphoepithelial nodules, *J. Cell Biol.* **13**:127.

Ackerman, G. A., and Knouff, R. A., 1959, Lymphocytopoiesis in the bursa of Fabricius, *Am. J. Anat.* **104**:165.

Aitken, I. D., and Survashe, B. D., 1977, Lymphoid cells in avian paraocular glands and paranasal tissue, *Comp. Biochem. Physiol. A* **58**:235.

Albini, B., and Wick, G., 1973, Immunoglobulin determinants on the surface of chicken lymphoid cells, *Int. Arch. Allergy Appl. Immunol.* **44**:804.

Albini, B., and Wick, G., 1974, Delineation of B and T lymphoid cells in the chicken, *J. Immunol.* **112**:444.

Albini, B., Wick, G., Rose, M. E., and Orlans, E., 1974, Immunoglobulin production in chicken Harderian glands, *Int. Arch. Allergy Appl. Immunol.* **47**:23.

Alm, G. V., 1970, The *in vivo* spleen response to sheep erythrocytes in bursectomized irradiated chickens, *Acta Pathol. Microbiol. Scand.* **78**:641.

Altman, L. C., and Kirchner, H., 1972, The production of a monocyte chemotactic factor by agammaglobulinemic chicken spleen cells, *J. Immunol.* **109**:1149.

Atwal, O. S., and McFarland, L. Z., 1971, Carbon uptake response by Sertoli cells and epithelial cells lining the duct system of the Japanese quail, (*Coturnix coturnix japonica*), *Poult. Sci.* **50**:159.

Bang, B. G., and Bang, F. B., 1968, Localized lymphoid tissues and plasma cells in paraocular and paranasal organ systems in chickens, *Am. J. Pathol.* **53**:735.

Bang, F. B., Bang, G. B., and Foard, M., 1975, Acute Newcastle viral infection of the upper respiratory tract of the chicken, *Am. J. Pathol.* **78**:417.

Bienenstock, J., Gauldie, J., and Perey, D. Y. E., 1973, Synthesis of IgG, IgA, IgM by chicken tissues: Immunofluorescent and ^{14}C-amino acid incorporation studies, *J. Immunol.* **111**:1112.

Biggs, P. M., 1957, The association of lymphoid tissue with the lymph vessels in the domestic chicken (*Gallus domesticus*), *Acta Anat.* **29**:36.

Blaese, R. M., Weiden, P. C., Koski, I., and Dooley, N., 1974, Infectious agammaglobulinemia: Transmission of immunodeficiency with grafts of agammaglobulinemic cells, *J. Exp. Med.* **140**:1097.

Bockman, D., 1979, Differentiation of epithelium, lymphocytes, granulocytes and erythrocytes in bursa of Fabricius transplanted to chorioallantoic membrane, *Dev. Comp. Immunol.* **3**:117.

Bockman, D. C., and Cooper, M. D., 1973, Pinocytosis by epithelium associated with lymphoid follicles in the bursa of Fabricius, appendix, and Peyer's patches: An electron microscopic study, *Am. J. Anat.* **136**:455.

Bockman, D., and Stevens, W., 1976, Gut-associated lymphoepithelial tissue: Bidirectional transport of tracer by specialized epithelial cells associated with lymphoid follicles, *J. Reticuloendothelial Soc.* **21**:245.

Bollum, F. J., 1974, Terminal deoxynucleotidyl transferase, in *The Enzymes*, Vol. 10 (P. D. Boyer, ed.), p. 145, Academic Press, New York.

Boyden, E. A., 1922, The development of the cloaca in birds with special reference to the origin of the bursa of Fabricius, the function of the urodaea sinus, and the regular occurrence of a cloacal fenestra, *Am. J. Anat.* **30**:163.

Burns, R. B., 1975, Plasma cells in the avian Harderian gland and the morphology of the gland in the rook, *Can. J. Zool.* **53**:1258.

Burns, R. B., 1976, Specific antibody production against a soluble antigen in the Harderian gland of the domestic fowl, *Clin. Exp. Immunol.* **26**:371.

Burns, R. B., 1979, Histological and immunological studies on the fowl lacrimal gland following surgical excision of Harder's gland, *Res. Vet. Sci.* **27**:69.

Burns, R. B., and Maxwell, M. H., 1979, The structure of the Harderian and lacrimal gland ducts of the turkey, fowl, and duck: A light microscope study, *J. Anat.* **128**:285.

Butler, D. G., Wilson, J. X., and Youson, J. H., 1978, Transformation of corticosterone to 11-dehydrocorticosterone by nasal, Harderian, and lacrimal glands of the domestic duck (*Anas platyrhynchos*) in vitro, *Gen. Comp. Endocrinol.* **34**:475.

Calder, E. A., Aitken, R. M., Penhale, W. J., McLeman, O., and Irvine, W. J., 1974, Lymphoid cell-mediated antibody-dependent cytotoxicity in untreated and bursectomized chickens, *Clin. Exp. Immunol.* **16**:137.

Calhoun, L. M., 1933, The microscopic anatomy of the digestive tract of *Gallus domesticus*, *Iowa State Coll. J. Sci.* **7**:261.

Carey, J., and Warner, N., 1964, Gamma globulin synthesis in hormonally bursectomized chickens, *Nature (London)* **203**:198.

Chiao, J. W., Lifter, J., Good, R. A., and Choi, Y., 1978, The Fc receptors on chicken T lymphocytes, *Cell. Immunol.* **40**:336.

Choi, Y., 1976, Biosynthesis of membrane bound Ig and secretion of Ig by chicken lymphoid cells, *Biochemistry* **15**:1037.

Choi, Y. S., 1977, Biosynthesis and secretion of antibodies by chicken lymphoid cells. II. Cooperation between carrier-specific T and hapten-specific B lymphocytes for the development of anti-hapten immune response, *Immunochemistry* **14**:59.

Claflin, A. J., Smithies, O., and Meyer, R. K., 1966, Antibody response in bursa-deficient chickens. *J. Immunol.* **97**:693.

Cogburn, L. A., and Glick, B., 1979a, Ontogenesis of pineal lymphocytes in chickens: A histological and functional study, *Poul. Sci.* **58**:1004.

Cogburn, L. A., and Glick, B., 1979b, The pineal gland: A prominent lymphoid tissue in young chickens, *Physiologist* **22**:20.

Cooper, M. D., Peterson, R. D. A., South, M. A., and Good, R. A., 1966, The functions of the thymus system and bursa system in the chicken, *J. Exp. Med.* **123**:75.

Cooper, M. D., Cain, W. A., Van Alten, P. J., and Good, R. A., 1969, Development and function of the immunoglobulin producing system. I. Effect of bursectomy at different stages of development on germinal centers, plasma cells, immunoglobulins, and antibody production, *Int. Arch. Allergy* **35**:242.

Cooper, M. D., Lawton, A. R., and Kincade, P. W., 1972, A two-stage model for development of antibody-producing cells, *Clin. Exp. Immunol.* **11**:143.

Cordier, A., 1969, Study of the innervation of the bursa of Fabricius in the embryonic and adult chick, *Acta Anat.* **73**:38.

Danielson, J. R., and Van Alten, P. J., 1974, Lymphocyte proliferation inhibited by cells and by effector substances obtained from bursal lymphocytes, *Prog. Exp. Tumor Res.* **19**:194.

Davelaar, F. G., and Kouwenhoven, B., 1976, Changes in the Harderian gland of the chicken following conjunctival and intranasal infection with infectious bronchitis virus in one- and 20-day-old chickens, *Avian Pathol.* **5**:39.

de Kruyff, R., Ponzio, N., and Thorbecke, G., 1977, Evaluation of the possible role of B-cell receptors in the tendency of B-cells to migrate into follicles in mice and chickens, *Eur. J. Immunol.* **7**:237.

Dransfield, J. W., 1945, The lymphatic system of the fowl, *Br. Vet. Sci.* **101**:171.

Droege, W., 1976, The antigen-inexperienced thymic suppressor cells: Class of lymphocytes in the young chicken thymus that inhibits antibody production and cell-mediated responses, *Eur. J. Immunol.* **6**:279.

Droege, W., Malchow, D., Strominger, J. L., and Linna, T. J., 1973, Cellular heterogeneity in the thymus: Graft-vs-host activity of fractionated thymus cells in the chicken, *Proc. Soc. Exp. Biol. Med.* **143**:249.

Droege, W., Zucker, R., and Hanning, K., 1974, Developmental changes in the cellular composition of the chicken thymus, *Cell. Immunol.* **12**:186.

Duncan, R., and McArthur, W. P., 1978, Partial characterization and the distribution of chicken mononuclear cells bearing the Fc receptor. *J. Immunol.* **120**:1014.

Ewald, S., and Sanders, B. G., 1977, EA rosette-forming cells in chickens: Presence in spleens of bursectomized chickens and relationship to B-lymphocytes, *Dev. Comp. Immunol.* **1**:363.

Ewald, S., Freedman, L., and Sanders, B. G., 1976a, EA rosette-forming lymphoid cells in chickens: Specificity of the Fc receptor and its relationship to other surface antigens receptor, *Immunology* **31**:1.

Ewald, S., Freedman, L., and Sanders, B. C., 1976b, EA rosette-forming lymphoid cells: Distribution as to cell types and organs from chickens of different developmental stages, *Cell. Immunol.* **23**:158.

Eyckmans, L., Schonne, E., and Eyssen, H., 1968, Antibody production in bursectomized chickens after injection of bacteriophage ØX174, *Life Sci.* **7**:161.

Fitzimmons, R. C., Garrod, E. M. F., and Garnett, L., 1973, Immunological responses following early embryonic surgical bursectomy, *Cell. Immunol.* **9**:377.

Fitzimmons, R. C., Dixon, D. K., and Kocal, E. M. F., 1977, The bursal–thymic interrelationship and ontogeny of the immune response in the chick embryo, in: *Developmental Immunobiology* (J. B. Solomon and J. D. Horton, eds.), p. 387, Elsevier/North-Holland, Amsterdam.

Glick, B., 1963, The effect of surgical and chemical bursectomy in the white Peking duck, *Poul. Sci.* **42**:1106.

Glick, B., 1968a, The immune response of bursaless birds as influenced by antibiotics nd age, *Proc. Soc. Exp. Biol. Med.* **127**:1054.

Glick, B., 1968b, Serum protein electrophoresis patterns in acrylamide gel: Patterns from normal and bursaless birds, *Poul. Sci.* **47**:807.

Glick, B., 1970, Hypertrophy of accessory spleen in splenectomized chickens, *Folio Biol. (Prague)* **16**:74.

Glick, B., 1976, Lymphocyte lifespan in chickens, in: *Phylogeny of Thymus and Bone Marrow-Bursa Cells* (R. K. Wright and E. L. Cooper, eds.), p. 237, Elsevier/North-Holland, Amsterdam.

Glick, B., 1977a, The bursa of Fabricius and immunoglobulin synthesis, *Int. Rev. Cytol.* **48**:345.

Glick, B., 1977b, Lymphocyte life span and migration of bursal lymphocytes, in: *Developmental Immunobiology* (J. B. Solomon and J. D. Horton, eds.), p. 371, Elsevier/North-Holland, Amsterdam.

Glick, B., 1978, The immune response in the chicken: Lymphoid development of the bursa of Fabricius and thymus and an immune response role for the gland of Harder, *Poul. Sci.* **57**:1441.

Glick, B., and McDuffie, F. C., 1974, Embryonic bursal development in the presence of testosterone, *Proc. World's Poul. Cong.* **15**:519.

Glick, B., and McDuffie, F. C., 1975, Immunoglobulin and the bursa of Fabricius, *J. Reticuloendothelial Soc.* **17**:119.

Glick, B., and Sadler, C. R., 1961, The elimination of the bursa of Fabricius and reduction of antibody production in birds from eggs dipped in hormone solutions, *Poul. Sci.* **40**:185.

Glick, B., and Sato, K., 1964, Accessory spleens in the chicken, *Poul. Sci.* **43**:1610.

Glick, B., and Whatley, S., 1967, The presence of immunoglobulin in the bursa of Fabricius, *Poul. Sci.* **46**:1587.

Glick, B., Chang, T. S., and Jaap, R. G., 1956, The bursa of Fabricius and antibody production, *Poul. Sci.* **35**:224.

Glick, B., Sato, K., and Cohenour, F., 1964, Comparison of the phagocytic ability of normal and bursectomized birds, *J. Reticuloendothelial Soc.* **1**:442.

Glick, B., Holbrook, K. A., and Perkins, W. D., 1977a, Scanning electron microscopy of the bursa of Fabricius from normal and testosterone-treated embryos, *Dev. Comp. Immunol.* **1**:41.

Glick, B., Subba, Rao, D. S. V., Stinson, R., and McDuffie, F. C., 1977b, Immunoglobulin-positive cells from the gland of Harder and bone marrow of the chicken, *Cell. Immunol.* **31**:177.

Glick, B., Holbrook, K. A., Oláh, I., Perkins, W. D., and Stinson, R., 1978, A scanning electron microscope study of the caecal tonsil: The identification of a bacterial attachment to the villi of the caecal tonsil and possible presence of lymphatics in the caecal tonsil, *Poul. Sci.* **57**:1408.

Good, R. A., and Finstad, J., 1967, The phylogenetic development of immune responses and the germinal center system, in: *Germinal Centers in Immune Response* (H. Cottier, N. Odartchenko, R. Schindler, and C. C. Congdon, eds.), pp. 4–27, Springer-Verlag, Berlin.

Granlund, D. J., and Loan, R. W., 1974, Cell-mediated immunity in the chicken: Cytotoxicity induced by specific soluble antigens, *J. Reticuloendothelial Soc.* **15**:503.

Granlund, D. J., Buening, G. M., and Loan, R. W., 1974, Cell-mediated immunity in the chicken. I. Development of a target cell destruction assay, *Cell. Immunol.* **11**:99.

Grebenau, M. D., Lerman, S. P., Palladino, M. A., and Thorbecke, G. J., 1976, Suppression of adoptive responses by addition of spleen cells from agammaglobulinemic chickens "immunized" with histocompatibility bursa cells, *Nature (London)* **260**:46.

Grebenau, M. D., Chi, D. S., and Thorbecke, G. J., 1979, T-Cell tolerance in the chicken. II. Lack of evidence for suppressor cells in tolerant agammaglobulinemic and normal chickens, *Eur. J. Immunol.* **9**:477.

Grossi, C. E., Lydard, P. M., and Cooper, M. D., 1976, B-Cell ontogeny in the chicken, *Ann. Immunol.* **127**:931.

Grossi, C., Lydard, P., and Cooper, M. D., 1977, Ontogeny of B-cells in the chicken. II. Changing patterns of cytoplasmic IgM expression and of modulation requirements for surface IgM by anti-μ antibodies, *J. Immunol.* **119**:749.

Hammond, W. S., 1954, Origin of thymus in chick embryo, *J. Morphol.* **95**:501.

Hayami, M., Hellstrom, I., Hellstrom, K. E., and Yamanouchi, K., 1972, Cell-mediated destruction of Rous sarcomas in Japanese quails, *Int. J. Cancer* **10**:507.

Hirota, Y., and Bito, Y., 1978, The role of the thymus for maturation of transferred bursa cells into competent B cells in chickens treated with cyclophosphamide, *Immunology* **35**:889.

Hoffman-Fezer, G., and Losch, U., 1973, Comparative histological and immunological studies on immunologically defective hens. I. Hormonally bursectomized hens, *Zentralbl. Veterinaermed. Reihe A* **20**:586.

Hoffman-Fezer, G., Rodt, H., Götze, D., and Thierfelder, S., 1977, Anatomical distribution of T and B lymphocytes identified by immunohistochemistry in the chicken spleen, *Int. Arch. Allergy Appl. Immunol.* **55**:86.

Holbrook, K. A., Perkins, W. D., and Glick, B., 1974, The fine structure of the bursa of Fabricius: "B" cell surface configuration and lymphoepithelial organization as revealed by scanning and transmission electron microscopy, *J. Reticuloendothelial Soc.* **16**:300.

Holbrook, K. A., Perkins, W. D., and Glick, B., 1977, Contact sites between lymphoid cells of the bursa of Fabricius, in vivo and in vitro, *Anat. Rec.* **189**:567.

Hoshi, H., and Mori, T., 1973, Identification of the bursa-dependent and thymus-dependent areas in the tonsilla caecalis of chickens, *Tohoku J. Exp. Med.* **111**:309.

Hudson, L., and Roitt, I. M., 1973, Immunofluorescent detection of surface antigens specific to T and B lymphocytes in the chicken, *Eur. J. Immunol.* **3**:63.

Ivanyi, J., and Evans, H., 1979, Analysis of the immunoglobulin isotype expression by peripheral blood rosette forming cells in chickens, *Immunology* **35**:947.

Ivanyi, J., Marvanova, H., and Skamene, E., 1969, Immunoglobulin synthesis and lymphocyte transformation by anti-immunoglobulin sera in bursectomized chickens, *Immunology* **17**:325.

Jankovic, B. D., 1968, The development and function of immunologically reactive tissue in the chicken, *Wiss. Z. Friedrich Schiller Univ. Jena Math. Naturwiss. Reihe* **17**:137.

Jankovic, B. D., and Isakovic, K., 1964, Role of the thymus and bursa of Fabricius in immune reactions in chickens. I. Changes in lymphoid tissues of chickens surgically thymectomized at hatching. *Int. Arch. Allergy Appl. Immunol.* **24**:278.

Jankovic, B. D., and Isakovic, K., 1966, Antibody production in bursectomized chickens given repeated injections of antigen, *Nature (London)* **211**:202.

Jankovic, B. D., and Isakovic, K., 1967, Natural haemagglutinins in surgically bursectomized and thymectomized chickens, *Folia Biol. (Prague)* **13**:401.

Jankovic, B. D., Mitrovic, K., Popeskovic, L., and Milosevic, D., 1966, Tonsilla caecalis: An immunologically active tissue in the chicken, *Iugoslav. Physiol. Pharmacol. Acta* **2**:71.

Jankovic, B. D., Knezevic, Z., Isakovic, K., Mitrovic, K., Markovic, B. M., and Rajcevic, M., 1975, Bursa lymphocytes and IgM containing cells in chicken embryos bursectomized at 52–64 hours of incubation, *Eur. J. Immunol.* **5**:656.

Jankovic, B. D., Isakovic, K., Markovic, B. M., Rajcevic, M., and Knezevic, Z., 1976, Nonbursal origin of humoral immunity: Immune capacity and cytomorphological changes in chickens bursectomized as 52 to 64 hour old embryos, *Exp. Hematol. (Copenhagen)* **4**:246.

Jankovic, B. D., Isakovic, K., Markovic, B. M., and Rajcevic, M., 1977, Immunological capacity of the chicken embryo. II. Humoral immune responses in embryos and young chickens bursectomized and sham-bursectomized at 52–64 h of incubation, *Immunology* **32**:689.

Jolly, J., 1915, La Bourse de Fabricius et les organes lympho-epitheliaux, *Arch. Anat. Microsc. Morphol. Exp.* **16**:363.

Kampmeier, O. F., 1969, *Evolution and Comparative Morphology of the Lymphatic System*, pp. 376–411, Thomas, Springfield, Ill.

Kennedy, G. Y., 1970, Harderoporphyrin: A new porphyrin from the Harderian glands of the rat, *Comp. Biochem. Physiol.* **36**:21.

Kermani-Arab, V., and Leslie, G. A., 1977, Suppression of immunoglobulin synthesis by transplantation of T-cells from anti-μ bursectomized chickens into normal recipients, *J. Immunol.* **119**:530.

Kincade, P., and Cooper, M. D., 1971, Development and distribution of immunoglobulin containing cells in the chicken: An immunofluorescence analysis using purified antibodies to μ, γ, and light chains, *J. Immunol.* **105**:371.

Kincade, P., and Cooper, M. D., 1973, Immunoglobulin A: Site and sequence of expression in developing chicks, *Science* **179**:398.

Kincade, P., Lawton, A. R., Bockman, D. E., and Cooper, M. D., 1970, Suppression of immunoglobulin G synthesis as a result of antibody-mediated suppression of immunoglobulin M synthesis in chickens, *Proc. Natl. Acad. Sci. USA* **67**:1918.

Kirchner, H., and Blaese, R. M., 1973, Pokeweed mitogen-, concanavalin A- and phytohemagglutinin-induced development of cytotoxic effector lymphocytes, *J. Exp. Med.* **138**:812.

Kirchner, H., Altman, L. C., Fridberg, A. P., and Oppenheim, J. J., 1974, Dissociation of *in vitro* lymphocyte transformation from production of a mononuclear leukocyte chemotactic factor in agammaglobulinemic chickens, *Cell. Immunol.* **10**:68.

Kondo, M., 1937, Die lymphatischen Gebilde in Lymphgefassystemdes Huhnes, *Folia Anat. Jpn.* **15**:309.

Kuhnel, W., and Beier, H. M., 1973, Morphologie und Cytochemie der Harderschen Drüse von Anatiden, *Z. Zellforsch. Mikrosk. Anat.* **141**:255.

Lam, K. M., and Linna, T. J., 1979, Natural cytotoxicity against Marek's disease tumor cells by non-immune chicken spleen cells, *Fed. Proc.* **38**:1278.

Lerner, K. G., Glick, B., and McDuffie, F. C., 1971, Role of the bursa of Fabricius in IgG and IgM production in the chicken: Evidence for the role of a nonbursal site in the development of humoral immunity, *J. Immunol.* **107**:493.

Lifter, J., and Choi, Y., 1977, Detergent solubilization of B-lymphocyte immunoglobulin, in: *Avian Immunology* (A. A. Benedict, ed.), pp. 100–107, Plenum Press, New York.

Lifter, J., Kincade, P., and Choi, Y., 1976, Subpopulations of chicken B lymphocytes, *J. Immunol.* **117**:2220.

Lind, L. R. (transl., ed.), 1963, *Aldrovandi on Chickens*, University of Oklahoma Press, Norman.

Long, P. L., and Pierce, A. E., 1963, Role of cellular factors in the mediation of immunity to avian coccidiosis (*Eimeria tenella*), *Nature (London)* **200**:426.

Looper, J. B., and Looper, M. H., 1929, A histological study of the colic caeca in the bantam fowl, *J. Morphol. Physiol.* **48**:585.

Lucas, A. M., and Jamroz, C., 1961, *Atlas of Avian Hematology*, Agric. Monogr. 25, U.S. Department of Agriculture, Superintendent of Documents, Washington, D.C.

Marinkovich, V. A., and Baluda, M. A., 1966, *In vitro* synthesis of M-like globulin by various chick embryonic cells, *Immunology* **10**:383.

Marvanova, H., and Hayek, P., 1969, The influence of bursectomy and thymectomy on the primary antibody formation to various antigens, *Folia Microbiol. (Prague)* **14**:171.

Matos, A., and Sousa, R., 1978, Tubuloreticular structures in chicken bursa Fabricii lymphocytes, *Experientia* **34**:1218.

Matsuda, H., Baba, T., and Bito, Y., 1976, The augmentation of antibody responses by preliminary intrabursal priming in the chicken, *Immunology* **31**:119.

Maxwell, M. H., and Burns, R. B., 1979, The ultrastructure of the epithelium of the diets of the Harderian and lacrimal glands of the turkey, fowl, and duck, *J. Anat.* **128**:445.

McArthur, W. P., Gilmour, D. G., and Thorbecke, G. J., 1972, Immunocompetent cells in the chicken, *Cell. Immunol.* **8**:103.

McCorkle, F., Stinson, R. S., Oláh, I., and Glick, B., 1979, The chicken's femoral-lymph nodules: T & B cells and the immune response, *J. Immunol.* **123**:667.

Medina, E., and Pedernera, E., 1976, Effect of early bursectomy on allografts survival in chicken, *Experientia* **33**:274.

Merkenschlager, M., Riedel, G., Kirchner, B., and Losch, U., 1966, Die immunoglobuline bursektomierter Küken in der immunelektrophorose (IE), *Naturwissenschaften* **16**:408.

Mitlin, N., and Glick, B., 1980, Nucleic acid and protein synthesis in lymphomyeloid organs of testosterone treated embryos of the chicken, (*Gallus domesticus*), *Comp. Biochem. Physiol.* **66A**:271.

Mizejewski, G. J., and Ramm, G. M., 1969, Phagocytosis as related to the reticuloendothelial system (RES) in the developing chick, *Growth* **33**:47.

Morgan, W., and Glick, B., 1972, A quantitative study of serum proteins in bursectomized and irradiated chickens, *Poul. Sci.* **51**:771.

Moticka, E. J., 1977, The presence of immunoregulatory cells in chicken thymus: Function in B & T cell responses, *J. Immunol.* **119**:987.

Moticka, E. J., and Van Alten, P. J., 1972a, Cellular haemolysin response of chickens bursectomized as embryos, *Folia Biol.* (*Prague*) **18**:331.

Moticka, E. J., and Van Alten, P. J., 1972b, Natural antibody to sheep erythrocytes in bursectomized chickens, *Proc. Soc. Exp. Biol. Med.* **141**:295.

Mueller, A. P., Sato, K., and Glick, B., 1971, The chicken lacrimal gland, gland of Harder, caecal tonsil and accessory spleens as a source of antibody producing cells, *Cell. Immunol.* **140**:151.

Muthmann, E., 1913, Beiträge zur vergleichende Anatomie des Blinddarmes und der lymphoiden Organe des Darmkanals bei Säugetieren und Vögeln, *Anat. Hefte.* **48**:67.

Naukkarinen, A., Arstila, A., and Sorvari, T., 1978, Morphological and functional differentiation of the surface epithelium of the bursa of Fabricius in chicken, *Anat. Rec.* **191**:415.

Nazerian, K., Ackerson, A., and Hooper, G., 1976, Scanning electron microscopy in the study of chicken T and B cells and cells from Marek's disease tumours, *Avian Pathol.* **5**:135.

Neumann, U., 1975, Die Cherurgische Entfernung der Harderschen Drüse des Huhnes, *Zentralbl. Veterinaermed. Reihe A* **23**:323.

Neumann, V., and Kaleta, E. T., 1977, Untersuchungen zur immunologischen Funktion der Harderschen Drüse des Huhnes, *Zentralbl. Veterinaermed. Reihe B* **24**:331.

Norton, S., and Wolfe, H. R., 1949, The growth of the spleen in the chicken, *Anat. Rec.* **105**:83.

Oates, C. M., Bissenden, J. F., Maini, R. N., Payne, L. N., and Dumonde, D. C., 1972, Thymus and bursa dependence of lymphocyte mitogenic factor in the chicken, *Nature New Biol.* **239**:137.

Odend'hal, S., and Breazile, J., 1979, Diffusely infiltrated area of lymphoid cells in the cloacal bursa, *J. Reticuloendothelial Soc.* **25**:315.

Oláh, I., and Glick, B., 1978a, The number and size of the follicular epithelium (FE) and follicles in the bursa of Fabricius, *Poult. Sci.* **57**:1445.

Oláh, I., and Glick, B., 1978b, Secretory cells in the medulla of the bursa of Fabricius, *Experientia* **34**:1642.

Oláh, I., and Glick, B., 1979, Structure of the germinal centers in the chicken caecal tonsil: Light and electron microscopic and autoradiographic studies, *Poult. Sci.* **58**:195.

Oláh, I., Glick, B., McCorkle, F., and Stinson, R., 1979, Light and electron microscope structure of secretory cells in the medulla of bursal follicles of normal and cyclophosphamide treated chickens, *Dev. Comp. Immunol.* **3**:101.

Orlans, E., and Rose, M. E., 1970, Antibody formation by transferred cells in inbred fowls, *Immunology* **18**:473

Ortega, L. G., and Der, B. K., 1964, Studies of agammaglobulinemia induced by ablation of the bursa of Fabricius, *Fed. Proc.* **23**:546.

Palladino, M. A., Lerman, S. P., and Thorbecke, G. J., 1976, Transfer of hypogammaglobulinemia in two inbred chicken strains by spleen cells from bursectomized donors, *J. Immunol.* **116**:1673.

Parry, S. H., and Aitken, I. D., 1973, Immunoglobulin A in the respiratory trait of the chicken following exposure to New Castle disease virus, *Vet. Rec.* **93**:258.

Payne, L. N., 1971, The lymphoid system, in: *Physiology and Biochemistry of the Domestic Fowl* (D. J. Bell and B. M. Freeman, eds.), pp. 988–1037, Academic Press, New York.

Penit, C., and Chaperille, F., 1977, Developmental changes in terminal deoxynucleotidyl transferase of the chicken thymus, *Biochem. Biophys. Res. Commun.* **74**:1046.

Peterson, R. D. A., and Good, R. A., 1965, Morphologic and developmental differences between the cells of the chicken's thymus and bursa of Fabricius, *Blood* **26**:269.

Pierce, A., Chubb, R., and Long, P., 1966, The significance of the bursa of Fabricius in relation to the synthesis of 7S and 19S immune globulins and specific antibody activity in the fowl, *Immunology* **10**:321.

Pintea, V., Constantinescu, G. H. M., abd Radu, C., 1967, Vascular and nervous supply of bursa of Fabricius in the hen, *Acta Vet. Acad. Sci. Hung.* **17**:263.

Potworowski, E. F., 1972, T and B lymphocytes organ and age distribution in the chicken, *Immunology* **23**:199.

Radzichovskaja, R., 1967a, Effect of bursectomy on the Rous virus tumor induction in chickens, *Nature (London)* **213**:1259.

Radzichovskaja, R., 1967b, Effect of thymectomy on Rous virus tumor growth induced in chickens, *Proc. Soc. Exp. Biol. Med.* **126**:13.

Retterer, E., 1885, Contributions à l'étude du cloaque et de la bourse de Fabricius les Oiseaux, *J. Anat. Physiol.* **21**:369.

Ritter, M., and Lebacq, A., 1977, Embryonic development *in vitro*, *Eur. J. Immunol.* **7**:468.

Romanoff, A., 1960, *The Avian Embryo*, Macmillan Co., New York.

Rose, M. E., and Orlans, E., 1968, Normal immune responses of bursaless chickens to a secondary antigenic stimulus, *Nature (London)* **217**:231.

Rothwell, B., Wight, P. A. L., Burns, R. B., and MacKenzie, G. M., 1972, The Harderian glands of the domestic fowl. III. Ultrastructure, *J. Anat.* **112**:233.

Ruth, R. F., Allen, C. P., and Wolfe, H. R., 1964, The effect of thymus on lymphoid tissue, in: *The Thymus in Immunobiology* (R. A. Good and A. E. Gabrielsen, eds.), pp. 183–206, Harper & Row (Hoeber), New York.

Ruuskanen, O., Toivanen, A., and Raekallio, J., 1977, Histochemical characterization of chicken lymphoid tissue, *Dev. Comp. Immunol.* **1**:231.

Sachs, H. G., Beezhold, D. H., and Van Alten, P. J., 1979, The effect of cyclophosphamide on the structure and function of the bursal epithelium. *J. Reticuloendothelial Soc.* **26**:1.

Sadler, C. R., and Glick, B., 1961, The antigenicity of duck-red-blood cells in bursectomized chickens, *Poult. Sci.* **40**:1767.

Sato, K., and Glick, B., 1970, Are splenic plaque forming cells sensitive to heat *in vitro*?, *Life Sci.* **9**:175.

Schaffner, T., Mueller, J., Hess, M. W., Cottier, H., Sordat, B., and Ropke, C., 1974, The bursa of Fabricius: A central organ providing for contact between the lymphoid system and intestinal content, *Cell. Immunol.* **13**:304.

Schaffner, T., Herring, J., Gerber, H., and Cottier, H., 1976, Bursa of Fabricius: Uptake of radioactive particles and radiotoxic "sealing" of bursal follicles, *Adv. Exp. Med. Biol.* **66**:33.

Schauenstein, K., 1979, Surface characteristics of chicken peripheral B and T lymphocytes defined by specific heteroantisera, *Dev. Comp. Immunol.* **3**:231.

Sharma, J. M., and Coulson, B. D., 1979, Presence of natural killer cells in specific pathogen-free chickens, *J. Natl. Cancer Inst.* **63**:527.

Sherman, J., and Auerbach, R., 1966, Quantitative characterization of chick thymus and bursa development, *Blood* **27**:371.

Sorvari, R., and Sorvari, T., 1977, Bursa Fabricii as a peripheral lymphoid organ: Transport of various materials from the anal lips to the bursal lymphoid follicles with reference to its immunological importance, *Immunology* **32**:499.

Sorvari, R., Naukkarinen, A., and Sorvari, T., 1977, Anal sucking-like movements in the chicken and chick embryo followed by the transportation of environmental material to the bursa of Fabricius, caeca, and caecal tonsils, *Poult. Sci.* **56**:1426.

Sorvari, T., Sorvari, R., Routsalbainen, P., Toivanen, A., and Toivanen, P., 1975, Uptake of environmental antigens by the bursa of Fabricius, *Nature (London)* **253**:217.

Stinson, R., and Glick, B., 1977, Scanning electron microscopy of chicken lymphocytes: A comparative study of thymic, bursal and splenic lymphocytes, *Dev. Comp. Immunol.* **2**:311.

Stinson, R. S., Mashaly, M. M., and Glick, B., 1979, Thrombocyte migration and the release of thrombocyte inhibitory factor (ThrIF) by T and B cells in the chicken, *Immunology* **36**:769.

Stutman, O., Yunis, E. S., and Good, R. A., 1970a, Studies on thymus function. I. Cooperative effect of thymus function and lymphopoietic cells in restoration of neonatally thymectomized mice, *J. Exp. Med.* **132**:583.

Stutman, O., Yunis, E. S., and Good, R. A., 1970b, Studies on thymus function. II. Cooperative effect of newborn and embryonic hemopoietic liver cells with thymus function, *J. Exp. Med.* **132**:601.

Subba Rao, D. S. V., and Glick, B., 1977, The production of a lymphocyte inhibitory factor (LyIF) by bursal and thymic lymphocytes, in: *Avian Immunology* (A. A. Benedict, ed.), pp. 87–98, Plenum Press, New York.

Subba Rao, D. S. V., McDuffie, F. C., and Glick, B., 1978, The regulation of IgM production in the chick: Roles of the bursa of Fabricius, Environmental antigens, and plasma IgG, *J. Immunol.* **120**:783.

Sugimoto, M., and Bollum, F. J., 1979, Terminal deoxynucleotidyl transferase (TdT) in chick embryo lymphoid tissue, *J. Immunol.* **122**:392.

Sugimoto, M., Yasuda, T., and Egashira, Y., 1977a, Development of the embryonic chicken thymus. I. Characteristic synchronous morphogenesis of lymphocytes accompanied by the appearance of an embryonic thymus-specific antigen, *Dev. Biol.* **56**:281.

Sugimoto, M., Yasuda, T., and Egashira, Y., 1977b, Development of the embryonic chicken thymus. I. Differentiation of the epithelial cells studied by electron microscopy, *Dev. Biol.* **56**:293.

Sundick, R. S., Albini, B., and Wick, G., 1973, Chicken Harder's gland: Evidence for relatively pure bursa-dependent lymphoid cell population, *Cell. Immunol.* **7**:332.

Survashe, B. D., Aitken, I. D., and Powell, J. R., 1979, The response of the Harderian gland of the fowl to antigen given by the ocular route. I. Histological changes, *Avian Pathol.* **8**:77.

Taliaffero, W. H., and Taliaffero, L. G., 1955, Reactions of the connective tissue in chickens to *Plasmodium gallinaceum* and *Plasmodium lophurae*. I. Histopathology during initial infections and superinfections, *J. Infect. Dis.* **97**:99.

Thorbecke, G. J., Warner, N. L., Hochwald, G. M., and Ohanian, S. H., 1968, Immune globulin production by the bursa of Fabricius of young chickens, *Immunology* **15**:123.

Toivanen, P., Toivanen, A., and Good, R. A., 1972, Ontogeny of bursal function in chicken. I. Embryonic stem cell for humoral immunity, *J. Immunol.* **109**:1058.

Van Alten, P. J., and Meuwissen, H. J., 1972, Production of specific antibody by lymphocytes of the bursa of Fabricius, *Science* **176**:45.

Van Alten, P. J., Cain, W. A., Good, R. A., and Cooper, M. D., 1968, Gamma globulin production and antibody synthesis in chickens bursectomized as embryos, *Nature (London)* **217**:358.

Van Alten, P. J., Waltenbaugh, C., Jaesson, L., and Danielson, J., 1976, Suppression of lymphocyte mitogenesis by a factor derived from bursal lymphocytes, in: *Neoplasm Immunity: Mechanisms* (R. G. Crispen, ed.), pp. 1–14, Proc. Chicago Symposium.

Van Meter, R., Good, R. A., and Cooper, M. D., 1969, Ontogeny of circulating immunoglobulins in normal, bursectomized and irradiated chicks, *J. Immunol.* **102**:370.

Waltenbaugh, C. R., Allen, C., Molnar, J., Sabet, T. Y., and Van Alten, P. J., 1976, Impairment of clearance by the reticuloendothelial system of bursectomized chickens, *J. Reticuloendothelial Soc.* **19**:3.

Warner, N. L., and Burnet, F. M., 1961, The influence of testosterone treatment on the development of the bursa of Fabricius in the chick embryo, *Aust. J. Biol. Sci.* **14**:380.

Weber, W. T., 1975, Avian B lymphocyte subpopulations: Origins and functional capacities, *Transplant. Rev.* **24**:8.

Weinbaum, F. I., Gilmour, D. G., and Thorbecke, G. J., 1973, Immunocompetent cells of the chicken. III. Cooperation of carrier sensitized T cells from agammaglobulinemic donors with hapten immune B cells, *J. Immunol.* **110**:1434.

Wenckebach, K., 1896, Die follickel der bursa Fabriciis, *Anat. Anz.* **11**:159.

Whitlock, D. R., Lushbaugh, W. B., Danforth, H. D., and Ruff, M. D., 1975, Scanning electron microscopy of the caecal mucosa in *Eimeria tenella* infected and uninfected chickens, *Avian Dis.* **19**:293.

Wight, P. A. L., and MacKenzie, G. M., 1974, Microsubstances in the Harderian gland of the domestic duck, *Res. Vet. Sci.* **17**:124.

Wight, P. A. L., Burns, R. B., Rothwell, B., and MacKenzie, G. M., 1971a, The Harderian gland of the domestic fowl. I. Histology with reference to the genesis of plasma cells and Russell bodies, *J. Anat.* **110**:307.

Wight, P. A. L., MacKenzie, G. M., Rothwell, B., and Burns, R. B., 1971b, The Harderian gland of the domestic fowl. II. Histochemistry, *J. Anat.* **110**:323.

Wolfe, H. R., Sheridan, S. A., Bilstad, N. P., and Johnson, M. H., 1962, The growth of lymphoidal organs and the testes of chickens, *Anat. Rec.* **142**:485.

Yamanouchi, K., Hayami, M., Kakura, S., Fukuda, A., and Kobune, F., 1971, Cellular immunity induced by Rous sarcoma virus in Japanese quail, *Jpn. J. Med. Sci. Biol.* **24**:1.

Yarborough, J. D., Wells, M., and Glick, B., 1971, Phagocytosis: The metabolism of thrombocytes from intact and bursaless birds, *J. Reticuloendothelial Soc.* **9**:248.

14

Molecular Basis of Self/Non-self Discrimination in the Ectothermic Vertebrates

GREGORY W. WARR and JOHN J. MARCHALONIS

1. INTRODUCTION

The ability to discriminate self from non-self is fundamental to the organization and survival of metazoan life. This ability resides primarily with the individual cells, although it can be mediated also by specific products secreted by the cells, as will be discussed later. Self/non-self discrimination is essential for the orderly processes of development, differentiation, and organogenesis; in addition, it is required so that an animal can maintain its integrity and prevent invasion and damage by foreign entities, either animate (viruses, bacteria, fungi, metazoan parasites) or inanimate (for example, bacterial exotoxins).

The immune system of all vertebrates can be considered to consist of two distinct but cooperating systems: the phagocytic and the lymphocytic. The phagocytic cells, such as the microphages or polymorphonuclear leukocytes, and the macrophages, or mononuclear phagocytes, can recognize and engulf many foreign substances of a particulate nature. This process of recognition by phagocytes can involve a variety of interactions, which are considered elsewhere in this volume. Although phagocytes are often able to recognize antigens by virtue of their possession of membrane receptors for other components of the immune system, such as antibody and complement, it seems that they also possess the capacity to recognize foreign materials directly by an interaction with their plasma membrane. Possible underlying mechanisms for this type of recognition are hydrophobic interactions, recognition of carbohydrate moieties by glycosyl transferases or by lectins, and other as yet unspecified mechanisms (Weir and Ogmundsdottir, 1977; Wilkinson, 1976). Perhaps the importance of this type of recognition may be appreciated by considering the invertebrate

---

GREGORY W. WARR and JOHN J. MARCHALONIS • Department of Biochemistry, Medical University of South Carolina, Charleston, South Carolina 29425.

phyla, which have survived (presumably potentially hostile) microbial contact for millions of years with phagocytic cells but none of the adaptive immune mechanisms (especially antibodies) that are the hallmark of the vertebrate response to non-self.

It is the lymphocyte system that enables the vertebrates to respond adaptively to foreign, non-self antigenic stimuli. By adaptive is meant the ability to alter specifically the degree of response to a non-self stimulus. A typical example is immunization, where prior contact with a microorganism can prevent subsequent infection with an identical or closely related organism, but not with an unrelated type. Although contact with an antigen usually heightens the response to a subsequent challenge, in some cases a depressed or absent response (tolerance) to subsequent challenge may result.

Before considering the molecular basis for the recognition of non-self by the lymphocytes of ectothermic vertebrates, it will be useful to describe briefly the phenomena of immunity. In addition, it will be necessary to refer constantly to the situation as it exists in the mammals; not because it provides a useful phylogenetic prototype (in fact, quite the opposite), but because it has been investigated much more thoroughly and is understood better than its counterpart in the ectothermic vertebrates. Consequently, we will attempt to interpret the information in ectotherms in a manner that might allow some evolutionary conclusions to be drawn.

Immunity in all vertebrates is manifest as either (1) a cellular reaction in which lymphocytes come into contact with and react to the antigen, or (2) by the presence in the circulation (and other sites) of soluble macromolecules, the antibodies, which have the ability to react specifically with antigens. Antibodies are produced by certain lymphocytes and the plasma cells into which they differentiate; and, in birds and mammals, the lymphocytes capable of producing antibody have been shown to derive from the bursa of Fabricius (in birds) or its equivalent (the bone marrow?) in mammals (Greaves et al., 1973). On the other hand, those cells that can react directly with non-self entities to destroy them are derived from the thymus, both in birds and in mammals (Greaves et al., 1973). Now it is accepted almost universally that the antibody-producing lymphocytes (B cells) are stimulated to produce immunoglobulins when antigen reacts with specific receptors on the plasma membrane. These receptors are also believed to be immunoglobulin in nature. The thymus-derived lymphocytes (T cells), which are capable of lysing certain types of cells recognized as foreign, mediate an inflammatory reaction to antigen, which is characterized by delayed onset, and an infiltration of lymphocytes and macrophages (Turk, 1975), the latter almost certainly being recruited in response to nonspecific mediators (lymphokines) released by antigen-activated T cells. T cells may also (in birds and mammals) give "help" to enable B cells to produce antibody most effectively (Miller, 1972). In addition, they can be shown experimentally to mediate reactions that have no natural counterpart. These reactions include in vitro blastogenic response to foreign lymphocytes, the ability to reject skin or other organs grafted to an animal from genetically different members of the same species (allograft) or a different species (xenograft), and the ability to react against the cells of a genetically distinct animal (usually the semiallogeneic F_1) into which they are injected [the so-called graft-versus-host (GVH) reaction]. T cells in mammals have been divided, on the basis of various criteria including surface antigenic markers, into a number of functional subsets (Cantor and Boyse, 1977; Chess and Schlossman, 1977).

It should be said here that some of the reactions of T cells, especially those related to the recognition of major histocompatibility antigens (MLC, GVH), may represent evolutionary carry-overs from invertebrate responses and possibly result from more primitive recognition systems than those which appear in the mammals. Such reactions have been termed quasi-immune by some investigators (Hildemann and Reddy, 1973).

One of the purposes of this review is to examine the evidence for lymphoid heterogeneity akin to T and B types in the ectothermic vertebrates; another is to examine evidence relating to their surface molecules involved in the recognition of antigens.

2. IMMUNOGLOBULINS

The typical response of a vertebrate to natural or experimental contact with many antigens is the production of circulating antibodies. In this section, we will work backwards from the properties of these observed molecules to what is known or can be inferred about the cellular basis of the phenomenon and the molecular nature of the receptor that allows the potential antibody-producing cell to recognize and respond to the stimulating antigen.

The immunoglobulins are a group of secreted molecules found in the blood and tissue fluids and are characterized by a variety of properties. They show electrophoretic mobilities and solubilities typical of the γ-globulins; they also possess detectable antibody activity (i.e., the ability to combine specifically with the appropriate antigen), characteristic subunit structure, and a heterogeneity of primary amino acid sequence reflected, for example, in a diversity of isoelectric points (Figure 1).

Intense investigation of Ig structure during the 1960s has resulted in the elucidation of the complete covalent structure of several Ig molecules, in nearly every case human or murine, and usually using the monoclonal, homogeneous product of a neoplastic lymphocyte (plasmacytoma or myeloma).

A diagrammatic representation of IgG from the human is shown in Figure 2. The molecule consists of four disulfide-bonded polypeptide chains, two heavy chains, designated γ and bearing the class-specific antigenic determinants, and two light chains, which are of one of two antigenic types, κ or λ, and which are shared by Igs of all the classes in humans (i.e., IgG, IgM, IgD, IgA, IgE). The light chains are primarily, but not totally, responsible for antigenic cross-reactions between Ig classes. For example, a rabbit antiserum to human IgG will usually cross-react, via the light chains, with *all* Ig classes, unless anti-light-chain activity is removed, by appropriate absorption, to make the antiserum γ-specific.

As result of sequence analysis of a variety of polypeptide chains, again generally from animal or human myelomas, it has become clear that the Ig polypeptide chains can be considered to consist of domains (Figure 2) approximately 110 amino acids in length. Thus, light chains consist of two domains; and although the NH_2-terminal domain shows considerable variability from myeloma to myeloma, the COOH-terminal domain shows little variation from myeloma to myeloma, within a class (κ or λ). Similarly, the γ chain consists of an NH_2-terminal variable region (V_H) and three constant-region domains ($C_{\gamma1-3}$). Homology (but not identity) between the three C γ domains suggests that they arose from tandem duplication of a primordial gene encoding one domain. Genetic polymorphism in the constant region leads to the definition of allotypes, both by amino acid sequence and by antigenic analysis. These can sometimes be traced to a single amino acid interchange, although the existence of complex allotypes (i.e., showing multiple amino acid substitutions) suggests that, in some cases, polymorphism of regulatory (not structural) genes may be the underlying cause of the observations. Variations in the primary amino acid sequence of the constant region of the heavy chain lead to antigenic differences, which allow for the distinction of classes and subclasses. The variable regions of the heavy and light chain are responsible for the heterogeneity of normal Igs and, as might be expected, confer the antigen specificity on the molecule. The

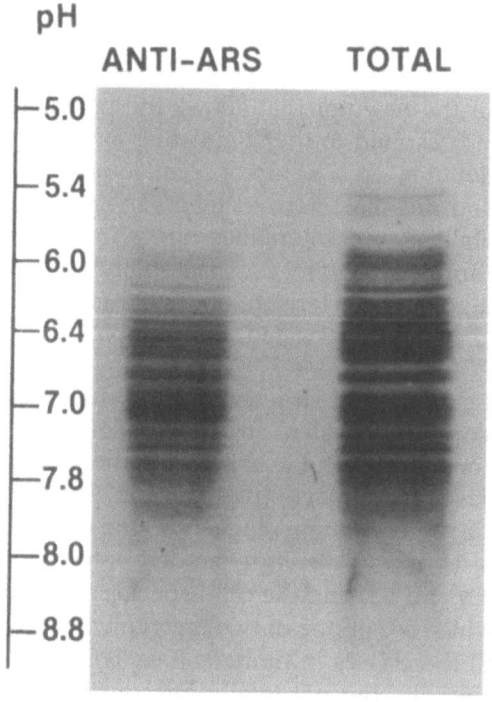

FIGURE 1. Isoelectric focusing patterns of immunoglobulins. A comparison is shown between total rabbit serum IgG (right) and purified specific antibody to the arsonate (ARS) hapten isolated from this total pool. The marked heterogeneity in isoelectric points of both total IgG and specific anti-ARS IgG should be noted. Using this system, a monoclonal antibody (e.g., a myeloma protein) would be expected to show no more than three or four bands.

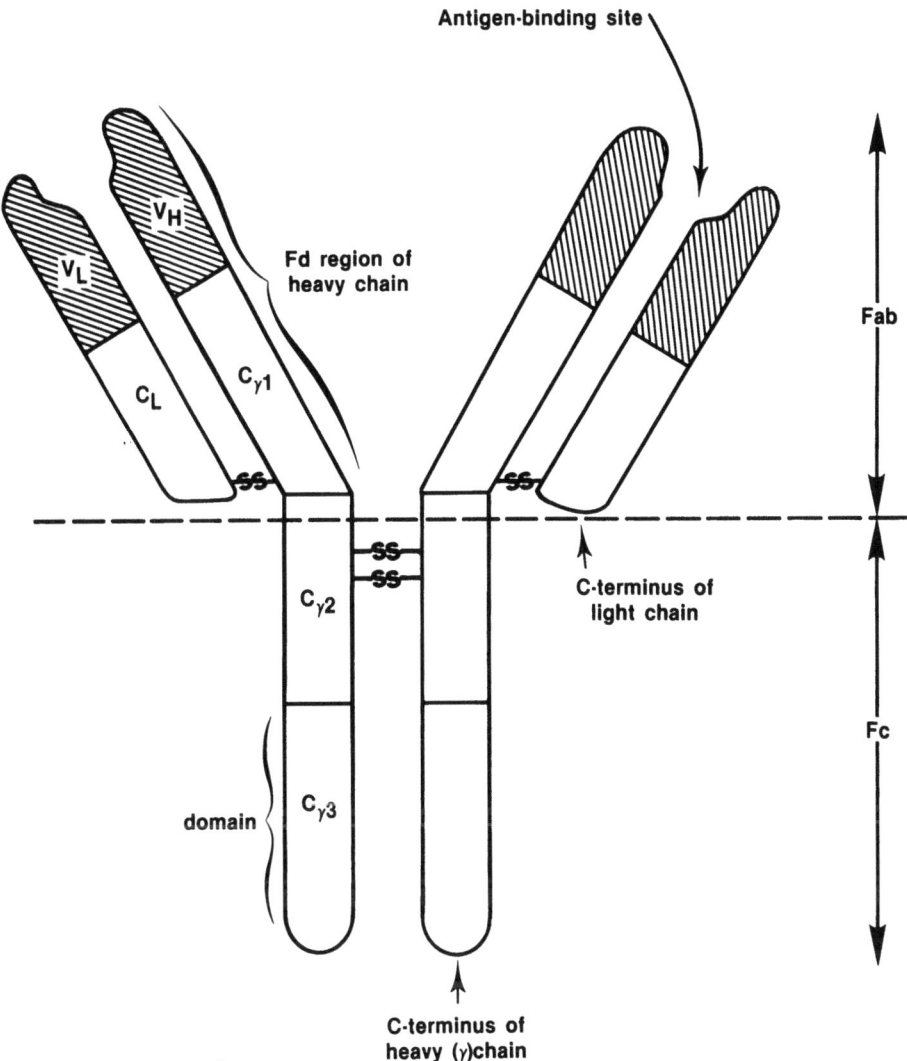

FIGURE 2. A diagrammatic representation of the human IgG molecule.–SS–, disulfide bonds.

binding site is formed from one V_H and one V_L domain, and the amino acids at certain hypervariable regions (three in the human κ chain; Wu and Kabat, 1970) are probably those intimately involved in forming the antigen binding pocket. Igs are glycoproteins, and the carbohydrate which is associated primarily with the heavy chain varies in amount from class to class (see, e.g., the data compiled by Marchalonis, 1977). Besides the heavy-chain-specified class-specific antigenic determinants, physicochemical differences exist between classes, although the H_2L_2 tetramer is the basic unit of all the Ig molecules. For example, mammalian IgM has a heavy (μ) chain composed of five domains (one variable, four constant), and the circulating IgM molecule is a pentamer, $(\mu_2L_2)_5$, which also contains a J or joining chain (molecular weight about 20,000) that is involved in the polymerization process (Koshland, 1975). Human IgA is a dimer, $(\alpha_2L_2)_2$, which, when present in secretions (e.g., of gut mucosa), contains a J chain and also a "secretory piece" (molecular weight 60,000).

Although the variable regions determine antigen-binding specificity, the constant regions of the molecule (especially heavy chain) specify other, secondary properties; for example, binding to macrophage Fc receptors (typical of IgG and mediated by the terminal or subterminal C-region domain) (Abramson et al., 1970; Hay et al., 1972) and binding to mast cell membranes (IgG subclasses, Muller-Eberhard, 1968).

2.1. IMMUNOGLOBULINS OF LOWER VERTEBRATES

Any description of Igs immediately raises the problem of nomenclature. What criteria can we use in defining homologies and postulating phylogenetic relationships? The best criterion is almost certainly primary amino acid sequence, preferably translated into the nucleotide sequence encoding it. However, while even this information can prove difficult to translate into firm rules for nomenclature, a discussion of this point with reference to the Igs of ectothermic vertebrates is rather academic, since sequence data are not available in abundance. The discussion on vertebrate Igs that follows below makes use of the sequence data that do exist, but it relies primarily on coarser properties, such as degree of polymerization, polypeptide chain size, degree of glycosylation, and antigenic cross-reaction, not to mention the traditional tools of the taxonomist, intuition and prejudice.

2.2. IgM OF FISH

Although, at first acquaintance, IgM would not appear to be a good candidate for early evolutionary appearance or wide phylogenetic distribution, it in fact seems to be the only class of Ig found in every vertebrate class. The reason that IgM might not be expected to appear early in evolution is that its heavy-chain constant region is probably encoded by a gene which has arisen by a series of tandem duplications, and, in addition, IgM frequently exists in various complex polymeric forms. It would have been satisfying to find somewhere in the "lower" vertebrates the products of the hypothetical unduplicated heavy-chain constant-region gene.

All living vertebrates appear to be capable of forming antibodies in response to immunization. This generalization applies to the cyclostomes, which, despite early technical problems with investigations, have been shown to be capable of producing antibodies, albeit in low amount and of poor quality (Boffa et al., 1967; Litman et al., 1970; Marchalonis and Edelman, 1968; Raison et al., 1978). This antibody appears to consist of a basic monomeric unit consisting of two heavy chains of a size similar to that of mammalian µ chains (molecular weight 70,000) and two light chains slightly larger than typical mammalian light chains (molecular weight 23,000).

There are two other interesting features to this cyclostome antibody: (1) individual polypeptide chains are not disulfide-bonded together (a property shared by some mammalian IgA molecules and T-cell surface Igs; Abel and Grey, 1968; Moseley et al., 1977) and (2) the basic monomer apparently can associate (noncovalently) to produce molecules of higher apparent molecular weight (Marchalonis and Edelman, 1968). Because of the difficulty of obtaining reasonable amounts of the antibody produced by lampreys and hagfish, the detailed chemical analyses required to establish the structure and evolutionary relationships of the molecules are lacking.

The major Ig produced by chondrichthean and osteichthean fish appears to be IgM. Some of these fish possess, in varying amounts, monomeric forms of IgM, whereas all of them possess higher-molecular-weight polymers, e.g., dimers, tetramers, and pentamers (Table 1).

The Igs of fish have been readily accepted as "classical" IgMs because their structure (especially that of the pentameric shark IgM studied initially) is so similar to that of mammalian IgM. Chondrichthean IgM typically exists as pentamer and monomer (but see Table 1 for exceptions) of structure $(\mu_2L_2)_5$ or μ_2L_2, the molecular weights of the heavy and light chain being similar to those of their mammalian counterparts. In addition, a J chain is claimed to have been isolated from leopard shark pentameric IgM (Klaus et al., 1971b), although others have reported such a chain to be absent from shark IgM (Weinheimer et al., 1971). Other properties, such as high carbohydrate content, amino acid composition, five-armed structure as visualized by electron microscopy, and relatively high degree of β-pleated sheet structure as assessed by circular dichroic spectroscopy (Frommel et al., 1971; Litman et al., 1971a,b), all lend support to the conclusion that the chondrichthean Ig can justifiably be called an IgM. The monomeric and polymeric forms of IgM in the shark have been shown to be identical by antigenic analysis and biochemical criteria such as amino acid composition and peptide mapping (Marchalonis and Edelman, 1965). It has been suggested that the monomeric IgM in the shark does *not* represent a precursor or breakdown product of the pentameric molecule; also suggested is that the monomer may carry out some of the physiological reactions of the distinct lower-molecular-weight Ig classes seen in the anuran amphibians, reptiles, and mammals (see below).

The work of Gitlin et al. (1973) has suggested that in sharks (16 species were investigated), between two and four antigenically distinguishable forms of the IgM exist. In every case, the monomeric and pentameric forms were found to be antigenically identical, and it seems likely that this antigenic heterogeneity has resulted from one or more duplications of the gene encoding the μ chain, thus giving rise to subclasses of IgM.

A possible candidate for a distinct low-molecular-weight Ig in elasmobranchs comes from the work of Fuller et al. (1978), who have reported that in nurse shark serum, there is a monomeric Ig characterized by a heavy chain of

TABLE 1. THE IMMUNOGLOBULINS OF FISH

	Molecular weights of constituent chains		Polymeric forms	Comments	References
	heavy	light			
Cyclostomata					
Sea lamprey (*Petromyzon marinus*)	70,000	25,000	Monomer (H_2L_2) and a higher polymeric form, $(H_2L_2)n$	Noncovalent H–L chain bonds, and noncovalent bonds in the polymeric form	Boffa *et al.* (1967), Marchalonis and Edelman (1968)
Pacific hagfish (*Eptatretus stoutii*)	—	—	Monomer and a higher polymeric form	Analogous molecule to the Ig of the sea lamprey, especially with regard to larger light chain	Raison *et al.* (1978)
Chondrichthyes					
Dogfish (*Mustelus canis*)	70,000	20,000	Monomer and pentamer $(H_2L_2)_5$	All the major Igs of the Chondrichthyes are tentatively classified as IgM. In general, the monomeric and polymeric forms of this IgM are identical antigenically and structurally, one exception being the Ig with low-molecular-weight H chain (50,000) in the nurse shark, which has not yet been assessed antigenically	Marchalonis and Edelman (1965, 1966)
Lemon shark (*Negaprion brevirostris*)	70,000	22,000	Monomer and pentamer		Clem and Small (1967)
Nurse shark (*Ginglymostoma cirratum*)	72,000 (50,000)	23,000	Monomer and pentamer. The 50,000-dalton H chain was found only in the monomeric form		Fuller *et al.* (1978)
Horned shark (*Heterodontus francisci*)			Monomer and pentamer		Frommel *et al.* (1971)
Stingray (*Dasyatis centroura*)	72,000	22,000	Predominantly dimer $(H_2L_2)_2$		Marchalonis and Schonfeld (1970)
Stingray (*Dasyatis americana*)	72,000	23,000	Predominantly pentamer		Johnston *et al.* (1971)
Sharks (16 species)			Monomer and pentamer		Gitlin *et al.* (1973)

Osteichthyes					
Paddlefish (*Polyodon spathula*)	75,000 (58,000)	23,500	Tetramer $(H_2L_2)_4$		Pollara et al. (1968), Acton et al. (1971)
Gar (*Lepisosteus platyrhincus*)	70,000	22,000	Tetramer	The polymeric Igs of bony fish are almost universally tetrameric molecules, considered to be analogous to IgM of mammals. Low-molecular-weight forms of IgM exist as the monomer. Monomeric Igs with a low-molecular-weight H chain also exist in some fish, such as grouper, bowfin, and lungfish. In the grouper and bowfin, this low-molecular-weight H chain is considered to be closely related to the high-molecular-weight H chain, of which it may represent a deletion mutation	Bradshaw et al. (1971), Acton et al. (1971)
Bowfin (*Amia calva*)	70,000 52,000	24,000	The 70,000-dalton H chain exists in a tetrameric Ig, the 50,000-dalton H chain in a monomeric Ig		Litman et al. (1971a)
Grouper (*Epinephelus itaira*)	70,000 40,000	22,000	The 70,000-dalton H chain exists in a tetrameric Ig, the 40,000-dalton H chain in a monomeric Ig		Clem (1971)
Margate	—	—	High- and low-molecular-weight forms believed to be monomeric and tetrameric forms showing antigenic identity		
Carp (*Cyprinus carpio*)	71,000	24,000	Tetramer		Shelton and Smith (1970), Marchalonis (1971)
Goldfish (*Carassius auratus*)	72,500	23,000	Tetramer		Marchalonis (1971), Trump (1970)
Plaice (*Pleuronectes platessa*)	70,000	22,000	Tetramer		Fletcher and Grant (1969)
Australian lungfish (*Neoceratodus forsteri*)	70,000 40,000	23,000	The 70,000-dalton H chain exists in a polymer $(H_2L_2)_n$, where n is probably 4 or 5. The 40,000-dalton H chain exists probably only as the monomer (H_2L_2)		Marchalonis (1969)
Catfish (*Ictalurus punctatus*)	70,000	23,000	Tetramer		Acton et al. (1971)

lower molecular weight (50,000) than that typical of μ chains. This Ig occurs at a higher concentration in the serum of immune animals, but, as yet, insufficient antigenic and structural information is available to allow an assessment of its relationship to other Ig classes. The predominant Ig found in the serum of bony fish is an IgM-like polymer, which consists of four subunits, as shown both by analytical ultracentrifugation and by direct electron microscopic observation. Some of the available information is summarized in Table 1. The IgM-like tetramers of bony fish are comprised of heavy chains and light chains which resemble, in size and other properties such as carbohydrate content, μ and light chains, respectively, of mammals.

Although J chains have been observed in the tetrameric IgM of catfish (Mestecky et al., 1975), they have been reported to be absent from the tetrameric IgM of gar and paddlefish (Weinheimer et al., 1971; Zikan, 1974).

There is a body of data which supports the contention that in some bony fish, in addition to the tetrameric IgM, there exist

1. Subclasses of the tetrameric IgM.
2. Low-molecular-weight monomeric Igs which are distinct physicochemically from IgM, but may be related to it nevertheless.
3. Monomeric forms of IgM.

Considering these points in turn, evidence for subclasses of teleost IgM comes from the studies of Trump (1970), who described a single molecular weight class of tetrameric IgM in the goldfish but was able to define antigenic and electrophoretic differences between molecules in this population. Evidence for the existence of low-molecular-weight monomeric Igs physicochemically distinguishable from IgM comes from the studies of the bowfin (Holostei) (Litman et al., 1971a) and the grouper (Teleostii) (Clem, 1971). The low-molecular-weight Igs in these species had heavy chains with a molecular weight of 40,000–52,000, and, primarily on the basis of antigenic analysis, both groups of investigators postulated that these heavy chains represented μ chains with a mass deletion. Whether this deletion is present in the DNA or occurs posttranscriptionally or posttranslationally has not been determined.

The occurrence of a monomeric form of IgM in teleosts is apparently a variable phenomenon. A monomeric IgM occurs in the margate (Clem and McLean, 1975) in amounts similar to or slightly greater than those of the antigenically indistinguishable tetramer. However, monomeric IgM appears to be absent from the serum of certain teleosts; for example, the Cyprinidae (Marchalonis, 1971; Trump, 1970), although some reports to the contrary eixst (Uhr et al., 1962). It is possible that a low-molecular-weight form of IgM exists in the serum of many, if not all, teleosts in a relatively low amount which may be elevated by certain immunization schedules and the ambient water temperature (Clem and McLean, 1975). Such appearance might reflect shedding of lymphocyte surface monomeric IgM molecules (Warr et al., 1976; Clem et al., 1977; Fiebig et al al., 1980).

2.3. IgM OF AMBHIBIANS AND REPTILES

All ectothermic tetrapod vertebrates that have been studied possess immune macroglobulins which appear to be similar to mammalian IgM on the basis of such properties as degree of polymerization, molecular weights of H and L chains, degree of glycosylation, and possession of J chain. The properties of IgM in some representative species are shown in Table 2. A few points are worthy of note; for example, although a pentameric form of the IgM is apparently typical of amphibians and reptiles, the IgM of *Xenopus laevis* is hexameric (Parkhouse et al., 1970), as visualized by electron microscopy. As will be discussed below, all amphibians and reptiles, except possibly for the urodeles (Marchalonis and Cohen, 1973), appear to possess at least one class of low-molecular-weight monomeric Ig distinguishable from IgM by both antigenic and physicochemical criteria. These immunoglobulins will be considered below.

2.4. NON-IgM IMMUNOGLOBULINS OF AMPHIBIANS AND REPTILES

Two low-molecular-weight Igs occur in reptiles and are distinguishable by their sedimentation coefficients in the ultracentrifuge (7 S and 5.7 S, respectively). The molecular weights of their heavy chains are approximately 62,000 and 40,000, respectively. The Ig with the approximately 62,000-molecular-weight heavy chain occurs in amphibians, reptiles, and birds; this has sometimes been called IgG. This point illustrates the problems of classification of Igs in lower vertebrates, because the heavy chain, on the basis of protein mass alone, is clearly not mammalian γ chain and has been variously called IgY (in the birds; Leslie and Clem, 1969), Ig RRA (for Reptilia, Amphibia, Aves; Atwell and Marchalonis, 1976), or intermediate-molecular-weight Ig.

The other low-molecular-weight Ig with a heavy chain of 40,000 is found not only in reptiles but also in some (but not all) birds (Benedict and Yamaga, 1976), in the Dipnoi (lungfish), and also apparently in some mammals (for example, the rabbit). This class can represent a substantial proportion of the antibody response of species in which it occurs and has been designated IgN, or low-molecular-weight Ig.

2.5. IMMUNOGLOBULIN LIGHT CHAINS

The molecules of all Ig classes of all the placoderm-derived vertebrates consist, as far as is known, of equivalent numbers of heavy and light chains, the variable regions of each heavy/light pair forming one antigen-binding site. In mammals, two antigenic forms of light chain, designated κ and λ, exist in varying proportions from species to species (Marchalonis, 1977). Attempts to assign light chains of ectothermic vertebrate Igs to either of these two classes have been made using criteria such as partial NH_2-terminal sequence or the presence of a

TABLE 2. IgM IMMUNOGLOBULINS OF TETRAPOD VERTEBRATES

Species	Class	Molecular weight		Polymeric form	Reference
		Heavy chain	Light chain		
Mudpuppy (*Necturus maculosus*)	Urodele amphibian	70,000	22,000	$(H_2L_2)_5$	Marchalonis and Cohen (1973)
Cane toad (*Bufo marinus*)	Anuran amphibian	67,000	22,500	$(H_2L_2)_5$	Acton et al. (1972a)
Clawed toad (*Xenopus laevis*)	Anuran amphibian	70,000	22,500	$(H_2L_2)_6$	Parkhouse et al. (1970)
Sleepy lizard (*Tiliqua rugosa*)	Reptile	77,000	23,000	$(H_2L_2)_5$	Acton et al. (1972b)
Turtle (*Pseudemys scripta*)	Reptile	67,000	23,000	$(H_2L_2)_5$	Acton et al. (1972b)
Chicken (*Gallus domesticus*)	Avian	70,000	22,000	$(H_2L_2)_5$	Leslie and Clem (1969)
Echidna (*Tachyglossus aculeatus*)	Monotreme mammal	69,000	22,500	$(H_2L_2)_5$	Atwell et al. (1973)
Human	Eutherian mammal	70,000	23,000	$(H_2L_2)_5$	From data compiled by Edelman and Gall (1969)

blocked NH$_2$-terminal residue (typical of many mammalian λ chains). These criteria, in combination with the small amount of information at present available, are not sufficiently exact to make definitive statements about class designations of light chains in the ectothermic vertebrates (see, e.g., Klaus et al., 1971a). However, tentative identifications (on an antigenic basis) of κ and λ type light chains have been made in the alligator (Saluk et al., 1970).

2.6. THE ANTIGEN-COMBINING SITE OF IMMUNOGLOBULINS OF ECTOTHERMIC VERTEBRATES

Information on the binding site of antibodies has come from a variety of sources: sequence analysis of V regions of both heavy and light chains; studies of the binding constants of the reaction between antigen (usually a small hapten) and antibody; determination of the thermodynamic characteristics of the reaction between antigen and antibody; and studies of the size and shape of the antigen-binding site, using, for example, X-ray crystallography or paramagnetic resonance spectroscopy with spin-labeled haptens.

Unfortunately, most investigators have applied these techniques predominantly to antibodies of the mammal, but what emerges from those limited studies that have been carried out on the antibodies of ectothermic vertebrates is that there has been a remarkable conservation of binding-site structure and function during vertebrate evolution.

It has been a general observation that the affinity of IgM produced by ectothermic vertebrates following immunization does not increase significantly with time or repeated immunization, a phenomenon which is similar to that reported for mammals. Interestingly enough, however, IgM antibodies in carp can show a significant increase in *functional* affinity (a property of the whole IgM molecule) upon reimmunization, while the intrinsic affinity of each binding site remains relatively unaffected (Fiebig et al., 1977).

Conversely, all the low-molecular-weight non-IgM classes in amphibians and reptiles show a rapid increase in binding affinity during immunization, a response typical of that of mammalian IgG (Siskind and Benacerraf, 1969).

Why progressive antigen-dependent selection of higher-affinity antibodies should favor classes other than IgM is a question not easily answered. Current immunological dogma (for the mammals) is that *all* the classes share a common pool of V_H genes which can be translocated to the appropriate heavy-chain constant-region genes for transcription and translation at the appropriate time.

Limited NH$_2$-terminal sequence analysis of vertebrate heavy chains shows clear similarities and the emergence of V_H subgroups as early as the elasmobranch fish. Similarities suggesting evolutionary relationships among mammalian V_H subgroups (especially V_HIII) and those seen in other species, e.g., birds and sharks, have been documented (Wasserman et al., 1974). Sequence analysis of the V regions of ectothermic vertebrate Ig chains has not been carried through the hypervariable region, so possible homologies in the binding-site-determining residues of antibodies directed to the same antigen but from different species remain to be investigated. Studies on the dimensions of the binding site of trout IgM antibodies and rabbit antibodies (both to dinitrophenyl) suggested that they are very similar (Roubal et al., 1974). Evaluation of the changes in the thermodynamic parameters (e.g., entropy, enthalpy, and free energy) during the interaction between antigen (dinitrophenyl) and antibody in the grouper and the rabbit showed that they were very similar in these two species, and again suggested that the binding site of antibody may have been highly conserved during vertebrate evolution (Clem and Small, 1970). With the use of available sequence data for the first 22 NH_2-terminal residues of the variable regions of heavy chains of selachian (Sledge et al., 1974; Kehoe et al., 1978) and avian (Kehoe and Capra, 1974) species, direct phylogenetic comparison of the first set of framework residues (Table 3) can be made. A striking degree of homology as manifested by identity of residue positions occurs. Five positions the expression of effector function (which, in the case of B cells, is the secretion

TABLE 3. COMPARISONS AMONG N-TERMINAL SEQUENCES OF HEAVY CHAINS OF MAMMALIAN, AVIAN, AND SELACHIAN SPECIES

Residue	Human[a] V_HIII	Mouse[b] V_HI	Dog[c] V_H	Cat[d] V_H	Pig[e] V_H	Duck[f] V_H	Chicken[f] V_H	Nurse shark[g] V_H	Horned shark[h] V_H
1	Glu	Asp	Glu	Asp	Glu	Ala	Ala	Glu	Asp
2	Val	Val	Val	Val	Glu	Ala	Val	Val	Val
3	Gln	Gln	Gln	Gln	Gln	Thr	Thr	Thr	Val
4	Leu	Leu	Leu	Leu	Leu	Leu	Leu	Leu	Leu
5	Val	Gln	Val	Val	Val	Asp	Asp	Thr	Thr
6	Glu	Glu	Glu	Glu	Glu	Glu	Glu	Glx	Gln
7	Ser	Ser	Ser	Ser	Ser	Ser	Ser	Pro	Pro
8	Gly	Gly	Gly	Gly	Gly	Gly	Gly	Glx	Glu
9	Gly	Pro	Gly	Gly	Gly	Gly	Gly	Ala	Ala
10	Gly	Gly	Asp	Asp	Gly	Gly	Gly	Glx	Glu
11	Leu	Leu	Leu	Leu	Leu	Leu	Leu	Asp	Thr
12	Val	Val	Val	Val	Val	Val	Gln	Ser	Gly
13	Glu	Lys	Glu	Gln	Gln	Gly	Thr	Glx	Lys
14	Pro	Pro	Pro	Pro	Pro	Pro	Pro	Pro	Pro
15	Gly	Ser	Ala	Gly	Gly	Gly	Gly	Gly	Gly
16	Gly	Glu	Gly	Asp	Gly	Val	Gly	Gly	Gly
17	Ser	Ser	Ser	Ser	Ser	Ser	Ala	Ala	Ser
18	Leu	Leu	Leu	Leu	Leu	Leu	Leu	Leu	Leu
19	Arg	Ser	Arg	Arg		Arg	Arg	Thr	Arg
20	Leu	Leu	Leu	Leu	Leu	Leu	Leu	Leu	Leu
21	Ser	Thr	Ser	Thr	Ser	Val	Val	Thr	Thr
22	Cys	Cys	Cys	Cys	Cys	Cys	Cys	Cys	Cys

[a]Data taken from Gally (1973).
[b]MOPC 315. Data tabulated in Kabat et al. (1976).
[c]GA. Data tabulated in Kabat et al. (1976).
[d]DI. Data tabulated in Kabat et al. (1976).
[e]Data tabulated in Kabat et al. (1976).
[f]Data from Kehoe and Capra (1974).
[g]Sledge et al. (1974).
[h]Kehoe et al. (1978).

(Leu^4, Pro^{14}, Leu^{18}, Leu^{20}, and Cys^{22}) are invariant throughout the 300 million years of evolutionary time which witnessed the emergence of vertebrate classes, genera, and species. A number of other residues, notably Val^2, Glu^6, Ser^7, Gly^{15}, and Arg^{19}, show remarkable conservation. Variation correlated with class and species evolution is also noted; for example, birds have Asp at position 5, whereas sharks have Thr, and mammals, predominantly Val. A species difference within birds is shown at residue 13 where ducks have Gly and chickens Thr. These data indicate that the V_H framework sequences are strongly conserved in evolution, and it is likely that the hypervariable regions which account for the combining-site specificity are likewise strongly conserved. When sharks and man, for example, produce antibodies to the same chemically defined structure, it is reasonable to predict that they will produce at least some antibodies with identical or similar combining sites. The minimal argument is that both species would be expressing the products of the same ancestral V_H genes which have been conserved in evolution, rather than having each species reinvent these genes (convergence). This hypothesis is, of course, clouded by the variability which would result from the somatic generation of V-region diversity. Although no direct sequence data are available on the broad evolutionary conservation of V_H, including framework and hypervariable portions, recent studies using anti-idiotype antibodies, i.e., antibodies directed against particular V-region determinants, show that rodent (many strains of mice and the guinea pig) antibodies to the synthetic terpolymer poly (L-Glu^{60}-LAla^{30}-L-Tyr^{10}) express the same idiotype (Ju et al., 1978; Ju and Dorf, 1979). It has also been shown that human and murine monoclonal Igs which bind phosphorylcholine express cross-reactive idiotype determinants (Riesen, 1979). Evidence of the evolutionary conservation of variable-region combining sites for antigen strongly suggests that these structures would also serve in the recognition of antigen by lymphocytes.

3. ANTIGEN RECEPTORS AND THE CELLULAR BASIS OF THE IMMUNE RESPONSE

Despite the sometimes bewildering picture of cellular interactions which dominate current concepts of mammalian cellular immunology, the basic aspects of the immune response are clear: lymphocytes are stimulated to divide, differentiate, and express their appropriate (preprogrammed?) effector functions upon contact with antigen. Most antigens are large, complex molecular or supramolecular structures, and both observation and intuition support the idea that a lymphocyte is triggered to react when it comes in contact with antigen at its plasma membrane. A corollary of this hypothesis is that the lymphocyte bears on its surface specific receptors which enable it to react with, and hence recognize, antigen. We now accept that most lymphocytes are preprogrammed to respond to one antigen (i.e., bear receptors specific for only one antigen) and that the appropriate antigen selects the lymphocyte(s) with complementary receptors, causes magnification of the response by inducing division, and leads to of antibody with the same antigen-specificity as existed for membrane-bound receptors). This concept of "clonal selection" was formulated by Burnet (1959) and has become a central dogma of cellular immunology.

The particular questions that we are going to consider for the ectothermic vertebrates are: (1) what is the nature of the receptor by which a lymphocyte is enabled to recognize antigen, (2) what is the evidence for functional heterogeneity in the responding lymphocytes, and is this heterogeneity reflected in the structure of the receptor for antigen on the lymphocytes?

In considering the first point, the simplest hypothesis appears to be that the receptor should equal the secreted product of the cells, i.e., be a typical secretory Ig. This, as we shall see later, is an oversimplification; but on teleological grounds, this system would clearly be economical for the antibody-producing (B-type) cell. The "minimal requirement" would be that the part of the surface receptors which shows specificity in antigen recognition is shared with the corresponding regions of antibodies. A complication arises when we consider the lymphocyte involved in cell-mediated immunity (T-type) which does not secrete (at least in large amounts) a readily detectable conventional Ig. This begs the second question posed above, regarding the heterogeneity of cell types in the ectothermic vertebrates, and we will briefly digress to consider this issue before returning to the nature of the antigen receptors.

The evidence relating to T- and B-type heterogeneity of lymphocytes in ectothermic vertebrates has been discussed in recent reviews (Ruben and Edwards, 1979; Warr and Marchalonis, 1978), and we will consider it very briefly here. The evidence for heterogeneity comes from: (1) lack of antibody production by thymocytes of certain species of amphibians and fish; (2) impairment of classical "T-cell-type functions" by thymus ablation in amphibians; (3) definition of lymphocyte subpopulations by surface markers, physical properties, and functional responsiveness; and (4) heterogeneity in expression of surface Ig, and, in some cases, physicochemical differences between surface Igs obtained from lymphocytes from different sites.

Against this can be set observations demonstrating antibody production in the thymus of some fish and amphibians (Moticka et al., 1973; Ortiz-Muniz and Sigel, 1971). There are clearly unresolved issues here, and possible explanations which would resolve them might include: (1) real distinctions between very diverse species in terms of the development of lymphoid heterogeneity; or (2) the existence of real functional differences in lymphocyte subpopulations, but an anatomical "scrambling" obscuring their organ partitioning normally seen in mammals (Cuchens, and Clem, 1977).

Our tentative conclusions do not exclude either of these possibilities, but we believe that there is a significant body of evidence pointing to a functional heterogeneity of lymphocytes detectable in many ectothermic vertebrates.

What then is known of the molecules which enable lymphocytes, of all types, to recognize antigen in the ectothermic vertebrates? Practical approaches to this problem are limited, as the small amount of protein associated with a lymphocyte membrane ($< 10^{-12}$ g) and the clonally restricted nature of the receptors mean that very little material is available for detailed immunochemical analysis. However, let it be stated clearly that the only class of molecules known to possess the intrinsic heterogeneity to cope with binding to the very large number of antigens encountered by an animal is the Igs.

Hence, the "minimal hypothesis" is that Ig, in some form, acts as receptor for antigen at the lymphocyte surface. Evidence relating to this postulated function of Ig will be presented, along with such molecular characterization of these molecules as exists at present.

4. IMMUNOGLOBULIN AS ANTIGEN RECEPTOR ON LYMPHOCYTES

Investigators have approached the question of whether or not Ig is the antigen receptor in several ways. First, can Ig be demonstrated at the lymphocyte surface? Second, does this Ig appear to act in the primary binding of antigen? Third, does the union of antigen with receptor lead to the "triggering" of the cell to divide and express effector functions, and if so, by what molecular mechanisms? This last point has hardly been investigated in the ectothermic vertebrates, and will not be considered further except to note several recent reviews of the available information derived from mammalian systems (Ling and Kay, 1975; Wedner and Parker, 1976; Cooper, 1973; Edelman, 1976; Decker and Marchalonis, 1978).

To the question, "can Ig be demonstrated at the surface of lymphocytes?," no short answer can be given because the reply depends markedly upon the nature of the Ig determinant under consideration (e.g., V or C region and the methods used for detection). Using direct-binding methods to demonstrate Ig *in situ* on the plasma membrane (e.g., by the use of fluorochrome- or radioisotope-tagged anti-Ig antibodies), the results differ from species to species. In Table 4 are shown results for anuran and urodele amphibians, teleosts, and elasmobranchs, along with results for the mouse, a commonly used mammal.

One summary of these results might be: all lymphocytes in fish and urodele amphibians bear readily detectable Ig; in mammals, birds, and adult anuran amphibians, thymocytes (and peripheral T cells?) are negative for Ig, but in the larval anuran amphibian the situation is the same as for the fish; i.e., all lymphocytes are positive for Ig. However, we consider this view to be grossly misleading, for a number of reasons. First, immunofluorescent or similar analyses should not be definitive criteria for establishing the existence of Ig on the membrane of a lymphocyte, for both misleading positive and misleading negative results can occur. For example, although thymocytes and T cells of mammals are negative using anti-Ig antibodies raised in other mammalian species, antisera raised in chickens can recognize membrane Ig on these cells (Hämmerling *et al.*, 1976; Jones *et al.*, 1976; Szenberg *et al.*, 1977). Another example of a misleading fluorescent reaction has been reported both in frogs and in fish. It appears that some antisera to IgM may contain anticarbohydrate antibodies which react with non-Ig glycoproteins on the surface of lymphocytes. This problem has been raised particularly in the case of thymocytes of certain ectothermic vertebrates, the larval amphibian *Rana catesbeiana* (Mattes and Steiner, 1978a,b) and the trout

TABLE 4. LYMPHOCYTE SURFACE IMMUNOGLOBULINS DETECTED BY STANDARD IMMUNOFLUORESCENT PROCEDURES

Species	Organ	% positive lymphocytes	Ig class(es) detected[a]	References
Mouse				
Mus musculus	Thymus	0–4		From data compiled by Warr (1979)
	Spleen	21–39	IgM, IgD (IgG)	
	Lymph node	9–52	IgM, IgD (IgG, IgA)	
	Blood	5–34	IgM, IgD (IgG)	
Clawed toad (Anura)				
Xenopus laevis	Thymus	9–70[b]	IgM	Jurd (1977), Jurd and Stevenson (1976), Du Pasquier et al. (1972), Du Pasquier (1976)
	Spleen	40–92	IgM (IgY)	
	Gut-associated lymphoid tissue	51–68	IgM (IgY)	
	Peripheral blood	25–82	IgM (IgY)	
Salamander (Urodele)				
Pleurodeles waltlii	Thymus	98	IgM	Charlemagne and Tournefier (1975)
	Spleen	65–85	IgM	
Skate (Elasmobranch)				
Raja naevus	Thymus	82		Ellis and Parkhouse (1975)
	Spleen	80		
	Blood	80		
Teleostii				
Pooled data for *Carassius auratus, C. carassius, Salmo gairdneri, Cyprinus carpio, Lepomis macrochirus*	Thymus	95–100	IgM	Clem et al. (1977), Emmrich et al. (1975), Etlinger et al. (1977), Warr et al. (1976, 1977, 1979), DeLuca et al. (1978)
	Spleen	25–99		
	Head nephros	85–100		
	Blood	25–90		

[a] The major class(es), where known, is (are) cited. Classes in parentheses are observed on a minority of the cells (generally less than 10%).
[b] The value declines, with developmental age, to a minimum in adult life (Du Pasquier et al., 1972; Du Pasquier, 1976).

Salmo gairdneri (Yamaga *et al.*, 1978a,b). Our opinion of these observations of anticarbohydrate reactivity in anti-IgM sera is that they are potential problems which ought to be recognized and controlled for in all immunofluorescent studies. Not only is caution necessary in the use of immunofluorescent systems, but we believe that biochemical criteria for the occurrence of Ig on lymphocytes (especially thymocytes or T-type cells) should be applied. These criteria might include detection of Ig in lysates of cells using radioimmunoassay, or the isolation of molecules sharing physicochemically the properties of Ig (e.g., polypeptide chain structure with disulfide-bonded heavy and light chains and appropriate molecular weight estimates for the Ig polypeptides). For this latter approach, it is usually necessary to isotopically label lymphocyte membrane proteins; e.g., by the use of the lactoperoxidase-catalyzed reaction to introduce radioisotopes of iodine into exposed tyrosines (Marchalonis *et al.*, 1971). Using these sorts of approaches (immunofluorescence, radioimmunoassay, and surface labeling), we (Warr *et al.*, 1979) have been able to confirm the presence of Ig molecules on the membrane of trout thymic lymphocytes, and Fiebig *et al.* (1980) have been able to demonstrate that the detection of Ig on carp thymocytes is not due to non-Ig carbohydrate cross-reactions. These investigators (Fiebig *et al.*, 1980) demonstrated Ig on carp thymocytes by fluorescence, radioimmunoassay, and immunochemical characterization.

It is our opinion that the proper application of biochemical techniques has established the existence of Ig-like molecules on virtually all lymphocytes of all studied vertebrates, both ectothermic and endothermic, including thymocytes of mammals, birds, and teleost fish such as the bream, goldfish, and trout.

For the ectothermic vertebrates, we have listed in Table 5 representative studies on the presence or absence of Ig on lymphocytes from various organs in animals so far studied, and we have noted the particular techniques used.

We should point out that other investigators may not agree with us about the universal occurrence of Ig on lymphocytes (especially thymocytes or T lymphocytes; Paul and Benacerraf, 1977). However, it appears that many, if not all, investigators now believe that T lymphocytes express at least the antigen-binding (variable) region of the Ig heavy-chain molecule as detected by anti-idiotypic reagents (Rajewsky and Eichmann, 1977; Binz and Wigzell, 1977), so we will proceed to outline what we consider the current state of our knowledge about the molecular structure of lymphocyte surface Ig (both T and B) in ectothermic vertebrates and try to point out technical or other problems which can lead to difficulties in identifying Ig on thymocytes or T-type cells.

In mammals, the surface Igs of B-type lymphocytes are typically either IgM or IgD, and, more rarely, IgG or another isotype. The problems which beset classification of serum isotypes in the ectothermic vertebrates occur also for the lymphocyte surface Igs. In all ectotherms, an IgM-like molecule occurs, and in those species with a low-molecular-weight Ig (e.g., IgY in anuran amphibians), this also appears on the membrane of some lymphocytes. In contrast to the polymeric IgM found in the serum of all vertebrates, the IgM which occurs on the membrane of lymphocytes is invariably monomeric; i.e., it consists of two heavy chains of μ-like properties and two light chains, typically disulfide-bond-

TABLE 5. LYMPHOCYTE MEMBRANE IMMUNOGLOBULINS OF ECTOTHERMIC VERTEBRATES DETECTED BY MEANS OTHER THAN DIRECT ANTIGLOBULIN BINDING

Species	Organ	Class observed	Method used[a]	References
Reptilia				
Agrionemys horsfieldii (tortoise)	Spleen, peripheral blood	IgM, IgY	A	Fiebig and Ambrosius (1975)
Alligator mississippiensis	Peripheral blood	IgM, IgY	A	Cuchens et al. (1975)
Amphibia	No data available			
Pisces				
Carassius auratus (goldfish)	Thymus, spleen, head nephros	IgM[b]	A	Warr et al. (1976), Ruben et al. (1977)
Cyprinus carpio (carp)	Thymus, spleen	IgM[b]	A, B	Fiebig and Ambrosius (1975), Fiebig et al. (1980)
Salmo gairdneri (trout)	Thymus, spleen	IgM	A, B	Warr et al. (1979)
Lepomis macrochirus (bluegill)	Thymus, spleen	IgM	A	Clem et al. (1977)

[a]Method A: Surface radioiodination, extraction, immune precipitation, and polyacrylamide gel electrophoresis. Method B: Radioimmunoassay.
[b]The thymocyte surface Ig in cyprinid fish is related to IgM but is structurally and antigenically distinct from the B cell or serum IgM molecule (Warr et al., 1976; Fiebig et al., 1980).

ed together into a four-chain structure. It has been suggested for the mammals that the μ chain of the B-lymphocyte surface IgM differs from that of the serum IgM in such properties as apparent mass and buoyant density (Melcher and Uhr, 1977; Melcher et al., 1975). While little comparable work has been done with ectothermic vertebrates, it appears that in the goldfish (but not trout or bream), there is a large difference in size between serum and splenic lymphocyte membrane "μ" chains (amounting to some 10,000 daltons nominal mass; Warr and Marchalonis, 1977a). The basis or significance of this observation is unknown.

The Ig associated with the membrane of thymocytes in ectothermic vertebrates has been studied biochemically, primarily in the teleost fish, having been isolated from these cells in the carp, goldfish, bream, and trout (Fiebig and Ambrosius, 1975; Fiebig et al., 1980; Clem et al., 1977; Warr et al., 1976, 1979). In the goldfish, the thymocyte surface Ig differs from the splenocyte Ig in certain properties, such as having an apparently smaller heavy chain, and its ease of solubilization with nonionic detergent (a property shared with Ig from mammalian thymocytes or T cells; Warr et al., 1976; Cone and Brown, 1976; Warr and Marchalonis, 1977b). Such differences may not be apparent in other teleost fish such as the trout (Warr et al., 1979). In the carp, Fiebig et al. (1980) have reported that thymocyte Ig differs from splenocyte or serum Ig in having a heavy chain similar to but clearly distinct from μ, and in the apparent lack of a light chain. It

TABLE 6. ANTIGEN-BINDING LYMPHOCYTES IN THE ECTOTHERMIC VERTEBRATES

	Antigens	Animals immunized	Lymphoid organ	Frequency of binding cells in unimmunized animals (%)	Inhibition by antiglobulin	References
Anuran Amphibia						
Grass frog (*Rana pipiens*)	Sheep and horse erythrocytes, trinitrophenyl hapten	Yes	Thymus, spleen	0.04	Yes	Edwards et al. (1975), Ruben and Edwards (1977)
Clawed toad (*Xenopus laevis*)	Sheep and horse erythrocytes, trinitrophenyl hapten	Yes	Spleen	—	NT[a]	Ruben and Edwards (1977)
Cane toad (*Bufo marinus*)	Flagellin of *Salmonella adelaide*	Yes	Spleen	—	NT	Azzolina (1975)
	Horse erythrocytes	Yes	Spleen	0.005	Yes	Cone and Marchalonis (1972)
Urodele Amphibia						
Newt (*Notophthalmus viridescens*)	Trinitrophenyl hapten, horse erythrocytes	Yes	Spleen	—	NT	Ruben and Edwards (1977)
Pisces (Teleostii)						
Goldfish (*Carassius auratus*)	Flagellin of *S. adelaide*	Yes	Spleen	—	NT	Azzolina (1975, 1978)
	Horse erythrocytes, trinitrophenyl hapten	Yes	Spleen, thymus, head nephros	—	Yes	Ruben et al. (1977)
	Keyhole limpet hemocyanin, myoglobin, horse spleen ferritin	Yes	Spleen, thymus, head nephros	0.3–3.4	NT	DeLuca et al. (1978)
Crucian carp (*Carassius carassius*)	Keyhole limpet hemocyanin, myoglobin, horse spleen ferritin	No	Spleen, thymus, head nephros	0.7–5.0	NT	DeLuca et al. (1978)
Trout (*Salmo gairdneri*)	Keyhole limpet hemocyanin, myoglobin, horse spleen ferritin	No	Spleen, thymus	0.6–3.0	NT	Warren et al. (1979)

[a] NT, not tested.

is, at present, difficult to conclude whether or not technical difficulties in isolating Ig from thymocytes or T cells may account for some problems in the ectothermic vertebrates as it has certainly done in the mammals (Cone, 1976).

A further way in which the functional role of lymphocyte membrane Ig has been investigated has been to look at the binding of antigen to cells (as well as splenocytes). Antigen-binding thymocytes or putative peripheral T cells have been described in immunized or unimmunized fish and amphibians (Table 5). In all cases where it has been investigated, the pretreatment of any kind of antigen-binding lymphocyte with anti-Ig antibody has inhibited subsequent binding of the antigen (Table 6). Furthermore, when the antigen and the anti-Ig are applied to the cell under conditions in which they do not prevent the binding of one or the other, they can, if tagged with fluorochromes emitting light of different colors, be seen to be coincident or codistributed in the plane of the membrane (DeLuca *et al.*, 1978; Warr *et al.*, 1979).

However, although these sorts of observations do not prove beyond all doubt that Ig variable regions act as antigen receptor on all lymphocytes in the animal studies, they can at least be taken as *prima facie* evidence for this hypothesis. It will be of considerable interest to see how this previously relatively understudied area of immunology expands our views of immune recognition in the next few years.

REFERENCES

Abel, C. A., and Grey, H. M., 1968, Studies on the structure of mouse A myeloma proteins, *Biochemistry* **7**:2682.

Abramson, N., Gelfand, E. W., Jandl, J. H., and Rosen, F. S., 1970, The interaction between human monocytes and red cells: Specificity for IgG subclasses and IgG fragments, *J. Exp. Med.* **132**:1207.

Acton, R. T., Weinheimer, P. F., Hall, S. J., Niedermeier, W., Shelton, E., and Bennett, J. C., 1971, Tetrameric immune macroglobulins in three orders of bony fish, *Proc. Natl. Acad. Sci. USA* **68**:107.

Acton, R. T., Evans, E. E., Weinheimer, P. F., Niedermeier, W., and Bennett, J. C., 1972a, Purification and characterization of two classes of immunoglobulins from the marine toad, *Bufo marinus*, *Biochemistry* **11**:2751.

Acton, R. T., Weinheimer, P. F., Shelton, E., Niedermeier, W., and Bennett, J. C., 1972b, Phylogeny of immunoglobulins—Purification and physicochemical characterization of the immune macroglobulin from the turtle, *Pseudemys scripta*, *Immunochemistry* **9**:421.

Atwell, J. L., and Marchalonis, J. J., 1976, Immunoglobulin classes of lower vertebrates distinct from IgM immunoglobulin, in: *Comparative Immunology* (J. J. Marchalonis, ed.), pp. 276–297, Blackwell, Oxford.

Atwell, J. L., Marchalonis, J. J., and Ealey, E. H. M., 1973, Major immunoglobulin classes of echidna, *Tachyglossus aculeatus*, *Immunology* **25**:835.

Azzolina, L. S., 1975, Differentiation of antigen-binding cells in the teleost *Carassius auratus* and in the anuran, *Bufo marinus*, *Haematologica* **60**:409.

Azzolina, L. S., 1978, Antigen recognition and immune response in goldfish *Carassius auratus* at different temperatures, *Dev. Comp. Immunol.* **2**:77.

Benedict, A. A., and Yamaga, K., 1976, Immunoglobulins and antibody production in avian species, in: *Comparative Immunology* (J. J. Marchalonis, ed.), pp. 335–375, Blackwell, Oxford.

Binz, H., and Wigzell, H., 1977, Antigen-binding, idiotypic T-lymphocyte receptors, *Contemp. Top. Immunobiol.* **7**:113.

Boffa, G. A., Fine, J. M., Drilhon, A., and Amouch, P., 1967, Immunoglobulins and transferrin in marine lamprey sera, *Nature (London)* **214**:700.

Bradshaw, C. M., Clem, L. W., and Sigel, M. M., 1971, Immunologic and immunochemical studies on the gar, *Lepisosteus platyrhincus*. II. Purification and characterization of immunoglobulin, *J. Immunol.* **106**:1480.

Burnet, F. M., 1959, *The Clonal Selection Theory of Acquired Immunity*, Vanderbilt University Press, Nashville, Tenn.

Cantor, H., and Boyse, E., 1977, Regulation of the immune response by T-cell subclasses, *Contemp. Top. Immunobiol.* **7**:47.

Charlemagne, J., and Tournefier, A., 1975, Cell surface immunoglobulins of thymus and spleen lymphocytes in urodele amphibian *Pleurodeles waltlii* (Salamandridae), *Adv. Exp. Med. Biol.* **64**:251.

Chess, L., and Schlossman, S. F., 1977, Functional analysis of distinct human T-cell subsets bearing unique differentiation antigens, *Contemp. Top. Immunobiol.* **7**:363.

Clem, L. W., 1971, Phylogeny of immunoglobulin structure and function. IV. Immunoglobulins of the giant grouper, *Epinephelus itaira*, *J. Biol. Chem.* **246**:9.

Clem, L. W., and McLean, W. E., 1975, Phylogeny of immunoglobulin structure and function. VII. Monomeric and tetrameric immunoglobulins of the margate, a marine teleost fish, *Immunology* **29**:791.

Clem, L. W., and Small, P. A., 1967, Phylogeny of immunoglobulin structure and function. I. Immunoglobulins of the lemon shark, *J. Exp. Med.* **125**:893.

Clem, L. W., and Small, P. A., 1970, Phylogeny of immunoglobulin structure and function. V. Valences and association constants of teleost antibodies to a haptenic determinant, *J. Exp. Med.* **132**:385.

Clem, L. W., McLean, W. E., Shankey, V. T., and Cuchens, M. A. 1977, Phylogeny of lymphocyte heterogeneity. I. Membrane immunoglobulins of teleost lymphocytes, *Dev. Comp. Immunol.* **1**:105.

Cone, R. E., 1976, Factors influencing the isolation of membrane immunoglobulins from T and B lymphocytes. I. Detergent effects and iodination conditions, *J. Immunol.* **116**:847.

Cone, R. E., and Brown, W. C., 1976, Isolation of membrane associated immunoglobulins from T lymphocytes by nonionic detergents, *Immunochemistry* **13**:571.

Cone, R. E., and Marchalonis, J. J., 1972, Cellular and humoral aspects of the influence of environmental temperature on the immune response of poikilothermic vertebrates, *J. Immunol.* **108**:952.

Cooper, H. L., 1973, Effects of mitogens on the mitotic cycle: A biochemical evaluation of lymphocyte activation, In: *Drugs and the Cell Cycle* (A. M. Zimmerman, G. M. Padille, and I. L. Cameron, eds.), pp. 137–194, Academic Press, New York.

Cuchens, M. A., and Clem, L. W., 1977, Phylogeny of lymphocyte heterogeneity. II. Differential effects of temperature on fish T-like and B-like cells, *Cell. Immunol.* **34**:219.

Cuchens, M. A., McLean, W. E., and Clem, L. W., 1975, Lymphocyte heterogeneity in fish and reptiles, in: *Phylogeny of Thymus and Bone Marrow-Bursa Cells* (R. K. Wright and E. L. Cooper, eds.), pp. 205–213, Elsevier/North-Holland, Amsterdam.

Decker, J. M., and Marchalonis, J. J., 1978, Molecular events in lymphocyte activation: Role of nonhistone chromosomal proteins in regulating gene expression, *Contemp. Top. Mol. Immunol.* **7**:365.

DeLuca, D., Warr, G. W., and Marchalonis, J. J., 1978, Phylogenetic origins of immune recognition: Lymphocyte surface immunoglobulins and antigen-binding in the genus *Carassius* (Teleostii), *Eur. J. Immunol.* **8**:525.

Du Pasquier, L., 1976, Amphibian models for the study of the ontogeny of immunity, in: *Comparative Immunology* (J. J. Marchalonis, ed.), pp. 390–418, Blackwell, Oxford.

Du Pasquier, L., Weiss, N., and Loor, F., 1972, Direct evidence for immunoglobulins on the surface of thymus lymphocytes of amphibian larvae, *Eur. J. Immunol.* **2**:366.

Edelman, G. M., 1976, Surface modulation in cell recognition and growth, *Science* **192**:218.

Edelman, G. M., and Gall, W. E., 1969, The antibody problem, *Annu. Rev. Biochem.* **38**:415.

Edwards, B. F., Ruben, L. N., Marchalonis, J. J., and Hylton, G., 1975, Surface characteristics of spleen cell–erythrocyte rosette formation in the grass frog, *Rana pipiens*, *Adv. Exp. Med. Biol.* **64:**397.

Ellis, A. E., and Parkhouse, R. M. E., 1975, Surface immunoglobulins on the lymphocytes of the skate, *Raja naevus*, *Eur. J. Immunol.* **5:**726.

Emmrich, F., Richter, R. F., and Ambrosius, H., 1975, Immunoglobulin determinants on the surface of lymphoid cells of carps, *Eur. J. Immunol.* **5:**76.

Etlinger, H. M., Hodgins, H. O., and Chiller, J. M., 1977, Evolution of the lymphoid system. II. Evidence for immunoglobulin determinants on all rainbow trout lymphocytes and demonstration of mixed leucocyte reaction, *Eur. J. Immunol.* **7:**881.

Fiebig, H., and Ambrosius, H., 1975, Cell surface immunoglobulin of lymphocytes in lower vertebrates, in: *Phylogeny of Thymus and Bone Marrow-Bursa Cells* (R. K. Wright and E. L. Cooper, eds.), pp. 195–203, Elsevier/North-Holland, Amsterdam.

Fiebig, H., Gruhn, R., and Ambrosius, H., 1977, Studies on the control of IgM antibody synthesis. III. Preferential formation of anti-DNP antibodies of high functional affinity in the course of the immune response in carp, *Immunochemistry* **14:**721.

Fiebig, H., Scherbaum, I., and Ambrosius, H., 1980, Evolutionary origin of the T lymphocyte receptor. I. Immunochemical investigation of immunoglobulinlike cell surface protein of carp thymocytes, *Mol. Immunol.* **17:**971.

Fletcher, T. C., and Grant, P. T., 1969, Immunoglobulins in the serum and mucus of the plaice (*Pleuronectes platessa*), *Biochem. J.* **115:**658.

Frommel, D., Litman, G. W., Finstad, J., and Good, R. A., 1971, The evolution of the immune response. XI. The immunoglobulins of the horned shark, *Heterodontus francisci*: Purification, characterization and structural requirement for antibody activity, *J. Immunol.* **106:**1234.

Fuller, L., Murray, J., and Jensen, J. A., 1978, Isolation from nurse shark of immune 7S antibodies with two different molecular weight H chains, *Immunochemistry* **15:**251.

Gally, J. A., 1973, Structure of immunoglobulins, in: The Antigens (M. Sela, ed.), Vol. I, pp. 161–298, Academic Press, New York.

Gitlin, D., Pericelli, A., and Gitlin, J. D., 1973, Multiple immunoglobulin classes among sharks and their evolution, *Comp. Biochem. Physiol. B* **66:**225.

Greaves, M. F., Owen, J. J., and Raff, M. C., 1973, T and B lymphocytes: Origins, properties and roles in immune responses, Excerpta Medica, Amsterdam.

Hämmerling, U., Mack, C., and Pickel, H. G., 1976, Immunofluorescence analysis of Ig determinants on mouse thymocytes and T cells, *Immunochemistry* **13:**525.

Hay, F. C., Torrigiani, G., and Roitt, I. M., 1972, The binding of human IgG subclasses to human monocytes, *Eur. J. Immunol.* **2:**257.

Hildemann, W. H., and Reddy, A. L., 1973, Phylogeny of immune responsiveness: Marine invertebrates, *Fed. Proc.* **32:**2188.

Johnston, W. H., Acton, R. T., Weinheimer, P. F., Niedermeier, W., Evans, E. E., Shelton, E., and Bennett, J. C., 1971, Isolation and physicochemical characterization of the IgM-like immunoglobulin from the stingray *Dasyatis americana*, *J. Immunol.* **107:**782.

Jones, V. E., Graves, H. E., and Orlans, E., 1976, The detection of (Fab')$_2$—related surface antigens on the thymocytes of children, *Immunology* **30:**281.

Ju, S.-T., and Dorf, M. E., 1979, Idiotypic analysis of antibodies against the terpolymer L-glutamic acid^{60}-L-alanine^{30}-L-tyrosine^{10} (GAT). IV. Induction of CGAT idiotype following immunization with various synthetic polymers containing glutamic acid and tyrosine, *Eur. J. Immunol.* **9:** 553.

Ju, S.-T., Benacerraf, B., and Dorf, M. E., 1978, Idiotypic analysis of antibodies to poly-(Glu^{60}Ala^{30}Tyr^{10}): Interstrain and interspecies idiotypic cross-reactions, *Proc. Natl. Acad. Sci. USA* **75:**6192.

Jurd, R. D., 1977, Secretory immunoglobulins and gut-associated lymphoid tissue in *Xenopus laevis*, in: *Developmental Immunobiology* (J. B. Solomon and J. D. Horton, eds.), pp. 307–316, Elsevier/North-Holland, Amsterdam.

Jurd, R. D., and Stevenson, G. T., 1976, Surface immunoglobulins on *Xenopus laevis* lymphocytes, *Comp. Biochem. Physiol. A* **53:**381.

Kabat, E. A., Wu, T. T., and Bilofsky, H., 1976, Variable regions of immunoglobulin chains, *Medical Computer Systems*, Cambridge, Mass.

Kehoe, J. M., and Capra, J. D., 1974, Phylogenetic aspects of immunoglobulin variable region diversity, *Contemp. Top. Mol. Immunol.* **3**:143.

Kehoe, J. M., Sharon, J., Gerber-Jenson, B., and Litman, G. W., 1978, The structure of immunoglobulin variable regions in the horned shark, *Heterodontus francisci*, *Immunogenetics* **7**:35.

Klaus, G. G. B., Nitecki, D. E., and Goodman, J. W., 1971a, Amino acid sequences of free and blacked N-termini of leopard shark immunoglobulins, *J. Immunol.* **107**:1250.

Klaus, G. G. B., Halpern, M. S., Koshland, M. E., and Goodman, J. W., 1971b, A polypeptide chain from leopard shark 19S immunoglobulin analogous to mammalian J chain, *J. Immunol.* **107**:1785.

Koshland, M. E., 1975, Structure and function of the J chain, *Adv. Immunol.* **20**:41.

Leslie, G. A., and Clem, L. W., 1969, Phylogeny of immunoglobulin structure and function. III. Immunoglobulins of the chicken, *J. Exp. Med.* **130**:1377.

Ling, N. R., and Kay, J. E., 1975, *Lymphocyte Stimulation*, North-Holland, Amsterdam.

Litman, G. W., Frommel, D., Finstad, J., Howell, J., Pollara, B. W., and Good, R. A., 1970, The evolution of the immune response. VII. Structural studies of the lamprey immunoglobulin, *J. Immunol.* **105**:1278.

Litman, G. W., Frommel, D., Finstad, J., and Good, R. A., 1971a, The evolution of the immune response. IX. Immunoglobulins of the bowfin; purification and characterization, *J. Immunol.* **106**:747.

Litman, G. W., Rosenberg, A., Frommel, D., Pollara, B., Finstad, J., and Good, R. A., 1971b, Biophysical studies of the immunoglobulins. The circular dichroic spectra of the immunoglobulins—A phylogenetic comparison, *Int. Arch. Allergy Appl. Immunol.* **40**:551.

Marchalonis, J. J., 1969, Isolation and characterization of immunoglobulin-like proteins of the Australian lungfish (*Neoceratodus forsteri*), *Aust. J. Exp. Biol. Med. Sci.* **47**:405.

Marchalonis, J. J., 1971, Isolation and partial characterization of immunoglobulins of goldfish (*Carassius auratus*) and carp (*Cyprinus carpio*), *Immunology* **20**:161.

Marchalonis, J. J., 1977, *Immunity in Evolution*, p. 110, Harvard University Press, Cambridge, Mass.

Marchalonis, J. J., and Cohen, N., 1973, Isolation and partial characterization of immunoglobulin from a urodele amphibian (*Necturus maculosus*), *Immunology* **24**:395.

Marchalonis, J. J., and Edelman, G. M., 1965, Phylogenetic origins of antibody structure. I. Multichain structure of immunoglobulins in the smooth dogfish (*Mustelus canis*), *J. Exp. Med.* **122**:610.

Marchalonis, J. J., and Edelman, G. M., 1966, Polypeptide chains of immunoglobulins from the smooth dogfish (*Mustelus canis*), *Science* **154**:1567.

Marchalonis, J. J., and Edelman, G. M., 1968, Phylogenetic origins of antibody structure. III. Antibodies in the primary immune response of the sea lamprey, *Petromyzon marinus*, *J. Exp. Med.* **127**:891.

Marchalonis, J. J., and Schonfeld, S. A., 1970, Polypeptide chain structure of stingray immunoglobulin, *Biochim. Biophys. Acta* **221**:604.

Marchalonis, J. J., Cone, R. E., and Santer, V. E., 1971, Enzymic iodination: A probe for accessible surface proteins of normal and neoplastic lymphocytes, *Biochem. J.* **124**:921.

Mattes, M. J., and Steiner, L. A., 1978a, Antisera to frog immunoglobulins cross-react with a periodate-sensitive cell surface determinant, *Nature (London)* **273**:761.

Mattes, M. J., and Steiner, L. A., 1978b, Surface immunoglobulin on frog lymphocytes: Identification of two lymphocyte populations, *J. Immunol.* **121**:1116.

Melcher, U., and Uhr, J. W., 1977, Density differences between membrane and secreted immunoglobulins of murine splenocytes, *Biochemistry* **16**:145.

Melcher, U., Eidels, L., and Uhr, J. W., 1975, Are immunoglobulins integral membrane proteins?, *Nature (London)* **258**:434.

Mestecky, J., Kulhavy, R., Schrohenloher, R. E., Tomana, M., and Wright, G. P., 1975, Identification and properties of J chain isolated from catfish macroglobulin, *J. Immunol.* **115**:993.

Miller, J. F. A. P., 1972, Lymphocyte interactions in antibody responses, *Int. Rev. Cytol.* **33**:77.

Moseley, J. M., Marchalonis, J. J., Harris, A. W., and Pye, J., 1977, Molecular properties of T lymphoma immunoglobulin. I. Serological and general physicochemical properties, *J. Immunogenet.* **4**:233.

Moticka, E. J., Brown, B. A., and Cooper, E. L., 1973, Immunoglobulin synthesis in bullfrog larvae, *J. Immunol.* **110**:855.

Muller-Eberhard, H. J., 1968, Chemistry and reaction mechanisms of complement, *Adv. Immunol.* **8**:1.

Ortiz-Muniz, G., and Sigel, M. M., 1971, Antibody synthesis in lymphoid organs of two marine teleosts, *J. Reticuloendothelial Soc.* **9**:42.

Parkhouse, R. M. E., Askonas, B. A., and Dourmashkin, R. R., 1970, Electron microscopic studies of mouse immunoglobulin M: Structure and reconstitution following reduction, *Immunology* **18**:575.

Paul, W. E., and Benacerraf, B., 1977, Functional specificity of thymus-derived lymphocytes, *Science* **195**:1293.

Pollara, B., Swan, A., Finstad, J., and Good, R. A., 1968, N-Terminal amino acid sequences of immunoglobulin chains in *Polyodon spathula*, *Proc. Natl. Acad. Sci. USA* **59**:1307.

Raison, R. L., Hull, C. J., and Hildemann, W. H., 1978, Characterization of immunoglobulin from the Pacific hagfish, a primitive vertebrate, *Proc. Natl. Acad. Sci. USA* **75**:5679.

Rajewsky, K., and Eichmann, K., 1977, Antigen receptors of T helper cells, *Contemp. Top. Immunobiol.* **7**:69.

Riesen, W. F., 1979, Idiotypic cross-reactivity of human and murine phosphorylcholine-binding immunoglobulins, *Eur. J. Immunol.* **9**:421.

Roubal, W. T., Etlinger, H. M., and Hodgins, H. O., 1974, Spin-label studies of a hapten combining site of rainbow trout antibody, *J. Immunol.* **113**:309.

Ruben, L. N., and Edwards, B. F., 1977, Phenotypic restriction of antigen-binding specificity on immunized amphibian spleen cells, *Cell. Immunol.* **33**:437.

Ruben, L. N., and Edwards, B. F., 1979, The phylogeny of the emergence of "T–B" collaboration, *Contemp. Top. Immunobiol.* **9**:55–89.

Ruben, L. N., Warr, G. W., Decker, J. M., and Marchalonis, J. J., 1977, Phylogenetic origins of immune recognition: Lymphoid heterogeneity and the hapten/carrier effect in the goldfish, *Carassius auratus Cell. Immunol.* **31**:266.

Saluk, P. H., Drauss, J., and Clem, L. W., 1970, The presence of two antigenically distinct light chains (κ and λ) in alligator immunoglobulins, *Proc. Soc. Exp. Biol. Med.* **113**:365.

Shelton, E., and Smith, M., 1970, The ultrastructure of carp (*Cyprinus carpio*) immunoglobulin: A tetrameric macroglobulin, *J. Mol. Biol.* **54**:615.

Siskind, G. W., and Benacerraf, B., 1969, Cell selection in the immune response, *Adv. Immunol.* **10**:1.

Sledge, C., Clem, L. W., and Hood, L., 1974, Antibody structure: Amino terminal sequence of nurse shark light and heavy chains, *J. Immunol.* **112**:941.

Szenberg, A., Marchalonis, J. J., and Warner, N. L., 1977, Direct demonstration of endogenous murine thymus-dependent cell surface immunoglobulin, *Proc. Natl. Acad. Sci. USA* **74**:2113.

Trump, G. N., 1970, Goldfish immunoglobulins and antibodies to bovine serum albumin, *J. Immunol.* **104**:1267.

Turk, J. L., 1975, *Delayed Hypersensitivity*, North-Holland, Amsterdam.

Uhr, J. W., Finkelstein, M. S., and Franklin, E. L., 1962, Antibody response to bacteriophage φX 174 in nonmammalian vertebrates, *Proc. Soc. Exp. Biol. Med.* **111**:13.

Warr, G. W., 1979, Membrane immunoglobulins of vertebrate lymphocytes, *Contemp. Top. Immunobiol.* **9**:141.

Warr, G. W., and Marchalonis, J. J., 1977a, Lymphocyte surface immunoglobulin of the goldfish differs from its serum counterpart, *Dev. Comp. Immunol.* **1**:15.

Warr, G. W., and Marchalonis, J. J., 1977b, Lymphocyte surface immunoglobulins: Detection, characterization and occurrence in diseases of the lymphoid system, *Crit. Rev. Clin. Lab. Sci.* **7**:185.

Warr, G. W., and Marchalonis, J. J., 1978, Specific immune recognition by lymphocytes: An evolutionary perspective, *Q. Rev. Biol.* **53**:225.

Warr, G. W., DeLuca, D., and Marchalonis, J. J., 1976, Phylogenetic origins of immune recognition: Lymphocyte surface immunoglobulins in the goldfish, *Carassius auratus*, *Proc. Natl. Acad. Sci. USA* **73**:2476.

Warr, G. W., DeLuca, D., Decker, J. M., Marchalonis, J. J., and Ruben, L. N., 1977, Lymphoid heterogeneity in teleost fish: Studies on the genus *Carassius*, in: *Developmental Immunobiology* (J. B. Solomon and J. D. Horton, eds.), pp. 241–248, Elsevier/North-Holland, Amsterdam.

Warr, G. W., DeLuca, D., and Griffin, B. R., 1979, Membrane immunoglobulin is present on thymic and splenic lymphocytes of the trout *Salmo gairdneri, J. Immunol.* **123**:910.

Wasserman, R. L., Kehoe, J. M., and Capra, J. D., 1974, The V_H III subgroup of immunoglobulin heavy chains: Phylogenetically associated residues in several avian species, *J. Immunol.* **13**:954.

Wedner, H. J., and Parker, C. W., 1976, Lymphocyte activation, *Prog. Allergy* **20**:195.

Weinheimer, P. F., Mestecky, J., and Acton, R. T., 1971, Species distribution of J chain, *J. Immunol.* **107**:1211.

Weir, D. W., and Ogmundsdottir, H. M., 1977, Nonspecific recognition mechanisms by mononuclear phagocytes, *Clin. Exp. Immunol.* **30**:323.

Wilkinson, P. C., 1976, Recognition and response in mononuclear and granular phagocytes, *Clin. Exp. Immunol.* **25**:355.

Wu, T. T., and Kabat, E. A., 1970, An analysis of the sequences of the variable regions of Bence–Jones proteins and myeloma light chains and their implications for antibody complementarity, *J. Exp. Med.* **132**:211.

Yamaga, K. M., Kubo, R. T., and Etlinger, H. M., 1978a, Studies on the question of conventional immunoglobulin on thymocytes from primitive vertebrates. I. Presence of anti-carbohydrate antibodies in rabbit anti-trout Ig sera, *J. Immunol.* **120**:2068.

Yamaga, K. M., Kubo, R. T., and Etlinger, H. M., 1978b, Studies on the question of conventional immunoglobulin on thymocytes from primitive vertebrates. II. Delineation between Ig-specific and cross-reactive membrane components, *J. Immunol.* **120**:2076.

Zikan, J., 1974 Workshop report, in: *Progress in Immunology. II. Immunochemical Aspects* (L. Brent and J. Holborow, eds.), Vol. 1, p. 246, American Elsevier, New York.

15

Ontogeny of Amphibian Hemopoietic Cells

JAMES B. TURPEN, NICHOLAS COHEN, PIERRE DEPARIS, ANDRÉ JAYLET, ROBERT TOMPKINS, and E. PETER VOLPE

1. INTRODUCTION

The developmental origin of lymphoid cells of vertebrates continues to be a fascinating, perplexing problem. Historically, the disputable issues have revolved around two fundamental aspects: the origin of the precursors of the various types of differentiated blood cells and the lineal relationships among these different cell types. There are special features of amphibians that make them suitable for clarifying the embryogenesis of hemopoietic cells. An array of microsurgical techniques from experimental embryology can be used to marked advantage. In this chapter, we will review the contributions made by investigators who have used the tools of experimental embryology to gain insight into the ontogeny of vertebrate blood cells.

The starting point for most discussions on hemopoiesis is the hematogeneous theory as set forth by Moore and Owen (1967) and Metcalf and Moore (1971). In essence, this theory states that pluripotent hemopoietic stem cells are formed in the extraembryonic blood islands of the yolk sac. Subsequently, these stem cells migrate via the circulation and colonize receptive intraembryonic

---

JAMES B. TURPEN • Department of Biology, Pennsylvania State University, University Park, Pennsylvania 16802. NICHOLAS COHEN • Department of Microbiology, Division of Immunology, University of Rochester School of Medicine and Dentistry, Rochester, New York 14642. PIERRE DEPARIS and ANDRÉ JAYLET • Laboratoire de Biologie générale, Université Paul Sabatier, 31077 Toulouse Cedex, France. ROBERT TOMPKINS • Department of Biology, Tulane University, New Orleans, Louisiana 70118. E. PETER VOLPE • Department of Basic Medical Sciences, Mercer University, School of Medicine, Macon, Georgia 31207. Research work and manuscript preparation supported by National Institutes of Health Grant HD-11179 to J.B.T.; HD-07901 to N.C.; by National Science Foundation Grant PCM-7903827 to R.T. and E.P.V.; and by the program of the Equipe de Recherche Associea no 327 (CNRS) to P.D. and A.J. J.B.T. is a recipient of Research Career Development Award HD-00319 from the USPHS.

hemopoietic organs where they are induced to differentiate along specific developmental pathways by local reticuloendothelial microenvironments. Sequential changes in sites of hemopoietic activity during development are the result of hematogeneous metastases arising from progeny of the initial stem cells.

The yolk-laden ventral blood islands in amphibians are the counterpart of the yolk sac blood islands in birds and mammals, a site where the initial red blood cell population differentiates. Additional hemopoietic organs include thymus, lymph nodes, spleen, liver, mesonephros, gut-associated lymphoid tissue, ventral cavity bodies, and bone marrow; each represents a hemopoietic microenvironment.

2. VENTRAL BLOOD ISLANDS AND ERYTHROPOIESIS

The mesodermal ventral blood islands (VBI) of the amphibian embryo are derived from the segment of the early gastrula that lies opposite the blastopore. Following involution during gastrulation, this mesoderm lies on the ventral surface of the embryo and extends from the ventral lip of the blastopore anteriorly into the free edges of the lateral plate in the region of the liver diverticulum (Yamada, 1937). By the time the embryo has reached the stage of muscular response, the blood islands extend along the midventral line from the cloaca to the region of the liver where the mesoderm splits to form a "Y" whose "arms" extend on each side of the liver (Fernald, 1943). Cellular differentiation begins within the blood islands at the time of muscular response.

In earliest experimental studies, the presumptive and definitive blood islands were extirpated from a variety of amphibian species (Federici, 1926; Goss, 1928; Slonimski, 1931; Stohr, 1931; Storti, 1935). The most notable experiments were performed by Federici (1926) on the frog *Rana fusca* and by Goss (1928) on the salamander *Ambystoma punctatum*. Removal of the blood islands prior to the establishment of circulation resulted in either reduction or total elimination of circulating RBCs. Goss was able to correlate a reduction in RBC numbers with a incomplete removal of the blood island primordium. Histological examinations of serial sections revealed a normal vasculature and normal endothelial linings in the experimental animals. However, Goss noted an overall reduction in the size of several major organs, particularly the liver and kidney.

Fernald (1940) extended these early observations to another anuran, *Hyla regilla*. Following the extirpation of blood island ectoderm and mesoderm from preneurula and neurula stages, there was a reduction in the number of circulating blood cells. In no case, however, was there a complete absence of cells, either red or white. In larvae that survived for at least 3 weeks, there was a gradual increase in the cellular components of the blood. Additionally, explants of blood island were cultured alone or in combination with other anlagen. Specifically, 71 explants of presumptive blood islands from the ventral marginal zone of early gastrulae were enclosed within an ectodermal jacket obtained from the ventral area of mid and late gastrulae; 58 explants survived for 8 days in culture and were examined histologically. Typically, these explants differentiated into mesenchyme and RBCs. Control explants of ectodermal jackets cultured alone developed only into epidermis. Thus, cells of ventral marginal zone are capable of self-differentiation into RBCs. Explants of presumptive blood islands from early neurulae, placed in ectodermal enclosures, likewise differentiated into RBCs (24 of 27 cases).

In a second series of experiments, presumptive blood islands from early neurulae were combined with notochordal material from the same stage and then wrapped in ectodermal jackets. Chordal material explanted alone differentiated into neural, chordal, and muscular tissue, but not into blood cells or pronephric tissues. Of 29 combined explants (chordal + VBI), 29 differentiated into neural and chordal tissue, 26 developed into muscle, 22 developed into pronephric tissue, and 10 developed into blood cells. These experiments suggest that chordal material induces ventral mesoderm to give rise to pronephros. They also suggest that the blood islands are not determined in the early neurula and that ventral mesoderm at this stage is multipotent.

These early reports allow us to draw the following conclusions about the VBI: First, this region gives rise to a population of differentiated RBCs quite early in amphibian embryogenesis. Second, explanation experiments suggest that VBI mesoderm is multipotent in the early neurula stage but probably becomes determined for erythropoiesis by the tail bud stage of development.

Subsequently, Finnegan (1953) analyzed the parameters involved in VBI determination and differentiation as assessed by the production of hemoglobin. Explants of VBI were compressed between glass slides. The cells that developed had the morphological appearance of RBCs but they did not appear to synthesize hemoglobin. Moreover, macrophagelike cells, capable of phagocytosing contaminating bacteria, also developed in these explants. If explants of VBI ectoderm and mesoderm were enclosed in an ectodermal cover (derived from late gastrulae), ovoid blood cells developed that also did not appear to synthesize hemoglobin. When VBI ectoderm, mesoderm, and endoderm were explanted, the results were the same as if VBI ectoderm and mesoderm had been wrapped in ectodermal coverings, i.e., RBC-shaped cells but no hemoglobin synthesis. In contrast, when all three VBI germ layers were enclosed in an ectodermal covering, differentiation of RBCs that synthesized hemoglobin resulted. From these experiments, it appears that morphological determination of RBCs can be separated from the expression of hemoglobin synthesis. To achieve complete differentiation, all three germ layers must be confined in an ectodermal ball. VBI ectoderm and mesoderm unconfined in Holtfreter's solution did not give rise to RBCs, but under these conditions, Finnegan reported phagocytosis by mesenchymal cells.

The results thus far discussed are consistent with the hypothesis of Moore and Owen—that stem cells, capable of differentiating into all classes of definitive blood cells, initially arise in the yolk sac, or its phylogenetic equivalent. In 1966, however, Hollyfield reported experimental results that could be interpreted along different lines. He studied the origin of circulating RBCs in larval *Rana pipiens* by using differences in the size of diploid and triploid RBCs as a cell marker. That is, presumptive hemopoietic anlagen were reciprocally exchanged between embryos of different chromosomal constitutions. Following exchange of VBI mesoderm at Shumway (1940) embryonic stage 15–16, the circulating RBC population of midlarval tadpoles [stages VIII to XXV of Taylor and Kollros (1946)] did not contain cells derived from the grafted material. In a second series of experiments, pronephric anlagen were exchanged between stage 15–16 embryos. Of the four surviving larvae analyzed between larval stages IX to XV, two animals contained circulating RBCs of graft origin and two did not. However, following an exchange of mesonephric anlagen, 13 of 14 larvae, analyzed between stages IX and XV, contained circulating RBCs derived from grafted mesoderm. Based on these results, Hollyfield suggested that the VBI contributes a population of short-lived RBCs, but does not produce stem cells responsible for erythropoiesis in other areas. Rather, the argument was made that erythroblasts residing in the intertubular spaces of the pronephros and mesonephros develop *in situ* (from the mesoderm of the transplanted anlagen). Thus, with regard to the development of circulating RBCs, Hollyfield's data indicate that the RBCs of the larval frog arise from the developing kidney region.

3. EMBRYONIC ORIGIN OF LYMPHOCYTES DURING NORMAL AMPHIBIAN DEVELOPMENT

3.1. ONTOGENY OF THYMOCYTES: STUDIES WITH ANURANS

Historically, the origin of lymphocytes within the thymus during vertebrate embryogenesis has been a complex and controversial issue involving two mutually exclusive viewpoints. One argument has been that thymocytes arise *de novo* from precursors located within the thymic anlage itself. An opposing argument is that thymocytes are derived from precursors that have migrated into the rudiment at some point during its histogenesis.

The problem of lymphocyte ontogeny in frogs was opened in the early 1970s by the studies of Turpen *et al.* (1973, 1975). The transplantation of thymic primordia between chromosomally distinct (diploid and triploid) embryos offered an approach to determining the relative importance during development of factors intrinsic and extrinsic to the differentiating thymus. It was concluded from transplantation experiments on the leopard frog (*R. pipiens*) that the lymphocytes of the thymus arise by the direct transformation of cells within the thymic anlage itself. It is now clear, however, that an error was made in the assertion (Turpen *et al.*, 1973, 1975; Volpe and Turpen, 1975) that the thymic lymphocytes of the frog arise *in situ* from elements in the thymic rudiment.

Recent studies (Nagata, 1977; Tochinai, 1978; Volpe *et al.*, 1977, 1979; Tompkins *et al.*, 1979, 1980; Turpen *et al.*, 1981) show unequivocally that in frogs, like in salamanders (see Section 3.2), the lymphocytes which initially populate the thymus originate from extrinsic hemocytoblasts that migrate into the thymic rudiment. The evidence bearing on the colonization of the thymic rudiment by extrinsic stem cells is based on experimental studies in *R. pipiens* and the African clawed frog (*Xenopus laevis*). It bears emphasizing, however, that there is unanimity in all studies (Turpen *et al.*, 1973, 1975; Turpen and Cohen, 1976b; Volpe *et al.*, 1980) that during normal development the migrant stem cells entering the thymic primordium are *not* derived from the VBI.

An instructive experiment in anurans involves the production of a chimeric embryo (Volpe *et al.*, 1979). In this procedure (Figure 1), two embryos are cut in half transversely and the front halves are interchanged. Such a chimeric embryo can develop into a frog of normal form. The use of chromosomally distinct embryonic halves permits an analysis of the extent to which the histogenesis of organs in each half is influenced by cells migrating from one half to the other. In the representative case illustrated in Figure 1, the chimera was diploid anteriorly and triploid posteriorly. In postmetamorphic life, the cells of both the brain and the thymus were assayed cytophotometrically. Although the anterior half of the chimera was diploid (as evidenced by the diploid brain cells), the lymphocytes of the anteriorly located thymus gland were exclusively triploid. There is no doubt that the anteriorly located thymus is colonized by lymphocytoblasts that originate in the posterior half of the chimera.

In the most recent investigations on chimeras, Flajnik, Horan, and Cohen (unpublished) produced chimeras in *X. laevis* by joining the anterior portions of diploid embryos (cut just behind the gill buds) to the posterior portions (cut just anterior to the pronephros) of triploid embryos of the same developmental age. The ploidy of the cell populations in the thymus and spleen of postmetamorphic chimeras was evaluated by flow cytometry rather than by microdensitometry.

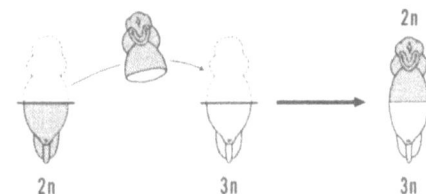

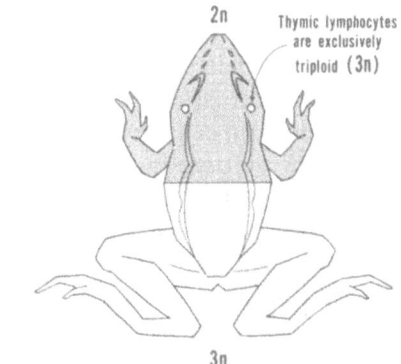

FIGURE 1. In the chimeric frog, the developing thymus is invaded by migratory, blood-borne lymphopoietic cells from the posterior region. The ventral blood islands are apparently not the source of the migratory stem cells.

This highly sensitive analytical procedure confirmed the posterior origin of stem cells that migrate into the thymus.

The production of a chimera differs from the more usual transplantation experiment in at least one important respect. In a transplantation study, a given part (the thymic primordium, for example) is excised and transplanted into the milieu of another embryo. It is conceivable that some controlling agency in the immediate vicinity of the thymic primordium may be important in the differentiation of the thymus, and that the controlling mechanism may be disrupted when the thymic rudiment alone is transplanted. On the other hand, when two halves are grafted together in a chimera, the thymic primordia themselves remain undisturbed in their relations to the whole organism. There is no experimental perturbation of the thymus gland. Accordingly, the presence of posteriorly derived lymphocytes in an undisturbed thymus gland in the anterior region of a chimera is strong evidence that cells migrate into the thymus during ontogeny.

Having established that the development of the thymus is dependent on the inflow of stem cells, we may inquire as to the source of the colonizing lymphopoietic cells. Once again, one might be inclined initially to the view that the colonizing cells are derived from the VBI of the posterior embryonic region. The absence of cellular chimerism in the thymus of chimeric embryos tends to argue against this view. The VBI of a chimeric embryo would be constituted of *both* diploid and triploid hemocytoblasts. If stem cells from all regions of the blood islands were to invade the developing thymus, one should expect to find representative populations of *both* diploid and triploid thymic lymphocytes. This expectation was not realized. To sustain the notion that the VBI do make a cellular contribution to the thymus, one would have to postulate a selective migration of mesenchymal blood elements strongly in favor of only one ploidy of cells.

Additional data substantiating that the VBI are *not* the primary source of thymic stem cells have been obtained by transplanting the VBI (Turpen *et al.*, 1973, 1975, 1979, 1981; Volpe *et al.*, 1977, 1979). In a series of experiments of similar design, presumptive blood islands from a diploid embryo were removed and transplanted orthotopically to a triploid host as well as the reciprocal combination, i.e., diploid hosts with triploid VBI. In both larval and postmetamorphic life, there was no evidence that the thymus derives its lymphoid elements from the orthotopically placed VBI. The thymocytes of the experimental animals were solely of host origin. The absence of donor VBI derivatives in the thymuses indicates that the differentiation of thymic lymphocytes is *not* dependent on cells from the VBI. The thymocytes evidently originate from a region of the embryo that is dorsal to the VBI.

Modern electron microscopic studies reinforce the notion that thymic lymphocytes originate from extrinsic stem cells that migrate into the developing thymic rudiment (Nagata, 1977). Nagata's investigation involved a developmental stage-by-stage analysis of the presence of lymphocyte precursors in the thymic region of *Xenopus*. Essentially, the number of cells in the mesenchyme adjacent to the thymus increased from day 2½ to day 4 postfertilization and decreased by day 6, a time that corresponds with an increase in the number of cells within the rudiment itself.

3.2. ONTOGENY OF THYMOCYTES: STUDIES WITH URODELES

In newts, the thymus gland, on each side, is dorsally situated with respect to the jaw joint. In representatives of various families of salamanders, the thymic buds first appear as thickenings of the gill epithelium that enlarge and subsequently separate from the surrounding mesenchyme. During larval development, the organ progressively acquires the structure and position of the adult thymus. In different families, variable numbers of buds are involved in the formation of the mature gland. In *A. punctatum*, for example, five thymic buds form from the endoderm of the pharyngeal pouches. The first two buds degenerate, and the other three develop and partially fuse to give the final three-lobed thymus (Baldwin, 1918). In *Necturus maculosus*, four thymic buds appear. The first never separates from the endoderm and ultimately degenerates; the adult thymus is formed from the three initial buds (Webster, 1934). In *Salamandra maculosus*, Maurer (1888) described the formation of three thymic buds, whereas in *S. perspicillata*, Livini (1902) observed five. Four of these degenerate, the fifth forming the thymus. In the genus *Triturus*, Maurer (1888) examined the species *T. taeniatus* where three thymic buds are formed. It is not clear, however, whether the adult thymus results from the development of the three buds or merely from one of them. In *T. alpestris*, only one thymic bud arises from the gill epithelium at the junction of the second and third pairs of gills. Migration of the thymus terminates during the 10th week of larval life with the gland in its final position, at which time two essential components are present: epithelial cells and lymphocytes (which appear during the 9th week; Tournefier, 1973). In *Pleurodeles waltlii*, the thymus is also formed from a single bud that arises from the epithelium of the third pair of gills (Desvaux, 1974). After becoming separated from the gill epithelium, the endodermal cell mass is invaded by stem cells, which are found in the neighboring mesenchyme from the 12th to the 16th day of embryonic life (Charlemagne, 1977). It can be seen, therefore, that the gill epithelium gives rise to a variable number of thymic buds in different species. Variability, however, does not coincide with the phylogenetic classification. In the family Salamandridae, for example, the number of buds varies from one in *T. alpestris* and *P. waltlii*, to five in *S. perspicillata*.

To clarify the ontogeny of thymocytes, cellularly and/or chromosomally marked embryonic grafts were transplanted between normal and marked animals before cellular colonization of the thymic rudiment. The species used in these experiments was *P. waltlii*, for which two types of marker are available. One marker, namely tetraploidy (Jaylet, 1972), serves as both a cellular and a chromosomal marker. Since the volume of tetraploid cells is trice that of the diploid cells (Deparis *et al.*, 1975), the cells at interphase are clearly distinguishable by size. Moreover, since the chromosomal complement is also different, $4n$ mitoses can easily be distinguished from $2n$ mitoses at metaphase. One might

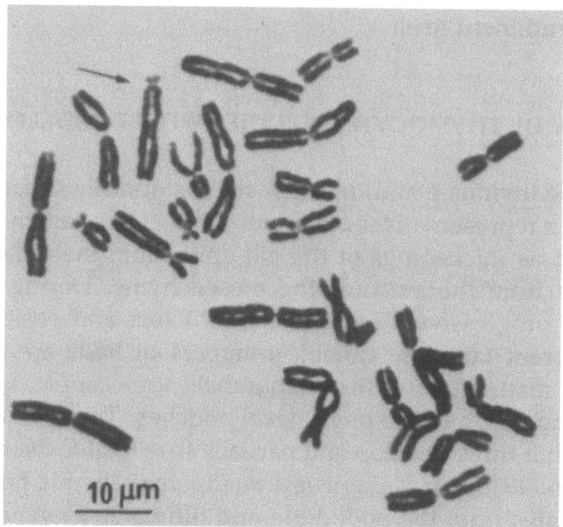

FIGURE 2. Metaphases of diploid *Pleurodeles waltlii* cells. Arrow points to the marker chromosome 6. (From Jaylet and Deparis, 1979; reproduced with permission from *Developmental and Comparative Immunology*.)

argue, however, that if there is any sort of competition between tetraploid and diploid cells, tetraploid cells are at a disadvantage either because of a slower division rate or because of restricted migration due to their size. The use of diploid embryos with a different chromosomal marker (Figure 2) (Jaylet, 1971) invalidates this contention. Accordingly, cell size as well as ploidy proved to be useful and valuable in the analysis.

The thymic bud in *Pleurodeles* develops from the epithelium of the third pair of gills. Therefore, the gill bud (containing the thymic primordium) was reciprocally exchanged, at the tail bud stage, between diploid and tetraploid embryos as well as between diploid embryos and diploid embryos carrying the chromosomal marker. The gill bud grafts were either orthotopically or heterotopically positioned. In the case of gill bud grafts between normal diploid and tetraploid embryos, the differentiated thymuses were sectioned and the diameters of the thymocyte nuclei were measured. Histograms (Figure 3) clearly reveal the host origin of thymocytes. The diameters of the $2n$ nuclei were between 6 and 10 μm, whereas those of the $4n$ cells were between 10 and 16 μm. These values agreed with those obtained from control animals of both ploidies. An overlap did exist, however, for the value of 10 μm. The number of cells found in this zone was low for either ploidy and it was impossible to classify the cells with any certainty in one ploidy group or the other.

In another series of experiments involving reciprocal gill bud grafts between $2n$ and $4n$ embryos, the ploidy of the thymocytes was determined by examining mitotic figures ($2n = 24$). The results (Table 1) again show unequivocally that the thymocytes originate from the host. Finally, the same kind of grafting was carried out but this time it involved normal diploid embryos with and without the chromosomal marker. The data confirmed the results obtained using the tetraploidy marker. Six thymuses were examined and 20 and 30 mitoses were studied in each. In all instances, 100% of the mitoses were from the graft recipient.

The foregoing series of experiments demonstrates an exogenous origin of thymocytes in *P. waltlii* (Deparis and Jaylet, 1976). The thymic stem cells do not originate from perithymic mesenchyme, since mesenchyme was introduced with the graft and no mitoses of the grafted cells were ever observed. To analyze the process of colonization and to study lymphocyte traffic in and out of the young thymus, normal embryos were parabiosed with chromosomally labeled diploid embryos at the tail bud stage. The two thymuses of each parabiont were

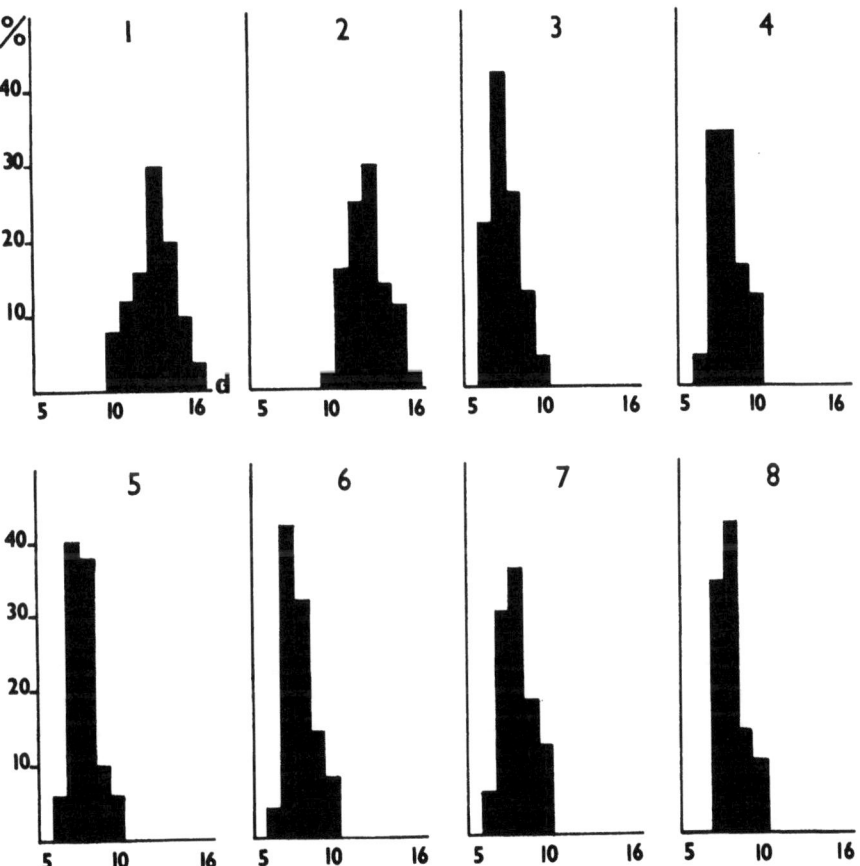

FIGURE 3. Gill buds were grafted between diploid and tetraploid *Pleurodeles* embryos. The differentiated thymuses were sectioned at 8 μm and histograms were plotted from measurements of the diameters of the thymocytes' nuclei. (1) 4n host with 2n graft in orthotopic position; (2) 4n host with 2n graft in heterotopic position; (3–6) 2n host with 4n graft in orthotopic position; (7, 8) 2n host with 4n graft in heterotopic position. % = cell percentage for a given diameter; d = diameter of nuclei (μm). [From Deparis and Jaylet, 1976; reproduced with permission from *Annales d'Immunologie (Institut Pasteur)*.]

TABLE 1. CHROMOSOMAL ANALYSIS OF CELLS PROLIFERATING IN THYMUSES WHOSE ANLAGEN HAD BEEN EXCHANGED BETWEEN $2n$ AND $4n$ PLEURODELES EMBRYOS AT STAGE 50[a]

Graft position	Host ploidy	Donor ploidy	Metaphases counted	Percentage of	
				$2n$ cells	$4n$ cells
Orthotopic	$2n$	$4n$	40	100	0
Orthotopic	$2n$	$4n$	150	100	0
Heterotopic	$2n$	$4n$	70	100	0
Orthotopic	$4n$	$2n$	100	0	100
Orthotopic	$4n$	$2n$	30	0	100
Orthotopic	$4n$	$2n$	50	0	100

[a]From Deparis and Jaylet (1976).

examined separately at Gallien and Durocher (1957) stage 52 (when the fourth digit of the hind leg has formed). This stage was chosen because it is a larval period of rapid growth of the thymus (Desvaux, 1974). The results show that the cell populations of each thymus of one animal develop independently. Indeed, very different cell compositions were frequently found in the two thymuses of the same animal. It was even recorded that one thymus contained a majority of cells from its parabiont partner, whereas the other contained only the cells of its own chromosomal constitution (Jaylet and Deparis, 1979). It is not known whether this observation only applied to the particular development time that was evaluated, namely the beginning of thymic development. It is certainly possible that later in life, cell traffic between the thymuses may become extensive.

The origin of the stem cells that colonize the newt thymus was studied by grafting the embryonic liver bud between $2n$ and $4n$ animals. When such grafts are carried out at the tail bud stage, the host subsequently acquires a large number of blood cells of donor origin (Deparis and Jaylet, 1975). The lymphoid component of the thymuses of these animals consists of a diploid and a tetraploid population that can easily be distinguished by the difference in size of the nucleus. It is probable, therefore, that the mesoderm of the graft (i.e., the anterior region of the embryo's trunk) gives rise to the subsequent blood cells and, in particular, to the stem cells that colonize the epithelial thymic bud.

4. EMBRYONIC ORIGIN OF OTHER HEMOPOIETIC CELLS DURING NORMAL ANURAN DEVELOPMENT

The information reviewed above argues against the doctrine that the lymphopoietic cells of amphibians owe their origin to the VBI (the counterpart of the yolk sac blood islands in amniotes). Recent studies by investigators in different laboratories have been directed toward localizing the primary source of hemocytoblasts to a blood-forming center or district within the embryo. Attention has focused on the liver, pronephros, and mesonephros.

4.1. HEPATIC HEMOPOIESIS

Differential counts of Wright–Giemsa-stained suspensions of *R. pipiens* liver cells have demonstrate that, like mammals, hepatic hemopoiesis involves lymphoid, myeloid, and erythroid cell lines (Turpen *et al.*, 1979). Hemopoietic activity, which appears to be initiated near the end of embryonic development, reaches a substantial level by 30 days of development. Erythropoiesis is the predominant hemopoietic activity in the larval liver. Examination of plastic-embedded tissue sections demonstrated that hemopoietic activity is localized in discrete foci within the endothelium-lined sinusoids. Following transplantation of chromosomally labeled presumptive VBI, derivatives of this region were readily apparent in the circulating RBC population and in the passenger RBC population of the liver. The WBC population, both of the circulation and of the resident hemopoietic cells of the liver, however, did *not* contain derivatives of the transplanted VBI. Since VBI-derived cells were not evident in the larval liver, an anterior–posterior chimera was produced in an attempt to localize the embryonic origin of these hemopoietic cells and to assess whether these cells are derived from a source extrinsic to the liver. The anterior portion of the animal, containing anlagen of both the thymus and the pronephros, was derived from a diploid embryo; the posterior portion (in this particular experiment), containing the liver anlage as well as the VBI, was derived from a triploid animal. Microdensitometric analysis demonstrated that the hepatocyte population was primarily triploid but that the hemopoietic cell population was predominantly diploid. Moreover, the lymphoid population of the thymus and the hemopoietic population of the pronephros were exclusively diploid.

These experiments with the liver are interpretable along several lines. First, extensive liver hemopoiesis appears to be the function of colonizing stem cells. Second, these stem cells do not originate in the VBI. Third, they do come from the dorsal region of the embryo.

4.2. PRONEPHRIC HEMOPOIESIS

Additional studies (Carpenter and Turpen, 1979) have explored the embryogenesis of hemopoietic cells in the pronephros of the leopard frog during embryonic and larval development. Differential cell counts of Wright–Giemsa-stained cell suspensions demonstrate that granulopoiesis is a predominant hemopoietic activity in this organ. Transplantation of the pronephric anlagen between chromosomally labeled embryos followed by microdensitometric analysis, demonstrated that the pronephric tubular cells were derived from the transplanted tissue but that the hemopoietic cells of the pronephros were derived from the host. VBI were first suspected as the extrinsic source of these cells. However, transplantation of VBI did not result in labeling of the resident granulopoietic population of the pronephros. It was intriguing that, whereas hemopoietic cells of the transplanted pronephros were clearly derived from the host in seven of nine instances, the hemopoietic cells of the remaining two individual recipients were of mixed donor and host origin. Thus, either colonization of the pronephric anlagen had occurred prior to the time the anlagen were removed from the donors, or cells from a putative dorsal stem cell compartment were included along with the pronephric anlagen transplants.

4.3. A DORSALLY LOCATED HEMOPOIETIC PRECURSOR CELL COMPARTMENT

The report by Carpenter and Turpen (1979) revealed that granulopoiesis is associated with the pronephros of the frog larva. Independently, in another laboratory, the results by Volpe et al. (1980) showed that lymphoid cells in the thymus as well as erythroid cells have a common developmental origin in the developing kidney region of *Xenopus*. Thus, both studies focused on a blood-forming region located in the dorsal portion of the embryo—particularly in the developing kidney area. Interestingly, there were indications in the earlier literature that the kidney region play an important role in hemopoiesis. In 1966, Hollyfield transplanted cytogenetically labeled kidney anlagen and found that the circulating RBCs of the host were of donor origin.

Turpen et al. (1981) transplanted cytogenetically labeled mesonephric anlagen, both unilaterally and bilaterally, between stage 15–16 *R. pipiens* embryos. This anlage lies posterior to the pronephros and ventral to the developing nervous system, somites, and notochord. Analysis of the hemopoietic organs from 13 stage IV to VI larvae (30–40 days postfertilization) demonstrated that the lymphoid populations of the thymuses contained mixtures of donor- and host-derived cells; the granulocyte populations of the pronephros contained mixtures of donor and host cells; the mesonephroses contained mixed cellular populations; and livers, spleen, and circulating blood also contained mixed populations. Additionally, bone marrow from one juvenile frog contained mixed populations of all types of hemopoietic cells. Thus, transplantation of a single embryonic region resulted in the labeling of representatives of all classes of differentiated hemopoietic cells. Transplantation of ectoderm from this region, without underlying mesoderm, did not result in the labeling of any hemopoietic cells. This indicates a mesodermal original of hemopoietic stem cells.

Similar experiments by Volpe et al. (1981) confirm the developing kidney region as a source of lymphoid precursor cells. As depicted in Figure 4, the regions of presumptive mesonephros from *both* lateral surfaces of a host embryo were removed and replaced with the corresponding right and left presumptive mesonephric tissues of a donor embryo. As shown in the stereodiagram, the transplant of presumptive mesonephros included closely adjacent tissues, notably the dorsal mesentery and lateral dorsal aorta.

In a representative experiment, the donor embryo was diploid; the host embryo was triploid. Cytophotometric analyses in late larval life showed that the lymphocytes of the host thymus gland were predominantly diploid (donor). Thus, the thymus gland is colonized by lymphocytoblasts that originate in the region of the developing mesonephros. Additionally, the experimental data show that RBCs are largely derived from the transplanted tissue of the developing mesonephric region. Approximately 75% of the circulating RBCs were of donor origin ($2n$).

We have independently confirmed in two different laboratories that the nephrogenic region is involved in some manner in the formation of the primary hemopoietic cells. Do the stem cells arise from the presumptive mesonephric tissue per se, or is their presence in this region during early development independent of the mesonephric tissue? Once again, the early experiments of Fernald (1943) and Finnegan (1953) provide insight into these questions as well as additional observations supporting the origin of hemopoietic stem cells in the mesonephric region. Presumptive pronephros and mesonephros from early neurulae of *Hyla* were explanted and cultured in ectodermal bags (Fernald, 1943). Of 31 explants cultured for 8 days, 26 were characterized by the presence of blood cells alone and 3 contained both pronephric tissue and blood cells. However, when presumptive pronephros was combined with notochordal material, the explants differentiated into neural, chordal, muscular, and pronephric tissue, but not blood cells. Finnegan (1953) reported the development of cells resembling embryonic lymphocytes in explants of the lateral hypomere from neurulae to tail-bud-stage *Ambystoma* embryos. In some, but not all cases, abortive pronephric tubules and ciliated nephrostomes also developed. In the experiments of Turpen and colleagues, the transplanted anlage consisted of a large region encompassing the nephrogenic mesoderm, traditionally thought to give rise to the mesonephros, as well as elements of the dorsal, lateral hypomere. Following unilateral transplantation, the hemopoietic cells in the contralateral mesonephros were of mixed donor and host origin. This finding suggests that the mesonephros itself is colonized by stem cells of extrinsic origin. Moreover, a considerable period of time separates the presence of stem cells in this region (67 hr) from the subsequent histogenesis of the organ at 10–15 days of development (Horton, 1970). These considerations argue against the direct involvement of mesonephric tissue per se and point to an independent development of the stem cells.

What we have established is that a population of hemopoietic precursor cells arises within a specific region on the lateral aspect of the amphibian embryo. As seen in the stereodiagram (Figure 4), this region includes the lateral mesoderm, the intermediate mesoderm, the dorsal mesentery, the lateral aorta, and the genital ridge. The blood vessels in this region merit attention. Turpen *et al.* (1981) have examined serially sectioned plastic-embedded stage 16 through stage 20 *Rana* embryos. The most intriguing finding was seen in the anterior pronephric region of stage 19–20 embryos. At stage 19, groups of cells were observed adhering to the outside of the developing dorsal aorta and to the outside of the tubular pronephros. By stage 20, these cells were seen both inside

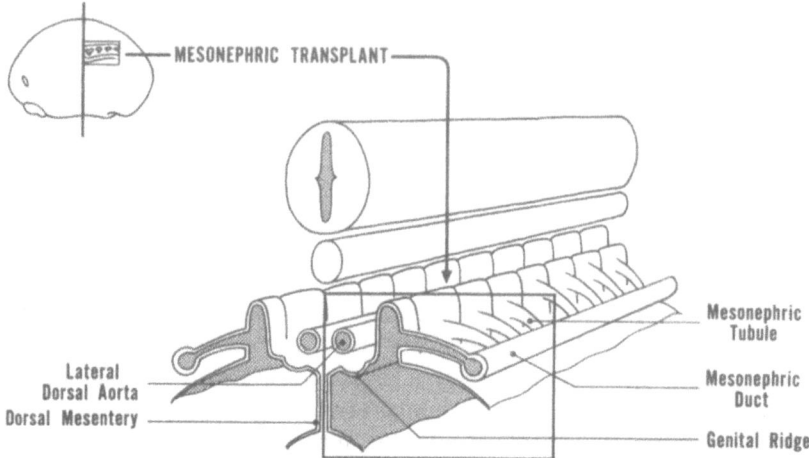

FIGURE 4. The site of the mesonephric transplant in *Xenopus* embryos and a highly schematic stereodiagram of the relationship of structures which may eventually develop within the region.

and outside the aortae and the pronephros. By tracing serial sections, it was shown that the posterior cardinal vein develops in close apposition to the pronephros, traversing the organ between the outer capsule and the inner tubular region. These observations suggest that the pronephric region and adjacent aortae may be sites where hemopoietic stem cells initially invade the circulatory network for subsequent distribution to the hemopoietic organs. Additionally, since the transplanted mesonephric anlage (containing the precursor cells) is located posterior to the pronephric region, these observations imply a posterior-to-anterior migratory pattern.

A series of experiments have been designed to answer the following two questions: Is there a posterior-to-anterior migration? Do cells derived from the dorsal, lateral hypomere invade the pronephros and aortae at stages 19–20 (Turpen and Knudson, 1982)? Anterior/posterior "half and half" chimeras have been constructed at stages 14, 15, 16, and 17 between $2n$ and $3n$ embryos. All embryos were "harvested" within a 1-week period at larval stage IV–VI. The line of bisection between the $2n$ and $3n$ halves of the embryo was at the posterior margin of the pronephros, a location that was confirmed by measuring the cytogenetic profile of the pronephric parenchymal cells. In all experiments except one, hemopoietic cells derived from the anterior portion of the chimeras were present in the hemopoietic cell populations of the various organs (thymus, pronephros, mesonephros, liver, and blood). However, the number of anterior-derived cells observed in the stage 14 chimeras was very low ($< 5\%$). The number of anterior-derived cells increased at stage 15–16 (to $\sim 10\%$) and increased again at stage 17 (to $\sim 22\%$). These data suggest that, with time, the number of precursor cells normally present in the anterior regions of the embryo increases. The following complementary experiment demonstrates that, by stage 19–20, these precursor cells are associated with the pronephros and aortae. Dorsal lateral mesoderm was reciprocally transplanted (either unilaterally or bilaterally) between cytogenetically labeled embryos at stages 14, 15, and 16. At stages 19 and 20, the pronephros and aortae were dissected from the embryos, dissociated into cell suspensions, and analyzed for their DNA content. At these stages, cells were readily observed adhering to the outside of the organs and vessels, as well as adhering to the outside of the developing pronephric duct. Feulgen-stained DNA analyses clearly showed that these adhering, invading cells were derived from the region that was transplanted at stages 14–16. Microscopic analysis of cells harvested from the aortae and pronephros showed that the cells had the morphological features of hemopoietic cells.

In a series of very preliminary experiments, Turpen (unpublished observations) exchanged skin grafts between reciprocal pairs of larvae that received stem cell transplants at 67 hr. Within the time limits of the experiment (45 days), reciprocal skin grafts from reciprocal stem cell donors survived (4 of 4), but these same animals either rejected (2 cases, rejection time 15 days) or tolerated (2 cases) skin grafts from third-party sibling donors. Control animals from within the same sibship destroyed skin grafts by a typical first-set rejection reaction (median survival time = 16 ± 4; $n = 13$). One control animal tolerated a sibling graft through metamorphosis but eventually rejected it. These experiments, then, support the prediction that transplantation of stem cells should result in tolerance of alloantigens of the stem cell donor.

5. DISCUSSION AND CONCLUSIONS

Our evidence in amphibians argues against the origin of internal blood-forming centers from migratory cells wandering in from the yolk sac. Rather, an internal blood-forming center arises from cells that appear to differentiate from local mesoderm. Specifically, the primary hemopoietic center of the embryo develops within the dorsal lateral region, in the vicinity of the presumptive mesonephros.

Based on several lines of experimental evidence, a picture of the early developmental relationships of hemopoietic cells in amphibians seems to be emerging (detailed by Turpen and Knudson, 1982). As seen in Figure 5, following gastrulation, two hematogeneous regions can be depicted in the embryo, VBI mesoderm and dorsal lateral plate mesoderm. Constituents of VBI differentiate into primitive (embryonic) RBCs in association with the developing vitelline veins. As this venous network becomes connected to the heart, RBCs are released into the circulation. These initial RBCs comprise a transient population which declines during the larval period. Dorsal lateral plate mesoderm, within the area described as the mesonephric anlage, appears to give rise to the precursors of definitive hemopoiesis. Following a postulated determinative interaction, these precursor cells (stem cells) migrate interstitially toward the anterior of the embryo. By stage 19, these migratory cells are in the anterior half of the embryo where they are associated with pronephros and the aortae. Access to the circulation is via the posterior cardinal vein and dorsal aortae at stage 20. This is commensurate with heartbeat and the completed circulatory system. Coloniza-

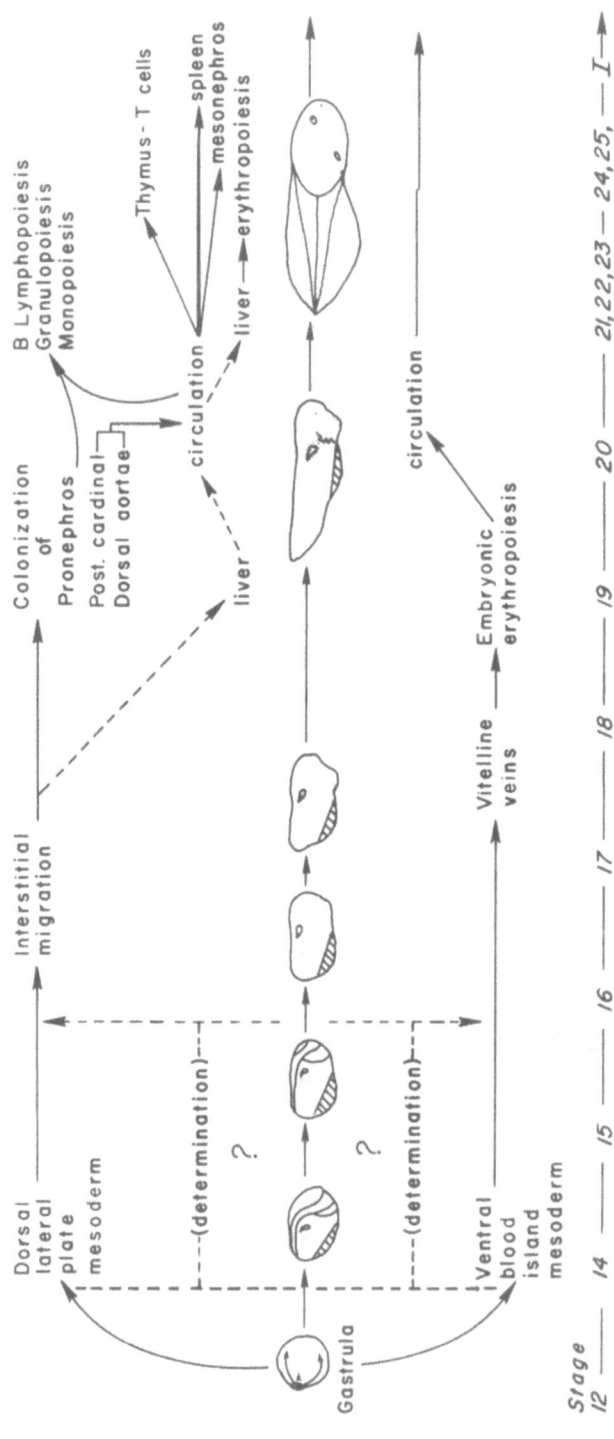

FIGURE 5. Relationships and migratory patterns of hemopoietic precursor cells during embryonic development in *Rana pipiens*. Dashed lines represent postulated events that have not been experimentally documented as yet.

tion of the thymus, spleen, and mesonephros appears to occur by a circulatory route and colonization of the pronephros appears to initially be by an interstitial route, although a return of stem cells to the pronephros by the circulatory route may also take place. Quite interestingly, the pronephros has been shown to be an early, if not the first, site of B-cell lymphopoiesis in the leopard frog (Zettergren et al., 1980) as well as a site of granulopoiesis (Carpenter and Turpen, 1979) and monocytopoiesis (Turpen and Pepe, unpublished observations). The liver may be colonized by an interstitial migration, as suggested by the data of Deparis and Jaylet, and/or by a circulatory route. Moreover, within this hypothetical framework, both the BVI and the dorsal stem cell compartment can be viewed as derivatives of lateral plate mesoderm. Clearly, the VBI are a product of the hypomere. Ample experimental evidence indicates that an interaction between endoderm and lateral-plate-derived splanchnopleure is involved in the differentiation of the VBI. As a working hypothesis, it seems reasonable to suggest that the dorsal stem cell compartment (perhaps dorsal blood islands?) may exhibit analogous development.

The stage 14 to 16 period of embryogenesis marks the occurrence of several interactions that result in the determination of various organ rudiments. Heterotopic transplantation of gill buds demonstrates that by stage 17 (84 hr), the thymus has already been determined (Turpen et al., 1977). Pronephric determination has already occurred by 72 hr at stage 16 (Fales, 1935). Presumably, the prior interactions leading to determination of these organ anlagen occur during stages 15–16. Several aspects of the experiments of Turpen and colleagues suggest that mesonephric determination, and perhaps initial stem cell determination, also occur during this period. For example, several embryos that received mesonephric anlagen transplants did not develop mesonephroses on the transplanted side. A consistent feature of these defective embryos was the slow healing of the transplanted tissue and the mispositioning of the anlagen as healing progressed. This suggests that the determinative interactions may have been interrupted. In these particular defective animals, very few transplant-derived hemopoietic cells were found. The nature of the interactions leading to the determination of hemopoietic stem cells is presently unknown, but it is the subject of intensive investigation.

An alternative hypothesis has been detailed by Volpe et al. (1980). The primary hemopoietic stem cells of the amphibian may be derived from the same intermediate cell mesodermal mass that produces the kidney system. These stem cells are capable of migrating interstitially toward the anterior region of the embryo. Indeed, perceptive views on this matter were expressed by Fraser more than 30 years ago. In 1950, Fraser wrote (p. 160): "It is important to remember that the entire intermediate cell mass does not participate in the formation of the excretory organ. Throughout the whole development and growth of the embryo, cells pass out from along the median wall of the intermediate cell mass. These ameboid cells are destined to form the *walls of the aorta* . . ." [italics ours].

As envisioned by Volpe et al. (1980), the nephrogenic band can be subdivided into sectional hemopoietic centers, each normally responsible for a particular population of stem cells. The primary site of erythropoiesis is the anterior part of the nephrogenic band. The mesonephric portion of the nephrogenic band is the definitive source of T cells; B cells are postulated to arise from the more posterior regions of the nephrogenic band. These regional hemopoietic centers may be viewed as embryonic fields which exhibit labile organization and may actually overlap. In essence, a given field may contain more than one type of blood cell precursor, and a heterogeneous population of precursors experimentally removed from a given field would give the impression that the selected cells are of a single kind only, and hence pluripotent. The foregoing views are speculative but they are of heuristic value in stimulating additional experimentation.

Thus, fundamental contributions from the students of amphibian (and avian) development are altering and refining our basic perspectives concerning the early establishment of the vertebrate hemopoietic system. Numerous exciting and important new questions are now being raised and experimentally addressed. As a small sample: Could thymic epithelial instruction, adaptive differentiation (Zinkernagel *et al.*, 1978), and adaptive modification of hemopoietic stem cells (Chertkov *et al.*, 1979) be related to fundamental processes occurring during hemopoietic embryogenesis? The answers to such probing questions should continue to enhance our understanding of the complex events of embryonic hemopoiesis.

REFERENCES

Baldwin, F. M., 1918, Pharyngeal derivatives of *Amblystoma*, *J. Morphol.* **30**:605.

Carpenter, K. L., and Turpen, J. B., 1979, Experimental studies on hemopoiesis in the pronephros of *Rana pipiens*, *Differentiation* **14**:167.

Charlemagne, J., 1977, Thymus development in amphibians: Colonization of thymic endodermal rudiments by lymphoid stem-cells of mesenchymal origin in the urodele *Pleurodeles waltlii* Michah, *Ann. Immunol. (Inst. Pasteur)* **128C**:897.

Chertkov, J. L., Gelfand, I. M., Gurevitch, O. A., Lemeneva, L. N., and Udalov, G. A., 1979, Hybrid resistance to parental bone marrow transplantation—Adaptive modification of hemopoietic stem cell in a non-syngeneic environment, *Proc. Natl. Acad. Sci. USA* **76**:2955.

Deparis, P., and Jaylet, A., 1975, Recherches sur l'origine des differentes lingnées de cellales sanguines chez l'amphibien *Pleurodeles waltlii*, *J. Embryol. Exp. Morphol* **33**:665.

Deparis, P., and Jaylet, A., 1976, Thymic lymphocyte origin in the newt *Pleurodeles waltlii* studied by embryonic grafts between diploid and tetraploid embryos. *Ann. Immunol. (Paris)* **127C**:827.

Deparis, P., Beetschen, J. C., and Jaylet, A., 1975, Red blood cells and hemoglobin concentration in normal diploid and several types of polyploid Salamanders, *Comp. Biochem. Physiol.* **50A**:263.

Desvaux, M., 1974, Etude de la morphogenese du thymus chez *Pleurodeles waltlii* Michah, *Bull. Soc. Zool.* **99**:259.

Fales, D. E., 1935, Experiments on the development of the pronephros of *Amblystoma punctatum*, *J. Exp. Zool.* **72**:147.

Federici, H., 1926, Recherches expérimentales sur les pótentialités de l'îlot sanguin ches l'embryon de *Rana* fusca, *Arch. Biol.* **36**:466.

Fernald, R. L., 1943, The origin and development of the blood islands of *Hyla regilla*, *Univ. Berkeley Calif. Publ. Zool.* **51**:129.

Finnegan, C. V., 1953, Studies of erythropoesis in Salamander embryos, *J. Exp. Zool.* **123**:379.

Fraser, E. A., 1950, *The Development of the Vertebrate Excretory System*, *Biol. Rev. Cambridge Philos. Soc.* **25**:159.

Gallien, L., and Dorocher, M., 1957, Table chronologique du développement chez *Pleurodeles waltlii* Michah, *Bull. Biol. Fr. Belg.* **91**:97.

Goss, C. M., 1928, Experimental removal of the blood island of *Amblystoma punctatum* embryos, *J. Exp. Zool.* **52**:45.

Hollyfield, J. G., 1966, The origin of erythroblasts in *Rana pipiens* tadpoles, *Dev. Biol.* **14**:461.

Horton, J. D., 1970, Histogenesis of the lymphomyeloid complex in the larval leopard frog, *Rana pipiens*, *J. Morphol.* **134**:1.

Jaylet, A., 1971, Modification du caryotype par une inversion péricentrique à l'état homozygote chez l'Amphibien Urodele *Pleurodeles waltlii* Michah, *Chromosoma* **35**:288.

Jaylet, A., 1972, Tétraploidie expérimentale chez le Triton *Pleurodeles waltlii* Michah, *Chromosoma* **38**:173.

Jaylet, A., and Deparis, P., 1979, An investigation into the origin of the thymocytes of the newt *Pleurodeles waltlii* using grafts between normal animals with a marker chromosome, *Dev. Comp. Immunol.* **3**:175.

Livini, F., 1902, Organi del sistema timo-tiroidea nella *Salamandrina perspicillata*, *Arch. Ital. Anat. Embriol.* **1**:3.

Maurer, F., 1888, Schilddrüse, Thymus and Kiemenreste der Amphibien, *Morphol. Jahrb.* **13**:3.

Metcalf, D., and Moore, M. A. S., 1971, Embryonic aspects of haemopoesis in: *Haemopoietic Cells* (A. Neuberger and E. L. Tatum, eds.), pp. 172–271. North-Holland, Amsterdam.

Moore, M. A. S., and Owen, J. J. T., 1967, Stem cell migration in developing myeloid and lymphoid systems, *Lancet* **1**:658.

Nagata, S., 1977, Electron microscopic study on the early histogenesis of thymus in the toad, *Xenopus laevis*, *Cell Tissue Res.* **179**:87.

Shumway, W., 1940, Stages in the normal development of *Rana pipiens*. I. External form, *Anat. Rec.* **78**:139.

Slonimski, P., 1931, Recherches experimentales sur la genèse du sang chez les Amphibiens, *Arch. Biol.* **42**:415.

Stohr, P., 1931, Beobachtungen zur Organentwicklung bei erythroztentrei Amphibien larven, *Arch. Entwicklungsmech. Org.* **124**:707.

Storti, E., 1935, Studio sull'ematopoiesi nella vita embrionale. I. Il periodo preepatico, *Arch. Zool. Ital.* **21**:241.

Taylor, A. C., and Kollros, J. J., 1946, Stages in the normal development of *Rana pipiens* larvae, *Anat. Rec.* **94**:7.

Tochinai, S., 1978, Thymocyte stem cell inflow in *Xenopus laevis* after grafting diploid thymic rudiments into triploid tadpoles, *Dev. Comp. Immunol.* **2**:627.

Tompkins, R., Reinschmidt, D., and Volpe, E. P., 1979, Colonization of the thymic primordium by migratory lymphocytoblasts in amphibian embryos, *Dev. Comp. Immunol.* **3**:635.

Tompkins, R., Volpe, E. P., and Reinschmidt, D., 1980, Origin of hemopoietic stem cells in amphibian ontogeny, *Development and Differentiation of Vertebrate Lymphocytes* (J. D. Horton, ed.), pp. 25–34, Elsevier/North Holland, Amsterdam.

Tournefier, A., 1973, Développment des organes lymphoides chez l'Amphibien Urodèle *Triturus alpestris* Laur.; tolérance des allograffes après la thymectomie larvaire, *J. Embryol. Exp. Morphol.* **29**:383.

Turpen, J. B., and Cohen, N., 1976b, Alternative sites of lymphopoiesis in the amphibian embryo, *Ann. Immunol. (Inst. Pasteur)* **127C**:841.

Turpen, J. B., and Knudson, C. M., 1982, Ontogeny of hematopoietic cells in *Rana pipiens*: Precursor cell migration during embryogenesis, *Dev. Biol.*, **89**:135.

Turpen, J. B., Volpe, E. P., and Cohen, N., 1973, Ontogeny and peripheralization of thymic lymphocytes, *Science* **182**:931.

Turpen, J. B., Volpe, E. P., and Cohen, N., 1975, On the origin of thymic lymphocytes, *Am. Zool.* **15**:51.

Turpen, J. B., Volpe, E. P., and Cohen, N., 1977, Stem cell influx following heterotropic transplantation of the thymic primordia between frog embryos, *Dev. Comp. Immunol.* **1**:255.

Turpen, J. B., Turpen, C. J., and Flajnik, M., 1979, Experimental analysis of hematopoietic cell development in the liver of larval *Rana pipiens*, *Dev. Biol.* **69**:466.

Turpen, J. B., Knudson, C. M., and Hoefen, P. S., 1981, The early ontogeny of hematopoietic cells studied by grafting cytogenetically-labeled tissue anlagen: Localization of a prospective stem cell compartment, *Dev. Biol.* **85**:99.

Volpe, E. P., and Turpen, J. B., 1975, Thymus: Central role in the immune system of the frog, *Science* **190**:1101.

Volpe, E. P., Tompkins, R., and Reinschmidt, D., 1977, Experimental studies on embryonic derivation of thymic lymphocytes, in: *Developmental Immunobiology* (J. B. Solomon and J. D. Horton, eds.), pp. 109–114, Elsevier/North-Holland, Amsterdam.

Volpe, E. P., Tompkins, R., and Reinschmidt, D., 1979, Clarification of studies on the origin of thymic lymphocytes, *J. Exp. Zool.* **208**:57.

Volpe, E. P., Tompkins, R., and Reinschmidt, D. C., 1981, Evolutionary modifications of nephrogenic mesoderm to establish the embryonic centers of hemopoiesis, in: *Aspects of Developmental and Comparative Immunology*, (J. B. Solomon, ed.), pp. 192–201, Pergamon Press, Elmsford, N.Y.

Webster, W. D., 1934, The development of the thymus bodies in *Necturus maculosus J. Morphol.* **56**:295.

Yamada, T., 1937, Der determinationszustand des rumpfmesoderms in molchkeim nach der gastrulation, *Wilhelm Roux Arch. Entwicklungsmech. Org.* **137**:151.

Zettergren, L. D., Kubagawa, H., and Cooper, M. D., 1980, Development of B cells in *Rana pipiens*, in: *Phylogeny of Immunological Memory* (M. J. Manning, ed.), p. 177, Elsevier/North-Holland, Amsterdam.

Zinkernagel, R. M., Callahan, G. N., Althage, A., Cooper, S., Klein, P. A., and Klein, J., 1978, On the thymus in the differentiation of "H-2 self-recognition" by T cells: Evidence for dual recognition?, *J. Exp. Med.* **147**:882.

16

Ontogeny of Avian Lymphocytes

N. LE DOUARIN, F. JOTEREAU, E. HOUSSAINT,
C. MARTIN, and F. DIETERLEN-LIÈVRE

1. INTRODUCTION

The chicken model has been extremely useful for investigating the development of the immune system, since birds are the only class of vertebrates in which a particular organ has been identified where differentiation of lymphocytes responsible for humoral immunity occurs.

Bruce Glick first recognized that a thymuslike organ called the bursa of Fabricius—located at the caudal end of the gastrointestinal tract of birds—is the site of B-cell-line development. Glick (1955) showed that injecting high doses of androgen to young chickens resulted in considerable reduction of the bursa of Fabricius together with a severe impairment of the development of the immune mechanisms. The complete absence of the bursa in the posthatched animal could even be obtained by administration of testosterone into the egg during early embryonic life (Meyer *et al.*, 1959; Papermaster *et al.*, 1962; Glick, 1964). This was followed by total agammaglobulinemia.

Around the same time, the function of the thymus in the acquisition of cell-mediated immunity was demonstrated in mammals (Miller, 1961). It was thereafter recognized, particularly through the work of Max Cooper and colleagues (1965), that a clear-cut separation between immunological functions exists in the chicken. As in mammals, the thymus is the organ of the body's recognition system, whereas the bursa of Fabricius is the site where the antibody-producing cells are produced.

The developmental processes through which the thymus and the bursa of Fabricius, both mesenchymoepithelial organs which arise as appendages of the digestive tract, undergo lymphoid differentiation were the subject of active debate long before the fundamental physiological role of lymphocytes was discovered. The controversy dealt with the embryonic origin of lymphocytes and

---

N. LE DOUARIN, F. JOTEREAU, E. HOUSSAINT, C. MARTIN, and F. DIETERLEN-LIÈVRE • Laboratoire d'Embryologie, Faculté des Sciences, 44 Nantes, France.

particularly on whether they were derived from the endodermal epithelial component of the rudiment or from the mesenchymal cells surrounding it (for reviews see Le Douarin, 1977, 1978, 1979).

These two possibilities, which are referred to respectively as the *transformation* theory and the *substitution* theory, were based essentially on descriptive embryological studies. The use of cell markers revealed that the developing thymus and bursa were the sites of cell immigration mediated through the bloodstream, and these invading cells were thought to be the precursors of both thymic and bursal lymphocytes. Exchanges of cells were observed between parabiosed chick embryos using sex chromosomes as markers (Moore and Owen, 1965). These experiments led to the formulation of the *hematogenous* theory of blood-forming organ development with its corollary that the yolk sac is the primary site of production of hemopoietic stem cells. According to this theory, yolk sac stem cells are the progenitors of lymphocytes and all other blood cells of the embryo and adult (see Metcalf and Moore, 1971, for a review).

Discrepancies in the various data available in the early 1970s suggested that the whole question of the ontogeny of the primary lymphoid organs should be reinvestigated using a technique which could mark all cells and trace back the origin of the various cell components of the primary lymphoid organs to the early primordium. Such a natural cell marker technique was devised by one of us (Le Douarin, 1969, 1973a,b). It is based on chimeric combinations of embryonic cells of the quail (*Coturnix coturnix japonica*) and the chick (*Gallus gallus*). Quail cells were found to have a particularly large nucleolus due to the presence of heterochromatin associated with the nucleolar ribonucleoproteins. Owing to the stability of this natural labeling, various chimeric tissues and even embryos could be devised in which both the origin and the differentiated cell type achieved could be readily recognizable. The ontogeny of the thymic and bursal primordia were investigated by this technique, and it was established that all lymphocytes developing in these organs were of extrinsic origin and derived from lymphocyte precursor cells (LPC) which invaded them at a definite embryonic stage. Two questions were thereafter investigated. One concerns the mechanisms which ensure the seeding of thymus and bursa rudiments by LPC; another deals with the origin of the hemopoietic cells which invade the intraembryonic blood-forming organs. The main features of this work will be reported in the present chapter.

2. ONTOGENY OF THE THYMIC PRIMORDIUM STUDIED IN THE QUAIL–CHICK CHIMERA SYSTEM

2.1. CONTRIBUTION OF THE ENDODERM AND MESENCHYME

In all vertebrates, the thymus originates from the primitive pharynx as an epitheliomesenchymal rudiment. It is generally accepted that the epithelium is endodermal in origin, although some ectodermal participation has been proposed (Fraser and Hill, 1915; Hammond, 1954) and could not be completely excluded. However, in birds, the ventrolateral part of the pharyngeal endoderm of the third and fourth branchial pouches can be separated by trypsinization from the underlying mesenchyme between the 6- and 30-somite stages. If taken from embryos before the opening of the branchial slit, the endoderm can be dissected out without any contamination by ectodermal cells. If it is associated with an appropriate mesenchyme, thymic histogenesis proceeds; thus, no intervention of ectoderm is required for thymic differentiation (Le Douarin, 1967; Le Douarin et al., 1968).

The embryonic origin of the mesenchyme that participates in thymic histogenesis was investigated in birds by means of the quail–chick marker system (Le Douarin, 1969, 1973a,b, 1974).

Experiments carried out essentially in amphibians had shown that cells of ectodermal origin, derived from the cephalic neural crest, take part in the formation of the branchial mesenchyme. However, the extent of the contribution of "mesectodermal" cells to the visceral arch structures was not clearly delineated due to the limitations of experimental approaches used. We have systematically investigated this question by transplanting isotopically and isochronically the mesencephalic and rhombencephalic primordia between quail and chick embryos (Le Douarin, 1974, 1976; Le Lièvre and Le Douarin, 1975). Neural crest cells originating from these regions of the neural axis were shown to migrate ventrally and give rise to the mesenchymal components of the maxillary and mandibulary buds, as well as of the other branchial arches. The so-called muscle plate, which forms a central core of mesenchymal tissue in the visceral arches, was found, however, to arise from the mesoderm. In the quail–chick chimeras examined at various developmental stages, the third and fourth endodermal pouches and, subsequently, the thymic epithelial cord were surrounded by mesenchymal cells derived from the grafted neural crest. When thymic lobulation occurred, the mesectodermal cells penetrated the epithelium, together with the blood vessels the endothelium of which was made up of host-type cells. The contribution of the mesenchymal components to thymic histogenesis appeared restricted to the interlobular connective tissue in the cortex and to strands of connective cells lining the blood vessels in the medulla. Therefore, although its initial mesenchymal component is of ectodermal origin, the vascularization of the thymus is ensured by mesodermal vascular buds invading the rudiment during its development. A similar mechanism operates for the other tissues (e.g., the dermis of the face and neck or the thyroid and parathyroid glands), the mesenchyme of which also originates from the mesectoderm.

The origin of the thymic lymphoid cell population was investigated with several kinds of techniques. By grafting (on the chorioallantoic membrane) chimeric combinations of prelymphoid mouse thymic epithelium with chick mesenchyme, Auerbach (1960, 1961) succeeded in obtaining lymphoid tissue in which the lymphocytes were of mouse type. His conclusion that lymphocytes arose from the epithelial thymic component provided, at that time, a powerful argument for the transformation theory. This view was strongly challenged, however, by experiments carried out in avian embryos by Moore and Owen (1965, 1967a), who used sex chromosomes as cell markers to analyze lymphocyte development. Chick embryos were joined by vascular anastomosis of either yolk sac or chorioallantoic blood vessels, according to the stage considered. Chromosomal analysis, following yolk sac parabiosis at day 4 or 5 of incubation, revealed high levels of chimerism in the thymus (44–70%). In contrast, little chimerism was found in parabionts joined at later stages. Since at the time of karyotype analysis (day 14 to 20 of incubation) the lymphoid cells were by far the major dividing cellular component of the thymus, one could assume that the chimerism involved the lymphocytes. An inflow of blood-borne lymphoid stem cells, invading the thymus after day 4 or 5 and before day 9 of incubation, was therefore proposed to account for these observations. This view was later supported by the results of culture experiments of chick or mouse thymic rudiments in cell-impermeable diffusion chambers on the chorioallantoic membrane: 10-day mouse or 7-day chick thymic anlagen failed to become lymphoid, whereas older thymuses were able to sustain lymphopoiesis in the same conditions (Owen and Ritter, 1969).

Careful histological examination of early mouse thymuses showed that, in fact, the large basophilic cells [which had already been suggested by Maximow (1909) to be the lymphocyte precursors] could not be detected in the epitheliomesenchymal thymic primordium before day 11 of gestation. At this time, they were first seen in the mesenchyme and appeared later in the epithelium where they could easily be recognized at day 12 (Metcalf and Moore, 1971). Such an observation explains Auerbach's previously mentioned results and places them in agreement with Moore, Owen and Metcalf's hematogenous theory: at day 12, the mouse thymic endoderm had already been populated by extrinsic stem cells before explantation.

However, since the chromosome marker system provided information only about dividing cells, a contribution of cells of the primary thymic rudiment to lymphopoiesis could not be ruled out by this experimental approach. From this point of view, the quail–chick marker technique offered more insight into the problem. To investigate the origin of the various differentiated cell types, several appropriate chimeric patterns were devised. In the experiment reported above, where the mesenchymal component of the rudiment was selectively labeled, the lymphocytes never had the same nuclear type as the connective cells. For instance, when a quail neural tube had been grafted into a chick, the thymic lymphocytes were always of chick type even though the thymic mesenchyme was made up of quail cells. This excluded the mesenchymal origin of LPC advocated by several authors (Hammar, 1905; Maximow, 1909; Venzke, 1952). The potential of the endoderm to give rise to lymphocytes was tested by transplanting the third and fourth pharyngeal pouch endoderm of a 15- to 30-somite quail embryo into the somatopleure of a chick (Figure 1) (Le Douarin and Jotereau, 1973). The ectopic thymus which developed in the body wall showed quail reticular cells but the connective and endothelial cells and lymphocytes were of the chick host type. Similar experiments carried out with chick thymic endoderm grafted into a quail embryo resulted in a reciprocal assortment, namely the lymphocytes never had the same nuclear type as the endoderm.

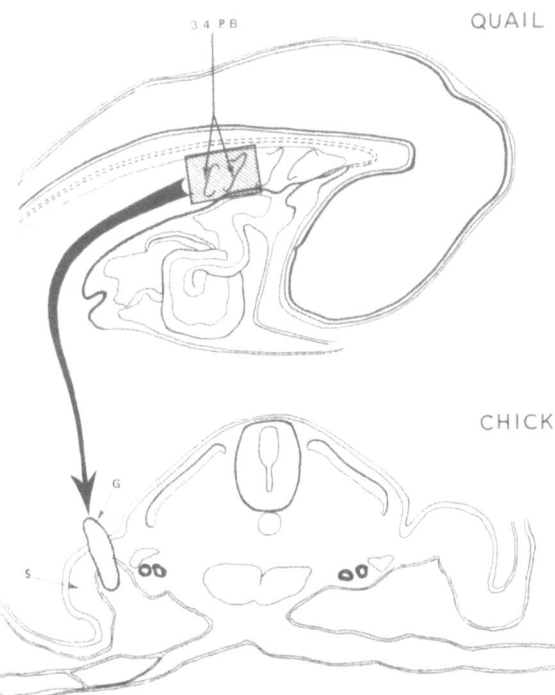

FIGURE 1. Schematic representation of the graft of the 3rd and 4th branchial pouches (3,4 P.B.) of a quail embryo into the somatopleure (S) of a 3-day chick host. G, graft.

These results confirmed the extrinsic origin of the lymphoid cell population that arises in the thymus and showed that in birds the endodermal cells are not able to differentiate into lymphocytes.

2.2. CHRONOLOGY OF HOMING OF HEMOPOIETIC STEM CELLS TO THE EMBRYONIC THYMUS

Since neither the endoderm nor the mesenchyme of the early thymic rudiment normally gives rise to lymphocytes, the LPC have to colonize the thymic epithelium sometime during development. Whether this colonization and its timing are continuous or discontinuous phenomena was investigated in the quail–chick system (Le Douarin and Jotereau, 1973, 1975).

First, quail thymic rudiments of various ages (between the 30-somite stage, i.e. day 2, and day 9 of incubation) were grafted into the somatopleure of 3-day chick host embryos. The graft was allowed to develop until it reached the total age of 14 days (age at grafting time plus duration of the graft). The lymphocytes which developed in the graft were:

1. Exclusively of host type, when the thymuses were taken from the quail before the end of the fifth day of incubation.

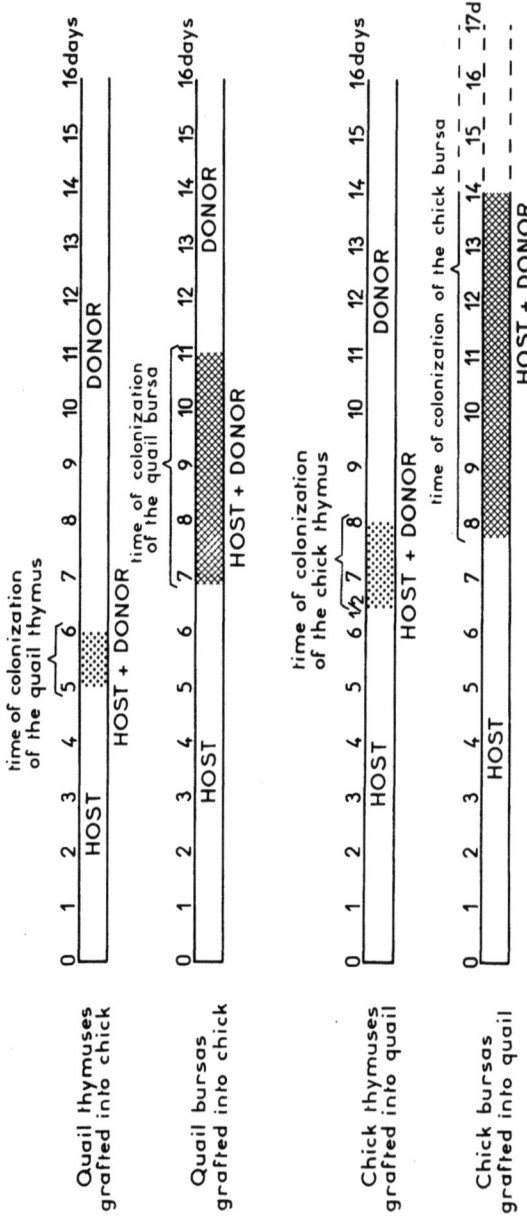

FIGURE 2. Diagram showing the timing of colonization of bursal and thymic rudiments by hemocytoblasts in quail and chick embryos.

2. A mixture of host and donor cells, when the thymuses originated from 5- to 6-day quail embryos.
3. Exclusively of donor type, when the thymuses originated from quails at day 6 of incubation onwards.

Therefore, in the quail embryo, the invasion of the thymic primordium by the lymphoid stem cells can be assumed to occur during the sixth day of incubation and to last about 24 hr (Figure 2).

Similar experiments in which chick thymuses were grafted into quails showed that, in the chick embryo, the stem cell inflow lasts approximately 36 hr and occurs from day 6½ to day 8 of incubation.

The first inflow of LPC is followed by a period during which no (or very few) hemocytoblasts penetrate the thymic rudiments. Research is now in progress to evaluate the duration of this nonreceptive period since we have already demonstrated that it is followed by a new inflow of LPC, the multiplication of which renews totally the lymphoid population of the thymus around hatching time in both quail and chick species (Le Douarin and Jotereau, 1975).

The existence of a thymus-regulated shutoff of stem cell influx following the first invasion, which occurs from day 6½ to day 8 in the chick embryo, is in line with previous observations by Moore and Owen (1967a). These investigators noticed that a chimerism in the parabiont thymuses was observed only when the vascular anastomosis had been established through the yolk sacs at day 4 or 5 of incubation. In contrast, parabioses established at later stages (day 9 or 10) through the chorioallantoic membrane resulted in few, if any, mixtures of cells in the respective thymus of joined embryos.

Moreover, the recovery of stem cell receptivity by the thymus, observed in our experiments during the second half of embryonic life, was also shown by Weber and Mausner (1977) by a different experimental approach. These investigators injected sex-chromosome-marked bone marrow cells into recipient embryos, previously exposed to 750 rads irradiation from a ^{137}Cs source. Since donor and host were of the SC line homozygous for the major histocompatibility complex (genotype B_2B_2), the injected cells settled permanently and could be detected in the hosts several weeks after hatching. Donor cells from 14-, 17-, and 19-day embryos as well as from 2-day posthatched birds, injected into 14-day embryos, were able to home to the thymus. Con A- and PHA-responsive T cells of donor origin were also detected in the spleens of the recipients, showing that donor bone marrow cells could differentiate into T cells in the host.

2.3. MECHANISMS CONTROLLING THE IMMIGRATION OF LPC INTO THE THYMUS

One striking result of quail–chick experiments is that hemocytoblasts begin to seed the thymic primordium at a precise stage of development. This could mean that, before this particular stage, there is either a lack of competent stem cells in the blood or an incapacity of the thymus itself to retain the hemocytoblasts.

By transplanting a 4-day quail thymic primordium into a 3-day chick host for 2 days and then into a quail host (until a total age of 14 days, as previously), we were able to show that stem cells ready to colonize the thymus are present in the chick circulation as early as day 3 or 4 of incubation (Le Douarin and Jotereau, 1975). Therefore, stem cells are available in the embryo at least 2½ days before they enter the thymic epithelium. It is thus clear that the onset of stem cell immigration is regulated by an intrinsic thymic property and does not depend on the availability of hemocytoblasts.

As a working hypothesis, we have assumed that this property, which appears in the thymus at day 5 in the quail and day 6½ in the chick, is realized by the production of a chemotactic factor to which the circulating stem cells are sensitive and which makes them stop in the bloodstream and migrate along an increasing gradient of concentration toward the thymic rudiment. The location of the thymus in the embryo does not influence this process, since an ectopically grafted thymus is seeded by stem cells in the same way as it is under normal conditions.

Thus, the receptive period of the thymus to stem cells is very short. This raises the issue of the mechanisms regulating the shutoff of hemocytoblast influx. Thymic rudiments taken before the start of colonization (at day 4 in the quail and day 6 in the chick) were cultured for 4 to 6 days either *in vitro* or in a diffusion chamber on the chorioallantoic membrane of 8-day chick embryos for 4 to 6 days. No stem cells could penetrate the primordium under these conditions, and the normal colonization period was over at the end of the culture time. The explants were then transplanted heterospecifically into the somatopleure of 3-day chick hosts and observed after 10 days' grafting. In all cases, normal thymic histogenesis occurred and the lymphoid population that developed in the explants was of host type.

This experiment shows that the arrest of stem cell influx into the thymus is not stage dependent, since attractivity can be maintained for several days if the epitheliomesenchymal primordium remains empty. Most probably, when a certain number of cells have seeded the rudiment, it becomes unreceptive for some time. Membrane-to-membrane site recognition between hemocytoblasts and thymic epithelial cells could be of importance in the mechanism by which the thymus captures circulating cells. Close membrane contacts between these two cellular components of the early thymus can indeed be observed during and soon after the invasion process. One can conceive that once all the available membrane recognition-sites are "saturated," no further stem cells can be retained in the rudiment.

Recovering of the homing capacity, observed in our experiments as well as in Weber and Mausner's, could be explained by the considerable growth of the thymus that occurs following the first stem cell influx. Therefore, new recognition sites appear on the epithelial cells and a seeding of stem cells can take place again. In addition, during the second half of incubation, stem cells from the first influx differentiate into lymphocytes which begin to become peripheralized. As a result, space becomes available for new stem cells in the thymus. Such a cell turnover in this organ has been shown to take place during its whole functional period (Weissman, 1967).

2.4. CHEMOTACTIC HYPOTHESIS AND POTENTIALITIES OF LPC

The chemotactic hypothesis, put forward as an element of the mechanisms regulating the onset of LPC seeding in the early thymic anlage, has been subjected to an experimental analysis in transfilter cultures. A method of culture, derived from that of Trowell, was used in these experiments (Jotereau et al., 1980).

A quail embryonic organ, chosen as a potential donor of LPC, was placed on the inferior side of a filter (Nuclepore, 5-μm pore diameter) and maintained in situ by a drop of agar; on the opposite side was positioned either a chick thymus at its receptive stage (day 6½ of incubation) or a control tissue from chick embryo. The quail bursa at the height of its receptive period (day 11 of incubation) was used as an LPC *donor explant*. The number of quail cells crossing the filter and seeding the chick organ were counted at various intervals after the beginning of the culture. The results indicated in Table 1 and Figure 3 show: (1) that a significant traffic of cells is initiated soon after the beginning of the culture from the lower to the upper compartment and (2) that this traffic is much higher when the upper explant is a thymus at its receptive period. The results strongly suggest the intervention of a chemotactic substance produced by the "attractive thymus," the release of which is shut off when the thymus has already been colonized by LPC (for instance, when the receptor explant is a 10-day chick thymus).

This experiment was complemented by transplanting the 6½-day thymus into a 3-day chick host after 3–4 days of *in vitro* association with the 11-day quail bursa. This allowed proliferation and differentiation of the lymphoid cells. It was indeed necessary to see whether the majority of cells crossing the filter were LPC since other cell types, e.g., fibroblasts of the bursal mesenchyme, were possible candidates. In fact, the thymus developed a chimeric lymphoid population in the chick host. The quail component of it was undoubtedly derived from LPC emigrated from the bursa during culture, whereas the chick lymphocytes originated from host LPC homing to the grafted thymus. The thymic explant retained competence to receive stem cells because the quail bursa did not provide the upper explant with enough LPC to shut off its attractivity.

Experiments carried out by Pyke and Bach (1979), using a technique of migration under agar derived from the one described by Nelson et al. (1975), have provided results in agreement with those reported above. They show, in particular, that a specifically oriented migration of hemopoietic cells from 12-day mouse fetal liver is induced by the lymphocyte-depleted mouse thymus.

Although our experiments strongly suggest that a chemotactic mechanism is responsible for stem cell seeding of the primary lymphoid organs, the results have to be correlated with events which occur during normal development. In

TABLE 1. NUMBER OF QUAIL BURSAL CELLS CROSSING THE FILTER AND FOUND IN THE UPPER COMPARTMENT OF THE CULTURES 5, 10, 16, AND 72 HR AFTER THE ONSET OF THE CULTURE

Duration	6½-day chick thymus/ 11-day quail bursa	6½-day chick liver/ 11-day quail bursa	6½-day chick mesonephros/ 11-day quail bursa	10-day chick thymus/ 11-day quail bursa
5 hr				
Mean ± S.E.	58.10 ± 9.20	3.33 ± 1.28	1.00 ± 0.35	
N[a]	10	6	5	ND[c]
C.V.[b]	47.50%	86.3%	70.7%	
10 hr				
Mean ± S.E.	111.75 ± 50.5			
N	4	ND	ND	ND
C.V.	78.2%			
16 hr				
Mean ± S.E.	268.33 ± 39.78	5.60 ± 2.39	2.50 ± 1.37	22.50 ± 8.17
N	9	5	4	8
C.V.	41.9%	85.3%	95.2%	96.1%
72 hr				
Mean ± S.E.	432.00 ± 68.28	29.40 ± 7.43	30.00 ± 3.86	26.55 ± 6.53
N	10	5	5	11
C.V.	47.4%	50.6%	25.7%	77.8%

[a]N, number of observations.
[b]C.V., coefficient of variation.
[c]ND, not determined.

this respect, it is interesting to note that in chick and quail embryos, large basophilic cells, considered to be LPC, can be seen in the blood vessels around the thymic anlage at the time of its receptivity. Similar observations can be made in the blood vessels of the bursal mesenchyme during the very period of its colonization. The first step of stem cell seeding is very likely the adhesion of LPC to the endothelium of the blood vessels irrigating (for the bursa) or lining (for the thymus) the primary lymphoid organ rudiments. Such an adhesion mediated by specific membrane molecules has been demonstrated to occur between the lymphocytes and the specialized lymph node venules (Butcher et al., 1979). This type of mechanism would be a general feature of the lymphoid (and perhaps hemopoietic) system and could be an important aspect of the specific homing of the circulating blood cells to the various compartments of the hemopoietic system. As to the colonization of the early lymphoid organ rudiments, once the LPC are stabilized by the endothelium of the vessels neighboring the thymus, they are induced to migrate along the gradient of the attractant produced by the thymus. However, before this hypothesis of a chemotactic gradient can be fully acceptable, it must be confirmed by the identification of both the factors responsible for the attraction and the receptors specific for the chemotactic substance on the LPC.

Our experiments further show that homing of cells to the bursa and to the thymus obey a similar mechanism, since cells which have already homed to a bursa can be induced to move into a thymus. This suggests that at these early stages of development, a common precursor could be responsible for colonization of both the primary lymphoid organs (Le Douarin et al., 1977; Le Douarin, 1978). Such a view has been challenged by Weber and Alexander (1978), who investigated by another method the thymus-seeding potential of hemopoietic cells recently immigrated into a 13- to 14-day embryonic chick bursa. Chromosomally marked cell populations obtained from 13- and 14-day embryonic bursas were transferred intravenously to γ-irradiated (780 rads) chick embryos of equivalent age. When the chimeras were examined at 4 to 12 weeks after cell transfer, donor cells were found to have proliferated primarily in the bursa, whereas significant donor cell influx into the thymus was not detected.

This was interpreted to mean that the LPC recently immigrated into the bursa during embryonic development are not pluripotential cells, but rather cells already precommitted to B-cell-line differentiation. Such a conclusion, however, seems unlikely in view of additional results of our group, i.e., of the bursal cell population which, in our experiments, migrates to the thymus, at least a portion acquire T-cell markers in the thymic environment (unpublished). It is important to stress the great complexity of the *in vivo* system used by Weber and Alexander and the numerous factors that can interfere with the seeding process of hemopoietic organs in the radiation chimeras. Furthermore, in these authors' experiments, the age of the thymus was 14 days. Thus, the first wave of colonization was not studied.

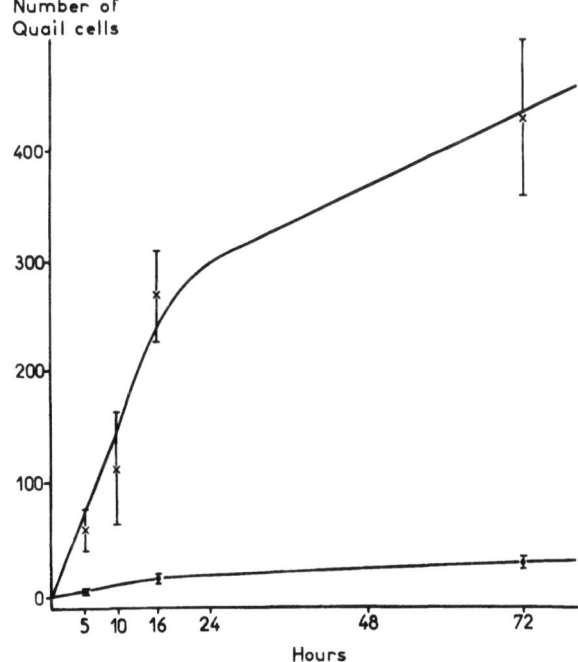

FIGURE 3. Comparison of the number of quail cells which cross the filter and seed the chick organ in transfilter cultures after 5, 10, 16, and 72 hr in culture. Upper curve: experiment in which the recipient organ is a 6½-day chick thymus. Lower curve: the results (pooled) obtained in the three other experimental series (chick recipient explant is either 6½-day liver, 6½-day mesonephros, or 10-day thymus; see Table 1).

3. ONTOGENY OF BURSA OF FABRICIUS

3.1. SEEDING BY EXTRINSIC LPC

Before lymphoid transformation of the bursal rudiment occurs, large undifferentiated cells with basophilic cytoplasm appear in the blood vessels and in the mesenchyme. Later on, similar cells, which are considered to be the progenitors of the lymphoid population, are present in the epithelium. Stem cells, characterized by their highly basophilic cytoplasm, first home to the bursal mesenchyme and begin to penetrate the epithelium at the apex of the bursal pouch from 10 days of incubation in the chick (Houssaint et al., 1976). From day 12, the number of these cells increases rapidly, first in the mesenchyme and later (day 13) in the epithelium. Some basophilic cells remain in the mesenchyme where they differentiate into granulocytes. The most conspicuous granulopoietic activity in the bursal mesenchyme takes place between days 8 and 12 in both quail and chick. At day 11 or 12 of incubation, follicular primordia start to develop from the epithelium. Later on, lymphocyte differentiation and follicular growth proceed rapidly, whereas granulopoiesis decreases in the mesenchyme.

According to Jolly (1915), the so-called primary hemocytoblasts are derived from the transformation of mesenchymal cells that invade the endoderm and differentiate there into lymphocytes. In contrast, an endodermal origin of the bursal hemopoietic cells has been suggested (Retterer, 1885). Extensive histological, histochemical, and ultrastructural studies of bursal development led Ackerman and Knouff (1959) to claim that medullary lymphocytes of the follicles have a purely epithelial origin but that cortical lymphocytes could arise both from epithelial cells migrating into the mesenchyme and from transformation of stellate mesenchymal cells. Moore and Owen (1965, 1966), using the chromosomal marker technique, concluded that most, if not all, of the lymphoid cells of the bursa are derived from an extrabursal source.

Bursal histogenesis was investigated in our laboratory, using the quail–chick marker system (Houssaint et al., 1976). Epitheliomesenchymal rudiments of the bursa were taken from 5- to 11-day quail embryos, grafted into the 3-day-old chick somatopleure, and fixed at a total age of 18 or 19 days. The species of the grafted bursal lymphoid population appeared to depend on the stage at which the organ was transferred from donor to host embryo. In bursas transplanted before the end of the seventh day of embryonic development, the whole lymphoid population was of host origin. Grafts of 7- to 11-day bursal anlagen contained a mixture of host and donor lymphocytes. When taken from quail embryos older than 11 days, the bursas contained only lymphoid cells of donor type.

This observation confirms that migration of LPC occurs during bursal development and shows that hemopoietic differentiation depends entirely on the seeding by stem cells of extrinsic origin. The migration, which results in differentiation of the first lymphocytes, takes place from days 7 to 11 of incubation in the quail embryo (Figure 4).

The chimerism observed in the bursas grafted during this time varied from one follicle to another. In some areas of the graft, the follicles contained lympho-

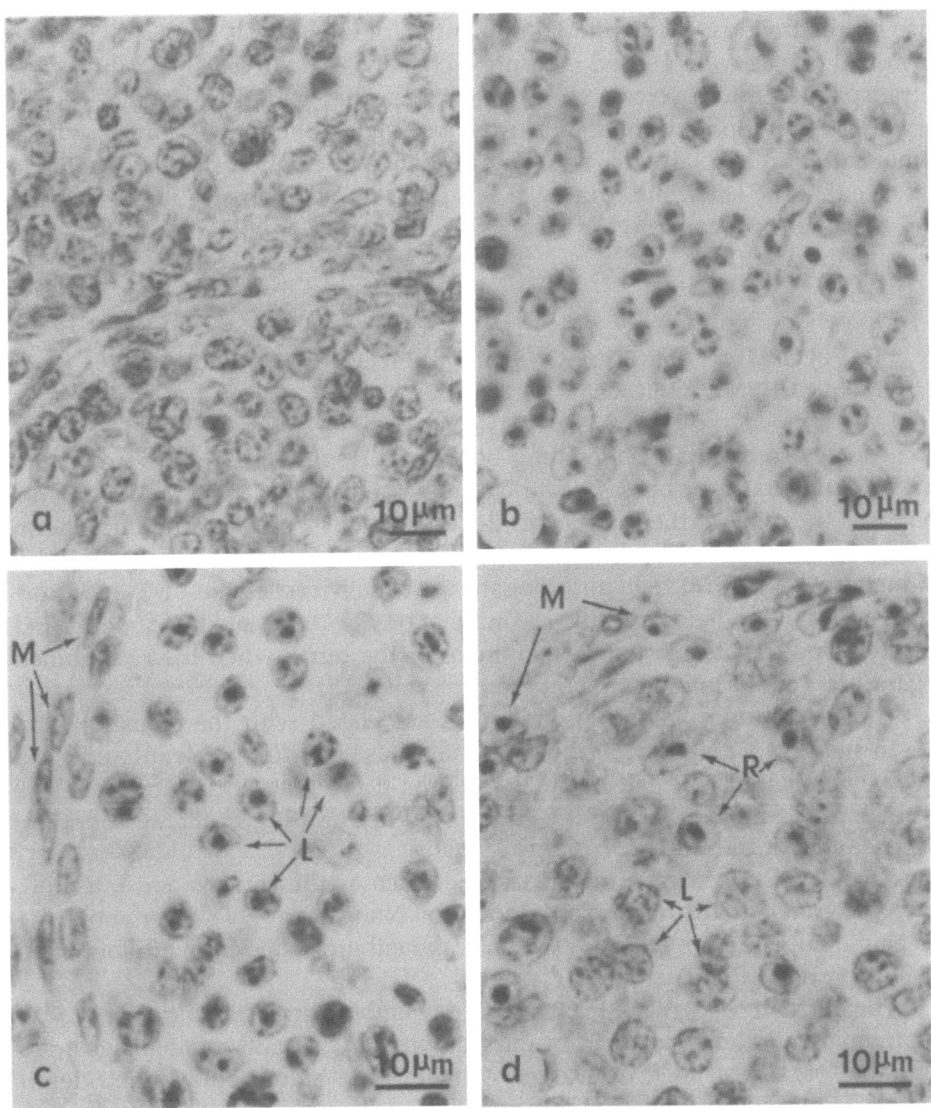

FIGURE 4. (a,b) Bursa of Fabricius of a chick (a) and a quail (b), 1 day after hatching. Feulgen staining. Note the condensations of heterochromatin in the quail nuclei. (c) Bursa of a 7-day chick embryo grafted for 13 days into a quail: the lymphoid population is of quail type. (d) Bursa of a 5-day quail embryo grafted for 13 days into a chick: lymphocytes of chick type (L), reticular cells of quail type (R). M, mesenchymal cells.

cytes exclusively of one type (host or donor), whereas in others both quail and chick lymphoid cells coexisted.

Reverse grafts of chick bursal primordium into quail embryos were also carried out. As in the quail, hemopoietic differentiation in the chick bursa entirely depends on blood-borne stem cells. The latter begin to invade the organ at the end of day 8 of incubation and the inflow of stem cells stops at 14 days.

Long-term grafting experiments have been conducted in order to see if, as is the case for the thymus, the lymphoid population of the bursa is renewed. Twelve-day quail bursas were grafted first into a chick host for 10 days, and then were retransplanted into a second chick host for another period of 10 days. No renewal of the lymphoid population was observed in the grafted bursas, which remained entirely quail after a 20-day residence in chick hosts. This seems to indicate that the embryonic inflow of stem cells into the bursa is a unique event and that these progenitor cells give rise to the entire population of B lymphocytes of the adult.

3.2. DEVELOPMENTAL RELATIONSHIPS BETWEEN ENDODERM AND MESENCHYME IN BURSAL ONTOGENY

The developmental relationships between the epithelial and the mesenchymal components of the bursa of Fabricius appeared to obey stricter requirements than was the case for the thymus. Endoderm can be separated from mesoderm in the bursa by enzymatic digestion at any stage in both the quail and the chick (Houssaint et al., 1976). The dissociation was usually carried out on 5- to 12-day bursas. Homo- or heterospecific recombinations of dissociated endoderm and mesenchyme lead to normal development of the bursa when the recombined organs are grafted into a host embryo.

On the contrary, if the endoderm is associated with the somatopleural mesenchyme of a 3-day host, it survives but does not differentiate into follicles. The basophilic cells present in the endoderm at the time of grafting leave the epithelium. Some can be found 2 to 3 days after grafting in the surrounding mesenchyme of the host.

Several other sources of mesenchyme, such as the somites and intestine, were associated with the bursal endoderm. None were able to promote the development of the epithelium, which was neither seeded by lymphoid stem cells nor formed follicular buds.

The bursal mesenchyme from 5- to 7-day embryos was also associated with the thymic endoderm taken from the pharynx at 25- to 30-somite stages. No lymphoid organs resulted from these chimeras; the thymic epithelium, identified by its nuclei, developed poorly and did not give rise to reticular cells.

Inasmuch as the bursal epithelium arises from the posterior gut, it seemed interesting to investigate the ability of the mesoderm of the bursa to induce bursal differentiation in the digestive endoderm. Associations of the endoderm isolated from the rectum of 5-day embryos with the mesenchyme of the bursa at the same developmental stage did not result in bursal histogenesis. The gut epithelium did not develop follicles and was not colonized by lymphoid stem cells.

The developmental potential of the bursal mesenchyme, when separated from the endoderm, was also studied. This was done by isolating the mesenchyme of 10- to 12-day quail bursas and grafting it into the 3-day chick somatopleure, the latter tissue being considered as a culture medium for the graft. The hemopoietic stem cells present in the explant at the time of grafting began to disperse into the surrounding mesenchyme soon after the implantation and later disappeared. Whether they died or penetrated the host blood vessels could not be determined due to their small number. When granulopoiesis was in progress in the bursas at the time of grafting, granulocytes could be seen for several days inside and in the vicinity of the explant.

Thus, tissue interactions between the mesenchymal and the endodermal components of the bursal primordium are of critical importance for lymphoid differentiation of this organ. The differentiating signals exchanged between the epithelium and the mesenchyme appear to be very specific, since none of the heterologous combinations designed resulted in bursal development. Such a situation is exceptional in epitheliomesenchymal organ histogenesis in which the effect exerted by the mesenchyme is usually "permissive" in nature and shows a high degree of nonspecificity. In most cases, the mesenchyme of the organ can actually be replaced by mesenchymal tissues taken from nearly any other rudiment at a similar developmental stage (for reviews see Saxen et al., 1976). The morphogenetic signal(s) coming from the mesenchyme provides conditions that enable the target epithelial cells to express preexisting differentiative bias. In the case of the bursa, one of the striking consequences of the separation of the epithelium from its mesenchyme was that neither one nor the other remained capable of retaining lymphoid stem cells when morphogenetic tissue interactions were prevented. Therefore, the cellular mechanisms that control stem cell homing depend upon reciprocal tissue interactions occurring between the two components of the rudiment.

It is important to stress that thymic histogenesis does not involve such strictly specific tissue interactions between the epithelial and the mesenchymal parts of the rudiment. Any stroma originating from the lateral plate mesoderm is able to promote the development of thymic endoderm. However, neither bursal mesoderm nor somite-derived mesenchymal tissues support the differentiation of the thymic epithelium (Le Douarin et al., 1968).

The effect of testosterone on bursal development has been extensively described (Kirkpatrick and Andrews, 1944; Glick, 1957, 1964; Meyer et al., 1959; Glick and Sadler, 1961; Warner and Burnet, 1961; Papermaster et al., 1962). Treatment of embryos with high doses of testosterone at any developmental stage stops the development of the bursa and inhibits B-cell differentiation. When testosterone propionate is injected into the egg, the lymphoid population of the bursa disappears completely within a few days; only granulocytes remain abundant in the mesenchyme when the injection is done from day 9 of incubation onwards. The follicles either do not grow or regress, depending upon the age at which the hormone was administered.

In a series of grafting experiments, Moore and Owen (1966) showed that normal bursas transplanted into testosterone-treated host embryos underwent lymphoid differentiation, although the host bursa completely failed to develop. This indicated that bursal lymphoid precursor cells were not injured by testosterone treatment but that it was stem cell homing that was suppressed in the androgen-treated rudiment. Inasmuch as this latter process depends on tissue interactions between the epithelium and the mesenchyme, it was of interest to learn whether testosterone prevents only one or both of the bursal tissue components from playing a role in bursal ontogeny.

Various tissue combinations between quail and chick bursal endoderm and mesenchyme, taken either from normal or from testosterone-treated embryos, showed that the androgen irreversibly damages the bursal epithelium: it could not be seeded by hemopoietic stem cells and undergo follicle formation, even if associated with normal mesenchyme. On the contrary, associations of testosterone-treated mesenchyme with normal endoderm resulted in normal bursal histogenesis. Early testosterone treatment (at 5 and 6 days in the chick) totally inhibited stem cell inflow in the bursa. If treated at later stages, the endoderm lost its ability to retain LPC or differentiating lymphocytes. The conspicuous effect of testosterone could be related to the levels of androgen receptors found in the bursa. Typical cytosol androgen receptors were demonstrated in both endoderm and mesoderm, although in higher amounts in the former. The concentration of binding sites in the bursa was found to be more than 10 times higher than in other organs (lungs, small intestine), contrasting with glucocorticoid receptors which were found in equivalent concentrations in all these organs (Le Douarin et al., 1980).

4. EMBRYONIC ORIGIN OF LPC

All the data analyzed in the previous sections point to the existence of extrinsic stem cells which have to seed the lymphopoietic organ rudiments at an early stage of development. When and where in the embryo do these stem cells arise? Since the yolk sac is the first blood-forming organ of amniote embryos and the only one with stem cells formed *in situ*, Moore and Owen (1967b) ascribed a central role of the extraembryonic blood islands as sole initiators of hemopoietic cell lines. They assumed that "the formation of stem cells from more primitive precursor cells, therefore, may be a developmental event unique to the early embryo." According to their hypothesis, all other sites of hemopoiesis are colonized by stem cells which form in the yolk sac and migrate through the circulation. The arguments advocated were the chimerism observed in parabiosed chick embryos (see Section 2.1) and the possibility of implanting 7-day yolk sac cells in the hemopoietic organs of irradiated 13-day chick embryos (Moore and Owen, 1967c). The points really demonstrated by these experiments are that stem cells travel through the blood during embryonic life and that totipotent stem cells are present in 7-day yolk sac. On the other hand, one can advance cogent arguments against a unique role of the yolk sac. For example, in the parabiosis experiment, traveling stem cells could have arisen from the parabiosed embryos rather than from their yolk sacs. In the reconstitution experiment, the stem cells present in 7-day yolk sac could also have originated from the embryo and emigrated out to the yolk sac. Indeed, such traffic has been demonstrated in a particular type of quail–chick chimera, which will be described next.

4.1. DEMONSTRATION OF INTRAEMBRYONIC STEM CELLS IN YOLK SAC CHIMERAS

These are quail embryos deprived of their own yolk sac and grafted *in ovo* on a chick yolk sac, according to a technique first devised by Martin (1972). These yolk sac chimeras have yielded a body of data that has led us to challenge the idea of a unique yolk sac origin of hemopoietic stem cells (Dieterlen-Lièvre, 1975). The chimeras were surgically constructed between stage-matched quail and chick blastoderms, at a time preceding the establishment of circulation (Figure 5); they were sacrificed at various ages between 5 and 13 days of incuba-

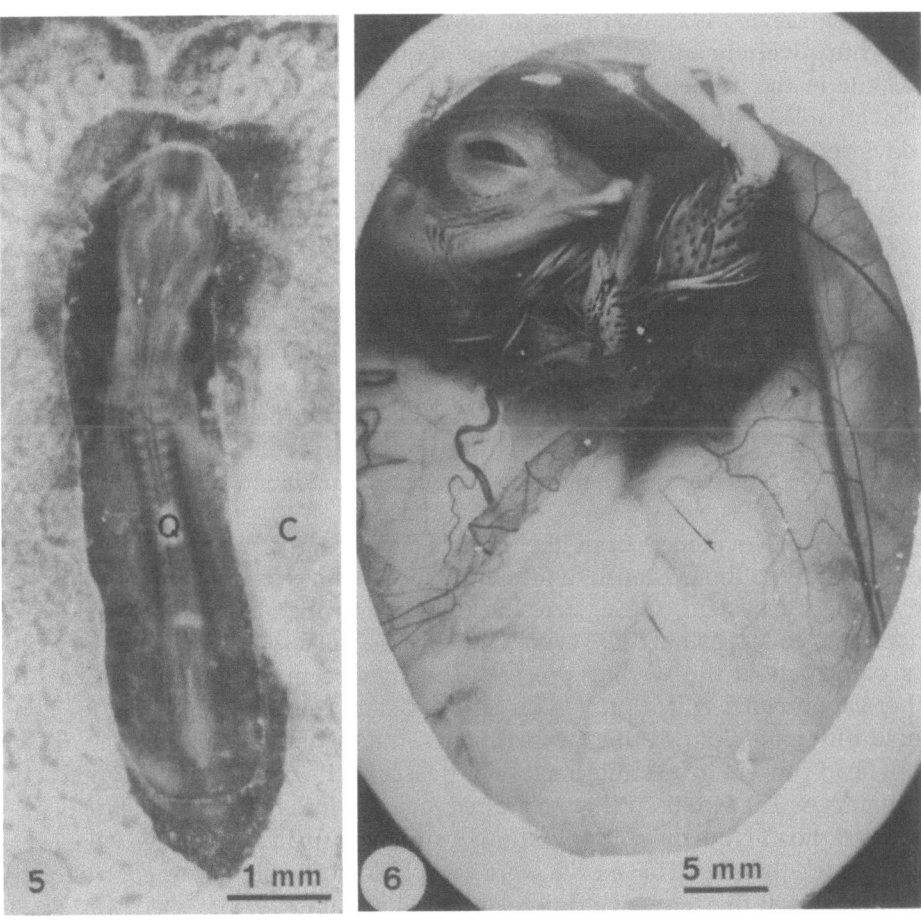

FIGURE 5. A yolk sac chimera *in ovo* shortly after the operation. The white line surrounding the embryonic body is the suture between the two components of the graft. Q, quail; C, chick. (Reproduced with permission of Elsevier/North-Holland from *Cell Differentiation* 7:115.)

FIGURE 6. A yolk sac chimera at day 13 of incubation. The pigmented quail embryo, grafted on a White Leghorn blastoderm (notice the white shell of the egg), has developed normal vascular connections with the chick yolk sac. The chorioallantois has been drawn to the side to allow full view of the embryo.

tion (Figure 6). The development of their hemopoietic system was monitored by several parameters, which all reveal the existence of an intraembryonic site of formation of hemopoietic stem cells. The species of the cells (quail or chick) making up the hemopoietic organs was analyzed in Feulgen–Rossenbeck-stained sections; the proportions of chick and quail erythrocytes in the blood were measured by an immune hemolysis technique using rabbit antibodies against either quail or chick red blood cells (Dieterlen-Lièvre et al., 1976).

The hemopoietic organs of the chimeras were predominantly populated by quail hemopoietic cells. Thymus and bursa contained only quail lymphocytes or lymphoblasts. The spleen showed a transitory and inconstant colonization by chick hemopoietic cells at 11–12 days. The bone marrow, when it began to differentiate in 12- to 13-day chimeras, contained only quail hemopoietic cells (Dieterlen-Lièvre, 1975; Martin et al., 1978).

Thus, overwhelming predominance of quail hemopoietic cells was the general rule in the intraembryonic organs of the chimeras. The quail hemopoietic cells can only be derived from the embryo proper. Moreover, in most cases, from 7 days on, the yolk sac blood islands of the chimeras contained a mixture of quail and chick hemocytoblasts or erythroblasts; often, quail erythroid cells were the most numerous (Figures 7 and 8). Thus, after a certain time of incubation, the yolk sac functions with stem cells from the embryo (Martin et al., 1978).

The limited ability of the yolk sac as a stem cell progenitor was further substantiated by the evolution of the blood composition. Until 5 days of incubation, 95% or more circulating erythrocytes were chick. From 6 days onwards, quail red blood cells increased in the blood and reached a mean of 43% at 13 days (Dieterlen-Lièvre et al., 1976; Beaupain et al., 1979). It is particularly striking that the thymus of these chimeras harbored exclusively quail lymphoid cells, although it was colonized at 5–6 days when circulating blood contained 95–100% chick erythrocytes.

Thus, the evidence from the chick yolk sac–quail embryo experiments points clearly to the existence of intraembryonic stem cells. Furthermore, in this heterospecific system, these were the only ones which colonized the thymus and bursa; yolk sac stem cells, even if they were available at the proper time, did not participate in the seeding process. This may seem paradoxical, in view of other experiments which prove that yolk sac stem cells are competent for lymphoid differentiation. In these experiments, fragments of quail yolk sac, taken from head-process to 2-somite stages were associated with 6-day chick embryonic thymuses. The two tissues were grafted together into the somatopleure of a 3-day chick host. Nine days after grafting, numerous quail thymocytes were seen in the chick thymus. These cells were morphologically similar to control quail lymphocytes at the same age (Jotereau and Houssaint, 1977).

Thus, it can be wondered whether in normal development intraembryonic and yolk sac stem cells may coparticipate in colonizing the rudiments. In interspecific chimeras, quail cells might have an advantage over chick cells, due, for instance, to their more rapid development. Chick–chick chimeras were devised to clarify this point. Blastoderms from the Leghorn histocompatible Iowa V strain were used. The markers identifying the components of the graft were

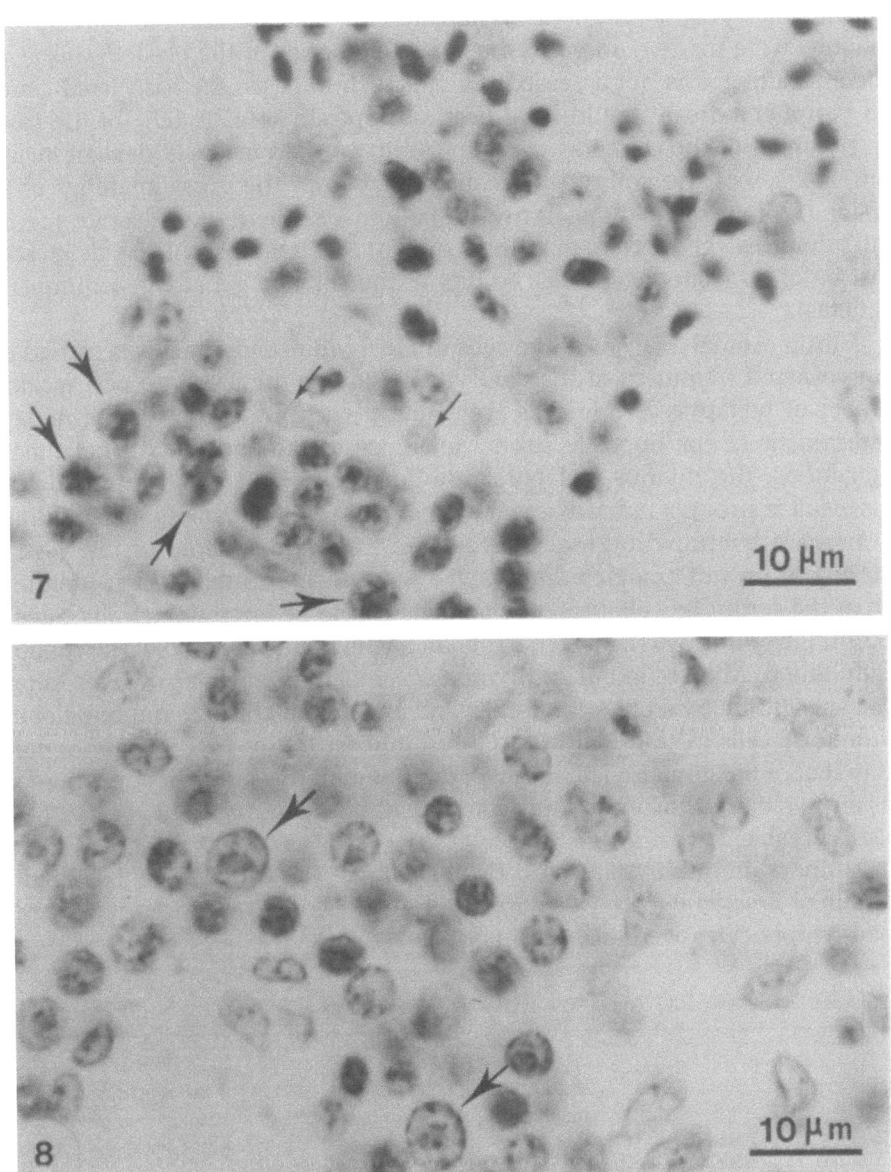

FIGURE 7. Blood island in the yolk sac of a yolk sac chimera. Erythropoietic cells have quail nuclei. The most immature cells are the most typical (left-hand corner, large arrows). Mesodermal cells of the yolk sac stroma are chick (small arrows). 5-μm section; Feulgen stain.

FIGURE 8. Control chick yolk sac blood island. Immature erythropoietic cell nuclei have a large, pale-staining nucleolus (arrows). 5-μm section; Feulgen stain.

either the sex-chromosome marker or two differing immunoglobulin allotypes (Lassila et al., 1978; Martin et al., 1979). The pattern of the graft is identical to that in quail–chick yolk sac chimeras. As regards the sex-chromosome marker, combinations were made randomly, since the genetic sex of the blastoderms at the time of grafting was not assessed. On a statistical basis, the associated central and peripheral areas should have been of opposite sexes in 50% of the cases. Development of the chimeras was allowed to proceed to 16–19 days of incubation. The sex of the embryos was diagnosed from the gross anatomy of the gonads (no intersexual appearance of the gonads was found, despite the fact that primordial germ cells migrating from the extraembryonic Swift crescent must be of opposite genetic sex to that of the somatic gonads in a number of chimeras).

Chromosomal analysis of the cells in the thymus and bursa was carried out after colcemid treatment of the embryos, to ensure the presence of a maximal number of metaphases. No cells were found with the chromosomal makeup characteristic of the opposite sex by reference to the gonads. At that time of ontogenesis, the thymus and bursa are densely packed with lymphoid cells. This result is strongly indicative that no cells from the yolk sac colonized durably the primary lymphoid organs. Blastoderms from two sublines of the V strain differing by an immunoglobulin allotype were also recombined. For identification of the serum IgG allotype, the animals had to be immunologically mature. Accordingly, effort was achieved to obtain hatching and survival of the chimeras to adulthood. Mitoses in bursa, thymus, spleen, and bone marrow were stimulated specifically by Con A for T cells and sheep anti-chicken immunoglobulin serum for B cells in 48-hr cultures. Only 4 mitoses out of 2631 were of opposite sex to that of the gonads. The IgG allotype found in the serum of adult chimeras was always that of the embryonic component of the graft and never that of the yolk sac (Table 2).

Humoral immunity, assessed by challenging the chimeras with human γ-globulin or *Brucella abortus,* was normal. PHA and Con A responses of peripheral blood lymphocytes were also normal.

TABLE 2. SERUM IgG ALLOTYPE EXPRESSION IN CHICK–CHICK YOLK SAC CHIMERA

Yolk sac	Embryo	Age at analyses (weeks)	Number of chickens	Number of sera[a] reacting with	
				Anti-Gl^a	Anti-Gl^b
Gl^{a}Gl^a	Gl^{b}Gl^b	3	5	0	5
		7	5	0	5
		20	5	0	5
Gl^{b}Gl^b	Gl^{a}Gl^a	3	2	2	0
		7	6	6[b]	0
		20	2	2	0

[a]The sera were tested against anti-Gl^a by double diffusion in agar.
[b]Four chimeras were tested only once at 7 weeks and sacrificed for chromosomal analysis.

4.2. THE PARA-AORTIC FOCI

To summarize these concurrent results, intraembryonic stem cells can be held as solely responsible for seeding lymphoid organ rudiments in avian embryos. Where then do they arise in the embryo? Their origin and history prior to their homing in hemopoietic organs is the subject of current investigations (Dieterlen-Lièvre and Martin, 1981). Diffuse hemopoiesis is a prominent occurrence in the chick embryo (Miller, 1913). Young hemopoietic cells, characterized by a round shape and a densely basophilic cytoplasm, are scattered among the meshes of the mesoderm in the dorsal mesentery. They are especially numerous at the level of the aortic roots and heart between days 6 and 8 of incubation and normally extend into the neck (Figure 9). They are numerous enough in fact to constitute well-delineated groups of cells. The shape, location, and size of these groups are highly variable between individual chick embryos. In quail embryos, two symmetrical foci are regularly present at the angle of the duct of Cuvier and the anterior cardinal vein (Figure 10). We propose to call them para-aortic foci, indicating their location and possible origin. Their maximal extent is at day 7, compared to day 8 in the chick. They decline from day 9 onwards. They are usually close to the rudiments of the developing thoracic ducts and jugular lymph sacs. Miller (1913) argued from this proximity that developing blood cells in these foci become enclosed within lymphatic spaces and eventually reach the general circulation (jugular vein) through the lymph ducts.

At earlier stages, i.e., days 4 and 5 of incubation, only isolated basophilic cells are present in the dorsal mesentery. At day 3, the ventral wall of the aorta appears thickened by basophilic cells densely aggregated toward the mesentery in the chick (Dantschakoff, 1909) and in the quail as well (Dieterlen-Lièvre and Martin, 1981). In the mesentery of yolk sac chimeras, these basophilic cells (i.e., young hemopoietic cells which may well be the stem cells themselves) are quail, which proves their intraembryonic origin.

The para-aortic foci can be considered as hemopoietic organs, the activity of which takes place during a transitory period before the definitive organ rudiments become functional. With time, cells in the foci can be seen to differentiate either to erythrocytes or to granulocytes. In the neighborhood of the foci or among them, small vessels packed with erythroblasts are often seen. It is not possible presently to identify these vessels as lymph ducts or venules. These foci function as several fundamental events occur in the developing hemopoietic system: these are the onset of definitive erythropoiesis, the primordial colonization of the thymic rudiment, the beginning of bursal and splenic colonization and, slightly later, bone marrow. During this same period, erythropoietic stem cells, capable of colony formation in the marrow of irradiated chicks, are present in the blood: in the 6-day embryo they are three times more numerous than in the yolk sac and they reach a maximum number at 13 days (Samarut and Nigon, 1976). Furthermore, stem cells capable of populating hemopoietic organs of irradiated 14-day chick embryos or of replenishing the bursa of cyclophosphamide-treated 18-day embryos have been detected in the mesoderm of 7-day chick embryos (Lassila *et al.*, 1979, 1980). In the cyclophosphamide

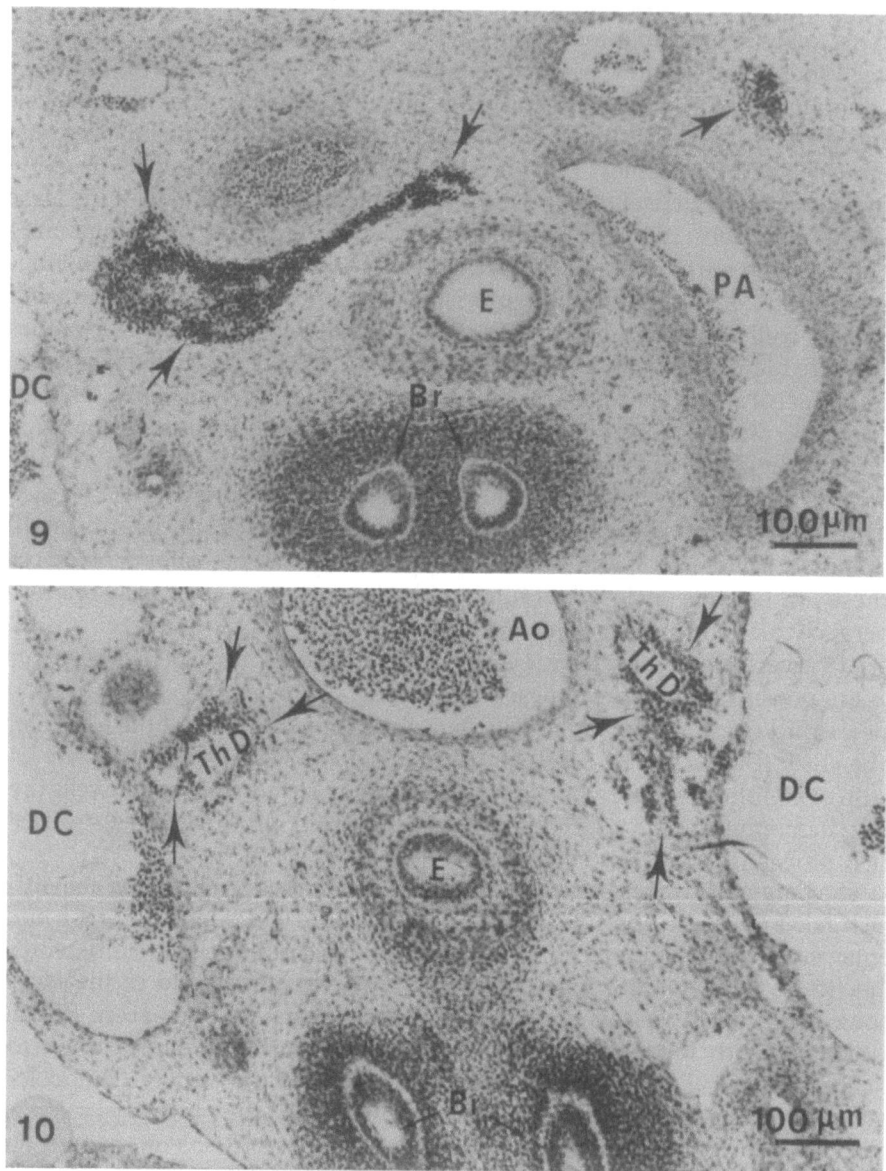

FIGURE 9. Two groups of hemopoietic cells (arrows) within the dorsal mesentery of a 7-day chick embryo. Br, bronchi; DC, duct of Cuvier; E, esophagus; PA, pulmonary artery.

FIGURE 10. The two symmetrical foci of hemopoietic cells (arrows) in the dorsal mesentery of a 6-day quail embryo. These foci are very constant in location, contrary to the variability seen in the chick. Ao, aorta; Br, bronchi; DC, duct of Cuvier; E, esophagus; ThD, thoracic duct.

model, the bursa, depleted of its lymphoid population, retains its capacity to be seeded a second time and to regenerate follicles. Cells from 9- to 18-day yolk sac (Eskola, 1977), from 15- to 18-day embryonic spleen, bone marrow, or thymus (Toivanen *et al.*, 1972; Eskola and Toivanen, 1976) were incapable of restoring the bursa. Thus, at these stages, totipotent LPC seem no longer present in these organs.

The para-aortic foci thus appear a likely source of intraembryonic stem cells. Our current hypothesis is that the cells are first emitted by the aortic endothelium, as suggested by Dantschakoff (1909), and that they find in the dorsal mesentery a favorable microenvironment for their multiplication, before they home to the hemopoietic organ rudiments.

Whether the hemopoietic cells only multiply in the mesentery and reenter blood vessels to migrate toward the organ rudiments or whether they can also migrate through the mesoderm is an open question. The existence of short circuits—as opposed to conveyance through the general circulation—is indicated by the regionalization observed in "partial chimeras" (Martin *et al.*, 1980).

These have a composite quail–chick body. They were obtained by grafting *in ovo* part of the central area of the quail blastoderm in place of the corresponding area of a stage-matched chick blastoderm (Martin, 1972). In that fashion, the

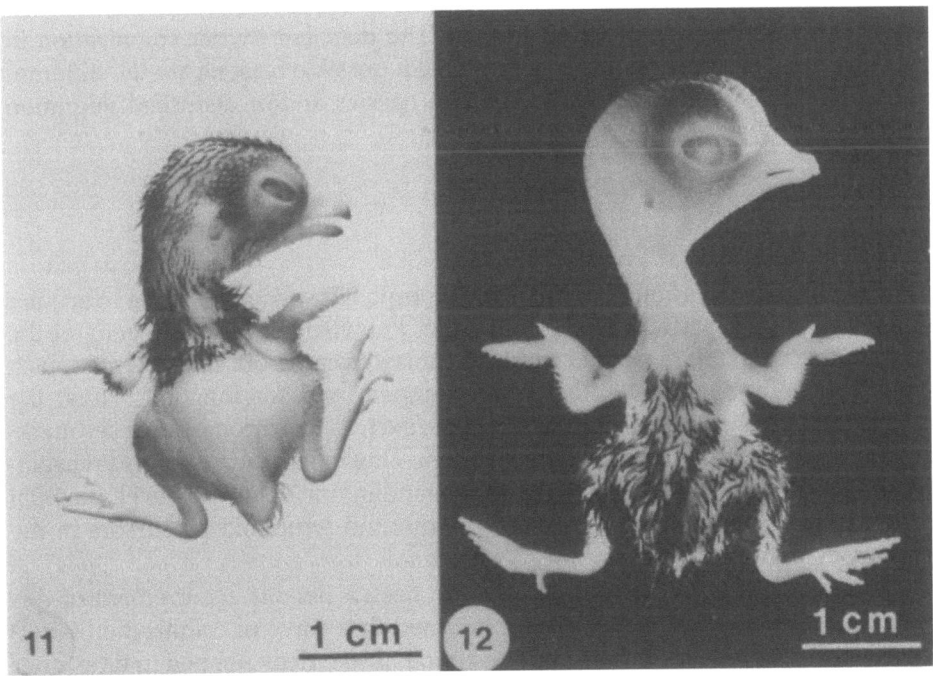

FIGURE 11. A quail head–neck–wing chimera. (In this particular specimen, the right wing does not have pigmented feathers, indicating an oblique suture.) In this type of chimera, the thymus is all quail (stromal cells and lymphoid population) while spleen and bursa are all chick.

FIGURE 12. A chick head–neck–wing chimera in which the thymus is all chick, while spleen and bursa are all quail.

quail component could be cephalic extending to the 10th or 17th somite, or caudal extending from the 10th or 17th somite. All chimeras developed on a chick yolk sac. In final appearance, at days 11 to 13 of incubation, they either had a quail head and neck on a chick thorax and abdomen, or a quail head, neck, and thorax on chick abdomen, or inversely (Figures 11 and 12). The makeup of their hemopoietic organs depended in striking fashion on the level of the suture. Quail head–neck embryos had a heterospecific thymus (quail stromal cells, chick hemopoietic cells) and all-chick spleen and bursa. Quail head–neck–wings embryos had an all-quail thymus and all-chick spleen and bursa. Thus, adding the wing level from the quail introduced a source of LPC capable of colonizing the thymus. This is precisely the level where para-aortic foci are present. However, the spleen and bursa were seeded by chick hemopoietic precursors. Since previous experiments excluded a yolk sac origin, the site of formation of stem cells within the embryo must be extensive. Furthermore, the different rudiments appear strictly regionalized, and in accordance with the nature of the region in which they are located. The mechanism of this regionalization is an intriguing problem now under investigation. A paradoxical observation was made in the reverse type of partial chimera: chick head–neck embryos retained an empty thymic primordium reduced to its chick stromal frame, while their spleen and bursa were all quail. When the chick contribution extended to the wing level, the hemopoietic organs were symmetrical to that of the reverse chimeras, i.e., all-chick thymus, all-quail spleen and bursa. The deficient thymic colonization in the chick head–neck chimeras is unexplained: possible reasons are the different locations of the para-aortic foci in the two species and/or disturbed migration pathways.

5. CONCLUSION

The unique accessibility of the avian embryo has made it possible to acquire in a few years a precise, though still partial, knowledge of the ontogeny of the lymphoid system. An essential feature of this system is cell movements which are extensive and critical both for development and for function. Thus, the investigations relied heavily on cell markers, such as the chromosomal sex marker, and more recently the quail–chick system. Due to its practicality and reliability, the latter has made it possible to understand in fine detail the developmental relationships between endoderm, mesoderm, and lymphoid precursors in the primary lymphoid organs.

The chronology of homing of LPC has been a notable achievement of this technique. It was shown that a sharply timed first wave of colonization is followed in the thymus by a nonreceptive period, also strictly defined in time, after which the process of colonization is resumed. The bursa, on the contrary, seems colonized by a unique wave of LPC.

An exclusive intraembryonic origin of LPC has been demonstrated. The identification of their sites of origin and history prior to their homing is well under way. Why they colonize the nearest rudiments and how the yolk sac stem cells are excluded, despite their capacity for lymphoid differentiation, are unknown.

Some of these features are already known to occur in other classes of vertebrates. In particular, intraembryonic stem cells have been demonstrated in amphibians (see Chapter 15), so that it can be seriously considered whether intraembryonic stem cells exist in mammals. Many of the other findings in the avian embryo may also serve as guidelines for future investigations in the mammalian embryo or fetus.

Very fundamental points still have to be solved, among them the mechanism of attractivity of LPC by the rudiments, the nature of the eventual chemotactic substance, the membrane recognition systems responsible for retention of LPC by endoderm, the acquisition of differentiation, the mechanism of release of differentiated lymphocytes. Finally, rather little is known in the avian embryo about the nature of LPC. Some data indicate that a common precursor may give rise to B and T cells but no experimental evidence obtains yet for the existence of a totipotential stem cell capable of entering all hemopoietic cell lines.

Certainly, the avian model will provide adequate approaches for investigating many of these issues. Refined markers will be needed, particularly markers of differentiation, some of which are already available.

REFERENCES

Ackerman, G. A., and Knouff, R. A., 1959, Lymphocytopoiesis in the bursa of Fabricius, *Am. J. Anat.* **104**:163.

Auerbach, R., 1960, Morphogenetic interactions in the development of the mouse thymus gland, *Dev. Biol.* **2**:271.

Auerbach, R., 1961, Experimental analysis of the origin of cell types in the development of the mouse thymus, *Dev. Biol.* **3**:336.

Beaupain, D., Martin, C., and Dieterlen-Lièvre, F., 1979, Are developmental hemoglobin changes related to the origin of stem cells and site of erythropoiesis?, *Blood* **53**:212.

Butcher, E., Scollay, R., and Weissman, I., 1979, Evidence of continuous evolutionary change in structure mediating adherence of lymphocytes to specialized venules, *Nature (London)* **280**:496.

Cooper, M. D., Peterson, R. D. A., and Good, R. A., 1965, Delineation of the thymic and bursal lymphoid systems in the chicken, *Nature (London)* **205**:143.

Dantschakoff, W., 1909, Untersuchungen über die Entwicklung von Blut and Bindegewebe bei Vögeln, *Arch. Mikrosk. Anat. Entwicklungsmech.* **73**:117.

Dieterlen-Lièvre, F., 1975, On the origin of haemopoietic stem cells in the avian embryo: An experimental approach, *J. Embryol. Exp. Morphol.* **33**:607.

Dieterlen-Lièvre, F., and Martin, C., 1981, Diffuse intraembryonic hemopoiesis in normal and chimeric avian development, *Dev. Biol.* **88**:180.

Dieterlen-Lièvre, F., Beaupain, D., and Martin, C., 1976, Origin of erythropoietic stem cells in avian development: Shift from the yolk sac to an intraembryonic site, *Ann. Immunol. (Inst. Pasteur)* **127C**:857.

Eskola, J., 1977, Cell transplantation into immunodeficient chicken embryos: Reconstituting capacity of cells from the yolk sac at different stages of development and from the liver, thymus, bursa of Fabricius, spleen and bone marrow of 15-day embryos, *Immunology* **32**:467.

Eskola, J., and Toivanen, A., 1976, Cell transplantation into immunodeficient chicken embryos: Reconstituting capacity of cells from the bursa of Fabricius, spleen, bone marrow, thymus and liver of 18-day-old embryos, *Cell. Immunol.* **26**:68.

Fraser, E. A., and Hill, J. P., 1915, The development of the thymus, epithelial bodies, and thyroid in the Marsupialia. Pt. I. Trichosurus vulpecula, *Philos. Trans. R. Soc. London Ser. B* **207**:1.

Glick, B., 1955, Growth and function of the bursa of Fabricius in the domestic fowl, Ph.D. dissertation, Ohio State University, Colombus.

Glick, B., 1957, Experimental modification of the growth of the bursa of Fabricius, *Poult. Sci.* **36**:18.

Glick, B., 1964, The bursa of Fabricius and the development of immunologic competence, in: *The Thymus in Immunobiology* (R. A. Good and A. E. Gabrielsen, eds.), pp. 345–358, Harper & Row (Hoeber), New York.

Glick, B., and Sadler, C., 1961, The elimination of the bursa of Fabricius and reduction of antibody production in birds from eggs dipped in hormone solutions, *Poul. Sci.* **40**:185.

Hammar, J. A., 1905, Zur Histogenese und Involution der Thymusdrüse, *Anat. Anz.* **27**:23.

Hammond, W. S., 1954, Origin of thymus in the chick embryo, *J. Morphol.* **95**:501.

Houssaint, E., Belo, M., and Le Douarin, N. M., 1976, Investigations on cell lineage and tissue interactions in the developing bursa of Fabricius through interspecific chimeras, *Dev. Biol.* **53**:250.

Jolly, J., 1915, La bourse de Fabricius et les organes lympho-épithéliaux, *Arch. Anat. Microsc. Morphol. Exp.* **16**:363.

Jotereau, F. V., and Houssaint, E., 1977, Experimental studies on the migration and differentiation of primary lymphoid stem cells in the avian embryo, in: *Developmental Immunobiology* (J. B. Solomon and J. D. Horton, eds.), pp. 123–130, Elsevier/North-Holland, Amsterdam.

Jotereau, F. V., Houssaint, E., and Le Douarin, N. M., 1980, Lymphoid stem cell homing to the early thymic primordium of the avian embryo, *Eur. J. Immunol.* **10**:620.

Kirkpatrick, C. M., and Andrews, F. N., 1944, The influence of the sex hormones on the bursa of Fabricius and the pelvis in the ring-necked pheasant, *Endocrinology* **34**:340.

Lassila, O., Eskola, J., Toivanen, P., Martin, C., and Dieterlen-Lièvre, F., 1978, The origin of lymphoid stem cells studied in chick yolk sac—embryo chimeras, *Nature (London)* **272**:353.

Lassila, O., Eskola, J., and Toivanen, P., 1979, Prebursal stem cells in the intraembryonic mesenchyme of the chick embryo at 7 days of incubation, *J. Immunol.* **123**:2091.

Lassila, O., Eskola, J., Toivanen, P., and Dieterlen-Lièvre, F., 1980, Lymphoid stem cells in the intraembryonic mesenchyme of the chicken, *Scand. J. Immunol.* **11**:445.

Le Douarin, N., 1967, Détermination précoce des ébauches de la thyroïde et du thymus chez l'embryon de Poulet, *C.R. Acad. Sci.* **264**:940.

Le Douarin, N., 1969, Particularités du noyau interphasique chez la Caille japonaise (Coturnix coturnix japonica): Utilisation de ces particularités comme "Marquage biologique" dans les recherches sur les interactions tissulaires et les migrations cellulaires au cours de l'ontogenèse, *Bull. Biol. Fr. Belg.* **103**:435.

Le Douarin, N., 1973a, A biological cell labelling technique and its use in experimental embryology, *Dev. Biol.* **30**:217.

Le Douarin, N., 1973b, A Feulgen-positive nucleolus, *Exp. Cell Res.* **77**:459.

Le Douarin, N. M., 1974, Cell recognition based on natural morphological nuclear markers, *Med. Biol.* **52**:281.

Le Douarin, N. M., 1976, Cell migration in early vertebrate development studied in interspecific chimeras, in: *Embryogenesis in Mammals*, Ciba Foundation Symposium No. 40, pp. 71–101, Elsevier/Excerpta Medica/North-Holland, Amsterdam.

Le Douarin, N. M., 1977, Ontogeny of primary lymphoid organs, in: *B and T Cells in Immune Recognition* (F. Loor and G. E. Roelants, eds.), pp. 1–19, Wiley, New York.

Le Douarin, N. M., 1978, Ontogeny of hematopoietic organs studied in avian embryo interspecific chimeras, in: *Differentiation of Normal and Neoplastic Hematopoietic Cells*, pp. 5–31, Cold Spring Harbor Laboratory, New York.

Le Douarin, N. M., 1979, Dependence of myeloid and lymphoid organ development on stem-cell seeding: Investigations on mechanisms in cell-marker analysis, in: *Mechanisms of Cell Change* (J. E. Ebert and T. S. Okada, eds.), pp. 293–326, Wiley, New York.

Le Douarin, N., and Jotereau, F., 1973, Recherches sur l'origine embryologique des lymphocytes du thymus chez l'embryon d'Oiseau, *C.R. Acad. Sci.* **276**:629.

Le Douarin, N., and Jotereau, F., 1975, Tracing of cells of the avian thymus through embryonic life in interspecific chimeras, *J. Exp. Med.* **142**:17.

Le Douarin, N., Bussonnet, C., and Chaumont, F., 1968, Etude des capacités de différenciation et du rôle morphogène de l'endoderme pharyngien chez l'embryon d'Oiseau, *Ann. Embryol. Morphog.* **1**:29.

Le Douarin, N. M., Houssaint, E., and Jotereau, F., 1977, Differentiation of the primary lymphoid organs in avian embryos: Origin and homing of the lymphoid stem cells, in: *Avian Immunology* (A. A. Benedict, ed.), pp. 29–37, Plenum Press, New York.

Le Douarin, N. M., Michel, G., and Baulieu, E. E., 1980, Studies of testosterone-induced involution of the bursa of Fabricius, *Dev. Biol.* **75**:288.

Le Lièvre, C. S., and Le Douarin, N. M., 1975, Mesenchymal derivatives of the neural crest: Analysis of chimaeric quail and chick embryos, *J. Embryol. Exp. Morphol.* **34**:125.

Martin, C., 1972, Technique d'explantation *in ovo* de blastodermes d'embryons d'Oiseaux, *C.R. Soc. Biol.* **166**:283.

Martin, C., Beaupain, D., and Dieterlen-Lièvre, F., 1978, Developmental relationships between vitelline and intra embryonic haemopoiesis studied in avian "yolk sac chimaeras," *Cell Differ.* **7**:115.

Martin, C., Lassila, O., Nurmi, T., Eskola, J., Dieterlen-Lièvre, F., and Toivanen, P., 1979, Intra-embryonic origin of lymphoid stem cells in the chicken: Studies with sex chromosome and IgG allotype markers in histocompatible yolk sac-embryo chimaeras, *Scand. J. Immunol.* **10**:333.

Martin, C., Beaupain, D., and Dieterlen-Lièvre, F., 1980, A study of the development of the hemopoietic system using quail–chick chimeras obtained by blastoderm recombination, *Dev. Biol.* **75**:303.

Maximow, A., 1909, Untersuchungen über Blut- und Bindegewebe: Uber die Histogenese des Thymus bei Säugetieren, *Arch. Mikrosk. Anat. Entwicklungsmech.* **74**:525.

Metcalf, D., and Moore, M. A. S., 1971, Embryonic aspects of haemopoiesis, in: *Haemopoietic Cells* (A. Neuberger and E. L. Tatum, eds.), pp. 172–271, North-Holland, Amsterdam.

Meyer, R. K., Rao, M. A., and Aspinall, R. L., 1959, Inhibition of the development of the bursa of Fabricius in the embryos of the common fowl by 19-nortestosterone, *Endocrinology* **64**:890.

Miller, A. M., 1913, Histogenesis and morphogenesis of the thoracic duct in the chick; development of blood cells and their passage to the blood stream via the thoracic duct, *Am. J. Anat.* **15**:131.

Miller, J. F. A. P., 1961, Immunological functions of the thymus, *Lancet* **2**:748.

Moore, M. A. S., and Owen, J. J. T., 1965, Chromosome marker studies on the development of the haemopoietic system in the chick embryo, *Nature (London)* **208**:956, 989.

Moore, M. A. S., and Owen, J. J. T., 1966, Experimental studies on the development of the bursa of Fabricius, *Dev. Biol.* **14**:40.

Moore, M. A. S., and Owen, J. J. T., 1967a, Experimental studies on the development of the thymus, *J. Exp. Med.* **126**:715.

Moore, M. A. S., and Owen, J. J. T., 1967b, Stem cell migration in developing myeloid and lymphoid systems, *Lancet* **ii**:658.

Moore, M. A. S., and Owen, J. J. T., 1967c, Chromosome marker studies in the irradiated chick embryo, *Nature (London)* **215**-1081.

Nelson, R. D., Quie, P. G., and Simmons, R. L. 1975, Chemotaxis under agarose: A new and simple method for measuring chemotaxis and spontaneous migration of human polymorphonuclear leukocytes and monocytes, *J. Immunol.* **115**:1650.

Owen, J. J. T., and Ritter, M. A. 1969, Tissue interaction in the development of thymus lymphocytes, *J. Exp. Med.* **129**:431.

Papermaster, B. W., Friedman, D. J., and Good, R. A., 1962, Relationships of the bursa of Fabricius to immunologic responsiveness and homograft immunity in the chicken, *Proc. Soc Exp. Biol. Med.* **110**:62.

Pyke, K. W., and Bach, J. F., 1979, In vitro migration of murine fetal liver cells to thymic rudiments, *Eur. J. Immunol.* **9**:317.

Retterer, E. J., 1885, Contribution à l'étude du cloaque et de la bourse de Fabricius chez les Oiseaux, *J. Anat. Physiol.* **21**:369.

Samarut, J., and Nigon, V., 1976, Properties and development of erythropoietic stem cells in the chick embryo, *J. Embryol. Exp. Morphol.* **36:**247.

Saxen, L., Karkinen-Jääskeläinen, M., Lehtonen, E., Nordling, S., and Wartiovaara, J., 1976, Inductive tissue interactions, in: *The Cell Surface in Animal Embryogenesis and Development* (G. Poste and G. L. Nicolson, eds.), pp. 331–407, North-Holland, Amsterdam.

Toivanen, P., Toivanen, A., and Good, R. A., 1972, Ontogeny of bursal function in chicken. I. Embryonic stem cell for humoral immunity, *J. Immunol.* **109:**1058.

Venzke, W. G., 1952, Morphogenesis of the thymus of chicken embryos, *Am. J. Vet. Res.* **13:**395.

Warner, N. L., and Burnet, F. M., 1961, The influence of testosterone treatment on the development of the bursa of Fabricius in the chick embryo, *Aust. J. Biol. Sci.* **14:**580.

Weber, W. T., and Alexander, J. E., 1978, The potential of bursa-immigrated hematopoietic precursor cells to differentiate to functional B and T cells, *J. Immunol.* **121:**653.

Weber, W. T., and Mausner, R., 1977, Migration patterns of avian embryonic bone marrow cells and their differentiation to functional T and B cells, in: *Avian Immunology* (A. A. Benedict, ed.), pp. 47–59, Plenum Press, New York.

Weissman, I. L., 1967, Thymus cell migration, *J. Exp. Med.* **126:**291.

17

Ontogeny of Mammalian Lymphocytes

J. J. T. OWEN

1. INTRODUCTION

Interest in the ontogeny of mammalian lymphocytes goes back long before the discovery of the importance of lymphocyte populations in immune responses. It is remarkable that scientists were prepared to spend so much time and effort in studying the origins of lymphocytes in the developing thymus for example, when the function of the thymus was completely unknown (Hammar, 1905). Perhaps it is even more remarkable that these studies generated such strong feelings. The main techniques used were essentially histological and, therefore, it was inevitable that there would be disagreement about the interpretation of the results. The issues must have been purely academic since lymphocytes were of unknown function. People working in this field had an interest which extended beyond the question of the origin of lymphocytes and in general they were concerned about the origins and interrelationships of all blood cell types. Indeed, it had already been postulated that all blood cells were derived from the same basic cell type—the hemopoietic stem cell (Maximow, 1924). This notion has been confirmed in recent years and, in retrospect, it can be seen that these early studies were rightly directed to the broader issues of hemopoiesis in general.

It is interesting to note that these early studies were directed to blood cell production in ontogeny rather than to renewal of blood cells in the adult. The thrust of this research reflected the enormous interest in embryology and the extensive efforts put into the reconstruction of early developmental processes by means of histological techniques. Again, subsequent events have amply confirmed the desirability of this approach since we can now see that many of the basic problems of immunobiology, such as the question of the origin of diversity of immune responsiveness, the acquisition of tolerance to self antigens, as well as the basis for congenital defects of immune responsiveness are most effectively

---

J. J. T. OWEN • Department of Anatomy, Medical School, Birmingham, England.

being studied at an ontogenetic level. In this context, studies on the ontogeny of lymphocyte populations in a variety of vertebrate species have contributed to our knowledge and understanding of these processes. Other chapters in this volume are concerned with some of these matters. In this chapter, I will attempt to review research on mammalian embryos, but by far the greatest amount of information is available on the mouse embryo. Hopefully, studies on a variety of vertebrate species will complement each other and produce information which is valid to the human situation.

I will discuss the ontogeny of mammalian lymphocytes in the broad context of the major subdivision of lymphocytes into T and B cells. This division, first elucidated a decade ago, still remains valid today, and provides a basis for the investigation of the subsets of T and B lymphocytes which have been demonstrated more recently. I will not attempt to summarize all of the available literature, but will indicate wherever possible what I consider to be key papers and reviews which have provided the basis for our current concepts. We remain woefully ignorant of many important issues and in most instances our knowledge does not extend to a biochemical and certainly not to a molecular level. I will not hesitate to indicate what I consider to be areas of great uncertainty, although my opinion may well not be shared by other investigators. It also follows from what I have said previously that I will need to consider blood cell types other than lymphocytes in order to provide a coherent account.

2. HEMOPOIETIC STEM CELLS AND THE ONTOGENY OF MAMMALIAN LYMPHOCYTES

Hemopoietic stem cells are cells which are capable of renewing their own kind, as well as proliferation and differentiation to mature blood cells. The mechanism by which they do so is poorly understood, but their capabilities are dramatically demonstrated by experiments in which it has been shown that all of the blood cells of a lethally irradiated mouse (in which the hemopoietic system has been destroyed) may be derived from a single stem cell of a normal donor mouse (Edwards et al., 1970). Thus, there are hemopoietic stem cells which not only have considerable proliferative capabilities, but which are also pluripotential in the sense that they can give rise to all of the various blood cell types.

The first assay for hemopoietic stem cells was introduced in 1961 (Till and McCullouch, 1961). Injection of limited numbers of normal bone marrow cells into lethally irradiated mice results in the production of discrete splenic nodules. These nodules have been shown to be colonies, each derived from a single donor stem cell. Thus, estimates of the number of splenic nodules provide an assay for the number of hemopoietic stem cells in the inoculum. These stem cells [known as colony-forming spleen (CFU-S) stem cells] are pluripotential. Thus, it is likely that lymphoid cells are derived from this stem cell type. CFU-S stem cells are first found in the embryonic yolk sac and thereafter in fetal liver (Metcalf and Moore, 1971). Finally, they appear within the developing bone marrow where they persist throughout life. The morphological identity of these cells has proved elusive. Consequently, it has proved difficult to study them other than by the CFU-S assay, and although they are ancestors of lymphocytes, their relationship with immature lymphocytes remains largely uncertain (Abramson *et al.*, 1977). In particular, intervening stages have not been identified with confidence.

3. MAMMALIAN LYMPHOCYTES: FUNCTIONAL TYPES AND THEIR ONTOGENY

The functional complexity of the immune system has only been fully demonstrated in recent years. Concepts of immune function and control are now based on the notion that there is an interactive system involving lymphocyte populations at a variety of levels. Control depends not only on the major subdivision into T and B lymphocytes and the interaction of these two populations, but on the existence of functional subsets of these major populations which have been shown to interact with one another. Our knowledge of this system is still at an elementary level, but I will attempt to describe its emergence in ontogenetic terms. In part, the emerging system depends upon a major lymphoid organ present in all mammalian embryos, namely the thymus, and this is where I will begin my description.

4. THE THYMUS AND T-LYMPHOCYTE ONTOGENY

The thymus is present in all mammals, although its final form is rather variable. In most instances, it exists as a pair of lobes in the upper thorax, but in some species (e.g., sheep) there is a cervical portion in the neck as well (Jordan, 1976). In all cases, however, the thymic primordium is initially derived from epithelium. Careful histological studies carried out many years ago indicate that the bulk of this epithelium is derived from the third pharyngeal pouch (Venzke, 1952). The epithelial primordium is formed at the point of junction of endoderm and ectoderm, and it has not been easy to distinguish between the relative contributions of the two portions. However, it is generally agreed that both endoderm and ectoderm do contribute to the epithelial thymus. The epithelial portion of the thymus, therefore, is initially a simple downgrowth of pharyngeal pouch material. The connection of primordium to the pharynx is soon lost and the central cavity is normally obliterated at an early stage.

While these events are taking place, two major types of cells are beginning to migrate into the thymic rudiment. On the one hand, septae of mesenchymal tissue penetrate the epithelium, dividing it into lobes and lobules. These septae are also accompanied by blood vessels so that the thymus becomes vascularized. In part at least, the mesenchyme invading the thymus at this stage of development is of neural crest origin (see Chapter 16). Perhaps more importantly, the thymic epithelium is invaded by lymphoid stem cells (Moore and Owen, 1967; Le Douarin and Joterau, 1975). These are large rounded cells with prominent nuclei and nucleoli. Their cytoplasm is basophilic, a property which has been shown to be due to the presence of large numbers of ribosomes. The cells are actively ameboid and leave small vessels in the tissue surrounding the thymic primordium and migrate through the mesenchyme and enter the thymus by penetrating the outer basement membrane. The precise origins of these stem cells and the factors which promote their migratory behavior are still unknown, especially in mammals. Some information is available in birds (see Chapter 16) which suggests that chemotactic factors liberated by thymic epithelial cells may be responsible (Pyke and Bach, 1979). With regard to the origin of these cells, it is known that they exist in a variety of embryonic hemopoietic sites such as yolk sac and fetal liver; however, their primary source has not been determined with certainty. Again, there is evidence in birds to suggest that they are of intra- rather than extraembryonic (yolk sac) origin (Dieterlen-Lièvre, 1975; see Chapter 16).

Once they are within the thymus, the stem cells proliferate and large numbers of small lymphocytes are produced. At the same time, the epithelial component of the thymus is proliferating and differentiating and well-defined cortical and medullary areas are generated. Thymic growth proceeds rapidly, so that the thymus is a well-developed organ at the time of birth. Indeed, in most mammals, the major events of thymic maturation are completed during ontogeny. Stem cell migration into the embryonic thymus proceeds at an extremely early stage of histogenesis (Moore and Owen, 1967). Thus, stem cells enter the epithelial rudiment while it is still connected to the pharyngeal region as a simple hollow tube. Lymphocyte maturation may be achieved well before the thymus is fully developed and some of the earliest lymphocytes to be formed may seed to developing peripheral lymphoid organs before birth. Subsequently, waves of maturing lymphocytes may add to the complement of peripheral lymphoid cells, but neonatal or even fetal thymectomy experiments suggest that the first T cells seeded may have considerable proliferative capabilities, and for thymectomy to produce severe T-cell deficiency in mice, it must be performed within 24 hr of birth (Miller and Osoba, 1967).

Much of our current interest in the events within the thymus, center on the respective roles of the epithelial cells in lymphocyte maturation, and in the lymphoid stem cells in the generation of the various T-lymphocyte subsets. I will therefore deal with these two items separately, although, of course, inevitably they are interlinked in the full functional maturation of the thymus.

5. ONTOGENETIC EVENTS IN THE MATURATION OF THYMIC EPITHELIUM

As mentioned earlier, thymic epithelium is derived from pharyngeal pouch ectoderm and endoderm. In the mouse embryo, where most ontogenetic studies have been carried out, the epithelium by downward migration loses its connection with the pharyngeal pouches and, at the 13th day of gestation, appears as a pear-drop structure with an obvious tail to which the parathyroid glands are attached. With the further migration of the thymic primordium down to the upper thoracic region, the parathyroid glands are left behind and become attached to the thyroid gland. It has been suggested that the more rapidly advancing rounded part of the primordium is mainly ectodermal, while the tail is endodermal. In the absence of cellular markers for ectoderm and endoderm, it is impossible to be sure about this. However, we have found that if the thymus is bisected into upper and lower halves at this stage, and the halves are then separately placed into organ culture, the upper half consistently shows cystic development which might indicate an endodermal origin (see discussion later about the nature of the defect in nude mice) (Jenkinson and Owen, unpublished observations).

Whatever the relative contributions of ectoderm and endoderm, rapid proliferation and morphological differentiation occur in the epithelial cells of the rudiment. The ultrastructural changes have been described by Mandel and principally consist of the development of vesicles and vacuoles within the cells; some of the vesicles contain small electron-dense bodies, similar in type to those seen in polypeptide hormone-secreting glands (Mandel and Russell, 1971). However, the significance of these ultrastructural changes is far from clear, and it is interesting to note that stem cells are migrating into the thymic primordium before these changes take place.

Recent studies have shown that there are important surface changes occurring on thymic epithelial cells during ontogeny which might be crucial for lymphocyte maturation (Jenkinson et al., 1980). These studies have been stimulated by evidence that T lymphocytes may learn to see foreign antigen in association with self MHC (major histocompatibility complex) components during their intrathymic maturation (MHC restriction) (Zinkernagel et al., 1978). Thus, those lymphocytes which have receptors for self MHC antigens may, by a process of selection, proliferate and be exported to peripheral tissues. This selection might be either positive, in the sense that only cells with receptors for self MHC will be driven to proliferation (Williamson, 1980), or negative, in the sense that cells lacking in self receptor will die. In either case, the expression of MHC antigens by thymic epithelial cells assumes special significance. Evidence has been presented that epithelial cells of the adult mouse thymus do express various MHC antigens. Most cells express Ia antigen, but there is doubt as to whether or not all cells express K antigen. Indeed, it has been proposed that there may be microenvironments within the thymus each of which is involved in the selective maturation of different T-cell subsets (Rouse et al., 1979).

However, the differences may be quantitative rather than qualitative, and it is relevant that these studies were carried out on frozen tissue sections where loss of antigenicity might be expected to occur. In recent studies on the ontogeny of MHC antigen expression on embryonic mouse thymus, we have found that Ia and K antigens are detectable on epithelial cells of thymus of 13-day embryos cultured overnight (Jenkinson et al., 1980). However, the intensity of staining in an indirect immunofluorescence assay is weak at this stage, but is followed by far greater intensity of staining on epithelial cells of 14- and 15-day thymus. In other words, a pattern emerges of limited MHC antigen expression occurring in the thymus during the 14th to 16th days of gestation.

The expression of K paralleled the expression of Ia in these studies which were carried out on the unfixed cell outgrowths of thymic cultures, and no evidence of microenvironmental differences was noted. However, studies on sections of embryonic thymus do indicate a pattern of staining for K antigen which is restricted to the medulla as seen in adult thymus (Jenkinson et al., 1981). However, again this may reflect quantitative differences in expression between cells in various regions of the thymus, together with a possible masking effect of greater proportions of lymphoid cells in the cortex. Whatever the explanation for the patterns of MHC antigen expression on thymic epithelial cells, these studies do show clearly that lymphoid cells are exposed to MHC antigens during intrathymic lymphopoiesis and that this is the case not only in the adult but during ontogeny. The situation merits further analysis; for example, it should be possible to study the effect of blocking antibodies directed against MHC components on epithelial cells on the development of lymphocytes in organ cultures of embryonic thymus. There might be a gross effect on lymphopoiesis or, alternatively, it may be possible in functional studies to show the absence of MHC restriction in specific populations.

It is well established that nude mice possess thymic stem cells as well as an epithelial rudiment, although normal thymic lymphopoiesis does not occur (Wortis et al., 1971). There have been reports that under certain conditions lymphopoiesis may take place in the nude thymic remnant in vivo or in vitro (Chakravarty et al., 1975); these reports have not been substantiated in other studies. Careful histological examination of the thymic remnant found in embryos derived from homozygous nude × nude matings has not supported the notion that any lymphopoiesis occurs, and organ culture studies on these remnants have also failed to provide evidence for lymphocyte differentiation (Jordan et al., 1977). Indeed, the basophilic stem cells which are normally seen in and around the thymus from an early stage of development (11 days gestation) are totally absent in the nude thymus. In recent studies, we have found that the nude thymic epithelial remnant, while expressing K antigen, does not express Ia antigen (Jenkinson et al., 1981). This observation suggests that there is a gross anomaly in the development of the epithelial primordium which might result from failure of Ia expression in a component of the thymus or, alternatively, might indicate that a component is totally absent. Support for the latter idea comes from suggestions that the ectodermal contribution to the normal thymus is absent in the nude embryo (Cordier and Heremans, 1975). This result taken with our observation on the absence of Ia expression in the nude thymus might suggest that it is the ectodermal component which normally becomes Ia positive and that the endoderm portion, left on its own, reverts to a cystic and disorganized structure. Large cysts are characteristic of the nude thymus both in embryogenesis and in the adult, and, as mentioned previously, can be found in organ cultures of the upper portion of normal thymus. Thus, it seems unlikely that the nude thymic remnant is directly involved either in the production of lymphocytes or in MHC restriction, at least in terms of Ia. However, there are a number of reports that, by the addition of suitable growth factors to cells of the nude mouse, it is possible to generate T-lympocyte responses, albeit at a much reduced level (Gillis *et al.*, 1979). These results might indicate that prethymic stem cells can undergo maturation outside of the thymus to a limited extent, provided that suitable growth factors are available. Whether or not these cells show MHC restriction is unknown (Hünig and Bevan, 1980; Gillis and Watson, 1981).

Before leaving this section, it is important to note that the nonlymphoid components of the thymus are not totally epithelial in origin. As mentioned previously, mesenchymal elements invade the thymus at an early stage of ontogeny. Some of these components go to make up the connective tissue stroma carrying the vascular supply. However, a variety of nonlymphoid blood cells migrate into the thymus, and of these, perhaps the macrophage component has received the most attention because of a possible role in lymphocyte maturation. Macrophages are known to be important in antigen presentation to T cells. Some antigen-presenting cells express Ia, and it has been suggested that some cells of this type are present in the thymus (Beller and Unanue, 1980). However, few macrophages are present in the early thymic primordium when lymphopoiesis is initiated and MHC antigens are already expressed, and it seems likely, therefore, that the epithelial components are mainly responsible for Ia expression. There is little doubt that macrophages play an important part in the overall function of the thymus, especially with regard to the removal of dead lymphoid cells (see discussion on turnover of thymic lymphocytes in the next section).

6. ONTOGENETIC EVENTS IN THE MATURATION OF THYMIC LYMPHOCYTES

It is not known whether the large basophilic cells which migrate into the thymus from the bloodstream are pluripotential hemopoietic stem cells or whether they are already committed to T-cell maturation before they enter the thymus. The main evidence that pre-T cells are committed to T-cell differentiation before they enter the thymus comes from studies which have shown that treatment of fetal liver cells (fetal liver is known to contain T-cell precursors) with agents which elevate intracellular cAMP results in the expression on some cells of Thy-1 and other surface alloantigens which are typically present on mature T cells (Scheid *et al.*, 1973; Komuro *et al.*, 1975). However, final proof that these cells are pre-T cells must depend upon studies which show that they migrate to and proliferate in the thymus. Certainly, the basophilic cells in the early thymus express only low amounts of T-cell antigens (e.g., Thy-1, Tl etc.) (Owen and Raff, 1970). Little is known about their general properties, for example whether they are in active cell cycle. It seems likely that they are indeed proliferating and that the considerable expansion of basophilic cells seen during the first few days after the migration process has begun, is not due solely to the recruitment of additional migrant stem cells. By day 14 of gestation in the mouse embryo, large basophilic cells are still very much in evidence. Few cells could be categorized as small lymphocytes. However, during the following 2 days of gestation, there is a dramatic change from predominantly large basophilic cells to small thymocytes. This change is accompanied by the strong expression of T-cell alloantigens (Owen and Raff, 1970; Kamarck and Gottlieb, 1977) and is coincident with the major quantitative change in expression of MHC antigens on the thymic epithelium.

During this period of time, there is considerable growth in the whole organ, and there may well be the beginnings of cell turnover which is such a feature of thymocyte production in the adult. Evidence has been reported in the adult that the vast majority of newly formed thymocytes have a short life-span and a large proportion actually die within the thymus (Shortman, 1977). The extent to which this is the case for the embryo is unknown, but it is likely that there is major export of newly formed cells to peripheral organs, during the embryonic phase.

During the process of proliferation, two distinct ontogenetic events are occurring. First of all, functional subsets of T cells are generated, e.g., helper cells, suppressor cells, etc. However, in addition, these cells have receptors for foreign antigen as well as receptors for self recognition in terms of MHC restriction.

With regard to the ontogeny of functional subsets of T cells, it has been shown that cells responsive in mixed lymphocyte culture (Robinson and Owen, 1977), as well as cells capable of cooperating with B cells in immune responses, appear in the thymus at about the time of birth (Chiscon and Golub, 1972). Precursors of cytotoxic T lymphocytes also are generated at about this time (Widmer and Cooper, 1979), and there is evidence that suppressor T cells are similarly generated (Mosier and Johnson, 1975). Cells responsive to mitogens appear somewhat earlier and have been detected from about the 17th day of gestation in the mouse embryo (Robinson and Owen, 1976). Although there are no extensive studies on the appearance of T-lymphocyte subsets in other lymphoid organs of the embryo, the evidence that is available suggests that functional subsets appear first in the thymus after which they are presumably exported to other lymphoid organs. However, it should be noted that there are some data which suggest that thymocytes might be exported before they reach full functional maturation. Coinciding with the appearance of these functional subsets, there are surface antigen changes evident in the thymic population as a whole. Most of the newly formed thymocytes express alloantigens such as Thy-1 and T1, and although a majority of cells express Lyt-1,2, and 3 antigens together (Kamarck and Gottlieb, 1977), it is likely that there are other cells which express Lyt-1 or Lyt-2 and 3 alone. Presumably, these correspond to helper/cytotoxic T-cell subsets (Cantor and Boyse, 1976).

The nature of receptors for antigen recognition on T lymphocytes has been a matter of considerable debate (Lindahl and Rajewsky, 1979). Current data support the notion that antigens are recognized by variable-region portions of immunoglobulin molecules comparable to those found on B cells. However, the manner by which T lymphocytes recognize self MHC antigens is still unclear. It has been suggested that whatever the recognition unit on the maturing thymocytes, a process of cell selection is involved (Williamson, 1980). The situation is open to experimental analysis and the final outcome must await the results of such experiments.

The full role of the thymic stroma in the maturation of the lymphoid component is still far from clear. In addition to a possible role in antigen presentation, there remains the question of the factors, hormonal or otherwise, which result in lymphocyte proliferation and maturation. It is outside the scope of this chapter to deal with the question of thymic hormones, but suffice it to say that there is evidence that such factors are involved in T-cell maturation both within the thymus and in other peripheral lymphoid organs (Trainin et al., 1977). However, thymocyte differentiation is probably regulated by a variety of factors, and it has been shown that hormones and other agents can influence differentiation of thymocytes probably by alterations in cyclic nucleotide metabolism (Singh et al., 1979).

7. MAMMALIAN B-LYMPHOCYTE ONTOGENY

The importance of the bursa of Fabricius in the ontogeny of avian B lymphocytes led to a search for specific lymphopoietic sites in mammals which might fulfill a comparable role (Fichtelius, 1967). However, attempts to define such a specific organ were unsuccessful, and it is now generally accepted that primary B-lymphocyte differentiation in mammals takes place in sites of general hemopoiesis along side the production of other blood cell types. Evidence for this view comes from two major sources. In the adult animal, the pioneering work of Osmond and his colleagues has shown that the bone marrow is a site of lymphopoiesis involving the production of large numbers of B lymphocytes on a scale comparable to the production of T cells in the thymus (Osmond and Nossal, 1974). In the embryo, evidence has come from organ culture studies which have shown that isolated fragments of fetal liver, maintained in vitro for a number of days, are capable of generating considerable numbers of B lymphocytes (Owen et al., 1975). Of course, the liver of the mammalian embryo is a major hemopoietic organ, and it seems likely that B cells are generated in it along side other blood cells in extravascular spaces. The fetal spleen is also active in hemopoiesis and again this organ has been shown to be a site of B-cell production. Although the first site of B-lymphocyte differentiation in the embryo has not been defined with accuracy, it has been suggested that the placenta might be the initial site of B lymphopoiesis (Melchers, 1979). However, B-lymphocyte precursors are found in embryonic blood as early as they are found in organs; consequently, their precise origin remains uncertain.

In many of these studies, B cells were identified by the presence of readily detectable surface immunoglobulin (sIg) which is thought to act as antigen recognition receptor. B lymphocytes bearing sIgM have been detected in mouse fetal liver at 17 days of gestation and at 11 weeks of gestation in the human fetus (Gathings et al., 1977). However, the question might rightly be asked as to whether cells with sIg are the first antibody-forming cells in the embryo, and the results discussed in the following section support the view that they are not.

8. PRE-B CELLS

By using purified antibodies against components of mouse immunoglobulin molecules conjugated with fluorochromes, μ chains have been demonstrated within the cytoplasm of cells of fetal liver as early as the 12th day of gestation (Raff et al., 1976). These cells lack sIgM as shown by double-labeling techniques. They are heterogeneous in size, and at least some of the larger cells are in cell cycle (Owen et al., 1977). The fact that these cells predate the appearance of cells with sIg and that they are found in fetal hemopoietic organs as well as in adult bone marrow, but not in peripheral lymphoid organs, has led to the notion that they are pre-B cells.

Pre-B cells have been detected in early fetal tissues by other assays. For example, in studies on Ig synthesis in fetal liver cells, IgM synthesis has been detected as early as the 11th day of gestation in the mouse (Melchers et al., 1975). In functional assays, it has been shown that early fetal cells contain pre-B cells which can readily mature *in vitro* to B cells and, in the presence of mitogens, to antibody-secreting cells (Melchers et al., 1977). These results point to the presence of pre-B cells in fetal tissues which synthesize μ chains, but do not insert Ig into their membranes. The question as to whether pre-B cells synthesize light chains is still unanswered (Burrows et al., 1979).

The nature of the immediate precursors of pre-B cells is unclear, and their relationship to pluripotential (CFU-S) stem cells has not been defined, although there is evidence for their origin from such a cell type. Although tissue culture conditions have now been defined which allow proliferation of CFU-S stem cells, primary production of B lymphocytes in long-term cultures has not been obtained (Dexter et al., 1977; Schrader and Schrader, 1978).

9. NEWLY DIFFERENTIATED B LYMPHOCYTES IN FETAL TISSUES

Newly formed B lymphocytes in fetal organs and, indeed, in adult bone marrow are a population of small cells, homogeneous in size. They are not in cell cycle as is the case for some pre-B cells, and they initially express sIgM (Owen et al., 1977). They express Fc receptors, and at least a proportion of them express Ia antigen (Kearney et al., 1977). C3 receptors also appear on a proportion of these cells shortly after their formation (Osmond, 1979).

Newly formed B lymphocytes can respond to mitogens such as lipopolysaccharide and will form colonies in soft agar (Johnson et al., 1976). However, it has been suggested that they may respond to antigenic stimuli in a manner different from that shown by mature B lymphocytes in peripheral lymphoid organs (Nossal and Pike, 1975). Thus, newly formed B cells may be rendered tolerant by amounts of antigen which produce responses in mature cells. However, this tolerance occurs only under certain experimental conditions, and whether or not these results point to a mechanism for the establishment and maintenance of tolerance to self antigens as suggested by Nossal and Pike remains an open question (Szewczuk and Siskind, 1977).

Paralleling this susceptibility to tolerance induction, newly formed B lymphocytes have sIg which can be readily modulated by anti-Ig antibodies (Sidman and Unanue, 1975). Indeed, continued incubation with anti-Ig antibodies results in complete suppression of sIg expression on immature B cells (Raff *et al.*, 1975). The functional significance of these observations remains unclear, but since they parallel ease of tolerance induction by antigen they should be investigated further.

Some newly formed B lymphocytes migrate from fetal liver to peripheral lymphoid organs. However, it is likely that the rate of production of cells, especially in the adult bone marrow, is so large (Osmond, 1975) as to suggest that many newly formed cells must die *in situ*, or shortly after migration. Indeed, it has been calculated that the half-life of the virgin B lymphocyte in the mouse is of the order of 7 days (Elson *et al.*, 1976). These observations suggest that many newly produced B lymphocytes are expressing antibody specificities which are irrelevant to the needs of the animal, either because the animal has not experienced the relevant antigens or because the specificities are reactive with self antigens.

Recently, a variety of differentiation antigens have been identified on murine B lymphocytes and it has been suggested that these antigens identify subsets of B cells (Ahmed and Scher, 1979). Further evidence for the existence of distinct B-cell subsets has been obtained in studies on congenitally immunodeficient mice of the CBA/N strain which lack certain types of B cells (Kincade, 1977). The manner in which the variety of functional subsets of B cells appear during ontogeny has not yet been examined.

There is very little information about the general factors which influence the primary production of B lymphocytes. A role for cyclic nucleotides in primary B-lymphocyte differentiation has been adduced from studies in which treatment of bone marrow cells with agents known to influence levels of intracellular cAMP has been shown to produce phenotypic conversion of cells from Ia negative to Ia positive, from sIg negative to sIg positive, and from complement receptor negative to complement receptor positive (Hammerling *et al.*, 1976). However, apart from these results, no hormonal or microenvironmental factors have been defined which are crucial to primary B-lymphocyte differentiation.

10. THE EXPRESSION OF V-REGION GENES DURING PRIMARY B-LYMPHOCYTE DIFFERENTIATION IN FETAL ORGANS

The question arises as to whether or not during primary differentiation of B lymphocytes in fetal tissues all antibody specificities appear simultaneously or whether some appear before others. There is evidence for a sequential appearance of immune reactivity to various antigens during ontogeny (Sterzl and Silverstein, 1967; Sherwin and Rowlands, 1975). However, immune responsiveness depends upon the functions of a number of interacting cells and it is difficult to evaluate the repertoire of V-region genes expressed by B lymphocytes in ontogeny from this type of data.

One approach is to test the ability of newly differentiated B cells to bind antigen. Antigen binding studies performed at various times during ontogeny may provide an estimate of the extent to which the antibody repertoire has developed. In this way, Edelman and his colleagues have found that coincident with the differentiation of small sIg-positive B lymphocytes, a repertoire of specificities is generated which is comparable in breadth of specificity and in range of binding affinities to those found in the adult (Cohen et al., 1977). However, Klinman and his colleagues in functional studies, where the abilities of newly formed B cells to produce antibodies have been tested by transfer to irradiated recipients, found evidence for a restricted repertoire in fetal mice (Klinman et al., 1976). It may be argued that ability to bind antigen does not imply functional capability and therefore antigen binding studies may give misleading results. Alternatively, of course, it may be argued that the complete antibody repertoire is not sampled in cell transfer studies. Thus, the matter remains unresolved, although technical improvements in lymphocyte culture methods in which every murine B lymphocyte can be stimulated by mitogen to proliferate and mature might allow more accurate sampling of the antibody repertoire in newly formed B lymphocytes in the future (Anderson et al., 1977). Whatever the final conclusion, it is clear that some antibody specificities are already present on newly formed B lymphocytes as they emerge in the fetal liver. The antibody specificities of pre-B cells have not as yet been characterized. In summary, there is disagreement as to the manner in which the antibody repertoire appears during primary B-lymphocyte differentiation in the fetal liver. It seems likely that this problem will only be resolved by studies examining the molecular mechanisms involved in the generation of antibody diversity in immature pre-B cells.

11. THE EXPRESSION OF C-REGION GENES DURING PRIMARY B-LYMPHOCYTE DIFFERENTIATION IN FETAL ORGANS

There is general agreement that the first immunoglobulin class synthesized during primary lymphocyte differentiation in ontogeny is IgM (Cooper et al., 1976a). Subsequently, lymphocytes expressing sIg of other classes are detected. Indeed, simultaneous expression of three Ig classes on one cell has been reported (Cooper et al., 1976b). These observation together with other evidence has supported the notion that diversification of Ig classes may be brought about by B lymphocytes switching the class of antibody they synthesize while maintaining the same antibody specificity. This concept of a switch within a single cell line has received considerable support from studies in birds and mammals; the strongest evidence has come from studies where multiple injections of anti-IgM antibodies during the neonatal period have been shown to suppress B lymphocytes expressing not only IgM but also other Ig classes (Lawton et al., 1972).

The functional significance of the Ig class expressed on newly differentiated B cells is the subject of much recent analysis; in particular, the role of sIgD has received considerable attention. It has been suggested that its presence may be necessary for lymphocyte triggering by antigen (Kettman *et al.*, 1979). However, further work is required before this problem can be resolved. In summary, IgM is the first class of antibody synthesized during ontogeny. There is evidence that other Ig classes appear on the progeny of cells which initially synthesize IgM. The relationship of the class of Ig on newly formed B cells to subsequent antibody production is under active investigation.

12. GENERAL SUMMARY

This chapter has focused on the first phases of lymphocyte differentiation during mammalian ontogeny. It is generally agreed that these earliest stages of differentiation proceed independently of antigen stimulation as normal developmental events. T and B lymphocytes emerge via separate developmental pathways in the thymus and hemopoietic tissues, respectively. Various stages in the development of both cell types can now be identified. During ontogeny of the thymus, MHC antigen expression can be detected on stromal cells during the earliest phase of lymphopoiesis and this observation may have considerable relevance to MHC restriction of mature T cells. The tissue distribution and properties of pre-B and newly formed B cells are now well worked out. Studies on factors controlling primary B-cell differentiation are at a preliminary stage. There is little doubt that the ontogeny of mammalian lymphocytes will be an important area for study in the future.

ACKNOWLEDGMENTS. I would like to thank Mrs. J. Lidstone for excellent help with the preparation of the manuscript.

REFERENCES

Abramson, S., Miller, R. S., and Phillips, R. A., 1977, The identification in adult bone marrow of pluripotent and restricted stem cells of myeloid and lymphoid systems, *J. Exp. Med.* **145**:1567.

Ahmed, A., and Scher, I., 1979, Murine B-cell heterogeneity defined by anti-Lyb5, an alloantiserum specific for a late appearing B-lymphocyte subpopulation, in: *B-lymphocytes in the Immune Response* (M. Cooper, D. E. Mosier, I. Scher, and E. S. Vitetta, eds.), Elsevier/North-Holland, Amsterdam.

Anderson, J., Cotinho, A., and Melchers, F., 1977, Frequencies of mitogen-reactive B-cells in the mouse, *J. Exp. Med.* **145**:1511.

Beller, D. I., and Unanue, E. R., 1980, Ia antigens and antigen-presenting function of thymic macrophages, *J. Immunol.* **124**:1433.

Burrows, P., Le Jeune, M., and Kearney, J. F., 1979, Evidence that murine pre-B-cells synthesize µ heavy chains but no light chains, *Nature (London)* **280**:838.

Cantor, H., and Boyse, E. A., 1976, Functional subclasses of T-lymphocytes bearing different Ly antigens. 1. The generation of functionally distinct T-cell subclasses is a differentiative process independent of antigen, *J. Exp. Med.* **141**:1376.

Chakravarty, A., Kubai, L., Sidky, Y., and Auerbach, R., 1975, Ontogeny of thymus function, *Ann. N.Y. Acad. Sci.* **249**:34.

Chiscon, M. O., and Golub, E. S., 1972, Functional development of the interacting cells in the immune response. 1. Development of T-cell and B-cell function, *J. Immunol.* **108**:1379.

Cohen, J. E., D'Eustachio, P., and Edelman, G. M., 1977, The specific antigen-binding cell populations of individual fetal mouse spleens: Repertoire composition, size, and genetic control, *J. Exp. Med.* **146**:394.

Cooper, M. D., Kearney, J. F., Lawton, A. R., Abney, E. R., Parkhouse, R. M. E., Preud'homme, J. L., and Seligman, M., 1976a, Generation of immunoglobulin class diversity in B-cells: A discussion with emphasis on IgD development, *Ann. Immunol. (Inst. Pasteur)* **127C**:573.

Cooper, M. D., Kearney, J. F., Lydyard, P. M., Grossi, C. E., and Lawton, A. R., 1976b, Studies of generation of B-cell diversity in mouse, man, and chicken, *Cold Spring Harbor Symp. Quant. Biol.* **41**:139.

Cordier, A. C., and Heremans, J. F., 1975, Nude mouse embryo: Ectodermal nature of the primordial thymic defect, *Scand. J. Immunol.* **4**:193.

Dexter, T. M., Allen, T. D., and Lajtha, L. S., 1977, Conditions controlling the proliferation of hemopoietic stem cells *in vitro*, *J. Cell. Physiol.* **91**:335.

Dieterlen-Lièvre, F., 1975, On the origin of haemopoietic stem cells in the avian embryo: An experimental approach, *J. Embryol. Exp. Morphol.* **33**:607.

Edwards, S. E., Miller, R. S., and Phillips, R. A., 1970, Differentiation of rosette-forming cells from myeloid stem cells, *J. Immunol.* **105**:719.

Elson, C. J., Jablonska, K. F., and Taylor, R. B., 1976, Functional half-life of virgin and primed B-lymphocytes, *Eur. J. Immunol.* **6**:634.

Fichtelius, K. E., 1967, The mammalian equivalent to bursa Fabricii of birds, *Exp. Cell Res.* **46**:231.

Gathings, W. E., Lawton, A. R., and Cooper, M. D., 1977, Immunofluorescent studies of the development of pre-B cells, B-lymphocytes, and immunoglobulin isotype diversity in humans, *Eur. J. Immunol.* **7**:804.

Gillis, S., and Watson, J., 1981, Interleukin-2 induction of hapten-specific cytolytic T cells in nude mice, *J. Immunol.* **126**:1245.

Gillis, S., Union, N. A., Baker, P. E., and Smith, K. A., 1979, The in vitro generation and sustained culture of nude mouse cytolytic T-lymphocytes, *J. Exp. Med.* **149**:1460.

Hammar, J. A., 1905, Zur Histogenese und Involution der Thymusdrüse, *Anat. Anz.* **27**:23.

Hammerling, V., Chin, A. F., and Abbott, J., 1976, Ontogeny of murine B-lymphocytes: Sequence of B-cell differentiation from surface-immunoglobulin-negative precursors to plasma cells, *Proc. Natl. Acad. Sci. USA* **73**:2008.

Hünig, T., and Bevan, M. J., 1980, Specificity of cytotoxic T cells from athymic mice, *J. Exp. Med.* **152**:688.

Jenkinson, E. J., Owen, J. J. T., and Aspinall, R., 1980 Lymphocyte differentiation and major histocompatibility complex antigen expression in the embryonic thymus, *Nature (London)* **284**:177.

Jenkinson, E. J., van Evijk, W., and Owen, J. J. T., 1981, Major histocompatibility complex antigen expression on the epithelium of the developing thymus in normal and nude mice, *J. Exp. Med.* **153**:280.

Johnson, G. R., Metcalf, D., and Wilson, J. W., 1976, Development of B-lymphocyte colony-forming cells in foetal mouse tissues, *Immunology* **30**:907.

Jordan, R. K., 1976, Development of sheep thymus in relation to in utero thymectomy experiments, *Eur. J. Immunol.* **6**:693.

Jordan, R. K., Owen, J. J. T., and Raff, M. C., 1977, Organ culture studies of nude mouse thymus, *Eur. J. Immunol.* **7**:736.

Kamarck, M. E., and Gottlieb, P. D., 1977, Expression of thymocyte surface alloantigens in the fetal mouse thymus in vivo and in organ culture, *J. Immunol.* **119**:407.

Kearney, J. F., Cooper, M. D., Klein, J., Abney, E. R., Parkhouse, R. M. E., and Lawton, A. R., 1977, Ontogeny of Ia and IgD on IgM-bearing B-lymphocytes in mice, *J. Exp. Med.* **146**:297.

Kettman, J. R., Cambier, J. C., Uhr, J. W., Ligler, F., and Vitetta, E. S., 1979, The role of receptor IgM and IgD in determining triggering and induction of tolerance in murine B-cells, *Immunol. Rev.* **43**:69.

Kincade, P. W., 1977, Defective colony formation by B-lymphocytes from CBA/N and C3H/H&J mice, *J. Exp. Med.* **145**:249.

Klinman, N. R., Sigal, N. H., Metcalf, E. S., Pierce, S. K., and Gearhart, P. J., 1976, The interplay of evolution and environment in B-cell diversification, *Cold Spring Harbor Symp. Quant. Biol.* **41**:165.

Komuro, K., Goldstein, G., and Boyse, E. A., 1975, Thymus-repopulating capacity of cells that can be induced to differentiate to T-cells in vitro, *J. Immunol.* **115**:195.

Lawton, A. R., Asotsky, R., Hylton, M. B., and Cooper, M. D., 1972, Suppression of immunoglobulin class synthesis in mice. 1. Effects of treatment with antibody to µ chain, *J. Exp. Med.* **135**:277.

Le Douarin, N., and Jotereau, F. V., 1975, Tracing of cells of the avian thymus through embryonic life in interspecific chimaeras, *J. Exp. Med.* **142**:17.

Lindahl, K. F., and Rajewsky, K., 1979, T-Cell recognition: Genes, molecules and functions, in: *International Review of Biochemistry: Defense and Recognition* (E. S. Lennox, ed.), IIA, Vol. 22, p. 97, University Park Press, Baltimore.

Mandel, T., and Russell, P. J., 1971, Differentiation of foetal mouse thymus: Ultrastructure of organ cultures and of subcapsular grafts, *Immunology* **21**:659.

Maximow, A. A., 1924, Relation of blood cells to connective tissue and endothelium, *Physiol. Rev.* **4**:533.

Melchers, F., 1979, Murine embryonic B-lymphocyte development in the placenta, *Nature (London)* **277**:219.

Melchers, F., von Boehmer, H., and Phillips, R. A., 1975, B-Lymphocyte subpopulations in the mouse: Organ distribution and ontogeny of immunoglobulin-synthesizing and antigen-sensitive cells, *Transplant. Rev.* **25**:26.

Melchers, F., Anderson, J., and Phillips, R. A., 1977, Ontogeny of murine B-lymphocytes: Development of immunoglobulin synthesis and of reactivities to mitogens and to anti-Ig antibodies, *Cold Spring Harbor Symp. Quant. Biol.* **41**:147.

Metcalf, D., and Moore, M. A. S., 1971, *Haemopoietic Cells* (A. Neuberger and E. L. Tatum, eds.), in: *Frontiers of Biology*, Vol. 24, pp. 172–271, North-Holland, Amsterdam.

Miller, J. F. A. P., and Osoba, D., 1967, Current concepts of the immunological function of the thymus, *Physiol. Rev.* **47**:437.

Moore, M. A. S., and Owen, J. J. T., 1967, Experimental studies on the development of the thymus, *J. Exp. Med.* **126**:715.

Mosier, D. E., and Johnson, B. M., 1975, Ontogeny of mouse lymphocyte function. II. Development of the ability to produce antibody is modulated by T-lymphocytes, *J. Exp. Med.* **141**:216.

Nossal, G. J. V., and Pike, B. L., 1975, Evidence for the clonal abortion theory of B-lymphocyte tolerance, *J. Exp. Med.* **141**:904.

Osmond, D. S., 1975, Formation and maturation of bone marrow lymphocytes, *J. Reticuloendothelial Soc.* **17**:99.

Osmond, D. S., 1979, Generation of B-lymphocytes in the bone marrow, in: B-lymphocytes in the Immune Response (M. Cooper, D. E. Mosier, I. Scher, and E. S. Vitetta, eds.), Elsevier/North-Holland, Amsterdam.

Osmond, D. S., and Nossal, G. J. V., 1974, Differentiation of lymphocytes in mouse bone marrow. II. Kinetics of maturation and renewal of antiglobulin-binding cells studied by double labeling, *Cell. Immunol.* **13**:132.

Owen, J. J. T., and Raff, M. C., 1970, Studies on the differentiation of thymus-derived lymphocytes, *J. Exp. Med.* **132**:1216.

Owen, J. J. T., Raff, M. C., and Cooper, M. D., 1975, Studies on the generation of B-lymphocytes in the mouse embryo, *Eur. J. Immunol.* **5**:468.

Owen, J. J. T., Wright, D. E., Habu, S., Raff, M. C., and Cooper, M. D., 1977, Studies on the generation of B-lymphocytes in fetal liver and bone marrow, *J. Immunol.* **118**:2067.

Pyke, K. W., and Bach, J.-F., 1979, The in vitro migration of murine fetal liver cells to thymic rudiments, *Eur. J. Immunol.* **9**:317.

Raff, M. C., Owen, J. J. T., Cooper, M. D., Lawton, A. R., Megson, M., and Gathings, W. E., 1975, Differences in susceptibility of mature and immature mouse B-lymphocytes to anti-immunoglobulin induced immunoglobulin suppression in vitro: Possible implications for B-cell tolerance to self, *J. Exp. Med.* **142**:1052.

Raff, M. C., Megson, M., Owen, J. J. T., and Cooper, M. D., 1976, Early production of intracellular IgM by B-lymphocyte precursor in the mouse, *Nature (London)* **259**:224.

Robinson, J. H., and Owen, J. J. T., 1976, Generation of T-cell function in organ culture of foetal mouse thymus. 1. Mitogen responsiveness, *Clin. Exp. Immunol.* **23**:347.

Robinson, J. H., and Owen, J. J. T., 1977, Generation of T-cell function in organ culture of foetal mouse thymus. II. Mixed lymphocyte culture reactivity, *Clin. Exp. Immunol.* **27**:322.

Rouse, R. V., van Ewijk, W., Jones, P. P., and Weissman, I. L., 1979, Expression of MHC antigens by mouse thymic dendritic cells, *J. Immunol.* **122**:2508.

Scheid, M. P., Hoffman, M. K., Komuro, K., Hammerling, V., Abbott, J., Boyse, E. A., Cohen, G. H., Hooper, J. A., Schulof, R. S., and Goldstein, A. L., 1973, Differentiation of T-cells induced by preparations from thymus and by non-thymic agents, *J. Exp. Med.* **138**:1027.

Schrader, J. W., Schrader, S., 1978, In vitro studies on lymphocyte differentiation. I. Long term in vitro culture of cells giving rise to functional lymphocytes in irradiated mice, *J. Exp. Med.* **148**:823.

Sherwin, W. K., and Rowlands, D. T., 1975, Determinants of the hierarchy of humoral, immune responsiveness during ontogeny. *J. Immunol.* **115**:1549.

Shortman, K., 1977, The pathway of T-cell development within the thymus, in: *Progress in Immunology III* (T. E. Mandel, C. Cheers, C. S. Hosking, I. F. C. McKenzie, and G. J. V. Nossal, eds.), p. 197, Australian Academy of Science.

Sidman, C. L., and Unanue, E., 1975, Receptor-mediated inactivation of early B-lymphocytes, *Nature (London)* **257**:149.

Singh, U., Milson, D. S., Smith, P. A., and Owen, J. J. T., 1979, Identification of B adrenoceptors during thymocyte ontogeny in mice, *Eur. J. Immunol.* **9**:31.

Sterzl, J., and Silverstein, A. M., 1967, Developmental aspects of immunity, *Adv. Immunol.* **6**:337.

Szewczuk, M. R., and Siskind, S. W., 1977, Ontogeny of B-lymphocyte function. III. In vivo and in vitro studies on the ease of tolerance induction in B-lymphocytes from fetal, neonatal and adult mice, *J. Exp. Med.* **145**:1590.

Till, J. E., and McCullouch, E. A., 1961, A direct measurement of the radiation sensitivity of normal and mouse bone marrow cells, *Radiat. Res.* **14**:213.

Trainin, N., Small, M., and Kook, A. I., 1977, The role of thymic hormones in regulation of the lymphoid system in B- and T-cells, in: *Immune Recognition* (F. Loor and G. E. Roelants, eds.), p. 83, Wiley, New York.

Venzke, W. G., 1952, Morphogenesis of the thymus of chicken embryo, *Am. J. Vet. Res.* **13**:395.

Widmer, M. B., and Cooper, E. L., 1979, Ontogeny of cell-mediated cytotoxicity: Induction of CTL in early postnatal thymocytes, *J. Immunol.* **122**:291.

Williamson, A., 1980, Three receptor, clonal expansion model for selection of self-recognition in the thymus, *Nature (London)* **283**:527.

Wortis, H. H., Nehlsen, S., and Owen, J. J. T., 1971, Abnormal development of the thymus in "nude" mice, *J. Exp. Med.* **134**:681.

Zinkernagel, R. M., Callihan, G. N., Althage, A., Cooper, S., Klein, P. A., and Klein, J., 1978, On the thymus in the differentiation of H-2 self recognition by T-cells: Evidence for dual recognition?, *J. Exp. Med.* **147**:882.

18

Ontogeny of Immunological Functions in Amphibians

LOUIS DU PASQUIER

1. INTRODUCTION

The theoretical basis for studying the ontogeny of immunological function in amphibians is linked to aspects of amphibian embryonic and larval development. The free-swimming larval period, the absence of maternal–fetal interactions, the relative simplicity of the larval immune system, and metamorphosis with its associated differentiation of adult antigens, all pose a variety of interesting problems to the developing immune system and to developmental immunologists. Answers to these problems can provide information in at least three major areas of research in immunology: the origin of antibody diversity, the role of the thymus, and the generation of tolerance to self. Moreover, the comparison between the development of the strikingly different immune systems of urodeles and anurans can reveal the phylogenetic aspects of some components of the immune system. During the past decade, amphibian models have been significantly refined largely due to the introduction: of biologically defined strains both in urodeles and in anurans (De Lanney and Blackler, 1969; Charlemagne and Tournefier, 1974; Nace and Richards, 1969; Kobel and Du Pasquier, 1977; Tochinai and Katagiri, 1975); of clones of isogenic and histocompatibility-defined *Xenopus* (Kobel and Du Pasquier, 1975, 1977); of natural (Tymowska and Fischberg, 1973) and laboratory-made polyploid species of *Xenopus* (Du Pasquier et al., 1977); and finally, of hyperdiploid *Xenopus* hybrids convenient for gene mapping (Kobel and Du Pasquier, 1979; Du Pasquier and Kobel, 1979). Thus, it is worthwhile reviewing the data concerning the ontogeny of immunological functions in amphibians, with special emphasis on results obtained with these new models.

In this chapter, Section 2 is primarily devoted to the ontogeny of the antibody production capacity (i.e., the B-cell system) as it relates to the thymus. Section 3 deals with the ontogeny of allogeneic recognition, a classical T-cell

---

LOUIS DU PASQUIER • Basel Institute for Immunology, Basel, Switzerland.

phenomenon in higher vertebrates. Graft rejection and mixed leukocyte reactions (MLR) will be analyzed according to their relationships to the thymus. Section 4 deals with the generation of tolerance to self and will contain a speculative paragraph on the possible relationships between self tolerance, metamorphosis, and the expression of the B-cell repertoire of recognition structures.

2. ANTIBODY PRODUCTION

2.1. DIFFERENTIATION OF B CELLS

2.1.1. The Amphibian B Cell

In mammals, the B lymphocyte is often characterized as a cell that expresses immunoglobulin (Ig) molecules at the surface of its membrane and that can be stimulated to divide and differentiate into an Ig-producing cell upon the action of certain substances like bacterial endotoxins (for reference on this subject, see *Cold Spring Harbor Symposa on Quantitative Biology*, Vol. 41, 1977, *Origins of Lymphocyte Diversity*). Although this definition may not strictly apply to B lymphocytes of all amphibians, it is still the most convenient one and will be used throughout this chapter. Although it is a fully differentiated cell, the amphibian B cell retains most, if not all, of its genome intact, since nuclei from *Xenopus* adult B lymphocytes used in nuclear transplantation experiments, can promote tadpole development (Wabl *et al.*, 1975).

2.1.2. Early Appearance of B Cells

The ontogeny of anuran B lymphocytes has been studied primarily in *Xenopus*. Strongly Ig-positive lymphocytes, detected by immunofluorescence, appear in the blood and spleen of tadpoles at early stages of development (Figure 1) and persist in these organs with approximately the same percentages throughout the entire life span of the frog (Du Pasquier *et al.*, 1972; Jurd and Stevenson, 1976). Studies performed in adults exclusively (Nagata and Katagiri, 1978) confirmed the general finding that in the spleen there exist two approximately equal populations of lymphocytes, one Ig positive, the other Ig negative. Similar findings have been reported in *Rana catesbeiana* (Mattes and Steiner, 1978). After capping or proteolytic treatment of the membrane, these Ig molecules were resynthesized by the cell. These molecules mostly belong to the IgM class (95%) and rarely to the low-molecular-weight (LMW Ig) type 5%) (Jurd and Stevenson, 1976), although it has been suggested that *Xenopus* lymphocytes can synthesize these two categories of Ig at the same time (Hadji-Azimi and Parrinello, 1978). Thymectomy at stage 45–48 results in a relative increase of the population of the strongly Ig-positive cells (up to 85–91%) (Weiss *et al.*, 1973; Nagata and Katagiri, 1978). Moreover, thymectomy does not abrogate the ability of *Xenopus* lymphocytes to respond to *E. coli* lipopolysaccharide (Manning *et al.*, 1976).

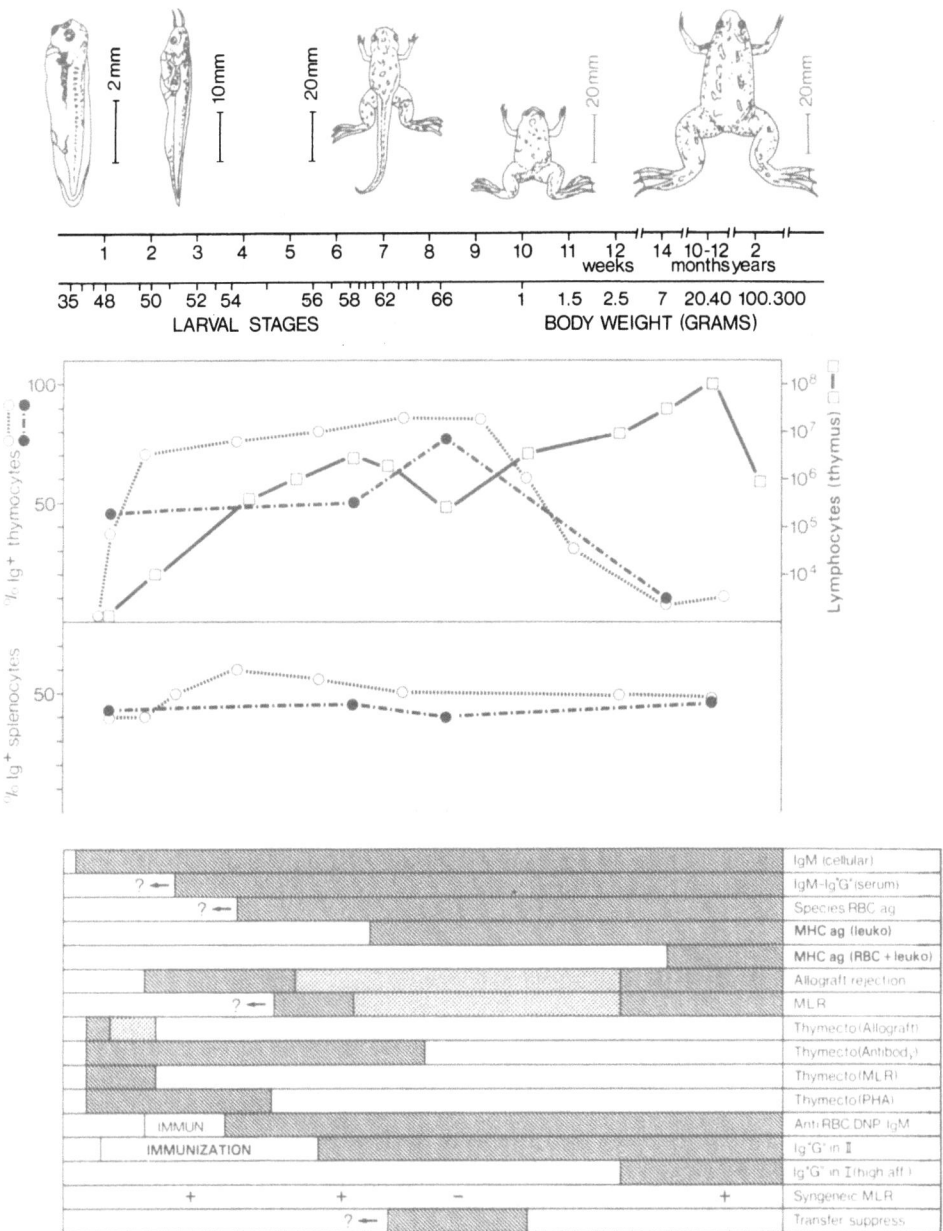

FIGURE 1. Recapitulation of the ontogeny of immunity in *Xenopus*. Thymecto (allograft, antibody, MLR, PHA): Period during which thymectomy is efficient in preventing allograft rejection, antibody production, MLR, and PHA responsiveness. Anti-RBC, DNP, IgM: Occurrence of IgM antibody response to RBC and DNP–KLH. Ig"G" in II: Detection of LMW antibodies in secondary response after priming early in larval life. Ig"G" in I: Production of LMW antibodies in primary response, of better affinity than the larval one. Transfer suppress.: Period during which thymic cell transfer into adult will suppress the response to allografts differing by minor histocompatibility antigens. ?←: Not investigated earlier. Striped areas indicate the occurrence of the function shown in the box at the right. Dotted areas indicate that there may be modification of this function during the particular period. (○) Data from Du Pasquier *et al.* (1972); (●) data from Jurd and Stevenson (1976).

2.2. ONSET OF IMMUNOGLOBULIN PRODUCTION AND OF SPECIFIC ANTIBODY RESPONSES

2.2.1. Immunoglobulins

Until recently, it was thought that the earliest production of Ig occurred concomittantly with the appearance of lymphocytes in the thymus, i.e., in *Xenopus* at stage 47 when young thymocytes express IgM on their surface. Recent studies in *Xenopus* (Moyer *et al.*, 1977; Leverone *et al.*, 1979), however, suggest that the onset of Ig production might occur as early as stage 35. Using a sensitive radioimmunoassay in conjunction with a biochemical characterization, the authors suggest that IgM is produced from stage 35 onwards. Although further biochemical characterization is necessary to make this date entirely convincing, one can already wonder about the nature of these early "B" cells, and whether the Ig remains cytoplasmic or is secreted. From the published curves of rate of synthesis, there even seems to be an increased synthesis 100 hr after fertilization. This roughly corresponds to the period of thymic differentiation.

As far as humoral Ig is concerned, the two classes of Ig in *R. catesbeiana* (Geczy *et al.*, 1973; Green and Steiner, 1976) are present in the serum of larvae. The younger Ig-positive *Xenopus*, tested by immunoprecipitation techniques, were at stage 50–51 of development. The sites of synthesis can be the spleen, the blood leukocytes, and the liver, as shown by biosynthesis experiments involving cell culture in the presence of radiolabeled amino acids and autoradiographic analysis of the supernatants (Du Pasquier and Feinstein, unpublished). The occurrence of LMW Ig in the serum is highly thymus dependent (Manning, 1975). No LMW Ig can be detected in the serum of thymectomized *Xenopus*, whereas the level of IgM is normal or higher than in sham thymectomized (Manning, 1975; Weiss *et al.*, 1973).

2.2.2. Onset of Specific Antibody Responses

That an embryo produces Ig or that larval thymocytes express Ig on their surface does not mean that the immune system of the animal is ready to be triggered by external antigens. Studies with specific ligands aimed at detecting the first antigen-binding cells and then the antibody-producing cells have been performed with a few species of anurans (*Alytes obstetricans*, *X. laevis*, *R. catesbeina*). They indicated that immune competence appears early and that larval amphibians express a relatively well-diversified repertoire of Ig recognition sites. The first signs of the existence of specific receptors recognizing xenoantigens were provided by analyses of natural antigen-binding cells (rosettes) during the development of the spleen. In *Alytes* and *Xenopus*, such cells can be detected very early, i.e., when the spleen has fewer than 5000 lymphocytes (Du Pasquier, 1970; Kidder *et al.*, 1973). An interesting observation in both species is that the proportion of cells binding a given antigen decreases with the growth of the spleen (Figure 2). Does this phenomenon reflect diversification of the B-cell repertoire, or the colonization of the spleen by a different population of cells?

The fine specificity of these natural rosette-forming cells is unknown. The incidence of early specific recognition is also indicated by the experiments of Jurd et al. (1975), showing that an injection of antigen at a stage where normally no response can be detected (stage 48) is, however, able to prime the animal for a specific secondary IgM and LMW Ig response.

That old amphibian larvae can mount an immune response to various antigens has been well demonstrated both in anurans (Cooper et al., 1963, 1964; Cooper and Hildemann, 1965a; reviewed by Du Pasquier, 1973, 1976) and in urodeles where the axolotl is a natural model of the permanently larval state (Ching and Wedgwood, 1967; Amirante and Parisi, 1967; Fougereau and Houdayer, 1968; Ambrosius et al., 1970). Precise analysis of the larval responses has been made more recently. In Alytes and Xenopus, the ability to mount an immune response to RBC antigens (measured by the rosette or plaque tech-

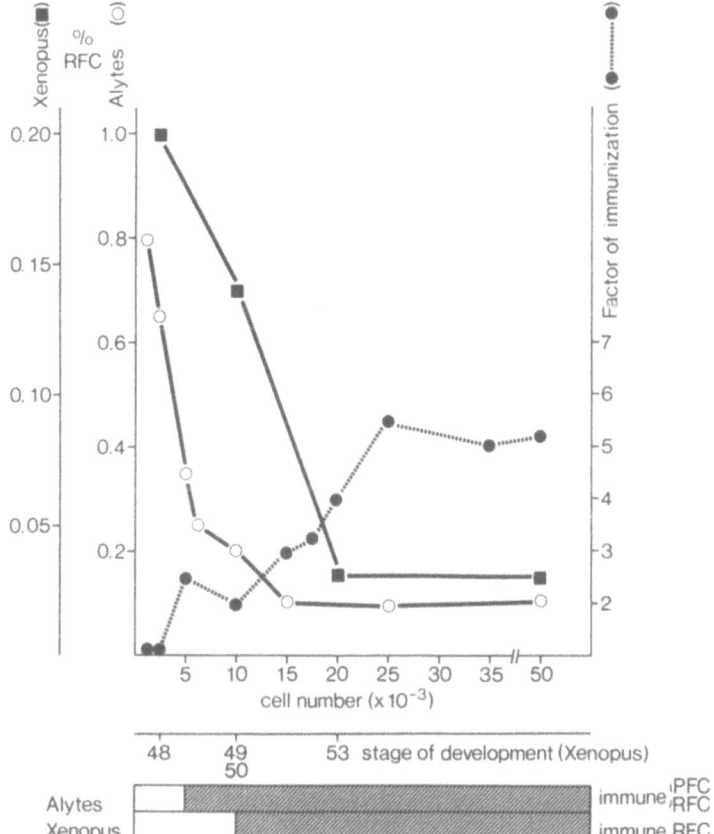

FIGURE 2. Evolution of the percentages of natural antigen-binding cells in Alytes (○) and Xenopus (■) in connection with the onset of immune reactivity estimated by the factor of immunization [i.e., (No. of rosette-forming cells in immunized animal)/(No. of rosette-forming cells in nonimmunized animal)]. See also the onset of the immune responses for Xenopus (data for Xenopus taken from Kidder et al., 1973; data for Alytes taken from Du Pasquier, 1970).

nique) appears at the time when the immune system contains about 10^6 lymphocytes precisely at the moment when the proportion of natural rosette-forming cells reaches a stable level (Figure 2). The larval response can be quantitatively as high as that of mammals if proper doses of antigen are used, but the anatomical location of plaque-forming cells can differ from that of mammals. For example, *R. catesbeiana* thymus can yield a fair number of plaque-forming cells (ca. 2000 per 10^6; Moticka et al., 1973). Blood and spleen are normally good sources of plaque-forming cells. The kinetics of larval and adult responses are similar. The class of the antibody produced by lymphocytes may not be easy to determine. IgM plaques are 2-ME sensitive, but there have been claims that frog LMW Ig is also 2-ME sensitive (Marchalonis and Edelman, 1966). Moreover, LMW IG plaque-forming cells of frogs need no enhancing antisera since these antibodies are spontaneously hemolytic in the presence of guinea pig complement.

The specificity of the larval response has been demonstrated by testing the cell suspensions immunized to one RBC against another type, without high degree of cross-reactivity (Du Pasquier, 1970, 1976; Kidder et al., 1973). The fine specificity of anti-DNP antibodies made during ontogeny has been analyzed in *R. catesbeiana* and *Xenopus*. Tadpoles exhibit IgM and LMW Ig anti-DNP responses but IgM is predominant. IgM antibodies can increase in affinity (10-fold) and reach an affinity and specificity identical to those of IgM antibodies made by mammals. LMW Ig are of low affinity in larvae but can reach higher affinities in adults, although never as high as those of mammals (Haimovich and Du Pasquier, 1973; Du Pasquier, 1973; Du Pasquier and Haimovich, 1974, 1976). These studies made on total serum or peritoneal fluid did not investigate the actual heterogeneity of the larval response. From the slopes of the curves of inhibition of inactivation of modified bacteriophages, one could only deduce that the response was not homogeneous. The heterogeneity of the IgM and LMW Ig larval response to bacteriophage has been demonstrated in *R. catesbeiana* (Pross and Rowlands, 1975). After isoelectric focusing in liquid, up to eight peaks of neutralizing antibodies could be found in the larval serum; adults did not show a more heterogeneous response. Similar results were obtained in the case of the anti-DNP response (Du Pasquier and Wabl, 1976). In *Xenopus*, the same technique has been applied but in gel, which only allows the detection of LMW antibodies. As already suggested from the small number of peaks detected in *Rana*, *Xenopus* was found to exhibit, in its primary or secondary response, a spectrum of anti-DNP antibodies that is much less heterogeneous than that shown by the mouse (Du Pasquier and Wabl, 1976). The number of antigen-binding bands after IEF separation is on the order of only 8–12 per individual. The larvae of *Xenopus*, whose LMW antibodies can be analyzed in the course of a secondary response, also mount a heterogeneous response, but one that is qualitatively different from the adult. This conclusion was reached by comparing the IEF patterns of anti-DNP sera obtained in a clone of isogenic individuals that differed only with regard to age. In fact, adult isogenic *Xenopus* show almost identical IEF patterns and it is relatively easy to establish a chart of adult antibody spectrotypes. The larvae also show a heterogeneous response (i.e., several spectrotypes for a given antigen) and again with most of the spectrotypes shared by the individuals of a clone. However, within the same clone, larval and adult pattern can differ significantly, even up to the point that no overlap exists between the larval and the adult antibody patterns. It is not impossible that a larval spectrotype is reexpressed in adult life since some overlap was indicated in other clones. On the other hand, if the larva has been primed during larval life and boosted in adult life, then the reexpression of larval clonotypes becomes the rule (Du Pasquier et al., 1979). This confirms previous findings that immunological memory can pass through metamorphosis (Du Pasquier and Haimovich, 1976). In *Xenopus*, metamorphosis does not seem to have a strong effect on the quantitative aspects of the anti-rabbit RBC response (Nagata, 1976). This contrasts with an apparently strong impairment of the anti-sheep RBC response reported for metamorphosing *Rana* (Moticka et al., 1973).

In summary, a diversified B-cell repertoire is expressed early in larval life (ca. 20 days after hatching), and a specific larval response with the two categories of IgM and LMW Ig can be detected after various antigenic stimulations in tadpoles. Although heterogeneous, the larval response is not identical to that of the adult of the same genotype. This suggests that the expression of antibody diversity depends very much on some ontogenetic steps occurring during metamorphosis.

2.3. THYMUS DEPENDENCY OF THE ANTIBODY RESPONSE

The thymus dependency of the LMW Ig in the serum has already been mentioned. Thymectomy during larval life up to a relatively advanced stage [stage 52–54 of Nieuwkoop and Faber (1967) according to Manning and Collie, 1975; stage 57 according to Horton et al., 1977] will affect antibody production. In the case of antigens that are T dependent in mammals, thymectomy in *Xenopus* abrogates the IgG response but leaves the IgM response untouched. For T-independent antigens, such as PVP or LPS, thymectomy does not abrogate the response which consists essentially of IgM antibody production (Turner and Manning, 1974; Collie et al., 1975; Tochinai, 1976). These observations for *Xenopus* are in agreement with previous preliminary work done with *Rana*, where late larval thymectomy also impaired the antibody response to soluble proteins (Cooper et al., 1964).

The effect of late thymectomy on the antibody response, in contrast to the absence of an effect on allograft immunity, suggests that the thymus itself is necessary to provide "help" during almost the entire larval period. There is no evidence whether this "help" is similar to that demonstrated recently in adults by *in vitro* collaboration between adult T and B cells responding to a hapten–carrier system (Blomberg et al., 1980). In fact, the characteristics of the normal larval response (i.e., poor LMW synthesis, lack of pronounced increase in affinity in their LMW Ig class) suggest that, although thymic function is present in larvae, it is not fully differentiated. Perhaps in larvae, mature antigen-specific T helper cells have not yet seeded lymphoid organs; the changes at the surface of thymocytes (Figure 1) may reflect a maturation step. Thymic function in larval stages might be restricted to production of humoral factors without much specificity. Another possible explanation of the difference between larval and adult responses is suggested by the data presented in the section dealing with the ontogeny of histocompatibility antigens (3.2). We know from experiments of MHC restriction in mammals that a specific collaboration between T and B cells occurs only when the two cells bear the same MHC antigens or share at least one haplotype. This also seems to be the situation in *Xenopus* (Bernard *et al.*, 1981). Since, to some extent, larval lymphocytes appear to lack such markers, it is tempting to suggest that T–B collaboration cannot be optimal in larval stages. The end of the period when thymectomy efficiently prevents antibody response to T-dependent antigens (stage 57) is practically coincident with the onset of histocompatibility antigen expression at the surface of lymphocytes (see Figure 1). In urodeles that appear to lack at least some elements of the MHC (Cohen, 1976), thymus dependency of antibody production could be interpreted in this context. Urodeles that respond exclusively to T-independent antigens and with production of IgM antibodies (Fougereau and Houdayer, 1968) behave a little like thymectomized *Xenopus*. They might have a defect in the specific helper mechanisms. Thymectomy in *Pleurodeles* had practically no effect on the antibody response to somatic antigens of *Salmonella*, but was followed by an increase of the antiflagellar antibody titer. Similarly, the anti-horse RBC response of thymectomized *Ambystoma* was higher than that of controls (Charlemagne and Tournefier, 1977a,b). Such an effect is reminiscent of the enhanced response to T-independent antigens observed in rabbits thymectomized as adults (Kerbel and Eidinger, 1972).

A similar effect had been found at the level of normal IgM in thymectomized *Xenopus* or *Alytes* (Du Pasquier, 1968; Weiss *et al.*, 1973). The high level of natural antibodies, the poor factor of immunization noticed in all immunized urodeles, suggests that the B-cell system expresses a restricted antibody repertoire and is permanently activated. It could lack the efficient control existing in anurans and probably exerted by the thymus. In some respects, urodeles may behave immunologically like the motheaten mouse mutant (Sidman *et al.*, 1978), where it has been proposed that the B-cell system is overactivated.

3. ALLOGENEIC RECOGNITION

3.1. DIFFERENTIATION OF THE T LYMPHOCYTE

Some of the properties of mammalian T cells are found in anuran and urodele amphibians. These include *in vitro* responses to PHA, Con A, and involvement in MLR (Goldshein and Cohen, 1972; Goldstein *et al.*, 1975; Collins *et al.*, 1975). Moreover, cytotoxic and helper T cells have recently been demonstrated in *Xenopus*. These cells pass through nylon wool and have no surface Ig detectable by immunofluorescence (Blomberg *et al.*, 1980). Such cells can be found in the spleen. In the thymus, the situation is more complex. In *Xenopus*, the thymus goes through a transient stage where up to 80–85% of its lymphocytes express membrane Ig (Du Pasquier et al., 1972; Jurd and Stevenson, 1976) (Figure 1). Similar observations have been made in urodeles (Charlemagne and Tournefier, 1975). After metamorphosis, either the thymocytes express less Ig or they no longer have Ig, since by conventional immunofluorescence tests, only a small percentage of adult thymocytes appear to be surface Ig positive (Figure 1). This ontogenetic change contrasts with the stability of the splenic cell population where one finds that throughout life, approximately 50% of the splenic lymphocytes are strongly Ig positive and 50% are Ig negative. Biosynthetic and immunochemical experiments showed that the membrane Ig of the thymocyte was of endogenous origin and of the IgM isotype (Du Pasquier et al., 1972; Du Pasquier, 1976; Jurd and Stevenson, 1976). Recently, it was reported that what was detected as surface Ig in *R. catesbeiana* larval thymus was, in fact, a non-Ig determinant that cross-reacted with Ig via a carbohydrate (Mattes and Steiner, 1978). It is, therefore, quite possible that antisera used by different authors have different properties and that it is, in fact, quite difficult to compare data. However, several points can be restated here which make us believe that larval thymocytes may actually express Ig on their surface. Antisera we used in our *Xenopus* studies were heavily adsorbed on RBC and ultracentrifuged before they were used at high dilution (1:256 routinely, but up to 1:1000 if necessary). Recent studies with peroxidase-labeled antibody also revealed Ig on the surface of 91% of thymocytes of young adults (Nagata and Katagiri, 1978) without any cross-reaction with RBC. This provides some support to a previous report arguing that adult thymocytes actually express surface Ig, but less than during the larval stage (Du Pasquier and Weiss, 1973). Finally, one does not know what the staining of the larval thymus in *Rana* corresponds to, since no comparison was made with the adult thymus, as was done with *Xenopus*. Despite such minor discrepancies, all the groups working on surface Ig in amphibians agree that, in these animals, there exists a heterogeneous population of lymphocytes containing surface-Ig-positive and surface-Ig-negative cells.

3.2. EARLY APPEARANCE OF T-CELL FUNCTION

3.2.1. Graft Rejection

Many experiments with diverse species have revealed that larval amphibians can reject allografts during larval life (*R. catesbeiana*, Hildemann and Haas, 1959; *Discoglossus pictus*, Dupuy, 1964; *Alytes obstetricans*, Du Pasquier, 1965; *R. pipiens*, Bovbjerg, 1966; *X. laevis*, Horton, 1969; reviewed by Du Pasquier, 1973; Cooper, 1973). Urodele larvae are also able to reject allografts and xenografts (Tournefier, 1968; Cohen, 1969; Orfila and Deparis, 1970). More precise data in *Xenopus* (Horton, 1969; reviewed by Horton, 1971) indicate that the onset of graft rejection capacity is at stage 49 when the lymphoid system of the animal contains as few as 0.5×10^6 lymphocytes (Horton and Manning, 1972). The genetic control of larval graft rejection has not been studied, however, and it cannot be assumed, a priori, that larval rejection is governed exactly by the same laws that govern histocompatibility reactions in adults (see next section). Nevertheless, anuran tadpoles (*R. catesbeiana*) show segregation of many alleles at various histocompatibility loci (Hildemann and Haas, 1961). The mechanisms and kinetics of graft rejection are very similar in adult and larvae (Baculi and Cooper, 1970; Rimmer and Horton, 1977). Depending on the protocol, grafting during early embryonic life may result in either tolerance or sensitization. The induction of tolerance, by creating chimeras, is routinely seen in *Xenopus* (Clark and Newth, 1972). Sensitization, which seems to be a regular phenomenon in the case of xenografts (reviewed by Du Pasquier, 1973), has also been reported in *R. pipiens* (Volpe and Gebhardt, 1965; reviewed by Volpe, 1971). The apparent dose effect recorded in these experiments (i.e., one neural fold graft = sensitization; two neural fold grafts = tolerance) may, in fact, depend on the nature of transplanted tissue, as previously found by Davison (1966) who formulated the hypothesis "no blood chimerism = no tolerance." Many of the experiments involving embryonic explant transfers should now be reexamined or repeated in the context of those MHC restriction phenomena that tell us that it is in the thymus that lymphocytes "learn" which MHC alleles to recognize on antigen-presenting cells, in the case of T_r-B collaboration, or in the generation of killer cells that recognize targets with modified self cell surfaces. Apparently, lymphoid cells expressing receptors for MHC products are selected by these antigens expressed on the thymic epithelium (Zinkernagel *et al.*, 1978). What will be the recognition capacities of the chimeric frog lymphocytes that were educated in an allogeneic thymus? Some experiments (à la Volpe or Davison) may have involved transfer of thymic primordia and may have, in fact, generated chimeras where the lymphocyte population has resided in two genetically different thymuses. Analyses of histocompatibility reactions and antibody responses of such animals could be most informative.

3.2.2. MLR

An *in vitro* correlate of graft rejection, MLR has also been studied during ontogeny, although for technical reasons, not at the very early stages of development. MLR between lymphocytes from stage 55–56 tadpoles of *Xenopus* could be detected by standard thymidine incorporation protocols. In *R. catesbeiana*, positive MLR was also detected by counting blasts in mixed suspensions of tadpole cells (Du Pasquier and Weiss, 1973; unpublished observations). The genetic control of the larval MLR has not been fully elucidated, because the analysis is complicated by the existence of a syngeneic MLR when larval cells are mixed with stimulators from syngeneic adults. Shortly before metamorphosis, the MLR classes determined in larvae seem to correlate well with the classes determined in adults. This question can now be approached in a better way with isogenic tadpoles, where it is possible to pool the cells and work with all the animals at exactly the same stage of development. In the neotenic "larval" urodele, *Ambystoma*, significant MLR indices (up to 2.8) were detected between members of different strains (De Lanney *et al.*, 1975).

Responses to T mitogens can also be detected in larval amphibians like *R.*

catesbeiana whose lymphocytes respond to PHA (Pross and Rowland, 1975). The axolotl's lymphocytes also are stimulated by PHA and Con A (Collins *et al.*, 1975). Other functions, such as generation of killer cells and helper cells, remain to be studied during ontogeny. The production of a thymic hormone (molecular weight < 50,000), detected in the serum of some species of urodeles (*Pleurodeles, Ambystoma,* and *Triturus*) (Dardenne *et al.*, 1973), has not yet been followed during early ontogeny.

3.3. APPEARANCE OF HISTOCOMPATIBILITY ANTIGENS

The study of allogeneic recognition would not be complete without a survey of the expression of the structures that are recognized by effector cells. Adult and embryonic or larval frogs share some histocompatibility antigens. This can be demonstrated by two types of experiments: in one, an adult individual is sensitized against an adult first graft and rejects a secondary embryonic graft in an accelerated way (Simnett, 1966); in the other, a larva, immunized against larval skin, subsequently rejects a secondary adult graft in an accelerated fashion (Volpe and Gebhardt, 1965; Volpe, 1964; Chardonnens, 1976). However, this type of experiment does not prove that all histocompatibility antigens are shared by larval and adult frogs. Actually, there is evidence for the presence of stage-specific antigens in anuran and urodele embryos (Spar, 1953; Clayton, 1957; Romanovsky, 1964). These studies concerned developmental steps in the embryo, but it is likely that equivalents can be found in larval stages. For instance, larval RBC differ from those of adults in *R. pipiens* (Hollyfield, 1966) and each type of RBC can be characterized by a few specific antigens.

In recent years, antisera have been raised against MHC determinants of *X. laevis*, (Du Pasquier *et al.*, 1975), as well as against other cell surface antigens. Two antisera, anti-XLA_1 and anti-XLA_2, have been used to type the cells of *Xenopus* strains during ontogeny. Other species-specific markers have been used as technical controls. MHC antigens detected by these antisera are absent (i.e., undetectable by immunofluorescence) from the surface of RBC and lymphocytes until stage 59–60. At this time, lymphocytes from the spleen and 30% of the thymic lymphocytes become positive. RBC MHC antigens appear later, $1\frac{1}{2}$ months after the completion of metamorphosis. These antigens, referred to as RBC ag, can also be present on lymphocytes and may, in fact, represent the MHC ubiquitous antigens. Species-specific non-MHC antigens, detected by similar techniques, were present on the youngest RBC tested (stage 53). This observation makes one wonder what kind of histocompatibility antigens are expressed in larvae. Is the MHC weakly expressed or is it present only on certain tissues? Should we admit that no MHC antigens at all are expressed before stage 59? Is there a relationship between the change in expression of surface Ig in thymic cells and the appearance of lymphocyte-specific MHC antigens (see Figure 1)? Could this lymphocyte antigen be the equivalent of Ia or a product of another MHC subregion? One could actually envisage that during ontogeny, the thymic cell Ig become less and less detectable because of the addition of other surface components such as those encoded by MHC. It could even be possible that MHC products participate in the thymic cell ot T-cell receptor together with Ig, as has been proposed in mammals.

Ontogenetic studies with cytotoxicity are necessary to determine where the functional equivalents of *K* or *D* ends of the MHC are expressed at the surface of larval cells. Antigenicity of the tissues may also change during ontogeny, although in *Xenopus*, the antigenicity of newly metamorphosed skin is the same as that of the adult (Bernardini *et al.*, 1970). Not much is known about larval skin. In urodeles (*P. waltlii*), antigenicity of the skin seems to change between larval and adult life (Orfila and Deparis, 1970). In *Ambystoma*, many of the grafts made from the skin of young individuals (20 days posthatching) exhibited indefinite survival when compared with the older transplants (80 days posthatching) (Cohen, 1969), which suggests a lower antigenicity of the young skin.

3.4. THYMUS DEPENDENCY OF MLR, GRAFT REJECTION, AND PHA RESPONSIVENESS

Because easy access to the larval thymus facilitates its surgical extirpation, amphibians have long been used to study the effect of early thymectomy (Cooper and Hildemann, 1965b, Du Pasquier, 1965, 1968; Horton and Manning, 1972; Tochinai and Katagiri, 1975; Curtis and Volpe, 1971). All these experiments proved that thymectomy, during anuran larval life, has a profound effect on allograft rejection. Moreover, although frogs are more resistant to deleterious effects of thymectomy than mice, they may eventually die from a runtinglike disease. Similar conclusions were reached with urodeles (Cohen, 1969; Charlemagne and Houillon, 1968; Fache and Charlemagne, 1975; Charlemagne, 1974). In this vertebrate order, histogenesis of the lymphoid system takes a longer time than in anurans. For example, in *Triturus alpestris*, the thymus (on which allograft reactivity depends) appears a few weeks after fertilization (Tournefier, 1973).

The effect of thymectomies performed at various stages of development has shed some light on the differentiation of various T-dependent functions. There is not complete agreement on whether thymectomy at stage 45–48 abrogates or impairs allograft rejection capacity. According to Tochinai and Katagiri (1975), thymectomy must be performed as early as stage 45 to completely prevent allograft rejection. Horton and his co-workers, who found that thymectomy at stage 48 only impaired allograft rejection, tend to agree that early thymectomy cannot suppress allograft rejection and that there is a T-independent component to it. However, both groups agree that after stage 48, thymectomy has no effect whatsoever on the fate of a subsequent allograft. The discrepancy between the two groups may relate to the stage at which thymectomy has been performed. It may also involve the genetic relationships between the donors and the hosts used in each laboratory. Such experiments should now be performed in genetically defined strains.

In this regard, unpublished observations of Nagata and Cohen (personal communication) seem relevant. They have observed chronic rejection of skin allografts by *Xenopus* that had been thymectomized at stage 45. These recipients were progeny of the MHC-homozygous animals used in Katagiri's laboratory; the donors were unrelated to those used in the studied by the Japanese group.

Similar minor contradictions can be found in the experiments concerning the thymus dependency of MLR and PHA responsiveness; some investigators report total abrogation of both functions (Du Pasquier and Horton, 1976), others, only impairment [Manning et al., 1976; Green, Donnelly, and Cohen, 1979]. Taking for granted that both experiments clearly indicate the thymus dependency of the phenomena studied, it would be interesting to know whether the residual activity detected by some groups is due to T cells that had already migrated out or to cells that have followed an alternative pathway of differentiation (see also Turpen and Cohen, 1976, for this point) or even to B cells.

Sequential thymectomy has indicated that allograft rejection, MLR, PHA responsiveness, and T helper functions do not mature at the same time and may be achieved by various subsets of T cells. There is a possibility that MLR and allograft rejection are connected, since it has been reported (Horton unpublished, mentioned in Horton and Sherif, 1977) that up to stage 51, thymectomy could have an effect on first-set allograft rejection. This stage is also the limit for obtaining an effect on PHA responsiveness (see Figure 1). Some attempts at reconstitution of thymectomized *Xenopus* have been performed. Adult thymus implanted into histocompatible thymusless toadlets (thymectomized at stage 45–46) can restore the capacity to reject allografts (although not fully) and the capacity to make antibody against a T-dependent antigen such as rabbit RBC (Tochinai et al., 1976). Other experiments involving the complication of the implantation of an allogeneic thymus (Horton and Horton, 1975; Ruben et al., 1977) will not be discussed here.

3.5. METAMORPHOSIS: ITS EFFECT ON ALLOGENEIC RECOGNITION

3.5.1. The Phenomenon

Experiments of skin grafting during the metamorphosis of urodeles (Orfila and Deparis, 1970) and anurans (Bernardini et al., 1969a,b, 1970) showed that the reactivity of the animals or the antigenicity of the graft may change during this period. This results in prolonged survival of some grafts, relative to survival of the same grafts on adult hosts. This early observation was the basis for further experiments in genetically defined strains of *Xenopus*, which suggested that grafts are most easily tolerated during the metamorphic period if they differ from the host only by minor (non-MHC) histocompatibility differences (Du Pasquier and Chardonnens, 1975; Chardonnens, 1975). To reach this conclusion, a panel of F_1 individuals of an MHC-typed family (family Y) provided skin grafts for many recipients from the same family (another spawning) that were in the process of metamorphosing. The screening of graft rejection and tolerance allowed us to distribute the animals into four classes of mutually tolerant animals. These four classes corresponded to the four MHC classes of the family as demonstrated by serology on the same animals later in development. Thus, *in this family*, tolerance seemed to be restricted to minor histocompatibility antigens. Some exceptions, however, were noted. In some cases, an animal of one MHC class did tolerate grafts from individuals whose MHC genotype differed by one haplotype. That differences in one MHC haplotype were not always sufficient to elicit graft rejection during metamorphosis was then frequently observed in other families and in many cases of grafts from outbred parents to F_1 sibling hosts. Tolerance to two MHC haplotype differences, however, was observed only once during all the experiments. Comparable results have recently been reported from Cohen's laboratory (Cohen *et al.*, 1980; Barlow *et al.*, 1981).

The specificity of this tolerance is far from well understood. A tolerant animal remains able to reject the grafts that show MHC differences with the original donor. On the other hand, a tolerant animal seems to be tolerant to all the minor histocompatibility antigens of the family. This was found in *X. laevis* families (Chardonnens, 1976) and in LG hybrids (Du Pasquier, unpublished). In the latter case, animals of a clone LG_{17} were made tolerant to an MHC-identical donor LG_{15}, and were later grafted with LG_7 of the same MHC genotype again, but with minor antigens different from LG_{17} and LG_{15}. This graft was permanently accepted. However, a graft from an MHC-nonidentical donor, LG_3, was rejected acutely. Is this poor specificity an indication of cross-reaction between various minor antigens, or of a nonspecific blockade of the effector mechanism? If a larva has rejected a graft of a given genotype, it will not be possible to induce tolerance to an identical graft later at metamorphosis. In other words, the perimetamorphic period does not affect secondary responses to allografts.

3.5.2. The Mechanisms

Is tolerance due to the antigenicity of the graft or to the immune system of the host? Earlier experiments done in urodeles indicated that the skin from juvenile *Pleurodeles* was less antigenic than that of adults and larvae. The authors suggested that the latent phase was more influenced by the size and the age of the graft than by the age of the host. In *Xenopus*, skin grafts from donors of different ages indicated that in this species, the state of tolerance did not depend on the antigenicity of the graft but on some characteristics of the host during its metamorphosis (Bernardini *et al.*, 1970). The perimetamorphic period suitable for induction of tolerance is characterized by several changes in the immune system: reduction of cell numbers in the thymus, change in the expression of surface Ig on thymic cells, second histogenesis of the thymus, appearance of MHC antigens, depression of the reactivity in the intrafamilial MLR test (Du Pasquier and Weiss, 1973; Du Pasquier and Chardonnens, 1975). A precise correlation of these phenomena with endocrine gland functions during development remains to be carried out. It is still not known whether the tolerance was induced because of a lack of effector cells (clonal deletion) or because of active suppression. Recently, however, the possibility of transferring the state of tolerance to adult animals via repeated injections of isogenic lymphocytes from metamorphosing animals clearly indicated that an active suppression (the specificity of which is not known) was involved (Du Pasquier and Bernard, 1980). On the other hand, it was impossible to prevent the induction of tolerance in metamorphosing animals by transferring adult lymphocytes. The possibility of observing tolerance of the grafts disappeared when the injections of cells from metamorphosing animals were stopped. These data are consistent with observations (Barlow, 1980) that the establishment of perimetamorphic tolerance is thymus dependent. That is, thymectomy performed during mid to late larval life is associated with graft rejection in donor–host combinations (MHC disparate) in which tolerance of intact metamorphic animals is routinely observed. Therefore, tolerance dected at metamorphosis in *Xenopus* and in other species is, at least in part, mediated by an active suppression by some lymphocytes. That active suppression mediates other manifestations of tolerance in *Xenopus* was already shown by the following experiment of Clark and Newth (1972). Tolerance was induced in embryos. The spleen from an animal immunized against A, when transferred into a compatible host tolerant to A, was unable to promote rejection of the A graft.

In summary, a study of the graft rejection capacity during ontogeny reveals that animals pass through a transient stage when they cannot reject certain grafts (either at all or as rapidly as can adults). The duration of this period depends on the immunogenetic disparity between donor and host, on the size of the graft, and on an MHC haplotype dose (DiMarzo and Cohen, 1979). The tolerance is due, at least partially, to an active suppression exerted by some lymphocytes during this period.

It would be unfair to leave the reader with this oversimplified view of the phenomenon. Actually, increasing numbers of experiments make the picture of perimetamorphic tolerance less and less clear. In concluding this section, I shall try to stress the ambiguities of the system. Some problems of interpretation arise from our poor knowledge of some aspects of larval graft rejection. If the observation of the late appearance of MHC antigens also applies to skin, then one wonders which antigens elicit graft rejection in larvae: are they minor antigens common to adults or are they larval-specific minor antigens (presumably there are both)? Some experiments by Chardonnens can be interpreted as an indication for the existence of larval-specific antigens. Some larvae grafted with larval skin rejected their graft during larval life, but could become tolerant to a new graft coming from an adult (Chardonnens, 1976). This point can now be clarified since MHC-homozygous strains have become available, as well as MHC-defined strains of gynogenetic hybrids. Another ambiguity arises from the existence of a syngeneic MLR detected when isogeneic lymphocytes from animals of different ages are mixed together. This complicates the anlysis of the genetic control of MLR during ontogeny. A further complication is due to the qualitative differences between alleles of the MHC in the process of tolerance induction and to the possibility of obviating these differences by changing graft size for instance (Bernardini *et al.*, 1970; Chardonnens and Du Pasquier, 1973; Cohen *et al.*, 1980).

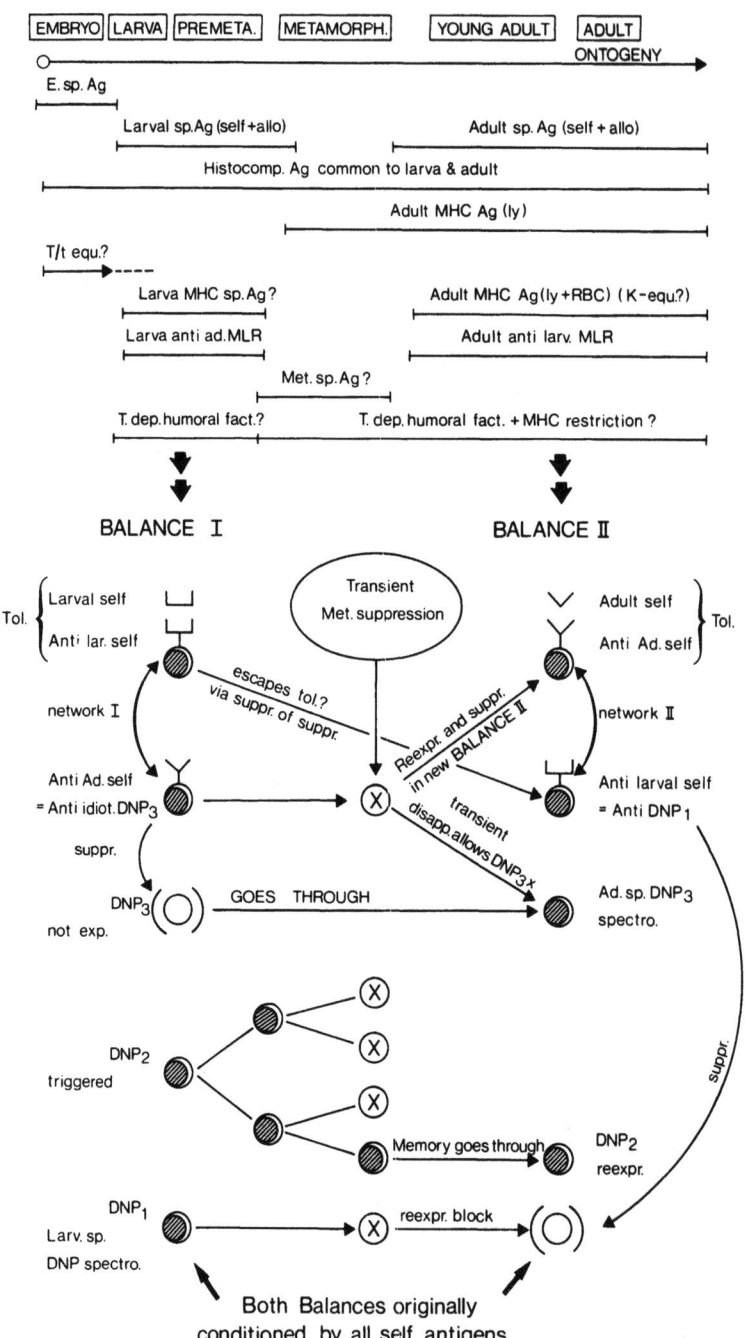

FIGURE 3. A recapitulation of the ontogeny of an amphibian (e.g., *Xenopus*) with a hypothetical schema for explaining the change in antibody diversity expression after metamorphosis. ad.: Adult. Adult, larval sp. Ag (self + allo): Polymorphic and nonpolymorphic adult- or larva-specific antigens. Anti idiot. DNP$_3$: Anti-idiotype of the antibody produced by the clone DNP$_3$. DNP$_3$x: Proliferation of the clone DNP$_3$. equ.: Equivalent. E. sp. Ag: Embryo-specific group of antigens. Larv. sp. DNP

All this makes it necessary to rigorously standardize experiments touching the field of perimetamorphic tolerance. A last difficulty is that metamorphosis extends over weeks of development, from the differentiation of hindlimbs to adulthood. Some of the parameters that one has to deal with are presented in the upper part of Figure 3 where it will be obvious that the transition between larva and adult is by no means an abrupt one.

4. SELF TOLERANCE

The biological significance of the previous findings will constitute the basis of this section. Several experiments have shown that the larval immune system could be stimulated by adult-specific antigens, and this gives strong support to the existence of a natural immunosuppression at metamorphosis: the genesis of tolerance to adult-specific self antigens.

4.1. EVIDENCE FOR ANTIADULT SELF REACTIVITY IN AMPHIBIANS

4.1.1. Early Experiments

The experiments of Triplett (1962) showed that tolerance to adult pituitary antigens was the result of the interaction of these antigens and the immune system. In *Hyla regilla* (a tree frog), the pituitary anlage from a tail bud embryo was transplanted into an allogeneic 2-week-old larva without rejection. The organ grew in the recipient. When at a later stage (beginning of metamorphosis) the pituitary was grafted back to its original donor, this autograft was rejected in most of the cases (10 of 13). Rejection was demonstrated by the modification of the skin pigmentation controlled by the pituitary pars intermedia. A control was included to show that the rejection was not due to alloantigens acquired by the pituitary during its transit in the host: the graft of half a pituitary should be rejected when grafted back to its donor if it had acquired some of the temporary-host alloantigens. Actually, such grafts were tolerated.

Similar experiments showing that tolerance to adult self structures is the result of interactions between these antigens and the larval immune system were done by Maniatis *et al.* (1969). Larval *R. catesbeiana*, when injected with hemoglobin of adults of the same species, can make specific antibodies against it, whereas during their normal development, tadpoles become tolerant to adult hemoglobin. The presence of such antibodies in metamorphosing animals was followed by the lack of expression of adult hemoglobin.

---

spectro.: Larva-specific anti-DNP antibody spectrotype. ly: Lymphocyte. Met.: Metamorphosis. MHC: Major histocompatibility complex. Reexpr. and suppr.: Reexpression and suppression conditioned by the new balance of adult self antigens. suppr.: Suppression. suppr. of suppr.: Suppression of suppressor during metamorphosis. T. dep. humoral fact.: Thymus dependency of the immune response under the control of humoral nonspecific factors. Transient Met. suppression: Lymphocyte-mediated suppression occurring during metamorphosis.

In *Xenopus*, a peculiar disease observed in animals that have received grafts of adult skin during their larval life has also been interpreted by Chardonnens as an autoimmune disease (Chardonnens, 1976). Tadpoles that have rejected several grafts during larval life suffer intradermal hermorrhages when they reach metamorphosis. The hemorrhages are located essentially in the belly and eye regions. Those that recover, later have a poor dorsal pigmentation, and do not grow as quickly as controls. This "red disease" could be the result of an immunization to adult antigens during larval stages and of the challenge occurring naturally at metamorphosis when the similar adult antigens appear by themselves.

4.1.2. Experiments in Isogenic Strains

All these experiments, although quite elegant, can be criticized because they deal, in one way or another, with allogeneic differences. It was thus necessary to reevaluate these possible antiadult reactions in isogenic strains. In *Xenopus*, this has been done by studying the MLR between adult and larval lymphocytes and by trying to immunize larvae with adult RBC known to express an adult-specific antigen.

4.1.2a. MLR. Larval thymic or splenic lymphocytes can be activated in unidirectional MLR by adult irradiated lymphocytes. Unlike in mammals (Von Boehmer *et al.*, 1972), the stimulation in this syngeneic MLR is not restricted to the B-cell population, but is possible with thymic, splenic, or peripheral blood lymphocytes. Vice versa, adults are able to recognize larval lymphocytes as nonself since they can be induced to proliferate by irradiated larval lymphocytes. No such reaction occurs when lymphocytes from metamorphosing animals are used as responders. This observation is interesting in connection with the ontogeny of histocompatibility antigens. Does the larval cell recognize the MHC determinants at the surface of adult cells? What does the adult lymphocyte recognize on larval cells? It is tempting to speculate that larval cells can express antigens that disappear in adult stages and escape the generation of self tolerance that occurs at metamorphosis, leaving the adult immune system able to respond to it later.

There may exist in frog larvae a phenomenon similar to the expression of the *T/t* locus in mouse embryos versus the expression of the *H-2* locus at later stages of development (review by Jacob 1977).

4.1.2b. RBC Antigens. Adult *Xenopus* RBC express a protein on their surface, that they have probably absorbed from the serum. It is completely lacking in larval RBC (Du Pasquier and Wabl, unpublished). Adult RBC also express MHC antigens whereas larval RBC do not. Injections of small doses (10^4–10^5) of adult isogenic RBC into larvae during early larval life (up to stage 51) resulted in the death of several experimental animals (8 or 12) when they reached the time of metamorphosis. Death was accompanied by symptoms that very much resemble those of "red disease," namely, intradermal hemorrhages in the tail and near the gill region. Since a normal animal does not suffer from autoimmune disease during its metamorphosis, there must exist mechanisms which prevent antiadult reactions from occurring. It is possible that tolerance to alloantigens observed at metamorphosis is the reflection of the activity of such mechanisms. The characteristics of perimetamorphic tolerance to alloantigens (see previous section) apply to what we know about tolerance to self antigens. It concerns only the primary response to antigens in both cases; therefore, if the tadpole is immunized against its own adult RBC, the tolerance mechanisms are not efficient.

The situation cannot, however, persist as it is at metamorphosis, and somehow the maintenance of tolerance during later life and induction of tolerance during a privileged period are not governed by the same mechanisms. During metamorphosis, all responses can be "weakened," including antibody response (Moticka et al., 1973), whereas in old tolerant animals, only responses to minor histocompatibility antigens are modified.

4.2. POSSIBLE INFLUENCE ON THE ANTIBODY REPERTOIRE

All this ensemble of phenomena points to the expectation that, qualitatively and functionally, the antibody repertoire of the adult animal may not be the same as that of the larvae, and that the adult repertoire has developed as a function of (1) what was in the genome of the cell, and (2) which antigen was on the surface of the surrounding cells. It is conceivable that during the general suppression, some clones of cells, antagonistic to the larval antiadult lymphocytes, can expand and thus prevent the comeback of these autoreactive cells. In all circumstances, the change in repertoire will vary according to the self antigens that develop. This is why it is possible that the change in antibody repertoire to external antigens observed from the larva to the adult is indirectly the reflection of the generation of tolerance to self that occurred during metamorphosis (see hypothetical scheme in Figure 3).

A direct approach to the question "How do self antigens modify the expression of the B-cell repertoire?" is now possible with the help of the aneuploid hyperdiploid *Xenopus*. These animals contain a normal diploid genome of *X. laevis* and an aneuploid set of *X. gilli* chromosomes (Kobel and Du Pasquier, 1979). Such animals will thus vary in the expression of some self *X. gilli* antigens. The effect of this variation has already been studied. As an example, an interesting relationship was found between the expression of an anti-DNP antibody IEF spectrotype of *X. gilli* and that of an *X. gilli* cell surface antigen coded by chromosome 9. In the hybrid animal, a few given *X. gilli* IEF spectrotypes were always missing when this above-mentioned antigen was present. One could interpret the finding as due to an influence of this self antigen on the proliferation of the various clones of B cells during ontogeny (Du Pasquier and Kobel, 1979).

In summary, several experiments prove that larval immunocytes can recognize adult antigens as foreign and vice versa. The state of perimetamorphic immunosuppression discovered in the case of certain histocompatibility antigens seems necessary to allow the establishment of tolerance by these self antigens. The differentiation of such antigens is likely to play a role in the change in the expression of antibody repertoire noticed during ontogeny. Some new models, such as the hyperdiploid *Xenopus* hybrids, should allow a direct study of the role of self antigens on B-cell diversification.

5. GENERAL CONCLUSIONS

One can summarize the development of immunological functions in amphibians as follows:

1. Early differentiation of the B-cell compartment with the production of a diversified immune response to practically all antigens tested, a finding recently confirmed in higher vertebrates.
2. Early differentiation of the allorecognition capacity by thymus-derived cells.
3. Sequential maturation of other T-cell functions (e.g., helper function) may be associated with the late appearance of MHC antigens on lymphocytes.
4. The differentiation of the animal occurs over a long period of time. The embryonic and very early larval period take place in the absence of an immune system. This period may be longer in urodeles than in anurans.
5. The perimetamorphic period takes place when the larval immune system has differentiated. An active immunosuppression, contemporary with profound changes in the thymus and at the surface of thymic cells, permits a harmonious transition between larval and adult life, without losing the benefit of immunizations that have occurred during larval life.
6. Qualitatively, larval and adult antibody responses are different. The repertoire of V regions expressed in larvae differs from that of adults. Several models now permit us to investigate whether and how these changes are modulated by self antigens.

Further experiments will also show, in the near future, whether the T–B cooperation that has recently been demonstrated in adult *Xenopus* (Blomberg *et al.*, 1980) can occur across the barriers of age, i.e., whether larval and adult cells can cooperate.

Another domain, the role of MHC in this cooperation and all the questions linked to the phenomenon of MHC restriction, can be particularly well approached in amphibians since at least in some species, there must be either a very low expression of a lack of MHC antigens on lymphocytes over a long period of time. This should allow interesting sequential experiments to help in determining what step of the normal immune response is governed by this restriction. It could thus help clarify the various aspects of thymic functions, not only in amphibians but in vertebrates in general, since after all, there are enough homologies between amphibians and mammals to permit reciprocal extrapolations.

REFERENCES

Ambrosius, H., Hemmerling, J., Richter, R., and Schimke, R., 1970, Immunoglobulins and the dynamics of antibody formation in poikilothermic vertebrates, in: *Developmental Aspects of Antibody Formation and Structure* (J. Sterzl and I. Riha, eds.), pp. 727–749, Academia, Prague.

Amirante, G. A., and Parisi, V., 1967, Anticorpopoeisi e carateri serologici negli anfibi, *Accad. Nazio. Lincei. Fasc. 1* **13**:88.

Baculi, B. S., and Cooper, E. L., 1970, Histopathology of skin allograft rejection in larval *Rana catesbeiana, J. Exp. Zool.* **173**:329.

Barlow, E. H., 1980, Thymus dependency and *in vitro* correlates of perimetamorphic tolerance in *Xenopus laevis*, Ph.D. thesis, University of Rochester.

Barlow, E. H., DiMarzo, S. J., and Cohen, N., 1981, Prolonged survival of MHC disparate skin allografts transplanted to the metamorphosing toad, *Xenopus laevis, Transplantation* **32**:51.

Bernard, C. C. A., Bordman, G., Blomberg, B., and Du Pasquier, L., 1981, Genetic control of T helper cell function in the clawed toad, *Xenopus laevis, Eur. J. Immunol.* **11**:151.

Bernardini, N., Chardonnens, X., and Simon, D., 1969a, Etude du comportement immunologique chez *Xenupus laevis*, en présence de deux greffes cutanées différentes, *C.R. Acad. Sci. Ser. D.* **269**:1107.

Bernardini, N., Chardonnens, X., and Simon, D., 1969b, Développement après la métamorphose de compétence immunologiques envers les homogreffes cutanées chez *Xenopus laevis* Daudin, *C.R. Acad. Sci.* **269**:1011.

Bernardini, N., Chardonnens, X., and Simon, D., 1970, Tolérance des allogreffes cutanées chez *Xenopus laevis:* Influence de la taille et de l'âge du greffon, *C.R. Acad. Sci.* **270**:2351.

Blomberg, B., Bernard, C. C. A., and Du Pasquier, L., 1980, *In vitro* evidence for T–B lymphocyte collaboration in the clawed toad, *Xenopus, Eur. J. Immunol.* **10**:869.

Bovbjerg, A. M., 1966, Rejection of skin homografts in larvae of *Rana pipiens, J. Exp. Zool.* **161**:69.

Chardonnens, X., 1975, Tissue typing by skin grafting during metamorphosis of the toad *Xenopus laevis* (Daudin), *Experientia* **31**:237.

Chardonnens, X., 1976, La tolérance aux antigènes d'histocompatibilif pendant la métamorphose de l'Amphibien Anoure, *Xenopus laevis:* Un modèle pour l'étude de la tolérance au self, Thèse Université de Genève, No. 1748.

Chardonnens, X., and Du Pasquier, L., 1973, Induction of skin allograft tolerance during metamorphosis of the toad *Xenopus laevis:* A possible model for studying generation of self tolerance to histocompatibility antigens, *Eur. J. Immunol.* **3**:569.

Charlemagne, J., 1974, Larval thymectomy and transplantation immunity in the urodele *Pleurodeles waltlii* (Michah) (Salamandridae), *Eur. J. Immunol.* **4**:390.

Charlemagne, J., and Houillon, C., 1968, Effects de la thymectomie larvaire chez l'Amphibien Urodèle, *Pleurodeles waltilii* Michah: Production à l'état adulte d'une tolérance aux homogreffes cutanées, *C.R. Acad Sci.* **267**:253.

Charlemagne, J., and Tournefier, A., 1974, Obtention of histocompatible strains in the Urodele Amphibian *Pleurodeles waltlii* Michah. (*Salamandridae*), J. Immunogenet. **1**:125.

Charlemagne, J., and Tournefier, A., 1975, Cell surface immunoglobulins of thymus and spleen lymphocytes in urodele amphibian *Pleurodeles waltlii* Salamandridae, *Adv. Exp. Med. Biol.* **64**:251.

Charlemagne, J., and Tournefier, A., 1977a, Humoral response to *Salmonella typhimurium* antigens in normal and thymectomized urodele amphibian *Pleurodeles waltlii* Michah, *Eur. J. Immunol.* **7**:500.

Charlemagne, J., and Tournefier, A., 1977b, Anti-horse red blood cells antibody synthesis in the Mexican axolotl (*Ambystoma mexicanum*): Effect of thymectomy, in: *Developmental Immunobiology* (J. B. Solomon and J. D. Horton, eds.), pp. 267–276, Elsevier/North-Holland, Amsterdam.

Ching, Y. C., and Wedgwood, R. J., 1967, Immunologic responses in the Axolotl, *Siredon mexicanum, J. Immunol.* **99**:191.

Clark, J. C., and Newth, D. R., 1972, Immunological activity of transplanted spleens in *Xenopus laevis, Experientia* **28**:951.

Clayton, R. M., 1957, Antigens in the developing newt embryo, *Nature (London)* **168**:120.

Cohen, N., 1969, Immunogenetic and developmental aspects of tissue transplantation immunity in urodele amphibians, in: *Biology of Amphibian Tumors* (M. Mizell, ed.), pp. 153–158, Springer-Verlag, Berlin.

Cohen, N., 1976, Phylogeny of the major histocompatibility complex: Theoretical implications of studies with urodele amphibians, in: *Phylogeny of Thymus and Bone Marrow-Bursa Cells* (R. K. Wright and E. L. Cooper, eds.), pp. 169–182, Elsevier–North-Holland, Amsterdam.

Cohen, N., DiMarzo, S. J., and Hailparn-Barlow, E., 1980, Induction of tolerance to alloantigens of the major histocompatibility complex in the metamorphosing frog, *Xenopus laevis*, in: Phylogeny of Immunological Memory (M. J. Manning, ed.), pp. 225–231, Elsevier/North-Holland, Amsterdam.

Collie, H., Turner, R. J., and Manning, M. J., 1975, Antibody production to lipopolysaccharide in thymectomixed *Xenopus, Eur. J. Immunol.* **5**:426.

Collins, N. H., Manickavel, V., and Cohen, N., 1975, In vitro responses of urodele lymphoid cells: Mitogenic and mixed lymphocyte culture reactivities, *Adv. Exp. Med. Biol.* **64**:305.

Cooper, E. L., 1973, The thymus and lymphomyeloid system in poikilothermic vertebrates, in: *Contemporary Topics in Immunobiology*, Volume 2, *Thymus Dependency* (A. J. S. Davis and R. C. Carter, eds.), pp. 13–38, Plenum Press, New York.

Cooper, E. L., and Hildemann, W. H., 1965a, The immune response of larval bullfrog *Rana catesbeiana* to diverse antigens, *Ann. N.Y. Acad. Sci.* **126**:647.

Cooper, E. L., and Hildemann, W. H., 1965b, Allograft reactions in bullfrog larvae in relation to thymectomy, *Transplantation* **3**:446.

Cooper, E. L., Hildemann, W. H., and Pinkerton, W., 1963, Serum antibody synthesis and skin homograft survival in larvae of the bullfrog *Rana catesbeiana*: Role of the thymus gland, *Immunogenet. Lett.* **3**:63.

Cooper, E. L., Pinkerton, W., and Hildemann, W. H., 1964, Serum antibody synthesis in larvae of the bullfrog, *Rana catesbeiana, Biol. Bull.* **127**:232.

Curtis, S. K., and Volpe, E. P., 1971, Modification of responsiveness to allografts in larvae of the leopard frog by thymectomy, *Dev. Biol.* **25**:177.

Dardenne, M., Tournefier, A., Charlemagne, J., and Bach, J.-F., 1973, Studies on thymus products. VII. Presence of thymic hormone in urodele serum, *Ann. Immunol. (Inst. Pasteur)* **124C**:465.

Davison, J., 1966, Chimeric and ex-parabiotic frogs (*Rana pipiens*): Specificity of tolerance, *Science* **152**:1250.

De Lanney, L. E., and Blackler, K., 1969, Acceptance and regression of a strain specific lymphosarcoma in Mexican exolotls, in: *Biology of Amphibian Tumors* (M. Mizell, ed.), pp. 399–408, Springer-Verlag, Berlin.

De Lanney, L. E., Collins, N. H., Cohen, N., and Reid, R., 1975, Transplantation immunogenetics and MLC reactivities of partially inbred strains of salamanders (*A. mexicanum*): Preliminary studies, *Adv. Exp. Med. Biol.* **64**:315.

Di Marzo, S. J., and Cohen, N., 1979, Ontogeny of alloimmunity to major histocompatibility antigens in the frog, *Xenopus, Am. Zool.* **19**:856.

Du Pasquier, L., 1965, Aspects cellulaires et humoraux de l'intolérance aux homogreffes de tissu musculaire chez le têtard d' *Alytes obstetricans*: Rôle du thymus, *C.R. Acad. Sci. Ser. D* **261**:1144.

Du Pasquier, L., 1968, Les protéines sériques et le complexe lympho-myéloïde chez le têtard d' *Alytes obstetricans* normal et thymectomisé, *Ann. Inst. Pasteur (Paris)* **114**:490.

Du Pasquier, L., 1970, Ontogeny of the immune response in animals having less than one million lymphocytes: The larvae of the toad *Alytes obstetricans, Immunology* **19**:353.

Du Pasquier, L., 1973, Ontogeny of Immune Response in Cold-Blooded Vertebrates, in: *Current Topics in Microbiology and Immunology*, Vol. 61, pp. 37–88, Springer-Verlag, Berlin.

Du Pasquier, L., 1976, Amphibian models for study of the ontogeny of immunity, in: *Comparative Immunology* (J. J. Marchalonis, ed.), pp. 390–418, Dlackwell, Oxford.

Du Pasquier, L., and Bernard, C. C. A., 1980, Active suppression of the allogeneic histocompatibility reactions during the metamorphosis of the clawed toad *Xenopus, Differentiation* **16**:1.

Du Pasquier, L., and Chardonnens, X., 1975, Genetic aspect of the tolerance to allografts induced at metamorphosis in the toad *Xenopus laevis, Immunogenetics* **2**:431.

Du Pasquier, L., and Haimovich, J., 1974, Changes in affinity of IgM antibodies in amphibian larvae, *Eur. J. Immunol.* **4**:580.

Du Pasquier, L., and Haimovich, J., 1976, The antibody response during amphibian ontogeny, *Immunogenetics* **3**:381.

Du Pasquier, L., and Horton, J. D., 1976, The effect of thymectomy on the mixed leukocyte reaction and phytohemagglutinin responsiveness in the clawed toad *Xenopus laevis, Immunogenetics* **3**:105.

Du Pasquier, L., and Kobel, H. R., 1979, Histocompatibility antigens and immunoglobulin genes in the clawed toad: Expression and linkage analysis in recombinant and hyperdiploid *Xenopus* hybrids, *Immunogenetics* **8**:299.

Du Pasquier, L., and Wabl, M. R., 1976, Antibody diversity studied in amphibians, in: The generation of Antibody Diversity: A New Look (A. J. Cunningham, ed.), pp. 151–164, Academic Press, New York.

Du Pasquier, L., and Weiss, N., 1973, The thymus during the ontogeny of the toad *Xenopus laevis*: Growth, membrane-bound immunoglobulins and mixed lymphocyte reaction, *Eur. J. Immunol.* **3**:773.

Du Pasquier, L., Weiss, N., and Loor, F., 1972, Direct evidence for immunoglobulins on the surface of thymus lymphocytes of amphibian larave, *Eur. J. Immunol.* **2**:366.

Du Pasquier, L., Chardonnens, X., and Miggiano, V. C., 1975, A major histocompatibility complex in the toad *Xenopus laevis* (Daudin), *Immunogenetics* **1**:482.

Du Pasquier, L., Miggiano, V. C., Kobel, H. R., and Fischberg, H., 1977, The genetic control of histocompatibility reactions in natural and laboratory-made polyploid individuals of the clawed toad *Xenopus*, *Immunogenetics* **5**:129.

Du Pasquier, L., Blomberg, B., and Bernard, C. C. A., 1979, Ontogeny of immunity in amphibians: Changes in antibody repertoires and appearance of adult major histocompatibility antigens in *Xenopus*, *Eur. J. Immunol.* **9**:900.

Dupuy, G., 1964, Les autogreffes, homogreffes et hétérogreffes de peau chez les têtards de *Discoglossus pictus* et d'*Alytes obstetricans*, Thèse 3eme Cycle Ens. Sup. No. 287, Bordeaux.

Fache, B., and Charlemagne, J., 1975, Influence on allograft rejection of thymectomy at different stages of larval development in urodele amphibian *Pleurodeles waltlii* Michah (Salamandridae), *Eur. J. Immunol.* **5**:155.

Fougereau, M., and Houdayer, M., 1968, Immunoglobulines et réponse immunitaire chez l'Axolotl (*Abystoma mexicanum*), *Ann. Inst. Pasteur* (Paris) **115**:968.

Geczy, C. L., Green, P. C., and Steiner, L., 1973, Immunoglobulins in the developing amphibian *Rana catesbeiana*, *J. Immunol.* **111**:1261.

Goldshein, S. J., and Cohen, N., 1972, Phylogeny of immunocompetent cells. I. In vitro blastogenesis and mitosis of toad (*Bufo marinus*) splenic lymphocytes in response to phytohemagglutinin and in mixed lymphocyte cultures, *J. Immunol.* **108**:1025.

Goldstine, S. N., Collins, N. H., and Cohen, N., 1975, Mitogens as probes of lymphocyte heterogeneity in anuran amphibians, *Adv. Exp. Med. Biol.* **64**:1.

Green, C., and Steiner, L. A., 1976, Low molecular weight immunoglobulins in *Rana catesbeiana* tadpoles, *J. Immunol.* **117**:375.

Green (Donnelly), N., and Cohen, N., 1979, Phylogeny of immunocompetent cells. III. Mitogen response characteristics of lymphocyte subpopulations from normal and thymectomized frogs (*Xenopus laevis*), *Cell. Immunol.* **48**:59.

Hadji-Azimi, I., and Parrinello, N., 1978, The simultaneous production of two classes of cytoplasmic immunoglobulins by single cells in *Xenopus laevis*, *Cell. Immunol.* **39**:316.

Haimovich, J., and Du Pasquier, L., 1973, Specificity of antibodies in amphibian larvae possessing a small number of lymphocytes, *Proc. Natl. Acad. Sci. USA* **70**:1898.

Hildemann, W. H., and Haas, R., 1959, Homotransplantation immunity and tolerance in the bullfrog, *J. Immunol.* **83**:478.

Hildemann, W. H., and Haas, R., 1961, Histocompatibility genetics of bullfrog populations, *Evolution* **15**:267.

Hollyfield, J. G., 1966, Erythrocyte replacement at metamorphosis in the frog *Rana pipiens*, *J. Morphol.* **119**:1.

Horton, J. D., 1969, Ontogeny of the immune response to skin allografts in relation to lymphoid organ development in the amphibian *Xenopus laevis* Daudin, *J. Exp. Zool.* **170**:449.

Horton, J. D., 1971, Ontogeny of the immune system in amphibians, *Am. Zool.* **11**:219.

Horton, J. D., and Horton, T. L., 1975, Development of transplantation immunity and restoration experiments in the thymectomized amphibian, *Am. Zool.* **15**:73.

Horton, J. D., and Manning, M. J., 1972, Response to skin allografts in *Xenopus laevis* following thymectomy at early stage of lymphoid organ maturation, *Transplantation* **14**:141.

Horton, J. D., and Sherif, N. E. H. S., 1977, Sequential thymectomy in the clawed toad: Effect on mixed leucocyte reactivity and phytohaemagglutinin responsiveness, in: Developmental Immunobiology (J. B. Solomon and J. D. Horton, eds.), pp. 238–290, Elsevier/North-Holland, Amsterdam.

Horton, J. D., Rimmer, J. J., and Horton, T. L., 1977, Critical role of the thymus in establishing humoral immunity in amphibians: studies on Xenopus thymectomized in larval and adult life, *Dev. Comp. Immunol.* **1**:119.

Jurd, R. D., and Stevenson, G. T., 1976, Surface immunoglobulins on *Xenopus laevis* lymphocytes, *Comp. Biochem. Physiol. A* **53**:381.

Jurd, R. D., Luther-Davies, S. M., and Stevenson, G. T., 1975, Humoral antibodies to soluble antigens in larvae of *Xenopus laevis*, *Comp. Biochem. Physiol. B* **50**:65.

Kerbel, R. S., and Eidinger, D., 1972, Enhanced immune responsiveness to a thymus independent antigen early after thymectomy: Evidence for a short-lived inhibitory thymus-derived cell, *Eur. J. Immunol.* **2**:114.

Kidder, G. M., Ruben, L. N., and Stevens, J. M., 1973, Cytodynamics and ontogeny of the immune response of *Xenopus laevis* against sheep erythrocytes, *J. Embryol. Exp. Morphol.* **29**:73.

Kobel, H. R., and Du Pasquier, L., 1975, Production of large clones of histocompatible fully identical clawed toads (*Xenopus*), *Immunogenetics* **2**:87.

Kobel, H. R., and Du Pasquier, L., 1977, Strains and species of *Xenopus* for immunological research, in: *Developmental Immunobiology* (J. B. Solomon and J. D. Horton, eds.), pp. 299–306, Elsevier/North-Holland, Amsterdam.

Kobel, H. R., and Du Pasquier, L., 1979, Hyperdiploid species hybrids for gene mapping in *Xenopus*, *Nature (London)* **279**:157.

Leverone, L., Suhar, T., Brown, R. D., and Armentrout, R. W., 1979, Characteristics of immunoglobulin synthesized early in the development of *Xenopus laevis*, *Dev. Biol.* **68**:319.

Maniatis, G. M., Steiner, L. A., and Ingram, V. M., 1969, Tadpole antibodies against hemoglobin and their effect on development, *Science* **165**:67.

Manning, M. J., 1975, The phylogeny of thymic dependence, *Am. Zool.* **15**:63.

Manning, M. J., and Collie, M. H., 1975, Thymic function in amphibians, *Adv. Exp. Med. Biol.* **64**:353.

Manning, M. J., Donnelly, N., and Cohen, N., 1976, Thymus-dependent and thymus-independent components of the amphibian immune system, in: *Phylogeny of Thymus and Bone Marrow-Bursa cells* (R. K. Wright and E. L. Cooper, eds.), pp. 123–132, Elsevier/North-Holland, Amsterdam.

Marchalonis, J. J., and Edelman, G. M., 1966, Phylogenetic origins of antibody structure. II. Immunoglobulins in the primary immune response of the bullfrog, *Rana catesbeiana*, *J. Exp. Med.* **124**:901.

Mattes, M. J., and Steiner, L. A., 1978, Surface immunoglobulin on frog lymphocytes: Identification of two lymphocyte populations, *J. Immunol.* **121**:1116.

Moticka, E. J., Brown, B. A., and Cooper, E. L., 1973, Immunoglobulin synthesis in bullfrog larvae, *J. Immunol.* **110**:855.

Moyer, C., Armentrout, R. W., and Brown, R. D., 1977, The onset of immunoglobulin synthesis during the development of *Xenopus laevis*, *Dev. Biol.* **61**:338.

Nace, G. W., and Richards, C. M., 1969, Development of biologically defined strains of amphibians, in: *Biology of Amphibian Tumors* (M. Mizell, ed.), pp. 409–418, Springer-Verlag, Berlin.

Nagata, S., 1976, Immune response against skin allograft and rabbit red blood cells in metamorphosing and metamorphosis-inhibited *Xenopus laevis*, *J. Fac. Sci. Hokkaido Univ. Serv. 6* **20**:183.

Nagata, S., and Katagiri, C., 1978, Lymphocyte surface immunoglobulin in *Xenopus laevis*. Light and electron microscopic demonstration by immuno peroxidase method, *Dev. Comp. Immunol.* **2**:277.

Nieuwkoop, P. D., and Faber, J., 1967, *Normal Table of Xenopus laevis Daudin*, 2nd ed., North-Holland, Amsterdam.

Orfila, C., and Deparis, P., 1970, Influance de l'âge du donneur et du receveur sur l'évolution des homogreffes cutanées chez les larves du triton "*Pleurodeles waltlii* Michah," *Pathol. Biol.* **18**:1033.

Pross, S. H., and Rowland, D. T., Jr., 1975, Immunity in the developing amphibians, *Adv. Exp. Med. Biol.* **64**:373.

Rimmer, J. J., and Horton, J. D., 1977, Allograft rejection in larval and adult *Xenopus* following early thymectomy, *Transplantation* **23**:142.

Romanovsky, A., 1964, Studies on antigenic differentiation in the embryonic development of *Rana temporaria*, *Folia Biol. (Prague)* **10**:1.

Ruben, L. N., Clothier, R., Hodgson, R., and Balls, M., 1977, The *in vitro* reconstitution of a thymus cell dependent humoral immune response in spleens of thymectomized *Xenopus laevis* with allogeneic thymocytes, in: *Developmental Immunobiology* (J. B. Solomon and J. D. Horton, eds.), pp. 227–283, Elsevier/North-Holland, Amsterdam.

Sidman, C. L., Shultz, L. D., and Unanue, E. R., 1978, The mouse mutant motheaten. 1. Development of lymphocyte populations, *J. Immunol.* **121**:2392.

Simnett, J. D., 1966, Factors influencing the differentiation of amphibian embryos implanted into homologous immunologically competent hosts (*Xenopus laevis*), *Dev. Biol.* **13**:112.

Spar, I. L., 1953, Antigenic differences among early developmental stages of *Rana pipiens*, *J. Exp. Zool.* **123**:467.

Tochinai, S., 1976, Demonstration of thymus-independent immune system in *Xenopus laevis*: Response to polyvinylpyrrolidone, *Immunology* **31**:125.

Tochinai, S., and Katagiri, C., 1975, Complete abrogation of immune response to skin allografts and rabbit erythrocytes in the early thymectomized *Xenopus*, *Dev. Growth Differ.* **17**:383.

Tochinai, S., Nagata, S., and Katagiri, C., 1976, Restoration of immune responsiveness in early thymectomized *Xenopus* by implantation of histocompatible adult thymus, *Eur. J. Immunol.* **6**:711.

Tournefier, A., 1973, Développemnt des organes lymphoïdes chez l'Amphibien Urodèle *Triturus alpestris* Laur.: Tolérance des allogreffes après la thymectomie larvaire, *J. Embryol. Exp. Morphol.* **29**:382.

Tournefier, A., 1968, Etude histologique des hétérogreffes embryonnaires de tégument chez les Amphibiens Urodèles, *Bull. Soc. Zool. France* **93**:99.

Triplett, E. L., 1962, On the mechanism of immunologic self recognition, *J. Immunol.* **89**:505.

Turner, R. J., and Manning, M. J., 1974, Thymic dependence of amphibian antibody response, *Eur. J. Immunol.* **4**:343.

Turpen, J. B., and Cohen, N., 1976, Alternate sites of lymphopoiesis in the amphibian embryo, *Ann. Immunol. (Inst. Pasteur)* **127C**:841.

Tymowska, J., and Fischberg, M., 1973, Chromosome complements of the genus *Xenopus*, *Chromosoma* **44**:335.

Volpe, E. P., 1964, Fate of neural crest homotransplants in pattern mutants of the leopard frog, *J. Exp. Zool.* **157**:179.

Volpe, E. P., 1971, Immunological tolerance in amphibians, *Am. Zool.* **11**:207.

Volpe, E. P., and Gebhardt, B. M., 1965, Effect of dosage on the survival of embryonic homotransplants in the leopard frog *Rana pipiens*, *J. Exp. Zool.* **160**:11.

Von Boehmer, H., Shortman, K., and Adams, P., 1972, Nature of the stimulating cell in the syngeneic and the allogeneic mixed lymphocyte reaction in mice, *J. Exp. Med.* **136**:1648.

Wabl, M. R., Brun, R. B., and Du Pasquier, L., 1975, Lymphocytes of the toad *Xenopus laevis* have the gene set for promoting tadpole development, *Science* **190**:1310.

Weiss, N., Horton, J. D., and Du Pasquier, L., 1973, The effect of thymectomy on cell surface associated and serum immunoglobulin in the toad *Xenopus laevis* (Daudin): A possible inhibitory role of the thymus on the expression of immunoglobulins, in: *L'Etude phylogénique et ontogénique de la réponse immunitaire et son apport à la théorie immunologique* (J. Panijel and P. Liacopoulos, eds.), pp. 165–174, INSERM, Paris.

Zinkernagel, R. M., Callahan, G. N., Klein, J., and Dennert, G., 1978, Cytotoxic T cells learn specificity for self H-2 during differentiation in the thymus, *Nature (London)* **271**:251.

19

Ontogeny of RES Function in Birds

PIERSON J. VAN ALTEN

1. INTRODUCTION

The chicken has served as an exquisite experimental animal model for investigating the functional ontogeny of the immune system. A major advance in immunobiology, to which the developing chicken has contributed to our present understanding, is that there are two distinct pathways for the differentiation of lymphoid cells. The first insight into this began with the fortuitous observations by Glick and his colleagues (Chang et al., 1955) showing that the removal of the bursa of Fabricius shortly after hatching resulted in profound impairment of the chicken's ability to produce antibody following antigenic stimulation. Likewise, Cooper et al. (1965) demonstrated that, as had been seen in rodents (Good et al., 1962; Miller 1961), there was a loss in the ability of chickens to generate cell-mediated immunity following thymectomy at hatching. Thymectomy, however, had little effect on antibody production. Extension of these observations led Warner and Szenberg (1962) and Cooper et al. (1965, 1966) to propose that the lymphoid system could be dissociated on the basis of ontogenetic criteria into a bursa-dependent compartment which gives rise to antibody-secreting cells, and a thymus-dependent compartment which gives rise to cellular immune functions. Based upon this experimental and developmental model and upon many other studies too numerous to recount in this review, Good and colleagues (Peterson et al., 1965) were motivated to the insightful hypothesis that the immune system was composed of two compartments and on this basis the pathogenesis of a large number of human immunological deficiency diseases could be better understood. Subsequently, it was found that the cells comprising

---

PIERSON J. VAN ALTEN • Department of Anatomy, University of Illinois at the Medical Center, Chicago, Illinois 60680. This review and some of the previously unpublished data were supported by Grant CA-20172 from the National Cancer Institute and a grant from the University of Illinois Campus Research Board.

these two populations could be identified by distinct cell membrane components and were given the shorthand names of T and B lymphocytes (Roitt et al., 1969).

It soon became apparent, particularly from experiments using mammals, that functionally the two compartments of the immune system were interdependent and modulated each other. It was Claman et al. (1966) who clearly demonstrated that for some antigens, the interaction between T and B lymphocytes was necessary for the induction of an antibody response. Subsequently, the T lymphocytes were called "helper" cells, whereas B lymphocytes were considered responsible for antibody secretion. In addition to the helper effect, T cells were also found to exert an important negative regulatory role on B cells which was called "suppressor effect" (Gershon et al., 1972). In addition to modulation by T lymphocytes, the reticuloendothelial system (RES), through its monocytic macrophages, was implicated in the induction of an antibody response by B lymphocytes (Halpern, 1959). Thus, for the induction of a primary antibody response to complex, multideterminant antigens, it now is apparent that interactions are required between at least three cell populations, that is, monocytic macrophages with T and B lymphocytes (Pierce et al., 1974; Claman and Mosier, 1972; Katz and Benacerraf, 1972).

Although the developing chicken as an experimental animal model has not had a major role in our understanding of how the RES develops functionally, it presents a unique system for investigating how the developing RES may modulate or be modulated by the lymphoid system which together will form a fully functional host defense system. That the immaturity of the macrophage system may be a limiting step in the maturation of immunocompetence has been shown to be true at least for the rat (Blaese, 1975). The main objective of the present discussion is to review pertinent literature which will give us insight into our present knowledge of the ontogeny of RES function in developing birds. In addition, we should keep in mind some of the advantages of using the developing chick as an experimental model for gaining insight into the maturation of the RES functions. First of all, the chick embryo constitutes one of the classical developmental systems, and therefore much is already known about its detailed morphological epigenesis. Second, it has separate organs for the differentiation of B and T lymphocytes which can be experimentally altered and thus their possible influence on the ontogeny of the RES determined. Third, the chick embryo as an organism is separated from the mother so that much of the maternal influence, which might otherwise overshadow the developing RES, is absent. Fourth, very early stages are available, that is, periods when organogenesis is just beginning. Fifth, at almost all stages of development, infusions of foreign substances can be made directly into the vascular system, making the evaluation of clearance possible. Sixth, any of the developing tissues or organs can be manipulated by experimental embryological techniques and the effect evaluated on subsequent functional deletions. Seventh, the various cell populations, tissues, or organs comprising the differentiating RES can be studied *in situ* or compared with those placed either in a heterotopic environment or in *in vitro* culture.

2. DEVELOPMENT OF COMPETENCE OF ENDOCYTOSIS AND CLEARANCE

2.1. UPTAKE OF COLLOIDAL MATERIALS

A major difficulty in considering the ontogeny of the RES is that it is composed of a series of interrelated cells and their fibrillar derivatives which may be either localized in the fixed macrophages lining the blood sinusoids of the liver, spleen, bone marrow, lungs, adrenal glands, and microglia or seen as wandering monocytic macrophages which are referred to as histiocytes of the connective tissues. To these organs may be added the thymus, bursa, cecal tonsils, lymphoid aggregates, and the diffuse lymphoid tissue found throughout the body. In fact, the RES has such a ubiquitous distribution that no organ is completely devoid of RE elements. Evaluation of the functional activity of the RES has been done almost entirely by means of bioassays. The usual criterion for a functional RES is that specific cells phagocytose dyes, colloidal materials, or particulates such as bacteria or foreign erythrocytes. This criterion, however, can pose a problem when one is dealing with embryonic development. These substances routinely have been used by developmental biologists for the marking of cells in the construction of embryonic fate maps (Rudnick, 1955). Such marked cells have been useful in following morphogenetic movements of presumptive cells prior to their differentiation into tissues and organs. For example, Spratt (1946) used colloidal carbon to identify cells in explants of blastocytes from unincubated chick eggs and followed their movements. He observed that carbon was taken up by the cells comprising the epiblast which were about ready to invaginate into the primitive streak. Although these cells were ectodermal at the time they took up the carbon, they were at a developmental stage when they would be considered presumptive mesoderm (Spratt, 1955). Such uptake of materials by even presumptive mesoderm can hardly be characterized as genuine RES activity. Therefore, it is difficult to precisely delineate in the early chick embryo what is true cellular phagocytosis or, for that matter, clearance of foreign material or sites of uptake. It cannot always be concluded that because particulate matter is taken into a cell, that cell is a phagocytic cell. It may very well be that the cell is not capable of processing, degrading, inactivating, or exocytosing the particulates and thus not functioning in the manner of a mature phagocytic cell.

One way of assessing the maturation of the RES is by introducing foreign colloids into the embryo's circulation and determining when cells arise which are capable of taking up such complex substances. Kent (1961) injected Thorotrast, a colloidal suspension of 25% thorium dioxide and 25% aqueous dextran by volume, into chicks ranging in age from 3 days of incubation until after hatching. Thorotrast was injected into the vitelline vein of 3- to 5-day embryos and into the allantoic vein of embryos incubated for 6 days or more. The injected vessels were sealed by cautery and incubation was continued for various periods of time. Thorotrast was identified in histological sections of the chick tissues and the cells which had taken it up had a characteristic "bubbly" or alveolar appearance. In this study it was observed in the very early embryos (3 days of incubation) that relatively little uptake occurred except for the wall of the yolk sac and then only by macrophages found in the mesodermal portion of the wall. These yolk sac cells continued to show uptake even until 7 days after hatching. When, however, the organs commonly associated with the RES were differentiated, rapid and extensive uptake of Thorotrast was seen by phagocytic cells resident in them. For example, in the liver, by the fourth day of incubation Thorotrast was clearly identifiable in the Kupffer cells lining the sinusoids. In the spleen, Thorotrast was taken up at 5 days of incubation although to a lesser extent than that by the liver. As these two organs continued to develop (embryos of 6 to 9 days of incubation), there was a decrease in the amount of Thorotrast taken up by the yolk sac. When 14-day embryos were injected and allowed to hatch, Thorotrast was also observed in the lungs and bone marrow as well as in the liver, spleen, and yolk sac. As is indicated by Kent, the decrease in uptake observed in the yolk sac may be due to migration of the macrophages from the wall of the yolk sac as soon as the mesodermal portion of its wall was formed and by the liver and spleen when they were differentiated into recognizable organs.

In contrast to the uptake of Thorotrast, when Perez del Castillo (1957) injected 1.6 mg of a carbon suspension into chick embryos at various stages of development, no phagocytosis was found prior to the 12th day of incubation. In the 14-day embryo, carbon was seen in a few Kupffer cells but it was not until day 16 that carbon was extensively phagocytosed by these sinusoidal cells of the liver. Perez del Castillo concluded that phagocytosis by the RES is not an embryonic characteristic. More recently, Edmonds (1970) injected 7- to 9-day chick embryos with India ink into the allantoic circulation. At the same time, he injured other allantoic vessels which resulted in hemostasis with the entrapment of thrombocytes. It was observed that the thrombocytes in the clot had taken up carbon, and carbon uptake also was seen in the liver, spleen, and kidneys. The greatest amount of carbon was found in the liver but was restricted to carbon-laden thrombocytes. Electron microscopic studies clearly demonstrated that the carbon was within membrane-bound cytoplasmic structures of thrombocytes. Although this study seems to contradict that of Perez del Castillo, the carbon uptake in these younger embryonic stages was limited to the thrombocytes and was not seen in the sinusoidal lining cells.

From these investigations into the development of the RES in the chick embryo, the quite different conclusions reached evidently may be ascribed to a selective process by which colloids are phagocytosed by Kupffer cells of the embryonic liver. Thus, although Thorotrast was taken up by Kupffer cells present in the liver of the 4-day embryo, it was not until the 14th day that carbon was phagocytosed. Thus, it seems likely that in these 10 days of incubation, some further functional differentiation occurred. One must realize that the situation in the embryo assumes a totally different aspect than that in the adult. One must keep in mind that even though embryonic cells or tissues can take up complex substances, the admission of such can occur both prior to and during the time when the basic differentiating functional capacity is occurring. In addition, by investigating different cell populations it became apparent that thrombocytes of 7 to 9 days were capable of taking up carbon, which was well before the 14th day when the Kupffer and other sinusoidal lining cells were first capable.

2.2. UPTAKE OF MICROORGANISMS

Another method of evaluating the ontogeny of the RES is to test how various particles are handled by the developing embryo. Goodpasture and Anderson (1937), who were not investigating the functional development of the RES, observed marked differences in the way the chick embryo was capable of reacting to various strains of bacteria. In some instances a strain of bacteria was phagocytosed and killed; in others there was uptake but the bacteria continued to grow within the cells; and for some there was no recognition of the bacteria as foreign so there was no uptake. Their method was to place bacteria from a fresh 24-hr culture on the chorioallantoic membrane (CAM) and evaluate the response made by the CAM and the embryo after further incubation. The precise age of the embryo is not given, but based upon the citation of the method of Goodpasture and Buddingh (1935), it seems safe to assume the age of the embryos to have been 12–15 days. When *Staphylococcus aureus* was applied to the CAM, it grew in a very circumscribed area and both polymorphonuclear and mononuclear leukocytes phagocytosed and destroyed the cocci and none were found within the embryonic organs. Introduction of *Streptococcus haemolyticus* resulted in necrosis of the membrane with extensive invasion of the spleen, liver, and blood vessels of the embryo. These organisms were not phagocytosed by the polymorphonuclear cells although monocytes phagocytosed and degraded them. It was seen that both monocytes and Kupffer cells phagocytosed *Streptococcus viridans*; however, the cocci continued to rapidly multiply within these cells. When *Eberthella typhi* was applied to the CAM, even though mononuclear cells were attracted to the site, no phagocytosis occurred. Finally, the mononuclear cells were not even attracted to the CAM when *Corynebacterium diphtheriae* was applied. Thus, there appears to be a large array of reactions to various species of bacteria by cells comprising the developing RES of the chicken. These reactions ranged from the complete destruction of the bacteria, to a failure to even mount an inflammatory response. Although these findings are interesting and appear to indicate a sequential onset of development for the RES, a full explanation cannot readily be made because comparable studies were not done showing how these same bacteria are handled in fully developed chickens. Regardless of this, it is not possible from this and the other studies cited (Kent and Perez del Castillo) to assign a precise point in time at which the mononuclear phagocytic cells of the chick embryo can be said to be fully differentiated.

2.3. CLEARANCE OF FOREIGN MATERIALS FROM THE VASCULAR SYSTEM

Until the 1950s, evaluating RES macrophage activity was mainly by bioassay methods which were limited to producing qualitative data. A new method, giving quantitative results, was introduced by Biozzi et al. (1953). This consisted of following the intravascular clearance of a particulate substance which was specifically phagocytosed by the intravascular sinusoidal macrophages lining the liver, spleen, lungs, and bone marrow. The particulate matter was placed in suspension and then given intravenously as an initial bolus and then at periodic intervals blood samples were taken and evaluated for the amount of the material still present. From this it was possible to calculate a clearance half-time, and assign an index which could be corrected for weight and other possible differences. Using a modification of this method, Karthigasu and Jenkin (1963) investigated the functional development of the RES of the chick embryo. They labeled two rough strains of *Escherichia coli*, one smooth strain of *E. coli*, and two strains of *Salmonella gallinarum* with ^{32}P. Chick embryos of various ages were injected via the CAM vein with a labeled bacterial suspension and at different time intervals a blood sample was removed. The radioactivity in the blood was assayed and the phagocytic index K was calculated for various aged embryos. It was observed that when the K value was calculated for clearance of one of the rough strains of *E. coli* and compared with that for a strain of *S. gallinarum*, the chick embryos throughout the period of 10 through 18 days of incubation were able to clear the *E. coli* at a much faster rate. Likewise, the K value for the *E. coli* increased with the increasing age of the embryo while that for the *Salmonella* strain remained relatively constant. The K values for other bacterial strains showed that only the other rough *E. coli* strain had a rapid rate of clearance and it also increased in the older embryos. These investigators also studied various organs of the embryos for the uptake of the bacteria. This was done by isotopically labeling a rough strain of *E. coli* and injecting it intravenously into the embryo. Sixty minutes following the injection of bacteria, the embryos were perfused with warm saline. Tissue samples from membranes of the amnion, allantois, and yolk sac, and also from the liver and spleen, were taken and analyzed for radioactivity. It was found that in 10-, 12-, and 14-day embryos, the greatest percentage of uptake occurred in the membranes while at 16, 18, and 20 days the highest percentage was in the liver. Throughout, there was *no* uptake seen in the spleen. The authors also investigated if the inability of the chick embryo to phagocytose the smooth strains of bacteria might be due to a lack of opsonins in the embryonic serum. Therefore, prior to injection into the embryo, the bacteria were incubated with either pig or chicken serum. The results obtained clearly showed that the clearance rates for the smooth strains of bacteria treated with serum were comparable to those for the rough strains, thus indicating that serum opsonins were the limiting factor. It was also interesting that the relative rate of rapid clearance for the rough strains of *E. coli* was not increased by prior incubation of the bacteria with serum. In order to examine if the faster rate of clearance for the rough strains of *E. coli* was attributable to phagocytosis, and not just mechanical trapping, RES blockading experiments of 14-day embryos were performed. Blockading of the RES was carried out by injecting a large dose of *E. coli* intravenously and at various intervals following this with a second dose of isotopically labeled *E. coli*, and then determining the phagocytic index. The data showed that RES blockade could be induced by the *E. coli;* that incubation of the labeled bacteria with either chicken or pig serum could overcome the blockade; and that *S. gallinarum* (which had a low clearance rate) was unable to function as a blockading agent for *E. coli*. From these data it was concluded that the clearance of rough strains of *E. coli* was an active phagocytic process and that specific serum opsonins directed against these organisms were present in the chick embryo and have an important role in RES clearance. It is quite possible, however, that the opsonin present in the chick embryo may not have been produced by the embryo itself but transferred there from the hen through the yolk (Brambell, 1970). To investigate this, Karthigasu and Jenkin (1963) immunized hens with heat-killed *S. gallinarum*, a smooth strain of bacteria which had a low phagocytic index. The eggs from these chickens were incubated and the embryos were tested for clearance of this bacteria at 14 and 17 days. The results of these experiments showed that the embryos from immunized hens had a markedly enhanced phagocytic index for the *Salmonella* compared with the embryos from untreated adults. In reviewing the above data, one must remain somewhat ambivalent in reaching any firm conclusion concerning a specific time as to when the chick embryo has developed an RES that can be considered functional. Although it was shown that 10-day embryos were able to clear rough strains of bacteria from their circulation, they were not able at even 20 days to do this for smooth strains. Again, it was clearly shown that uptake of bacteria was dependent on the presence of opsonic factors, yet it is not known if the opsonins were produced by the embryo or had been passed from the hen to the embryo.

2.4. THE ROLE OF OPSONINS IN CLEARANCE AND UPTAKE

Certainly, it cannot be questioned that at least some components of the RES are present and functional in the chick embryo. The method of testing for phagocytosis or clearance did not give a clear definition of competence. Another approach which has been used is to determine if the tissues phagocytosing the bacteria also are able to destroy the bacteria (Jenkin, 1963). Therefore, Karthigasu *et al.* (1965) investigated if chick embryos of various developmental stages possessed bactericidal activity. They did this by injecting a rough strain of *E. coli* into a vein of the CAM and then at various time intervals determining the numbers of viable bacteria in both embryonic organs and extraembryonic membranes. The results showed that the potential of the phagocytic cells increased with increasing age; that is, the 11- and 13-day embryos killed very few of the organisms while the 15- and 17-day embryos destroyed almost all the injected bacteria by 3 hr after injection. They also investigated the ability of the embryo and neonatal chick to phagocytose a smooth strain (*S. gallinarum*) which was shown to be phagocytosed poorly unless opsonins were present. Prior to intravenous injection into the embryo, the bacteria were treated with adult chicken serum as a source of opsonin. They then observed that *S. gallinarum* was not destroyed by either the 17-day embryo or by the 1-day hatched chick, but that the 7-day hatched chick was able to kill substantial numbers of organisms. In conclusion then, it appears from this study that the capacity of the RES to deal with ingesting and bactericidal potential at one particular embryonic or neonatal time may vary according to the type of organism which is tested. Also, although Jenkin (1963) had shown that opsonins may have an effect on intracellular killing, as well as ingestion, in the case of *S. gallinarum*, opsonin did not appear to be the limiting factor.

The possible role of opsonin in the onset of resistance to infection by *S. gallinarum* was investigated in chicks from hatching to 5 weeks of age (Solomon, 1968). He compared the virulence for both a so-called virulent and an avirulent strain by measuring the $LD_{50}/100$ g body wt. He also evaluated the ability of the liver and spleen to destroy these organisms by counting the outgrowth of colonies from homogenates of these organs. The results showed that resistance to infection as measured by an increasing $LD_{50}/100$ g body wt increased about 10,000-fold for the avirulent strain between the first and fifth days following hatching, while it took about 5 weeks for this to be attained by the chick against the virulent strain. It also was found that both strains multiplied at about the same rate in the liver and spleen of 1-day chicks, while there was about a 30-fold decrease for the avirulent compared with the virulent strain in these organs of the 4-day chick. The clearance rate for the virulent strain remained almost unchanged throughout the 5 weeks studied, while the avirulent strain was cleared with increasing rapidity between 2 and 5 weeks; at 5 weeks it was being cleared 7.5-fold faster than the virulent strain. To test what function opsonins might serve, either heat-killed avirulent *S. gallinarum* was injected intravenously into 14- and 15-day embryos or low doses of live organisms were injected into hatched chickens of 1 day to 5 weeks of age. In using the phagocytic index as a measurement of clearance, no significant differences were observed when compared with uninjected controls. On the basis of these results, Solomon concluded that opsonins, at best, play only a minor role in the onset of resistance to infection to this organism. However, he did not show by *in vitro* methods that serum opsonins had been produced in the injected chickens. Nevertheless, this study has clearly demonstrated that for two strains of *Salmonella*, which antigenically were closely related and to which 5-week chickens were highly resistant, there were dramatic differences in their rates of clearance and their ability to grow in the liver and spleen. Thus, the capacity of the developing RES as measured by several parameters again was shown to vary markedly, depending upon the nature of the organism being utilized in the testing procedures.

In another study, Solomon (1966) observed a situation where "natural" antibodies appear to have an important function as the chick embryo develops competence to clear mammalian erythrocytes from its circulation. Clearance rates of goat erythrocytes were observed in chick embryos between 9 days and hatching and in hatched chickens from 1 day through 5 weeks. During this time period, because there is a marked change in the liver and spleen weights relative to the body weight, a correction of the phagocytic index as developed by Benacerraf *et al.* (1957) had to be used. The results showed there was no increase in the phagocytic index of the embryo from 9 to 11 days of incubation but a threefold increase occurred between 11 and 19 days. The phagocytic index decreased for the first 6 days after hatching and then rose slightly and leveled off for the next 5 weeks. Similar results were found when guinea pig erythrocytes were used. From these results, Solomon postulated that the peak clearance values were correlated with the time of maximum incorporation of maternal antibody into the embryonic circulation (Jungherr and Terrell, 1948). By use of agglutinin techniques, very low levels of antibody to goat erythrocytes were found in the pooled serum of 1-day hatched chicks. To determine if serum from adult chickens could function as an effective opsonin *in vivo*, both immune and nonimmune sera were injected intravenously into 16- and 17-day embryos which previously had received goat erythrocytes. It was observed that both immune and nonimmune sera caused an increase in the phagocytic index, with immune sera being far more effective. The normal serum's opsonic ability was decreased following heating at 56°C and the immune serum's ability was reduced by treatment with 2-mercaptoethanol. The rather marked changes in the phagocytic index of the chick embryo for mammalian erythrocytes correlated with the passage of maternal serum components from the yolk into the embryo's circulation (Brambell, 1970).

In this section it has been shown that the development of the RES in the chick is very complex and does not occur as a simple all-or-none phenomenon or even necessarily in a clear sequential fashion. Even though 4- to 5-day embryos were able to phagocytose colloidal particles, it was also seen that for some strains of bacteria, phagocytosis did not occur until after hatching. In some instances, even the ability to recognize organisms as foreign differed widely depending on the strain or form of the bacteria. Part of this inability to react to some of the organisms could be correlated with the absence of opsonins while for others even the addition of serum did not promote either uptake or killing. Therefore, in defining RES competence, one is confronted with having to develop an operational definition dependent upon the methods used.

3. RES FUNCTION IN IMMUNOLOGICALLY DEFICIENT CHICKENS

3.1. CLEARANCE OF ANTIGENS AND COLLOIDS FOLLOWING BURSECTOMY

It has been shown in the previous section that macrophages were functional at very early stages in the developing chick embryo. The extent, variety of materials used, and efficiency of endocytosis, however, were not comparable to those of hatched chicks. Several of the studies indicated that serum factors appeared to be needed for recognition of foreignness to certain of the organisms by the embryonic macrophages. The importance of serum components for the promotion of phagocytosis was probably already recognized by Wright and Douglas (1904). They observed that bacteria had to interact with serum, which they termed "opsonins," before significant phagocytosis by white blood cells from mature animals would occur. The best studied of the serum components are antibodies formed in response to a specific antigen (Rowley, 1962). Certain classes of the immunoglobulins have been shown to be cytophilic for the macrophage plasma membrane (Boyden, 1964). Although it is not yet completely understood how the macrophage interacts with the antibody–antigen complex, Phillips-Quagliata et al. (1969) proposed that the Fc portion of the IgG molecule attaches to the macrophage membrane, which in turn functions to initiate the uptake of the foreign material. For the chicken to develop its capacity to produce immunoglobulin, it is well known that the bursa of Fabricius plays a major role (Glick et al., 1956; Glick, 1977; Cooper et al., 1965, 1969; Van Alten et al., 1968). Thus, various investigators have used a variety of *in vivo* and *in vitro* studies to evaluate the RES activity of bursectomized chickens.

Mueller et al. (1960) investigated the clearance of BSA of 6-week-old chickens which had been either surgically bursectomized at 1 or 2 weeks after hatching or hormonally bursectomized by treatment of the embryo with 19-nor-testosterone. Following a single intravenous injection of 40 mg BSA/kg body wt, antigen clearance was measured as the last day that the antigen could be detected within the circulation. The elimination of BSA for the normal chickens occurred on day 5 and for both groups of those surgically bursectomized it occurred on day 6. For the group of chickens injected with 19-nortestosterone at 5 days of incubation, the clearance of antigen did not occur until between 8 and 10 days. All of the bursectomized chickens showed a reduction in their antibody response to BSA and the testosterone-treated group showed the greatest reduction. Mueller et al. (1964) reported that clearance of BSA following a primary intravenous injection at 12 weeks of age in other groups of similarly treated chicks followed the same pattern as at 6 weeks. Thus, these reductions in the ability of bursectomized chickens to clear BSA from their circulations did not appear to be transient and probably correlate best with the reduction in their ability to produce specific antibody. Another approach used by Mueller and his associates was to compare the elimination of homologous erythrocytes by bursectomized and normal chickens. They determined the rate of elimination from the circulation of chromium-labeled erythrocytes of another strain of chicken. Mitchison (1962) had shown that accelerated elimination of such homologous erythrocytes represented an immune response. The chicks were surgically bursectomized between 1 and 7 days following hatching. Three separate experiments were carried out using groups of chicks injected at 8, 11, and 26 days posthatching. In none of these experiments were differences observed between the bursectomized and the normal chickens in their ability to clear the chromium-labeled cells. The differences in clearance seen between the study using the erythrocytes and BSA might in part be due to the difference between a particulate and a soluble antigen. Another probable explanation is that for the homologous erythrocytes there was persistence of serum antibodies derived from the yolk well beyond the time at which these tests were carried out (Brambell, 1970). In using the clearance time as a measure for RES function, it was seen that for some foreign proteins there was an increase following both surgical and hormonal bursectomy. Although Mueller et al. (1964) believed the reduction of antibody was responsible for the diminished clearance rate for the bursectomized chicken, their experiments did not allow them to determine if there might be reduction either in the antigen uptake by the tissues comprising the RES or in the functional ability of the macrophages.

The effects of thymectomy and bursectomy were assessed by determining the clearance rate for colloidal gold in 1-month-old chickens (Cooper et al., 1966). In these studies, either the thymus or the bursa was removed surgically on the day of hatching and then the chicks were given irradiation to destroy B and T lymphocytes which had already been processed by the thymus or bursa and now were in the peripheral tissues. Following this treatment, they observed that neither the thymectomized nor the bursectomized chickens had a detectable alteration in the clearance of colloidal gold. They also reported that under similar conditions there were no detectable changes in the clearance of radioiodine-labeled hemocyanin from the circulation in bursectomized and irradiated chickens when compared to controls. Their results suggest that neither the thymus nor the bursa seems to directly influence the RES. All of the clearance studies, including the controls, were carried out using irradiated chickens. Although not highly probable, it is possible that the effect of the irradiation of the RES may still be persisting and that any influence that the bursa or thymus might have had, therefore, may have been overshadowed.

Studies which monitored only antigen or colloid elimination from the blood, at best, gave contradictory information as to a possible effect that the bursa of Fabricius had on the function of the full development of the RES. Thus, experiments described by Glick et al. (1964) investigated the ability of bursectomized and normal chickens to clear carbon from the blood and for the heterophils to phagocytose bacteria. For the clearance of carbon they used 10-week-old chickens which had been surgically bursectomized at hatching. Their results showed that the clearance rate of the bursectomized chickens was not significantly different from that of normal chickens. Likewise, no differences were observed for the clearance of carbon by bursectomized and normal chickens at 7 days following an intravenous injection of a sheep erythrocyte suspension. It was thought that the injection of the erythrocytes might generate nonspecific opsonins and thus promote the phagocytosis of carbon in normal but not in bursectomized chickens. One hour after injection of carbon, when it was no longer detectable in the blood, autopsies were carried out on some control chickens. They observed in these chickens that extensive uptake had occurred in the liver, spleen, bone marrow, and accessory spleens. Only slight uptake was seen in the kidneys and lungs, and none in the thymus or bursa of Fabricius. The method for the *in vitro* assessment of phagocytosis was to combine equal portions of *Staphylococcus aureus* were oxalated whole blood and incubate this mixture with agitation for 30 min at 37°C. At the end of the incubation a blood smear was made, air dried, and stained with Wright's stain. Heterophils were counted and the numbers containing bacteria to the total numbers counted were recorded. At 1 and 3 weeks of age, there were no apparent differences between heterophils of normal and bursectomized chickens. At 8 weeks, 30% of the heterophils of normal chickens had phagocytosed the bacteria compared to only 21% for bursectomized chickens. Although these authors state that there were no differences in phagocytic ability between the heterophils of normal and bursectomized chickens, no statistical support for this conclusion was given. In this same study, they reported that significant numbers of thrombocytes of normal chickens phagocytosed bacteria. Their data show that 15% of the thrombocytes of 1-week-old chickens contained bacteria and that 40% of the thrombocytes of 3- and 8-week-old chickens did. From this it appears that the thrombocytes of adult chickens are more active than those of young chickens. That the chicken's thrombocytes have a very important function as a circulating phagocyte has recently been shown by Chang and Hamilton (1979). They also used the *in vitro* method of direct counting with whole blood and counted the percentage of thrombocytes, heterophils, and monocytes which phagocytosed formalinized *Enterobacter cloaca*. Their results showed a statistically significant reduction (ca. 50%) in uptake at 2.5 min of incubation for the thrombocytes of 3-week-old chickens as compared with those 6 to 8 weeks of age. After 10 min of incubation, however, no differences were seen between the percentages of thrombocytes phagocytosing bacteria from either age group. This evidence also suggests that the thrombocytes of older chickens were more active phagocytically than those of younger chickens.

In the above review a clear functional deficiency of the RES was not consistently demonstrated following bursectomy. This may have been in part due to the experimental design of the various investigations. For example, antigen clearance, as a criterion for RES function, was not used in several studies, and as was pointed out, using young chicks, which possibly still retained serum proteins of the hen, would not show possible opsonin deficiencies. It should not be surprising, therefore, to observe an alteration in RES function when changes are found in the antibody production system. Subba Rao and Glick (1975) reported a depression in the ability of the chicken to clear colloidal carbon following both positive and negative alterations in the antibody response. Their method consisted of injecting phytohemagglutinin (PHA) either intravenously or intraperitoneally into 4-week-old chickens and testing them for antibody production to sheep erythrocytes and for clearance of carbon. Following intravenous injections of PHA they observed an 18% decrease in the clearance rate of carbon and a significant enhancement of the antibody reponse. When PHA was administered intraperitoneally, they observed a decrease of from 35 to 43% in the clearance rate and a significant decrease in antibody production. From this study it is not possible to determine if the perturbation of RES function was due to a direct effect of PHA on macrophages, alterations in serum opsonins, or an indirect modulating effect by lymphocytes responding to the mitogen.

3.2. ALTERATION IN THE NONIMMUNE OPSONINS FOLLOWING BURSECTOMY

In addition to the immune opsonizing system, which functions primarily as the recognition system for the removal of foreign pathogens, there are the nonimmune opsonins which are involved in the phagocytosis of nonantigenic colloids. One such colloid, which has been extensively used *in vivo* and *in vitro*, is a gelatinized lipid emulsion (Saba and DiLuzio, 1969). Such a nonimmune opsonin which promotes the uptake of gelantinized lipid colloids has been purified from both rat (Molnar et al., 1977) and human sera (Blumenstock et al., 1978). To investigate if such an opsonin might be altered in the bursectomized chicken, Waltenbaugh et al. (1976) carried out a series of in vivo experiments. In 4- to 6-week-old chickens, an ^{3}H-labeled gelatin-coated lipid emulsion was injected intravenously; blood samples were collected at 2-min intervals; and half-time clearance rates were calculated. They reported that in bursectomized chickens the half-time clearance was 6.3 min, in contrast to only 1.8 min in normal chickens. Thus, surgical bursectomy on the day of hatching significantly altered the ability of the chicken to clear the gelatinized lipid colloid from the circulation. To determine if this defect in RES function might be due to serum opsonins, RES-blockaded bursectomized and RES-blockaded normal chickens were investigated. Following blockading of the RES, the chickens were injected intravenously with [^{3}H]thymidine-labeled lipid emulsion which had been incubated with fresh serum of either normal or bursectomized chickens. They observed that clearance in both bursectomized and normal chickens was substantially restored when the test emulsion was incubated with serum from normal chickens prior to injection. The RES blockade, however, was not overcome in either normal or busectomized chickens when the emulsion had been incubated with sera from bursectomized chickens. The decrease in clearance ability of the bursectomized chickens, therefore, appears to be in the sera which was shown to be unable to overcome the RES blockade. Furthermore, it was also shown that recovery from blockade in bursectomized chickens also occurred when the lipid emulsion was incubated with rat α_2-macroglobulin. This material, found in the serum of rats, had been shown by Allen et al. (1973) to promote phagocytic activity for the gelatinized lipid emulsion in rats. These investigators also determined the uptake of the radioactively labeled emulsion by the liver, lungs, and spleen of normal and bursectomized chickens (Allen, Waltenbaugh, Molnar, Sabet, and Van Alten, unpublished observations, 1976). When the test emulsion had been incubated with fresh serum prior to intravenous injection into RES-blockaded normal and RES-blockaded bursectomized chickens, over 60% of the initially injected material was found in the liver, 2–3% in the lungs, and 1.5–2% in the spleen. If the lipid emulsion had been incubated with serum from bursectomized chickens, only about 30% of the initially injected material was present in the liver, and the lungs and spleen again had taken up between 1 and 3%. The studies of the uptake by these organs show that reduced clearance by the bursectomized chicken is not due to a defect in organs which clear foreign substances from the blood. From all of these studies the evidence indicates that the depressed clearance by the RES in bursectomized chickens was due to a serum factor deficiency rather than cellular impairment.

To further support these in vivo findings, in vitro experiments were undertaken and some of the preliminary results were described by Sabet et al. (1975). One of the major difficulties encountered was the ability to consistently obtain functional peritoneal macrophages to carry out these in vitro studies. Such a technique, however, was developed by Sabet et al. (1977) in which Sephadex suspensions were used as an irritant to elicit the peritoneal exudate cells. They found that monolayers of the glass-adherent peritoneal exudate cells were ac-

TABLE 1. PHAGOCYTIC UPTAKE OF SHEEP ERYTHROCYTES BY
PERITONEAL MACROPHAGES OF NORMAL CHICKENS WHEN
INCUBATED WITH VARIOUS SERA

Source of serum	Percentage of cells phagocytosing erythrocytes
None	8
Normal chicken	78
Bursectomized chicken	38
Human	13
Rabbit	15
Mouse	12
Fetal calf	1

tively phagocytic for either latex particles or antibody-sensitized erythrocytes, or both. Using such monolayers of peritoneal exudate cells, derived from 6- to 8-week-old chickens, Sabet et al. (1980) carried out an investigation of the impairment of RES function in the bursectomized chicken. To the monolayers of peritoneal macrophages, prepared in Hanks' medium, were added sheep erythrocytes alone or with 5% fresh serum of either normal or bursectomized chickens. All sera were obtained from chickens of more than 6 weeks of age to avoid the possibility of serum components of maternal origin. The phagocytic cells were incubated in one of the various media for 1 hr at 37°C, after which the medium was removed; they were treated with distilled water for 1 min to lyse erythrocytes which had not been internalized; and then they were placed in fresh Hanks' medium. Up to 1000 macrophages were counted and the percentage of cells containing erythrocytes was calculated. From the data in Table 1 it is seen that only about one-half as many cells (38%) of normal chickens phagocytosed erythrocytes in the presence of sera from bursectomized chickens as when in the presence of sera from normal chickens (78%). In addition, it is evident that fresh human, rabbit, and mouse sera and commercial fetal calf serum were ineffective in stimulating phagocytosis. From these results the im-

TABLE 2. PERCENTAGE OF PERITONEAL MACROPHAGES FROM
NORMAL AND BURSECTOMIZED CHICKENS CAPABLE OF
PHAGOCYTOSING ERYTHROCYTES WHEN INCUBATED WITH
VARIOUS SERA

Source of serum	Source of peritoneal cells	
	Normal	Bursectomized
None	3%	2%
Normal chicken	91%	96%
Normal chicken	83%	87%
Bursectomized chicken	58%	53%
Normal chicken heated at 56°C for 30 min	9%	—

pairment in phagocytic ability by the bursectomized chicken appears to be due to a decrease in the level of serum opsonin(s). To further test this, the macrophages of normal and bursectomized chickens were compared. In Table 2 it is shown that macrophages from bursectomized chickens were as effective at phagocytosing erythrocytes as were the cells from normal chickens when cultured in the presence of sera from normal chickens. For cells from both normal and bursectomized chickens there was a marked decrease in the percentage of uptake when sera from bursectomized chickens were used. Evidence is also shown in Table 2 that heating normal serum at 56°C for 30 min greatly decreased its phagocytosis-promoting ability. From these experiments it can be concluded that the defect in RES function in the bursectomized chicken is due to some change in a component in the serum rather than alteration in the monocytic macrophage cells. To gain further information concerning the possible defect in the phagocytosis-promoting activity of the serum from bursectomized chickens, isolated chicken IgG and C1 were evaluated for their ability to promote phagocytosis. Previously, Cooper et al. (1969) found that the major defect in chickens surgically bursectomized at hatching was a decrease in serum IgG. When IgG was used in the incubation media there was no enhancement in the percentage of macrophages of normal chickens to phagocytose erythrocytes. As is shown in Table 2, the serum factor for promoting phagocytosis was heat labile. Thus, C1 was added to the incubation media and failed to enhance the ability of the macrophages to phagocytose the erythrocytes. The preparation of C1 was according to the method of Stolfi et al. (1971) as modified by Gabrielson (1975). From these two studies it would seem reasonable to conclude that the serum opsonic defect is not due to either a diminished IgG or possible alteration in C1, neither of which promoted *in vitro* phagocytosis of erythrocytes by peritoneal macrophages.

Using both *in vivo* and *in vitro* methods showed that bursectomized chickens had a defect in RES function. Clearly, this altered RES response was not cellular but was due to some serum component. A possible component of the serum which may be reduced or altered is the α_2-macroglobulin which has been shown to be required for phagocytosis of gelatin-coated colloids in mammals (Molnar et al., 1977). It will be recalled that even though it was of rat origin, such an α_2 opsonic protein was able to overcome RES blockade in bursectomized chickens (Waltenbaugh et al., 1976). In human plasma such a nonimmune opsonin, antigenically similar to α_2-macroglobulin, has been demonstrated to be cold-insoluble globulin or fibrinonectin (Molnar et al., 1979). Recently, from chicken plasma, Marquette et al. (1981) have been able to isolate fibrinonectin by three different methods. The purity of these preparations was demonstrated by SDS-polyacrylamide gel electrophoresis, and in the presence of mercaptoethanol showed a major band at 210,000 daltons. The phagocytosis-promoting activity of fibrinonectin was tested by incubating chicken peritoneal macrophages with [^{125}I]gelatin-coated latex particles. It was shown that uptake of the latex increased linearly with added concentrations of fibrinonectin. This study shows that normal chicken plasma fibrinonectin acts as a nonimmune phagocytosis-promoting opsonin for gelatin-coated colloids. So far, such an investigation has not been undertaken to compare the plasma fibrinonectin from normal and bursectomized chickens to determine if it is quantitatively decreased or altered in some other way in bursectomized chickens. Such a study, however, is presently being organized (Molnar and Van Alten).

In this section it was shown that there was a depression of RES function in chickens following both surgical and hormonal bursectomy. The methods used to evaluate RES function were quite variable and thus the results obtained seem contradictory. Evidence was presented showing that in addition to a decrease in immunologically dependent phagocytosis there was also a defect in the nonimmune opsonic system. Although at this time the defect in the nonimmune opsonic system of the bursectomized chicken has not been fully clarified, it appears that it might be related to a possible quantitative decrease or some other alteration in plasma fibrinonectin. In none of the papers reviewed was any evidence found that either cells or tissues comprising the RES of bursectomized chickens were deficient in their phagocytic ability.

4. RES FUNCTIONS OF THE EPITHELIUM COVERING THE LYMPHOID FOLLICLES OF THE BURSA OF FABRICIUS

4.1. MORPHOLOGY AND FUNCTION OF THE BURSAL EPITHELIUM

In addition to tissues which are regularly considered part of the RES, there appear to be epithelial areas of the gastrointestinal tract which are particularly adapted for handling environmental materials, including nondigestible particles. These areas are in contrast to most portions of the digestive tract, in which an as yet poorly understood barrier function is present. At these modified sites the epithelium appears both structurally and functionally to be peculiarly adapted to and in direct contact with the underlying lymphatic tissue. In chickens, this epithelial–lymphatic arrangement is found only in the bursa of Fabricius. That the bursal epithelium is unique was first reported by Bockman and Cooper (1973) and subsequently by Holbrook et al. (1974). A striking characteristic of this epithelium was that there were two types present and was most clearly shown with the electron microscope. It was observed that overlying the lymphoid follicles, the epithelium was pseudostratified and displayed evenly distributed microvilli, extensive intercellular spaces, and few smooth membrane vesicles; it has been called follicle-associated epithelium (FAE). The interfollicular epithelium, by contrast, was described as a simple epithelium showing pitlike depressions on its surface, having unevenly distributed, irregularly shaped microvilli, and an abundance of smooth membrane vesicles.

It was Bockman and Cooper (1973) who provided new insight into the possible functional role of the bursal FAE. They reported that 1 hr following intraluminal administration of either colloidal carbon or ferritin, the FAE had pinocytosed these colloids. In addition to their presence in the FAE, these colloids also were found in the intercellular spaces and in the lymphoid tissue. That the bursal FAE was unique in its ability to take up inert particles was demonstrated by Schaffner et al. (1974). They reported that following oral administration of carbon, the bursa was the only area along the entire avian digestive tract which was capable of taking it up. They also observed that following cloacal administration of carbon, it was present in the FAE within 15 min, in the subepithelial areas at 2 hr, and in the medullary portions of the follicles, mainly as aggregates within the macrophages, at 16 hr. They also introduced latex spheres and Thorotrast into the cloaca and 24 hr later observed that these particles were present within both FAE and subepithelial macrophages and between the lymphoid cells. In addition, they introduced suspensions containing either *E. coli* or *Staphylococcus albus* into the bursal duct. Although no intact microorganisms were observed in either the epithelium or the bursal follicles, myelinlike cytoplasmic inclusions were numerous in both FAE and subepithelial macrophages of chicks which were given bacteria at 1 to 16 days of age. From these investigations it is apparent that the epithelium overlying the bursal lymphoid follicles is uniquely adapted for the entry and passage of particulates introduced into the intestinal lumen. In addition to small inert particles, Bockman and Stevens (1977) observed that the glycoprotein horseradish peroxidase (HRP) was taken up and transported by the FAE. An advantage of HRP over particulates is that it is a more uniform colloid and by histochemical techniques can be visualized by light as well as electron microscopy. Recently, Naukkarinen *et al.* (1978) investigated both the structural and the functional differentiation of the bursal epithelium. They observed that in the bursa of 13-day embryos there were morphological changes in the epithelium indicating two cell types, that is, an FAE and an interfollicular epithelium. It was not until the 19th day of incubation, however, that they found the FAE capable of endocytosing carbon particles. With other colloids, it now appears that the FAE may be capable of pinocytosis at an even earlier stage of development. Bockman (1979) has reported that after transplantation of bursae of 10-day embryos into a heterotopic site (CAM), the bursae continued to differentiate for 5 days and then the FAE was found capable of taking up ferritin. Thus, bursae comparable to that of 15-day embryos showed endocytosis. Unfortunately, there are no reports showing that the bursae under normal conditions have the ability to take up colloids other than carbon prior to the 19th day of incubation. Nevertheless, substantial evidence has accumulated (Bockman and Cooper, 1973; Schaffner *et al.*, 1974; Sorvari *et al.*, 1975; Bockman and Stevens, 1977; Sorvari and Sorvari, 1977; Naukkarinen *et al.*, 1978; Oláh and Glick, 1978) showing that the FAE of the bursa is modified for the function of transmitting various materials from the lumen into the lymphoid follicles.

4.2. BURSAL FUNCTION AS A PERIPHERAL LYMPHOID ORGAN IN LOCALIZED IMMUNITY

The question now arises as to what possible meaning this transmission of substances by the FAE can have for the bursa. It is well known that the bursa has the unique importance of being a central lymphoid organ which functions as a site for the development of immunological competence of B lymphocytes (Glick, 1977). Van Alten and Meuwissen (1972) were the first to report that administration of sheep red blood cells (SRBC) directly into the bursal lumen resulted in the formation of significant numbers of antibody-specific plaque-forming cells (PFC) by the bursal follicular lymphocytes. No increase in the numbers of PFC present in the spleen resulted following the application of SRBC into the bursal lumen. When they injected the SRBC directly into the bursal parenchyma, no PFC were induced in the bursa. Further, they reported that on the fourth day after hatching and at all subsequent times, lymphocytes were present in the bursa which bound *E. coli* to their surfaces. In a subsequent study, Waltenbaugh and Van Alten (1974) observed that binding of *E. coli* could be blocked by specific antibody to chicken immunoglobulin and that significant numbers of bursal lymphocytes secrete specific antibody to *E. coli* flagella. They were not able, however, to detect significant antibody titers to *E. coli* in the serum even when there were significant numbers of lymphocytes binding these bacteria in the bursa. On the basis of these observations they concluded that the bursa possibly plays a significant role in localized host defense. They also proposed that passage of antigen through the FAE was necessary for bursal lymphocytes to respond and become localized antibody-secreting cells. Evidence supporting the idea that there was a functional significance between the association of the FAE and the underlying lymphoid follicles was shown by Waltenbaugh et al. (1977) using organ-cultured bursae. They observed that bursal lymphocytes in suspension culture were immunologically unresponsive to SRBC, while under similar culture conditions splenic cells responded by forming significant numbers of PFC. In contrast, bursal tissue incubated in organ culture with either SRBC or BSA for 72 hr resulted in a significant increase in the numbers of PFC to both antigens. In using organ cultures, the bursal pieces were placed on the grids so that the epithelium was exposed and the antigen applied directly on it. Thus, it appears that a critical component of the pathway to bring about induction of a local bursal response to antigen is the surface epithelium which overlies the lymphoid follicles. That the FAE may function in the processing, as well as in the passage of antigen, can be inferred from *in vitro* studies using bursal cell suspension cultures. Van Alten et al. (1977) showed that bursal lymphocytes were able to produce significant numbers of PFC in suspension cultures if a cell-free supernatant of cultured splenic cells from bursectomized chickens was added to the cultured bursal cells along with the SRBC. The cell-free supernatant was obtained from splenic cells cultured with SRBC for 72 hr prior to its addition to the bursal cell cultures. They further characterized the splenic cell fraction responsible for producing this bursal cell-activating factor by showing that only splenic cells adherent to plastic were necessary. All of these studies confirm that the bursa of Fabricius functions as a peripheral lymphoid organ in localized antibody production and that the unique relationship between the FAE and the lymphatic tissue is necessary for this immunological function.

4.3. BURSAL FUNCTION AS A PERIPHERAL LYMPHOID ORGAN IN SYSTEMIC IMMUNITY

That the bursa also may function in systemic immunization has been shown by several reports in which various antigens were administered via the cloacal orifice. Sorvari *et al.* (1975) observed that when barium sulfate was placed on the anal lips of 6-week-old chickens and examined with X-rays, almost all the contrast medium was taken up into the bursa. They then immunized 4-week-old chickens with *Brucella abortus* and SRBC on 6 consecutive days. On the day following the final administration of the antigens, a significant increase in the serum antibody titer to *Brucella* was observed while that for SRBC was not significantly different from unimmunized chickens. In subsequent experiments, Sorvari and Sorvari (1977) immunized 1-day-old, 4-week-old, and 10-week old chickens on 5 consecutive days by applying SRBC and *Brucella* to the anal lips. Following immunization of the 1-day-old and 4-week old groups, no significant antibody responses were detected to SRBC, but to *Brucella* there was a substantial antibody titer by 3 days following the last application of the antigens to the 4-week-old chickens. Ten-week-old chickens responded to both antigens with serum antibody, although for SRBC, a significant increase in the serum titer was not detected until after the second week of antigen administration. For all ages of chickens, continued antigen administration *per anum* over the following 3 weeks resulted in a marked decline in the antibody titers. In addition, these investigators also studied serum antibody production following cloacal administration of antigens after performing bursectomies on 10-week-old chickens. They observed that the serum agglutinin response for SRBC applied to the cloaca was no different at 14 days after the beginning of immunization than that of intraperitoneally injected bursectomized chickens. For *Brucella*, however, no significant serum antibody titer was elicited subsequent to *per anum* application of this antigen to the bursectomized chicken, while a normal antibody response was present following intraperitoneal injections. From these studies they concluded that for some antigens the bursa appears to have an important role for the chicken to gain part of its immunity.

Others also have demonstrated that antigen given via the cloaca may contribute to the state of the systemic humoral immunity of the chicken. When Hughes and Henderson (1977) placed 0.5 ml of 10 mg human serum albumin (HSA) on the cloacal orifice of 4- to 6-week-old chickens, they reported that they could not detect the presence of serum antibody to this antigen between 3 and 12 days following its administration. On the other hand, they did detect such antibody following either intravenous or intramuscular administration of HSA. When, however, they challenged these various groups of chickens 28 days after the primary administration of HSA intravenously, they observed significant levels of antibody by the group which was given primary administration of antigen cloacally. The titer for this group was comparable to the groups given an initial injection of the antigen intravenously or intramuscularly. They investigated the numbers of rosette-forming antigen-binding peripheral blood lymphocytes and found significantly fewer rosetting cells in chickens receiving a primary cloacally administered antigen than those given an intravenous injection. They concluded that cloacally administered antigen resulted in a state of immunological memory without inducing an appreciable immediate production of serum antibody in 4- to 6-week-old chickens. They proposed that cloacal antigen administration leads to the generation of an enlarged population of antigen-specific cells by the bursa which will produce an amplified systemic response when subsequently challenged by parenteral administration of the antigen. There is no evidence that this author is aware of which shows that following antigenic stimulation of the avian bursa, a common membrane-associated immunity results as has been proposed for the mammalian gut-associated lymphoid tissue (Craig and Cebra, 1971). Recently, Muehleman and Van Alten (1979) used the cloaca of the newly hatched chick as a route for the establishment of systemic immunological tolerance. The antigen used was BSA and serum antibody production was determined by passive hemagglutination. For the establishment of tolerance they administered deaggregated BSA either cloacally or intramuscularly on the day of hatching and weekly for the following 4 weeks. At 5 weeks the chickens that had received BSA via the cloaca were given the antigen in Freund's complete adjuvant, also cloacally. Following this they were divided into two groups and on the sixth week one group was given antigen in the adjuvant cloacally and the other group, intramuscularly. Ten days following this last injection, the chickens were bled and it was found that the serum antibody titer was no greater than that of uninjected chickens. In contrast, groups of chickens that had not been given the tolerizing antigen as newly hatched chicks, produced significant amounts of antibody when they received antigen in adjuvant cloacally on the fifth week and intramuscularly on the sixth week. On the other hand, when aggregated BSA was given cloacally in the neonatal period, tolerance was not induced. Thus, the deaggregated form of the antigen was necessary for the establishment of immunological tolerance by cloacal administration of antigen. From these investigations, then, it appears that the bursa, in addition to having a role in localized host defense, also contributes to systemic immunity, production of immunological memory, and establishment of immunological tolerance. In the last two studies cited (Hughes and Henderson, 1977; Muehleman and Van Alten, 1979), it must be remembered that neither has definitively demonstrated that the bursa was responsible for the altered systemic responses described; it is possible that some other gut-associated lymphoid organ might be involved.

4.4. EPITHELIAL–LYMPHOID INTERACTION IN THE BURSA AND CONSEQUENT FUNCTIONING AS BOTH A CENTRAL AND A PERIPHERAL LYMPHOID ORGAN

From the investigations reviewed above it now seems well established that the bursa functions as a peripheral as well as a central lymphoid organ. For it to function as a peripheral lymphoid organ, it appears that the epithelium, particularly the FAE portion, is important for both transporting and processing antigens. The close association of epithelium and lymphocytes occurs early in development with the first development of the lymphoid follicles between 9 and 10 days of incubation. The follicles are first seen as a proliferation of solid epithelial buds at various points along the epithelium which then push into the tunica propria of the bursal plicae (Bockman and Cooper, 1976). At this same period in development, Edwards *et al.* (1974) observed the first blood-derived lymphoid precursor cells migrating through the basement membrane into the epithelium where they proliferated and then accumulated as subepithelial foci. That there was an interdependency between the epithelium and the lymphocytes was evident following chemical bursectomy of either embryonic or newly hatched chicks. Sachs et al. (1979) studied the influence of cyclophosphamide on bursal FAE of newly hatched chicks. Although a number of investigators (Glick, 1971; Linna et al., 1972; Toivanen et al., 1972; Hoffmann-Fezer et al., 1977) have described the devastating effects of this drug on bursal morphology, most of the studies have focused on the lymphoid elements and largely ignored the epithelium. Rouse and Szenberg (1974), however, suggested that cyclophosphamide treatment resulted in chronic degenerative changes in the bursal epithelium. Histological changes seen with both light and electron microscopy by Sachs et al. (1979) included decreased size of the plicae, thinning of the FAE, hyperplasia of the interfollicular epithelium, decrease in the numbers of medullary lymphocytes, and virtual absence of cortical lymphocytes. They examined the function of the FAE by cloacally administering carbon particles and observed that bursae from treated chickens showed practically no uptake of carbon, in contrast to the effective transport seen by this epithelium in age-matched untreated normal chickens. In contrast to cyclophosphamide, which produced degenerative changes in already formed bursae, Beezhold et al. (1980) examined the effect of testosterone treatment on the FAE when given to 5-day embryos. This drug has been shown to inhibit the development of the bursa (Rao et al., 1960; Meyer et al., 1959; Glick and Sadler, 1961). Although treatment with testosterone resulted in a variable inhibition, in general, both the numbers and the size of the follicles were markedly decreased. When the follicles were present, they often were small and appeared embedded within the epithelium, similar to those described by Hoffmann-Fezer (1975) for normal 18- or 19-day embryos. Beezhold et al. (1980) reported that the interfollicular epithelium exhibited areas of marked hypertrophy and that heterophilic granulocytes often were observed within this epithelium. When present, the FAE was histologically similar to that in bursae of normal chickens. The function of the FAE was assessed at 1 week following hatching by cloacally administering HRP. When the FAE was found associated with follicles containing a substantial lymphoid component, uptake of HRP occurred. This uptake appeared normal at the electron microscopic level. In bursae completely devoid of follicles, no uptake by any of the epithelia was observed. These investigators also treated a group of newly hatched chicks with cyclophosphamide and evaluated the ability of the bursal FAE to pinocytose HRP. Again, it was observed that if a sufficient number of follicular lymphocytes were still asssociated with the FAE, uptake of HRP occurred. These results suggested that for the FAE to carry out endocytosis of HRP, a degree of lymphoid integrity within the follicle underlying the FAE is necessary. Thus, using two methods of chemical bursectomy, both of which spared portions of the bursa, it was shown that the bursal FAE was capable of endocytosing some materials present in the bursal lumen but only when there was an underlying follicle populated with lymphocytes.

From the above review it was clearly demonstrated that the avian bursa of

Fabricius functions as a peripheral lymphoid organ in addition to being a central lymphoid organ for the differentiation of B lymphocytes. In functioning as a peripheral lymphoid organ, it appears to carry out RES functions, such as clearing foreign materials from the gut lumen. The tissue responsible for this RES function was the specialized epithelium overlying the lymphoid follicles. It also was seen that a close interrelationship exists between the epithelium and the underlying lymphoid tissue. For example, during embryonic development of the bursa, it was shown that the epithelium had an inductive function in the differentiation of the lymphoid follicles. Thus, for the bursa to function as a central lymphoid organ, the lymphoid precursor cells were observed interacting with the epithelial cells before undergoing differentiation into B lymphocytes which then traffic to various peripheral tissues and become antibody producers. In contrast, for the bursa to function as a peripheral lymphoid organ, it was demonstrated that the epithelium had to remain in intimate contact with the underlying lymphoid follicle for it to have the capacity to take up, transport, and process luminal antigens. Therefore, the relationship between these two tissues results in a unique interaction which is necessary for the bursa to be fully functional.

5. CONCLUSIONS

Sufficient evidence has been presented in this review to support the conclusion that the differentiation of the RES, like any other developing system, is epigenetic in character. Thus, the various components, cells, fibers, and plasma proteins, are end products of differentiation and not causal components of the mechanisms of differentiation. It is apparent that differentiation of the RES is a very complex phenomenon, and because cells demonstrate endocytosis of some foreign substance does not necessarily indicate that they are either part of the RES or have their full functional competence. Similarly, clearance ability of one particular material cannot be used as a definitive endpoint of maturation. Also, uptake and even degradation of foreign substances are dependent upon components in the plasma proteins (opsonins) which, although they may be present in the embryo or neonatal chicken, may not have been synthesized by the developing organism but obtained from the hen through the yolk.

The resulting depression of RES function that was observed following surgical and chemical bursectomy was shown to be due to changes in the plasma opsonic proteins and not to alterations in macrophage functions. The deficiency in opsonins, due to bursectomy, was not confined to only a decrease in the immunoglobulins but also to some nonimmune opsonins which have been demonstrated to promote the phagocytosis of nonantigenic colloids. One such opsonin appears to be an α_2-macroglobulin which is probably the same as plasma fibrinonectin.

In reviewing the role of the bursa in clearing foreign materials from the gut lumen, it was seen that the RES function was carried out by the follicle-associated epithelium. This epithelium appeared to be especially adapted for the uptake of such substances. By means of chemical bursectomy, which did not totally destroy the bursa, it was seen that the epithelium was able to pinocytose some colloids if it remained in contact with a morphologically identifiable lymphoid follicle. The unique relationship between the epithelium and its underlying lymphoid tissue has not been fully clarified.

From this review of the functional ontogeny of the avian RES it is evident that most of the present knowledge is based primarily upon observations of phenomenological events. Thus, there is no clear understanding of what cellular or tissue interactions must take place in order to initiate differentiation of the RES. Also, little or no data as yet are available concerning how the developing RES may influence other maturing host defense systems. For example, from recent studies with mammalian cells there now is evidence that cells comprising the RES, in addition to clearing foreign substances and presenting antigens to lymphocytes, also function as modulators of immunocompetent cells. Apparently, the way in which the monocytic macrophages function in this capacity is by the secretion of biologically active materials (Unanue, 1976). Although most of these materials appear to be lymphoregulatory factors, they may have broader potential and affect also the vascular system (Polverini *et al.*, 1977). Of particular interest in considering maturation of RES function would be to evaluate the ability of the developing monocytic macrophage to secrete both thymocyte-differentiating factor and B-cell-activating factor as has been demonstrated *in vitro* to be a property of macrophages from adult mammals (Beller *et al.*, 1978). Likewise, the reciprocal situation should be evaluated whereby lymphocytes also produce factors which have been shown in *in vitro* cultures to cause activation and proliferation of macrophages (Hadden *et al.*, 1978). Thus far, neither monokines nor lymphokines have been studied in a developing system. It would seem plausible that cultures of embryonic chick cells or tissues would be advantageous for evaluating the regulatory role of macrophages on lymphocyte differentiation and of lymphocytes on macrophage ontogeny.

REFERENCES

Allen, C., Saba, T. M., and Molnar, J., 1973, Isolation, purification and characterization of opsonic protein, *J. Reticuloendothelial Soc.* **13**:410.

Beezhold, D. H., Sachs, H. G., and Van Alten, P. J., 1980, The influence of embryonic testosterone treatment on bursal epithelial pinocytotic activity, *Dev. Comp. Immunol.* **6**:121–130.

Beller, D. I., Farr, A. G., and Unanue, E. R., 1978, Regulation of lymphocytic proliferation and differentiation by macrophages, *Fed. Proc.* **37**:91.

Benacerraf, B., Biozzi, G., Halpern, B. N., and Stiffel, C., 1957, Physiology of phagocytosis of particles by the reticuloendothelial system, in: *Physiopathology of the Reticuloendothelial System* (B. N. Halpern, B. Benacerraf, and J. F. Delafresnaye, eds.), pp. 52–79, Thomas, Springfield, Ill.

Biozzi, G., Benacerraf, B., and Halpern, B. N., 1953, Quantitative study of granulopectic activity of the reticulo-endothelial system. II. A study of the kinetics of the granulopectic activity of the R.E.S. in relation to the dose of carbon injected, *Br. J. Exp. Pathol.* **34**:441.

Blaese, R. M., 1975, Macrophages and the development of immunocompetence, in: *The Phagocytic Cell in Host Resistance* (J. A. Bellanti and D. H. Dayton, eds.), pp. 309–319. Raven Press, New York.

Blumenstock, F. A., Saba, T. M., Weber, P. B., and Laffin, R., 1978, Physiological function of cold insoluble globulin: Identity with human opsonic alpha-2-SB glycoprotein, *J. Biol. Chem.* **253**:4287.

Bockman, D. E., 1979, Differentiation of epithelium, lymphocytes, granulocytes and erythrocytes in bursa of Fabricius transplanted to chorioallantoic membrane, *Dev. Comp. Immunol.* **3**:117.

Bockman, D. E., and Cooper, M. D., 1973, Pinocytosis by epithelium associated with lymphoid follicles in the bursa of Fabricius, appendix and Peyer's patches: An electron microscopic study, *Am. J. Anat.* **136**:455.

Bockman, D. E., and Cooper, M. D., 1976, Participation of follicle associated epithelium in differentiation of B-cells on avian and mammalian lymphoepithelial tissues, in: *Phylogeny of Thymus and Bone Marrow-Bursa Cells* (A. R. Wright and E. L. Cooper, eds.), pp. 277–286, Elsevier/North-Holland, Amsterdam.

Bockman, D. E., and Stevens, W., 1977, Gut associated lymphoepithelial tissues: Bidirectional transport of tracer by specialized epithelial cells associated with lymphoid follicles, *J. Reticuloendothelial Soc.* **21**:245.

Boyden, S. V., 1964, Cytophilic antibody in guinea pigs with delayed type hypersensitivity, *Immunology* **7**:474.

Brambell, F. W. R., 1970, *The Transmission of Passive Immunity from Mother to Young*, Elsevier/North-Holland, Amsterdam.

Chang, C.-F., and Hamilton, P. B., 1979, The thrombocyte as the primary circulating phagocyte in chickens, *J. Reticuloendothelial Soc.* **25**:585.

Chang, T. S., Glick, B., and Winter, R. A., 1955, Significance of the bursa of Fabricius of chickens in antibody production, *Poult. Sci.* **34**:1187.

Claman, H. N., and Mosier, D. E., 1972, Cell–cell interactions in antibody production, *Prog. Allergy* **16**:40.

Claman, H. N., Chaperon, E. A., and Triplett, R. F., 1966, Thymus–marrow cell combinations: Synergism in antibody production, *Proc. Soc. Exp. Biol. Med.* **122**:1167.

Cooper, M. D., Peterson, R., and Good, R. A., 1965, Delineation of the thymic and bursal lymphoid systems in the chicken, *Nature (London)* **205**:143.

Cooper, M. D., Peterson, R., South, M., and Good, R. A., 1966, The functions of the thymus system and the bursa system in the chicken, *J. Exp. Med.* **123**:75.

Cooper, M. D., Cain, W. A., Van Alten, P. J., and Good, R. A., 1969, Development and function of the immunoglobulin producing system, *Int. Arch. Allergy Appl. Immunol.* **35**:242.

Craig, S. W., and Cebra, J. J., 1971, Peyer's patches: An enriched source of precursors for IgA-producing immunocytes in the rabbit, *J. Exp. Med.* **134**:188.

Edmonds, R. H., 1970, Electron microscope studies on hemostatic process in bird embryos. II. *In vivo* phagocytosis by nucleated thrombocytes, *J. Ultrastruct. Res.* **30**:184.

Edwards, J. L., Murphy, R. C., and Cho, Y., 1974, On the development of the lymphoid follicles of the bursa of Fabricius, *Anat. Rec.* **181**:735.

Gabrielson, A. E., Pickering, R. J., Linna, T. J., and Good, R. A., 1973, Haemolysis in chicken serum II. Ontogenetic development, *Immunology* **25**:179.

Gabrielson, A. E., Linna, T. J., Weitekamp, D. P., and Pickering, R. J., 1974, Reduced hemolytic C1 activity in serum of hypogammaglobulinemic chickens, *Immunology* **27**:463.

Gershon, R. K., Cohen, P., Henchin, R., and Liebhaber, S. A., 1972, Suppressor T cells, *J. Immunol.* **108**:586.

Glick, B., 1971, Morphological changes and humoral immunity in cyclophosphamide-treated chicks, *Transplantation* **11**:433.

Glick, B., 1977, The bursa of Fabricius and immunoglobulin synthesis, *Int. Rev. Cytol.* **48**:345.

Glick, B., and Sadler, C. R., 1961, The elimination of the bursa of Fabricius and reduction of antibody production in birds from eggs dipped in hormone solutions, *Poult. Sci.* **40**:185.

Glick, B., Chang, T. S., and Jaap, R. G., 1956, The bursa of Fabricius and antibody production, *Poult. Sci.* **35**:224.

Glick, B., Sato, K., and Cohenour, F., 1964, Comparison of the phagocytic ability of normal and bursectomized birds, *J. Reticuloendothelial Soc.* **1**:442.

Good, R. A., Dalmasso, A. P., Martinez, C., Archer, O. K., Pierce, J. C., and Papermaster, B. W., 1962, The role of the thymus in development of immunologic capacity in rabbits and mice, *J. Exp. Med.* **116**:773.

Goodpasture, E. W., and Anderson, K., 1937, Problem of infection as presented by bacterial invasion of the chorioallantoic membrane of chick embryos, *Am. J. Pathol.* **13**:149.

Goodpasture, E. W., and Buddingh, G. J., 1935, The preparation of antismallpox vaccine by culture of the virus in the chorioallantoic membrane of chick embryos, and its use in human immunization, *Am. J. Hyg.* **21**:319.

Hadden, J. W., Sadlik, J. R., and Hadden, E. M., 1978, The induction of macrophage of proliferation *in vitro* by a lymphocyte-produced factor, *J. Immunol.* **121**:231.

Halpern, B. N., 1959, The role and function of the reticulo-endothelial system in immunological processes, *J. Pharm. Pharmacol.* **11**:321.

Hoffmann-Fezer, G., 1975, Der Einfluss der hormonellen Bursektomie auf die embryonale Entwicklung der follikel in der Bursa Fabricii des Huhnes, *Avian Pathol.* **4**:205.

Hoffmann-Fezer, G., Hoffmann, R., Fiedler, H., and Losch, V., 1977, Recovery of the bursa of Fabricius in chickens after cyclophosphamide treatment, *Int. Arch. Allergy Appl. Immunol.* **53**:206.

Holbrook, K. A., Perkins, W. D., and Glick, B., 1974, The fine structure of the bursa of Fabricius: "B" cell surface configuration and lymphoepithelial organization as revealed by scanning and transmission electron microscopy, *J. Reticuloendothelial Soc.* **16**:300.

Hughes, C. L., and Henderson, D. C., 1977, Induction of avian immunological responsiveness following cloacal drinking of immunogen, *Immunol. Commun.* **6**:195.

Jenkin, C. R., 1963, The effect of opsonins on intracellular survival of bacteria, *Br. J. Exp. Pathol.* **44**:47.

Jungherr, E. L., and Terrell, N. L., 1948, Naturally acquired passive immunity to infectious bronchitis in chicks, *Am. J. Vet. Res.* **9**:201.

Karthigasu, K., and Jenkin, C., 1963, The functional development of the reticulo-endothelial system of the chick embryo, *Immunology* **6**:255.

Karthigasu, K., Reade, P., and Jenkin, C., 1965, The functional development of the reticulo-endothelial system. III. The bactericidal capacity of fixed macrophages of foetal and neonatal chicks and rats, *Immunology* **9**:67.

Katz, D. H., and Benacerraf, B., 1972, The regulatory influence of activated T cells on B cell responses to antigen, *Adv. Immunol.* **15**:1.

Kent, R., 1961, The development of the phagocytic activity of the reticulo-endothelial system in the chick, *J. Embryol. Exp. Morphol.* **9**:128.

Linna, T. J., Frommel, D., and Good, R. A., 1972, Effects of early cyclophosphamide treatment on the development of lymphoid organs and immunoglobulin function in the chicken, *Int. Arch. Allergy Appl. Immunol.* **42**:20.

Marquette, D., Molnar, J., Yamada, K., Darby, S., Van Alten, P., and Schlesinger, D., 1979, Opsonic activity of chicken plasma and fibroblast cell surface fibronectins, *J. Reticuloendothelial Soc.* **26**:35a.

Marquette, D., Molnar, J., Yamada, K., Schlesinger, D., Darby, S., and Van Alten, P., 1981, Phagocytosis-promoting activity of avian plasma and fibroblastic cell surface fibronectins, *Mol. Cell. Biochem.* **36**:147.

Meyer, R. K., Rao, R. A., and Aspinall, R. L., 1959, Inhibition of the development of the bursa of Fabricius in the embryos of the common fowl by 19-nortestosterone, *Endocrinology* **64**:890.

Miller, J. F. A. P., 1961, Immunological function of the thymus, *Lancet* **2**:748.

Mitchison, N. A., 1962, Tolerance of erythrocytes in poultry: Induction and specificity, *Immunology* **5**:341.

Molnar, J., McLain, S., Allen, C., Gara, A., and Gelder, F., 1977, The role of an α_2-macroglobulin of rat serum in the phagocytosis of colloidal particles, *Biochim. Biophys. Acta* **493**:37.

Molnar, J., Gudewicz, P., Ming-Zong, L., Credo, B., Siefring, G., and Lorand, L., 1979, Role of plasma fibronectin in phagocytosis of gelatin-coated colloids, *Fed. Proc.* **38**:303.

Muehleman, C., and Van Alten, P. J., 1979, Induction of immunologic tolerance in newly hatched chicks following cloacal administration of antigen, *Fed. Proc.* **38**:1009.

Mueller, A. P., 1964, Antibody studies in hormonally and surgically bursectomized chickens, in: *The Thymus in Immunobiology* (R. A. Good and A. E. Gabrielsen, eds.), pp. 359–373, Harper & Row (Hoeber), New York.

Mueller, A. P., Wolfe, H. R., and Meyer, R. K., 1960, Precipitin production in chickens. XXI. Antibody production in bursectomized chickens and in chickens injected with 19-nortestosterone on the fifth day of incubation, *J. Immunol.* **85**:172.

Mueller, A. P., Wolfe, H. R., and Cote, W. P., 1964, Antibody studies in hormonally and surgically bursectomized chickens, in: *The Thymus in Immunobiology* (R. A. Good and A. E. Gabrielson, eds), pp. 359–373, Harper and Row, New York.

Naukkarinen, A., Arstila, A., and Sorvari, T., 1978, Morphological and functional differentiation of the surface epithelium of the bursa fabricii in chicken, *Anat. Rec.* **191**:415.

Oláh, I., and Glick, B., 1978, The number and size of the follicular epithelium (FE) and follicles in the bursa of Fabricius, *Poult. Sci.* **57**:1445.

Perez del Castillo, C. E., 1957, Symposium, in: *Physiopathology of the Reticuloendothelial System* (B. N. Halpern, B. Benacerraf, and J. F. Delafresnaye, eds.), p. 312, Thomas, Springfield, Ill.

Peterson, R., Cooper, M. D., and Good, R. A., 1965, The pathogenesis of immunologic deficiency diseases, *Am. J. Med.* **38**:579.

Phillips-Quagliata, J. M., Levine, B. B., and Uhr, J. W., 1969, Studies on the mechanism of binding of immune complexes to phagocytes, *Nature (London)* **222**:1290.

Pierce, C. W., Kapp, J. A., Wood, D. D., and Benacerraf, B., 1974, Immune responses *in vitro*. X. Function of macrophages, *J. Immunol.* **112**:1181.

Polverini, P. J., Cotran, R. S., Gimbrone, M. A., Jr., and Unanue, E. R., 1977, Activated macrophages induce vascular proliferation, *Nature (London)* **269**:804.

Rao, M. A., Aspinall, R. L., and Meyer, R. K., 1960, Effect of dose and time of administration of 19-nortestosterone on the differentiation of lymphoid tissue in the bursa Fabricii of chick embryos, *Endocrinology* **70**:159.

Roitt, I. M., Greaves, M. F., Torrigianti, G., Brostoff, J., and Playfair, J. H. L., 1969, The cellular basis of immunological responses, *Lancet* **2**:367.

Rouse, B. T., and Szenberg, A., 1974, Functional and morphological observations on the effect of cyclophosphamide on the immune response of the chicken, *Aust. J. Exp. Biol. Med. Sci.* **52**:873.

Rowley, D., Phagocytosis, in: *Advances in Immunology* (W. H. Taliaferro and J. H. Humphrey, eds.), pp. 241–264, Academic Press, New York.

Rudnick, D., 1955, Teleosts and birds, in: *Analysis of Development* (B. H. Willier, P. A. Weiss, and V. Hamburger, eds.), pp. 297–314, Saunders, Philadelphia.

Saba, T. M., and DiLuzio, N. R., 1969, Reticuloendothelial blockade and recovery as a function of opsonic activity, *Am. J. Physiol.* **216**:197.

Sabet, T., Hsia, W., Waltenbaugh, C., Stanisz, M., Van Alten, P. J., and El-Domeiri, A., 1975, Impairment of phagocytosis in bursectomized chickens, *in vitro* study, *J. Reticuloendothelial Soc.* **18**:1a.

Sabet, T., Hsia, W., Stanisz, M., El-Domeiri, A., and Van Alten, P. J., 1977, A simple method for obtaining peritoneal macrophages from chickens, *J. Immunol. Methods* **14**:103.

Sachs, H. G., Beezhold, D. H., and Van Alten, P. J., 1979, The effect of cyclophosphamide on structure and function of the bursal epithelium, *J. Reticuloendothelial Soc.* **19**:3.

Schaffner, T., Hess, M. W., and Cottier, H., 1974, A reappraisal of bursal functions, *Ser. Haematol.* **VII**:568.

Solomon, J., 1966, The appearance and nature of opsonins for goat erythrocytes during the development of the chicken, *Immunology* **11**:79.

Solomon, J., 1968, Immunity to *Salmonella gallinarum* during ontogeny of the chicken. I. Onset of resistance to infection; the minor role of opsonins, *Immunology* **15**:197.

Sorvari, R., and Sorvari, T. E., 1977, Bursa Fabricii as a peripheral lymphoid organ: Transport of various materials from the anal lips to the bursal lymphoid follicles with reference to its immunological importance, *Immunology* **32**:499.

Sorvari, T., Sorvari, R., Ruotsalainen, P., Toivanen, A., and Toivanen, P., 1975, Uptake of environmental antigens by the bursa of Fabricius, *Nature (London)* **253**:217.

Spratt, N. T., Jr., 1946, Formation of the primitive streak in the explanted chick blastoderm marked with carbon particles, *J. Exp. Zool.* **103**:259.

Spratt, N. T., Jr., 1955, Studies on the organizer centers of the early chick embryo, in: Aspects of Synthesis and Order in Growth (D. Rudnick, ed.), pp. 209–231, Princeton University Press, Princeton, N.J.

Stolfi, R. L., Fugmann, R. A., Jensen, J. J., and Sigel, M. M., 1971, A C1-fixation method for the measurement of chicken antiviral antibody, *Immunology* **20**:299.

Subba Rao, D. S. V., and Glick, B., 1975, Antibody and cell-mediated immunity in phytohemagglutinin-treated chickens, *Int. Arch. Allergy Appl. Immunol.* **48**:30.

Toivanen, P., Toivanen, A., and Good, R. A., 1972, Ontogeny of bursal function in chicken. I. Embryonic stem cell for humoral immunity, *J. Immunol.* **109**:1058.

Unanue, E. R., 1976, Secretory function of mononuclear phagocytes, *Am. J. Pathol.* **83**:395.

Van Alten, P. J., and Meuwissen, H. J., 1972, Production of specific antibody by lymphocytes of the bursa of Fabricius, *Science* **176**:45.

Van Alten, P. J., Cain, W. A., Good, R. A., and Cooper, M. D., 1968, Gamma globulin production and antibody synthesis in chickens bursectomized as embryos, *Nature (London)* **217**:358.

Van Alten, P. J., Waltenbaugh, C. R., and Sachs, H. G., 1977, Regulation of bursal lymphocyte antibody production *in vitro*, in: *Developmental Immunobiology* (J. B. Solomon and J. D. Horton, eds.), pp. 403–409, Elsevier/North-Holland, Amsterdam.

Waltenbaugh, C. R., and Van Alten, P. J., 1974, The production of antibody by bursal lymphocytes, *J. Immunol.* **113**:1079.

Waltenbaugh, C. R., Allen, C., Molnar, J., Sabet, T. Y., and Van Alten, P. J., 1976, Impairment of clearance by the reticuloendothelial system of bursectomized chickens, *J. Reticuloendothelial Soc.* **19**:3.

Waltenbaugh, C. R., Sachs, H. G., and Van Alten, P. J., 1977, Antibody production in organ culture by bursae of mature chickens and the development of immunological competence, *Dev. Comp. Immunol.* **1**:353.

Warner, N. L., and Szenberg, A., 1962, Dissociation of immunological responsiveness in fowls with a hormonally arrested development of lymphoid tissues, *Nature (London)* **194**:146.

Wright, A. E., and Douglas, S. R., 1904, An experimental investigation of the role of blood fluids in connection with phagocytosis, *Proc. R. Soc. London* **72**:357.

20

Ontogeny of Cellular Immune Reactivity in the Mouse

ROBERT AUERBACH, AMIELA GLOBERSON,
and TEHILA UMIEL

1. INTRODUCTION

Developmental immunology, once a peripheral area of investigation, has become a central arena for studies directed at understanding the nature of molecular diversification, for analysis of the expansion of specific antibody-forming clones, for investigations into the sequential changes accompanying immunoglobulin production following antigenic challenge, and indeed for research aimed at resolving the basic steps of gene replication, transcription, and translation as these are influenced during the maturation of the cells that comprise the B-cell system responsible for antibody production. The literature dealing with the ontogeny of B cells is overwhelming as is the information that is becoming available concerning the postnatal and poststimulatory changes that occur in B cells even after functional maturation has been largely completed.

This review, however, will not address the question of B-cell ontogeny, and the reader is referred instead to the several excellent recent symposia and reviews that together can provide a reasonable coverage of this important aspect of developmental immunology (Seidman and Leder, 1978; Siskind et al., 1979; Horton, 1980). In the context of the present volume, which attempts to bring together the concepts of phylogeny and ontogeny, an analysis of the development of cell-mediated rather than antibody-mediated immune reactivity seems more appropriate. If indeed a case is made for grouping together phylogeny and ontogeny (beyond the legacy of the Haeckelian school of philosophy), it is that

---

ROBERT AUERBACH • Department of Zoology, University of Wisconsin, Madison, Wisconsin 53706. AMIELA GLOBERSON and TEHILA UMIEL • Department of Cell Biology, The Weizmann Institute of Science, Rehovot, Israel. Original studies reported in this paper have been supported in part by grants from the US–Israel Binational Science Foundation (A.G. and R.A.). Additional support has been received from the National Institutes of Health (R.A.) and the Ministry of Health, Israel (T.U.).

cell recognition systems have either shown a remarkable type of conservation through evolution or that analogous systems have been required to deal with the basic problem of recognition of self and defense against non-self.

If one were to begin conceptually with a teleological approach to the study of T-cell functions in the embryo, one would ask what activities are required, immune-type in nature, to permit the embryo to pass through the sequence of events leading to maturation of the definitive immune system. Just as different hemoglobins are best suited for embryonic, fetal, or adult oxygen transport, and just as different isoenzymes may be of greater or lesser advantage as the type of glycolysis changes during ontogeny, so too the defense mechanisms required to operate at various stages of embryogenesis may call sequentially for a variety of functional effector systems.

The embryo, even within the uterus, must be able to either overcome or suppress maternally derived lymphoid cell activities directed against embryonic or paternal alloantigens; it may need to recognize and respond to placentally transmitted infectious agents, and it may be required to cope with foreign material introduced through the uterine tissues or the maternal blood to which the extraembryonic membranes are exposed. Toxins may need to be neutralized; immune-type recognition may be a necessary prelude to or adjunct of induction of specific unresponsiveness or clonal deletion; "natural" cytotoxic reactions may be critical in protection against transplacental carcinogenesis; immune-mediated vasculogenesis may be an integral part of the sequence of responses needed to permit scavenger-cell access to sites of environmental insult at various stages of development. And it has yet to be determined what pattern of ontogenetic evolution is associated with the initiation of the sequence of selective events that collectively are responsible for the establishment of the mature B-cell and T-cell repertoire.

Thus, although this chapter deals with the ontogeny of T-cell functions, we have of necessity broadened our approach to include cell-mediated immune functions wherever they can be detected in the embryo. We will discuss effector cell activities that include mixed leukocyte culture responses (MLC), cell-mediated lympholysis (CML), and graft-versus-host (GVH) reactions; we will examine the development of the ability of effector cells to induce neovascularization (lymphocyte-induced angiogenesis, LIA) following allogeneic stimulation; we will survey the capacity of the embryo to generate suppressor cells or factors; and we will review to the extent information exists the ontogeny of "natural" immunity as this impinges on the development of cellular immune reactivity.

2. EMBRYONIC EFFECTOR ORGANS

2.1. THE EMBRYONIC YOLK SAC: FIRST EFFECTOR ORGAN OF THE IMMUNE SYSTEM

There is always an intrinsic interest in determining how early a particular developmental or functional event can be traced back in ontogeny. The problem in the case of cell-mediated immune-type activity is to a large extent one of definition. Recognition of self vs. non-self, for example, is a key event throughout the process of fertilization. Sperm–egg recognition follows many of the operational rules of immune recognition and response: specificity, a limited kind of memory, cell activation, and induction of cell division all accompany the fertilization reaction as seen in species specificity, blocks to polyspermy, fertilization membrane and cortical changes accompanying sperm contact with the egg surface, and induce cleavage (cf. Auerbach, 1970; cf. also Ginsburg as cited in Auerbach, 1977). Similarly, cell recognition, selective cell adhesions, and specific cellular interactions accompany all of the major processes taking place during early embryogenesis.

On the other hand, if we restrict our definition to include only the functional activity of differentiated lymphocytes, then by that very definition we eliminate consideration of cell-mediated immunity as it might be expressed in the embryo prior to the formation of structurally acceptable lymphoid cells as these become recognizable entities within defined effector organs. A useful middle course between these two extremes seems to us to be a consideration of the embryonic yolk sac, which is a readily identifiable, organized structure, whose cells collectively might well comprise the first effector organ for immune-type functional activity and whose time of appearance as well as location make it a prime candidate as the embryo's first line of defense as postulated by the teleological arguments presented earlier.

In looking at the yolk sac as an effector organ, we should at once exclude from discussion the question of yolk sac vs. non-yolk sac origin of thymic or bursal-type lymphocytes. Whether the yolk sac gives rise to any (Moore and Owen, 1965) or all (Moore and Owen, 1967; Le Douarin and Jotereau, 1975) lymphoid stem cells, or whether there is a prior (Dieterlen-Lièvre et al., 1980; Martin et al., 1980) or totally exclusive (Tompkins et al., 1980, Turpen, 1980) embryonic origin of such cells has no bearing on the question of whether the yolk sac at various times during embryogenesis serves a *functional* role as measured by assays for cell-mediated immune-type responses. We will look, in turn, simply at this latter aspect and ask whether cells contained within the embryonic yolk sac can, in fact, generate GVH reactions, whether they can induce angiogenesis following allogeneic stimulation, whether they can manifest proliferative activity in response to such stimulation, and whether they can carry out, without overt stimulation, natural effector cell functions such as antitumor-directed cytotoxicity.

Studies of Hofman and Globerson (1973) have shown that yolk sac cells can induce a GVH reaction *in vitro*, as measured by induction *in vitro* of enlargement of allogeneic but not of syngeneic neonatal splenic fragments. This ability, determined initially by physical measurements, to establish splenomegaly was confirmed by Auerbach (1976) who used a radioisotope incorporation assay to permit more precise quantitation of the *in vitro* GVH reaction. Hofman and Globerson (1976) also demonstrated that yolk sac cells could induce GVH reactions *in vivo* as determined by their ability to induce regional lymph node enlargement in allogeneic but not syngeneic combinations. In contrast to this activity, a response of yolk sac cells in MLC assays, or following mitogen stimulation, was not observed (Hofman, 1975), but this inability should not be considered altogether conclusive in that yolk sac cells *in vitro* are at any rate considerably more active mitotically than are adult splenocytes or lymph node cells.

Another manifestation of cell-mediated immunity is the ability of stimulated lymphocytes to induce neovascularization or angiogenesis. LIA has been demonstrated to be evoked by T lymphocytes in response to stimulation by allogeneic cells (Sidky and Auerbach, 1975). Specifically, the reaction has been determined to be based on recognition primarily of I-region-associated determinants (Auerbach and Sidky, 1979a), and to be carried out by Thy-1^+, Lyt-$1^+,2,3^-$, mature lymphocytes (Roehm, Sidky, and Auerbach, unpublished observations). LIA has been measured by three assay methods: induction of neovascularization on the chick chorioallantoic membrane, eliciting of newly visualizable blood vessels following intradermal inoculation into adult, irradiated mice, and the induction of new capillaries from the limbal vasculature following implantation of lymphoid tissue into the cornea of adult mice (Sidky and Auerbach, 1975; Auerbach *et al.*, 1976; Muthukkaruppan and Auerbach, 1979). The latter two assays are carried out by implantation of parental strain cells into semiallogeneic or irradiated, allogeneic host animals.

The results of LIA assays involving yolk sac cells are summarized in Table 1. Yolk sac cells from 10-day embryos can elicit LIA responses following allogeneic stimulation, and the reaction is manifested in a manner comparable to that seen for adult splenic cells. Thus, LIA as well as the induction of splenomegaly attest to the ability of embryonic yolk sac cells to carry out effector cell functions following stimulation by allogeneic or semiallogeneic cells.

Once again, further experiments are needed. For example, while syngeneic controls serve to prove that allogeneic stimulation is prerequisite to LIA induction by yolk sac cells, it has not been determined whether allogeneic stimulation is I-region dependent for yolk sac cells as it is for mature T cells, nor has it been determined whether prior exposure of yolk sac cells *in vitro*, or possibly by

TABLE 1. ANGIOGENESIS INDUCTION BY MOUSE YOLK SAC GRAFTS[a]

Expt	Tissue source	Method of assay	No. positive/total
1	10-day yolk sac	Intracorneal graft into allogeneic hosts	5/5
		Intracorneal graft into syngeneic hosts	0/4[b]
2	12-day yolk sac	CAM graft	6/8
	12-day brain		0/2
	12-day spinal cord		0/2
3[c]	9-day yolk sac	CAM graft	0/9
		Intradermal injection into allogeneic hosts	15/15[d]

[a]Table reproduced from Dahl *et al.* (1980).
[b]3/4 grafts showed transient reactions (2 days) in contrast to positive reactions which always persist for at least 5 days.
[c]Data from Auerbach and Sidky (1979b).
[d]11/12 syngeneic controls showed a weak vascular reaction.

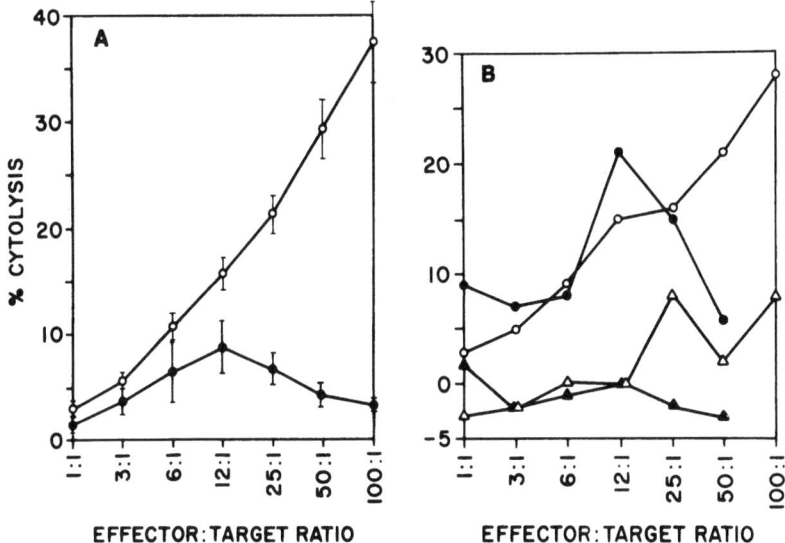

FIGURE 1. Natural cytotoxicity of embryonic mouse yolk sac cells. (A) Natural cytotoxicity of BALB/c-derived effector cell populations against YAC-1 target cells. Data points represent the mean ± S.E. of seven experiments. Effector cells obtained from adult spleen (○) or 10-day yolk sac (●). (B) Natural cytotoxicity of adult BALB/c spleen and BALB/c 10-day embryonic yolk sac against target cell lines of differing susceptibilities to NK lysis. Adult splenic cells against YAC-1 targets (○) or P815 targets (△). Ten-day yolk-sac-derived cells against YAC-1 targets (●) or P815 targets (▲). (From Dahl et al., 1980. For further control data see Dahl, 1980).

selection of appropriate maternal phenotype, alters LIA responsiveness as has been shown for adult lymphocytes (Sidky and Auerbach, 1979a). The response of yolk sac cells following irradiation or mitomycin C treatment has also not been tested and such tests could provide additional useful correlates (cf. Auerbach and Shalaby, 1973; Sidky and Auerbach, 1979b).

The question of the relation between natural cytotoxicity and immune reactions is a complex one, for a wide range of cytotoxic reactions may be obtained depending on the test cells employed and the assay systems utilized. Dahl et al. (1980) have approached the question of possible expression of yolk-sac-mediated natural cytotoxicity by employing the test procedures developed by Herberman et al. (1975; Herberman and Holden, 1978; cf. also Kiessling et al., 1975) to delineate the nature of NK cells. Specifically, ^{51}Cr-labeled YAC-1 tumor cells were combined with yolk sac cells and tested in a chromium release assay. In analogy to the adult NK system, P815 mastocytoma cells, insensitive to killing by NK cells, were used as negative controls for the targets, while embryonic-type teratoma cells were used as negative controls for the effector yolk sac cells. The results are reproduced in Figure 1: Embryonic yolk sac cells were as capable of killing YAC-1 cells as were splenic cells, up to an effector : target cell ratio of 12 : 1 or 25 : 1. At this point, yolk sac cells reached their peak efficiency, while adult splenic cell cytotoxicity continued to increase until a 200 : 1 effector : target cell ratio was obtained. The fact that yolk sac cells could kill YAC-1 but not P815 cells suggests that the observed functional activity of the yolk sac was "NK-like" (Dahl et al., 1980; Dahl, 1980), and the fact that neither teratoma cells nor embryonic liver or thymic cells showed such NK-like activity attests to the active role played quite uniquely by yolk sac cells.

It has been suggested that NK activity may be due to "prothymocyte" action, in that NK activity is high in *nu/nu* mice known to have T-cell precursors but lacking mature T cells, and the accompanying finding that NK activity declines in *nu/nu* mice following thymus-induced maturation of T-cell systems (Herberman et al., 1979). In terms of a primitive immune response, one might well argue that natural cytotoxicity provides a first innate rather than adaptive response mechanism to deal with unwanted cells (cf. Auerbach et al., 1980). Similarly, both the GVH reaction (Roehm, 1977) and the LIA reaction (cf. Auerbach and Sidky, 1979a) may in large part be mediated by soluble factors released from activated effector cells (lymphokines). It may well be that the earliest manifestations of adaptive immunity both phylogenetically and ontogenetically may be seen in these types of cell-mediated immune reactions (Auerbach, 1982).

2.2. THE EMBRYONIC THYMUS AS AN EFFECTOR ORGAN

Although the role of the thymus in the generation and maturation of T cells has long been accepted, and although thymic humoral activity has been demonstrated repeatedly as a major function of the thymus in the development of immunocompetence by stem cells, little attention has been placed on the thymus as an effector organ in its own right. Neonatal thymectomy has been seen primarily as a means of abrogating subsequent (adult) immune functions of the peripheral lymphoid tissues, the thymic microenvironment has been considered primarily to act as a means of permitting or restricting maturation of T cells that would be expected to function only secondarily after emigration, and even thymic restitution experiments have been employed primarily to provide stem cells that could differentiate or be influenced by antigens to differentiate selectively to permit the evolution of functional T cells.

That the generative and regulatory roles of the thymus are of major importance is not argued at all. Yet it is also important to consider separately the question of thymic effector functions. For example, the neonatal thymus has long been known to contain cells that are competent to initiate GVH reactions, and indeed the newborn mouse thymus is as efficient at inducing GVH reactions as is the adult thymus (Cohen et al., 1963). However, the thymus was found to be only 5–10% as efficient in GVH–inducing capacity as is the adult spleen (cf. Elkins, 1971; Umiel & Auerbach, 1973). A priori this could have been explained on the basis of lower activity of each individual thymic cell, or the existence of a minor population of cells that are as active as those of the spleen. The studies of Reisner et al. (1976) have clearly demonstrated that a specific minor subset of thymic cells, defined by binding of the PNA lectin is highly active. In a similar manner, assays by serological markers (Cantor & Weissman, 1976) also point to a minority cell population for GVH effector activity. Yet, the neonatal spleen does not have the capacity to induce a GVH reaction *in vivo*, and even the spleen

TABLE 2. ABILITY OF SEMIALLOGENEIC EMBRYONIC, NEWBORN, OR ADULT THYMIC CELLS TO INDUCE SPLENOMEGALY (GRAFT-VERSUS-HOST REACTION) IN VITRO[a]

Source of thymic cells	Number of cells	Number of experiments	Number of cultures	Spleen index
Adult	1.0×10^6	24	145	1.25[b]
Newborn	1.0×10^6	28	212	1.25[b]
19-day embryo	1.0×10^6	4	41	1.01
18-day embryo	1.0×10^6	10	107	0.99
17-day embryo	1.0×10^6	5	47	1.12
16-day embryo	1.0×10^6	5	38	0.99
13-day embryo, after 7 days in organ culture	1.0×10^6	1	4	1.45[b]
Adult	0.2×10^6			(1.00)[c]
Newborn	0.2×10^6			(0.97)[c]
13-day embryo, after 7 days in organ culture	0.2×10^6	5	20	1.22[b]
13-day embryo, 850 R X-rays, then 7 days in vitro	0.2×10^6	7	27	1.33[b]
13-day nu/nu embryo, 7 days in vitro	0.2×10^6	3	4	1.29[d]
13-day embryo, 1–3 days in hydrocortisone, 4–6 days further culture	0.2×10^6	5	18	1.18[d]

[a]Table reproduced from Chakravarty et al. (1975).
[b]Highly significant; see Auerbach and Shalaby (1973) for validation and experimental detail.
[c]From previous experiments.
[d]$0.05 > p < 0.10$.

of the young mouse at 1 week of age is only marginally active (Simonsen, 1962a; cf. Umiel and Auerbach, 1973). Thus, thymic GVH reactivity should be considered within the context of the neonatal mouse as representing the *major* source of immunoreactive GVH-inducing cells of the neonate.

One may wonder whether this thymic reactivity is due to a transient sojourn of competent cells obtained from peripheral sources, although such a source in fact would not negate the effector organ role of the thymus. On the other hand, *in vitro* studies have established (Chakravarty et al., 1975; cf. Robinson, 1980) that differentiation of thymic cells to immunocompetence can occur *in vitro* following explantation of thymic rudiments from 13-day embryos (Table 2).

Given the fact that thymic cells capable of developing immunocompetence for GVH reactivity are already present in the early thymic rudiment, we may ask at what stage in development the functional maturation of such stem cells has actually occurred. When GVH reactivity of thymic cells was tested at various times during development, it was found that reactivity is first manifest in the 24-hr period preceding birth, i.e., at about 19 days after fertilization, or on the day of birth. Before that time, thymic GVH reactivity is either not present (Chakravarty et al., 1975) or is compromised by the simultaneous presence of suppressor cells (see below).

Wu (1978; Wu et al., 1975) carried out extensive studies on the ontogeny of CML and MLC responsiveness in the mouse thymus (cf. Auerbach, 1976, for discussion of earlier results). These studies, subsequently confirmed by Widmer and Cooper (1979) and Pilarski (1977), indicated that the thymus on the day of birth already has the capability of responding well in MLC reactions; moreover, the MLC reactivity could be seen with H-2 congenic mouse strains, indicating that the major histocompatibility complex (MHC) was involved in the MLC recognition by thymic cells of the neonate. CML reactivity was found to be established almost concurrently with the development of MLC competence. Once again, maturation of thymic effector cells occurred prior to splenic effector cell development, and, moreover, the number of such cells in the thymus further emphasizes the potential role of the thymus as an effector organ in the neonatal mouse.

Chakravarty (1977), in his study of ontogeny of thymic cell function, assessed the ability of Con A to stimulate thymic helper cell activity, as expressed by the ability of Con A-treated cells to help B cells in the response to SRBC *in vitro*. Overnight incubation of thymic cells with Con A had previously been shown to be adequate for generating adult thymic helper cell activity in the SRBC system; Chakravarty found that not only were neonatal thymic cells as capable of furnishing T-cell help as were adult thymic cells, but that such help could be generated with embryonic thymic cells as well. Beginning with 16-day embryonic thymic cells, corresponding to the first appearance of small lymphocytes in the thymus, helper cell function could be demonstrated (Chakravarty et al., 1975; Chakravarty, 1977).

In a sense, the requirement of thymic cells of all ages for activation by Con A argues for a precursor rather than an effector cell in this system. Chakravarty had shown that Con A activation leads to a change in expression (maturation?) of Thy-1 and H-2 antigens similar to that seen when thymic cells become peripheralized. On the other hand, the experiments at the very least argue for a mitogen responsiveness associated with T-cell functions, and the fact that Con A activation involves rapid alterations (6 hr) seems to support the idea that cells from the thymus are, at least, ready to act as helper cells with little in the way of additional steps of maturation even as early as the 16th day of embryonic development.

Two tests for thymic effector cell activity have been negative. Dahl (1980) has not been able to detect NK activity in thymic cells from embryonic or neonatal rudiments. Sidky and Auerbach (1979b) have failed to detect LIA activity of thymocytes from embryonic or neonatal rudiments tested on the chorioallantoic membrane or in the intradermal LIA assay system. The LIA test might logically have been extended to use Con A-activated thymocytes, but the additional problems of Con A carry-over in the face of a highly Con A-sensitive endothelial cell response has made this type of experiment difficult to interpret.

When one compares yolk sac and thymic effector cell functions, the presence of NK-like and LIA activity in the former and MLC and CML responsiveness in the latter suggest that the pattern of effector organ activity is quite different. It would be most useful to obtain information on other lymphokine systems such as elaboration of allogeneic effect factor, migration inhibition factor, and leukocyte adherence factor, for it may well be that during the course of ontogeny the type of lymphokinelike substances produced may be changing as the functional requirements of the developing embryo change and as different cell lineages become mature.

2.3. THE EMBRYONIC LIVER AND THE DEVELOPMENT OF T-CELL-TYPE IMMUNE FUNCTIONS

The mammalian embryonic liver has been recognized as a B-cell organ analogous to the fowl bursa of Fabricius, since it is the major organ in the embryo in which B-cell precursors first appear. Hence, embryonic liver cells are the first to express surface immunoglobulin (Nossal and Pike, 1973; Spear et al., 1973). Furthermore, even prior to the expression of Ig-bearing cells, the embryonic liver contains precursor B cells that can mature to Ig-bearing cells in vitro (Owen et al., 1974), and can restore B-cell function to lethally irradiated animals (Umiel and Globerson, 1974; Phillips and Melchers, 1976; Rosenberg and Cunningham, 1977). However, the question of B-cell maturation in relation to the mouse embryonic liver is not one to be addressed in this chapter.

On the other hand, the question of whether the liver serves in ontogeny only as a B-cell developing site or whether it also contributes to T-cell development or engages in T-cell-type functions needs consideration. Since the concept of a liver contribution to the T-cell lineage has not been unequivocally established, we will discuss the evidence that the embryonic liver can give rise to mature cells capable of carrying out cell-mediated effector cell functions, suggesting that the liver contains precursors of T cells, and we will consider the role of the liver as an effector organ in its own right.

The idea that the fetal liver can contribute to the establishment of immune reactivity, including antibody production and allograft rejection, was in fact shown in experiments employing fetal liver radiation chimeras before the dichotomy of T- and B-cell lineages was fully recognized in mammals (cf. Van Bekkum and de Vries, 1967). A precise evaluation of the development of the T-cell compartment requires a different methodological approach, and we shall focus in this discussion on studies in which assessment of cells clearly characterized as T lymphocytes was attempted. One should distinguish in this respect between direct observation of T-cell functions produced by fetal liver cell populations and studies in which the fetal liver cells do not manifest T-cell properties, but rather acquire them during the course of the experiment.

Observations of the first category were reported on human fetal liver cells. Accordingly, the fetal liver cells could react in an MLC system (Carr et al., 1973; Stiles et al., 1974; Rabinowich and Klajman, 1982). However, mouse fetal liver cells seem to lack such T-cell properties (Umiel, unpublished). Studies on the mouse system thus fall into the second category, namely in following the appearance of T-cell properties in cells derived from the fetal liver.

The most direct line of search for T-cell precursors in the liver involved a study of the induction of Thy-1 membrane antigens (Scheid et al., 1975). Thus, whereas 16-day fetal liver cells did not initially seem to express this antigen,

Thy-1$^+$ cells became overt after treatment with thymosin (Goldstein and White, 1973). The relevance of the thymosin effect was questionable, though, in suggesting a thymus-dependent process, since other agents, not derived from the thymus, were also capable of inducing exposure of the Thy-1 marker on the membrane (Cohen and Patterson, 1975).

Most of the studies concentrated on T-cell functions rather than on cell membrane markers. It has been shown that embryonic liver cells (12–14 days' gestation) could be induced to elicit a GVH response by coculturing of the liver cells with thymic explants (Umiel et al., 1968). The assay (Auerbach and Globerson, 1966) was based on measuring the *in vitro* correlate of the splenomegaly component of the GVH reaction (Simonsen, 1962a,b) as already discussed in Sections 2.1 and 2.2. Immunogenetic analysis pointed out that indeed the effector cells were of hepatic origin. In subsequent studies (Globerson et al., 1975a), 17- to 19-day fetal or newborn mouse liver cells acquired the ability to produce such a GVH response upon incubation with thymic extracts (THF) prepared according to the method of Trainin and Small (1970). It may be recalled that 9-day embryonic yolk sac cells can elicit this type of GVH reaction even without THF or thymic treatment, and the observation that fetal liver cells require thymic induction is thus most intriguing. It might be useful to expand these studies now employing more sensitive methods for analysis of the GVH reaction (cf. Auerbach and Shalaby, 1973) as well as other assays, to determine whether the thymus plays an obligatory or only an enhancing role in the *in vitro* model systems.

In considering potential functional activities of embryonic and fetal liver cells it seemed well to ask whether natural cytotoxic reactions, analogous to those seen in early yolk sac, might be found in liver cells. Preliminary observations (C. Dahl, unpublished) suggest that 16-day fetal liver cells manifest NK-like activity when tested on the standard YAC-1 target cells. Whereas further studies are still required to characterize these cells in the fetal liver and to test for the range of target specificities, it does appear that the NK-like function is the first to be expressed by fetal lymphohemapoietic organs, requiring neither inductive interactions, hormonal triggering or antigen–specific sensitization.

Although these studies demonstrate that the potential to give rise to T-cell-type properties exists in the embryonic and fetal liver, one may wonder whether these *in vitro* observations reflect differentiation processes that also occur *in vivo*. In fact, studies involving *in vivo* experimental systems do lead to similar conclusions. A systematic survey toward this end has recently been carried out by grafting 14-day embryonic liver cells into lethally irradiated adult mice (Rabinowich, 1980; Umiel et al., 1981; Rabinowich et al., 1982). It was found that the chimeras obtained indeed developed a wide spectrum of T-cell functions. To ensure that the reactive cells were of donor liver origin, C3H/Disn liver cells and F_1(C3H/Disn × C3H/SW) recipients were employed and the cells of the chimeras were treated with C3H/Disn (H-2k) anti-C3H/SW(H-2b) congenic antisera to eliminate the host cell contribution prior to assay of the various functions. The results demonstrated that cells exhibiting T-cell functions developed in the liver chimeras and that the majority of the reaction was attributable to the donor rather than to the host cells.

TABLE 3. DAY OF ONSET OF IMMUNE RESPONSES BY MOUSE FETAL LIVER-DERIVED CELLS[a] IN LYMPHOID ORGANS OF RADIATION CHIMERAS AS MEASURED IN VITRO[b]

Response		Spleen	Lymph nodes	Thymus
B-Cell functions				
Proliferative responses to mitogens	D × S	14	—	—
	LPS	21	26	—
Antibodies to α-DNP-PLL[c]		17		
T-Cell functions				
Proliferative responses to mitogens	Con A	30	18–26	16
	PMA	30	26	24–33
MLR		30	30	21–30
CML (alloantigens)		35–42	21	
CML (TNP-modified syngeneic cells)		14	14	

[a] Liver cells were obtained from 14-day embryos.
[b] Data summarized from Rabinowich (1982); Umiel et al. (1981); and Rabinowich et al. (1982).
[c] Data from Umiel and Globerson (1974).

That the results reflected a long-term maturational sequence rather than a transfer of already competent cells was reflected in the hierarchy of reactivities seen at various times and in various organs following cell transfer (Table 3). It is of interest to note that the first types of responses detected were of the B-cell lineage. Thus, first a response to dextran sulfate was seen (14 days) followed by a response to bacterial lipopolysaccharide (LPS) at 21 days, as has also been reported by Gronowicz et al. (1974). T-Cell functions first were manifested after 21–30 days, the CML response to TNP-modified syngeneic splenic cells being detectable before any of the other T-cell functions (Rabinowich et al., 1979a; Umiel et al., 1981). The significance of this observation is discussed below (Section 3.4).

These cell transfer experiments suggest that while the embryonic liver does contain precursor cells capable of differentiation into T-cell lineages, the cells only acquire these types of T-cell properties after transfer rather than within the liver itself. Ultimately, T-cell differentiation seems to take place within the thymic tissue or is conferred at least in part by thymic hormonal effects. Indeed, Stutman (Stutman et al., 1970; Stutman, 1975, 1978) has proposed that differentiation of early embryonic liver cells requires the integrity of thymic tissue but that later in development thymic hormonal factors may be sufficient to induce differentiation.

A certain level of development of T-cell functions of fetal liver cells can be observed even in the absence of the thymus altogether. Fetal liver cells could induce a GVH reaction in vivo, in 1-day-old mice thymectomized at birth (Umiel, 1971). The splenomegaly reaction was as pronounced as it was in unoperated normal control mice. Since an effect of residual thymic hormone could not be ruled out under this procedure, passage of the cells through adult, irradiated recipients was attempted. Accordingly 14- to 17-day fetal liver cells were transferred to adult thymectomized or unoperated, irradiated mice, and the spleens of the recipients were subsequently examined, at various time intervals, for their ability to induce a GVH response in vitro. It was found (Umiel, 1971) that cells capable of a response did appear in the spleens of the thymectomized mice, but later than in the intact ones (after 21 days, compared to 2–6 days). Analysis of the origin of the reactive cells involving appropriate immunogenetic combinations and the use of anti-H-2 alloantisera demonstrated that the reactive cells were of liver origin (Umiel, 1973). In a similar manner, 19-day fetal liver cells were shown to give rise to cytotoxic effector cells after passage through adult, thymectomized, irradiated recipients (Galli and Dröge, 1980).

The immediate message from the information gained so far is that T-cell precursors are indeed resident in the embryonic and fetal liver. The question now raised is whether the subsequent differentiation observed after passage in thymectomized recipients should be interpreted to suggest that the cells represent a distinct subpopulation that is independent or less dependent on thymic influence. It may well be argued that in studies employing liver from fetuses in the third trimester of gestation, when the thymus is already a lymphoid (Auerbach, 1964) and hormone-secreting (Clark, 1966) organ, the cells acquiring T-cell-type properties have actually been derived from and/or influenced by the embryonic thymus.

It is possible that the embryonic liver contains cells which can contribute to the T-cell compartment even before the thymus itself is established as a lymphoid or secretory organ. The fact that 12- and 14-day embryonic liver explants acquire the potential to elicit a GVH response *in vitro* in cultures which also include a thymic explant (Umiel *et al.*, 1968) suggests that this may be the case. On the other hand, it could be argued that the liver, a priori, may contain potentially active T cells that are unable to function because the liver may be rich in regulatory cells that operate as effective suppressors of cell-mediated immune reactions. Indeed, recent experiments all point to the fetal liver as a major source of suppressor cells during ontogeny. Since this suppression represents yet another arm related to the T-cell type of immunoregulatory function and since it appears to use to be of such major importance as a factor regulating the *manifestation* of immunocompetence during development (in contradistinction to the *existence* of immunocompetent cells), we will devote the next section of this chapter to a detailed analysis of the ontogeny of suppressor cells.

3. ONTOGENY OF SUPPRESSOR CELLS

3.1. EMBRYONIC LIVER AS A SOURCE OF CELLS CAPABLE OF SUPPRESSING *IN VITRO* IMMUNE REACTIONS

The question as to whether suppressor cells exist in the fetal liver was investigated by attempting to find out whether fetal liver cells are capable of interfering with immune reactivity of mature lymphocytes. Indeed, cells of the liver as early as the 12th to 13th day of gestation, when added to MLC or CML systems, led to reduced responses (Globerson *et al.*, 1975b). Such interference with the response of adult lymphoid cells was also manifested when liver cells were added to cultures of adult spleens stimulated *in vitro* to produce antibodies to SRBC (Globerson *et al.*, 1977) and to α-DNP-PLL (Globerson *et al.*, 1975b,

1980). Similarly, suppression of the reaction of lymphocytes to Con A, PHA, LPS, and dextran sulfate was seen (Globerson *et al.*, 1979, 1981b; Rabinowich *et al.*, 1979b). The ability of embryonic liver cells to depress lymphocyte functions extended through fetal life until birth.

The obvious question was then raised as to whether this decreased response was due to a function of liver suppressor cells or whether the embryonic liver cells exerted their effect by emitting factors interfering with cell metabolism and function through mechanisms other than those of an immune type. We examined the possibility that α-fetoprotein produced by the liver cells (Abelev, 1971; Murgita and Tomasi, 1975a,b) might interfere with immune reactions. However, anti-α-fetoprotein antibodies added to liver cells in the MLC system did not prevent suppression (Umiel, unpublished observations). Furthermore, suppression was seen only if the liver cells were added within a short time of antigen stimulation in both the GVH system (Umiel and Globerson, 1979) and the *in vitro* antibody formation assay (Globerson *et al.*, 1981b). The implication of the series of experiments is that embryonic liver cells can actively interfere with the immunological functions of adult, competent lymphocytes.

TABLE 4. SUPPRESSION OF IMMUNE RESPONSES BY CELLS OF EMBRYONIC AND FETAL MOUSE ORGANS

Organ	Gestation day	Response suppressed	Reference
Liver	12 to birth	MLC	Globerson *et al.* (1975b), Globerson and Umiel (1978)
	12 to birth	CML	Globerson *et al.* (1975b), Umiel *et al.* (1981)
	14 to birth	PHA, Con A, LPS, dextran sulfate	Globerson *et al.* (1977, 1981), Rabinowich *et al.* (1979b)
	14 to birth	*In vitro* antibody response to SRBC (in organ and cell cultures) and to α-DNP-PLL (in organ culture)	Globerson *et al.* (1975b, 1977, 1981)
	16 to birth	GVH mortality	Globerson and Umiel (1978)
Spleen	17–18 to birth	MLC; *in vitro* antibody responses to SRBC and to α-DNP-PLL	Globerson *et al.* (1977)
	18 to birth	Local GVH response	Skowron-Cendrzak and Ptak (1976)
	19	Antibody response to SRBC in cell culture	Ptak *et al.* (1979)
Thymus	14, after 7–12 days in culture	*In vitro* antibody response to SRBC in cell culture (and not the response to NIP-POL)	Luckenbach *et al.* (1978)
	16–17 only	*In vitro* antibody response to SRBC in organ and cell cultures and to α-DNP-PLL	Globerson *et al.* (1977)
	16–17 only	MLC	Globerson *et al.* (1977), Globerson and Umiel (1978)
	19 only	Antibody response to SRBC in cell cultures	Ptak *et al.* (1979)

3.2. OTHER EMBRYONIC ORGANS AS SOURCES OF SUPPRESSOR CELLS

Having established that the embryonic and fetal liver contain suppressor cells, it seemed important to determine whether other embryonic organs also contained suppressor cells. A survey was made of embryonic spleen and thymus from the earliest stages at which they could be harvested to the early postnatal animal (Globerson and Umiel, 1978). The experiments indicated that the thymus contains cells which can interfere with MLC and with production of antibodies *in vitro*, but that these cells can only be found during a short period (16–17 days of gestation; cf. also Robinson, 1980). On the other hand, the spleen contains suppressor cells beginning on the 18th day of gestation and continuing into postnatal life (Tables 4 and 5). Thus, during the first 6 days after birth, suppression of MLC reactions was noted (Rodriguez *et al.*, 1979), and for the first 2 weeks, cells interfering with antibody responses *in vitro* were detectable (Mosier and Johnson, 1975) as were cells that reduced NK cell activity (Savary and Lotzova, 1978; Cudkowicz and Hochman, 1979). Furthermore, cells interfering with the GVH splenomegaly reaction were observed for extended periods after birth (Hardin *et al.*, 1973).

TABLE 5. SUPPRESSION OF IMMUNE RESPONSES BY CELLS OF NEONATAL MOUSE ORGANS

Organ	Age	Response suppressed	Reference
Liver	0–24 hr	Antibody response to SRBC in cell and organ cultures and to α-DNP-PLL in organ culture	Globerson *et al.*, 1977; 1981
	0–24 hr	Response to mitogens	Rabinowich, 1982; Globerson *et al.*, 1981
	0–24 hr	GVH mortality	Globerson and Umiel, 1978
	0–24 hr	MLR, CML	Globerson *et al.*, 1975b; 1977; Globerson and Umiel, 1978; Umiel *et al.*, 1981
	<7 days[a]	MLR	Rollwagen and Stutman, 1979
	5 days	GVH splenomegaly in vivo	Bortin *et al.*, 1969
Spleen	0–24 hr; and 5 days	GVH, local lymph node enlargement	Skowron–Cendrzak and Ptak, 1976
	0–24 hr	MLR, CTL	Argyris, 1979
	0–6 days	MLR	Bassett *et al.*, 1977; Rodriguez *et al.*, 1979
	0–7 days	MLR	Rollwagen and Stutman, 1979
	>7 days	CTL	Rollwagen and Stutman, 1979
	0–2 wk	Antibody response to SRBC in cell culture	Mosier and Johnson, 1975
	12 days	Antibody response to SRBC in cell culture; con–A response	Ptak *et al.*, 1979
	8–13 days	NK activity	Savary and Lotzova, 1978; Cudkowicz and Hochman, 1979
	6 wk	GVH splenomegaly in vivo	Hardin *et al.*, 1973

[a] After preculture.

Since the embryonic liver already contains suppressor cells at earlier times during ontogeny, the finding that there are transient suppressor cells in the thymus and that such cells later are found in the spleen may be interpreted in terms of migration of suppressor cells from the liver via the thymus to the spleen. However, it is equally possible that entirely different cell populations are involved in the thymus and spleen and that they may develop independently of suppressor cells of the liver. It is of interest to note in this respect that 14-day embryonic thymus cultivated in organ cultures for 7–12 days could subsequently suppress the *in vitro* antibody response to SRBC (Luckenbach *et al.*, 1978). It may be suggested that while the thymus is kept in a closed system, without the ability for the cells to migrate out, the suppressor cells are maintained for extended periods in this organ.

3.3. *IN VIVO* FUNCTION OF EMBRYONIC SUPPRESSOR CELLS

The question as to whether these suppressor cells can function *in vivo* as well as *in vitro* led to experiments in which the possible effect of the embryonic liver cells was assessed in an *in vivo* GVH response system, measured by mortality (Globerson *et al.*, 1977; Globerson and Umiel, 1978; Umiel and Globerson, 1979). It was found that when embryonic liver cells were administered to F_1 irradiated adult or neonatal mice, together with parental type adult splenic cells, the level of GVH mortality was reduced. The optimal liver : spleen ratio was 1 : 1 to 2 : 1. It was of interest to observe that the splenomegaly manifestation of the GVH reaction was less readily suppressed than was the mortality (Globerson and Umiel, 1978). However, other reports have indicated a reduction of spleen index in neonatal F_1 mice injected with adult parental type splenic cells and neonatal liver cells (Bortin *et al.*, 1969). A reduction of lymph node enlargement was also observed when neonatal splenic cells were employed as suppressors (Skowron-Cendrzak and Ptak, 1976). Reduction of GVH mortality rate was observed only when the embryonic liver and effector splenic cells were syngeneic to each other. Thus C57BL/6 liver cells suppressed GVH mortality induced in either (BALB/c × C57BL/6)F_1 or (C3H/3b × C57BL/6)F_1 mice by adult C57BL/6 splenic cells, while liver cells originating in C3H/eb, BALB/c, or the F_1-type embryos had no effect in this system. In a similar manner, a response induced by C3H/eb splenic cells was suppressed by C3H/eb liver but not by C57BL/6 or any of the other strains mentioned above (Globerson *et al.*, 1977; Umiel *et al.*, 1977). This suggests that the liver cells express specific H-2 restriction. Initial analysis of this possibility was performed using *H-2* congenic mice (Umiel and Globerson, 1979). The experiments showed that embryonic liver cells of C3H/Disn (*H-2k*) mice reduced the mortality rates induced by C3H/Disn but not by C3H/SW (*H-2b*) liver cells, whereas C3H/SW cells interfered with the reaction of C57BL/6 (*H-2b*) splenic cells. Further assessment of this requirement for *H-2* identity is needed, however, before definitive conclusions concerning the details of the observed histocompatibility restrictions can be drawn.

3.4. DEVELOPMENTAL CONSIDERATIONS

Since embryonic suppressor cells become manifest at an early stage in the development of the liver, one may ask whether these suppressor cells represent early, less mature reactive cells or whether they belong to a distinct, independent cell lineage. While ideally one would distinguish between these alternatives by cloning suppressor cells and applying optimal conditions for their further differentiation, we do not yet have the methodology necessary for carrying out such experiments. A more conventional way could be based on analysis of the cell phenotype by employing reagents to various cell membrane markers. However, at the early stage when fetal liver cells manifest suppression, the cells do not seem to express any of the membrane markers characterizing adult-type T suppressor cells (Umiel and Globerson, preliminary observations). Furthermore, even after birth (0–6 days), suppressor cells interfering with the MLC reaction were described as lacking T- or B-cell markers, and they could be characterized as NK cells (Rodriguez et al., 1979). On the other hand, postnatal splenic suppressor cells were characterized as T-cells when antibody responses in cell cultures (Mosier and Johnson, 1975) and local GVH reactions (Skowron-Cendrzak and Ptak, 1976) were measured.

As an alternative approach, experiments were carried out to examine whether the embryonic suppressor cells have detectable cell surface properties characteristic of immature lymphoid cells. Peanut agglutinin (PNA) was chosen, since it was known to agglutinate the less mature cell population in the thymus (Reisner et al., 1976). Under experimental conditions similar to those previously employed for analysis of thymic cell populations, suppressor cells of the liver were found to be present in the PNA-agglutinated fraction, whereas the nonagglutinated cells (PNA$^-$), poor in suppressor function, expressed reactivity to dextran sulfate and LPS (Rabinowich et al., 1979b). These studies indicate that the suppressor cell population is distinct from the reactive B-cell type associated with the liver, but do not yet permit more extensive conclusions concerning the cell lineage of this population.

An immediate outcome of these experiments relates to the question of whether potentially reactive T cells are present in the liver which do not function because of the coexistence of suppressor cells. It was found that the PNA-unagglutinable cells remaining after removal of PNA$^+$ suppressor cells did not manifest a response to T-cell mitogens Con A and PHA (Rabinowich et al., 1979b). That such uncoupling of reactive from suppressor cells is feasible has recently been demonstrated by employing the same methodological approach to aging mouse splenic cells (Globerson et al., 1980, 1981a). On the other hand, the possible presence in fetal liver of PNA$^+$ reactive cells cannot be excluded.

Another developmental aspect of liver suppressor cells concerns the question of the role of the thymus in determining histocompatibility requirements for functional manifestation *in vivo*, especially since H-2 restriction does not appear to be essential in the *in vitro* test systems. *In vivo* experiments carried out in thymectomized mice indicated that in the absence of the thymus, no liver cell suppression could be obtained (Umiel et al., 1977). However, since THF could restore the liver cell suppressor function, the MHC restriction appeared to be related to the liver cells rather than to the genetic constitution of the thymus itself. These observations thus suggest that the fetal liver-derived suppressor cells, at least in these experiments, do not acquire the education to *H-2* self recognition in the thymus, in contrast to the situation with cytotoxic T-cell development (Zinkernagel *et al.*, 1978). Whether the case of these suppressor cells is similar in this respect to the potentially reactive T cells in the liver, in which a distinction was made between early embryonic thymic tissue dependent and later stages inducible by thymic humoral factors (Stutman, 1978), is an interesting possibility that warrants further critical analysis.

Finally, the emergence of suppressor cells in ontogeny prior to manifestations of immune reactivity may suggest that the fetal suppressor cells play a role in immunosurveillance, preventing reactivity to self. The recent observations that the fetal liver contains precursors of cells capable of a response to TNP-modified self splenic cells on the one hand, and suppressor cells interfering with such a reaction on the other hand, seem most intriguing in this respect.

Furthermore, fetal liver derived cells can be induced by hapten-carrier conjugates to differentiate into suppressor cells interfering specifically with the antibody response to the haptenic determinants (Ritterband & Globerson, 1982). It thus seems inviting to speculate that the establishment of tolerance to self in ontogeny is mediated, at least in part, by embryonic suppressor cells.

3.5. THE POSSIBLE RELEVANCE OF EMBRYONIC LIVER-DERIVED SUPPRESSOR CELLS FOR THE ESTABLISHMENT OF TOLERANCE TO SELF

The idea that tolerance to self is an "educational" developmental process and not a genetic property derives from the fact that an F_1 hybrid, although genetically P1 × P2, has the potential of reactivity against both parental type antigens, yet in practice does not manifest such a response. Various mechanisms may be proposed to account for this phenomenon. One of these is based on the notion that tolerance is mediated by suppressor cells. This led to the investigation of whether suppressor cells emerging during early ontogenetic development can indeed be involved in the establishment of a state of tolerance. The embryonic liver chimera seemed an appropriate model to examine this possibility. It was demonstrated that lethally irradiated F_1 mice repopulated with parental embryonic liver cells do not suffer from a GVH syndrome (Barnes *et al.*, 1958; Uphoff, 1958) to the same extent as is the case when parental type adult bone marrow cells are applied (Trentin, 1956). The idea that the liver chimera is repopulated with donor-type lymphopoietic cells and that tolerance to the other parent type is manifest has been shown unequivocally (cf. Van Bekkum and de Vries, 1967). The question as to whether this tolerance is based on suppressor mechanisms was investigated by reexposing the chimera to total-body irradiation (500 rads) and challenging with adult competent splenic cells of the liver donor type. Under these conditions, no GVH response developed, although a GVH response expressed in mortality was seen when the challenging splenic cells were of the second parent type or were genetically unrelated to either the donor or the recipient (Umiel, 1975, 1976). Further studies revealed that the lymph nodes of the chimera contained suppressor cells of the liver donor type directed against the second parent type antigens (Umiel, 1977).

In this case, as in the studies described earlier, suppressor cells were not manifest in thymectomized mice, thus emphasizing the thymic dependency. Furthermore, here also specific H-2 recognition was not "dictated" by thymic tissue, since specificity was expressed according to the liver origin and not the thymic graft types and was also conferred by THF without any intact thymic tissue (Umiel and Trainin, 1976). The thymic influence was found to be necessary for the establishment of suppressor cells, during a limited critical period, after which it could be removed without interference with the function of suppressor cells (Umiel and Trainin, 1976).

4. GENERAL DISCUSSION

Perhaps most critical in the evaluation of the several observations made that cells capable of carrying out various cell-mediated functions exist in the yolk sac, thymus, and liver of the developing embryo is the question of whether that existence bears any meaningful relationship to expression of immunological functions during that time, or whether, rather, it simply means that the maturation of immunocompetent cells precedes by a considerable time the actual expression of immunoreactivity by those cells.

For the embryonic yolk sac, the question is probably the most difficult to answer experimentally. Surveillance, if it exists at this stage, is difficult to simulate in the laboratory, and chimeras, so useful in the study of cell lineages, cannot readily be made between extraembryonic membranes and the early mammalian embryo. On the other hand, there is no a priori reason to believe that surveillance and defense mechanisms would not exist at this stage, and the evolutionary conservatism of embryonic and extraembryonic aspects of amniote development leads to the suggestion that the question of yolk sac function may be answered by the use of avian material. The ability of making extraembryonic/embryonic chimeras (e.g., Dieterlen-Lièvre et al., 1980; Martin et al., 1980) combined with the ability to manipulate the external environment of avian embryos, make such studies of great potential importance.

The presence of functional immunocompetent cells in the thymus is in a certain sense likely to represent a laboratory phenomenon rather than a meaningful one for the embryo itself. The fact that cells within the thymus *can* respond to mitogens or *can* induce GVH reactions may simply reflect the fact that the cells are close to being released into the circulating pool of lymphocytes and that it is only peripherally, rather than in the thymus itself, that the immunocompetence of the thymic cells finds biological relevance. In terms of actual *in situ* function, we may not be examining the "right" reactions. The cell-mediated functions seen in our test systems certainly indicate that cell recognition systems are present and that effector cell functions can be carried out. Within the thymus itself it may well be that various aspects of H-2 restriction, of establishment of conditions for tolerance or self recognition, require a set of functional differentiations related to the ones studied in the *in vitro* test systems. For example, the ability of thymic cells to cause a GVH reaction may be related to the ability to eliminate unwanted cells within the thymic environment during the development of the later, functionally restricted, but highly active T cells destined to emigrate to the lymph nodes and spleen of the postnatal mouse.

Similarly, one may wonder whether the finding that liver suppressor cells are actually "educated" in the chimeric environment in experimental situations reflects a process normally present during embryogenesis. While a direct test of this question has not been possible in the mouse embryo, circumstantial evidence indeed indicates that there is an *in vivo* significance to the chimera studies. Suppressor cells are detectable in the embryo at a stage before immune reactivity is overt, embryonic tissues at an early stage of development already express H-2 and Ia antigens, and the response, for example, in a CML assay for response to TNP-modified syngeneic splenic cells is expressed by embryonic liver cells sooner than is the response to alloantigens (cf. Table 3).

If one speculates that the meaning of the early appearance of suppressor cells in the embryonic liver is to prevent the response of cells potentially capable of a reaction to self antigens, then the yolk sac might, because of its appearance prior to the development of the liver, be expected to exert nondiscriminatory (anti-self as well as anti-non-self) types of reactions. The observation that yolk sac cells exert NK-like functions supports this idea, as does the finding that there is a weak but detectable LIA in syngeneic systems (Auerbach and Sidky, 1979b). The autoreactive yolk sac cell population should represent no threat to the developing embryo both because the organ itself is separated from the embryo proper, and because there is now considerable doubt that the lymphoid cell precursors destined to enter the various embryonic lymphoid organs ever originate from the yolk sac (cf. Dieterlen-Lièvre *et al.*, 1980).

This then brings us to an attempt to synthesize from the ontogenetic studies presented a picture of the emerging immune system that is not altogether conventional but which appears to us to represent fairly the experimental findings:

1. The yolk sac functions as a primitive effector organ. Its role may be defensive in nature, for surveillance against external undesirable agents and possibly against maternal cells that could otherwise be damaging to the developing embryo.
2. The primitive yolk sac system exerts its functions by secretion of active factors (lymphokinelike substances) in response to allogeneic or antigenic stimulation, and by carrying out cytotoxic reactions against unwanted cells. The observed GVH activity of yolk sac cells may reflect both types of functional activity.
3. There is no compelling reason to think that the yolk sac gives rise to the lymphoid cell precursors destined to develop in thymus, liver, or spleen, any more than that it is reasonable to suggest that cells of the pronephros or mesonephros necessarily must be the precursors of cells of the metanephros. Sequential appearance of functional activity does not imply direct lineage.
4. The embryonic liver is also an effector organ during development, replacing and/or supplementing yolk sac functions. In line with the need of specifically controlling the immunological repertoire for preventing unwanted immune reactions while the processes of self recognition and tolerance are becoming established, the primary immunological activity recognized in the liver is one of immunosuppression.

5. There is good evidence to suggest that the fetal liver serves the embryo by providing a source of T as well as of B precursor cells, analogous therefore to the later developing bone marrow. Thus, we conceive of the liver as a transitional effector organ on the one hand, and as a generative organ on the other. A major reason for the lack of clarity in defining effector cell functions for the liver beyond that of suppression is that the suppressor activity itself is enough to interfere with assessment of positive immunological functions of liver cells.
6. The embryonic thymus contains stem cells that are capable of developing into immunocompetent cells *in vitro*, and contains, as well, both positive (helper) and negative (suppressor) cells. Whether these cells function directly in the thymus or only on release into the circulation is not really at issue, for what seems clear is that thymus-generated cells are available to carry out T-cell activities beginning with the terminal phases of murine intrauterine development. At birth, they probably form the predominant T-cell pool available for T-cell activities.
7. Lyt-1^+,$2,3^+$ cells alone make up the T-cell population at birth, with Lyt-1^+,$2,3^-$ and Lyt-1^-,$2,3^+$ cells first appearing only about 1 week after birth. The presence at birth of MLC-positive cells, of helper cells, and of suppressor cells, and the appearance of CML-positive cells within 48 hr in the thymus and spleen, combined with the new and conflicting data concerning the absence vs. presence of Lyt-1 antigens on virtually all T cells throughout life (Roehm, 1980; Roehm *et al.*, 1981; Ledbetter *et al.*, 1980), argue for a conservative approach to the question of postnatal T-cell development. Moreover, the view that the Lyt $1^+2,3^+$ cell is the obligatory precursor phenotype for all T–cell lineages must itself be tempered by the observation (Mathieson *et al.*, 1981) that Lyt 1^+,$2,3^-$ cells were found at an early time in T–cell development. An analysis of this aspect of developmental immunology is not, however, within the scope of the present chapter.

Finally, we wish to return briefly once more to the question of ontogeny and phylogeny. Our review stresses a developmental sequence for cells in the yolk sac, liver, thymus, and spleen that reflects in turn the functional requirements of the developing mouse embryo and fetus. In a sense, the sequence of functional activities that we have observed may also be seen in the phylogenetic evolution within the animal kingdom, as groups of animals are challenged to respond to various and different types of environmentally or maternally originated foreign stimuli. Basic recognition of self is a fundamental attribute of all living cells. The production of simple effector molecules and the evolution, in turn, of increasingly sophisticated response patterns is a general feature of the evolutionary process. And while it may no longer be fashionable to state directly that "ontogeny repeats phylogeny," parallels between evolutionary and developmental histories are nonetheless fascinating and may as well be most informative.

ACKNOWLEDGMENTS. The assistance of B. Houser and B. V. Nguyen in the preparation of the manuscript is acknowledged with thanks.

REFERENCES

Abelev, G. I., 1971, Alpha-fetoprotein in ontogenesis and its association with malignant tumors, *Adv. Cancer Res.* **14**:295.

Argyris, B. F., 1979, Further studies on suppressor cell activity in the spleen of neonatal mice, *Cell. Immunol.* **48**:398.

Auerbach, R., 1964, On the function of the embryonic thymus, in: *The Thymus*, Wistar Institute Monograph No. 2, pp. 1–8.

Auerbach, R., 1970, Toward a developmental theory of antibody formation: The germinal theory of immunity, in: *Developmental Aspects of Antibody Formation and Structure*, pp. 23–33, Academic Press, New York.

Auerbach, R., 1976, Ontogeny of immune responsiveness in the mouse, *Ann. Immunol. (Inst. Pasteur)* **127C**:983.

Auerbach, R., 1977, Toward a developmental theory of immunity: Selective differentiation of teratoma cells, in: *Cell and Tissue Interactions* (J. W. Lash and M. M. Burger, eds.), pp. 47–55, Raven Press, New York.

Auerbach, R., 1982, Developmental aspects of immunoregulation: Embryonic effector cells and the emergence of functional restrictions during fetal development, in: *The Biological Significance of Immune Regulation* (L. N. Ruben and M. E. Gershwin, eds.), Dekker, New York, in press.

Auerbach, R., and Globerson, G., 1966, In vitro induction of the graft-versus-host reaction, *Exp. Cell Res.* **42**:31.

Auerbach, R., and Shalaby, M. R., 1973, Graft-vs.-host reaction in tissue culture, *J. Exp. Med.* **138**:1506.

Auerbach, R., and Sidky, Y. A., 1979a, Nature of the stimulus leading to lymphocyte-induced angiogenesis, *J. Immunol.* **123**:751.

Auerbach, R., and Sidky, Y. A., 1979b, Studies on the maturation of immune responsiveness in the mouse. III. Ontogeny of immunocompetence as measured by the ability to embryonic yolk sac cells to evoke angiogenesis in allogeneic hosts, in: *Animal Models of Comparative and Developmental Aspects of Immunity and Disease* (M. E. Gershwin and E. L. Cooper, eds.), pp. 166–174, Pergamon Press, Elmsford, N.Y.

Auerbach, R., Kubai, L., and Sidky, Y., 1976, Angiogenesis induction by tumors, embryonic tissues, and lymphocytes, *Cancer Res.* **36**:3435.

Auerbach, R., Cairns, J. S., and Roehm, N., 1980, Development of the mouse immune system, *Fortschr. Zool.* **26**:287–305.

Barnes, D. W. H., Ilbery, P. L. T., and Loutit, J. F., 1958, Avoidance of "secondary disease" in radiation chimeras, *Nature (London)* **181**:488.

Bassett, M., Coons, T. A., Wallis, W., Goldberg, E. A., and Williams, R. C., 1977, Suppression of stimulation in mixed leukocyte cultures by newborn splenic lymphocytes in the mouse, *J. Immunol.* **119**:1855.

Bortin, M. M., Rimm, A. A., and Saltzstein, E. C., 1969, Graft versus host inhibition. I. Incubated parental strain spleen and liver cells administered to F_1 mice, *J. Immunol.* **102**:1042.

Cantor, H., and Weissman, I., 1976, Development and function of subpopulations of thymocytes and T lymphocytes, *Progr. Allergy* **20**:1.

Carr, M. C., Stites, D. P., and Fudenberg, H. H., 1973, Dissociation of responses to phytohaemagglutinin and adult allogeneic lymphocytes in humal foetal lymphoid tissues, *Nature New Biol.* **241**:279.

Chakravarty, A. K., 1977, An in vitro study of functional maturation of murine thymus cells, *Differentiation* **8**:21.

Chakravarty, A., Kubai, L., Sidky, Y., and Auerbach, R., 1975, Ontogeny of thymus cell function, *Ann. N.Y. Acad. Sci.* **249**:34.

Clark, S. L., 1966, Cytological evidences of secretion in the thymus, in: *The Thymus: Experimental and Clinical Studies* (G. Wolstenholme and R. Porter, eds.), pp. 3–57, CIBA Foundation Symposium.

Cohen, J. J., and Patterson, C. K., 1975, Induction of theta-positive lymphocytes in mouse bone marrow by mitogens, *J. Immunol.* **114**:374.

Cohen, M. W., Thorbeck, G. J., Hochwald, G. M., and Jacobson, E. B., 1963, Induction of graft-versus-host reaction in newborn mice by injection of newborn or adult homologous thymus cells, *Proc. Soc. Exp. Biol. Med.* **114**:242.

Cudkowicz, G., and Hochman, P. S., 1979, Regulation of natural killer activity by macrophage-like and other types of suppressor cells, in: *Developmental Immunobiology* (G. W. Siskind, S. D. Litwin, and M. E. Weksler, eds.), pp. 1–24, Grune & Stratton, New York.

Dahl, C. A., 1980, Natural cytotoxicity in the mouse embryo. 1. Evidence for the presence of natural cytotoxic function in the yolk sac of the 10-day mouse embryo, *J. Immunol.* **125**:1924.

Dahl, C. A., Kahan, B. W., and Auerbach, R., 1980, Studies with mouse embryonic yolk sac cells: An approach to understanding the pattern of ontogeny of cell-mediated immune functions, in: *Development and Differentiation of Vertebrate Lymphocytes* (J. D. Horton, ed.), pp. 241–253, Elsevier/North-Holland, Amsterdam.

Dieterlen-Lièvre, F., Beaupain, D., Lassila, O., Tiovanen, P., and Martin, C., 1980, Embryonic origin of lymphoid stem cells investigated in avian inter- and intra-specific chimeras, in: *Development and Differentiation of Vertebrate Lymphocytes* (J. D. Horton, ed.), pp. 35–43, Elsevier/North-Holland, Amsterdam.

Elkins, W. L., 1971, Cellular immunology and the pathogenesis of graft versus host reaction, *Prog. Allergy* **15**:78.

Galli, P., and Dröge, W., 1980, Development of cytotoxic T lymphocyte precursors in the absence of the thymus, *Eur. J. Immunol.* **10**:87.

Globerson, A., and Umiel, T., 1978, Ontogeny of suppressor cells. II. Suppression of graft-versus-host and mixed leukocyte culture responses by embryonic cells, *Transplantation* **26**:438.

Globerson, A., Umiel, T., and Friedman, D., 1975a, Activation of immune competence by thymus factors, *Ann. N.Y. Acad. Sci.* **249**:248.

Globerson, A., Zinkernagel, R. M., and Umiel, T., 1975b, Immunosuppression by embryonic liver cells, *Transplantation* **20**:480.

Globerson, A., Rabinowich, H., and Umiel, T., 1977, Ontogeny of suppressor cells, in: *Developmental Immunobiology* (J. B. Solomon and J. D. Horton, eds.), pp. 331–337, Elsevier/North-Holland, Amsterdam.

Globerson, A., Rabinowich, H., Umiel, T., Reisner, Y., and Sharon, N., 1979, Characterization of reactive and suppressive cells in the mouse embryonic liver by peanut agglutinin (PNA), in: *Function and Structure of the Immune System* (W. Müller-Ruchholtz and H. K. Müller-Hermelink, eds.), pp. 345–350, Plenum Press, New York.

Globerson, A., Abel, L., and Umiel, T., 1980, Immunosuppression in ageing, expressed by PNA-agglutinated cells, Abstracts, *IVth International Congress on Immunology*, Paris, 3.8.16.

Globerson, A., Abel, L., and Umiel, T., 1981a, Immune reactivity during ageing. 3. Removal of peanut-agglutinin binding cells from ageing mouse spleen cells leads to increased reactivity to mitogens, *Mech. Ageing Dev.* **16**:275.

Globerson, A., Rabinowich, H., and Umiel, T., 1981b, Ontogeny of suppressor cells in the mouse and their relation to potential immune reactivity in the fetal liver, in: *Developmental and Comparative Immunology* (J. B. Solomon, ed.), pp. 415–421. Pergamon Press, Oxford.

Goldstein, A. L., and White, A., 1973, Thymosin and other thymic hormones: Their nature and roles in the thymic dependency of immunological phenomena, in: *Contemporary Topics in Immunobiology*, Vol. II (A. J. S. Davies and R. L. Carter, eds.), pp. 339–350, Plenum Press, New York.

Gronowicz, E., Coutinho, A., and Möller, G., 1974, Differentiation of B cells: Sequential appearance of responsiveness to polyclonal activators, *Scand. J. Immunol.* **3**:413.

Hardin, J. A., Chused, T. M., and Steinberg, A. D., 1973, Suppressor cells in the graft versus host reaction, *J. Immunol.* **11**:650.

Herberman, R. B., and Holden, H. T., 1978, Natural cell-mediated immunity, *Adv. Cancer Res.* **27**:305.

Herberman, R. B., Myrthel, E. N., and Lavrin, D. H., 1975, Natural cytotoxic reactivity of mouse lymphoid cells against syngeneic and allogeneic tumors. 1. Distribution of reactivity and specificity, *Int. J. Cancer* **16**:216.

Herberman, R. B., Djeu, J. Y., Kay, H. D., Ortaldo, J. R., Riccardi, C., Bonnard, G. D., Holden, H. T., Fagnani, R., Santoni, A., and Pucetti, P., 1979, Natural killer cells: Characteristics and regulation of activity, *Immunol. Rev.* **44**:43.

Hofman, F., 1975, The development of immunocompetence in cells of yolk cell origin, Ph.D. dissertation, Weizmann Institute.

Hofman, F., and Globerson, A., 1973, Graft-versus-host response induced in vitro by mouse yolk sac cells, *Eur. J. Immunol.* **3**:179.

Hofman, F., and Globerson, A., 1976, Immunological potential of yolk sac cells, in: *Immune Reactivity of Lymphocytes: Development, Expression, and Control* (M. Feldman and A. Globerson, eds.), pp. 51–57, Plenum Press, New York.

Horton, J. D. (ed.), 1980, *Development and Differentiation of Vertebrate Lymphocytes*, Elsevier/North-Holland, Amsterdam.

Kiessling, R., Klein, E., and Wigzell, H., 1975, "Natural" killer cells in the mouse. I. Cytotoxic cells with specificity for mouse Moloney leukemia cells. Specificity and distribution according to genotype, *Eur. J. Immunol.* **5**:112.

Ledbetter, J. A., Rouse, R. V., Micklem, H. S., and Herzenberg, L. A., 1980, T cell subsets defined by expression of Lyt-1,2,3 and Thy-1 antigens, *J. Exp. Med.* **152**:280.

LeDouarin, N., and Jotereau, F., 1975, Tracing of cells of the avian thymus through embryonic life in interspecific chimeras, *J. Exp. Med.* **142**:17.

Luckenbach, G. A., Kennedy, M. M., Kelly, A., and Mandel, T. E., 1978, Suppression of an *in vitro* humoral immune response by cultured fetal thymus cells, *Eur. J. Immunol.* **8**:8.

Martin, C., Beaupain, D., and Dieterlen-Lièvre, F., 1980, A study of the development of the hemopoietic system using quail–chick chimeras obtained by blastoderm recombination, *Dev. Biol.* **75**:303.

Mathieson, B. J., Sharrow, J. O., Rosenberg, Y., and Hammerling, U., 1981, Ly 1^+23^- cells appear in the thymus before Ly 123^+ cells, *Nature (London)* **289**:179.

Moore, M. A. S., and Owen, J. J. T., 1965, Chromosome marker studies on the development of the haemopoietic system in the chick embryo, *Nature (London)* **208**:956, 989.

Moore, M. A. S., and Owen, J. J. T., 1967, Experimental studies on the development of the thymus, *J. Exp. Med.* **126**:715.

Mosier, D. E., and Johnson, B. M., 1975, Ontogeny of mouse lymphocyte function. II. Development of the ability to produce antibody is modulated by T-lymphocytes, *J. Exp. Med.* **141**:216.

Murgita, R., and Tomasi, T. B., 1975a, Suppression of the immune response by α-fetoprotein. I. The effect of mouse α-fetoprotein on the primary and secondary antibody response, *J. Exp. Med.* **141**:269.

Murgita, R., and Tomasi, T. B., 1975b, Suppression of the immune response by α-fetoprotein. II. The effect of mouse α-fetoprotein on mixed lymphocyte reactivity and mitogen induced lymphocyte transformation, *J Exp. Med.* **141**:440.

Muthukkaruppan, VR., and Auerbach, R., 1979, Angiogenesis in the mouse cornea, *Science* **205**:1416.

Nossal, G. J. V., and Pike, B. L., 1973, Studies on the differentiation of B lymphocytes in the mouse, *Immunology* **25**:33.

Owen, J. J. T., Cooper, M. D., and Raff, M. C., 1974, In vitro generation of B lymphocytes in mouse fetal liver, a mammalian 'bursa equivalent,' *Nature (London)* **249**:361.

Phillips, R. A., and Melchers, F., 1976, Appearance of functional lymphocytes in fetal liver, *J. Immunol.* **117**:1099.

Pilarski, L. M., 1977, Ontogeny of cell-mediated immunity: Early development of alloantigen-specific cytotoxic T-cell precursors in postnatal mice, *J. Exp. Med.* **146**:887.

Ptak, W., Naidorf, K. F., Strzyewska, J., and Gershon, R. K., 1979, Ontogeny of cells involved in the suppressor circuit of the immune response, *Eur. J. Immunol.* **9**:495.

Rabinowich, H., 1980, Immunological functions of fetal liver cells, Ph.D. Dissertation, Weizmann Institute.

Rabinowich, H., and Klajman, A., 1982, Cellular and humoral suppressor activity induced by concanaval in A–stimulated human fetal liver cells, *Isr. J. Med. Sci.* in press.

Rabinowich, H., Umiel, T., and Globerson, A., 1982, T cell progenitors in the mouse fetal liver, *Transplantation.* in press.

Rabinowich, H., Umiel, T., Droege, W., and Globerson, A., 1979a, Precursors of cells reactive to foreign and to modified-self antigens in the fetal liver, *Israel J. Med. Sci.* **15**:873.

Rabinowich, H., Umiel, T., Reisner, Y., Sharon, N., and Globerson, A., 1979b, Characterization of embryonic liver suppressor cells by peanut agglutinin, *Cell. Immunol.* **47**:347.

Reisner, Y., Linker-Israeli, M., and Sharon, N., 1976, Separation of mouse thymocytes into two subpopulations by the use of peanut agglutinin, *Cell. Immunol.* **25**:129.

Ritterband, M., and Globerson, A., 1982, Developmental aspects of T-suppressor cells induced by hapten-carrier conjugates, *7th Conference on Lymphatic Tissues and Germinal Centers in Immune Reactions.* in press.

Robinson, J. H., 1980, A review of thymus in organ culture, in: *Development and Differentiation of Vertebrate Lymphocytes* (J. D. Horton, ed.), pp. 111–126, Elsevier/North-Holland, Amsterdam.

Rodriguez, G., Andersson, G., Wigzell, H., and Peck, A. B., 1979, Non-T cell nature of the naturally occurring, spleen-associated suppressor cells present in the newborn mouse, *Eur. J. Immunol.* **9:**737.

Roehm, N., 1977, Induction of splenomegaly with stimulated leukocyte culture supernatants, *Transplantation* **23:**49.

Roehm, N. W., 1980, Lyt phenotype analyses of T cells generating and expressing cell mediated cytotoxicity in allogeneic mixed leukocyte cultures, Ph.D. dissertation, University of Wisconsin, Madison.

Roehm, N. W., Sidky, Y. A., and Auerbach, R., 1981, Lyt phenotype analysis of the effector cells responsible for evoking lymphocyte induced angiogenesis (LIA), *Cell Immunol.* **63:**272.

Rollwagen, F. M. and Stutman, O., 1979, Ontogeny of culture-generated suppressor cells, *J. Exp. Med.* **150:**1359.

Rosenberg, Y. J., and Cunningham, A. J., 1977, Ontogeny of the antibody-forming cell line in mice. III. The generation of mature anti-sheep red blood cell-specific B cells is antigen-dependent, *Eur. J. Immunol.* **7:**257.

Savary, C. A., and Lotzova, E., 1978, Suppression of natural killer cell cytotoxicity by splenocytes from *Corynebacterium parvum*-injected bone-marrow-tolerant infant mice, *J. Immunol.* **120:**239.

Scheid, M. P., Goldstein, G., Hämmerling, U., and Boyse, E. A., 1975, Lymphocyte differentiation from precursor cells in vitro, *Ann. N.Y. Acad. Sci.* **249:**351.

Seidman, G. J., and Leder, P., 1978, The arrangement and rearrangement of antibody genes, *Nature (London)* **276:**790.

Sidky, Y. A., and Auerbach, R., 1975, Lymphocyte-induced angiogenesis: A quantitative and sensitive assay of the graft-vs.-host reaction, *J. Exp. Med.* **141:**1084.

Sidky, Y. A., and Auerbach, R., 1979a, Response of the host vascular system to immunocompetent lymphocytes: Effect of preimmunization of donor or host animals, *Proc. Soc. Exp. Biol. Med.* **161:**174.

Sidky, Y., and Auerbach, R., 1979b, Studies on the maturation of immune responsiveness in the mouse. III. Ontogeny of immunocompetence as measured by the ability of embryonic yolk sac cells to evoke angiogenesis in allogeneic hosts, in: *Animal Models of Comparative and Developmental Aspects of Immunity and Disease* (M. E. Gershwin and E. L. Cooper, eds.), pp. 166–174, Pergamon Press, Elmsford, N.Y.

Simonsen, M., 1962a, The factor of immunization: Clonal selection theory investigated by spleen assays of graft-versus-host reaction, *Transplant. CIBA Found. Symp.*, pp. 185–215.

Simonsen, M., 1962b, Graft-versus-host reactions: Their natural history, and applicability as tools of research, *Prog. Allergy* **6:**349.

Siskind, G. W., Litwin, S. D., and Weksler, M. E. (eds.), 1979, *Developmental Immunobiology*, 260 pp. Grune & Stratton, New York.

Skowron-Cendrzak, A., and Ptak, W., 1976, Suppression of local graft-versus-host reactions by mouse fetal and newborn spleen cells, *Eur. J. Immunol.* **6:**451.

Spear, P. G., Wang, A. L., Rutishauser, U., and Edelman, G. M., 1973, Characterization of lymphoid cells in fetal and newborn mice, *J. Exp. Med.* **138:**557.

Stites, D., Carr, M. C., and Fudenberg, H. H., 1974, Ontogeny of cellular immunity in the human fetus: Development of responses to phytohemagglutinin and to allogeneic cells, *Cell. Immunol.* **11:**257.

Stutman, O., 1975, Humoral thymic factors influencing postthymic cells, *Ann. N.Y. Acad. Sci.* **249:**89.

Stutman, O., 1978, Intrathymic and extrathymic T cell maturation, *Immunol. Rev.* **42:**138.

Stutman, O., Yunis, E. G., and Good, R. A., 1970, Studies on thymus function. II. Cooperative effect of newborn and embryonic hemopoietic liver cells with thymus function, *J. Exp. Med.* **132:**601.

Tompkins, R., Volpe, E. P., and Reinschmidt, D. C., 1980, Origin of hemopoietic stem cells in amphibian ontogeny, in: *Development and Differentiation of Vertebrate Lymphocytes* (J. D. Horton, ed.), pp. 25–34, Elsevier/North-Holland, Amsterdam.

Trainin, N., and Small, M., 1970, Studies on some physicochemical properties of a thymus humoral factor conferring immunocompetence on lymphoid cells, *J. Exp. Med.* **132**:885.

Trentin, J. J., 1956, Mortality and skin transplantability in X-irradiated mice receiving isologous, homologous or heterologous bone marrow, *Proc. Soc. Exp. Biol. Med.* **92**:688.

Turpen, J. B., 1980, Early embryogenesis of hematopoietic cells in *Rana pipiens*, in: *Development and Differentiation of Vertebrate Lymphocytes* (J. D. Horton, ed.), pp. 15–24, Elsevier/North-Holland, Amsterdam.

Umiel, T., 1971, Thymus influenced immunological maturation of embryonic liver cells, *Transplantation* **11**:531.

Umiel, T., 1973, Requirements for development of immunocompetence of embryonic liver cells: The graft-versus-host response, *Differentiation* **1**:295.

Umiel, T., 1975, Specific inhibition of the graft-versus-host reaction in fetal or neonatal liver chimeras, *Transplantation* **19**:485.

Umiel, T., 1976, Immunosuppression by fetal liver as a model for tolerance to self, in: *Immune Reactivity of Lymphocytes* (M. Feldman and A. Globerson, eds.), pp. 565–569, Plenum Press, New York.

Umiel, T., 1977, Development of specific suppressor cells of embryonic or neonatal liver origin and their possible role in immunological tolerance, in: *Developmental Immunobiology* (J. B. Solomon and J. D. Horton, eds.), pp. 323–330, Elsevier/North-Holland, Amsterdam.

Umiel, T., and Auerbach, R., 1973, Studies on the development of immunity: The graft-versus-host reaction, *Pathobiol. Annu.* **32**:27.

Umiel, T., and Globerson, A., 1974, Analysis of lymphoid cell types developing in mouse fetal liver, *Differentiation* **2**:169.

Umiel, T., and Globerson, A., 1979, Suppression of GvH mortality by mouse neonatal liver cells, *Transplant. Proc.* **11**:1497.

Umiel, T., and Trainin, N., 1976, The role of the thymus in the establishment of suppressor cells in liver chimeras, *Cell. Immunol.* **23**:232.

Umiel, T., Globerson, A., and Auerbach, R., 1968, Role of the thymus in the development of immunocompetence of embryonic liver cells in vitro, *Proc. Soc. Exp. Biol. Med.* **129**:598.

Umiel, T., Globerson, A., and Trainin, N., 1977, Development of immunosuppressor cells: Exertion of specific suppression by neonatal liver cells, *Transplantation* **24**:282.

Umiel, T., Rabinowich, H., and Globerson, A., 1981, Suppression of a response to modified self cells by mouse fetal liver cells, in: *Developmental and Comparative Immunology* (J. B. Solomon, ed.), pp. 423–429, Pergamon Press, Elmsford, N.Y.

Uphoff, D. E., 1958, Preclusion of secondary phase of irradiation syndrome by inoculation of fetal hematopoietic tissue following lethal total body X-irradiation, *J. Natl. Cancer Inst.* **20**:625.

Van Bekkum, D. W., and de Vries, M. J., 1967, *Radiation Chimeras*, Logos Press/Academic Press, New York.

Widmer, M. B., and Cooper, E. L., 1979, Ontogeny of cell-mediated cytotoxicity: Induction of CTL in early postnatal thymocytes, *J. Immunol.* **122**:291.

Wu, S., 1978, Ontogeny of cell-mediated immunity of murine thymocytes and spleen cells: In vitro mixed leukocyte culture and cell-mediated lympholysis reactions, *Differentiation* **11**:169.

Wu, S., Bach, F. H., and Auerbach, R., 1975, Cell-mediated immunity: Differential maturation of mixed leukocyte reaction and cell-mediated lympholysis, *J. Exp. Med.* **142**:1301.

Zinkernagel, R. M., Callahan, G. N., Althage, A., Cooper, S., Klein, P. A., and Klein, J., 1978, On the thymus in the differentiation of "H-2 self-recognition" by T cells: Evidence for dual recognition?, *J. Exp. Med.* **147**:882.

21

Aging and Functions of the RES

RICHARD H. WEINDRUCH and ROY L. WALFORD

1. INTRODUCTION

Among the most striking alterations occurring with aging are those involving the immune apparatus. These alterations are of two different categories and affect thymus-dependent functions to a greater extent than thymus-independent functions. A *decreased* proliferative and/or functional response capacity to a wide array of exogenous stimuli occurs postpubertally together with an age-dependent *increase* in autoimmune phenomena. The first category has been amply documented via standard tests including lymphocyte proliferation following mitogen stimulation, mixed lymphocyte reaction (MLR), humoral immune responses, and T-cell-mediated cytolysis. The second category, though clinically appreciated in its later and grosser expressions, has not been as well studied immunologically due perhaps to a lack of assay systems to measure the apparently subtle losses in self tolerance brought on by normal aging.

Investigative interest in immunogerontology sharply increased and diversified during the late 1970s. Studies in this area have almost exclusively used either rodents (usually mice) or humans. As discussed elsewhere, the most gerontologically relevant murine studies have employed long-lived strains as opposed to short-lived strains prone to develop a specific disease (Gottesman and Walford, 1982). A representative picture of age changes in survival, immune response capacity, and autoimmunity for a long-lived mouse strain is shown in Figure 1. We herein review this field beginning with a discussion of immunological theories of aging. The subject is next treated at the organismic level via an examination of immunosenescent contributions to mortality and to certain diseases of old age. A brief review of age alterations in the RES at the organ level follows. We next survey the diverse senescent influences at the cellular/subcellular level, which comprises the main focus of this review. Finally, strategies

---

RICHARD H. WEINDRUCH and ROY L. WALFORD • Department of Pathology, School of Medicine, University of California, Los Angeles, California 90024. Supported by USPHS Research Grant AG 00424 and National Research Service Postdoctoral Fellowship CA 9030 from the National Cancer Institute.

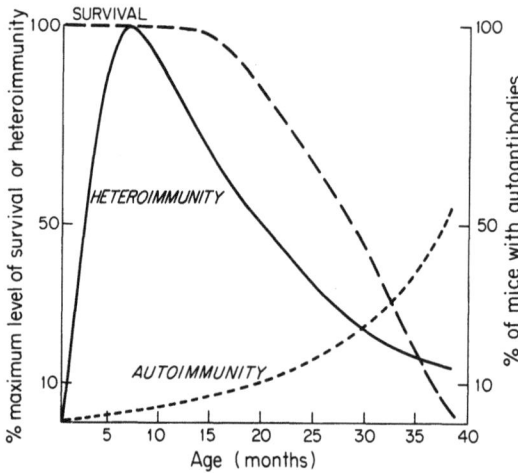

FIGURE 1. Characteristic age-related changes in survivorship, immune response capacity, and autoimmunity for long-lived mouse strains.

which decelerate the aging process or show potential of doing so will be examined for influences or senescent changes in the RES.

2. IMMUNOLOGICAL THEORIES OF AGING

2.1. IMMUNE DECLINE, AUTOIMMUNITY, AND AGING

How might age-related decreases in exogenous response capacities and increases in autoimmunity relate to each other and to aging of the whole animal? Early notions about the role of autoimmunity in the aging process (Walford, 1962; Burnet, 1959) concerned the possible emergence of forbidden clones of autoreactive cells. A more recent view (Walford, 1969, 1974) holds that autoimmune phenomena of aging may reflect a disturbance in tolerance homeostasis, suggesting a deregulation of the immune system rather than a *de novo* arising of forbidden clones. Further, this disturbance in self tolerance is postulated to cause a chronic graft-versus-host-like reaction (GVHR) in the animal. Table 1 lists features common to the chronic GVHR and aging. This view places the key role of immune phenomena in aging at the level of pathogenesis and does not necessarily explain the etiology of aging, which probably lies at a deeper level and could involve a number of other phenomena (e.g., DNA repair defects, error catastrophe, somatic mutation, hypothalamic dysfunction). It is noteworthy that recent demonstrations of immunorecognition systems in all major groups of multicellular animals (Hildemann, 1978) do much to dispel a key objection against an important general role for immune phenomena in aging.

Immune functional decline likely contributes to age-associated increases in susceptibility to infectious, neoplastic, and autoimmune diseases. Whether decline in immune function or increase in autoimmunity is the primary event has been discussed (Kay, 1976; Walford, 1974). It may well be that declining immune function and autoimmunity interact continually and generate a progressively dysfunctional state both immunologically and generally in the aging animal.

TABLE 1. FEATURES COMMON TO THE CHRONIC GVHR AND AGING

Lymphoid depletion and hypoplasia, with increased incidence of lymphoma or leukemia.
Thymic atrophy
Increase of plasma cells in lymphoid organs
Changes in skin (dermatitis, atrophy) and hair (graying)
Amyloidosis
Weight loss—"shriveling up" of body
Changes in ratio of soluble/insoluble collagen
Increase in renal lysozyme
Activation of latent tumor viruses?
Autoantibodies:
    To red blood cells, lymphocytes, nuclear antigens
Humoral immune response capacity:
    Markedly decreased to T-dependent antigens
    Less or no effect on T-independent antigens
Hyper- or dysgammaglobulinemia
Cellular immunity:
    Variable effect on suppressor T cells
Strain-specific relation to main histocompatibility complex

2.2. ROLE OF THE MAJOR HISTOCOMPATIBILITY COMPLEX

The immune system of all vertebrate species so far examined is largely controlled by a cluster of genes, the major histocompatibility complex (MHC) located on chromosome No. 6 in humans (the HLA system) and No. 17 in mice (the H-2 system). The best studied of the MHCs is *H-2*, a region large enough to house many more genes than the dozen or so now identified. The genes within *H-2* (reviewed by Klein, 1979; Paul and Benacerraf, 1977; Shreffler, 1977) mainly code for cell surface proteins displayed on almost all body cells; however, the products of certain regions are displayed only on T cells, B cells, and macrophages. Immunologically, *H-2* loci chiefly affect T-dependent functions including the development of: the MLR, the cell-mediated lymphocytotoxic reaction (CML), an immune response to certain antigens, suppressor T lymphocytes. Also *H-2* dependent is susceptibility vs. resistance to some spontaneous neoplasms and virally induced leukemias. Klein (1978) lists 59 traits reported as associated with or linked to *H-2*. Similarly, a large number of traits have been listed by Amos and Kostyu (1980) for HLA. The notion that the ultimate biological function of MHC-coded antigens lies in allowing lymphocytes to recognize self from non-self is strongly supported (see Meredith and Walford, 1979). Therefore, an influence of MHC on aging would certainly be predicted by the immune/aging theories just described. As elaborated elsewhere (Walford, 1979), in addition to MHC genes per se, genes concerned with other gerontologically important processes (e.g., DNA repair, superoxide dismutase, mixed-function oxidases) are linked to the MHC. Chromosome No. 17 in the mouse and No. 6 in man might thus contain one or more multigene families concerned with processes of recognition, protection, and repair, in short with homeostatic life maintenance processes. Existence of these "supergene" regions would accord with conclusions of Sacher (1975) and Cutler (1975) that the primary aging processes may be controlled by a relatively small number of genes or gene regulatory systems.

The strongest, most direct evidence supporting a major immunological component in aging comes from study of mouse strains congenic for *H-2* (i.e., whose genomes differ from one another only within this relatively short region of chromosome No. 17. Smith and Walford (1978) compared seven strains congenic for *H-2* on a C57BL/10 background, three on a C3H background, and four on an A strain background. They reported that this region significantly influenced mean and maximum life spans, aging rate, and both the frequency and age-specific incidences of diverse spontaneous cancers, Popp (1978) found that insertion of the $H-2^n$ allele into a C57BL/10 background produced premature graying of the hair and reduced longevity. Meredith and Walford (1977) measured lymphocyte proliferation in response to B- and T-cell mitogens as a function of age in the *H-2*-congenic lines just described. They reported that *H-2* strongly influences the age-specific development and loss of proliferative potential. *H-2* effects became more striking with advancing age. The long-lived strains tended to have the best preservation of T-cell mitogenic function in old age. In sum, these data suggest that *H-2* influences aging, its pathologies, peak response levels, and the rates of change in response capacity during growth and aging.

Evidence for an MHC effect on aging also comes from human studies. The HLA data do not provide the direct support offered by the mouse studies since HLA congenic humans are, of course, unavailable. Nevertheless, associations exist between some highly immunologically-influenced diseases (certain autoimmune diseases, lymphoma, leukemias, and other cancers) and HLA antigenic makeup (Amos and Kostyu, 1980; Naeim and Walford, 1981; Dausset and Svejgaard, 1977). HLA effects on longevity have also been observed. An increased heterozygosity at the HLA-B locus with age has been reported (Converse and Williams, 1978; Macurova *et al.*, 1975). Macurova *et al.* (1975) also observed a significant age-related increase in W10 (now called HLA-Bw40). A decrease with age in incidence of HLA-B8 and A1-B8 was reported in females (Greenberg and Yunis, 1978), which might relate to the correlation of HLA-B8 to autoimmune susceptibility. A significant increase in HLA-Bw40 was reported in people over 70 years (Yarnell *et al.*, 1979), but these authors denied the importance of this observation because the frequency did not continue to increase in the older over-70 cohorts.

3. ORGANISMIC LEVEL: IMMUNOLOGICAL ASPECTS OF MORTALITY AND THE DISEASES OF AGING

The "diseases of aging" are expressions of basic aging. These contribute to an exponentially increasing risk of death with advancing age and include certain cancers, arteriosclerosis, diabetes, some autoimmune diseases, hypertension, renal disease, and infections. Other age-associated diseases such as amyloidosis and rheumatoid arthritis, while not usually the cause of death, probably place the elderly in a less adaptable state. In this regard it is noteworthy that many of the elderly do not die of overt disease but are killed by accidents (which also increase exponentially in incidence with age).

Various features of immunological aging are associated with a greater risk of death. Older subjects with autoantibodies in their sera may exhibit a decreased expectancy of survival (Mathews et al., 1973). The mortality rates of individuals over 80 years of age showing low response profiles in tests of delayed hypersensitivity were higher over the next 2 years than in age-matched nonhyporesponsive individuals (Roberts-Thomson et al., 1974). Buckley et al. (1974) observed an increased mortality rate for aged persons with low (but not abnormal) serum levels of IgG. In an aged population of 71- to 99-year-old humans, 2-year survivorship was highly related to performance in a Con A-activated suppressor cell assay in which the initial suppressor capacity of cells from survivors averaged 20.6% versus 0.1% for nonsurvivors (Hallgren and Yunis, 1981).

Immunological inputs into some major pathologies of old age are summarized in Table 2. For certain diseases (e.g., rheumatoid arthritis, amyloidosis), immune system links are clear-cut, whereas for others (e.g., vascular disease), present evidence for a pathogenic involvement of immune mechanisms must be regarded as substantial but inconclusive. Clearly, many questions remain unanswered regarding the pathogenesis of these conditions. Since these are age-related diseases of variable immunological dependency, it seems likely that insights into pathogenesis could be gained by applying observations from the rapidly growing body of knowledge concerning cellular and molecular immunosenescent changes.

Recent studies point toward hastened immunological senescence in certain models of accelerated human aging including Down's syndrome (Walford et al., 1981a) and other chromosomal instability syndromes (Gatti and Walford, 1981). These findings are discussed in some detail elsewhere (Walford et al., 1981b).

TABLE 2. IMMUNOLOGICAL ASSOCIATIONS WITH CERTAIN DISEASES OF OLD AGE

Disease	Immunological association	Reference
Rheumatoid arthritis	Strong HLA association	Stastny (1978)
	High incidence of Ig complexes in bone and joint fluids	Pope et al. (1974)
Amyloidosis	Senile amyloid consists largely of Ig light chains	Wisniewski and Terry (1976)
	Associated with both immunodeficiency and immune hyperactivity	Anderson et al. (1977)
Diabetes (onset at maturity)	Increased incidence of autoantibodies	Whittingham et al. (1971)
CNS pathology	Increased incidence of brain-reactive autoantibodies in aging mice	Nandy et al. (1975)
	HLA association with Alzheimer's disease	Henschke et al. (1978), Cohen et al. (1979)
	Presence of Ig in amyloid fibrils of senile plaques in Alzheimer's disease	Ishii and Haga (1976)

(continued)

TABLE 2. (Continued)

Disease	Immunological association	Reference
Vascular disease	Immunosuppressive drugs inhibit development of rabbit atherosclerosis	Poston and Davies (1974)
	Rabbit atherosclerotic lesions produced via immunization with elastin and aortic glycoproteins	Robert and Robert (1973)
	Arteries in transplanted organs subject to rejection are prone to develop arteriosclerotic lesions	Finn et al. (1976)
	HLA association with ischemic heart disease	Pandey et al. (1978)
Neoplasia	Humans and animals bearing tumors often show a generalized immunodeficiency	Kamo and Friedman (1977)
	T cells from old mice enhance the growth of a transplantable lung carcinoma	Gozes and Trainin (1977)
	H-2-congenic mice differ in susceptibility to spontaneous cancers	Smith and Walford (1978)
	Some human cancers show HLA associations	Murphy et al. (1977)

4. ORGAN LEVEL

Aging strikingly alters the sizes and histological patterns of two RES organs, the thymus and appendix (Walford, 1969). Any structural changes in the spleen, lymph nodes, and bone marrow occur later in life and are subtler than the involution of these two central lymphoid organs. Mounting evidence suggests that thymic involution plays a key role in the genesis of aging changes in the RES. On the other hand, little is known of the role of the appendix. Two immunological effects relating to the thymus are likely operative. First, thymic involution is associated with a decline in serum thymic hormone levels (Bach and Beaurain, 1979; Lewis et al., 1978; Bach et al., 1975) presumably leading to age-dependent, function-specific losses in T-cell differentiative capacities (Hirokawa and Sado, 1978; Hirokawa and Makinodan, 1975). Second, there is strong evidence that T cells learn to distinguish self from non-self in the thymus (Zinkernagel, 1978). Clearly, disruptions in this process could directly contribute to the autoimmune aspect of aging.

It has long been appreciated that the human thymus undergoes a marked loss of mass beginning at about puberty (Boyd, 1932). Microscopically, thymic involution occurs chiefly via a loss of cortical lymphocytes (Hirokawa, 1978). In CBA mice, the number of T cells per thymus decreases by about a factor of 10 from 3 to 24 months of age (Kay, 1979). With regard to the epithelial cell component, three morphological types of thymic epithelial cells have been described in mice, and these are not lost at equal rates with aging (Hirokawa, 1978). In one rat strain (but not in others), proliferation of epithelial cells suggestive of a true hyperplasia was observed in 2- to 4-year-old thymuses (Meihuizen and Burek, 1978).

Thymic tissue shows autonomy in its growth/involutional behavior as shown by the fact that weights of syngeneic mouse thymic grafts were most strongly determined by the age of the graft and not the age of the host (Metcalf *et al.*, 1961). Further, the growth of each graft was not affected by the presence of as many as 23 other grafts in the same host (Metcalf, 1965). Observations of this sort led Burnet (1974) to suggest that the thymus serves as a biological clock. Interestingly, some very long-lived mouse strains such as CBA reach a peak thymic weight at 3 months of age compared to 1 month for much shorter-lived strains, such as A and NZB (Yunis *et al.*, 1973). Other observations likewise point toward an important role of the thymus in aging. Many similarities exist between young genetically thymusless mice and old genetically normal mice (Pantelouris, 1974). Also, thymectomy of adult mice has been found to decrease longevity (Jeejeebhoy, 1971), hasten the appearance of antinuclear antibodies (Teague *et al.*, 1970), and accelerate both age-associated declines in immunological vigor (Pachciarz and Teague, 1976; Miller, 1965) and increases in autoreactive lymphoid cells (Carnaud *et al.*, 1977).

Although mostly thought of in terms of immunology, the thymus is deeply involved in extraimmunological activities. It influences endocrine/neuroendocrine development of mice as illustrated by the finding that genetically athymic mice and neonatally thymectomized normal mice show derangements in thyroidal and gonadal functions very early in life, which persist (Pierpaoli and Besedovsky, 1975). Thymectomy of 2-day-old but not 10-day-old mice caused about a 2-week delay in vaginal opening time (an index of puberty) (Besedovsky and Sorkin, 1974). The involvement of the thymus in both endocrine and immunological functions is noteworthy in view of mounting evidence for a complex network of immune system–endocrine system interactions (Besedovsky and Sorkin, 1977). Describing and explaining these interactions and how they might be altered by age will form an important investigative area.

Immunosenescent alterations occur in peripheral blood mononuclear cell populations. Some cross-sectional studies suggested that the absolute number of human peripheral blood lymphocytes (PBL) decreases in the elderly (Kishimoto *et al.*, 1978; MacKinney, 1978; Diaz-Jouanen *et al.*, 1975), whereas other investigators detected no age effects (Gupta and Good, 1979; Weiner *et al.*, 1978; Hallgren *et al.*, 1978). An ongoing longitudinal study of humans demonstrated constancy in total white blood cells, monocytes, granulocytes, and lymphocytes in peripheral blood over a 10-year observation period (Adler and Nagel, 1981). Changes in representation of certain lymphoid cell subtypes have been observed. These data are discussed in Section 5. Longitudinal data regarding age effects on human serum Ig levels are contradictory. Levels of IgG and IgA have been reported to increase with age without fluctuation in IgM (Buckley *et al.*, 1974). Another study failed to show altered levels of those Ig classes with age (Adler and Nagel, 1981). A cross-sectional study also found no changes (Czlonkowska and Korlak, 1979). Sera from humans over 80 years showed a restriction if Ig heterogeneity (Riesen *et al.*, 1976). In CBA mice, IgG1 and IgG2 levels increased from adulthood (6 months) into early senescence (24 months), but systematic changes later in life were not observed (Haaijman *et al.*, 1977a). Thus, it is difficult to account for the immunodeficiency of aging on the basis of serum Ig.

The frequency of sera positive for autoantibodies clearly increases in old humans and mice (Riesen et al., 1976; Hallgren et al., 1973; Siegal et al., 1972; Walford, 1969) as does the frequency in sera of immune complexes (Rosenthal, 1978). Another common alteration in the sera of old humans and mice is the appearance of an idiopathic paraproteinemia (benign monoclonal gammopathy) which attains a strain-dependent incidence of 10–70% in mice by age 27–30 months (Radl et al., 1978). These paraprotein-producing cells die out after three or four *in vivo* transfers, which contrasts to the situation for B-cell malignancies (Radl et al., 1979).

5. CELLULAR/SUBCELLULAR LEVEL

5.1. STEM CELLS

Deficiencies in lymphohemopoietic stem cell numbers accompany aging in some but not all strains of mice. The direction of age-associated changes in bone marrow cellularity were found to be highly strain dependent in mice (Kay, 1978). In a long-lived hybrid mouse strain, about 90% of the stem cells were located in the bone marrow (Chen, 1971). Although the concentration of marrow stem cells fell with age in these mice, the number of stem cells remained fairly constant due to an age increase in total marrow cellularity (Chen, 1971). Similarly, another strain (RFM/Un) showed a decrease in marrow stem cell concentration with almost all of the deficit developing between 50 and 80 weeks of age (Davis et al., 1971). A different result obtained in SAS/4 mice, which displayed no major changes in marrow stem cell concentration even in the very old (Coggle and Proukakis, 1970). A more thorough study (Coggle et al., 1975) confirmed this observation and also provided evidence for age-related decreases in splenic stem cell numbers. No major decreases in the number of stem cells per femur occurred in adult-thymectomized or control mice of three inbred strains during the first 16 months of life (Pachciarz and Teague, 1976).

As discussed by Popp and Popp (1979), the pool of bone marrow stem cells is quite heterogeneous and most methods employed to assay aging effects have not differentiated between specific cellular changes versus shifts in distribution of cell types.

Evidence is conflicting about possible age deficits in stem cell capacities for both growth and the production of functional progeny. In one study, the proliferative capacity of bone marrow stem cells from old mice decreased with age and continued to display this defect even after a sojourn within a young recipient (Albright and Makinodan, 1976). The same workers also reported a deleterious effect of an "aged" environment on the growth potential of young stem cells, as did Coggle et al. (1975). An age-related decline has been reported in the ability of mouse bone marrow to generate B cells (Kishimoto et al., 1976; Farrar et al., 1974; Price and Makinodan, 1972a). An age-associated decrease in mouse bone marrow T-cell progenitors has been reported, based on decreased thymic regeneration occurring in lethally irradiated syngeneic recipients injected with marrow from old compared to young mice (Tyan, 1977). In contrast to the above, the findings of Harrison and his associates do not support the existence of an age deficit in stem cell growth and function:

1. Marrow cell lines from old and young control mice functioned equally well for 15–21 months in maintaining the health of lethally irradiated recipients (Harrison, 1975). Also, marrow lines from some old donors functioned normally for periods of time much greater than the life span of the donors. A later study (Harrison, 1979a) showed that erythropoietic stem cell lines from old mice produced erythrocytes for 100 months, and any eventual functional decline could be attributed to the number of transplantations. Harrison (1979b) has presented a lucid review of aging studies which examined the proliferative capacities of erythropoietic stem cells.
2. Grafts of marrow or spleen cells from old mice produced antibody-forming cells as effectively as did grafts from younger controls if tested 5–10 months after transplantation (Harrison and Doubleday, 1975). A subsequent study indicated that little or none of the age decrease in either the generation of antibody-secreting cells in response to injected sheep red blood cells (SRBC) or in the proliferative response of splenic lymphocytes to phytohemagglutinin (PHA) is intrinsic to old lymphoid stem cells (Harrison et al., 1977).
3. Marrow stem cell lines from old and young donors showed equally rapid growth rates in the spleens of lethally irradiated syngeneic recipients (Harrison et al., 1978). It was also observed that the stress of serial transplantation acutely lowers stem cell proliferative powers to a far greater extent than does a lifetime of function in an undisturbed mouse.

These discordant observations about stem cells could in part reflect the time period separating cell transfer and functional assay. Harrison's group allowed for a longer interval than did those groups reporting an age-associated defect, and maintain that to assay shortly after grafting may test for responses of differentiated donor cells already immunocompetent when transferred. In sum, the collective data make it difficult to ascribe deficits in stem cell number or proliferative capacities as a major contributor to age changes in the RES.

5.2. MACROPHAGES

Most findings point toward normal or increased numbers and activity of "old macrophages." No major changes in blood monocyte counts were observed in elderly humans (Munan and Kelly, 1979). The total number of peritoneal macrophages harvestable from aged mice was significantly greater than from young adult mice and the phagocytic activities of cells from both ages were either comparable or showed an age increase in some tests (Perkins, 1971). Young and old mice did not differ in the percentage of splenic latex-engulfing cells (Gozes and Trainin, 1977). The activity of three lysosomal hydrolases of mouse peritoneal macrophages increased slightly with age, and more free enzyme (as opposed to membrane-bound enzyme) was found in macrophages from old mice (Heidrick, 1972). The age deficit in mouse lymphocyte proliferation (see below) to PHA, lipopolysaccharide (LPS), and in the MLR could not be attributed to old macrophages (Callard, 1978; Callard and Basten, 1977). Likewise, the lower number of plaque-forming cells (PFC) generated in splenic cell cultures from old mice stimulated by SRBC was not due to a defect in cells adherent to plastic (70% macrophages) but appeared to be caused by nonadherent cells (94% lymphocytes) (Heidrick and Makinodan, 1973). Cantrell and Elko (1973) found no age effect in carbon clearance in young, growing rats (all less than 1 year old) when allowances were made for differences in body sizes. On the other hand, the maximal velocity of carbon clearance fell sharply in rats between 1 and 14 months of age (Bilder and Denckla, 1977). Neither of these studies tested truly old rats.

Three reports describe impaired functioning of macrophages from aging or old rodents. Old mice showed a delayed onset of macrophage activation during acute *Toxoplasma gondii* infection, whereas no age defect in lymphocyte transformation to this agent occurred (Gardner and Remington, 1978). Peritoneal cells from old mice were not as efficient as cells from young mice in supporting an *in vitro* PFC response to a T-independent antigen (Nordin and Buchholz, 1981). Macrophage-mediated natural cytotoxicity was reduced in 12- to 16-month-old rats compared to rats 2–5 months of age (Keller, 1978).

Even though the above data do not reveal a major macrophage contribution to immunosenescence, further study appears warranted since rapidly mounting evidence attests to the diversity and importance of macrophage function.

5.3. B CELLS

It remains unresolved if numbers of proportions of B cells are altered in old age. Some investigators have found that human PBL bearing either surface Ig or complement receptors show age-related increases in relative numbers (Reddy and Goh, 1979; Hallgren *et al.*, 1978; Girard *et al.*, 1977; Diaz-Jouanen *et al.*, 1975). Clot *et al.* (1978) reported increases in relative and absolute numbers of PBL bearing surface Ig in the aged, but a decrease in those with complement receptors, perhaps indicating a differential effect of age on B-cell subtypes. By contrast, other investigators did not find any effects of age on human B-cell levels (Adler and Nagel, 1981; Gupta and Good, 1979; Kishimoto *et al.*, 1978; Ben-Zwi *et al.*, 1977; Davey and Huntington, 1977; Weksler and Hutteroth, 1974). In mice, there is good evidence for a moderate rise occurring with aging in numbers and proportions of B cells in the spleen, lymph nodes, and bone marrow (Kay, 1978; Callard *et al.*, 1977). Quantitation of mouse cells containing cytoplasmic immunoglobulin indicated that the major source of these cells early in life is the spleen, changing later in life to bone marrow (Haaijman *et al.*, 1977b).

The proliferative capacity of mouse splenic lymphocytes (measured by [^{3}H]thymidine incorporation into DNA) responding to B-cell stimulants such as purified protein derivative (PPD), LPS, and pokeweed mitogen (PWM, which also stimulates cortisone-resistant T cells) declined in advanced age in most (Abraham *et al.*, 1977; Callard *et al.*, 1977; Gerbase-DeLima *et al.*, 1974; Mathies *et al.*, 1973) but not all (Nariuchi and Adler, 1979; Kay, 1978) studies. LPS-stimulated B-cell colony formation seemed identical in splenocytes from young and old mice (Duwe *et al.*, 1979). In general, B-cell proliferative deficits do not appear as severe or to occur as early in life as T-cell deficits. The frequency of LPS-reactive splenic cells in mice declined with age (Abraham *et al.*, 1977; Andersson *et al.*, 1977; Callard *et al.*, 1977; Price and Makinodan, 1972a), but those old cells still capable of responding did not appear to differ from young responding cells in either total cell cycle time or number of division cycles undergone (Abraham *et al.*, 1977). A study of H-2-congenic mice showed that H-2 influenced patterns of aging changes for splenic cell responses to PPD, LPS, and PWM (Meredith and Walford, 1977). Not much information is available about human B-cell proliferative responses, although an increased response by PBL from elderly persons to anti-Ig antibodies has been reported (Weiner *et al.*, 1978).

Primary responses (As judged of PFC) to a variety of T-independent antigens were reduced in aged mice (Nordin and Buchholz, 1981; Friedman and Globerson, 1978a,b; Callard *et al.*, 1977; Smith, 1976; Gerbase-DeLima *et al.*, 1974). Some mechanistic insights to these deficits are being gained. Callard *et al.* (1977) studied the response to pneumococcal polysaccharide type III and could demonstrate neither T-cell suppression nor age changes in levels of antigen-binding splenic cells. Studying the response to polyvinyl pyrrolidone, Friedman and Globerson (1978a,b) also could not detect T-cell suppression. These data point toward intrinsic B-cell defects. Along these lines, age changes in T-cell–B-cell cooperation in the allogenic effect point toward problems in recruiting old B cells (Popp and Francis, 1979). It is noteworthy that results of most gerontological studies of T-cell dependent humoral immunity (discussed below) do not rule out possible involvement of intrinsic B-cell lesions.

5.4. T CELLS

The influences of aging on T cells and T-cell-dependent functions are better characterized than for B cells, and are probably more gerontologically significant. It is now believed that, in essence, T cells function via cell surface antigen recognition. Paul and Benacerraf (1977) argue that T cells accomplish their most basic function, self vs. non-self recognition, by focusing in on MHC-coded surface antigens on living cells and seeking variants. This idea regarding the major role of T cells coincides with the key function predicted to go awry with advancing age (Walford, 1969, 1974). The notion that T-cell-like functions are evolutionarily older than those of B cells is well supported (Hildemann, 1974) and also bolsters the importance of T cells to RES aging. Age changes in T-cell numbers and functions will now be discussed.

5.4.1 T-Cell Numbers

Human T-cell levels have been measured in adulthood and old age by the E rosette assay. Most studies have detected lower numbers and/or lower percentages of T cells in the aged (Czlonkowska and Korlak, 1979; Reddy and Goh,

1979; Clot et al., 1978; Kishimoto et al., 1978; Weiner et al., 1978; Ben-Zwi et al., 1977; Girard et al., 1977; Diaz-Jouanen et al., 1975; Augener et al., 1974; Carosella et al., 1974; Smith et al., 1974), but others have not found a decrease (Adler and Nagel, 1981; Gupta and Good, 1979; Portaro et al., 1978; Davey et al., 1977; Weksler and Hutteroth, 1974). An increase in the number of circulating T cells in very old people (9th and 10th decades) has been reported (Hallgren et al., 1978) as well as an age-related rise in the proportion of T cells with receptors for the Fc portion of IgG (Gupta and Good, 1979; Kishimoto et al., 1978). It is perhaps noteworthy that this subclass contains T cells capable of suppressing B-cell differentiation to plasma cells after PWM stimulation.

Mouse T-cell levels have been quantitated via expression of surface Thy-1 antigen on splenic and lymph node lymphocytes. Most studies using long-lived, non-NZB-type strains did not observe decreases in T cells in aging mice (Gozes and Trainin, 1977; Callard and Basten, 1977; Fernandes et al., 1976a; Gerbase-Delima et al., 1974; Stutman, 1974; Hori et al., 1973; Kishimoto et al., 1973). An exception was the report of Brennan and Jaroslow (1975) describing a linear 50% fall in splenic T cells from 6 to 22 months of age along with a decrease in the amount of Thy-1 per cell. The percent of Thy-1$^+$ lymphocytes in mouse blood decreased between 6 weeks and 1 year of age (Olsson and Claesson, 1973). These last workers also observed a decrease in the ratio of short-lived to long-lived Thy-1$^+$ cells in blood, spleen, and lymph node during this same time span. Thus, major changes may occur within T-cell compartments not grossly changing in size in the aging RES.

Studies using more functionally informative T-cell surface markers such as the Ly system (Cantor and Boyse, 1977) are needed. Cantor et al. (1978) provide the only report of this sort, describing changes during the first year of life in the splenic makeup of Ly-123$^+$, Ly-1$^+$, and Ly-23$^+$ subtypes in "normal" mice (BALB/c) and NZB mice. NZB mice are born with a deficit of Ly-123$^+$ cells; however, BALB/c mice take a year to gradually reach the same low level.

Thus, age-related decreases in T-cell numbers are more clearly documented in humans than in mice, but the difference may well be due to the source of the cells (peripheral blood vs. spleen, lymph node) and the marker measured (SRBC receptor vs. Thy-1 antigen).

5.4.2. Persistence of Thymic Function

How does age (and involution) affect the differentiative capacities of the thymus? Hirokawa and Makinodan (1975) implanted thymic lobes from neonatal, 3-, 24-, or 33-month-old mice into 3-month-old irradiated recipients reconstituted with syngeneic young bone marrow cells, and which had been thymectomized 1.5 to 2 months previously. T-Cell emergence was assessed over the ensuing 3 months by several criteria. Only neonatal thymic grafts led to repopulation of T-cell-dependent lymph node areas and brought splenic cell PHA and Con A responses to control levels. Thymuses from neonatal and 3-month-old donors reconstituted responsiveness to injected SRBC equally well and restored total number of Thy-1$^+$ cells per spleen. Recipients of 24- or 33-month-old thymuses had levels of splenic Thy-1+ cells only slightly different from those receiving younger tissue. The MLR response of the four recipient groups were equal to each other and to age-matched controls. Thus, very old thymus retains some function. Hirokawa and Sado (1978) further defined differences between mouse thymuses ranging in age from 1 day to 11 weeks. The mitogenic responsiveness of lymph node and splenic cells declined when thymus grafts were from donors 1 month or older. A decline of the *in vivo* SRBC PFC response occurred in mice grafted with thymus from 11-week-old donors. It appears that the thymus loses sequentially during life the ability to influence some but not all T-cell functions.

It is of interest that young recipients were used in the preceding studies. Bach and Beaurain (1979) have observed that newborn thymus is less efficient in restoring serume thymic factor levels in middle-age- than in young-adult-thymectomized recipients, thereby implicating extrinsic factors in older mice involved in age-related decreases in thymic factor levels.

Adult thymectomy constitutes another model useful in evaluating the extent of thymic function during adulthood. Thymectomy of 2- to 4-month-old mice lessened the allogeneic GVH activity of lymph node lymphocytes beginning about 4 months later (Miller, 1965; Taylor, 1965). Thymectomy of mice at 2–3 months of age led to a deficit in the *in vivo* PFC response to SRBC appearing at 4 to 10 months postthymectomy (Miller, 1965). Splenic cells from 30-month-old mice thymectomized at 12–15 months of age exhibited greatly reduced responsiveness to SRBC (Perkins *et al.*, 1975). Similarly, 24- to 25-month-old mice thymectomized at 2–3 or 12–15 months of age displayed T-cell deficits such as increased sensitivity to *in vivo* allogeneic tumor cell challenge and drastically reduced PHA responsiveness (Peterson *et al.*, 1975). Thymectomy of mice at 6–8 weeks of age accelerated the normal decline in splenic cell responses to two T-cell mitogens (Pachciarz and Teague, 1976).

The collective findings suggest that thymic functions may decline selectively with age, some diminishing sooner due perhaps to reduced thymic hormone production, or alternatively, aging may cause preexisting, intrinsic defects in T-cell precursors, rendering them unresponsive to reduced but functionally adequate thymic hormone levels.

5.4.3. Declines in T-Cell Proliferative Capacities

While negative reports exist (e.g., Agarwal *et al.*, 1978; Portaro *et al.*, 1978; Halbrecht *et al.*, 1977; Stutman, 1974), most studies agree that T cells from old humans and mice respond less vigorously than cells from young animals to proliferative stimuli such as plant mitogens (PHA, Con A, PWM) or alloantigens (MLR, GVHR). Effects of age on these responses are discussed here with emphasis on cellular/molecular explanations of the repeatedly described declines.

The early work on age effects on T-cell mitogen responses (e.g., Weksler and Hutteroth, 1974; Hori *et al.*, 1973; Mathies *et al.*, 1973) described deficiencies in old mouse and hamster splenic cells or old human PBL in the incorporation of [^{3}H]thymidine into DNA following *in vitro* mitogen stimulation. Responses in old age usually ranged from 0 to 50% of youthful peak values. More recent mouse data (Kay, 1978; Meredith and Walford, 1977; Fernandes et al., 1976a) indicate that the timing of maturational and senescent changes in proliferative abilities is strain specific, lymphoid organ specific, mitogen specific, and H-2 influenced. Interestingly, strains which are slow to mature immunologically (e.g., CBA) or H-2 congenics (e.g., C57BL/10.RIII/Sn) are long-lived, whereas very-short-lived strains such as NZB and (NZB × NZW)F_1 mice attain immunological peaks much earlier in life. For human PBL, recent findings have continued to describe age-lowered PHA and Con A responses (Adler and Nagel, 1981; Czlonkowska and Korlak, 1979; Hallgren et al., 1978; Kishimoto et al., 1978).

These deficits in T-cell proliferation after PHA or Con A stimulation seem to relate to defects *intrinsic* to old T cells. Old mouse splenic cells did not differ from young cells in optimal mitogen dosage or in the timing of peak response (Meredith and Walford, 1977; Callard and Basten, 1977), suggesting that different *in vitro* requirements for old cells were probably not contributing to the low responses. Likewise, the low PHA response of old splenocytes could not be explained by reduced numbers of T cells, by macrophage regulation, or by the release of soluble inhibitors during culture (Callard and Basten, 1977). The chromatin of PHA-stimulated PBL from old humans did not bind as much [^{3}H]actinomysin as young PBL, indicating that older chromatin may be more condensed than young chromatin (Preumont et al., 1978). Also, the effects of aging on cyclic nucleotide levels in unstimulated and mitogen-stimulated mouse splenic cells suggest intrinsic defects (Tam and Walford, 1978).

Strong evidence suggests that age-related depression of PHA and Con A reactivity is at least partially due to decreases in numbers of responding cells. This loss of responding cells was not detectable in mouse spleen by measuring the frequency of cells binding ^{125}I-PHA (Callard and Basten, 1977; Hung et al., 1975a). Colchicine inhibition of mitogenesis indicated that fewer old CBA splenic cells (Callard and Basten, 1977) and old human PBL (Inkeles et al., 1977) respond to PHA. The mitotic index of PHA-stimulated PBL fell in old age (Tice et al., 1979; Barbaruk, 1974). Autoradiographic studies revealed that a lower percentage of splenocytes from old mice responded to PHA and Con A (Abraham et al., 1977; Hung et al., 1975b). It remains unresolved if those cells from the aged still capable of responding do so in "youthful" way. Studies using human PBL showed that old cells have longer mean cell cycle times and are less capable of repeated divisions (Tice et al., 1979; Inkeles et al., 1977), whereas Con A (or LPS)-stimulated splenic cells from young and old mice did not differ in cell cycle time or reentry into mitosis following an initial cycle (Abraham et al., 1977).

Alterations in T cell mitogenesis may be involved in senescent changes in Con A-induced suppressor activity of human PBL. Old Con A-activated PBL did not suppress Con A responses of control cells as potently as did young Con A-activated PBL (Hallgren and Yunis, 1977, 1981). An age decrease was also observed for Con A-induced suppression of Ig production by PWM-stimulated PBL (Kishimoto et al., 1979). On the other hand, Rice et al. (1979) found no effect of age on Con A-induced suppression of the MLR, but did detect age-related increases in both an "adherent cell suppressor system" and "prostaglandin-related suppressor system." Con A-activated PBL from the elderly produced less suppressor factor than young PBL but were more susceptible to influences of this uncharacterized factor (Antel and Arnason, 1979; Antel et al., 1978). This recalls the observation that the sensitivity of PBL to prostaglandin E_2-mediated inhibition of the PHA response increased in old subjects (Goodwin and Messner, 1979). The *in vivo* relevance of these findings is unclear.

The MLR also declines in old age, both for lymph node lymphocytes (Merhav and Gershon, 1977; Meredith et al., 1975b) and splenocytes (Konen et al., 1973). Autoradiographic studies of the MLR using lymph node cells suggested that old mice had fewer responding cells but those still stimulatable showed the same cell cycle time and mitotic potential as cells from young mice (Merhav and Gershon, 1977). Maturational peaks for the MLR occur quite late in life for some mouse strains (Meredith et al., 1975b; Walters and Claman, 1975). The MLR also falls in aged humans (Hallgren et al., 1978; Kishimoto et al., 1978; Weksler and Hutteroth, 1974).

The MLR has been used to test for age shifts in mouse T-cell subpopulations. Synergy between lymph node cells and syngeneic thymocytes declined with age in two mouse hybrids, suggesting a particular deficiency in the recirculating lymphoid pool (probably T_2, Ly-23$^+$ cells) (Meredith et al., 1975b). On the other hand, splenic cells from old mice suppressed the MLR of young syngeneic lymph node cells (Gerbase-DeLima et al., 1975a). The MLR of thymocytes from young mice increased by a factor of 12 when the mice were injected with cortisone 2 days before assay, whereas injection of old mice did not alter their MLR, suggesting that cirtisone-resistant thymocytes lost with age may be suppressor T cells (mostly T_2, Ly-23$^+$) (Gerbase-DeLima and Walford, 1975). Thus, based on MLR manipulations, mice appear especially susceptible to age effects on T_2-like cells. Along these lines, a greater age effect on T_2-like cells was inferred from the declining PHA/Con A stimulatory ratio in aging mice (Meredith et al., 1975a).

The GVHR (the *in vivo* counterpart of the MLR) also declines in aging mice as judged by proliferation of splenic or lymph node cells within neonatal (immunologically compromised) hosts (Perkins and Cacheiro, 1977; Walters and Claman, 1975; Stutman, 1974; Kishimoto et al., 1973). It is not known which cell subtypes contribute to this decline.

H-2-congenic mice have been used to measure age effects on proliferative responses of lymph node cells to total H-2 differences and also to D-end and K-end incompatibilities (Popp, 1975). Lymphocytes from young and old mice responded to the same extent to total H-2 differences. Responses to D-end antigens were depressed when lymphocytes from old mice were tested (Popp, 1978), but responses by old cells to K-end differences did not differ from young cells (Popp, 1975). The D-end responding lymph node lymphocytes from 4-month-old but not 20-month old mice were sensitive to *in vivo* cortisone treatment (Popp, 1977). It was concluded that a population of immunocompetent cells exists in old mice which is not detectable in young mice.

5.4.4. Cytotoxic T Cells

The ability of mice to survive injection with allogeneic tumor cells is severely compromised with age (Perkins and Cacheiro, 1977; Goodman and Makinodan, 1975). This probably relates to deficits in cytolysis of allogeneic target cells by immune T cells (a T_2', Ly-23$^+$ function directed mostly to *H-2K* and *H-2D* region determinants). Splenic cells from middle-aged mice immunized with a suboptimal dose of allogeneic tumor cells showed about sixfold less cytolysis than splenic cells from young mice (Menon *et al.*, 1974). When sensitization occurred *in vitro* via the MLR, only one-half of the middle-aged spleens displayed a lower cytolytic response than young spleens (Shigemoto *et al.*, 1975). Lymph node cells from the two age groups generated killer cells *in vitro* equally well. Bach (1979) observed a slight fall in MLR-generated cytotoxicity by middle age and attributed this to coincident alterations in different T-cell subsets. Using a limiting-dilution *in vivo* transfer system, Goodman and Makinodan (1975) observed a fourfold decrease in the ability of mouse splenic cells to generate cytolytic cells. About one-half of this deficiency appeared due to decreases in relative numbers of immunocompetent precursors and the remainder to a fall in proliferative capacities of antigen-stimulated precursor cells. However, the final products of differentiation were equally effective at killing target cells. In addition to responses to allogeneic tumor cells, the cytotoxic T-cell response to lymphocytic choriomeningitis virus was delayed in onset and lower in magnitude in old mice (Doherty, 1977). Cell-mediated cytotoxicity also was reduced in elderly humans (Kishimoto *et al.*, 1978).

The activity of other non-T-cell cytolytic effectors declines with age. Natural killer (NK) cell activity in mice peaked between 1 and 3 months of age and declined thereafter (Herberman *et al.*, 1975; Kiessling *et al.*, 1975). NK cells appear to be involved in tumor surveillance (Welsh, 1978), and deficits in this function in old age could contribute to cancer susceptibility. Marginal deficits in NK activity have been reported for PBL from aged humans (Adler and Nagel, 1981; Sato *et al.*, 1979). Antibody-dependent cellular cytotoxicity decreased in mice after 3 months of age (Becker *et al.*, 1979), but did not decrease in aging humans (Edwards and Avis, 1979).

5.4.5. T-Cell-Dependent Antibody Responses: T Helper and Suppressor Activity

Primary and secondary T-dependent humoral responses have been extensively studied gerontologically in mice using both *in vivo* and *in vitro* sensitization. These responses are depressed in the aged as most often assessed by PFC numbers and less often by serum antibody titers. Primary T-dependent responses require interaction of T cells, macrophages, and B cells. Although B-cell-derived PFC are usually tallied, it is clear that age-depressed PFC numbers could arise via alterations in any of these cell types in addition to contributions from an aged environment. Age-related increases in T suppressor function, age-related decreases in T helper function, and/or qualitative B-cell defects may all be involved in the decrement.

A striking decline occurs in old mice in the splenic PFC responses to injected SRBC. For example, old mice sensitized *in vivo* produced about 1/40th as many IgM-producing PFC (direct PFC or DPFC) per 10^6 splenic cells as mice 1 year old (the age peak for this response) (Friedman and Globerson, 1978a). Numbers of IgG-producing PFC (indirect PFC or IPFC) were even more reduced by aging than DPFC. Limiting-dilution and dose–response analyses of the *in vivo* SRBC response of young and old mice (Price and Makinodan, 1972a,b) indicated that about 90% of the decreased response was attributable to intrinsic cellular deficiencies and the remaining 10% to extrinsic defects (Makinodan and Adler, 1975). Other studies also employing *in vivo* transfer of young or old splenic cells have led to similar conclusions regarding intrinsic cellular deficiencies in the SRBC response (Friedman and Globerson, 1978b; Kishimoto and Yamamura, 1971). Low responses by old splenic cells were partially restored via co-injection of young thymoctyes, suggesting a T helper cell defect among old splenic cells (Friedman and Globerson, 1978b).

The primary *in vitro* response of mouse splenocytes to SRBC also falls in old age. Makinodan *et al.* (1976) observed a 30-fold decrease due to at least three classes of alterations: an increase in numbers and/or activity of suppressor cells (the commonest change), a decrease in number of one type of immunocompetent cell occurring in excess in young mice, and a decrease in the functional efficiencies and/or numbers of immunocompetent cells in the aged. Thus, the decreased response may be for different reasons in different mice, and it is important in such studies not to pool spleens from old mice. Roder *et al.* (1978) detected an increased activity of non-T suppressor cells for the SRBC response in old spleens and bone marrow. Along these lines, non-T-cell-mediated suppression of a secondary T-dependent response has been reported to increase with age in two or three mouse strains tested (Segre *et al.*, 1981).

Primary responses to haptenic determinants coupled to xenogeneic carriers such as SRBC, bovine serum albumin (BSA), or bovine γ-globulin (BGG) require T helper function and have been studied in variously aged mice. Pooled splenic cells from middle-aged mice primed *in vitro* to trinitrophenylated SRBC (TNP-SRBC) generated about 4-fold fewer TNP-specific PFC than cells from young mice (Kishimoto *et al.*, 1976). T-Cell suppression could not be detected. The TNP-directed antibodies made by old cells were, on the average, 30-fold less avid than those made by young cells, suggesting a preferential loss with age of B cells with high-affinity receptors for TNP. Age-related decreases in antibody avidity to TNP were also reported by Doria *et al.* (1978). The primary IgE response to dinitrophenol (DNP) was reduced in magnitude and avidity in old mice (Fujiwara and Kishimoto, 1979). Goidl *et al.* (1976) studied the primary response to DNP in young and old mice immunized with DNP-BGG. Old mice generated 5-fold fewer PFC per spleen and the deficit was most severe for IPFC. Aging preferentially reduced those PFC secreting high-avidity antibody to DNP. Adoptive transfer of pooled old splenic cells and DNP-BGG into lethally irradiated, thymectomized 2-month-old hosts produced a lower number of PFC and an older avidity profile than when 2-month splenic cells were injected. The addition of young thymocytes to old splenic cells formed an inoculum which generated a more youthful avidity profile. When 2- and 24-month-old splenic cells were mixed and injected, the old cells suppressed the PFC and avidity responses of the young cells. These data suggested therefore that old mice show impaired helper and increased suppressor cell activity.

Secondary responses to xenogeneic RBC are not as severely compromised as primary responses. The secondary antibody response to rat RBC in mice primed at 1 month of age peaked at about 4 months and declined gradually thereafter (Makinodan and Peterson, 1966). However, much of this decline may be due to the time interval between priming and testing since the secondary PFC response to SRBC was quite similar in young and old mice when both were primed 1 month before boosting (Finger et al., 1972; Makinodan et al., 1971).

Evaluation of depressed secondary responses to haptens in old mice indicates that major senescent alterations occur in T cells. Krogsrud and Perkins (1977) studied the ability of carrier (BGG)-primed T cells harvested from young and old mice to interact with hapten (DNP)-primed B cells from young mice in generating DNP-specific PFC in young, irradiated recipients. The carrier-primed splenic cells to be injected were enriched for T cells. Pooled spleens were always used. They observed that primed T cells from old mice were sixfold less capable of generating DNP PFC with DNP-primed, young cells than were primed T cells from young mice. These workers viewed their findings as suggestive of age-related increases in suppressor T-cell activity and/or age deficits in the capacity of old T helper cells to activate B cells.

M. Segre and Segre (1976) also studied the secondary response to DNP. They injected young and old mice with DNP conjugated to keyhole limpet hemocyanin (KLH) and sacrificed the animals 1 month later. Splenic cell suspensions from pooled old and young spleens were mixed with DNP-KLH in a diffusion chamber, placed within a young, irradiated recipient, and assayed for DNP PFC. The peak response of cells from old mice was less than 1/10th that of young mice. When young and old DNP-KLH-primed splenic cells were mixed in the diffusion chamber together with DNP-KLH and placed in the recipient, the number of secondary PFC was significantly less than when young and old cells were cultured separately (D. Segre and Segre, 1976). Old spleens were not pooled in these studies. The suppressive effect by the old cells appeared to be dependent on T cells. Also, suppression did not occur when cells from unprimed old spleens were cocultured with primed young cells and DNP-KLH. Like Makinodan et al. (1976), Segre and Segre (1977) have observed that old mice show a variety of lesions when tested via this same system. Preliminary findings suggested that with aging, increased suppressor T-cell activity occurred first, followed by helper T-cell defects and finally by B-cell lesions.

Recent work by Callard and co-workers also points toward multiple changes reducing humoral immunity in aged mice. A variety of T-dependent antibody responses were studied via adoptive transfer of T and B cells from young or old mice into young recipients (Callard and Basten, 1978). Old B cells showed defects in both primary and secondary responses since young T cells failed to restore responsiveness of old B cells. Decreases in numbers of antigen-binding cells were not observed, suggesting qualitative B-cell defects. Old T cells did not collaborate with young B cells as well as did young T cells, indicating a loss of T helper activity. Suppressor T-cell activity was also shown by old splenic cells via mixing studies. These suppressor cells were the subject of a more recent study (Callard *et al.*, 1980). T-Cell suppression of primary responses to T-dependent and T-independent antigens occurred, as did suppression of a secondary T-dependent response. The suppressor effect was mediated by a T-cell–T-cell interaction between Ly-1$^+$ and Ly-2$^+$ cells, with the final effector cell most likely being a Ly-1$^-$, 2$^+$, Ia$^+$ T cell. These suppressor cells were widely distributed (spleen, lymph nodes, bone marrow).

5.5. BIOCHEMICAL CHANGES IN LYMPHOID CELLS WITH AGE

Relatively few reports describe biochemical alterations in lymphoid cells harvested from old animals or humans. Clearly, these effects might be expected to provide molecular explanations for the oft-described functional changes. We herein discuss three areas in which old lymphoid cells have been examined: cyclic nucleotide content, DNA repair, and cell surface/cytoskeletal alterations.

Our laboratory has studied age alterations in cyclic nucleotide levels in lymphocytes of young and old humans and mice, and in Down's syndrome, which shows many features of accelerated aging (Tam and Walford, 1980; Tam and Walford, 1978). With age the levels of cAMP were greatly decreased and cGMP increased. The activities of adenylate cyclase and guanylate cyclase were increased and decreased, respectively. Results from Down's syndrome subjects were characteristic of humans of chronologically much greater age. The changes in cyclic nucleotides with aging are paradoxical in that old lymphocytes have low cAMP levels and proliferative capacity but, in general, an increase in cAMP is believed to inhibit proliferation.

DNA repair capacity was shown to relate to aging by the studies of Hart and Setlow (1974) who demonstrated a correlation between DNA excision repair and maximum life span in a number of mammalian species. DNA repair in human PBL was reported to decline with age as measured by repair of UV-induced lesions (Lambert *et al.*, 1979) and as sensitivity to bleomycin (Seshadri *et al.*, 1979). Some evidence exists possibly linking the MHC and DNA repair capacity. DNA repair is defective in systemic lupus erythematosus (Beighlie and Teplitz, 1975) and this is an HLA-linked condition (reviewed by Walford and Bergmann, 1979) showing features of accelerated aging (Walford *et al.*, 1981b). Likewise, reheumatoid arthritis is a highly HLA-linked malady (Stastny, 1978) and may show decreased DNA repair (Matsuura *et al.*, 1978). In preliminary studies, we have observed what appear to be significant differences in DNA repair capacity in lymphocytes of mice congenic at the *H-2* locus (Walford and Bergmann, 1979).

Lymphocytes from old rodents show alterations in cell membrane composition and in the mobility and density of cell surface molecules. The density of "antigen B," Ia, and Fc receptors was decreased on the surface of splenic cells from old rats (Woda *et al.*, 1979). Old cells also showed slower capping rates for three membrane proteins and decreases in membrane fluidity and lateral diffusion. Likewise, fewer splenic B cells from old rats showed a high density of surface immunoglobulin, and capping rates were slower in old cells, as was shedding of cross-linked surface receptors (Woda and Feldman, 1979). Colchicine increased capping rates only in cells from young animals. Results in lymphocytes from old mice concur in that plasma membrane viscosity was about 20% higher than that of young mice, due probably to an elevation in the cholesterol/phospholipid ratio of the membrane (Rivnay et al., 1979). Naeim and Walford (1980) have shown that the degree of capping and patching of human lymphocyte declines greatly with age, and also in Down's syndrome. It seems clear that molecularly oriented studies will provide explanations for immunosenescent dysfunction.

5.6. AUTOREACTIVITY/TOLERANCE

Age effects on self tolerance may represent an important aspect of the pathogenesis of aging. Data providing cellular explanations for the maintenance and loss of self tolerance in young mice have only recently been published. Not surprisingly, little aging work has yet been done. The demonstration of age-dependent changes in self reactivity would seem to be a biologically relevant measure. While the functional tests usually done in aging studies (e.g., mitogenesis, MLR, SRBC) provide indices of immunosenescence, assessment of autoreactive status probably is more directly informative of immunological processes affecting life-maintenance.

Self-reactive potential clearly exists in young mice. CBA mice generated large numbers of DPFC to injected bromelain-treated syngeneic RBC (Br-MRBC) (Cunningham, 1974). (Bromelain probably exposes hidden autoantigens.) This same response increased about 20-fold following antilymphocyte serum treatment, indicating the presence of active cellular suppressor mechanisms involved in preserving tolerance (Cunningham, 1975). Young C57BL/6J splenic cells exposed in vitro to bromelain-treated isologous RBC generated not only PFC but, upon transfer to normal C57BL/6J mice, generated T cells mediating delayed-type hypersensitivity (Ramshaw and Eidinger, 1977). A nylon-nonadherent, non-T splenic cell of medium density was shown to exhibit autocytotoxicity against syngeneic targets, and this response was suppressed by a light-density T cell (Osband and Parkman, 1978).

What observations have been made regarding cellular alterations contributing to increasing autoimmune tendencies in old mice from long-lived strains or in aged humans?

1. Tolerance induction: Immunological tolerance to foreign antigens was more easily induced in young than in old mice and the most striking age-related changes involved induction of B-cell (not T-cell) tolerance (Habicht and Jerrard, 1981; Weksler et al., 1981: McIntosh and Segre, 1976; Smith, 1976). The increased difficulty in inducing B-cell tolerance in aged mice appeared intrinsic to B cells (Weksler et al., 1981).

2. Autologous RBC binding by T cells: Human PBL (Fournier and Charreire, 1977) and long-lived mouse strains' splenocytes and thymocytes (Hiramoto et al., 1979; Charreire and Bach, 1975) showed age-dependent increases in levels of T cells which bind autologous RBC. It is presently unclear if and how these cells contribute to the autoimmune aspect of aging.
3. PFC to altered self antigens: Old mice responded more vigorously than young mice in a primary *in vivo* PFC response to TNP-modified syngeneic erythrocytes (Naor et al., 1976). Similarly, LPS induction of PFC to Br-MRBC was greatest in old mice (Goidl et al., 1981; Meredith et al., 1979). The low response found only in young mice was boosted by an injection of cyclophosphamide, which suggests the possibility of cellular suppression of autoimmunity in young and deregulation in old age (Meredith et al., 1979).
4. Cell-mediated immunity to self: Lymph node cells from old rats showed increased *in vitro* reactivity to MHC-coded autoantigens on testicular cell monolayers, and this autoreactivity occurred earlier in life than the decrease in Con A and MLR responses (Kyewski and Wekerle, 1978). Splenic cells from old mice could induce an *in vivo* syngeneic GVHR in both young and old recipients (Gozes et al., 1978). This reaction appeared to depend on T cells. This observation may accord with evidence by Callard et al. (1979) for the expression of new or altered antigenic determinants on surfaces of splenic lymphocytes from old mice. They observed a syngeneic MLR with young cells responding to old cells and vice versa. Response by young cells to old suggested antigenic diversification with age. Response by old to young cells was not explained by the authors, but again points toward a disruption in self recognition.
5. Autoantibodies to T cells in old humans: Serum samples from 30 of 83 people 60–99 years old reacted with a subpopulation of self T cells (Strelkauskas et al., 1981). Interestingly, the autoantibodies in the aged were directed to similar T-cell subsets as antibodies in juvenile rheumatoid arthritis.

6. METHODS OF DECELERATING IMMUNOLOGICAL AGING

Senescent alterations in RES functions and the immunopathology of old age are both ameliorated to a variable degree by certain procedures, including core body temperature reduction, drug or hormone therapy, immunological reconstitution, and dietary restriction. These methods are important for both theoretical and practical concerns. From the standpoint of theory, it has been repeatedly maintained that the only valid test of an aging hypothesis requires extension of maximum life span (Walford, 1974, 1969). Both the two procedures (core body temperature reduction and dietary restriction) which do this affect the immune system, as well as other systems. The more practical side of decelerating immunosenescence involves possible benefits to humankind of improved health in late-life and/or life extension. We reviewed this topic previously (Walford et al., 1977) and herein focus on information postdating that review.

6.1. BODY TEMPERATURE REDUCTION

The life span of poikilothermic vertebrates can be increased by body temperature reduction. The annual fish *Cynolebias bellottii* maintained in 15°C water lived longer than fish kept at 20°C, but not as long as fish kept at 20°C until 8 months of age and then transferred to 15°C until death (Liu and Walford, 1975). Since temperature reduction generally suppresses immune functions (reviewed by Liu and Walford, 1972), it may well be that temperature suppression of age-induced autoimmunity may override the disadvantage of simultaneously accentuating the immunodeficiency of aging. The life span-prolonging effect of temperature reduction cannot be attributed simply to metabolic slowdown (Liu and Walford, 1972).

A chronic, nontoxic hypothermia has not been established in homeotherms; therefore, life span or immune effects remain uninvestigable in them. Progress is being made as small brain peptides have been shown to transiently reduce body temperature. These include substance P (Uyeno et al., 1979), neurotensin (Loosen et al., 1978), and histidyl-proline diketopiperazine (a metabolite of thyrotropin-releasing hormone) (Prasad et al., 1978). These should provide molecular insights into thermoregulatory mechanisms and thereby contribute to safe ways to control body temperature in homeotherms. On thermodynamic grounds, a mild temperature reduction in humans might significantly extent life span (Rosenberg et al., 1973).

6.2. HORMONE AND DRUG THERAPY

An assortment of hormones and drugs have been reported to ameliorate senescent changes in the RES. Some also inhibit late-life diseases.

Treatment with diverse hormones of aging and truly old rodents has yielded encouraging results. Late-middle-age mice given 15 daily injections of L-thyroxine displayed a fivefold higher PFC response to injected SRBC than age-matched controls (Piantanelli and Fabris, 1978). Injection of old mice with thymopoietin$_{32-36}$ (a pentapeptide fragment of a thymic hormone) partially restored a T-dependent PFC response and corrected declines in antibody affinity (Weksler et al., 1978). Splenic cells from old mice showed an increased ability to elicit a GVHR if pretreated in vitro with thymic humoral factor (another thymic hormone) (Friedman et al., 1974). Another promising hormone, though not yet studied immunologically, is the adrenal steroid dehydroepiandrosterone. Long-term treatment of mice with this hormone inhibits spontaneous breast cancer in a susceptible strain (Schwartz, 1979). Denckla and co-workers, hypothesizing that immunological aging may relate to the pituitary's postpubertal release of a substance which inhibits the peripheral action of thyroid hormones (Denckla, 1974), tested immunological effects of hypophysectomy. Hypophysectomy of adult rats (followed by appropriate hormone replacement) led to "younger" responses in allograft rejection, carbon clearance (Bilder and Denckla, 1977), and to injected SRBC (Scott et al., 1979).

Various nonhormonal drug therapies also favorably influence senescent immunological alterations. The sulfhydryl compound 2-mercaptoethanol raised the impaired primary SRBC response of old mice, being effective both in vitro (Makinodan and Albright, 1979a) and in vivo (Makinodan and Albright, 1979b).

Free-radical inhibitors such as vitamin E or Santoquin added to diets of mice variably boosted Con A and SRBC responses (Harman et al., 1977) and inhibited experimentally induced amyloidosis (Harman et al., 1976). Long-term injection of mice with the antibiotic bestatin inhibited spontaneous cancers and restored a variety of diminished immune responses (Bruley-Rosset et al., 1979). Similarly, long-term treatment of mice with levamisole resulted in younger-acting immune systems (Morimoto et al., 1979; Bruley-Rosset et al., 1978). Treatment of 1-year-old mice for 6 months with a weekly levamisole injection reduced the incidence of spontaneous tumors when evaluated at 2 years of age (Bruley-Rosset et al., 1978). Injection of middle-aged mice with the synthetic polynucleotide polyadenylate-polyuridylate boosted PFC responses to SRBC (Han and Johnson, 1976). Administration of polyadenylate-polyuridylate to mice during the first week of life inhibited spontaneous mammary tumors in late life (Lacour et al., 1975). Old mice injected with coenzyme Q_{10} four days after SRBC immunization showed higher antibody titers than controls (Bliznakov, 1978).

6.3. IMMUNOLOGICAL RECONSTITUTION

Young thymuses, lymphocytes, and/or stem cells have been transferred into old syngeneic mouse recipients. The results in long-lived mouse strains indicate that these manipulations can positively influence age-induced immunological changes and perhaps survivorship as well. Grafting both young bone marrow and newborn thymus into late-middle-age and old mice clearly boosted the PFC response to SRBC as well as PHA reactivity above control levels (Hirokawa et al., 1976). Transfer of splenic cells from young mice previously immunized with *Salmonella typhimurium* to old mice protected the latter against subsequent *in vivo* challenge with this bacterium (Perkins et al., 1972). Mice receiving monthly injections of young lymph node cells starting at 112 weeks of age and 200 R of X-irradiation at 150 weeks appeared to survive slightly longer than noninjected mice or animals receiving either treatment alone (Walford et al., 1977). Grafting a neonatal thymus into old mice corrected the impaired isoproterenol-induced DNA-synthetic response of the submandibular gland, and normalized age changes in serum levels of triiodothyronine and insulin (Piantanelli et al., 1978).

These transfer strategies could be refined for future human application. Lymphoid cells and tissue fragments from a young person could be cryopreserved for use in self reconstitution in late life.

6.4. DIETARY RESTRICTION

Many studies with rodents have shown that restricted feeding regimens initiated early in life (about the time of weaning, i.e., 3–4 weeks of age) and continued until death inhibit late-life diseases and extend life span (Tucker, 1979; Nolen, 1972; Ross, 1961, 1969; Berg and Simms, 1961; McCay et al., 1935, 1939). The collective data point toward reductions of food intake of 20–60%

being associated with extensions of mean survivorship of about 15–40%. These animals are undernourished (not malnourished) because the diets are limited in calories but enriched in salts and vitamins (and occasionally protein). Lifelong underfeeding of this sort is the only method proven capable of decreasing the actuarial rate of aging in homeotherms, whereas pharmacological therapies boost average life expectancies (not maximal life span) via a reduction in disease vulnerability and not by decreasing the aging rate (Sacher, 1977).

Postweaning-initiated dietary restriction strikingly affects aging of the immune system. Six- to eight-month-old C57BL/6J female mice underfed since weaning responded less vigorously than controls to five T- and B-cell mitogens, to injected SRBC, and to allogeneic skin grafts, but by midlife the restricted mice generally outperformed controls in these tests and carried their superior response capacities into late life (Gerbase-DeLima et al., 1975b). These data suggested that RES maturation was slowed by underfeeding but that the immune system stayed younger longer in diet-restricted mice. This immunological preservation was associated with greater longevity. The early life effect of underfeeding on splenic T-cell mitogen responses appears strain dependent as higher responses have been reported for diet-restricted 3-month-old female C3H mice (Fernandes et al., 1976b) and 9-month-old female (C57BL/10Sn × C3H/HeDiSn)F_1 mice (Weindruch et al., 1979). Ten-month-old diet-restricted (NZB × NZW)F_1 mice showed a more vigorous immune system than controls as tested by T-cell cytotoxicity to alloantigens and PFC response to SRBC (Fernandes et al., 1978). Inhibition by underfeeding of the spontaneous development of suppressor lymphocytes was also observed. Late-life increases in immune complex levels were reduced by underfeeding long-lived (Barnett et al., 1981). and short-lived (Fernandes et al., 1979) mouse strains. Antinuclear antibody incidence was also reduced by food restriction of a long-lived mouse strain (Barnett et al., 1981). These various observations may relate causally to influences of underfeeding on the thymus. As judged by organ weight, mouse thymic growth was dampened by underfeeding (Weindruch et al., 1979) but at 6 months of age, the thymuses of diet-restricted mice appeared to have undergone less histological involution changes than controls (Weindruch and Suffin, 1980).

A more relevant but less studied question involves life span, disease, and immunological effects of food restriction begun at middle age in rodents previously fed *ad libitum*. Some findings point toward survival benefits following adult-onset underfeeding (Weindruch et al., 1979; Barrows and Kokkonen, 1975; Nolen, 1972; Ross, 1972; Miller and Payne, 1968). Spontaneous diseases were also observed to be inhibited by food restriction in adulthood (Friend et al., 1978; Tucker et al., 1976; Tannenbaum, 1942). Immunological effects have only recently been investigated. The gradually imposed restriction of calories or of calories and protein on previously *ad libitum*-fed 1-year-old mice resulted in more vigorous splenic lymphocyte proliferative responses to T-cell mitogens at 16 months of age (Weindruch et al., 1979). Similarly, restricted feeding begun at 17 months raised mitogen responses when assayed at 23 months of age (Mann, 1978). These results are encouraging in terms of potential human application. This model of adult-onset underfeeding is also more attractive than diet restriction initiated early in life in that adult underfeeding does not tamper with growth nor alter the timing of sexual maturity.

7. CONCLUDING COMMENTS

Present findings point toward lesions intrinsic to T cells and B cells as characterizing age changes in the RES. Alterations within T-cell subsets have also been documented. T-Cell problems could relate directly to age-associated deficits in thymic differentiative capacities. Sources of B-cell changes are less obvious. Age effects on stem cells could contribute to all these alterations but are difficult to evaluate from the present conflicting data. Studies at the molecular level are providing descriptions and explanations of T- and B-cell alterations. Investigative interest in both human immunogerontology and in the autoimmune aspect of aging has recently accelerated with cellular insights into the latter at last being gained. The same is not true for the diseases of aging. For example, the spontaneous cancers of old age have not been looked at in the framework of modern immunology because tumor immunologists have largely studied virally and chemically induced tumors. Mounting evidence indicates that genes linked to the MHC play a central role in senescent immunological alterations. Keeping pace with these advances is the development of methods to inhibit aging of the RES, which could elevate the quality if not the quantity of the later years.

REFERENCES

Abraham, C., Tal, Y., and Gershon, H., 1977, Reduced *in vitro* response to concanavalin A and lipopolysaccharide in senescent mice: A function of reduced number of responding cells, *Eur. J. Immunol.* **7**:301.

Adler, W. H., and Nagel, J. E., 1981, Studies of immune function in a human population, in: *Immunological Aspects of Aging* (D. Segre and L. Smith, eds.), pp. 295–311, Dekker, New York.

Agarwal, S. S., Tuffner, M., and Loeb, L. A., 1978, DNA replication in human lymphocytes during aging, *J. Cell. Physiol.* **96**:235.

Albright, J. W., and Makinodan, T., 1976, Decline in the growth potential of spleen-colonizing bone marrow stem cells of long-lived aging mice, *J. Exp. Med.* **144**:1204.

Amos, D. B., and Kostyu, D. D., 1980, HLA—A central immunological agency of man, *Adv. Hum. Genet.* **10**:137.

Anderson, R. E., Doughty, W. E., and Troup, G. M., 1977, Immunological responsiveness and aging phenomena in germfree mice, in: *Immunology and Aging* (T. Makinodan and E. Yunis, eds.), pp. 151–170, Plenum Press, New York.

Andersson, J., Coutinho, A., and Melchers, F., 1977, Frequencies of mitogen reactive B cells in the mouse. I. Distribution in different lymphoid organs from different inbred strains of mice at different ages, *J. Exp. Med.* **145**:1511.

Antel, J. P., and Arnason, B. G. W., 1979, Suppressor cell function in man: Evidence for altered sensitivity of responder cells with age, *Clin. Immunol. Immunopathol.* **13**:119.

Antel, J. P., Weinrich, M., and Arnason, B. G. W., 1978, Circulating suppressor cells in man as a function of age, *Clin. Immunol. Immunopathol.* **9**:134.

Augener, W., Cohnen, G., Reuter, A., and Brittinger, G., 1974, Decrease of T-lymphocytes during ageing, *Lancet* **1**:1164.

Bach, J.-F., Dardenne, M., Pleau, J.-M., and Bach, M.-A., 1975, Isolation, biochemical characteristics and biological activity of a circulating thymic hormone in the mouse and in the human, *Ann. N.Y. Acad. Sci.* **249**:186.

Bach, M.-A., 1979, Influence of aging on T-cell subpopulations involved in the *in vitro* generation of allogeneic cytotoxicity, *Clin. Immunol. Immunopathol.* **13**:220.

Bach, M.-A., and Beaurain, G., 1979, Respective influence of extrinsic and intrinsic factors on the age-related decrease of thymic secretion, *J. Immunol.* **122**:2505.

Barbaruk, L. G., 1974, Blast transformation and mitotic activity in cultured peripheral-blood lymphocytes from persons of different ages, *Cytol. Genet.* **8**:28.

Barnett, E. V., Chia, D., Knutson, D., Van Lancker, J., Cheney, K., Weindruch, R., and Walford, R. L. 1981, SLE, an accelerated form of aging, in: *Immunological Aspects of Aging* (D. Segre and L. Smith, eds.), pp. 467–474, Dekker, New York.

Barrows, C. H., Jr., and Kokkonen, G., 1975, Protein synthesis, development, growth and life span, *Growth* **39**:525.

Becker, M. J., Roubinian, J., Feldman, J. L., Blackman, M. A., Klajman, A., and Talal, N., 1979, Age-related changes in antibody-dependent cell-mediated cytotoxicity in mouse spleen, *Isr. J. Med. Sci.* **15**:147.

Beighlie, D. J., and Teplitz, R. L., 1975, Repair of UV damaged DNA in systemic lupus erythematosus, *J. Rheumatol.* **2**:149.

Ben-Zwi, A., Galili, U., Russell, A., and Schlesinger, M., 1977, Age-associated changes in subpopulations of human lymphocytes, *Clin. Immunol. Immunopathol.* **7**:139.

Berg, B. N., and Simms, H. S., 1961, Nutrition and longevity in the rat. III. Food restriction beyond 800 days, *J. Nutr.* **74**:23.

Besedovsky, H., and Sorkin, E., 1974, Thymus involvement in female sexual maturation, *Nature (London)* **249**:356.

Besedovsky, H., and Sorkin, E., 1977, Network of immune–neuroendocrine interactions, *Clin. Exp. Immunol.* **27**:1.

Bilder, G. E., and Denckla, W. D., 1977, Restoration of ability to reject xenografts and clear carbon after hypophysectomy of adult rats, *Mech. Ageing Dev.* **6**:153.

Bliznakov, E. G., 1978, Immunological senescence in mice and its reversal by coenzyme Q_{10}, *Mech. Ageing Dev.* **7**:189.

Boyd, E., 1932, The weight of the thymus gland in health and disease, *Am. J. Dis. Child.* **43**:1162.

Brennan, P. C., and Jaroslow, B. N., 1975, Age-associated decline in theta antigen on spleen thymus-derived lymphocytes of $B6CF_1$ mice, *Cell. Immunol.* **15**:51.

Bruley-Rosset, M., Florentin, I., Kiger, N., Davigny, M., and Mathe, G., 1978, Effects of Bacillus Calmette-Guerin and levamisole on immune responses in young adult and age-immunodepressed mice, *Cancer Treat. Rep.* **62**:1641.

Bruley-Rosset, M., Florentin, I., Kiger, N., Schulz, J., and Mathe, G., 1979, Restoration of impaired immune functions of aged animals by chronic bestatin treatment, *Immunology* **38**:75.

Buckley, C. E., III, Buckley, E. G., and Dorsey, F. C., 1974, Longitudinal changes in serum immunoglobulin levels in older humans, *Fed. Proc.* **33**:2036.

Burnet, F. M., 1959, Autoimmune disease. II. Pathology of the immune response, *Br. Med. J.* **2**:720.

Burnet, F. M., 1974, *Intrinsic Mutagenesis: A Genetic Approach to Ageing*, Wiley, New York.

Callard, R. E., 1978, Immune function in aged mice. III. Role of macrophages and effect of 2-mercaptoethanol in the response of spleen cells from old mice to phytohemagglutinin, lipopolysaccharide and allogeneic cells, *Eur. J. Immunol.* **8**:697.

Callard, R. E., and Basten, A., 1977, Immune function in aged mice. I. T-Cell responsiveness using phytohaemagglutinin as a functional probe, *Cell. Immunol.* **31**:13.

Callard, R. E., and Basten, A., 1978, Immune function in aged mice. IV. Loss of T cell and B cell function in thymus-dependent antibody responses, *Eur. J. Immunol.* **8**:552.

Callard, R. E., Basten, A., and Waters, L. K., 1977, Immune function in aged mice. II. B-Cell function, *Cell Immunol.* **31**:26.

Callard, R. E., Basten, A., and Blanden, R. V., 1979, Loss of immune competence with age may be due to a qualitative abnormality in lymphocyte membranes, *Nature (London)* **281**:218.

Callard, R. E., De St. Groth, B. F., Basten, A., and McKenzie, I. F. C., 1980, Immune function in aged mice. V. Role of suppressor cells, *J. Immunol.* **124**:52.

Cantor, H., and Boyse, E. A., 1977, Lymphocytes as models for the study of mammalian cellular differentiation, *Immunol. Rev.* **33**:105.

Cantor, H., McVay-Broudreau, L., Hugenberger, J., Naidorf, K., Shen, F. W., and Gershon, R. K., 1978, Immunoregulatory circuits among T cell sets. II. Physiologic role of feedback inhibition *in vivo*: Absence in NZB mice, *J. Exp. Med.* **147**:1116.

Cantrell, W., and Elko, E. E., 1973, Effect of age on phagocytosis of carbon in the rat, *Exp. Parasitol.* **34**:337.

Carnaud, C., Charreire, J., and Bach, J.-F., 1977, Adult thymectomy promotes the manifestation of autoreactive lymphocytes, *Cell. Immunol.* **28**:274.

Carosella, E. D., Mochanko, K., and Braun, M., 1974, Rosette-forming T cells in human peripheral blood at different ages, *Cell. Immunol.* **12**:323.

Charreire, J., and Bach, J.-F., 1975, Binding of autologous erythrocytes to immature T-cells, *Proc. Natl. Acad. Sci. USA* **72**:3201.

Chen, M. G., 1971, Age-related changes in hematopoietic stem cell populations of a long-lived hybrid mouse, *J. Cell. Physiol.* **78**:225.

Clot, J., Charmasson, E., and Brochier, J., 1978, Age-dependent changes of human blood lymphocyte subpopulations, *Clin. Exp. Immunol.* **32**:346.

Coggle, J. E., and Proukakis, C., 1970, The effect of age on the bone marrow cellularity of the mouse, *Gerontologia* **16**:25.

Coggle, J. E., Gordon, M. Y., Proukakis, C., and Bogg, C. E., 1975, Age-related changes in the bone marrow and spleen of SAS/4 mice, *Gerontologia* **21**:1.

Cohen, D., Zeller, E., Eisdorfer, C., and Walford, R. L., 1979, Alzheimer's disease and the main histocompatibility complex, *Gerontologist* **19**(2):57.

Converse, P. J., and Williams, D. R. R., 1978, Increased HLA-B heterozygosity with age, *Tissue Antigens* **12**:275.

Cunningham, A. J., 1974, Large numbers of cells in normal mice produce antibody components of isologous erythrocytes, *Nature (London)* **252**:749.

Cunningham, A. J., 1975, Active suppressor mechanisms maintaining tolerance to some self components, *Nature (London)* **254**:143.

Cutler, R. G., 1975, Evolution of human longevity and the genetic complexity governing aging, *Proc. Natl. Acad. Sci. USA* **72**:4664.

Czlonkowska, A., and Korlak, J., 1979, The immune response during aging, *J. Gerontol.* **34**:9.

Dausset, J., and Svejgaard, A. (eds.), 1977, *HLA and Disease*, Munksgaard, Copenhagen.

Davey, F. R., and Huntington, S., 1977, Age-related variation in lymphocyte subpopulations, *Gerontologia* **23**:381.

Davis, M. L., Upton, A. C., and Satterfield, L. C., 1971, Growth and senescence of the bone marrow stem cell pool in RFM/Un mice, *Proc. Soc. Exp. Biol. Med.* **137**:1452.

Denckla, W. D., 1974, Role of the pituitary and thyroid glands in the decline of minimal O_2 consumption with age, *J. Clin. Invest.* **53**:572.

Diaz-Jouanen, E., Strickland, R. G., and Williams, R. C., Jr., 1975, Studies on human lymphocytes in the newborn and aged, *Am. J. Med.* **58**:620.

Doherty, P. C., 1977, Diminished T cell surveillance function in old mice infected with lymphocyte choriomeningitis virus, *Immunology* **32**:751.

Doria, G., D'agostaro, G., and Poretti, A., 1978, Age-dependent variations of antibody avidity, *Immunology* **35**:601.

Duwe, A. K., Roder, J. C., and Singhal, S. K., 1979, Immunological senscence. II. Normal *in vitro* colony formation by B cells from old mice. *Immunology* **37**:293.

Edwards, D. L., and Avis, F. P., 1979, Antibody-dependent cellular cytotoxicity effector cell capability among normal individuals, *J. Immunol.* **123**:1887.

Farrar, J. J., Loughman, B. E., and Nordin, A. A., 1974, Lymphopoietic potential of bone marrow cells from aged mice: Comparison of the cellular constituents of bone marrow from young and aged mice, *J. Immunol.* **112**:1244.

Fernandes, G., Yunis, E. J., and Good, R. A., 1976a, Age and genetic influence on immunity in NZB and autoimmune-resistant mice, *Clin. Immunol. Immunopathol.* **6**:318.

Fernandes, G., Yunis, E. J., and Good, R. A., 1976b, Suppression of adenocarcinoma by the immunological consequences of caloric restriction, *Nature (London)* **263**:504.

Fernandes, G., Friend, P., Yunis, E. J., and Good, R. A., 1978, Influence of dietary restriction on immunologic function and renal disease in (NZB × NZW)F$_1$ mice, *Proc. Natl. Acad. Sci. USA* **75**:1500.

Fernandes, G., West, A., and Good, R. A., 1979, Nutrition, immunity and cancer—A review. Part III. Effects of diet on the diseases of aging, *Clin. Bull.* **9**:91.

Finger, H., Beneke, G., Emmerling, P., Bertz, R., and Plager, L., 1972, Secondary antibody-forming potential of aged mice, with special reference to the influence of adjuvant on priming, *Gerontologia* **18**:77.

Finn, R., Nichol, F. E., and Coates, P., 1976, Immunological factors in accelerated atheroma associated with renal disease, *Lancet* **2**:1141.

Fournier, C., and Charreire, J., 1977, Increase in autologous erythrocyte binding by T cells with ageing in man, *Clin. Exp. Immunol.* **29**:468.

Friedman, D., and Globerson, A., 1978a, Immune reactivity during aging. I. T-Helper dependent and independent antibody responses to different antigens, *in vivo* and *in vitro*, *Mech. Ageing Dev.* **7**:289.

Friedman, D., and Globerson, A., 1978b, Immune reactivity during aging. II. Analysis of cellular mechanisms involved in the deficient antibody response in old mice, *Mech. Ageing Dev.* **7**:299.

Friedman, D., Keiser, V., and Globerson, A., 1974, Reactivation of immunocompetence in spleen cells of aged mice, *Nature (London)* **251**:545.

Friend, P. S., Fernandes, G., Good, R. A., Michael, A. F., and Yunis, E. J., 1978, Dietary restrictions early and late: Effects on the nephropathy of the NZB × NZW mouse, *Lab. Invest.* **38**:629.

Fujiwara, M., and Kishimoto, S., 1979, IgE antibody formation and aging. I. Age-related changes in IgE antibody formation and avidity for the DNP determinant in mice, *J. Immunol.* **123**:263.

Gardner, I. D., and Remington, J. S., 1978, Aging and the immune response. II. Lymphocyte responsiveness and macrophage activation in *Toxoplasma gondii*-infected mice, *J. Immunol.* **120**:944.

Gatti, R. A., and Walford, R. L., 1981, Immune function and features of aging in chromosomal instability syndromes, in: *Immunological Aspects of Aging* (D. Segre and L. Smith, eds.), pp. 449–464, Dekker, New York.

Gerbase-DeLima, M., and Walford, R. L., 1975, Effect of cortisone in delineating thymus cell subsets in advanced age, *Proc. Soc. Exp. Biol. Med.* **149**:562.

Gerbase-DeLima, M., Wilkinson, J., Smith, G. S., and Walford, R. L., 1974, Age-related decline in thymic-independent immune function in a long-lived mouse strain, *J. Gerontol.* **29**:261.

Gerbase-DeLima, M., Meredith, P., and Walford, R. L., 1975a, Age-related changes, including synergy and suppression, in the mixed lymphocyte reaction in long-lived mice, *Fed. Proc.* **34**:159.

Gerbase-DeLima, M., Liu, R. K., Cheney, K. E., Mickey, R., and Walford, R. L., 1975b, Immune function and survival in a long-lived mouse strain subjected to undernutrition, *Gerontologia* **21**:184.

Girard, J. P., Paychere, M., Cuevas, M., and Fernandes, B., 1977, Cell-mediated immunity in an ageing population, *Clin. Exp. Immunol.* **27**:85.

Goidl, E. A., Innes, J. B., and Weksler, M. E., 1976, Immunological studies of aging. II. Loss of IgG and high avidity plaque-forming cells and increased suppressor cell activity in aging mice, *J. Exp. Med.* **144**:1037.

Goidl, E. A., Kim, Y. T., and Weksler, M. E., 1981, Studies on autoimmunity in aging, in: *Immunological Aspects of Aging* (D. Segre and L. Smith, eds.), pp. 165–173, Dekker, New York.

Goodman, S. A., and Makinodan, T., 1975, Effect of age on cell-mediated immunity in long-lived mice, *Clin. Exp. Immunol.* **19**:533.

Goodwin, J. S., and Messner, R. P., 1979, Sensitivity of lymphocytes to prostaglandin E$_2$ increases in subjects over age 70, *J. Clin. Invest.* **64**:434.

Gottesman, S. R. S., and Walford, R. L., 1982, Autoimmunity and aging, in: *Methods in Aging Research*, Vol. 14 (R. C. Adelman and G. S. Roth, eds.), CRC Press, West Palm Beach.

Gozes, Y., and Trainin, N., 1977, Enhancement of Lewis lung carcinoma in a syngeneic host by spleen cells of C57BL/6 old mice, *Eur. J. Immunol.* **7**:159.

Gozes, Y., Umiel, T., Meshorer, A., and Trainin, N., 1978, Syngeneic GVH induced in popliteal lymph nodes by spleen cells of old C57BL/6 mice, *J. Immunol.* **121**:2199.

Greenberg, L. J., and Yunis, E. J., 1978, Histocompatibility determinants, immune responsiveness and aging in man, *Fed. Proc.* **37**:1258.

Gupta, S., and Good, R. A., 1979, Subpopulations of human T lymphocytes. X. Alterations in T, B, third population cells, and T cells with receptors for immunoglobulin M (Tμ) or G (Tγ) in aging humans, *J. Immunol.* **122**:1214.

Haaijman, J. J., Van Den Berg, P., and Brinkhof, J., 1977a, Immunoglobulin class and subclass levels in the serum of CBA mice throughout life, *Immunology* **32**:923.

Haaijman, J. J., Schuit, H. R. E., and Hijmans, W., 1977b, Immunoglobulin-containing cells in different lymphoid organs of the CBA mouse during its life-span, *Immunology* **32**:427.

Habicht, G. S., and Jerrard, D. A., 1981, Tolerance to HGG in mice of different age groups, in: *Immunological Aspects of Aging* (D. Segre and L. Smith, eds.), pp. 127–138, Dekker, New York.

Halbrecht, I., Komlos, L., and Strauss, Z., 1977, Lymphocyte reactivity in the aged, *J. Am. Geriatr. Soc.* **25**:354.

Hallgren, H. M., and Yunis, E. J., 1977, Suppressor lymphocytes in young and aged humans, *J. Immunol.* **118**:2004.

Hallgren, H. M., and Yunis, E. J., 1981, Immune function, immune regulation, and survival in an aging human population, in: *Immunological Asects of Aging* (D. Segre and L. Smith, eds.), pp. 281–293, Dekker, New York.

Hallgren, H. M., Buckley, C. E., III, Gilbertsen, V. A., and Yunis, E. J., 1973, Lymphocyte phytohemagglutinin responsiveness, immunoglobulins and autoantibodies in aging humans, *J. Immunol.* **111**:1101.

Hallgren, H. M., Kersey, J. H., Dubey, D. P., and Yunis, E. J., 1978, Lymphocyte subsets and integrated immune function in aging humans, *Clin. Immunol. Immunopathol.* **10**:65.

Han, I. H., and Johnson, A. G., 1976, Regulation of the immune system by synthetic polynucleotides. VI. Amplification of the immune response in young and aging mice, *J. Immunol.* **117**:423.

Harman, D., Eddy, D. E., and Noffsinger, J., 1976, Free radical theory of aging: Inhibition of amyloidosis in mice by antioxidants; possible mechanism, *J. Am. Geriatr. Soc.* **24**:203.

Harman, D., Heidrick, M. L., and Eddy, D. E., 1977, Free radical theory of aging: Effect of free-radical-reaction inhibitors on the immune response, *J. Am. Geriatr. Soc.* **25**:400.

Harrison, D. E., 1975, Normal function of transplanted marrow cell lines from aged mice, *J. Gerontol.* **30**:279.

Harrison, D. E., 1979a, Mouse erythropoietic stem cell lines function normally 100 months: Loss related to number of transplantations, *Mech. Ageing Dev.* **9**:427.

Harrison, D. E., 1979b, Proliferative capacity of erythropoietic stem cell lines and aging: An overview, *Mech. Ageing Dev.* **9**:409.

Harrison, D. E., and Doubleday, J. W., 1975, Normal function of immunologic stem cells from aged mice, *J. Immunol.* **114**:1314.

Harrison, D. E., Astle, C. M., and Doubleday, J. W., 1977, Stem cell lines from old immunodeficient donors give normal responses in young recipients, *J. Immunol.* **118**:1223.

Harrison, D. E., Astle, C. M., and Delaittre, J., 1978, Loss of proliferative capacity in immunohemopoietic stem cells caused by serial transplantation rather than aging, *J. Exp. Med.* **147**:1526.

Hart, R. W., and Setlow, R. B., 1974, Correlation between deoxyribonucleic acid excision-repair and lifespan in a number of mammalian species, *Proc. Natl. Acad. Sci. USA* **71**:2169.

Heidrick, M. L., 1972, Age-related changes in hydrolase activity of peritoneal macrophages, *Gerontologist* **12**(2):28.

Heidrick, M. L., and Makinodan, T., 1973, Presence of impairment of humoral immunity in nonadherent spleen cells of old mice, *J. Immunol.* **111**:1502.

Henschke, P. J., Bell, D. A., and Cape, R. D. T., 1978, Alzheimer's disease and HLA, *Tissue Antigens* **12**:132.

Herberman, R. B., Nunn, M. E., and Larvin, D. H., 1975, Natural cytotoxic reactivity of mouse lymphoid cells against syngeneic and allogeneic tumors. I. Distribution of reactivity and specificity, *Int. J. Cancer* **16**:216.

Hildemann, W. H., 1974, Some new concepts in immunological phylogeny, *Nature (London)* **250**:116.

Hildemann, W. H., 1978, Phylogenetic and immunogenetic aspects of aging, in: *Genetic Effects on Aging* (D. Bergsma and D. E. Harrison, eds.), pp. 97–107, Liss, New York.

Hiramoto, R. N., Ghanta, V. K., Hsu, L., and Davis, D. W., 1979, Age-related changes in autoerythrocyte rosettes in the C57BL/6J and NZB/BINJ mice, *Mech. Ageing Dev.* **11**:45.

Hirokawa, K., 1978, Age-related changes of thymus—Morphological and functional aspects, *Acta Pathol. Jpn.* **28**:843.

Hirokawa, K., and Makinodan, T., 1975, Thymic involution: Effect on T cell differentiation, *J. Immunol.* **114**:1659.

Hirokawa, K., and Sado, T., 1978, Early decline of thymic effect on T cell differentiation, *Mech. Ageing Dev.* **7**:89.

Hirokawa, K., Albright, J. W., and Makinodan, T., 1976, Restoration of impaired immune functions in aging animals. I. Effect of syngeneic thymus and bone marrow grafts, *Clin. Immunol. Immunopathol.* **5**:371.

Hori, Y., Perkins, E. H., and Halsall, M. K., 1973, Decline in phytohemagglutinin responsiveness of spleen cells from aging mice, *Proc. Soc. Exp. Biol. Med.* **144**:48.

Hung, L. Y., Perkins, E. H., and Yang, W. K., 1975a, Age-related refractoriness of PHA-induced lymphocyte transformation. II. ^{125}I-PHA binding to spleen cells from young and old mice, *Mech. Ageing Dev.* **4**:103.

Hung, L. Y., Perkins, E. H., and Yang, W. K., 1975b, Age-related refractoriness of PHA-induced lymphocyte transformation. I. Comparable sensitivity of spleen cells from young and old mice to culture conditions, *Mech. Ageing Dev.* **4**:29.

Inkeles, B., Innes, J. B., Kuntz, M. M., Kadish, A. S., and Weksler, M. E., 1977, Immunological studies of aging. III. Cytokinetic basis for the impaired response of lymphocytes from aged humans to plant lectins, *J. Exp. Med.* **145**:1176.

Ishii, T., and Haga, S., 1976, Immuno-electron microscopic localization of immunoglobulins in amyloid fibrils of senile plaques, *Acta Neuropathol.* **36**:243.

Jeejeebhoy, H. F., 1971, Decreased longevity of mice following thymectomy in adult life, *Transplantation* **12**:525.

Kamo, I., and Friedman, H., 1977, Immunosuppression and the role of suppressive factors in cancer, *Adv. Cancer Res.* **25**:271.

Kay, M. M. B., 1976, Autoimmune disease: The consequence of deficient T cell function, *J. Am. Geriatr. Soc.* **24**:253.

Kay, M. M. B., 1978, Immunologic aging patterns: Effect of parainfluenza Type 1 virus infection on aging mice of eight strains and hybrids, in: *Genetic Effects on Aging* (D. Bergsma and D. E. Harrison, eds.), pp. 213–240, Liss, New York.

Kay, M. M. B., 1979, Parainfluenza infection of aged mice results in autoimmune disease, *Clin. Immunol. Immunopathol.* **12**:301.

Keller, R., 1978, Macrophage-mediated natural cytotoxicity against various target cells *in vitro*. II. Macrophages from rats of different ages, *Br. J. Cancer* **37**:742.

Kiessling, R., Klein, E., Pross, H., and Wigzell, H., 1975, "Natural" killer cells in the mouse. II. Cytotoxic cells with specificity for mouse Moloney leukemia cells. Characteristics of the killer cell, *Eur. J. Immunol.* **5**:117.

Kishimoto, S., and Yamamura, Y., 1971, Immune responses in aged mice: Changes of antibody-forming cell precursors and antigen-reactive cells with ageing, *Clin. Exp. Immunol.* **8**:957.

Kishimoto, S., Shigemoto, S., and Yamamura, Y., 1973, Immune response in aged mice, *Transplantation* **15**:455.

Kishimoto, S., Takahama, T., and Mizumachi, H., 1976, *In vitro* immune response to the 2,4,6-trinitrophenyl determinant in aged C57BL/6J mice: Changes in the humoral immune response to, avidity for the TNP determinant and responsiveness to LPS effect with aging, *J. Immunol.* **116**:294.

Kishimoto, S., Tomino, S., Inomata, K., Kotegawa, S., Saito, T., Kuroki, M., Mitsuya, H., and Hisamitsu, S., 1978, Age-related changes in the subsets and functions of human T lymphocytes, *J. Immunol.* **121**:1773.

Kishimoto, S., Tomino, S., Mitsuya, H., and Fujiwara, H., 1979, Age-related changes in suppressor functions of human T cells, *J. Immunol.* **123**:1586.

Klein, J., 1978, H-2 mutations: Their genetics and effect on immune functions, *Adv. Immunol.* **26**:56.

Klein, J., 1979, The major histocompatibility complex of the mouse, *Science* **203**:516.

Konen, T. G., Smith, G. S., and Walford, R. L., 1973, Decline in mixed lymphocyte reactivity of spleen cells from aged mice of a long-lived strain, *J. Immunol.* **110**:1216.

Krogsrud, R. L., and Perkins, E. H., 1977, Age-related changes in T cell function, *J. Immunol.* **118**:1607.

Kyewski, B., and Wekerle, H., 1978, Increase of T lymphocyte self-reactivity in aging inbred rats: In vitro studies with a model of experimental autoimmune orchitis, *J. Immunol.* **120**:1249.

Lacour, F., Delage, G., and Chianale, C., 1975, Reduced incidence of spontaneous mammary tumors in C3H/He mice after treatment with polyadenylate-polyuridylate, *Science* **187**:256.

Lambert, B., Ringborg, U., and Skoog, L., 1979, Age-related decrease of ultraviolet light-induced DNA repair synthesis in human peripheral leukocytes, *Cancer Res.* **39**:2792.

Lewis, V. M., Twomey, J. J., Bealmear, P., Goldstein, G., and Good, R. A., 1978, Age, thymic involution, and circulating thymic hormone activity, *J. Clin. Endocrinol. Metab.* **47**:145.

Liu, R. K., and Walford, R. L., 1972, The effect of lowered body temperature on lifespan and immune and non-immune processes, *Gerontologia* **18**:363.

Liu, R. K., and Walford, R. L., 1975, Mid-life temperature-transfer effects on life-span of annual fish, *J. Gerontol.* **30**:129.

Loosen, P. T., Nemeroff, C. B., Bisette, G., Burnet, G. B., Prange, A. J., Jr., and Lipton, M. A., 1978, Neurotensin-induced hypothermia in the rat: Structure–activity studies, *Neuropharmacology* **17**:109.

MacKinney, A. A., Jr., 1978, Effect of aging on the peripheral blood lymphocyte count, *J. Gerontol.* **33**:213.

Macurova, H., Ivanyi, P., Sajdlova, H., and Trojan, J., 1975, HL-A antigens in aged persons, *Tissue Antigens* **6**:269.

Makinodan, T., and Adler, W. H., 1975, Effects of aging on the differentiation and proliferation potentials of cells of the immune system, *Fed. Proc.* **34**:153.

Makinodan, T., and Albright, J. W., 1979a, Restoration of impaired immune functions in aging animals. II. Effect of mercaptoethanol in enhancing the reduced primary antibody responsiveness in vitro, *Mech. Ageing Dev.* **10**:325.

Makinodan, T., and Albright, J. W., 1979b, Restoration of impaired immune functions in aging animals. III. Effect of mercaptoethanol in enhancing the reduced primary antibody responsiveness in vivo, *Mech. Ageing Dev.* **11**:1.

Makinodan, T., and Peterson, W. J., 1966, Secondary antibody-forming potential of mice in relation to age—Its significance in senescence, *Dev. Biol.* **14**:96.

Makinodan, T., Chino, F., Lever, W. E., and Brewen, B. S., 1971, The immune systems of mice reared in clean and in dirty conventional laboratory farms. III. Ability of old mice to be sensitized to undergo a secondary antibody response, *J. Gerontol.* **26**:515.

Makinodan, T., Albright, J. W., Good, P. I., Peter, C. P., and Heidrick, M. L., 1976, Reduced humoral immune activity in long-lived old mice: An approach to elucidating its mechanisms, *Immunology* **31**:903.

Mann, P. L., 1978, The effect of various dietary restricted regimes on some immunological parameters of mice, *Growth* **42**:87.

Mathews, J. D., Hooper, B. M., Wittingham, S., Mackay, I. R., and Stenhouse, N. S., 1973, Association of autoantibodies with smoking, cardiovascular morbidity, and death in the Busselton population, *Lancet* **2**:754.

Mathies, M., Lipps, L., Smith, G. S., and Walford, R. L., 1973, Age-related decline in response to phytohemagglutinin and pokeweed mitogen by spleen cells from hamsters and a long-lived mouse strain, *J. Gerontol.* **28**:425.

Matsuura, M., Shimada, H., and Namba, K., 1978, Deficient scheduled and unscheduled DNA synthesis in the cultured fibroblasts derived from patients with rheumatoid arthritis (RA), XIth International Congress Gerontology, Tokyo, p. 39 (abstract).

McCay, C. M., Crowell, M. F., and Maynard, L. A., 1935, The effect of retarded growth upon the length of the life span and upon the ultimate body size, *J. Nutr.* **10**:63.

McCay, C. M., Maynard, L. A., Sperling, G., and Barnes, L. L., 1939, Retarded growth, life span, ultimate body size and age changes in the albino rat after feeding diets restricted in calories, *J. Nutr.* **18**:1.

McIntosh, K. R., and Segre, D., 1976, B and T cell tolerance induction in young-adult and old mice, *Cell. Immunol.* **27**:230.

Meihuizen, S. P., and Burek, J. D., 1978, The epithelial cell components of the thymuses of aged female BN/Bi rats. A light microscopic, electron microscopic, and autoradiographic study, *Lab. Invest.* **39**:613.

Menon, M., Jaroslow, B. N., and Koesterer, R., 1974, The decline of cell-mediated immunity in aging mice, *J. Gerontol.* **29**:499.

Meredith, P. J., and Walford, R. L., 1977, Effect of age on response to T and B cell mitogens in mice congenic at the H-2 locus, *Immunogenetics* **5**:109.

Meredith, P. J., and Walford, R. L., 1979, Autoimmunity, histocompatibility and aging, *Mech. Ageing Dev.* **9**:61.

Meredith, P., Gerbase-DeLima, M., and Walford, R. L., 1975a, Age-related changes in the PHA:Con A stimulatory ratios of cells from spleens of a long-lived mouse strain, *Exp. Gerontol.* **10**:247.

Meredith, P., Tittor, W., Gerbase-DeLima, M., and Walford, R. L., 1975b, Age-related changes in the cellular immune response of lymph node and thymus cells in long-lived mice, *Cell. Immunol.* **18**:324.

Meredith, P. J., Kristie, J. A., and Walford, R. L., 1979, Aging increases expression of LPS-induced autoantibody-secreting B cells, *J. Immunol.* **123**:87.

Merhav, S., and Gershon, H., 1977, The mixed lymphocyte response of senescent mice: Sensitivity to alloantigen and cell replication time, *Cell. Immunol.* **34**:354.

Metcalf, D., 1965, Delayed effect of thymectomy in adult life on immunological competence, *Nature (London)* **208**:1336.

Metcalf, D., Sparrow, N., Nakamura, K., and Ishidate, M., 1961, The behaviour of thymus grafts in high and low leukaemia strains of mice, *Aust. J. Exp. Biol. Med. Sci.* **39**:441.

Miller, D. S., and Payne, P. R., 1968, Longevity and protein intake, *Exp. Gerontol.* **3**:231.

Miller, J. F. A. P., 1965, Effect of thymectomy in adult mice on immunological responsiveness, *Nature (London)* **208**:1337.

Morimoto, C., Abe, T., and Homma, M., 1979, Restoration of T-cell function in aged mice with long-term administration of levamisole, *Clin. Immunol. Immunopathol.* **12**:316.

Munan, L., and Kelly, A., 1979, Age-dependent changes in blood monocyte populations in man, *Clin. Exp. Immunol.* **35**:161.

Murphy, G. P., Cohen, E., Fitzpatrick, J. F., and Pressman, D. (eds.), 1977, *HLA and Malignancy*, Volume 16 of *Progress in Clinical and Biological Research*, Liss, New York.

Naeim, F., and Walford, R. L., 1980, Disturbance of redistribution of surface membrane receptors of peripheral mononuclear cells of patients with Down's syndrome and of aged individuals, *J. Gerontol.* **35**:640.

Naeim, F., and Walford, R. L., 1981, Human Ia-like alloantigens and their medical significances, in: *Current Trends in Histocompatibility* (R. A. Reisfeld and S. Ferrone, eds.), pp. 227–256, Plenum Press, New York.

Nandy, K., Fritz, R. B., and Threatt, J., 1975, Specificity of brain-reactive antibodies in serum of old mice, *J. Gerontol.* **30**:269.

Naor, D., Bonavida, B., and Walford, R. L., 1976, Autoimmunity and aging: The age-related response of mice of a long-lived strain to trinitrophenylated syngeneic mouse red blood cells, *J. Immunol.* **117**:2204.

Nariuchi, H., and Adler, W. H., 1979, Dissociation between proliferation and antibody formation by old mouse spleen cells in response to LPS stimulation, *Cell. Immunol.* **45**:295.

Nolen, G. A., 1972, Effects of various restricted dietary regimens on the growth, health and longevity of albino rats, *J. Nutr.* **102**:1477.

Nordin, A. A., and Buchholz, M., 1981, The effect of age on the *in vitro* immune response of C57BL/6 mice to a T-independent antigen, in: *Immunological Aspects of Aging* (D. Segre and L. Smith, eds.), pp. 91–105, Dekker New York.

Olsson, L., and Claesson, M. H., 1973, Studies on subpopulations of theta-bearing lymphoid cells, *Nature New Biol.* **244**:50.

Osband, M., and Parkman, R., 1978, The control of autoreactivity. I. Lack of autoreactivity in murine spleens is due to concomitant presence of suppressor and autocytotoxic lymphocytes, *J. Immunol.* **121**:179.

Pachciarz, J. A., and Teague, P. O., 1976, Age-associated involution of cellular immune function. I. Accelerated decline of mitogen reactivity of spleen cells in adult thymectomized mice, *J. Immunol.* **116**:982.

Pandey, J. P., Fudenberg, H. H., Root, B., and Loadholt, C. B., 1978, Association of HLA antigens, ischaemic heart disease, and serum calcium levels: A preliminary report, *Age* **1**:21.

Pantelouris, E. M., 1974, Common parameters in genètic athymia and senescence, *Exp. Gerontol.* **9**:161.

Paul, W. E., and Benacerraf, B., 1977, Functional specificity of thymus-dependent lymphocytes, *Science* **195**:1293.

Perkins, E. H., 1971, Phagocytic activity of aged mice, *J. Reticuloendothelial Soc.* **9**:642.

Perkins, E. H., and Cacheiro, L. H., 1977, A multiple-parameter comparison of immunocompetence and tumor resistance in aged BALB/c mice, *Mech. Ageing Dev.* **6**:15.

Perkins, E. H., Makinodan, T., and Seibert, C., 1972, Model approach to immunological rejuvenation of the aged, *Infect. Immun.* **6**:518.

Perkins, E. H., Peterson, W. J., Gottlieb, C. F., Halsall, M. K., Cacheiro, L. H., and Makinodan, T., 1975, The late effects of selected immunosuppressants on immunocompetence, disease incidence and mean lifespan. I. Humoral immune activity, *Mech. Ageing Dev.* **4**:231.

Peterson, W. H., Perkins, E. H., Goodman, S. A., Hori, Y., Halsall, M. K., and Makinodan, T., 1975, The late effects of selected immunosuppressants on immunocompetence, disease incidence and mean lifespan. II. Cell-mediated immune activity, *Mech. Ageing Dev.* **4**:241.

Piantanelli, L., and Fabris, N., 1978, Hypopituitary dwarf and athymic nude mice and the study of the relationships among thymus, hormones and aging, in: *Genetic Effects on Aging* (D. Bergsma and D. E. Harrison, eds.), pp. 315–333, Liss, New York.

Piantanelli, L., Basso, A., Muzzioli, M., and Fabris, N., 1978, Thymus-dependent reversibility of physiological and isoproterenol evoked age-related parameters in athymic (nude) and old normal mice, *Mech. Ageing Dev.* **7**:171.

Pierpaoli, W., and Besedovsky, H. O., 1975, Role of the thymus in programming of neuroendocrine function, *Clin. Exp. Immunol.* **20**:323.

Pope, R. M., Teller, D. C., and Mannik, M., 1974, The molecular basis of self association of antibodies of IgG (rheumatoid factors) in rheumatoid arthritis, *Proc. Natl. Acad. Sci. USA* **71**:517.

Popp, D. M., 1975, The effect of age on antigen-sensitive cells, *Mech. Ageing Dev.* **4**:221.

Popp, D. M., 1977, Qualitative changes in immunocompetent cells with age: Reduced sensitivity to cortisone acetate, *Mech. Ageing Dev.* **6**:355.

Popp, D. M., 1978, Use of congenic mice to study the genetic basis of degenerative disease in: *Genetic Effects on Aging* (D. Bergsma and D. E. Harrison, eds.), pp. 261–279, Liss, New York.

Popp, D. M., and Francis, M., 1979, Age-associated changes in T–B cell cooperation demonstration by the allogeneic effect, *Mech. Ageing Dev.* **10**:341.

Popp, D. M., and Popp, R. A., 1979, Hemopoietic stem cell heterogeneity: Use of cell cycle-specific drugs to look for age-associated alterations, *Mech. Ageing Dev.* **9**:441.

Portaro, J. K., Glick, G. I., and Zighelboim, J., 1978, Population immunology: Age and immune cell parameters, *Clin. Immunol. Immunopathol.* **11**:339.

Poston, R. N., and Davies, D. F., 1974, Immunity and inflammation in pathogenesis of atherosclerosis, *Atherosclerosis*, **19**:353.

Prasad, C., Matsui, T., Williams, J., and Peterkofsky, A., 1978, Thermoregulation in rats: Opposing effects of thyrotropin releasing hormone and its metabolite histidyl-proline diketopiperazine, *Biochem. Biophys. Res. Commun.* **85**:1582.

Preumont, A. M., Van Gansen, P., and Brachet, J., 1978, Cytochemical study of human lymphocytes stimulated by PHA in function of donor age, *Mech. Ageing Dev.* **7**:25.

Price, G. B., and Makinodan, T., 1972a, Immunologic deficiencies in senescence. I. Characterization of intrinsic deficiencies, *J. Immunol.* **108**:403.

Price, G. B., and Makinodan, T., 1972b, Immunologic deficiencies in senscence. II. Characterization of extrinsic deficiencies, *J. Immunol.* **108**:413.

Radl, J., Hollander, C. F., Van Den Berg, P., and De Glopper, E., 1978, Idiopathic paraproteinaemia. Studies in an animal model—the ageing C57BL/KaLwRij mouse, *Clin. Exp. Immunol.* **33**:395.

Radl, J., De Glopper, E., Schuit, H. R. E., and Zurcher, C., 1979, Idiopathic paraproteinemia. II. Transplantation of the paraprotein-producing clone from old to young C57BL/KaLwRij mice, *J. Immunol.* **122**:609.

Ramshaw, I. A., and Eidinger, D., 1977, T-Cell mediated immunity towards antigen(s) on isologous erythrocytes, *Nature (London)* **267**:441.

Reddy, M. M., and Goh, K., 1979, B and T lymphocytes in man. IV. Circulating B, T and "null" lymphocytes in aging population, *J. Gerontol.* **34**:5.

Rice, L., Laughter, A. H., and Twomey, J. J., 1979, Three suppressor systems in human blood that modulate lymphoproliferation, *J. Immunol.* **122**:991.

Riesen, W., Keller, H., Skvaril, F., Morell, A., and Barandun, S., 1976, Restriction of immunoglobulin heterogeneity, autoimmunity and serum protein levels in aged people, *Clin. Exp. Immunol.* **26**:280.

Rivnay, B., Globerson, A., and Shinitzky, M., 1979, Viscosity of lymphocyte plasma membrane in aging mice and its possible relation to serum cholesterol, *Mech. Ageing Dev.* **10**:71.

Robert, L., and Robert, B., 1973, Immunology and aging, *Gerontologia* **19**:330.

Roberts-Thomson, I. C., Whittingham, S., Youngchaiyud, U., and MacKay, I. R., 1974, Ageing, immune response and mortality, *Lancet* **2**:368.

Roder, J. C., Duwe, A. K., Bell, D. A., and Singhal, S. K., 1978, Immunological senescence. I. The role of suppressor cells, *Immunology* **35**:837.

Rosenberg, B., Kemeny, G., Smith, L. G., Skurnick, I. D., and Bandurski, M. J., 1973, The kinetics and thermodynamics of death in multicellular organisms, *Mech. Ageing Dev.* **2**:275.

Rosenthal, M., 1978, Age and immunity. III. Circulating immune complexes in different age groups, *Blut* **37**:271.

Ross, M. H., 1961, Length of life and nutrition in the rat, *J. Nutr.* **75**:197.

Ross, M. H., 1969, Aging, nutrition, and hepatic enzyme activity patterns in the rat, *J. Nutr.* **97**:565.

Ross, M. H., 1972, Length of life and caloric intake, *Am. J. Clin. Nutr.* **25**:834.

Sacher, G. A., 1975, Maturation and longevity in relation to cranial capacity in hominid evolution, in: *Primates: Functional Morphology and Evolution*, Vol. 1 (R. Tuttle, ed.), pp. 417–441, Mouton, The Hague.

Sacher, G. A., 1977, Life table modification and life prolongation, in: *Handbook of the Biology of Aging* (C. E. Finch and L. Hayflick, eds.), pp. 582–638, Van Nostrand–Reinhold, Princeton, N.J.

Sato, T., Fuse, A., and Kuwata, T., 1979, Enhancement by interferon of natural cytotoxic activities of lymphocytes from human cord blood and peripheral blood of aged persons, *Cell. Immunol.* **45**:458.

Schwartz, A. G., 1979, Inhibition of spontaneous breast cancer formation in female C3H (A^{vy}/a) mice by long-term treatment with dehydroepiandrosterone, *Cancer Res.* **39**:1129.

Scott, M., Bolla, R., and Denckla, W. D., 1979, Age-related changes in immune function of rats and the effect of long-term hypophysectomy, *Mech. Ageing Dev.* **11**:127.

Segre, D., and Segre, M., 1976, Humoral immunity in aged mice. II. Increased suppressor T cell activity in immunologically deficient old mice, *J. Immunol.* **116**:735.

Segre, D., and Segre, M., 1977, Age-related changes in B and T lymphocytes and decline of humoral immune responsiveness in aged mice, *Mech. Ageing Dev.* **6**:115.

Segre, D., Greeley, E. H., and Segre, M., 1981, Cyclophosphamide induction of non-T suppressor cells and their possible role in self-tolerance, in: *Immunological Aspects of Aging* (D. Segre and L. Smith, eds.), pp. 151–164, Dekker New York.

Segre, M., and Segre, D., 1976, Humoral immunity in aged mice. I. Age-related decline in the secondary response to DNP of spleen cells propagated in diffusion chambers, *J. Immunol.* **116**:731.

Seshadri, R. S., Morley, A. A., Trainor, K. J., and Sorrell, J., 1979, Sensitivity of human lymphocytes to bleomycin increases with age, *Experientia* **35**:233.

Shigemoto, S., Kishimoto, S., and Yamamura, Y., 1975, Change of cell-mediated cytotoxicity with aging, *J. Immunol.* **115**:307.

Shreffler, D. C., 1977, The H-2 model: Genetic control of immune functions, in: *HLA and Disease* (J. Dausset and A. Svejgaard, eds.), pp. 32–45, Munksgaard, Copenhagen.

Siegal, B. V., Braun, M., and Morton, J. J., 1972, Detection of antinuclear antibodies in NZB and other mouse strains, *Immunology* **22**:457.

Smith, A. M., 1976, The effects of age on the immune response to type III pneumococcal polysaccharide (SIII) and bacterial lipopolysaccharide (LPS) in BALB/c, SJL/J and C3H mice, *J. Immunol.* **116**:469.

Smith, G. S., and Walford, R. L., 1978, Influence of the H-2 and H-1 histocompatibility systems upon lifespan and spontaneous cancer incidence in congenic mice, in: *Genetic Effects on Aging* (D. Bergsma and D. E. Harrison, eds.), pp. 281–312, Liss, New York.

Smith, M. A., Evans, J., and Steel, C. M., 1974, Age-related variation in proportion of circulating T cells, *Lancet* **2**:922.

Stastny, P., 1978, Association of the B-cell alloantigen DRw4 with rheumatoid arthritis, *N. Engl. J. Med.* **298**:869.

Strelkauskas, A. J., Andrew, J. A., Hallgren, H. M., and Yunis, E. J., 1981, Autoantibodies to a regulatory T-cell subset in human aging, in: *Immunological Aspects of Aging* (D. Segre and L. Smith, eds.), pp. 267–280, Dekker, New York.

Stutman, O., 1974, Cell-mediated immunity and aging, *Fed. Proc.* **33**:2028.

Tam, C. F., and Walford, R. L., 1978, Cyclic nucleotide levels in resting and mitogen-stimulated spleen cell suspensions from young and old mice, *Mech. Ageing Dev.* **7**:309.

Tam, C. F., and Walford, R. L., 1980, Alterations in cyclic nucleotides and cyclase specific activities in T-lymphocytes of aging normal humans and patients with Down's Syndrome, *J. Immunol.* **125**:1665.

Tannenbaum, A., 1942, The genesis and growth of tumors. II. Effects of caloric restriction *per se*, *Cancer Res.* **2**:460.

Taylor, R. B., 1965, Decay of immunological responsiveness after thymectomy in adult life, *Nature (London)* **208**:1334.

Teague, P. O., Yunis, E. J., Rodey, G., Fish, A. J., Stutman, O., and Good, R. A., 1970, Autoimmune phenomena and renal disease in mice: Role of thymectomy, aging, and involution of immunologic capacity, *Lab. Invest.* **22**:121.

Tice, R. R., Schneider, E. L., Kram, D., and Thorne, P., 1979, Cytokinetic analysis of the impaired proliferative response of peripheral lymphocytes from aged humans to phytohemagglutinin, *J. Exp. Med.* **149**:1029.

Tucker, M. J., 1979, The effect of long-term food restriction on tumours in rodents, *Int. J. Cancer* **23**:803.

Tucker, S. M., Mason, R. L., and Beauchene, R. E., 1976, Influence of diet and feed restriction on kidney function of aging male rats, *J. Gerontol.* **31**:264.

Tyan, M., 1977, Age-related decrease in mouse T-cell progenitors, *J. Immunol.* **118**:846.

Uyeno, E. T., Chang, D., and Folkers, K., 1979, Substance P found to lower body temperature and aggression, *Biochem. Biophys. Res. Commun.* **86**:837.

Walford, R. L., 1962, Auto-immunity and aging, *J. Gerontol.* **17**:281.

Walford, R. L., 1969, *The Immunologic Theory of Aging*, Munksgaard, Copenhagen.

Walford, R. L., 1974, Immunologic theory of aging: Curent status, *Fed. Proc.* **33**:2020.

Walford, R. L., 1979, Multigene families, histocompatibility systems, transformation, meiosis, stem cells, and DNA repair, *Mech. Ageing Dev.* **9**:19.

Walford, R. L., and Bergmann, K., 1979, Influence of genes associated with the main histocompatibility complex on deoxyribonucleic acid excision repair capacity and bleomycin sensitivity in mouse lymphocytes, *Tissue Antigens* **14**:336.

Walford, R. L., Meredith, P. J., and Cheney, K. E., 1977, Immunoengineering: prospects for correction of age-related immunodeficiency states, in: *Immunology and Aging* (T. Makinodan and E. Yunis, eds.), pp. 183–201, Plenum Press, New York.

Walford, R. L., Barnett, E. V., Chia, D., Fahey, J. L., Gatti, R. A., Gossett, T. C., Grossman, H., Medici, M. A., Mutola, M., Naeim, F., Sparkes, R. S., Spina, C., Tam, C. F., Tomura, T., and Van Lancker, J., 1981a, Immunological and biochemical studies of Down's Syndrome as a model for accelerated aging, in: *Immunological Aspects of Aging* (D. Segre and L. Smith, eds.), pp. 479–532, Dekker, New York.

Walford, R. L., Weindruch, R. H., Gottesman, S. R. S., and Tam, C. F., 1981b, Immunopathology of aging, in: *Annual Review of Gerontology and Geriatrics*, Vol. 2 (C. Eisdorfer, B. Starr, and V. Cristofalo, eds.), pp. 3–48, Springer, New York.

Walters, C. S., and Claman, H. N., 1975, Age-related changes in cell-mediated immunity in BALB/c mice, *J. Immunol.* **115**:1438.

Weindruch, R. H., and Suffin, S. C., 1980, Quantitative histologic effects on mouse thymus of controlled dietary restriction, *J. Gerontol.* **34**:525.

Weindruch, R. H., Kristie, J. A., Cheney, K. E., and Walford, R. L., 1979, Influence of controlled dietary restriction on immunologic function and aging, *Fed. Proc.* **38**:2007.

Weiner, H. L., Scribner, D. J., Schocket, A. L., and Moorhead, J. W., 1978, Increased proliferative responses of human peripheral blood lymphocytes to anti-immunoglobulin antibodies in elderly people, *Clin. Immunol. Immunopathol.* **9**:356.

Weksler, M. E., and Hutteroth, T. H., 1974, Impaired lymphocyte function in aged humans, *J. Clin. Invest.* **53**:99.

Weksler, M. E., Innes, J. B., and Goldstein, G., 1978, Immunological studies of aging. IV. The contribution of thymic involution to the immune deficiencies of aging mice and reversal with thymopoietin$_{32-36}$, *J. Exp. Med.* **148**:996.

Weksler, M. E., Dekruyff, R., Dobken, J., and Siskind, G. W., 1981, Ease of tolerance induction in mice of different ages, in: *Immunological Aspects of Aging* (D. Segre and L. Smith, eds.), pp. 119–126, Dekker, New York.

Welsh, R. M., Jr., 1978, Mouse natural killer cells: Induction, specificity and function, *J. Immunol.* **121**:1631.

Whittingham, S., Mathews, J. D., MacKay, I. R., Stocks, A. E., Ungar, B., and Martin, F. I. R., 1971, Diabetes mellitus, autoimmunity and ageing, *Lancet* **1**:763.

Wisniewski, H. M., and Terry, R. D., 1976, Neuropathology of the aging brain in: *Neurobiology of Aging* (R. D. Terry and S. Gershon, eds.), pp. 265–280, Raven Press, New York.

Woda, B. A., and Feldman, J. D., 1979, Density of surface immunoglobulin and capping on rat B lymphocytes. I. Changes with aging, *J. Exp. Med.* **149**:416.

Woda, B. A., Yguerabide, J., and Feldman, J. D., 1979, Mobility and density of AgB, "Ia", and Fc receptors on the surface of lymphocytes from young and old rats, *J. Immunol.* **123**:2162.

Yarnell, J. W. G., St. Leger, A. S., Balfour, I. C., and Russell, R. B., 1979, The distribution, age effects and disease associations of HLA antigens and other blood group markers in a random sample of an elderly population, *J. Chronic Dis.* **32**:555.

Yunis, E. J., Fernandes, G., Smith, J., Stutman, O., and Good, R. A., 1973, Involution of the thymus dependent lymphoid system, in: *Microenvironmental Aspects of Immunity* (B. D. Jankovic and K. Iskovic, eds.), pp. 301–306, Plenum Press, New York.

Zinkernagel, R. M., 1978, Thymus and lymphohemopoietic cells: Their role in T cell maturation in selection of T cells' H-2-restriction-specificity in H-2 linked Ir gene control, *Immunol. Rev.* **42**:224.

Index

Acoelomate, 91
Acrorhagi, 65, 66
Adipohemocyte, of insects, 209
Aging
 B cell changes, 722–723
 biochemical changes in lymphocytes, 731–732
 bone marrow changes, 720–721
 diet, 735–737
 diseases during, 717
 drug therapy of, 734–735
 immunological theories of, 714
 macrophage function, 721–722
 mixed leukocyte reactions, 727
 reticuloendothelial system, 713
 T-cells, 723–724
 temperature effects on, 733
 thymus changes, 719
Agglutinin, see also Hemagglutinins
 of ascidins, 290, 299
 of earthworms, 130–121
 of horseshoe crab, 173, 174
 of insects, 233
 of molluscs, 158, 159
 of sponges, 42
 of sipunculids, 130
Aggregation factor, in sponges, 43–47
Albumen gland, gastropods, 334, 335, 337, 344
Allograft, see also Histocompatibility; Transplantation; Xenografts
 in annelids, 115, 119, 120
 in ascidians, 308–314
 in coelenterates, 72–82
 in fish, 404, 405, 415
 in invertebrates, 14
 in reptiles, 486
 in sea stars, 274–275
 in sea urchins, 275–279
 in sponges, 48–54
α2 macroglobulin, 673

Amebocyte, see also Coelomocyte; Hemocyte; Phagocytosis; Encapsulation
 of ascidians, 286, 287, 289–291
 of coelenterates, 61–63, 68, 73–75
 in hemolymph clotting, 171–173
 of molluscs, 147
 of starfish, 258–260
 of worms, 105
Anaphylaxis, in reptiles, 479, 480
Angiogenesis, see also Lymphocyte-induced angiogenesis
 sea star factor, 24, 25
Antibiotic
 from coelenterates, 71
 from sponges, 40, 41
Antibody
 cytophilic, in birds, 668
 natural, in chickens, 666
Antibody forming cell
 in amphibian
  spleen, 436–437
  thymus, 428
 in reptiles, 480–481
Antibody response
 in amphibians
  affinity, 638
  heterogeneity, 638
  impairment in metamorphosis, 639, 651
  ontogeny, 636
 in reptiles, 476–477, 481–482
  seasonal effects on, 500
Antigen binding, see also Rosette forming cells
 by B cells during ontogeny, 628, 636
Antigen clearance, see also Phagocytosis in crustaceans, 355–357
Antithymocyte antibody
 anti-fish, 412
 anti-reptile, 474, 491, 499
Archeocyte, see also Phagocytosis
 of sponges, 39, 42

749

Autoimmunity, *see also* Tolerance
  and aging, 714, 732–733
  in amphibians
    reactions against adult, 650
    repertoire expression, 650

Bacterial clearance, *see also* Phagocytosis
  by crustacean hemocytes, 191–195
Bacteriolysin
  in ascidians, 297
  in crustaceans, 195
  in earthworm, 129, 130
  in sipunculids, 130
B cell, *see* B lymphocyte
Blood group, *see also* Hemagglutinin
  specificity for, gastropod hemagglutinins, 335
B lymphocyte, *see also* sIg positive cells
  in aging, 722–723
  of amphibians
    surface Ig positive, 635
    effects of thymectomy on, 634
  of birds, 589, 599, 603, 613
  colonies, 626
  C-region genes, 628–629
  differentiation antigens, 626–627
  of fish, 410
    mitogen reactive, 411
    surface Ig positive, 410, 415, 416
    in thymus, 411–412, 415
  ontogeny of
    in frogs, 635
    in mammals, 625–627
  of reptiles
    sIg positive, 474–475
    ultrastructure, 472–474
  tolerance, 626–627
  V region genes, 627–628
Bone marrow
  in aging, 720–721
  in amphibians, 447
  B cell source in mammals, 625, 626
  in birds, 524
Bursa of Fabricius, 589–590, 602–604, 625, 659
  antigen localization in, 513–514
  follicle-associated epithelium, 674–675
  histology, 511–512
  innervation, 515
  local immunity in, 675–676
  lumen of, 675–676
  lymphocytes, life span, 524
  ontogeny, 509–515, 600–604
    endomesodermal relationships, 602
    seeding by precursor cells, 600–602
  systemic immunity, 676–678

Bursectomy, 516
  effects on
    immunoglobulin synthesis, 526–527
    reticuloendothelial system, 673
    phagocytosis, 668, 670
  testosterone-induced, 603–604, 679

Cancer
  in coelenterates, 75–76
  in evolution, 4–5
  in molluscs, 144–145
Cecal tonsil, in birds, 517–520
Cell cooperation, *see also* MHC, Restriction; T-cells, Helper
  in amphibians, 430, 431, 639
  in fish, 409, 410, 412, 414
  in reptiles, 482–484, 491, 493
Cell marker, *see also* Chimera
  in hemopoiesis studies
    immunoglobulin allotypes, 608
    ploidy in amphibians, 571, 575
    quail–chick system, 590–91, 593–596, 600, 605, 612
    sex chromosome in birds, 590, 592, 595, 599, 608, 612
Choanocyte, in sponges, 37
Chemotaxis
  in avian hemopoiesis, 597–600, 612
  in crustaceans, 198
  by earthworm coelomocytes
  by insect hemocytes
Chimera, *see also* Cell marker
  amphibian hemopoiesis, 573–575, 578
  quail–chick system, 590–591, 593–596, 600, 605, 612
  radiation, fetal liver grafts, 699
  tolerance, in frogs, 642
  yolk sac, in birds, 606–609
Chloragogue cell of annelids, 99, 104
Chorioallantoic membrane, 663, 665
  culture in diffusion chambers, 592, 596
  grafting on, 591–592
  lymphocyte-induced angiogenesis on, 690
Circulatory system
  of arachnids, 175
  of birds, 663–664
  of crustaceans, 177
  of insects, 207
  of myriapods, 201
  of onychophorans, 201
  of tunicates, 284
Cloaca, in birds as route of immunization, 676–678
Clotting reaction
  in arachnid hemolymph, 175

Clotting reaction *(cont.)*
    in echinoderms, 271–272
    in Holothuroidea, 265
    in horseshoe crab hemolymph, 171
    in sea urchins, 262
Coagulogen, 171–172
Coelomate, 92
Coelomocyte, *see also* Amebocyte; Hemocyte
    of annelids, 99–10
    of Crinoidea, 266
    of Holothuroidea, 263–265
    of Ophiuroidea, 267
    of sea stars, 258–260
    of sea urchins, 278
Colony forming cells, spleen, 618, 626
Complement
    in birds, 673
    homology with invertebrate humoral substances, 19–20
Concanavalin A, *see also* Mitogens
    amphibian lymphocytes, 438
    fish lymphocytes, 408, 411
    mouse helper T cells, 694
    worm coelomocytes, 107
Cytotoxicity, *see also* Natural killer cells
    in arthropods, 23
    in earthworms, 126–127
    in fish, 406–408
    in sipunculids, 22, 123–126
    in sponges, 53
    T cells, aging, 728
    by thymocytes, during ontogeny, of mouse, 693

Delayed type hypersensitivity
    in amphibians, 22
    in reptiles, 488
Dendritic cell
    in frog spleen, 441
Deuterostome, 5, 6, 10, 89, 257
Differentiation antigen
    on B cells, 627
    on fetal liver cells, 695
    on T cells, aging, 724
Diseases, *see also* Cancer; Parasites; Pathogen
    of crustaceans, 178
    of insects, 207, 208
    of myriapods, 201
    of onychophorans, 201

Ectoderm
    in coelenterates, 62, 72–74, 79
Encapsulation defense reaction
    by ascidians, 302–307
    by crustaceans, 196–197

Encapsulation defense reaction *(cont.)*
    by diplopods, 206
    by insects, 224–228
    by molluscs, 147–149, 150, 151
Endoderm, 61, 69, 74
Epigonal organ, cartilaginous fish, 395–396
Epithelial cell
    follicle-associated, bursa, 674, 675
    of thymus, mouse, 621
Erythropoiesis
    in amphibians, 439, 570–572
Evolution
    of echinoderms, 257–258
    of immunoglobulin, 554–555
    of lymphocytes, 267–268
    of nonself-self recognition, 142
    theories of
        invertebrate, 16–18, 21
        major histocompatibility complex, 27–30
        V region genes, 29–30

Fc receptor
    in birds, 523
    in opsonization, 668
Fetal lymphoid organs
    site of B cell generation, 625
    suppressor cells in
        liver, 695, 698, 701
        spleen, 701
        thymus, 700
Fibrinonectin, in birds, 673

Gene duplication, 546
Germinal center
    absence of
        in birds
            cecal tonsil
            lymph nodules
            spleen
    in amphibians, 450
    in reptiles, 469
Gill bud, transplantation, 576
Graft versus host reaction
    in aging, 714
    in frogs, 689
    in mice, induced by
        fetal liver, 690, 697
        neonatal thymus, 692
        yolk sac cells, 689
    in reptiles, 487–488
Granulopoiesis
    of fish, 399–401
    in frog pronephros, 580

Gut associated lymphoid tissue
  in amphibians
    larvae, 443
    adults, 443
  in reptiles, 462
    ontogeny of, 467–468
Hapten-carrier, see cell cooperation
Harderian gland, 521–522, 530
Hemagglutinins, see also Agglutinins; Opsonins; Phagocytes
  biochemistry of
    in Arachnida, 358–360
    in Asteroidea, 360–362
    in Bivalvia, 345–349
    in Cephalopoda, 349–350
    in Crustacea, 350–357
    in Holothuroidea, 364–365
    in Insecta, 357–358
    in Porifera, 323–325
    in Protochordata, 367–379
Hemocyte, see also Amebocyte; Coelomocyte
  of arachnids, 175
  of crustaceans, 178–183
    clumping in, 189–194
  of horseshoe crab, 169–173
  of insects
    free, 208–211
    phagocytosis by, 218–222
    sessile, 211
  of myriapods, 201–205
  of onychophorans, 201–205
  of pycnogonids, 176
Hemolymph
  of arachnids, 175
  coagulation in
    crustaceans, 197
    horseshoe crab, 170
    insects, 228–229
    myriapods, 206
    onychophorans, 206
  of molluscs, 330
Hematogeneous theory, 569
Hemoglobin
  transition during amphibian metamorphosis, 649
  in ventral blood island transplants, 371
Hemolysin
  in earthworms, 127–128
  in molluscs, 158
  in sea urchin, 362
  in sipunculids, 129
Hemopoietic organ, tissue
  of cartilaginous fish, 395
  of crustaceans, 183–185
  of echinoderms, 267–268

Hemopoietic organ, tissue (cont.)
  of insects, 212–214
  phagocytic cells, 219–224
  of onychophoran, 205
Hemopoietic stem cell, hemocytoblast
  in ascidians, 286–28
  in birds, 595, 596, 600, 606
  in amphibians, 581–583
  in mammals, 618, 626
Hemopoiesis, see also Lymphopoiesis
  in amphibians
    hepatic, 448, 579
    mesonephros, 579–580
    pronephros, 580–581
  in ascidians, 293–296
  in fish
    elasmobranchs, 395–396
    teleost kidney, 397
    lamprey, protovertebral arch, 394
  in placenta, 625
Hepatopancreas, horseshoe crab, 169
Histocompatibility, see also Allografts; Memory; Tolerance; Transplantation
  antigens in amphibians
    larval versus adult, 650
    ontogeny of, 643
    minor, 646
  gene dose
    effects on tolerance, 645–646
  genetics of
    annelids, 120–122
    ascidians colony fusion, 310–314
    insects, 234–235
    Nemerteans, 120
    sponges, 49–51
  reactions
    in annelids, 111, 123
    in ascidians, 308–314
    in coelenterates, 76–82
    in frogs, metamorphosis, 645–646
    in fish, 404–405, 415
    in Holothuroidea, 272–274
    in insects, 234–236
    in invertebrates, 3, 26
    in platyhelminthes, 110–123
    in reptiles, 456, 490
    in sea stars, 274–275
    in sea urchins, 2, 8, 275–279
    in sponges, 49
Hyaline cell
  of crustaceans, 179–180
Hypodermal gland, 170–173

Immunodeficiency
  of CBA/N mice, 627

Immunoglobulin
  combining site, 553
  cytoplasmic, 524–525
    in birds, 625
    in mice, 544
  isotypes IgM
    in amphibians, 638
    in fish, 546
    in mammals, 628
    in reptiles, 478, 479, 551
  low molecular weight
    in amphibians, 551, 634, 636
    in reptiles, 551
  light chains, 551, 553
    B cell differentiation, 626
  receptor, 557
  structure, 543–544, 553
  surface Ig positive cells
    B cells in mammals, 625, 626
    Harderian gland cells, 531
    fish cells, 410, 415, 416
    frog cells, 635
    thymocytes, 560
  switch during ontogeny, 628–629
  synthesis in birds, during ontogeny, 525–52
  thymectomy, effects on IgM, 634–638
Implants, in insects, 234–236
Inflammation, in coelenterates, 74
Interferon, in fish, 405–406
Interstitial cell, in coelenterates, 63, 77, 82
Irradiation
  effects on
    bone marrow reconstitution, 448
    insects, 212
    transplantation in amphibians, 437, 443

Kidney
  amphibian lymphocytes, 440
Kupffer cell
  in chick liver, 662

Lectins
  from Geodia, 43, 46
  on Helix hemocytes, 328, 330
Leukocyte
  of annelids, 99–101
  cytotoxic effectors
    in sipunculids, 123–126
    in worms, 126
  of fish, 397–403
  of nemerteans, 95, 96
  of platyhelminthes, 95–96
  of sipunculids, 96–99
Leydig organ, 395
LIA, see Lymphocyte induced angiogenesis

Limulin, 358–360
Lipopolysaccharide, see also Mitogens
  mitogenic for
    amphibian thymocytes, 431
    worm leukocytes, 108–109
Liver
  lymphopoiesis in amphibian, 449, 579
  fetal, mouse
    site of B cell generation, 625
    suppressor cells in, 695, 698, 701
Lymph nodes
  in amphibians, 441
  in birds, 522–523
  in reptiles, 462
Lymphocyte, see also B cells; T cells
  of ascidians, 286, 288, 289
  of birds, 514, 515, 517
  of reptiles, 471, 474
Lymphocyte induced angiogenesis during
      ontogeny
  by thymocytes, 694
  by yolk sac cells, 690
Lymphokines, in birds, 529–530
Lymphopoiesis
  in amphibian
    adult kidney, 447
    adult liver, 449, 579
    larval lymph gland, 443
Lymphocyte precursor cells
  in birds, 590, 592, 593, 600, 604, 611, 612
  chemotactic hypothesis, 597–600, 612
  colonization of thymus, 593–595
  embryogenesis of, 604–612
  immigration into thymus, 596

Macrophages, see also Hemocyte; Phagocytosis
  and aging, 721–722
  in fish, 398–399
  in mouse thymus, 623
  in reptiles, 470
Major histocompatibility complex, see also
      Histocompatibility
  aging, 715–716
  evolution of, 27–30
  frogs, 643, 646
    ontogeny of, 643–644
    restriction, 640, 642
  invertebrates, 25, 26
  restriction, 621–623, 629
    in suppression, 702
    thymic epithelium, 621–623
Melanization
  in crustaceans, 190, 199
Memory
  annelids, 115–120

Memory (cont.)
  coelenterates, 23, 80–82
  crustaceans, 200
  Holothuroidea, 272–273
  insects, 236
  invertebrates, 22
  Nemerteans, 114, 115
  Platyhelminthes, 114
  reptiles, 477
  sea stars, 274–275
  sea urchins, 278–280
  sipunculids, 14
  sponges, 51, 52
Mesoglea, coelenterates, 61, 63, 75
Mesonephros
  stem cell source in frogs, 580
Mesophyl, of sponges, 38
Metamorphosis of amphibians
  antibody response, 639
  bone marrow, 447
  hemoglobin, 649
  histocompatibility antigens, 643–644
  mixed leukocyte reaction, 646
  suppressor cells, 646–647
  tolerance, 645–648
Migration inhibition
  in birds, 530
  in reptiles, 487–489
Mitogens
  aging
    B cells, 722–723
    T cells, 725–727
  ascidians, 308
  amphibians, 438, 640, 643–645
    bone marrow, 448
    liver, 449
    thymectomized, 644, 645
  echinoderm, 275
  fish, 410, 411
  reptiles, 475
  worms, 105, 109
Mixed leukocyte reaction
  during aging, 727–728
  in amphibians, 431, 438, 642, 644
    during metamorphosis, 646
    syngeneic, 650
    thymectomy, 644
  in ascidians, 307–309
  in fish
  in mice
    fetal liver, 615
    ontogeny, 624
    suppressor cell assay, 700
    thymocytes, 695
MLC, see mixed leukocyte reaction

Monocyte, see Macrophage
Morula cells
  of Holothuroidea, 265–266
  of sea urchins, 262
Mucus
  in coelenterates, 67, 70
Myelopoiesis
  in amphibians
    kidney, 440
    liver, 448
    spleen, 437

Nacrezation, 151
Natural killer cells
  and aging, 728
  in birds, 529
  in fish, 409
  in mouse yolk sac, 691
Nephrocyte
  of coelenterates, 65–68
  of crustaceans, 185–186, 195–196
  of insects, 214–215, 230–231
  of myriapodans, 205
  of onychophorans, 205
Nodules
  formation
    in crustaceans, 189–194
    in insects, 222–224
Normal lymphocyte transfer reaction
  in reptiles, 489–490
Nude mice, 621, 622
Nutrition
  effects on
    aging, 735–737
    immunity in reptiles, 496–497

Oenocytes, of insects, 215
Opsonins
  of birds, 665–670
  of crustaceans, 198, 199, 215, 354–355
  history of, 321–323
  of insects, 232–233
  of molluscs, 153–155, 327
  of tunicates, 299–300

Paraaortic foci
  in hemopoiesis
    birds, 609–612
    frogs, 581
Parasites, see also Disease; Pathogen
  of coelenterates, 68
  of crustaceans, 178
  of insects, 208
    defense against, 224–228
  of molluscs, 149, 150
    defense against, 24

Pathogens, *see also* Disease; Parasites
  of coelenterates, 70–71
  of molluscs, 143–146
  of sponges, 41
Peanut agglutinin, 702
Pericardial cells, *see* Nephrocyte
Peritoneal exudate
  in birds, 671–672
Phagocytosis, by cells, 1, 2, 7, 27–29
  of amphibians, 449
  of annelids, 101
  of arachnids, 175–176
  of ascidians, 300–302
  of birds, 527, 661–662, 664
  of coelenterates, 61, 69, 73, 74
  of earthworms, 103–104
  of echinoderms, 187–189, 198–199, 354
  of fish, 403–404
  of horseshoe crab, 173
  of insects, 215–222
  of molluscs, 146–147, 151, 152, 328, 330–332
  of Myriapoda, 205–206
  of Nemerteans, 95
  of Onychophora, 205–206
  of Platyhelminthes, 95
  of sea urchins, 269–271
  of sipunculids, 96
  of sponges, 39, 41–42
Phytohemagglutinin, *see also* Mitogens
  responses of
    earthworm cells, 108
    fish cells, 408–411
Pinacoderm, 37, 42
Pinacocytes, of sponges, 39, 42
Pinocytosis
  in follicle-associated epithelium, 674–695
  in molluscs, 157–158
Placenta
  B cell hemopoiesis, 625
Plasma cells
  in amphibian
    bone marrow, 448
    kidney, 440
    liver, 449
    lymph nodes, 441
    thymus, 431
    spleen, 438
Plasmacytes, of insects, 789
Polyploidy
  in amphibians, 633
Posterior cardinal vein
  in amphibian hemopoiesis, 583
Pre-B cells, mammals, 625–626
Pre-T cells, mammals, 623

Pronephros
  amphibians, 579–580
  fish, 397, 416
Protostomes, 89, 92, 257
Receptor
  of aggregation factor, 45, 46
  on B cells, 542
  on crustacean hemocytes, 198
  self-nonself, 142
    in invertebrates, 381, 382
    protosomes, 9, 13
  on T cells, 624–625
Red disease
  in frogs, 649
Regeneration
  of coelenterates, 71–74
  of worms, 105
Rosette-forming cells
  in birds, 523
  in fish, 479
  in frogs, 636–638
  in reptiles, 487
Season, effects on
  amphibian
    spleen, 429, 440
    thymus, 429
  reptile
    antibody response, 500–501
    lymphoid tissue, 453, 497–499
Spiral valve, 395
Spleen
  in amphibians
    antigen localization, 435
    histology, 431
    follicular dendritic cell, 437
    pyroninophilia, 437
    red and white pulp, 432
  in birds, 517
  in cartilaginous fish, 395
Splenectomy in amphibians, 439
Stem cells, birds intraembryonic, 606–609, 612
Superoxide dismutase in aging, 715
Suppressor cells, *see also* T-cells
  in fetal liver, 698–699
  in frogs, adoptive transfer, 646–647
  ontogeny, 701
  self tolerance, 703–705
Symbiosis
  of coelenterates, 67–70
  of sponges, 43, 44

T-cell, *see* T-lymphocyte
Temperature, effects
  on aging, 733

Temperature, effects (cont.)
  on antibody production in fish, 410, 414, 415
  on cultured cells in fish, 409, 411–412
  on alloimmunity
    in coelenterates, 77–78, 81–82
    in fish, 415
    in reptiles, 486, 488
    in sponges, 53
  on lymphoid tissue in reptiles, 464
Testosterone
  for bursectomy, 679
Tetraploidy, see Cell marker
Thrombocytes
  of chickens, 670
  of fish, 401–402
Thymus hormones, 428, 625, 695
Thymectomy
  in adults, aging, 725
  in amphibians, 430, 432, 438
  in birds, 669
  in frogs, effects on
    antibody production, 634
    mitogen responses, 644
    mixed leukocyte reaction, 644
    ontogeny, of immunity, 645
  in reptiles, 490–491, 494
  in salamanders, 640
Thymic lymphocytes
  of amphibians
    ontogeny of, 572–578
    surface Ig positive, 429
  of birds, 515
  cyclic nucleotides, 625
  receptors on, 560
  of reptiles, 475
  surface alloantigens, 623–624
Thymus
  and aging, 718–719
  in amphibians, 423, 574
    antibody, 639, 556
    colonization during ontogeny, 527–528
    cortex, 425, 428
    cysts, 428
    dependent areas in spleen, 435
    embryology of, 423, 574
    function, 430
    Hassall's corpuscle, 427
    medulla, 425
    metamorphic changes, 429, 641
    myoid cells, 426
    ontogeny of, 574–578
    plasma cells in, 429, 431
    seasonal effects on, 429

Thymus (cont.)
  in birds, 589–590, 605–607, 609
    ontogeny, 515, 590–600
    mesectoderm, 591
  in bony fish, 396–397
    antibody formation in, 556
  in hagfish?, 394
  in lamprey?, 394
  in mammals
    MHC antigen, 622
    ontogeny, 619–622, 692–693
    stroma, 625
    stem cells, 620
  in nude mice, 622
  in reptiles
    dependent areas of spleen, 465
    histology, 463
    ontogeny, 466–467
Tiedemann bodies, 267
T-lymphocyte
  in aging, 723–724, 728–731
  in birds, 595, 599, 613
  cytotoxic
    and aging, 728
    in birds, 529
  differentiation antigens, 623, 624
  helpers
    in aging, 728
    in amphibians, 430, 639, 645
    in birds, 528
    in fish, 415
    in ontogeny, 624
    in reptiles, 482
  mitogen reactive in aging, 725–727
  receptor, 559–561
  suppressors
    in birds, 528–529
Tolerance
  and aging, 714, 732–733
  in birds, 678
  in frogs
    in chimeras, 583
    in embryos, 642
    during metamorphosis, 626, 627
    to self, 649–651
  in mammals
    of B cells, 626–627
    to self, 703–705
Transplantation, see also Allograft; Histocompatibility; MHC; Xenograft
  in ascidians, 308–314
  in coelenterates, 77–83
  as criterion of invertebrate immunity, 76
  in echinoderms, 272–279

Transplantation (cont.)
　in fish, 404–405, 415
　in frogs
　　antigen dose, 646
　　during metamorphosis, 645
　　during ontogeny, 641
　in reptiles, 486
Transfilter cultures, 597
Triploidy, see Cell marker

Urn
　of sipunculids, 96

Vacuolated cells
　of ascidians, 291–292
Ventral blood islands
　in amphibians erythropoiesis
　　explants, 570-71
　　extirpation, 570
　　transplantation, 570, 572
V region genes
　diversity, 553, 555
　in evolution, 39, 30
　repertoire during ontogeny, 627

White body, 147

Wound healing
　in coelenterates, 71–74
　in crustaceans, 197
　in insects, 228–230
　in sea urchins, 202
　in sponges, 39
　in worms, 105

Xenografts, see also Allografts; Histocompatibility; Transplantation
　in annelids, 14, 114–120
　in coelenterates, 77–82
　in sponges, 40, 47

Yolk sac, 569, 571, 583, 618, 662
　of birds, 590, 592, 595, 04, 605, 609, 611
　　chimeras, 605–609
　　parabiosis, 592
　cells functioning in
　　graft versus host reaction, 688
　　natural cytotoxicity, 91
　of reptiles, 467

Zooanthellae, 44, 69–71
Zoochlorellae, 68–70

If you have any concerns about our products,
you can contact us on
ProductSafety@springernature.com

In case Publisher is established outside the EU,
the EU authorized representative is:
**Springer Nature Customer Service Center GmbH
Europaplatz 3, 69115 Heidelberg, Germany**

Printed by Libri Plureos GmbH
in Hamburg, Germany